Business Statistics

Business Statistics

Norean R. Sharpe
Babson College

Richard D. De Veaux
Williams College

Paul F. Velleman
Cornell University

Addison Wesley

Boston San Francisco New York
London Toronto Sydney Tokyo Singapore Madrid
Mexico City Munich Paris Cape Town Hong Kong Montreal

Editor in Chief	Deirdre Lynch
Vice President/Executive Director, Development	Carol Trueheart
Senior Development Editor	Elaine Page
Senior Project Editor	Chere Bemelmans
Associate Editor	Sara Oliver Gordus
Assistant Editor	Christina Lepre
Editorial Assistant	Dana Jones
Senior Managing Editor	Karen Wernholm
Senior Production Supervisor	Peggy McMahon
Cover Design	Barbara T. Atkinson
Interior Design	The Davis Group
Cover Photo	Chris Cheadle / All Canada Photos / Getty Images
Senior Market Development Manager	Dona Kenly
Executive Marketing Manager	Roxanne McCarley
Marketing Manager	Alex Gay
Marketing Assistant	Kathleen DeChavez
Photo Researcher	Beth Anderson
Media Producer	Christine Stavrou
Software Development	Edward Chappell and Marty Wright
Senior Author Support/Technology Specialist	Joe Vetere
Rights and Permissions Advisor	Tim Nicholls
Senior Manufacturing Buyers	Carol Melville and Ginny Michaud
Production Coordination, Illustration, and Composition	Pre-Press PMG

About the cover: The building on the cover is the Royal Bank Plaza Building in Toronto, Canada. Its distinctive gold color is the result of the layer of 24-karat gold coating its 14,000 windows. Besides serving as insulation, the gold acts as a stunning reflector of light.

Library of Congress Cataloging-in-Publication Data
Sharpe, Norean Radke.
  Business statistics / Norean R. Sharpe, Richard D. De Veaux, Paul F. Velleman.—1st ed.
    p. cm.
  Includes index.
  ISBN 0-321-42659-2
  1. Commercial statistics.   I. De Veaux, Richard D.   II. Velleman, Paul F., 1949-   III. Title.
    HF1017.S467 2010
    650.01′5195—dc22

                              2007060101

Addison-Wesley
is an imprint of

www.pearsonhighered.com

ISBN-10: 0-321-42659-2 ISBN-13: 978-0-321-42659-8
2 3 4 5 6 7 8 9 10—CRK—11 10 09

*To my husband Peter, for his encouragement and support,
and to my children, Katrina and PJ, whose own inquisitive and
exploratory natures continue to impress and astound me.*
—Norean

To my parents.
—Dick

*To my father, who taught me about ethical business practice by
his constant example as a small businessman and parent.*
—Paul

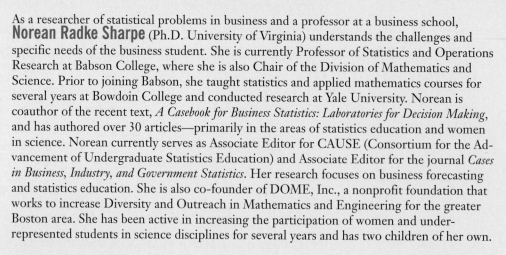

Meet the Authors

As a researcher of statistical problems in business and a professor at a business school, **Norean Radke Sharpe** (Ph.D. University of Virginia) understands the challenges and specific needs of the business student. She is currently Professor of Statistics and Operations Research at Babson College, where she is also Chair of the Division of Mathematics and Science. Prior to joining Babson, she taught statistics and applied mathematics courses for several years at Bowdoin College and conducted research at Yale University. Norean is coauthor of the recent text, *A Casebook for Business Statistics: Laboratories for Decision Making*, and has authored over 30 articles—primarily in the areas of statistics education and women in science. Norean currently serves as Associate Editor for CAUSE (Consortium for the Advancement of Undergraduate Statistics Education) and Associate Editor for the journal *Cases in Business, Industry, and Government Statistics*. Her research focuses on business forecasting and statistics education. She is also co-founder of DOME, Inc., a nonprofit foundation that works to increase Diversity and Outreach in Mathematics and Engineering for the greater Boston area. She has been active in increasing the participation of women and under-represented students in science disciplines for several years and has two children of her own.

Richard D. De Veaux (Ph.D. Stanford University) is an internationally known educator, consultant, and lecturer. Dick has taught statistics at a business school (Wharton), an engineering school (Princeton), and a liberal arts college (Williams). While at Princeton, he won a Lifetime Award for Dedication and Excellence in Teaching. Since 1994, he has been a Professor of Statistics at Williams College. Dick holds degrees from Princeton University in Civil Engineering and Mathematics and from Stanford University in Dance Education and Statistics, where he studied with Persi Diaconis. His research focuses on the analysis of large data sets and data mining in science and industry. Dick has won both the Wilcoxon and Shewell awards from the American Society for Quality and is a Fellow of the American Statistical Association. Dick is well known in industry, having consulted for such Fortune 500 companies as American Express, Hewlett-Packard, Alcoa, DuPont, Pillsbury, General Electric, and Chemical Bank. He was named the "Statistician of the Year" for 2008 by the Boston Chapter of the American Statistical Association for his contributions to teaching, research, and consulting. In his spare time he is an avid cyclist and swimmer. He also is the founder and bass for the doo-wop group "Diminished Faculty" and is a frequent soloist with various local choirs and orchestras. Dick is the father of four children.

Paul F. Velleman (Ph.D. Princeton University) has an international reputation for innovative statistics education. He designed the Data Desk® software package and is also the author and designer of the award-winning ActivStats® multimedia software, for which he received the EDUCOM Medal for innovative uses of computers in teaching statistics and the ICTCM Award for Innovation in Using Technology in College Mathematics. He is the founder and CEO of Data Description, Inc. (www.datadesk.com), which supports both of these programs. He also developed the Internet site, *Data and Story Library* (DASL) (www.dasl.datadesk.com), which provides data sets for teaching Statistics. Paul coauthored (with David Hoaglin) the book *ABCs of Exploratory Data Analysis*. Paul has taught Statistics at Cornell University on the faculty of the School of Industrial and Labor Relations since 1975. His research often focuses on statistical graphics and data analysis methods. Paul is a Fellow of the American Statistical Association and of the American Association for the Advancement of Science and baritone of the barbershop quartet *Alchemy*. Paul's experience as a professor, entrepreneur, and business leader brings a unique perspective to the book.

Dick De Veaux and Paul Velleman have authored successful books in the introductory college and AP High School market with Dave Bock, including *Intro Stats*, Third Edition (Pearson, 2009), *Stats: Modeling the World*, Third Edition (Pearson, 2010), and *Stats: Data and Models*, Second Edition (Pearson, 2008).

Contents

*Indicates an optional topic

Part IV   Building Models for Decision Making   637

Appendixes

From the Classroom...

Providing Real Business Context

Chapter Openers

Each chapter opens with an interesting business example. The stories of companies such as Amazon.com, The Home Depot, and KEEN Inc. enhance and illustrate the message of each chapter, showing students how and why statistical thinking is vital to modern business decision-making. We analyze data from these examples throughout the chapter.

CHAPTER **4**

Displaying and Describing Categorical Data

KEEN Inc.

KEEN Inc. was started to create a sandal designed for a variety of water activities. The sandals quickly became popular due to their unique patented toe protection—a black bumper to protect the toes when adventuring out on rivers and trails. Today the KEEN brand offers over 100 different outdoor performance and outdoor inspired casual footwear styles.

Few companies experience th[e] KEEN did in less than four years done this with relatively little adv ing primarily to specialty footwea in addition to online outlets.

After the 2004 Tsunami disast advertising budget almost comple $1 million to help the victims and Foundation to s[...]

In-Text Examples

Real business examples motivate the discussions, often returning to the chapter-opening company.

Table 4.2 A relative frequency table for the same data.

100.01%?
If you are careful to add the percentages in Table 4.2, you will notice the total is 100.01%. Of course the real total has to be 100.00%. The discrepancy is due to individual percentages being rounded. You'll often see this in tables of percents, sometimes with explanatory footnotes.

4.3 Charts

The Area Principle

Now that we have a frequency table, we're ready to follow the three rules of data analysis and make a picture of the data. But we can't make just any picture; a bad picture can distort our understanding rather than help it. For example, here's a graph of the frequencies of Table 4.1. What impression do you get of the relative frequencies of visits from each source?

While it's true that the majority of people came to KEEN's website from Google, in Figure 4.2 it looks like nearly all did. That doesn't seem right. What's wrong? The lengths of the sandals *do* match the totals in the table. But our eyes tend to be more impressed by the *area* (or perhaps even the *volume*) than by other aspects of each

Figure 4.2 Although the length of each sandal corresponds to the correct number, the impression we get is all wrong because we perceive the entire area of the sandal. In fact, only a little more than 50% of all visitors used Google to get to the website.

Applying the Concepts

The Sharpe Edge: Plan, Do, Report

There are three simple steps to doing Statistics right: *Plan, Do,* and *Report.*

We lead students through the process of making business decisions with data. The first step is planning how to tackle a problem, the second is doing the calculations, and the third is reporting the results and conclusions. In each chapter, we apply the new concepts learned in *Guided Examples*. Examples are structured to reflect the way statisticians approach and solve problems. These step-by-step examples show students how to produce the kind of solutions and reports that clients expect to see.

PLAN first. Know where you're headed and why. Clearly defining and understanding your objective will save you a lot of work. What do you know? What do you hope to learn? Are the assumptions and conditions satisfied?

DO is the mechanics of calculating statistics. This is what most people think Statistics is about. But the computations don't tell the whole story.

REPORT what you've learned. Until you've explained your results in the context of the business question in your Plan, the job isn't done. We present the report step as a memo to emphasize the decision aspect of each example.

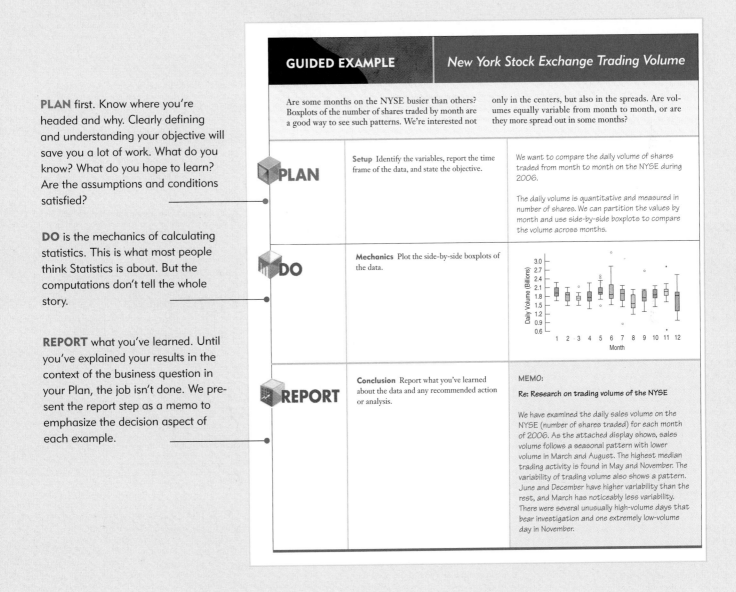

GUIDED EXAMPLE — *New York Stock Exchange Trading Volume*

Are some months on the NYSE busier than others? Boxplots of the number of shares traded by month are a good way to see such patterns. We're interested not only in the centers, but also in the spreads. Are volumes equally variable from month to month, or are they more spread out in some months?

PLAN

Setup Identify the variables, report the time frame of the data, and state the objective.

We want to compare the daily volume of shares traded from month to month on the NYSE during 2006.

The daily volume is quantitative and measured in number of shares. We can partition the values by month and use side-by-side boxplots to compare the volume across months.

DO

Mechanics Plot the side-by-side boxplots of the data.

REPORT

Conclusion Report what you've learned about the data and any recommended action or analysis.

MEMO:

Re: Research on trading volume of the NYSE

We have examined the daily sales volume on the NYSE (number of shares traded) for each month of 2006. As the attached display shows, sales volume follows a seasonal pattern with lower volume in March and August. The highest median trading activity is found in May and November. The variability of trading volume also shows a pattern. June and December have higher variability than the rest, and March has noticeably less variability. There were several unusually high-volume days that bear investigation and one extremely low-volume day in November.

Promoting Understanding

What Can Go Wrong?

The most common mistakes for those new to statistical analysis usually involve misusing a method, not miscalculating a statistic. We acknowledge these mistakes with *What Can Go Wrong?*, found at the end of each chapter. Our goal is to arm students with the tools to detect statistical errors and offer practice in debunking misuses of statistics.

Math Box

The mathematical underpinnings of statistical methods and concepts are set apart to avoid interrupting the explanation of the topic at hand. We use these derivations to increase students' understanding of the underlying mathematics, but they can be skipped by less mathematically inclined students.

By Hand

Although we encourage using technology to perform statistical calculations, we recognize the benefits of knowing how to compute by hand. *By Hand* boxes explain formulas and help students through the calculation of a worked example.

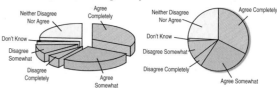

WHAT CAN GO WRONG?

- **Don't violate the area principle.** This is probably the most common mistake in a graphical display. Violations of the area principle are often made for the sake of artistic presentation. Here, for example, are two versions of the same pie chart for the *Regional Preference* data.

The one on the left looks interesting, doesn't it? But showing the pie three dimensionally on a slant violates the area principle and makes it much more difficult to compare fractions of the whole made up of each category of the response—the principal feature that a pie chart ought to show.

- **Keep it honest.** Here's a pie chart that displays data on the percentage of high ... e in specified dangerous behaviors as reported by ... ntrol. What's wrong with this plot?

... ges. Or look at the 50% slice. Does it look right?
... e percentages of? Is there a "whole" that has been
... e proportions shown by each slice of the pie must
... n individual must fall into only one category. Of
... a slant makes it even harder to detect the error.
... le. This bar chart shows the number of airline
... rity screening.

(continued)

MATH BOX

Standardizing the variables first gives us an easy to understand expression for the correlation.

$$r = \frac{\sum z_x z_y}{n - 1}$$

But sometimes you'll see other formulas. Remembering how standardizing works gets us from one formula to the other.

Since
$$z_x = \frac{x - \bar{x}}{s_x}$$

and
$$z_y = \frac{y - \bar{y}}{s_y},$$

we can substitute these and get

$$r = \left(\frac{1}{n-1}\right)\sum z_x z_y = \left(\frac{1}{n-1}\right)\sum \frac{(x - \bar{x})}{s_x}\frac{(y - \bar{y})}{s_y} = \sum \frac{(x - \bar{x})(y - \bar{y})}{(n-1)s_x s_y}.$$

That's one version. And since we know the formula for standard deviation,

$$s_y = \sqrt{\frac{\sum (y - \bar{y})^2}{n - 1}},$$

we could use substitution to write:

$$r = \left(\frac{1}{n-1}\right)\sum \frac{(x - \bar{x})}{s_x}\frac{(y - \bar{y})}{s_y}$$

$$= \left(\frac{1}{n-1}\right)\frac{\sum (x - \bar{x})(y - \bar{y})}{\sqrt{\frac{\sum (x - \bar{x})^2}{n-1}}\sqrt{\frac{\sum (y - \bar{y})^2}{n}}}$$

$$= \left(\frac{1}{n-1}\right)\frac{\sum (x - \bar{x})}{\left(\frac{1}{n-1}\right)\sqrt{\sum (x - }}$$

$$= \frac{\sum (x - \bar{x})(y - \bar{y})}{\sqrt{\sum (x - \bar{x})^2 \sum (y - \bar{y})^2}}.$$

That's the other common version. If you ever h... tion by hand, it's easier to start with one of thes... ber how correlation works, stick with the first f...

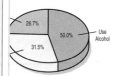

Finding the correlation coefficient by hand

To find the correlation coefficient by hand, we'll use a formula in original units, rather than z-scores. This will save us the work of having to standardize each individual data value first. Start with the summary statistics for both variables: $\bar{x}$, $\bar{y}$, s_x, and s_y. Then find the deviations as we did for the standard deviation, but now in both x and y: $(x - \bar{x})$ and $(y - \bar{y})$. For each data pair, multiply these deviations together: $(x - \bar{x}) \times (y - \bar{y})$. Add the products up for all data pairs. Finally, divide the sum by the product of $(n - 1) \times s_x \times s_y$ to get the correlation coefficient.

Here we go.

Suppose the data pairs are:

x	6	10	14	19	21
y	5	3	7	8	12

Then $\bar{x} = 14$, $\bar{y} = 7$, $s_x = 6.20$, and $s_y = 3.39$

Deviations in x	Deviations in y	Product
$6 - 14 = -8$	$5 - 7 = -2$	$-8 \times -2 = 16$
$10 - 14 = -4$	$3 - 7 = -4$	16
$14 - 14 = 0$	$7 - 7 = 0$	0
$19 - 14 = 5$	$8 - 1 = 1$	5
$21 - 14 = 7$	$12 - 7 = 5$	35

Add up the products: $16 + 16 + 0 + 5 + 35 = 72$
Finally, we divide by $(n - 1) \times s_x \times s_y = (5 - 1) \times 6.20 \times 3.39 = 84.07$
The ratio is the correlation coefficient:

$$r = 72/84.07 = 0.856$$

Checking Understanding

Just Checking

Once or twice per chapter, *Just Checking* asks students to stop and think about what they've read. These questions are designed to check student understanding and involve little calculation. Answers are provided at the end of the chapter so students can easily check their work.

JUST CHECKING

So that they can balance their inventory, an optometry shop collects the following data for customers in the shop.

		Eye Condition			
		Near Sighted	Far Sighted	Need Bifocals	Total
Sex	Males	6	20	6	32
	Females	4	16	12	32
	Total	10	36	18	64

1 What percent of females are far-sighted?
2 What percent of near-sighted customers are female?
3 What percent of all customers are far-sighted females?
4 What's the distribution of *Eye Condition*?
5 What's the conditional distribution of *Eye Condition* for males?
6 Compare the percent who are female among near-sighted customers to the percent of all customers who are female.
7 Does it seem that *Eye Condition* and *Sex* might be dependent? Explain.

What have we learned?

We've learned that we can summarize categorical data by counting the number of cases in each category, sometimes expressing the resulting distribution as percents. We can display the distribution in a bar chart or a pie chart. When we want to see how two categorical variables are related, we put the counts (and/or percentages) in a two-way table called a contingency table.

- We look at the marginal distribution of each variable (found in the margins of the table).
- We also look at the conditional distribution of a variable within each category of the other variable.
- We can display these conditional and marginal distributions using bar charts or pie charts.
- If the conditional distributions of one variable are (roughly) the same for every category of the other, the variables are independent.

What Have We Learned?

These chapter-ending summaries highlight the concepts introduced in the chapter, define new terms, and list the skills presented in the chapter. If students understand all these parts, they're probably ready for the exam.

Terms

Area principle	A principle that helps to interpret ___ that in a statistical display, each da___
Bar chart (relative frequency bar chart)	A chart that represents the count ___ variable as a bar, allowing easy vi___
Column percent	The proportion of each column ___
Conditional distribution	The distribution of a variable res___ of individuals.
Contingency table	A contingency table displays cou___ falling into named categories on ___ individuals on all variables at on___ may be contingent on the catego___
Distribution	The distribution of a variable is ___ • all the possible values of the v___ • the relative frequency of each ___
Frequency table (relative frequency table)	A table that lists the categories in a categorical variable and gives the number (the percentage) of observations for each category. The row percent is the proportion of each row contained in the cell of a frequency table, while the column percent is the proportion of each column contained in the cell of a frequency table.
Independent variables	Variables for which the conditional distribution of one variable is the same for each category of the other.
Marginal distribution	In a contingency table, the distribution of either variable alone. The counts or percentages are the totals found in the margins (usually the right-most column or bottom row) of the table.

Skills

PLAN
- Recognize when a variable is categorical and choose an appropriate display for it.
- Understand how to examine the association between categorical variables by comparing conditional and marginal percentages.

DO
- Summarize the distribution of a categorical variable with a frequency table.
- Display the distribution of a categorical variable with a bar chart or pie chart.
- Construct and examine a contingency table.
- Construct and examine displays of the conditional distributions of one variable for two or more groups.

REPORT
- Describe the distribution of a categorical variable in terms of its possible values and relative frequencies.
- Describe any anomalies or extraordinary features revealed by the display of a variable.
- Describe and discuss patterns found in a contingency table and associated displays of conditional distributions.

Integrating Technology

Technology Help

In business, Statistics is practiced with computers. We offer specific guidance for several of the most common Statistics software (Excel® 2007 and 2003, Minitab®, SPSS®, JMP®, and DataDesk®), often with an annotated example to help students get started with the technology of their choice.

Technology Help: Confidence Intervals for Proportions

Confidence intervals for proportions are so easy and natural that many statistics packages don't offer special commands for them. Most statistics programs want the "raw data" for computations. For proportions, the raw data are the "success" and "failure" status for each case. Usually, these are given as 1 or 0, but they might be category names like "yes" and "no." Often we just know the proportion of successes, $\hat{p}$, and the total count, n. Computer packages don't usually deal with summary data like this easily, but the statistics routines found on many graphing calculators allow you to create confidence intervals from summaries of the data—usually all you need to enter are the number of successes and the sample size.

In some programs you can reconstruct variables of 0's and 1's with the given proportions. But even when you have (or can reconstruct) the raw data values, you may not get exactly the same margin of error from a computer package as you would find working by hand. The reason is that some packages make approximations or use other methods. The result is very close but not exactly the same. Fortunately, Statistics means never having to say you're certain, so the approximate result is good enough.

EXCEL

Inference methods for proportions are not part of the standard Excel tool set.

Comments

For summarized data, type the calculation into any cell and evaluate it. Confidence intervals for a proportion are available in the DDXL add-in. Select the range of data holding the variable. Then choose **Confidence Intervals** from the **DDXL** menu. Choose **1 Var Prop Interval** from the menu, indicate the variable, and click **OK**.

MINITAB

Choose **Basic Statistics** from the **Stat** menu.

- Choose **1 Proportion** from the Basic Statistics submenu.
- If the data are category names in a variable, assign the variable from the variable list box to the **Samples in columns** box. If you have summarized data, click the **Summarized Data** button and fill in the number of trials and the number of successes.
- Click the **Options** button and specify the remaining details.

- If you have a large sample, check **Use test and interval based on normal distribution**. Click the **OK** button.

Comments

When working from a variable that names categories, MINITAB treats the last category as the "success" category. You can specify how the categories should be ordered.

SPSS

SPSS does not find confidence intervals for proportions.

JMP

For a **categorical** variable that holds category labels, the **Distribution** platform includes tests and intervals for proportions. For summarized data, put the category names in one variable and the frequencies in an adjacent variable. Designate the frequency column to have the **role** of **frequency**. Then use the **Distribution** platform.

Comments

JMP uses slightly different methods for proportion inferences than those discussed in this text. Your answers are likely to be slightly different, especially for small samples.

DATA DESK

...nts

...marized data, open a Scratchpad to compute the ... deviation and margin of error by typing the calcula-... en use z-interval for individual μs.

Mini Case Study Projects

*Fuel Efficiency

With the ever increasing price of gasoline, both drivers and auto companies are motivated to raise the fuel efficiency of cars. Recent information posted by the U.S. government proposes some simple ways to increase fuel efficiency (see www.fueleconomy.gov): avoid rapid acceleration, avoid driving over 60 mph, reduce idling, and reduce the vehicle's weight. An extra 100 pounds can reduce fuel efficiency (mpg) by up to 2%. A marketing executive is studying the relationship between the fuel efficiency of cars (as measured in miles per gallon) and their weight to design a new compact car campaign. In the data set **ch07_MCSP_Fuel_Efficiency** you'll find data on the variables below.

- Model of Car
- Engine Size (L)
- Cylinders
- MSRP (Manufacturer's Suggested Retail Price in $)
- City (mpg)
- Highway (mpg)
- Weight (pounds)
- Type and Country of manufacturer

Describe the relationship of Weight, MSRP, and Engine Size with fuel efficiency (both City and Highway) in a written report. Be sure to transform the variables if necessary.

Mini Case Study Projects

Each chapter includes one or two *Mini Case Study Projects* that use real data and ask students to investigate a question or make a decision. Students define the objective, plan the process, complete the analysis, and report their conclusion. Data for the *Mini Case Study Projects* are available on the DVD and website, formatted for various technologies.

Tackling Problems

Exercises

We have worked hard to ensure that exercises contain relevant and modern questions with real data. The exercises generally start with a straightforward application of the chapter ideas, then tackle larger problems. Many break a problem into several parts to help guide the student through the logic of a complete analysis. Finally, there are exercises that ask the student to synthesize and incorporate their ideas with less guidance. Data for exercises marked **T** are available on the DVD and website, formatted for various technologies.

EXERCISES

40. Selling fuel economy 2007. In 2006, a study by *Consumer Reports* found that 37% of nationwide respondents reported that they were considering replacing their current car with one with greater fuel economy. Here are advertised horsepower ratings and expected gas mileage for several 2007 vehicles (www.kbb.com/KBB/ReviewsAndRating).

Vehicle	Horsepower	Highway Gas Mileage (mpg)
Audi A4	200	32
BMW 328	230	30
Buick LaCrosse	200	30
Chevy Cobalt	148	32
Chevy TrailBlazer	291	22
Ford Expedition	300	20
GMC Yukon	295	21
Honda Civic	140	40
Honda Accord	166	34
Hyundai Elantra	138	36
Lexus IS 350	306	28
Lincoln Navigator	300	18
Mazda Tribute	212	25
Toyota Camry	158	34
Volkswagen Beetle	150	30

a) Make a scatterplot for these data.
b) Describe the direction, form, and strength of the plot.
c) Find the correlation between horsepower and miles per gallon.
d) Write a few sentences telling what the plot says about fuel economy.

T 41. Pizza sales. Here is a scatterplot for the weekly sales (in pounds) for every fourth week of a brand of frozen pizza versus the unit price of the pizza for a sample of stores in the Dallas area.

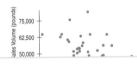

T 42. Housing costs. Concern over the possibility of a "home cost bubble" has lead many economists to examine housing costs. The Office of Federal Housing Enterprise Oversight (www.ofheo.gov) collects data on various aspects of housing costs around the United States. Here is a scatterplot of the *Housing Cost Index* versus the *Median Family Income* for each of the 50 states. The correlation is 0.65.

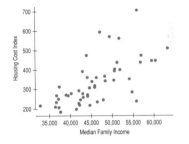

a) Describe the relationship between the *Housing Cost Index* and the *Median Family Income* by state.
b) If we standardized both variables, what would the correlation coefficient between the standardized variables be?
c) If we had measured *Median Family Income* in thousands of dollars instead of dollars, how would the correlation change?
d) Washington, DC, has a *Housing Cost Index* of 548 and a median income of about $45,000. If we were to include DC in the data set, how would that affect the correlation coefficient?
e) Do these data provide proof that by raising the median income in a state, the *Housing Cost Index* will rise as a result? Explain.

T 43. Mutual funds. Here is a scatterplot showing the association between money flowing into mutual funds (fund flows in $M) and a specific type of market return (Wilshire Index) for each month from 1990 to 2002.

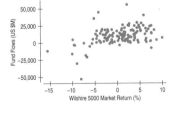

a) Is it appropriate to calculate a correlation? Explain.
b) Identify the largest outlier in the scatterplot. Find and discuss

---

Ethics in Action

Our ethics vignettes in each chapter illustrate some of the judgments needed in statistical analysis, identify possible errors, link the issues to the ASA's Ethical Guidelines, and then propose ethically and statistically sound alternatives.

Preface

We set out to write a book for business students that answers the simple question: "How can I make better decisions?" As entrepreneurs and consultants, we know that Statistics is essential to survive and thrive in today's competitive environment. As educators, we've seen a disconnect between the way Statistics is taught to business students and the way it is used in making business decisions. In *Business Statistics*, we try to close the gap between theory and practice by presenting statistical methods so they are both relevant and interesting to students.

The data that inform a business decision have a story to tell, and the role of Statistics is to help us hear that story clearly. Like other textbooks, *Business Statistics* teaches how to calculate a particular statistic or test and highlights definitions and formulas. But, unlike other textbooks, *Business Statistics* also teaches the "why" and insists that results be reported in the context of business decisions. Students will come away knowing how to think statistically to make better business decisions and how to effectively communicate the analysis that led to the decision to others.

Business Statistics is written with the understanding that today Statistics is practiced with technology. This insight informs everything from our choice of forms for equations (favoring intuitive forms over calculation forms) to our extensive use of real data. But most important, understanding the value of technology allows us to focus on teaching statistical thinking rather than calculation. The questions that motivate each of our hundreds of examples are not "how do you find the answer?" but "how do you think about the answer, and how does it help you make a better decision?"

Our focus on statistical thinking ties the chapters of the book together. An introductory Business Statistics course covers an overwhelming number of new terms, concepts, and methods. But they have a central core: how we can understand more about the world and make better decisions by understanding what the data tell us. From this perspective, students can see that the many ways to draw inferences from data are several applications of the same core concepts.

Our Goal: Read This Book!

The best textbook in the world is of little value if students don't read it. Here are some of the ways we made *Business Statistics* more approachable:

- **Readability.** You'll see immediately that this book doesn't read like other Statistics texts. We strive for a conversational, approachable style, and we introduce anecdotes to maintain interest. In class tests, instructors report their amazement that students are voluntarily reading ahead of their assignments. Students write to tell us (to their amazement) that they actually enjoyed the book.

- **Focus on assumptions and conditions.** More than any other textbook, *Business Statistics* emphasizes the need to verify assumptions to use statistical procedures. We reiterate this focus throughout the examples and exercises. We make every effort to provide students with templates that reinforce the practice of checking these assumptions and conditions, rather than rushing through the computations of a real-life problem.

- **Emphasis on graphing and exploring data.** Our consistent emphasis on the importance of displaying data is evident from the first chapters on understanding data to the complex model-building chapters at the end. Examples always include data displays and often illustrate the value of examining data visually, and the Exercises reinforce this. When we graph data, we are able to see structures, or patterns, that we can't see otherwise. These patterns often raise new questions and guide our statistical analysis and case analysis process. Emphasizing graphics throughout the book helps the student see that the simple structures we look for when we graph data illustrate the concepts we call on for our most sophisticated analyses.

- **Consistency.** We work hard to avoid the "do what we say, not what we do" trap. Having taught the importance of plotting data and checking assumptions and conditions, we are careful to model that behavior throughout the book. (Check the Exercises in the chapters on multiple regression or time series and you'll find us still requiring and demonstrating the plots and checks that were introduced in the early chapters.) This consistency helps reinforce these fundamental principles.

- **The need to read.** Students who plan just to skim the book may find that important concepts, definitions, and sample solutions are not always set aside in boxes. This book needs to be read, so we've tried to make the reading experience enjoyable.

Coverage

The topics covered in a Business Statistics course are generally consistent and mandated by our students' needs in their studies and in their future professions. But the *order* of these topics and the relative emphasis given to each is not well established. In *Business Statistics*, you may encounter some topics sooner or later than you expected. Although we have written many chapters specifically so they can be taught in a different order, we urge you to consider the order we have chosen.

We've been guided in the order of topics by the fundamental principle that this should be a coherent course in which concepts and methods fit together to give students a new understanding of how reasoning with data can uncover new and important truths. We have tried to ensure that each new topic fits into the growing structure of understanding that we hope students will build. For example, we teach inference concepts with proportions first and then with means. Students have a wider experience with proportions, seeing them in polls and advertising. And by starting with proportions, we can teach inference with the Normal model and then introduce inference for means with the Student's t distribution.

We introduce the concepts of association, correlation, and regression early in *Business Statistics*. Our experience in the classroom shows that introducing students to these fundamental ideas early motivates them at the beginning of the course. Later in the semester, when we discuss inference, students recall what they have learned and find it natural and relatively easy to build on the fundamental concepts they experienced by exploring data with these methods.

In an introductory course, the relative emphasis placed on topics requires planning by the instructor. Introductory Business Statistics courses often have limited time, so it can be hard to get to the important topics of multiple regression and model building with enough time to treat them thoroughly. We've moved discussions of risk and probability later in the topic sequence so students can get to practical applications sooner and have time for an adequate emphasis on these essential skills. We've been guided in our choice of what to emphasize by the GAISE (Guidelines for Assessment and Instruction in Statistics Education) Report,[1] which emerged from extensive studies of how students best learn Statistics. Those recommendations, now officially adopted and recommended by the American Statistical Association, urge (among other detailed suggestions) that Statistics education should:

1. Emphasize statistical literacy and develop statistical thinking;
2. Use real data;
3. Stress conceptual understanding rather than mere knowledge of procedures;
4. Foster active learning;
5. Use technology for developing conceptual understanding and analyzing data;
6. Make assessment a part of the learning process

In this sense, this book is thoroughly modern.

But to be effective, a course must fit comfortably with the instructor's preferences. There are several equally effective pathways through this material depending on the emphasis a particular instructor wants to make. We describe some alternative orders that also work comfortably with these materials.

Flexible Syllabus. In *Business Statistics*, we have chosen to follow the GAISE Guidelines and the suggestions of the most innovative educators in Statistics education. These experts agree that it is best to expose students to real data early and often in the course and to emphasize real-world interpretations of analyses throughout. Because we are acutely aware of the challenge that Statistics instructors face in covering essential material in a limited period of time—often in a one-semester course—we have made two departures from the traditional Business Statistics sequence.

1. We have placed an introductory (exploratory, noninference-based) section on regression earlier in the text (Chapters 7 and 8).
2. We have shifted detailed discussion of probability and decision-making (Chapters 21 and 22) to later in the text. We realize that this order may not suit everyone's needs, so we have made these chapters as modular as possible—they may be picked up and covered at different points in the course.

Here are a few topic-sequencing options.

Focus on data and early regression coverage (with added emphasis on probability). Covering Chapter 21 before Chapter 9 allows the instructor to reference the binomial model when explaining the sampling distribution of the sample proportion. Note that this will introduce the Normal model theoretically; calculations using normal tables are in Chapter 9.

[1] www.amstat.org/education/gaise

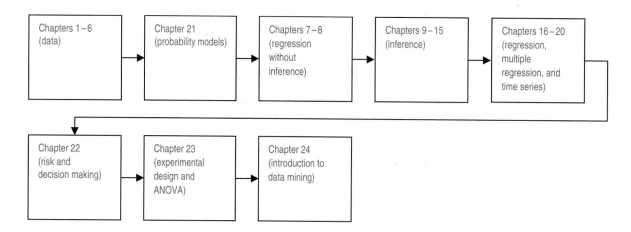

For this approach in a one-semester course, we recommend Chapters 1–6, Chapter 21 (probability models), Chapters 7–8 (regression without inference), Chapters 9–15, and Chapters 16–17 (regression with inference). Chapter 21 can also be covered after Chapter 8, just before the inference chapters, instead of directly after Chapter 6.

Focus on probability with later coverage of regression.   To focus on probability and cover regression later in the course, Chapters 21 and 22 can be covered after Chapters 1–6. One possible sequence is:

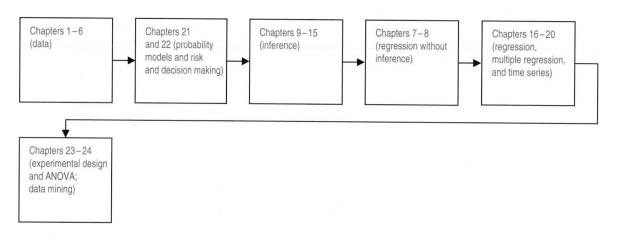

For this approach in a one-semester course, we recommend Chapters 1–6, Chapter 21 (probability models), Chapters 9–15, Chapters 7–8 (regression without inference), and Chapters 16–17 (regression with inference).

Focus on data and early regression coverage without emphasis on probability.   This is our standard sequence. To cover this in a one-semester course, we recommend the following sequence.

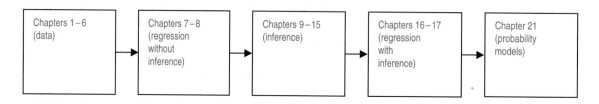

Features

A textbook isn't just words on a page. A textbook is many features that come together to form a big picture. The features in *Business Statistics* provide a real-world context for concepts, help students apply these concepts, promote problem-solving, and integrate technology—all of which help students understand and see the big picture of Business Statistics.

Motivating Examples. Each chapter opens with a motivating example, often taken from the authors' consulting experiences. These companies—such as Amazon.com, Zillow.com, Keen Inc., and Whole Foods Market—enhance and illustrate the story of each chapter and show students how and why statistical thinking is so vital to modern business decision-making. We analyze the data from those companies throughout the chapter.

Step-by-Step Guided Examples. The ability to clearly communicate statistical results is crucial to helping Statistics contribute to business decision-making. To that end, some examples in each chapter are presented as Guided Examples. A good solution is modeled in the right column while commentary appears in the left column. The overall analysis follows our innovative **Plan, Do, Report** template. That template begins each analysis with a clear question about a decision and ends with a report that answers that question. To emphasize the decision aspect of each example, we present the **Report** step as a business memo that summarizes the results in the context of the example and states a recommendation if the data are able to support one. In addition, whenever possible we include limitations of the analysis or models in the concluding memo.

PLAN

DO

REPORT

Mini Case Study Projects. Each chapter includes one or two *Mini Case Study Projects* that use real data and ask students to investigate a question or make a decision. Students define the objective, plan the process, complete the analysis, and report a conclusion. Data for the *Mini Case Study Projects* are available on the DVD and website, formatted for various technologies.

What Can Go Wrong? Each chapter contains an innovative section called *What Can Go Wrong?* which highlights the most common statistical errors and the misconceptions people have about Statistics. The most common mistakes for the new user of statistical methods involve misusing a method, not miscalculating a statistic. Most of the mistakes we discuss have been experienced by the authors in a business context rather than a classroom situation. One of our goals is to arm students with the tools to detect statistical errors and to offer practice in debunking misuses of Statistics, whether intentional or not. In this spirit, some of our exercises probe the understanding of such errors.

By Hand. Even though we encourage the use of technology to calculate statistical quantities, we realize the pedagogical benefits of doing a calculation by hand. The *By Hand* boxes break apart the calculation of some of the simpler formulas and help the student through the calculation of a worked example.

Reality Check. We regularly remind students that Statistics is about understanding the world and making decisions with data. Results that make no sense are probably wrong, no matter how carefully we think we did the calculations. Mistakes are often easy to spot with a little thought, so we ask students to stop for a reality check before interpreting results.

Notation Alert. Throughout this book, we emphasize the importance of clear communication. Proper notation is part of the vocabulary of Statistics, but it can be daunting. Students who know that in Algebra n can stand for any variable may be surprised to learn that in Statistics n is always and only the sample size. Statisticians dedicate many letters and symbols for specific meanings (b, e, n, p, q, r, s, t, and z, along with many Greek letters all carry special connotations). Students learn more effectively when they are clear about the letters and symbols statisticians use.

Just Checking. To help students check their understanding of material they've just read, we ask questions at points throughout the chapter. These questions are a quick check; most involve very little calculation. The answers are at the end of the exercise sets in each chapter so students can easily check themselves to be sure they understand the key ideas. The questions can also be used to motivate class discussion.

MATH BOX

Math Boxes. In many chapters, we present the mathematical underpinnings of the statistical methods and concepts. Different students learn in different ways, and even the same student can understand the material by more than one path. By setting these proofs, derivations, and justifications apart from the narrative, we allow the student to continue to follow the logical development of the topic at hand, yet also make available the underlying mathematics for greater depth.

What Have We Learned? These chapter-ending summaries highlight new concepts, define new terms introduced in the chapter, and list the skills that the student should have acquired. Students can think of these as study guides. If they understand the concepts in the summary, know the terms, and have the skills, they're probably ready for the exam.

ETHICS IN ACTION

Ethics in Action. Students are often surprised to learn that Statistics is not just plugging numbers into formulas. Most statistical analyses require a fair amount of judgment. The best guidance for these judgments is that we make an honest and ethical attempt to learn the truth. Anything less than that can lead to poor and even dangerous decisions. Our *Ethics in Action* vignettes in each chapter illustrate some of the judgments needed in statistical analyses, identify possible errors, link the issues to the American Statistical Association's Ethical Guidelines, and then propose ethically and statistically sound alternative approaches.

Exercises. We've worked hard to make sure that exercises contain relevant, modern, and real-world questions. Many come from news stories; some come from recent research articles. Whenever possible, the data are on the disk and website (always in a variety of formats) so students can explore them further. The exercises marked with a ⓣ indicate that the data are provided. Sometimes, because of the size of the data set, the data are only available electronically. Throughout, we pair the exercises so that each odd-numbered exercise (with answer in the back of the book) is followed by an even-numbered exercise on the same Statistics topic. Exercises are roughly ordered within each chapter by both topic and by level of difficulty.

Data Sources. Most of the data used in examples and exercises are from real-world sources, and we list many sources in this edition. Whenever we can, we include references to the Internet data sources used, often in the form of URLs. As Internet users (and thus, our students) know well, URLs can "break" as websites evolve. To minimize the impact of such changes, we point as high in the address tree as is practical. Moreover, the data themselves often change as more recent values become available. The data we use are usually on the DVD packaged with new copies of *Business Statistics* and on the companion website, www.aw.com/sharpe. If you seek the data—or an updated version of the data—on the Internet, we try to direct you to a good starting point.

Videos on DVD with Optional Captioning. These videos, featuring *Business Statistics* authors, help students review the high points of each chapter, and concept videos (Business Insight Videos) focus on statistical concepts as they pertain to the real world. The presentations feature the same student-friendly style and emphasis on critical thinking as the textbook. The DVD format makes it easy and convenient to watch the videos from a computer at home or on campus. Videos are available with captioning. They also can be viewed from within the online MyStatLab course.

Technology Help. In business, Statistics is practiced with computers, but not with a single software platform. Instead of emphasizing a particular Statistics program, at the end of each chapter, we summarize what students can find in the most common packages, often with annotated output. We then offer specific guidance for several of the most common packages (Excel 2007 and 2003, Minitab, SPSS, JMP, and Data Desk) to help students get started with the software of their choice.

Supplements

Student Supplements

Business Statistics, for-sale student edition (ISBN-13: 978-0-321-42659-8; ISBN-10: 0-321-42659-2)

Student's Solutions Manual, by Linda Dawson, University of Washington, provides detailed, worked-out solutions to odd-numbered exercises. (ISBN-13: 978-0-321-50691-7; ISBN-10: 0-321-50691-X)

Excel Manual, by Jim Zimmer, Chattanooga State University (ISBN-13: 978-0-321-57135-9; ISBN-10: 0-321-57135-5)

Minitab Manual, by Robert H. Carver, Stonehill College (ISBN-13: 978-0-321-57059-8; ISBN-10: 0-321-57059-6)

SPSS Manual, by Rita Akin, Westminster College (ISBN-13: 978-0-321-57136-6; ISBN-10: 0-321-57136-3)

Instructor Supplements

Instructor's Edition contains answers to all exercises. (ISBN-13: 978-0-321-50566-8; ISBN-10: 0-321-50566-2)

Online Test Bank (download only), by Rose Sebastianelli, University of Scranton, includes chapter quizzes and part level tests. The Test Bank is available at www.pearsonhighered.com/irc.

Instructor's Solutions Manual, by Linda Dawson, University of Washington, contains detailed solutions to all of the exercises. (ISBN-13: 978-0-321-50686-3; ISBN-10: 0-321-50686-3)

TestGen® (www.pearsonhighered.com/testgen) enables instructors to build, edit, print, and administer tests using a computerized bank of questions developed to cover all the objectives of the text. TestGen is algorithmically based, allowing instructors to create multiple but equivalent versions of the same question or test with the click of a button. Instructors can also modify test bank questions or add new questions. Tests can be printed or administered online. The software and test bank are available for download from Pearson Education's online catalog.

PowerPoint Lecture Slides provide an outline to use in a lecture setting, presenting definitions, key concepts, and figures from the text. These slides are available within MyStatLab or at www.pearsonhighered.com/irc.

Pearson Math Adjunct Support Center. The Pearson Math Adjunct Support Center (www.pearsontutorservices.com/math-adjunct.html) is staffed by qualified instructors with more than 50 years of combined experience at both the community college and university levels. Assistance is provided for faculty in the following areas:

- Suggested syllabus consultation
- Tips on using materials packed with your book
- Book-specific content assistance
- Teaching suggestions, including advice on classroom strategies

Technology Resources

A companion DVD is bound in new copies of *Business Statistics*. The DVD holds a number of supporting materials, including:

- **Data** for exercises marked **T** are available on the DVD and website formatted for DataDesk, Excel, JMP, MINITAB, SPSS, and as text files suitable for these and virtually any other statistics software.
- **DDXL,** an Excel add-in, adds sound statistics and statistical graphics capabilities to Excel. DDXL adds, among other capabilities, boxplots, histograms, statistical scatterplots, normal probability plots, and statistical inference procedures not available in Excel's Data Analysis pack.

ActivStats® for Business Statistics (Mac and PC). The award-winning ActivStats multimedia program supports learning chapter by chapter with the book. It complements the book with videos of real-world stories, worked examples, animated expositions of each of the major Statistics topics, and tools for performing simulations, visualizing inference, and learning to use statistics software. ActivStats includes 15 short video clips; 183 animated activities and teaching applets; 260 data sets; interactive graphs, simulations, visualization tools, and much more. ActivStats (Mac and PC) is available in an all-in-one version for Data Desk, Excel, JMP, MINITAB, and SPSS. (ISBN-13: 978-0-321-57719-1; ISBN-10: 0-321-57719-1)

MathXL® for Statistics is a powerful online homework, tutorial, and assessment system that accompanies Pearson textbooks in Statistics. With MathXL for Statistics, instructors can create, edit, and assign online homework and tests using algorithmically generated exercises correlated at the objective level to the textbook. They can also create and assign their own online exercises and import TestGen tests for added flexibility. All student work is tracked in MathXL's online gradebook. Students can take chapter tests in MathXL and receive personalized study plans based on their test results. The study plan diagnoses weaknesses and links students directly to tutorial exercises for the objectives they need to study and retest. Students can also access supplemental animations directly from selected exercises. MathXL for Statistics is available to qualified adopters. For more information, visit our website at www.mathxl.com, or contact your sales representative.

MyStatLab™ —part of the MyMathLab® product family— is a text-specific, easily customizable online course that integrates interactive multimedia instruction with textbook content. MyStatLab gives you the tools you need to deliver all or a portion of your course online, whether your students are in a lab setting or working from home.

- **Interactive Tutorial Exercises:** A comprehensive set of exercises—correlated to your textbook at the objective level—are algorithmically generated for unlimited practice and mastery. Most exercises are free-response and provide guided solutions, sample problems, and learning aids for extra help at point-of-use.
- **Personalized Study Plan:** When students complete a test or quiz in MyStatLab, the program generates a personalized study plan for each student that indicates which topics have been mastered and links students directly to tutorial exercises for topics they need to study and retest
- **Multimedia Learning Aids:** Students can use online learning aids, such as video lectures, animations, and a complete multimedia textbook, to help them independently improve their understanding and performance.
- **Statistics Tools:** MyStatLab includes built-in tools for Statistics, including statistical software called StatCrunch. Students also have access to statistics animations and applets that illustrate key ideas for the course. For those who use technology in their course, technology manual PDFs are included.
- **Assessment Manager:** An easy-to-use assessment manager lets instructors create online homework, quizzes, and tests that are automatically graded and correlated directly to your textbook. Assignments can be created using a mix of questions from the MyStatLab exercise bank, instructor-created custom exercises, and/or TestGen test items.
- **Gradebook:** Designed specifically for mathematics and statistics, the MyStatLab gradebook automatically tracks students' results and gives you control over how to calculate final grades. You can also add offline (paper-and-pencil) grades to the gradebook.
- **Math Exercise Builder:** You can use the MathXL Exercise Builder to create static and algorithmic exercises for your online assignments. A library of sample exercises provides an easy starting point for creating questions, and you can also create questions from scratch.
- **Pearson Tutor Center** (www.pearsontutorservices.com): Access is automatically included with MyStatLab. The Tutor Center is staffed by qualified mathematics instructors who provide textbook-specific tutoring for students via toll-free phone, fax, e-mail, and interactive Web sessions.

MyStatLab is powered by CourseCompass™, Pearson Education's online teaching and learning environment, and by MathXL®, our online homework, tutorial, and assessment system. MyStatLab is available to qualified adopters. For more information about MyStatLab, visit our website at www.mystatlab.com or contact your Pearson sales representative.

StatCrunch is a powerful online tool that provides an interactive environment for doing statistics. You can use StatCrunch for both numerical and graphical data analysis, taking advantage of interactive graphics to help you see the connection between objects selected in a graph and the underlying data. In MyStatLab, the data sets from your textbook are preloaded into StatCrunch. StatCrunch is also available as a tool from the online homework and practice exercises in MyStatLab and in MathXL for Statistics. Also available is Statcrunch.com, web-based software that allows students to perform complex statistical analysis in a simple manner.

Videos on DVD with Optional Captioning feature the textbook authors reviewing the high points of each chapter, and **Business Insight Videos** focus on statistical concepts as they pertain to the real world. The DVD format makes it easy and convenient to watch the videos from a computer at home or on campus. The videos can also be downloaded to Video iPods® from within MyStatLab. Contact your Pearson representative for details. (ISBN 13: 978-0-321-57134-2; ISBN-10: 0-321-57134-7)

PowerPoint Lecture Slides provide an outline to use in a lecture setting, presenting definitions, key concepts, and figures from the text. These slides are available within MyStatLab or at www.pearsonhighered.com/irc.

Companion Website (www.aw.com/sharpe) provides additional resources for instructors and students.

Active Learning Questions. Prepared in PowerPoint®, these questions are intended for use with classroom response systems. Several multiple-choice questions are available for each chapter of the book, allowing instructors to quickly assess mastery of material in class. The Active Learning Questions are available to download from within MyStatLab® and from Pearson Education's online catalog.

The **Student Edition of Minitab** is a condensed edition of the Professional release of Minitab statistical software that offers the full range of statistical methods and graphical capabilities, along with worksheets that can include up to 10,000 data points. Individual copies of the software can be bundled with the text. (ISBN 13: 978-0-321-11313-9; ISBN-10: 0-321-11313-6)

JMP Student Edition is an easy-to-use, streamlined version of JMP desktop statistical discovery software from SAS, Institute, Inc. and is available for bundling with the text. (ISBN 13: 978-0-321-51738-8; ISBN-10: 0-321-51738-5)

SPSS, a statistical and data management software package, is also available for bundling with the text. (ISBN 13: 978-0-13-605348-4; ISBN-10: 0-13-605348-3)

Acknowledgements

Many people have contributed to this book from the first day of its conception to its publication. *Business Statistics* would have never seen the light of day without the assistance of the incredible team at Pearson. Our Editor in Chief, Deirdre Lynch, was central to the support, development, and realization of the book from day one. Sara Oliver Gordus, Associate Editor, and Chere Bemelmans, Senior Project Editor, kept us on task as much as humanly possible. Peggy McMahon, Senior Production Supervisor, and Laura Hakala, Senior Project Manager at Pre-Press PMG, worked miracles to get the book out the door. We are indebted to them. Christina Lepre, Assistant Editor; Dana Jones, Editorial Assistant; Alex Gay, Marketing Manager; Kathleen DeChavez, Marketing Associate; and Dona Kenly, Senior Market Development Manager, were essential in managing all of the behind-the-scenes work that needed to be done. Christine Stavrou, Media Producer, put together a top-notch media package for this book. Barbara Atkinson, Senior Designer, and Geri Davis are responsible for the wonderful way the book looks. Evelyn Beaton, Manufacturing Manager, and Ginny Michaud, Senior Manufacturing Buyer, worked miracles to get this book and DVD in your hands, and Greg Tobin, President, was supportive and good-humored throughout all aspects of the project.

Special thanks go out to Pre-PressPMG, the compositor, for the wonderful work they did on this book and in particular to Laura Hakala, the project manager, for her close attention to detail.

We'd also like to thank our accuracy checkers whose monumental task was to make sure we said what we thought we were saying: Eugene Allevato, Woodbury University; Dave Bregenzer, Utah State University; Ann Cannon, Cornell College; Joan Donohue, University of South Carolina; David Doorn, University of Minnesota, Duluth; David Hudgins, University of Oklahoma, Norman; John Lawrence, California State University, Fullerton; Joe Kupresanin, Cecil College; Monnie McGee, Southern Methodist University; Jackie Miller, The Ohio State University; Doug Morris, University of New Hampshire; Michael Polomsky, Cleveland State University; Gary Smith, Florida State University; Joe Sullivan, Mississippi State University; Dirk Tempelaar, Maastricht University; William Warde, Oklahoma State University; and Jim Zimmer, Chattanooga State University.

We wish to thank the following individuals who joined us for a weekend to discuss business statistics education, emerging trends, technology, and business ethics. These individuals made invaluable contributions to *Business Statistics*:

Dr. Taiwo Amoo, CUNY Brooklyn
Dave Bregenzer, Utah State University
Joan Donohue, University of South Carolina
Soheila Fardanesh, Towson University
Chun Jin, Central Connecticut State University
Brad McDonald, Northern Illinois University
Amy Luginbuhl Phelps, Duquesne University
Michael Polomsky, Cleveland State University
Robert Potter, University of Central Florida
Rose Sebastianelli, University of Scranton
Debra Stiver, University of Nevada, Reno
Minghe Sun, University of Texas—San Antonio
Mary Whiteside, University of Texas—Arlington

We also thank those who provided feedback through focus groups, class tests, and reviews:

Alabama: Nancy Freeman, Shelton State Community College; Rich Kern, Montgomery County Community College; Robert Kitahara, Troy University; Tammy Prater, Alabama State University **Arizona:** Kathyrn Kozak, Coconino Community College; Robert Meeks, Pima Community College; Philip J. Mizzi, Arizona State University; Yvonne Sandoval, Pima Community College; Alex Sugiyama, University of Arizona **California:** Eugene Allevato, Woodbury University; Randy Anderson, California State University, Fresno; Paul Baum, California State University, Northridge; Giorgio Canarella, California State University, Los Angeles; Natasa Christodoulidou, California State University, Dominguez Hills; Abe Feinberg, California State University, Northridge; Bob Hopfe, California State University, Sacramento; John Lawrence, California State University, Fullerton; Elaine McDonald-Newman, Sonoma State University; Khosrow Moshirvaziri, California State University; Sunil Sapra, California State University, Los Angeles; Carlton Scott, University of California, Irvine; Yeung-Nan Shieh, San Jose State University; Dr. Rafael Solis, California State University, Fresno; T. J. Tabara, Golden Gate University; Dawit Zerom, California State University, Fullerton **Canada:** Jianan Peng, Acadia University; Brian E. Smith, McGill University **Colorado:** Sally Hay, Western State College; Rutilio Martinez, University of Northern Colorado; Gerald Morris, Metropolitan State College of Denver; Charles Trinkel, DeVry University, Colorado **Connecticut:** Judith Mills, Southern Connecticut State University; William Pan, University of New Haven; Frank Bensics, Central Connecticut State University; Lori Fuller, Tunxis Community College; Chun Jin, Central Connecticut State University; Jason Molitierno, Sacred Heart University **Florida:** David Afshartous, University of Miami; Dipankar Basu, Miami University; Ali Choudhry, Florida International University; Nirmal Devi, Embry Riddle Aeronautical University; Dr. Chris Johnson, University of North Florida; Robert Potter, University of Central Florida; Gary Smith, Florida State University; Roman Wong, Barry University **Georgia:** Dr. Michael Deis, Clayton University; Swarna Dutt, State University of West Georgia; John Grout, Berry College; Michael Parzen, Emory University; Barbara Price, Georgia Southern University **Idaho:** Craig Johnson, Brigham Young University; Teri Peterson, Idaho State University; Dan Petrak, Des Moines Area Community College **Illinois:** Lori Bell, Blackburn College; Jim Choi, DePaul University; David Gordon, Illinois Valley Community College; John Kriz, Joliet Junior College; Constantine Loucopoulos, Northeastern Illinois University; Brad McDonald, Northern Illinois University; Ozgur Orhangazi, Roosevelt University **Indiana:** H. Lane David, Indiana University South Bend; Ting Liu, Ball State University; Constance McLaren, Indiana State University; Dr. Ceyhun Ozgur, Valparaiso University; Hedayeh Samavati, Indiana University, Purdue; Mary Ann Shifflet, University of Southern Indiana; Cliff Stone, Ball State University; Sandra Strasser, Valparaiso University **Iowa:** Ann Cannon, Cornell College; Timothy McDaniel, Buena Vista University; Dan Petrack, Des Moines Area Community College; Mount Vernon, Iowa; Osnat Stramer, University of Iowa; Bulent Uyar, University of Northern Iowa; Blake Whitten, University of Iowa **Kansas:** John E. Boyer, Jr., Kansas State University **Louisiana:** Zhiwei Zhu, University of Louisiana at Lafayette **Maastricht, The Netherlands** Dirk Tempelaar, Maastricht University **Maryland:** John F. Beyers, University of Maryland University College; Deborah Collins, Anne Arundel Community College; Frederick W. Derrick, Loyola College in Maryland; Soheila Fardanesh, Towson University; Dr. Jeffery Michael, Towson University; Dr. Timothy Sullivan, Towson University **Massachusetts:** Elaine Allen, Babson College; Paul D. Berger, Bentley College; Scott Callan, Bentley College; Ken Callow, Bay Path College; Robert H. Carver, Stonehill College; Richard Cleary, Bentley College; Ismael Dambolena,

Babson College; Steve Erikson, Babson College; Elizabeth Haran, Salem State College; David Kopcso, Babson College; Supriya Lahiri, University of Massachusetts, Lowell; John MacKenzie, Babson College; Dennis Mathaisel, Babson College; Abdul Momen, Framingham State University; Ken Parker, Babson College; John Saber, Babson College; Ahmad Saranjam, Bridgewater State College; Daniel G. Shimshak, University of Massachusetts, Boston; Erl Sorensen, Bentley College; Denise Sakai Troxell, Babson College; Janet M. Wagner, University of Massachusetts, Boston; Elizabeth Wark, Worcester State College; Fred Wiseman, Northeastern University **Michigan:** Sheng-Kai Chang, Wayne State University, **Minnesota:** Daniel G. Brick, University of St. Thomas; Dr. David J. Doorn, University of Minnesota Duluth; Howard Kittleson, Riverland Community College; Craig Miller, Normandale Community College **Mississippi:** Dal Didia, Jackson State University; J. H. Sullivan, Mississippi State University; Wenbin Tang, The University of Mississippi **Missouri:** Emily Ross, University of Missouri, St. Louis **Nevada:** Debra K. Stiver, University of Nevada, Reno; Grace Thomson, Nevada State College **New Hampshire:** Parama Chaudhury, Dartmouth College; Doug Morris, University of New Hampshire **New Jersey:** Kunle Adamson, DeVry University; Dov Chelst, DeVry University—New Jersey; Leonard Presby, William Paterson University; Subarna Samanta, The College of New Jersey **New York:** Dr. Taiwo Amoo, City University of New York, Brooklyn; Bernard Dickman, Hofstra University; Mark Marino, Niagara University **North Carolina:** Margaret Capen, East Carolina University; Warren Gulko, University of North Carolina, Wilmington; Geetha Vaidyanathan, University of North Carolina **Ohio:** David Booth, Kent State University, Main Campus; Arlene Eisenman, Kent State University; Michael Herdlick, Tiffin University; Joe Nowakowski, Muskingum College; Jayprakash Patankar, The University of Akron; Michael Polomsky, Cleveland State University; Anirudh Ruhil, Ohio University; Bonnie Schroeder, Ohio State University; Gwen Terwilliger, University of Toledo; Yan Yu, University of Cincinnati **Oklahoma:** Anne M. Davey, Northeastern State University; Damian Whalen, St. Gregory's University; David Hudgins, University of Oklahoma—Norman; Dr. William D. Warde, Oklahoma State University—Main Campus **Oregon:** Jodi Fasteen, Portland State University **Pennsylvania:** Dr. Deborah Gougeon, University of Scranton; Rose Sebastianelli, University of Scranton; Jack Yurkiewicz, Pace University; Rita Akin, Westminster College; H. David Chen, Rosemont College; Laurel Chiappetta, University of Pittsburgh; Burt Holland, Temple University; Ronald K Klimberg, Saint Joseph's University; Amy Luginbuhl Phelps, Duquesne University; Sherryl May, University of Pittsburg—KGSB; Dr. Bruce McCullough, Drexel University; Tracy Miller, Grove City College; Heather O'Neill, Ursinus College; Tom Short, Indiana University of Pennsylvania; Keith Wargo, Philadelphia Biblical University **Rhode Island:** Paul Boyd, Johnson & Wales University; Jeffrey Jarrett, University of Rhode Island **South Carolina:** Karie Barbour, Lander University; Joan Donohue, University of South Carolina; Woodrow Hughes, Jr., Converse College; Willis Lewis, Lander University; M. Patterson, Midwestern State University; Kathryn A. Szabat, LaSalle University **Tennessee:** Ferdinand DiFurio, Tennessee Technical University; Farhad Raiszadeh, University of Tennessee—Chattanooga; Scott J. Seipel, Middle Tennessee State University; Han Wu, Austin Peay State University; Jim Zimmer, Chattanooga State University **Texas:** Raphael Azuaje, Sul Ross State University; Mark Eakin, University of Texas—Arlington; Betsy Greenberg, University of Texas—Austin; Daniel Friesen, Midwestern State University; Erin Hodgess, University of Houston—Downtown; Joyce Keller, St. Edward's University; Gary Kelley, West Texas A&M University; Monnie McGee, Southern Methodist University; John M. Miller, Sam Houston State University; Carolyn H. Monroe, Baylor University; Ranga Ramasesh, Texas Christian University; Plamen Simeonov, University of Houston—Downtown; Lynne Stokes, Southern Methodist University; Minghe Sun, University of Texas—San Antonio; Rajesh Tahiliani. University of Texas—El Paso; Mary Whiteside, University of Texas—Arlington; Stuart Warnock, Tarleton State University **Utah:** Dave

Bregenzer, Utah State University; Camille Fairbourn, Utah State University **Virginia:** Sidhartha R. Das, George Mason University; Quinton J. Nottingham, Virginia Polytechnic & State University; Ping Wang, James Madison University **Washington:** Nancy Birch, Eastern Washington University; Mike Cicero, Highline Community College; Fred DeKay, Seattle University; Stergios Fotopoulous, Washington State University; Teresa Ling, Seattle University **West Virginia:** Clifford Hawley, West Virginia University **Wisconsin:** Nancy Burnett University of Wisconsin—Oshkosh; Thomas Groleau, Carthage College; Patricia Ann Mullins, University of Wisconsin, Madison.

Finally, want to thank our families. This has been a long project, and it has required many nights and weekends. Our families have sacrificed so that we could write the book we envisioned.

Norean Sharpe
Richard De Veaux
Paul Velleman

Index of Applications

IE = In-Text Example; BE = Boxed Example; GE = Guided Example; EIA = Ethics in Action; E = Exercises; JC = Just Checking; P = Project

Accounting

Administrative and Training Costs (E) 78, 367–368, 431–432
Annual Reports (E) 75–76
Audits and Tax Returns (E) 106, 278, 343, 368, 431–432, 740
Bankruptcy (E) 399
Bookkeeping (IE) 12; (E) 50, 308, 310
Budgets (E) 254, 338
Company Assets, Profit, and Revenue (IE) 4, 9, 15, 159, 194, 226, 346, 489–491, 548; (E) 75, 77–78, 81, 153–154, 188, 217, 398–399, 465, 469, 470, 541, 543, 579, 626, 670; (BE) 553
Cost Cutting (E) 430, 433
CPAs (E) 106, 343
Earnings per Share Ratio (E) 371
Expenses (IE) 12, 16; (E) 501–502
Financial Close Process (E) 372
IT Costs and Policies (E) 431–432
Legal Accounting and Reporting Practices (E) 431–432
Purchase Records (IE) 11, 12–13, 194; (E) 50, 152–153, 255

Advertising

Ads (E) 109, 274, 308, 311, 313, 372–374, 389, 539; (IE) 267, 376
Advertising in Business (IE) 4, 13, 161, 257; (E) 22, 77–78, 82–83, 109, 218, 372, 399–400, 467, 504, 539, 695–696; (GE) 93; (EIA) 177, 573; (BE) 286
Branding (E) 372–373; (IE) 375, 474, 708, 719–721; (GE) 723–727
Coupons (EIA) 302; (IE) 705, 710, 712–713, 717–718
Free Products (IE) 288, 328, 710, 712–713, 717–718; (E) 341
International Advertising (E) 108
Jingles (IE) 376
Product Claims (JC) 120; (E) 252, 254, 340–342, 374–375, 395–396, 398, 427, 429–430, 737; (BE) 347–348
Sexual Images in Advertising (E) 389, 397
Target Audience (E) 108–109, 217–218, 279, 309, 343, 370–371, 429–430, 632, 672, 737–738; (JC) 321; (IE) 376, 708; (EIA) 761
Truth in Advertising (E) 311, 393

Agriculture

Agricultural Discharge (EIA) 43; (E) 51
Beef and Livestock (E) 336–337, 542–543
Coffee Farmers (EIA) 136
Dairy Farmers (E) 393
Drought and Crop Losses (E) 390
Farmers' Markets (E) 671
Fruit Growers (E) 507
Lawn Equipment (E) 696–697
Lobster Fishing Industry (E) 505–506, 508, 541–542, 582–583, 586–587
Lumber (E) 25, 507
Poultry Farmers (E) 253, 389
Seeds (E) 254, 311

Banking

Annual Percentage Rate (P) 668–669; (IE) 709
ATMs (IE) 345
Bank Tellers (E) 739
Certificates of Deposit (CDs) (P) 668–669
Collection Agencies (E) 341
Credit Card Charges (GE) 123–124, 295–297, 382–385; (E) 144, 217, 279, 337, 471–472; (IE) 226, 479–480; (P) 389
Credit Card Companies (IE) 17, 86, 225–227, 268, 345–347, 479–480, 699–701, 747–749; (P) 22; (GE) 123–124, 295–297, 350–355; (E) 217, 278–279, 313, 337, 341; (BE) 269; (JC) 295, 299
Credit Card Customers (GE) 123–124, 295–297, 350–355, 382–385; (E) 217, 278, 337, 341, 432, 672; (IE) 225–227, 237, 268, 346–347, 479–480, 700–701; (BE) 269; (JC) 295, 299; (P) 389
Credit Card Debt (JC) 295, 299; (E) 341, 373
Credit Card Offers (IE) 17, 86–87, 92, 268, 346–347, 479–480, 702–703, 709, 719–721; (P) 22; (GE) 169–170, 295–297, 350–355, 706–707, 723–727; (BE) 269; (E) 279, 313
Credit Unions (EIA) 271
Credit Scores (IE) 85–86
Federal Reserve Bank (E) 104; (P) 181
Interest Rates (E) 104, 182, 184–185, 502–503, 630, 635–636; (P) 182, 668–669; (IE) 171, 225, 705
Investment Banks (E) 695

Liquid Assets (E) 626
Loan Approval (E) 254
Maryland Bank National Association (IE) 225–226
Mortgages (E) 24, 182, 184–185; (GE) 238–239
Subprime Loans (IE) 17, 86
World Bank (E) 154, 215

Business (General)

Attracting New Business (E) 342–343
Best Places to Work (IE) 127–129; (E) 434, 468
Business Planning (IE) 9, 297
Chief Executives (E) 109–110, 152, 186, 252, 337, 432; (IE) 132–133, 152, 160, 320–321
Company Case Reports and Lawyers (IE) 5; (GE) 233
Company Databases (IE) 15, 17, 258
Contract Bids (E) 670, 698
Ease of Doing Business (E) 154
Elder Care Business (EIA) 456
Enterprise Resource Planning (E) 372, 434–435
Entrepreneurial Skills (E) 432
Forbes 500 Companies (BE) 488
Fortune 500 Companies (IE) 132, 320–321, 699; (E) 337, 465
Franchises (EIA) 208, 456; (BE) 553
Industry Sector (E) 433
International Business (IE) 28; (P) 47; (E) 48, 74–75, 83, 105, 276
Job Growth (E) 434, 468
Outside Consultants (IE) 68
Outsourcing (E) 433
Research and Development (IE) 9–10; (E) 78; (JC) 382
Small Business (IE) 10, 675–676; (E) 76, 78, 107, 144, 277, 338, 393, 399, 432–433, 670, 696–697; (EIA) 136; (E) 501–502, 539–540, 583
Start-Up Companies (E) 23, 147–148, 154, 279, 311, 695; (EIA) 208
Trade Secrets (IE) 437
Women-Led Businesses (E) 75, 81–82, 277, 311

Company Names

Adair Vineyards (E) 144–146
Alpine Medical Systems, Inc. (EIA) 532
Amazon.com (IE) 9–11, 14–15
American Express (IE) 345

Consumers

Demographics

PART I

Exploring and Collecting Data

CHAPTER 1

Statistics and Variation

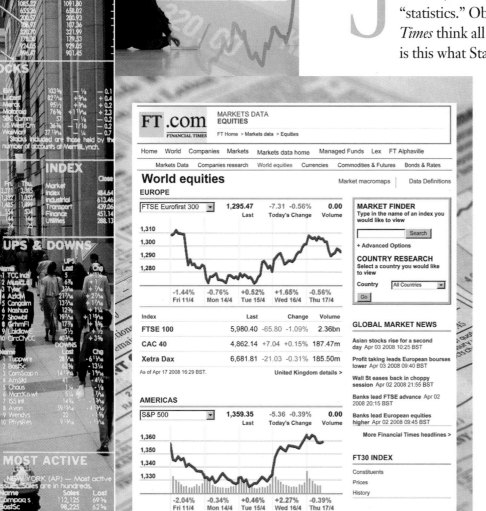

J ust look at a page from the *Financial Times* website, like the one shown here. It's full of "statistics." Obviously, the writers of the *Financial Times* think all this information is important, but is this what Statistics is all about? Well, yes and no. This page may contain a lot of facts, but as we'll see, the subject is much more interesting and rich than just spreadsheets and tables.

"Why should I learn Statistics?" you might ask. "After all, I don't plan to do this kind of work. In fact, I'm going to hire people to do this stuff." That's fine. But the decisions you make based on data are too important to delegate. You'll want to be able to interpret the data that surrounds you and to come to your own conclusions. And you'll find that studying Statistics is much more important and enjoyable than you thought.

So, What Is Statistics?

Q: What is Statistics?

A: Statistics is a way of reasoning, along with a collection of tools and methods, designed to help us understand the world.

Q: What are statistics?

A: Statistics (plural) are quantities calculated from data.

Q: So what is data?

A: You mean, "what *are* data?" Data is the plural form. The singular is datum.

Q: So, what are data?

A: Data are values along with their context.

It seems every time we turn around, someone is collecting data on us, from every purchase we make in the grocery store to every click of our mouse as we surf the Web. The United Parcel Service (UPS) tracks every package it ships from one place to another around the world and stores these records in a giant database. You can access part of it if you send or receive a UPS package. The database is about 17 terabytes—about the same size as a database that contained every book in the Library of Congress would be. (But, we suspect, not quite as interesting.) What can anyone hope to do with all these data?

Statistics plays a role in making sense of our complex world. Statisticians assess the risk of genetically engineered foods or of a new drug being considered by the Food and Drug Administration (FDA). Statisticians predict the number of new cases of AIDS by regions of the country or the number of customers likely to respond to a sale at the supermarket. And statisticians help scientists, social scientists, and business leaders understand how unemployment is related to environmental controls, whether enriched early education affects the later performance of school children, and whether vitamin C really prevents illness. Whenever you have data and a need to understand the world, you need Statistics.

If we want to analyze student perceptions of business ethics (a question we'll come back to in a later chapter), should we administer a survey to every single university student in the United States—or, for that matter, in the world? Well, that wouldn't be very practical or cost effective. What should we do instead? Give up and abandon the survey? Maybe we should try to obtain survey responses from a smaller, representative group of students. Statistics can help us make the leap from the data we have at hand to an understanding of the world at large. We talk about the specifics of sampling in Chapter 3, and the theme of generalizing from the specific to the general is one that we revisit throughout this book. We hope this text will empower *you* to draw conclusions from data and make valid business decisions in response to such questions as:

◆ Do university students from different parts of the world perceive business ethics differently?

◆ What is the effect of advertising on sales?

◆ Do aggressive, "high-growth" mutual funds really have higher returns than more conservative funds?

◆ Is there a seasonal cycle in your firm's revenues and profits?

◆ What is the relationship between shelf location and cereal sales?

◆ How reliable are the quarterly forecasts for your firm?

◆ Are there common characteristics about your customers and why they choose your products?—and, more importantly, are those characteristics the same among those who aren't your customers?

Our ability to answer questions such as these and draw conclusions from data depends largely on our ability to understand *variation*. That may not be the term you expected to find at the end of that sentence, but it is the essence of Statistics. The key to learning from data is understanding the variation that is all around us.

Data vary. People are different. So are economic conditions from month to month. We can't see everything, let alone measure it all. And even what we do measure, we measure imperfectly. So the data we wind up looking at and basing our decisions on provide, at best, an imperfect picture of the world. Variation lies at the heart of what Statistics is all about. How to make sense of it is the central challenge of Statistics.

How Will This Book Help?

A fair question. Most likely, this book will not turn out to be what you expect. It emphasizes graphics and understanding rather than computation and formulas. Instead of learning how to plug numbers in formulas you'll learn the process of model development and come to understand the limitations both of the data you analyze and the methods you use. Every chapter uses real data and real business scenarios so you can see how to use data to make decisions.

Graphs

Close your eyes and open the book at random. Is there a graph or table on the page? Do it again, say, ten times. You probably saw data displayed in many ways, even near the back of the book and in the exercises. Graphs and tables help you understand what the data are saying. So, each story and data set and every new statistical technique will come with graphics to help you understand both the methods and the data.

Process

To help you use Statistics to make business decisions, we'll lead you through the entire process of thinking about a problem, finding and showing results, and telling others what you have discovered. The three simple steps to doing Statistics for business right are: **Plan**, **Do**, and **Report**.

 Plan first. Know where you're headed and why. Clearly defining and understanding your objective will save you a lot of work.

Do is what most students think Statistics is about. The mechanics of calculating statistics and making graphical displays are important, but the computations are usually the least important part of the process. In fact, we usually turn the computations over to technology and get on with understanding what the results tell us.

 **Report** what you've learned. Until you've explained your results in a context that someone else can understand, the job is not done.

> "Get your facts first, and then you can distort them as much as you please. (Facts are stubborn, but statistics are more pliable.)"
> —Mark Twain

Guided Example

Each chapter applies the new concepts taught in worked examples called **Guided Examples**. These examples model how you should approach and solve problems using the Plan, Do, Report framework. They illustrate how to plan an analysis, the appropriate techniques to use, and how to report what it all means. These step-by-step examples show you how to produce the kind of solutions and case study reports that instructors and managers or, better yet, clients expect to see. You will find a model solution in the right-hand column and background notes and discussion in the left-hand column.

JUST CHECKING

Sometimes, in the middle of the chapter, you'll find sections called **Just Checking**, which pose a few short questions you can answer without much calculation. Use them to check that you've understood the basic ideas in the chapter. You'll find the answers at the end of the chapter exercises.

Ethics in Action

Statistics often requires judgment, and the decisions based on statistical analyses may influence people's health and even their lives. Decisions in government can affect policy decisions about how people are treated. In science and industry, interpretations of data can influence consumer safety and the environment. And in business, misunderstanding what the data say can lead to disastrous decisions. The central guiding principle of statistical judgment is the ethical search for a true understanding of the real world. In all spheres of society it is vitally important that a statistical analysis of data be done in an ethical and unbiased way. Allowing preconceived notions, unfair data gathering, or deliberate slanting to affect statistical conclusions is harmful to business and to society.

At various points throughout the book, you will encounter a scenario under the title **Ethics in Action** in which you'll read about an ethical issue. Think about the issue and how you might deal with it. Then read the summary of the issue and one solution to the problem, which follow the scenario. We've related the ethical issues to guidelines that the American Statistical Association has developed.[1] These scenarios can be good topics for discussion. We've presented one solution, but we invite you to think of others.

What Can Go Wrong?

One of the interesting challenges of Statistics is that, unlike some math and science courses, there can be more than one right answer. This is why two statisticians can testify honestly on opposite sides of a court case. And it's why some people think that you can prove anything with statistics. But that's not true. People make mistakes using statistics, and sometimes people misuse statistics to mislead others. Most of the mistakes are avoidable. We're not talking about arithmetic. Mistakes usually involve using a method in the wrong situation or misinterpreting results. So each chapter has a section called **What Can Go Wrong?** to help you avoid some of the most common mistakes that we've seen in our years of consulting and teaching experience.

Mini Case Study Projects

At the end of each chapter you'll find an extended problem or two that use real data and ask you to investigate a question or make a decision. These mini case studies are a good way to test your ability to attack an open-ended (and thus more realistic) problem. You'll be asked to define the objective, plan your process, complete the analysis, and report your conclusion. These are good opportunities to apply the template provided by the **Guided Examples**. And they provide an opportunity to practice reporting your conclusions in written form to refine your communication skills where statistical results are involved. Data sets for these case studies can be found on the disk included with this text.

Technology Help: Using the Computer

Although we show you all the formulas you need to understand the calculations, you will most often use a calculator or computer to perform the mechanics of a statistics problem. And the easiest way to calculate statistics with a computer is with a statistics package. Several different statistics packages are used widely. Although they differ in the details of how to use them, they all work from the same basic information and find the same results. Rather than adopt one package for this

> *"Far too many scientists have only a shaky grasp of the statistical techniques they are using. They employ them as an amateur chef employs a cookbook, believing the recipes will work without understanding why. A more cordon bleu attitude . . . might lead to fewer statistical soufflés failing to rise."*
> —The Economist, June 3, 2004, "Sloppy stats shame science."

---

[1] www.amstat.org/profession/index.cfm?fuseaction=ethicalstatistics

You'll find all sorts of stuff in margin notes, such as stories and quotations. For example:

"Computers are useless. They can only give you answers."

—Pablo Picasso

While Picasso underestimated the value of good statistics software, he did know that creating a solution requires more than just *Doing*—it means you have to *Plan* and *Report*, too!

book, we present generic output and point out common features that you should look for. We also give a table of instructions to get you started on five packages: Excel, Minitab, SPSS, JMP, and Data Desk.

> From time to time we'll take time out to discuss an interesting or important side issue. We indicate these by setting them apart like this.[2]

What Have We Learned?

At the end of each chapter, you'll see a brief summary of the important concepts in a section called **What Have We Learned?** That section includes a list of the **Terms** and a summary of the important **Skills** you've acquired in the chapter. You won't be able to learn the material from these summaries, but you can use them to check your knowledge of the important ideas in the chapter. If you have the skills, know the terms, and understand the concepts, you should be well prepared—and ready to use Statistics!

Exercises

Beware: No one can learn Statistics just by reading or listening. The only way to learn it is to do it. So, at the end of each chapter (except this one) you'll find **Exercises** designed to help you learn to use the Statistics you've just read about. Some exercises are marked with a red ⓣ. You'll find the data for these exercises on the book's website, www.aw-bc.com/sharpe or on the book's disk, so you can use technology as you work the exercises.

We've paired up and grouped the exercises, so if you're having trouble doing an exercise, you'll find a similar exercise either just before or just after it. You'll find answers to the odd-numbered exercises at the back of the book. But these are only "answers" and not complete solutions. What's the difference? The answers are sketches of the complete solutions. For most problems, your solution should follow the model of the **Guided Examples**. If your calculations match the numerical parts of the answer and your argument contains the elements shown in the answer, you're on the right track. Your complete solution should explain the context, show your reasoning and calculations, and state your conclusions. Don't worry too much if your numbers don't match the printed answers to every decimal place. Statistics is more than computation—it's about getting the reasoning correct—so pay more attention to how you interpret a result than to what the digit in the third decimal place is.

*Optional Sections and Chapters

Some sections and chapters of this book are marked with an asterisk (*). These are optional in the sense that subsequent material does not depend on them directly. We hope you'll read them anyway, as you did this section.

Getting Started

It's only fair to warn you: You can't get there by just picking out the highlighted sentences and the summaries. This book is different. It's not about memorizing definitions and learning equations. It's deeper than that. And much more interesting. But . . .

You have to read the book!

---

[2] Or in a footnote.

2

Data

Amazon.com

Amazon.com opened for business in July 1995, billing itself even then as "Earth's Biggest Bookstore," with an unusual business plan: They didn't plan to turn a profit for four to five years. Although some shareholders complained when the dotcom bubble burst, Amazon continued its slow, steady growth, becoming profitable for the first time in 2002. Since then, Amazon has remained profitable and has continued to grow. By 2004, they had more than 41 million active customers in over 200 countries and were ranked the 74th most valuable brand by *Business Week*. Their selection of merchandise has expanded to include almost anything you can imagine, from $400,000 necklaces, to yak cheese from Tibet, to the largest book in the world. In 2006, profits were $190 million—even after a $662 million charge for Research and Development.

Amazon R&D is constantly monitoring and evolving their website to best serve their customers and maximize their sales performance. To make changes to the site, they experiment by collecting data and analyzing what works best. As Ronny Kohavi, former director of Data Mining and Personalization, said, "Data trumps intuition. Instead of using our intuition, we experiment on the live site and let our customers tell us what works for them."

Amazon.com has recently stated "many of the important decisions we make at Amazon.com can be made with data. There is a right answer or a wrong answer, a better answer or a worse answer, and math tells us which is which. These are our favorite kinds of decisions."[1] While we might prefer that Amazon refer to these methods as Statistics instead of math, it's clear that data analysis, forecasting, and statistical inference are the core of the decision making tools of Amazon.com.

Many years ago, stores in small towns knew their customers personally. If you walked into the hobby shop, the owner might tell you about a new bridge that had come in for your Lionel train set. The tailor knew your dad's size, and the hairdresser knew how your mom liked her hair. There are still some stores like that around today, but we're increasingly likely to shop at large stores, by phone, or on the Internet. Even so, when you phone an 800 number to buy new running shoes, customer service representatives may call you by your first name or ask about the socks you bought six weeks ago. Or the company may send an e-mail in October offering new head warmers for winter running. This company has millions of customers, and you called without identifying yourself. How did the sales rep know who you are, where you live, and what you had bought?

The answer to all these questions is data. Collecting data on their customers, transactions, and sales lets companies track inventory and know what their customers prefer. These data can help them predict what their customers may buy in the future so they know how much of each item to stock. The store can use the data and what they learn from the data to improve customer service, mimicking the kind of personal attention a shopper had 50 years ago.

2.1 What *Are* Data?

THE W'S:
WHO
WHAT
WHEN
WHERE
WHY

We bet you thought you knew this instinctively. Think about it for a minute. What exactly *do* we mean by "data"? Do data even have to be numbers? The amount of your last purchase in dollars is numerical data, but some data record names or other labels. The names in Amazon.com's database are data, but are not numerical.

Sometimes, data can have values that look like numerical values but are just numerals serving as labels. This can be confusing. For example, the ASIN

---

[1] Amazon.com 2005 Annual Report

10

(Amazon Standard Item Number) of a book may have a numerical value, such as 978-0321426592, but it's really just another *name* for the book *Business Statistics*.

Data values, no matter what kind, are useless without their context. Newspaper journalists know that the lead paragraph of a good story should establish the "Five W's": *Who, What, When, Where,* and (if possible) *Why*. Often, we add *How* to the list as well. Answering these questions can provide a **context** for data values. The answers to the first two questions are essential. If you can't answer *Who* and *What*, you don't have data, and you don't have any useful information.

Here's an example of some of the data Amazon might collect:

10675489	B000001OAA	10.99	Chris G.	902	Boston	15.98	Kansas	Illinois
Samuel P.	Orange County	10783489	12837593	N	B000068ZVQ	Bad Blood	Nashville	Katherine H.
Canada	Garbage	16.99	Ohio	N	Chicago	N	11.99	Massachusetts
B000002BK9	312	Monique D.	Y	413	B00000I5Y6	440	15783947	Let Go

Table 2.1 *An example of data with no context. It's impossible to say anything about what these values might mean without knowing their context.*

Try to guess what they represent. Why is that hard? Because these data have no *context*. We can make the meaning clear if we add the context of *Who* and *What* and organize the values into a **data table** such as this one.

Purchase Order Number	Name	Ship to State/Country	Price	Area Code	Previous CD Purchase	Gift?	ASIN	Artist
10675489	Katherine H.	Ohio	10.99	440	Nashville	N	B00000I5Y6	Kansas
10783489	Samuel P.	Illinois	16.99	312	Orange County	Y	B000002BK9	Boston
12837593	Chris G.	Massachusetts	15.98	413	Bad Blood	N	B000068ZVQ	Chicago
1578397	Monique D.	Canada	11.99	902	Let Go	N	B000001OAA	Garbage

Table 2.2 *Example of a data table. The variable names are in the top row. Typically, the Who of the table are found in the leftmost column.*

Now we can see that these are four purchase records, relating to CD orders from Amazon. The column titles tell *What* has been recorded. The rows tell us *Who*. But, be careful. Look at all the variables to see *Who* the variables are about. Even if people are involved, they may not be the *Who* of the data. For example, the *Who* here are the purchase orders (not the people who made the purchases) because each row refers to a different purchase order, not necessarily a different *person*. A common place to find the *Who* of the table is the leftmost column. The other W's might have to come from the company's database administrator.[2]

In general, the rows of a data table correspond to individual **cases** about *Whom* (or about which—if they're not people) we record some characteristics. These cases go by different names, depending on the situation. Individuals who answer a survey are referred to as **respondents**. People on whom we experiment are **subjects** or (in an attempt to acknowledge the importance of their role in the experiment) **participants,** but animals, plants, website, and other inanimate subjects are often called **experimental units.** In a database, rows are called **records**—in this example, purchase records. Perhaps the most generic term is **cases.** In the table, the cases are the individual CD purchase orders.

Sometimes people refer to data values as *observations*, without being clear about the *Who*. Be sure you know the *Who* of the data, or you may not know what the

---

[2] In database management, this kind of information is called "metadata" or data about data.

data say. The *characteristics* recorded about each individual or case are called **variables.** These are usually shown as the columns of a data table, and they should have a name that identifies *What* has been measured.

A general term for a data table like this is a **spreadsheet,** a name that comes from bookkeeping ledgers of financial information. The data were typically spread across facing pages of a bound ledger, the book used by an accountant for keeping records of expenditures and sources of income. For the accountant, the columns were the types of expenses and income, and the cases were transactions, typically invoices or receipts. Since the advent of computers, use of spreadsheet programs has become a common skill, and the programs have become some of the most successful applications in the computer industry. It is usually easy to move a data table from a spreadsheet program to a program designed for statistical graphics and analysis, either directly or by copying the data table and pasting it into the statistics program.

Although data tables and spreadsheets are great for relatively small data sets, they are cumbersome for the complex data sets that companies must maintain on a day-to-day basis. Various other database architectures are used to store data. The most common is a relational database. In a **relational database**, two or more separate data tables are linked together so that information can be merged across them. Each data table is a *relation* because it is about a specific set of cases with information about each of these cases for all (or at least most) of the variables ("fields" in database terminology). For example, a table of customers, along with demographic information on each, is such a relation. A data table with information about a different collection of cases is a different relation. For example, a data table of all the items sold by the company, including information on price, inventory, and past history, is a relation as well (for example, as in Table 2.3). Finally, the day-to-day transactions may be held in a third database where each purchase of an item by a customer is listed as a case. In a relational database, these three relations can be

Customers

Customer Number	Name	City	State	Zip Code	Customer since	Gold Member?
473859	R. De Veaux	Williamstown	MA	01267	2007	No
127389	N. Sharpe	Wellesley	MA	02481	2000	Yes
335682	P. Velleman	Ithaca	NY	14580	2003	No
...						

Items

Product ID	Name	Price	Currently in Stock?
SC5662	Sliver Cane	43.50	Yes
TH2839	Top Hat	29.99	No
RS3883	Red Sequined Shoes	35.00	Yes
...			

Transactions

Transaction Number	Date	Customer Number	Product ID	Quantity	Shipping Method	Free Ship?
T23478923	9/15/08	473859	SC5662	1	UPS 2nd Day	N
T23478924	9/15/08	473859	TH2839	1	UPS 2nd Day	N
T63928934	10/20/08	335473	TH2839	3	UPS Ground	N
T72348299	12/22/08	127389	RS3883	1	Fed Ex Ovnt	Y

Table 2.3 *A relational database shows all the relevant information for three separate relations linked together by customer and product numbers.*

linked together. For example, you can look up a customer to see what he or she purchased or look up an item to see which customers purchased it.

In statistics, all analyses are performed on a single data table. But often the data must be retrieved from a relational database. Retrieving data from these databases often requires specific expertise with that software. In the rest of the book, we'll assume that all data have been downloaded to a data table or spreadsheet with variables listed as columns and cases as the rows.

It is wise to be careful. The *What* and *Why* of area codes are not as simple as they may first seem. When area codes were first introduced, AT&T was still the source of all telephone equipment, and phones had dials.

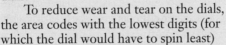

To reduce wear and tear on the dials, the area codes with the lowest digits (for which the dial would have to spin least) were assigned to the most populous regions—those with the most phone numbers and thus the area codes most likely to be dialed. New York City was assigned 212, Chicago 312, and Los Angeles 213, but rural upstate New York was given 607, Joliet was 815, and San Diego 619. For that reason, at one time, the numerical value of an area code could be used to guess something about the population of its region. Since the advent of push-button phones, area codes have finally become just categories.

2.2 Variable Types

Variables play different roles, and knowing the variable's *type* is crucial to knowing what to do with it and what it can tell us. When a variable names categories and answers questions about how cases fall into those categories, we call it a **categorical variable**.[3] When a variable has measured numerical values with *units* and the variable tells us about the quantity of what is measured, we call it a **quantitative variable.** Sometimes, the variable may be viewed as categorical or quantitative depending on the situation. It's more a decision about what we hope to learn from a variable than a quality of the variable itself. It's the questions we ask of a variable (the *Why* of our analysis) that shape how we think about it and how we treat it.

Descriptive responses to questions are often categories. For example, the responses to the questions "What type of mutual fund do you invest in?" or "What kind of advertising does your firm use?" yield categorical values. An important special case of categorical variables is one that has only two possible responses (usually "yes" or "no"), which arise naturally from questions like "Do you invest in the stock market?" or "Do you make online purchases from this website?"

Be careful. If you treat a variable as quantitative, be sure the values have units and measure a quantity of something. For example, area codes are numbers, but do we use them that way? Is 610 twice 305? Of course it is, but that's not really the point. We don't care that Allentown, PA (area code 610), is twice Key West, FL (305). The numerical values of area codes are completely arbitrary (well, not quite—see the side bar). The numbers assigned by the area codes are codes that *categorize* the phone number into a geographical area. So, we treat area code as a categorical variable.

For quantitative variables, the **units** tell how each value has been measured. Even more important, units such as yen, cubits, carats, angstroms, nanoseconds, miles per hour, or degrees Celsius tell us the *scale* of measurement. The units tell us how much of something we have or how far apart two values are. Without units, the values of a measured variable have no meaning. It does little good to be promised a raise of 5,000 a year if you don't know whether it will be paid in euros, dollars, yen, or Estonian krooni. An essential part of a quantitative variable is its units.

Sometimes the type of the variable is clear. Some variables can answer questions *only* about categories. If the values of a variable are words rather than numbers, it's a good bet that it is categorical. But

One tradition that hangs on in some quarters is to name variables with cryptic abbreviations written in uppercase letters. This can be traced back to the 1960s, when the very first statistics computer programs were controlled with instructions punched on cards. The earliest punch card equipment used only uppercase letters, and the earliest statistics programs limited variable names to six or eight characters, so variables were called things like PRSRF3. Modern programs do not have such restrictive limits, so there is no reason for variable names that you wouldn't use in an ordinary sentence.

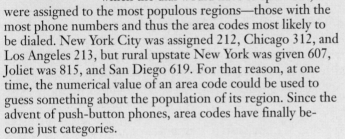

---

[3] You may also see them called *qualitative* variables.

some variables can answer both kinds of questions. For example, Amazon could ask for your *Age* in years. That seems quantitative, and would be if they want to know the average age of those customers who visit their site after 3 A.M. But suppose they want to decide which CD to offer you in a special deal—one by Raffi, Blink182, Carly Simon, or Mantovani—and need to be sure they have adequate supplies on hand to meet the demand. Then thinking of your age in one of the categories child, teen, adult, or senior might be more useful. If it isn't clear whether to treat a variable as categorical or quantitative, think about *Why* you are looking at it and what you want it to tell you.

A typical course evaluation survey asks:

"How valuable do you think this course will be to you?"

1 = Not valuable;   2 = Slightly valuable;   3 = Moderately valuable;   4 = Extremely valuable.

Is this variable categorical or quantitative? Once again, we'll look to the *Why*. A teacher might simply count the number of students who gave each response for her course, treating *Educational Value* as a categorical variable. When she wants to see whether the course is improving, she might treat the responses as the *amount* of perceived value—in effect, treating the variable as quantitative. But what are the units? There is certainly an *order* of perceived worth; higher numbers indicate higher perceived worth. A course that averages 3.5 seems more valuable than one that averages 2, but we should be careful about treating *Educational Value* as purely quantitative. To treat it as quantitative, she'll have to imagine that it has "educational value units" or some similar arbitrary construction. Because there are no natural units, she should be cautious.

Question	Categories or Responses
Do you invest in the stock market?	___ Yes ___ No
What kind of advertising do you use?	___ Newspapers ___ Internet ___ Direct mailings
What is your class at school?	___ Freshman ___ Sophomore ___ Junior ___ Senior
I would recommend this course to another student.	___ Strongly Disagree ___ Slightly Disagree ___ Slightly Agree ___ Strongly Agree
How satisfied are you with this product?	___ Very Unsatisfied ___ Unsatisfied ___ Satisfied ___ Very Satisfied

Table 2.4 *Some examples of categorical variables*

Counts

In Statistics, we often count things. When Amazon considers a special offer of free shipping to customers, they might first analyze how purchases have been shipped in the recent past. They might start by counting the number of purchases shipped in each category: ground transportation, second-day air, and overnight air. Counting is a natural way to summarize the categorical variable *Shipping Method*. So *every* time we see counts, does that mean that the associated variable is categorical? Actually, no.

We also use counts to *measure* the amounts of things. How many songs are on your digital music player? How many classes are you taking this semester? To measure these quantities, we'd naturally count. The variables (*Songs, Classes*) are quantitative, whose units are the "number of," or generically, just "counts" for short.

So we use counts in two different ways. When we have a categorical variable, we count the cases in each category to summarize what the variable tells us. The counts themselves are not the data, but are something we use to summarize the data. For example, Amazon counts the number of purchases in each category of the categorical variable *Shipping Method*.

Shipping Method	Number of Purchases
Ground	20,345
Second-day	7,890
Overnight	5,432

Table 2.5 *A summary of the categorical variable* Shipping Method *that shows the counts, or number of cases for each category.*

At other times, our focus is on the amount of something, which we measure by counting. Amazon might track the growth in the number of teenage customers each month to forecast CD sales. *Teen* was a category when we looked at the categorical variable *Age*. But now it's a quantitative variable in its own right whose amount is measured by counting the number of customers as shown in Table 2.6.

Month	Number of Teenage Customers
January	123,456
February	234,567
March	345,678
April	456,789
May	...
...	...

Table 2.6 *A summary of the quantitative variable* Teenage Customers *that shows the counts, or how many teenage customers made purchases, for each month of the year.*

Identifiers

What's your student ID number? It may be numerical, but is it a quantitative variable? No, it doesn't have units. Is it categorical? Yes, but a special kind. Look at how many categories there are and at how many individuals there are in each category. There are exactly as many categories as individuals and only one individual in each category. While it's easy to count the totals for each category, it's not very interesting. This is an **identifier variable**. Amazon wants to know who you are when you sign in again and doesn't want to confuse you with some other customer. So they assign you a unique identifier.

Identifier variables themselves don't tell us anything useful about the categories because we know there is exactly one individual in each. However, they are crucial in this era of large data sets because by uniquely identifying the cases, they make it possible to combine data from different sources, protect confidentiality, and provide unique labels. Most company databases are, in fact, relational databases. The identifier is crucial to linking one data table to another in a relational database. The identifiers in Table 2.3 are the *Customer Number*, *Product ID*, and *Transaction Number*. Variables like *UPS Tracking Number*, *Social Security Number*, and Amazon's *ASIN* are other examples of identifiers.

You'll want to recognize when a variable is playing the role of an identifier so you won't be tempted to analyze it. Knowing that Amazon's average ASIN number increased 10% from 2007 to 2008 doesn't really tell you anything—any more than analyzing any categorical variable as if it were quantitative would.

Be careful not to be inflexible in your typing of variables. Variables can play different roles, depending on the question we ask of them, and classifying variables rigidly into types can be misleading. For example, in their annual reports, Amazon refers to its database and looks at the variables *Sales* and *Year*. When analysts ask how many books Amazon sold in 2005, what role does *Year* play? There's only one row for 2005, and *Year* identifies it, so it plays the role of an identifier variable. In its role as an identifier, you might match other data from Amazon, or the economy in general, for the same year. But analysts also track sales growth over time. In this role, *Year* measures time. Now it's being treated as a quantitative variable with units of years. The difference lies in the consideration of the *Why* of our question.

Other Data Types

Categorical variables used only to name categories are sometimes called **nominal variables.** Sometimes all we want to know about a variable is the order of its values. For example, we may want to pick out the first, the last, or the middle value. In such cases, we can say that our variable has **ordinal** values. Values can be individually ordered (e.g., the ranks of employees based on the number of days they've worked for the company) or ordered in classes (e.g., Freshman, Sophomore, Junior, Senior). But the ordering always depends on our purpose. Are the categories Infant, Youth, Teen, Adult, and Senior ordinal? Well, if we are ordering on age, they surely are. But if we are ordering (as Amazon might) on purchase volume, it is likely that either Teen or Adult will be the top group.

Some people differentiate quantitative variables according to whether their measured values have a defined value for zero. This is a technical distinction and usually not one we'll need to make. (For example, it isn't correct to say that a temperature of 80°F is twice as hot as 40°F because 0° is an arbitrary value. On the Celsius scale those temperatures are 26.6°C and 4.44°C—a ratio of 6.) The term *interval scale* is sometimes applied to data such as these, and the term *ratio scale* is applied to measurements for which such ratios are appropriate.

Cross-Sectional and Time Series Data

The quantitative variable *Teenage Customers* in Table 2.6 is an example of a **time series** because we have the same variable measured at regular intervals over time. Time series are common in business. Typical measuring points are months, quarters, or years, but virtually any consistently-spaced time interval is possible. Variables collected over time hold special challenges for statistical analysis, and Chapter 20 discusses these in more detail. By contrast, most of the methods in this book are better suited for **cross-sectional data**, where several variables are measured at the same time point.

For example, if we collect data on sales revenue, number of customers, and expenses for last month at each Starbucks (more than 13,000 locations as of 2007), this would be cross-sectional data. If we expanded our data collection process to include sales revenue and expenses each day over a time span of several months, we would now have a time series for sales and expenses. Because different methods are used to analyze these different types of data, it is important to be able to identify both time series and cross-sectional data sets.

2.3 Where, How, and When

We must know *Who, What,* and *Why* to analyze data. Without knowing these three, we don't have enough to start. Of course, we'd always like to know more. The more we know, the more we'll understand. If possible, we'd like to know the *When* and *Where* of data as well. Values recorded in 1803 may mean something different than similar values recorded last year. Values measured in Tanzania may differ in meaning from similar measurements made in Mexico.

How the data are collected can make the difference between insight and nonsense. As we'll see later, data that come from a voluntary survey on the Internet are almost always worthless. In a recent Internet poll, 84% of respondents said "no" to the question of whether subprime borrowers should be bailed out. While it may be true that 84% of those 23,418 respondents did say that, it's dangerous to assume that that group is representative of any larger group. Chapter 3 discusses sound methods for collecting data from surveys and polls so that you can make inferences from the data you have at hand to the world at large.

You may also collect data by performing an experiment in which you actively manipulate variables (called factors) to see what happens. Most of the "junk mail" credit card offers that you receive are actually experiments done by marketing groups in those companies. Chapter 23 discusses both the design and the analysis of experiments like these.

Sometimes, the answer to the question you have may be found in data that someone, or more typically, some organization has already collected. Companies, nonprofit organizations, and government agencies collect a vast amount of data that is becoming increasingly easy to access via the Internet, although some organizations may charge a fee for accessing or downloading their data. The U.S. government through its various agencies collects information on nearly every aspect of life in the United States, both social and economic (see for example www.census.gov), as the European Union does for Europe (see ec.europa.eu/eurostat). International organizations such as the World Health Organization (www.who.org) and polling agencies such as Gallup (www.gallup.com) offer information on a variety of topics as well.

There's a world of data on the Internet

These days, one of the richest sources of data is the Internet. With a bit of practice, you can learn to find data on almost any subject. Many of the data sets we use in this book were found in this way. The Internet has both advantages and disadvantages as a source of data. Among the advantages are the fact that often you'll be able to find even more current data than we present. The disadvantage is that references to Internet addresses can "break" as sites evolve, move, and die.

Our solution to these challenges is to offer the best advice we can to help you search for the data, wherever they may be residing. We usually point you to a website. We'll sometimes suggest search terms and offer other guidance.

Some words of caution, though: Data found on Internet sites may not be formatted in the best way for use in statistics software. Although you may see a data table in standard form, an attempt to copy the data may leave you with a single column of values. You may have to work in your favorite statistics or spreadsheet program to reformat the data into variables. You will also probably want to remove commas from large numbers and such extra symbols as money indicators ($, ¥, £, €); few statistics packages can handle these.

Throughout this book, whenever we introduce data, we will provide a margin note listing the W's of the data. It's a habit we recommend. The first step of any data analysis is to know why you are examining the data (what you want to know), whom each row of your data table refers to, and what the variables (the columns of the table) record. These are the *Why*, the *Who*, and the *What*. Identifying them is a key part of the *Plan* step of any analysis. Make sure you know all three before you spend time analyzing the data.

JUST CHECKING

An insurance company that specializes in commercial property insurance has a separate database for their policies that involve churches and schools. Here is a small portion of that database.

Policy Number	Years Claim Free	Net Property Premium ($)	Net Liability Premium ($)	Total Property Value ($1,000)	Median Age of Zip Code	School?	Territory	Coverage
4000174699	1	3107	503	1036	40	FALSE	AL580	BLANKET
8000571997	2	1036	261	748	42	FALSE	PA192	SPECIFIC
8000623296	1	438	353	344	30	FALSE	ID60	BLANKET
3000495296	1	582	339	270	35	TRUE	NC340	BLANKET
5000291199	4	993	357	218	43	FALSE	OK590	BLANKET
8000470297	2	433	622	108	31	FALSE	NV140	BLANKET
1000042399	4	2461	1016	1544	41	TRUE	NJ20	BLANKET
4000554596	0	7340	1782	5121	44	FALSE	FL530	BLANKET
3000260397	0	1458	261	1037	42	FALSE	NC560	BLANKET
8000333297	2	392	351	177	40	FALSE	OR190	BLANKET
4000174699	1	3107	503	1036	40	FALSE	AL580	BLANKET

1 List as many of the W's as you can for this data set.

2 Classify each variable as to whether you think it should be treated as categorical or quantitative (or both); if quantitative, identify the units.

WHAT CAN GO WRONG?

- **Don't label a variable as categorical or quantitative without thinking about the data and what they represent.** The same variable can sometimes take on different roles.

- **Don't assume that a variable is quantitative just because its values are numbers.** Categories are often given numerical labels. Don't let that fool you into thinking they have quantitative meaning. Look at the context.

- **Always be skeptical.** One reason to analyze data is to discover the truth. Even when you are told a context for the data, it may turn out that the truth is a bit (or even a lot) different. The context colors our interpretation of the data, so those who want to influence what you think may slant the context. A survey that seems to be about all students may in fact report just the opinions of those who visited a fan website. The question that respondents answered may be posed in a way that influences responses.

ETHICS IN ACTION

Sarah Potterman, a doctoral student in educational psychology, is researching the effectiveness of various interventions recommended to help children with learning disabilities improve their reading skills. Among the approaches examined is an interactive software system that uses analogy-based phonics. Sarah contacted the company that developed this software, RSPT Inc., in order to obtain the system free of charge for use in her research. RSPT Inc. expressed interest in having her compare their product with other intervention strategies and was quite confident that their approach would be the most effective. Not only did the company provide Sarah with free software, but RSPT Inc. also generously offered to fund her research with a grant to cover her data collection and analysis costs.

ETHICAL ISSUE *Both the researcher and company should be careful about the funding source having a vested interest in the research result (related to Item H, ASA Ethical Guidelines).*

ETHICAL SOLUTION *RSPT Inc. should not pressure Sarah Potterman to obtain a particular result. Both parties should agree on paper before the research is begun that the research results can be published even if they show that RSPT's interactive software system is not the most effective.*

Jim Hopler is operations manager for a local office of a top-ranked full service brokerage firm. With increasing competition from both discount and online brokers, Jim's firm has redirected attention to attaining exceptional customer service through its client-facing staff, namely brokers. In particular, they wish to emphasize the excellent advisory services provided by their brokers. Results from surveying clients about the advice received from brokers at the local office revealed that 20% rated it *poor*, 5% rated it *below average*, 15% rated it *average*, 10% rated it *above average*, and 50% rated it *outstanding*. With corporate approval, Jim and his management team instituted several changes in an effort to provide the best possible advisory services at the local office. Their goal was to increase the percentage of clients who viewed their advisory services as *outstanding*. Surveys conducted after the changes were implemented showed the following results: 5% *poor*, 5% *below average*, 20% *average*, 40% *above average*, and 30% *outstanding*. In discussing these results, the management team expressed concern that the percentage of clients who considered their advisory services *outstanding* fell from 50% to 30%. One member of the team suggested an alternative way of summarizing the data. By coding the categories on a scale from 1 = poor to 5 = outstanding and computing the average, they found that the average rating increased from 3.65 to 3.85 as a result of the changes implemented. Jim was delighted to see that their changes were successful in improving the level of advisory services offered at the local office. In his report to corporate, he only included average ratings for the client surveys.

ETHICAL ISSUE *By taking an average, Jim is able to show improved customer satisfaction. However, their goal was to increase the percentage of outstanding ratings. Jim redefined his study after the fact to support a position (related to Item A, ASA Ethical Guidelines).*

ETHICAL SOLUTION *Jim should report the percentages for each rating category. He can also report the average. He may wish to include in his report a discussion of what those different ways of looking at the data say and why they appear to differ. He may also want to explore with the survey participants the perceived differences between "above average" and "outstanding."*

What have we learned?

We've learned that data are information in a context.

- The W's help nail down the context: *Who, What, Why, Where, When.*
- We must know at least the *Who, What,* and *Why* to be able to say anything useful based on the data. The *Who* are the *cases.* The *What* are the *variables.* A variable gives information about each of the cases. The *Why* helps us decide which way to treat the variables.

We treat variables in two basic ways, as *categorical* or *quantitative.*

- Categorical variables identify a category for each case. Usually we think about the counts of cases that fall in each category. (An exception is an identifier variable that just names each case.)
- Quantitative variables record measurements or amounts of something; they must have *units.*
- Sometimes we treat a variable as categorical or quantitative depending on what we want to learn from it, which means some variables can't be pigeonholed as one type or the other. That's an early hint that in Statistics we can't always pin things down precisely.

Terms

Case	A case is an individual about whom or which we have data.
Categorical variable	A variable that names categories (whether with words or numerals) is called categorical.
Context	The context ideally tells *Who* was measured, *What* was measured, *How* the data were collected, *Where* the data were collected, and *When* and *Why* the study was performed.
Cross-sectional data	Data taken from situations that vary over time but measured at a single time instant is said to be a cross-section of the time series.
Data	Systematically recorded information, whether numbers or labels, together with its context.
Data table	An arrangement of data in which each row represents a case and each column represents a variable.
Experimental unit	An individual in a study for which or for whom data values are recorded. Human experimental units are usually called subjects or participants.
Identifier variable	A categorical variable that records a unique value for each case, used to name or identify it.
Nominal variable	The term "nominal" can be applied to data whose values are used only to name categories.

Ordinal variable	The term "ordinal" can be applied to data for which some kind of order is available but for which measured values are not available.
Participant	A human experimental unit. Also called a subject.
Quantitative variable	A variable in which the numbers are values of measured quantities with units.
Record	Information about an individual in a database.
Relational database	A relational database stores and retrieves information. Within the database, information is kept in data tables that can be "related" to each other.
Respondent	Someone who answers, or responds to, a survey.
Spreadsheet	A spreadsheet is layout designed for accounting that is often used to store and manage data tables. Excel is a common example of a spreadsheet program.
Subject	A human experimental unit. Also called a participant.
Time series	Data measured over time. Usually the time intervals are equally-spaced (e.g., every week, every quarter, or every year).
Units	A quantity or amount adopted as a standard of measurement, such as dollars, hours, or grams.
Variable	A variable holds information about the same characteristic for many cases.

Skills

 PLAN

- Be able to identify the *Who*, *What*, *When*, *Where*, *Why*, and *How* of data, or recognize when some of this information has not been provided.
- Be able to identify the cases and variables in any data set.
- Know how to treat a variable as categorical or quantitative depending on its use.
- For any quantitative variable, be able to identify the units in which the variable has been measured (or note that they have not been provided).

 REPORT

- Be sure to describe a variable in terms of its *Who*, *What*, *When*, *Where*, *Why*, and *How* (and be prepared to remark when that information is not provided).

Technology Help

Most often we find statistics on a computer using a program, or *package*, designed for that purpose. There are many different statistics packages, but they all do essentially the same things. If you understand what the computer needs to know to do what you want and what it needs to show you in return, you can figure out the specific details of most packages pretty easily.

For example, to get your data into a computer statistics package, you need to tell the computer:

- Where to find the data. This usually means directing the computer to a file stored on your computer's disk or to data on a database. Or it might just mean that you have copied the data from a spreadsheet program or Internet site and it is currently on your computer's clipboard. Usually, the data should be in the form of a data table. Most computer statistics packages prefer the *delimiter* that marks the division between elements of a data table to be a *tab* character and the delimiter that marks the end of a case to be a *return* character.

- Where to put the data. (Usually this is handled automatically.)
- What to call the variables. Some data tables have variable names as the first row of the data, and often statistics packages can take the variable names from the first row automatically.

Mini Case Study Project

Credit Card Bank

Like all credit and charge card companies, this company makes money on each of its cardholders' transactions. Thus, its profitability is directly linked to card usage. To increase customer spending on its cards, the company sends many different offers to its cardholders, and market researchers analyze the results to see which offers yield the largest increases in the average amount charged.

On your disk (in the file **ch02_MCSP_Credit_Card_Bank**) is a small part of a database like the one used by the researchers. For each customer, it contains several variables in a spreadsheet.

Examine the data in the data file. List as many of the W's as you can for these data and classify each variable as categorical or quantitative. If quantitative, identify the units.

EXERCISES

For each description of data in Exercises 1 to 26, identify the W's, name the variables, specify for each variable whether its use indicates it should be treated as categorical or quantitative, and for any quantitative variable identify the units in which it was measured (or note that they were not provided).

1. The news. Find a newspaper or magazine article in which some data are reported (e.g., see *The Wall Street Journal, Financial Times, Business Week,* or *Fortune*). For the data discussed in the article, answer the questions above. Include a copy of the article with your report.

2. Investments. In the U.S., 401(k) plans permit employees to shift part of their before-tax salaries into investments such as mutual funds. One company, concerned with what it believed was a low employee participation rate in its 401(k) plan, sampled 30 other companies with similar plans and asked for their 401(k) participation rates.

3. Oil spills. After several major ocean oil spills by oil tankers, Congress passed the 1990 Oil Pollution Act, which requires all tankers to have thicker hulls. Further improvements in the structural design of a tanker have been proposed since then, each with the objective of reducing the likelihood of an oil spill and decreasing the amount of outflow in the event of a hull puncture. Infoplease (www.infoplease.com) reports the date, the spillage amount and cause of puncture for 50 recent major oil spills from tankers and carriers.

4. Sales. A major U.S. company is interested in seeing how various promotional activities are related to domestic sales. Analysts decide to measure the money spent on different forms of advertising ($ thousand) and sales ($ million) on a monthly basis for three years (2004–2006).

5. Food store. A food retailer that specializes in selling organic food has decided to open a new store. To help determine the best location in the United States for the new store, researchers decide to examine data from existing stores, including weekly sales ($), town population (thousands), median age of town, median income of town ($), and whether or not the store sells wine and beer.

6. Sales II. The company in Exercise 4 is also interested in the impact of national indicators on their sales. It decides to obtain measurements for unemployment rate (%) and inflation rate (%) on a quarterly basis to compare to their quarterly sales ($ million) over the same time period (2004–2006).

7. Arby's menu. A listing posted by the Arby's restaurant chain gives, for each of the sandwiches it sells, the type of meat in the sandwich, number of calories, and serving size in ounces. The data might be used to assess the nutritional value of the different sandwiches.

8. MBA admissions. A school in the northeastern United States is concerned with the recent drop in female students in its MBA program. It decides to collect data from the admissions office on each applicant, including: sex of each applicant, age of each applicant, whether or not they were accepted, whether or not they attended, and the reason for not attending (if they did not attend). The school hopes to find commonalities among the female accepted students who have decided not to attend the business program.

9. Climate. In a study appearing in the journal *Science*, a research team reports that plants in southern England are flowering earlier in the spring. Records of the first flowering dates for 385 species over a period of 47 years indicate that flowering has advanced an average of 15 days per decade, an indication of climate warming according to the authors.

10. MBA admissions II. An internationally recognized MBA program in London intends to also track the GPA of the MBA students and compares MBA performance to standardized test scores over a 5-year period (2000–2005).

11. Schools. A State Education Department requires local school districts to keep records on all students, recording: age, race or ethnicity, days absent, current grade level, standardized test scores in reading and mathematics, and any disabilities or special educational needs the student may have.

12. Pharmaceutical firm. Scientists at a major pharmaceutical firm conducted an experiment to study the effectiveness of an herbal compound to treat the common cold. They exposed volunteers to a cold virus, then gave them either the herbal compound or a sugar solution known to have no effect on colds. Several days later they assessed each patient's condition using a cold severity scale ranging from 0–5. They found no evidence of the benefits of the compound.

13. Start-up company. A start-up company is building a database of customers and sales information. For each customer, it records name, ID number, region of the country (1 = East, 2 = South, 3 = Midwest, 4 = West), date of last purchase, amount of purchase, and item purchased.

14. Cars. A survey of autos parked in executive and staff lots at a large company recorded the make, country of origin, type of vehicle (car, van, SUV, etc.), and age.

15. Vineyards. Business analysts hoping to provide information helpful to grape growers compiled these data about vineyards: size (acres), number of years in existence, state, varieties of grapes grown, average case price, gross sales, and percent profit.

16. Environment. As research for an ecology class, students at a college in upstate New York collect data on streams each year to study the impact of the environment. They record a number of biological, chemical, and physical variables, including the stream name, the substrate of the stream (limestone, shale, or mixed), the acidity of the water (pH), the temperature (°C), and the BCI (a numerical measure of biological diversity).

17. Gallup Poll. The Gallup Poll conducted a representative telephone survey of 1180 American voters. Among the reported results were the voter's region (Northeast, South, etc.), age, political party affiliation, whether the respondent owned any shares of stock, and their attitude (on a scale of 1 to 5) toward unions.

18. FAA. The Federal Aviation Administration (FAA) monitors airlines for safety and customer service. For each flight the carrier must report the type of aircraft, number of passengers, whether or not the flights departed and arrived on schedule, and any mechanical problems.

19. EPA. The Environmental Protection Agency (EPA) tracks fuel economy of automobiles. Among the data EPA analysts collect from the manufacturer are the manufacturer (Ford, Toyota, etc.), vehicle type (car, SUV, etc.), weight, horsepower, and gas mileage (mpg) for city and highway driving.

20. Consumer Reports. In 2002, Consumer Reports published an article evaluating refrigerators. It listed 41 models, giving the brand, cost, size (cu ft), type (such as top-freezer), estimated annual energy cost, overall rating (good, excellent, etc.), and repair history for that brand (percentage requiring repairs over the past 5 years).

21. Lotto. A study of state-sponsored Lotto games in the United States (*Chance*, Winter 1998) listed the names of the states and whether or not the state had Lotto. For states that did, the study indicated the number of numbers in the lottery, the number of matches required to win, and the probability of holding a winning ticket.

22. L.L. Bean. L.L. Bean is a large U.S. retailer that depends heavily on its catalog sales. It collects data internally and tracks the number of catalogs mailed out, the number of square inches in each catalog, and the sales ($ thousands) in the 4 weeks following each mailing. The company is interested in learning more about the relationship (if any) among the timing and space of their catalogs and their sales.

23. Stock market. An online survey of students in a large MBA Statistics class at a business school in the northeastern United States asked them to report their total personal investment in the stock market ($), total number of different stocks currently held, total invested in mutual funds ($), and the name of each mutual fund in which they have invested. The data were used in the aggregate for classroom illustrations.

24. Theme park sites. A study on the potential for developing theme parks in various locations throughout Europe in 2008 collects the following information: the country where the proposed site is located, estimated cost to acquire site (in euros), size of population within a one hour drive of the site, size of the site (in hectares), mass transportation within 5 minutes of the site. The data will be used to present to prospective developers.

25. Indy. The 2.5-mile Indianapolis Motor Speedway has been the home to a race on Memorial Day nearly every year since 1911. Even during the first race there were controversies. Ralph Mulford was given the checkered flag first but took three extra laps just to make sure he'd completed 500 miles. When he finished, another driver, Ray Harroun, was being presented with the winner's trophy, and Mulford's protests were ignored. Harroun averaged 74.6 mph for the 500 miles. Here are the data for the first few and three recent Indianapolis 500 races.

Year	Winner	Car	Time (hrs)	Speed (mph)	Car #
1911	Ray Harroun	Marmon Model 32	6.7022	74.602	32
1912	Joe Dawson	National	6.3517	78.719	8
1913	Jules Goux	Peugeot	6.5848	75.933	16
...					
...					
2005	Dan Wheldon	Dallara/ Honda	3.1725	157.603	26
2006	Sam Hornish, Jr.	Dallara/ Honda	3.1830	157.085	6
2007	Dario Franchitti	Dallara/ Honda	2.7343	151.774	27

26. Kentucky Derby. The Kentucky Derby is a horse race that has been run every year since 1875 at Churchill Downs, Louisville, Kentucky. The race started as a 1.5-mile race, but in 1896 it was shortened to 1.25 miles because experts felt that 3-year-old horses shouldn't run such a long race that early in the season. (It has been run in May every year but one—1901—when it took place on April 29.) The table at the bottom of the page shows the data for the first few and a few recent races.

When you organize data in a spreadsheet, it is important to lay it out as a data table. For each of these examples in Exercises 27 to 30, show how you would lay out these data. Indicate the headings of columns and what would be found in each row.

27. Mortgages. For a study of mortgage loan performance: amount of the loan, the name of the borrower.

28. Employee performance. Data collected to determine performance-based bonuses: employee ID, average contract closed (in $), supervisor's rating (1–10), years with the company.

29. Company performance. Data collected for financial planning: weekly sales, week (week number of the year), sales predicted by last year's plan, difference between predicted sales and realized sales.

30. Command performance. Data collected on investments in Broadway shows: number of investors, total invested, name of the show, profit/loss after one year.

For the following examples in Exercises 31 to 34, indicate whether the data are a time series or a cross section.

31. Car sales. Number of cars sold by each salesperson in a dealership in September.

Date	Winner	Margin (lengths)	Jockey	Winner's Payoff ($)	Duration (min:sec)	Track Condition
May 17, 1875	Aristides	2	O. Lewis	2850	2:37.75	Fast
May 15, 1876	Vagrant	2	B. Swim	2950	2:38.25	Fast
May 22, 1877	Baden-Baden	2	W. Walker	3300	2:38.00	Fast
May 21, 1878	Day Star	1	J. Carter	4050	2:37.25	Dusty
May 20, 1879	Lord Murphy	1	C. Shauer	3550	2:37.00	Fast
...						
May 5, 2001	Monarchos	4 3/4	J. Chavez	812000	1:59.97	Fast
May 4, 2002	War Emblem	4	V. Espinoza	1875000	2:01.13	Fast
May 3, 2003	Funny Cide	1 3/4	J. Santos	800200	2:01.19	Fast
May 1, 2004	Smarty Jones	2 3/4	S. Elliott	854800	2:04.06	Sloppy

32. Motorcycle sales. Number of motorcycles sold by a dealership in each month of 2008.

33. Cross sections. Average diameter of trees brought to a sawmill in each week of a year.

34. Series. Attendance at the third World Series game recording the age of each fan.

JUST CHECKING ANSWERS

1 Who—policies on churches and schools
What—policy number, years claim free, net property premium ($), net liability premium ($), total property value ($000), median age in zip code, school?, territory, coverage
How—company records
When—not given

2 Policy number: identifier (categorical)
Years claim free: quantitative
Net property premium: quantitative ($)
Net liability premium: quantitative ($)
Total property value: quantitative ($)
School?: categorical (true/false)
Territory: categorical
Coverage: categorical

3

Surveys and Sampling

Roper Polls

Public opinion polls are a relatively new phenomenon. In 1948, as a result of telephone surveys of likely voters, all of the major organizations—Gallup, Roper, and Crossley—consistently predicted, throughout the summer and into the fall, that Thomas Dewey would defeat Harry Truman in the November presidential election. By October the results seemed so clear that *Fortune* magazine declared, "Due to the overwhelming evidence, *Fortune* and Mr. Roper plan no further detailed reports on change of opinion in the forthcoming presidential campaign...."

Of course, Harry Truman went on to win the 1948 election, and the picture of Truman in the early morning after the election holding up the *Chicago Tribune* (printed the night before), with its headline declaring Dewey the winner, has become legend.

The public's faith in opinion polls plummeted after the election, but Elmo Roper vigorously defended the pollsters. Roper was a principal and founder of one of the first market research firms, Cherington, Wood, and Roper, and director of the *Fortune Survey*, which was the first national poll to use scientific sampling techniques. He argued that rather than abandoning polling, business leaders should learn what had gone wrong in the 1948 polls so that market research could be improved. His frank admission of the mistakes made in those polls helped to restore confidence in polling as a business tool.

For the rest of his career, Roper split his efforts between two projects, commercial polling and public opinion. He established the Roper Center for Public Opinion Research at Williams College as a place to house public opinion archives, convincing fellow polling leaders Gallup and Crossley to participate as well. Now located at the University of Connecticut, the Roper Center is one of the world's leading archives of social science data. Roper's market research efforts started as Roper Research Associates and later became the Roper Organization, which was acquired in 2005 by GfK. Founded in Germany in 1934 as the Gesellschaft für Konsumforschung (literally, "Society for Consumption Research"), GfK now stands for "growth from knowledge." It is the fourth largest international market research organization, with over 130 companies in 70 countries and more than 7700 employees worldwide.

GfK Roper Consulting conducts a yearly, global study to examine cultural, economic, and social information that may be crucial to companies that wish to do business worldwide. These companies use the information to help make marketing and advertising decisions in different markets around the world.

How do the researchers at GfK Roper Consulting know that the responses they get reflect the real attitudes of consumers? After all, they don't ask everyone, but they don't want to limit their conclusions to just the people they surveyed. Generalizing from the data at hand to the world at large is something that market researchers, investors, and pollsters do every day. To do it wisely, they need three fundamental ideas.

3.1 Three Ideas of Sampling

Idea 1: Examine a Part of the Whole

The first idea is to draw a sample. We'd like to know about an entire population of individuals, but examining all of them is usually impractical, if not impossible. So we settle for examining a smaller group of individuals—a sample—selected from the population. The whole world is the population the Roper researchers are interested in, but it's not practical, cost-effective, or feasible to survey the entire population. So they examine a sample selected from the population.

You take samples of a larger population every day. For example, if you want to know how the vegetable soup you're cooking for dinner tonight is going to taste, you blow on a spoonful and try it. You certainly don't consume the whole pot. You trust that the taste will *represent* the flavor of the entire pot. The idea of tasting is that a small sample, if selected properly, can represent the entire population.

The GfK Roper Reports® Worldwide poll is an example of a **sample survey**, designed to ask questions of a small group of people in the hope of learning something about the entire population. Most likely, you've never been selected to be part of a national opinion poll. That's true of most people. So how can the pollsters claim that a sample is representative of the entire population? Professional researchers like those who run the Roper survey work hard to ensure that the "taste"—the sample that they take—represents the population fairly. If they are not careful, the sample can produce misleading information about the population.

Selecting a sample to represent the population fairly is more difficult than it sounds. Polls or surveys most often fail because the sample fails to represent part of the population. The way the sample is drawn may overlook subgroups that are hard to find. For example, a telephone survey may get no responses from people with caller ID and may favor other groups, such as the retired or the homebound, who would be more likely to be near their phones when the interviewer calls. Samples that over- or underemphasize some characteristics of the population are said to be biased. When a sample is **biased**, the summary characteristics of a sample differ from the corresponding characteristics of the population it is trying to represent. Conclusions based on biased samples are inherently flawed. There is usually no way to fix bias after the sample is drawn and no way to salvage useful information from it.

What are the basic techniques for making sure that a sample is representative? To make the sample as representative as possible, you might be tempted to hand-pick the individuals included in the sample. But the best strategy is to do something quite different: We should select individuals for the sample *at random*.

Idea 2: Randomize

Think back to our example of sampling soup. Suppose you add some salt to the pot. If you sample it from the top before stirring, you'll get the misleading idea that the whole pot is salty. If you sample from the bottom, you'll get the equally misleading idea that the whole pot is bland. But by stirring the soup, you *randomize* the amount of salt throughout the pot, making each taste more typical of the saltiness of the whole pot. Deliberate randomization is one of the great tools of Statistics.

Randomization can protect against factors that you aren't aware of, as well as those you know are in the data. Suppose, while you aren't looking, a friend adds a handful of peas to the soup. The peas sink to the bottom of the pot, mixing with the other vegetables. If you don't randomize the soup by stirring, your test spoonful from the top won't have any peas. By stirring in the salt, you *also* randomize the

---

The W's and Sampling

The population we are interested in is usually determined by the *Why* of our study. The participants or cases in the sample we draw will be the *Who*. *When* and *How* we draw the sample may depend on what is practical.

peas throughout the pot, making your sample taste more typical of the overall pot *even though you didn't know the peas were there.* So randomizing protects us by giving us a representative sample even for effects we were unaware of.

How do we "stir" people in our survey? We select them at random. Randomizing protects us from the influences of *all* the features of our population by making sure that *on average* the sample looks like the rest of the population.

We all think we know what it means for something to be random. Rolling dice, spinning spinners, and shuffling cards all produce random outcomes. What's the most important aspect of the randomness in these games? Randomness makes them fair.

Two things make **randomization** seem fair. First, nobody can guess the outcome before it happens. Second, when we want things to be fair, usually some underlying set of outcomes will be equally likely (although in many games, some combinations of outcomes are more likely than others). We'll soon see how to use randomness to ensure that the sample we draw is representative of the population we want to study.

Truly random values are surprisingly hard to get. Computers are a popular way to generate random numbers. Even though they often do much better than humans, computers can't generate truly random numbers either. Computers follow programs. Start a computer from the same place, and, all things being equal, it will follow the same path every time. So numbers generated by a computer program are not truly random. Technically, "random" numbers generated by computer are *pseudorandom.* Fortunately, pseudorandom values are good enough for most purposes because they are virtually indistinguishable from truly random numbers.

There *are* ways to generate random numbers that are both equally likely and truly random. If you want to select subjects for a survey at random from a list of potential respondents, you can get as many random numbers as you need online, from a source such as www.random.org, match them up with your list, sort the numbers while carrying along the respondent IDs, and start from the top of the sorted list, selecting as many respondents as you need, now at random.

◆ **Why not match the sample to the population?**   Rather than randomizing, we could try to design our sample to include every possible, relevant characteristic: income level, age, political affiliation, marital status, number of children, place of residence, etc. Clearly we can't possibly think of all the things that might be important. Even if we could, we wouldn't be able to match our sample to the population for all these characteristics.

How well does a sample represent the population from which it was selected? Here's an example using the database of the Paralyzed Veterans of America, a philanthropic organization with a donor list of about 3.5 million people. We've taken two samples, each of 8000 individuals at random from the population. Table 3.1 shows how the means and proportions match up on seven variables.

	Age (yr)	White (%)	Female (%)	# of Children	Income Bracket (1–7)	Wealth Bracket (1–9)	Homeowner? (% Yes)
Sample 1	61.4	85.12	56.2	1.54	3.91	5.29	71.36
Sample 2	61.2	84.44	56.4	1.51	3.88	5.33	72.30

Table 3.1   *Means and proportions for seven variables from two samples of size 8000 from the Paralyzed Veterans of America data. The fact that the summaries of the variables from these two samples are so similar gives us confidence that either one would be representative of the entire population.*

Notice that the two samples match closely in every category. This shows how well randomizing has stirred the population. We didn't preselect the samples for these variables, but randomizing has matched the results closely. We can reasonably assume that since the two samples don't differ too much from each other, they don't differ much from the rest of the population either.

Even if a survey is given to multiple random samples, the samples will differ from each other and so, therefore, will the responses. These sample-to-sample differences are referred to as **sampling error** even though no error has occurred.

Idea 3: The Sample Size Is What Matters

You probably weren't surprised by the idea that a sample can represent the whole. And the idea of sampling randomly makes sense when you stop to think about it, too. But the third important idea of sampling often surprises people. The third idea is that the *size of the sample* determines what we can conclude from the data *regardless of the size of the population*. Many people think that we need a large percentage, or *fraction*, of the population, but in fact all that matters is the size of the sample. The size of the *population* doesn't matter at all.[1] A random sample of 100 students in a college represents the student body just about as well as a random sample of 100 voters represents the entire electorate of the United States. This is perhaps the most surprising idea in designing surveys.

To understand how this works, let's return one last time to your pot of soup. If you're cooking for a banquet rather than just for a few people, your pot will be bigger, but you don't need a bigger spoon to decide how the soup tastes. The same size spoonful is probably enough to make a decision about the entire pot, no matter how large the pot. What *fraction* of the population you sample doesn't matter. It's the *sample size* itself that's important. This idea is of key importance to the design of any sample survey, because it determines the balance between how well the survey can measure the population and how much the survey costs.

How big a sample do you need? That depends on what you're estimating, but too small a sample won't be representative of the population. To get an idea of what's really in the soup, you need a large enough taste to be a *representative* sample from the pot, including, say, a selection of the vegetables. For a survey that tries to find the proportion of the population falling into a category, you'll usually need at least several hundred respondents.[2]

◆ **What do the professionals do?** How do professional polling and market research companies do their work? The most common polling method today is to contact respondents by telephone. Computers generate random telephone numbers for telephone exchanges known to include residential customers; so pollsters can contact people with unlisted phone numbers. The person who answers the phone will be invited to respond to the survey—if that person qualifies. (For example, only adults are usually surveyed, and the respondent usually must live at the residence phoned.) If the person answering doesn't qualify, the caller will ask for an appropriate alternative. When they conduct the interview, the pollsters often list possible responses (such as product names) in randomized orders to avoid biases that might favor the first name on the list.

---

[1] Well, that's not exactly true. If sample is more than 10% of the whole population, it *can* matter. It doesn't matter whenever, as usual, our sample is a very small fraction of the population.

[2] Chapter 9 gives the details behind this statement and shows how to decide on a sample size for a survey.

Do these methods work? The Pew Research Center for the People and the Press, reporting on one survey, says that

> *Across five days of interviewing, surveys today are able to make some kind of contact with the vast majority of households (76%), and there is no decline in this contact rate over the past seven years. But because of busy schedules, skepticism and outright refusals, interviews were completed in just 38% of households that were reached using standard polling procedures.*

Nevertheless, studies indicate that those actually sampled can give a good snapshot of larger populations from which the surveyed households were drawn.

3.2 A Census—Does It Make Sense?

Why bother determining the right sample size? If you plan to open a store in a new community, why draw a sample of residents to understand their interests and needs? Wouldn't it be better to just include everyone and make the "sample" be the entire population? Such a special sample is called a **census**. Although a census would appear to provide the best possible information about the population, there are a number of reasons why it might not.

First, it can be difficult to complete a census. There always seem to be some individuals who are hard to locate or hard to measure. Do you really need to contact the folks away on vacation when you collect your data? How about those with no telephone or mail address? The cost of locating the last few cases may far exceed the budget. It can also be just plain impractical to take a census. The quality control manager for Hostess® Twinkies® doesn't want to census *all* the Twinkies on the production line to determine their quality. Aside from the fact that nobody could eat that many Twinkies, it would defeat their purpose: There would be none left to sell.

Second, the population we're studying may change. For example, in any human population, babies are born, people travel, and folks die during the time it takes to complete the census. News events and advertising campaigns can cause sudden shifts in opinions and preferences. A sample, surveyed in a shorter time frame, may actually generate more accurate information.

Finally, taking a census can be cumbersome. A census usually requires a team of pollsters and/or the cooperation of the population. Because it tries to count everyone, the U.S. Census records too many college students. Many are included both by their families and in a report filed by their schools. Errors of this sort, of both under- and overcounting can be found throughout the U.S. Census.

3.3 Populations and Parameters

GfK Roper Reports Worldwide reports that 60.5% of people over 50 worry about food safety, but only 43.7% of teens do. What does this claim mean? We can be sure the Roper researchers didn't take a census. So they can't possibly know *exactly* what percentage of teenagers worry about food safety. So what does "43.7%" mean?

To generalize from a sample to the world at large, we need a model of reality. Such a model doesn't need to be complete or perfect. Just as a model of an airplane in a wind tunnel can tell engineers what they need to know about aerodynamics even though it doesn't include every rivet of the actual plane, models of data can give us summaries that we can learn from and use even though they don't fit each data value exactly. It's important to remember that they're only models of reality and not reality itself. But without models, what we can learn about the world at large is limited to only what we can say about the data we have at hand.

Any quantity that we calculate from data could be called a "statistic." But in practice, we usually obtain a statistic from a sample and use it to estimate a population parameter.

> Population model parameters are not just unknown—usually they are *unknowable*. We have to settle for sample statistics.

Models use mathematics to represent reality. We call the key numbers in those models **parameters**. All kinds of models have parameters, so sometimes a parameter used in a model for a population is called (redundantly) a **population parameter**.

But let's not forget about the data. We use the data to try to estimate values for the population parameters. Any summary found from the data is a **statistic**. Those statistics that estimate population parameters are particularly interesting. Sometimes—and especially when we match statistics with the parameters they estimate—we use the term **sample statistic**.

We draw samples because we can't work with the entire population. We hope that the statistics we compute from the sample will estimate the corresponding parameters accurately. A sample that does this is said to be **representative**.

JUST CHECKING

1 Various claims are often made for surveys. Why is each of the following claims not correct?

   a) It is always better to take a census than to draw a sample.

   b) Stopping customers as they are leaving a restaurant is a good way to sample opinions about the quality of the food.

   c) We drew a sample of 100 from the 3000 students in a school. To get the same level of precision for a town of 30,000 residents, we'll need a sample of 1000.

   d) A poll taken at a popular website (www.statsisfun.org) garnered 12,357 responses. The majority of respondents said they enjoy doing Statistics. With a sample size that large, we can be sure that most Americans feel this way.

   e) The true percentage of all Americans who enjoy Statistics is called a "population statistic."

3.4 Simple Random Sample (SRS)

How would you select a representative sample? It seems fair to say that every individual in the population should have an equal chance to be selected, but that's not sufficient. There are many ways to give everyone an equal chance that still wouldn't give a representative sample. Consider, for example, a manufacturer that samples customers by drawing at random from product registration forms, half of which arrived by mail and half by online registration. They flip a coin. If it comes up heads, they draw 100 mail returns; tails, they draw 100 electronic returns. Each customer has an equal chance of being selected, but if tech-savvy customers are different, then the samples are hardly representative.

We need to do better. Suppose we insist that every possible *sample* of the size we plan to draw has an equal chance of being selected. This ensures that situations like the all tech-savvy (or not) samples are not likely to occur and still guarantees that each person has an equal chance of being selected. With this method each *combination* of individuals has an equal chance of being selected as well. A sample drawn in this way is called a **simple random sample,** usually abbreviated **SRS**. An SRS is the standard against which we measure other sampling methods, and the sampling method on which the theory of working with sampled data is based.

To select a sample at random, we first need to define a **sampling frame,** a list of individuals from which the sample will be drawn. For example, to draw a random sample of regular customers, a store might sample from its list of all "frequent buyers." In defining the sampling frame, we must deal with the details of defining the population. Are former frequent buyers who have moved away included? How about those who still live in the area but haven't shopped at our store in over a year? The answers to these questions may depend on the purpose of the survey.

Once we have a sampling frame, the easiest way to choose an SRS is with random numbers. We can assign a sequential number to each individual in the sampling frame. We then draw random numbers to identify those to be sampled. Let's look at an example.

> We want to select 5 students from the 80 enrolled in a Business Statistics class. We start by numbering the students from 00 to 79. Now we get a sequence of random digits from a table (such as the table in the back of this book), technology (most statistics packages and spreadsheets can generate random numbers), or the Internet (e.g., a site like www.random.org). For example, we might get 051662930577482. Taking those random numbers two digits at a time gives us 05, 16, 62, 93, 05, 77, and 48. We ignore 93 because no one had a number that high. And to avoid picking the same person twice, we also skip the repeated number 05. Our simple random sample consists of students with the numbers 05, 16, 62, 77, and 48.

Often the sampling frame is so large that it would be awkward to search through the list to locate each randomly selected individual. An alternative method is to generate random numbers of several digits in length, assigning one to each member of the sampling frame. Then *sort* the random numbers, *carrying along* the identities of the individuals in the sampling frame. (Spreadsheets and statistics programs typically can do this kind of sort.) Now you can pick a random sample of any size you like off the top of the sorted list.

Samples drawn at random generally differ one from another. Each draw of random numbers selects *different* people for our sample. These differences lead to different values for the variables we measure. We call these sample-to-sample differences **sampling variability**. Surprisingly, sampling variability isn't a problem; it's an opportunity. If different samples from a population vary little from each other, then most likely the underlying population harbors little variation. If the samples show much sampling variability, the underlying population probably varies a lot. In the coming chapters, we'll spend much time and attention working with sampling variability to better understand what we are trying to measure.

Sampling Errors vs. Bias
We referred to sample-to-sample variability earlier in this chapter as *sampling error*, making it sound like it's some kind of mistake. It's not. We understand that samples will vary, so "sampling errors" are to be expected. It's *bias* we must strive to avoid. Bias means our sampling method distorts our view of the population. Of course, bias leads to mistakes. Even more insidious, bias introduces errors that we cannot correct with subsequent analysis.

3.5 Other Sample Designs

Simple random sampling is not the only fair way to sample. More complicated designs may save time or money or avert sampling problems. All statistical sampling designs have in common the idea that chance, rather than human choice, is used to select the sample.

Stratified Sampling

Designs that are used to sample from large populations—especially populations residing across large areas—are often more complicated than simple random samples. Sometimes we slice the population into homogeneous groups, called **strata**, and then use simple random sampling within each stratum, combining the results at the end. This is called **stratified random sampling**.

Why would we want to stratify? Suppose we want to survey how shoppers feel about a potential new anchor store at a large suburban mall. The shopper population is 60% women and 40% men, and we suspect that men and women have different views on their choice of anchor stores. If we use simple random sampling to select 100 people for the survey, we could end up with 70 men and 30 women or

35 men and 65 women. Our resulting estimates of the attractiveness of a new anchor store could vary widely. To help reduce this sampling variability, we can force a representative balance, selecting 40 men at random and 60 women at random. This would guarantee that the proportions of men and women within our sample match the proportions in the population, and that should make such samples more accurate in representing population opinion.

You can imagine that stratifying by race, income, age, and other characteristics can be helpful, depending on the purpose of the survey. When we use a sampling method that restricts by strata, additional samples are more like one another, so statistics calculated for the sampled values will vary less from one sample to another. This reduced sampling variability is the most important benefit of stratifying.

Cluster and Multistage Sampling

Sometimes dividing the sample into homogeneous strata isn't practical, and even simple random sampling may be difficult. For example, suppose we wanted to assess the reading level of a product instruction manual based on the length of the sentences. Simple random sampling could be awkward; we'd have to number each sentence and then find, for example, the 576th sentence or the 2482nd sentence, and so on. Doesn't sound like much fun, does it?

We could make our task much easier by picking a few *pages* at random and then counting the lengths of the sentences on those pages. That's easier than picking individual sentences and works if we believe that the pages are all reasonably similar to one another in terms of reading level. Splitting the population in this way into parts or **clusters** that each represent the population can make sampling more practical. We select one or a few clusters at random and perform a census within each of them. This sampling design is called **cluster sampling**. If each cluster fairly represents the population, cluster sampling will generate an unbiased sample.

What's the difference between cluster sampling and stratified sampling? We stratify to ensure that our sample represents different groups in the population, and sample randomly within each stratum. This reduces the sample-to-sample variability. Strata are homogeneous, but differ from one another. By contrast, clusters are more or less alike, each heterogeneous and resembling the overall population. We cluster to save money or even to make the study practical.

Sometimes we use a variety of sampling methods together. In trying to assess the reading level of our instruction manual, we might worry that the "quick start" instructions are easy to read, but the "troubleshooting" chapter is more difficult. If so, we'd want to avoid samples that selected heavily from any one chapter. To guarantee a fair mix of sections, we could randomly choose one section from each chapter of the manual. Then we would randomly select a few pages from each of those sections. If altogether that made too many sentences, we might select a few sentences at random from each of the chosen pages. So, what is our sampling strategy? First we stratify by the chapter of the manual and randomly choose a section to represent each stratum. Within each selected section, we choose pages as clusters. Finally, we consider an SRS of sentences within each cluster. Sampling schemes that combine several methods are called **multistage samples**. Most surveys conducted by professional polling organizations and market research firms use some combination of stratified and cluster sampling as well as simple random samples.

Systematic Samples

Sometimes we draw a sample by selecting individuals systematically. For example, a **systematic sample** might select every tenth person on an alphabetical list of employees. To make sure our sample is random, we still must start the systematic selection with a randomly selected individual—not necessarily the first person on

Strata or Clusters?
We create strata by dividing the population into groups of similar individuals so that each stratum is different from the others. (For example, we often stratify by age, race, or sex.) By contrast, we create clusters that all look pretty much alike, each representing the wide variety of individuals seen in the population.

JUST CHECKING

2 We need to survey a random sample of the 300 passengers on a flight from San Francisco to Tokyo. Name each sampling method described below.

a) Pick every tenth passenger as people board the plane.

b) From the boarding list, randomly choose five people flying first class and 25 of the other passengers.

c) Randomly generate 30 seat numbers and survey the passengers who sit there.

d) Randomly select a seat position (right window, right center, right aisle, etc.) and survey all the passengers sitting in those seats.

the list. When there is no reason to believe that the order of the list could be associated in any way with the responses measured, systematic sampling can give a representative sample. Systematic sampling can be much less expensive than true random sampling. When you use a systematic sample, you should justify the assumption that the systematic method is not associated with any of the measured variables.

Think about the reading level sampling example again. Suppose we have chosen a section of the manual at random, then three pages at random from that section, and now we want to select a sample of 10 sentences from the 73 sentences found on those pages. Instead of numbering each sentence so we can pick a simple random sample, it would be easier to sample systematically. A quick calculation shows 73/10 = 7.3, so we can get our sample by picking every seventh sentence on the page. But where should you start? At random, of course. We've accounted for $10 \times 7 = 70$ of the sentences, so we'll throw the extra three into the starting group and choose a sentence at random from the first 10. Then we pick every seventh sentence after that and record its length.

GUIDED EXAMPLE   *Market Demand Survey*

In a course at a business school in the United States, the students form business teams, propose a new product, and use seed money to launch a business to sell the product on campus.

Before committing funds for the business, each team must complete the following assignment: "Con-

duct a survey to determine the potential market demand on campus for the product you are proposing to sell." Suppose your team's product is a 500-piece jigsaw puzzle of the map of your college campus. Design a marketing survey and discuss the important issues to consider.

PLAN

Setup State the goals and objectives of the survey (the *Why*).

Our team designed a study to find out how likely students at our school are to buy our proposed product—a 500-piece jigsaw puzzle of the map of our college campus.

Population and Parameters Identify the population to be studied and the associated sampling frame. The *What* identifies the parameters of interest and the variables measured. The *Who* is the sample of people we draw.

The population studied will be students at our school. We have obtained a list of all students currently enrolled to use as the sampling frame. The parameter of interest is the proportion of students likely to buy this product. We'll also collect some demographic information about the respondents.

Sampling Plan Specify the sampling method and the sample size, *n*. Specify how the sample was actually drawn. What is the sampling frame?

The description should, if possible, be complete enough to allow someone to replicate the procedure, drawing another sample from the same population in the same manner. A good description of the procedure is essential, even if it could never practically be repeated. The question you ask is important, so state the wording of the question clearly. Be sure that the question is useful in helping you with the overall goal of the survey.

We will select a simple random sample of students. We decided against stratifying by sex or class because we thought that students were all more or less alike in their likely interest in our product.

We will ask the students we contact:

Do you solve jigsaw puzzles for fun?

Then we will show them a prototype puzzle and ask:

If this puzzle sold for $10, would you purchase one?

We will also record the respondent's sex and class.

DO

Sampling Practice Specify *When, Where,* and *How* the sampling will be performed. Specify any other details of your survey, such as how respondents were contacted, any incentives that were offered to encourage them to respond, how nonrespondents were treated, and so on.

The survey will be administered in the middle of the fall semester during October. We have a master list of registered students, which we will randomize by matching it with random numbers from www.random.org and sorting on the random numbers, carrying the names. We will contact selected students by phone or email and arrange to meet with them. If a student is unwilling to participate, the next name from the randomized list will be substituted until a sample of 200 participants is found.

We will meet with students in an office set aside for this purpose so that each will see the puzzle under similar conditions.

REPORT

Summary and Conclusion This report should include a discussion of all the elements needed to design the study. It's good practice to discuss any special circumstances or other issues that may need attention.

MEMO:

Re: Survey Plans

Our team's plans for the puzzle market survey call for a simple random sample of students. Because subjects need to be shown the prototype puzzle, we must arrange to meet with selected participants. We have arranged an office for that purpose.

We will also collect demographic information so we can determine whether there is in fact a difference in interest level among classes or between men and women.

3.6 Defining the Population

The *Who* of a survey can refer to different groups, and the resulting ambiguity can tell you a lot about the success of a study. To start, you should think about the population of interest. Often, this is not a well-defined group. For example, who, exactly, is a mall "shopper": only the hurrying couples already carrying a purchase, or should we include people eating at the food court? How about teenagers outside the mall's video store, who may be carrying purchases or just hanging out, or both? Even when the population is clear, it may not be a practical group to study. For example, election polls want to sample from all those who will vote in the next election—a population that is particularly tricky to identify before election day.

Second, you must specify the sampling frame. Usually, the sampling frame is not the group you *really* want to know about, and sometimes it's actually much smaller. The sampling frame limits what your survey can find out.

Then there's your target sample. These are the individuals for whom you *intend* to measure responses. You're not likely to get responses from all of them. ("I know it's dinner time, but I'm sure you wouldn't mind answering a few questions. It'll only take 20 minutes or so. Oh, you're busy?") Nonresponse is a problem in many surveys.

Finally, there is your sample—the actual respondents. These are the individuals about whom you *do* get data and can draw conclusions. Unfortunately, they might not be representative of either the sampling frame or the population.

At each step, the group we can study may be constrained further. The *Who* keeps changing, and each constraint can introduce biases. A careful study should address the question of how well each group matches the population of interest. One of the main benefits of simple random sampling is that it never loses its sense of who's *Who*. The *Who* in an SRS is the population of interest from which we've drawn a representative sample. That's not always true for other kinds of samples.

When people (or committees!) decide on a survey, they often fail to think through the important questions about who are the *Who* of the study and whether they are the individuals about whom the answers would be interesting or have meaningful business consequences. This is a key step in performing a survey and should not be overlooked.

> The population is determined by the *Why* of the study. Unfortunately, the sample is just those we can reach to obtain responses—the *Who* of the study. This difference could undermine even a well-designed study.

CALVIN AND HOBBES © 1993 Watterson. Reprinted with permission of Universal Press Syndicate. All rights reserved.

3.7 The Valid Survey

It isn't sufficient to draw a sample and start asking questions. You want to feel confident your survey can yield the information you need about the population you are interested in. We want a *valid survey*.

To help ensure a valid survey, you need to ask four questions:

- ◆ What do I want to know?
- ◆ Who are the right respondents?
- ◆ What are the right questions?
- ◆ What will be done with the results?

These questions may seem obvious, but there are a number of specific pitfalls to avoid:

Know what you want to know.   Far too often, decisionmakers decide to perform a survey without any clear idea of what they hope to learn. Before considering a survey, you must be clear about what you hope to learn and about whom you hope to learn it. If you don't know that, you can't even judge whether you have a valid survey. The survey *instrument*—the questionnaire itself—can be a source of errors. Perhaps the most common error is to ask unnecessary questions. The longer the survey, the fewer people will complete it, leading to greater nonresponse bias. For each question on your survey, you should ask yourself whether you really want to know this and what you would do with the responses if you had them. If you don't have a good use for the answer to a question, don't ask it.

Use the right sampling frame.   A valid survey obtains responses from appropriate respondents. Be sure you have a suitable sampling frame. Have you identified the population of interest and sampled from it appropriately? A company looking to expand its base might survey customers who returned warrantee registration cards—after all, that's a readily available sampling frame—but if the company wants to know how to make its product more attractive, it needs to survey customers who rejected its product in favor of a competitor's product. This is the population that can tell the company what about its product needs to change to capture a larger market share. The errors in the presidential election polls of 1948 were likely due to the use of telephone samples in an era when telephones were not affordable by the less affluent—who were the folks most likely to vote for Truman.

It is equally important to be sure that your respondents actually know the information you hope to discover. Your customers may not know much about the competing products, so asking them to compare your product with others may not yield useful information.

Ask specific rather than general questions.   It is better to be specific. "Do you usually recall TV commercials?" won't be as useful as "How many TV commercials can you recall from last night?" or better, yet, "Please describe for me all the TV commercials you can recall from your viewing last night."

Watch for biases.   Even with the right sampling frame, you must beware of bias in your sample. If customers who purchase more expensive items are less likely to respond to your survey, this can lead to **nonresponse bias**. Although you can't expect all mailed surveys to be returned, if those individuals who don't respond have common characteristics, your sample will no longer represent the population you hope to learn about. Surveys in which respondents volunteer to participate, such as online surveys, suffer from **voluntary response bias**. Individuals with the strongest feelings on either side of an issue are more likely to respond; those who don't care may not bother.

Be careful with question phrasing.   Questions must be carefully worded. A respondent may not understand the question—or may not understand the question the way the researcher intended it. For example, "Does anyone in your family own a Ford truck?" leaves the term "family" unclear. Does it include only spouses and children or parents and siblings, or do in-laws and second cousins count too? A question like "Was your Twinkie fresh?" might be interpreted quite differently by different people.

Be careful with answer phrasing.   Respondents and survey-takers may also provide inaccurate responses, especially when questions are politically or sociologically sensitive. This also applies when the question does not take into account all possible answers, such as a true-false or multiple-choice question to which there may be other answers. Or the respondent may not know the correct answer to the question on the survey. In 1948, there were four major candidates for President,[3] but some survey respondents might not have been able to name them all. A survey question that just asked "Who do you plan to vote for?" might have underrepresented the less prominent candidates. And one that just asked "What do you think of Wallace?" might yield inaccurate results from voters who simply didn't know who he was. We refer to inaccurate responses (intentional or unintentional) as **measurement errors.** One way to cut down on measurement errors is to provide a range of possible responses. But be sure to phrase them in neutral terms.

The best way to protect a survey from measurement errors is to perform a pilot test. In a **pilot test,** a small sample is drawn from the sampling frame, and a draft form of the survey instrument is administered. A pilot test can point out flaws in the instrument. For example, during a staff cutback at one of our schools, a researcher surveyed faculty members to ask how they felt about the reduction in staff support. The scale ran from "It's a good idea" to "I'm very unhappy." Fortunately, a pilot study showed that everyone was very unhappy or worse. The scale was re-tuned to run from "unhappy" to "ready to quit."

WHAT CAN GO WRONG?—OR, HOW TO SAMPLE BADLY

Bad sample designs yield worthless data. Many of the most convenient forms of sampling can be seriously biased. And there is no way to correct for the bias from a bad sample. So it's wise to pay attention to sample design—and to beware of reports based on poor samples.

Voluntary Response Sample

One of the most common dangerous sampling methods is the voluntary response sample. In a **voluntary response sample,** a large group of individuals is invited to respond, and all who do respond are counted. This method is used by call-in shows, 900 numbers, Internet polls, and letters written to members of

---

[3] Harry Truman, Thomas Dewey, Strom Thurmond, and Henry Wallace.

Congress. Voluntary response samples are almost always biased, and so conclusions drawn from them are almost always wrong.

It's often hard to define the sampling frame of a voluntary response study. Practically, the frames are groups such as Internet users who frequent a particular website or viewers of a particular TV show. But those sampling frames don't correspond to the population you are likely to be interested in.

Even if the sampling frame is of interest, voluntary response samples are often biased toward those with strong opinions or those who are strongly motivated—and especially from those with strong negative opinions. A request that travelers who have used the local airport visit a survey site to report on their experiences is much more likely to hear from those who had long waits, cancelled flights, and lost luggage than from those whose flights were on time and carefree. The resulting voluntary response bias invalidates the survey.

Convenience Sampling

Another sampling method that doesn't work is convenience sampling. As the name suggests, in **convenience sampling** we simply include the individuals who are convenient. Unfortunately, this group may not be representative of the population. A survey of 437 potential home buyers in Orange County, California, found, among other things, that

> all but 2 percent of the buyers have at least one computer at home, and 62 percent have two or more. Of those with a computer, 99 percent are connected to the Internet (Jennifer Hieger, "Portrait of Homebuyer Household: 2 Kids and a PC," Orange County Register, July 27, 2001).

Later in the article, we learn that the survey was conducted via the Internet. That was a convenient way to collect data and surely easier than drawing a simple random sample, but perhaps home builders shouldn't conclude from this study that *every* family has a computer and an Internet connection.

Many surveys conducted at shopping malls suffer from the same problem. People in shopping malls are not necessarily representative of the population of interest. Mall shoppers tend to be more affluent and include a larger percentage of teenagers and retirees than the population at large. To make matters worse, survey interviewers tend to select individuals who look "safe," or easy to interview.

Convenience sampling is not just a problem for beginners. In fact, convenience sampling is a widespread problem in the business world. When a company wants to find out what people think about its products or services, it may turn to the easiest people to sample: its own customers. But the company will never learn how those who *don't* buy its product feel about it.

Bad Sampling Frame?

An SRS from an incomplete sampling frame introduces bias because the individuals included may differ from the ones not in the frame. It may be easier to sample workers from a single site, but if a company has many sites and they differ in worker satisfaction, training, or job descriptions, the resulting sample can be biased. There is serious concern among professional pollsters that the increasing numbers of people who can be reached only by cell phone may bias telephone-based market research and polling.

(continued)

Do you use the Internet?
Click here ⬤ for yes
Click here ⬤ for no

Internet convenience surveys are often worthless. As voluntary response surveys, they have no well-defined sampling frame (all those who use the Internet and visit their site?) and thus report no useful information. Do not use them.

Undercoverage

Many survey designs suffer from **undercoverage, in which some portion of the population is not sampled at all or has a smaller representation in the sample than it has in the population.** Undercoverage can arise for a number of reasons, but it's always a potential source of bias. Are people who use answering machines to screen callers (and are thus less available to blind calls from market researchers) different from other customers in their purchasing preferences?

WHAT *Else* CAN GO WRONG?

- **Nonrespondents.** No survey succeeds in getting responses from everyone. The problem is that those who don't respond may differ from those who do. And if they differ on just the variables we care about, the lack of response will bias the results. Rather than sending out a large number of surveys for which the response rate will be low, it is often better to design a smaller, randomized survey for which you have the resources to ensure a high response rate.

- **Long, dull surveys.** Surveys that are too long are more likely to be refused, reducing the response rate and biasing all the results. Keep it short.

- **Response bias.** Response bias includes the tendency of respondents to tailor their responses to please the interviewer and the consequences of slanted question wording.

THE WIZARD OF ID   parker and hart

Push polls, which masquerade as surveys, present one side of an issue before asking a question. For example, a question like

> Would the fact that the new store that just opened by the mall sells mostly goods made overseas by workers in sweatshop conditions influence your decision to shop there rather than in the downtown store that features American-made products?

is designed not to gather information, but to spread ill-will toward the new store.

How to Think about Biases

- **Look for biases in any survey.** If you design a survey of your own, ask someone else to help look for biases that may not be obvious to you. Do this *before* you collect your data. There's no way to recover from a biased sample or a survey that asks biased questions.

   A bigger sample size for a biased study just gives you a bigger useless study. A really big sample gives you a really big useless study.

- **Spend your time and resources reducing biases.** No other use of resources is as worthwhile as reducing the biases.

- **If you possibly can, pretest or pilot your survey.** Administer the survey in the exact form that you intend to use it to a small sample drawn from the population you intend to sample. Look for misunderstandings, misinterpretation, confusion, or other possible biases. Then redesign your survey instrument.

- **Always report your sampling methods in detail.** Others may be able to detect biases where you did not expect to find them.

ETHICS IN ACTION

The Lackawax River Group is interested in applying for state funds in order to continue their restoration and conservation of the Lackawax River, a river that has been polluted from years of industry and agricultural discharge. While they have managed to gain significant support for their cause through education and community involvement, the executive committee is now interested in presenting the state with more compelling evidence. They decided to survey local residents regarding their attitudes toward the proposed expansion of the river restoration and conservation project. With limited time and money (the deadline for the grant application was fast approaching), the executive committee was delighted that one of its members, Harry Greentree, volunteered to undertake the project. Harry owned a local organic food store and agreed to have a sample of his shoppers interviewed during the next one-week period. The only concern that the committee had was that the shoppers be selected in a systematic fashion, for instance, by interviewing every fifth person who entered the store. Harry had no problem with this request and was eager to help the Lackawax River Group.

ETHICAL ISSUE *Introducing bias into the results (even if not intentional). One might expect consumers of organic food to be more concerned about the environment than the general population (related to Item C, ASA Ethical Guidelines).*

ETHICAL SOLUTION *Harry is using a convenience sample from which results cannot be generalized. If the Lackawax River Group cannot improve their sampling scheme and survey design (for example, for lack of expertise or time), they should openly discuss the weaknesses of their sampling method when they disclose details of their study. When reporting the results, they should note that their findings are from a convenience sample and include an appropriate disclaimer.*

What have we learned?

We've learned that a representative sample can offer important insights about populations. It's the size of the sample—and not its fraction of the larger population—that determines the precision of the statistics it yields.

We've learned several ways to draw samples, all based on the power of randomness to make them representative of the population of interest:

- A simple random sample (SRS) is our standard. Every possible group of *n* individuals has an equal chance of being our sample. That's what makes it *simple*.

- Stratified samples can reduce sampling variability by identifying homogeneous subgroups and then randomly sampling within each.

- Cluster samples randomly select among heterogeneous subgroups that each resemble the population at large, making our sampling tasks more manageable.

- Systematic samples can work in some situations and are often the least expensive method of sampling. But we still want to start them randomly.

- Multistage samples combine several random sampling methods.

We've learned that bias can destroy our ability to gain insights from our sample:

- Nonresponse bias can arise when sampled individuals will not or cannot respond.

- Response bias arises when respondents' answers might be affected by external influences, such as question wording or interviewer behavior.

We've learned that bias can also arise from poor sampling methods:

- Voluntary response samples are almost always biased and should be avoided and distrusted.

- Convenience samples are likely to be flawed for similar reasons.

- Even with a reasonable design, sample frames may not be representative. Undercoverage occurs when individuals from a subgroup of the population are selected less often than they should be.

Finally, we've learned to look for biases in any survey we find and to be sure to report our methods whenever we perform a survey so that others can evaluate the fairness and accuracy of our results.

Terms

Biased	Any systematic failure of a sampling method to represent its population.
Census	An attempt to collect data on the entire population of interest.
Cluster	A representative subset of a population chosen for reasons of convenience, cost, or practicality.
Cluster sampling	A sampling design in which groups, or clusters, representative of the population are chosen at random and a census is then taken of each.

Convenience sampling	A sample that consists of individuals who are conveniently available.
Multistage sample	Sampling schemes that combine several sampling methods.
Nonresponse bias	Bias introduced to a sample when a large fraction of those sampled fails to respond.
Parameter	A numerically valued attribute of a model for a population. We rarely expect to know the value of a parameter, but we do hope to estimate it from sampled data.
Pilot test	A small trial run of a study to check that the methods of the study are sound.
Population	The entire group of individuals or instances about whom we hope to learn.
Population parameter	A numerically valued attribute of a model for a population.
Randomization	A defense against bias in the sample selection process, in which each individual is given a fair, random chance of selection.
Representative sample	A sample from which the statistics computed accurately reflect the corresponding population parameters.
Response bias	Anything in a survey design that influences responses.
Sample	A subset of a population, examined in hope of learning about the population.
Sample size	The number of individuals in a sample.
Sample survey	A study that asks questions of a sample drawn from some population in the hope of learning something about the entire population.
Sampling frame	A list of individuals from which the sample is drawn. Individuals in the population of interest but who are not in the sampling frame cannot be included in any sample.
Sampling variability (or sampling error)	The natural tendency of randomly drawn samples to differ, one from another. Sometimes called *sampling error*.
Simple random sample (SRS)	A sample in which each set of n elements in the population has an equal chance of selection.
Statistic, sample statistic	A value calculated for sampled data, particularly one that corresponds to, and thus estimates, a population parameter. The term "sample statistic" is sometimes used, usually to parallel the corresponding term "population parameter."
Strata	Subsets of a population that are internally homogeneous but may differ one from another.
Stratified random sample	A sampling design in which the population is divided into several homogeneous subpopulations, or strata, and random samples are then drawn from each stratum.
Systematic sampling	A sample drawn by selecting individuals systematically from a sampling frame.
Voluntary response bias	Bias introduced to a sample when individuals can choose on their own whether to participate in the sample.

Voluntary response sample	A sample in which a large group of individuals are invited to respond and decide individually whether or not to participate. Voluntary response samples are generally worthless.
Undercoverage	A sampling scheme that biases the sample in a way that gives a part of the population less representation than it has in the population.

Skills

 PLAN

- Know the basic concepts and terminology of sampling.
- Be able to recognize population parameters in descriptions of populations and samples.
- Understand the value of randomization as a defense against bias.
- Understand the value of sampling to estimate population parameters from statistics calculated on representative samples drawn from the population.
- Understand that the size of the sample (not the fraction of the population) determines the precision of estimates.

 DO

- Know how to draw a simple random sample from a master list of a population, using a computer or a table of random numbers.

 REPORT

- Know what to report about a sample as part of your account of a statistical analysis.
- Be sure to report possible sources of bias in sampling methods. Recognize voluntary response and nonresponse as sources of bias in a sample survey.

Technology Help: Random Sampling

Computer-generated pseudorandom numbers are usually quite good enough for drawing random samples. But there is little reason not to use the truly random values available on the Internet. Here's a convenient way to draw an SRS of a specified size using a computer-based sampling frame. The sampling frame can be a list of names or of identification numbers arrayed, for example, as a column in a spreadsheet, statistics program, or database:

1. Generate random numbers of enough digits so that each exceeds the size of the sampling frame list by several digits. This makes duplication unlikely.

2. Assign the random numbers arbitrarily to individuals in the sampling frame list. For example, put them in an adjacent column.

3. Sort the list of random numbers, *carrying* along the sampling frame list.

4. Now the first *n* values in the sorted sampling frame column are an SRS of *n* values from the entire sampling frame.

Mini Case Study Projects

Market Survey Research

You are part of a marketing team that needs to research the potential of a new product. Your team decides to e-mail an interactive survey to a random sample of consumers. Write a short questionnaire that will generate the information you need about the new product. Select a sample of 200 using an SRS from your sampling frame. Discuss how you will collect the data and how the responses will help your market research.

The GfK Roper Reports Worldwide Survey

GfK Roper Consulting conducts market research for multinational companies who want to understand attitudes in different countries so they can market and advertise more effectively to different cultures. Every year they conduct a poll worldwide, which asks hundreds of questions of people in approximately 30 different countries. Respondents are asked a variety of questions about food. Some of the questions are simply yes/no (agree/disagree) questions: Please tell me whether you agree or disagree with each of these statements about your appearance: (Agree = 1; Disagree = 2; Don't know = 9).

> The way you look affects the way you feel.
>
> I am very interested in new skin care breakthroughs.
>
> People who don't care about their appearance don't care about themselves.

Other questions are asked on a 5 point scale (Please tell me the extent to which you disagree or agree with it using the following scale: Disagree completely = 1; Disagree somewhat = 2; Neither disagree nor agree = 3; Agree somewhat = 4; Agree completely = 5; Don't know = 9).

Examples of such questions include:

> I read labels carefully to find out about ingredients, fat content, and/or calories.
>
> I try to avoid eating fast food.

and

> When it comes to food I'm always on the lookout for something new.

Think about designing a survey on such a global scale:

- What is the population of interest?
- Why might it be difficult to select an SRS from this sampling frame?
- What are some potential sources of bias?
- Why might it be difficult to ensure a representative number of men and women and all age groups in some countries?
- What might be a reasonable sampling frame?

EXERCISES

1. Roper. As discussed in the chapter, GfK Roper Consulting conducts a global consumer survey to help multinational companies understand different consumer attitudes throughout the world. In India, the researchers interviewed 1000 people aged 13–65. Their sample is designed so that they get 500 males and 500 females. (www.gfkamerica.com)

a) Are they using a simple random sample? How do you know?

b) What kind of design do you think they are using?

2. Coffee shop survey. For their class project, a group of Business students decide to survey the student body to assess opinions about a proposed new student coffee shop to judge how successful it might be. Their sample of 200 contained 50 first-year students, 50 sophomores, 50 juniors, and 50 seniors.

a) Do you think the group was using an SRS? Why?

b) What kind of sampling design do you think they used?

3. Software licenses. The website www.gamefaqs.com asked, as their question of the day to which visitors to the site were invited to respond, *"Do you ever read the end-user license agreements when installing software or games?"* Of the 98,574 respondents, 63.47% said they never read those agreements—a fact that software manufacturers might find important.

a) What kind of sample was this?

b) How much confidence would you place in using 63.47% as an estimate of the fraction of people who don't read software licenses?

4. Drugs in baseball. Major League Baseball, responding to concerns about their "brand," tests players to see whether they are using performance-enhancing drugs. Officials select a team at random, and a drug-testing crew shows up unannounced to test all 40 players on the team. Each testing day can be considered a study of drug use in Major League Baseball.

a) What kind of sample is this?

b) Is that choice appropriate?

5. Gallup. At its website (www.galluppoll.com) the Gallup Poll publishes results of a new survey each day. Scroll down to the end, and you'll find a statement that includes an explanation such as this:

Results are based on telephone interviews with 1,008 national adults, aged 18 and older, conducted April 2–5, 2007. . . . In addition to sampling error, question wording and practical difficulties in conducting surveys can introduce error or bias into the findings of public opinion polls.

a) For this survey, identify the population of interest.

b) Gallup performs its surveys by phoning numbers generated at random by a computer program. What is the sampling frame?

c) What problems, if any, would you be concerned about in matching the sampling frame with the population?

6. Defining the survey. At its website (www.gallupworldpoll. com) the Gallup World Poll reports results of surveys conducted in various places around the world. At the end of one of these reports, they describe their methods, including explanations such as the following:

Results are based on face-to-face interviews with randomly selected national samples of approximately 1,000 adults, aged 15 and older, who live permanently in each of the 21 sub-Saharan African nations surveyed. Those countries include Angola (areas where land mines might be expected were excluded), Benin, Botswana, Burkina Faso, Cameroon, Ethiopia, Ghana, Kenya, Madagascar (areas where interviewers had to walk more than 20 kilometers from a road were excluded), Mali, Mozambique, Niger, Nigeria, Senegal, Sierra Leone, South Africa, Tanzania, Togo, Uganda (the area of activity of the Lord's Resistance Army was excluded from the survey), Zambia, and Zimbabwe. . . . In all countries except Angola, Madagascar, and Uganda, the sample is representative of the entire population.

a) Gallup is interested in sub-Saharan Africa. What kind of survey design are they using?

b) Some of the countries surveyed have large populations. (Nigeria is estimated to have about 130 million people.) Some are quite small. (Togo's population is estimated at 5.4 million.) Nonetheless, Gallup sampled 1000 adults in each country. How does this affect the precision of its estimates for these countries?

7–16. Survey details. *For the following reports about statistical studies, identify the following items (if possible). If you can't tell, then say so—this often happens when we read about a survey.*

a) The population

b) The population parameter of interest

c) The sampling frame

d) The sample

e) The sampling method, including whether or not randomization was employed

f) Any potential sources of bias you can detect and any problems you see in generalizing to the population of interest

7. HR directors. A business magazine mailed a questionnaire to the human resources directors of all Fortune 500 companies, and received responses from 23% of them. Those responding reported that they did not find that such surveys intruded significantly on their workday.

8. Health insurance. A question posted on the Lycos website asked visitors to the site to say whether they thought that businesses should be required to pay for their employees' health insurance.

9. Alternative medicine. Consumers Union asked all subscribers whether they had used alternative medical treatments and, if so, whether they had benefited from them. For almost all of the treatments, approximately 20% of those responding reported cures or substantial improvement in their condition.

10. Global warming. The Gallup Poll interviewed 1007 randomly selected U.S. adults aged 18 and older, March 23–25, 2007. Gallup reports that when asked when (if ever) the effects of global warming will begin to happen, 60% of respondents said the effects had already begun. Only 11% thought that they would never happen.

11. At the bar. Researchers waited outside a bar they had randomly selected from a list of such establishments. They stopped every 10th person who came out of the bar and asked whether he or she thought drinking and driving was a serious problem.

12. Election poll. Hoping to learn what issues may resonate with voters in the coming election, the campaign director for a mayoral candidate selects one block from each of the city's election districts. Staff members go there and interview all the residents they can find.

13. Toxic waste. The Environmental Protection Agency took soil samples at 16 locations near a former industrial waste dump and checked each for evidence of toxic chemicals. They found no elevated levels of any harmful substances.

14. Housing discrimination. Inspectors send trained "renters" of various races and ethnic backgrounds, and of both sexes to inquire about renting randomly assigned advertised apartments. They look for evidence that landlords deny access illegally based on race, sex, or ethnic background.

15. Quality control. A company packaging snack foods maintains quality control by randomly selecting 10 cases from each day's production and weighing the bags. Then they open one bag from each case and inspect the contents.

16. Contaminated milk. Dairy inspectors visit farms unannounced and take samples of the milk to test for contamination. If the milk is found to contain dirt, antibiotics, or other foreign matter, the milk will be destroyed and the farm is considered to be contaminated pending further testing.

17. Pulse poll. A local TV station conducted a "PulsePoll" to predict the winner in the upcoming mayoral election. Evening news viewers were invited to phone in their votes, with the results to be announced on the late-night news. Based on the phone calls, the station predicted that Amabo would win the election with 52% of the vote. They were wrong: Amabo lost, getting only 46% of the vote. Do you think the station's faulty prediction is more likely to be a result of bias or sampling error? Explain.

18. Paper poll. Prior to the mayoral election discussed in Exercise 17, the newspaper also conducted a poll. The paper surveyed a random sample of registered voters stratified by political party, age, sex, and area of residence. This poll predicted that Amabo would win the election with 52% of the vote. The newspaper was wrong: Amabo lost, getting only 46% of the vote. Do you think the newspaper's faulty prediction is more likely to be a result of bias or sampling error? Explain.

19. Cable company market research. A local cable TV company, Pacific TV, with customers in 15 towns is considering offering high-speed Internet service on its cable lines. Before launching the new service they want to find out whether customers would pay the $50 per month that they plan to charge. An intern has prepared several alternative plans for assessing customer demand. For each, indicate what kind of sampling strategy is involved and what (if any) biases might result.
a) Put a big ad in the newspaper asking people to log their opinions on the PTV website.
b) Randomly select one of the towns and contact every cable subscriber by phone.
c) Send a survey to each customer and ask them to fill it out and return it.
d) Randomly select 20 customers from each town. Send them a survey, and follow up with a phone call if they do not return the survey within a week.

20. Cable company market research, part 2. Four new sampling strategies have been proposed to help PTV determine whether enough cable subscribers are likely to purchase high-speed Internet service. For each, indicate what kind of sampling strategy is involved and what (if any) biases might result.
a) Run a poll on the local TV news, asking people to dial one of two phone numbers to indicate whether they would be interested.
b) Hold a meeting in each of the 15 towns, and tally the opinions expressed by those who attend the meetings.
c) Randomly select one street in each town and contact each of the households on that street.
d) Go through the company's customer records, selecting every 40th subscriber. Send employees to those homes to interview the people chosen.

21. Churches. For your marketing class, you'd like to take a survey from a sample of all the Catholic Church members in your city to assess the market for a DVD about the pope's visit to the United States. A list of churches shows 17 Catholic churches within the city limits. Rather than try to obtain a list of all members of all these churches, you decide to pick 3 churches at random. For those churches, you'll ask to get a list of all current members and contact 100 members at random.
a) What kind of design have you used?
b) What could go wrong with the design that you have proposed?

22. Fishing in the Great Lakes. The U.S. Fish and Wildlife Service plans to study the fishing industry around Saginaw Bay. To do that, they decide to randomly select 5 fishing boats at the end of a randomly chosen fishing day and to count the numbers and types of all the fish on those boats.
a) What kind of design have they used?
b) What could go wrong with the design that they have proposed?

23. Amusement park riders. An amusement park has opened a new roller coaster. It is so popular that people are waiting for up to 3 hours for a 2-minute ride. Concerned about how patrons (who paid a large amount to enter the park and ride on the rides) feel about this, they survey every 10th person on the line for the roller coaster, starting from a randomly selected individual.

a) What kind of sample is this?

b) Is it likely to be representative?

c) What is the sampling frame?

24. Playground. Some people have been complaining that the children's playground at a municipal park is too small and is in need of repair. Managers of the park decide to survey city residents to see if they believe the playground should be rebuilt. They hand out questionnaires to parents who bring children to the park. Describe possible biases in this sample.

25. Survey wording. The intern designing the study of high speed Internet service for exercises 19 and 20 has proposed some questions that might be used in the surveys.

Question 1: If PTV offered state-of-the-art high-speed Internet service for $50 per month, would you subscribe to that service?

Question 2: Would you find $50 per month—less than the cost of a daily cappuccino—an appropriate price for high-speed Internet service?

a) Do you think these are appropriately worded questions? Why or why not?

b) Which one has more neutral wording? Explain.

26. More words. Here are more proposed survey questions.

Question 3: Do you find that the slow speed of dial-up Internet access reduces your enjoyment of web services?

Question 4: Given the growing importance of high-speed Internet access for your children's education, would you subscribe to such a service if it were offered?

a) Do you think these are appropriately worded questions? Why or why not?

b) Propose a question with more neutral wording.

27. Another ride. The survey of patrons waiting in line for the roller coaster in Exercise 23 asks whether they think it is worthwhile to wait a long time for the ride and whether they'd like the amusement park to install still more roller coasters. What biases might cause a problem for this survey?

28. Playground bias. The survey described in Exercise 24 asked, *Many people believe this playground is too small and in need of repair. Do you think the playground should be repaired and expanded even if that means raising the entrance fee to the park?*

Describe two ways this question may lead to response bias.

29. (Possibly) Biased questions. Examine each of the following questions for possible bias. If you think the question is biased, indicate how and propose a better question.

a) *Should companies that pollute the environment be compelled to pay the costs of cleanup?*

b) *Should a company enforce a strict dress code?*

30. More possibly biased questions. Examine each of the following questions for possible bias. If you think the question is biased, indicate how and propose a better question.

a) *Do you think that price or quality is more important in selecting an MP3 player?*

b) *Given humanity's great tradition of exploration, do you favor continued funding for space flights?*

31. Phone surveys. Anytime we conduct a survey, we must take care to avoid undercoverage. Suppose we plan to select 500 names from the city phone book, call their homes between noon and 4 P.M., and interview whoever answers, anticipating contacts with at least 200 people.

a) Why is it difficult to use a simple random sample here?

b) Describe a more convenient, but still random, sampling strategy.

c) What kinds of households are likely to be included in the eventual sample of opinion? Who will be excluded?

d) Suppose, instead, that we continue calling each number, perhaps in the morning or evening, until an adult is contacted and interviewed. How does this improve the sampling design?

e) Random-digit dialing machines can generate the phone calls for us. How would this improve our design? Is anyone still excluded?

32. Cell phone survey. What about drawing a random sample only from cell phone exchanges? Discuss the advantages and disadvantages of such a sampling method compared with surveying randomly generated telephone numbers from non–cell phone exchanges. Do you think these advantages and disadvantages have changed over time? How do you expect they'll change in the future?

33. Change. How much change do you have on you right now? Go ahead, count it.

a) How much change do you have?

b) Suppose you check on your change every day for a week as you head for lunch and average the results. What parameter would this average estimate?

c) Suppose you ask 10 friends to average *their* change every day for a week, and you average those 10 measurements. What is the population now? What parameter would this average estimate?

d) Do you think these 10 average change amounts are likely to be representative of the population of change amounts in your class? In your college? In the country? Why or why not?

34. Fuel economy. Occasionally, when I fill my car with gas, I figure out how many miles per gallon my car got. I wrote down those results after 6 fill-ups in the past few months. Overall, it appears my car gets 28.8 miles per gallon.

a) What statistic have I calculated?

b) What is the parameter I'm trying to estimate?

c) How might my results be biased?

d) When the Environmental Protection Agency (EPA) checks a car like mine to predict its fuel economy, what parameter is it trying to estimate?

35. Accounting. Between quarterly audits, a company likes to check on its accounting procedures to address any problems before they become serious. The accounting staff processes payments on about 120 orders each day. The next day, the supervisor rechecks 10 of the transactions to be sure they were processed properly.

a) Propose a sampling strategy for the supervisor.

b) How would you modify that strategy if the company makes both wholesale and retail sales, requiring different bookkeeping procedures?

36. Happy workers? A manufacturing company employs 14 project managers, 48 foremen, and 377 laborers. In an effort to keep informed about any possible sources of employee discontent, management wants to conduct job satisfaction interviews with a simple random sample of employees every month.

a) Do you see any danger of bias in the company's plan? Explain.

b) How might you select a simple random sample?

c) Why do you think a simple random sample might not provide the representative opinion the company seeks?

d) Propose a better sampling strategy.

e) Listed below are the last names of the project managers. Use random numbers to select two people to be interviewed. Be sure to explain your method carefully.

Barrett	Bowman	Chen
DeLara	DeRoos	Grigorov
Maceli	Mulvaney	Pagliarulo
Rosica	Smithson	Tadros
Williams	Yamamoto	

37. Quality control. Sammy's Salsa, a small local company, produces 20 cases of salsa a day. Each case contains 12 jars and is imprinted with a code indicating the date and batch number. To help maintain consistency, at the end of each day, Sammy selects three bottles of salsa, weighs the contents, and tastes the product. Help Sammy select the sample jars. Today's cases are coded 07N61 through 07N80.

a) Carefully explain your sampling strategy.

b) Show how to use random numbers to pick the three jars for testing.

c) Did you use a simple random sample? Explain.

38. Fish quality. Concerned about reports of discolored scales on fish caught downstream from a newly sited chemical plant, scientists set up a field station in a shoreline public park. For one week they asked fishermen there to bring any fish they caught to the field station for a brief inspection. At the end of the week, the scientists said that 18% of the 234 fish that were submitted for inspection displayed the discoloration. From this information, can the researchers estimate what proportion of fish in the river have discolored scales? Explain.

39. Sampling methods. Consider each of these situations. Do you think the proposed sampling method is appropriate? Explain.

a) We want to know what percentage of local doctors accept Medicaid patients. We call the offices of 50 doctors randomly selected from local Yellow Page listings.

b) We want to know what percentage of local businesses anticipate hiring additional employees in the upcoming month. We randomly select a page in the Yellow Pages and call every business listed there.

40. More sampling methods. Consider each of these situations. Do you think the proposed sampling method is appropriate? Explain.

a) We want to know if business leaders in the community support the development of an "incubator" site at a vacant lot on the edge of town. We spend a day phoning local businesses in the phone book to ask whether they'd sign a petition.

b) We want to know if travelers at the local airport are satisfied with the food available there. We go to the airport on a busy day and interview every 10th person in line in the food court.

JUST CHECKING ANSWERS

1 a) It can be hard to reach all members of a population, and it can take so long that circumstances change, affecting the responses. A well-designed sample is often a better choice.

 b) This sample is probably biased—people who didn't like the food at the restaurant might not choose to eat there.

 c) No, only the sample size matters, not the fraction of the overall population.

 d) Students who frequent this website might be more enthusiastic about Statistics than the overall population of Statistics students. A large sample cannot compensate for bias.

 e) It's the population "parameter." "Statistics" describe samples.

2 a) systematic

 b) stratified

 c) simple

 d) cluster

4

Displaying and Describing Categorical Data

KEEN Inc.

KEEN Inc. was started to create a sandal designed for a variety of water activities. The sandals quickly became popular due to their unique patented toe protection—a black bumper to protect the toes when adventuring out on rivers and trails. Today the KEEN brand offers over 100 different outdoor performance and outdoor inspired casual footwear styles.

Few companies experience the kind of growth that KEEN did in less than four years. Amazingly, they've done this with relatively little advertising and by selling primarily to specialty footwear and outdoor stores, in addition to online outlets.

After the 2004 Tsunami disaster, KEEN cut its advertising budget almost completely and donated over $1 million to help the victims and establish the KEEN Foundation to support environmental and social causes. Philanthropy and community projects continue to play an integral

part of the KEEN brand values. In fact, the KEEN website has three sections: 1. HybridLife, a section devoted to consumers living a balanced and outdoor lifestyle, 2. Product Showcase, a website featuring the products that carry the KEEN brand name, and 3. the KEEN Foundation, a philanthropic effort devoted to helping the environment, conservation, and social movements involving the outdoors.

KEEN Footwear, like most companies, collects data on visits to its website. Each visit to the site and each subsequent action the visitor takes (changing the page, entering data etc.) is recorded in a file called a usage, or access weblog. These logs contain a lot of potentially worthwhile information, but they are not easy to use. Here's one line from a log:

245.240.221.71 - - [03/Jan/2007:15:20:06-0800]" GET http://www.keenfootwear.com/ pdp_page.cfm?productID=148" 200 8788 "http://www.google. com/" "Mozilla/3.0 WebTV/1.2 (compatible; MSIE 2.0)"

Unless the company has the analytic resources to deal with these files, it must rely on a third party to summarize the data. KEEN, like many other small and midsized companies, uses *Google Analytics* to collect and summarize its log data.

Imagine a whole table of data like the one above—with a line corresponding to every visit. In September 2006 there were 93,173 visits to the KEEN site, which would be a table with 93,173 rows. The problem with a file like this—and in fact even with data tables—is that we can't see what's going on. And seeing is exactly what we want to do. We need ways to show the data so that we can see patterns, relationships, trends, and exceptions.

4.1 The Three Rules of Data Analysis

There are three things you should always do with data:

1. **Make a picture.** A display of your data will reveal things you are not likely to see in a table of numbers and will help you to *plan* your approach to the analysis and think clearly about the patterns and relationships that may be hiding in your data.

2. **Make a picture.** A well-designed display will *do* much of the work of analyzing your data. It can show the important features and patterns. A picture will also reveal things you did not expect to see: extraordinary (possibly wrong) data values or unexpected patterns.

3. **Make a picture.** The best way to *report* to others what you find in your data is with a well-chosen picture.

These are the three rules of data analysis. These days, technology makes drawing pictures of data easy, so there is no reason not to follow the three rules. Here are some displays showing various aspects of traffic on one of the authors' websites.

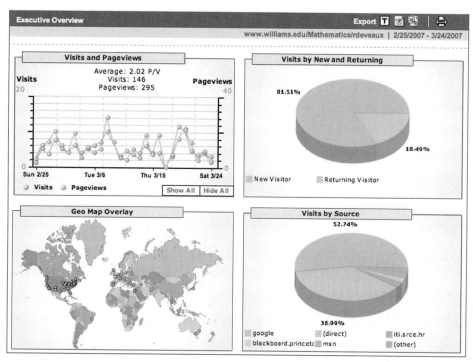

Figure 4.1 *Part of the output from Google Analytics (www.google.com) for the period Feb. 25 to March 24, 2007 displaying website traffic.*

Some displays communicate information better than others. We'll discuss some general principles for displaying information honestly in this chapter.

4.2 Frequency Tables

To make a picture of data, we start by putting the data into piles. We pile together things that seem to go together, so we can see how the cases distribute across different categories. For categorical data, piling is easy since we can just count the number of cases corresponding to each category.

For the Web data, one way to start might be to pile together all the visits that used the same search engine. KEEN Footwear might use that information in deciding where to advertise in the future. So, they might count how many visits come to the KEEN website from each search engine. All of these piles can be organized into a **frequency table** (Table 4.1) which records totals and category names. (The label "Direct" indicates that the visit resulted from directly typing the URL into a browser's address bar and not from a search engine query.)

The names of the categories label each row in the frequency table. For *Search Engine* these are "Google," "Direct," "Yahoo," and so on. Even with thousands of

Search Engine	Visits
Google	50,629
Direct	22,173
Yahoo	7,272
MSN	3,166
SnapLink	946
All Others	8,987
Total	**93,167**

Table 4.1 *A frequency table of the Search Engine used by visitors to the KEEN Footwear website.*

Search Engine	Visits by %
Google	54.34%
Direct	23.80%
Yahoo	7.80%
MSN	3.40%
SnapLink	1.02%
All Others	9.65%
Total	**100.00%**

Table 4.2   *A relative frequency table for the same data.*

100.01%?

If you are careful to add the percentages in Table 4.2, you will notice the total is 100.01%. Of course the real total has to be 100.00%. The discrepancy is due to individual percentages being rounded. You'll often see this in tables of percents, sometimes with explanatory footnotes.

cases, a variable that doesn't have too many categories produces a frequency table that is easy to read. A frequency table with dozens or hundreds of categories would be much harder to read. Notice the label of the last line of the table—"All Others." When the number of categories gets too large, we often lump together values of the variable into "Other." When to do that is a judgment call, but it's a good idea to have fewer than about a dozen categories.

Counts are useful, but sometimes we want to know the fraction or **proportion** of the data in each category, so we divide the counts by the total number of cases. Usually we multiply by 100 to express these proportions as **percentages.** A **relative frequency table** (Table 4.2) displays the *percentages*, rather than the counts, of the values in each category. Both types of tables show how the cases are distributed across the categories. In this way, they describe the **distribution** of a categorical variable because they name the possible categories and tell how frequently each occurs.

4.3 Charts

The Area Principle

Now that we have a frequency table, we're ready to follow the three rules of data analysis and make a picture of the data. But we can't make just any picture; a bad picture can distort our understanding rather than help it. For example, here's a graph of the frequencies of Table 4.1. What impression do you get of the relative frequencies of visits from each source?

While it's true that the majority of people came to KEEN's website from Google, in Figure 4.2 it looks like nearly all did. That doesn't seem right. What's wrong? The lengths of the sandals *do* match the totals in the table. But our eyes tend to be more impressed by the *area* (or perhaps even the *volume*) than by other aspects of each

Figure 4.2   *Although the length of each sandal corresponds to the correct number, the impression we get is all wrong because we perceive the entire area of the sandal. In fact, only a little more than 50% of all visitors used Google to get to the website.*

sandal image, and it's that aspect of the image that we notice. Since there were about two times as many people who came from Google as those who typed the URL in directly, the sandal depicting the number of Google visitors is about two times longer than the sandal below it, but it occupies about four times the area. As you can see from the frequency table, that just isn't a correct impression.

The best data displays observe a fundamental principle of graphing data called the **area principle,** which says that the area occupied by a part of the graph should correspond to the magnitude of the value it represents.

Bar Charts

Figure 4.3 gives us a chart that obeys the area principle. It's not as visually entertaining as the sandals, but it does give a more *accurate* visual impression of the distribution. The height of each bar shows the count for its category. The bars are the same width, so their heights determine their areas, and the areas are proportional to the counts in each class. Now it's easy to see that more than half the site hits came from places other than Google—not the impression that the sandals in Figure 4.2 conveyed. We can also see that there were a little more than twice as many visits that originated with a Google search as there were visits that came directly. Bar charts make these kinds of comparisons easy and natural.

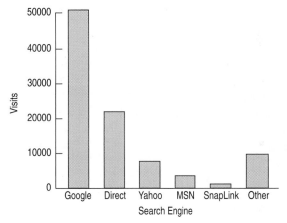

Figure 4.3 *Visits to the KEEN Footwear website by Search Engine choice. With the area principle satisfied, the true distribution is clear.*

A **bar chart** displays the distribution of a categorical variable, showing the counts for each category next to each other for easy comparison. Bar charts should have small spaces between the bars to indicate that these are freestanding bars that could be rearranged into any order. The bars are lined up along a common base.

Bar charts are usually drawn vertically in columns, 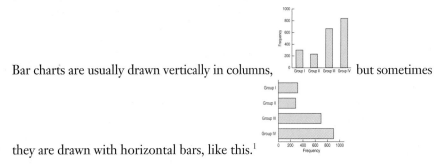 but sometimes they are drawn with horizontal bars, like this.[1]

---

[1] Excel refers to this display as a bar graph.

If we want to draw attention to the relative *proportion* of visits from each *Search Engine*, we could replace the counts with percentages and use a **relative frequency bar chart**, like the one shown in Figure 4.4.

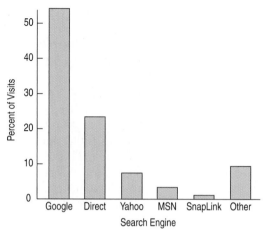

Figure 4.4 *The relative frequency bar chart looks the same as the bar chart (Figure 4.3) but shows the proportion of visits in each category rather than the counts.*

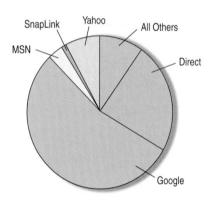

Figure 4.5 *Number of visits by Search Engine*

Pie Charts

Another common display that shows how a whole group breaks into several categories is a pie chart. **Pie charts** show the whole group of cases as a circle. They slice the circle into pieces whose size is proportional to the fraction of the whole in each category.

Pie charts give a quick impression of how a whole group is partitioned into smaller groups. Because we're used to cutting up pies into 2, 4, or 8 pieces, pie charts are good for seeing relative frequencies near 1/2, 1/4, or 1/8. For example, in Figure 4.5, you can easily see that the slice representing Google is just slightly more than half the total. Unfortunately, other comparisons are harder to make with pie charts. Were there more visits from Yahoo, or from All Others? It's hard to tell since the two slices look about the same. Comparisons such as these are usually easier in a bar chart. (Compare to Figure 4.4.)

◆ **Think before you draw.**   Our first rule of data analysis is *Make a picture*. But what kind of picture? We don't have a lot of options—yet. There's more to Statistics than pie charts and bar charts, and knowing when to use every type of display we'll discuss is a critical first step in data analysis. That decision depends in part on what type of data you have and on what you hope to communicate.

We always have to check that the data are appropriate for whatever method of analysis we choose. Before you make a bar chart or a pie chart, always check the **Categorical Data Condition:** that the data are counts or percentages of individuals in categories.

If you want to make a pie chart or relative frequency bar chart, you'll need to also make sure that the categories don't overlap, so that no individual is counted in two categories. If the categories do overlap, it's misleading to make a pie chart, since the percentages won't add up to 100%. For the *Search Engine* data, either kind of display is appropriate because the categories don't overlap—each visit comes from a unique source.

Throughout this course, you'll see that doing Statistics right means selecting the proper methods. That means you have to think about the situation at hand. An important first step is to check that the type of analysis you plan is appropriate. Our Categorical Data Condition is just the first of many such checks

4.4 Contingency Tables

WHO	Respondents in the GfK Roper Reports Worldwide Survey
WHAT	Responses to questions relating to perceptions of food and health
WHEN	Fall 2005; published in 2006
WHERE	Worldwide
HOW	Data collected by GfK Roper Consulting using a multistage design
WHY	To understand cultural differences in the perception of the food and beauty products we buy and how they affect our health

In Chapter 3 we saw how GfK Roper Consulting gathered information on consumers attitudes about health, food, and health care products. In order to effectively market food products across different cultures, it's essential to know how strongly people in different cultures feel about their food. One question in the Roper survey asked respondents whether they agreed with the following statement: "I have a strong preference for regional or traditional products and dishes from where I come from." Here is a frequency table (Table 4.3) of the responses.

Response to *Regional Food Preference Question*	Counts	Relative Frequency
Agree Completely	2346	30.51%
Agree Somewhat	2217	28.83%
Neither Disagree Nor Agree	1738	22.60%
Disagree Somewhat	811	10.55%
Disagree Completely	498	6.48%
Don't Know	80	1.04%
Total	**7690**	**100.00%**

Table 4.3 *A combined frequency and relative frequency table for the responses (from all 5 countries represented: China, France, India, the U.K., and the U.S.) to the statement "I have a strong preference for regional or traditional products and dishes from where I come from."*

The pie chart (Figure 4.6) shows clearly that more than half of all the respondents agreed with the statement.

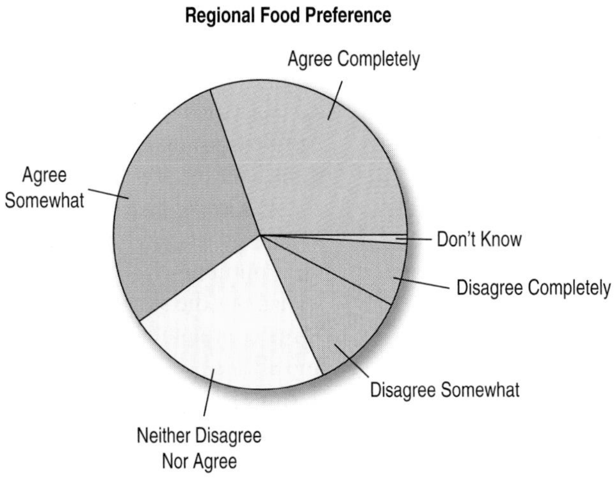

Regional Food Preference

Figure 4.6 *It's clear from the pie chart that the majority of respondents identify with their local foods.*

But if we want to target our marketing differently in different countries, wouldn't it be more interesting to know how opinions vary from country to country?

To find out, we need to look at the two categorical variables *Regional Preference* and *Country* together, which we do by arranging the data in a two-way table. Table 4.4 is a two-way table of *Regional Preference* by *Country*. Because the table shows how the individuals are distributed along each variable, depending on, or *contingent on*, the value of the other variable, such a table is called a **contingency table**.

		Agree Completely	Agree Somewhat	Neither Disagree Nor Agree	Disagree Somewhat	Disagree Completely	Don't Know	Total
		Regional Preference						
Country	China	518	576	251	117	33	7	**1502**
	France	347	475	400	208	94	15	**1539**
	India	960	282	129	65	95	4	**1535**
	U.K.	214	407	504	229	175	28	**1557**
	U.S.	307	477	454	192	101	26	**1557**
	Total	**2346**	**2217**	**1738**	**811**	**498**	**80**	**7690**

Table 4.4   *Contingency table of Regional Preference and Country. The bottom line "Totals" are the values that were in Table 4.3.*

The margins of a contingency table give totals. In the case of Table 4.4, these are shown in both the right-hand column (in bold) and the bottom row (also in bold). The totals in the bottom row of the table show the frequency distribution of the variable *Regional Preference*. The totals in the right-hand column of the table show the frequency distribution of the variable *Country*. When presented like this, at the margins of a contingency table, the frequency distribution of either one of the variables is called its **marginal distribution**.

Each **cell** of a contingency table (any intersection of a row and column of the table) gives the count for a combination of values of the two variables. If you look across the row in Table 4.4 for the United Kingdom, you can see that 504 people neither agreed nor disagreed. Looking down the Agree Completely column, you can see that the largest number of responses in that column (960) are from India. Are Britons less likely to agree with the statement than Indians or Chinese? Questions like this are more naturally addressed using percentages.

We know that 960 people from India agreed completely with the statement. We could display this number as a percentage, but as a percentage of what? The total number of people in the survey? (960 is 12.5% of the total.) The number of Indians in the survey? (960 is 62.5% of the row total.) The number of people who agree completely? (960 is 40.9% of the column total.) All of these are possibilities, and all are potentially useful or interesting. You'll probably wind up calculating (or letting your technology calculate) lots of percentages. Most statistics programs offer a choice of **total percent, row percent,** or **column percent** for contingency tables. Unfortunately, they often put them all together with several numbers in each cell of the table. The resulting table (Table 4.5) holds lots of information but is hard to understand.

Regional Preference

Country	Agree Completely	Agree Somewhat	Neither Disagree Nor Agree	Disagree Somewhat	Disagree Completely	Don't Know	Total
China	518	576	251	117	33	7	**1502**
% of Row	34.49	38.35	16.71	7.79	2.20	0.47	**100.00**
% of Column	22.08	25.98	14.44	14.43	6.63	8.75	**19.53**
% of Table	6.74	7.49	3.26	1.52	0.43	0.09	**19.53**
France	347	475	400	208	94	15	**1539**
% of Row	22.55	30.86	25.99	13.52	6.11	0.97	**100.00**
% of Column	14.79	21.43	23.01	25.65	18.88	18.75	**20.01**
% of Table	4.51	6.18	5.20	2.70	1.22	0.20	**20.01**
India	960	282	129	65	95	4	**1535**
% of Row	62.54	18.37	8.40	4.23	6.19	0.26	**100.00**
% of Column	40.92	12.72	7.42	8.01	19.08	5.00	**19.96**
% of Table	12.48	3.67	1.68	0.845	1.24	0.05	**19.96**
U.K.	214	407	504	229	175	28	**1557**
% of Row	13.74	26.14	32.37	14.71	11.24	1.80	**100.00**
% of Column	9.12	18.36	29.00	28.24	35.14	35.00	**20.24**
% of Table	2.78	5.29	6.55	2.98	2.28	0.36	**20.24**
U.S.	307	477	454	192	101	26	**1557**
% of Row	19.72	30.64	29.16	12.33	6.49	1.67	**100.00**
% of Column	13.09	21.52	26.12	23.67	20.28	32.50	**20.24**
% of Table	3.99	6.20	5.90	2.50	1.31	0.34	**20.24**
Total	**2346**	**2217**	**1738**	**811**	**498**	**80**	**7690**
% of Row	**30.51**	**28.83**	**22.60**	**10.55**	**6.48**	**1.04**	**100.00**
% of Column	**100.00**	**100.00**	**100.00**	**100.00**	**100.00**	**100.00**	**100.00**
% of Table	**30.51**	**28.83**	**22.60**	**10.55**	**6.48**	**1.04**	**100.00**

Table 4.5 *Another contingency table of Regional Preference and Country. This time we see not only the counts for each combination of the two variables, but also the percentages these counts represent. For each count, there are three choices for the percentage: by row, by column, and by table total. There's probably too much information here for this table to be useful.*

To simplify the table, let's first pull out the values corresponding to the percentages of the total.

Regional Preference

Country	Agree Completely	Agree Somewhat	Neither Disagree Nor Agree	Disagree Somewhat	Disagree Completely	Don't Know	Total
China	6.74	7.49	3.26	1.52	0.43	0.09	**19.53**
France	4.51	6.18	5.20	2.70	1.22	0.20	**20.01**
India	12.48	3.67	1.68	0.85	1.24	0.05	**19.96**
U.K.	2.78	5.29	6.55	2.98	2.28	0.36	**20.25**
U.S.	3.99	6.20	5.90	2.50	1.31	0.34	**20.25**
Total	**30.51**	**28.83**	**22.60**	**10.55**	**6.48**	**1.04**	**100.00**

Table 4.6 *A contingency table of Regional Preference and Country showing only the total percentages.*

These percentages tell us what percent of *all* respondents belong to each combination of column and row category. For example, we see that 3.99% of the respondents were Americans who agreed completely with the question, which is slightly more than the percentage of Indians who agreed somewhat. Is this fact useful? Is that really what we want to know?

> **Percent of what?** The English language can be tricky when we talk about percentages. If asked, "What percent of those answering 'I Don't Know' were from India?" it's pretty clear that you should focus only on the *Don't Know* column. The question itself seems to restrict the *Who* in the question to that column, so you should look at the number of those in each country among the 80 people who replied "I don't know." You'd find that in the column percentages, and the answer would be 4 out of 80 or 5.00%.
>
> But if you're asked, "What percent were Indians who replied 'I don't know?'" you'd have a different question. Be careful. The question really means "what percent of the entire sample were both from India and replied 'I don't know'?" So the *Who* is all respondents. The denominator should be 7690, and the answer is the table percent 4/7690 = 0.05%.
>
> Finally, if you're asked, "What percent of the Indians replied 'I don't know'?" you'd have a third question. Now the *Who* is Indians. So the denominator is the 1535 Indians, and the answer is the row percent, 4/1535 = 0.26%.

> Always be sure to ask "percent of what." That will help define the *Who* and will help you decide whether you want *row*, *column*, or *table* percentages.

Conditional Distributions

The more interesting questions are contingent on something. We'd like to know, for example, what percentage *of Indians* agreed completely with the statement and how that compares to the percentage *of Britons* who also agreed. Equivalently, we might ask whether the chance of agreeing with the statement depended on the *Country* of the respondent. We can look at this question in two ways. First, we could ask how the distribution of *Regional Preference* changes across *Country*. To do that we look at the row percentages.

		Regional Preference						
		Agree Completely	**Agree Somewhat**	**Neither Disagree Nor Agree**	**Disagree Somewhat**	**Disagree Completely**	**Don't Know**	**Total**
Country	**India**	960	282	129	65	95	4	**1535**
		62.54	18.37	8.40	4.23	6.19	0.26	**100%**
	U.K.	214	407	504	229	175	28	**1557**
		13.74	26.14	32.37	14.71	11.24	1.80	**100%**

Table 4.7 *The conditional distribution of Regional Preference conditioned on two values of Country: India and the United Kingdom. This table shows the row percentages.*

By focusing on each row separately, we see the distribution of *Regional Preference* under the condition of being in the selected *Country*. The sum of the percentages in each row is 100%, and we divide that up by the responses to the question. In effect, we can temporarily restrict the *Who* first to Indians and look at how their response are distributed. A distribution like this is called a **conditional distribution** because it shows the distribution of one variable for just those cases that satisfy a condition on another. We can compare the two conditional distributions with pie charts (Figure 4.7). Of course, we could also turn the question around. We could look at the distribution of *Country* for each category of *Regional Preference*. To do this, we would look at the column percentages.

Looking at how the percentages change across each row, it sure looks like the distribution of responses to the question is different in each *Country*. To make the

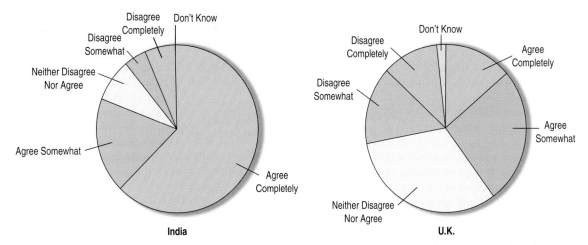

Figure 4.7 *Pie charts of the conditional distributions of* Regional Food Preference *importance for India and the United Kingdom. The percentage of people who agree is much higher in India than in the United Kingdom.*

differences more vivid, we could also display the conditional distributions. Figure 4.8 shows an example of a side-by-side bar chart, displaying the responses to the questions for India and the United Kingdom.

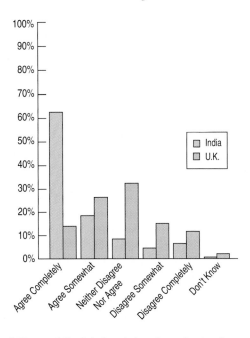

Figure 4.8 *Side-by-side bar charts showing the conditional distribution of* Regional Food Preference *for both India and the United Kingdom. It's easier to compare percentages within each country with side-by-side bar charts than pie charts.*

From Figure 4.8, it is clear that Indians have a stronger preference for their own cuisine than Britons have for theirs. For food companies, including GfK Roper's clients, that means Indians are less likely to accept a food product they perceive as foreign, and people in Great Britain are more accepting of "foreign" foods. This could be invaluable information for marketing products.

Variables can be associated in many ways and to different degrees. The best way to tell whether two variables are associated is to ask whether they are *not*.[2] In a

---

[2] This kind of "backwards" reasoning shows up surprisingly often in science—and in Statistics.

contingency table, when the distribution of one variable is the same for all categories of another, we say that the variables are **independent**. That tells us there's no association between these variables. We'll see a way to check for independence formally later in the book. For now, we'll just compare the distributions.

JUST CHECKING

So that they can balance their inventory, an optometry shop collects the following data for customers in the shop.

		Eye Condition			
		Near Sighted	**Far Sighted**	**Need Bifocals**	**Total**
Sex	**Males**	6	20	6	**32**
	Females	4	16	12	**32**
	Total	**10**	**36**	**18**	**64**

1 What percent of females are far-sighted?

2 What percent of near-sighted customers are female?

3 What percent of all customers are far-sighted females?

4 What's the distribution of *Eye Condition*?

5 What's the conditional distribution of *Eye Condition* for males?

6 Compare the percent who are female among near-sighted customers to the percent of all customers who are female.

7 Does it seem that *Eye Condition* and *Sex* might be dependent? Explain.

Segmented Bar Charts

We could display the Roper survey information by dividing up bars rather than circles as we did when making pie charts. The resulting **segmented bar chart** treats each bar as the "whole" and divides it proportionally into segments corresponding to the percentage in each group. We can see that the distributions of responses to the question are very different in the two countries, indicating again that *Regional Preference* is not independent of *Country*.

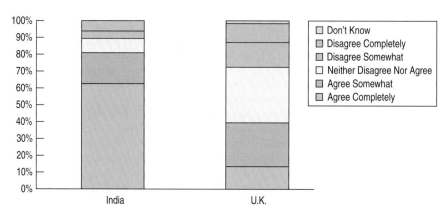

Figure 4.9 *Although the totals for India and the United Kingdom are different, the bars are the same height because we have converted the numbers to percentages. Compare this display with the side-by-side pie charts of the same data in Figure 4.7.*

GUIDED EXAMPLE | *Food Safety*

Food storage and food safety are major issues for multinational food companies. A client wants to know if people of all age groups have the same degree of concern so GfK Roper Consulting asked 1500 people in five countries whether they agree with the following statement: "I worry about how safe the food I buy is." We might want to report to a client who was interested in how concerns about food safety were related to age.

PLAN

Setup

- State the objectives and goals of the study.
- Identify and define the variables.
- Provide the time frame of the data collection process.

Determine the appropriate analysis for data type.

The client wants to examine the distribution of responses to the food safety question and see whether they are related to the age of the respondent. GfK Roper Consulting collected data on this question in the fall of 2005 for their 2006 Worldwide report. We will use the data from that study.

The variable is *Food Safety*. The responses are in nonoverlapping categories of agreement, from Agree Completely to Disagree Completely (and Don't Know). There were originally 12 Age groups, which we can combine into five:

Teen	13–19
Young Adult	20–29
Adult	30–39
Middle Aged	40–49
Mature	50 and older

Both variables, *Food Safety* and *Age*, are ordered categorical variables. To examine any differences in responses across age groups, it is appropriate to create a contingency table and a side-by-side bar chart. Here is a contingency table of "Food Safety" by "Age."

DO

Mechanics For a large data set like this, we rely on technology to make table and displays.

		Food Safety						
		Agree Completely	Agree Somewhat	Neither Disagree Nor Agree	Disagree Somewhat	Disagree Completely	Don't Know	Total
Age	**Teen**	16.19	27.50	24.32	19.30	10.58	2.12	**100%**
	Young Adult	20.55	32.68	23.81	14.94	6.98	1.04	**100%**
	Adult	22.23	34.89	23.28	12.26	6.75	0.59	**100%**
	Middle Aged	24.79	35.31	22.02	12.43	5.06	0.39	**100%**
	Mature	26.60	33.85	21.21	11.89	5.82	0.63	**100%**

A side-by-side bar chart is particularly helpful when comparing multiple groups.

A side-by-side bar chart shows the percent of each response to the question by Age group.

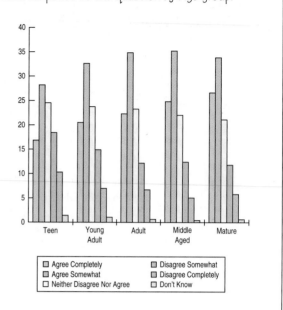

REPORT

Summary and Conclusions Summarize the charts and analysis in context. Make recommendations if possible and discuss further analysis that is needed.

MEMO:

Re: Food safety concerns by age

Our analysis of the GfK Roper Reports™ Worldwide survey data for 2006 shows a pattern of concern about food safety that generally increases from youngest to oldest.

Our analysis thus far has not considered whether this trend is consistent across countries. If it were of interest to your group, we could perform a similar analysis for each of the countries.

The enclosed tables and plots provide support for these conclusions.

WHAT CAN GO WRONG?

- **Don't violate the area principle.** This is probably the most common mistake in a graphical display. Violations of the area principle are often made for the sake of artistic presentation. Here, for example, are two versions of the same pie chart for the *Regional Preference* data.

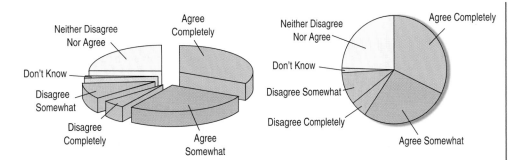

The one on the left looks interesting, doesn't it? But showing the pie three dimensionally on a slant violates the area principle and makes it much more difficult to compare fractions of the whole made up of each category of the response—the principal feature that a pie chart ought to show.

- **Keep it honest.** Here's a pie chart that displays data on the percentage of high school students who engage in specified dangerous behaviors as reported by the Centers for Disease Control. What's wrong with this plot?

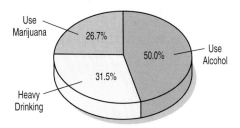

Try adding up the percentages. Or look at the 50% slice. Does it look right? Then think: What are these percentages of? Is there a "whole" that has been sliced up? In a pie chart, the proportions shown by each slice of the pie must add up to 100%, and each individual must fall into only one category. Of course, showing the pie on a slant makes it even harder to detect the error.

Here's another example. This bar chart shows the number of airline passengers searched by security screening.

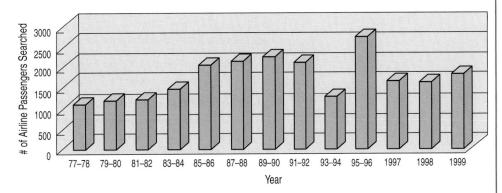

Looks like things didn't change much in the final years of the 20th century—until you read the bar labels and see that the last three bars represent single years, while all the others are for *pairs* of years. The false depth makes it even harder to see the problem.

(*continued*)

- **Don't confuse percentages.** Many percentages based on a conditional and joint distributions sound similar, but are different:
  - The percentage of French who answered "Agree Completely": This is 347/1539 or 22.5%.
  - The percentage of those who answered "Don't Know" who are French: This is 15/80 or 18.75%.
  - The percentage of those who were French *and* answered "Agree Completely": This is 347/7690 or 4.5%.

	Regional Preference						
Country	**Agree Completely**	**Agree Somewhat**	**Neither Disagree Nor Agree**	**Disagree Somewhat**	**Disagree Completely**	**Don't Know**	**Total**
China	518	576	251	117	33	7	**1502**
France	347	475	400	208	94	15	**1539**
India	960	282	129	65	95	4	**1535**
UK	214	407	504	229	175	28	**1557**
USA	307	477	454	192	101	26	**1557**
Total	**2346**	**2217**	**1738**	**811**	**498**	**80**	**7690**

In each instance, pay attention to the wording that makes a restriction to a smaller group (those who are French those who answered "Don't Know" and all respondents, respectively) before a percentage is found. This restricts the *Who* of the problem and the associated denominator for the percentage. Your discussion of results must make these differences clear.

- **Don't forget to look at the variables separately, too.** When you make a contingency table or display a conditional distribution, be sure to also examine the marginal distributions. It's important to know how many cases are in each category.

- **Be sure to use enough individuals.** When you consider percentages, take care that they are based on a large enough number of individuals (or cases). Take care not to make a report such as this one:

  We found that 66.67% of the companies surveyed improved their performance by hiring outside consultants. The other company went bankrupt.

- **Don't overstate your case.** Independence is an important concept, but it is rare for two variables to be *entirely* independent. We can't conclude that one variable has no effect whatsoever on another. Usually, all we know is that little effect was observed in our study. Other studies of other groups under other circumstances could find different results.

- **Don't use unfair or inappropriate percentages.** Sometimes percentages can be misleading. Sometimes they don't make sense at all. Be careful when finding percentages across different categories not to combine percentages inappropriately. The next section gives an example.

One famous example of Simpson's Paradox arose during an investigation of admission rates for men and women at the University of California at Berkeley's graduate schools. As reported in an article in *Science*, about 45% of male applicants were admitted, but only about 30% of female applicants got in. It looked like a clear case of discrimination. However, when the data were broken down by school (Engineering, Law, Medicine, etc.), it turned out that within each school, the women were admitted at nearly the same or, in some cases, much *higher* rates than the men. How could this be? Women applied in large numbers to schools with very low admission rates. (Law and Medicine, for example, admitted fewer than 10%.) Men tended to apply to Engineering and science. Those schools have admission rates above 50%. When the total applicant pool was combined and the percentages were computed, the women had a much lower *overall* rate, but the combined percentage didn't really make sense.

Simpson's Paradox

Here's an example showing that combining percentages across very different values or groups can give absurd results. Suppose there are two sales representatives, Peter and Katrina. Peter argues that he's the better salesperson, since he managed to close 83% of his last 120 prospects compared with Katrina's 78%. But let's look at the data a little more closely. Here (Table 4.8) are the results for each of their last 120 sales calls, broken down by the product they were selling.

| | Product | | |
Sales Rep	**Printer Paper**	**USB Flash Drive**	**Overall**
Peter	90 out of 100 90%	10 out of 20 50%	100 out of 120 83%
Katrina	19 out of 20 95%	75 out of 100 75%	94 out of 120 78%

Table 4.8 *Look at the percentages within each Product category. Who has a better success rate closing sales of paper? Who has the better success rate closing sales of Flash Drives? Who has the better performance overall?*

Look at the sales of the two products separately. For printer paper sales, Katrina had a 95% success rate, and Peter only had a 90% rate. When selling flash drives, Katrina closed her sales 75% of the time, but Peter only 50%. So Peter has better "overall" performance, but Katrina is better selling each product. How can this be?

This problem is known as **Simpson's Paradox**, named for the statistician who described it in the 1960s. Although it is rare, there have been a few well-publicized cases of it. As we can see from the example, the problem results from inappropriately combining percentages of different groups. Katrina concentrates on selling flash drives, which is more difficult, so her *overall* percentage is heavily influenced by her flash drive average. Peter sells more printer paper, which appears to be easier to sell. With their different patterns of selling, taking an overall percentage is misleading. Their manager should be careful not to conclude rashly that Peter is the better salesperson.

The lesson of Simpson's Paradox is to be sure to combine comparable measurements for comparable individuals. Be especially careful when combining across different levels of a second variable. It's usually better to compare percentages *within* each level, rather than across levels.

ETHICS IN ACTION

Lyle Erhart has been working in sales for a leading vendor of Customer Relationship Management (CRM) software for the past three years. He was recently made aware of a published research study that examined factors related to the successful implementation of CRM projects among firms in the financial services industry. Lyle read the research report with interest and was excited to see that his company's CRM software product was included. Among the results were tables reporting the number of projects that were successful based on type of CRM implementation (Operational versus Analytical) for each of the top leading CRM products of 2006. Lyle quickly found the results for his company's product and their major competitor. He summarized the results into one table as follows:

	His Company	**Major Competitor**
Operational	16 successes out of 20	68 successes out of 80
Analytical	90 successes out of 100	19 successes out of 20

At first he was a bit disappointed, especially since most of their potential clients were interested in Operational CRM. He had hoped to be able to disseminate the findings of this report among the sales force so they could refer to it when visiting potential clients. After some thought, he realized that he could combine the results. His company's overall success rate was 106 out of 120 (over 88%) and was higher than that of its major competitor. Lyle was now happy that he found and read the report.

ETHICAL ISSUE *Lyle, intentionally or not, has benefited from Simpson's Paradox. By combining percentages, he can present the findings in a manner favorable to his company (related to item A, ASA Ethical Guidelines).*

ETHICAL SOLUTION *Lyle should not combine the percentages as the results are misleading. If he decides to disseminate the information to his sales force, he must do so without combining.*

What have we learned?

We've learned that we can summarize categorical data by counting the number of cases in each category, sometimes expressing the resulting distribution as percents. We can display the distribution in a bar chart or a pie chart. When we want to see how two categorical variables are related, we put the counts (and/or percentages) in a two-way table called a contingency table.

- We look at the marginal distribution of each variable (found in the margins of the table).
- We also look at the conditional distribution of a variable within each category of the other variable.
- We can display these conditional and marginal distributions using bar charts or pie charts.
- If the conditional distributions of one variable are (roughly) the same for every category of the other, the variables are independent.

Terms

Area principle
A principle that helps to interpret statistical information with distortion by insisting that in a statistical display, each data value be represented by the same amount of area.

Bar chart (relative frequency bar chart)
A chart that represents the count (or percentage) of each category in a categorical variable as a bar, allowing easy visual comparisons across categories.

Column percent	The proportion of each column contained in the cell of a frequency table.
Conditional distribution	The distribution of a variable restricting the *Who* to consider only a smaller group of individuals.
Contingency table	A contingency table displays counts and, sometimes, percentages of individuals falling into named categories on two or more variables. The table categorizes the individuals on all variables at once, to reveal possible patterns in one variable that may be contingent on the category of the other.
Distribution	The distribution of a variable is a list of: • all the possible values of the variable • the relative frequency of each value
Frequency table (relative frequency table)	A table that lists the categories in a categorical variable and gives the number (the percentage) of observations for each category. The row percent is the proportion of each row contained in the cell of a frequency table, while the column percent is the proportion of each column contained in the cell of a frequency table.
Independent variables	Variables for which the conditional distribution of one variable is the same for each category of the other.
Marginal distribution	In a contingency table, the distribution of either variable alone. The counts or percentages are the totals found in the margins (usually the right-most column or bottom row) of the table.
Pie chart	Pie charts show how a "whole" divides into categories by showing a wedge of a circle whose area corresponds to the proportion in each category.
Row percent	The proportion of each row contained in the cell of a frequency table.
Simpson's paradox	A phenomenon that arises when averages, or percentages, are taken across different groups, and these group averages appear to contradict the overall averages.
Total percent	The proportion of the total contained in the cell of a frequency table.

Skills

 PLAN

- Recognize when a variable is categorical and choose an appropriate display for it.
- Understand how to examine the association between categorical variables by comparing conditional and marginal percentages.

 DO

- Summarize the distribution of a categorical variable with a frequency table.
- Display the distribution of a categorical variable with a bar chart or pie chart.
- Construct and examine a contingency table.
- Construct and examine displays of the conditional distributions of one variable for two or more groups.

 REPORT

- Describe the distribution of a categorical variable in terms of its possible values and relative frequencies.
- Describe any anomalies or extraordinary features revealed by the display of a variable.
- Describe and discuss patterns found in a contingency table and associated displays of conditional distributions.

Technology Help: Displaying Categorical Data on the Computer

Although every package makes a slightly different bar chart, they all have similar features:

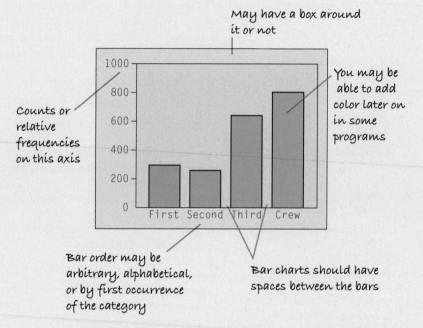

May have a box around it or not

You may be able to add color later on in some programs

Counts or relative frequencies on this axis

Bar order may be arbitrary, alphabetical, or by first occurrence of the category

Bar charts should have spaces between the bars

Sometimes the count or a percentage is printed above or on top of each bar to give some additional information. You may find that your statistics package sorts category names in annoying orders by default. For example, many packages sort categories alphabetically or by the order the categories are seen in the data set. Often, neither of these is the best choice.

EXCEL

First make a pivot table (Excel's name for a frequency table). From the **Data** menu, choose **Pivot Table** and **Pivot Chart Report**.

When you reach the Layout window, drag your variable to the row area and drag your variable again to the data area. This tells Excel to count the occurrences of each category.

Once you have an Excel pivot table, you can construct bar charts and pie charts.

Click inside the Pivot Table.

Click the Pivot Table Chart Wizard button. Excel creates a bar chart.

A longer path leads to a pie chart; see your Excel documentation.

Comments

Excel uses the pivot table to specify the category names and find counts within each category. If you already have that information, you can proceed directly to the Chart Wizard.

EXCEL 2007

To make a bar chart:
- Select the variable in Excel you want to work with.
- Choose the **Column** command from the Insert tab in the Ribbon.
- Select the appropriate chart from the drop down dialog.

To change the bar chart into a pie chart:
- Right-click the chart and select **Change Chart Type...** from the menu. The Chart type dialog opens.
- Select a pie chart type.
- Click the **OK** button. Excel changes your bar chart into a pie chart.

MINITAB

To make a bar chart, choose **Bar Chart** from the **Graph** menu.

Then select either a Simple, Cluster, or Stack chart from the options and click **OK**. To make a **Simple** bar chart, enter the name of the variable to graph in the dialog box. To make a

relative frequency chart, click **Chart Options**, and choose **Show Y as Percent**.

In the Chart dialog, enter the name of the variable that you wish to display in the box labeled "Categorical variables." Click **OK**.

SPSS

To make a bar chart, open the **Chart Builder** from the **Graphs** menu.

Click the **Gallery** tab.

Choose **Bar Chart** from the list of chart types.

Drag the appropriate bar chart onto the canvas.

Drag a categorical variable onto the x-axis drop zone.

Click **OK**

Comments

A similar path makes a pie chart by choosing **Pie chart** from the list of chart types.

DATA DESK

To make a bar chart or pie chart, select the variable. In the **Plot** menu, choose **Bar Chart** or **Pie Chart**.

To make a frequency table, in the **Calc** menu choose **Frequency Table**.

Comments

These commands treat the data as categorical even if they are numerals. If you select a quantitative variable by mistake, you'll see an error message warning of too many categories.

JMP

JMP makes a bar chart and frequency table together. From the **Analyze** menu, choose **Distribution**.

In the Distribution dialog, drag the name of the variable into the empty variable window beside the label "Y, Columns"; click **OK**.

To make a pie chart, choose **Chart** from the **Graph** menu.

In the Chart dialog, select the variable name from the Columns list, click on the button labeled "Statistics," and select "N" from the drop-down menu.

Click the "**Categories, X, Levels**" button to assign the same variable name to the X-axis.

Under Options, click on the **second** button—labeled "**Bar Chart**"—and select "Pie" from the drop-down menu.

Mini Case Study Project

KEEN Footwear

More of the data that KEEN Footwear obtained from *Google Analytics* are in the file **ch04_MCSP_KEEN**. Open the data file using a statistics package and find data on *Country of Origin, Top Keywords, Online Retailers, User Statistics,* and *Page Visits*. Create frequency tables, bar charts, and pie charts using your software. What might KEEN want to know about their Web traffic? Which of these tables and charts is most useful to address the question of where they should advertise and how they should position their products? Write a case report summarizing your analysis and results.

EXERCISES

1. Graphs in the news. Find a bar graph of categorical data from a business publication (e.g., *Business Week, Fortune, The Wall Street Journal*, etc.).

a) Is the graph clearly labeled?
b) Does it violate the area principle?
c) Does the accompanying article tell the W's of the variable?
d) Do you think the article correctly interprets the data? Explain.

2. Graphs in the news, part 2. Find a pie chart of categorical data from a business publication (e.g., *Business Week, Fortune, The Wall Street Journal*, etc.).

a) Is the graph clearly labeled?
b) Does it violate the area principle?
c) Does the accompanying article tell the W's of the variable?
d) Do you think the article correctly interprets the data? Explain.

3. Tables in the news. Find a frequency table of categorical data from a business publication (e.g., *Business Week, Fortune, The Wall Street Journal*, etc.).

a) Is it clearly labeled?
b) Does it display percentages or counts?
c) Does the accompanying article tell the W's of the variable?
d) Do you think the article correctly interprets the data? Explain.

4. Tables in the news, part 2. Find a contingency table of categorical data from a business publication (e.g., *Business Week, Fortune, The Wall Street Journal*, etc.).

a) Is it clearly labeled?
b) Does it display percentages or counts?
c) Does the accompanying article tell the W's of the variable?
d) Do you think the article correctly interprets the data? Explain.

5. U.S. market share. An article in the *The Wall Street Journal* (March 16, 2007) reported the 2006 U.S. market share of leading sellers of carbonated drinks, summarized in the following pie chart:

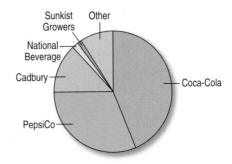

a) Is this an appropriate display for these data? Explain.
b) Which company had the largest share of the market?

6. World market share. *The Wall Street Journal* article described in Exercise 5 also indicated the 2005 world market share for leading distributors of total confectionery products. The following bar chart displays the values:

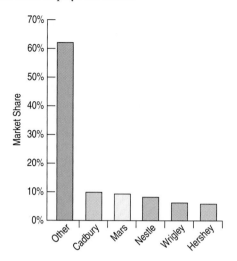

a) Is this an appropriate display for these data? Explain.
b) Which company had the largest share of the candy market?

7. Market share again. Here's a bar chart of the data in Exercise 5.

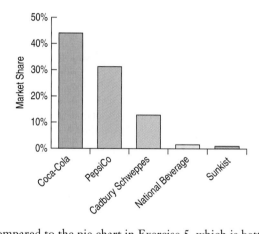

a) Compared to the pie chart in Exercise 5, which is better for displaying the relative portions of market share? Explain.
b) What is missing from this display that might make it misleading?

8. World market share again. Here's a pie chart of the data in Exercise 6.

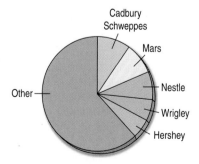

a) Which display of these data is best for comparing the market shares of these companies? Explain.

b) Does Cadbury Schweppes or Mars have a bigger market share?

9. Insurance company. An insurance company is updating its payouts and cost structure for their insurance policies. Of particular interest to them is the risk analysis for customers currently on heart or blood pressure medication. The Centers for Disease Control lists causes of death in the United States during one year as follows.

Cause of Death	Percent
Heart disease	30.3
Cancer	23.0
Circulatory diseases and stroke	8.4
Respiratory diseases	7.9
Accidents	4.1

a) Is it reasonable to conclude that heart or respiratory diseases were the cause of approximately 38% of U.S. deaths during this year?
b) What percent of deaths were from causes not listed here?
c) Create an appropriate display for these data.

10. Revenue growth. A 2005 study by Babson College and The Commonwealth Institute surveyed the top women-led businesses in the state of Massachusetts in 2003 and 2004. The study reported the following results for continuing participants with a 9% response rate.

2003–2004 Revenue Growth	
Decline	7%
Modest Decline	9%
Steady State	10%
Modest Growth	18%
Growth	54%

a) Describe the distribution of companies with respect to revenue growth.
b) Is it reasonable to conclude that 72% of all women-led businesses in the U.S. reported some level of revenue growth? Explain.

11. Web conferencing. Cisco Systems Inc. announced plans in March 2007 to buy WebEx Communications, Inc. for $3.2 billion, demonstrating their faith in the future of Web conferencing. The leaders in market share for the venders in the area of Web conferencing in 2006 are as follows: WebEx 58.4% and Microsoft 26.3%. Create an appropriate graphical display of this information and write a sentence or two that might appear in a newspaper article about the market share.

12. Mattel. In their 2006 annual report, Mattel Inc. reported that their domestic market sales were broken down as follows: 44.1% Mattel Girls and Boys brand, 43.0% Fisher-Price brand and the rest of the nearly $3.5 billion revenues were due to their American Girl brand. Create an appropriate graphical display of this information and write a sentence or two that might appear in a newspaper article about their revenue breakdown.

13. Small business productivity. The Wells Fargo/Gallup Small Business Index asked 592 small business owners in March 2004 what steps they had taken in the past year to increase productivity. They found that 60% of small business owners had updated their computers, 52% had made other (noncomputer) capital investments, 37% hired part-time instead of full-time workers, 24% had not replaced workers who left voluntarily, 15% had laid off workers, and 10% had lowered employee salaries.

a) What do you notice about the percentages listed? How could this be?

b) Make a bar chart to display the results and label it clearly.

c) Would a pie chart be an effective way of communicating this information? Why or why not?

d) Write a couple of sentences on the steps taken by small businesses to increase productivity.

14. Small business hiring. In 2004, the Wells Fargo/Gallup Small Business Index found that 86% of the 592 small business owners they surveyed said their productivity for the previous year had stayed the same or increased and most had substituted productivity gains for labor. (See Exercise 13.) As a follow-up question, the survey gave them a list of possible economic outcomes and asked if that would make them hire more employees. Here are the percentages of owners saying that they would "definitely or probably hire more employees" for each scenario: a substantial increase in sales—79%, a major backlog of sales orders—71%, a general improvement in the economy—57%, a gain in productivity—50%, a reduction in overhead costs—43%, and more qualified employees available—39%.

a) What do you notice about the percentages listed?

b) Make a bar chart to display the results and label it clearly.

c) Would a pie chart be an effective way of communicating this information? Why or why not?

d) Write a couple of sentences on the responses to small business owners about hiring given the scenarios listed.

15. Environmental hazard. Data from the International Tanker Owners Pollution Federation Limited (www.itopf.com)

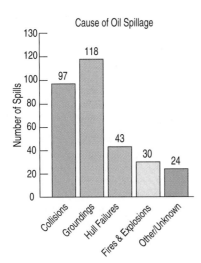

Cause of Oil Spillage

give the cause of spillage for 312 large oil tanker accidents from 1974–2006. Here are the displays. Write a brief report interpreting what the displays show. Is a pie chart an appropriate display for these data? Why or why not?

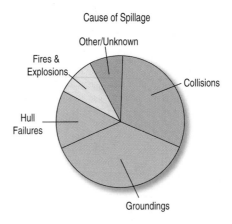

Cause of Spillage

16. Winter Olympics. Twenty-six countries won medals in the 2006 Winter Olympics. The following table lists them, along with the total number of medals each won.

Country	Medals	Country	Medals
Germany	29	Finland	9
United States	25	Czech Republic	4
Canada	24	Estonia	3
Austria	23	Croatia	3
Russia	22	Australia	2
Norway	19	Poland	2
Sweden	14	Ukraine	2
Switzerland	14	Japan	1
South Korea	11	Belarus	1
Italy	11	Bulgaria	1
China	11	Great Britain	1
France	9	Slovakia	1
Netherlands	9	Latvia	1

a) Try to make a display of these data. What problems do you encounter?

b) Can you find a way to organize the data so that the graph is more successful?

17. Importance of wealth. GfK Roper Reports Worldwide surveyed people in 2004, asking them "How important is acquiring wealth to you?" The percent who responded that it was of more than average importance were: 71.9% China, 59.6% France, 76.1% India, 45.5% UK, and 45.3% USA. There were about 1500 respondents per country. A report showed the following bar chart of these percentages.

a) How much larger is the proportion of those who said acquiring wealth was important in India than in the United States?

b) Is that the impression given by the display? Explain.

c) How would you improve this display?

d) Make an appropriate display for the percentages.

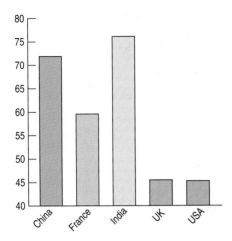

e) Write a few sentences describing what you have learned about attitudes toward acquiring wealth.

18. Importance of power. In the same survey as that discussed in Exercise 17, GfK Roper Consulting also asked "How important is having control over people and resources to you?" The percent who responded that it was of more than average importance are given in the following table:

China	49.1%
France	44.1%
India	74.2%
UK	27.8%
USA	36.0%

Here's a pie chart of the data:

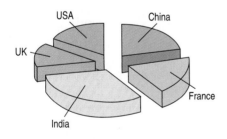

a) List the errors you see in this display.
b) Make an appropriate display for the percentages.
c) Write a few sentences describing what you have learned about attitudes toward acquiring power.

19. Google financials. Google Inc. derives revenue from three major sources: advertising revenue from their websites, advertising revenue from the thousands of third-party websites that comprise the Google Network, and licensing and miscellaneous revenue. The following table shows the percentage of all revenue derived from these sources for the period 2002 to 2006.

a) Are these row or column percentages?
b) Make an appropriate display of these data.
c) Write a brief summary of this information.

		Year			
	2002	**2003**	**2004**	**2005**	**2006**
Google websites	70%	54%	50%	55%	60%
Google network websites	24%	43%	49%	44%	39%
Licensing & other revenue	6%	3%	1%	1%	1%

(Revenue Source)

20. Real estate pricing. A study of a sample of 1057 houses in upstate New York reports the following percentages of houses falling into different Price and Size categories.

		Price		
	Low	**Med Low**	**Med High**	**High**
Small	61.5%	35.2%	5.2%	2.4%
Med Small	30.4%	45.3%	26.4%	4.7%
Med Large	5.4%	17.6%	47.6%	21.7%
Large	2.7%	1.9%	20.8%	71.2%

(Size)

a) Are these column, row, or total percentages? How do you know?
b) What percent of the highest priced houses were small?
c) From this table, can you determine what percent of all houses were in the low price category?
d) Among the lowest prices houses, what percent were small or medium small?
e) Write a few sentences describing the association between *Price* and *Size*.

21. Stock performance. The following table displays information for 40 widely held U.S. stocks, on how their one day change on March 15, 2007 compared with their previous 52-week change.

		Over prior 52 weeks	
		Positive Change	**Negative Change**
Positive Change	14	9
Negative Change	11	6

(March 15, 2007)

a) What percent of the companies reported a positive change in their stock price over the prior 52 weeks?
b) What percent of the companies reported a positive change in their stock price over both time periods?
c) What percent of the companies reported a negative change in their stock price over both time periods?
d) What percent of the companies reported a positive change in their stock price over one period and then a negative change in the other period?

e) Among those companies reporting a positive change in their stock price over the prior day what percentage also reported a positive change over the prior year?

f) Among those companies reporting a negative change in their stock price over the prior day what percentage also reported a positive change over the prior year?

g) What relationship, if any, do you see between the performance of a stock on a single day and its 52-week performance?

22. New product. A company started and managed by business students is selling campus calendars. The students have conducted a market survey with the various campus constituents to determine sales potential and identify which market segments should be targeted. (Should they advertise in the Alumni Magazine and/or the local newspaper?) The following table shows the results of the market survey.

	Buying Likelihood			
Campus Group	**Unlikely**	**Moderately Likely**	**Very Likely**	**Total**
Students	197	388	320	**905**
Faculty/Staff	103	137	98	**338**
Alumni	20	18	18	**56**
Town Residents	13	58	45	**116**
Total	**333**	**601**	**481**	**1415**

a) What percent of all these respondents are alumni?

b) What percent of these respondents are very likely to buy the calendar?

c) What percent of the respondents who are very likely to buy the calendar are alumni?

d) Of the alumni, what percent are very likely to buy the calendar?

e) What is the marginal distribution of the campus constituents?

f) What is the conditional distribution of the campus constituents among those very likely to buy the calendar?

g) Does this study present any evidence that this company should focus on selling to certain campus constituents?

23. Real estate. *The Wall Street Journal* reported in March 2007 that the real estate market in Nashville, Tennessee, slowed slightly from 2006 to 2007. The supporting data are summarized in the following table.

	Type of Sale				
Year	**Condos**	**Farms/ Land**	**Residential**	**Multi-family**	**Total**
2006	266	177	2119	48	**2610**
2007	341	190	2006	38	**2575**
Total	**607**	**367**	**4125**	**86**	**5185**

a) What percent of all sales in February, 2006 were condominiums (condos)? In February 2007?

b) What percent of the sales in February 2006 were multifamily? In February 2007?

c) Overall, what was the percentage change in real estate sales in Nashville, Tennessee, from February 2006 to February 2007?

24. Google financials, part 2. Google Inc. divides their total costs and expenses into five categories: cost of revenues, research and development, sales and marketing, general administrative, and miscellaneous. See the table at the bottom of the page.

a) What percent of all costs and expenses were cost of revenues in 2005? In 2006?

b) What percent of all costs and expenses were due to research and development in 2005? In 2006?

c) Have general administrative costs grown as a percentage of all costs and expenses over this time period?

25. Movie ratings. The movie ratings system is a voluntary system operated jointly by the Motion Picture Association of America (MPAA) and the National Association of Theatre Owners (NATO). The ratings themselves are given by a board of parents who are members of the Classification and Ratings Administration (CARA). The board was created in response to outcries from parents in the 1960s for some kind of regulation of film content, and the first ratings were introduced in 1968. Here

Cost and Expenses	**2002**	**2003**	**2004**	**2005**	**2006**
Cost of revenues	$132,575	$634,411	$1,468,967	$2,577,088	$4,225,027
Research and development	$40,494	$229,605	$385,164	$599,510	$1,228,589
Sales and marketing	$48,783	$164,935	$295,749	$468,152	$849,518
General administrative	$31,190	$94,519	$188,151	$386,532	$751,787
Miscellaneous	$0	$0	$201,000	$90,000	$0
Total Costs and Expenses	**$253,042**	**$1,123,470**	**$2,539,031**	**$4,121,282**	**$7,054,921**

is information on the ratings of 120 movies that came out in 2005, also classified by their genre.

Genre	Rating				
	G	**PG**	**PG-13**	**R**	**Total**
Action/Adventure	4	5	17	9	**35**
Comedy	2	12	20	4	**38**
Drama	0	3	8	17	**28**
Thriller/Horror	0	0	11	8	**19**
Total	**6**	**20**	**56**	**38**	**120**

a) Find the conditional distribution (in percentages) of movie ratings for action/adventure films.
b) Find the conditional distribution (in percentages) of movie ratings for thriller/horror films.
c) Create a graph comparing the ratings for the four genres.
d) Are *Genre* and *Rating* independent? Write a brief summary of what these data show about movie ratings and the relationship to the genre of the film.

26. Wireless access. The Pew Internet and American Life Project has had access to the Internet since the 1990s. Here is an income breakdown of 798 Internet users surveyed in December 2006, asking whether they have logged on to the Internet using a wireless device or not.

Income	Wireless Users	Other Internet Users	Total
Under $30K	34	128	**162**
$30K–50K	31	133	**164**
$50K–$75K	44	72	**116**
Over $75K	83	111	**194**
Don't know/ refused	51	111	**162**
Total	**243**	**555**	**798**

a) Find the conditional distribution (in percentages) of income distribution for the wireless users.
b) Find the conditional distribution (in percentages) of income distribution for nonwireless users.
c) Create a graph comparing the income distributions of the two groups.
d) Do you see any differences between the conditional distributions? Write a brief summary of what these data show about wireless use and its relationship to income.

27. MBAs. A survey of the entering MBA students at a university in the United States classified the country of origin of the students, as seen in the following table.

Origin	MBA Program		
	Two-Year MBA	**Evening MBA**	**Total**
Asia/Pacific Rim	31	33	**64**
Europe	5	0	**05**
Latin America	20	1	**21**
Middle East/Africa	5	5	**10**
North America	103	65	**168**
Total	**164**	**104**	**268**

a) What percent of all MBA students were from North America?
b) What percent of the Two-Year MBAs were from North America?
c) What percent of the Evening MBAs were from North America?
d) What is the marginal distribution of origin?
e) Obtain the column percentages and show the conditional distributions of origin by MBA Program.
f) Do you think that origin of the MBA student is independent of the MBA program? Explain.

28. MBAs, part 2. The same university as in Exercise 27 reported the following data on the gender of their students in their two MBA programs.

Sex	Type		
	Two-Year	**Evening**	**Total**
Men	116	66	**182**
Women	48	38	**86**
Total	**164**	**104**	**268**

a) What percent of all MBA students are women?
b) What percent of Two-Year MBAs are women?
c) What percent of Evening MBAs are women?
d) Do you see evidence of an association between the *Type* of MBA program and the percentage of women students? If so, why do you believe this might be true?

T 29. Top producing movies. The following table shows the Motion Picture Association of America (MPA) (www.mpaa.org) ratings for the top 20 grossing films in the United States for each of the 10 years from 1996 to 2005. (Data are number of films.)

	Rating				
Year	**G**	**PG**	**PG-13**	**R**	**Total**
2005	1	4	13	2	20
2004	1	6	10	3	20
2003	1	3	11	5	20
2002	1	6	13	0	20
2001	2	4	10	4	20
2000	0	3	12	5	20
1999	2	3	7	8	20
1998	3	3	9	5	20
1997	1	4	8	7	20
1996	2	4	5	9	20
Total	**14**	**40**	**98**	**48**	**200**

a) What percent of all these top 20 films are G rated?

b) What percent of all top 20 films in 2005 were G rated?

c) What percent of all top 20 films were PG-13 and came out in 1999?

d) What percent of all top 20 films produced in 2000 or later were PG-13?

e) What percent of all top 20 films produced from 1996 to 1999 were PG-13?

f) Compare the conditional distributions of the ratings for films produced in 2000 or later to those produced in 1996 to 1999. Write a couple of sentences summarizing what you see.

T **30. Movie admissions.** The following table shows attendance data collected by the Motion Picture Association of America during the period 2002 to 2006. Figures are in millions of movie admissions.

	Patron Age						
Year	**12 to 24**	**25 to 29**	**30 to 39**	**40 to 49**	**50 to 59**	**60 and Over**	**Total**
2006	485	136	246	219	124	124	**1334**
2005	489	135	194	216	125	122	**1281**
2004	567	132	265	236	145	132	**1477**
2003	567	124	269	193	152	118	**1423**
2002	551	158	237	211	119	130	**1406**
Total	**2659**	**685**	**1211**	**1075**	**665**	**626**	**6921**

a) What percent of all admissions during this period were bought by people between the ages of 12 and 24?

b) What percent of admissions in 2003 were bought by people between the ages of 12 and 24?

c) What percent of the admission were bought by people between the ages of 12 and 24 in 2006?

d) What percent of admissions in 2006 were bought by people over 60 years old?

e) What percent of the admissions bought by people 60 and over were in 2002?

f) Compare the conditional distributions of the age groups across years. Write a couple of sentences summarizing what you see.

31. Tattoos. A study by the University of Texas Southwestern Medical Center examined 626 people to see if there was an increased risk of contracting hepatitis C associated with having a tattoo. If the subject had a tattoo, researchers asked whether it had been done in a commercial tattoo parlor or elsewhere. Write a brief description of the association between tattooing and hepatitis C, including an appropriate graphical display.

	Tattoo done in commercial parlor	**Tattoo done elsewhere**	**No tattoo**
Has hepatitis C	17	8	18
No hepatitis C	35	53	495

32. Working parents. In July 1991 and again in April 2001, the Gallup Poll asked random samples of 1015 adults about their opinions on working parents. The following table summarizes responses to this question: *"Considering the needs of both parents and children, which of the following do you see as the ideal family in today's society?"* Based upon these results, do you think there was a change in people's attitudes during the 10 years between these polls? Explain.

		Year	
Response		**1991**	**2001**
	Both work full-time	142	131
	One works full-time, other part-time	274	244
	One works, other works at home	152	173
	One works, other stays home for kids	396	416
	No opinion	51	51

33. Revenue growth, last one. The study completed in 2005 and described in Exercise 10 also reported on education levels of the women chief executives. The column percentages for CEO education for each level of revenue are summarized in the following table. (Revenue is in $ million.)

	Graduate Education and Firm Revenue Size		
	< $10 M revenue	$10–$49.999 M revenue	≥ $50 M revenue
% with High School Education only	8%	4%	8%
% with College Education, but no Graduate Education	48%	42%	33%
% with Graduate Education	44%	54%	59%
Total	**100%**	**100%**	**100%**

a) What percent of these CEOs in the highest revenue category had only a high school education?

b) From this table, can you determine what percent of all these CEOs had graduate education? Explain.

c) Among the CEOs in the lowest revenue category, what percent had more than a high school education?

d) Write a few sentences describing the association between *Revenue* and *Education*.

T 34. Minimum wage workers. The U.S. Department of Labor (www.bls.gov) collects data on the number of U.S. workers who are employed at or below the minimum wage. Here is a contingency table of age and sex of the 2,013,000 workers in 2006 who were employed at or below $5.15 per hour, the minimum wage.

Age	Male	Female
16–24	14.7%	28.3%
25–34	6.1%	27.5%
35–44	3.3%	6.2%
45–54	2.0%	5.5%
55–64	1.5%	2.0%
65 and older	0.7%	2.1%

a) What percent of the women were ages 16 to 24?

b) Using graphical displays, compare the age distributions of the men and women who worked for minimum wage. Write a couple of sentences summarizing what you see.

35. Moviegoers and ethnicity. The Motion Picture Association of America studies the ethnicity of moviegoers to understand changes in the demographics of moviegoers over time. Here are the numbers of moviegoers (in millions) classified as to whether they were Hispanic, African-American, or Caucasian for the years 2002 to 2006.

		Year				
		2002	2003	2004	2005	2006	Total
Ethnicity	**Hispanic**	21	23	25	25	26	**120**
	African-American	21	20	22	21	20	**104**
	Caucasian	118	127	127	113	120	**605**
	Total	**160**	**170**	**174**	**159**	**166**	**829**

a) Find the marginal distribution *Ethnicity* of moviegoers.

b) Find the conditional distribution of *Ethnicity* for the year 2006.

c) Compare the conditional distribution of *Ethnicity* for all 5 years with a segmented bar graph.

d) Write a brief description of the association between *Year* and *Ethnicity* among these respondents.

36. Department store. A department store is planning its next advertising campaign. Since different publications are read by different market segments, they would like to know if they should be targeting specific age segments. The results of a marketing survey are summarized in the following table by *Age* and *Shopping Frequency* at their store.

		Age			
	Shopping	**Under 30**	**30–49**	**50 and Over**	**Total**
Frequency	**Low**	27	37	31	**95**
	Moderate	48	91	93	**232**
	High	23	51	73	**147**
	Total	**98**	**179**	**197**	**474**

a) Find the marginal distribution of *Shopping Frequency*.

b) Find the conditional distribution of *Shopping Frequency* within each age group.

c) Compare these distributions with a segmented bar graph.

d) Write a brief description of the association between *Age* and *Shopping Frequency* among these respondents.

e) Does this prove that customers ages 50 and over are more likely to shop at this department store? Explain.

37. Women's business centers. A study conducted in 2002 by Babson College and the Association of Women's Centers surveyed women's business centers in the United States. The data showing the location of established centers (at least 5 years old) and less established centers are summarized in the following table.

	Location	
	Urban	**Nonurban**
Less Established	74%	26%
Established	80%	20%

a) Are these percentages column percentages, row percentages, or table percentages?

b) Use graphical displays to compare these percentages of women's business centers by location.

38. Advertising. A company that distributes a variety of pet foods is planning their next advertising campaign. Since different publications are read by different market segments, they would like to know how pet ownership is distributed across different income segments. The U.S. Census Bureau reports the number of households owning various types of pets. Specifically, they keep track of dogs, cats, birds, and horses.

INCOME DISTRIBUTION OF HOUSEHOLDS OWING PETS (PERCENT)

		Pet			
		Dog	**Cat**	**Bird**	**Horse**
Income	**Under $12,500**	14	15	16	9
	$12,500 to $24,999	20	20	21	21
	$25,000 to $39,999	24	23	24	25
	$40,000 to $59,999	22	22	21	22
	$60,000 and over	20	20	18	23
	Total	**100**	**100**	**100**	**100**

a) Do you think the income distributions of the households who own these different animals would be roughly the same? Why or why not?

b) The table shows the percentages of income levels for each type of animal owned. Are these row percentages, column percentages, or table percentages?

c) Do the data support that the pet food company should not target specific market segments based on household income? Explain.

39. Worldwide toy sales. Around the world, toys are sold through different channels. For example, in some parts of the world toys are sold primarily through large toy store chains, while in other countries department stores sell more toys. The following table shows the percentages by region of the distribution of toys sold through various channels in Europe and America in

2003, accumulated by the International Council of Toy Industries (www.toy-icti.org).

a) Are these row percentages, column percentages, or table percentages?

b) Can you tell what percent of toys sold by mail order in both Europe and America are sold in Europe? Why or why not?

c) Use a graphical display to compare the distribution of channels between Europe and America.

d) Summarize the distribution of toy sales by channel in a few sentences. What are the biggest differences between these two continents?

40. Internet piracy. Illegal downloading of copyrighted movies is an international problem estimated to have cost the international movie industry more than $18 billion in 2005. The typical pirate worldwide is a 16 to 24-year old male living in an urban area, according to a study by the international strategy consulting firm LEK (www.mpaa.org/researchStatistics.asp). The following table compares the age distribution of the U.S. pirate to the rest of the world.

		Age			
		16–24	**25–29**	**30–39**	**Over 40**
Region	**United States**	71	11	7	11
	Rest of World	58	15	18	9

a) Are these row percentages, column percentages, or table percentages?

b) Can you tell what percent of pirates worldwide are in the 16 to 24 age group?

c) Use a graphical display to compare the age distribution of pirates in the United States to the distribution in the rest of the world.

d) Summarize the distribution of *Age* by *Region* in a few sentences. What are the biggest differences between these two regions?

41. Insurance company, part 2. An insurance company that provides medical insurance is concerned with recent data. They suspect that patients who undergo surgery at large hospitals have their discharges delayed for various reasons—which results in increased medical costs to the insurance company. The recent data

		Channel					
		General Merchandise	**Toy Specialists**	**Department Stores**	**Mass Merchant Discounters & Food Hypermarkets**	**Mail Order**	**Other**
Location	**America**	9%	25%	3%	51%	4%	8%
	Europe	13%	36%	7%	24%	5%	15%

for area hospitals and two types of surgery (major and minor) are shown in the following table.

Procedure		Discharge Delayed	
		Large Hospital	**Small Hospital**
	Major surgery	120 of 800	10 of 50
	Minor surgery	10 of 200	20 of 250

a) Overall, for what percent of patients was discharge delayed?
b) Were the percentages different for major and minor surgery?
c) Overall, what were the discharge delay rates at each hospital?
d) What were the delay rates at each hospital for each kind of surgery?
e) The insurance company is considering advising their clients to use large hospitals for surgery to avoid postsurgical complications. Do you think they should do this?
f) Explain, in your own words, why this confusion occurs.

42. Delivery service. A company must decide which of two delivery services they will contract with. During a recent trial period, they shipped numerous packages with each service and have kept track of how often deliveries did not arrive on time. Here are the data.

Delivery Service	Type of Service	Number of Deliveries	Number of Late Packages
Pack Rats	Regular	400	12
	Overnight	100	16
Boxes R Us	Regular	100	2
	Overnight	400	28

a) Compare the two services' overall percentage of late deliveries.
b) Based on the results in part a, the company has decided to hire Pack Rats. Do you agree they deliver on time more often? Why or why not? Be specific.
c) The results here are an instance of what phenomenon?

43. Graduate admissions. A 1975 article in the magazine *Science* examined the graduate admissions process at Berkeley for evidence of gender bias. The following table shows the number of applicants accepted to each of four graduate programs.

Program	Males Accepted (of Applicants)	Females Accepted (of Applicants)
1	511 of 825	89 of 108
2	352 of 560	17 of 25
3	137 of 407	132 of 375
4	22 of 373	24 of 341
Total	**1022 of 2165**	**262 of 849**

a) What percent of total applicants were admitted?
b) Overall, were a higher percentage of males or females admitted?
c) Compare the percentage of males and females admitted in each program.
d) Which of the comparisons you made do you consider to be the most valid? Why?

44. Simpson's Paradox. Develop your own table of data that is a business example of Simpson's Paradox. Explain the conflict between the conclusions made from the conditional and marginal distributions.

JUST CHECKING ANSWERS

1 50.0%
2 40.0%
3 25.0%
4 15.6% Near-sighted, 56.3% Far-sighted, 28.1% Need Bifocals
5 18.8% Near-sighted, 62.5% Far-sighted, 18.8% Need Bifocals
6 40% of the near-sighted customers are female, while 50% of customers are female.
7 Since near-sighted customers appear less likely to be female, it seems that they may not be independent. (But the numbers are small.)

Randomness and Probability

Credit Reports and the Fair Isaacs Corporation

You've probably never heard of the Fair Isaacs Corporation, but they probably know you. Whenever you apply for a loan, a credit card, or even a job, your credit "score" will be used to determine whether you are a good risk. And because the most widely used credit scores are Fair Isaacs' FICO® scores, the company may well be involved in the decision. The Fair Isaacs Corporation (FICO) was founded in 1956, with the idea that data, used intelligently, could improve business decision-making. Today, Fair Isaacs claims that their services provide companies around the world with information for more than 180 billion business decisions a year.

Your credit score is a number between 350 and 850 that summarizes your credit "worthiness." It's a snapshot of credit risk today based on your credit history and past behavior. Lenders of all kinds use credit scores to predict behavior, such as how likely you are to make your loan payments on time or to default on a loan. Lenders use the score to determine not only whether to give credit, but also the cost of the credit that they'll offer. There are no established boundaries, but generally scores over 750 are considered excellent, and applicants with those scores get the best rates. An applicant with a score below 620 is generally considered to be a poor risk. Those with very low scores may be denied credit outright or only offered "subprime" loans at substantially higher rates.

It's important that you be able to verify the information that your score is based on, but until recently, you could only hope that your score was based on correct information. That changed in 2000, when a California law gave mortgage applicants the right to see their credit scores. Today, the credit industry is more open about giving consumers access to their scores and the U.S. government, through the Fair and Accurate Credit Transaction Act (FACTA), now guarantees that you can access your credit report at no cost, at least once a year.[1]

Companies have to manage risk to survive, but by its nature, risk carries uncertainty. A bank can't know for certain that you'll pay your mortgage on time—or at all. What can they do with events they can't predict? They start with the fact that, although individual outcomes cannot be anticipated with certainty, random phenomena do, in the long run, settle into patterns that are consistent and predictable. It's this property of random events that makes Statistics practical.

5.1 Random Phenomena and Probability

When a customer calls the 800 number of a credit card company, he or she is asked for a card number before being connected with an operator. As the connection is made, the purchase records of that card and the demographic information of the customer are retrieved and displayed on the operator's screen. If the customer's FICO score is high enough, the operator may be prompted to "cross-sell" another service—perhaps a new "platinum" card for customers with a credit score of at least 750.

Of course, the company doesn't know which customers are going to call. Call arrivals are an example of a random phenomenon. With **random phenomena**, we can't predict the individual outcomes, but we can hope to understand characteristics of their long-run behavior. We don't know whether the *next* caller will qualify

---

[1] However, the score you see in your report will be an "educational" score intended to show consumers how scoring works. You still have to pay a "reasonable fee" to see your FICO score.

for the platinum card, but as calls come into the call center, the company will find that the percentage of platinum-qualified callers who qualify for cross-selling will settle into a pattern, like that shown in the graph in Figure 5.1.

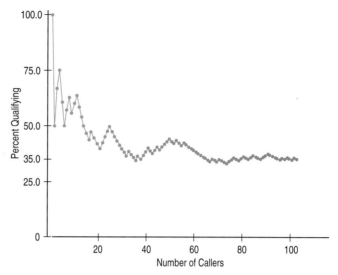

Figure 5.1 *The percentage of credit card customers who qualify for the premium card.*

As calls come into the call center, the company might record whether each caller qualifies. The first caller today qualified. Then the next five callers' qualifications were no, yes, yes, no, and no. If we plot the percentage who qualify against the call number, the graph would start at 100% because the first caller qualified (1 out of 1, for 100%). The next caller didn't qualify, so the accumulated percentage dropped to 50% (1 out of 2). The third caller qualified (2 out of 3, or 67%), then yes again (3 out of 4, or 75%), then no twice in a row (3 out of 5, for 60%, and then 3 out of 6, for 50%), and so on (Table 5.1). With each new call, the new datum is a smaller fraction of the accumulated experience, so, in the long run, the graph settles down. As it settles down, it appears that, in fact, the fraction of customers who qualify is about 35%.

Call	FICO Score	Qualify?	% Qualify
1	750	Yes	100
2	640	No	50
3	765	Yes	66.7
4	780	Yes	75
5	680	No	60
6	630	No	50
⋮	⋮		⋮

Table 5.1 *Data on the first six callers showing their FICO score, whether they qualified for the platinum card offer, and a running percentage of number of callers who qualified.*

> A **phenomenon** consists of **trials**. Each trial has an **outcome**. Outcomes combine to make **events**.

When talking about long-run behavior, it helps to define our terms. For any random phenomenon, each attempt, or **trial**, generates an **outcome**. For the call center, each call is a trial. Something happens on each trial, and we call whatever happens the outcome. Here the outcome is whether the caller qualifies or not. We use the more general term **event** to refer to outcomes or combinations of outcomes. For example, suppose we categorize callers into 6 risk categories and number these outcomes from 1 to 6 (of increasing credit worthiness). The three outcomes 4, 5, or 6 could make up the event "caller is at least a category 4."

We sometimes talk about the collection of *all possible outcomes*, a special event that we'll refer to as the **sample space**. We denote the sample space **S**; you may also see the Greek letter Ω used. But whatever symbol we use, the sample space is the set that contains all the possible outcomes. For the calls, if we let Q = qualified and N = not qualified, the sample space is simple: **S** = {Q, N}. If we look at two calls together, the sample space has four outcomes: **S** = {QQ, QN, NQ, NN}. If we were interested in at least one qualified caller from the two calls, we would be interested in the event (call it **A**) consisting of the three outcomes QQ, QN, and NQ, and we'd write **A** = {QQ, QN, NQ}.

Although we may not be able to predict a *particular* individual outcome, such as which incoming call represents a potential upgrade sale, we can say a lot about the long-run behavior. Look back at Figure 5.1. If you were asked for the probability that a random caller will qualify, you might say that it was 35% because, in the *long run*, the percentage of the callers who qualify is about 35%. And, that's exactly what we mean by **probability**.

That seems simple enough, but do random phenomena always behave this well? Couldn't it happen that the frequency of qualified callers never settles down, but just bounces back and forth between two numbers? Maybe it hovers around 45% for awhile, then goes down to 25%, and then back and forth forever. When we think about what happens with a series of trials, it really simplifies things if the individual trials are independent. Roughly speaking, **independence** means that the outcome of one trial doesn't influence or change the outcome of another. Recall, that in Chapter 4, we called two variables *independent* if the value of one categorical variable did not influence the value of another categorical variable. (We checked for independence by comparing relative frequency distributions across variables.) There's no reason to think that whether the one caller qualifies influences whether another caller qualifies, so these are independent trials. We'll see a more formal definition of independence later in the chapter.

Fortunately, for independent events, we can depend on a principle called the **Law of Large Numbers (LLN)**, which states that if the events are independent, then as the number of calls increases, over days or months or years, the long-run relative frequency of qualified calls gets closer and closer to a single value. This gives us the guarantee we need and makes probability a useful concept.

Because the LLN guarantees that relative frequencies settle down in the long run, we know that the value we called the probability is legitimate and the number it settles down to is called the probability of that event. For the call center, we can write *P*(qualified) = 0.35. Because it is based on repeatedly observing the event's outcome, this definition of probability is often called **empirical probability**.

> The **probability** of an event is its long-run relative frequency. A relative frequency is a fraction, so we can write it as $\frac{35}{100}$, as a decimal, 0.35, or as a percentage, 35%.

> **Law of Large Numbers**
> The *long-run relative frequency* of repeated, independent events eventually produces the *true relative frequency* as the number of trials increases.

5.2 The Nonexistent Law of Averages

The Law of Large Numbers says that the relative frequency of a random event settles down to a single number in the long run. But, it is often misunderstood to be a "law of averages," perhaps because the concept of "long run" is hard to grasp. Many people believe, for example, that an outcome of a random event that hasn't occurred in many trials is "due" to occur. The original "dogs of the Dow" strategy for buying stocks recommended buying the 10 worst performing stocks of the 30 that make up the Dow Jones Industrial Average, figuring that these "dogs" were bound to do better next year. After all, we know that in the long run, the relative

"*Slump? I ain't in no slump. I just ain't hittin'.*"

—*Yogi Berra*

You may think it's obvious that the frequency of repeated events settles down in the long run to a single number. The discoverer of the Law of Large Numbers thought so, too. The way he put it was: "*For even the most stupid of men is convinced that the more observations have been made, the less danger there is of wandering from one's goal.*"

—Jacob Bernoulli, 1713

"*In addition, in time, if the roulette-betting fool keeps playing the game, the bad histories [outcomes] will tend to catch up with him.*"

—Nassim Nicholas Taleb in *Fooled by Randomness*

frequency will settle down to the probability of that outcome, so now we have some "catching up" to do, right? Wrong. In fact, Louis Rukeyser (the former host of *Wall Street Week*) said of the "dogs of the Dow" strategy, "that theory didn't work as promised."

Actually, we know very little about the behavior of random events in the short run. The fact that we are seeing independent random events makes each individual result impossible to predict. Relative frequencies even out *only* in the long run. And, according to the LLN, the long run is really long (infinitely long, in fact). The "Large" in the law's name means *infinitely* large. Sequences of random events don't compensate in the short run and don't need to do so to get back to the right long-run probability. Any short-run deviations will be overwhelmed in the long run. If the probability of an outcome doesn't change and the events are independent, the probability of any outcome in another trial is always what it was, no matter what has happened in other trials.

Many people confuse the Law of Large numbers with the so-called Law of Averages that would say that things have to even out in the short run. But even though the Law of Averages doesn't exist at all, you'll hear people talk about it as if it does. Is a good hitter in baseball who has struck out the last six times due for a hit his next time up? If the stock market has been down for the last three sessions, is it due to increase today? No. This isn't the way random phenomena work. There is no Law of Averages for short runs—no "Law of Small Numbers." A belief in such a "law" can lead to poor business decisions.

Keno and the Law of Averages. Of course, sometimes an apparent drift from what we expect means that the probabilities are, in fact, *not* what we thought. If you get 10 heads in a row, maybe the coin has heads on both sides!

Keno is a simple casino game in which numbers from 1 to 80 are chosen. The numbers, as in most lottery games, are supposed to be equally likely. Payoffs are made depending on how many of those numbers you match on your card. A group of graduate students from a Statistics department decided to take a field trip to Reno. They (*very* discreetly) wrote down the outcomes of the games for a couple of days, then drove back to test whether the numbers were, in fact, equally likely. It turned out that some numbers were *more likely* to come up than others. Rather than bet on the Law of Averages and put their money on the numbers that were "due," the students put their faith in the LLN—and all their (and their friends') money on the numbers that had come up before. After they pocketed more than $50,000, they were escorted off the premises and invited never to show their faces in that casino again. Not coincidentally, the ringleader of that group currently makes his living on Wall Street.

You've just flipped a fair coin and seen six heads in a row. Does the coin "owe" you some tails? Suppose you spend that coin and your friend gets it in change. When she starts flipping the coin, should we expect a run of tails? Of course not. Each flip is a new event. The coin can't "remember" what it did in the past, so it can't "owe" any particular outcomes in the future. Just to see how this works in practice, we simulated 100,000 flips of a fair coin on a computer. In our 100,000 "flips," there were 2981 streaks of at least 5 heads. The "Law of Averages" suggests that the next flip after a run of 5 heads should be tails more often to even things out. Actually, the next flip was heads more often than tails: 1550 times to 1431 times. That's 51.9% heads. You can perform a similar simulation easily.

JUST CHECKING

1 It has been shown that the stock market fluctuates randomly. Nevertheless, some investors believe that they should buy right after a day when the market goes down because it is bound to go up soon. Explain why this is faulty reasoning.

5.3 Different Types of Probability

Model-Based (Theoretical) Probability

We've discussed *empirical probability*—the relative frequency of an event's occurrence as the probability of an event. There are other ways to define probability as well. Probability was first studied extensively by a group of French mathematicians who were interested in games of chance. Rather than experiment with the games and risk losing their money, they developed mathematical models of probability. To make things simple (as we usually do when we build models), they started by looking at games in which the different outcomes were equally likely. Fortunately, many games of chance are like that. Any of 52 cards is equally likely to be the next one dealt from a well-shuffled deck. Each face of a die is equally likely to land up (or at least it should be).

When outcomes are equally likely, their probability is easy to compute—it's just 1 divided by the number of possible outcomes. So the probability of rolling a 3 with a fair die is one in six, which we write as 1/6. The probability of picking the ace of spades from the top of a well-shuffled deck is 1/52.

It's almost as simple to find probabilities for events that are made up of several equally likely outcomes. We just count all the outcomes that the event contains. The probability of the event is the number of outcomes in the event divided by the total number of possible outcomes.

For example, Pew Research[2] reports that of 10,189 randomly generated working phone numbers called for a survey, the initial results of the calls were as follows:

Result	Number of Calls
No Answer	311
Busy	61
Answering Machine	1336
Callbacks	189
Other Non-Contacts	893
Contacted Numbers	7400

The phone numbers were generated randomly, so each was equally likely. To find the probability of a contact, we just divide the number of contacts by the number of calls: $7400/10,189 = 0.7263$.

> We can write:
>
> $$P(\mathbf{A}) = \frac{\# \text{ of outcomes in } \mathbf{A}}{\text{total } \# \text{ of outcomes}}$$
>
> and call this the **(theoretical) probability** of the event.

> In an attempt to understand why, an interviewer asked someone who had just purchased a lottery ticket, "What do you think your chances are of winning the lottery?" The reply was, "Oh, about 50–50." The shocked interviewer asked, "How do you get that?" to which the response was, "Well, the way I figure it, either I win or I don't!" The moral of this story is that events are not always equally likely.

---

[2] www.pewinternet.org/pdfs/PIP_Digital_Footprints.pdf.

But don't get trapped into thinking that random events are always equally likely. The chance of winning a lottery—especially lotteries with very large payoffs—is small. Regardless, people continue to buy tickets.

Personal Probability

What's the probability that gold will sell for more than $1000 an ounce at the end of next year? You may be able to come up with a number that seems reasonable. Of course, no matter how confident you feel about your prediction, your probability should be between 0 and 1. How did you come up with this probability? In our discussion of probability, we've defined probability in two ways: 1) in terms of the relative frequency—or the fraction of times—that an event occurs in the long run or 2) as the number of outcomes in the event divided by the total number of outcomes. Neither situation applies to your assessment of gold's chances of selling for more than $1000.

We use the *language* of probability in everyday speech to express a degree of uncertainty without basing it on long-run relative frequencies. Your personal assessment of an event expresses your uncertainty about the outcome. That uncertainty may be based on your knowledge of commodities markets, but it can't be based on long-run behavior. We call this kind of probability a subjective, or **personal probability**.

Although personal probabilities may be based on experience, they are not based either on long-run relative frequencies or on equally likely events. Like the two other probabilities we defined, they need to satisfy the same rules as both empirical and theoretical probabilities that we'll discuss in the next section.

5.4 Probability Rules

For some people, the phrase "50/50" means something vague like "I don't know" or "whatever." But when we discuss probabilities, 50/50 has the precise meaning that two outcomes are *equally likely*. Speaking vaguely about probabilities can get you into trouble, so it's wise to develop some formal rules about how probability works. These rules apply to probability whether we're dealing with empirical, theoretical, or personal probability.

Rule 1. If the probability of an event occurring is 0, the event can't occur; likewise if the probability is 1, the event *always* occurs. Even if you think an event is very unlikely, its probability can't be negative, and even if you're sure it will happen, its probability can't be greater than 1. So we require that:

> **A probability is a number between 0 and 1.**
> **For any event A, $0 \leq P(A) \leq 1$.**

"Baseball is 90% mental. The other half is physical."
—Yogi Berra

Rule 2. If a random phenomenon has only one possible outcome, it's not very interesting (or very random). So we need to distribute the probabilities among all the outcomes a trial can have. How can we do that so that it makes sense? For example, consider the behavior of a certain stock. The possible daily outcomes might be:

A: The stock price goes up.
B: The stock price goes down.
C: The stock price remains the same.

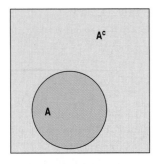

*The set **A** and its complement **A^c**. Together, they make up the entire sample space **S**.*

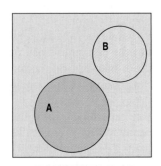

| *Two disjoint sets, **A** and **B**.*

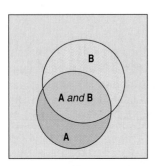

*Two sets **A** and **B** that are not disjoint. The event (**A** and **B**) is their intersection.*

! NOTATION ALERT:
You may see the event (**A** or **B**) written as (**A** ∪ **B**). The symbol ∪ means "union" and represents the outcomes in event **A** or event **B**. Similarly the symbol ∩ means intersection and represents outcomes that are in *both* event **A** and event **B**. You may see the event (**A** and **B**) written as (**A** ∩ **B**).

When we assign probabilities to these outcomes, we should be sure to distribute all of the available probability. Something always occurs, so the probability of *something* happening is 1. This is called the **Probability Assignment Rule**:

> **The probability of the set of all possible outcomes must be 1.**
>
> $$P(S) = 1$$

where S represents the set of all possible outcomes and is called the **sample space**.

Rule 3.   Suppose the probability that you get to class on time is 0.8. What's the probability that you don't get to class on time? Yes, it's 0.2. The set of outcomes that are *not* in the event **A** is called the "complement" of **A**, and is denoted **A^C**. This leads to the **Complement Rule**:

> **The probability of an event occurring is 1 minus the probability that it doesn't occur.**
>
> $$P(A) = 1 - P(A^C)$$

Rule 4.   Whether or not a caller qualifies for a platinum card is a random outcome. Suppose the probability of qualifying is 0.35. What's the chance that the next two callers qualify? The **Multiplication Rule** says that to find the probability that two independent events occur, we multiply the probabilities:

> **For two independent events A and B, the probability that both A *and* B occur is the product of the probabilities of the two events.**
> $$P(A \text{ and } B) = P(A) \times P(B), \text{ provided that A and B are independent.}$$

Thus if **A** = {customer 1 qualifies} and **B** = {customer 2 qualifies}, the chance that both qualify is:

$$0.35 \times 0.35 = 0.1225$$

Of course, to calculate this probability, we have used the assumption that the two events are independent. We'll expand the multiplication rule to be more general later in this chapter.

Rule 5.   Suppose the card center operator has more options. She can **A**: offer a special travel deal, **B**: offer a platinum card, or **C**: decide to send information about a new affinity card. If she can do one, but only one, of these, then these outcomes are **disjoint** (or **mutually exclusive**). To see whether two events are disjoint, we separate them into their component outcomes and check whether they have any outcomes in common. For example, if the operator can choose to both offer the travel deal and send the affinity card information, those would not be disjoint. The **Addition Rule** allows us to add the probabilities of disjoint events to get the probability that *either* event occurs:

> $$P(A \text{ or } B) = P(A) + P(B).$$

Thus the probability that the caller *either* is offered a platinum card or is sent the affinity card information is the sum of the two probabilities, since the events are disjoint.

Rule 6.   Suppose we would like to know the probability that either of the next two callers is qualified for a platinum card? We know $P(\mathbf{A}) = P(\mathbf{B}) = \mathbf{0.35}$, but $P(\mathbf{A} \text{ or } \mathbf{B})$ is not simply the sum $P(\mathbf{A}) + P(\mathbf{B})$ because the events **A** and **B** are not disjoint in this case. Both customers could qualify. So we need a new probability rule.

JUST CHECKING

2 MP3 players have relatively high failure rates for a consumer product, especially those models that contain a disk drive as opposed to those that have less storage but no drive. The worst failure rate for all iPod models was the 40GB Click wheel (as reported by MacIntouch.com) at 30%. If a store sells this model and failures are independent,

    a) What is the probability that the next one they sell will have a failure?

    b) What is the probability that there will be failures on *both* of the next two?

    c) What is the probability that the stores first failure problem will be with the third one they sell?

    d) What is the probability the store will have a failure problem with at least one of the next five that they sell?

We can't simply add the probabilities of **A** and **B** because that would count the outcome of *both* customers qualifying twice. So, if we started by adding the two probabilities, we could compensate by subtracting out the probability of that outcome. In other words,

$$P(\text{customer A } or \text{ customer B qualifies}) = P(\text{customer A qualifies}) + P(\text{customer B qualifies}) - P(\text{both customers qualify})$$

$$= (0.35) + (0.35) - (0.35 \times 0.35) \text{ (since events are independent)}$$

$$= (0.35) + (0.35) - (0.1225)$$

$$= 0.5775$$

It turns out that this method works in general. We add the probabilities of two events and then subtract out the probability of their intersection. This gives us the **General Addition Rule**, which does not require disjoint events:

$$P(\textbf{A or B}) = P(\textbf{A}) + P(\textbf{B}) - P(\textbf{A and B})$$

GUIDED EXAMPLE   M&M's Modern Market Research

In 1941, when M&M's® milk chocolate candies were introduced to American GIs in World War II, there were six colors: brown, yellow, orange, red, green, and violet. Mars®, the company that manufactures M&M's, has used the introduction of a new color as a marketing and advertising event several times in the years since then. In 1980, the candy went international adding 16 countries to their markets. In 1995, the company conducted a "worldwide survey" to vote on a new color. Over 10 million people voted to add blue. They even got the lights of the Empire State Building in New York City to glow blue to help announce the addition. In 2002, they used the Internet to help pick a new color. Children from over 200 countries were invited to respond via the Internet, telephone, or mail. Millions of voters chose among purple, pink, and teal. The global winner was purple, and for a brief time, purple M&M's could be found in packages worldwide (although in 2008, the colors were brown, yellow, red, blue, orange, and green). In the United States, 42% of those who voted said purple, 37% said teal, and only 19% said pink. But in Japan the percentages were 38% pink, 36% teal, and only 16% purple. Let's use Japan's percentages to ask some questions.

1. What's the probability that a Japanese M&M's survey respondent selected at random preferred either pink or teal?

2. If we pick two respondents at random, what's the probability that they *both* selected purple?

3. If we pick three respondents at random, what's the probability that *at least one* preferred purple?

PLAN

Setup The probability of an event is its long-term relative frequency. This can be determined in several ways: by looking at many replications of an event, by deducing it from equally likely events, or by using some other information. Here, we are told the relative frequencies of the three responses.

Make sure the probabilities are legitimate. Here, they're not. Either there was a mistake or the other voters must have chosen a color other than the three given. A check of other countries shows a similar deficit, so probably we're seeing those who had no preference or who wrote in another color.

The M&M's website reports the proportions of Japanese votes by color. These give the probability of selecting a voter who preferred each of the colors:

$$P(\text{pink}) = 0.38$$
$$P(\text{teal}) = 0.36$$
$$P(\text{purple}) = 0.16$$

Each is between 0 and 1, but these don't add up to 1. The remaining 10% of the voters must have not expressed a preference or written in another color. We'll put them together into "other" and add $P(\text{other}) = 0.10$.

With this addition, we have a legitimate assignment of probabilities.

Question 1. What's the probability that a Japanese M&M's survey respondent selected at random preferred either pink or teal?

PLAN

Setup Decide which rules to use and check the conditions they require.

The events "pink" and "teal" are individual outcomes (a respondent can't choose both colors), so they are disjoint. We can apply the General Addition Rule.

DO

Mechanics Show your work.

$$P(\text{pink or teal}) = P(\text{pink}) + P(\text{teal})$$
$$- P(\text{pink and teal})$$
$$= 0.38 + 0.36 - 0 = 0.74$$

The probability that both pink and teal were chosen is zero, since respondents were limited to one choice.

REPORT

Conclusion Interpret your results in the proper context.

The probability that the respondent said pink or teal is 0.74.

Question 2. If we pick two respondents at random, what's the probability that they both said purple?

PLAN

Setup The word "both" suggests we want $P(\mathbf{A} \text{ and } \mathbf{B})$, which calls for the Multiplication Rule. Check the required condition.

Independence
It's unlikely that the choice made by one respondent affected the choice of the other, so the events seem to be independent. We can use the Multiplication Rule.

DO	**Mechanics** Show your work. For both respondents to pick purple, each one has to pick purple.	$P(\text{both purple})$ $= P(\text{first respondent picks purple and second respondent picks purple})$ $= P(\text{first respondent picks purple})$ $\quad \times P(\text{second respondent picks purple})$ $= 0.16 \times 0.16 = 0.0256$
REPORT	**Conclusion** Interpret your results in the proper context.	The probability that both respondents pick purple is 0.0256.

Question 3. If we pick three respondents at random, what's the probability that at least one preferred purple?

PLAN	**Setup** The phrase "at least one" often flags a question best answered by looking at the complement, and that's the best approach here. The complement of "at least one preferred purple" is "none of them preferred purple." Check the conditions.	$P(\text{at least one picked purple})$ $\quad = P(\{\text{none picked purple}\}^c)$ $\quad = 1 - P(\text{none picked purple}).$ **Independence.** These are independent events because they are choices by three random respondents. We can use the Multiplication Rule.
DO	**Mechanics** We calculate $P(\text{none purple})$ by using the Multiplication Rule. Then we can use the Complement Rule to get the probability we want.	$P(\text{none picked purple}) = P(\text{first not purple})$ $\qquad\qquad \times P(\text{second not purple})$ $\qquad\qquad \times P(\text{third not purple})$ $\qquad\qquad = [P(\text{not purple})]^3.$ $P(\text{not purple}) = 1 - P(\text{purple})$ $\qquad\qquad = 1 - 0.16 = 0.84.$ So $P(\text{none picked purple}) = (0.84)^3 = 0.5927.$ $P(\text{at least 1 picked purple})$ $\quad = 1 - P(\text{none picked purple})$ $\quad = 1 - 0.5927 = 0.4073.$
REPORT	**Conclusion** Interpret your results in the proper context.	There's about a 40.7% chance that at least one of the respondents picked purple.

5.5 Joint Probability and Contingency Tables

As part of a Pick Your Prize Promotion, a chain store invited customers to choose which of three prizes they'd like to win (while providing name, address, phone number, and e-mail address). At one store, the responses could be placed in the contingency table in Table 5.2.

		Prize preference			
		MP3	**Camera**	**Bike**	**Total**
Sex	**Man**	117	50	60	**227**
	Woman	130	91	30	**251**
	Total	**247**	**141**	**90**	**478**

| **Table 5.2** *Prize preference for 478 customers.*

> A **marginal probability** uses a marginal frequency (from either the Total row or Total column) to compute the probability.

If the winner is chosen at random from these customers, the probability we select a woman is just the corresponding relative frequency (since we're equally likely to select any of the 478 customers). There are 251 women in the data out of a total of 478, giving a probability of:

$$P(\text{woman}) = 251/478 = 0.525$$

This is called a **marginal probability** because it depends only on totals found in the margins of the table. The same method works for more complicated events. For example, what's the probability of selecting a woman whose preferred prize is the camera? Well, 91 women named the camera as their preference, so the probability is:

$$P(\text{woman } and \text{ camera}) = 91/478 = 0.190$$

Probabilities such as these are called **joint probabilities** because they give the probability of two events occurring together.

The probability of selecting a customer whose preferred prize is a bike is:

$$P(\text{bike}) = 90/478 = 0.188$$

Since our sample space is these 478 customers, we can recognize the relative frequencies as probabilities. What if we are given the information that the selected customer is a woman? Would that change the probability that the selected customer's preferred prize is a bike? You bet it would! The pie charts show that women are much less likely to say their preferred prize is a bike than are men. When we restrict our focus to women, we look only at the women's row of the table, which gives the conditional distribution of preferred prizes given "woman." Of the 251 women, only 30 of them said their preferred prize was a bike. We write the probability that a selected customer wants a bike *given* that we have selected a woman as:

$$P(\text{bike}|\text{woman}) = 30/251 = 0.120$$

For men, we look at the conditional distribution of preferred prizes given "man" shown in the top row of the table. There, of the 227 men, 60 said their preferred prize was a bike. So, $P(\text{bike}|\text{man}) = 60/227 = 0.264$, more than twice the women's probability. (see Figure 5.1)

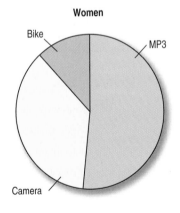

Women

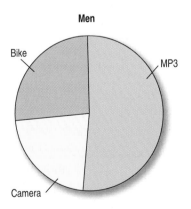

Men

Figure 5.1 Conditional distributions of *Prize Preference* for *Women* and for *Men*.

5.6 Conditional Probability

In general, when we want the probability of an event from a *conditional* distribution, we write $P(\mathbf{B}|\mathbf{A})$ and pronounce it "the probability of **B** *given* **A**." A probability that takes into account a given *condition* such as this is called a **conditional probability**.

Let's look at what we did. We worked with the counts, but we could work with the probabilities just as well. There were 30 women who selected a bike as a prize, and there were 251 women customers. So we found the probability to be 30/251. To find the probability of the event **B** *given* the event **A**, we restrict our attention to the outcomes in **A**. We then find in what fraction of *those* outcomes **B** also occurred. Formally, we write:

$$P(\mathbf{B}|\mathbf{A}) = \frac{P(\mathbf{A} \ and \ \mathbf{B})}{P(\mathbf{A})}$$

NOTATION ALERT:
$P(\mathbf{B}|\mathbf{A})$ is the conditional probability of **B** *given* **A**.

We can use the formula directly with the probabilities derived from the contingency table (Table 5.2) to find:

$$P(\text{bike}|\text{woman}) = \frac{P(\text{bike } and \text{ woman})}{P(\text{woman})} = \frac{30/478}{251/478} = \frac{0.063}{0.525} = 0.120 \text{ as before.}$$

The formula for conditional probability requires one restriction. The formula works only when the event that's given has probability greater than 0. The formula doesn't work if $P(\mathbf{A})$ is 0 because that would mean we had been "given" the fact that **A** was true even though the probability of **A** is 0, which would be a contradiction.

Rule 7. Remember the Multiplication Rule for the probability of **A** *and* **B**? It said

$$P(\mathbf{A} \text{ and } \mathbf{B}) = P(\mathbf{A}) \times P(\mathbf{B})$$

when **A** and **B** are independent. Now we can write a more general rule that doesn't require independence. In fact, we've already written it. We just need to rearrange the equation a bit.

The equation in the definition for conditional probability contains the probability of **A** *and* **B**. Rearranging the equation gives the **General Multiplication Rule** for compound events that does not require the events to be independent:

$$P(\mathbf{A} \ and \ \mathbf{B}) = P(\mathbf{A}) \times P(\mathbf{B}|\mathbf{A})$$

The probability that two events, **A** and **B**, both occur is the probability that event **A** occurs multiplied by the probability that event **B** *also* occurs—that is, by the probability that event **B** occurs given that event **A** occurs.

Of course, there's nothing special about which event we call **A** and which one we call **B**. We should be able to state this the other way around. Indeed we can. It is equally true that:

$$P(\mathbf{A} \ and \ \mathbf{B}) = P(\mathbf{B}) \times P(\mathbf{A}|\mathbf{B}).$$

Let's return to the question of just what it means for events to be independent. We said informally in Chapter 4 that what we mean by independence is that the outcome of one event does not influence the probability of the other. With our new notation for conditional probabilities, we can write a formal definition. Events **A** and **B** are **independent** whenever:

$$P(\mathbf{B}|\mathbf{A}) = P(\mathbf{B}).$$

> If we had to pick one key idea in this chapter that you should understand and remember, it's the definition and meaning of independence.

Now we can see that the Multiplication Rule for independent events is just a special case of the General Multiplication Rule. The general rule says

$$P(\mathbf{A} \ and \ \mathbf{B}) = P(\mathbf{A}) \times P(\mathbf{B}|\mathbf{A})$$

whether the events are independent or not. But when events **A** and **B** are independent, we can write $P(\mathbf{B})$ for $P(\mathbf{B}|\mathbf{A})$ and we get back our simple rule:

$$P(\mathbf{A} \ and \ \mathbf{B}) = P(\mathbf{A}) \times P(\mathbf{B}).$$

Sometimes people use this statement as the definition of independent events, but we find the other definition more intuitive. Either way, the idea is that the probabilities of independent events don't change when you find out that one of them has occurred.

Using our earlier example, is the probability of the event *choosing a bike* independent of the sex of the customer? We need to check whether

$$P(\text{bike}|\text{man}) = \frac{P(\text{bike} \ and \ \text{man})}{P(\text{man})} = \frac{0.126}{0.475} = 0.265$$

is the same as $P(\text{bike}) = 0.189$.

Because these probabilities aren't equal, we can say that prize preference is *not* independent of the sex of the customer. Whenever at least one of the joint probabilities in the table is *not* equal to the product of the marginal probabilities, we say that the variables are not independent.

◆ **Independent *vs.* Disjoint.**   Are disjoint events independent? Both concepts seem to have similar ideas of separation and distinctness about them, but in fact disjoint events *cannot* be independent.[3] Let's see why. Consider the two disjoint events {you get an A in this course} and {you get a B in this course}. They're disjoint because they have no outcomes in common. Suppose you learn that you *did* get an A in the course. Now what is the probability that you got a B? You can't get both grades, so it must be 0.

Think about what that means. Knowing that the first event (getting an A) occurred changed your probability for the second event (down to 0). So these events aren't independent.

Mutually exclusive events can't be independent. They have no outcomes in common, so knowing that one occurred means the other didn't. A common error is to treat disjoint events as if they were independent and apply the Multiplication Rule for independent events. Don't make that mistake.

5.7 Constructing Contingency Tables

Sometimes we're given probabilities without a contingency table. You can often construct a simple table to correspond to the probabilities.

A survey of real estate in upstate New York classified homes into two price categories (Low—less than \$175,000 and High—over \$175,000). It also noted whether the houses had at least 2 bathrooms or not (True or False). We are told that 56% of the houses had at least 2 bathrooms, 62% of the houses were Low

---

[3] Technically two disjoint events *can* be independent, but only if the probability of one of the events is 0. For practical purposes, we can ignore this case, since we don't anticipate collecting data about things that can't possibly happen.

priced, and 22% of the houses were both. That's enough information to fill out the table. Translating the percentages to probabilities, we have:

Price	At least 2 Bathrooms: True	False	Total
Low	0.22		**0.62**
High			
Total	**0.56**		**1.00**

The 0.56 and 0.62 are marginal probabilities, so they go in the margins. What about the 22% of houses that were both Low priced and had at least 2 bathrooms? That's a *joint* probability, so it belongs in the interior of the table.

Because the cells of the table show disjoint events, the probabilities always add to the marginal totals going across rows or down columns.

Price	At least 2 Bathrooms: True	False	Total
Low	0.22	0.40	**0.62**
High	0.34	0.04	**0.38**
Total	**0.56**	**0.44**	**1.00**

Now, finding any other probability is straightforward. For example, what's the probability that a high-priced house has at least 2 bathrooms?

$$P(\text{at least 2 bathrooms} \mid \text{high-priced})$$
$$= P(\text{at least 2 bathrooms } \textit{and} \text{ high-priced})/P(\text{high-priced})$$
$$= 0.34/0.38 = 0.895 \text{ or } 89.5\%.$$

JUST CHECKING

3 Suppose a supermarket is conducting a survey to find out the busiest time and day for shoppers. Survey respondents are asked 1) whether they shopped at the store on a weekday or on the weekend and 2) whether they shopped at the store before or after 5 p.m. The survey revealed that:

- 48% of shoppers visited the store before 5 p.m.
- 27% of shoppers visited the store on a weekday (Mon.–Fri.)
- 7% of shoppers visited the store before 5 p.m. on a weekday.

a) Make a contingency table for the variables *time of day* and *day of week*.

b) What is the probability that a randomly selected shopper who shops before 5 p.m. also shops on a weekday?

c) Are time and day of the week disjoint events?

d) Are time and day of the week independent events?

WHAT CAN GO WRONG?

- **Beware of probabilities that don't add up to 1.** To be a legitimate assignment of probability, the sum of the probabilities for all possible outcomes must total 1. If the sum is less than 1, you may need to add another category ("other") and assign the remaining probability to that outcome. If the sum is more than 1, check that the outcomes are disjoint. If they're not, then you can't assign probabilities by counting relative frequencies.

- **Don't add probabilities of events if they're not disjoint.** Events must be disjoint to use the Addition Rule. The probability of being under 80 *or* a female is not the probability of being under 80 *plus* the probability of being female. That sum may be more than 1.

- **Don't multiply probabilities of events if they're not independent.** The probability of selecting a customer at random who is over 70 years old *and* retired is not the probability the customer is over 70 years old *times* the probability the customer is retired. Knowing that the customer is over 70 changes the probability of his or her being retired. You can't multiply these probabilities. The multiplication of probabilities of events that are not independent is one of the most common errors people make in dealing with probabilities.

- **Don't confuse disjoint and independent.** Disjoint events *can't* be independent. If **A** = {you get a promotion} and **B** = {you don't get a promotion}, **A** and **B** are disjoint. Are they independent? If you find out that **A** is true, does that change the probability of **B**? You bet it does! So they can't be independent.

What have we learned?

We've learned that probability is based on long-run relative frequencies and that the Law of Large Numbers speaks only of long-run behavior. Because the long run is a very long time, we need to be careful not to misinterpret the Law of Large Numbers as a law of averages. Even when we've observed a string of heads, we shouldn't expect extra tails in subsequent coin flips.

Also, we've learned some basic rules for combining probabilities of outcomes to find probabilities of more complex events. These include:

1. Probability for any event is between 0 and 1
2. Probability of the sample space, **S**, the set of possible outcomes = 1
3. Complement Rule
4. Multiplication Rule for independent events
5. General Addition Rule
6. General Multiplication Rule

Terms

Addition Rule   If **A** and **B** are disjoint events, then the probability of **A** or **B** is

$$P(\mathbf{A} \text{ or } \mathbf{B}) = P(\mathbf{A}) + P(\mathbf{B}).$$

Complement Rule   The probability of an event occurring is 1 minus the probability that it doesn't occur:

$$P(\mathbf{A}) = 1 - P(\mathbf{A}^{\mathrm{C}}).$$

Conditional probability	$$P(\mathbf{B}|\mathbf{A}) = \frac{P(\mathbf{A} \text{ and } \mathbf{B})}{P(\mathbf{A})}.$$ $P(\mathbf{B}|\mathbf{A})$ is read "the probability of **B** *given* **A**."
Disjoint (or Mutually Exclusive) Events	Two events are disjoint if they share no outcomes in common. If **A** and **B** are disjoint, then knowing that **A** occurs tells us that **B** cannot occur. Disjoint events are also called "mutually exclusive."
Empirical probability	When the probability comes from the long-run relative frequency of the event's occurrence, it is an empirical probability.
Event	A collection of outcomes. Usually, we identify events so that we can attach probabilities to them. We denote events with bold capital letters such as **A**, **B**, or **C**.
General Addition Rule	For any two events, **A** and **B**, the probability of **A** *or* **B** is: $$P(\mathbf{A} \text{ or } \mathbf{B}) = P(\mathbf{A}) + P(\mathbf{B}) - P(\mathbf{A} \text{ and } \mathbf{B}).$$
General Multiplication Rule	For any two events, **A** and **B**, the probability of **A** and **B** is: $$P(\mathbf{A} \text{ and } \mathbf{B}) = P(\mathbf{A}) \times P(\mathbf{B}|\mathbf{A}).$$
Independence (informally)	Two events are *independent* if the fact that one event occurs does not change the probability of the other.
Independence (used formally)	Events **A** and **B** are independent when $P(\mathbf{B}|\mathbf{A}) = P(\mathbf{B})$.
Joint probabilities	The probability that two events both occur.
Law of Large Numbers (LLN)	The Law of Large Numbers states that the *long-run relative frequency* of repeated, independent events settles down to the *true relative frequency* as the number of trials increases.
Marginal probability	In a joint probability table a marginal probability is the probability distribution of either variable separately, usually found in the rightmost column or bottom row of the table.
Multiplication Rule	If **A** and **B** are independent events, then the probability of **A** *and* **B** is: $$P(\mathbf{A} \text{ and } \mathbf{B}) = P(\mathbf{A}) \times P(\mathbf{B}).$$
Outcome	The outcome of a trial is the value measured, observed, or reported for an individual instance of that trial.
Personal probability	When the probability is subjective and represents your personal degree of belief, it is called a personal probability.
Probability	The probability of an event is a number between 0 and 1 that reports the likelihood of the event's occurrence. A probability can be derived from a model (such as equally likely outcomes), from the long-run relative frequency of the event's occurrence, or from subjective degrees of belief. We write $P(\mathbf{A})$ for the probability of the event **A**.
Probability Assignment Rule	The probability of the entire sample space must be 1: $$P(S) = 1.$$

Random phenomenon	A phenomenon is random if we know what outcomes could happen, but not which particular values will happen.
Sample space	The collection of all possible outcome values. The sample space has a probability of 1.
Theoretical probability	When the probability comes from a mathematical model (such as, but not limited to, equally likely outcomes), it is called a theoretical probability.
Trial	A single attempt or realization of a random phenomenon.

Skills

 PLAN

- Be able to understand that random phenomena are unpredictable in the short term but show long-run regularity.
- Know how to recognize random outcomes in a real-world situation.
- Know that the relative frequency of an outcome of a random phenomenon settles down as we gather more random outcomes. Be able to state the Law of Large Numbers.
- Know the basic definitions and rules of probability.
- Be able to recognize when events are disjoint and when events are independent. Understand the difference and that disjoint events cannot be independent.

 DO

- Be able to use the facts about probability to determine whether an assignment of probabilities is legitimate. Each probability must be a number between 0 and 1, and the sum of the probabilities assigned to all possible outcomes must be 1.
- Know how and when to apply the General Addition Rule. Know when events are disjoint.
- Know how and when to apply the General Multiplication Rule. Be able to use the Multiplication Rule to find probabilities for combinations of both independent and nonindependent events.
- Know how to use the Complement Rule to make calculating probabilities simpler. Recognize that probabilities of "at least" are likely to be simplified in this way.

 REPORT

- Be able to use statements about probability in describing a random phenomenon. You will need this skill soon for making statements about statistical inference.
- Know and be able to use correctly the terms "sample space," "disjoint events," and "independent events."
- Be able to make a statement about a conditional probability that makes clear how the condition affects the probability.
- Avoid making statements that assume independence of events when there is no clear evidence that they are in fact independent.

Mini Case Study Project

Market Segmentation

The data from the "Chicago Female Fashion Study"[4] were collected using a self-administered survey of a sample of homes in the greater Chicago metropolitan area. The marketing manager for department store X wants to know how important quality is to her customers. A consultant reports that based on past research, 30% of all consumers nationwide are more interested in quantity than quality. The marketing manager suspects that customers from her store are different, and that customers of different ages might have different views as well. Using conditional probabilities, marginal probabilities, and joint probabilities constructed from the data in the file **ch05_MCSP_Market_Segmentation**,[5] write a report to the manager on what you find.

Keep in mind: The manager may be more interested in the opinions of "frequent" customers than those who never or hardly ever shop at her store. These "frequent" customers contribute a disproportionate amount of profit to the store. Keep that in mind as you do your analysis and write up your report.

Variable and Question	Categories
Age *Into which of the following age categories do you belong?*	18–24 yrs old 25–34 35–44 45–54 55–64 65 or over
Frequency *How often do you shop for women's clothing at [department store X]?*	Never–hardly ever 1–2 times per year 3–4 times per year 5 times or more
Quality *For the same amount of money, I will generally buy one good item than several of lower price and quality.*	1. Definitely Disagree 2. Generally Disagree 3. Moderately Disagree 4. Moderately Agree 5. Generally Agree 6. Definitely Agree

---

[4] Original *Market Segmentation Exercise* prepared by K. Matsuno, D. Kopcso, and D. Tigert, Babson College in 1997 (Babson Case Series #133-C97A-U).

[5] For a version with the categories coded as integers see **ch05_MCSP_Market_Segmentation_Coded**.

EXERCISES

1. What does it mean? part 1. Respond to the following questions:

a) A casino claims that its roulette wheel is truly random. What should that claim mean?

b) A reporter on *Market Place* says that there is a 50% chance that the Federal Reserve Bank will cut interest rates by a quarter point at their next meeting. What is the meaning of such a phrase?

2. What does it mean? part 2. Respond to the following questions:

a) After an unusually dry autumn, a radio announcer is heard to say, "Watch out! We'll pay for these sunny days later on this winter." Explain what he's trying to say, and comment on the validity of his reasoning.

b) A batter who had failed to get a hit in seven consecutive times at bat then hits a game-winning home run. When talking to reporters afterward, he says he was very confident that last time at bat because he knew he was "due for a hit." Comment on his reasoning.

3. Airline safety. Even though commercial airlines have excellent safety records, in the weeks following a crash, airlines often report a drop in the number of passengers, probably because people are afraid to risk flying.

a) A travel agent suggests that since the law of averages makes it highly unlikely to have two plane crashes within a few weeks of each other, flying soon after a crash is the safest time. What do you think?

b) If the airline industry proudly announces that it has set a new record for the longest period of safe flights, would you be reluctant to fly? Are the airlines due to have a crash?

4. Economic predictions. An investment newsletter makes general predictions about the economy to help their clients make sound investment decisions.

a) Recently they said that because the stock market had been up for the past three months in a row that it was "due for a correction" and advised their client to reduce their holdings. What "law" are they applying? Comment.

b) They advised buying a stock that had gone down in the past four sessions because they said that it was clearly "due to bounce back." What "law" are they applying? Comment.

5. Fire insurance. Insurance companies collect annual payments from homeowners in exchange for paying to rebuild houses that burn down.

a) Why should you be reluctant to accept a $300 payment from your neighbor to replace his house should it burn down during the coming year?

b) Why can the insurance company make that offer?

6. Casino gambling. Recently, the International Gaming Technology company issued the following press release:

(LAS VEGAS, Nev.)—Cynthia Jay was smiling ear to ear as she walked into the news conference at the Desert Inn Resort in Las Vegas today, and well she should. Last night, the 37-year-old cocktail waitress won the world's largest slot jackpot—$34,959,458—on a Megabucks machine. She said she had played $27 in the machine when the jackpot hit. Nevada Megabucks has produced 49 major winners in its 14-year history. The top jackpot builds from a base amount of $7 million and can be won with a 3-coin ($3) bet.

a) How can the Desert Inn afford to give away millions of dollars on a $3 bet?

b) Why did the company issue a press release? Wouldn't most businesses want to keep such a huge loss quiet?

7. Toy company. A toy company manufactures a spinning game and needs to decide what probabilities are involved in the game. The plastic arrow on the spinner stops rotating to point at a color that will determine what happens next. Knowing these probabilities will help determine how easy or difficult it is for a person to win the game and helps to determine how long the average game will last. Are each of the following probability assignments possible? Why or why not?

	Probabilities of . . .			
	Red	**Yellow**	**Green**	**Blue**
a)	0.25	0.25	0.25	0.25
b)	0.10	0.20	0.30	0.40
c)	0.20	0.30	0.40	0.50
d)	0	0	1.00	0
e)	0.10	0.20	1.20	−1.50

8. Store discounts. Many stores run "secret sales": Shoppers receive cards that determine how large a discount they get, but the percentage is revealed by scratching off that black stuff (what *is* that?) only after the purchase has been totaled at the cash register. The store is required to reveal (in the fine print) the distribution of discounts available. Are each of these probability assignments plausible? Why or why not?

	Probabilities of . . .			
	10% off	**20% off**	**30% off**	**50% off**
a)	0.20	0.20	0.20	0.20
b)	0.50	0.30	0.20	0.10
c)	0.80	0.10	0.05	0.05
d)	0.75	0.25	0.25	−0.25
e)	1.00	0	0	0

9. Quality control. A tire manufacturer recently announced a recall because 2% of its tires are defective. If you just bought a new set of four tires from this manufacturer, what is the probability that at least one of your new tires is defective?

10. Pepsi promotion. For a sales promotion, the manufacturer places winning symbols under the caps of 10% of all Pepsi bottles. If you buy a six-pack of Pepsi, what is the probability that you win something?

11. Auto warranty. In developing their warranty policy, an automobile company estimates that over a 1-year period 17% of their new cars will need to be repaired once, 7% will need repairs twice, and 4% will require three or more repairs. If you buy a new car from them, what is the probability that your car will need:

a) No repairs?
b) No more than one repair?
c) Some repairs?

12. Consulting team. You work for a large global management consulting company. Of the entire work force of analysts, 55% have had no experience in the telecommunications industry, 32% have had limited experience (less than 5 years), and the rest have had extensive experience (5 years or more). On a recent project, you and two other analysts were chosen at random to constitute a team. It turns out that part of the project involves telecommunications. What is the probability that the first teammate you meet has:

a) Extensive telecommunications experience?
b) Some telecommunications experience?
c) No more than limited telecommunications experience?

13. Auto warranty, part 2. Consider again the auto repair rates described in Exercise 11. If you bought two new cars, what is the probability that:

a) Neither will need repair?
b) Both will need repair?
c) At least one car will need repair?

14. Consulting team, part 2. You are assigned to be part of a team of three analysts of a global management consulting company as described in Exercise 12. What is the probability that of your other two teammates:

a) Neither has any telecommunications experience?
b) Both have some telecommunications experience?
c) At least one has had extensive telecommunications experience?

15. Auto warranty, again. You used the Multiplication Rule to calculate repair probabilities for your cars in Exercise 11.

a) What must be true about your cars in order to make that approach valid?
b) Do you think this assumption is reasonable? Explain.

16. Final consulting team project. You used the Multiplication Rule to calculate probabilities about the telecommunications experience of your consulting teammates in Exercise 12.

a) What must be true about the groups in order to make that approach valid?
b) Do you think this assumption is reasonable? Explain.

17. Real estate. Real estate ads suggest that 64% of homes for sale have garages, 21% have swimming pools, and 17% have both features. What is the probability that a home for sale has:

a) A pool or a garage?
b) Neither a pool nor a garage?
c) A pool but no garage?

18. Human resource data. Employment data at a large company reveal that 72% of the workers are married, 44% are college graduates, and half of the college grads are married. What's the probability that a randomly chosen worker is:

a) Neither married nor a college graduate?
b) Married but not a college graduate?
c) Married or a college graduate?

19. Market research on energy. A Gallup Poll in March 2007 asked 1005 U.S. adults whether increasing domestic energy production or protecting the environment should be given higher priority. Here are the results.

Response	Number
Increase Production	342
Protect the Environment	583
Equally Important	50
No Opinion	30
Total	**1005**

If we select a person at random from this sample of 1005 adults:

a) What is the probability that the person responded "Increase Production"?
b) What is the probability that the person responded "Equally Important" or had "No Opinion"?

20. More market research on energy. Exercise 19 shows the results of a Gallup Poll about energy. Suppose we select three people at random from this sample.

a) What is the probability that all three responded "Protect the Environment"?
b) What is the probability that none responded "Equally Important"?
c) What assumption did you make in computing these probabilities?
d) Explain why you think that assumption is reasonable.

21. Telemarketing contact rates. Marketing research firms often contact their respondents by sampling random telephone numbers. Although interviewers currently reach about 76% of selected U.S. households, the percentage of those contacted who agree to cooperate with the survey has fallen. Assume that the percentage of those who agree to cooperate in telemarketing surveys is now only 38%. Each household is assumed to be independent of the others.

a) What is the probability that the next household on the list will be contacted but will refuse to cooperate?
b) What is the probability of failing to contact a household or of contacting the household but not getting them to agree to the interview?
c) Show another way to calculate the probability in part b.

22. Telemarketing contact rates, part 2. According to Pew Research, the contact rate (probability of contacting a selected household) in 1997 was 69%, and in 2003, it was 76%. However, the cooperation rate (probability of someone at the contacted household agreeing to be interviewed) was 58% in 1997 and dropped to 38% in 2003.

a) What is the probability (in 2003) of obtaining an interview with the next household on the sample list? (To obtain an interview, an interviewer must both contact the household and then get agreement for the interview.)

b) Was it more likely to obtain an interview from a randomly selected household in 1997 or in 2003?

23. Mars product information. The Mars company says that before the introduction of purple, yellow made up 20% of their plain M&M candies, red made up another 20%, and orange, blue, and green each made up 10%. The rest were brown.

a) If you picked an M&M at random from a pre-purple bag of candies, what is the probability that it was:

   i) Brown?

   ii) Yellow or orange?

   iii) Not green?

   iv) Striped?

b) Assuming you had an infinite supply of M&M's with the older color distribution, if you picked three M&M's in a row, what is the probability that:

   i) They are all brown?

   ii) The third one is the first one that's red?

   iii) None are yellow?

   iv) At least one is green?

24. American Red Cross. The American Red Cross must track their supply and demand for various blood types. They estimate that about 45% of the U.S. population has Type O blood, 40% Type A, 11% Type B, and the rest Type AB.

a) If someone volunteers to give blood, what is the probability that this donor:

   i) Has Type AB blood?

   ii) Has Type A or Type B blood?

   iii) Is not Type O?

b) Among four potential donors, what is the probability that:

   i) All are Type O?

   ii) None have Type AB blood?

   iii) Not all are Type A?

   iv) At least one person is Type B?

25. More Mars product information. In Exercise 23, you calculated probabilities of getting various colors of M&M's.

a) If you draw one M&M, are the events of getting a red one and getting an orange one disjoint or independent or neither?

b) If you draw two M&M's one after the other, are the events of getting a red on the first and a red on the second disjoint or independent or neither?

c) Can disjoint events ever be independent? Explain.

26. American Red Cross, part 2. In Exercise 24, you calculated probabilities involving various blood types.

a) If you examine one donor, are the events of the donor being Type A and the donor being Type B disjoint or independent or neither? Explain your answer.

b) If you examine two donors, are the events that the first donor is Type A and the second donor is Type B disjoint or independent or neither?

c) Can disjoint events ever be independent? Explain.

27. Tax accountant. A recent study of IRS audits showed that, for estates worth less than $5 million, about 1 out of 7 of all estate tax returns are audited, but that probability increases to 50% for estates worth over $5 million. Suppose a tax accountant has three clients who have recently filed returns for estates worth more than $5 million. What are the probabilities that:

a) All three will be audited?

b) None will be audited?

c) At least one will be audited?

d) What did you assume in calculating these probabilities?

28. Casinos. Because gambling is big business, calculating the odds of a gambler winning or losing in every game is crucial to the financial forecasting for a casino. A standard slot machine has three wheels that spin independently. Each has 10 equally likely symbols: 4 bars, 3 lemons, 2 cherries, and a bell. If you play once, what is the probability that you will get:

a) 3 lemons?

b) No fruit symbols?

c) 3 bells (the jackpot)?

d) No bells?

e) At least one bar (an automatic loser)?

29. Information technology. A company has recently replaced their e-mail server because previously mail was interrupted on about 15% of workdays. To see how bad the situation was, calculate the probability that during a 5-day work week, there would be an e-mail interruption.

a) On Monday and again on Tuesday?

b) For the first time on Thursday?

c) Every day?

d) At least once during the week?

30. Information technology, part 2. At a mid-sized Web design and maintenance company, 57% of the computers are PCs, 29% are Macs, and the rest are Unix-based machines. Assuming that users of each of the machines are equally likely to call in to the information technology help line, what is the probability that of the next three calls:

a) All are Macs?

b) None are PCs?

c) At least one is a Unix machine?

d) All are Unix machines?

31. Casinos, part 2. In addition to slot machines, casinos must understand the probabilities involved in card games. Suppose you are playing at the blackjack table, and the dealer shuffles a deck of cards. The first card shown is red. So is the second and the third. In fact, you are surprised to see 5 red cards in a row. You start thinking, "The next one is due to be black!"

a) Are you correct in thinking that there's a higher probability that the next card will be black than red? Explain.

b) Is this an example of the Law of Large Numbers? Explain.

32. Inventory. A shipment of road bikes has just arrived at The Spoke, a small bicycle shop, and all the boxes have been placed in the back room. The owner asks her assistant to start bringing in the boxes. The assistant sees 20 identical-looking boxes and starts bringing them into the shop at random. The owner knows that she ordered 10 women's and 10 men's bicycles, and so she's surprised to find that the first six are all women's bikes. As the seventh box is brought in, she starts thinking, "This one is bound to be a men's bike."

a) Is she correct in thinking that there's a higher probability that the next box will contain a men's bike? Explain.

b) Is this an example of the Law of Large Numbers? Explain.

33. International food survey. A GfK Roper Worldwide survey in 2005 asked consumers in five countries whether they agreed with the statement "I am worried about the safety of the food I eat." Here are the responses classified by the age of the respondent.

		Agree	Neither Agree nor Disagree	Disagree	Don't Know/ No Response	Total
Age	**13–19**	661	368	452	32	**1513**
	20–29	816	365	336	16	**1533**
	30–39	871	355	290	9	**1525**
	40–49	914	335	266	6	**1521**
	50+	966	339	283	10	**1598**
	Total	**4228**	**1762**	**1627**	**73**	**7690**

If we select a person at random from this sample:

a) What is the probability that the person agreed with the statement?

b) What is the probability that the person is younger than 50 years old?

c) What is the probability that the person is younger than 50 *and* agrees with the statement?

d) What is the probability that the person is younger than 50 *or* agrees with the statement?

34. Cosmetics marketing. A GfK Roper Worldwide survey asked consumers in five countries whether they agreed with the statement "I follow a skin care routine every day." Here are the responses classified by the country of the respondent.

		Response			
		Agree	**Disagree**	**Don't know**	**Total**
Country	**China**	361	988	153	**1502**
	France	695	763	81	**1539**
	India	828	689	18	**1535**
	U.K.	597	898	62	**1557**
	USA	668	841	48	**1557**
	Total	**3149**	**4179**	**362**	**7690**

If we select a person at random from this sample:

a) What is the probability that the person agreed with the statement?

b) What is the probability that the person is from China?

c) What is the probability that the person is from China *and* agrees with the statement?

d) What is the probability that the person is from China *or* agrees with the statement?

35. E-commerce. Suppose an online business organizes an e-mail survey to find out if online shoppers are concerned with the security of business transactions on the Web. Of the 42 individuals who respond, 24 are concerned, and 18 are not concerned. Eight of those concerned about security are male and 6 of those not concerned are male. If a respondent is selected at random, find each of the following conditional probabilities:

a) The respondent is male, given that the respondent is not concerned about security.

b) The respondent is not concerned about security, given that it is female.

c) The respondent is female, given that the respondent is concerned about security.

36. Automobile inspection. Twenty percent of cars that are inspected have faulty pollution control systems. The cost of repairing a pollution control system exceeds $100 about 40% of the time. When a driver takes her car in for inspection, what's the probability that she will end up paying more than $100 to repair the pollution control system?

37. Pharmaceutical company. A U.S. pharmaceutical company is considering manufacturing and marketing a pill that will help to lower both an individual's blood pressure and cholesterol. The company is interested in understanding the demand for such a product. The joint probabilities that an adult American man has high blood pressure and/or high cholesterol are shown in the table.

Cholesterol	Blood Pressure: High	Blood Pressure: OK
High	0.11	0.21
OK	0.16	0.52

a) What's the probability that an adult American male has both conditions?
b) What's the probability that an adult American male has high blood pressure?
c) What's the probability that an adult American male with high blood pressure also has high cholesterol?
d) What's the probability that an adult American male has high blood pressure if it's known that he has high cholesterol?

38. International relocation. A European department store is developing a new advertising campaign for their new U.S. location, and their marketing managers need to better understand their target market. Based on survey responses, a joint probability table that an adult shops at their new U.S. store classified by their age is shown below.

Age	Shop: Yes	Shop: No	Total
< 20	0.26	0.04	0.30
20–40	0.24	0.10	0.34
> 40	0.12	0.24	0.36
Total	**0.62**	**0.38**	**1.00**

a) What's the probability that a survey respondent will shop at the U.S. store?
b) What is the probability that a survey respondent will shop at the store given that they are younger than 20 years old?
c) What is the probability that a survey respondent who is older than 40 years shops at the store?
d) What is the probability that a survey respondent is younger than 20 or will shop at the store?

39. Pharmaceutical company, again. Given the table of probabilities compiled for marketing managers in Exercise 37, are high blood pressure and high cholesterol independent? Explain.

Cholesterol	Blood Pressure: High	Blood Pressure: OK
High	0.11	0.21
OK	0.16	0.52

40. International relocation, again. Given the table of probabilities compiled for a department store chain in Exercise 38, are age and shopping at the department store independent? Explain.

41. International food survey, part 2. Look again at the data from the GfK Roper Worldwide survey on food attitudes in Exercise 33.

a) If we select a respondent at random, what's the probability we choose a person between 13 and 19 years old who agreed with the statement?
b) Among the 13- to 19-year-olds, what is the probability that a person responded "Agree"?
c) What's the probability that a person who agreed was between 13 and 19?
d) If the person responded "Disagree," what is the probability that they are at least 50 years old?
e) What's the probability that a person 50 years or older disagreed?
f) Are response to the question and age independent?

42. Cosmetics marketing, part 2. Look again at the data from the GfK Roper Worldwide survey on skin care in Exercise 34.

a) If we select a respondent at random, what's the probability we choose a person from the US who agreed with the statement?
b) Among those from the US, what is the probability that a person responded "Agree"?
c) What's the probability that a person who agreed was from the US?
d) If the person responded "Disagree," what is the probability that they are from the USA?
e) What's the probability that a person from the USA disagreed?
f) Are response to the question and Country independent?

43. Real estate, part 2. In the real estate research described in Exercise 17, 64% of homes for sale have garages, 21% have swimming pools, and 17% have both features.

a) What is the probability that a home for sale has a garage, but not a pool?
b) If a home for sale has a garage, what's the probability that it has a pool, too?
c) Are having a garage and a pool independent events? Explain.
d) Are having a garage and a pool mutually exclusive? Explain.

44. Employee benefits. Fifty-six percent of all American workers have a workplace retirement plan, 68% have health insurance, and 49% have both benefits. If we select a worker at random:

a) What's the probability that the worker has neither employer-sponsored health insurance nor a retirement plan?
b) What's the probability that the worker has health insurance if they have a retirement plan?
c) Are having health insurance and a retirement plan independent? Explain.
d) Are having these two benefits mutually exclusive? Explain.

45. Telemarketing. Telemarketers continue to attempt to reach consumers by calling landline phone numbers. According to estimates from a national 2003 survey, based on face-to-face interviews in 16,677 households, approximately 58.2% of U.S. adults have both a landline in their residence and a cell phone, 2.8% have only cell phone service but no land line, and 1.6% have no telephone service at all.

a) Polling agencies won't phone cell phone numbers because customers object to paying for such calls. What proportion of U.S. households can be reached by a landline call?

b) Are having a cell phone and having a landline independent? Explain.

46. Snoring. According to the British United Provident Association (BUPA), a major health care provider in the U.K., snoring can be an indication of sleep apnea which can cause chronic illness if left untreated. In the U.S.A., the National Sleep Foundation reports that 36.8% of the 995 adults they surveyed snored. Of the respondents, 81.5% were over the age of 30, and 32% were both over the age of 30 and snorers.

a) What percent of the respondents were 30 years old or younger and did not snore?

b) Is snoring independent of age? Explain.

47. Selling cars. A recent ad campaign for a major automobile manufacturer is clearly geared toward an older demographic. You are surprised, so you decide to conduct a quick survey of your own. A random survey of autos parked in the student and staff lots at your university classified the brands by country of origin, as seen in the table. Is country of origin independent of type of driver?

		Driver	
		Student	**Staff**
Origin	**American**	107	105
	European	33	12
	Asian	55	47

48. Fire sale. A survey of 1056 houses in the Saratoga Springs, New York area found the following relationship between price (in $) and whether the house had a fireplace in 2006. Is the price of the house independent of whether it has a fireplace?

		Fireplace	
		No	**Yes**
House Price	**Low—less than $112,000**	198	66
	Med Low ($112 to $152K)	133	131
	Med High ($152 to $207K)	65	199
	High—over $207,000	31	233

49. Used cars. A business student is searching for a used car to purchase, so she posts an ad to a website saying she wants to buy a used Jeep between $18,000 and $20,000. From Kelly's Blue-Book.com, she learns that there are 149 cars matching that description within a 30-mile radius of her home. If we assume that those are the people who will call her and that they are equally likely to call her:

a) What is the probability that the first caller will be a Jeep Liberty owner?

b) What is the probability that the first caller will own a Jeep Liberty that costs between $18,000 and $18,999?

c) If the first call offers her a Jeep Liberty, what is the probability that it costs less than $19,000?

d) Suppose she decides to ignore calls with cars whose cost is ≥ $19,000. What is the probability that first call she takes will offer to sell her a Jeep Liberty?

		Price		
	Make	**$18,000–$18,999**	**$19,000–$19,999**	**Total**
Car	**Commander**	3	6	**9**
	Compass	6	1	**7**
	Grand Cherokee	33	33	**66**
	Liberty	17	6	**23**
	Wrangler	33	11	**44**
	Total	**92**	**57**	**149**

50. CEO relocation. The CEO of a mid-sized company has to relocate to another part of the country. To make it easier, the company has hired a relocation agency to help purchase a house. The CEO has 5 children and so has specified that the house have at least 5 bedrooms, but hasn't put any other constraints on the search. The relocation agency has narrowed the search down to the houses in the table and has selected one house to showcase to the CEO and family on their trip out to the new site. The agency doesn't know it, but the family has its heart set on a Cape Cod house with a fireplace. If the agency selected the house at random, without regard to this:

		Fireplace?		
		No	**Yes**	**Total**
House Type	**Cape Cod**	7	2	**9**
	Colonial	8	14	**22**
	Other	6	5	**11**
	Total	**21**	**21**	**42**

a) What is the probability that the selected house is a Cape Cod?

b) What is the probability that the house is a Colonial with a fireplace?

c) If the house is a Cape Cod, what is the probability that it has a fireplace?

d) What is the probability that the selected house is what the family wants?

JUST CHECKING ANSWERS

1 The probability of going up on the next day is not affected by the previous day's outcome.

2 a) 0.30

b) $0.30(0.30) = 0.09$

c) $(1 - 0.30)^2(0.30) = 0.147$

d) $1 - (1 - 0.30)^5 = 0.832$

3 a)

		Weekday		
		Yes	**No**	**Total**
Before Five	**Yes**	0.07	0.41	**0.48**
	No	0.20	0.32	**0.52**
	Total	**0.27**	**0.73**	**1.00**

b) $P(\mathbf{WD}|\mathbf{BF}) = P(\mathbf{WD}\ and\ \mathbf{BF})/P(\mathbf{BF}) = .07/.48 = .146$

c) No, shoppers can do both (and 7% do).

d) To be independent, we'd need $P(\mathbf{BF}|\mathbf{WD}) = P(\mathbf{BF})$. $P(\mathbf{BF}|\mathbf{WD}) = 0.259$, but $P(\mathbf{BF}) = 0.48$. They do not appear to be independent.

Displaying and Describing Quantitative Data

Enron Corporation

Enron Corporation was once one of the world's biggest corporations. From its beginnings as an interstate natural gas supply company in 1985, it grew steadily throughout the 1990s, diversifying into nearly every form of energy transaction and eventually dominating the energy trading business. Its stock price followed this spectacular growth. In 1985, Enron stock sold for about $5 a share, but at the end of 2000, Enron stock closed at a 52-week high of $89.75, and the company's stock was worth more than $6 billion. Less than one year later it hit a low of $0.25 a share, having lost more than 99% of its value.

Many employees who had taken advantage of generous stock plans lost retirement packages worth hundreds of thousands of dollars. Portfolio managers typically examine stock prices (or changes in stock prices) over time to determine stock volatility and to help them decide which stocks to buy and sell. Were there warning signs in the data?

To learn more about the behavior and volatility of Enron's stock, let's start by looking at Table 6.1, which gives the monthly changes in stock price (in dollars) for the five years leading up to the company's failure.

	Jan.	Feb.	Mar.	Apr.	May	June	July	Aug.	Sept.	Oct.	Nov.	Dec.
1997	–$1.44	–0.75	–0.69	–0.88	0.12	0.75	0.81	–1.75	0.69	–0.22	–0.16	0.34
1998	0.78	0.62	2.44	–0.28	2.22	–0.50	2.06	–0.88	–4.50	4.12	1.16	–0.50
1999	3.28	3.34	–1.22	0.47	5.62	–1.59	4.31	1.47	–0.72	–0.38	–3.25	0.03
2000	5.72	21.06	4.50	4.56	–1.25	–1.19	–3.12	8.00	9.31	1.12	–3.19	–17.75
2001	14.38	–1.08	–10.11	–12.11	5.84	–9.37	–4.74	–2.69	–10.61	–5.85	–17.16	–11.59

| **Table 6.1** *Monthly stock price change in dollars of Enron stock for the period January 1997 to December 2001.*

It's hard to tell very much from tables of values like this. You might get a rough idea of how much the stock changed from month to month—usually less than $10 in either direction—but that's about it.

6.1 Displaying Distributions

Instead, let's follow the first rule of data analysis and make a picture. What kind of picture should we make? It can't be a bar chart or a pie chart. Those are only for categorical variables, and Enron's stock price change is a *quantitative* variable, whose units are dollars.

Histograms

Here are the monthly price changes of Enron stock displayed in a histogram.

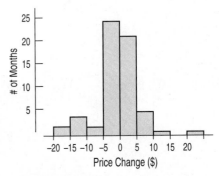

Figure 6.1 *Monthly price changes of Enron stock. The histogram displays the distribution of price changes by showing for each "bin" of price changes, the number of months having price changes in that bin.*

Like a bar chart, a **histogram** plots the bin counts as the heights of bars. In this histogram of monthly price changes, each bin has a width of $5, so, for example, the height of the second tallest bar says that there were about 20 monthly price changes of between $0 and $5. In this way, the histogram displays the entire distribution of price changes. Unlike a bar chart, which puts gaps between bars to separate the categories, there are no gaps between the bars of a histogram *unless* there are actual gaps in the data. Gaps indicate a region where there are no values, such as the gap between $15 and $20 in Figure 6.1. That can be important, so watch out for them.

For categorical variables, each category got its own bar. That was easy; there was no choice, except maybe to combine categories for ease of display. But for quantitative variables, we have to choose how to slice up all the possible values into bins. Once we have equal width bins, the histogram can count the number of cases that fall into each bin, represent the counts as bars, and plot them against the bin values. In this way, it displays the distribution at a glance.

◆ **How do histograms work?** If you were to make a histogram by hand or in Excel, you'd need to make some decisions about the bins. First, you would need to decide how wide to make the bins. Bin choice is important because some features of the distribution may appear more obvious at different bin width choices. With many statistics packages, you can easily vary the bin width interactively so you can make sure that a feature you think you see isn't just a consequence of a certain bin width choice.

Next you'd need to decide where to place the endpoints of the bins. (Statistics packages and graphing calculators make these choices for you automatically.) Bins are always equal in width and are typically multiples of five or ten. But what do you do with a value of $5 if one bin spans from $0 to $5 and the next bin spans from $5 to $10? The standard rule for a value that falls exactly on a bin boundary is to put it into the next higher bin, so you'd put a month with a change of $5 into the $5 to $10 bin instead of into the $0 to $5 bin.

From the histogram, we can see that these months typically have changes near $0. We can see that although they vary, most of the monthly price changes are less than $5 in either direction. Only in a very few months were the changes larger than $10 in either direction. There appear to be about as many positive as negative price changes—indicating that the stock went up as often as it went down.

Does the distribution look as you expected? It's often a good idea to imagine what the distribution might look like before making the display. That way you're less likely to be fooled by errors either in your display or in the data themselves.

If our focus is on the overall pattern of how the values are distributed rather than on the counts themselves, it can be useful to make a relative frequency histogram, replacing the counts on the vertical axis with the percentage of the total number of cases falling in each bin. The shape of the histogram is exactly the same; only the labels are different. A **relative frequency histogram** is faithful to the area principle by displaying the *percentage* of cases in each bin instead of the count.

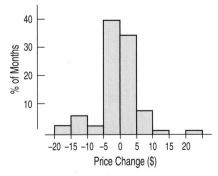

Figure 6.2 A relative frequency histogram looks just like a frequency histogram except that the y-axis now shows the percentage of months in each bin.

Stem-and-Leaf Displays

Histograms provide an easy-to-understand summary of the distribution of a quantitative variable, but they don't show the data values themselves. **Stem-and-leaf displays** are like histograms, but they also give the individual values. They are easy to make by hand for data sets that aren't too large, so they're a great way to look at a small batch of values quickly.[1] Here's a stem-and-leaf display for just the first three years of the Enron price change data, alongside a histogram of the same data.

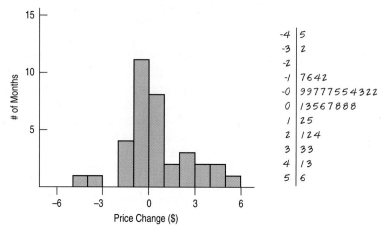

Figure 6.3　*The first 36 months of Enron monthly stock price changes displayed both by a histogram (left) and stem-and-leaf display (right). Stem-and-leaf displays are typically made by hand, so we are most likely to use them for small data sets. For much larger data sets, we use a histogram.*

◆ **How do stem-and-leaf displays work?**　Stem-and-leaf displays use part of each number (called the stem) to name the bins. To make the "bars," they use the next digit of the number. For example, if we had a monthly price change of $21, we could write 2|1, where 2 serves as the stem and 1 as the leaf. To display the changes 21, 22, 24, 33, and 33 together, we would write

$$2|124$$
$$3|33$$

Often we put the higher numbers on top, but either way is common. Higher numbers on top is often natural, but putting the higher numbers on the bottom keeps the direction of the histogram the same when you tilt your head to look at it—otherwise the histogram appears reversed.

When you make a stem-and-leaf display by hand, make sure you give each digit about the same width, in order to satisfy the area principle. (That can lead to some fat 1s and thin 8s—but it keeps the display honest.)

There are both positive and negative values in the price changes. Values of $0.3 and $0.5 are displayed as leaves of "3" and "5" on the "0" stem. But values of −$0.3 and −$0.5 must be plotted below zero. So the stem-and-leaf display has a "−0" stem to hold them—again with leaves of "3" and "5." It may seem a little

---

[1] The authors like to make stem-and-leaf displays whenever data are presented (without a suitable display) at committee meetings or working groups. The insights from just that quick look at the distribution are often quite valuable.

strange to see two zero stems, one labeled "−0." But, if you think about it, you'll see that it's a sensible way to deal with negative values.

Unlike most other displays discussed in this book, stem-and-leaf displays are great pencil-and-paper constructions. They are well-suited to moderate amounts of data—say, between 10 and a few hundred values. For larger data sets, histograms do a better job.

In Chapter 4, you learned to check the Categorical Data Condition before making a pie chart or a bar chart. Now, by contrast, before making a stem-and-leaf display, or a histogram, you need to check the **Quantitative Data Condition**: The data are values of a quantitative variable whose units are known.

Although a bar chart and a histogram may look similar, they're not the same display. You can't display categorical data in a histogram or quantitative data in a bar chart. Always check the condition that confirms what type of data you have before making your display.

6.2 Shape

Once you've displayed the distribution in a histogram or stem-and-leaf display, what can you say about it? When you describe a distribution, you should pay attention to three things: its shape, its center, and its spread.

> We describe the **shape** of a distribution in terms of its modes, its symmetry, and whether it has any gaps or outlying values.

Mode

Does the histogram have a single, central hump (or peak) or several, separated humps? These humps are called **modes**.[2] Formally, the mode is the single, most frequent value, but we rarely use the term that way. Sometimes we talk about the mode as being the value of the variable at the center of this hump. The Enron stock price changes have a single mode at just below $0 (Figure 6.1). We often use modes to describe the shape of the distribution. A distribution whose histogram has one main hump, such as the one for the Enron price changes, is called **unimodal**; distributions whose histograms have two humps are **bimodal**, and those with three or more are called **multimodal**. For example, here's a bimodal distribution.

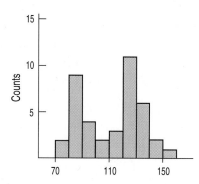

Figure 6.4 *A bimodal distribution has two apparent modes.*

The **mode** is typically defined as the single value that appears most often. That definition is fine for categorical variables because we need only to count the number of cases for each category. For quantitative variables, the meaning of *mode* is more ambiguous. For example, what's the mode of the Enron data? No price change occurred more than twice, but two months had drops of $0.50. Should that be the mode? Probably not. For quantitative data, it makes more sense to use the word *mode* in the more general sense of "peak in a histogram," rather than as a single summary value.

You've heard of pie à la mode. Is there a connection between pie and the mode of a distribution? Actually, there is! The mode of a distribution is a *popular* value near which a lot of the data values gather. And à la mode means "in style"—*not* "with ice cream." That just happened to be a *popular* way to have pie in Paris around 1900.

---

[2] Technically, the mode is the value on the *x*-axis of the histogram below the highest peak, but informally we often refer to the peak or hump itself as a mode.

A bimodal histogram is often an indication that there are two groups in the data. It's a good idea to investigate when you see bimodality.

A distribution whose histogram doesn't appear to have any mode and in which all the bars are approximately the same height is called **uniform**.

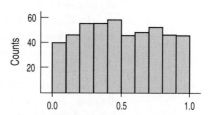

Figure 6.5 In a uniform distribution, bars are all about the same height. The histogram doesn't appear to have a mode.

Symmetry

Could you fold the histogram along a vertical line through the middle and have the edges match pretty closely, as in Figure 6.6, or are more of the values on one side, as in the histograms in Figure 6.7? A distribution is **symmetric** if the halves on either side of the center look, at least approximately, like mirror images.

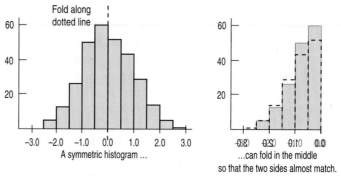

Figure 6.6 A symmetric histogram can fold in the middle so that the two sides almost match.

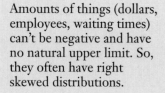

Amounts of things (dollars, employees, waiting times) can't be negative and have no natural upper limit. So, they often have right skewed distributions.

The (usually) thinner ends of a distribution are called the **tails**. If one tail stretches out farther than the other, the distribution is said to be **skewed** to the side of the longer tail.

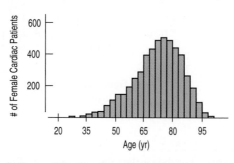
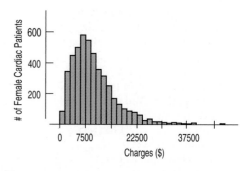

Figure 6.7 Two skewed histograms showing the age (on left) and hospital charges (on right) for all female heart attack patients in New York State in one year. The histogram of Age (in blue) is skewed to the left, while the histogram of Charges (in purple) is skewed to the right.

Outliers

Do any features appear to stick out? Often such features tell us something interesting or exciting about the data. You should always point out any stragglers or **outliers** that stand off away from the body of the distribution. For example, if you're studying the personal wealth of Americans and Bill Gates is in your sample, he would be an outlier. Because his wealth would be so obviously atypical, you'd want to point it out as a special feature.

Outliers can affect almost every method we discuss in this book, so we'll always be on the lookout for them. An outlier can be the most informative part of your data, or it might just be an error. Either way, you shouldn't throw it away without comment. Treat it specially and discuss it when you report your conclusions about your data. (Or find the error and fix it if you can.) We'll soon learn a rule of thumb for how we can decide if and when a value might be considered to be an outlier and some advice for what to do when you encounter them.

◆ **Using Your Judgement.** How you characterize a distribution is often a judgment call. Does the gap you see in the histogram really reveal that you have two subgroups, or will it go away if you change the bin width slightly? Are those observations at the high end of the histogram truly unusual, or are they just the largest ones at the end of a long tail? These are matters of judgment on which different people can legitimately disagree. There's no automatic calculation or rule of thumb that can make the decision for you. Understanding your data and how they arose can help. What should guide your decisions is an honest desire to understand what is happening in the data.

Looking at a histogram at several different bin widths can help to see how persistent some of the features are. If the number of observations in each bin is small enough so that moving a couple of values to the next bin changes your assessment of how many modes there are, be careful. Be sure to think about the data, where they came from and what kinds of questions you hope to answer from them.

6.3 Center

Look again at the Enron price changes in Figure 6.1. If you had to pick one number to describe a *typical* price change, what would you pick? When a histogram is unimodal and symmetric, most people would point to the center of the distribution, where the histogram peaks. The typical price change is around $0.

If we want to be more precise and *calculate* a number, we can *average* the data. In the Enron example, the average price change is −$0.37, about what we might expect from the histogram. You already know how to average values, but this is a good place to introduce notation that we'll use throughout the book. We'll call a generic variable y, and use the Greek capital letter sigma, Σ, to mean "sum" (sigma is "S" in Greek), and write:[3]

> **NOTATION ALERT:**
> A bar over any symbol indicates the mean of that quantity.

$$\bar{y} = \frac{Total}{n} = \frac{\Sigma y}{n}.$$

---

[3] You may also see the variable called x and the equation written $\bar{x} = \dfrac{Total}{n} = \dfrac{\Sigma x}{n}.$

According to this formula, we add up all the values of the variable, y, and divide that sum (*Total*, or Σy) by the number of data values, n. We call this value the **mean** of y.[4]

Although the mean is a natural summary for unimodal, symmetric distributions, it can be misleading for skewed data or for distributions with gaps or outliers. For example, Figure 6.7 showed a histogram of the total charges for hospital stays of female heart attack patients in one year in New York State. The mean value is $10,260.70. Locate that value on the histogram. Does it seem a little high as a summary of a typical cost? In fact, about two thirds of the charges are lower than that value. It might be better to use the **median**—the value that splits the histogram into two *equal* areas. We find the median by counting in from the ends of the data until we reach the middle value. So the median is resistant; it isn't affected by unusual observations or by the shape of the distribution. Because of its resistance to these effects, the median is commonly used for variables such as cost or income, which are likely to be skewed. For the female heart attack patient charges, the median cost is $8,619, which seems like a more appropriate summary.

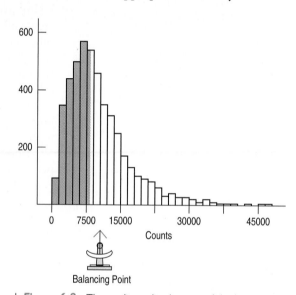

Balancing Point

Figure 6.8   *The median splits the area of the histogram in half at $8,619. Because the distribution is skewed to the right, the mean $10,260 is higher than the median. The points at the right have pulled the mean toward them, away from the median.*

Finding the median by hand

Finding the median of a batch of n numbers is easy as long as you remember to order the values first. If n is odd, the median is the middle value. Counting in from the ends, we find this value in the $\frac{n+1}{2}$ position.

When n is even, there are two middle values. So, in this case, the median is the average of the two values in positions $\frac{n}{2}$ and $\frac{n}{2} + 1$.

Here are two examples:

Suppose the batch has the values 14.1, 3.2, 25.3, 2.8, −17.5, 13.9, and 45.8. First we order the values: −17.5, 2.8, 3.2, 13.9, 14.1, 25.3, and 45.8. Since there are 7 values, the median is the $(7 + 1)/2 = $ 4th value counting from the top or bottom: 13.9.

Suppose we had the same batch with another value at 35.7. Then the ordered values are −17.5, 2.8, 3.2, 13.9, 14.1, 25.3, 35.7, and 45.8. The median is the average of the 8/2, or 4th, and the (8/2) + 1, or 5th, values. So the median is (13.9 + 14.1)/2 = 14.0.

Does it really make a difference whether we choose a mean or a median? The mean price change for the Enron stock is −$0.37. Because the distribution of the price changes is roughly symmetric, we'd expect the mean and median to be close. In fact, we compute the median to be −$0.25. But for variables with skewed distributions, the story is quite different. For a right skewed distribution like the hospital charges in Figure 6.8, the mean is larger than the median: $10,260 compared to $8,619. The difference is due to the overall shape of the distributions.

---

[4] Once you've averaged the data, you might logically expect the result to be called the *average*. But average is used too colloquially as in the "average" home buyer, where we don't sum up anything. Even though average *is* sometimes used in the way we intend, as in the Dow Jones Industrial Average (which is actually a weighted average) or a batting average, we'll often use the more precise term *mean* throughout the book.

An easy way to find the quartiles is to first split the sorted data at the median. (If *n* is odd, include the median with each half.) Then find the median of each of these halves and use them as the quartiles. Even though quartiles are easy to find, there are at least six rules for finding them. The rules all give pretty much the same values when we have a lot of data. But when *n* is small, they can differ. If your calculator, statistics package, or friend in a different class gets a slightly different value, don't worry about it. Usually the exact value isn't important.

The mean is the point at which the histogram would balance. Just like a child who moves away from the center of a see-saw, a bar of the histogram far from the center has more leverage, pulling the mean in its direction. It's hard to argue that a summary that's been pulled aside by only a few outlying values or by a long tail is what we mean by the center of the distribution. That's why the median is usually a better choice for skewed data.

However, when the distribution is unimodal and symmetric, the mean offers better opportunities to calculate useful quantities and draw more interesting conclusions. It will be the summary value we work with much more throughout the rest of the book.

6.4 Spread of the Distribution

We know that the typical price change of the Enron stock is around $0, but knowing the mean or median alone doesn't tell us about the entire distribution. A stock whose price change doesn't move away from $0 isn't very interesting. The more the data vary, the less a measure of center can tell us. We need to know how spread out the data are as well.

One simple measure of spread is the **range**, defined as the difference between the extremes:

$$\text{Range} = max - min.$$

For the Enron data, the range is $21.06 - (-$17.75) = $38.81. Notice that the range is *a single number* that describes the spread of the data, not an interval of values—as you might think from its use in common speech. If there are any unusual observations in the data, the range is not resistant and will be influenced by them. Concentrating on the middle of the data avoids this problem. The **quartiles** are the values that frame the middle 50% of the data. One quarter of the data lies below the lower quartile, Q1, and one quarter of the data lies above the upper quartile, Q3. The **interquartile range (IQR)** summarizes the spread by focusing on the middle half of the data. It's defined as the difference between the two quartiles:

$$\text{IQR} = Q3 - Q1.$$

For the Enron data, there are 30 values on either side of the median. After ordering the data, we average the 15th and 16th values to find Q1 = −$1.68. We average the 45th and 46th values to find Q3 = $2.14. So the IQR = Q3 − Q1 = $2.14 − (−$1.68) = $3.82.

The IQR is usually a reasonable summary of spread, but because it uses only the two quartiles of the data, it ignores much of the information about how individual values vary.

A more powerful measure of spread—and the one we'll use most often—is the standard deviation, which, as we'll see, takes into account how far each value is from the mean. Like the mean, the standard deviation is appropriate only for symmetric data and can be influenced by outlying observations.

As the name implies, the standard deviation uses the *deviations* of each data value from the mean. If we tried to average these deviations, the positive and negative differences would cancel each other out, giving an average deviation of 0—not very useful. Instead, we square each deviation. The average of the *squared* deviations is called the **variance** and is denoted by s^2:

$$s^2 = \frac{\sum(y - \bar{y})^2}{n - 1}.$$

Why do banks favor a single line that feeds several teller windows rather than separate lines for each teller? The average waiting time is the same. But the time you can expect to wait is less variable when there is a single line, and people prefer consistency.

✓ JUST CHECKING

Thinking About Variation

1   The U.S. Census Bureau reports the median family income in its summary of census data. Why do you suppose they use the median instead of the mean? What might be the disadvantages of reporting the mean?

2   You've just bought a new car that claims to get a highway fuel efficiency of 31 miles per gallon. Of course, your mileage will "vary." If you had to guess, would you expect the IQR of gas mileage attained by all cars like yours to be 30 mpg, 3 mpg, or 0.3 mpg? Why?

3   A company selling a new MP3 player advertises that the player has a mean lifetime of 5 years. If you were in charge of quality control at the factory, would you prefer that the standard deviation of lifespans of the players you produce be 2 years or 2 months? Why?

The variance plays an important role in statistics, but as a measure of spread, it has a problem. Whatever the units of the original data, the variance is in *squared* units. We want measures of spread to have the same units as the data, so we usually take the square root of the variance. That gives the **standard deviation**.

$$s = \sqrt{\frac{\sum (y - \bar{y})^2}{n - 1}}.$$

For the Enron stock price changes, $s = \$6.29$.

Finding the standard deviation by hand

To find the standard deviation, start with the mean, $\bar{y}$. Then find the *deviations* by taking $\bar{y}$ from each value: $(y - \bar{y})$. Square each deviation: $(y - \bar{y})^2$.

Now you're nearly home. Just add these up and divide by $n - 1$. That gives you the variance, s^2. To find the standard deviation, s, take the square root.

Suppose the batch of values is 4, 3, 10, 12, 8, 9, and 3.

The mean is $\bar{y} = 7$. So find the deviations by subtracting 7 from each value:

Original Values	Deviations	Squared Deviations
4	$4 - 7 = -3$	$(-3)^2 = 9$
3	$3 - 7 = -4$	$(-4)^2 = 16$
10	$10 - 7 = 3$	9
12	$12 - 7 = 5$	25
8	$8 - 7 = 1$	1
9	$9 - 7 = 2$	4
3	$3 - 7 = -4$	16

Add up the squared deviations:
$9 + 16 + 9 + 25 + 1 + 4 + 16 = 80$.
Now, divide by $n - 1$: $80/6 = 13.33$.
Finally, take the square root: $s = \sqrt{13.33} = 3.65$

6.5 Shape, Center, and Spread—A Summary

What should you report about a quantitative variable? Report the shape of its distribution, and include a center and a spread. But which measure of center and which measure of spread? The guidelines are pretty easy.

◆ If the shape is skewed, point that out and report the median and IQR. You may want to include the mean and standard deviation as well, explaining why the mean and median differ. The fact that the mean and median do not agree is a sign that the distribution may be skewed. A histogram will help you make the point.

◆ If the shape is unimodal and symmetric, report the mean and standard deviation and possibly the median and IQR as well. For unimodal symmetric data, the IQR is usually a bit larger than the standard deviation. If that's not true for your data set, look again to make sure the distribution isn't skewed or mutimodal and that there are no outliers.

◆ If there are multiple modes, try to understand why. If you can identify a reason for separate modes, it may be a good idea to split the data into separate groups.

◆ If there are any clearly unusual observations, point them out. If you are reporting the mean and standard deviation, report them computed with and without the unusual observations. The differences may be revealing.

◆ Always pair the median with the IQR and the mean with the standard deviation. It's not useful to report one without the other. Reporting a center without a spread can lead you to think you know more about the distribution than you do. Reporting only the spread omits important information.

6.6 Five-Number Summary and Boxplots

The volume of shares traded on the New York Stock Exchange (NYSE) is important to investors, research analysts, and policymakers. It can predict market volatility, and has been used in models for predicting price fluctuations. How many shares are typically traded in a day on the NYSE? One good way to summarize a distribution with just a few values is with a five-number summary. The **five-number summary** of a distribution reports its median, quartiles, and extremes (maximum and minimum). For example, the five-number summary of NYSE volume during the entire year 2006 looks like this (in billions of shares).

Max	3.287
Q3	1.972
Median	1.824
Q1	1.675
Min	0.616

Table 6.2 The five-number summary of a NYSE daily volume (in billions of shares) for the year 2006.

How to Build a Boxplot

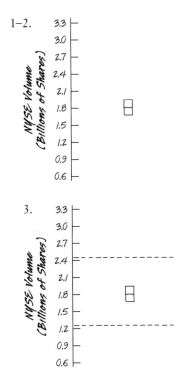

The five-number summary provides a good overall look at the distribution. For example, because the quartiles frame the middle half of the data, we can see that on half of the days the volume was between 1.675 and 1.972 billion shares. We can also see the extremes of over 3 billion shares on the high end and just over half a billion shares on the low end. Were those days extraordinary for some reason or just the busiest and quietest days? To answer that, we'll need to work with the summaries a bit more.

Once we have a five-number summary of a (quantitative) variable, we can display that information in a **boxplot**. To make a boxplot of the daily volumes, follow these steps:

1. Draw a single vertical axis spanning the extent of the data.

2. Draw short horizontal lines at the lower and upper quartiles and at the median. Then connect them with vertical lines to form a box. The width isn't important unless you plan to show more than one group.

3. Now erect (but don't show in the final plot) "fences" around the main part of the data, placing the upper fence 1.5 IQRs above the upper quartile and the

4.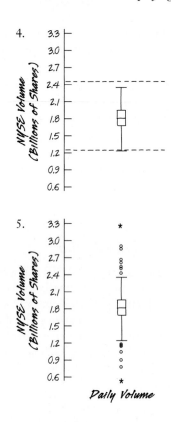

lower fence 1.5 IQRs below the lower quartile. For the NYSE share volume data, compute:

$$\textit{Upper fence} = Q3 + 1.5\,IQR = 1.97 + 1.5 \times 0.29 = 2.405 \text{ billion shares}$$

and

$$\textit{Lower fence} = Q1 - 1.5\,IQR = 1.68 - 1.5 \times 0.29 = 1.245 \text{ billion shares}$$

4. Grow "whiskers." Draw lines from each end of the box up and down to *the most extreme data values found within the fences.* If a data value falls outside one of the fences, do *not* connect it with a whisker.

5. Finally, add any outliers by displaying data values that lie beyond the fences with special symbols. Here there are about 15 such values. (We often use one symbol for outliers that lie less than 3 IQRs from the quartiles and a different symbol for "far outliers"—data values more than 3 IQRs from the quartiles.)

Now that you've drawn the boxplot, let's summarize what it shows. The center of a boxplot is (remarkably enough) a box that shows the middle half of the data, between the quartiles. The height of the box is equal to the IQR. If the median is roughly centered between the quartiles, then the middle half of the data is roughly symmetric. If it is not centered, the distribution is skewed. The whiskers show skewness as well if they are not roughly the same length. Any outliers are displayed individually, both to keep them out of the way for judging skewness and to encourage you to give them special attention. They may be mistakes, or they may be the most interesting cases in your data.

> The prominent statistician John W. Tukey, the originator of the boxplot, was asked (by one of the authors) why the outlier nomination rule cut at 1.5 IQRs beyond each quartile. He answered that the reason was that 1 IQR would be too small and 2 IQRs would be too large.

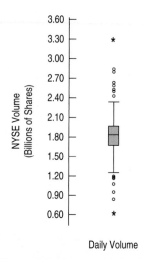

Figure 6.9 *Boxplot of daily volume of shares traded on NYSE in 2006 (in billions of shares).*

The boxplot for NYSE volume shows the middle half the days—those with average volume between 1.676 and 1.970 billion shares—as the central box. From the shape of the box, it looks like the central part of the distribution of volume is roughly symmetric, and the similar length of the two whiskers shows the outer parts of the distribution to be roughly symmetric as well. We also see several high-volume and low-volume days. Boxplots are particularly good at exhibiting outliers. We also see two extreme outliers, one on each side. These extreme days may deserve more attention. (When and why did they occur?)

GUIDED EXAMPLE | *Credit Card Bank Customers*

In order to focus on the needs of particular customers, companies often segment their customers into groups with similar needs or spending patterns. A major credit card bank wanted to see how much a particular group of cardholders charged per month on their cards in order to understand the potential growth in their card use. The data for each customer was the amount he or she spent using the card during a three-month period in 2008. Boxplots are especially useful for one variable when combined with a histogram and numerical summaries. Let's summarize the spending of this segment.

PLAN

Setup Identify the *variable*, the time frame of the data, and the objective of the analysis.

We want to summarize the average monthly charges (in dollars) made by 500 cardholders from a market segment of interest during a three-month period in 2008. The data are quantitative, so we'll use histograms and boxplots, as well as numerical summaries.

DO

REALITY CHECK

Mechanics Select an appropriate display based on the nature of the data and what you want to know about it.

It is always a good idea to think about what you expected to see and to check whether the histogram is close to what you expected. Are the data about what you might expect for customers to charge on their cards in a month? A typical value is a few hundred dollars. That seems like the right ballpark.

Note that outliers are often easier to see with boxplots than with histograms, but the histogram provides more details about the shape of the distribution. This computer program "jitters" the outliers in the boxplot so they don't lie on top of each other, making them easier to see.

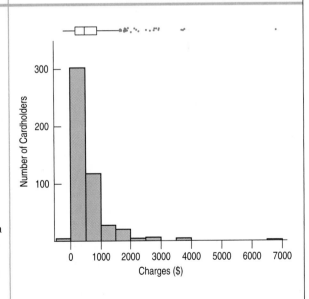

Both graphs show a distribution that is highly skewed to the right with several outliers and an extreme outlier near $7000.

Summary of Monthly Charges	
Count	500
Mean	544.749
Median	370.65
StdDev	661.244
IQR	624.125
Q1	114.54
Q3	738.665

The mean is much larger than the median. The data do not have a symmetric distribution.

REPORT

Interpretation   Describe the shape, center, and spread of the distribution. Be sure to report on the symmetry, number of modes, and any gaps or outliers.

Recommendation   State a conclusion and any recommended actions or analysis.

MEMO:

Re: Report on segment spending.

The distribution of charges for this segment during this time period is unimodal and skewed to the right. For that reason, we recommend summarizing the data with the median and interquartile range (IQR).

The median amount charged was $370.65. Half of the cardholders charged between $114.54 and $738.67.

In addition, there are several high outliers, with one extreme value at $6745.

There are also a few negative values. We suspect that these are people who returned more than they charged in a month, but because the values might be data errors, we suggest that they be checked.

Future analyses should look at whether charges during these three months in 2008 was similar to charges in the rest of the year. We would also like to investigate if there is a seasonal pattern and, if so, whether it can be explained by our advertising campaigns or by other factors.

6.7 Comparing Groups

As we saw earlier, the volume on the NYSE can vary greatly from day to day, but if we step back a bit, we may be able to find patterns that can help us understand, model, and predict it. We might be interested not only in individual daily values, but also in looking for patterns in the volume when we group the days into time periods such as weeks, months, or seasons. Such comparisons of distributions can reveal patterns, differences, and trends.

Let's start with the "big picture." We'll split the year into halves: January through June and July through December. Here are histograms of the NYSE volume for 2006.

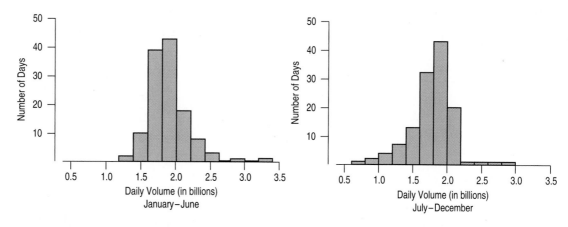

| *Figure 6.10* Daily volume on the NYSE split into two halves of the year. How do the two distributions differ?

The centers and spreads are not too different, but the shape appears to be slightly right skewed in the first half, while the second half of the year appears to be left skewed with more days on the lower end. There are several noticeable outlying values on the high side in both graphs.

Histograms work well for comparing two groups, but what if we want to compare the volume across four quarters? Or twelve months? Histograms are best at displaying one or two distributions. When we compare several groups, boxplots usually do a better job. Boxplots offer an ideal balance of information and simplicity, hiding the details while displaying the overall summary information. And we can plot them side by side, making it easy to compare multiple groups or categories.

When we place boxplots side by side, we can easily see which group has the higher median, which has the greater IQR, where the central 50% of the data is located, and which has the greater overall range. We can also get a general idea of symmetry from whether the medians are centered within their boxes and whether the whiskers extend roughly the same distance on either side of the boxes. Equally important, we can see past any outliers in making these comparisons because they've been displayed separately. We can also begin to look for trends in the medians and in the IQRs.

GUIDED EXAMPLE | *New York Stock Exchange Trading Volume*

Are some months on the NYSE busier than others? Boxplots of the number of shares traded by month are a good way to see such patterns. We're interested not only in the centers, but also in the spreads. Are volumes equally variable from month to month, or are they more spread out in some months?

PLAN

Setup Identify the variables, report the time frame of the data, and state the objective.

We want to compare the daily volume of shares traded from month to month on the NYSE during 2006.

The daily volume is quantitative and measured in number of shares. We can partition the values by month and use side-by-side boxplots to compare the volume across months.

DO	**Mechanics** Plot the side-by-side boxplots of the data.	
REPORT	**Conclusion** Report what you've learned about the data and any recommended action or analysis.	MEMO: **Re: Research on trading volume of the NYSE** We have examined the daily sales volume on the NYSE (number of shares traded) for each month of 2006. As the attached display shows, sales volume follows a seasonal pattern with lower volume in March and August. The highest median trading activity is found in May and November. The variability of trading volume also shows a pattern. June and December have higher variability than the rest, and March has noticeably less variability. There were several unusually high-volume days that bear investigation and one extremely low-volume day in November.

6.8 Identifying Outliers

When we looked at a boxplot for volumes of the entire year, there were 15 outliers. Now, when we group the days by *Month*, the boxplots display fewer days as outliers, and identifies different days as the extraordinary ones. This change occurs because our outlier nomination rule for boxplots depends on the quartiles of the data being displayed. Days that may have seemed ordinary when placed against the entire year's data can look like outliers for the month they're in and *vice versa*. That high-volume day in March certainly wouldn't stand out in May or June, but for March it was remarkable, and that very low-volume day in November really stands out now. What should we do with such outliers?

Cases that stand out from the rest of the data deserve our attention. Boxplots have a rule for nominating extreme cases to display as outliers, but that's just a rule of thumb—not a definition. The rule doesn't tell you what to do with them.

So, what *should* we do with outliers? The first thing to do is to try to understand them in the context of the data. Look back at the boxplot in the Guided Example. The boxplot for November (month 11) shows a fairly symmetric body of data with one low-volume and one high-volume day set clearly apart from the other days. Such a large gap suggests that the volume really is quite different.

Once you've identified likely outliers, you should always investigate them. Some outliers are unbelievable and may simply be errors. A decimal point may have been misplaced, digits transposed, or digits repeated or omitted. Or, the units may be wrong. If you saw the number of shares traded on the NYSE listed as 2 shares for a particular day, you'd know something was wrong. It could be that it was meant as 2 billion shares, but you'd have to check to be sure. Sometimes a number is transcribed incorrectly, perhaps copying an adjacent value on the original data sheet. If you can identify the error, then you should certainly correct it.

Many outliers are not wrong; they're just different. These are the cases that often repay your efforts to understand them. You may learn more from the extraordinary cases than from summaries of the overall dataset.

What about that low November day? It was November 24, 2006, the Friday after Thanksgiving, a day when most likely, traders would have rather stayed home.

The high-volume day, September 15, was a "triple witching day," a day when during the final trading hour, options and futures contracts expire. Such days often experience large trading volume and price fluctuations.

14-year-old widowers?

Careful attention to outliers can often reveal problems in data collection and management. Two researchers, Ansley Coale and Fred Stephan, looking at data from the 1950 census noticed that the number of widowed 14-year-old boys had increased from 85 in 1940 to a whopping 1600 in 1950. The number of divorced 14-year-old boys had increased, too, from 85 to 1240. Oddly, the number of teenaged widowers and divorcees *decreased* for every age group after 14, from 15 to 19. When Coale and Stephan also noticed a large increase in the number of young Native Americans in the Northeast United States, they began to look for data problems. Data in the 1950 census were recorded on computer cards. (For a picture of a computer card, see p. 13.) Cards are hard to read and mistakes are easy to make. It turned out that data punches had been shifted to the right by one column on hundreds of cards. Because each card column meant something different, the shift turned 43-year-old widowed males into 14-year-olds, 42-year-old divorcees into 14-year-olds, and children of white parents into Native Americans. Not all outliers have such a colorful (or famous) story, but it is always worthwhile to investigate them. And, as in this case, the explanation is often surprising. (A. Coale and F. Stephan, "The case of the Indians and the teen-age widows." *J. Am. Stat. Assoc.* 57 [Jun 1962]: 338–347.)

6.9 Standardizing

The data we compared by groups in previous sections were all the same variable. It was easy to compare volume on the NYSE in July to volume on the NYSE in December because the data had the same units. Sometimes, however, we want to compare very different variables—apples to oranges, so to speak. For example, the Great Place to Work Institute measures more than 50 aspects of companies and publishes through *Fortune Magazine* a ranking of the top places to work. In 2007, the top honor was captured by Google.

What was the key to Google's winning? Was it the free food offered to all employees? Maybe the on-site day care? How about the salaries—do they compare favorably with other companies? Were they better on all 50 variables? Probably not, but it isn't obvious how to combine and balance all these different aspects to come up with a single number. The variables don't even have the same units; for example, average salary is in dollars, perceptions are often measured on a seven-point scale, and diversity measures are in percentages.

The trick to comparing very different-looking values is to standardize the values. Rather than working with the original values, we ask "how far is this value from the mean?" Then—and this is the key—we measure that distance with the standard deviation. The result is the standardized value which records how many standard deviations each value is above or below the overall mean. The standard deviation provides a ruler, based on the underlying variability of all the values, against which we can compare values that otherwise have little in common.

It turns out that statisticians do this all the time. Over and over during this course (and in any additional Statistics courses you may take), questions such as "How far is this value from the mean?" or "How different are these two values?" will be answered by measuring the distance or difference in standard deviations.

In order to see how standardizing works, we'll focus on just two of the 50 variables that the Great Places to Work Institute reports: the number of *New Jobs* created during the year and the reported *Average Pay* for salaried employees for two companies. We'll choose two companies that were farther down the list to show how standardization works: Starbucks and the Wrigley Company (the company that makes Wrigley's chewing gum among other things).[5]

When we compare two variables, it's always a good idea to start with a picture. Here we'll use stem-and-leaf displays (Figure 6.11), so we can see the individual distances, highlighting Starbucks in red and Wrigley in blue. The mean number of new jobs created for all the companies was 305.8. Starbucks with over 2,000 jobs is well above average, as we can see from the stem-and-leaf display. Wrigley, with only 16 jobs (rounded to 0 in the stem-and-leaf), is closer to the center. On the other hand, Wrigley's average salary was $56,350 (rounded to 6), compared with Starbuck's $44,790 (represented as 4), so even though both are below average, Wrigley is closer to the center (see margin).

Variable	Mean	SD
New Jobs	305.8	1508.0
Avg Pay	$73,229.42	$34,055.24

New Jobs

```
  4 |
  3 | 67
  2 | 25
  1 | 01234567
  0 | 111111122222223333333444555666667778888
 -0 | 65444332110000
 -1 | 1
 -2 |
 -3 | 3
 -4 |
 -5 |
 -6 |
 -7 |
 -8 |
 -9 | 1
```
3|6 represents 3600

Average Pay

```
  2 | 5
  2 |
  2 |
  1 |
  1 |
  1 | 45
  1 | 222
  1 | 000001
  0 | 88889999999999
  0 | 66666666667777777777
  0 | 4444444455555555555
  0 | 3
  0 | 1
```
2|5 represents 250000

Figure 6.11 *Stem-and-leaf displays for both the number of new jobs created and the average pay of salaried employees at the top 100 companies to work for in 2005 from Fortune magazine. Starbucks (in red) had more jobs created, but Wrigley (in blue) did better in average pay. Which company did better for both variables combined?*

When we compare scores from different variables, our eye naturally looks at how far from the center of each distribution the value lies. We adjust naturally for the fact that these variables have very different scales. Starbucks did better on *New Jobs*, and Wrigley did better on *Average Pay*. To quantify *how much* better each one did and to combine the two scores, we'll ask how many standard deviations they each are from the variable means.

---

[5] The data we analyze here are actually from 2005, the last year for which we have data and the year Wegman's Supermarkets was the number one company to work for.

To find how many standard deviations a value is from the mean we find:

$$z = \frac{y - \bar{y}}{s}.$$

We call the resulting value a **standardized value** and denote it with the letter z. Usually, we just call it a **z-score**.

A z-score of 2.0 indicates that a data value is two standard deviations above the mean. Data values below the mean have negative z-scores, so a z-score of -0.84 means that the data value is 0.84 standard deviations *below* the mean.

	New Jobs	Average Pay
Mean (All companies) **SD**	305.9 1507.97	$73,299.42 $34,055.25
Starbucks **z-score**	2193 **1.25** = (2193 − 305.9)/1507.97	$44,790 **−0.84** = (44790 − 73299.42)/34055.25
Wrigley **z-score**	16 **−0.19** = (16 − 305.9)/1507.97	$56,351 **−0.50** = (56351 − 73299.42)/34055.25
Total z Score **Starbucks** **Wrigley**	**1.25 − 0.84 = 0.41** **−0.19 − 0.50 = −0.69**	

Table 6.3 *For each variable, the z-score for each observation is found by subtracting the mean from the value and then dividing that difference by the standard deviation. By adding the two z-scores, we see that even though Starbucks has a lower average salary than Wrigley, it is compensated for by the number of new jobs they offered.*

Starbucks offered more new jobs than Wrigley, but Wrigley had a higher average salary. The z-score for Starbucks' *New Jobs* (2193 − 305.9)/1507.97 = 1.25 is 1.44 higher than Wrigley's −0.19 (see Table 6.3 for details). By comparison, Wrigley's *Average Pay* z-score of −0.50 is only 0.34 better than Starbucks' −0.84. So in terms of standardized scores, Starbucks' *New Jobs* performance dominates Wrigley's *Average Pay*.

Is this the result we wanted? Adding z-scores together doesn't always make sense, or give the answer we want. Maybe we should put more weight on salary than on number of new jobs created. Our combined z-score added the two variables equally, but we could weight the variables differently. To determine the best company to work for, the Great Places to Work Institute had to combine scores from 50 different variables into one ranking, a task best accomplished by working with standardized scores. By using the standard deviation as a ruler to measure statistical distance from the mean, we compare values that are measured on different variables, with different scales, with different units, or for different populations.

> **Standardizing into z-scores:**
> - Shifts the mean to 0.
> - Changes the standard deviation to 1.
> - Does not change the shape.
> - Removes the units.

6.10 Time Series Plots

The volume on the NYSE is reported daily. Earlier, we grouped the days into months and half years, but we could simply look at the volume day by day. Whenever we have time series data, it is a good idea to look for patterns by plotting the data in time order. Figure 6.12 shows the *daily volumes* plotted over time for 2006.

A display of values against time is sometimes called a **time series plot**. This plot reflects the pattern that we saw when we plotted the daily volume by month, but without the arbitrary divisions between months we can see periods of relative calm contrasted with periods of greater activity. We can also see that the volume both became more variable and increased during certain parts of the year.

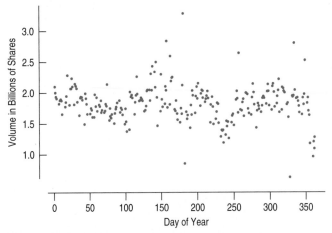

Figure 6.12   *A time series plot of daily volume shows the overall pattern and changes in variation.*

Time series plots often show a great deal of point-to-point variation, as Figure 6.12 does, and you'll often see time series plots drawn with all the points connected, especially in financial publications.

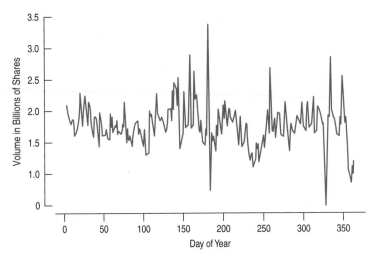

Figure 6.13   *The daily volumes of Figure 6.12, drawn by connecting all the points. Sometimes this can help us see the underlying pattern.*

Often it is better to try to smooth out the local point-to-point variability. After all, we usually want to see past this variation to understand any underlying trend and think about how the values vary around that trend—the time series version of center and spread. There are many ways for computers to run a smooth trace through a time series plot. Some follow local bumps, others emphasize long-term trends. Some provide an equation that gives a typical value for any given time point, others just offer a smooth trace.

A smooth trace can highlight long-term patterns and help us see them through the more local variation. Here are the daily volumes of Figures 6.12 and 6.13 with a typical smoothing function, available in many statistics programs.[6] With the smooth trace, it's a bit easier to see a pattern. The trace helps our eye follow the main trend and alerts us to points that don't fit the overall pattern.

---

[6] We'll discuss the most common ways to smooth data in Chapter 20.

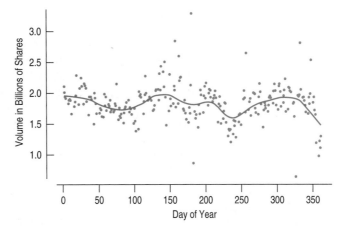

Figure 6.14 *The daily volumes of Figure 6.12, with a smooth trace added to help your eye see the long-term pattern.*

It is always tempting to try to extend what we see in a timeplot into the future. Sometimes that makes sense. Most likely, the NYSE volume follows some regular patterns throughout the year. It's probably safe to predict more volume on triple witching days and less activity in the week between Christmas and New Year's Day. But we certainly wouldn't predict a record every June 30.

Other patterns are riskier to extend into the future. If a stock's price has been rising, how long will it continue to go up? No stock has ever increased in value indefinitely, and no stock analyst has consistently been able to forecast when a stock's value will turn around. Stock prices, unemployment rates, and other economic, social, or psychological measures are much harder to predict than physical quantities. The path a ball will follow when thrown from a certain height at a given speed and direction is well understood. The path interest rates will take is much less clear.

Unless we have strong (nonstatistical) reasons for doing otherwise, we should resist the temptation to think that any trend we see will continue indefinitely. Statistical models often tempt those who use them to think beyond the data. We'll pay close attention later in this book to understanding when, how, and how much we can justify doing that.

Let's return to the Enron data we saw at the beginning of the chapter. The stock price changes are a time series from January 1997 to December 2001 with frequency one month. The histogram (Figure 6.1) showed a symmetric, possibly unimodal distribution for the most part concentrated between −$5 and +$5 with a second group near −$15. The time series plot shows a different story.

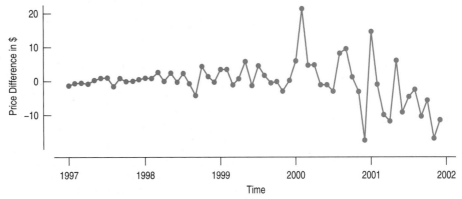

Figure 6.15 *A time series plot of monthly Enron stock price changes shows the increased volatility that began in early 2000.*

The time series plot shows that price changes were relatively small from 1997 to mid 1998, then increased to a new level until early 2000, when the prices became quite volatile. All the large price changes came after January 2000. Curiously, most stock analysts weren't concerned about this volatility until some time in 2001, when most of the changes became negative. Perhaps they weren't looking at the time series plot.

The histogram fails to summarize this distribution well because of the change in the behavior of the series over time. When a time series is **stationary**[7] (without a strong trend or change in variability), then a histogram can provide a useful summary, especially in conjunction with a time series plot. However, when the time series is not stationary, a histogram is unlikely to capture much of interest. Then, a time series plot is the best graphical display to use in describing the behavior of the data.

*6.11 Transforming Skewed Data

When a distribution is skewed, it can be hard to summarize the data simply with a center and spread, and hard to decide whether the most extreme values are outliers or just part of the stretched-out tail. How can we say anything useful about such data? The secret is to apply a simple function to each data value. One such function that can change the shape of a distribution is the logarithmic function. Let's examine an example in which a set of data is severely skewed.

In 1980, the average CEO made about 42 times the average worker's salary. In the two decades that followed, CEO compensation soared when compared with the average worker's pay; by 2000, that multiple had jumped to 525.[8] What does the distribution of the Fortune 500 companies' CEOs look like? Figure 6.16 shows a histogram of the 2005 compensation.

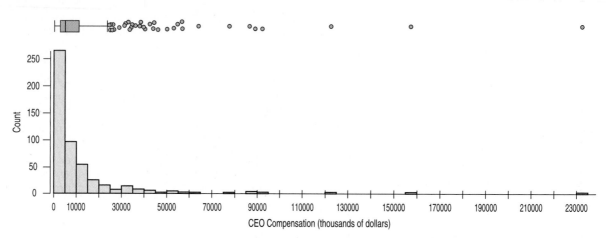

Figure 6.16   *The total compensation for CEOs (in $000) of the 500 largest companies is skewed and includes some extraordinarily large values.*

These values are reported in *thousands* of dollars. The boxplot indicates that some of the 500 CEOs received extraordinarily high compensation. The first bin of the histogram, containing about half the CEOs, covers the range $0 to $5,000,000. The reason that the histogram seems to leave so much of the area blank is that the

---

[7] Sometimes we separate out the properties and say the series is stationary with respect to the mean (if there is no trend) or stationary with respect to the variance (if the spread doesn't change), but unless otherwise noted, we'll assume that *all the statistical properties* of a stationary series are constant over time.

[8] Sources: United for a Fair Economy, *Business Week* annual CEO pay surveys, Bureau of Labor Statistics, "Average Weekly Earnings of Production Workers, Total Private Sector." Series ID: EEU00500004.

largest observations are so far from the bulk of the data, as we can see from the boxplot. Both the histogram and boxplot make it clear that this distribution is *very* skewed to the right.

Total compensation for CEOs consists of their base salaries, bonuses, and extra compensation, usually in the form of stock or stock options. Data that add together several variables, such as the compensation data, can easily have skewed distributions. It's often a good idea to separate the component variables and examine them individually, but we don't have that information for the CEOs.

Skewed distributions are difficult to summarize. It's hard to know what we mean by the "center" of a skewed distribution, so it's not obvious what value to use to summarize the distribution. What would you say was a typical CEO total compensation? The mean value is $10,307,000, while the median is "only" $4,700,000. Each tells something different about how the data are distributed.

One way to make a skewed distribution more symmetric is to **re-express**, or **transform**, the data by applying a simple function to all the data values. Variables with a distribution that is skewed to the right often benefit from a re-expression by logarithms or square roots. Those skewed to the left may benefit from squaring the data values. It doesn't matter what base you use for a logarithm.

◆ **Dealing with logarithms** You probably don't encounter logarithms every day. In this book, we use them to make data behave better by making model assumptions more reasonable. Base 10 logs are the easiest to understand, but natural logs are often used as well. (Either one is fine.) You can think of base 10 logs as roughly one less than the number of digits you need to write the number. So 100, which is the smallest number to require 3 digits, has a $\log_{10}$ of 2. And 1000 has a $\log_{10}$ of 3. The $\log_{10}$ of 500 is between 2 and 3, but you'd need a calculator to find that it's approximately 2.7. All salaries of "six figures" have $\log_{10}$ between 5 and 6. Logs are incredibly useful for making skewed data more symmetric. Fortunately, with technology, remaking a histogram or other display of the data is as easy as pushing a button.

The histogram of the logs of the total CEO compensations in Figure 6.17 is much more symmetric, so we can see that a typical *log compensation* is between 6.0 and 7.0, which means that it lies between $1 million and $10 million. To be more precise, the mean log_{10} value is 6.73, while the median is 6.67 (that's $5,370,317 and $4,677,351, respectively). Note that nearly all the values are between 6.0 and 8.0—in other words, between $1,000,000 and $100,000,000 per year. Logarithmic transformations are common, and because computers and calculators are available to do the calculating, you should consider transformation as a helpful tool whenever you have skewed data.

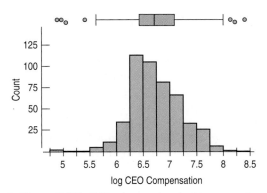

Figure 6.17 *Taking logs makes the histogram of CEO total compensation nearly symmetric.*

WHAT CAN GO WRONG?

A data display should tell a story about the data. To do that it must speak in a clear language, making plain what variable is displayed, what any axis shows, and what the values of the data are. And it must be consistent in those decisions.

The task of summarizing a quantitative variable requires that we follow a set of rules. We need to watch out for certain features of the data that make summarizing them with a number dangerous. Here's some advice:

- **Don't make a histogram of a categorical variable.** Just because the variable contains numbers doesn't mean it's quantitative. Here's a histogram of the insurance policy numbers of some workers. It's not very informative because the policy numbers are categorical. A histogram or stem-and-leaf display of a categorical variable makes no sense. A bar chart or pie chart may do better.

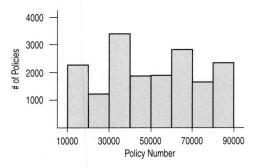

Figure 6.18 *It's not appropriate to display categorical data like policy numbers with a histogram.*

- **Choose a scale appropriate to the data.** Computer programs usually do a pretty good job of choosing histogram bin widths. Often, there's an easy way to adjust the width, sometimes interactively. Here is the Enron price change histogram with two other choices for the bin size.

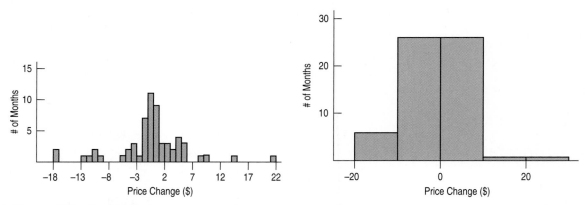

Figure 6.19 *Changing the bin width changes how the histogram looks. The Enron stock price changes look very different with these two choices.*

- **Avoid inconsistent scales.** Parts of displays should be mutually consistent—no fair changing scales in the middle or plotting two variables on different scales but on the same display. When comparing two groups, be sure to draw them on the same scale.

- **Label clearly.** Variables should be identified clearly and axes labeled so a reader knows what the plot displays.

Here's a remarkable example of a plot gone wrong. It illustrated a news story about rising college costs. It uses time series plots, but it gives a misleading impression. First, think about the story you're being told by this display. Then try to figure out what has gone wrong.

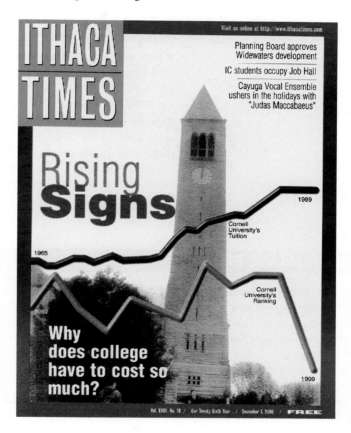

What's wrong? Just about everything.

- The horizontal scales are inconsistent. Both lines show trends over time, but for what years? The tuition sequence starts in 1965, but rankings are graphed from 1989. Plotting them on the same (invisible) scale makes it seem that they're for the same years.
- The vertical axis isn't labeled. That hides the fact that it's using two different scales. Does it graph dollars (of tuition) or ranking (of Cornell University)?

This display violates three of the rules. And it's even worse than that. It violates a rule that we didn't even bother to mention. The two inconsistent scales for the vertical axis don't point in the same direction! The line for Cornell's rank shows that it has "plummeted" from 15th place to 6th place in academic rank. Most of us think that's an *improvement*, but that's not the message of this graph.

- **Do a reality check.** Don't let the computer (or calculator) do your thinking for you. Make sure the calculated summaries make sense. For example, does the mean look like it is in the center of the histogram? Think about the spread. An IQR of 50 mpg would clearly be wrong for a family car. And no measure of

(continued)

spread can be negative. The standard deviation can take the value 0, but only in the very unusual case that all the data values equal the same number. If you see the IQR or standard deviation equal to 0, it's probably a sign that something's wrong with the data.

- **Don't compute numerical summaries of a categorical variable.** The mean zip code or the standard deviation of Social Security numbers is not meaningful. If the variable is categorical, you should instead report summaries such as percentages. It is easy to make this mistake when you let technology do the summaries for you. After all, the computer doesn't care what the numbers mean.

- **Watch out for multiple modes.** If the distribution—as seen in a histogram, for example—has multiple modes, consider separating the data into groups. If you cannot separate the data in a meaningful way, you should not summarize the center and spread of the variable.

- **Beware of outliers.** If the data have outliers but are otherwise unimodal, consider holding the outliers out of the further calculations and reporting them individually. If you can find a simple reason for the outlier (for instance, a data transcription error), you should remove or correct it. If you cannot do either of these, then choose the median and IQR to summarize the center and spread.

ETHICS IN ACTION

Beth Tully owns Zenna's Café, an independent coffee shop located in a small mid-western city. Since opening Zenna's in 2002, she has been steadily growing her business and now distributes her custom coffee blends to a number of regional restaurants and markets. She operates a microroaster that offers specialty grade Arabica coffees recognized as some as the best in the area. In addition to providing the highest quality coffees, Beth also wants her business to be socially responsible. Toward that end, she pays fair prices to coffee farmers and donates funds to help charitable causes in Panama, Costa Rica, and Guatemala. In addition, she encourages her employees to get involved in the local community. Recently, one of the well-known multinational coffeehouse chains announced plans to locate shops in her area. This chain is one of the few to offer Certified Free Trade coffee products and work toward social justice in the global community. Consequently, Beth thought it might be a good idea for her to begin communicating Zenna's socially responsible efforts to the public, but with an emphasis on their commitment to the local community. Three months ago she began collecting data on the number of volunteer hours donated by her employees per week. She has a total of 12 employees, of whom 10 are full time. Most employees volunteered less than 2 hours per week, but Beth noticed that one part-time employee volunteered more than 20 hours per week. She discovered that her employees collectively volunteered an average of 15 hours per month (with a median of 8 hours). She planned to report the average number and believed most people would be impressed with Zenna's level of commitment to the local community.

ETHICAL ISSUE *The outlier in the data affects the average in a direction that benefits Beth Tully and Zenna's Café (related to Item C, ASA Ethical Guidelines).*

ETHICAL SOLUTION *Beth's data are highly skewed. There is an outlier value (for a part-time employee) that pulls the average number of volunteer hours up. Reporting the average is misleading. In addition, there may be justification to eliminate the value since it belongs to a part-time employee (10 of the 12 employees are full time). It would be more ethical for Beth to: (1) report the average but discuss the outlier value; (2) report the average for only full-time employees, or (3) report the median instead of the average.*

What have we learned?

We've learned how to display and summarize quantitative data to help us see the story the data have to tell.

- We can display the distribution of quantitative data with a histogram or a stem-and-leaf display.
- We report what we see about the distribution by talking about shape, center, spread, outliers, and any unusual features.

We've learned how to summarize distributions of quantitative variables numerically.

- Measures of center for a distribution include the median and the mean.
- Measures of spread include the range, IQR, and standard deviation.
- We'll report the median and IQR when the distribution is skewed. If it's symmetric, we'll summarize the distribution with the mean and standard deviation (and possibly the median and IQR as well). Always pair the median with the IQR and the mean with the standard deviation.

We've learned to think about the type of variable we're summarizing.

- All the methods in this chapter assume that the data are quantitative.
- The Quantitative Data Condition serves as a check that the data are, in fact, quantitative. One good way to be sure is to know the measurement units.

We've learned the value of comparing groups and looking for patterns among groups and over time.

- We've seen that boxplots are very effective for comparing groups graphically.
- When we compare groups, we discuss their shape, center, spreads, and any unusual features.

We've experienced the value of identifying and investigating outliers, and we've seen that when we group data in different ways, it can allow different cases to emerge as possible outliers.

- We've graphed data that have been measured over time against a time axis and looked for trends both by eye and with a data smoother.

We've learned the power of standardizing data.

- Standardizing uses the standard deviation as a ruler to measure distance from the mean, creating z-scores.
- Using these z-scores, we can compare apples and oranges—values from different distributions or values based on different units.
- A z-score can identify unusual or surprising values among data.

Terms

Bimodal	Distributions with two modes.
Boxplot	A boxplot displays the 5-number summary is a central box with whiskers that extend to the non-outlying values. Boxplots are particularly effective for comparing groups.
Center	The middle of the distribution, usually summarized numerically by the mean or the median.
Distribution	The distribution of a variable gives: • possible values of the variable • frequency or relative frequency of each value
Five-number summary	A five-number summary for a variable consists of: • The minimum and maximum • The quartiles Q1 and Q3 • The median
Histogram (relative frequency histogram)	A histogram uses adjacent bars to show the distribution of values in a quantitative variable. Each bar represents the frequency (relative frequency) of values falling in an interval of values.
Interquartile range (IQR)	The difference between the first and third quartiles. $IQR = Q3 - Q1$.
Mean	A measure of center found as $\Sigma y/n$.
Median	The middle value with half of the data above it and half below it.
Mode	A peak or local high point in the shape of the distribution of a variable. The apparent location of modes can change as the scale of a histogram is changed.
Multimodal	Distributions with more than two modes.
Outliers	Extreme values that don't appear to belong with the rest of the data. They may be unusual values that deserve further investigation or just mistakes; there's no obvious way to tell.
Quartile	The lower quartile (Q1) is the value with a quarter of the data below it. The upper quartile (Q3) has a quarter of the data above it. The median and quartiles divide data into four equal parts.
Range	The difference between the lowest and highest values in a data set: Range = *max* − *min*.
Re-express or **transform**	We re-express or transform data by taking the logarithm, square root, reciprocal or some other mathematical operation on all values of the data set.

Shape	The visual appearance of the distribution. To describe the shape, look for:
	• single vs. multiple modes
	• symmetry vs. skewness
Skewed	A distribution is skewed if one tail stretches out farther than the other.
Spread	The description of how tightly clustered the distribution is around its center. Measures of spread include the IQR and the standard deviation.
Standard deviation	A measure of spread found as $s = \sqrt{\dfrac{\Sigma(y - \bar{y})^2}{n - 1}}$.
Standardized value	We standardize a value by subtracting the mean and dividing by the standard deviation for the variable. These values, called z-scores, have no units.
Stationary	A time series is said to be stationary if its statistical properties don't change over time.
Stem-and-leaf display	A stem-and-leaf display shows quantitative data values in a way that sketches the distribution of the data. It's best described in detail by example.
Symmetric	A distribution is symmetric if the two halves on either side of the center look approximately like mirror images of each other.
Tail	The tails of a distribution are the parts that typically trail off on either side.
Time series plot	Displays data that change over time. Often, successive values are connected with lines to show trends more clearly.
Uniform	A distribution that's roughly flat is said to be uniform.
Unimodal	Having one mode. This is a useful term for describing the shape of a histogram when it's generally mound-shaped.
Variance	The standard deviation squared.
z-score	A standardized value that tells how many standard deviations a value is from the mean; z-scores have a mean of 0 and a standard deviation of 1.

Skills

PLAN

- Be able to identify an appropriate display for any quantitative variable.
- Be able to select a suitable measure of center and a suitable measure of spread for a variable based on information about its distribution.
- Know the basic properties of the median: The median divides the data into the half of the data values that are below the median and the half that are above the median.
- Know the basic properties of the mean: The mean is the point at which the histogram balances.
- Know that the standard deviation summarizes how spread out all the data are around the mean.
- Know that standardizing uses the standard deviation as a ruler.

DO

- Know how to display the distribution of a quantitative variable with a stem-and-leaf display or a histogram.

- Know how to make a time series plot of data that are collected at regular time intervals.

- Know how to compute the mean and median of a set of data and know when each is appropriate.

- Know how to compute the standard deviation and IQR of a set of data and know when each is appropriate.

- Know how to compute a five-number summary of a variable.

- Know how to construct a boxplot by hand from a five-number summary.

- Know how to calculate the z-score of an observation.

 REPORT

- Be able to describe and compare the distributions of quantitative variables in terms of their shape, center, and spread.

- Be able to discuss any outliers in the data, noting how they deviate from the overall pattern of the data.

- Be able to describe summary measures in a sentence. In particular, know that the common measures of center and spread have the same units as the variable that they summarize and that they should be described in those units.

- Be able to compare two or more groups by comparing their boxplots.

- Be able to discuss patterns in a time series plot, both in terms of the general trend and any changes in the spread of the distribution over time.

Technology Help: Displaying and Summarizing Quantitative Variables

Almost any program that displays data can make a histogram, but some will do a better job of determining where the bars should start and how they should partition the span of the data (see the art on the next page).

Many statistics packages offer a prepackaged collection of summary measures. The result might look like this:

```
Variable: Weight
N = 234
Mean = 143.3      Median = 139
St. Dev = 11.1    IQR = 14
```

Alternatively, a package might make a table for several variables and summary measures:

Variable	N	mean	median	stdev	IQR
Weight	234	143.3	139	11.1	14
Height	234	68.3	68.1	4.3	5
Score	234	86	88	9	5

It is usually easy to read the results and identify each computed summary. You should be able to read the summary statistics produced by any computer package.

Packages often provide many more summary statistics than you need. Of course, some of these may not be appropriate when the data are skewed or have outliers. It is your responsibility to check a histogram or stem-and-leaf display and decide which summary statistics to use.

It is common for packages to report summary statistics to many decimal places of "accuracy." Of course, it is rare to find data that have such accuracy in the original measurements. The ability to calculate to six or seven digits beyond the decimal point doesn't mean that those digits have any meaning. Generally, it's a good idea to round these values, allowing perhaps one more digit of precision than was given in the original data.

Displays and summaries of quantitative variables are among the simplest things you can do in most statistics packages.

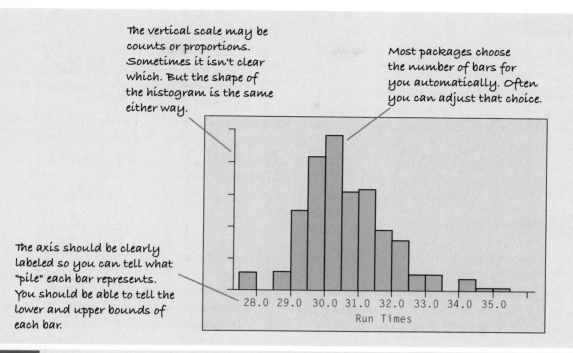

The vertical scale may be counts or proportions. Sometimes it isn't clear which. But the shape of the histogram is the same either way.

Most packages choose the number of bars for you automatically. Often you can adjust that choice.

The axis should be clearly labeled so you can tell what "pile" each bar represents. You should be able to tell the lower and upper bounds of each bar.

EXCEL

To calculate summaries, Click on an empty cell. Type an equals sign and choose "**Average**" from the popup list of functions that appears to the left of the text editing box. Enter the data range in the box that says "**Number 1**. " Click the **OK** button. To compute the standard deviation of a column of data directly, use the **STDEV** from the popup list of functions in the same way.

Excel cannot make histograms, boxplots, or dotplots without a third-party add-in. To make a histogram, boxplot, or dotplot using the DDXL add-in on the CD, select the range of data you wish to display. Include the variable name if it is the first row of your column of data. Then choose **Charts**

and Plots from the **DDXL** menu. From the DDXL dialog, choose **Dotplot**, **Boxplot**, or **Histogram**. Indicate the variable by dragging it into the Quantitative Variables area, uncheck First row is variable names if there isn't a name in the first row, and click OK.

Comments

Excel's Data Analysis add-in does offer something called a histogram, but it isn't a statistically appropriate histogram. Excel's STDEV function should not be used for data values larger in magnitude than 100,000 or for lists of more than a few thousand values. It is fine for smaller datasets.

EXCEL 2007

In Excel 2007 there is another way to find some of the standard summary statistics. For example, to compute the mean:

- Click on an empty cell.
- Go to the Formulas tab in the Ribbon. Click on the drop down arrow next to "AutoSum" and choose "**Average**".
- Enter the data range in the formula displayed in the empty box you selected earlier.
- Press **Enter**. This computes the mean for the values in that range.

To compute the standard deviation:

- Click on an empty cell.
- Go to the Formulas tab in the Ribbon and click the drop down arrow next to "AutoSum" and select "**More functions...**"
- In the dialog window that opens, select "**STDEV**" from the list of functions and click **OK**. A new dialog window opens. Enter a range of fields into the text fields and click **OK**.

Excel 2007 computes the standard deviation for the values in that range and places it in the specified cell of the spreadsheet.

MINITAB

To make a histogram:

- Choose **Histogram** from the **Graph** menu.
- Select "Simple" for the type of graph and click **OK**.
- Enter the name of the quantitative variable you wish to display in the box labeled "Graph variables." Click **OK**.

To make a boxplot:

- Choose **Boxplot** from the **Graph** menu and specify your data format.

To calculate summary statistics:

- Choose **Basic statistics** from the **Stat** menu. From the **Basic Statistics** submenu, choose **Display Descriptive Statistics**.
- Assign variables from the variable list box to the Variables box. MINITAB makes a Descriptive Statistics table.

SPSS

To make a histogram or boxplot in SPSS open the Chart Builder from the Graphs menu.

- Click the **Gallery** tab
- Choose **Histogram** or **Boxplot** from the list of chart types.
- Drag the icon of the plot you want onto the canvas.
- Drag a scale variable to the y-axis drop zone.
- Click **OK**.

To make side-by-side boxplots, drag a categorical variable to the x-axis drop zone and click **OK**.

To calculate summary statistics:

- Choose **Explore** from the **Descriptive Statistics** submenu of the **Analyze** menu. In the Explore dialog, assign one or more variables from the source list to the Dependent List and click the **OK** button.

JMP

To make a histogram and find summary statistics:

- Choose **Distribution** from the **Analyze** menu.
- In the **Distribution** dialog, drag the name of the variable that you wish to analyze into the empty window beside the label "**Y, Columns**."
- Click **OK**. JMP computes standard summary statistics along with displays of the variables.

To make boxplots:

- Choose **Fit y by x**. Assign a continuous response variable to **Y, Response** and a nominal group variable holding the group names to **X, Factor,** and click **OK**. JMP will offer (among other things) dotplots of the data. Click the red triangle and, under **Display Options**, select Boxplots. Note: If the variables are of the wrong type, the display options might not offer boxplots.

DATA DESK

To make a histogram:

- Select the variable to display.
- In the **Plot** menu, choose **Histogram**.

To make boxplots:

- If the data are in separate variables, select the variables and choose **Boxplot side by side** from the **Plot** menu.
- If the data are a single quantitative variable and a second variable holding group names, select the quantitative variable as Y and the group variable as X. Then choose **Boxplot y by x** from the **Plot** menu.

To calculate summaries:

- In the **Calc** menu, open the **summaries** submenu. **Options** offer separate tables, a single unified table, and other formats.

Mini Case Study Projects

Hotel Occupancy Rates

Many properties in the hospitality industry experience strong seasonal fluctuations in demand. To be successful in this industry it is important to anticipate such fluctuations and to understand demand patterns. The file **ch06_MCSP_Occupancy_Rates** contains data on monthly *Hotel Occupancy Rates* (in % capacity) for Honolulu, Hawaii from January 2000 to December 2004.

Examine the data and prepare a report for the manager of a hotel chain in Honolulu on patterns in *Hotel Occupancy* during this period. Include both numerical summaries and graphical displays and summarize the patterns that you see. Discuss any unusual features of the data and explain them if you can, including a discussion of whether the manager should take these features into account for future planning.

Value and Growth Stock Returns

Investors in the stock market have choices of how aggressive they would like to be with their investments. To help investors, stocks are classified as "growth" or "value" stocks. Growth stocks are generally shares in high quality companies that have demonstrated consistent performance and are expected to continue to do well. Value stocks on the other hand are stocks whose prices seem low compared to their inherent worth (as measured by the book to price ratio). Managers invest in these hoping that their low price is simply an over reaction to recent negative events.[9]

In the data set **ch06_MCP_Returns**[10] are the monthly returns of 2500 stocks classified as Growth and Value for the time period January 1975 to June 1997. Examine the distributions of the two types of stocks and discuss the advantages and disadvantages of each. Is it clear which type of stock offers the best investment? Discuss briefly.

---

[9] The cynical statistician might say that the manager who invests in growth funds puts his faith in extrapolation, while the value manager is putting his faith in the Law of Averages.

[10] Source: Independence International Associates, Inc. maintains a family of international style indexes covering 22 equity markets. The highest book-to-price stocks are selected one by one from the top of the list. The top half of these stocks become the constituents of the "value index," and the remaining stocks become the "growth index."

EXERCISES

1. Statistics in business. Find a histogram that shows the distribution of a variable in a business publication (e.g., *The Wall Street Journal, Business Week,* etc.).

a) Does the article identify the W's?
b) Discuss whether the display is appropriate for the data.
c) Discuss what the display reveals about the variable and its distribution.
d) Does the article accurately describe and interpret the data? Explain.

2. Statistics in business, part 2. Find a graph other than a histogram that shows the distribution of a quantitative variable in a business publication (e.g., *The Wall Street Journal, Business Week,* etc.).

a) Does the article identify the W's?
b) Discuss whether the display is appropriate for the data.
c) Discuss what the display reveals about the variable and its distribution.
d) Does the article accurately describe and interpret the data? Explain.

3. Two-year college tuition. The histogram shows the distribution of average tuitions charged by each of the fifty U.S. states for public two-year colleges in the 2007–2008 academic year. Write a short description of this distribution (shape, center, spread, unusual features).

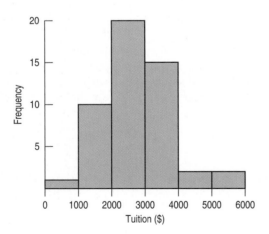

4. Gas prices. The website MSN auto (www.autos.msn.com) provides prices of gasoline at stations all around the United States. This histogram shows the price of regular gas (in $/gallon) for 57 stations in the Los Angeles area during the week before Christmas 2007. Describe the shape of this distribution (shape, center, spread, unusual features).

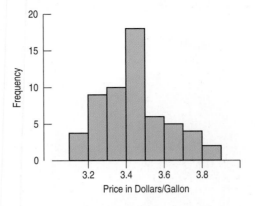

5. Credit card charges. The histogram shows the December charges (in $) for 5000 customers from one marketing segment from a credit card company. (Negative values indicate customers who received more credits than charges during the month.)

a) Write a short description of this distribution (shape, center, spread, unusual features).
b) Would you expect the mean or the median to be larger? Explain.
c) Which would be a more appropriate summary of the center, the mean or the median? Explain.

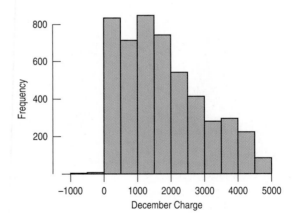

6. Vineyards. Adair Vineyard is a 10-acre vineyard in New Paltz, New York. The winery itself is housed in a 200-year-old historic Dutch barn, with the wine cellar on the first floor and the tasting room and gift shop on the second. Since they are relatively small and considering an expansion, they are curious about how their size compares to that of other vineyards. The histogram shows the sizes (in acres) of 36 wineries in upstate New York.

a) Write a short description of this distribution (shape, center, spread, unusual features).

b) Would you expect the mean or the median to be larger? Explain.

c) Which would be a more appropriate summary of the center, the mean or the median? Explain.

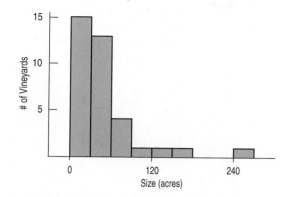

T 7. Mutual funds. The histogram displays the 12-month returns (in percent) for a collection of mutual funds in 2007. Give a short summary of this distribution (shape, center, spread, unusual features).

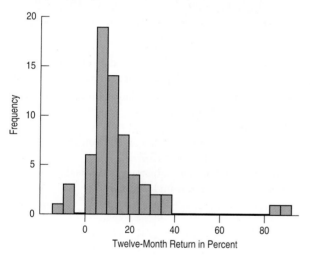

T 8. Car discounts. A researcher, interested in studying gender differences in negotiations, collects data on the prices that men and women pay for new cars. Here is a histogram of the discounts (the amount in $ below the list price) that men and women received at one car dealership for the last 100 transactions. (54 men and 46 women). Give a short summary of this distribution (shape, center, spread, unusual features). What do you think might account for this particular shape?

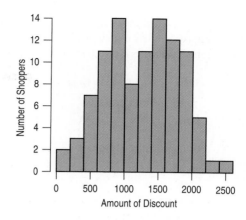

T 9. Mutual funds, part 2. Use the dataset of Exercise 7 to answer the following questions.

a) Find the five-number summary for these data.

b) Find appropriate measures of center and spread for these data.

c) Create a boxplot for these data.

d) What can you see, if anything, in the histogram that isn't clear in the boxplot?

T 10. Car discounts, part 2. Use the dataset of Exercise 8 to answer the following questions.

a) Find the five-number summary for these data.

b) Create a boxplot for these data.

c) What can you see, if anything, in the histogram of Exercise 8 that isn't clear in the boxplot?

11. Vineyards, part 2. Here are summary statistics for the sizes (in acres) of upstate New York vineyards from Exercise 6.

Variable	N	Mean	StDev	Minimum	Q1	Median	Q3	Maximum
Acres	36	46.50	47.76	6	18.50	33.50	55	250

a) Would you describe this distribution as symmetric or skewed? Explain.

b) Are there any outliers? Explain.

c) Using these data, sketch a boxplot. What additional information would you need to complete the boxplot?

12. Graduation. A survey of major universities asked what percentage of incoming freshmen usually graduate "on time" in 4 years. Use the summary statistics given to answer these questions.

	% on time
Count	48
Mean	68.35
Median	69.90
StdDev	10.20
Min	43.20
Max	87.40
Range	44.20
25th %tile	59.15
75th tile	74.75

a) Would you describe this distribution as symmetric or skewed?
b) Are there any outliers? Explain.
c) Create a boxplot of these data.

T 13. Vineyards, again. The data set provided contains the data from Exercises 6 and 11. Create a stem-and-leaf display of the sizes of the vineyards in acres. Point out any unusual features of the data that you can see from the stem-and-leaf.

T 14. Gas prices, again. The data set provided contains the data from Exercise 4 on the price of gas for 57 stations around Los Angeles in December 2007. Round the data to the nearest penny (e.g., 3.459 becomes 3.46) and create a stem-and-leaf display of the data. Point out any unusual features of the data that you can see from the stem-and-leaf.

15. Hockey. During his 20 seasons in the National Hockey League, Wayne Gretzky scored 50% more points than anyone else who ever played professional hockey. He accomplished this amazing feat while playing in 280 fewer games than Gordie Howe, the previous record holder. Here are the number of games Gretzky played during each season:

79, 80, 80, 80, 74, 80, 80, 79, 64, 78, 73, 78, 74, 45, 81, 48, 80, 82, 82, 70

a) Create a stem-and-leaf display.
b) Sketch a boxplot.
c) Briefly describe this distribution.
d) What unusual features do you see in this distribution? What might explain this?

16. Baseball. In his 17-year career as a player in major league baseball, Mark McGwire hit 583 home runs, placing him eighth on the all-time home run list (as of 2008). Here are the number of home runs that McGwire hit for each year from 1986 through 2001:

3, 49, 32, 33, 39, 22, 42, 9, 9, 39, 52, 34, 24, 70, 65, 32, 29

a) Create a stem-and-leaf display.
b) Sketch a boxplot.
c) Briefly describe this distribution.
d) What unusual features do you see in this distribution? What might explain this?

17. Gretzky returns. Look once more at data of hockey games played each season by Wayne Gretzky, seen in Exercise 15.

a) Would you use the mean or the median to summarize the center of this distribution? Why?
b) Without actually finding the mean, would you expect it to be lower or higher than the median? Explain.
c) A student was asked to make a histogram of the data in Exercise 15 and produced the following. Comment.

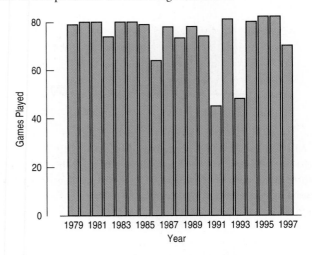

18. McGwire, again. Look once more at data of home runs hit by Mark McGwire during his 17-year career as seen in Exercise 16.

a) Would you use the mean or the median to summarize the center of this distribution? Why?
b) Find the median.
c) Without actually finding the mean, would you expect it to be lower or higher than the median? Explain.
d) A student was asked to make a histogram of the data in Exercise 16 and produced the following. Comment.

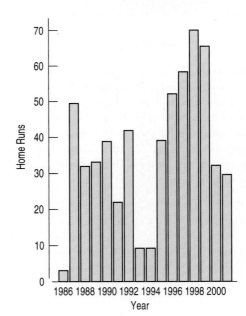

T 19. Pizza prices. The weekly prices of one brand of frozen pizza over a three-year period in Dallas are provided in the data file. Use the price data to answer the following questions.

a) Find the five-number summary for these data.
b) Find the range and IQR for these data.
c) Create a boxplot for these data.
d) Describe this distribution.
e) Describe any unusual observations.

T 20. Pizza prices, part 2. The weekly prices of one brand of frozen pizza over a three-year period in Chicago are provided in the data file. Use the price data to answer the following questions.

a) Find the five-number summary for these data.
b) Find the range and IQR for these data.
c) Create a boxplot for these data.
d) Describe the shape (center and spread) of this distribution.
e) Describe any unusual observations.

T 21. Gasoline usage. The U.S. Department of Transportation collects data on the amount of gasoline sold in each state and the District of Columbia. The following data show the per capita (gallons used per person) consumption in the year 2005. Write a report on the gasoline usage by state in the year 2005, being sure to include appropriate graphical displays and summary statistics.

State	Gasoline Usage	State	Gasoline Usage
Alabama	556.91	Montana	486.15
Alaska	398.99	Nebraska	439.46
Arizona	487.52	Nevada	484.26
Arkansas	491.85	New Hampshire	521.45
California	434.11	New Jersey	481.79
Colorado	448.33	New Mexico	482.33
Connecticut	441.39	New York	283.73
Delaware	514.78	North Carolina	491.07
District of Columbia	209.47	North Dakota	513.16
Florida	485.73	Ohio	434.65
Georgia	560.90	Oklahoma	501.12
Hawaii	352.02	Oregon	415.67
Idaho	414.17	Pennsylvania	402.85
Illinois	392.13	Rhode Island	341.67
Indiana	497.35	South Carolina	570.24
Iowa	509.13	South Dakota	498.36
Kansas	399.72	Tennessee	509.77
Kentucky	511.30	Texas	505.39
Louisiana	489.84	Utah	409.93
Maine	531.77	Vermont	537.94
Maryland	471.52	Virginia	518.06
Massachusetts	427.52	Washington	423.32
Michigan	470.89	West Virginia	444.22
Minnesota	504.03	Wisconsin	440.45
Mississippi	539.39	Wyoming	589.18
Missouri	530.72		

T 22. GDP growth. Established in Paris in 1961, the Organisation for Economic Co-operation and Development (OECD) (www.oced.org) collects information on many economic and social aspects of countries around the world. Here are the growth rates of 30 industrialized countries (in percentages) of their gross domestic products (GDP) in 2005. Write a brief report on the GDP growth rates of these countries in 2005, being sure to include appropriate graphical displays and summary statistics.

Country	Growth Rate
Turkey	0.074
Czech Republic	0.061
Slovakia	0.061
Iceland	0.055
Ireland	0.055
Hungary	0.041
Korea, Republic of (South Korea)	0.040
Luxembourg	0.040
Greece	0.037
Poland	0.034
Spain	0.034
Denmark	0.032
United States	0.032
Mexico	0.030
Canada	0.029
Finland	0.029
Sweden	0.027
Japan	0.026
Australia	0.025
New Zealand	0.023
Norway	0.023
Austria	0.020
Switzerland	0.019
United Kingdom	0.019
Belgium	0.015
Netherlands	0.015
France	0.012
Germany	0.009
Portugal	0.004
Italy	0.000

23. Start-up. A start-up company is planning to build a new golf course. For marketing purposes, the company would like to be able to advertise the new course as one of the more difficult courses in the state of Vermont. One measure of the difficulty of a golf course is its length: the total distance (in yards) from tee to hole for all 18 holes. Here are the histogram and summary statistics for the lengths of all the golf courses in Vermont.

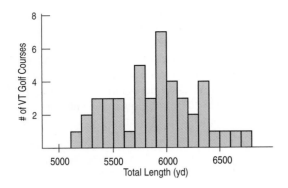

Count	45
Mean	5892.91 yd
StdDev	386.59
Min	5185
Q1	5585.75
Median	5928
Q3	6131
Max	6796

a) What is the range of these lengths?
b) Between what lengths do the central 50% of these courses lie?
c) What summary statistics would you use to describe these data?
d) Write a brief description of these data (shape, center, and spread).

24. Real estate. A real estate agent has surveyed houses in twenty nearby zip codes in an attempt to put together a comparison for a new property that she would like to put on the market. She knows that the size of the living area of a house is a strong factor in the price, and she'd like to market this house as being one of the biggest in the area. Here is a histogram and summary statistics for the sizes of all the houses in the area.

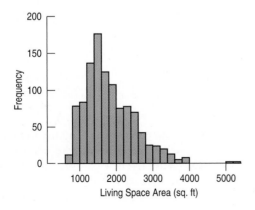

Count	1057
Mean	1819.498 sq. ft
Std Dev	662.9414
Min	672
Q1	1342
Median	1675
Q3	2223
Max	5228
Missing	0

a) What is the range of these sizes?
b) Between what sizes do the central 50% of these houses lie?
c) What summary statistics would you use to describe these data?
d) Write a brief description of these data (shape, center, and spread).

T 25. Food sales. Sales (in $) for one week were collected for 18 stores in a food store chain in the northeastern United States. The stores and the towns they are located in vary in size.

a) Make a suitable display of the sales from the data provided.
b) Summarize the central value for sales for this week with a median and mean. Why do they differ?
c) Given what you know about the distribution, which of these measures does the better job of summarizing the stores' sales? Why?
d) Summarize the spread of the sales distribution with a standard deviation and with an IQR.
e) Given what you know about the distribution, which of these measures does the better job of summarizing the stores' sales? Why?
f) If we were to remove the outliers from the data, how would you expect the mean, median, standard deviation, and IQR to change?

T 26. Insurance profits. Insurance companies don't know whether a policy they've written is profitable until the policy matures (expires). To see how they've performed recently, an analyst looked at mature policies and investigated the net profit to the company (in $).

a) Make a suitable display of the profits from the data provided.
b) Summarize the central value for the profits with a median and mean. Why do they differ?
c) Given what you know about the distribution, which of these measures might do a better job of summarizing the company's profits? Why?
d) Summarize the spread of the profit distribution with a standard deviation and with an IQR.
e) Given what you know about the distribution, which of these measures might do a better job of summarizing the company's profits? Why?
f) If we were to remove the outliers from the data, how would you expect the mean, median, standard deviation, and IQR to change?

T **27.** iPod failures. MacInTouch (www.macintouch.com/reliability/ipodfailures.html) surveyed readers about the reliability of their iPods. Of the 8926 iPods owned, 7510 were problem-free while the other 1416 failed. From the data on the disk, compute the failure rate for each of the 17 iPod models. Produce an appropriate graphical display of the failure rates and briefly describe the distribution.

T **28.** Unemployment. The data set provided contains 2008 unemployment rates for 23 developed countries (www.oecd.org). Produce an appropriate graphical display and briefly describe the distribution of unemployment rates.

29. Sales. Here are boxplots of the weekly sales (in $ U.S.) over a two-year period for a regional food store for two locations. Location #1 is a metropolitan area that is known to be residential where shoppers walk to the store. Location #2 is a suburban area where shoppers drive to the store. Assume that the two towns have similar populations and that the two stores are similar in square footage. Write a brief report discussing what these data show.

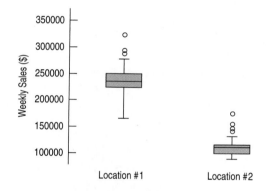

30. Sales, part 2. Recall the distributions of the weekly sales for the regional stores in Exercise 29. Following are boxplots of weekly sales for this same food store chain for three stores of similar size and location for two different states: Massachusetts (MA) and Connecticut (CT). Compare the distribution of sales for the two states and describe in a report.

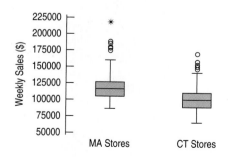

31. Gas prices, part 2. Here are boxplots of weekly gas prices at a service station in the Midwest United States (prices in $ per gallon).

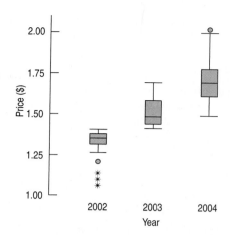

a) Compare the distribution of prices over the three years.
b) In which year were the prices least stable (most volatile)? Explain.

32. Fuel economy. American automobile companies are becoming more motivated to improve the fuel efficiency of the automobiles they produce. It is well known that fuel efficiency is impacted by many characteristics of the car. Describe what these boxplots tell you about the relationship between the number of cylinders a car's engine has and the car's fuel economy (mpg).

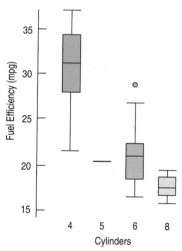

33. Wine prices. The boxplots display case prices (in dollars) of wines produced by vineyards along three of the Finger Lakes in upstate New York.

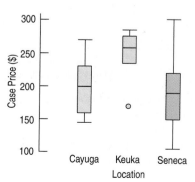

a) Which lake region produces the most expensive wine?
b) Which lake region produces the cheapest wine?
c) In which region are the wines generally more expensive?
d) Write a few sentences describing these prices.

34. Ozone. Ozone levels (in parts per billion, ppb) were recorded at sites in New Jersey monthly between 1926 and 1971. Here are boxplots of the data for each month (over the 46 years) lined up in order (January = 1).

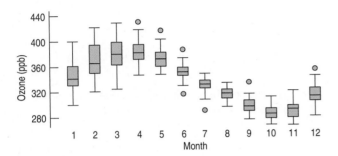

a) In what month was the highest ozone level ever recorded?
b) Which month has the largest IQR?
c) Which month has the smallest range?
d) Write a brief comparison of the ozone levels in January and June.
e) Write a report on the annual patterns you see in the ozone levels.

35. Derby speeds. How fast do horses run? Kentucky Derby winners top 30 miles per hour, as shown in the graph. This graph shows the percentage of Kentucky Derby winners that have run *slower* than a given speed. Note that few have won running less than 33 miles per hour, but about 95% of the winning horses have run less than 37 miles per hour. (A cumulative frequency graph like this is called an **ogive**.)

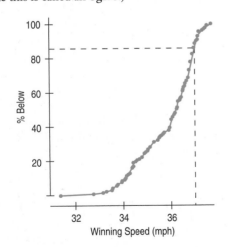

a) Estimate the median winning speed.
b) Estimate the quartiles.
c) Estimate the range and the IQR.
d) Create a boxplot of these speeds.
e) Write a few sentences about the speeds of the Kentucky Derby winners.

36. Mutual fund, part 3. Here is an ogive of the distribution of monthly returns for a group of aggressive (or high growth) mutual funds over a period of 22 years from 1975 to 1999. (Recall from Exercise 35 that an ogive, or cumulative relative frequency graph, shows the percent of cases at or below a certain value. Thus this graph always begins at 0% and ends at 100%.)

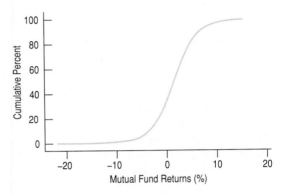

a) Estimate the median.
b) Estimate the quartiles.
c) Estimate the range and the IQR.
d) Create a boxplot of these returns.

37. Test scores. Three Statistics classes all took the same test. Here are histograms of the scores for each class.

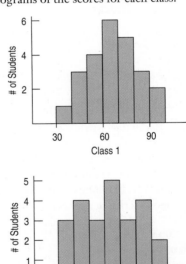

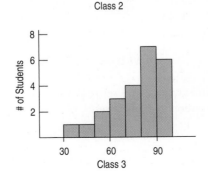

a) Which class had the highest mean score?

b) Which class had the highest median score?

c) For which class are the mean and median most different? Which is higher? Why?

d) Which class had the smallest standard deviation?

e) Which class had the smallest IQR?

38. Test scores, again. Look again at the histograms of test scores for the three Statistics classes in Exercise 37.

a) Overall, which class do you think performed better on the test? Why?

b) How would you describe the shape of each distribution?

c) Match each class with the corresponding boxplot.

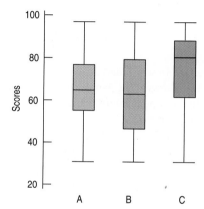

T 39. Quality control. Engineers at a computer production plant tested two methods for accuracy in drilling holes into a PC board. They tested how fast they could set the drilling machine by running 10 boards at each of two different speeds. To assess the results, they measured the distance (in inches) from the center of a target on the board to the center of the hole. The data and summary statistics are shown in the table.

	Fast	Slow
	0.000101	0.000098
	0.000102	0.000096
	0.000100	0.000097
	0.000102	0.000095
	0.000101	0.000094
	0.000103	0.000098
	0.000104	0.000096
	0.000102	0.975600
	0.000102	0.000097
	0.000100	0.000096
Mean	0.000102	0.097647
StdDev	0.000001	0.308481

Write a report summarizing the findings of the experiment. Include appropriate visual and verbal displays of the distributions, and make a recommendation to the engineers if they are most interested in the accuracy of the method.

T 40. Fire sale. A real estate agent notices that houses with fireplaces often fetch a premium in the market and wants to assess the difference in sales price of 60 homes that recently sold. The data and summary are shown in the table.

No Fireplace	Fireplace
142,212	134,865
206,512	118,007
50,709	138,297
108,794	129,470
68,353	309,808
123,266	157,946
80,248	173,723
135,708	140,510
122,221	151,917
128,440	235,105,000
221,925	259,999
65,325	211,517
87,588	102,068
88,207	115,659
148,246	145,583
205,073	116,289
185,323	238,792
71,904	310,696
199,684	139,079
81,762	109,578
45,004	89,893
62,105	132,311
79,893	131,411
88,770	158,863
115,312	130,490
118,952	178,767
	82,556
	122,221
	84,291
	206,512
	105,363
	103,508
	157,513
	103,861
Mean 116,597.54	**7,061,657.74**
Median 112,053	**136,581**

Write a report summarizing the findings of the investigation. Include appropriate visual and verbal displays of the distributions, and make a recommendation to the agent about the average premium that a fireplace is worth in this market.

41. Customer database. A philanthropic organization has a database of millions of donors that they contact by mail to raise money for charities. One of the variables in the database, *Title*, contains the title of the person or persons printed on the address label. The most common are Mr., Ms., Miss, and Mrs., but there are also Ambassador and Mrs., Your Imperial Majesty, and Cardinal, to name a few others. In all there are over 100 different titles, each with a corresponding numeric code. Here are a few of them.

Code	Title
000	MR.
001	MRS.
1002	MR. and MRS.
003	MISS
004	DR.
005	MADAME
006	SERGEANT
009	RABBI
010	PROFESSOR
126	PRINCE
127	PRINCESS
128	CHIEF
129	BARON
130	SHEIK
131	PRINCE AND PRINCESS
132	YOUR IMPERIAL MAJESTY
135	M. ET MME.
210	PROF.
⋮	⋮

An intern who was asked to analyze the organization's fundraising efforts presented these summary statistics for the variable *Title*.

Mean	54.41
StdDev	957.62
Median	1
IQR	2
n	94649

a) What does the mean of 54.41 mean?
b) What are the typical reasons that cause measures of center and spread to be as different as those in this table?
c) Is that why these are so different?

42. **CEOs.** For each CEO, a code is listed that corresponds to the industry of the CEO's company. Here are a few of the codes and the industries to which they correspond.

Industry	Industry Code	Industry	Industry Code
Financial services	1	Energy	12
Food/drink/tobacco	2	Capital goods	14
Health	3	Computers/communications	16
Insurance	4	Entertainment/information	17
Retailing	6	Consumer non-durables	18
Forest products	9	Electric utilities	19
Aerospace/defense	11		

A recently hired investment analyst has been assigned to examine the industries and the compensations of the CEOs. To start the analysis, he produces the following histogram of industry codes.

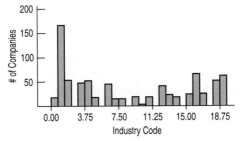

a) What might account for the gaps seen in the histogram?
b) What advice might you give the analyst about the appropriateness of this display?

43. **Mutual funds types.** The 64 mutual funds of Exercise 7 are classified into three types: U.S. Domestic Large Cap Funds, U.S. Domestic Small/Mid Cap Funds, and International Funds. Compare the 3-month return of the three types of funds using an appropriate display and write a brief summary of the differences.

44. **Car discounts, part 3.** The discounts negotiated by the car buyers in Exercise 8 are classified by whether the buyer was Male (code = 0) or Female (code = 1). Compare the discounts of men vs. women using an appropriate display and write a brief summary of the differences.

45. **Houses for sale.** Each house listed on the multiple listing service (MLS) is assigned a sequential ID number. A recently hired real estate agent decided to examine the MLS numbers in a recent random sample of homes for sale by one real estate agency in nearby towns. To begin the analysis, the agent produces the following histogram of ID numbers.

a) What might account for the distribution seen in the histogram?
b) What advice might you give the analyst about the appropriateness of this display?

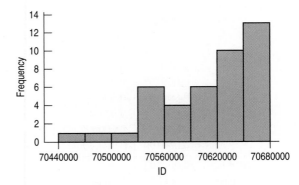

46. **Zip codes.** Holes-R-Us, an Internet company that sells piercing jewelry, keeps transaction records on its sales. At a recent sales meeting, one of the staff presented the following histogram and summary statistics of the zip codes of the last 500 customers, so that the staff might understand where sales are coming from. Comment on the usefulness and appropriateness of this display.

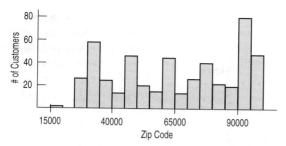

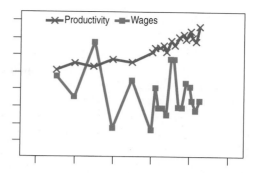

47. Hurricanes. Buying insurance for property loss from hurricanes has become increasingly difficult since hurricane Katrina caused record property loss damage. Many companies have refused to renew policies or write new ones. The data set provided contains the total number of hurricanes by every full decade from 1851 to 2000 (from the National Hurricane Center). Some scientists claim that there has been an increase in the number of hurricanes in recent years.

a) Create a histogram of these data.
b) Describe the distribution.
c) Create a time series plot of these data.
d) Discuss the time series plot. Does this graph support the claim of these scientists, at least up to the year 2000?

48. Hurricanes, part 2. Using the hurricanes data set, examine the number of major hurricanes (category 3, 4, or 5) by every full decade from 1851 to 2000.

a) Create a histogram of these data.
b) Describe the distribution.
c) Create a timeplot of these data.
d) Discuss the timeplot. Does this graph support the claim of scientists that the number of major hurricanes has been increasing (at least up through the year 2000)?

49. Productivity study. The National Center for Productivity releases information on the efficiency of workers. In a recent report, they included the following graph showing a rapid rise in productivity. What questions do you have about this?

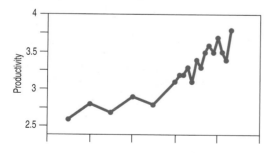

50. Productivity study revisited. A second report by the National Center for Productivity analyzed the relationship between productivity and wages. Comment on the graph they used.

51. Assets. Here is a histogram of the assets (in millions of dollars) of 79 companies chosen from the *Forbes* list of the nation's top corporations.

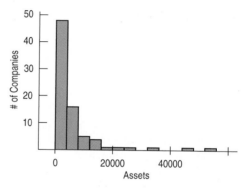

a) What aspect of this distribution makes it difficult to summarize, or to discuss, center and spread?
b) What would you suggest doing with these data if we want to understand them better?

52. Assets, again. Here are the same data you saw in Exercise 51 after re-expressions as the square root of assets and the logarithm of assets.

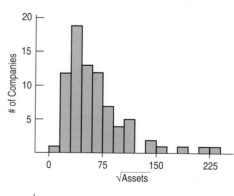

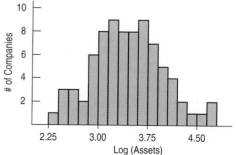

a) Which re-expression do you prefer? Why?

b) In the square root re-expression, what does the value 50 actually indicate about the company's assets?

53. Real estate, part 2. The 1057 houses described in Exercise 24 have a mean price of $167,900, with a standard deviation of $77,158. The mean living area is 1819 sq. ft., with a standard deviation of 663 sq. ft. Which is more unusual, a house in that market that sells for $400,000 or a house that has 4000 sq. ft of living area? Explain.

T 54. Tuition. The data set provided contains the average tuition of private four-year colleges and universities as well as the average 2007–2008 tuitions for each state seen in Exercise 3. The mean tuition charged by a public two-year college was $2763, with a standard deviation of $988. For private four-year colleges the mean was $21,259, with a standard deviation of $6241. Which would be more unusual: a state whose average public two-year college is $700 or a state whose average private four-year college tuition was $10,000? Explain.

T 55. Food consumption. FAOSTAT, the Food and Agriculture Organization of the United Nations, collects information on the production and consumption of more than 200 food and agricultural products for 200 countries around the world. Here are two tables, one for meat consumption (per capita in kg per year) and one for alcohol consumption (per capita in gallons per year). The United States leads in meat consumption with 267.30 pounds, while Ireland is the largest alcohol consumer at 55.80 gallons.

Using z-scores, find which country is the larger consumer of both meat and alcohol together.

Country	Alcohol	Meat	Country	Alcohol	Meat
Australia	29.56	242.22	Luxembourg	34.32	197.34
Austria	40.46	242.22	Mexico	13.52	126.50
Belgium	34.32	197.34	Netherlands	23.87	201.08
Canada	26.62	219.56	New Zealand	25.22	228.58
Czech Republic	43.81	166.98	Norway	17.58	129.80
Denmark	40.59	256.96	Poland	20.70	155.10
Finland	25.01	146.08	Portugal	33.02	194.92
France	24.88	225.28	Slovakia	26.49	121.88
Germany	37.44	182.82	South Korea	17.60	93.06
Greece	17.68	201.30	Spain	28.05	259.82
Hungary	29.25	179.52	Sweden	20.07	155.32
Iceland	15.94	178.20	Switzerland	25.32	159.72
Ireland	55.80	194.26	Turkey	3.28	42.68
Italy	21.68	200.64	United Kingdom	30.32	171.16
Japan	14.59	93.28	United States	26.36	267.30

56. World Bank. The World Bank, through their Doing Business project (www.doingbusiness.org), ranks nearly 200 economies on the ease of doing business. One of their rankings measures the ease of starting a business and is made up (in part)

of the following variables: number of required start-up procedures, average startup time (in days), and average start-up cost (in % of per capita income). The following table gives the mean and standard deviations of these variables for 95 economies.

	Procedures (#)	Time (Days)	Cost (%)
Mean	7.9	27.9	14.2
SD	2.9	19.6	12.9

Here are the data for three countries.

	Procedures	Time	Cost
Spain	10	47	15.1
Guatemala	11	26	47.3
Fiji	8	46	25.3

a) Use z-scores to combine the three measures.

b) Which country has the best environment after combining the three measures? Be careful—a lower rank indicates a better environment to start up a business.

T 57. Gasoline prices. The data set provided contains U.S. regular retail gasoline prices (cents/gallon) from August 20, 1990 to May 28, 2007, from a national sample of gasoline stations obtained from the U.S. Department of Energy.

a) Create a histogram of the data and describe the distribution.

b) Create a time series plot of the data and describe the trend.

c) Which graphical display seems the more appropriate for these data? Explain.

T 58. House prices. Standard and Poor's Case-Shiller® Home Price Index measures the residential housing market in 20 metropolitan regions across the United States. The national index is a composite of the 20 regions and can be found in the data set provided.

a) Create a histogram of the data and describe the distribution.

b) Create a time series plot of the data and describe the trend.

c) Which graphical display seems the more appropriate for these data? Explain.

59. Unemployment rate. The histogram shows the monthly U.S. unemployment rate from June 1995 to June 2004.

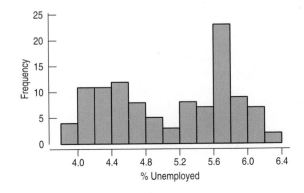

Here is the time series plot for the same data.

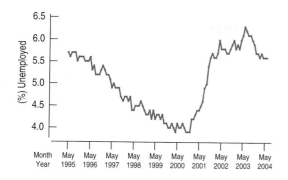

a) What features of the data can you see in the histogram that aren't clear in the time series plot?
b) What features of the data can you see in the time series plot that aren't clear in the histogram?
c) Which graphical display seems the more appropriate for these data? Explain.
d) Write a brief description of unemployment rates over this time period in the United States.

60. Mutual fund performance. The following histogram displays the monthly returns for a group of mutual funds considered aggressive (or high growth) over a period of 22 years from 1975 to 1997.

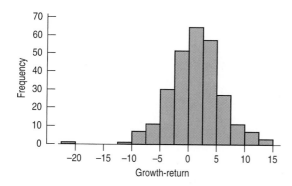

Here is the time series plot for the same data.

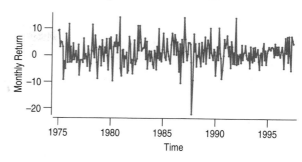

a) What features of the data can you see in the histogram that aren't clear from the time series plot?
b) What features of the data can you see in the time series plot that aren't clear in the histogram?
c) Which graphical display seems the more appropriate for these data? Explain.
d) Write a brief description of unemployment rates over this time period in the United States.

 JUST CHECKING ANSWERS

1 Incomes are probably skewed to the right and not symmetric, making the median the more appropriate measure of center. The mean will be influenced by the high end of family incomes and not reflect the "typical" family income as well as the median would. It will give the impression that the typical income is higher than it is.

2 An IQR of 30 mpg would mean that only 50% of the cars get gas mileages in an interval 30 mpg wide. Fuel economy doesn't vary that much. 3 mpg is reasonable. It seems plausible that 50% of the cars will be within about 3 mpg of each other. An IQR of 0.3 mpg would mean that the gas mileage of half the cars varies little from the estimate. It's unlikely that cars, drivers, and driving conditions are that consistent.

3 We'd prefer a standard deviation of 2 months. Making a consistent product is important for quality. Customers want to be able to count on the MP3 player lasting somewhere close to 5 years, and a standard deviation of 2 years would mean that lifespans were highly variable.

PART II

Understanding Data and Distributions

Scatterplots, Association, and Correlation

CHAPTER 7

The Home Depot

Handy Dan was a successful chain of home improvement stores in the 1970s, thanks in part to the efforts of two of its executives, Bernie Marcus and Arthur Blank. However, on a spring day in 1978, after an argument with their boss, the two were summarily fired. Rather than search the *Help Wanted* section, they formed MB Associates. Their idea was to open a chain of warehouse-style home improvement stores. After raising some venture capital, they managed to open three stores in their first year, hiring 200 sales associates and generating $7 million in sales. They changed the name of the stores to The Home Depot. Five years later they had 10 times as many stores, 20 times as many associates, and over 60 times the sales. The extraordinary growth continued throughout the 1980s and 1990s. The Home Depot reached the milestones of $30, $40, $50, and $60 billion in sales, faster than any other retailer in history. By 2005, it was the second largest retailer in the United States, behind only Wal-Mart, and the third largest retailer in the world behind Wal-Mart and the French supermarket giant Carrefour.

In 2000, the original founders, Marcus and Blank, retired to pursue philanthropic activities. Robert (Bob) L. Nardelli, who had spent almost 30 years at General Electric, where he had risen to president and CEO of GE Power Systems, became chairman, president, and CEO of The Home Depot. After Nardelli took over, the company continued to grow, nearly doubling again in those first five years, but the stock price languished. In September 2006, Nardelli was reelected as chairman at the annual stockholders' meeting, but nearly a third of the voting shareholders withheld their support, citing disappointment with the stock price and questioning his compensation, which over a five-year period totaled $123.7 million. Nardelli abruptly announced his departure on January 3, 2007, saying that it was a "mutual decision," and left with a severance package estimated to be worth over $200 million. Frank Blake, who joined The Home Depot in 2002 as executive vice president, became CEO following Nardelli's departure.

WHO	Economic Quarters
WHAT	Seasonally Adjusted Quarterly *Housing Starts* in the United States
UNITS	Thousands of units
WHEN	1995–2004
WHERE	United States
WHY	To analyze trends in home building

The decade from 1995 to 2005 saw extraordinary growth for The Home Depot. Naturally, The Home Depot is interested in trends in the housing market. What has been the trend in housing starts? Here is a plot showing the number of new *Housing Starts* (seasonally adjusted by quarter, in thousands of units) plotted by *Quarter* for that decade. If you were asked to summarize the trend in housing starts over this decade what would you say?

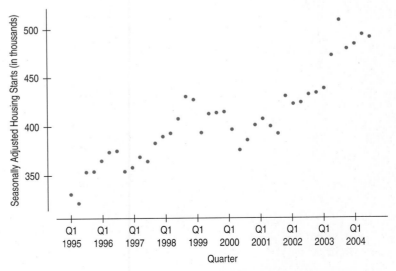

Figure 7.1 *Seasonally Adjusted Quarterly Housing Starts (in thousands) from 1995 to 2005.*

Clearly, U.S. housing starts grew between 1995 and 2005, starting at below 350,000 per quarter and ending at close to 500,000. The plot shows fairly steady growth from 1995 to the end of 1999 when starts decreased and then picked up again from 2001 onward.

This time series plot is an example of a more general kind of display called a scatterplot. A **scatterplot**, which plots one quantitative variable against another, can be an effective display for data. Whenever you want to understand the relationship between two quantitative variables, you should make a scatterplot. Just by looking at a scatterplot, you can see patterns, trends, relationships, and even the occasional unusual values standing apart from the others. Scatterplots are the best way to start observing the relationship between two *quantitative* variables.

Relationships between variables are often at the heart of what we'd like to learn from data.

◆ Is consumer confidence related to oil prices?

◆ What happens to customer satisfaction as sales increase?

◆ Is an increase in money spent on advertising related to sales?

◆ What is the relationship between a stock's sales volume and its price?

Questions such as these relate two quantitative variables and ask whether there is an **association** between them. Scatterplots are the ideal way to *picture* such associations.

7.1 Looking at Scatterplots

The Texas Transportation Institute, which studies the mobility provided by the nation's transportation system, issues an annual report on traffic congestion and its costs to society and business. Figure 7.2 shows a scatterplot of the annual *Congestion Cost* Per Person of traffic delays (in dollars) in 65 cities in the United States against the Peak Period *Freeway Speed* (mph).

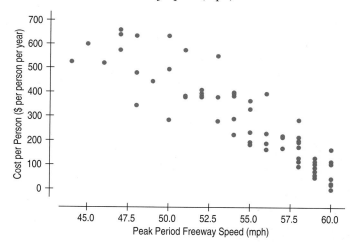

> **Figure 7.2** *Congestion Cost Per Person ($ per year) of traffic delays against Peak Period Freeway Speed (mph) for 65 U.S. cities.*

Everyone looks at scatterplots. But, if asked, many people would find it hard to say what to look for in a scatterplot. What do *you* see? Try to describe the scatterplot of *Congestion Cost* against *Freeway Speed*.

You might say that the **direction** of the association is important. As the peak freeway speed goes up, the cost of congestion goes down. A pattern that runs from the upper left to the lower right is said to be **negative**. A pattern running the other way is called **positive**.

WHO	Cities in the United States
WHAT	*Congestion Cost* Per Person and Peak Period *Freeway Speed*
UNITS	*Congestion Cost* Per Person ($ per person per year); Peak Period *Freeway Speed* (mph)
WHEN	2000
WHERE	Across the United States
WHY	To examine the relationship between congestion on the highways and its impact on society and business

Look for **Direction**: What's the sign—positive, negative, or neither?

The second thing to look for in a scatterplot is its **form**. If there is a straight line relationship, it will appear as a cloud or swarm of points stretched out in a generally consistent, straight form. For example, the scatterplot of traffic congestion has an underlying **linear** form, although some points stray away from it.

Scatterplots can reveal many different kinds of patterns. Often they will not be straight, but straight line patterns are both the most common and the most useful for statistics.

If the relationship isn't straight, but curves gently, while still increasing or decreasing steadily, we can often find ways to straighten it out. But if

> Look for **Form**: Straight, curved, something exotic, or no pattern?

it curves sharply—up and then down, for example,   —then you'll need more advanced methods.

The third feature to look for in a scatterplot is the **strength** of the relationship.

At one extreme, do the points appear tightly clustered in a single stream (whether straight, curved, or bending all over the place)? Or, at the other extreme, do the points seem to be so variable and spread out that we can barely discern any

trend or pattern?   The traffic congestion plot shows moderate scatter around a generally straight form. That indicates that there's a moderately strong linear relationship between cost and speed.

> Look for **Strength**: How much scatter?

Finally, always look for the unexpected. Often the most interesting discovery in a scatterplot is something you never thought to look for. One example of such a surprise is an unusual observation, or **outlier**, standing away from the overall pattern of the scatterplot. Such a point is almost always interesting and deserves special attention. You may see entire clusters or subgroups that stand away or show a trend in a different direction than the rest of the plot. That should raise questions about why they are different. They may be a clue that you should split the data into subgroups instead of looking at them all together.

> Look for **Unusual Features**: Are there unusual observations or subgroups?

7.2 Assigning Roles to Variables in Scatterplots

Scatterplots were among the first modern mathematical displays. The idea of using two axes at right angles to define a field on which to display values can be traced back to René Descartes (1596–1650), and the playing field he defined in this way is formally called a *Cartesian plane*, in his honor.

The two axes Descartes specified characterize the scatterplot. The axis that runs up and down is, by convention, called the *y*-axis, and the one that runs from side to side is called the *x*-axis. These terms are standard. If someone refers to the *y*-axis, you may be sure they mean the vertical, up-and-down axis, and similarly with the *x*-axis.[1]

Descartes was a philosopher, famous for his statement cogito ergo sum: I think, therefore I am.

---

[1] The axes are also called the "ordinate" and the "abscissa"—but we can never remember which is which because statisticians don't generally use these terms. In Statistics (and in all statistics computer programs) the axes are generally called "*x*" (abscissa) and "*y*" (ordinate) and are usually labeled with the names of the corresponding variables.

To make a scatterplot of two quantitative variables, assign one to the *y*-axis and the other to the *x*-axis. As with any graph, be sure to label the axes clearly, and indicate the scales of the axes with numbers. Scatterplots display *quantitative* variables. Each variable has units, and these should appear with the display—usually near each axis. Each point is placed on a scatterplot at a position that corresponds to values of these two variables. Its horizontal location is specified by its *x*-value, and its vertical location is specified by its *y*-value variable. Together, these are known as *coordinates* and written (*x*, *y*).

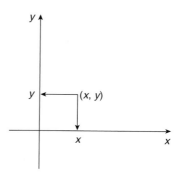

Scatterplots made by computer programs (such as the two we've seen in this chapter) often do not—and usually should not—show the *origin*, the point at $x = 0$, $y = 0$ where the axes meet. If both variables have values near or on both sides of zero, then the origin will be part of the display. If the values are far from zero, though, there's no reason to include the origin. In fact, it's far better to focus on the part of the Cartesian plane that contains the data. In our example about freeways, none of the speeds was anywhere near 0 mph, so the computer drew the scatterplot in Figure 7.2 with axes that don't quite meet.

Which variable should go on the *x*-axis and which on the *y*-axis? What we want to know about the relationship can tell us how to make the plot. We often have questions such as:

♦ Is The Home Depot employee satisfaction related to productivity?

♦ Will increased sales at The Home Depot be reflected in the stock price?

♦ What other economic factors besides housing starts are related to The Home Depot sales?

In all of these examples, one variable plays the role of the **explanatory** or **predictor variable**, while the other takes on the role of the **response variable**. We place the explanatory variable on the *x*-axis and the response variable on the *y*-axis. When you make a scatterplot, you can assume that those who view it will think this way, so choose which variables to assign to which axes carefully.

The roles that we choose for variables have more to do with how we *think* about them than with the variables themselves. Just placing a variable on the *x*-axis doesn't necessarily mean that it explains or predicts *anything*, and the variable on the *y*-axis may not respond to it in any way. We plotted *Congestion Cost* Per Person against peak *Freeway Speed*, thinking that the slower traffic moves, the more it costs in delays. But maybe *spending* $500 per person in freeway improvement would increase speed. If we were examining that option, we might choose to plot *Congestion Cost* Per Person as the explanatory variable and *Freeway Speed* as the response.

The *x*- and *y*-variables are sometimes referred to as the **independent** and **dependent** variables, respectively. The idea is that the *y*-variable *depends* on the

NOTATION ALERT:

So *x* and *y* are reserved letters as well, but not just for labeling the axes of a scatterplot. In Statistics, the assignment of variables to the *x*- and *y*-axes (and choice of notation for them in formulas) often conveys information about their roles as predictor or response.

WHO	Economic Quarters
WHAT	Quarterly *Sales* at The Home Depot and the (unadjusted) Quarterly U.S. *Housing Starts*
UNITS	*Sales* ($ billion) and *Housing Starts* (thousands)
WHEN	May 1995–June 2004
WHERE	United States
WHY	To examine association between sales and housing starts

x-variable and the *x*-variable acts *independently* to make *y* respond. These names, however, conflict with other uses of the same terms in Statistics. Instead, we'll sometimes use the terms "explanatory" or "predictor variable" and "response variable" when we're discussing roles, but we'll often just say *x-variable* and *y-variable*.

7.3 Understanding Correlation

Generally, a stronger economy accompanies greater consumer spending. Does this also apply to the home improvement industry? We've seen that during the period 1995 to 2005 seasonally adjusted housing starts increased dramatically (Figure 7.1). Let's examine a scatterplot of the unadjusted data to see if there's an association between the growth in *Housing Starts* and *Sales* at The Home Depot. It should be no great surprise to discover that there is a positive association between the two. As you might suspect, the greater the number of new houses built, the higher the sales at Home Depot.

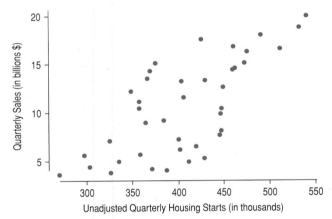

Figure 7.3   *The Home Depot quarterly sales ($ billion) and U.S. Housing Starts (thousands) from 1995 to 2004.*

There is clearly a positive association, and the scatterplot looks reasonably straight, but how strong is the association? If you had to put a number (say, between 0 and 1) on the strength, what would it be? Your measure shouldn't depend on the choice of units for the variables. After all, if sales had been recorded in euros instead of dollars or housing starts in millions of units instead of thousands, the scatterplot would look the same. The direction, form, and strength won't change, so neither should our measure of the association's strength.

Since the units don't matter, why not just remove them altogether? One way we can do that is by standardizing both variables, turning the coordinates of each point into a pair of *z*-scores, z_x and z_y. Remember from Chapter 6 that to standardize values we subtract the mean of each variable and then divide by its standard deviation:

$$(z_x, z_y) = \left(\frac{x - \bar{x}}{s_x}, \frac{y - \bar{y}}{s_y} \right).$$

The resulting scatterplot does look roughly the same (if you don't read the axis labels).

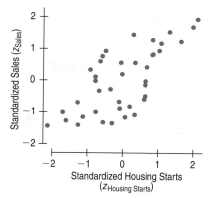

Figure 7.4 *Standardized Quarterly Sales (z_{Sales}) against Standardized Housing Starts ($z_{Housing\ Starts}$).*

Because standardizing makes the means of both variables 0, the center of the new scatterplot is now at the origin, and the scales on both axes are now in standard deviation units.

But wait. The two plots aren't *exactly* the same. Can you see the differences? For one thing, the underlying linear pattern seems steeper in the standardized plot. That's because standardizing makes the scales of the axes the same. Now it's natural to make the length of one standard deviation the same vertically and horizontally. When we worked in the original units, we were free to make the plot as tall and thin

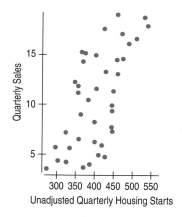

or as squat and wide

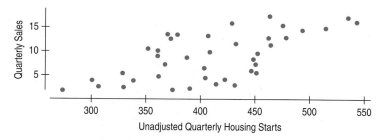

as we liked. Typically, we make the *x*-axis longer than the *y*-axis for aesthetic reasons,[2] but equal scaling gives a neutral way of drawing the scatterplot and a fairer impression of the strength of the association.

---

[2] The aesthetic choice for the ratio between the lengths of the two axes has a long history and is related to the golden ratio of the ancient Greeks.

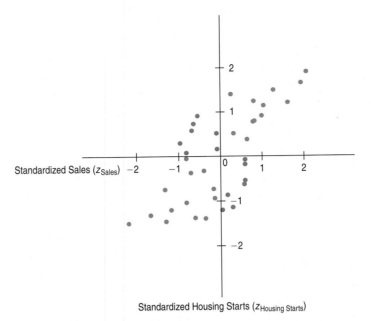

Standardized Sales (z_{Sales})

Standardized Housing Starts ($z_{Housing\ Starts}$)

Figure 7.5 *In this scatterplot of z-scores, points are colored by how they affect the association: green for positive and red for negative.*

Since standardizing the variables doesn't change the strength of the association, we can work with the z-scores to understand how to measure the strength. Which points in the scatterplot of the z-scores (Figure 7.5) give the impression of a positive association? In a positive association, y tends to increase as x increases. So, the points in the upper right and lower left of Figure 7.5 strengthen that impression. For these points, z_x and z_y have the same sign. If we multiplied them together, the product, $z_x z_y$, would be positive. Points far from the origin (which make the association look more positive) have a bigger product. The points in the upper left and lower right quadrants of Figure 7.5 tend to weaken the positive association (or support a negative association). For these points, z_x and z_y have opposite signs, so the product $z_x z_y$ for these points is negative. Points far from the origin (which make the association look more negative) have a more negative product.

Points with z-scores of zero on either variable don't vote either way because $z_x z_y = 0$. These points would be found on the axes in Figure 7.5. In order to turn these products into a measure of the strength of the association, we can just add up the $z_x z_y$ products for every point in the scatterplot.

$$\sum z_x z_y$$

This summarizes the direction *and* strength of the association for all the points. If most of the points are in the quadrants where the z-scores have the same signs, the sum will be positive. If most are in the quadrants where the z-scores have opposite signs, it will be negative.

But the *size* of this sum gets bigger the more data we have. To adjust for this we divide the sum by $n - 1$.[3] This ratio is called the **correlation coefficient**:

$$r = \frac{\sum z_x z_y}{n - 1}.$$

---

[3] Yes, the same $n - 1$ we used for calculating the standard deviation, which we will explain in later chapters.

For The Home Depot Quarterly *Sales* and *Housing Starts*, the correlation coefficient is 0.70.

There are alternative formulas for the correlation in terms of the variables x and y. Here are two of the more common:

$$r = \frac{\sum (x - \bar{x})(y - \bar{y})}{\sqrt{\sum (x - \bar{x})^2 \sum (y - \bar{y})^2}} = \frac{\sum (x - \bar{x})(y - \bar{y})}{(n - 1)s_x s_y}.$$

These formulas can be more convenient for calculating correlation by hand, but the form given using z-scores is best for understanding what correlation means. If you want to see how to go from the formula using z-scores to the other two formulas, look in the Math Box for details.

MATH BOX

Standardizing the variables first gives us an easy to understand expression for the correlation.

$$r = \frac{\sum z_x z_y}{n - 1}$$

But sometimes you'll see other formulas. Remembering how standardizing works gets us from one formula to the other.

Since

$$z_x = \frac{x - \bar{x}}{s_x}$$

and

$$z_y = \frac{y - \bar{y}}{s_y},$$

we can substitute these and get

$$r = \left(\frac{1}{n - 1}\right) \sum z_x z_y = \left(\frac{1}{n - 1}\right) \sum \frac{(x - \bar{x})}{s_x} \frac{(y - \bar{y})}{s_y} = \sum \frac{(x - \bar{x})(y - \bar{y})}{(n - 1)s_x s_y}.$$

That's one version. And since we know the formula for standard deviation,

$$s_y = \sqrt{\frac{\sum (y - \bar{y})^2}{n - 1}},$$

we could use substitution to write:

$$r = \left(\frac{1}{n - 1}\right) \sum \frac{(x - \bar{x})}{s_x} \frac{(y - \bar{y})}{s_y}$$

$$= \left(\frac{1}{n - 1}\right) \frac{\sum (x - \bar{x})(y - \bar{y})}{\sqrt{\dfrac{\sum (x - \bar{x})^2}{n - 1}} \sqrt{\dfrac{\sum (y - \bar{y})^2}{n - 1}}}$$

$$= \left(\frac{1}{n - 1}\right) \frac{\sum (x - \bar{x})(y - \bar{y})}{\left(\dfrac{1}{n - 1}\right) \sqrt{\sum (x - \bar{x})^2} \sqrt{\sum (y - \bar{y})^2}}$$

$$= \frac{\sum (x - \bar{x})(y - \bar{y})}{\sqrt{\sum (x - \bar{x})^2 \sum (y - \bar{y})^2}}.$$

That's the other common version. If you ever have to compute the correlation by hand, it's easier to start with one of these. But if you want to remember how correlation works, stick with the first formula in this Math Box.

Finding the correlation coefficient by hand

To find the correlation coefficient by hand, we'll use a formula in original units, rather than z-scores. This will save us the work of having to standardize each individual data value first. Start with the summary statistics for both variables: $\bar{x}$, $\bar{y}$, s_x, and s_y. Then find the deviations as we did for the standard deviation, but now in both x and y: $(x - \bar{x})$ and $(y - \bar{y})$. For each data pair, multiply these deviations together: $(x - \bar{x}) \times (y - \bar{y})$. Add the products up for all data pairs. Finally, divide the sum by the product of $(n - 1) \times s_x \times s_y$ to get the correlation coefficient.

Here we go.

Suppose the data pairs are:

x	6	10	14	19	21
y	5	3	7	8	12

Then $\bar{x} = 14$, $\bar{y} = 7$, $s_x = 6.20$, and $s_y = 3.39$

Deviations in x	Deviations in y	Product
$6 - 14 = -8$	$5 - 7 = -2$	$-8 \times -2 = 16$
$10 - 14 = -4$	$3 - 7 = -4$	16
$14 - 14 = 0$	$7 - 7 = 0$	0
$19 - 14 = 5$	$8 - 1 = 1$	5
$21 - 14 = 7$	$12 - 7 = 5$	35

Add up the products: $16 + 16 + 0 + 5 + 35 = 72$
Finally, we divide by $(n - 1) \times s_x \times s_y = (5 - 1) \times 6.20 \times 3.39 = 84.07$
The ratio is the correlation coefficient:

$$r = 72/84.07 = 0.856$$

Correlation Conditions

Correlation measures the strength of the *linear* association between two *quantitative* variables. Before you use correlation, you must check three *conditions*:

◆ **Quantitative Variables Condition:** Correlation applies only to quantitative variables. Don't apply correlation to categorical data masquerading as quantitative. Check that you know the variables' units and what they measure.

◆ **Linearity Condition:** Sure, you can *calculate* a correlation coefficient for any pair of variables. But correlation measures the strength only of the *linear* association and will be misleading if the relationship is not straight enough. What is "straight enough"? This question may sound too informal for a statistical condition, but that's really the point. We can't verify whether a relationship is linear or not. Very few relationships between variables are perfectly linear, even in theory, and scatterplots of real data are never perfectly straight. How nonlinear looking would the scatterplot have to be to fail the condition? This is a judgment call that you just have to think about. Do you think that the underlying relationship is curved? If so, then summarizing its strength with a correlation would be misleading.

◆ **Outlier Condition:** Unusual observations can distort the correlation and can make an otherwise small correlation look big or, on the other hand, hide a large correlation. It can even give an otherwise positive association a negative correlation coefficient (and vice versa). When you see an outlier, it's often a good idea to report the correlation both with and without the point.

Each of these conditions is easy to check with a scatterplot. Many correlations are reported without supporting data or plots. You should still think about the conditions. You should be cautious in interpreting (or accepting others' interpretations of) the correlation when you can't check the conditions for yourself.

✓ JUST CHECKING

For the years 1992 to 2002, the quarterly stock price of the semiconductor companies Cypress and Intel have a correlation of 0.86.

1 Before drawing any conclusions from the correlation, what would you like to see? Why?

2 If your coworker tracks the same prices in euros, how will this change the correlation? Will you need to know the exchange rate between euros and U.S. dollars to draw conclusions?

3 If you standardize both prices, how will this affect the correlation?

4 In general, if on a given day the price of Intel is relatively low, is the price of Cypress likely to be relatively low as well?

5 If on a given day the price of Intel stock is high, is the price of Cypress stock definitely high as well?

GUIDED EXAMPLE

Customer Spending

A major credit card company sends an incentive to its best customers in hope that the customers will use the card more. They wonder how often they can offer the incentive. Will repeated offerings of the incentive result in repeated increased credit card use? To examine this question, an analyst took a random sample of 184 customers from their highest use segment and investigated the charges in the two months in which the customers had received the incentive.

PLAN

Setup State the objective. Identify the quantitative variables to examine. Report the time frame over which the data have been collected and define each variable. (State the W's.)

Our objective is to investigate the association between the amount that a customer charges in the two months in which they received an incentive. The customers have been randomly selected from among the highest use segment of customers. The variables measured are the total credit card charges (in $) in the two months of interest.

✓ **Quantitative Variable Condition.** Both variables are quantitative. Both charges are measured in dollars.

Make the scatterplot and clearly label the axes to identify the scale and units.

Because we have two quantitative variables measured on the same cases, we can make a scatterplot.

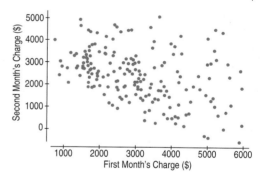

Check the conditions.

✓ **Linearity Condition.** The scatterplot is straight enough.

✓ **Outlier Condition.** There are no obvious outliers.

DO

Mechanics Once the conditions are satisfied, calculate the correlation with technology.

The correlation is −0.391.

The negative correlation coefficient confirms the impression from the scatterplot.

REPORT

Conclusion Describe the direction, form, and the strength of the plot, along with any unusual points or features. Be sure to state your interpretation in the proper context.

MEMO:

Re: Credit Card Spending

We have examined some of the data from the incentive program. In particular, we looked at the charges made in the first two months of the program. We noted that there was a negative association between charges in the second month and charges in the first month. The correlation was −0.391, which is only moderately strong, and indicates substantial variation.

We've concluded that while the observed pattern is negative, these data do not allow us to find the causes of this behavior. It is likely that some customers were encouraged by the offer to increase their spending in the first month, but then returned to former spending patterns. It is possible that others didn't change their behavior until the second month of the program, increasing their spending at that time. Without data on the customers' pre-incentive spending patterns it would be hard to say more.

We suggest further research, and we suggest that the next trial extend for a longer period of time to help determine whether the patterns seen here persist.

Correlation Properties

Because correlation is so widely used as a measure of association it's a good idea to remember some of its basic properties. Here's a useful list of facts about the correlation coefficient:

- **The sign of a correlation coefficient gives the direction of the association.**
- **Correlation is always between −1 and +1.** Correlation *can* be exactly equal to −1.0 or +1.0, but watch out. These values are unusual in real data because they mean that all the data points fall *exactly* on a single straight line.
- **Correlation treats *x* and *y* symmetrically.** The correlation of *x* with *y* is the same as the correlation of *y* with *x*.
- **Correlation has no units.** This fact can be especially important when the data's units are somewhat vague to begin with (customer satisfaction, worker efficiency, productivity, and so on).
- **Correlation is not affected by changes in the center or scale of either variable.** Changing the units or baseline of either variable has no effect on the correlation coefficient because the correlation depends only on the *z*-scores.
- **Correlation measures the strength of the *linear* association between the two variables.** Variables can be strongly associated but still have a small correlation if the association is not linear.
- **Correlation is sensitive to unusual observations.** A single outlier can make a small correlation large or make a large one small.

How strong is strong? Be careful when using the terms "weak," "moderate," or "strong," because there's no agreement on exactly what those terms mean. The same numerical correlation might be strong in one context and weak in another. You might be thrilled to discover a correlation of 0.7 between an economic index and stock market prices, but finding "only" a correlation of 0.7 between a drug treatment and blood pressure might be viewed as a failure by a pharmaceutical company. Using general terms like "weak," "moderate," or "strong" to describe a linear association can be useful, but be sure to report the correlation and show a scatterplot so others can judge for themselves.

Correlation Tables

Sometimes you'll see the correlations between each pair of variables in a data set arranged in a table. The rows and columns of the table name the variables, and the cells hold the correlations.

	Closing Price	Interest Rate	Unemployment Rate
Closing Price	1.000		
Interest Rate	−0.214	1.000	
Unemployment Rate	−0.445	−0.679	1.000

Table 7.1 A correlation table for some other variables measured monthy during the period 1995 to 2005. Closing Price = The Home Depot Stock price at end of month, Interest Rate = Prevailing Bank Prime Interest Rate, and Unemployment Rate is in percent.

Correlation tables are compact and give a lot of summary information at a glance. They can be an efficient way to start to look at a large data set. The diagonal cells of a correlation table always show correlations of exactly 1.000, and the upper half of the table is symmetrically the same as the lower half (can you see why?), so traditionally only the lower half is shown. A table like this can be convenient, but be sure to check for linearity and unusual observations or the correlations in the table may be misleading or meaningless. Can you be sure, looking at Table 7.1, that the variables are linearly associated? Correlation tables are often produced by statistical software packages. Fortunately, these same packages often offer simple ways to make all the scatterplots you need to look at.[4]

*7.4 Straightening Scatterplots

After the Dow Jones Industrial Average, the S&P 500 is the most widely-watched index of U.S. stocks. Since its introduction in 1957, the S&P Index, which is comprised of large publicly held companies, has seen a period of remarkable growth. On January 2, 1957, the S&P 500 stood at 46.2, and it reached a high of 1527.46 on March 24, 2000 (see Figure 7.6).

If you heard that the correlation between *Time* and *S&P 500 Index* was 0.76, you might think there was a strong linear association. But the time series plot of the data tells quite a different story. The growth was relatively modest until around 1980, when the index started growing at a much faster rate until its peak in March 2000. (It is also interesting to see that the "crash" of 1987 now appears as just a small blip in the overall growth.)

---

[4] A table of scatterplots arranged just like a correlation table is sometimes called a *scatterplot matrix*, or SPLOM, and is easily created using a statistics package.

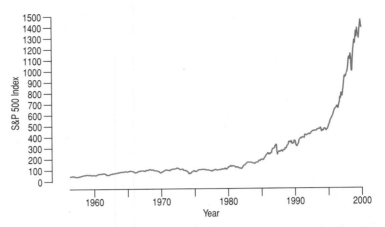

| **Figure 7.6** *A time series plot of the S&P 500 Index shows a curved relationship.*

Remember, correlation measures only the strength of a "linear" association. In the time series plot of the *S&P 500 Index*, it's clear that the Index is not increasing linearly. What if we look at the *logarithm* of the *S&P 500* over *Time*?

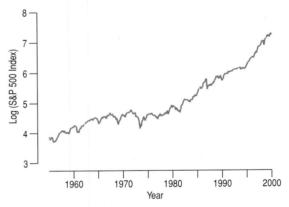

Figure 7.7   *Re-expressing the S&P 500 by finding the logarithm helps to straighten the plot. Because the plot is straighter, the correlation is now a more appropriate measure of association.*

This plot looks straighter, so the correlation is now a more appropriate measure of association. Another benefit of this plot is that the periods of different growth are clearer. What appeared to be a period of low growth in the early 1960s is now revealed as a period of typical growth—something that was hidden in the original plot because of the enormous increases later. The period from about 1970 to 1980, which suffered from both high unemployment and high inflation, shows little growth. Finally, the bull market that lasted from the early 1980s to its peak in March 2000 shows nearly steady growth on the log scale. The "crash" of 1987 now appears even less significant when viewed in the context of this extraordinary 20-year run. Indices like the *S&P 500* are often plotted on a log scale in order to make it easier to see what's going on.

Simple transformations such as the logarithm, square root, or reciprocal can sometimes straighten a scatterplot's form. Future chapters discuss simple ways to find a good re-expression of data.

7.5 Lurking Variables and Causation

An educational researcher finds a strong association between height and reading ability among elementary school students in a nationwide survey. Taller children tend to have higher reading scores. Does that mean that students' heights *causes* their reading scores to go up? No matter how strong the correlation is between two variables, there's no simple way to show from observational data that one variable causes the other. A high correlation just increases the temptation to think and to say that the *x*-variable *causes* the *y*-variable. Just to make sure, let's repeat the point again.

No matter how strong the association, no matter how large the *r* value, no matter how straight the form, there is no way to conclude from a high correlation *alone* that one variable causes the other. There's always the possibility that some third variable—a **lurking variable**—is affecting both of the variables you have observed. In the reading score example, you may already have guessed that the lurking variable is the age of the child. Older children tend to be taller and have stronger reading skills. But even when the lurking variable isn't as obvious, resist the temptation to think that a high correlation implies causation. Here's another example.

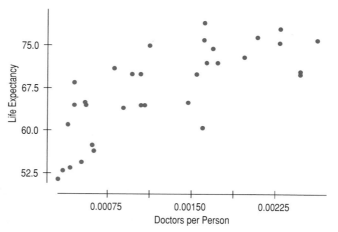

Figure 7.8 *Life Expectancy and numbers of Doctors Per Person in 34 countries shows a fairly strong, positive linear relationship with a correlation of 0.74.*

The scatterplot shows the *Life Expectancy* (average of men and women, in years) for each of 34 countries of the world, plotted against the number of *Doctors per Person* in each country. The strong positive association ($r = 0.74$) seems to confirm our expectation that more *Doctors per Person* improves health care, leading to longer lifetimes and a higher *Life Expectancy*. Perhaps we should send more doctors to developing countries to increase life expectancy.

If we increase the number of doctors, will the life expectancy increase? That is, would adding more doctors *cause* greater life expectancy? Could there be another explanation of the association? Here's another scatterplot. *Life Expectancy* is still the response, but this time the predictor variable is not the number of doctors, but the number of *Televisions per Person* in each country (see Figure 7.9). The positive association in this scatterplot looks even *stronger* than the association in the previous plot. If we wanted to calculate a correlation, we should straighten the plot first, but even from this plot, it's clear that higher life expectancies are associated with more televisions per person. Should we conclude that increasing the number of televisions extends lifetimes? If so, we should send televisions instead of doctors to

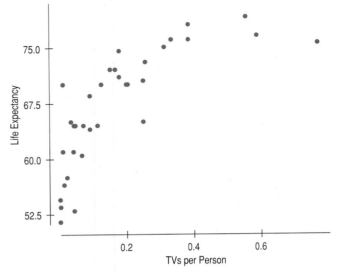

Figure 7.9 *Life Expectancy and number of Televisions per Person shows a strong, positive (although clearly not linear) relationship.*

developing countries. Not only is the association with life expectancy stronger, but televisions are cheaper than doctors. What's wrong with this reasoning? Maybe we were a bit hasty earlier when we concluded that doctors *cause* greater life expectancy. Maybe there's a lurking variable here. Countries with higher standards of living have both longer life expectancies *and* more doctors. Could higher living standards cause changes in the other variables? If so, then improving living standards might be expected to prolong lives, increase the number of doctors, and increase the number of televisions. From this example, you can see how easy it is to fall into the trap of mistakenly inferring causality from a correlation. For all we know, doctors (or televisions) *do* increase life expectancy. But we can't tell that from data like these no matter how much we'd like to. Resist the temptation to conclude that *x* causes *y* from a correlation, no matter how obvious that conclusion seems to you.

WHAT CAN GO WRONG?

- **Don't say "correlation" when you mean "association."** How often have you heard the word "correlation"? Chances are pretty good that when you've heard the term, it's been misused. It's one of the most widely misused Statistics terms, and given how often Statistics are misused, that's saying a lot. One of the problems is that many people use the specific term *correlation* when they really mean the more general term *association*. Association is a deliberately vague term used to describe the relationship between two variables.

  Correlation is a precise term used to describe the strength and direction of a linear relationship between quantitative variables.

- **Don't correlate categorical variables.** Be sure to check the Quantitative Variables Condition. It makes no sense to compute a correlation of categorical variables.

- **Make sure the association is linear.** Not all associations between quantitative variables are linear. Correlation can miss even a strong nonlinear association. A company, concerned that customers might use ovens with imperfect temperature controls, performed a series of experiments[5] to assess the effect of baking temperature on the quality of brownies made from their freeze-dried reconstituted brownies. The company wants to understand the sensitivity of brownie quality to variation in oven temperatures around the recommended baking temperature of 325°F. The lab reported a correlation of −0.05 between the scores awarded by a panel of trained taste-testers and baking temperature and told management that there is no relationship. Before printing directions on the box telling customers not to worry about the temperature, a savvy intern asks to see the scatterplot.

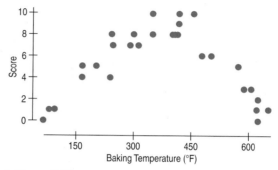

Figure 7.10 *The relationship between brownie taste score and baking temperature is strong, but not linear.*

The plot actually shows a strong association—but not a linear one. Don't forget to check the Linearity Condition.

- **Beware of outliers.** You can't interpret a correlation coefficient safely without a background check for unusual observations. Here's an example. The relationship between IQ and shoe size among comedians shows a surprisingly strong positive correlation of 0.50. To check assumptions, we look at the scatterplot.

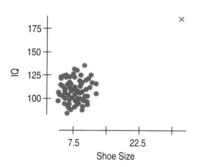

| Figure 7.11 *IQ vs. Shoe Size.*

From this "study," what can we say about the relationship between the two? The correlation is 0.50. But who *does* that point in the upper right-hand corner belong to? The outlier is Bozo the Clown, known for his large shoes and widely acknowledged to be a comic "genius." Without Bozo the correlation is near zero.

(continued)

---

[5] Experiments designed to assess the impact of environmental variables outside the control of the company on the quality of the company's products were advocated by the Japanese quality expert Dr. Genichi Taguchi starting in the 1980s in the United States.

Even a single unusual observation can dominate the correlation value. That's why you need to check the Unusual Observations Condition.

• **Don't confuse correlation with causation.** Once we have a strong correlation, it's tempting to try to explain it by imagining that the predictor variable has *caused* the response to change. Humans are like that; we tend to see causes and effects in everything. Just because two variables are related does not mean that one *causes* the other.

> **Does cancer cause smoking?**   Even if the correlation of two variables is due to a causal relationship, the correlation itself cannot tell us what causes what.
>
> Sir Ronald Aylmer Fisher (1890–1962) was one of the greatest statisticians of the 20th century. Fisher testified in court (paid by the tobacco companies) that a causal relationship might underlie the correlation of smoking and cancer:
>
> "Is it possible, then, that lung cancer . . . is one of the causes of smoking cigarettes? I don't think it can be excluded . . . the pre-cancerous condition is one involving a certain amount of slight chronic inflammation . . .
>
> A slight cause of irritation . . . is commonly accompanied by pulling out a cigarette, and getting a little compensation for life's minor ills in that way. And . . . is not unlikely to be associated with smoking more frequently."
>
> Ironically, the proof that smoking indeed is the cause of many cancers came from experiments conducted following the principles of experiment design and analysis that Fisher himself developed.

Scatterplots and correlation coefficients *never* prove causation. This is, for example, partly why it took so long for the U.S. Surgeon General to get warning labels on cigarettes. Although there was plenty of evidence that increased smoking was *associated* with increased levels of lung cancer, it took years to provide evidence that smoking actually *causes* lung cancer. (The tobacco companies used this to great advantage.)

• **Watch out for lurking variables.** A scatterplot of the damage (in dollars) caused to a house by fire would show a strong correlation with the number of firefighters at the scene. Surely the damage doesn't cause firefighters. And firefighters actually do cause damage, spraying water all around and chopping holes, but does that mean we shouldn't call the fire department? Of course not. There is an underlying variable that leads to both more damage and more firefighters—the size of the blaze. A hidden variable that stands behind a relationship and determines it by simultaneously affecting the other two variables is called a **lurking variable.** You can often debunk claims made about data by finding a lurking variable behind the scenes.

An ad agency hired by a well known manufacturer of dental hygiene products (electric toothbrushes, oral irrigators, etc.) put together a creative team to brainstorm ideas for a new ad campaign. Trisha Simes was chosen to lead the team as she has had the most experience with this client to date. At their first meeting, Trisha communicated to her team the client's desire to differentiate themselves from their competitors by not focusing their message on the cosmetic benefits of good dental care. As they brainstormed ideas, one member of the team, Brad Jonns, recalled a recent CNN broadcast that reported a "correlation" between flossing teeth and reducing the risk of heart disease. Seeing potential in promoting the health benefits of proper dental care, the team agreed to pursue this idea further. At their next meeting several team members commented on how surprised they were to find so many articles, medical, scientific, and popular, that seemed to claim good dental hygiene resulted in good health. One member noted that he found articles that linked gum disease not only to heart attacks and strokes but to diabetes and even cancer. While Trisha puzzled over why their client's competitors had not yet capitalized on these research findings, her team was on a roll and had already begun to focus on designing the campaign around this core message.

ETHICAL ISSUE *Correlation does not imply causation. The possibility of lurking variables is not explored. For example, it is likely that those who take better care of themselves would floss regularly and also have less risk of heart disease (related to Item C, ASA Ethical Guidelines).*

ETHICAL SOLUTION *Refrain from implying cause and effect from correlation results.*

What have we learned?

In previous chapters we learned how to listen to the story told by data from a single variable. Now we've turned our attention to the more complicated (and more interesting) story we can discover in the association between two quantitative variables.

We've learned to begin our investigation by looking at a scatterplot. We're interested in the *direction* of the association, the *form* it takes, and its *strength*.

We've learned that, although not every relationship is linear, when the scatterplot is straight enough, the *correlation coefficient* is a useful numerical summary.

- The sign of the correlation tells us the direction of the association.
- The magnitude of the correlation tells us of the *strength* of a linear association. Strong associations have correlations near +1 or −1, and very weak associations have correlations near 0.
- Correlation has no units, so shifting or scaling the data, standardizing, or even swapping the variables has no effect on the numerical value.

We've learned that to use correlation we have to check certain conditions for the analysis to be valid.

- Before finding or talking about a correlation, we'll always check the Linearity Condition.
- And, as always, we'll watch out for unusual observations!

Finally, we've learned not to make the mistake of assuming that a high correlation or strong association is evidence of a cause-and-effect relationship. Beware of lurking variables!

Terms

Association	• **Direction:** A positive direction or association means that, in general, as one variable increases, so does the other. When increases in one variable generally correspond to decreases in the other, the association is negative.
	• **Form:** The form we care about most is straight, but you should certainly describe other patterns you see in scatterplots.
	• **Strength:** A scatterplot is said to show a strong association if there is little scatter around the underlying relationship.
Correlation coefficient	A numerical measure of the direction and strength of a linear association.

$$r = \frac{\sum z_x z_y}{n - 1}$$

Explanatory or independent variable (*x*-variable)	The variable that accounts for, explains, predicts, or is otherwise responsible for the *y*-variable.
Lurking variable	A variable other than *x* and *y* that simultaneously affects both variables, accounting for the correlation between the two.
Outlier	A point that does not fit the overall pattern seen in the scatterplot.
Response or dependent variable (*y*-variable)	The variable that the scatterplot is meant to explain or predict.
Scatterplot	A graph that shows the relationship between two quantitative variables measured on the same cases.

Skills

 PLAN

- Recognize when interest in the pattern of a possible relationship between two quantitative variables suggests making a scatterplot.
- Be able to identify the roles of the variables and to place the response variable on the *y*-axis and the explanatory variable on the *x*-axis.
- Know the conditions for correlation and how to check them.
- Know that correlations are between −1 and +1 and that each extreme indicates a perfect linear association.
- Understand how the magnitude of the correlation reflects the strength of a linear association as viewed in a scatterplot.
- Know that the correlation has no units.
- Know that the correlation coefficient is not changed by changing the center or scale of either variable.
- Understand that causation cannot be demonstrated by a scatterplot or correlation.

 DO

- Be able to make a scatterplot by hand (for a small set of data) or with technology.
- Know how to compute the correlation of two variables.
- Know how to read a correlation table produced by a statistics program.

 REPORT

- Be able to describe the direction, form, and strength of a scatterplot.
- Be prepared to identify and describe points that deviate from the overall pattern.
- Be able to use correlation as part of the description of a scatterplot.

- Be alert to misinterpretations of correlation.
- Understand that finding a correlation between two variables does not indicate a causal relationship between them. Beware the dangers of suggesting causal relationships when describing correlations.

Technology Help: Scatterplots and Correlation

Statistics packages generally make it easy to look at a scatterplot to check whether the correlation is appropriate. Some packages make this easier than others.

Many packages allow you to modify or enhance a scatterplot, altering the axis labels, the axis numbering, the plot symbols, or the colors used. Some options, such as color and symbol choice, can be used to display additional information on the scatterplot.

EXCEL

To make a Scatterplot with the Excel Chart Wizard,

- Click on the **Chart Wizard** Button in the menu bar. Excel opens the Chart Wizard's Chart Type Dialog window.
- Make sure the **Standard Types** tab is selected, and select **XY (Scatter)** from the choices offered.
- Specify the **scatterplot without lines** from the choices offered in the Chart subtype selections. The **Next** button takes you to the Chart Source Data dialog.
- If it is not already frontmost, click on the **Data Range** tab, and enter the data range in the space provided.
- By convention, we always represent variables in columns. The Chart Wizard refers to variables as Series. Be sure the **Column** option is selected.
- Excel places the leftmost column of those you select on the x-axis of the scatterplot. If the column you wish to see on the x-axis is not the leftmost column in your spreadsheet, click on the **Series** tab and edit the specification of the individual axis series.
- Click the **Next** button. The Chart Options dialog appears.
- Select the **Titles** tab. Here you specify the title of the chart and names of the variables displayed on each axis.
- Type the chart title in the **Chart title:** edit box.
- Type the x-axis variable name in the **Value (X) Axis:** edit box. Note that you must name the columns correctly here. Naming another variable will not alter the plot, only mislabel it.
- Type the y-axis variable name in the **Value (Y) Axis:** edit box.
- Click the **Next** button to open the chart location dialog.

- Select the **As new sheet:** option button.
- Click the **Finish** button.

Often, the resulting scatterplot will require rescaling. By default, Excel includes the origin in the plot even when the data are far from zero. You can adjust the axis scales. To change the scale of a plot axis in Excel,

- Double-click on the axis. The **Format Axis Dialog** appears.
- If the **scale tab** is not the frontmost, select it.
- Enter new minimum or new maximum values in the spaces provided. You can drag the dialog box over the scatterplot as a straightedge to help you read the maximum and minimum values on the axes.
- Click the **OK** button to view the rescaled scatterplot.
- Follow the same steps for the x-axis scale.

Compute a correlation in Excel with the **CORREL** function from the drop-down menu of functions. If CORREL is not on the menu, choose **More Functions** and find it among the statistical functions in the browser.

In the dialog that pops up, enter the range of cells holding one of the variables in the space provided.

Enter the range of cells for the other variable in the space provided.

To make a scatterplot using the DDXL add-in, select the two variables to display. They should be columns. If the first row holds column labels, include it. From the **DDXL** menu choose **Charts and Plots.** From the dialog's Function menu choose **Scatterplot.** Drag the x-variable into the **X-Axis Variable** area and the y-variable into the **Y-Axis Variable** area. If you have a column that names each case, drag it into the **Label Variable** area. Click **OK.**

EXCEL 2007

To make a scatterplot in Excel 2007:

- Select the columns of data to use in the scatterplot. You can select more than one column by holding down the control key while clicking.

- In the Insert tab, click on the **Scatter** button and select the **Scatter with only Markers** chart from the menu.

To make the plot more useful for data analysis, adjust the display as follows:

- With the chart selected click on the **Gridlines** button in the Layout tab to cause the Chart Tools tab to appear.

- Within Primary Horizontal Gridlines, select **None.** This will remove the gridlines from the scatterplot.

- To change the axis scaling, click on the numbers of each axis of the chart, and click on the **Format Selection** button in the Layout tab.

- Select the **Fixed** option instead of the Auto option, and type a value more suited for the scatterplot. You can use the popup dialog window as a straightedge to approximate the appropriate values.

Excel 2007 automatically places the leftmost of the two columns you select on the x-axis, and the rightmost one on the y-axis. If that's not what you'd prefer for your plot, you'll want to switch them.

To switch the X and Y-variables:

- Click the chart to access the **Chart Tools** tabs.

- Click on the **Select Data** button in the Design tab.

- In the popup window's Legend Entries box, click on **Edit.**

- Highlight and delete everything in the Series X Values line, and select new data from the spreadsheet. (Note that selecting the column would inadvertently select the title of the column, which would not work well here.)

- Do the same with the Series Y Values line.

- Press **OK,** then press **OK** again.

To make a scatterplot using the DDXL add-in, select the two variables to display. They should be columns. If the first row holds column labels, include it. From the **DDXL** menu choose **Charts and Plots.** From the dialog's Function menu choose **Scatterplot.** Drag the x-variable into the **X-Axis Variable** area and the y-variable into the **Y-Axis Variable** area. If you have a column that names each case, drag it into the **Label Variable** area. Click **OK.**

MINITAB

To make a scatterplot, choose **Scatterplot** from the **Graph** menu. Choose "Simple" for the type of graph. Click **OK.** Enter variable names for the Y-variable and X-variable into the table. Click **OK.**

To compute a correlation coefficient, choose **Basic Statistics** from the **Stat** menu. From the Basic Statistics submenu, choose **Correlation.** Specify the names of at least two quantitative variables in the "Variables" box. Click **OK** to compute the correlation table.

SPSS

To make a scatterplot in SPSS, open the Chart Builder from the Graphs menu. Then

- Click the Gallery tab.

- Choose Scatterplot from the list of chart types.

- Drag the scatterplot onto the canvas.

- Drag a scale variable you want as the response variable to the y-axis drop zone.

- Drag a scale variable you want as the factor or predictor to the x-axis drop zone.

- Click OK.

To compute a correlation coefficient, choose **Correlate** from the **Analyze** menu. From the Correlate submenu, choose **Bivariate.** In the Bivariate Correlations dialog, use the arrow button to move variables between the source and target lists. Make sure the **Pearson** option is selected in the Correlation Coefficients field.

JMP

To make a scatterplot and compute correlation, choose **Fit Y by X** from the **Analyze** menu.
In the Fit Y by X dialog, drag the Y variable into the "**Y, Response**" box, and drag the X variable into the "**X, Factor**" box. Click the **OK** button.

Once JMP has made the scatterplot, click on the red triangle next to the plot title to reveal a menu of options. Select **Density Ellipse** and select .95. JMP draws an ellipse around the data and reveals the **Correlation** tab. Click the blue triangle next to Correlation to reveal a table containing the correlation coefficient.

DATA DESK

To make a scatterplot of two variables, select one variable as Y and the other as X and choose **Scatterplot** from the **Plot** menu. Then find the correlation by choosing **Correlation** from the scatterplot's HyperView menu.
Alternatively, select the two variables and choose **Pearson Product-Moment** from the **Correlations** submenu of the **Calc** menu.

Comments
We prefer that you look at the scatterplot first and then find the correlation. But if you've found the correlation first, click on the correlation value to drop down a menu that offers to make the scatterplot.

Mini Case Study Projects

Fuel Efficiency

With the ever increasing price of gasoline, both drivers and auto companies are motivated to raise the fuel efficiency of cars. Recent information posted by the U.S. government proposes some simple ways to increase fuel efficiency (see www.fueleconomy.gov): avoid rapid acceleration, avoid driving over 60 mph, reduce idling, and reduce the vehicle's weight. An extra 100 pounds can reduce fuel efficiency (mpg) by up to 2%. A marketing executive is studying the relationship between the fuel efficiency of cars (as measured in miles per gallon) and their weight to design a new compact car campaign. In the data set **ch07_MCSP_Fuel_Efficiency** you'll find data on the variables below.

- Model of Car
- Engine Size (L)
- Cylinders
- MSRP (Manufacturer's Suggested Retail Price in $)
- City (mpg)
- Highway (mpg)
- Weight (pounds)
- Type and Country of manufacturer

Describe the relationship of Weight, MSRP, and Engine Size with fuel efficiency (both City and Highway) in a written report. Be sure to transform the variables if necessary.

The U.S. Economy and Home Depot Stock Prices

The file **ch07_MCSP_Home_Depot** contains economic variables, as well as stock market data for The Home Depot, Inc. Economists, investors, and corporate executives use measures of the U.S. economy to evaluate the impact of inflationary pressures and employment fluctuations on the stock market. Inflation is often tracked through interest rates. While there are many different types of interest rates, here we include the monthly values for the bank prime loan rate, where the rate is posted by a majority of the top 25 (based on assets) insured U.S.-chartered commercial banks. The prime rate is often used by banks to price short-term business loans. In addition, we provide the interest rates on 6-month CDs, the unemployment rates (seasonally adjusted), and the rate on Treasury bills. Investigate the relationships between *Closing Price* for Home Depot stock and the following variables from 2006 to 2008:[6]

- Unemployment Rate (%)
- Bank Prime Rate (Interest Rate in %)
- CD Rate (%)
- Treasury Bill Rate (%)

Describe the relationship of each of these variables with Home Depot *Closing Price* in a written report. Be sure to use scatterplots and correlation tables in your analysis and transform variables, if necessary.

EXERCISES

1. Association. Suppose you were to collect data for each pair of variables. You want to make a scatterplot. Which variable would you use as the explanatory variable and which as the response variable? Why? What would you expect to see in the scatterplot? Discuss the likely direction and form.

a) Cell phone bills: number of text messages, cost.

b) Automobiles: Fuel efficiency (mpg), sales volume (number of autos).

c) For each week: Ice cream cone sales, air conditioner sales.

d) Product: Price ($), demand (number sold per day).

2. Association, part 2. Suppose you were to collect data for each pair of variables. You want to make a scatterplot. Which variable would you use as the explanatory variable and which as the response variable? Why? What would you expect to see in the scatterplot? Discuss the likely direction and form.

a) T-shirts at a store: price each, number sold.

b) Real estate: house price, house size (square footage).

c) Economics: Interest rates, number of mortgage applications.

d) Employees: Salary, years of experience.

3. Scatterplots. Which of the scatterplots show:

a) Little or no association?

b) A negative association?

c) A linear association?

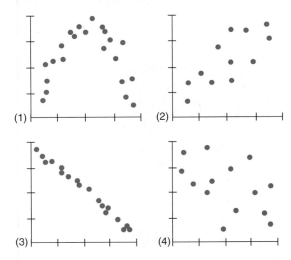

(1) (2) (3) (4)

d) A moderately strong association?

e) A very strong association?

4. Scatterplots, part 2. Which of the scatterplots on the next page show:

a) Little or no association?

b) A negative association?

c) A linear association?

d) A moderately strong association?

e) A very strong association?

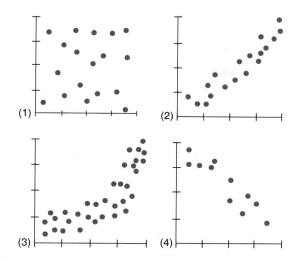

(1)   (2)

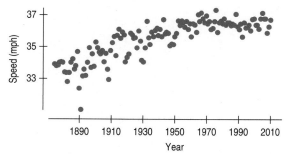

(3)   (4)

T 5. Kentucky Derby 2007. The fastest horse in Kentucky Derby history was Secretariat in 1973. The scatterplot shows speed (in miles per hour) of the winning horses each year.

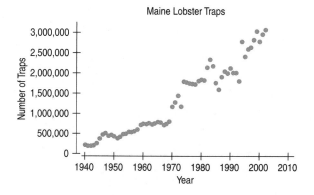

What do you see? In most sporting events, performances have improved and continue to improve, so surely we anticipate a positive direction. But what of the form? Has the performance increased at the same rate throughout the last 125 years?

T 6. Lobster traps. In many parts of the world lobster fishing is big business. The graph shows the growth in the number of lobster traps set (legally) in the state of Maine in the United States since 1940.

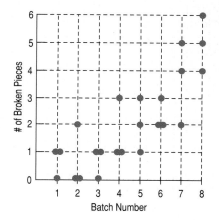

a) What do you see? Although we would expect the trend to be positive, what can you say about the form?

b) Can you see an impact of introducing wire-mesh traps and electronic equipment on lobster boats in the early 1970s? What effect, if any, did this appear to have?

7. Manufacturing. A ceramics factory can fire eight large batches of pottery a day. Sometimes a few of the pieces break in the process. In order to understand the problem better, the factory records the number of broken pieces in each batch for 3 days and then creates the scatterplot shown.

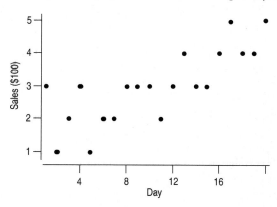

a) Make a histogram showing the distribution of the number of broken pieces in the 24 batches of pottery examined.
b) Describe the distribution as shown in the histogram. What feature of the problem is more apparent in the histogram than in the scatterplot?
c) What aspect of the company's problem is more apparent in the scatterplot?

8. Coffee sales. Owners of a new coffee shop tracked sales for the first 20 days and displayed the data in a scatterplot (by day).

a) Make a histogram of the daily sales since the shop has been in business.
b) State one fact that is obvious from the scatterplot, but not from the histogram.
c) State one fact that is obvious from the histogram, but not from the scatterplot.

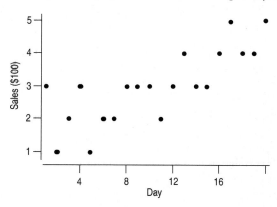

9. Matching. Here are several scatterplots. The calculated correlations are −0.923, −0.487, 0.006, and 0.777. Which is which?

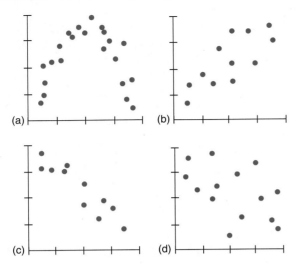

(a) (b)

(c) (d)

10. Matching, part 2. Here are several scatterplots. The calculated correlations are −0.977, −0.021, 0.736, and 0.951. Which is which?

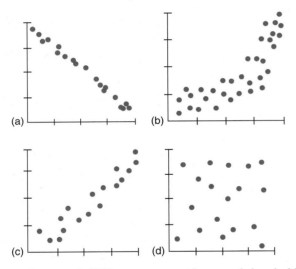

(a) (b)

(c) (d)

11. Packaging. A CEO announces at the annual shareholders meeting that the new see-through packaging for the company's flagship product has been a success. In fact, he says, "There is a strong correlation between packaging and sales." Criticize this statement on statistical grounds.

12. Insurance. Insurance companies carefully track claims histories so that they can assess risk and set rates appropriately. The National Insurance Crime Bureau reports that Honda Accords, Honda Civics, and Toyota Camrys are the cars most frequently reported stolen, while Ford Tauruses, Pontiac Vibes, and Buick LeSabres are stolen least often. Is it reasonable to say that there's a correlation between the type of car you own and the risk that it will be stolen?

13. Book sales. An analyst considers the correlation between book sales (number of books) at a college book store and day of

the year (1 = Jan 1, . . . , 365 = Dec 31). What might you expect the correlation between *Book Sales* and *Day Number* to be? Do you think there is an association between these variables? Explain.

14. Internet sales. An article in a business magazine reported that commerce over the Internet has exploded recently, doubling nearly every three years. It then stated that there was a high correlation between sales made on the Internet and *Year*. Do you think this is an appropriate summary? Explain.

T 15. Diamond prices. The price of a diamond depends on its color, cut, clarity, and carat weight. Here are data from a quality diamond merchant (so we can assume good cut) for diamonds of the best color (D) and high clarity (VS1).

Carat	Price	Carat	Price
0.33	1079	0.62	3116
0.33	1079	0.63	3165
0.39	1030	0.64	2600
0.40	1150	0.70	3080
0.41	1110	0.70	3390
0.42	1210	0.71	3440
0.42	1210	0.71	3530
0.46	1570	0.71	4481
0.47	2113	0.72	4562
0.48	2147	0.75	5069
0.51	1770	0.80	5847
0.56	1720	0.83	4930
0.61	2500		

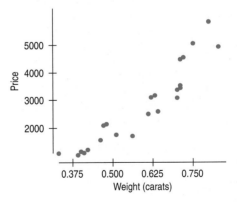

a) Are the assumptions and conditions met for finding a correlation?

b) The correlation is 0.937. Using that information, describe the relationship.

T 16. Interest rates and mortgages. Since 1980, average mortgage interest rates have fluctuated from a low of under 6% to a high of over 14%. Is there a relationship between the amount of money people borrow and the interest rate that's offered? Here is a scatterplot of *Total Mortgages* in the United States (in millions of 2005 dollars) versus *Interest Rate* at various times over the past 26 years. The correlation is −0.84.

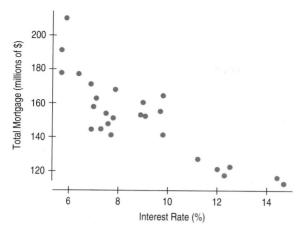

a) Describe the relationship between *Total Mortgages* and *Interest Rate*.

b) If we standardized both variables, what would the correlation coefficient between the standardized variables be?

c) If we were to measure *Total Mortgages* in thousands of dollars instead of millions of dollars, how would the correlation coefficient change?

d) Suppose in another year, interest rates were 11%, and mortgages totaled $250 million. How would including that year with these data affect the correlation coefficient?

e) Do these data provide proof that if mortgage rates are lowered, people will take out more mortgages? Explain.

17. Carbon footprint. The scatterplot shows, for 2008 cars, the carbon footprint (tons of CO_2 per year) vs. the new Environmental Protection Agency (EPA) highway mileage for 82 family sedans as reported by the U.S. government (www.fueleconomy.gov/feg/byclass.htm). The car with the highest highway mpg and lowest carbon footprint is the Toyota Prius.

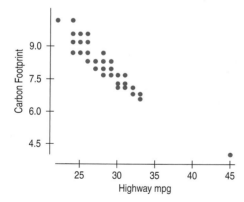

a) The correlation is −0.947. Describe the association.

b) Are the assumptions and conditions met for computing correlation?

c) Using technology, find the correlation of the data when the Prius is not included with the others. Can you explain why it changes in that way?

18. EPA mpg. In 2008, the EPA revised their methods for estimating the fuel efficiency (mpg) of cars—a factor that plays an increasingly important role in car sales. How do the new highway

and city estimated mpg values relate to each other? Here's a scatterplot for 83 family sedans as reported by the U.S. government. These are the same cars as in Exercise 17 except that the Toyota Prius has been removed from the data and two other hybrids, the Nissan Altima and Toyota Camry, are included in the data (and are the cars with highest city mpg.)

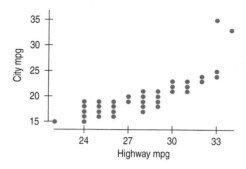

a) The correlation of these two variables is 0.823. Describe the association.

b) If the two hybrids were removed from the data, would you expect the correlation to increase, decrease, or stay the same? Try it using technology. Report and discuss what you find.

19. Oil consumption. The scatterplot shows the relationship between *Life Expectancy* and the logarithm of *Oil Consumption* for the 137 countries of the world for which both variables are available.

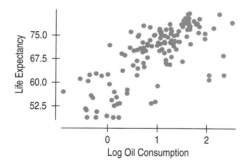

a) Is it appropriate to compute a correlation? Explain

b) In fact, the correlation is 0.80. Describe the association.

20. Antidepressants. Fourteen years after Eli Lilly introduced Prozac, antidepressants have grown into a $12 billion industry, second only to heart medications among sales of prescription drugs. But their effectiveness is still debated. A study compared the effectiveness of several antidepressants by examining the experiments in which they had passed the Food and Drug Administration (FDA) requirements. Each of those experiments compared the active drug with a placebo, an inert pill given to some of the subjects. In each experiment some patients treated with the placebo improved, a phenomenon called the *placebo effect*. Patients' depression levels were evaluated on the standard scale for quantitatively rating depression, called the Hamilton Depression Rating Scale. Positive changes in Hamilton scale ratings recorded improvements in patients' conditions. The Hamilton scale is a widely accepted standard that was used in each of the independently run studies. The scatterplot compares mean improvement levels for the antidepressants and placebos for these independently run trials.

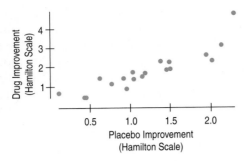

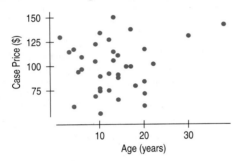

a) Is it appropriate to calculate the correlation? Explain.
b) The correlation is 0.898. Explain what we have learned about the results of these experiments.

T **21. Vineyards.** Here is the scatterplot and correlation for *Case Price* of wines from 36 vineyards in the Finger Lakes region of New York State and the *Age* of those vineyards.

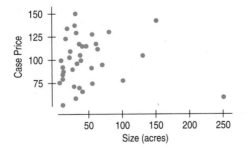

Correlation = 0.16
a) Check the assumptions and conditions for correlation.
b) Does it appear that older vineyards get higher prices for their wines? Explain.
c) What does this analysis tell us about vineyards in the rest of the world?

T **22. Vineyards, again.** Instead of the age of a vineyard, considered in Exercise 21, perhaps the *Size* of the vineyard (in acres) is associated with the price of the wines. Look at the scatterplot.

a) The correlation is −0.022. Does price get lower with increasing vineyard size? Explain.
b) If the point corresponding to the largest vineyard were removed, what effect would that have on the correlation?

T **23. Real estate.** Using a random sample of homes for sale, a prospective buyer is interested in examining the relationship

between price and number of bedrooms. The graph shows the scatterplot for price versus number of bedrooms. The correlation is 0.723.

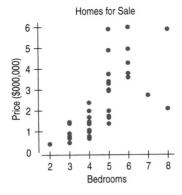

a) Check the assumptions and conditions for correlation.
b) Describe the relationship.

T **24. Real estate, again.** Maybe the number of total rooms in the house is associated with the price of a house. Here is the scatterplot for the same homes as we examined in Exercise 23:

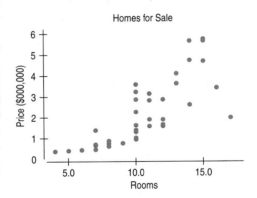

a) Is there an association?
b) Check the assumptions and conditions for correlation.

25. Regional sales. The head of the sales force for a retail clothing chain is analyzing whether the company does better in some parts of the country than others. She examines a scatterplot of last year's total *Sales* by *State*, where the states are numbered in alphabetical order, Alaska = 01 , . . . , Wyoming = 50. The correlation is only 0.045 from which she concludes that there are no differences in sales among the 50 states. Comment.

26. Human resources. At a small company, the CFO is concerned about absenteeism among the employees and asks the head of human resources to investigate. The jobs are coded from 01 to 99 with 01 = Stockroom Clerk and 99 = President. The human resource manager plots number of days absent last year by job type and finds a correlation of −0.034 and no obvious trend. He then reports to the CFO that there seems to be no relationship between absenteeism and job type. Comment.

27. Public financing of education. All 50 states of the United States offer public higher education through four-year colleges and universities and two-year colleges. Tuition charges by different states vary widely for both types. But would you expect there to be a relationship between the tuition states charge for the two types? (The data for the year 2007–2008 are found in the variables *Public.2yr* and *Public.4yr* in the data set Ch06_Tuition.)

a) Examine a scatterplot of the average tuition for four-year colleges against the tuition charged for two-year colleges. Describe the relationship.

b) Is the direction of the relationship what you expected?

c) Is a correlation an appropriate numerical summary of the strength of the relationship? Explain. If so, find it.

28. Public and private higher education. In Exercise 27, we examined the relationship between the average tuition charged by the 50 U.S. states for four-year colleges vs. two-year colleges. Now we'll look at the relationship between average tuition charged by private four-year colleges and universities (*Private.4yr*) against that of public four-year institutions (*Public.4yr*). The data are in Ch06_Tuition.

a) Would you expect the relationship between tuition charged by private and public four-year colleges and universities to be as strong as the relationship between public four-year and two-year institutions?

b) Examine a scatterplot and describe the relationship.

c) Is a correlation an appropriate numerical summary of the strength of the relationship? Explain. If so, find it.

29. Colombian peso. In 2007, the Colombian peso (COP) was the one of the global currencies that appreciated the most. A student noticed that during that year the exchange rate of the peso seemed to move in a direction opposite to the Dow Jones Industrial Average (DJIA) and calculated the correlation to be −0.81. The student concluded that the DJIA was driving down the value of the peso. Here is a scatterplot of the COP value (in pesos/dollar) vs. the DJIA (www.measuringworth.co).

a) Describe the relationship.

b) Is a correlation an appropriate numerical summary of the strength of the relationship?

c) Comment on the student's conclusion.

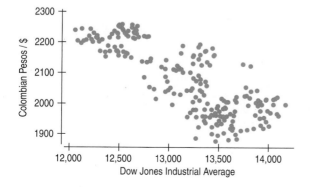

30. Colombian peso, part 2. In Exercise 29, a student commented on the negative correlation between the exchange rate of the Colombian peso and the Dow Jones Industrial Average during the year 2007. Another student in the class decided to research the history of the relationship and examined the scatterplot of the yearly average of the exchange rate and the DJIA from 1928 to 2006 seen below (www.measuringworth.com). (The DJIA reached 9000 for the first time in history in 1998.) She concluded that, in fact, the relationship is positive because the correlation is 0.97.

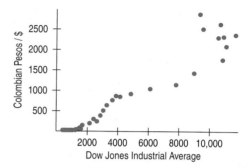

a) Describe the relationship.

b) Is a correlation an appropriate numerical summary of the strength of the relationship?

c) Comment on the student's conclusion and explain how the two students could reach such different conclusions.

31. Housing industry. A student following the housing industry compared quarterly sales at The Home Depot from 1995 to 2004 to quarterly U.S. housing starts and created the scatterplot we saw in Figure 7.3. He calculates the correlation to be 0.70 and writes a report that claims that an increase in housing starts will result in higher sales for The Home Depot.

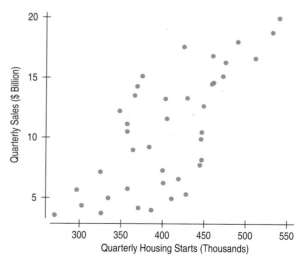

a) Describe the relationship.

b) Is a correlation an appropriate numerical summary of the strength of the relationship?

c) Comment on the student's conclusion.

T **32. Economic analysis.** An economics student is studying the American economy and finds that the correlation between the inflation adjusted Dow Jones Industrial Average and the Gross Domestic Product (GDP) (also inflation adjusted) is 0.77 (www.measuringworth.com). From that he concludes that there is a strong linear relationship between the two series and predicts that a drop in the GDP will make the stock market go down. Here is a scatterplot of the adjusted DJIA against the GDP (in year 2000 $). Describe the relationship and comment on the student's conclusions.

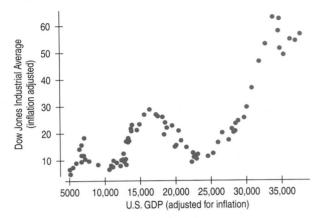

33. International economics correlation errors. The instructor in your International Economics course assigns your class to investigate factors associated with the Gross Domestic Product of nations. Each student examines a different factor (such as *Life Expectancy, Literacy Rate*, etc.) for a few countries and reports to the class. Apparently, some of your classmates do not understand Statistics very well because several of their conclusions are incorrect. Explain the mistakes in their statements.

a) "My correlation of −0.772 shows that there is almost no association between *GDP* and *Infant Mortality Rate*."

b) "There was a correlation of 0.44 between *GDP* and *Continent.*"

34. More international economics correlation errors. Students in the class discussed in Exercise 33 also wrote these conclusions. Explain the mistakes they made.

a) "There was a very strong correlation of 1.22 between *Life Expectancy* and *GDP.*"

b) "The correlation between *Literacy Rate* and *GDP* was 0.83. This shows that countries wanting to increase their standard of living should invest heavily in education."

35. Investments. An investment analyst looking at the association between sales and assets of companies was surprised when she calculated the correlation. She had expected to find a fairly strong association, yet the correlation was near 0. Explain how a scatterplot could still reveal the strong associations he anticipated.

36. Used cars. A customer shopping for a used car believes that there should be a negative association between the mileage a used car has on it and the price of the car. Yet, when she runs a correlation, it is near 0, and she is surprised. Explain how a scatterplot could help her understand the relationship.

37. Oil consumption, again. In Exercise 19, we found there was a strong association between the logarithm of oil consumption and life expectancy across many countries of the world.

a) Does this mean that consuming oil is good for the health?

b) What might explain the strong correlation?

38. Age and income. The correlations between *Age* and *Income* as measured on 100 people is $r = 0.75$. Explain whether or not each of these possible conclusions is justified:

a) When *Age* increases, *Income* increases as well.

b) The form of the relationship between *Age* and *Income* is straight.

c) There are no outliers in the scatterplot of *Income* vs. *Age*.

d) Whether we measure *Age* in years or months, the correlation will still be 0.75.

T **39. Reducing truck shipping costs.** Regulators must keep an eye on the weights of trucks on major highways, but making trucks stop to be weighed is costly both for the regulators and the truckers. The Minnesota Department of Transportation hoped that they could keep costs down by measuring the weights of big trucks without actually stopping the vehicles and instead using a newly developed "weight-in-motion" scale. To see if the new device was accurate, they conducted a calibration test. They weighed several trucks when stopped (static weight), assuming that this weight was correct. Then they weighed the trucks again while they were moving to see how well the new scale could estimate the actual weight. Their data are given in the table.

WEIGHT OF A TRUCK (THOUSANDS OF POUNDS)	
Weight-in-Motion	**Static Weight**
26.0	27.9
29.9	29.1
39.5	38.0
25.1	27.0
31.6	30.3
36.2	34.5
25.1	27.8
31.0	29.6
35.6	33.1
40.2	35.5

a) Make a scatterplot for these data.

b) Describe the direction, form, and strength of the plot.

c) Write a few sentences telling what the plot says about the data. (*Note:* The sentences should be about weighing trucks, not about scatterplots.)

d) Find the correlation.

e) If the trucks were weighed in kilograms (1 kilogram = 2.2 pounds), how would this change the correlation?

f) Do any points deviate from the overall pattern? What does the plot say about a possible recalibration of the weight-in-motion scale?

40. Selling fuel economy 2007. In 2006, a study by *Consumer Reports* found that 37% of nationwide respondents reported that they were considering replacing their current car with one with greater fuel economy. Here are advertised horsepower ratings and expected gas mileage for several 2007 vehicles (www.kbb.com/KBB/ReviewsAndRating).

Vehicle	Horsepower	Highway Gas Mileage (mpg)
Audi A4	200	32
BMW 328	230	30
Buick LaCrosse	200	30
Chevy Cobalt	148	32
Chevy TrailBlazer	291	22
Ford Expedition	300	20
GMC Yukon	295	21
Honda Civic	140	40
Honda Accord	166	34
Hyundai Elantra	138	36
Lexus IS 350	306	28
Lincoln Navigator	300	18
Mazda Tribute	212	25
Toyota Camry	158	34
Volkswagen Beetle	150	30

a) Make a scatterplot for these data.
b) Describe the direction, form, and strength of the plot.
c) Find the correlation between horsepower and miles per gallon.
d) Write a few sentences telling what the plot says about fuel economy.

T 41. Pizza sales. Here is a scatterplot for the weekly sales (in pounds) for every fourth week of a brand of frozen pizza versus the unit price of the pizza for a sample of stores in the Dallas area.

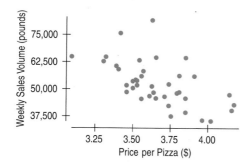

a) Check the assumptions and conditions for correlation.
b) Compute the correlation between sales and price.
c) Does this graph support the theory that as prices drop, demand for the product will increase?
d) If we assume that the number of pounds of pizza per box is consistent and we measure sales in the number of pizza boxes sold instead of pounds, will the correlation change? Explain.

T 42. Housing costs. Concern over the possibility of a "home cost bubble" has lead many economists to examine housing costs. The Office of Federal Housing Enterprise Oversight (www.ofheo.gov) collects data on various aspects of housing costs around the United States. Here is a scatterplot of the *Housing Cost Index* versus the *Median Family Income* for each of the 50 states. The correlation is 0.65.

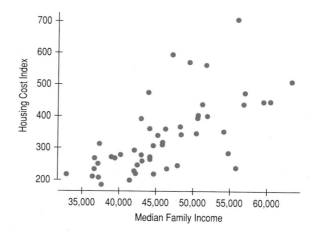

a) Describe the relationship between the *Housing Cost Index* and the *Median Family Income* by state.
b) If we standardized both variables, what would the correlation coefficient between the standardized variables be?
c) If we had measured *Median Family Income* in thousands of dollars instead of dollars, how would the correlation change?
d) Washington, DC, has a *Housing Cost Index* of 548 and a median income of about $45,000. If we were to include DC in the data set, how would that affect the correlation coefficient?
e) Do these data provide proof that by raising the median income in a state, the *Housing Cost Index* will rise as a result? Explain.

T 43. Mutual funds. Here is a scatterplot showing the association between money flowing into mutual funds (fund flows in $M) and a specific type of market return (Wilshire Index) for each month from 1990 to 2002.

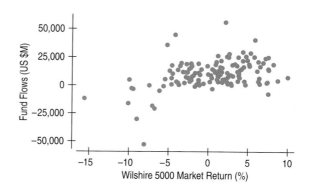

a) Is it appropriate to calculate a correlation? Explain.
b) Identify the largest outlier in the scatterplot. Find and discuss the correlation after setting this outlier aside.

44. Football salaries. Payrolls for the 32 different teams in the National Football League (NFL) vary widely. Do higher payrolls lead to more victories? Here is a scatterplot of number of wins vs. team payroll for 2006.

a) Check the assumptions and conditions for correlation.
b) Is there support for the NFL coaches who claim that if only they had more resources, they could win more games?

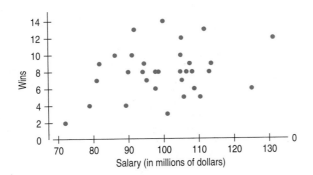

45. Attendance 2006. American League baseball games are played under the designated hitter rule, meaning that weak-hitting pitchers do not come to bat. Baseball owners believe that the designated hitter rule means more runs scored, which in turn means higher attendance. Is there evidence that more fans attend games if the teams score more runs? Data collected from American League games during the 2006 season have a correlation of 0.667 between *Runs Scored* and the number of people at the game (www.mlb.com).

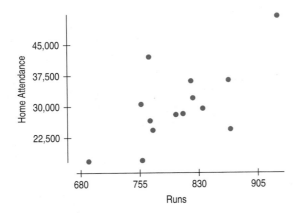

a) Does the scatterplot indicate that it's appropriate to calculate a correlation? Explain.
b) Describe the association between attendance and runs scored.
c) Does this association prove that the owners are right that more fans will come to games if the teams score more runs?

46. Second inning 2006. Perhaps fans are just more interested in teams that win. The displays are based on American League

teams for the 2006 season (espn.go.com). Are the teams that win necessarily those that score the most runs?

CORRELATION			
	Wins	**Runs**	**Attend**
Wins	1.000		
Runs	0.605	1.000	
Attend	0.697	0.667	1.000

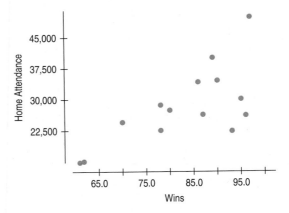

a) Do winning teams generally enjoy greater attendance at their home games? Describe the association.
b) Is attendance more strongly associated with winning or scoring runs? Explain
c) How strongly is scoring more runs associated with winning more games?

47. Fundraising. Analysts at a philanthropic organization want to predict who is most likely to give to their next fundraising campaign. They considered the potential donors' *Marital Status* (single = 1, married = 2, divorced = 3, widowed = 4) and *Giving* (No = 0, Yes = 1). They found a correlation of 0.089 between the two variables. Comment on their conclusion that this shows that marital status has no association with whether the person will respond to the campaign. What should the organization have done with these data?

48. Merging data sources. A business is checking its database to see if the two data sources they used sampled the same zip codes. The variable *data source* = 1 if the data source is Metro-Media, 2 if the data source is DataQwest, and 3 if it's RollingPoll. The organization finds that the correlation between five-digit zip code and *data source* is −0.0229. It concludes that the correlation is low enough to state that there is no dependency between *zip code* and *data source*. Comment.

49. Oil production. This table shows the oil production of the United States from 1949 to 2005 (in thousands of barrels per year).

Year	Production (thousands of barrels)	Year	Production (thousands of barrels)
1949	1,841,940	1978	3,178,216
1950	1,973,574	1979	3,121,310
1951	2,247,711	1980	3,146,365
1952	2,289,836	1981	3,128,624
1953	2,357,082	1982	3,156,715
1954	2,314,988	1983	3,170,999
1955	2,484,428	1984	3,249,696
1956	2,617,283	1985	3,274,553
1957	2,616,901	1986	3,168,252
1958	2,448,987	1987	3,047,377
1959	2,574,590	1988	2,979,126
1960	2,574,933	1989	2,778,772
1961	2,621,758	1990	2,684,689
1962	2,676,189	1991	2,707,039
1963	2,752,723	1992	2,624,631
1964	2,786,822	1993	2,499,033
1965	2,848,514	1994	2,431,476
1966	3,027,763	1995	2,394,269
1967	3,215,742	1996	2,366,016
1968	3,329,042	1997	2,354,831
1969	3,371,751	1998	2,281,920
1970	3,517,450	1999	2,146,732
1971	3,453,914	2000	2,130,706
1972	3,455,368	2001	2,117,512
1973	3,360,903	2002	2,097,124
1974	3,202,585	2003	2,073,454
1975	3,056,779	2004	1,983,300
1976	2,976,180	2005	1,890,107
1977	3,009,265		

a) Find the correlation of production and year.

b) A reporter concludes that a low correlation between *Year* and *Production* shows that oil production has remained steady over the 57-year period. Do you agree with this interpretation? Explain.

 50. Airline industry. The airline industry grew rapidly during the decade from 1995 to 2005. The table shows the number of flights flown in each of these years.

Year	Flights
1995	5,327,435
1996	5,351,983
1997	5,411,843
1998	5,384,721
1999	5,527,884
2000	5,683,047
2001	5,967,780
2002	5,271,359
2003	6,488,539
2004	7,129,270
2005	7,140,596

a) Find the correlation of *Flights* and *Year*.

b) Make a scatterplot.

c) Note two reasons that the correlation you found in a) is not a suitable summary of the strength of the association. Can you account for these violations of the conditions?

✓ JUST CHECKING ANSWERS

1 We know the scores are quantitative. We should check to see if the *Linearity Condition* and the *Outlier Condition* are satisfied by looking at a scatterplot of the two scores.

2 It won't change.

3 It won't change.

4 They are more likely to do poorly. The positive correlation means that low closing prices for Intel are associated with low closing prices for Cypress.

5 No, the general association is positive, but daily closing prices may vary.

Linear Regression

CHAPTER 8

Best Buy Co., Inc.

In 1966, Richard Schulze opened a small music store called Sound of Music in St. Paul, Minnesota, to sell audio component systems. By 1980, he had expanded to nine locations, but in 1981, his most successful store was destroyed by a tornado. Capitalizing on misfortune, Schulze turned the event into a highly successful "tornado sale" that combined a large selection with low prices. That same year, the company expanded into consumer electronics, and in 1983, it officially became known as Best Buy. The company, which went public in 1985, pioneered the retail concept of putting all the inventory on the sales floor, with salaried (noncommissioned) product specialists as sales assistants, dressed in a "uniform" of blue polo shirts, khaki pants, and black shoes.

In 2001, Best Buy acquired Future Shop, Canada's largest consumer electronics retailer, and the following year bought Geek Squad®, which offered their customers round-the-clock technical support. In 2006, Best Buy purchased a majority interest in China's fourth-largest appliance and consumer electronics retailer.

193

Today, Best Buy is a retailer of consumer electronics, home office products, appliances, and entertainment software, with nearly a thousand retail stores across the United States and Canada. In 2006, Best Buy reported revenue of over $27 billion and profit of nearly $1 billion.

Companies like Best Buy depend on large computer systems to manage and store millions of customer transactions, inventory records, payroll information, and other types of company data. So the company must have enough computing power to be able to process and retrieve that data quickly and efficiently. For a growing company, assuring enough computing capacity and speed is critical.

Each year Best Buy purchases mainframe computing, measured in MIPS (Millions of Instructions Per Second). For planning and budgeting purposes they also want to forecast the number of MIPS needed the following year. Before 2001, Best Buy did not use Statistics to predict their computer needs; they only looked at the numbers and then guessed at the amount of MIPS needed in the coming year. But we can do better than that. Figure 8.1 shows monthly mainframe computing use and the number of stores Best Buy had between August 1996 and July 2000.

WHO	Months
WHAT	*Monthly* Computer Use at Best Buy Number of *Stores*
UNITS	Millions of instructions per second (MIPS)
WHEN	August 1996 to July 2000
WHY	To predict computer capacity needs

Extrapolation
A prediction just one year ahead doesn't seem like an unusual request. But whenever we reach beyond the range of the data, such an *extrapolation* can be dangerous. The model can provide a prediction for any value, but management should be cautious when using any model to make predictions for values that lie far beyond the data on which the model was built.

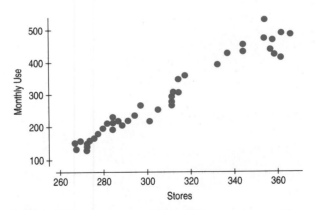

Figure 8.1 *We want to model monthly computer use with the number of stores, so we'll put Monthly Use on the y-axis and Stores on the x-axis.*

From the scatterplot, you can see that the relationship between computer capacity and number of stores is positive and linear. The high correlation of 0.979 attests to its strength. But the strength of the relationship is only part of the picture. In 2000, management might have wanted to predict how many MIPS they'd need to support the 419 stores they projected they'd have by the end of fiscal 2001. That's a reasonable business question, but we can't read the answer directly from the scatterplot. We need a model for the trend. The correlation says "there seems to be a strong linear association between these two variables," but it doesn't tell us *what the line is.*

8.1 The Linear Model

"Statisticians, like artists, have the bad habit of falling in love with their models."

—George Box, famous statistician

Of course, we can say more. We can model the relationship with a line and give the equation. For the Best Buy case, we use a linear model to describe the relationship between computer use and number of stores. A **linear model** is just an equation of a straight line through the data. The points in the scatterplot don't all line up, but a straight line can summarize the general pattern with only a few parameters. This model can help us understand how the variables are associated.

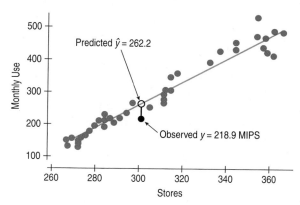

Figure 8.2 *A linear model for computer Monthly Use versus Stores at Best Buy, 1996–2000.*

Residuals

> **NOTATION ALERT:**
> "Putting a hat on it" is standard Statistics notation to indicate that something has been predicted by a model. Whenever you see a hat over a variable name or symbol, you can assume it is the predicted version of that variable or symbol.

A *negative* residual means the predicted value is too big— an overestimate. A *positive* residual shows the model makes an underestimate. These may seem backwards at first.

We know the model won't be perfect. No matter what line we draw, it won't go through many of the points. The best line might not even hit any of the points. Then how can it be the "best" line? We want to find the line that somehow comes *closer* to all the points than any other line. Some of the points will be above the line and some below. A linear model can be written as $\hat{y} = b_0 + b_1 x$, where b_0 and b_1 are numbers estimated from the data and $\hat{y}$ (pronounced *y*-hat) is the **predicted value**. We use the *hat* to distinguish the predicted value from the observed value *y*. The difference between these two is called the **residual**:

$$e = y - \hat{y}.$$

The residual value tells us how far the model's prediction is from the observed value at that point. For example, the residual for 301 stores is $y - \hat{y} = 218.9 - 262.2 = -43.3$ MIPS. Figure 8.2 shows the observed (black dot) value, the predicted (hollow dot) value and the residual (the difference, as shown by the vertical line).

To find the residuals, we always subtract the predicted values from the observed ones. The negative residual -43.3 tells us that the actual computer use was about 43 MIPS *less* than the model predicts when there were 301 U.S. Best Buy stores in operation.

Our question now is how to find the right line.

The Line of "Best Fit"

When we draw a line through a scatterplot, some residuals are positive, and some are negative. We can't assess how well the line fits by adding up all the residuals—the positive and negative ones would just cancel each other out. We need to find the line that's closest to all the points, and to do that, we need to make all the distances positive. We faced the same issue when we calculated a standard deviation to measure spread. And we deal with it the same way here: by squaring the residuals to make them positive. The sum of all the squared residuals tells us how well the line we drew fits the data—the smaller the sum, the better the fit. A different line will produce a different sum, maybe bigger, maybe smaller. The **line of best fit** is the line for which the sum of the squared residuals is smallest—often called the **least squares line**.

This line has the special property that the variation of the data around the model, as seen in the residuals, is the smallest it can be for any straight line model for these data. No other line has this property. Speaking mathematically, we say that this line minimizes the sum of the squared residuals. You might think that finding this "least squares line" would be difficult. Surprisingly, it's not, although it was an exciting mathematical discovery when Legendre published it in 1805.

> **Who Was First?**
>
> French mathematician Adrien-Marie Legendre was the first to publish the "least squares" solution to the problem of fitting a line to data when the points don't all fall exactly on the line. The main challenge was how to distribute the errors "fairly." After considerable thought, he decided to minimize the sum of the squares of what we now call the residuals. After Legendre published his paper in 1805, Carl Friedrich Gauss, the German mathematician and astronomer, claimed he had been using the method since 1795 and, in fact, had used it to calculate the orbit of the asteroid Ceres in 1801. Gauss later referred to the "least squares" solution as *"our method"* (principium *nostrum*), which certainly didn't help his relationship with Legendre.

8.2 Correlation and the Line

Any straight line can be written as:

$$y = b_0 + b_1 x.$$

If we were to plot all the (x, y) pairs that satisfy this equation, they'd fall exactly on a straight line. We'll use this form for our linear model. Of course, with real data, the points won't all fall on the line. So, we write our model as $\hat{y} = b_0 + b_1 x$, using $\hat{y}$ for the predicted values, because it's the predicted values (not the data values) that fall on the line. If the model is a good one, the data values will scatter closely around it.

For the Best Buy data, the line is:

$$\widehat{Monthly\ Use} = -833.4 + 3.64\ Stores.$$

What does this mean? The **slope** 3.64 says that each store is associated with (perhaps because it requires) an additional 3.64 MIPS, on average. Slopes are always expressed in *y*-units per *x*-units. They tell you how the response variable changes for a one unit step in the predictor variable. So we'd say that the slope is 3.64 MIPS per store.

The **intercept** −833.4 is the value of the line when the *x*-variable is zero. What does it mean here? The intercept often serves just as a starting value for our predictions. We don't interpret it unless a 0 value for the predictor variable would really mean something under the circumstances. Here, it's unlikely that a linear model that fits for a company with hundreds of stores would be the right model for one with only a few stores, so it isn't of much concern that the intercept value has no meaning.

How do we find the slope and intercept of the least squares line? The formulas are simple. The model is built from the summary statistics we've used before. We'll

JUST CHECKING

A scatterplot of sales per month (in thousands of dollars) vs. number of employees for all the outlets of a large computer chain shows a relationship that is straight, with only moderate scatter and no outliers. The correlation between *Sales* and *Employees* is 0.85, and the equation of the least squares model is:

$$\widehat{Sales} = 9.564 + 122.74 \, Employees$$

1 What does the slope of 122.74 mean?
2 What are the units of the slope?
3 The outlet in Dallas, Texas, has 10 more employees than the outlet in Cincinnati. How much more *Sales* do you expect it to have?

need the correlation (to tell us the strength of the linear association), the standard deviations (to give us the units), and the means (to tell us where to locate the line).

The slope of the line is computed as:

$$b_1 = r \frac{s_y}{s_x}.$$

We've already seen that the correlation tells us the sign and the strength of the relationship, so it should be no surprise to see that the slope inherits this sign as well. If the correlation is positive, the scatterplot runs from lower left to upper right, and the slope of the line is positive.

Correlations don't have units, but slopes do. How x and y are measured—what units they have—doesn't affect their correlation, but does change the slope. The slope gets its units from the ratio of the two standard deviations. Each standard deviation has the units of its respective variable. So, the units of the slope are a ratio, too, and are always expressed in units of y per unit of x.

How do we find the intercept? If you had to predict the y-value for a data point whose x-value was average, what would you say? The best fit line predicts $\bar{y}$ for points whose x-value is $\bar{x}$. Putting that into our equation and using the slope we just found gives:

$$\bar{y} = b_0 + b_1\bar{x}$$

and we can rearrange the terms to find:

$$b_0 = \bar{y} - b_1\bar{x}.$$

It's easy to use the estimated linear model to predict the amount of computer use (in MIPS) that we'll need for any number of stores. For example, to estimate the *Monthly Use* needed for 419 stores, we substitute 419 for *Stores* in the equation:

$$\widehat{Monthly\ Use} = -833.4 + 3.64 \, (Stores)$$

and find:

$$\widehat{Monthly\ Use} = -833.4 + 3.64\,(419) = 691.76 \text{ MIPS.}$$

Best Buy
Summary statistics:

$$Monthly\ Use: \bar{y} = 276.49; \, s_y = 117.09$$

$$Stores: \bar{x} = 304.65; \, s_x = 31.468$$

Correlation $r = 0.979$

So, $b_1 = r \dfrac{s_y}{s_x} = (0.979)\dfrac{117.09}{31.468} = 3.64$ MIPS/Store and

$$b_0 = \bar{y} - b_1\bar{x} = 276.49 - 3.643(304.65) = -833.3 \text{ MIPS}$$

(The difference in the final digit is due to rounding error.)

Least squares lines are commonly called **regression** lines. Although this name is an accident of history (as we'll soon see), "regression" almost always means "the linear model fit by least squares." Clearly, regression and correlation are closely related. We'll need to check the same condition for regression as we did for correlation:

1. **Quantitative Variables Condition**
2. **Linearity Condition**
3. **Outlier Condition**

A little later in the chapter we'll add a fourth.

Getting from Correlation to the Line

We've seen the equation of the least squares line, but we can gain more insight by looking back at a plot of standardized variables to see what the line means in this context. Here are the Best Buy data again after standardizing both variables. The scatterplot is essentially the same, except for the change in the axes.

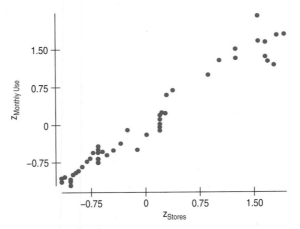

Figure 8.3   *The scatterplot of standardized variables looks just like the original plot except for the change in the axes.*

If we look at the regression for these standardized variables, we can gain some insight into how regression works. We can find the regression slope for any two variables x and y from:

$$b_1 = r \frac{s_y}{s_x}.$$

However, here these are standardized variables, z_x and z_y, both of which have standard deviations equal to one. So we have:

$$b_1 = r \frac{s_{z_y}}{s_{z_x}} = r \frac{1}{1} = r.$$

(Of course, the standard deviations cancel like this whenever they are the same for the two variables.) What about the intercept? We know that:

$$b_0 = \bar{y} - b_1 \bar{x}$$

but when we standardize the variables, we have $b_0 = \bar{z}_y - b_1\bar{z}_x = 0 - r0 = 0$. So the regression has an equation that's about as simple as we could possibly hope for:

$$\hat{z}_y = rz_x.$$

This equation for the line is simple and it tells us a lot, not only about these standardized variables, but also about how regression works in general. For example, it says that if you have an observation 1 SD above the mean in x (with a z_x score of 1), you'd expect y to have a z-score of r. Now we can see that the correlation is more than just a vague measure of strength of association: It's a great way to think about what the model tells us.

Let's be more specific. For the computer *Monthly Use*, the correlation is 0.979. So, we know immediately that:

$$\hat{z}_{Monthly\ Use} = rz_{Stores} = 0.979\ z_{Stores}.$$

But we don't have to standardize the two variables to get the benefit of this equation. It tells us about the original variables as well, saying that for every standard deviation above (or below) the mean we are in *Stores*, we'll predict that the computer usage is 0.979 standard deviations above (or below) the mean *Monthly Use*.

Recall from Chapter 7 that if $r = 0$, there's no linear relationship. No matter how many standard deviations you move in x, the predicted value for y doesn't change. On the other hand, if $r = 1.0$ or -1.0, there's a perfect linear association. In that case, moving any number of standard deviations in x moves exactly the same number of standard deviations in y. In general, moving any number of standard deviations in x moves our prediction r times that number of standard deviations in y.

JUST CHECKING

Let's go back to our regression of sales (in thousands of dollars) vs. employees. The correlation between *Sales* and *Employees* is 0.85, and the equation of the regression model is:

$$\widehat{Sales} = 9.564 + 122.74\ Employees.$$

4 If an outlet were 1 SD above the mean in number of *Employees*, how many SDs above the mean would you predict its *Sales* to be?

5 What would you predict about the sales of an outlet that's 2 SDs below average in number of employees?

8.3 Regression to the Mean

Suppose you were told that a new male student was about to join the class and you were asked to guess his height in inches. What would be your guess? A good guess would be the mean height of male students. Now suppose you are also told that this student had a grade point average (GPA) of 3.9—about 2 SDs above the mean GPA. Would that change your guess? Probably not. The correlation between GPA and height is near 0, so knowing the GPA value doesn't tell you anything and doesn't move your guess. (And the standardized regression equation, $\hat{z}_y = rz_x$, tells us that as well, since it says that we should move 0×2 SDs from the mean.)

Sir Francis Galton was the first to speak of "regression," although others had fit lines to data by the same method.

The First Regression

Sir Francis Galton related the heights of sons to the heights of their fathers with a regression line. The slope of his line was less than 1. That is, sons of tall fathers were tall, but not as much above the average height as their fathers had been above their mean. Sons of short fathers were short, but generally not as far from their mean as their fathers. Galton interpreted the slope correctly as indicating a "regression" toward the mean height—and "regression" stuck as a description of the method he had used to find the line.

On the other hand, if you were told that, measured in centimeters, the student's height was 2 SDs above the mean, you'd know his height in inches. There's a perfect correlation between *Height* in inches and *Height* in centimeters ($r = 1$), so you know he's 2 SDs above mean height in inches as well. (The standardized regression equation would tell us to move 1.0×2 SDs from the mean.)

What if you were told that the student was 2 SDs above the mean in shoe size? Would you still guess that he's of average height? You might guess that he's taller than average, since there's a positive correlation between height and shoe size. But would you guess that he's 2 SDs above the mean? When there was no correlation, we didn't move away from the mean at all. With a perfect correlation, we moved our guess the full 2 SDs. Any correlation between these extremes should lead us to move somewhere between 0 and 2 SDs above the mean. (To be exact, the standardized regression equation tells us to move $r \times 2$ standard deviations away from the mean.)

Notice that if x is 2 SDs above its mean, we won't ever move more than 2 SDs away for y, since r can't be bigger than 1.0. So, each predicted y tends to be closer to its mean (in standard deviations) than its corresponding x was. This property of the linear model is called **regression to the mean**. This is why the line is called the regression line.

MATH BOX

Where does the equation of the line of best fit come from? To write the equation of any line, we need to know a point on the line and the slope. It's logical to expect that an average x will correspond to an average y, and, in fact, the line does pass through the point $(\bar{x}, \bar{y})$. (This is not hard to show as well.)

To think about the slope, we look once again at the z-scores. We need to remember a few things.

1. The mean of any set of z-scores is 0. This tells us that the line that best fits the z-scores passes through the origin (0, 0).

2. The standard deviation of a set of z-scores is 1, so the variance is also 1. This means that $\dfrac{\sum(z_y - \bar{z}_y)^2}{n-1} = \dfrac{\sum(z_y - 0)^2}{n-1} = \dfrac{\sum z_y^2}{n-1} = 1$, a fact that will be important soon.

3. The correlation is $r = \dfrac{\sum z_x z_y}{n-1}$, also important soon.

Remember that our objective is to find the slope of the best fit line. Because it passes through the origin, the equation of the best fit line will be of the form $\hat{z}_y = m z_x$. We want to find the value for m that will minimize the sum of the squared errors. Actually we'll divide that sum by $n - 1$ and minimize this mean squared error (MSE). Here goes:

Minimize:	$MSE = \dfrac{\sum(z_y - \hat{z}_y)^2}{n-1}$
Since $\hat{z}_y = m z_x$:	$MSE = \dfrac{\sum(z_y - m z_x)^2}{n-1}$
Square the binomial:	$= \dfrac{\sum(z_y^2 - 2m z_x z_y + m^2 z_x^2)}{n-1}$
Rewrite the summation:	$= \dfrac{\sum z_y^2}{n-1} - 2m\dfrac{\sum z_x z_y}{n-1} + m^2\dfrac{\sum z_x^2}{n-1}$

4. Substitute from (2) and (3): $= 1 - 2mr + m^2$

This last expression is a quadratic. A parabola in the form $y = ax^2 + bx + c$ reaches its minimum at its turning point, which occurs when $x = \dfrac{-b}{2a}$. We can minimize the mean of squared errors by choosing $m = \dfrac{-(-2r)}{2(1)} = r$.

The slope of the best fit line for z-scores is the correlation, r. This fact leads us immediately to two important additional results, listed here.

A slope with value r for z-scores means that a difference of 1 standard deviation in z corresponds to a difference of r standard deviations in $\hat{z}_y$. Translate that back to the original x and y values: "Over one standard deviation in x, up r standard deviations in $\hat{y}$."

The slope of the regression line is $b = \dfrac{rs_y}{s_x}$.

We know choosing $m = r$ minimizes the sum of the squared errors (SSE), but how small does that sum get? Equation (4) told us that the mean of the squared errors is $1 - 2mr + m^2$. When $m = r$, $1 - 2mr + m^2 = 1 - 2r^2 + r^2 = 1 - r^2$. This is the percentage of variability *not* explained by the regression line. Since $1 - r^2$ of the variability is *not* explained, the percentage of variability in y that is explained by x is r^2. This important fact will help us assess the strength of our models.

And there's still another bonus. Because r^2 is the percent of variability explained by our model, r^2 is at most 100%. If $r^2 \leq 1$, then $-1 \leq r \leq 1$, proving that correlations are always between -1 and $+1$.

Why _r_ for correlation?

In his original paper on correlation, Galton used r for the "index of correlation"—what we now call the correlation coefficient. He calculated it from the regression of y on x or of x on y after standardizing the variables, just as we have done. It's fairly clear from the text that he used r to stand for (standardized) regression.

8.4 Checking the Model

The linear regression model is perhaps the most widely used model in all of Statistics. It has everything we could want in a model: two easily estimated parameters, a meaningful measure of how well the model fits the data, and the ability to predict new values. It even provides a self-check in plots of the residuals to help us avoid all kinds of mistakes. Most models are useful only when specific **assumptions** are true. Of course, assumptions are hard—often impossible—to check. That's why we *assume* them. But we should check to see whether the assumptions are *reasonable*. Fortunately, we can often check *conditions* that provide information about the assumptions. For the linear model, we start by checking the same ones we checked in Chapter 7 for using correlation.

Linear models only make sense for quantitative data. The **Quantitative Data Condition** is pretty easy to check, but don't be fooled by categorical data recorded as numbers. You don't want to try to predict zip codes from credit card account numbers.

The regression model *assumes* that the relationship between the variables is, in fact, linear. If you try to model a curved relationship with a straight line, you'll usually get what you deserve. We can't ever verify that the underlying relationship between two variables is truly linear, but an examination of the scatterplot will let you decide whether the **Linearity Assumption** is reasonable. The **Linearity Condition** we used in Chapter 7 is designed to do precisely that and is satisfied if the scatterplot looks reasonably straight. If the scatterplot is not straight enough, stop. You can't use a linear model for just *any* two variables, even if they are related.

> **Make a Picture**
> Check the scatterplot. The shape must be linear, or you can't use regression for the variables in their current form. And watch out for outliers.

The two variables must have a *linear* association, or the model won't mean a thing. Some nonlinear relationships can be saved by re-expressing the data to make the scatterplot more linear.

Watch out for outliers. The linearity assumption also requires that no points lie far enough away to distort the line of best fit. Check the **Outlier Condition** to make sure no point needs special attention. Outlying values may have large residuals, and squaring makes their influence that much greater. Outlying points can dramatically change a regression model. Unusual observations can even change the sign of the slope, misleading us about the direction of the underlying relationship between the variables.

8.5 Learning More from the Residuals

We always check conditions with a scatterplot of the data, but we can learn even more after we've fit the regression model. There's extra information in the residuals that we can use to help us decide how reasonable our model is and how well the model fits. So, we plot the residuals and check the conditions again.

The residuals are the part of the data that *hasn't* been modeled. We can write

$$Data = Predicted + Residual$$

or, equivalently,

$$Residual = Data - Predicted.$$

Or, as we showed earlier, in symbols:

$$e = y - \hat{y}.$$

> **Why e for *residual*?**
> The easy answer is that *r* is already taken for correlation, but the truth is that *e* stands for "error." It's not that the data point is a mistake but that statisticians often refer to variability not explained by a model as error.

Residuals help us to see whether the model makes sense. When a regression model is appropriate, it should model the underlying relationship. Nothing interesting should be left behind. So after we fit a regression model, we usually plot the residuals in the hope of finding . . . nothing.

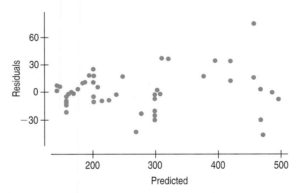

Figure 8.4 *A scatterplot of residuals against predicted values for the Best Buy data shows only that there are several values with the same predicted value. Otherwise, the plot is appropriately boring.*

We check the Linearity Condition and Outlier Condition in this plot. A scatterplot of the residuals versus the *x*-values should be a plot without patterns.[1] It shouldn't have any interesting features—no direction, no shape. It should stretch

---

[1] Most computer statistics packages plot the residuals as we did in Figure 8.4, against the predicted values, rather than against *x*. When the slope is positive, the scatterplots are virtually identical except for the axes labels. When the slope is negative, the two versions are mirror images. Since all we care about is the patterns (or, better, lack of patterns) in the plot, either plot is useful.

horizontally, showing no bends, and it should have no outliers. If you see nonlinearities, outliers, or clusters in the residuals, find out what the regression model missed.

Not only can the residuals help check the conditions, but they can also tell us how well the model performs. The better the model fits the data, the less the residuals will vary around the line. The standard deviation of the residuals, s_e, gives us a measure of how much the points spread around the regression line. Of course, for this summary to make sense, the residuals should all share the same underlying spread. So we must *assume* that the standard deviation around the line is the same wherever we want the model to apply.

This new assumption about the standard deviation around the line gives us a new condition, called the **Equal Spread Condition**. The associated question to ask is does the plot thicken—or fan out? We check to make sure that the spread is about the same throughout. We can check that either in the original scatterplot of y against x or in the scatterplot of residuals (or, preferably, in both plots). We estimate the **standard deviation of the residuals** in almost the way you'd expect:

$$s_e = \sqrt{\frac{\sum e^2}{n - 2}}.$$

> **Equal Spread Condition**
> This condition requires that the scatter is about equal for all x-values. It's often checked using a plot of residuals against predicted values. The underlying assumption of equal variance is also called **homoscedasticity**.

We don't need to subtract the mean of the residuals because $\bar{e} = 0$. Why divide by $n - 2$ rather than $n - 1$? We used $n - 1$ for s when we estimated the mean. Now we're estimating both a slope and an intercept. Looks like a pattern—and it is. We subtract one more for each parameter we estimate.

When we predicted use for 301 stores, we saw a residual of -43.3 MIPS. For the Best Buy data, the residual standard deviation (s_e) is 24.07 MIPS, so our prediction is about $-43.3/24.07 = -1.8$ standard deviations away from the actual value, a fairly typical size for residual since it's within 2 standard deviations.

8.6 Variation in the Model and R^2

The variation in the residuals is the key to assessing how well the model fits. *Monthly Use* has a standard deviation of 117.0 MIPS. If we had to guess how much capacity we'll need without knowing the number of stores, we might guess the mean of 276.5 MIPS. The SD would be around 117.0 MIPS. We might expect to be wrong by roughly twice the SD—plus or minus 234.0 MIPS—probably not accurate enough for planning. But after fitting the line, the residuals have a standard deviation of only 24.07 MIPS, so knowing the number of stores allows us to make much better predictions. If the correlation were 1.0 and the model predicted the capacity perfectly, the residuals would all be zero and have no variation. We couldn't possibly do any better than that.

How well have we done? Look at the boxplots.

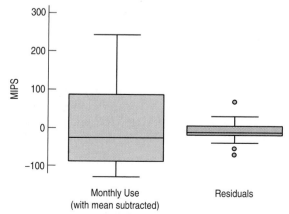

Figure 8.5 *The boxplots show how much the variation is reduced by fitting the linear regression model.*

If the correlation had been zero, the model would simply predict the mean (as we might do, if we didn't know the number of stores). The residuals from that prediction would just be the observed values minus their mean. These residuals would have the same spread as the original data because, as we know, just subtracting the mean doesn't change the spread.

How well does the Best Buy model do? The variation in the residuals is certainly a lot smaller than in the data, but still bigger than zero. How much of the variation is left in the residuals? If you had to put a number between 0% and 100% on the fraction of the variation left in the residuals, what would you say?

All regression models fall somewhere between the two extremes of zero correlation and perfect correlation ($r = \pm 1$). We'd like to gauge where our model falls. Can we use the correlation to do that? Well, a regression model with correlation -0.5 is doing as well as one with correlation $+0.5$. They just have different directions. But if we *square* the correlation coefficient, we'll get a value between 0 and 1, and the direction won't matter. It turns out, as we showed you in the Math Box, this works perfectly. The squared correlation, r^2, gives the fraction of the data's variation accounted for by the model, and $1 - r^2$ is the fraction of the original variation left in the residuals. For the Best Buy model, $r^2 = (0.979)^2 = 0.959$, and $1 - r^2$ is 0.041, so only 4.1% of the variability in monthly use has been left in the residuals.

All regression analyses include this statistic; although by tradition, it is written with a capital letter, R^2, pronounced "R-squared." An R^2 of 0 means that none of the variance in the data is in the model; all of it is still in the residuals. It would be hard to imagine using that model for anything. Because R^2 is a fraction of a whole, it is often given as a percentage.[2]

When interpreting a regression model, you need to report what R^2 means. According to our linear model, 95.9% of the variation in *Monthly Use* is accounted for by the number of stores.

◆ **How can we see that R^2 is really the fraction of variance accounted for by the model?** It's a simple calculation. The variance of *Monthly Use* is $117.09^2 = 13,710.7$. If we treat the residuals as data, the variance of the residuals is 566.8.[3] As a fraction of the variance of *Monthly Use*, that's 0.0413 or 4.13%. That's the fraction of the variance that is *not* accounted for by the model. The fraction that *is* accounted for is 100% − 4.13% = 95.9%, just the value we got for R^2.

<aside>
Sum of Squares

The sum of the squared residuals $\Sigma(y - \hat{y})^2$ is sometimes written as SSE (sum of squared errors). If we call $\Sigma(y - \bar{y})^2$ SST (for total sum of squares) then

$$R^2 = 1 - \frac{\text{SSE}}{\text{SST}}.$$
</aside>

<aside>
Is a correlation of 0.80 twice as strong as a correlation of 0.40? Not if you think in terms of R^2. A correlation of 0.80 means an R^2 of $0.80^2 = 64\%$. A correlation of 0.40 means an R^2 of $0.40^2 = 16\%$—only a quarter as much of the variability accounted for. A correlation of 0.80 gives an R^2 *four* times as strong as a correlation of 0.40 and accounts for four times as much of the variability.
</aside>

JUST CHECKING

Let's go back to our regression of sales ($000) on number of employees again.

$$\widehat{Sales} = 9.564 + 122.74 \, Employees$$

The R^2 value is reported as 71.4%.

6 What does the R^2 value mean about the relationship of *Sales* and *Employees*?

7 Is the correlation of *Sales* and *Employees* positive or negative? How do you know?

8 If we measured the *Sales* in thousands of euros instead of thousands of dollars, would the R^2 value change? How about the slope?

---

[2] By contrast, we give correlation coefficients as decimal values between −1.0 and 1.0.
[3] This isn't quite the same as squaring s_e which we discussed previously, but it's very close.

How Big Should R^2 Be?

The value of R^2 is always between 0% and 100%. But what is a "good" R^2 value? The answer depends on the kind of data you are analyzing and on what you want to do with it. Just as with correlation, there is no value for R^2 that automatically determines that the regression is "good." Data from scientific experiments often have R^2 in the 80% to 90% range and even higher. Data from observational studies and surveys, though, often show relatively weak associations because it's so difficult to measure reliable responses. An R^2 of 30% to 50% or even lower might be taken as evidence of a useful regression. The standard deviation of the residuals can give us more information about the usefulness of the regression by telling us how much scatter there is around the line.

As we've seen, an R^2 of 100% is a perfect fit, with no scatter around the line. The s_e would be zero. All of the variance would be accounted for by the model with none left in the residuals. This sounds great, but it's too good to be true for real data.[4]

8.7 Reality Check: Is the Regression Reasonable?

Statistics don't come out of nowhere. They are based on data. The results of a statistical analysis should reinforce common sense. If the results are surprising, then either you've learned something new about the world or your analysis is wrong.

Whenever you perform a regression, think about the coefficients and ask whether they make sense. Is the slope reasonable? Does the direction of the slope seem right? The small effort of asking whether the regression equation is plausible will be repaid whenever you catch errors or avoid saying something silly or absurd about the data. It's too easy to take something that comes out of a computer at face value and assume that it makes sense.

Always be skeptical and ask yourself if the answer is reasonable.

GUIDED EXAMPLE | *Home Size and Price*

Real estate agents know the three most important factors in determining the price of a house are *location, location,* and *location.* But what other factors help determine the price at which a house should be listed? Number of bathrooms? Size of the yard? A student amassed publicly available data on thousands of homes in upstate New York. We've drawn a random sample of 1057 homes to examine house pricing. Among the variables she collected were the total living area (in square feet), number of bathrooms, number of bedrooms, size of lot (in acres), and age of house (in years). We will investigate how well the size of the house, as measured by living area, can predict the selling price.

[4] If you see an R^2 of 100%, it's a good idea to investigate what happened. You may have accidentally regressed two variables that measure the same thing.

PLAN

Setup State the objective of the study.

Identify the variables and their context.

Model We need to check the same conditions for regression as we did for correlation. To do that, make a picture. Never fit a regression without looking at the scatterplot first.

Check the Linearity, Equal Spread, and Outlier Conditions.

We want to find out how well the living area of a house in upstate NY can predict its selling price.

We have two quantitative variables: the living area (in square feet) and the selling price ($). These data come from public records in upstate New York in 2006.

✓ **Quantitative Variables Condition**

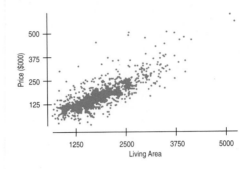

✓ **Linearity Condition** The scatterplot shows two variables that appear to have a fairly strong positive association. The plot appears to be fairly linear.

✓ **Outlier Condition** There appear to be a few possible outliers, especially among large, relatively expensive houses. A few smaller houses are expensive for their size. We will check their influence on the model later.

✓ **Equal Spread Condition** The scatterplot shows a consistent spread across all the x's we are modeling.

We have two quantitative variables that appear to satisfy the conditions, so we will model this relationship with a regression line.

DO

Mechanics Find the equation of the regression line using a statistics package. Remember to write the equation of the model using meaningful variable names.

Our software produces the following output.

```
Dependent variable is: Price
1057 total cases
R squared = 62.43%
s = 57930 with 1000 − 2 = 998 df
Variable          Coefficient
Intercept         6378.08
Living Area       115.13
```

Once you have the model, plot the residuals and check the Equal Spread Condition again.

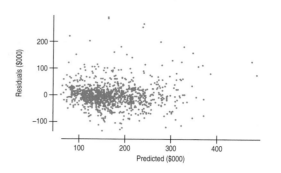

The residual plot appears generally patternless. The few relatively expensive small houses are evident, but setting them aside and refitting the model did not change either the slope or intercept very much so we left them in. There is a slight tendency for cheaper houses to have less variation, but the spread is roughly the same throughout.

REPORT

Conclusion Interpret what you have found in the proper context.

MEMO:

Re: Report on housing prices.

We examined how well the size of a house could predict its selling price. Data were obtained from recent sales of 1057 homes in upstate New York. The model is:

$$\widehat{Price} = \$6376.08 + 115.13 \times Living\ Area$$

In other words, from a base of $6376.08, houses cost about $115.13 per square foot in upstate NY.

This model appears reasonable from both a statistical and real estate perspective. While we know that size is not the only factor in pricing a house, the model accounts for 62.4% of the variation in selling price.

As a reality check, we checked with several real estate pricing sites (www.realestateabc.com, www.zillow.com) and found that houses in this region were averaging $100 to $150 per square foot, so our model is plausible.

Of course, not all house prices are predicted well by the model. We computed the model without several of these houses, but their impact on the regression model was small. We believe that this is a reasonable place to start to assess whether a house is priced correctly for this market. Future analysis might benefit by considering other factors.

WHAT CAN GO WRONG?

Regression analyses can be more subtle than they seem at first. Here are some guidelines to help you use this powerful method effectively.

- **Don't fit a straight line to a nonlinear relationship.** Linear regression is suited only to relationships that are, in fact, linear. Fortunately, we can often improve the linearity easily by using re-expression (see Chapter 17).

- **Beware of extraordinary points.** Data values can be extraordinary or unusual in a regression in two ways. They can have y-values that stand off from the linear pattern suggested by the bulk of the data. These are what we have been calling outliers; although with regression, a point can be an outlier by being far from the linear pattern even if it is not the largest or smallest y-value. Points can also be extraordinary in their x-values. Such points can exert a strong influence on the line. Both kinds of extraordinary points require attention.

- **Don't extrapolate far beyond the data. A linear model will often do a reasonable job of summarizing a relationship in the range of observed x-values.** Once we have a working model for the relationship, it's tempting to use it. But beware of predicting y-values for x-values that lie too far outside the range of the original data. The model may no longer hold there, so such extrapolations too far from the data are dangerous.

- **Don't infer that x causes y just because there is a good linear model for their relationship.** When two variables are strongly correlated, it is often tempting to assume a causal relationship between them. Putting a regression line on a scatterplot tempts us even further, but it doesn't make the assumption of causation any more valid.

- **Don't choose a model based on R^2 alone.** Although R^2 measures the *strength* of the linear association, a high R^2 does not demonstrate the *appropriateness* of the regression. A single unusual observation, or data that separate into two groups, can make the R^2 seem quite large when, in fact, the linear regression model is simply inappropriate. Conversely, a low R^2 value may be due to a single outlier. It may be that most of the data fall roughly along a straight line, with the exception of a single point. Always look at the scatterplot.

ETHICS IN ACTION

Jill Hathway is looking for a career change and is interested in starting a franchise. After spending the last 20 years working as a mid-level manager for a major corporation, Jill wants to indulge her entrepreneurial spirit and strike out on her own. She currently lives in a small southwestern city and is considering a franchise in the health and fitness industry. She is considering several possibilities including *Pilates One*, for which she requested a franchise packet. Included in the packet information were data showing how various regional demographics (age, gender, income) related to franchise success (revenue, profit, return on investment). *Pilates One* is a relatively new franchise with only a few scattered locations. Nonetheless, the company reported various graphs and data analysis results to help prospective franchisers in their decision-making process. Jill was particularly interested in the graph and the regression analysis that related the proportion of women over the age of 40 within a 20-mile radius of a *Pilates One* location to return on investment for the franchise. She noticed that there was a positive relationship. With a little research, she discovered that the proportion of women over the age of 40 in her city was higher than for any other *Pilates One* location (attributable, in part, to the large number of retirees relocating to the southwest). She then used the regression equation to project return on investment for a *Pilates One* located in her city and was very pleased with the result. With such objective data, she felt confident that *Pilates One* was the franchise for her.

ETHICAL ISSUE *Pilates One is reporting analysis based on only a few observations. Jill is extrapolating beyond the range of x-values (related to Item C, ASA Ethical Guidelines).*

ETHICAL SOLUTION *Pilates One should include a disclaimer that the analysis was based on very few observations and that the equation should not be used to predict success at other locations or beyond the range of x-values used in the analysis.*

What have we learned?

We've learned that when the relationship between quantitative variables is linear, a linear model can help summarize that relationship and give us insights about it.

- The regression (best fit) line doesn't pass through all the points, but it is the best compromise in the sense that the sum of squares of the residuals is the smallest possible.

We've learned several things the correlation, r, tells us about the regression:

- The slope of the line is based on the correlation, adjusted for the standard deviations of x and y. We've learned to interpret that slope in context.
- For each SD that a case is away from the mean of x, we expect it to be r SDs in y away from the y mean.
- Because r is always between -1 and $+1$, each predicted y is fewer SDs away from its mean than the corresponding x was, a phenomenon called *regression to the mean*.
- The square of the correlation coefficient, R^2, gives us the fraction of the variation of the response accounted for by the regression model. The remaining $1 - R^2$ of the variation is left in the residuals.

Terms

Intercept	The intercept, b_0, gives a starting value in y-units. It's the $\hat{y}$ value when x is 0. $$b_0 = \bar{y} - b_1\bar{x}$$
Least squares	A criterion that specifies the unique line that minimizes the variance of the residuals or, equivalently, the sum of the squared residuals.
Linear model (Line of best fit)	The linear model of the form $\hat{y} = b_0 + b_1 x$ fit by least squares. Also called the regression line. To interpret a linear model, we need to know the variables and their units.
Predicted value	The prediction for y found for each x-value in the data. A predicted value, $\hat{y}$, is found by substituting the x-value in the regression equation. The predicted values are the values on the fitted line; the points $(x, \hat{y})$ lie exactly on the fitted line.
Residual	The difference between the actual data value and the corresponding value predicted by the regression model—or, more generally, predicted by any model.
Regression line	The particular linear equation that satisfies the least squares criterion, often called the line of best fit.
Regression to the mean	Because the correlation is always less than 1.0 in magnitude, each predicted y tends to be fewer standard deviations from its mean than its corresponding x is from its mean.
R^2	• The square of the correlation between y and x • The fraction of the variability of y accounted for by the least squares linear regression on x • An overall measure of how successful the regression is in linearly relating y to x
Standard deviation of the residuals	s_e is found by: $$s_e = \sqrt{\frac{\sum e^2}{n - 2}}.$$
Slope	The slope, b_1, is given in y-units per x-unit. Differences of one unit in x are associated with differences of b_1 units in predicted values of y: $$b_1 = r\frac{s_y}{s_x}.$$

Skills

 PLAN

- Know how to identify response (y) and explanatory (x) variables in context.
- Understand how a linear equation summarizes the relationship between two variables.
- Recognize when a regression should be used to summarize a linear relationship between two quantitative variables.
- Know how to judge whether the slope of a regression makes sense.
- Examine a scatterplot of your data for violations of the Linearity, Equal Spread, and Outlier Conditions that would make it inappropriate to compute a regression.

- Understand that the least squares slope is easily affected by extreme values.
- Define residuals as the differences between the data values and the corresponding values predicted by the line and that the Least Squares Criterion finds the line that minimizes the sum of the squared residuals.

- Know how to find the slope and intercept values of a regression.
- Be able to use regression to predict a value of *y* for a given *x*.
- Know how to compute the residual for each data value and how to compute the standard deviation of the residuals.
- Be able to evaluate the Equal Spread Condition with a scatterplot of the residuals after computing the regression.

 REPORT

- Write a sentence explaining what a linear equation says about the relationship between *y* and *x*, basing it on the fact that the slope is given in *y*-units per *x*-unit.
- Understand how the correlation coefficient and the regression slope are related. Know that R^2 describes how much of the variation in *y* is accounted for by its linear relationship with *x*.
- Be able to describe a prediction made from a regression equation, relating the predicted value to the specified *x*-value.

Technology Help: Regression

All statistics packages make a table of results for a regression. These tables may differ slightly from one package to another, but all are essentially the same—and all include much more than we need to know for now. Every computer regression table includes a section that looks something like this:

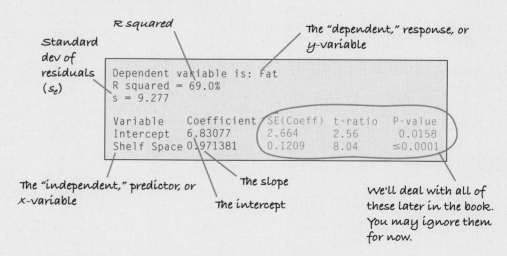

The slope and intercept coefficient are given in a table such as this one. Usually the slope is labeled with the name of the *x*-variable, and the intercept is labeled "Intercept" or "Constant." So the regression equation shown here is

$$\widehat{Sales} = 6.83077 + 0.97138 \; Shelf \; Space.$$

It is not unusual for statistics packages to give many more digits of the estimated slope and intercept than could possibly be estimated from the data. (The original data were reported to the nearest gram.) Ordinarily, you should round most of the reported numbers to one digit more than the precision of the data, and the slope to two. We will learn about the other numbers in the regression table later in the book. For now, all you need to be able to do is find the coefficients, the s_e and the R^2 value.

EXCEL

Make a scatterplot of the data. With the scatterplot frontmost, select **Add Trendline ...** from the **Chart** menu. Click the **Options** tab and select **Display Equation on Chart.** Click **OK.**

Comments

The computer section for Chapter 7 shows how to make a scatterplot. We don't repeat those steps here.

EXCEL 2007

- Click on a blank cell in the spreadsheet.
- Go to the **Formulas** tab in the Ribbon and click **More Functions** → **Statistical.**
- Choose the **CORREL** function from the drop-down menu of functions.
- In the dialog that pops up, enter the range of one of the variables in the space provided.
- Enter the range of the other variable in the space provided.
- Click **OK.**

Comments

The correlation is computed in the selected cell. Correlations computed this way will update if any of the data values are changed.

Before you interpret a correlation coefficient, always make a scatterplot to check for nonlinearity and outliers. If the variables are not linearly related, the correlation coefficient cannot be interpreted.

MINITAB

Choose **Regression** from the **Stat** menu. From the Regression submenu, choose **Fitted Line Plot.** In the Fitted Line Plot dialog, click in the **Response Y** box, and assign the y-variable from the Variable list. Click in the **Predictor X** box, and assign the x-variable from the Variable list. Make sure that the Type of Regression Model is set to Linear. Click the **OK** button.

SPSS

From the **Analyze** menu, choose **Regression > Linear ...** In the Linear Regression dialog, specify the **Dependent** (y), and **Independent** (x) variables. Click the **Plots** button to specify plots and Normal Probability Plots of the residuals. Click **OK.**

JMP

Choose **Fit Y by X** from the **Analyze** menu. Specify the y-variable in the Select Columns box and click the **"Y, Response"** button. Specify the x-variable and click the **"X, Factor"** button. Click **OK** to make a scatterplot. In the scatterplot window, click on the red triangle beside the heading labeled "Bivariate Fit . . ." and choose **"Fit Line."** JMP draws the least squares regression line on the scatterplot and displays the results of the regression in tables below the plot.

DATA DESK

Select the y-variable and the x-variable. In the **Plot** menu choose **Scatterplot.** From the scatterplot HyperView menu, choose **Add Regression Line** to display the line. From the HyperView menu, choose **Regression** to compute the regression.

Comments

Alternatively, find the regression first with the **Regression** command in the **Calc** menu. Click on the x-variable's name to open a menu that offers the scatterplot.

Mini Case Study Projects

Cost of Living

The Mercer Human Resource Consulting website (www.mercerhr.com) lists prices of certain items in selected cities around the world. They also report an overall cost-of-living index for each city compared to the costs of hundreds of items in New York City. For example, London at 110.6 is 10.6% more expensive than New York. You'll find the 2006 data for 16 cities in the data set **ch08_MCSP_Cost_of_Living**. Included are the 2006 cost of living index, cost of a luxury apartment (per month), price of a bus or subway ride, price of a compact disc, price of an international newspaper, price of a cup of coffee (including service), and price of a fast-food hamburger meal. All prices are in U.S. dollars.

Examine the relationship between the overall cost of living and the cost of each of these individual items. Verify the necessary conditions and describe the relationship in as much detail as possible. (Remember to look at direction, form, and strength.) Identify any unusual observations.

Based on the correlations and linear regressions, which item would be the best predictor of overall cost in these cities? Which would be the worst? Are there any surprising relationships? Write a short report detailing your conclusions.

Mutual Funds

According to the U.S. Securities and Exchange Commission (SEC), a mutual fund is a professionally-managed collection of investments for a group of investors in stocks, bonds, and other securities. The fund manager manages the investment portfolio and tracks the wins and losses. Eventually the dividends are passed along to the individual investors in the mutual fund. The first group fund was founded in 1924, but the spread of these types of funds was slowed by the stock market crash in 1929. Congress passed the Securities Act in 1933 and the Securities Exchange Act in 1934 to require that investors be provided disclosures about the fund, the securities, and the fund manager. The SEC drafted the Investment Company Act, which provided guidelines for registering all funds with the SEC. By the end of the 1960s, funds reported $48 billion in assets and, by October 2007 there were over 8,000 mutual funds with combined assets under management of over $12 trillion.

Investors often choose mutual funds on the basis of past performance, and many brokers, mutual fund companies, and other websites offer such data. In the file **ch08_MCSP_Mutual_Funds**, you'll find the 3-month Return, the annualized 1 yr, 5 yr, and 10 yr returns, and the return since inception of 64 funds of various types. Which data from the past provides the best predictions of the recent 3 months? Examine the scatterplots and regression models for predicting 3-month returns and write a short report containing your conclusions.

EXERCISES

T 1. Pizza sales and price. A linear model fit to predict weekly *Sales* of frozen pizza (in pounds) from the average *Price* ($/unit) charged by a sample of stores in the city of Dallas in 39 recent weeks is:

$$\widehat{Sales} = 141{,}865.53 - 24{,}369.49\ Price.$$

a) What is the explanatory variable?
b) What is the response variable?
c) What does the slope mean in this context?
d) What does the *y*-intercept mean in this context? Is it meaningful?
e) What do you predict the sales to be if the average price charged was $3.50 for a pizza?
f) If the sales for a price of $3.50 turned out to be 60,000 pounds, what would the residual be?

2. Used Saab prices. A linear model to predict the *Price* of a 2004 Saab 9-3 (in $) from its *Mileage* (in miles) was fit to 38 cars that were available during the week of January 11, 2008 (Kelly's Blue Book, www.kbb.com). The model was:

$$\widehat{Price} = 24{,}356.15 - 0.0151\ Mileage.$$

a) What is the explanatory variable?
b) What is the response variable?
c) What does the slope mean in this context?
d) What does the *y*-intercept mean in this context? Is it meaningful?
e) What you do predict the price to be for a car with 100,000 miles on it?
f) If the price for a car with 100,000 miles on it was $24,000, what would the residual be?

T 3. Football salaries. Is there a relationship between total team salary and the performance of teams in the National Football League (NFL)? For the 2006 season, a linear model predicting *Wins* (out of 16 regular season games) from the total team *Salary* ($M) for the 32 teams in the league is:

$$\widehat{Wins} = 1.783 + 0.062\ Salary.$$

a) What is the explanatory variable?
b) What is the response variable?
c) What does the slope mean in this context?
d) What does the *y*-intercept mean in this context? Is it meaningful?
e) If one team spends $10 million more than another on salary, how many more games on average would you predict them to win?
f) If a team spent $50 million on salaries and won 8 games, would they have done better or worse than predicted?
g) What would the residual of the team in part f be?

T 4. Baseball salaries. In 2007, the Boston Red Sox won the World Series and spent $143 million on salaries for their players (benfry.com/salaryper). Is there a relationship between salary and team performance in Major League Baseball? For the 2007 season, a linear model fit to the number of *Wins* (out of 162 regular

season games) from the team *Salary* ($M) for the 30 teams in the league is:

$$\widehat{Wins} = 70.097 + 0.132\ Salary.$$

a) What is the explanatory variable?
b) What is the response variable?
c) What does the slope mean in this context?
d) What does the *y*-intercept mean in this context? Is it meaningful?
e) If one team spends $10 million more than another on salaries, how many more games on average would you predict them to win?
f) If a team spent $110 million on salaries and won half (81) of their games, would they have done better or worse than predicted?
g) What would the residual of the team in part f be?

T 5. Pizza sales and price, revisited. For the data in Exercise 1, the average *Sales* was 52,697 pounds (SD = 10,261 pounds), and the correlation between *Price* and *Sales* was = −0.547.

If the *Price* in a particular week was one SD higher than the mean *Price*, how much pizza would you predict was sold that week?

6. Used Saab prices, revisited. The 38 cars in Exercise 2 had an average *Price* of $23,847 (SD = $923), and the correlation between *Price* and *Mileage* was = −0.169.

If the *Mileage* of a 2004 Saab was 1 SD below the average number of miles, what *Price* would you predict for it?

7. Sales by region. A sales manger for a major pharmaceutical company analyzes last year's sales data for her 96 sales representatives, grouping them by region (1 = East Coast U.S.; 2 = Mid West U.S.; 3 = West U.S.; 4 = South U.S.; 5 = Canada; 6 = Rest of World). She plots *Sales* (in $1000) against *Region* (1–6) and sees a strong negative correlation.

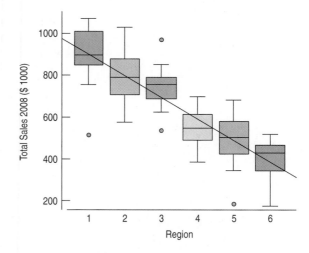

She fits a regression to the data and finds:

$$\widehat{Sales} = 1002.5 - 102.7\ Region.$$

The R^2 is 70.5%.

Write a few sentences interpreting this model and describing what she can conclude from this analysis.

8. Salary by job type. Recall the human resource manager from Chapter 7 Exercise 26 who was trying to understand the relationship between job type and absenteeism. Now he wants to examine salary in order to prepare annual reviews. He selects 28 employees at random with job types ranging from 01 = Stocking clerk to 99 = President. He plots *Salary* ($) against *Job Type* and finds a strong linear relationship with a correlation of 0.96.

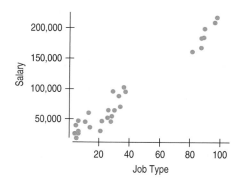

The regression output gives:

$$\widehat{Salary} = 15827.9 + 1939.1 \; Job \; Type$$

Write a few sentences interpreting this model and describing what he can conclude from this analysis.

T 9. GDP growth. Is economic growth in the developing world related to growth in the industrialized countries? Here's a scatterplot of the growth (in % of Gross Domestic Product) of the developing countries vs. the growth of developed countries for 180 countries as grouped by the World Bank (www.ers.usda.gov/data/macroeconomics). Each point represents one of the years from 1970 to 2007. The output of a regression analysis follows.

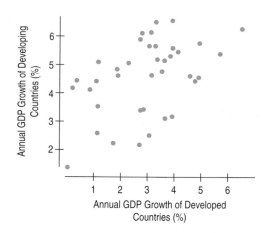

Dependent variable: GDP Growth Developing Countries
R^2 = 20.81%

Variable	Coefficient
Intercept	3.46
GDP Growth Developed Countries	0.433

$s = 1.244$

a) Check the assumptions and conditions for the linear model.
b) Explain the meaning of R^2 in this context.
c) What are the cases in this model?

T 10. European GDP growth. Is economic growth in Europe related to growth in the United States? Here's a scatterplot of the average growth in 25 European countries (in % of Gross Domestic Product) vs. the growth in the United States. Each point represents one of years from 1970 to 2007.

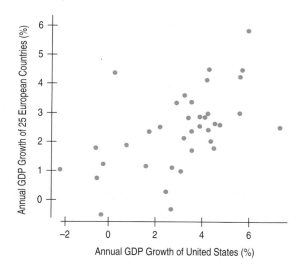

Dependent variable: 25 European Countries GDP Growth
R^2 = 29.65%

Variable	Coefficient
Intercept	1.330
U.S. GDP Growth	0.3616

$s = 1.156$

a) Check the assumptions and conditions for the linear model.
b) Explain the meaning of R^2 in this context.

T 11. GDP growth part 2. From the linear model fit to the data on GDP growth of Exercise 9.

a) Write the equation of the regression line.
b) What is the meaning of the intercept? Does it make sense in this context?
c) Interpret the meaning of the slope.
d) In a year in which the developed countries grow 4%, what do you predict for the developing world?
e) In 2007, the developed countries experienced a 2.65% growth, while the developing countries grew at a rate of 6.09%. Is this more or less than you would have predicted?
f) What is the residual for this year?

T 12. European GDP growth part 2. From the linear model fit to the data on GDP growth of Exercise 10.
a) Write the equation of the regression line.
b) What is the meaning of the intercept? Does it make sense in this context?
c) Interpret the meaning of the slope.
d) In a year in which the United States grows at 0%, what do you predict for European growth?
e) In 2007, the United States experienced a 3.20% growth, while Europe grew at a rate of 2.16%. Is this more or less than you would have predicted?
f) What is the residual for this year?

T 13. Tuition. All 50 states offer public higher education through four-year colleges and universities and two-year colleges (often called community colleges). Tuition charges by different states vary widely for both types. Would you expect to find a relationship between the tuition states charge for the two types?
a) Using the data on the disk, make a scatterplot of the average tuition for four-year colleges against the tuition charged for two-year colleges. Describe the relationship.
b) Is the direction of the relationship what you expected?
c) What is the regression equation for predicting the tuition at a four-year college from the tuition at a two-year college in the same state?
d) Is a linear model appropriate?
e) How much more do states charge on average in yearly tuition for four-year colleges compared to two-year colleges according to this model?
f) What is the R^2 value for this model? Explain what it says.

T 14. Tuition, public versus private. Exercise 13 examined the relationship between the tuition charged by states for four-year colleges and universities compared to the tuition for two-year colleges. Now, examine the relationship between private and public four-year colleges and universities in the states.
a) Would you expect the relationship between tuition ($ per year) charged by private and public four-year colleges and universities to be as strong as the relationship between public four-year and two-year institutions?
b) Using the data on the disk, examine a scatterplot of the average tuition for four-year private institutions against the tuition charged for four-year public institutions. Describe the relationship.
c) What is the regression equation for predicting the tuition at a four-year private institution from the tuition at a four-year public institution in the same state?
d) Is a linear model appropriate?
e) Interpret the regression equation. How much more is the tuition for four-year private institutions compared to four-year public institutions in the same state according to this model?
f) What is the R^2 value for this model? Explain what it says.

15. Mutual funds. As the nature of investing shifted in the 1990's, (more day traders and faster flow of information using technology), the relationship between mutual fund monthly performance (*Return*) in percent and money flowing (*Flow*) into

mutual funds ($ million) shifted. Using only the values for the 1990s (we'll examine later years in later chapters), answer the following questions. (You may assume that the assumptions and conditions for regression are met.)

The least squares linear regression is:

$$\widehat{Flow} = 9747 + 771\ Return.$$

a) Interpret the intercept in the linear model.
b) Interpret the slope in the linear model.
c) What is the predicted fund *Flow* for a month that had a market *Return* of 0%?
d) If during this month, the recorded fund *Flow* was $5 billion, what is the residual using this linear model? Did the model provide an underestimate or overestimate for this month?

16. Online clothing purchases. An online clothing retailer examined their transactional database to see if total yearly *Purchases* ($) were related to customers' *Incomes* ($). (You may assume that the assumptions and conditions for regression are met.)

The least squares linear regression is:

$$\widehat{Purchases} = -31.6 + 0.012\ Income.$$

a) Interpret the intercept in the linear model.
b) Interpret the slope in the linear model.
c) If a customer has an *Income* of $20,000, what is his predicted total yearly *Purchases*?
d) This customer's yearly *Purchases* were actually $100. What is the residual using this linear model? Did the model provide an underestimate or overestimate for this customer?

T 17. Home Depot. Analysts at The Home Depot want to predict quarterly sales from U.S. housing starts. They use *Quarterly Sales* at The Home Depot from 1995 to 2004 and U.S. *Housing Starts* and find the correlation is 0.70. They then examine the scatterplot and decide it is appropriate to fit a regression model to predict *Sales* ($B) from *Housing Starts* (in thousands).
a) What units does the slope have?
b) What is the R^2 value for the model?
c) What would you predict about the *Sales* for a quarter that has housing starts one standard deviation below average in *Housing Starts*?

T 18. House prices. House prices are subject to a variety of economic factors but are, to some extent, based on the living area of the house. Analysts examined the recent sales of 1000 homes and found the correlation to be 0.79. After examining a scatterplot, they decide a linear model is appropriate and fit a regression model to predict *House Price* ($) from *Living Area* (sq. ft).
a) What units does the slope have?
b) What is the R^2 value for the model?
c) What would you predict about the *Price* of a house that is two standard deviations larger in *Living Area* than the mean?

19. Retail sales. Sales are often related to economic indicators. One possible indicator is the unemployment rate. Data for a large retail store were used to obtain a linear regression model to predict quarterly *Sales* ($B) in the United States based on the U.S. unemployment *Rate* (in %) over a period of four years. This regression model produced an $R^2 = 88.3\%$ and a slope of -2.99.

a) Interpret the meaning of R^2.
b) What is the correlation of *Sales* and unemployment *Rate*?
c) If a quarter has an unemployment *Rate* 1% larger than another, what is the predicted impact on *Sales*?

T 20. Pizza sales and price, part 3. The linear model in Exercise 1 predicting *Sales* of frozen pizza (in pounds) from *Price* ($/unit) has an R^2 of 32.9% and slope of $-24,369.5$.

a) Interpret the meaning of R^2.
b) What is the correlation of *Sales* and *Price*?
c) If in one week the *Price* is $0.50 higher than another, what is the predicted difference in *Sales*?

21. Residual plots. Tell what each of the following residual plots indicates about the appropriateness of the linear model that was fit to the data.

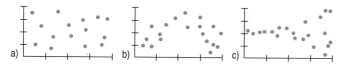

22. Residual plots, again. Tell what each of the following residual plots indicates about the appropriateness of the linear model that was fit to the data.

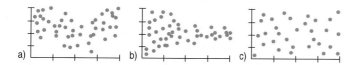

23. More Home Depot. Consider the quarterly Home Depot *Sales* in Exercise 17 again. The regression analysis gives the model:

$$\widehat{Sales} = -11.5 + 0.0535 \; Housing \; Starts.$$

a) Explain what the slope of the line says.
b) What would you predict for quarterly sales when housing starts are 500,000 units?
c) If quarterly sales are $3 billion higher than predicted given the reported housing starts during a quarter, what is this difference called?

24. Last retail sales. Consider the regression described in Exercise 19 again. The regression analysis gives the model:

$$\widehat{Sales} = 20.91 - 2.994 \; Rate.$$

a) Explain what the slope of the line says.
b) If the unemployment *Rate* is 6.0%, how much do you predict *Sales* will be?

c) If the unemployment *Rate* next quarter is 4.0% and *Sales* are reported as $8.5 billion, is this less than or more than you would predict? By how much? What is that called?

T 25. Consumer spending. An analyst at a large credit card bank is looking at the relationship between customers' charges to the bank's card in two successive months. He selects 150 customers at random, regresses charges in *March* ($) on charges in *February* ($), and finds an R^2 of 79%. The intercept is $730.20, and the slope is 0.79. After verifying all the data with the company's CPA, he concludes that the model is a useful one for predicting one month's charges from the other. Examine the data on the disk and comment on his conclusions.

T 26. Insurance policies. An actuary at a mid-sized insurance company is examining the sales performance of the company's sales force. She has data on the average size of the policy ($) written in two consecutive years by 200 salespeople. She fits a linear model and finds the slope to be 3.00 and the R^2 is 99.92%. She concludes that the predictions for next year's policy size will be very accurate. Examine the data on the disk and comment on her conclusions.

T 27. Supermarket sales. A regional high-end specialty supermarket is considering opening a new store and is curious about the relationship between demographic data and store sales for their existing stores. For example, are store sales related to the population in the town where the store is located? Data for 10 stores in the northeastern United States in 2000 produced this scatterplot and regression. (Population numbers are from the 2000 Census.)

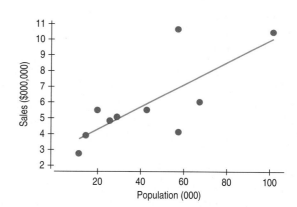

Predictor	Coef
Intercept	2.924
Population	0.0703
$s = 1.842$	R-Sq = 56.9%

a) Do you think a linear model is appropriate here? Explain.
b) What is the correlation between store sales and town population?
c) Explain the meaning of R^2 in this context.

T **28. More supermarket sales.** Take another look at the regression analysis of *Sales* and *Population* in Exercise 27.

a) Estimate the *Sales* of a store located in a town with a population of 80,000.

b) Interpret the meaning of the slope of the regression line in this context.

c) What does the intercept mean? Does this make sense?

29. What slope? If you create a regression model for predicting the sales ($ million) from money spent on advertising the prior month ($ thousand), is the slope most likely to be 0.03, 300 or 3000? Explain.

30. What slope, part 2? If you create a regression model for estimating a student's business school GPA (on a scale of 1–5) based on his math SAT (on a scale of 200–800), is the slope most likely to be 0.01, 1, or 10? Explain.

31. Misinterpretations. An advertising agent who created a regression model using amount spent on *Advertising* to predict annual *Sales* for a company made these two statements. Assuming the calculations were done correctly, explain what is wrong with each interpretation.

a) My R^2 of 93% shows that this linear model is appropriate.

b) If this company spends $1.5 million on advertising, then annual sales will be $10 million.

32. More misinterpretations. An economist investigated the association between a country's *Literacy Rate* and *Gross Domestic Product (GDP)* and used the association to draw the following conclusions. Explain why each statement is incorrect. (Assume that all the calculations were done properly.)

a) The *Literacy Rate* determines 64% of the *GDP* for a country.

b) The slope of the line shows that an increase of 5% in *Literacy Rate* will produce a $1 billion improvement in *GDP*.

33. Business admissions. An analyst at a business school's admissions office claims to have developed a valid linear model predicting success (measured by starting salary ($) at time of graduation) from a student's undergraduate performance (measured by GPA). Describe how you would check each of the four regression conditions in this context.

34. School rankings. A popular magazine annually publishes rankings of both U.S. business programs and international business programs. The latest issue claims to have developed a linear model predicting the school's ranking (with "1" being the highest ranked school) from its financial resources (as measured by size of the school's endowment). Describe how you would apply each of the four regression conditions in this context.

T **35. SAT scores.** The SAT is a test often used as part of an application to college. SAT scores are between 200 and 800 but have no units. Tests are given in both math and verbal areas. Doing the SAT math problems also involves the ability to read and understand the questions, but can a person's verbal score be used to predict the math score? Verbal and Math SAT scores of a high school graduating class are displayed in the scatterplot, with the regression line added.

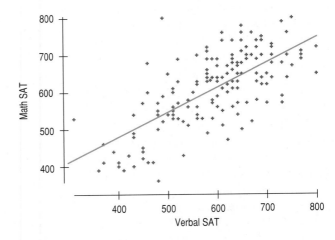

a) Describe the relationship.

b) Are there any students whose scores do not seem to fit the overall pattern?

c) For these data, $r = 0.685$. Interpret this statistic.

d) The verbal scores averaged 596.3, with a standard deviation of 99.5, and the math scores averaged 612.1, with a standard deviation of 98.1. Write the equation of the regression line.

e) Interpret the slope of the line.

f) Predict the math score of a student with a verbal score of 500.

g) Every year some student scored a perfect 1600. Based on this model, what would that student's residual be for her math score?

36. Success in college. Colleges use SAT scores in the admissions process because they believe these scores provide some insight into how a high school student will perform at the college level. Suppose the entering freshmen at a certain college have mean combined *SAT scores* of 1222, with a standard deviation of 83. In the first semester, these students attained a mean GPA of 2.66, with a standard deviation of 0.56. A scatterplot showed the association to be reasonably linear, and the correlation between SAT score and GPA was 0.47.

a) Write the equation of the regression line.

b) Explain what the *y*-intercept of the regression line indicates.

c) Interpret the slope of the regression line.

d) Predict the GPA of a freshman who scored a combined 1400.

e) Based upon these statistics, how effective do you think SAT scores would be in predicting academic success during the first semester of the freshman year at this college? Explain.

f) As a student, would you rather have a positive or a negative residual? Explain.

37. SAT scores, part 2. Suppose we wanted instead to use SAT math scores to estimate verbal scores based on the data in Exercise 35.

a) What is the correlation?

b) Write the equation of the regression of predicted *Verbal* scores on *Math* scores.

c) What would a positive residual mean in this context?

d) Predict the *Verbal* score of someone with a *Math* score of 500.

e) Using the predicted value and the equation you created in Exercise 35, predict her *Math* score.

f) Why doesn't the result in part e come out to 500?

38. Success in college, part 2. Suppose we wanted to use data from Exercise 36 and use a student's first semester GPA to predict what their SAT score was. This new regression produces an R^2 of 22.1%.

a) What is the correlation?

b) Compare this to the correlation in Exercise 36.

c) What would the predicted SAT score be among freshmen who attained a first semester GPA of 3.0?

39. Used BMW prices. A business student needs cash, so he decides to sell his car. The car is a valuable BMW 840 that was only made over the course of a few years in the late 1990s. He would like to sell it on his own, rather than through a dealer so he'd like to predict the price he'll get for his car's model year.

a) Make a scatterplot for the data on used BMW 840's provided.

b) Describe the association between year and price.

c) Do you think a linear model is appropriate?

d) Computer software says that $R^2 = 57.4\%$. What is the correlation between year and price?

e) Explain the meaning of R^2 in this context.

f) Why doesn't this model explain 100% of the variability in the price of a used BMW 840?

40. More used BMW prices. Use the advertised prices for BMW 840s given in Exercise 39 to create a linear model for the relationship between a car's *Year* and its *Price*.

a) Find the equation of the regression line.

b) Explain the meaning of the slope of the line.

c) Explain the meaning of the intercept of the line.

d) If you want to sell a 1997 BMW 840, what price seems appropriate?

e) You have a chance to buy one of two cars. They are about the same age and appear to be in equally good condition. Would you rather buy the one with a positive residual or the one with a negative residual? Explain.

41. Supermarket sales, revisited. The following table shows the total sales for 10 high-end supermarket stores in the northeastern United States. These are the same stores that you examined in Exercises 27 and 28. In addition to the population of the towns in which the stores are located, supermarket executives believe their store products appeal to a younger generation. Here are the data for total annual *Sales* and *Median Age* of residents of the town in 2000.

Sales ($M)	5.540	10.700	10.532	5.995	5.090
Median Age	39.5	34.5	30.4	36.2	40.8

Sales ($M)	3.955	2.774	4.828	5.511	4.195
Median Age	41.5	34.7	41.4	38.0	40.0

a) Examine a scatterplot and describe the general relationship between *Sales* and *Median Age* of the town residents.

b) Do you think a linear model is appropriate?

c) The marketing manager fit the regression line. What is the equation of that line?

d) Using this model, predict the annual sales of a supermarket of this type in a town whose median age is 32 years.

e) Do you have any reservations about the prediction you found in part d? Explain.

42. Supermarket sales, final visit. Executives at the same high-end specialty supermarket in Exercise 27 are trying to determine the best location for a new store. They believe that sales are also related to the total number of housing units in the towns. Based on 2000 Census data, they would like to develop a model to describe this relationship.

a) Examine a scatterplot and describe the general relationship between *Sales* and *Housing Units* in the town.

b) Do you think a linear model is appropriate?

c) The marketing manager fit the regression line. What is the equation of that line?

d) Using this model, predict the annual sales of a supermarket of this type in a town that has 100,000 housing units.

e) Do you have any reservations about the prediction you found in part d? Explain.

43. Cost of living. The *Worldwide Cost of Living Survey City Rankings* determine the cost of living in the most expensive cities in the world as an index. This index scales New York City as 100 and expresses the cost of living in other cities as a percentage of the New York cost. For example, in 2007, the cost of living index in Tokyo was 122.1, which means that it was 22% higher than New York. The scatterplot shows the index for 2007 plotted against the 2006 index for the 15 most expensive cities of 2007.

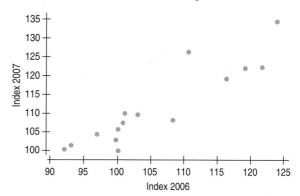

a) Describe the association between cost of living indices in 2007 and 2006.

b) The R^2 for the regression equation is 0.837. Interpret the value of R^2.

c) Using the data provided, find the correlation.

d) Predict the 2007 cost of living of Moscow and find its residual.

T 44. Lobster prices. Over the past few decades both the demand for lobster and the price of lobster have continued to increase. The scatterplot shows this increase in the *Price* of Maine lobster (*Price*/pound) since 1990.

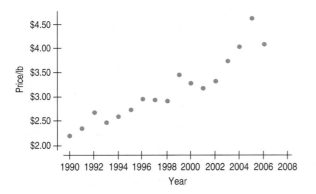

a) Describe the increase in the *Price* of lobster since 1990.
b) The R^2 for the regression equation is 88.5%. Interpret the value of R^2.
c) Find the correlation.
d) Find the linear model and examine the plot of residuals versus predicted values. Is the Equal Spread Condition satisfied? (Use time starting at 1990 so that 1990 = 0.)

T 45. Wal-Mart sales. The table shows the increase in Wal-Mart sales for the past few years. Using these data, find a linear model to predict *Sales*.

Wal-Mart Net Sales ($ billion)	Year
180.787	2001
204.011	2002
226.479	2003
252.792	2004
281.488	2005
308.945	2006

a) Using technology, obtain a regression equation for these data. (Enter *Year* as 0, 1, 2, etc.)
b) Interpret the meaning of the slope.
c) Interpret the intercept in your equation.

T 46. Lobster value. Maine catches and sells the most lobster of any state in the United States. The *Value* of the total lobster catch ($ million) for the state of Maine since 1990 is shown in the scatterplot.

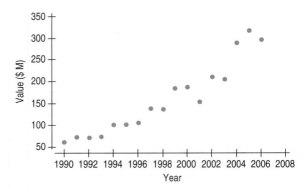

a) Find the linear model for the increase in the *Value* of the total lobster catch in the state of Maine since 1990. (Use the time variable, where 1990 = 0.)
b) Interpret the meaning of the slope and intercept in your equation.
c) Examine the plot of residuals versus predicted values. Is the Equal Spread Condition satisfied?

T 47. Wal-Mart sales, again. The table in Exercise 45 shows the increase in Wal-Mart sales for the past few years.

a) Use your model obtained in Exercise 45 to make a prediction for Wal-Mart *Sales* in 2007.
b) Discuss the danger in using your model from Exercise 45 to make a prediction of Wal-Mart *Sales* in 2010.

T 48. Lobster value, again. The plot in Exercise 46 shows the increase in the *Value* of the lobster catch in the state of Maine since 1990.

a) Use your model obtained in Exercise 46 to make a prediction for the *Value* of the lobster catch in 2007.
b) Discuss the danger in using your model from Exercise 46 to make a prediction for 2010.

49. El Niño. Concern over the weather associated with El Niño has increased interest in the possibility that the climate on Earth is getting warmer. The most common theory relates an increase in atmospheric levels of carbon dioxide (CO_2), a greenhouse gas, to increases in temperature. Here is a scatterplot showing the mean annual CO_2 concentration in the atmosphere, measured in parts per million (ppm) at the top of Mauna Loa in Hawaii, and the mean annual air temperature over both land and sea across the globe, in degrees Celsius (C).

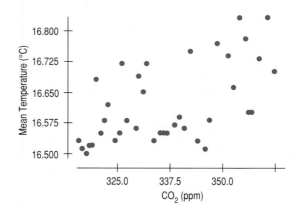

A regression predicting *Mean Temperature* from CO_2 produces the following output table (in part).

Dependent variable: Temperature
R-squared = 33.4%

Variable	Coefficient
Intercept	15.3066
CO2	0.004

a) What is the correlation between CO_2 and *Mean Temperature?*
b) Explain the meaning of *R*-squared in this context.
c) Give the regression equation.
d) What is the meaning of the slope in this equation?
e) What is the meaning of the intercept of this equation?
f) Here is a scatterplot of the residuals vs. CO_2. Does this plot show evidence of the violations of any of the assumptions of the regression model? If so, which ones?
g) CO_2 levels may reach 364 *ppm* in the near future. What *Mean Temperature* does the model predict for that value?

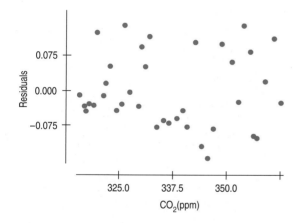

T 50. U.S. birthrates. The table shows the number of live births per 1000 women aged 15–44 years in the United States, starting in 1965. (National Center for Health Statistics, www.cdc.gov/nchs/)

Year	1965	1970	1975	1980	1985
Rate	19.4	18.4	14.8	15.9	15.6

Year	1990	1995	2000	2005
Rate	16.4	14.8	14.4	14.0

a) Make a scatterplot and describe the general trend in *Birthrates*. (Enter *Year* as years since 1900: 65, 70, 75, etc.)
b) Find the equation of the regression line.
c) Check to see if the line is an appropriate model. Explain.
d) Interpret the slope of the line.
e) The table gives rates only at 5-year intervals. Estimate what the rate was in 1978.
f) In 1978, the birthrate was actually 15.0. How close did your model come?
g) Predict what the *Birthrate* will be in 2010. Comment on your faith in this prediction.
h) Predict the *Birthrate* for 2025. Comment on your faith in this prediction.

T 51. Heptathlon. The heptathlon is an Olympic event for women combining scores on seven events. The table on the next page shows the results from the high jump, 800-meter run, and long jump for the 26 women who successfully completed all three events in 2004 (www.espn.com).

Let's examine the association among these events. Perform a regression to predict high jump performance from the 800-meter results.

a) What is the regression equation? What does the slope mean?
b) What is the R^2 value?
c) Do good high jumpers tend to be fast runners? (Be careful—low times are good for running events, and high distances are good for jumps.)
d) What does the residual plot reveal about the model?
e) Do you think this is a useful model? Would you use it to predict high jump performance? (Compare the residual standard deviation to the standard deviation of the high jumps.)

Name	Country	High Jump (m)	800-meter (sec)	Long Jump (m)
Carolina Klüft	SWE	1.91	134.15	6.51
Austra Skujyte	LIT	1.76	135.92	6.30
Kelly Sotherton	GBR	1.85	132.27	6.51
Shelia Burrell	USA	1.70	135.32	6.25
Yelena Prokhorova	RUS	1.79	131.31	6.21
Sonja Kesselschlaeger	GER	1.76	135.21	6.42
Marie Collonville	FRA	1.85	133.62	6.19
Natalya Dobrynska	UKR	1.82	137.01	6.23
Margaret Simpson	GHA	1.79	137.72	6.02
Svetlana Sokolova	RUS	1.70	133.23	5.84
JJ Shobha	IND	1.67	137.28	6.36
Claudia Tonn	GER	1.82	130.77	6.35
Naide Gomes	POR	1.85	140.05	6.10
Michelle Perry	USA	1.70	133.69	6.02
Aryiro Strataki	GRE	1.79	137.90	5.97
Karin Ruchstuhl	NED	1.85	133.95	5.90
Karin Ertl	GER	1.73	138.68	6.03
Kylie Wheeler	AUS	1.79	137.65	6.36
Janice Josephs	RSA	1.70	138.47	6.21
Tiffany Lott Hogan	USA	1.67	145.10	6.15
Magdalena Szczepanska	POL	1.76	133.08	5.98
Irina Naumenko	KAZ	1.79	134.57	6.16
Yuliya Akulenko	UKR	1.73	142.58	6.02
Soma Biswas	IND	1.70	132.27	5.92
Marsha Mark-Baird	TRI	1.70	141.21	6.22
Michaela Hejnova	CZE	1.70	145.68	5.70

T 52. Heptathlon, again. We saw the data for the women's 2004 Olympic heptathlon in Exercise 51. Are the two jumping events associated? Perform a regression to predict the long jump results from the high jump results.

a) What is the regression equation? What does the slope mean?

b) What percentage of the variation in long jumps can be accounted for by high jump performances?

c) Do good high jumpers tend to be good long jumpers?

d) What does the residuals plot reveal about the model?

e) Do you think this is a useful model? Would you use it to predict long jump performance? (Compare the residual standard deviation to the standard deviation of the long jumps.)

JUST CHECKING ANSWERS

1 For each additional employee, monthly sales increase, on average, $122,740.

2 Thousands of $ per employee.

3 $1,227,400 per month.

4 About 0.85 SDs.

5 About 1.7 SDs below the mean sales.

6 Differences in the number of employees account for about 71.4% of the variation in the monthly sales.

7 It's positive. The correlation and the slope have the same sign.

8 R^2, No. Slope, Yes.

CHAPTER 9

Sampling Distributions and the Normal Model

Marketing Credit Cards: The MBNA Story

When Delaware substantially raised its interest rate ceiling in 1981, banks and other lending institutions rushed to establish corporate headquarters there. One of these was the Maryland Bank National Association, which established a credit card branch in Delaware using the acronym MBNA. Starting in 1982 with 250 employees in a vacant supermarket in Ogletown, Delaware, MBNA grew explosively in the next two decades.

One of the reasons for this growth was MBNA's use of affinity groups—issuing cards endorsed by alumni associations, sports teams, interest groups, and labor unions, among others. MBNA sold the idea to these groups by letting them share a small percentage of the profit. By 2006, MBNA had become Delaware's largest private employer. At its peak, MBNA had more than 50 million cardholders and had outstanding credit card loans of $82.1 billion, making MBNA the third-largest U.S. credit card bank.

"In American corporate history, I doubt there are many companies that burned as brightly, for such a short period of time, as MBNA," said Rep. Mike Castle, R-Del.[1] MBNA was bought by Bank of America in 2005 for $35 billion. Bank of America kept the brand briefly before issuing all cards under its own name in 2007.

Unlike the early days of the credit card industry when MBNA established itself, the environment today is intensely competitive, with companies constantly looking for ways to attract new customers and to maximize the profitability of the customers they already have. Many of the large companies have millions of customers, so instead of trying out a new idea with all their customers, they almost always conduct a pilot study or trial first, conducting a survey or an experiment on a sample of their customers.

Credit card companies make money on their cards in three ways: they earn a percentage of every transaction, they charge interest on balances that are not paid in full, and they collect fees (yearly fees, late fees, etc.). To generate all three types of revenue, the marketing departments of credit card banks constantly seek ways to encourage customers to increase the use of their cards.

A marketing specialist at one company had an idea of offering double air miles to their customers with an airline-affiliated card if they increased their spending by at least $800 in the month following the offer. In order to forecast the cost and revenue of the offer, the finance department needed to know what percent of customers would actually qualify for the double miles. The marketer decided to send the offer to a random sample of 1000 customers to find out. In that sample, she found that 211 (21.1%) of the cardholders increased their spending by more than the required $800. But, another analyst drew a different sample of 1000 customers of whom 202 (20.2%) of the cardholders exceeded $800.

The two samples don't agree. We know that observations vary, but how much variability among samples should we expect to see?

Why do sample proportions vary at all? How can two samples of the same population measuring the same quantity get different results? The answer is fundamental to statistical inference. Each proportion is based on a *different* sample of cardholders. The proportions vary from sample to sample because the samples are comprised of different people.

WHO	Cardholders of a bank's credit card
WHAT	Whether cardholders increased their spending by at least $800 in the subsequent month
WHEN	February 2008
WHERE	United States
WHY	To predict costs and benefits of a program offer

9.1 Modeling the Distribution of Sample Proportions

We'd like to know how much proportions can vary from sample to sample. We've talked about *Plan*, *Do*, and *Report*, but to learn more about the variability, we have to add *Imagine*. When we sample, we see only the results from the actual sample that we draw, but we can *imagine* what we might have seen had we drawn *all*

---

[1] Delaware *News Online*, January 1, 2006.

Imagine

We see only the sample we actually drew, but if we *imagine* the results of all the other possible samples we could have drawn (by modeling or simulating them), we can learn more.

other possible random samples. What would the histogram of all those sample proportions look like?

If we could take many random samples of 1000 cardholders, we would find the proportion of each sample who spent more than $800 and collect all of those proportions into a histogram. Where would you expect the center of that histogram to be? Of course, we don't *know* the answer, but it is reasonable to think that it will be at the true proportion in the population. We probably will never know the value of the true proportion. But it is important to us, so we'll give it a label, p for "true proportion."

9.2 Simulations

In fact, we can do better than just imagining. We can *simulate*. We can't really take all those different random samples of size 1000, but we can use a computer to pretend to draw random samples of 1000 individuals from some population of values over and over. In this way, we can model the process of drawing many samples from a real population. A *simulation* can help us understand how sample proportions vary due to random sampling.

When we have only two possible outcomes for an event, the convention in Statistics is to arbitrarily label one of them "success" and the other "failure." Here, a "success" would be that a customer increases card charges by at least $800, and a "failure" would be that the customer didn't. In the simulation, we'll set the true proportion of successes to a known value, draw random samples, and then record the sample proportion of successes, which we'll denote by $\hat{p}$, for each sample.

The proportion of successes in each of our simulated samples will vary from one sample to the next, but the *way* in which the proportions vary shows us how the proportions of real samples would vary. Because we can specify the true proportion of successes, we can see how close each sample comes to estimating that true value. Here's a histogram of the proportions of cardholders who increased spending by at least $800 in 2000 independent samples of 1000 cardholders, when the true proportion, $p = 0.21$. (We know this is the true value of p because in a simulation we can control it.)

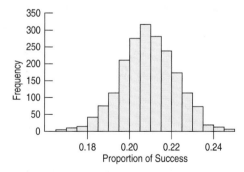

Figure 9.1 *The distribution of 2000 sample values of $\hat{p}$, from simulated samples of size 1000 drawn from a population in which the true p is 0.21.*

It should be no surprise that we don't get the same proportion for each sample we draw, even though the underlying true value, p, stays the same at $p = 0.21$. Since each $\hat{p}$ comes from a random sample, we don't expect them to all be equal to p. And since each comes from a *different* independent random sample, we don't expect them to be equal to each other, either. The remarkable thing is that even though the $\hat{p}$'s vary from sample to sample, they do so in a way that we can model and understand.

9.3 The Normal Distribution

The collection of $\hat{p}$'s may be better behaved than you expected. The histogram in Figure 9.1 is unimodal and symmetric. It is also bell-shaped—something that we'll see again and again in Statistics. In order to make general statements about how often values occur in histograms like this, statisticians make models for distributions. The model for symmetric, unimodal histograms like this one is called the **Normal model**. You've probably seen Normal models before, and if you've seen a "bell-shaped curve," chances are it was a Normal model. Normal models are defined by two parameters, a mean and a standard deviation. By convention, we denote parameters with Greek letters. For example, we denote the mean of such a model with the Greek letter μ, which is the Greek equivalent of "m," for *m*ean, and the standard deviation with the Greek letter σ, the Greek equivalent of "s," for *s*tandard deviation. So we write $N(\mu, \sigma)$ to represent a Normal model with mean μ and standard deviation σ.

There's a different Normal model for every combination of μ and σ, but if we standardize our data first as we did in Chapter 6, creating z-scores by subtracting the mean to make the mean 0 and dividing by the standard deviation to make the standard deviation 1, then we'll need only the model with mean 0 and standard deviation 1. We call this the **standard Normal model** (or the **standard Normal distribution**).

Of course, we shouldn't use a Normal model for every data set. If the histogram isn't unimodal and symmetric to begin with, the z-scores won't be well modeled by the Normal model. And standardizing won't help because standardizing doesn't change the shape of the distribution. So always check the histogram of the data before using the Normal model.

The 68-95-99.7 Rule

Normal models are useful because they can give us an idea of how extreme a value is by telling us how likely we are to find one that far from the mean. We'll soon see how to find these values for any z-score, but for now, there's a simple rule, called the **68-95-99.7 Rule**, that tells us roughly how the values are distributed.

In bell-shaped distributions, about 68% of the values fall within one standard deviation of the mean, about 95% of the values fall within two standard deviations of the mean, and about 99.7%—almost all—of the values fall within three standard deviations of the mean (Figure 9.2).[2]

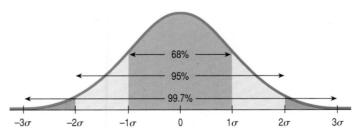

Figure 9.2   *Reaching out one, two, and three standard deviations in a bell-shaped distribution gives the 68-95-99.7 Rule.*

> [2] This rule was first recognized by the mathematician Abraham De Moivre in 1733, based on empirical observations of data, so it is sometimes called the **Empirical Rule**. But it's a better mnemonic to call it the 68-95-99.7 Rule, for the three numbers that define it.

z-scores

$$z = \frac{y - \bar{y}}{s}$$

for data

$$z = \frac{y - \mu}{\sigma}$$

for models

Finding Other Percentiles

Finding the probability that a proportion is at least 1 SD above the mean is easy. We know that 68% of the values lie within 1 SD of the mean, so 32% lie farther away. Since the Normal model is symmetric, half of those 32% (or 16%) are more than 1 SD above the mean. But what if we want to know the percentage of observations that fall more than 1.8 SD above the mean? We already know that no more than 16% of observations have z-scores above 1. By similar reasoning, no more than 2.5% of the observations have a z-score above 2. Can we be more precise with our answer than "between 16% and 2.5%"?

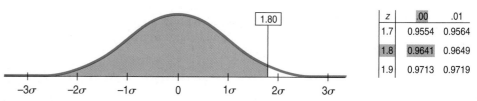

Figure 9.3 *A table of Normal percentiles (Table Z in Appendix C) lets us find the percentage of individuals in a standard Normal distribution falling below any specified z-score value.*

These days, finding percentiles from a Normal table is rarely necessary. Most of the time, we can use a calculator, a computer, or a website.

When the value doesn't fall exactly 0, 1, 2, or 3 standard deviations from the mean, we can look it up in a table of **Normal percentiles**.[3] Tables use the standard Normal model, so we'll have to convert our data to z-scores before using the table. If our data value was 1.8 standard deviations above the mean, we would standardize it to a z-score of 1.80, and then find the value associated with a z-score of 1.80. If we use a table, as shown in Figure 9.3, we find the z-score by looking down the left column for the first two digits (1.8) and across the top row for the third digit, 0. The table gives the percentile as 0.9641. That means that 96.4% of the z-scores are less than 1.80. Since the total area is always 1, and $1 - 0.9641 = 0.0359$ we know that only 3.6% of all observations from a Normal model have z-scores higher than 1.80. We can also find the probabilities associated with z-scores using technology such as calculators, statistical software, and various websites.

9.4 Practice with Normal Distribution Calculations

Finding the percentiles associated with any value in a Normal model isn't hard, but a little practice might be useful. Most of you have taken standardized tests of one kind or another, and you probably focused as much on the "percentile" of your performance as on the raw score. The scores from most standardized tests, such as SATs, GMATs, and LSATs, are well modeled by a Normal and often reported both as raw scores and percentiles. For practice, let's see how we can convert SAT scores to percentiles.

---

[3] See Table Z in Appendix C. Many calculators and statistics computer packages do this as well.

Example I

Problem: Each Scholastic Aptitude Test (SAT) has a distribution that is roughly unimodal and symmetric and is designed to have an overall mean of 500 and a standard deviation of 100. In any one year, the mean and standard deviation may differ from these target values by a small amount, but we can use these values as good overall approximations.

Suppose you earned a 600 on an SAT test. From that information and the 68-95-99.7 Rule, where do you stand among all students who took the SAT?

Solution: Because we're told that the distribution is unimodal and symmetric, we can model the distribution with a Normal model. We are also told the scores have a mean of 500 and an SD of 100. So, we'll use a $N(500,100)$ model. It's good practice at this point to draw the distribution. Find the score whose percentile you want to know and locate it on the picture. When you finish the calculation, you should check to make sure that it's a reasonable percentile from the picture.

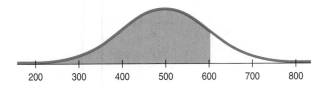

A score of 600 is 1 SD above the mean. That corresponds to one of the points in the 68-95-99.7% Rule. About 32% $(100\% - 68\%)$ of those who took the test were more than one standard deviation from the mean, but only half of those were on the high side. So about 16% (half of 32%) of the test scores were better than 600.

Example II

Problem: Assuming the SAT scores are nearly normal with $N(500,100)$, what proportion of SAT scores falls between 450 and 600?

Solution: *The first step is to find the z-scores associated with each value.* Standardizing the scores we are given, we find that for 600, $z = (600 - 500)/100 = 1.0$ and for 450, $z = (450 - 500)/100 = -0.50$. We can label the axis below the picture either in the original values or the z-scores or even use both scales as the following picture shows.

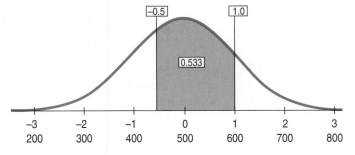

From Table Z, we find the area $z \le 1.0 = 0.8413$, which means that 84.13% of scores fall below 1.0, and the area $z \le -0.50 = 0.3085$, which means that 30.85% of the values fall below -0.5, so the proportion of z-scores *between* them is $84.13\% - 30.85\% = 53.28\%$. So, the Normal model estimates that about 53.3% of SAT scores fall between 450 and 600.

Finding areas from z-scores is the simplest way to work with the Normal model. But sometimes we start with areas and are asked to work backward to find the corresponding z-score or even the original data value. For instance, what z-score represents the first quartile, Q1, in a Normal model? In our first set of examples, we knew the z-score and used the table or technology to find the percentile. Now we want to find the cut point for the 25th percentile. Make a picture, shading the leftmost 25% of the area. Look in Table Z for an area of 0.2500. The exact area is not there, but 0.2514 is the closest number. That shows up in the table with -0.6 in the left margin and .07 in the top margin. The z-score for Q1, then, is approximately $z = -0.67$. Computers and calculators can determine the cut point more precisely (and more easily).[4]

Example III

Problem: Suppose a college says it admits only people with SAT scores among the top 10%. How high an SAT score does it take to be eligible?

Solution: The college takes the top 10%, so their cutoff score is the 90th percentile. Draw an approximate picture like this one.

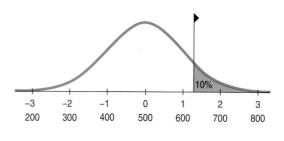

	0.07	0.08	0.09
1.0	0.8577	0.8599	0.8621
1.1	0.8790	0.8810	0.8830
1.2	0.8980	0.8997	0.9015
1.3	0.9147	0.9162	0.9177
1.4	0.9292	0.9306	0.9319

From our picture we can see that the z-value is between 1 and 1.5 (if we've judged 10% of the area correctly), and so the cutoff score is between 600 and 650 or so. Using technology, you may be able to select the 10% area and find the z-value directly. Using a table, such as Table Z, locate 0.90 (or as close to it as you can; here 0.8997 is closer than 0.9015) in the *interior* of the table and find the corresponding z-score (see table above). Here the 1.2 is in the left margin, and the .08 is in the margin above the entry. Putting them together gives 1.28. Now, convert the z-score back to the original units. From Table Z, the cut point is $z = 1.28$. A z-score of 1.28 is 1.28 standard deviations above the mean. Since the standard deviation is 100, that's 128 SAT points. The cutoff is 128 points above the mean of 500, or 628. Because the school wants SAT scores in the top 10%, the cutoff is 628. (Actually since SAT scores are reported only in multiples of 10, you'd have to score at least a 630.)

---

[4] We'll often use those more precise values in our examples. If you're finding the values from the table you may not get *exactly* the same number to all decimal places as your classmate who's using a computer package.

GUIDED EXAMPLE | *Cereal Company*

A cereal manufacturer has a machine that fills the boxes. Boxes are labeled "16 oz," so the company wants to have that much cereal in each box. But since no packaging process is perfect, there will be minor variations. If the machine is set at exactly 16 oz and the Normal model applies (or at least the distribution is roughly symmetric), then about half of the boxes will be underweight, making consumers unhappy and exposing the company to bad publicity and possible lawsuits. To prevent underweight boxes, the manufacturer has to set the mean a little higher than 16.0 oz. Based on their experience with the packaging machine, the company believes that the amount of cereal in the boxes fits a Normal model with a standard deviation of 0.2 oz. The manufacturer decides to set the machine to put an average of 16.3 oz in each box. Let's use that model to answer a series of questions about these cereal boxes.

Question 1. What fraction of the boxes will be underweight?

 PLAN

Setup State the variable and the objective.

The variable is weight of cereal in a box. We want to determine what fraction of the boxes risk being underweight.

Model Check to see if a Normal model is appropriate.

We have no data, so we cannot make a histogram. But we are told that the company believes the distribution of weights from the machine is Normal.

Specify which Normal model to use.

We use an N(16.3, 0.2) model.

 **DO**

Mechanics Make a graph of this Normal model. Locate the value you're interested in on the picture, label it, and shade the appropriate region.

REALITY CHECK

Estimate from the picture the percentage of boxes that are underweight. (This will be useful later to check that your answer makes sense.)

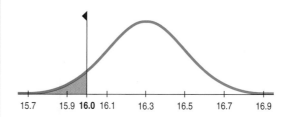

(It looks like a low percentage—maybe less than 10%.)

We want to know what fraction of the boxes will weigh less than 16 oz.

Convert your cutoff value into a z-score.

$$z = \frac{y - \mu}{\sigma} = \frac{16 - 16.3}{0.2} = -1.50.$$

Look up the area in the Normal table, or use your calculator, or software.

Area $(y < 16) =$ Area $(z < -1.50) = 0.0668.$

REPORT

Conclusion State your conclusion in the context of the problem.

We estimate that approximately 6.7% of the boxes will contain less than 16 oz of cereal.

Question 2. The company's lawyers say that 6.7% is too high. They insist that no more than 4% of the boxes can be underweight. So the company needs to set the machine to put a little more cereal in each box. What mean setting do they need?

 **PLAN**

Setup State the variable and the objective.

The variable is weight of cereal in a box. We want to determine a setting for the machine.

Model Check to see if a Normal model is appropriate.

We have no data, so we cannot make a histogram. But we are told that a Normal model applies.

Specify which Normal model to use. This time you are not given a value for the mean!

We don't know μ, the mean amount of cereal. The standard deviation for this machine is 0.2 oz. The model, then, is $N(\mu, 0.2)$.

REALITY CHECK We found out earlier that setting the machine to $\mu = 16.3$ oz made 6.7% of the boxes too light. We'll need to raise the mean a bit to reduce this fraction.

We are told that no more than 4% of the boxes can be below 16 oz.

 **DO**

Mechanics Make a graph of this Normal model. Center it at μ (since you don't know the mean) and shade the region below 16 oz.

The z-score that has 0.04 area to the left of it is $z = -1.75$.

Using the Normal table, a calculator, or software, find the z-score that cuts off the lowest 4%.

Use this information to find μ. It's located 1.75 standard deviations to the right of 16.

Since 16 must be 1.75 standard deviations below the mean, we need to set the mean at $16 + 1.75 \cdot 0.2 = 16.35$.

 REPORT

Conclusion State your conclusion in the context of the problem.

The company must set the machine to average 16.35 oz of cereal per box.

Question 3. The company president vetoes that plan, saying the company should give away less free cereal, not more. Her goal is to set the machine no higher than 16.2 oz and still have only 4% underweight boxes. The only way to accomplish this is to reduce the standard deviation. What standard deviation must the company achieve, and what does that mean about the machine?

PLAN	**Setup** State the variable and the objective.	The variable is weight of cereal in a box. We want to determine the necessary standard deviation to have only 4% of boxes underweight.
	Model Check that a Normal model is appropriate.	The company believes that the weights are described by a Normal model.
	Specify which Normal model to use. This time you don't know σ.	Now we know the mean, but we don't know the standard deviation. The model is therefore $N(16.2, \sigma)$.
REALITY CHECK	We know the new standard deviation must be less than 0.2 oz.	

DO	**Mechanics** Make a graph of this Normal model. Center it at 16.2, and shade the area you're interested in. We want 4% of the area to the left of 16 oz.	
	Find the *z*-score that cuts off the lowest 4%.	We already know that the *z*-score with 4% below it is $z = -1.75$.
	Solve for σ. (Note that we need 16 to be 1.75 σ's below 16.2, so 1.75σ must be 0.2 oz. You could just start with that equation.)	$$z = \frac{y - \mu}{\sigma}$$ $$-1.75 = \frac{16 - 16.2}{\sigma}$$ $$1.75\sigma = 0.2$$ $$\sigma = 0.114.$$

REPORT	**Conclusion** State your conclusion in the context of the problem.	The company must get the machine to box cereal with a standard deviation of only 0.114 oz. This means the machine must be more consistent (by nearly a factor of 2) in filling the boxes.
	As we expected, the standard deviation is lower than before—actually, quite a bit lower.	

9.5 The Sampling Distribution for Proportions

The distribution of proportions over many independent samples from the same population is called the **sampling distribution** of the proportions. Section 9.2 showed a simulation in which that distribution was bell-shaped and centered at the true proportion, *p*. If we knew that the sampling distribution of proportions always followed a bell-shaped distribution, we could use the Normal model to describe the behavior of proportions. With the Normal model, we could find the percentage of values falling between any two values. But to make that work, we need to

NOTATION ALERT:

We use p for the proportion in the population and $\hat{p}$ for the observed proportion in a sample. We'll also use q for the proportion of failures ($q = 1 - p$), and $\hat{q}$ for its observed value, just to simplify some formulas.

know one more thing. Normal models are determined by their mean and standard deviation, and we only know that the mean is p, the true proportion. What about the standard deviation?

An amazing fact about proportions is that (unlike quantitative data) once we know the mean, p, and the sample size, n, we also know the standard deviation of the sampling distribution as you can see from its formula:

$$SD(\hat{p}) = \sqrt{\frac{p(1 - p)}{n}} = \sqrt{\frac{pq}{n}}.$$

If the true proportion of credit cardholders who increased their spending by more than \$800 is 0.21, then for samples of size 1000, we expect the distribution of sample proportions to have a standard deviation of:

$$SD(\hat{p}) = \sqrt{\frac{p(1 - p)}{n}} = \sqrt{\frac{0.21(1 - 0.21)}{1000}} = 0.0129, \text{ or about } 1.3\%.$$

We have now answered the question raised at the start of the chapter. To discover how variable a sample proportion is, we need to know the proportion and the size of the sample. That's all.

Remember that the two samples of size 1000 had proportions of 21.1% and 20.2%. Since the standard deviation of proportions is 1.3%, these two proportions are not even a full standard deviation apart. In other words, the two samples don't really disagree. Proportions of 21.1% and 20.2% from samples of 1000 are both *consistent* with a true proportion of 21%. We know from Chapter 3 that this difference between sample proportions is referred to as **sampling error**. But it's not really an *error*. It's just the *variability* you'd expect to see from one sample to another. A better term might be *sampling variability*.

Look back at Figure 9.1 to see how well the model worked in our simulation. If $p = 0.21$, we now know that the standard deviation should be about 0.013. The 68-95-99.7 Rule from the Normal model says that 68% of the samples will have proportions within 1 SD of the mean of 0.21. How closely does our simulation match the predictions? The actual standard deviation of our 2000 *sample* proportions is 0.0129 or 1.29%. And, of the 2000 simulated samples, 1346 of them had proportions between 0.197 and .223 (one standard deviation on either side of 0.21). The 68-95-99.7 Rule predicts 68%—the actual number is 1346/2000 or 67.3%.

Effect of Sample Size

Because n is in the denominator of $SD(\hat{p})$, the larger the sample, the smaller the standard deviation. We need a small standard deviation to make sound business decisions, but larger samples cost more. That tension is a fundamental issue in statistics.

Now we know everything we need to know to model the sampling distribution. We know the mean and standard deviation of the sampling distribution of proportions: they're p, the true population proportion, and $\sqrt{\frac{pq}{n}}$. So the particular Normal model, $N\left(p, \sqrt{\frac{pq}{n}}\right)$, is a **sampling distribution model for the sample proportion**.

We saw this worked well in a simulation, but can we rely on it in all situations? It turns out that this model can be justified theoretically with just a little mathematics. It won't work for *all* situations, but it works for most situations that you'll encounter in practice. We'll provide conditions to check so you'll know when the model is useful.

The Sampling Distribution Model for a Proportion

Provided that the sampled values are independent and the sample size is large enough, the sampling distribution of $\hat{p}$ is modeled by a Normal model with mean $\mu(\hat{p}) = p$ and standard deviation $SD(\hat{p}) = \sqrt{\frac{pq}{n}}$.

JUST CHECKING

1 You want to poll a random sample of 100 shopping mall customers about whether they like the proposed location for the new coffee shop on the third floor, with a panoramic view of the food court. Of course, you'll get just one number, your sample proportion, $\hat{p}$. But if you imagined all the possible samples of 100 customers you could draw and imagined the histogram of all the sample proportions from these samples, what shape would it have?

2 Where would the center of that histogram be?

3 If you think that about half the customers are in favor of the plan, what would the standard deviation of the sample proportions be?

The sampling distribution model for $\hat{p}$ is valuable for a number of reasons. First, because it is known from mathematics to be a good model (and one that gets better and better as the sample size gets larger), we don't need to actually draw many samples and accumulate all those sample proportions, or even to simulate them. The Normal sampling distribution model tells us what the distribution of sample proportions would look like. Second, because the Normal model is a mathematical model, we can calculate what fraction of the distribution will be found in any region. You can find the fraction of the distribution in *any* interval of values using Table Z at the back of the book or with technology.

How Good Is the Normal Model?

We've seen that the simulated proportions follow the 68-95-99.7 Rule well. But do all sample proportions really work like this? Stop and think for a minute about what we're claiming. We've said that if we draw repeated random samples of the same size, n, from some population and measure the proportion, $\hat{p}$, we get for each sample, then the collection of these proportions will pile up around the underlying population proportion, p, in such a way that a histogram of the sample proportions can be modeled well by a Normal model.

There must be a catch. Suppose the samples were of size 2, for example. Then the only possible numbers of successes could be 0, 1, or 2, and the proportion values would be 0, 0.5, and 1. There's no way the histogram could ever look like a Normal model with only three possible values for the variable (Figure 9.4).

Well, there *is* a catch. The claim is only approximately true. (But, that's fine. Models are *supposed* to be only approximately true.) And the model becomes a better and better representation of the distribution of the sample proportions as the sample size gets bigger.[5] Samples of size 1 or 2 just aren't going to work very well, but the distributions of proportions of many larger samples do have histograms that are remarkably close to a Normal model.

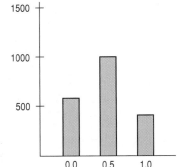

Figure 9.4 *Proportions from samples of size 2 can take on only three possible values. A Normal model does not work well here.*

---

[5] Formally, we say the claim is true in the limit as the sample size (n) grows.

9.6 Assumptions and Conditions

Most models are useful only when specific assumptions are true. In the case of the model for the distribution of sample proportions, there are two assumptions:

Independence Assumption: The sampled values must be *independent* of each other.

Sample Size Assumption: The sample size, *n*, must be *large* enough.

Of course, the best we can do with assumptions is to think about whether they are likely to be true, and we should do so. However, we often can check corresponding *conditions* that provide information about the assumptions as well. Think about the Independence Assumption and check the following corresponding conditions before using the Normal model to model the distribution of sample proportions:

Randomization Condition: If your data come from an experiment, subjects should have been randomly assigned to treatments. If you have a survey, your sample should be a simple random sample of the population. If some other sampling design was used, be sure the sampling method was not biased and that the data are representative of the population.

10% Condition: If sampling has not been made with replacement (that is, returning each sampled individual to the population before drawing the next individual), then the sample size, *n*, must be no larger than 10% of the population. If it is, you must adjust the size of the confidence interval with methods more advanced than those found in this book.

Success/Failure Condition: The Success/Failure condition says that the sample size must be big enough so that both the number of "successes," *np*, and the number of "failures," *nq*, are expected to be at least 10.[6] Expressed without the symbols, this condition just says that we need to expect at least 10 successes and at least 10 failures to have enough data for sound conclusions. For the bank's credit card promotion example, we labeled as a "success" a cardholder who increases monthly spending by at least $800 during the trial. The bank observed 211 successes and 789 failures. Both are at least 10, so there are certainly enough successes and enough failures for the condition to be satisfied.[7]

These two conditions seem to contradict each other. The Success/Failure condition wants a big sample size. How big depends on *p*. If *p* is near 0.5, we need a sample of only 20 or so. If *p* is only 0.01, however, we'd need 1000. But the 10% condition says that the sample size can't be too large a fraction of the population. Fortunately, the tension between them isn't usually a problem in practice. Often, as in polls that sample from all U.S. adults, or industrial samples from a day's production, the populations are much larger than 10 times the sample size.

---

[6] Why 10? We'll discuss this when we discuss confidence intervals in Chapter 10.

[7] The Success/Failure condition is about the number of successes and failures we *expect*, but if the number of successes and failures that *occurred* is $\geq$ 10, then you can use that.

GUIDED EXAMPLE   *Foreclosures*

An analyst at a home loan lender was looking at a package of 90 mortgages that the company had recently purchased in central California. The analyst was aware that in that region about 13% of the homeowners with current mortgages will default on their loans in the next year and the house will go into foreclosure. In deciding to buy the collection of mortgages, the finance department assumed that no more than 15 of the mortgages would go into default. Any amount above that will result in losses for the company. In the package of 90 mortgages, what's the probability that there will be more than 15 foreclosures?

PLAN

Setup State the objective of the study.	We want to find the probability that in a group of 90 mortgages, more than 15 will default. Since 15 out of 90 is 16.7%, we need the probability of finding more than 16.7% defaults out of a sample of 90, if the proportion of defaults is 13%.
Model Check the conditions.	✓ **Independence Assumption** If the mortgages come from a wide geographical area, one homeowner defaulting should not affect the probability that another does. However, if the mortgages come from the same neighborhood(s), the independence assumption may fail and our estimates of the default probabilities may be wrong.
	✓ **Randomization Condition.** The 90 mortgages in the package can be considered as a random sample of mortgages in the region.
	✓ **10% Condition.** The 90 mortgages are less than 10% of the population.
	✓ **Success/Failure Condition**
	$np = 90(0.13) = 11.7 \geq 10$
	$nq = 90(0.87) = 78.3 \geq 10$
State the parameters and the sampling distribution model.	The population proportion is $p = 0.13$. The conditions are satisfied, so we'll model the sampling distribution of $\hat{p}$ with a Normal model, with mean 0.13 and standard deviation
	$$SD(\hat{p}) = \sqrt{\frac{pq}{n}} = \sqrt{\frac{(0.13)(0.87)}{90}} \approx 0.035.$$
	Our model for $\hat{p}$ is $N(0.13, 0.035)$. We want to find $P(\hat{p} > 0.167)$.

Plot Make a picture. Sketch the model and shade the area we're interested in, in this case the area to the right of 16.7%.

| 0.167 |

| 0.145 |

| 0.025 | 0.06 | 0.095 | 0.130 | 0.165 | 0.2 | 0.235 |
| -3σ | -2σ | -1σ | p | 1σ | 2σ | 3σ |

DO

Mechanics Use the standard deviation as a ruler to find the z-score of the cutoff proportion. Find the resulting probability from a table, a computer program, or a calculator.

$$z = \frac{\hat{p} - p}{SD(\hat{p})} = \frac{0.167 - 0.13}{0.035} = 1.06$$

$$P(\hat{p} > 0.167) = P(z > 1.06) = 0.1446$$

REPORT

Conclusion Interpret the probability in the context of the question.

MEMO:

Re: Mortgage defaults

Assuming that the 90 mortgages we recently purchased are a random sample of mortgages in this region, there is about a 14.5% chance that we will exceed the 15 foreclosures that Finance has determined as the break-even point.

9.7 The Central Limit Theorem—The Fundamental Theorem of Statistics

Proportions summarize categorical variables. When we sample at random, the results we get will vary from sample to sample. The Normal model seems an incredibly simple way to summarize all that variation. Could something that simple work for means? We won't keep you in suspense. It turns out that means also have a sampling distribution that we can model with a Normal model. And it turns out that there's a theoretical result that proves it to be so. As we did with proportions, we can get some insight from a simulation.

Simulating the Sampling Distribution of a Mean

Here's a simple simulation with a quantitative variable. Let's start with one fair die. If we toss this die 10,000 times, what should the histogram of the numbers on the face of the die look like? Here are the results of a simulated 10,000 tosses:

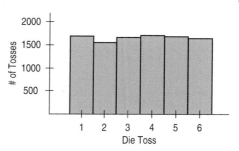

That's called the *uniform distribution*, and it's certainly not Normal. Now let's toss a *pair* of dice and record the average of the two. If we repeat this (or at least simulate repeating it) 10,000 times, recording the average of each pair, what will the histogram of these 10,000 averages look like? Before you look, think a minute. Is getting an average of 1 on *two* dice as likely as getting an average of 3 or 3.5? Let's see:

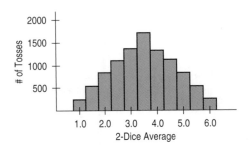

We're much more likely to get an average near 3.5 than we are to get one near 1 or 6. Without calculating those probabilities exactly, it's fairly easy to see that the *only* way to get an average of 1 is to get two 1s. To get a total of 7 (for an average of 3.5), though, there are many more possibilities. This distribution even has a name—the *triangular distribution*.

What if we average three dice? We'll simulate 10,000 tosses of three dice and take their average.

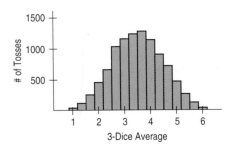

What's happening? First notice that it's getting harder to have averages near the ends. Getting an average of 1 or 6 with three dice requires all three to come up 1 or 6, respectively. That's less likely than for two dice to come up both 1 or both 6. The distribution is being pushed toward the middle. But what's happening to the shape?

Let's continue this simulation to see what happens with larger samples. Here's a histogram of the averages for 10,000 tosses of five dice.

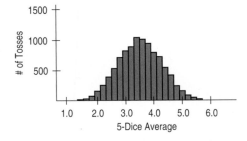

The pattern is becoming clearer. Two things are happening. The first fact we knew already from the Law of Large Numbers, which we saw in Chapter 5. It says that as the sample size (number of dice) gets larger, each sample average tends to become closer to the population mean. So we see the shape continuing to tighten around 3.5. But the shape of the distribution is the surprising part. It's becoming bell-shaped. In fact, it's approaching the Normal model.

| *Pierre-Simon Laplace, 1749–1827.*

"*The theory of probabilities is at bottom nothing but common sense reduced to calculus.*"

—Laplace, in Théorie Analytique des Probabilitiés, *1812*

Laplace was one of the greatest scientists and mathematicians of his time. In addition to his contributions to probability and statistics, he published many new results in mathematics, physics, and astronomy (where his nebular theory was one of the first to describe the formation of the solar system in much the way it is understood today). He also played a leading role in establishing the metric system of measurement.

His brilliance, though, sometimes got him into trouble. A visitor to the Académie des Sciences in Paris reported that Laplace let it be known widely that he considered himself the best mathematician in France. The effect of this on his colleagues was not eased by the fact that Laplace was right.

Are you convinced? Let's skip ahead and try 20 dice. The histogram of averages for throws 10,000 of 20 dice looks like this.

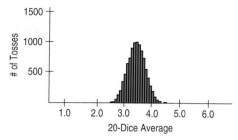

Now we see the Normal shape again (and notice how much smaller the spread is). But can we count on this happening for situations other than dice throws? What kinds of sample means have sampling distributions that we can model with a Normal model? It turns out that Normal models work well amazingly often.

The Central Limit Theorem

The dice simulation may look like a special situation. But it turns out that what we saw with dice is true for means of repeated samples for almost every situation. When we looked at the sampling distribution of a proportion, we had to check only a few conditions. For means, the result is even more remarkable. There are almost no conditions at all.

Let's say that again: The sampling distribution of *any* mean becomes Normal as the sample size grows. All we need is for the observations to be independent and collected with randomization. We don't even care about the shape of the population distribution![8] This surprising fact was proved in a fairly general form in 1810 by Pierre-Simon Laplace, and caused quite a stir (at least in mathematics circles) because it is so unintuitive. Laplace's result is called the **Central Limit Theorem**[9] (CLT).

Not only does the distribution of means of many random samples get closer and closer to a Normal model as the sample size grows, but *this is true regardless of the shape of the population distribution!* Even if we sample from a skewed or bimodal population, the Central Limit Theorem tells us that means of repeated random samples will tend to follow a Normal model as the sample size grows. Of course, you won't be surprised to learn that it works better and faster the closer the population distribution is to a Normal model. And it works better for larger samples. If the data come from a population that's exactly Normal to start with, then the observations themselves are Normal. If we take samples of size 1, their "means" are just the observations—so, of course, they have a Normal sampling distribution. But now suppose the population distribution is very skewed (like the CEO data from Chapter 5, for example). The CLT works, although it may take a sample size of dozens or even hundreds of observations for the Normal model to work well.

For example, think about a real bimodal population, one that consists of only 0s and 1s. The CLT says that even means of samples from this population will follow a Normal sampling distribution model. But wait. Suppose we have a categorical variable and we assign a 1 to each individual in the category and a 0 to each individual not in the category. Then we find the mean of these 0s and 1s. That's the same as counting the number of individuals who are in the category and dividing by *n*. That mean will be the *sample proportion*, $\hat{p}$, of individuals who are in the category (a "success"). So maybe it wasn't so surprising after all that proportions, like means, have Normal sampling distribution models; proportions are actually just a special case of Laplace's remarkable theorem. Of course, for such an extremely

---

[8] Technically, the data must come from a population with a finite variance.
[9] The word "central" in the name of the theorem means "fundamental." It doesn't refer to the center of a distribution, so our section title is really redundant.

bimodal population, we need a reasonably large sample size—and that's where the Success/Failure condition for proportions comes in.

> ### The Central Limit Theorem (CLT)
> The mean of a random sample has a sampling distribution whose shape can be approximated by a Normal model. The larger the sample, the better the approximation will be.

Be careful. We have been slipping smoothly between the real world, in which we draw random samples of data, and a magical mathematical-model world, in which we describe how the sample means and proportions we observe in the real world might behave if we could see the results from every random sample that we might have drawn. Now we have *two* distributions to deal with. The first is the real-world distribution of the sample, which we might display with a histogram (for quantitative data) or with a bar chart or table (for categorical data). The second is the math-world *sampling distribution* of the statistic, which we model with a Normal model based on the Central Limit Theorem. Don't confuse the two.

For example, don't mistakenly think the CLT says that the *data* are Normally distributed as long as the sample is large enough. In fact, as samples get larger, we expect the distribution of the data to look more and more like the distribution of the population from which it is drawn—skewed, bimodal, whatever—but not necessarily Normal. You can collect a sample of CEO salaries for the next 1000 years, but the histogram will never look Normal. It will be skewed to the right. The Central Limit Theorem doesn't talk about the distribution of the data from the sample. It talks about the sample *means* and sample *proportions* of many different random samples drawn from the same population. Of course, we never actually draw all those samples, so the CLT is talking about an imaginary distribution—the sampling distribution model.

The CLT does require that the sample be big enough when the population shape is not unimodal and symmetric. But it is still a very surprising and powerful result.

9.8 The Sampling Distribution of the Mean

The CLT says that the sampling distribution of any mean or proportion is approximately Normal. But which Normal? We know that any Normal model is specified by its mean and standard deviation. For proportions, the sampling distribution is centered at the population proportion. For means, it's centered at the population mean. What else would we expect?

What about the standard deviations? We noticed in our dice simulation that the histograms got narrower as the number of dice we averaged increased. This shouldn't be surprising. Means vary less than the individual observations. Think about it for a minute. Which would be more surprising, having *one* person in your Statistics class who is over 6′9″ tall or having the *mean* of 100 students taking the course be over 6′9″? The first event is fairly rare.[10] You may have seen somebody this tall in one of your classes sometime. But finding a class of 100 whose mean height is over 6′9″ tall just won't happen. Why? *Means have smaller standard deviations than individuals.*

That is, the Normal model for the sampling distribution of the mean has a standard deviation equal to $SD(\bar{y}) = \dfrac{\sigma}{\sqrt{n}}$ where σ is the standard deviation of the population. To emphasize that this is a standard deviation *parameter* of the sampling distribution model for the sample mean, $\bar{y}$, we write $SD(\bar{y})$ or $\sigma(\bar{y})$.

"The n's justify the means."
—Apocryphal statistical saying

---

[10] If students are a random sample of adults, fewer than 1 out of 10,000 should be taller than 6′9″. Why might college students not really be a random sample with respect to height? Even if they're not a perfectly random sample, a college student over 6′9″ tall is still rare.

> ### The Sampling Distribution Model for a Mean
>
> When a random sample is drawn from any population with mean μ and standard deviation σ, its sample mean, $\bar{y}$, has a sampling distribution with the same mean μ but whose standard deviation is $\dfrac{\sigma}{\sqrt{n}}$ $\left(\text{and we write } \sigma(\bar{y}) = SD(\bar{y}) = \dfrac{\sigma}{\sqrt{n}}\right)$. No matter what population the random sample comes from, the shape of the sampling distribution is approximately Normal as long as the sample size is large enough. The larger the sample used, the more closely the Normal approximates the **sampling distribution model for the mean.**

We now have two closely related sampling distribution models. Which one we use depends on which kind of data we have.

◆ When we have categorical data, we calculate a sample proportion, $\hat{p}$. Its sampling distribution follows a Normal model with a mean at the population proportion, p, and a standard deviation $SD(\hat{p}) = \sqrt{\dfrac{pq}{n}} = \dfrac{\sqrt{pq}}{\sqrt{n}}$.

◆ When we have quantitative data, we calculate a sample mean, $\bar{y}$. Its sampling distribution has a Normal model with a mean at the population mean, μ, and a standard deviation $SD(\bar{y}) = \dfrac{\sigma}{\sqrt{n}}$.

The means of these models are easy to remember, so all you need to be careful about is the standard deviations. Remember that these are standard deviations of the *statistics* $\hat{p}$ and $\bar{y}$. They both have a square root of n in the denominator. That tells us that the larger the sample, the less either statistic will vary. The only difference is in the numerator. If you just start by writing $SD(\bar{y})$ for quantitative data and $SD(\hat{p})$ for categorical data, you'll be able to remember which formula to use.

Assumptions and Conditions for the Sampling Distribution of the Mean

The CLT requires essentially the same assumptions as we saw for modeling proportions:

Independence Assumption: The sampled values must be independent of each other.

> **Randomization Condition:** The data values must be sampled randomly, or the concept of a sampling distribution makes no sense.

Sample Size Assumption: The sample size must be sufficiently large. We can't check these directly, but we can think about whether the Independence Assumption is plausible. We can also check some related conditions:

> **10% Condition:** When the sample is drawn without replacement (as is usually the case), the sample size, n, should be no more than 10% of the population.

> **Large Enough Sample Condition:** The CLT doesn't tell us how large a sample we need. The truth is, it depends; there's no one-size-fits-all rule. If the population is unimodal and symmetric, even a fairly small sample is okay. You may hear that 30 or 50 observations is always enough to guarantee Normality, but in truth, it depends on the shape of the original data distribution. For highly skewed distributions, it may require samples of several hundred for the sampling distribution of means to be approximately Normal. Always plot the data to check.

9.9 Sample Size—Diminishing Returns

The standard deviation of the sampling distribution declines only with the square root of the sample size. The mean of a random sample of 4 has half $\left(\dfrac{1}{\sqrt{4}} = \dfrac{1}{2}\right)$ the standard deviation of an individual data value. To cut it in half again, we'd need a sample of 16, and a sample of 64 to halve it once more. In practice, random sampling works well, and means have smaller standard deviations than the individual data values that were averaged. This is the power of averaging.

If only we could afford a much larger sample, we could get the standard deviation of the sampling distribution *really* under control so that the sample mean could tell us still more about the unknown population mean. As we shall see, that square root limits how much we can make a sample tell about the population. This is an example of something that's known as the **Law of Diminishing Returns**.

Example

Problem: Suppose that the mean weight of boxes shipped by a company is 12 lbs, with a standard deviation of 4 lbs. Boxes are shipped in palettes of 10 boxes. The shipper has a limit of 150 lbs for such shipments. What's the probability that a palette will exceed that limit?

Solution: Asking the probability that the total weight of a sample of 10 boxes exceeds 150 lbs is the same as asking the probability that the *mean* weight exceeds 15 lbs. First we'll check the conditions. We will assume that the 10 boxes on the palette are a random sample from the population of boxes and that their weights are mutually independent. And 10 boxes is surely less than 10% of the population of boxes shipped by the company.

Under these conditions, the CLT says that the sampling distribution of $\bar{y}$ has a Normal model with mean 12 and standard deviation

$$SD(\bar{y}) = \frac{\sigma}{\sqrt{n}} = \frac{4}{\sqrt{10}} = 1.26 \text{ and } z = \frac{\bar{y} - \mu}{SD(\bar{y})} = \frac{15 - 12}{1.26} = 2.38$$

$$P(\bar{y} > 150) = P(z > 2.38) = 0.0087$$

So the chance that the shipper will reject a palette is only .0087—less than 1%. That's probably good enough for the company.

9.10 How Sampling Distribution Models Work

Both of the sampling distributions we've looked at are Normal. We know for proportions, $SD(\hat{p}) = \sqrt{\dfrac{pq}{n}}$, and for means, $SD(\bar{y}) = \dfrac{\sigma}{\sqrt{n}}$. These are great if we know, or can pretend that we know, p or σ, and sometimes we'll do that.

Standard Error

Often we know only the observed proportion, $\hat{p}$, or the sample standard deviation, s. So of course we just use what we know, and we estimate. That may not seem like a big deal, but it gets a special name. Whenever we estimate the standard deviation of a sampling distribution, we call it a **standard error (SE)**.

For a sample proportion, $\hat{p}$, the standard error is:

$$SE(\hat{p}) = \sqrt{\frac{\hat{p}\hat{q}}{n}}.$$

For the sample mean, $\bar{y}$, the standard error is:

$$SE(\bar{y}) = \frac{s}{\sqrt{n}}.$$

You may see a "standard error" reported by a computer program in a summary or offered by a calculator. It's safe to assume that if no statistic is specified, what was meant is $SE(\bar{y})$, the standard error of the mean.

JUST CHECKING

4 The entrance exam for business schools, the GMAT, given to 100 students had a mean of 520 and a standard deviation of 120. What was the standard error for the mean of this sample of students?

5 As the sample size increases, what happens to the standard error, assuming the standard deviation remains constant?

6 If the sample size is doubled, what is the impact on the standard error?

To keep track of how the concepts we've seen combine, we can draw a diagram relating them. At the heart is the idea that *the statistic itself (the proportion or the mean) is a random quantity.* We can't know what our statistic will be because it comes from a random sample. A different random sample would have given a different result. This sample-to-sample variability is what generates the sampling distribution, the distribution of all the possible values that the statistic could have had.

We could simulate that distribution by pretending to take lots of samples. Fortunately, for the mean and the proportion, the CLT tells us that we can model their sampling distribution directly with a Normal model.

The two basic truths about sampling distributions are:

1. Sampling distributions arise because samples vary. Each random sample will contain different cases and, so, a different value of the statistic.

2. Although we can always simulate a sampling distribution, the Central Limit Theorem saves us the trouble for means and proportions.

Figure 9.5 diagrams the process.

Figure 9.5 *We start with a population model, which can have any shape. It can even be bimodal or skewed (as this one is). We label the mean of this model μ and its standard deviation, σ.*

We draw one real sample (solid line) of size n and show its histogram and summary statistics. We imagine (or simulate) drawing many other samples (dotted lines), which have their own histograms and summary statistics.

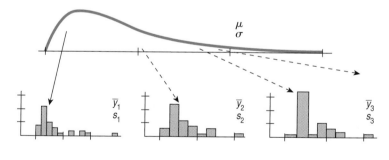

We (imagine) gathering all the means into a histogram.

The CLT tells us we can model the shape of this histogram with a Normal model. The mean of this Normal is μ, and the standard deviation is $SD(\bar{y}) = \dfrac{\sigma}{\sqrt{n}}$.

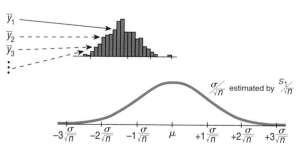

When we don't know σ, we estimate it with the standard deviation of the one real sample. That gives us the standard error, $SE(\bar{y}) = \dfrac{s}{\sqrt{n}}$.

WHAT CAN GO WRONG?

- **Don't use Normal models when the distribution is not unimodal and symmetric.** Normal models are so easy and useful that it is tempting to use them even when they don't describe the data very well. That can lead to wrong conclusions. Don't use a Normal model without first looking at a picture of the data to check that it is unimodal and symmetric. A histogram, or boxplot, can help you tell whether a Normal model is appropriate.

- **Don't use the mean and standard deviation when outliers are present.** Both means and standard deviations can be distorted by outliers, and no model based on distorted values will do a good job. So, it's a good idea to always check for outliers. How? Make a picture.

- **Don't confuse the sampling distribution with the distribution of the sample.** When you take a sample, you always look at the distribution of the values, usually with a histogram, and you may calculate summary statistics. Examining the distribution of the sample like this is wise. But that's not the sampling distribution. The sampling distribution is an imaginary collection of the values that a statistic might have taken for all the random samples—the one you got and the ones you didn't get. Use the sampling distribution model to make statements about how the statistic varies.

- **Beware of observations that are not independent.** The CLT depends crucially on the assumption of independence. Unfortunately, this isn't something you can check in your data. You have to think about how the data were gathered. Good sampling practice and well-designed randomized experiments ensure independence.

- **Watch out for small samples from skewed populations.** The CLT assures us that the sampling distribution model is Normal if n is large enough. If the population is nearly Normal, even small samples may work. If the population is very skewed, then n will have to be large before the Normal model will work well. If we sampled 15 or even 20 CEOs and used $\bar{y}$ to make a statement about the mean of all CEOs' compensation, we'd likely get into trouble because the underlying data distribution is so skewed. Unfortunately, there's no good rule to handle this.[11] It just depends on how skewed the data distribution is. Always plot the data to check.

---

[11] For proportions, there is a rule: the Success/Failure condition. That works for proportions because the standard deviation of a proportion is linked to its mean. You may hear that 30 or 50 observations is enough to guarantee Normality, but it really depends on the skewness of the original data distribution.

ETHICS IN ACTION

Home Illusions, a national retailer of contemporary furniture and home décor has recently experienced customer complaints about the delivery of its products. This retailer uses different carriers depending on the order destination. Its policy with regard to most items it sells and ships is to simply deliver to the customer's doorstep. However, its policy with regard to furniture is to "deliver, unpack, and place furniture in the intended area of the home." Most of their recent complaints have been from customers in the northeastern region of the United States who were dissatisfied because their furniture deliveries were not unpacked and placed in their homes. Since the retailer uses different carriers, it is important for them to label their packages correctly so the delivery company can distinguish between furniture and nonfurniture deliveries. Home Illusions sets as a target "1% or less" for incorrect labeling of packages. Joe Zangard, V.P. Logistics, was asked to look into the problem. The retailer's largest warehouse in the northeast prepares about 1000 items per week for shipping. Joe's initial attention was directed at this facility, not only because of its large volume, but also because he had some reservations about the newly hired warehouse manager, Brent Mossir. Packages at the warehouse were randomly selected and examined over a period of several weeks. Out of 1000 packages, 13 were labeled incorrectly. Since Joe had expected the count to be 10 or fewer, he was confident that he had now pinpointed the problem. His next step was to set up a meeting with Brent in order to discuss the ways in which he can improve the labeling process at his warehouse.

ETHICAL ISSUE *Joe is treating the sample proportion as if it were the true fixed value. By not recognizing that this sample proportion varies from sample to sample, he has unfairly judged the labeling process at Brent's warehouse. This is consistent with his initial misgivings about Brent being hired as warehouse manager (related to Item A, ASA Ethical Guidelines).*

ETHICAL SOLUTION *Joe Zangard needs to use the normal distribution to model the sampling distribution for the sample proportion. In this way, he would realize that the sample proportion observed is less than one standard deviation away from 1% (the upper limit of the target) and thus not conclusively larger than the limit.*

What have we learned?

In Chapter 1, we said that Statistics is about variation. We know that no sample fully and exactly describes the population; sample proportions and means will vary from sample to sample. That's sampling error (or, better, sampling variability). We know it will always be present—indeed, the world would be a boring place if variability didn't exist. You might think that sampling variability would prevent us from learning anything reliable about a population by looking at a sample, but that's just not so. The fortunate fact is that sampling variability is not just unavoidable—it's predictable!

We've learned how the Central Limit Theorem describes the behavior of sample proportions—shape, center, and spread—as long as certain conditions are met. The sample must be random, of course, and large enough that we expect at least 10 successes and 10 failures. Then:

- The sampling distribution (the imagined histogram of the proportions from all possible samples) is shaped like a Normal model.
- The mean of the sampling model is the true proportion in the population.
- The standard deviation of the sample proportions is $\sqrt{\dfrac{pq}{n}}$.

We've learned to describe the behavior of sample means as well, also based on the Central Limit Theorem—the Fundamental Theorem of Statistics. Again the sample must be random and needs to be larger if our data come from a population that's not roughly unimodal and symmetric. Then:

- Regardless of the shape of the original population, the shape of the distribution of the means of all possible samples can be described by a Normal model, provided the samples are large enough.
- The center of the sampling model will be the true mean of the population from which we took the sample.
- The standard deviation of the sample means is the population's standard deviation divided by the square root of the sample size, $\frac{\sigma}{\sqrt{n}}$.

Terms

68-95-99.7 Rule (or Empirical Rule)
In a Normal model, 68% of values fall within one standard deviation of the mean, 95% fall within two standard deviations of the mean, and 99.7% fall within three standard deviations of the mean.

Central Limit Theorem
The Central Limit Theorem (CLT) states that the sampling distribution model of the sample mean (and proportion) is approximately Normal for large n, regardless of the distribution of the population, as long as the observations are independent.

Normal model
A unimodal, symmetric, bell-shaped distribution with important applications in Statistics. Normal models are characterized by their mean and standard deviation, so they are commonly indicated as $N(\mu, \sigma)$.

Normal percentile
A percentile corresponding to a z-score that gives the percentage of values in a standard Normal distribution found at that z-score or below.

Parameter
A numerically valued attribute of a model, such as the values of μ and σ in a $N(\mu, \sigma)$ model.

Sampling distribution
The distribution of a statistic over many independent samples of the same size from the same population.

Sampling distribution model for a mean
If the independence assumption and randomization condition are met and the sample size is large enough, the sampling distribution of the sample mean is well modeled by a Normal model with a mean equal to the population mean, μ, and a standard deviation equal to $\frac{\sigma}{\sqrt{n}}$.

Sampling distribution model for a proportion
If the independence assumption and randomization condition are met and we expect at least 10 successes and 10 failures, then the sampling distribution of a proportion is well modeled by a Normal model with a mean equal to the true proportion value, p, and a standard deviation equal to $\sqrt{\frac{pq}{n}}$.

Sampling error
The variability we expect to see from sample to sample is often called the sampling error, although sampling variability is a better term.

Standard error
When the standard deviation of the sampling distribution of a statistic is estimated from the data, the resulting statistic is called a standard error (SE).

Standard Normal model or Standard Normal distribution
A Normal model, $N(\mu, \sigma)$, with mean $\mu = 0$ and standard deviation $\sigma = 1$.

Skills

 PLAN
- Understand that the variability of a statistic (as measured by the standard deviation of its sampling distribution) depends on the size of the sample. Statistics based on larger samples are less variable.
- Understand that the Central Limit Theorem gives the sampling distribution model of the mean for sufficiently large samples regardless of the underlying population.

 DO
- Be able to use a sampling distribution model to make simple statements about the distribution of a proportion or mean under repeated sampling.

 REPORT
- Be able to interpret a sampling distribution model as describing the values taken by a statistic in all possible realizations of a sample or randomized experiment under the same conditions.

Mini Case Study Project

Real Estate Simulation

Many variables important to the real estate market are skewed, limited to only a few values or considered as categorical variables. Yet, marketing and business decisions are often made based on means and proportions calculated over many homes. One reason these statistics are useful is the Central Limit Theorem.

Data on 1063 houses sold recently in the Saratoga, New York area, are in the file **ch09_MCSP_Real_Estate** on your disk. Let's investigate how the CLT guarantees that the sampling distribution of proportions approaches the Normal and that the same is true for means of a quantitative variable even when samples are drawn from populations that are far from Normal.

Part 1: Proportions

The variable *Fireplace* is a dichotomous variable where 1 = *has a fireplace* and 0 = *does not have a fireplace*.

- Calculate the proportion of homes that have fireplaces for all 1063 homes. Using this value, calculate what the standard error of the sample proportion would be for a sample of size 50.
- Using the software of your choice, draw 100 samples of size 50 from this population of homes, find the proportion of homes with fireplaces in each of these samples, and make a histogram of these proportions.
- Compare the mean and standard deviation of this (sampling) distribution to what you previously calculated.

Means

- Select one of the quantitative variables and make a histogram of the entire population of 1063 homes. Describe the distribution (including its mean and SD).
- Using the software of your choice, draw 100 samples of size 50 from this population of homes, find the means of these samples, and make a histogram of these means.
- Compare the (sampling) distribution of the means to the distribution of the population.
- Repeat the exercise with samples of sizes 10 and of 30. What do you notice about the effect of the sample size?

Some statistics packages make it easier than others to draw many samples and find means. Your instructor can provide advice on the path to follow for your package. If you are using Excel, you will need to use the DDXL add-in to make your histograms.

An alternative approach is to have each member of the class draw one sample to find the proportion and mean and then combine the statistics for the entire class.

EXERCISES

For Exercises 1–8, use the 68-95.99.7 Rule to approximate the probabilities rather than using technology to find the values more precisely. Answers given for probabilities or percentages from Exercise 9 and on assume that a calculator or software has been used. Answers found from using Z-tables may vary slightly.

T 1. Mutual fund returns. In the last quarter of 2007, a group of 64 mutual funds had a mean return of 2.4% with a standard deviation of 5.6%. If a Normal model can be used to model them, what percent of the funds would you expect to be in each region?

Be sure to draw a picture first.

a) Returns of 8.0% or more

b) Returns of 2.4% or less

c) Returns between −8.8% and 13.6%

d) Returns of more than 19.2%

2. Human resource testing. Although controversial and the subject of some recent law suits (e.g., *Satchell et al. vs. FedEx Express*), some human resource departments administer standard IQ tests to all employees. The Stanford-Binet test scores are well modeled by a Normal model with mean 100 and standard deviation 16. If the applicant pool is well modeled by this distribution, a randomly selected applicant would have what probability of scoring in the following regions?

a) 100 or below

b) Above 148

c) Between 84 and 116

d) Above 132

3. Mutual funds, again. From the 64 mutual funds in Exercise 1 with quarterly returns that are well modeled by a Normal model with a mean of 2.4% and a standard deviation of 5.6%, find the cutoff return value(s) that would separate the

a) highest 50%.

b) highest 16%.

c) lowest 2.5%.

d) middle 68%.

4. Human resource testing, again. For the IQ test administered by human resources and discussed in Exercise 2, what cutoff value would separate the

a) lowest 0.15% of all applicants?

b) lowest 16%?

c) middle 95%?

d) highest 2.5%?

5. Currency exchange rates. The daily exchange rates for the five-year period 2003 to 2008 between the euro (EUR) and the British pound (GBP) are well modeled by a Normal distribution with mean 1.459 euros (to pounds) and standard deviation 0.033 euros. Given this model, what is the probability that on a randomly selected day during this period, the pound was worth

a) less than 1.459 euros?
b) more than 1.492 euros?
c) less than 1.393 euros?
d) Which would be more unusual, a day on which the pound was worth less than 1.410 euros or more than 1.542 euros?

6. Stock prices. For the 900 trading days from January 2003 through July 2006, the daily closing price of IBM stock (in $) is well modeled by a Normal model with mean $85.60 and standard deviation $6.20. According to this model, what is the probability that on a randomly selected day in this period the stock price closed

a) above $91.80?
b) below $98.00?
c) between $73.20 and $98.00?
d) Which would be more unusual, a day on which the stock price closed above $93 or below $70?

7. Currency exchange rates, again. For the model of the EUR/GBP exchange rate discussed in Exercise 5, what would the cutoff rates be that would separate the

a) highest 16% of EUR/GBP rates?
b) lowest 50%?
c) middle 95%?
d) lowest 2.5%?

8. Stock prices, again. According to the model in Exercise 6, what cutoff value of price would separate the

a) lowest 16% of the days?
b) highest 0.15%?
c) middle 68%?
d) highest 50%?

9. Mutual fund probabilities. According to the Normal model $N(0.024, 0.056)$ describing mutual fund returns in the 4th quarter of 2007 in Exercise 1, what percent of this group of funds would you expect to have return

a) over 6.8%?
b) between 0% and 7.6%?
c) more than 1%?
d) less than 0%?

10. Normal IQs. Based on the Normal model $N(100, 16)$ describing IQ scores from Exercise 2, what percent of applicants would you expect to have scores

a) over 80?
b) under 90?
c) between 112 and 132?
d) over 125?

11. Mutual funds, once more. Based on the model $N(0.024, 0.056)$ for quarterly returns from Exercise 1, what are the cutoff values for the

a) highest 10% of these funds?
b) lowest 20%?
c) middle 40%?
d) highest 80%?

12. More IQs. In the Normal model $N(100, 16)$ for IQ scores from Exercise 2, what cutoff value bounds the

a) highest 5% of all IQs?
b) lowest 30% of the IQs?
c) middle 80% of the IQs?
d) lowest 90% of all IQs?

13. Mutual funds, finis. Consider the Normal model $N(0.024, 0.056)$ for returns of mutual funds in Exercise 1 one last time.

a) What value represents the 40th percentile of these returns?
b) What value represents the 99th percentile?
c) What's the IQR of the quarterly returns for this group of funds?

14. IQs, finis. Consider the IQ model $N(100, 16)$ one last time.

a) What IQ represents the 15th percentile?
b) What IQ represents the 98th percentile?
c) What's the IQR of the IQs?

15. Parameters. Every Normal model is defined by its parameters, the mean and the standard deviation. For each model described here, find the missing parameter. As always, start by drawing a picture.

a) $\mu = 20$, 45% above 30; $\sigma = ?$
b) $\mu = 88$, 2% below 50; $\sigma = ?$
c) $\sigma = 5$, 80% below 100; $\mu = ?$
d) $\sigma = 15.6$, 10% above 17.2; $\mu = ?$

16. Parameters, again. Every Normal model is defined by its parameters, the mean and the standard deviation. For each model described here, find the missing parameter. Don't forget to draw a picture.

a) $\mu = 1250$, 35% below 1200; $\sigma = ?$
b) $\mu = 0.64$, 12% above 0.70; $\sigma = ?$
c) $\sigma = 0.5$, 90% above 10.0; $\mu = ?$
d) $\sigma = 220$, 3% below 202; $\mu = ?$

17. SAT or ACT? Each year thousands of high school students take either the SAT or ACT, standardized tests used in the college admissions process. Combined SAT scores can go as high as 1600, while the maximum ACT composite score is 36. Since the two exams use very different scales, comparisons of performance are difficult. (A convenient rule of thumb is $SAT = 40 \times ACT + 150$; that is, multiply an ACT score by 40 and add 150 points to estimate the equivalent SAT score.) Assume that one year the combined SAT can be modeled by $N(1000, 200)$ and the ACT can be modeled by $N(27, 3)$. If an applicant to a university has taken the SAT and scored 1260 and another student has taken the ACT and scored 33, compare these students scores using z-values. Which one has a higher relative score? Explain.

18. Economics. Anna, a business major, took final exams in both Microeconomics and Macroeconomics and scored 83 on both. Her roommate Megan, also taking both courses, scored 77 on the Micro exam and 95 on the Macro exam. Overall, student scores on the Micro exam had a mean of 81 and a standard deviation of 5, and the Macro scores had a mean of 74 and a standard deviation of 15. Which student's overall performance was better? Explain.

19. Claims. Two companies make batteries for cell phone manufacturers. One company claims a mean life span of 2 years, while the other company claims a mean life span of 2.5 years (assuming average use of minutes/month for the cell phone).

a) Explain why you would also like to know the standard deviations of the battery life spans before deciding which brand to buy.

b) Suppose those standard deviations are 1.5 months for the first company and 9 months for the second company. Does this change your opinion of the batteries? Explain.

T 20. Car speeds. The police department of a major city needs to update its budget. For this purpose, they need to understand the variation in their fines collected from motorists for speeding. As a sample, they recorded the speeds of cars driving past a location with a 20 mph speed limit, a place that in the past has been known for producing fines. The mean of 100 readings was 23.84 mph, with a standard deviation of 3.56 mph. (The police actually recorded every car for a two-month period. These are 100 representative readings.)

a) How many standard deviations from the mean would a car going the speed limit be?

b) Which would be more unusual, a car traveling 34 mph or one going 10 mph?

21. CEOs. A business publication recently released a study on the total number of years of experience in industry among CEOs. The mean is provided in the article, but not the standard deviation. Is the standard deviation most likely to be 6 months, 6 years, or 16 years? Explain which standard deviation is correct and why.

22. Stocks. A newsletter for investors recently reported that the average stock price for a blue chip stock over the past 12 months was $72. No standard deviation was given. Is the standard deviation more likely to be $6, $16, or $60? Explain.

23. Fuel economy. Recent Environmental Protection Agency (EPA) fuel economy estimates for automobile models tested predicted a mean of 24.8 mpg and a standard deviation of 6.2 mpg for highway driving. Assume that a Normal model can be applied.

a) Draw the model for auto fuel economy. Clearly label it, showing what the 68-95-99.7 Rule predicts about miles per gallon.

b) In what interval would you expect the central 68% of autos to be found?

c) About what percent of autos should get more than 31 mpg?

d) About what percent of cars should get between 31 and 37.2 mpg?

e) Describe the gas mileage of the worst 2.5% of all cars.

24. Job satisfaction. Some job satisfaction assessments are standardized to a Normal model with a mean of 100 and a standard deviation of 12.

a) Draw the model for these job satisfaction scores. Clearly label it, showing what the 68-95-99.7 Rule predicts about the scores.

b) In what interval would you expect the central 95% of job satisfaction scores to be found?

c) About what percent of people should have job satisfaction scores above 112?

d) About what percent of people should have job satisfaction scores between 64 and 76?

e) About what percent of people should have job satisfaction scores above 124?

25. Low job satisfaction. Exercise 24 proposes modeling job satisfaction scores with $N(100, 12)$. Human resource departments of corporations are generally concerned if the job satisfaction drops below a certain score. What score would you consider to be unusually low? Explain.

26. Low return. Exercise 1 proposes modeling quarterly returns of a group of mutual funds with $N(0.024, 0.056)$. The manager of this group of funds would like to flag any fund whose return is unusually low for a quarter. What level of return would you consider to be unusually low? Explain.

27. Management survey. A survey of 200 middle managers showed a distribution of the number of hours of exercise they participated in per week with a mean of 3.66 hours and a standard deviation of 4.93 hours.

a) According to the Normal model, what percent of managers will exercise fewer than one standard deviation below the mean number of hours?

b) For these data, what does that mean? Explain.

c) Explain the problem in using the Normal model for these data.

28. Customer database. A large philanthropic organization keeps records on the people who have contributed to their cause. In addition to keeping records of past giving, the organization buys demographic data on neighborhoods from the U.S. Census Bureau. Eighteen of these variables concern the ethnicity of the neighborhood of the donor. Here is a histogram and summary statistics for the percentage of whites in the neighborhoods of 500 donors.

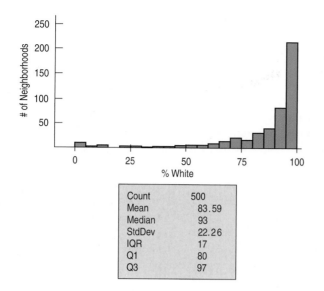

Count	500
Mean	83.59
Median	93
StdDev	22.26
IQR	17
Q1	80
Q3	97

a) Which is a better summary of the percentage of white residents in the neighborhoods, the mean or the median? Explain.
b) Which is a better summary of the spread, the IQR or the standard deviation? Explain.
c) From a Normal model, about what percentage of neighborhoods should have a percent white residents within one standard deviation of the mean?
d) What percentage of neighborhoods actually have a percent white within one standard deviation of the mean?
e) Explain the problem in using the Normal model for these data.

29. Drug company. Manufacturing and selling drugs that claim to reduce an individual's cholesterol level is big business. A company would like to market their drug to women if their cholesterol is in the top 15%. Assume the cholesterol levels of adult American women can be described by a Normal model with a mean of 188 mg/dL and a standard deviation of 24.

a) Draw and label the Normal model.
b) What percent of adult women do you expect to have cholesterol levels over 200 mg/dL?
c) What percent of adult women do you expect to have cholesterol levels between 150 and 170 mg/dL?
d) Estimate the interquartile range of the cholesterol levels.
e) Above what value are the highest 15% of women's cholesterol levels?

30. Tire company. A tire manufacturer believes that the tread-life of its snow tires can be described by a Normal model with a mean of 32,000 miles and a standard deviation of 2500 miles.

a) If you buy a set of these tires, would it be reasonable for you to hope that they'll last 40,000 miles? Explain.
b) Approximately what fraction of these tires can be expected to last less than 30,000 miles?
c) Approximately what fraction of these tires can be expected to last between 30,000 and 35,000 miles?

d) Estimate the IQR for these data.
e) In planning a marketing strategy, a local tire dealer wants to offer a refund to any customer whose tires fail to last a certain number of miles. However, the dealer does not want to take too big a risk. If the dealer is willing to give refunds to no more than 1 of every 25 customers, for what mileage can he guarantee these tires to last?

31. Fidelity funds. Fidelity offers its customers many different investment options, depending on the risk each investor would like to take. For example, the Fidelity Aggressive International Fund is known to invest in more volatile stocks. Over the past 10 years, the mean annual return for this high-growth fund is 10.47%, and the standard deviation is 25.90%. Assume that the distribution is normally distributed and find the probability that the annual return for this fund is

a) greater than 0%.
b) greater than 5%.
c) greater than 10%
d) less than −5%.

32. Fidelity funds, part 2. Fidelity also offers the Fidelity Mid-Cap Growth Fund, which specializes in mid-growth stocks. Many investors believe that both the average return and the volatility of a mid-growth fund will be less than that of a high-growth fund. The mean annual return for this fund over the past 10 years is reported to be 7.98%, with a standard deviation of 21.13%. Assume the distribution is normally distributed and find the probability that the annual return for this fund is

a) greater than 0%.
b) greater than 5%.
c) greater than 10%
d) less than −5%.
e) Compare your answers above to your answers in Exercise 31 for the high-growth fund.

33. Quality control. A farmer is concerned about the number of eggs he has been collecting that are "below weight" because this impacts his bottom line. Hens usually begin laying eggs when they are about 6 months old. Young hens tend to lay smaller eggs, often weighing less than the desired minimum weight of 54 grams.

a) The average weight of the eggs produced by the young hens is 50.9 grams, and only 28% of their eggs exceed the desired minimum weight. If a Normal model is appropriate, what would the standard deviation of the egg weights be?
b) By the time these hens have reached the age of 1 year, the eggs they produce average 67.1 grams, and 98% of them are above the minimum weight. What is the standard deviation for the appropriate Normal model for these older hens?
c) Are egg sizes more consistent for the younger hens or the older ones? Explain.
d) A certain poultry farmer finds that 8% of his eggs are underweight and that 12% weigh over 70 grams. Estimate the mean and standard deviation of his eggs.

34. **Selling tomatoes.** Agricultural scientists are working on developing an improved variety of Roma tomatoes. Marketing research indicates that customers are likely to bypass Romas that weigh less than 70 grams. The current variety of Roma plants produces fruit that average 74 grams, but 11% of the tomatoes are too small. It is reasonable to assume that a Normal model applies.

a) What is the standard deviation of the weights of Romas now being grown?

b) Scientists hope to reduce the frequency of undersized tomatoes to no more than 4%. One way to accomplish this is to raise the average size of the fruit. If the standard deviation remains the same, what target mean should they have as a goal?

c) The researchers produce a new variety with a mean weight of 75 grams, which meets the 4% goal. What is the standard deviation of the weights of these new Romas?

d) Based on their standard deviations, compare the tomatoes produced by the two varieties.

35. **Loans.** Based on past experience, a bank believes that 7% of the people who receive loans will not make payments on time. The bank has recently approved 200 loans.

a) What are the mean and standard deviation of the proportion of clients in this group who may not make timely payments?

b) What assumptions underlie your model? Are the conditions met? Explain.

c) What's the probability that over 10% of these clients will not make timely payments?

36. **Stock market.** Assume that 30% of all business students at a university invest in the stock market.

a) We randomly pick 100 students. Let $\hat{p}$ represent the proportion of students in this sample who buy and sell stocks. What's the appropriate model for the distribution of $\hat{p}$? Specify the name of the distribution, the mean, and the standard deviation. Be sure to verify that the conditions are met.

b) What's the approximate probability that more than one third of this sample invests in the stock market?

37. **Polling.** Just before a referendum on a school budget, a local newspaper polls 400 voters in an attempt to predict whether the budget will pass. Suppose that the budget actually has the support of 52% of the voters. What's the probability the newspaper's sample will lead them to predict defeat? Be sure to verify that the assumptions and conditions necessary for your analysis are met.

38. **Selling seeds.** Information on a packet of seeds claims that the germination rate is 92%. The manufacturer needs to understand the likelihood of this claim. What's the probability that more than 95% of the 160 seeds in the packet will germinate? Be sure to discuss your assumptions and check the conditions that support your model.

39. **Apples.** When a truckload of apples arrives at a packing plant, a random sample of 150 is selected and examined for bruises, discoloration, and other defects. The whole truckload will be rejected if more than 5% of the sample is unsatisfactory.

Suppose that in fact 8% of the apples on the truck do not meet the desired standard. What's the probability that the shipment will be accepted anyway?

40. **Equipment testing.** It's believed that 4% of children have a gene that may be linked to juvenile diabetes. Researchers at a firm would like to test new monitoring equipment for diabetes. Hoping to have 20 children with the gene for their study, the researchers test 732 newborns for the presence of the gene linked to diabetes. What's the probability that they find enough subjects for their study?

41. **Customer demand.** While some nonsmokers do not mind being seated in a smoking section of a restaurant, about 60% of the customers demand a smoke-free area. A new restaurant with 120 seats is being planned.

How many seats should be in the nonsmoking area in order to be very sure of having enough seating there?

Comment on the assumptions and conditions that support your model, and explain what "very sure" means to you.

42. **Customer demand, part 2.** A restaurateur anticipates serving about 180 people on a Friday evening and believes that about 20% of the patrons will order the chef's steak special.

How many of those meals should he plan on serving in order to be pretty sure of having enough steaks on hand to meet customer demand?

Justify your answer, including an explanation of what "pretty sure" means to you.

43. **Sampling.** A sample is chosen randomly from a population that can be described by a Normal model.

a) What's the sampling distribution model for the sample mean? Describe shape, center, and spread.

b) If we choose a larger sample, what's the effect on this sampling distribution model?

44. **Sampling, part 2.** A sample is chosen randomly from a population that was strongly skewed to the left.

a) Describe the sampling distribution model for the sample mean if the sample size is small.

b) If we make the sample larger, what happens to the sampling distribution model's shape, center, and spread?

c) As we make the sample larger, what happens to the expected distribution of the data in the sample?

45. **Home values.** Assessment records indicate that the value of homes in a small city is skewed right, with a mean of $140,000 and a standard deviation of $60,000. To check the accuracy of the assessment data, officials plan to conduct a detailed appraisal of 100 homes selected at random. Using the 68-95-99.7 Rule, draw and label an appropriate sampling model for the mean value of the homes selected.

46. **Fidelity funds, part 3.** Statistics for the closing price of the Fidelity Aggressive International Fund in 2006 indicate that the average closing price was $39.01, with a standard deviation of $2.86. Assume that this will be indicative of performance next year and assume that a Normal model applies. Using the

68-95-99.7 Rule, draw and label an appropriate sampling model for the mean closing price of 35 funds selected at random.

47. At work. Some business analysts estimate that the length of time people work at a job has a mean of 6.2 years and a standard deviation of 4.5 years.

a) Explain why you suspect this distribution may be skewed to the right.

b) Explain why you could estimate the probability that 100 people selected at random had worked for their employers an average of 10 years or more, but you could not estimate the probability that an individual had done so.

48. Store receipts. Grocery store receipts show that customer purchases have a skewed distribution with a mean of $32 and a standard deviation of $20.

a) Explain why you cannot determine the probability that the next customer will spend at least $40.

b) Can you estimate the probability that the next 10 customers will spend an average of at least $40? Explain.

c) Is it likely that the next 50 customers will spend an average of at least $40? Explain.

49. Quality control, again. The weight of potato chips in a medium-size bag is stated to be 10 ounces. The amount that the packaging machine puts in these bags is believed to have a Normal model with a mean of 10.2 ounces and a standard deviation of 0.12 ounces.

a) What fraction of all bags sold are underweight?

b) Some of the chips are sold in "bargain packs" of 3 bags. What's the probability that none of the 3 is underweight?

c) What's the probability that the mean weight of the 3 bags is below the stated amount?

d) What's the probability that the mean weight of a 24-bag case of potato chips is below 10 ounces?

50. Milk production. Although most of us buy milk by the quart or gallon, farmers measure daily production in pounds. Ayrshire cows average 47 pounds of milk a day, with a standard deviation of 6 pounds. For Jersey cows, the mean daily production is 43 pounds, with a standard deviation of 5 pounds. Assume that Normal models describe milk production for these breeds.

a) We select an Ayrshire at random. What's the probability that she averages more than 50 pounds of milk a day?

b) What's the probability that a randomly selected Ayrshire gives more milk than a randomly selected Jersey?

c) A farmer has 20 Jerseys. What's the probability that the average production for this small herd exceeds 45 pounds of milk a day?

JUST CHECKING ANSWERS

1 A Normal model (approximately).

2 At the actual proportion of all customers who like the new location.

3 $SD(\hat{p}) = \sqrt{\dfrac{(0.5)(0.5)}{100}} = 0.05$

4 $SE(\bar{y}) = 120/\sqrt{100} = 12$

5 Decreases

6 The standard error decreases by $1/\sqrt{2}$.

Confidence Intervals for Proportions

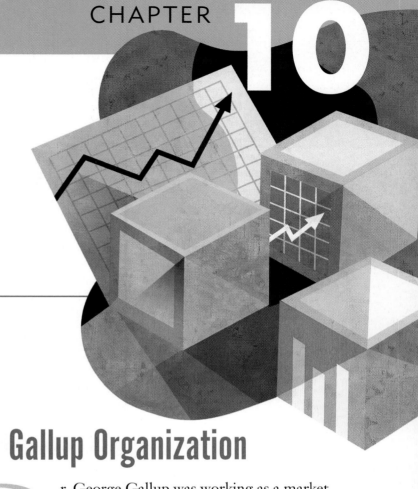

The Gallup Organization

Dr. George Gallup was working as a market research director at an advertising agency in the 1930s when he founded the Gallup Organization to measure and track the public's attitudes toward political, social, and economic issues. He gained notoriety a few years later when he defied common wisdom and predicted that Franklin Roosevelt would win the U.S. presidential election in 1936. Today, the Gallup Poll is a household name. During the late 1930s, he founded the Gallup International Research Institute to conduct polls across the globe. International businesses use the Gallup polls to track how consumers think and feel about such issues as corporate behavior, government policies, and executive compensation. During the late twentieth century, the Gallup Organization partnered with CNN and *USA Today* to conduct and publish public opinion polls. As Gallup once said, "If politicians and special interests have polls to guide them in pursuing their interests, the voters should have polls as well."[1]

[1] Source: The Gallup Organization, Princeton, NJ, www.gallup.com.

Gallup's Web-based data storage system now holds data from polls taken over the last 65 years on a variety of topics, including consumer confidence, household savings, stock market investment, and unemployment.

WHO	U.S. adults
WHAT	Proportion who think economy is getting better
WHEN	January 2008
WHY	To measure expectations about the economy

In order to plan their inventory and production needs, businesses use a variety of forecasts about the economy. One important attribute is consumer confidence in the overall economy. Tracking changes in consumer confidence over time can help businesses gauge whether the demand for their products is on an upswing or about to experience a downturn. The Gallup Poll periodically asks a random sample of U.S. adults whether they think economic conditions are getting better, getting worse, or staying about the same. When they polled 1023 respondents in January 2008, only 153 thought economic conditions in the United States were getting better—a sample proportion of $\hat{p} = 153/1023 = 15.0\%$.[2] We (and Gallup) hope that this observed proportion is close to the population proportion, p, but we know that a second sample of 1023 adults wouldn't have a sample proportion of exactly 15.0%. In fact, Gallup did sample another group of adults just a few days later and found a sample proportion of 13.0%.

From Chapter 9, we know it isn't surprising that two random samples give slightly different results. We'd like to say something, not about different random *samples*, but about the proportion of *all* adults who thought that economic conditions in the United States were getting better in January 2008. The sampling distribution will be the key to our ability to generalize from our sample to the population.

10.1 A Confidence Interval

What do we know about our sampling distribution model? We know that it's centered at the true proportion, p, of all U.S. adults who think the economy is improving. But we don't know p. It isn't 15.0%. That's the $\hat{p}$ from our sample. What we do know is that the sampling distribution model of $\hat{p}$ is centered at p, and we know that the standard deviation of the sampling distribution is $\sqrt{\dfrac{pq}{n}}$. We also know, from the Central Limit Theorem, that the shape of the sampling distribution is approximately Normal, when the sample is large enough.

We don't know p, so we can't find the true standard deviation of the sampling distribution model. But we'll use $\hat{p}$ and find the standard error:

$$SE(\hat{p}) = \sqrt{\frac{\hat{p}\hat{q}}{n}} = \sqrt{\frac{(0.15)(1 - 0.15)}{1023}} = 0.011$$

Since the Gallup sample of 1023 is large, we know that the sampling distribution model for $\hat{p}$ should look approximately like the one shown in Figure 10.1.

NOTATION ALERT:

Remember that $\hat{p}$ is our sample estimate of the true proportion p. Recall also that q is just shorthand for $1 - p$, and $\hat{q} = 1 - \hat{p}$.

---

[2] A proportion is a *number* between 0 and 1. In business it's usually reported as a percentage. You may see it written either way.

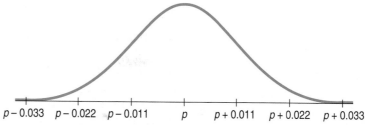

Figure 10.1 *The sampling distribution of sample proportions is centered at the true proportion, p, with a standard deviation of 0.011.*

The sampling distribution model for $\hat{p}$ is Normal with a mean of p and a standard deviation we estimate to be $\sqrt{\dfrac{\hat{p}\hat{q}}{n}}$. Because the distribution is Normal, we'd expect that about 68% of all samples of 1023 U.S. adults taken in January 2008 would have had sample proportions within 1 standard deviation of p. And about 95% of all these samples will have proportions within $p \pm 2$ SEs. But where is *our* sample proportion in this picture? And what value does p have? We still don't know!

We do know that for 95% of random samples, $\hat{p}$ will be no more than 2 SEs away from p. So let's reverse it and look at it from $\hat{p}$'s point of view. If I'm $\hat{p}$, there's a 95% chance that p is no more than 2 SEs away from me. If I reach out 2 SEs, or 2×0.011, away from me on both sides, I'm 95% sure that p will be within my grasp. Of course, I won't know, and even if my interval does catch p, I still don't know its true value. The best I can do is state a probability that I've covered the true value in our interval.

Figure 10.2 *Reaching out 2 SEs on either side of $\hat{p}$ makes us 95% confident we'll trap the true proportion, p.*

What Can We Say about a Proportion?

So what can we really say about p? Here's a list of things we'd like to be able to say and the reasons we can't say most of them:

1. **"15.0% of *all* U.S. adults thought the economy was improving."** It would be nice to be able to make absolute statements about population values with certainty, but we just don't have enough information to do that. There's no way to be sure that the population proportion is the same as the sample proportion; in fact, it almost certainly isn't. Observations vary. Another sample would yield a different sample proportion.

2. **"It is *probably* true that 15.0% of all U.S. adults thought the economy was improving."** No. In fact, we can be pretty sure that whatever the true proportion is, it's not exactly 15.0%, so the statement is not true.

3. **"We don't know exactly what proportion of U.S. adults thought the economy was improving, but we *know* that it's within the interval 15.0% ± 2 × 1.1%. That is, it's between 12.8% and 17.2%."** This is getting closer, but we still can't be certain. We can't know for sure that the true proportion is in this interval—or in any particular range.

4. **"We don't know exactly what proportion of U.S. adults thought the economy was improving, but the interval from 12.8% to 17.2% *probably* contains the true proportion."** We've now fudged twice—first by giving an interval and second by admitting that we only think the interval "probably" contains the true value.

"Far better an approximate answer to the right question, . . . than an exact answer to the wrong question."

—John W. Tukey

That last statement may be true, but it's a bit wishy-washy. We can tighten it up by quantifying what we mean by "probably." We saw that 95% of the time when we reach out 2 SEs from $\hat{p}$, we capture p, so *we can be 95% confident that this is one of those times*. After putting a number on the probability that this interval covers the true proportion, we've given our best guess of where the parameter is and how certain we are that it's within some range.

5. **"We are 95% confident that between 12.8% to 17.2% of U.S. adults thought the economy was improving."** This is now an appropriate interpretation of our confidence intervals. It's not perfect, but it's about the best we can do.

Each confidence interval discussed in the book has a name. You'll see many different kinds of confidence intervals in the following chapters. Some will be about more than *one* sample, some will be about statistics other than *proportions*, and some will use models other than the Normal. The interval calculated and interpreted here is an example of a **one-proportion z-interval**.[3] We'll lay out the formal definition in the next few pages.

What Does "95% Confidence" Really Mean?

What do we mean when we say we have 95% confidence that our interval contains the true proportion? Formally, what we mean is that "95% of samples of this size will produce confidence intervals that capture the true proportion." This is correct but a little long-winded, so we sometimes say "we are 95% confident that the true proportion lies in our interval." Our uncertainty is about whether the particular sample we have at hand is one of the successful ones or one of the 5% that fail to produce an interval that captures the true value. In Chapter 9, we saw how proportions vary from sample to sample. If other pollsters had selected their own samples of adults, they would have found some who thought the economy was getting better, but each sample proportion would almost certainly differ from ours. When they each tried to estimate the true proportion, they'd center their confidence intervals at the proportions they observed in their own samples. Each would have ended up with a different interval.

Figure 10.3 shows the confidence intervals produced by simulating 20 samples. The purple dots are the simulated proportions of adults in each sample who thought the economy was improving, and the orange segments show the confidence intervals found for each simulated sample. The green line represents the true percentage of adults who thought the economy was improving. You can see that most of the simulated confidence intervals include the true value—but one missed. (Note that it is the *intervals* that vary from sample to sample; the green line doesn't move.)

Of course, a huge number of possible samples *could* be drawn, each with its own sample proportion. This simulation approximates just some of them. Each sample can be used to make a confidence interval. That's a large pile of possible

---

[3] In fact, this confidence interval is so standard for a single proportion that you may see it simply called a "confidence interval for the proportion."

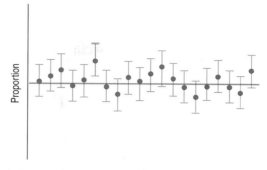

Figure 10.3 *The horizontal green line shows the true proportion of people in January 2008 who thought the economy was improving. Most of the 20 simulated samples shown here produced 95% confidence intervals that captured the true value, but one missed.*

confidence intervals, and ours is just one of those in the pile. Did *our* confidence interval "work"? We can never be sure because we'll never know the true proportion of all U.S. adults who thought in January 2008 that the economy was improving. However, the Central Limit Theorem assures us that 95% of the intervals in the pile are winners, covering the true value, and only 5% on average, miss the target. That's why we're 95% *confident* that our interval is a winner.

10.2 Margin of Error: Certainty vs. Precision

We've just claimed that at a certain confidence level we've captured the true proportion of all U.S. adults who thought the economy was improving in January 2008. Our confidence interval stretched out the same distance on either side of the estimated proportion with the form:

$$\hat{p} \pm 2\ SE(\hat{p}).$$

The *extent* of that interval on either side of $\hat{p}$ is called the **margin of error (ME)**. In general, confidence intervals look like this:

$$estimate \pm ME.$$

The margin of error for our 95% confidence interval was 2 SEs. What if we wanted to be more confident? To be more confident, we'd need to capture p more often, and to do that, we'd need to make the interval wider. For example, if we want to be 99.7% confident, the margin of error will have to be 3 SEs.

> **Confidence Intervals**
>
> We'll see many confidence intervals in this book. All have the form:
>
> estimate ± ME.
>
> For proportions at 95% confidence:
>
> ME ≈ $2SE(\hat{p})$.

$\hat{p} - 3\ SE$ $\hat{p}$ $\hat{p} + 3\ SE$

Figure 10.4 *Reaching out 3 SEs on either side of $\hat{p}$ makes us 99.7% confident we'll trap the true proportion p. Compare the width of this interval with the interval in Figure 10.2.*

The more confident we want to be, the larger the margin of error must be. We can be 100% confident that any proportion is between 0% and 100%, but that's not very useful. Or we could give a narrow confidence interval, say, from 14.98% to 15.02%. But we couldn't be very confident about a statement this precise. Every confidence interval is a balance between certainty and precision.

The tension between certainty and precision is always there. There is no simple answer to the conflict. Fortunately, in most cases we can be both sufficiently certain and sufficiently precise to make useful statements. The choice of confidence level is somewhat arbitrary, but you must choose the level yourself. The data can't do it for you. The most commonly chosen confidence levels are 90%, 95%, and 99%, but any percentage can be used. (In practice, though, using something like 92.9% or 97.2% might be viewed with suspicion.)

Garfield © 1999 Paws, Inc. Reprinted with permission of UNIVERSAL PRESS SYNDICATE. All rights reserved.

10.3 Critical Values

In our opening example, our margin of error was 2 SEs, which produced a 95% confidence interval. To change the confidence level, we'll need to change the *number* of SEs to correspond to the new level. A wider confidence interval means more confidence. For any confidence level the number of SEs we must stretch out on either side of $\hat{p}$ is called the **critical value**. Because it is based on the Normal model, we denote it z^*. For any confidence level, we can find the corresponding critical value from a computer, a calculator, or a Normal probability table, such as Table Z in the back of the book.

For a 95% confidence interval, the precise critical value is $z^* = 1.96$. That is, 95% of a Normal model is found within ±1.96 standard deviations of the mean. We've been using $z^* = 2$ from the 68-95-99.7 Rule because 2 is very close to 1.96 and is easier to remember. Usually, the difference is negligible, but if you want to be precise, use 1.96.[4]

Suppose we could be satisfied with 90% confidence. What critical value would we need? We can use a smaller margin of error. Our greater precision is offset by our acceptance of being wrong more often (that is, having a confidence interval that misses the true value). Specifically, for a 90% confidence interval, the critical value is only 1.645 because for a Normal model, 90% of the values are within 1.645 standard deviations from the mean. By contrast, suppose your boss demands more confidence. If she wants an interval in which she can have 99% confidence, she'll need to include values within 2.576 standard deviations, creating a wider confidence interval.

> ! **NOTATION ALERT:**
> We put an asterisk on a letter to indicate a critical value. We usually use "z" when we talk about Normal models, so z^* is always a critical value from a Normal model.

Some common confidence levels and their associated critical values:

CI	z^*
90%	1.645
95%	1.960
99%	2.576

---

[4] It's been suggested that since 1.96 is both an unusual value and so important in Statistics, you can recognize someone who's had a Statistics course by just saying "1.96" and seeing whether they react.

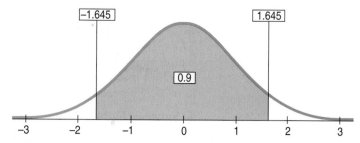

Figure 10.5 For a 90% confidence interval, the critical value is 1.645 because for a Normal model, 90% of the values fall within 1.645 standard deviations of the mean.

10.4 Assumptions and Conditions

The statements we made about what all U.S. adults thought about the economy were possible because we used a Normal model for the sampling distribution. But is that model appropriate?

As we've seen, all statistical models make assumptions. If those assumptions are not true, the model might be inappropriate, and our conclusions based on it may be wrong. Because the confidence interval is built on the Normal model for the sampling distribution, the assumptions and conditions are the same as those we discussed in Chapter 9. But, because they are so important, we'll go over them again.

You can never be certain that an assumption is true, but you can decide intelligently whether it is reasonable. When you have data, you can often decide whether an assumption is plausible by checking a related condition in the data. However, you'll want to make a statement about the world at large, not just about the data. So the assumptions you make are not just about how the data look, but about how representative they are.

Here are the assumptions and the corresponding conditions to check before creating (or believing) a confidence interval about a proportion.

Independence Assumption

You first need to think about whether the independence assumption is plausible. You can look for reasons to suspect that it fails. You might wonder whether there is any reason to believe that the data values somehow affect each other. (For example, might any of the adults in the sample be related?) This condition depends on your knowledge of the situation. It's not one you can check by looking at the data. However, now that you have data, there are two conditions that you can check:

◆ **Randomization Condition:** Were the data sampled at random or generated from a properly randomized experiment? Proper randomization can help ensure independence.

◆ **10% Condition:** Samples are almost always drawn without replacement. Usually, you'd like to have as large a sample as you can. But if you sample from a small population, the probability of success may be different for the last few individuals you draw than it was for the first few. For example, if most of the women have already been sampled, the chance of drawing a woman from the remaining population is lower. If the sample exceeds 10% of the population, the probability of a success changes so much during the sampling that a Normal model may no longer be appropriate. But if less than 10% of the population is sampled, it is safe to assume to have independence.

Sample Size Assumption

The model we use for inference is based on the Central Limit Theorem. So, the sample must be large enough for the Normal sampling model to be appropriate. It turns out that we need more data when the proportion is close to either extreme (0 or 1). This requirement is easy to check with the following condition:

◆ **Success/Failure Condition:** We must expect our sample to contain at least 10 "successes" and at least 10 "failures." Recall that by tradition we arbitrarily label one alternative (usually the outcome being counted) as a "success" even if it's something bad. The other alternative is then a "failure." So we check that both $n\hat{p} \geq 10$ and $n\hat{q} \geq 10$.

One-proportion z-interval

When the conditions are met, we are ready to find the confidence interval for the population proportion, p. The confidence interval is $\hat{p} \pm z^{*} \times SE(\hat{p})$, where the standard deviation of the proportion is estimated by $SE(\hat{p}) = \sqrt{\dfrac{\hat{p}\hat{q}}{n}}$.

GUIDED EXAMPLE	*Public Opinion*

Not long after the Enron debacle, the Gallup Poll asked 508 randomly sampled adults the question: "Do you think the problem of corporate corruption has gotten worse in the past few years, or has it always been like this?" Of these adults, 47% responded that they think corporate corruption has gotten worse. What can we conclude from this survey?

To answer this question, we'll build a confidence interval for the proportion of all U.S. adults who think that corporate corruption has gotten worse. As with other procedures, there are three steps to building and summarizing a confidence interval for proportions: Plan, Do, and Report.

WHO	Adults in the United States
WHAT	Proportion who think corporate corruption is worse
WHEN	July 29–31, 2002
WHERE	United States
HOW	508 adults were randomly sampled by the Gallup Poll
WHY	To investigate public opinion of corporate corruption

PLAN

Setup State the context of the question.

Identify the *parameter* you wish to estimate. Identify the *population* about which you wish to make statements.

Choose and state a confidence level.

Model Think about the assumptions and check the conditions to decide whether we can use the Normal model.

We want to find an interval that is likely with 95% confidence to contain the true proportion, p, of U.S. adults who think that corporate corruption has gotten worse. We have a random sample of 508 U.S. adults, with a sample proportion of 47%.

✓ **Independence Assumption:** Gallup phoned a random sample of U.S. adults. It is unlikely that any respondent influenced another.

✓ **Randomization Condition:** Gallup drew a random sample from all U.S. adults. We don't have details of their randomization but assume that we can trust it.

✓ **10% Condition:** Although sampling was necessarily without replacement, there are many more U.S. adults than were sampled. The sample is certainly less than 10% of the population.

✓ **Success/Failure Condition:**
$n\hat{p} = 508 \times 0.47 = 239 \geq 10$ and
$n\hat{q} = 508 \times 0.53 = 269 \geq 10$,
so the sample is large enough.

State the sampling distribution model for the statistic. Choose your method.

The conditions are satisfied, so I can use a Normal model to find a one-proportion z-interval.

DO

Mechanics Construct the confidence interval. First, find the standard error. (Remember: It's called the "standard error" because we don't know p and have to use $\hat{p}$ instead.)

$n = 508$, $\hat{p} = 0.47$, so

$$SE(\hat{p}) = \sqrt{\frac{0.47 \times 0.53}{508}} = 0.022$$

Next, find the margin of error. We could informally use 2 for our critical value, but 1.96 is more accurate.

Because the sampling model is Normal, for a 95% confidence interval, the critical value $z^* = 1.96$. The margin of error is:

$$ME = z^* \times SE(\hat{p}) = 1.96 \times 0.22 = 0.043$$

Write the confidence interval.

So the 95% confidence interval is:

$$0.47 \pm 0.043 \text{ or } (0.427, 0.513).$$

REALITY CHECK

Check that the interval is plausible. We may not have a strong expectation for the center, but the width of the interval depends primarily on the sample size—especially when the estimated proportion is near 0.5.

The confidence interval covers a range of nearly 10%, but that's about the width we might expect for a sample of 500 (when $\hat{p}$ is close to 0.5).

REPORT

Conclusion Interpret the confidence interval in the proper context. We're 95% confident that our interval captured the true proportion.

MEMO:

Re: Corruption Survey

The Gallup Poll surveyed 508 U.S. adults and asked their opinion about corporate corruption in 2002. Although we can't know the true proportion of U.S. adults who thought that corporate corruption has gotten worse, based on Gallup's results, we can be 95% confident that between 42.7% and 51.3% thought that corporate corruption had gotten worse. Because this is an ongoing concern for public relations, we may want to repeat the survey to obtain more current data. We may also want to keep these results in mind in planning advertising campaigns and corporate public relations.

JUST CHECKING

Think some more about the 95% confidence interval we just created for the proportion of U.S. adults who thought that corporate corruption had gotten worse.

1 If we wanted to be 98% confident, would our confidence interval need to be wider or narrower?

2 Our margin of error was about ±4%. If we wanted to reduce it to ± 3% without increasing the sample size, would our level of confidence be higher or lower?

3 If the Gallup Organization had polled more people, would the interval's margin of error have likely been larger or smaller?

*10.5 A Confidence Interval for Small Samples

When the Success/Failure condition fails, all is not lost. A simple adjustment to the calculation lets us make a confidence interval anyway. All we do is add four *synthetic* observations, two to the successes and two to the failures. So instead of the proportion $\hat{p} = \dfrac{y}{n}$, we use the adjusted proportion $\widetilde{p} = \dfrac{y + 2}{n + 4}$, and for convenience, we write $\widetilde{n} = n + 4$. We modify the interval by using these adjusted values for both the center of the interval *and* the margin of error. Now the adjusted interval is:

$$\widetilde{p} \pm z^* \sqrt{\frac{\widetilde{p}(1 - \widetilde{p})}{\widetilde{n}}}.$$

This adjusted form gives better performance overall[5] and works much better for proportions near 0 or 1. It has the additional advantage that we don't need to check the Success/Failure condition that $n\hat{p}$ and $n\hat{q}$ are greater than 10.

---

[5] By "better performance" we mean that the 95% confidence interval's actual chance of covering the true population proportion is closer to 95%. Simulation studies have shown that our original, simpler confidence interval covers the true population proportion less than 95% of the time when the sample size is small or the proportion is very close to 0 or 1. The original idea was E. B. Wilson's, but the simplified approach we suggest here appeared in A. Agresti and B. A. Coull, "Approximate Is Better Than 'Exact' for Interval Estimation of Binomial Proportions," *The American Statistician*, 52 (1998): 119–126.

Suppose a student in an advertising class is studying the impact of ads placed during the Super Bowl, and wants to know what proportion of students on campus watched it. She takes a random sample of 25 students and find that all 25 watched the Super Bowl for a $\hat{p}$ of 100%. A 95% confidence interval is

$$\hat{p} \pm 1.96\sqrt{\frac{\hat{p}\hat{q}}{n}} = 1.0 \pm 1.96\sqrt{\frac{1.0(0.0)}{25}} = (1.0, 1.0).$$ Does she really believe that *every* one of the 30,000 students on her campus students watched the Super Bowl? Probably not. And she realizes that the Success/Failure condition is severely violated because there are *no* failures.

Using the pseudo observation method described above, she adds two successes and two failures to the sample to get 27/29 successes, for $\tilde{p} = \dfrac{27}{29} = 0.931$. The standard error is no longer 0, but $SE(\tilde{p}) = \sqrt{\dfrac{\tilde{p}\tilde{q}}{\tilde{n}}} = \sqrt{\dfrac{(0.931)(0.069)}{29}} = 0.047$.

Now, a 95% confidence interval is $0.931 \pm 1.96(0.047) = (0.839, 1.023)$. In other words, she's 95% confident that between 83.9% and 102.3% of all students on campus watched the Super Bowl. Because any number greater than 100% makes no sense, she will report simply that with 95% confidence the proportion is at least 83.9%.

10.6 Choosing the Sample Size

Every confidence interval must balance precision—the width of the interval—against confidence. Although it is good to be precise and comforting to be confident, there is a trade-off between the two. A confidence interval that says that the percentage is between 10% and 90% wouldn't be of much use, although you could be quite confident that it covered the true proportion. An interval from 43% to 44% is reassuringly precise, but not if it carries a confidence level of 35%. It's a rare study that reports confidence levels lower than 80%. Levels of 95% or 99% are more common.

The time to decide whether the margin of error is small enough to be useful is when you design your study. Don't wait until you compute your confidence interval. To get a narrower interval without giving up confidence, you need to have less variability in your sample proportion. How can you do that? Choose a larger sample.

Consider a company planning to offer a new service to their customers. Product managers want to estimate the proportion of customers who are likely to purchase this new service to within 3% with 95% confidence. How large a sample do they need?

Let's look at the margin of error:

$$ME = z^*\sqrt{\frac{\hat{p}\hat{q}}{n}}$$

$$0.03 = 1.96\sqrt{\frac{\hat{p}\hat{q}}{n}}.$$

What $\hat{p}$ should we use?

Often you'll have an estimate of the population proportion based on experience or perhaps on a previous study. If so, use that value as $\hat{p}$ in calculating what size sample you need. If not, the cautious approach is to use $\hat{p} = 0.5$. That will determine the largest sample necessary regardless of the true proportion. It's the *worst case* scenario.

They want to find n, the sample size. To find n, they need a value for $\hat{p}$. They don't know $\hat{p}$ because they don't have a sample yet, but they can probably guess a value. The worst case—the value that makes the SD (and therefore n) largest—is 0.50, so if they use that value for $\hat{p}$, they'll certainly be safe.

The company's equation, then, is:

$$0.03 = 1.96\sqrt{\frac{(0.5)(0.5)}{n}}.$$

To solve for *n*, just multiply both sides of the equation by $\sqrt{n}$ and divide by 0.03:

$$0.03\sqrt{n} = 1.96\sqrt{(0.5)(0.5)}$$

$$\sqrt{n} = \frac{1.96\sqrt{(0.5)(0.5)}}{0.03} \approx 32.67$$

Then square the result to find *n*:

$$n \approx (32.67)^2 \approx 1067.1$$

That method will probably give a value with a fraction. To be safe, always round up. The company will need at least 1068 respondents to keep the margin of error as small as 3% with a confidence level of 95%.

Unfortunately, bigger samples cost more money and require more effort. Because the standard error declines only with the *square root* of the sample size, to cut the standard error (and thus the ME) in half, you must *quadruple* the sample size.

Generally a margin of error of 5% or less is acceptable, but different circumstances call for different standards. The size of the margin of error may be a marketing decision or one determined by the amount of financial risk you (or the company) are willing to accept. Drawing a large sample to get a smaller ME, however, can run into trouble. It takes time to survey 2400 people, and a survey that extends over a week or more may be trying to hit a target that moves during the time of the survey. A news event or new product announcement can change opinions in the middle of the survey process.

Keep in mind that the sample size for a survey is the number of respondents, not the number of people to whom questionnaires were sent or whose phone numbers were dialed. Also keep in mind that a low response rate turns any study essentially into a voluntary response study, which is of little value for inferring population values. It's almost always better to spend resources on increasing the response rate than on surveying a larger group. A complete or nearly complete response by a modest-size sample can yield useful results.

Surveys are not the only place where proportions pop up. Credit card banks sample huge mailing lists to estimate what proportion of people will accept a credit card offer. Even pilot studies may be mailed to 50,000 customers or more. Most of these customers don't respond. But in this case, that doesn't make the sample smaller. In fact, they did respond in a way—they just said "No thanks." To the bank, the response rate[6] is $\hat{p}$. With a typical success rate below 1%, the bank needs a very small margin of error—often as low as 0.1%—to make a sound business decision. That calls for a large sample, and the bank should take care when estimating the size needed. For our election poll example, we used $p = 0.5$, both because it's safe and because we honestly believed p to be near 0.5. If the bank used 0.5, they'd get an absurd answer. Instead they base their calculation on a value of p that they expect to find from their experience.

> Public opinion polls often use a sample size of 1000, which gives an ME of about 3% (at 95% confidence) when *p* is near 0.5. But businesses and nonprofit organizations often use much larger samples to estimate the response to a direct mail campaign. Why? Because the proportion of people who respond to these mailings is very low, often 5% or even less. An ME of 3% may not be precise enough if the response rate is that low. Instead, an ME like 0.1% would be more useful, and that requires a very large sample size.

---

[6] Be careful. In marketing studies like this *every* mailing yields a response—"yes" or "no"—and response rate means the success rate, the proportion of customers who accept the offer. That's a different use of the term response rate from the one used in survey response.

How Much of a Difference Can It Make?

A credit card company is about to send out a mailing to test the market for a new credit card. From that sample, they want to estimate the true proportion of people who will sign up for the card nationwide. To be within a tenth of a percentage point, or 0.001 of the true acquisition rate with 95% confidence, how big does the test mailing have to be? Similar mailings in the past lead them to expect that about 0.5% of the people receiving the offer will accept it. Using those values, they find:

$$ME = 0.001 = z^* \sqrt{\frac{\hat{p}\hat{q}}{n}} = 1.96 \sqrt{\frac{(0.005)(0.995)}{n}}$$

$$(0.001)^2 = 1.96^2 \frac{(0.005)(0.995)}{n} \Rightarrow n = \frac{1.96^2(0.005)(0.995)}{(0.001)^2}$$

$$= 19{,}111.96 \; or \; 19{,}112$$

That's a perfectly reasonable size for a trial mailing. But if they had used 0.50 for their estimate of p they would have found:

$$ME = 0.001 = z^* \sqrt{\frac{pq}{n}} = 1.96 \sqrt{\frac{(0.5)(0.5)}{n}}$$

$$(0.001)^2 = 1.96^2 \frac{(0.5)(0.5)}{n} \Rightarrow n = \frac{1.96^2(0.5)(0.5)}{(0.001)^2} = 960{,}400.$$

Quite a different result!

WHAT CAN GO WRONG?

Confidence intervals are powerful tools. Not only do they tell us what is known about the parameter value, but—more important—they also tell us what we *don't* know. In order to use confidence intervals effectively, you must be clear about what you say about them. Don't misstate what the interval means.

What *Can* I Say?

Confidence intervals are based on random samples, so the interval is random, too. The Central Limit Theorem tells us that 95% of the random samples will yield intervals that capture the true value. That's what we mean by being 95% confident.

Technically, we should say "I am 95% confident that the interval from 12.8% to 17.2% captures the true proportion of U.S. adults who thought the economy was improving in January 2008." That formal phrasing emphasizes that *our confidence (and our uncertainty) is about the interval, not the true proportion.* But you may choose a more casual phrasing like "I am 95% confident that between 12.8% and 17.2% of U.S. adults thought the economy was improving in January 2008." Because you've made it clear that the uncertainty is yours and you didn't suggest that the randomness is in the true proportion, this is OK. Keep in mind that it's the interval that's random. It's the focus of both our confidence and our doubt.

- **Don't suggest that the parameter varies.** A statement like "there is a 95% chance that the true proportion is between 12.8% and 17.2%" sounds as though you think the population proportion wanders around and sometimes happens to fall between 12.8% and 17.2%. When you interpret a confidence interval, make it clear that *you* know that the population parameter is fixed and that it is the interval that varies from sample to sample.

- **Don't claim that other samples will agree with yours.** Keep in mind that the confidence interval makes a statement about the true population proportion. An interpretation such as "in 95% of samples of U.S. adults the

(*continued*)

proportion who thought the economy was improving in January 2008 will be between 12.8% and 17.2%" is just wrong. The interval isn't about sample proportions but about the population proportion. There is nothing special about the sample we happen to have; it doesn't establish a standard for other samples.

- **Don't be certain about the parameter.** Saying "between 12.8% and 17.2% of U.S. adults thought the economy was improving in January 2008" asserts that the population proportion cannot be outside that interval. Of course, you can't be absolutely certain of that (just pretty sure).

- **Don't forget: It's about the parameter.** Don't say "I'm 95% confident that $\hat{p}$ is between 12.8% and 17.2%." Of course, you are—in fact, we calculated that our sample proportion was 15.0%. So we already *know* the sample proportion. The confidence interval is about the (unknown) population parameter, p.

- **Don't claim to know too much.** Don't say "I'm 95% confident that between 12.8% and 17.2% of all U.S. adults think the economy is improving." Gallup sampled adults during January 2008, and public opinion shifts over time.

- **Do take responsibility.** Confidence intervals are about *un*certainty. *You* are the one who is uncertain, not the parameter. You have to accept the responsibility and consequences of the fact that not all the intervals you compute will capture the true value. In fact, about 5% of the 95% confidence intervals you find will fail to capture the true value of the parameter. You *can* say "I am 95% confident that between 12.8% and 17.2% of U.S. adults thought the economy was improving in January 2008."

Violations of Assumptions

Confidence intervals and margins of error are often reported along with poll results and other analyses. But it's easy to misuse them and wise to be aware of the ways things can go wrong.

- **Watch out for biased sampling.** Don't forget about the potential sources of bias in surveys that we discussed in Chapter 3. Just because we have more statistical machinery now doesn't mean we can forget what we've already learned. A questionnaire that finds that 85% of people enjoy filling out surveys still suffers from nonresponse bias even though now we're able to put confidence intervals around this (biased) estimate.

- **Think about independence.** The assumption that the values in a sample are mutually independent is one that you usually cannot check. It always pays to think about it, though.

- **Be careful of sample size.** The validity of the confidence interval for proportions may be affected by sample size. Avoid using the confidence interval on "small" samples.

ETHICS IN ACTION

One of Tim Solsby's major responsibilities at MassEast Federal Credit Union is managing online services and website content. In an effort to better serve MassEast members, Tim routinely visits the sites of other financial institutions to get ideas on how he can improve MassEast's online presence. One of the features that caught his attention was a "teen network" that focused on educating teenagers about personal finances. He thought that this was a novel idea and one that could help build a stronger online community among MassEast's members. The executive board of MassEast was meeting next month to consider proposals for improving credit union services, and Tim was eager to present his idea for adding an online teen network. To strengthen his proposal, he decided to poll current credit union members. On the MassEast Federal Credit Union website, he posted an online survey. Among the questions he asked are "Do you have teenage children in your household?" and "Would you encourage your teenage children to learn more about managing personal finances?" Based on 850 responses, Tim constructed a 95% confidence interval and was able to estimate (with 95% confidence) that between 69% and 75% of MassEast members had teenage children at home and that between 62% and 68% would encourage their teenagers to learn more about managing personal finances. Tim believed these results would help convince the executive board that MassEast should add this feature to its website.

ETHICAL ISSUE *The sampling method introduces bias because it is a voluntary response sample and not a random sample. Customers who do have teenagers are more likely to respond than those that do not (related to Item A, ASA Ethical Guidelines).*

ETHICAL SOLUTION *Tim should revise his sampling methods. He might draw a simple random sample of credit union customers and try and contact them by mail or telephone. Whatever method he uses, Tim needs to disclose the sampling procedure to the Board and discuss possible sources of bias.*

What have we learned?

The first few chapters of the book explored graphical and numerical ways of summarizing and presenting sample data. We've learned to use the sample we have at hand to say something about the *world at large*. This process, called *statistical inference*, is based on our understanding of sampling models and will be our focus for the rest of the book.

As our first step in statistical inference, we've learned to use our sample to make a *confidence interval* that estimates what proportion of a population has a certain characteristic.

We've learned that:

- Our best estimate of the true population proportion is the proportion we observed in the sample, so we center our confidence interval there.

- Samples don't represent the population perfectly, so we create our interval with a *margin of error*. This method successfully captures the true population proportion most of the time, providing us with a level of confidence in our interval.

- For a given sample size, the higher the level of confidence we want, the *wider* our confidence interval becomes.

- For a given level of confidence, the larger the sample size we have, the *narrower* our confidence interval can be.

- When designing a study, we can calculate the sample size we'll need to enable us to reach conclusions that have a desired margin of error and level of confidence.

- There are important assumptions and conditions we must check before using this (or any) statistical inference procedure.

We've learned to interpret a confidence interval by reporting what we believe is true in the entire population from which we took our random sample. Of course, we can't be *certain*. We've learned not to overstate or misinterpret what the confidence interval says.

Terms

Confidence interval

An interval of values usually of the form

$$estimate \pm margin\ of\ error$$

found from data in such a way that a percentage of all random samples can be expected to yield intervals that capture the true parameter value.

Critical value

The number of standard errors to move away from the mean of the sampling distribution to correspond to the specified level of confidence. The critical value, denoted z^*, is usually found from a table or with technology.

Margin of error (ME)

In a confidence interval, the extent of the interval on either side of the observed statistic value. A margin of error is typically the product of a critical value from the sampling distribution and a standard error from the data. A small margin of error corresponds to a confidence interval that pins down the parameter precisely. A large margin of error corresponds to a confidence interval that gives relatively little information about the estimated parameter.

One-proportion z-interval

A confidence interval for the true value of a proportion. The confidence interval is

$$\hat{p} \pm z^*SE(\hat{p})$$

where z^* is a critical value from the Standard Normal model corresponding to the specified confidence level.

Skills

 PLAN

- Understand confidence intervals as a balance between the precision and the certainty of a statement about a model parameter.
- Understand that the margin of error of a confidence interval for a proportion changes with the sample size and the level of confidence.
- Know how to examine your data for violations of conditions that would make inference about a population proportion unwise or invalid.

 DO

- Be able to construct a one-proportion z-interval.

 REPORT

- Know how to interpret a one-proportion z-interval in a simple sentence or two. Be able to write such an interpretation so that it does not state or suggest that the parameter of interest is itself random, but rather that the bounds of the confidence interval are the random quantities about which we state our degree of confidence.

Technology Help: Confidence Intervals for Proportions

Confidence intervals for proportions are so easy and natural that many statistics packages don't offer special commands for them. Most statistics programs want the "raw data" for computations. For proportions, the raw data are the "success" and "failure" status for each case. Usually, these are given as 1 or 0, but they might be category names like "yes" and "no." Often we just know the proportion of successes, $\hat{p}$, and the total count, n. Computer packages don't usually deal with summary data like this easily, but the statistics routines found on many graphing calculators allow you to create confidence intervals from summaries of the data—usually all you need to enter are the number of successes and the sample size.

In some programs you can reconstruct variables of 0's and 1's with the given proportions. But even when you have (or can reconstruct) the raw data values, you may not get *exactly* the same margin of error from a computer package as you would find working by hand. The reason is that some packages make approximations or use other methods. The result is very close but not exactly the same. Fortunately, Statistics means never having to say you're certain, so the approximate result is good enough.

EXCEL

Inference methods for proportions are not part of the standard Excel tool set.

Comments

For summarized data, type the calculation into any cell and evaluate it. Confidence intervals for a proportion are available in the DDXL add-in. Select the range of data holding the variable. Then choose **Confidence Intervals** from the **DDXL** menu. Choose **1 Var Prop Interval** from the menu, indicate the variable, and click **OK.**

MINITAB

Choose **Basic Statistics** from the **Stat** menu.

- Choose **1Proportion** from the Basic Statistics submenu.
- If the data are category names in a variable, assign the variable from the variable list box to the **Samples in columns** box. If you have summarized data, click the **Summarized Data** button and fill in the number of trials and the number of successes.
- Click the **Options** button and specify the remaining details.

- If you have a large sample, check **Use test and interval based on normal distribution.**
  Click the **OK** button.

Comments

When working from a variable that names categories, MINITAB treats the last category as the "success" category. You can specify how the categories should be ordered.

SPSS

SPSS does not find confidence intervals for proportions.

JMP

For a **categorical** variable that holds category labels, the **Distribution** platform includes tests and intervals for proportions. For summarized data, put the category names in one variable and the frequencies in an adjacent variable. Designate the frequency column to have the **role** of **frequency**. Then use the **Distribution** platform.

Comments

JMP uses slightly different methods for proportion inferences than those discussed in this text. Your answers are likely to be slightly different, especially for small samples.

DATA DESK

Data Desk does not offer built-in methods for inference with proportions.

Comments

For summarized data, open a Scratchpad to compute the standard deviation and margin of error by typing the calculation. Then use **z-interval for individual μs**.

Mini Case Study Projects

Investment

During the period from June 27–29, 2003, the Gallup Organization asked stock market investors questions about the amount and type of their investments. The questions asked the investors were:

1. Is the total amount of your investments right now $10,000 or more, or is it less than $10,000?

2. If you had $1000 to invest, would you be more likely to invest it in stocks or bonds?

In response to the first question, 65% of the 692 investors reported that they currently have at least $10,000 invested in the stock market. In response to the second question, 48% of the 692 investors reported that they would be more likely to invest in stocks (over bonds). Compute the standard error for each sample proportion. Compute and describe the 95% confidence intervals in the context of the question. What would the size of the sample need to be for the margin of error to be 3%?

Find a recent survey about investment practices or opinions and write up a short report on your findings.

Forecasting Demand

Utilities must forecast the demand for energy use far into the future because it takes decades to plan and build new power plants. Ron Bears, who worked for a northeast utility company, had the job of predicting the proportion of homes that would choose to use electricity to heat their homes. Although he was prepared to report a confidence interval for the true proportion, after seeing his preliminary report, his management demanded a single number as his prediction.

Help Ron explain to his management why a confidence interval for the desired proportion would be more useful for planning purposes. Explain how the precision of the interval and the confidence we can have in it are related to each other. Discuss the business consequences of an interval that is too narrow and the consequences of an interval with too low a confidence level.

EXERCISES

1. Margin of error. A corporate executive reports the results of an employee satisfaction survey, stating that 52% of employees say they are either "satisfied" or "extremely satisfied" with their jobs, and then says "the margin of error is plus or minus 4%." Explain carefully what that means.

2. Margin of error, again. A market researcher estimates the percentage of adults between the ages of 21 and 39 who will see their television ad is 15%, adding that he believes his estimate

has a margin of error of about 3%. Explain what the margin of error means.

3. Conditions. Consider each situation described below. Identify the population and the sample, explain what p and $\hat{p}$ represent, and tell whether the methods of this chapter can be used to create a confidence interval.

a) Police set up an auto checkpoint at which drivers are stopped and their cars inspected for safety problems. They find that 14 of the 134 cars stopped have at least one safety violation. They want to estimate the proportion of all cars in this area that may be unsafe.

b) A CNN show asks viewers to register their opinions on corporate corruption by logging onto a website. Of the 602 people who voted, 488 thought corporate corruption was "worse" this year than last year. The show wants to estimate the level of support among the general public.

4. More conditions. Consider each situation described below. Identify the population and the sample, explain what p and $\hat{p}$ represent, and tell whether the methods of this chapter can be used to create a confidence interval.

a) A large company with 10,000 employees at their main research site is considering moving its day care center off-site to save money. Human resources gathers employees' opinions by sending a questionnaire home with all employees; 380 surveys are returned, with 228 employees in favor of the change.

b) A company sold 1632 MP3 players last month, and within a week, 1388 of the customers had registered their products online at the company website. The company wants to estimate the percentage of all their customers who enroll their products.

5. Conditions, again. Consider each situation described. Identify the population and the sample, explain what p and $\hat{p}$ represent, and tell whether the methods of this chapter can be used to create a confidence interval.

a) A consumer group hoping to assess customer experiences with auto dealers surveys 167 people who recently bought new cars; 3% of them expressed dissatisfaction with the salesperson.

b) A cell phone service provider wants to know what percent of U.S. college students have cell phones. A total of 2883 students were asked as they entered a football stadium, and 2243 indicated they had phones with them.

6. Final conditions. Consider each situation described. Identify the population and the sample, explain what p and $\hat{p}$ represent, and tell whether the methods of this chapter can be used to create a confidence interval.

a) A total of 240 potato plants in a field in Maine are randomly checked, and only 7 show signs of blight. How severe is the blight problem for the U.S. potato industry?

b) Concerned about workers' compensation costs, a small company decided to investigate on the job injuries. The company reported that 12 of their 309 employees suffered an injury on-the-job last year. What can the company expect in future years?

7. Catalog sales. A catalog sales company promises to deliver orders placed on the Internet within 3 days. Follow-up calls to a few randomly selected customers show that a 95% confidence interval for the proportion of all orders that arrive on time is 88% ± 6%. What does this mean? Are the conclusions in parts a–e correct? Explain.

a) Between 82% and 94% of all orders arrive on time.

b) 95% of all random samples of customers will show that 88% of orders arrive on time.

c) 95% of all random samples of customers will show that 82% to 94% of orders arrive on time.

d) The company is 95% sure that between 82% and 94% of the orders placed by the customers in this sample arrived on time.

e) On 95% of the days, between 82% and 94% of the orders will arrive on time.

8. Belgian euro. Recently, two students made worldwide headlines by spinning a Belgian euro 250 times and getting 140 heads—that's 56%. That makes the 90% confidence interval (51%, 61%). What does this mean? Are the conclusions in parts a–e correct? Explain your answers.

a) Between 51% and 61% of all euros are unfair.

b) We are 90% sure that in this experiment this euro landed heads on between 51% and 61% of the spins.

c) We are 90% sure that spun euros will land heads between 51% and 61% of the time.

d) If you spin a euro many times, you can be 90% sure of getting between 51% and 61% heads.

e) 90% of all spun euros will land heads between 51% and 61% of the time.

9. Confidence intervals. Several factors are involved in the creation of a confidence interval. Among them are the sample size, the level of confidence, and the margin of error. Which statements are true?

a) For a given sample size, higher confidence means a smaller margin of error.

b) For a specified confidence level, larger samples provide smaller margins of error.

c) For a fixed margin of error, larger samples provide greater confidence.

d) For a given confidence level, halving the margin of error requires a sample twice as large.

10. Confidence intervals, again. Several factors are involved in the creation of a confidence interval. Among them are the sample size, the level of confidence, and the margin of error. Which statements are true?.

a) For a given sample size, reducing the margin of error will mean lower confidence.

b) For a certain confidence level, you can get a smaller margin of error by selecting a bigger sample.

c) For a fixed margin of error, smaller samples will mean lower confidence.

d) For a given confidence level, a sample 9 times as large will make a margin of error one third as big.

11. Cars. A student is considering publishing a new magazine aimed directly at owners of Japanese automobiles. He wanted to estimate the fraction of cars in the United States that are made in Japan. The computer output summarizes the results of a random sample of 50 autos. Explain carefully what it tells you.

z-interval for proportion
With 90.00% confidence
0.29938661 < p(japan) < 0.46984416

12. Quality control. For quality control purposes, 900 ceramic tiles were inspected to determine the proportion of defective (e.g., cracked, uneven finish, etc.) tiles. Assuming that these tiles are representative of all tiles manufactured by an Italian tile company, what can you conclude based on the computer output?

```
z-interval for proportion
With 95.00% confidence
0.025 < p(defective) < 0.035
```

13. E-mail. A small company involved in e-commerce is interested in statistics concerning the use of e-mail. A poll found that 38% of a random sample of 1012 adults, who use a computer at their home, work, or school, said that they do not send or receive e-mail.

a) Find the margin of error for this poll if we want 90% confidence in our estimate of the percent of American adults who do not use e-mail.
b) Explain what that margin of error means.
c) If we want to be 99% confident, will the margin of error be larger or smaller? Explain.
d) Find that margin of error.
e) In general, if all other aspects of the situation remain the same, will smaller margins of error involve greater or less confidence in the interval?

14. Biotechnology. A biotechnology firm in Boston is planning its investment strategy for future products and research labs. A poll found that only 8% of a random sample of 1012 U.S. adults approved of attempts to clone a human.

a) Find the margin of error for this poll if we want 95% confidence in our estimate of the percent of American adults who approve of cloning humans.
b) Explain what that margin of error means.
c) If we only need to be 90% confident, will the margin of error be larger or smaller? Explain.
d) Find that margin of error.
e) In general, if all other aspects of the situation remain the same, would smaller samples produce smaller or larger margins of error?

15. Teenage drivers. An insurance company checks police records on 582 accidents selected at random and notes that teenagers were at the wheel in 91 of them.

a) Create a 95% confidence interval for the percentage of all auto accidents that involve teenage drivers.
b) Explain what your interval means.
c) Explain what "95% confidence" means.
d) A politician urging tighter restrictions on drivers' licenses issued to teens says, "In one of every five auto accidents, a teenager is behind the wheel." Does your confidence interval support or contradict this statement? Explain.

16. Advertisers. Direct mail advertisers send solicitations ("junk mail") to thousands of potential customers in the hope that some will buy the company's product. The response rate is usually quite low. Suppose a company wants to test the response to a new flyer and sends it to 1000 people randomly selected from their mailing list of over 200,000 people. They get orders from 123 of the recipients.

a) Create a 90% confidence interval for the percentage of people the company contacts who may buy something.
b) Explain what this interval means.
c) Explain what "90% confidence" means.
d) The company must decide whether to now do a mass mailing. The mailing won't be cost-effective unless it produces at least a 5% return. What does your confidence interval suggest? Explain.

17. Retailers. Some food retailers propose subjecting food to a low level of radiation in order to improve safety, but sale of such "irradiated" food is opposed by many people. Suppose a grocer wants to find out what his customers think. He has cashiers distribute surveys at checkout and ask customers to fill them out and drop them in a box near the front door. He gets responses from 122 customers, of whom 78 oppose the radiation treatments. What can the grocer conclude about the opinions of all his customers?

18. Local news. The mayor of a small city has suggested that the state locate a new prison there, arguing that the construction project and resulting jobs will be good for the local economy. A total of 183 residents show up for a public hearing on the proposal, and a show of hands finds 31 in favor of the prison project. What can the city council conclude about public support for the mayor's initiative?

19. Internet music. In a survey on downloading music, the Gallup Poll asked 703 Internet users if they "ever downloaded music from an Internet site that was not authorized by a record company, or not", and 18% responded "yes." Construct a 95% confidence interval for the true proportion of Internet users who have downloaded music from an Internet site that was not authorized.

20. Economy worries. In 2008, a Gallup Poll asked 2,335 U.S. adults, aged 18 or over, how they rated economic conditions. In a poll conducted from January 27–February 1, 2008, only 24% rated the economy as Excellent/Good. Construct a 95% confidence interval for the true proportion of Americans who rated the U.S. economy as Excellent/Good.

21. International business. In Canada, the vast majority (90%) of companies in the chemical industry are ISO 14001 certified. The ISO 14001 is an international standard for environmental management systems. An environmental group wished to estimate the percentage of U.S. chemical companies that are ISO 14001 certified. Of the 550 chemical companies sampled, 385 are certified.

a) What proportion of the sample reported being certified?
b) Create a 95% confidence interval for the proportion of U.S. chemical companies with ISO 14001 certification. (Be sure to check conditions.) Compare to the Canadian proportion.

22. Worldwide survey. In Chapter 4, Exercise 17, we learned that GfK Roper surveyed people worldwide asking them "how important is acquiring wealth to you." Of 1535 respondents in India, 1168 said that it was of more than average importance. In the United States of 1317 respondents, 596 said it was of more than average importance.

a) What proportion thought acquiring wealth was of more than average importance in each country's sample?

b) Create a 95% confidence interval for the proportion who thought it was of more than average importance in India. (Be sure to test conditions.) Compare that to a confidence interval for the U.S. population.

23. Business ethics. In a survey on corporate ethics, a poll split a simple random sample of 1076 faculty and corporate recruiters into two halves, asking 538 respondents the question, "Generally, do you believe that MBAs are more or less aware of ethical issues in business today than five years ago?" The other half were asked: "Generally, do you believe that MBAs are less or more aware of ethical issues in business today than five years ago?" These may seem like the same questions, but sometimes the order of the choices matters. In response to the first question, 53% thought MBA graduates were more aware of ethical issues, but when the question was phrased differently, this proportion dropped to 44%.

a) What kind of bias may be present here?

b) Each group consisted of 538 respondents. If we combine them, considering the overall group to be one larger random sample, what is a 95% confidence interval for the proportion of the faculty and corporate recruiters that believe MBAs are more aware of ethical issues today?

c) How does the margin of error based on this pooled sample compare with the margins of error from the separate groups? Why?

24. Media survey. In 2007, a Gallup Poll conducted face-to-face interviews with 1006 adults in Saudi Arabia, aged 15 and older, asking them questions about how they get information. Among them was the question: "Is international television very important in keeping you well-informed about events in your country?" Gallup reported that 82% answered "yes" and noted that at 95% confidence there was a 3% margin of error and that "in addition to sampling error, question wording and practical difficulties in conducting surveys can introduce error or bias into the findings of public opinion polls."

a) What kinds of bias might they be referring to?

b) Do you agree with their margin of error? Explain.

25. Gambling. A city ballot includes a local initiative that would legalize gambling. The issue is hotly contested, and two groups decide to conduct polls to predict the outcome. The local newspaper finds that 53% of 1200 randomly selected voters plan to vote "yes," while a college Statistics class finds 54% of 450 randomly selected voters in support. Both groups will create 95% confidence intervals.

a) Without finding the confidence intervals, explain which one will have the larger margin of error.

b) Find both confidence intervals.

c) Which group concludes that the outcome is too close to call? Why?

26. Casinos. Governor Deval Patrick of Massachusetts proposed legalizing casinos in Massachusetts although they are not currently legal, and he included the revenue from them in his latest state budget. The website www.boston.com conducted an Internet poll on the question: "Do you agree with the casino plan the governor is expected to unveil?" As of the end of 2007, there were 8663 votes cast, of which 63.5% of respondents said: "No. Raising revenues by allowing gambling is shortsighted."

a) Find a 95% confidence interval for the proportion of voters in Massachusetts who would respond this way.

b) Are the assumptions and conditions satisfied? Explain.

27. Pharmaceutical company. A pharmaceutical company is considering investing in a "new and improved" vitamin D supplement for children. Vitamin D, whether ingested as a dietary supplement or produced naturally when sunlight falls upon the skin, is essential for strong, healthy bones. The bone disease rickets was largely eliminated in England during the 1950s, but now there is concern that a generation of children more likely to watch TV or play computer games than spend time outdoors is at increased risk. A recent study of 2700 children randomly selected from all parts of England found 20% of them deficient in vitamin D.

a) Find a 98% confidence interval for the proportion of children in England who are deficient in vitamin D.

b) Explain carefully what your interval means.

c) Explain what "98% confidence" means.

d) Does the study show that computer games are a likely cause of rickets? Explain.

28. Wireless access. In Chapter 4, Exercise 26, we saw that the Pew Internet and American Life Project polled 798 Internet users in December 2006, asking whether they have logged on to the Internet using a wireless device or not and 243 responded "Yes."

a) Find a 98% confidence interval for the proportion of all U.S. Internet users who have logged in using a wireless device.

b) Explain carefully what your interval means.

c) Explain what "98% confidence" means.

29. Funding. In 2005, a survey developed by Babson College and the Association of Women's Business Centers (WBCs) was distributed to WBCs in the United States. Of a representative sample of 20 WBCs, 40% reported that they had received funding from the national Small Business Association (SBA).

a) Check the assumptions and conditions for inference on proportions.

b) If it's appropriate, find a 90% confidence interval for the proportion of WBCs that receive SBA funding. If it's not appropriate, explain and/or recommend an alternative action.

30. Real estate survey. A real estate agent looks over the 15 listings she has in a particular zip code in California and finds that 80% of them have swimming pools.

a) Check the assumptions and conditions for inference on proportions.

b) If it's appropriate, find a 90% confidence interval for the proportion of houses in this zip code that have swimming pools. If it's not appropriate, explain and/or recommend an alternative action.

31. Benefits survey. A paralegal at the Vermont State Attorney General's office wants to know how many companies in Vermont provide health insurance benefits to all employees. She chooses 12 companies at random and finds that all 12 offer benefits.

a) Check the assumptions and conditions for inference on proportions.

b) Find a 95% confidence interval for the true proportion of companies that provide health insurance benefits to all their employees.

32. Awareness survey. A telemarketer at a credit card company is instructed to ask the next 18 customers that call into the 800 number whether they are aware of the new Platinum card that the company is offering. Of the 18, 17 said they were aware of the program.

a) Check the assumptions and conditions for inference on proportions.

b) Find a 95% confidence interval for the true proportion of customers who are aware of the new card.

33. IRS. In a random survey of 226 self-employed individuals, 20 reported having had their tax returns audited by the IRS in the past year. Estimate the proportion of self-employed individuals nationwide who've been audited by the IRS in the past year.

a) Check the assumptions and conditions (to the extent you can) for constructing a confidence interval.

b) Construct a 95% confidence interval.

c) Interpret your interval.

d) Explain what "95% confidence" means in this context.

34. ACT, Inc. In 2004, ACT, Inc. reported that 74% of 1644 randomly selected college freshmen returned to college the next year. Estimate the national freshman-to-sophomore retention rate.

a) Check that the assumptions and conditions are met for inference on proportions.

b) Construct a 98% confidence interval.

c) Interpret your interval.

d) Explain what "98% confidence" means in this context.

35. Internet music, again. A Gallup Poll (exercise 19) asked Americans if the fact that they can make copies of songs on the Internet for free made them more likely—or less likely—to buy a performer's CD. Only 13% responded that it made them "less likely." The poll was based on a random sample of 703 Internet users.

a) Check that the assumptions and conditions are met for inference on proportions.

b) Find the 95% confidence interval for the true proportion of all U.S. Internet users who are "less likely" to buy CDs.

36. ACT, Inc., again. The ACT, Inc. study described in Exercise 34 was actually stratified by type of college—public or private. The retention rates were 71.9% among 505 students enrolled in public colleges and 74.9% among 1139 students enrolled in private colleges.

a) Will the 95% confidence interval for the true national retention rate in private colleges be wider or narrower than the 95% confidence interval for the retention rate in public colleges? Explain.

b) Find the 95% confidence interval for the public college retention rate.

c) Should a public college whose retention rate is 75% proclaim that they do a better job than other public colleges of keeping freshmen in school? Explain.

37. Politics. A poll of 1005 U.S. adults split the sample into four age groups: ages 18–29, 30–49, 50–64, and 65+. In the youngest age group, 62% said that they thought the U.S. was ready for a woman president, as opposed to 35% who said "no, the country was not ready." (3% were undecided.) The sample included 250 18- to 29-year-olds.

a) Do you expect the 95% confidence interval for the true proportion of all 18- to 29-year-olds who think the U.S. is ready for a woman president to be wider or narrower than the 95% confidence interval for the true proportion of all U.S. adults? Explain.

b) Find the 95% confidence interval for the true proportion of all 18- to 29-year-olds who believe the U.S. is ready for a woman president.

38. Wireless access, again. The survey in Exercise 28 asking about wireless Internet access also classified the 798 respondents by income.

Income	Wireless Users	Other Internet Users	Total
Under $30K	34	128	**162**
$30K–$50K	31	133	**164**
$50K–$75K	44	72	**116**
Over $75K	83	111	**194**
Don't know/ refused	51	111	**162**
Total	**243**	**555**	**798**

a) Do you expect the 95% confidence interval for the true proportion of all those making more than $75K who are wireless users to be wider or narrower than the 95% confidence interval for the true proportion among those who make between $50K and $75K. Explain briefly.

b) Find the 95% confidence interval for the true proportion of those making more than $75K who are wireless users.

39. More Internet music. A random sample of 168 students was asked how many songs were in their digital music library and what fraction of them was legally purchased. Overall, they reported having a total of 117,079 songs, of which 23.1% were legal. The music industry would like a good estimate of the proportion of songs in students' digital music libraries that are legal.

a) Think carefully. What is the parameter being estimated? What is the population? What is the sample size?

b) Check the conditions for making a confidence interval.

c) Construct a 95% confidence interval for the fraction of legal digital music.

d) Explain what this interval means. Do you believe that you can be this confident about your result? Why or why not?

40. Trade agreement. Results from a January 2008 telephone survey conducted by Gallup showed that 57% of urban Colombian adults support a free trade agreement (FTA) with the United States. Gallup used a cluster sample in which the cities of Bogota, Cali, Barranquilla, and Medellin provided a representative sample of 1000 urban Colombians aged 15 and older.

a) What is the parameter being estimated? What is the population? What is the sample size?

b) Check the conditions for making a confidence interval.

c) Construct a 95% confidence interval for the fraction of Colombians in agreement with the FTA.

d) Explain what this interval means. Do you believe that you can be this confident about your result? Why or why not?

41. CDs. A company manufacturing CDs is working on a new technology. A random sample of 703 Internet users were asked: "As you may know, some CDs are being manufactured so that you can only make one copy of the CD after you purchase it. Would you buy a CD with this technology, or would you refuse to buy it even if it was one you would normally buy?" Of these users, 64% responded that they would buy the CD.

a) Create a 90% confidence interval for this percentage.

b) If the company wants to cut the margin of error in half, how many users must they survey?

42. Internet music, last time. The research group that conducted the survey in Exercise 39 wants to provide the music industry with definitive information, but they believe that they could use a smaller sample next time. If the group is willing to have twice as big a margin of error, how many songs must be included?

43. Graduation. It's believed that as many as 25% of adults over age 50 never graduated from high school. We wish to see if this percentage is the same among the 25 to 30 age group.

a) How many of this younger age group must we survey in order to estimate the proportion of nongrads to within 6% with 90% confidence?

b) Suppose we want to cut the margin of error to 4%. What's the necessary sample size?

c) What sample size would produce a margin of error of 3%?

44. Hiring. In preparing a report on the economy, we need to estimate the percentage of businesses that plan to hire additional employees in the next 60 days.

a) How many randomly selected employers must we contact in order to create an estimate in which we are 98% confident with a margin of error of 5%?

b) Suppose we want to reduce the margin of error to 3%. What sample size will suffice?

c) Why might it not be worth the effort to try to get an interval with a margin of error of only 1%?

45. Graduation, again. As in Exercise 43, we hope to estimate the percentage of adults aged 25 to 30 who never graduated from high school. What sample size would allow us to increase our confidence level to 95% while reducing the margin of error to only 2%?

46. Better hiring info. Editors of the business report in Exercise 44 are willing to accept a margin of error of 4% but want 99% confidence. How many randomly selected employers will they need to contact?

47. Pilot study. A state's environmental agency worries that a large percentage of cars may be violating clean air emissions standards. The agency hopes to check a sample of vehicles in order to estimate that percentage with a margin of error of 3% and 90% confidence. To gauge the size of the problem, the agency first picks 60 cars and finds 9 with faulty emissions systems. How many should be sampled for a full investigation?

48. Another pilot study. During routine conversations, the CEO of a new start-up reports that 22% of adults between the ages of 21 and 39 will purchase her new product. Hearing this, some investors decide to conduct a large-scale study, hoping to estimate the proportion to within 4% with 98% confidence. How many randomly selected adults between the ages of 21 and 39 must they survey?

49. Approval rating. A newspaper reports that the governor's approval rating stands at 65%. The article adds that the poll is based on a random sample of 972 adults and has a margin of error of 2.5%. What level of confidence did the pollsters use?

50. Amendment. The Board of Directors of a publicly traded company says that a proposed amendment to their bylaws is likely to win approval in the upcoming election because a poll of 1505 stock owners indicated that 52% would vote in favor. The Board goes on to say that the margin of error for this poll was 3%.

a) Explain why the poll is actually inconclusive.

b) What confidence level did the pollsters use?

T 51. Customer spending. The data set provided contains last month's credit card purchases of 500 customers randomly chosen from a segment of a major credit card issuer. The marketing department is considering a special offer for customers who spend more than $1000 per month on their card. From these data construct a 95% confidence interval for the proportion of customers in this segment who will qualify.

T 52. Advertising. A philanthropic organization knows that its donors have an average age near 60 and is considering taking out an ad in the American Association of Retired People (AARP) magazine. An analyst wonders what proportion of their donors are actually 50 years old or older. He takes a random sample of the records of 500 donors. From the data provided, construct a 95% confidence interval for the proportion of donors who are 50 years old or older.

53. Health insurance. Based on a 2007 survey of U.S. house-holds (see www.census.gov), 87% (out of 3060) of males in Massachusetts (MA) have health insurance.

a) Examine the conditions for constructing a confidence interval for the proportion males in MA who had health insurance.

b) Find the 95% confidence interval for the percent of males who have health insurance.

c) Interpret your confidence interval.

54. Health insurance, part 2. Using the same survey and data as in Exercise 53, we find that 84% of those respondents in Massachusetts who identified themselves as Black/African-Americans (out of 440) had health insurance.

a) Examine the conditions for constructing a confidence interval for the proportion of Black/African-Americans in MA who had health insurance.

b) Find the 95% confidence interval.

c) Interpret your confidence interval.

JUST CHECKING ANSWERS

1 Wider

2 Lower

3 Smaller

Testing Hypotheses about Proportions

Dow Jones Industrial Average

More than a hundred years ago Charles Dow changed the way people look at the stock market. Surprisingly, he wasn't an investment wizard or a venture capitalist. He was a journalist who wanted to make investing understandable to ordinary people. Although he died at the relatively young age of 51 in 1902, his impact on how we track the stock market has been both long-lasting and far-reaching.

In the late 1800s, when Charles Dow reported on Wall Street, investors preferred bonds, not stocks. Bonds were reliable, backed by the real machinery and other hard assets the company owned. What's more, bonds were predictable; the bond owner knew when the bond would mature and so, knew when and how much the bond would pay. Stocks simply represented "shares" of ownership, which were risky and erratic. In May 1896, Dow and Edward Jones, whom he had known since their days as reporters for the *Providence Evening Press*, launched the now-famous Dow Jones Industrial Average

(DJIA) to help the public understand stock market trends. The original DJIA averaged 11 stock prices. Of those original industrial stocks, only General Electric is still in the DJIA.

Since then, the DJIA has become synonymous with overall market performance and is often referred to simply as the Dow. The index was expanded to 20 stocks in 1916 and to 30 in 1928 at the height of the roaring 20's bull market. That bull market peaked on September 3, 1929, when the Dow reached 381.17. On October 28 and 29, 1929, the Dow lost nearly 25% of its value. Then things got worse. Within four years, on July 8, 1932, the 30 industrials reached an all-time low of 40.65. The highs of September 1929 were not reached again until 1954.

Today the Dow is a weighted average of 30 industrial stocks, with weights used to account for splits and other adjustments. The "Industrial" part of the name is largely historic. Today's DJIA includes the service industry and financial companies and is much broader than just heavy industry. And it is still one of the most watched indicators of the state of the U.S. stock market and the global economy.

WHO	Days on which the stock market was open ("trading days")
WHAT	Closing price of the Dow Jones Industrial Average (*Close*)
UNITS	Points
WHEN	August 1982 to December 1986
WHY	To test theory of stock market behavior

How does the stock market move? Here are the DJIA closing prices for the bull market that ran from mid 1982 to the end of 1986.

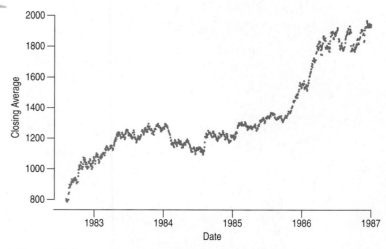

Figure 11.1 *Daily closing prices of the Dow Jones Industrials from mid 1982 to the end of 1986.*

The DJIA clearly increased during this famous bull market, more than doubling in value in less than five years. One common theory of market behavior says that on a given day, the market is just as likely to move up as down. Another way of phrasing this is that the daily behavior of the stock market is random. Can that be true during such periods of obvious increase? Let's investigate if the Dow is just as likely to move higher or lower on any given day. Out of the 1112 trading days in that period, the average increased on 573 days, a sample proportion of 0.5153 or 51.53%. That *is* more "up" days than "down" days, but is it far enough from 50% to cast doubt on the assumption of equally likely up or down movement?

11.1 Hypotheses

NOTATION ALERT:
Capital H is the standard letter for hypotheses. H_0 labels the null hypothesis, and H_A labels the alternative.

How can we state and test a hypothesis about daily changes in the DJIA? Hypotheses are working models that we adopt temporarily. To test whether the daily fluctuations are equally likely to be up as down, we assume that they are, and that any apparent difference from 50% is just random fluctuation. So, our starting hypothesis, called the null hypothesis, is that the proportion of days on which the DJIA increases is 50%. The **null hypothesis**, which we denote H_0, specifies a population model parameter and proposes a value for that parameter. We usually write down a null hypothesis about a proportion in the form $H_0: p = p_0$. This is a concise way to specify the two things we need most: the identity of the parameter we hope to learn about (the true proportion) and a specific hypothesized value for that parameter (in this case, 50%). We need a hypothesized value so we can compare our observed statistic to it. Which value to use for the hypothesis is not a statistical question. It may be obvious from the context of the data, but sometimes it takes a bit of thinking to translate the question we hope to answer into a hypothesis about a parameter. For our hypothesis about whether the DJIA moves up or down with equal likelihood, it's pretty clear that we need to test $H_0: p = 0.5$.

The **alternative hypothesis**, which we denote H_A, contains the values of the parameter that we consider plausible if we reject the null hypothesis. In our example, our null hypothesis is that the proportion, p, of "up" days is 0.5. What's the alternative? During a bull market, you might expect more up days than down, but we'll assume that we're interested in a deviation in either direction from the null hypothesis, so our alternative is $H_A: p \neq 0.5$.

What would convince you that the proportion of up days was not 50%? If on 95% of the days, the DJIA closed up, most people would be convinced that up and down days were not equally likely. But if the sample proportion of up days were only slightly higher than 50%, you'd be skeptical. After all, observations do vary, so we wouldn't be surprised to see some difference. How different from 50% must the proportion be before we *are* convinced that it has changed? Whenever we ask about the size of a statistical difference, we naturally think of the standard deviation. So let's start by finding the standard deviation of the sample proportion of days on which the DJIA increased.

We've seen 51.53% up days out of 1112 trading days. The sample size of 1112 is certainly big enough to satisfy the Success/Failure condition. (We expect $0.50 \times 1112 = 556$ daily increases.) We suspect that the daily price changes are random and independent. And we know what hypothesis we are testing. To test a hypothesis we (temporarily) *assume* that it is true so we can see whether that description of the world is plausible. If we assume that the Dow increases or decreases with equal likelihood, we'll need to center our Normal sampling model at a mean of 0.5. Then, we can find the standard deviation of the sampling model as

$$SD(\hat{p}) = \sqrt{\frac{pq}{n}} = \sqrt{\frac{(0.5)(1-0.5)}{1112}} = 0.015$$

◆ **Why is this a standard deviation and not a standard error?** This is a standard deviation because we haven't estimated anything. Once we assume that the null hypothesis is true, it gives us a value for the model parameter p. With proportions, if we know p then we also automatically know its standard deviation. Because we find the standard deviation from the model parameter, this is a standard deviation and not a standard error. When we found a confidence interval for p, we could not assume that we knew its value, so we estimated the standard deviation from the sample value, $\hat{p}$.

Now we know both parameters of the Normal sampling distribution model for our null hypothesis. For the mean, μ, we use $p = 0.50$, and for σ we use the standard deviation of the sample proportions $SD(\hat{p}) = 0.015$. We want to know how likely it would be to see the observed value $\hat{p}$ as far away from 50% as the value of 51.53% that we actually have observed. Looking first at a picture (Figure 11.2), we can see that 51.53% doesn't look very surprising. The more exact answer (from a calculator, computer program, or the Normal table) is that the probability is about 0.308. This is the probability of observing more than 51.53% up days (or more than 51.53% down days) if the null model were true. In other words, if the chance of an up day for the Dow is 50%, we'd expect to see stretches of 1112 trading days with as many as 51.53% up days about 15.4% of the time and with as many as 51.53% down days about 15.4% of the time. That's not terribly unusual, so there's really no convincing evidence that the market did not act randomly.

> To remind us that the parameter value comes from the null hypothesis, it is sometimes written as p_0 and the standard deviation as
>
> $$SD(\hat{p}) = \sqrt{\frac{p_0 q_0}{n}}$$

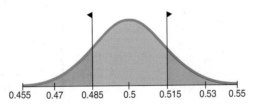

| 0.455 | 0.47 | 0.485 | 0.5 | 0.515 | 0.53 | 0.55 |

Figure 11.2 *How likely is a proportion of more than 51.5% or less than 48.5% when the true mean is 50%? This is what it looks like. Each red area is 0.154 of the total area under the curve.*

It may surprise you that even during a bull market, the direction of daily movements is random. But, the probability that any given day will end up or down appears to be about 0.5 regardless of the longer-term trends. It may be that when the stock market has a long run up (or possibly down, although we haven't checked that), it does so not by having more days of increasing or decreasing value, but by the actual amounts of the increases or decreases being unequal.

11.2 A Trial as a Hypothesis Test

We started by assuming that the probability of an up day was 50%. Then we looked at the data and concluded that we couldn't say otherwise because the proportion that we actually observed wasn't far enough from 50%. Does this reasoning of hypothesis tests seem backwards? That could be because we usually prefer to think about getting things right rather than getting them wrong. But, you've seen this reasoning before in a different context. This is the logic of jury trials.

Let's suppose a defendant has been accused of robbery. In British common law and those systems derived from it (including U.S. law), the null hypothesis is that the defendant is innocent. Instructions to juries are quite explicit about this.

The evidence takes the form of facts that seem to contradict the presumption of innocence. For us, this means collecting data. In the trial, the prosecutor presents evidence. ("If the defendant were innocent, wouldn't it be remarkable that the police found him at the scene of the crime with a bag full of money in his hand, a mask on his face, and a getaway car parked outside?") The next step is to judge the evidence. Evaluating the evidence is the responsibility of the jury in a trial, but it falls on your shoulders in hypothesis testing. The jury considers the evidence in light of the *presumption* of innocence and judges whether the evidence against the defendant would be plausible *if the defendant were in fact innocent.*

Like the jury, we ask: "Could these data plausibly have happened by chance if the null hypothesis were true?" If they are very unlikely to have occurred, then the evidence raises a reasonable doubt about the null hypothesis. Ultimately, *you* must make a decision. The standard of "beyond a reasonable doubt" is purposefully ambiguous because it leaves the jury to decide the degree to which the evidence contradicts the hypothesis of innocence. Juries don't explicitly use probability to help them decide whether to reject that hypothesis. But when you ask the same question of your null hypothesis, you have the advantage of being able to quantify exactly how surprising the evidence would be if the null hypothesis were true.

How unlikely is unlikely? Some people set rigid standards. Levels like 1 time out of 20 (0.05) or 1 time out of 100 (0.01) are common. But if *you* have to make the decision, you must judge for yourself in each situation whether the probability of observing your data is small enough to constitute "reasonable doubt."

11.3 P-Values

The fundamental step in our reasoning is the question: "Are the data surprising, given the null hypothesis?" And the key calculation is to determine exactly how likely the data we observed would be if the null hypothesis were the true model of the world. So we need a *probability*. Specifically, we want to find the probability of seeing data like these (or something even less likely) *given* the null hypothesis. This probability is the value on which we base our decision, so statisticians give this probability a special name. It's called the **P-value**.

A low enough P-value says that the data we have observed would be very unlikely if our null hypothesis were true. We started with a model, and now that same model tells us that the data we have are unlikely to have happened. That's surprising. In this case, the model and data are at odds with each other, so we have to make a choice. Either the null hypothesis is correct and we've just seen something remarkable, or the null hypothesis is wrong, (and, in fact, we were wrong to use it as the basis for computing our P-value). If you believe in data more than in assumptions, then, given that choice, when you see a low P-value you should reject the null hypothesis.

> **Beyond a Reasonable Doubt**
> We ask whether the data were unlikely beyond a reasonable doubt. We've just calculated that probability. The probability that the observed statistic value (or an even more extreme value) could occur if the null model were true—in this case, 0.308—is the P-value.

When the P-value is *high* (or just not low *enough*), what do we conclude? In that case, we haven't seen anything unlikely or surprising at all. The data are consistent with the model from the null hypothesis, and we have no reason to reject the null hypothesis. Events that have a high probability of happening happen all the time. So, when the P-value is high does that mean we've proved the null hypothesis is true? No! We realize that many other similar hypotheses could also account for the data we've seen. The most we can say is that it doesn't appear to be false. Formally, we say that we "fail to reject" the null hypothesis. That may seem to be a pretty weak conclusion, but it's all we can say when the P-value is not low enough. All that means is that the data are consistent with the model that we started with.

What to Do with an "Innocent" Defendant

"If the People fail to satisfy their burden of proof, you must find the defendant not guilty."
—New York State jury instructions

Let's see what that last statement means in a jury trial. If the evidence is not strong enough to reject the defendant's presumption of innocence, what verdict does the jury return? They do not say that the defendant is innocent. They say "not guilty." All they are saying is that they have not seen sufficient evidence to reject innocence and convict the defendant. The defendant may, in fact, be innocent, but the jury has no way to be sure.

Said statistically, the jury's null hypothesis is: innocent defendant. If the evidence is too unlikely (the P-value is low) then, given the assumption of innocence, the jury rejects the null hypothesis and finds the defendant guilty. But—and this is an important distinction—if there is *insufficient evidence* to convict the defendant (if the P-value is *not* low), the jury does not conclude that the null hypothesis is true and declare that the defendant is innocent. Juries can only *fail to reject* the null hypothesis and declare the defendant "not guilty."

In the same way, if the data are not particularly unlikely under the assumption that the null hypothesis is true, then the most we can do is to "fail to reject" our null hypothesis. We never declare the null hypothesis to be true. In fact we simply do not know whether it's true or not. (After all, more evidence may come along later.)

Don't We Want to Reject the Null?

Often the people who collect the data or perform the experiment hope to reject the null. They hope the new drug is better than the placebo; they hope the new ad campaign is better than the old one; or they hope their candidate is ahead of the opponent. But when we practice Statistics, we can't allow that hope to affect our decision. The essential attitude for a hypothesis tester is skepticism. Until we become convinced otherwise, we cling to the null's assertion that there's nothing unusual, nothing unexpected, no effect, no difference, etc. As in a jury trial, the burden of proof rests with the alternative hypothesis—innocent until proven guilty. When you test a hypothesis, you must act as judge and jury, but you are not the prosecutor.

Imagine a test of whether a company's new website design encourages a higher percentage of visitors to make a purchase (as compared to the site they've used for years). The null hypothesis is that the new site is no more effective at stimulating purchases than the old one. The test sends visitors randomly to one version of the website or the other. Of course, some will make a purchase, and others won't. If we compare the two websites on only 10 customers each, the results are likely *not to be clear*, and we'll be unable to reject the hypothesis. Does this mean the new design is a complete bust? Not necessarily. It simply means that we don't have enough evidence to reject our null hypothesis. That's why we don't start by assuming that the new design is *more* effective. If we were to do that, then we could test just a few customers, find that the results aren't clear, and claim that since we've been unable to reject our original assumption the redesign must be effective. The Board of Directors is unlikely to be impressed by that argument.

Conclusion

If the P-value is "low", reject H_0 and conclude H_A.

If the P-value is not "low enough" then fail to reject H_0 and the test is inconclusive.

JUST CHECKING

1 A pharmaceutical firm wants to know whether aspirin helps to thin blood. The null hypothesis says that it doesn't. The firm's researchers test 12 patients, observe the proportion with thinner blood, and get a P-value of 0.32. They proclaim that aspirin doesn't work. What would you say?

2 An allergy drug has been tested and found to give relief to 75% of the patients in a large clinical trial. Now the scientists want to see whether a new, "improved" version works even better. What would the null hypothesis be?

3 The new allergy drug is tested, and the P-value is 0.0001. What would you conclude about the new drug?

11.4 The Reasoning of Hypothesis Testing

Hypothesis tests follow a carefully structured path. To avoid getting lost as we navigate down it, we divide that path into four distinct sections: hypothesis, model, mechanics, and conclusion.

Hypotheses

"The null hypothesis is never proved or established, but is possibly disproved, in the course of experimentation. Every experiment may be said to exist only in order to give the facts a chance of disproving the null hypothesis."
—Sir Ronald Fisher, The Design of Experiments, 1931

First, state the null hypothesis. That's usually the skeptical claim that nothing's different. The null hypothesis assumes the default (often the status quo) is true (the defendant is innocent, the new method is no better than the old, customer preferences haven't changed since last year, etc.).

In statistical hypothesis testing, hypotheses are almost always about model parameters. To assess how unlikely our data may be, we need a null model. The null hypothesis specifies a particular parameter value to use in our model. In the usual notation, we write H_0: *parameter = hypothesized value*. The alternative hypothesis, H_A, contains the values of the parameter we consider plausible when we reject the null.

Model

When the Conditions Fail . . .
You might proceed with caution, explicitly stating your concerns. Or you may need to do the analysis with and without an outlier, or on different subgroups, or after re-expressing the response variable. Or you may not be able to proceed at all.

To plan a statistical hypothesis test, specify the *model* for the sampling distribution of the statistic you will use to test the null hypothesis and the parameter of interest. For proportions, we use the Normal model for the sampling distribution. Of course, all models require assumptions, so you will need to state them and check any corresponding conditions. For a test of a proportion, the assumptions and conditions are the same as for a one-proportion z-interval.

Your model step should end with a statement such as: *Because the conditions are satisfied, we can model the sampling distribution of the proportion with a Normal model.* Watch out, though. Your Model step could end with: *Because the conditions are not satisfied, we can't proceed with the test.* (If that's the case, stop and reconsider.)

Each test we discuss in this book has a name that you should include in your report. We'll see many tests in the following chapters. Some will be about more than one sample, some will involve statistics other than proportions, and some will use models other than the Normal (and so will not use z-scores). The test about proportions is called a **one-proportion z-test**.[1]

One-proportion z-test

The conditions for the one-proportion z-test are the same as for the one-proportion z-interval. We test the hypothesis $H_0: p = p_0$ using the statistic

$$z = \frac{(\hat{p} - p_0)}{SD(\hat{p})}.$$

We use the hypothesized proportion to find the standard deviation: $SD(\hat{p}) = \sqrt{\frac{p_0 q_0}{n}}$. When the conditions are met and the null hypothesis is true, this statistic follows the standard Normal model, so we can use that model to obtain a P-value.

---

[1] It's also called the "one-sample test for a proportion."

"They make things admirably plain,
But one hard question will remain:
If one hypothesis you lose,
Another in its place you choose . . ."
—*James Russell Lowell*, Credidimus
Jovem Regnare

Mechanics

Under "Mechanics" we perform the actual calculation of our test statistic from the data. Different tests we encounter will have different formulas and different test statistics. Usually, the mechanics are handled by a statistics program or calculator. The ultimate goal of the calculation is to obtain a P-value—the probability that the observed statistic value (or an even more extreme value) could occur if the null model were correct. If the P-value is small enough, we'll reject the null hypothesis.

Conclusions and Decisions

The primary conclusion in a formal hypothesis test is only a statement about the null hypothesis. It simply states whether we reject or fail to reject that hypothesis. As always, the conclusion should be stated in context, but your conclusion about the null hypothesis should never be the end of the process. You can't make a decision based solely on a P-value. Business decisions have consequences, with actions to take or policies to change. The conclusions of a hypothesis test can help *inform* your decision, but they shouldn't be the only basis for it.

Business decisions should always take into consideration three things: the statistical significance of the test, the *cost* of the proposed action, and the *effect size* of the statistic they observed. For example, a cellular telephone provider finds that 30% of their customers switch providers (or *churn*) when their two-year subscription contract expires. They try a small experiment and offer a random sample of customers a free $350 top-of-the-line phone if they renew their contracts for another two years. Not surprisingly, they find that the new switching rate is lower by a statistically significant amount. Should they offer these free phones to all their customers? Obviously, the answer depends on more than the P-value of the hypothesis test. Even if the P-value is statistically significant, the correct business decision also depends on the cost of the free phones and by how much the churn rate is lowered (the effect size). It's rare that a hypothesis test alone is enough to make a sound business decision.

11.5 Alternative Hypotheses

In our example about the DJIA, we were equally interested in proportions that deviate from 50% in *either* direction. So we wrote our alternative hypothesis as $H_A: p \neq 0.5$. Such an alternative hypothesis is known as a **two-sided alternative** because we are equally interested in deviations on either side of the null hypothesis value. For two-sided alternatives, the P-value is the probability of deviating in *either* direction from the null hypothesis value.

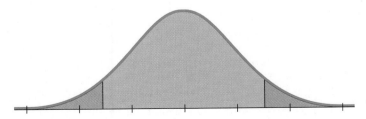

Figure 11.3 The P-value for a two-sided alternative adds the probabilities in both tails of the sampling distribution model outside the value that corresponds to the test statistic.

Suppose we want to test whether the proportion of customers returning merchandise has decreased under our new quality monitoring program. We know the quality has improved, so we can be pretty sure things haven't gotten worse. But have the customers noticed? We would only be interested in a sample proportion *smaller* than the null hypothesis value. We'd write our alternative hypothesis as $H_A: p < p_0$. An alternative hypothesis that focuses on deviations from the null hypothesis value in only one direction is called a **one-sided alternative**.

> **Alternative Hypotheses**
> Two-sided
> $H_0: p = p_0$
> $H_A: p \neq p_0$
> One-sided
> $H_0: p = p_0$
> $H_A: p < p_0$ or $p > p_0$

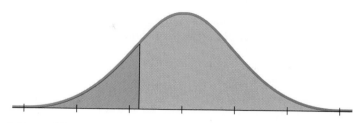

Figure 11.4 *The P-value for a one-sided alternative considers only the probability of values beyond the test statistic value in the specified direction.*

For a hypothesis test with a one-sided alternative, the P-value is the probability of deviating *only in the direction of the alternative* away from the null hypothesis value.

| **GUIDED EXAMPLE** | *Home Field Advantage* |

Major league sports are big business. And the fans are more likely to come out to root for the team if the home team has a good chance of winning. Anyone who follows or plays sports has heard of the "home field advantage." It is said that teams are more likely to win when they play at home. That *would* be good for encouraging the fans to come to the games. But is it true?

In the 2006 Major League Baseball (MLB) season, there were 2429 regular season games. (One rained-out game was never made up.) It turns out that the home team won 1327 of the 2429 games, or 54.63% of the time. If there were no home field advantage, the home teams would win about half of all games played. Could this deviation from 50% be explained just from natural sampling variability, or does this evidence suggest that there really is a home field advantage, at least in professional baseball?

To test the hypothesis, we will ask whether the observed rate of home team victories, 54.63%, is so much greater than 50% that we cannot explain it away as just chance variation.

Remember the four main steps to performing a hypothesis test—hypotheses, model, mechanics, and conclusion? Let's put them to work and see what this will tell us about the home team's chances of winning a baseball game.

PLAN

Setup State what we want to know.

Define the variables and discuss their context.

We want to know whether the home team in professional baseball is more likely to win. The data are all 2429 games from the 2006 Major League Baseball season. The variable is whether or not the home team won. The parameter of interest is the proportion of home team wins. If there is an advantage, we'd expect that proportion to be greater than 0.50. The observed statistic value is $\hat{p} = 0.5463$.

Hypotheses The null hypothesis makes the claim of no home field advantage.

We are interested only in a home field *advantage*, so the alternative hypothesis is one-sided.

$$H_0: p = 0.50$$

$$H_A: p > 0.50$$

Model Think about the assumptions and check the appropriate conditions.

✓ **Independence Assumption.** Generally, the outcome of one game has no effect on the outcome of another game. But this may not always be strictly true. For example, if a key player is injured, the probability that the team will win in the next couple of games may decrease slightly, but independence is still roughly true.

Consider the time frame carefully.

✓ **Randomization Condition.** We have results for all 2429 games of the 2006 season. But we're not just interested in 2006. While these games were not randomly selected, they may be reasonably representative of all recent professional baseball games.

✓ **10% Condition.** This is not a random sample, but these 2429 games are fewer than 10% of all games played over the years.

✓ **Success/Failure Condition.** Both $np_0 = 2429(0.50) = 1214.5$ and $nq_0 = 2429(0.50) = 1214.5$ are at least 10.

Specify the sampling distribution model.

Tell what test you plan to use.

Because the conditions are satisfied, we'll use a Normal model for the sampling distribution of the proportion and do a one-proportion z-test.

Mechanics The null model gives us the mean, and (because we are working with proportions) the mean gives us the standard deviation.

The null model is a Normal distribution with a mean of 0.50 and a standard deviation of:

$$SD(\hat{p}) = \sqrt{\frac{p_0 q_0}{n}} = \sqrt{\frac{(0.5)(1 - 0.5)}{2429}} = 0.01015$$

The observed proportion $\hat{p}$ is 0.5463.

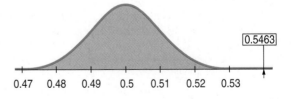

From technology, we can find the P-value, which tells us the probability of observing a value that extreme (or more).

The probability of observing a $\hat{p}$ of 0.5463 or more in our Normal model can be found by computer, calculator, or table to be <0.001.

The corresponding P-value is <0.001.

Conclusion State your conclusion about the parameter—in context.	MEMO: **Re: Home field advantage** Our analysis of outcomes during the 2006 Major League Baseball season showed a statistically significant advantage to the home team (P < 0.001). We can be quite confident that playing at home gives a baseball team an advantage.

11.6 Alpha Levels and Significance

Sometimes we need to make a firm decision about whether or not to reject the null hypothesis. A jury must *decide* whether the evidence reaches the level of "beyond a reasonable doubt." A business must *select* a Web design. You need to decide which section of a Statistics course to enroll in.

When the P-value is small, it tells us that our data are rare *given the null hypothesis*. As humans, we are suspicious of rare events. If the data are "rare enough," we just don't think that could have happened due to chance. Since the data *did* happen, something must be wrong. All we can do now is to reject the null hypothesis.

But how rare is "rare"? How low does the P-value have to be?

We can define "rare event" arbitrarily by setting a threshold for our P-value. If our P-value falls below that point, we'll reject the null hypothesis. We call such results *statistically significant*. The threshold is called an **alpha level**. Not surprisingly, it's labeled with the Greek letter α. Common α-levels are 0.10, 0.05, and 0.01. You have the option—almost the *obligation*—to consider your alpha level carefully and choose an appropriate one for the situation. If you're assessing the safety of air bags, you'll want a low alpha level; even 0.01 might not be low enough. If you're just wondering whether folks prefer their pizza with or without pepperoni, you might be happy with $\alpha = 0.10$. It can be hard to justify your choice of α, though, so often we arbitrarily choose 0.05.

Sir Ronald Fisher (1890–1962) was one of the founders of modern Statistics.

◆ **Where did the value 0.05 come from?** In 1931, in a famous book called *The Design of Experiments*, Sir Ronald Fisher discussed the amount of evidence needed to reject a null hypothesis. He said that it was *situation dependent*, but remarked, somewhat casually, that for many scientific applications, 1 out of 20 *might be* a reasonable value, especially in a *first* experiment—one that will be followed by confirmation. Since then, some people—indeed some entire disciplines—have acted as if the number 0.05 were sacrosanct.

! **NOTATION ALERT:**
The first Greek letter, α, is used in Statistics for the threshold value of a hypothesis test. You'll hear it referred to as the alpha level. Common values are 0.10, 0.05, 0.01, and 0.001.

The alpha level is also called the **significance level**. When we reject the null hypothesis, we say that the test is "significant at that level." For example, we might say that we reject the null hypothesis "at the 5% level of significance." You must select the alpha level *before* you look at the data. Otherwise you can be accused of finagling the conclusions by tuning the alpha level to the results after you've seen the data.

What can you say if the P-value does not fall below α? When you have not found sufficient evidence to reject the null according to the standard you have established, you should say: "The data have failed to provide sufficient evidence to reject the null hypothesis." Don't say: "We accept the null hypothesis." You certainly haven't proven or established the null hypothesis; it was assumed to begin

> **It could happen to you!**
> Of course, if the null hypothesis *is* true, no matter what alpha level you choose, you still have a probability α of rejecting the null hypothesis by mistake. When we do reject the null hypothesis, no one ever thinks that *this* is one of those rare times. As statistician Stu Hunter notes, "The statistician says 'rare events do happen—but not to me!'"

> **Conclusion**
> If the P-value $< \alpha$, then reject H_0.
>   If the P-value $\geq \alpha$, then fail to reject H_0.

with. You *could* say that you have *retained* the null hypothesis, but it's better to say that you've failed to reject it.

Look again at the home field advantage example. The P-value was <0.001. This is so much smaller than any reasonable alpha level that we can reject H_0. We concluded: "We reject the null hypothesis. There is sufficient evidence to conclude that there is a home field advantage over and above what we expect with random variation."

The automatic nature of the reject/fail-to-reject decision when we use an alpha level may make you uncomfortable. If your P-value falls just slightly above your alpha level, you're not allowed to reject the null. Yet a P-value just barely below the alpha level leads to rejection. If this bothers you, you're in good company. Many statisticians think it better to report the P-value than to choose an alpha level and carry the decision through to a final reject/fail-to-reject verdict. So when you declare your decision, it's always a good idea to report the P-value as an indication of the strength of the evidence.

◆ **It's in the stars.**   Some disciplines carry the idea further and code P-values by their size. In this scheme, a P-value between 0.05 and 0.01 gets highlighted by a single asterisk (*). A P-value between 0.01 and 0.001 gets two asterisks (**), and a P-value less than 0.001 gets three (***). This can be a convenient summary of the weight of evidence against the null hypothesis, but it isn't wise to take the distinctions too seriously and make black-and-white decisions near the boundaries. The boundaries are a matter of tradition, not science; there is nothing special about 0.05. A P-value of 0.051 should be looked at seriously and not casually thrown away just because it's larger than 0.05, and one that's 0.009 is not very different from one that's 0.011.

Sometimes it's best to report that the conclusion is not yet clear and to suggest that more data be gathered. (In a trial, a jury may "hang" and be unable to return a verdict.) In such cases, it's an especially good idea to report the P-value, since it's the best summary we have of what the data say or fail to say about the null hypothesis.

> **Practical vs. Statistical Significance**
> A large insurance company mined its data and found a statistically significant ($P = 0.04$) difference between the mean value of policies sold in 2001 and those sold in 2002. The difference in the mean values was $0.98. Even though it was statistically significant, management did not see this as an important difference when a typical policy sold for more than $1000. On the other hand, a marketable improvement of 10% in relief rate for a new pain medicine may not be statistically significant unless a large number of people are tested. The effect, which is economically significant, might not be statistically significant.

What do we mean when we say that a test is statistically significant? All we mean is that the test statistic had a P-value lower than our alpha level. Don't be lulled into thinking that "statistical significance" necessarily carries with it any practical importance or impact.

For large samples, even small, unimportant ("insignificant") deviations from the null hypothesis can be statistically significant. On the other hand, if the sample is not large enough, even large, financially or scientifically important differences may not be statistically significant.

It's good practice to report the magnitude of the difference between the observed statistic value and the null hypothesis value (in the data units) along with the P-value on which you have based your decision about statistical significance.

11.7 Critical Values

When building a confidence interval, we found a **critical value**, z^*, to correspond to our selected confidence level. Critical values can also be used as a shortcut for hypothesis tests. Before computers and calculators were common, P-values were hard to find. It was easier to select a few common alpha levels (0.05, 0.01, 0.001, for example) and learn the corresponding critical values for the Normal model (that is, the critical values corresponding to confidence levels 0.95, 0.99, and 0.999, respectively). Rather than find the probability that corresponded to your observed statistic, you'd just calculate how many standard deviations it was away from the hypothesized value and compare that value directly against these z^* values. (Remember that whenever we measure the distance of a value from the mean in standard deviations, we are finding a z-score.) Any z-score larger in magnitude (that is, more extreme) than a particular critical value has to be less likely, so it will have a P-value smaller than the corresponding alpha.

If we were willing to settle for a flat reject/fail-to-reject decision, comparing an observed z-score with the critical value for a specified alpha level would give a shortcut path to that decision. For the home field advantage example, if we choose $\alpha = 0.05$, then in order to reject H_0, our z-score has to be larger than the one-sided critical value of 1.645. The observed proportion was actually 4.78 standard deviations above 0.5, so we clearly reject the null hypothesis. This is perfectly correct and does give us a yes/no decision, but it gives us less information about the hypothesis because we don't have the P-value to think about. With technology, P-values are easy to find. And since they give more information about the strength of the evidence, you should report them.

Here are the traditional z^* critical values from the Normal model[2]:

If you need to make a decision on the fly with no technology, remember "2." That's our old friend from the 68-95-99.7 Rule. It's roughly the critical value for testing a hypothesis against a two-sided alternative at $\alpha = 0.05$. The exact critical value is 1.96, but 2 is close enough for most decisions.

α	1-sided	2-sided
0.05	1.645	1.96
0.01	2.33	2.576
0.001	3.09	3.29

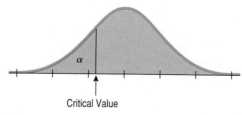

Figure 11.5 *When the alternative is one-sided, the critical value puts all of α on one side.*

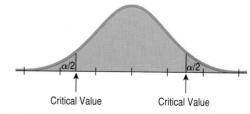

Figure 11.6 *When the alternative is two-sided, the critical value splits α equally into two tails.*

---

[2] In a sense, these are the flip side of the 68-95-99.7 Rule. There we chose simple statistical distances from the mean and recalled the areas of the tails. Here we select convenient tail areas (0.05, 0.01, and 0.001, either on one side or adding both together), and record the corresponding statistical distances.

11.8 Confidence Intervals and Hypothesis Tests

Confidence intervals and hypothesis tests are built from the same calculations. They have the same assumptions and conditions. As we have just seen, you can approximate a hypothesis test by examining the confidence interval. Just ask whether the null hypothesis value is consistent with a confidence interval for the parameter at the corresponding confidence level. Because confidence intervals are naturally two-sided, they correspond to two-sided tests. For example, a 95% confidence interval corresponds to a two-sided hypothesis test at $\alpha = 5\%$. In general, a confidence interval with a confidence level of C% corresponds to a two-sided hypothesis test with an α level of $100 - C\%$.

The relationship between confidence intervals and one-sided hypothesis tests gives us a choice. For a one-sided test with $\alpha = 5\%$, you could construct a one-sided confidence level of 95%, leaving 5% in one tail.

A one-sided confidence interval leaves one side unbounded. For example, in the home field example, we wondered whether the home field gave the home team an *advantage*, so our test was naturally one-sided. A 95% one-sided confidence interval would be constructed from one side of the associated two-sided confidence interval:

$$0.5463 - 1.645 \times 0.0101 = 0.530.$$

In order to leave 5% on one side, we used the z^* value 1.645 that leaves 5% in one tail. Writing the one-sided interval as $(0.530, \infty)$ allows us to say with 95% confidence that we know the home team will win, on average, at least 53.0% of the time. To test the hypothesis $H_0: p = 0.50$ we note that the value 0.50 is not in this interval. The lower bound of 0.53 is clearly above 0.50, showing the connection between hypothesis and confidence intervals.

As we saw in Chapter 10, it is not *exactly* true that hypothesis tests and confidence intervals are equivalent for proportions. For a confidence interval, we estimate the standard deviation of $\hat{p}$ from $\hat{p}$ itself, making it a *standard error*. For the corresponding hypothesis test, we use the model's *standard deviation* for $\hat{p}$ based on the null hypothesis value p_0. When $\hat{p}$ and p_0 are close, these calculations give similar results. When they differ, you're likely to reject H_0 (because the observed proportion is far from your hypothesized value). In that case, you're better off building your confidence interval with a standard error estimated from the data rather than rely on the model you just rejected.

For convenience, and to provide more information, however, we sometimes report a two-sided confidence interval even though we are interested in a one-sided test. For the home field example, we could report a 90% confidence interval:

$$0.5463 \pm 1.645 \times 0.0101 = (0.530, 0.563).$$

Notice that we *matched* the left end point by leaving α in *both* sides, which made the corresponding confidence level 90%. We can still see the correspondence that since the 95% (two-sided) confidence interval for $\hat{p}$ doesn't contain 0.50, we reject the null hypothesis, but it also tells us that the home team winning percentage is unlikely to be greater than 56.3%, an added benefit to understanding. You can see the relationship between the two confidence intervals in Figure 11.7.

There's another good reason for finding a confidence interval along with a hypothesis test. Although the test can tell us whether the observed statistic differs from the hypothesized value, it doesn't say by how much. Often, business decisions depend not only on whether there is a statistically significant difference, but also on whether the difference is meaningful. For the home field advantage, the corresponding confidence interval shows that over a full season, home field advantage adds an average of about two to six extra victories for a team. That could make a meaningful difference in both the team's standing and in the size of the crowd.

"Extraordinary claims require extraordinary proof."

   —Carl Sagan

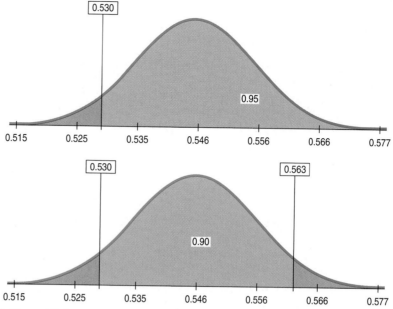

Figure 11.7 *The one-sided 95% confidence interval (top) leaves 5% on one side (in this case the left), but leaves the other side unbounded. The 90% confidence interval is symmetric and matches the one-sided interval on the side of interest. Both intervals indicate that a one-sided test of $p = 0.50$ would be rejected at $\alpha = 0.05$ for any value of $\hat{p}$ greater than 0.530.*

 JUST CHECKING

4 A bank is testing a new method for getting delinquent customers to pay their past-due credit card bills. The standard way was to send a letter (costing about $0.60 each) asking the customer to pay. That worked 30% of the time. The bank wants to test a new method that involves sending a DVD to the customer encouraging them to contact the bank and set up a payment plan. Developing and sending the DVD costs about $10.00 per customer. What is the parameter of interest? What are the null and alternative hypotheses?

5 The bank sets up an experiment to test the effectiveness of the DVD. The DVD is mailed to several randomly selected delinquent customers, and employees keep track of how many customers then contact the bank to arrange payments. The bank just got back the results on their test of the DVD strategy. A 90% confidence interval for the success rate is (0.29, 0.45). Their old send-a-letter method had worked 30% of the time. Can you reject the null hypothesis and conclude that the method increases the proportion at $\alpha = 0.05$? Explain.

6 Given the confidence interval the bank found in the trial of the DVD mailing, what would you recommend be done? Should the bank scrap the DVD strategy?

GUIDED EXAMPLE | *Credit Card Promotion*

A credit card company plans to offer a special incentive program to customers who charge at least $500 next month. The marketing department has pulled a sample of 500 customers from the same month last year and noted that the mean amount charged was $478.19 and the median amount was $216.48. The finance department says that the only relevant quantity is the proportion of customers who spend more than $500. If that proportion is less than 25%, the program will lose money.

Among the 500 customers, 148 or 29.6% of them charged $500 or more. Can we use a confidence interval to test whether the goal of 25% for all customers was met?

PLAN

Setup State the problem and discuss the variables and the context.

Hypotheses The null hypothesis is that the proportion qualifying is 25%. The alternative is that it is higher. It's clearly a one-sided test, so if we use a confidence interval, we'll have to be careful about what level we use.

Model Check the conditions.

State your method. Here we are using a confidence interval to test a hypothesis.

We want to know whether 25% or more of the customers will spend $500 or more in the next month and qualify for the special program. We will use the data from the same month a year ago to estimate the proportion and see whether the proportion was at least 25%.

The statistic is $\hat{p} = 0.296$, the proportion of customers who charged $500 or more.

$$H_0: p = 0.25$$
$$H_A: p > 0.25$$

✓ **Independence Assumption.** Customers are not likely to influence one another when it comes to spending on their credit cards.

✓ **Randomization Condition.** This is a random sample from the company's database.

✓ **10% Condition.** The sample is less than 10% of all customers.

✓ **Success/Failure Condition.** There were 148 successes and 352 failures, both at least 10. The sample is large enough.

Under these conditions, the sampling model is Normal. We'll create a one-proportion z-interval.

DO

Mechanics Write down the given information and determine the sample proportion.

To use a confidence interval, we need a confidence level that corresponds to the alpha level of the test. If we use $\alpha = 0.05$, we should construct a 90% confidence interval because this is a one-sided test. That will leave 5% on *each* side of the observed proportion. Determine the standard error of the sample proportion and the margin of error. The critical value is $z^* = 1.645$.

The confidence interval is estimate ± margin of error.

$n = 500$, so

$$\hat{p} = \frac{148}{500} = 0.296 \text{ and}$$

$$SE(\hat{p}) = \sqrt{\frac{\hat{p}\hat{q}}{n}} = \sqrt{\frac{(0.296)(0.704)}{500}} = 0.020$$

$$ME = z^* \times SE(\hat{p})$$
$$= 1.645(0.020) = 0.033$$

The 90% confidence interval is 0.296 ± 0.033 or (0.263, 0.329).

REPORT

Conclusion Link the confidence interval to your decision about the null hypothesis, then state your conclusion in context.

MEMO:

Re: Credit card promotion

Our study of a sample of customer records indicates that between 26.3% and 32.9% of customers charge $500 or more. We are 90% confident that this interval includes the true value. Because the minimum suitable value of 25% is below this interval, we conclude that it is not a plausible value, and so we reject the null hypothesis that only 25% of the customers charge more than $500 a month. The goal appears to have been met assuming that the month we studied is typical.

11.9 Two Types of Errors

Nobody's perfect. Even with lots of evidence, we can still make the wrong decision. In fact, when we perform a hypothesis test, we can make mistakes in *two* ways:

I. The null hypothesis is true, but we mistakenly reject it.

II. The null hypothesis is false, but we fail to reject it.

These two types of errors are known as **Type I** and **Type II errors** respectively. One way to keep the names straight is to remember that we start by assuming the null hypothesis is true, so a Type I error is the first kind of error we could make.

In medical disease testing, the null hypothesis is usually the assumption that a person is healthy. The alternative is that he or she has the disease we're testing for. So a Type I error is a *false positive*—a healthy person is diagnosed with the disease. A Type II error, in which an infected person is diagnosed as disease free, is a *false negative*. These errors have other names, depending on the particular discipline and context.

Which type of error is more serious depends on the situation. In a jury trial, a Type I error occurs if the jury convicts an innocent person. A Type II error occurs if the jury fails to convict a guilty person. Which seems more serious? In medical diagnosis, a false negative could mean that a sick patient goes untreated. A false positive might mean that the person must undergo further tests.

In business planning, a false positive result could mean that money will be invested in a project that turns out not to be profitable. A false negative result might mean that money won't be invested in a project that would have been profitable. Which error is worse, the lost investment or the lost opportunity? The answer always depends on the situation, the cost, and your point of view.

Here's an illustration of the situations:

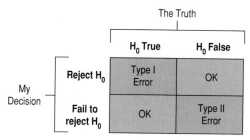

Figure 11.8 *The two types of errors occur on the diagonal where the truth and decision don't match. Remember that we start by assuming H_0 to be true, so an error made (rejecting it) when H_0 is true is called a Type I error. A Type II error is made when H_0 is false (and we fail to reject it).*

! NOTATION ALERT:

In Statistics, α is the probability of a Type I error and β is the probability of a Type II error.

The null hypothesis specifies a single value for the parameter. So it's easy to calculate the probability of a Type I error. But the alternative gives a whole range of possible values, and we may want to find a β for several of them.

We have seen ways to find a sample size by specifying the margin of error. Choosing the sample size to achieve a specified β (for a particular alternative value) is sometimes more appropriate, but the calculation is more complex and lies beyond the scope of this book.

How often will a Type I error occur? It happens when the null hypothesis is true but we've had the bad luck to draw an unusual sample. To reject H_0, the P-value must fall below α. When H_0 is true, that happens *exactly* with probability α. So when you choose level α, you're setting the probability of a Type I error to α.

What if H_0 is not true? Then we can't possibly make a Type I error. You can't get a false positive from a sick person. A Type I error can happen only when H_0 is true.

When H_0 is false and we reject it, we have done the right thing. A test's ability to detect a false hypothesis is called the **power** of the test. In a jury trial, power is a measure of the ability of the criminal justice system to convict people who are guilty. We'll have a lot more to say about power soon.

When H_0 is false but we fail to reject it, we have made a Type II error. We assign the letter β to the probability of this mistake. What's the value of β? That's harder to assess than α because we don't know what the value of the parameter really is. When H_0 is true, it specifies a single parameter value. But when H_0 is false, we don't have a specific one; we have many possible values. We can compute the probability β for any parameter value in H_A, but the choice of which one to pick is not always clear.

One way to focus our attention is by thinking about the *effect size*. That is, ask: "How big a difference would matter?" Suppose a charity wants to test whether placing personalized address labels in the envelope along with a request for a donation increases the response rate above the baseline of 5%. If the minimum response that would pay for the address labels is 6%, they would calculate β for the alternative $p = 0.06$.

Of course, we could reduce β for *all* alternative parameter values by increasing α. By making it easier to reject the null, we'd be more likely to reject it whether it's true or not. The only way to reduce *both* types of error is to collect more evidence or, in statistical terms, to collect more data. Otherwise, we just wind up trading off one kind of error against the other. Whenever you design a survey or experiment, it's a good idea to calculate β (for a reasonable α level). Use a parameter value in the alternative that corresponds to an effect size that you want to be able to detect. Too often, studies fail because their sample sizes are too small to detect the change they are looking for.

JUST CHECKING

7   Remember our bank that's sending out DVDs to try to get customers to make payments on delinquent loans? It is looking for evidence that the costlier DVD strategy produces a higher success rate than the letters it has been sending. Explain what a Type I error is in this context and what the consequences would be to the bank.

8   What's a Type II error in the bank experiment context and what would the consequences be?

9   If the DVD strategy *really* works well—actually getting 60% of the people to pay off their balances—would the power of the test be higher or lower compared to a 32% payoff rate? Explain briefly.

*11.10 Power

Remember, we can never prove a null hypothesis true. We can only fail to reject it. But when we fail to reject a null hypothesis, it's natural to wonder whether we looked hard enough. Might the null hypothesis actually be false and our test too weak to tell?

When the null hypothesis actually *is* false, we hope our test is strong enough to reject it. We'd like to know how likely we are to succeed. The power of the test gives us a way to think about that. The power of a test is the probability that it correctly rejects a false null hypothesis. When the power is high, we can be confident that we've looked hard enough. We know that β is the probability that a test *fails* to reject a false null hypothesis, so the power of the test is the complement, $1 - \beta$. We might have just written $1 - \beta$, but power is such an important concept that it gets its own name.

Whenever a study fails to reject its null hypothesis, the test's power comes into question. Was the sample size big enough to detect an effect had there been one? Might we have missed an effect large enough to be interesting just because we failed to gather sufficient data or because there was too much variability in the data we could gather? Might the problem be that the experiment simply lacked adequate power to detect their ability?

When we calculate power, we imagine that the null hypothesis is false. The value of the power depends on how far the truth lies from the null hypothesis value. We call the distance between the null hypothesis value, p_0, and the truth, p, the **effect size**. The power depends directly on the effect size. It's easier to see larger effects, so the further p_0 is from p, the greater the power.

How can we decide what power we need? Choice of power is more a financial or scientific decision than a statistical one because to calculate the power, we need to specify the "true" parameter value we're interested in. In other words, power is calculated for a particular effect size, and it changes depending on the size of the effect we want to detect.

Graph It!

It makes intuitive sense that the larger the effect size, the easier it should be to see it. Obtaining a larger sample size decreases the probability of a Type II error, so it increases the power. It also makes sense that the more we're willing to accept a Type I error, the less likely we will be to make a Type II error.

Figure 11.9 may help you visualize the relationships among these concepts. Suppose we are testing $H_0: p = p_0$ against the alternative $H_A: p > p_0$. We'll reject

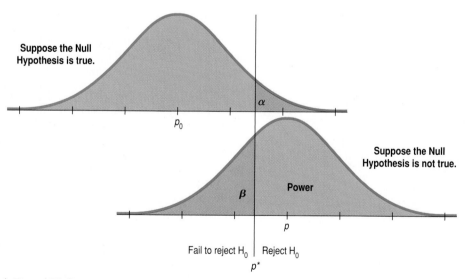

Figure 11.9 *The power of a test is the probability that it rejects a false null hypothesis. The upper figure shows the null hypothesis model. We'd reject the null in a one-sided test if we observed a value of in the red region to the right of the critical value, p*. The lower figure shows the true model. If the true value of p is greater than p_0, then we're more likely to observe a value that exceeds the critical value and make the correct decision to reject the null hypothesis. The power of the test is the green region on the right of the lower figure. Of course, even drawing samples whose observed proportions are distributed around p, we'll sometimes get a value in the red region on the left and make a Type II error of failing to reject the null.*

the null if the observed proportion, $\hat{p}$ is big enough. By *big enough*, we mean $\hat{p} > p^*$ for some critical value p^* (shown as the red region in the right tail of the upper curve). The upper model shows a picture of the sampling distribution model for the proportion when the null hypothesis is true. If the null were true, then this would be a picture of that truth. We'd make a Type I error whenever the sample gave us $\hat{p} > p^*$ because we would reject the (true) null hypothesis. Unusual samples like that would happen only with probability α.

In reality, though, the null hypothesis is rarely *exactly* true. The lower probability model supposes that H_0 is not true. In particular, it supposes that the true value is p, not p_0. It shows a distribution of possible observed $\hat{p}$ values around this true value. Because of sampling variability, sometimes $\hat{p} < p^*$ and we fail to reject the (false) null hypothesis. Then we'd make a Type II error. The area under the curve to the left of p^* in the bottom model represents how often this happens. The probability is β. In this picture, β is less than half, so most of the time we *do* make the right decision. The *power* of the test—the probability that we make the right decision—is shown as the region to the right of p^*. It's $1 - \beta$.

We calculate p^* based on the upper model because p^* depends only on the null model and the alpha level. No matter what the true proportion, p^* doesn't change. After all, we don't *know* the truth, so we can't use it to determine the critical value. But we always reject H_0 when $\hat{p} > p^*$.

How often we reject H_0 when it's *false* depends on the effect size. We can see from the picture that if the true proportion were further from the hypothesized value, the bottom curve would shift to the right, making the power greater.

We can see several important relationships from this figure:

◆ Power = $1 - \beta$.

◆ Moving the critical value, p^*, to the right, reduces α, the probability of a Type I error, but increases β, the probability of a Type II error. It correspondingly reduces the power.

◆ The larger the true effect size, the real difference between the hypothesized value, p_0, and the true population value, p, the smaller the chance of making a Type II error and the greater the power of the test.

If the two proportions are very far apart, the two models will barely overlap, and we would not be likely to make any Type II errors at all—but then, we are unlikely to really need a formal hypothesis testing procedure to see such an obvious difference.

Reducing Both Type I and Type II Errors

Figure 11.9 seems to show that if we reduce Type I error, we automatically must increase Type II error. But there is a way to reduce both. Can you think of it?

If we can make both curves narrower, as shown in Figure 11.10, then the probability of both Type I errors and Type II errors will decrease, and the power of the test will increase.

How can we do that? The only way is to reduce the standard deviations by increasing the sample size. (Remember, these are pictures of sampling distribution models, not of data.) Increasing the sample size works regardless of the true population parameters. But recall the curse of diminishing returns. The standard deviation of the sampling distribution model decreases only as the *square root* of the sample size, so to halve the standard deviations, we must *quadruple* the sample size.

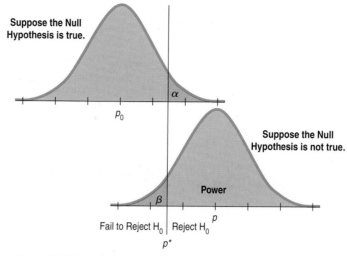

Figure 11.10 *Making the standard deviations smaller increases the power without changing the alpha level or the corresponding z-critical value. The means are just as far apart as in Figure 11.9, but the error rates are reduced.*

WHAT CAN GO WRONG?

- **Don't base your null hypotheses on what you see in the data.** You are not allowed to look at the data first and then adjust your null hypothesis so that it will be rejected. If your sample value turns out to be $\hat{p} = 51.8\%$ with a standard deviation of 1%, don't form a null hypothesis like H_0: $p = 49.8\%$, knowing that will enable you to reject it. Your null hypothesis describes the "nothing interesting" or "nothing has changed" scenario and should not be based on the data you collect.

- **Don't base your alternative hypothesis on the data either.** You should always think about the situation you are investigating and base your alternative hypothesis on that. Are you interested only in knowing whether something has *increased*? Then write a one-tail (upper tail) alternative. Or would you be equally interested in a change in either direction? Then you want a two-tailed alternative. You should decide whether to do a one- or two-tailed test based on what results would be of interest to you, not on what you might see in the data.

- **Don't make your null hypothesis what you want to show to be true.** Remember, the null hypothesis is the status quo, the nothing-is-strange-here position a skeptic would take. You wonder whether the data cast doubt on that. You can reject the null hypothesis, but you can never "accept" or "prove" the null.

- **Don't forget to check the conditions.** The reasoning of inference depends on randomization. No amount of care in calculating a test result can save you from a biased sample. The probabilities you compute depend on the independence assumption. And your sample must be large enough to justify your use of a Normal model.

- **Don't believe too strongly in arbitrary alpha levels.** There's not really much difference between a P-value of 0.051 and a P-value of 0.049, but sometimes it's regarded as the difference between night (having to retain H_0) and day (being able to shout to the world that your results are "statistically significant"). It may just be better to report the P-value and a confidence interval and let the world (perhaps your manager or client) decide along with you.

(continued)

- **Don't confuse practical and statistical significance.** A large sample size can make it easy to discern even a trivial change from the null hypothesis value. On the other hand, you could miss an important difference if your test lacks sufficient power.

- **Don't forget that in spite of all your care, you might make a wrong decision.** No one can ever reduce the probability of a Type I error (α) or of a Type II error (β) to zero (but increasing the sample size helps).

ETHICS IN ACTION

Many retailers have recognized the importance of staying connected to their in-store customers via the Internet. Retailers not only use the Internet to inform their customers about specials and promotions, but also to send them e-coupons redeemable for discounts. Shellie Cooper, long-time owner of a small organic food store, specializes in locally produced organic foods and products. Over the years Shellie's customer base has been quite stable, consisting mainly of health-conscious individuals who tend not to be very price sensitive, opting to pay higher prices for better-quality local, organic products. However, faced with increasing competition from grocery chains offering more organic choices, Shellie is now thinking of offering coupons. She needs to decide between the newspaper and the Internet. She recently read that the percentage of consumers who use printable Internet coupons is on the rise but, at 15%, is much less than the 40% who clip and redeem newspaper coupons. Nonetheless, she's interested in learning more about the Internet and sets up a meeting with Jack Kasor, a Web consultant. She discovers that for an initial investment and continuing monthly fee, Jack would design Shellie's website, host it on his server, and broadcast e-coupons to her customers at regular intervals. While she was concerned about the difference in redemption rates for e-coupons vs. newspaper coupons, Jack assured her that e-coupon redemptions are continuing to rise and that she should expect between 15% and 40% of her customers to redeem them. Shellie agreed to give it a try. After the first six months, Jack informed Shellie that the proportion of her customers who redeemed e-coupons was significantly greater than 15%. He determined this by selecting several broadcasts at random and found the number redeemed (483) out of the total number sent (3000). Shellie thought that this was positive and made up her mind to continue the use of e-coupons.

ETHICAL ISSUE *Statistical vs. practical significance. While it is true that the percentage of Shellie's customers redeeming e-coupons is significantly greater than 15% statistically, in fact, the percentage is just over 16%. This difference amounts to about 33 customers more than 15%, which may not be of practical significance to Shellie (related to Item A, ASA Ethical Guidelines). Mentioning a range of 15% to 40% may mislead Shellie into expecting a value somewhere in the middle.*

ETHICAL SOLUTION *Jack should report the difference between the observed value and the hypothesized value to Shellie, especially since there are costs associated with continuing e-coupons. Perhaps he should recommend that she reconsider using the newspaper.*

What have we learned?

We've learned to use what we see in a random sample to test a particular hypothesis about the world. This is our second step in statistical inference, complementing our use of confidence intervals.

We've learned that testing a hypothesis involves proposing a model and then seeing whether the data we observe are consistent with that model or so unusual that we must reject it. We do this by finding a P-value—the probability that data like ours could have occurred if the model is correct. If the data are out of line with the null hypothesis model, the P-value will be small, and we will reject the null hypothesis. If the data are consistent with the null hypothesis model, the P-value will be large, and we will not reject the null hypothesis.

We've learned that:

- We start with a *null hypothesis* specifying the parameter of a model we'll test using our data.
- Our *alternative hypothesis* can be one- or two-sided, depending on what we want to learn.
- We must check the appropriate *assumptions* and *conditions* before proceeding with our test.
- The *significance level* of the test establishes the level of proof we'll require. That determines the critical value of z that will lead us to reject the null hypothesis.
- *Hypothesis tests* and *confidence intervals* are really two ways of looking at the same question. The hypothesis test gives us the answer to a decision about a parameter; the confidence interval tells us the plausible values of that parameter.
- If the null hypothesis is really true and we reject it, that's a *Type I error*; the alpha level of the test is the probability that this happens.
- If the null hypothesis is really false but we fail to reject it, that's a *Type II error*.

*Optional Sections

- The *power* of the test is the probability that we reject the null hypothesis when it's false. The larger the size of the effect we're testing for, the greater the power of the test in detecting it.
- Tests with a greater likelihood of Type I error have more power and less chance of a Type II error. We can increase power while reducing the chances of both kinds of error by increasing the sample size.

Terms

Alpha level
The threshold P-value that determines when we reject a null hypothesis. Using an alpha level of α, if we observe a statistic whose P-value based on the null hypothesis is less than α, we reject that null hypothesis.

Alternative hypothesis
The hypothesis that proposes what we should conclude if we find the null hypothesis to be unlikely.

Critical value
The value in the sampling distribution model of the statistic whose P-value is equal to the alpha level. Any statistic value further from the null hypothesis value than the critical value will have a smaller P-value than α and will lead to rejecting the null hypothesis. The critical value is often denoted with an asterisk, as z^*, for example.

Effect size
The difference between the null hypothesis value and the true value of a model parameter.

Null hypothesis
The claim being assessed in a hypothesis test. Usually, the null hypothesis is a statement of "no change from the traditional value," "no effect," "no difference," or "no relationship." For a claim to be a testable null hypothesis, it must specify a value for some population parameter that can form the basis for assuming a sampling distribution for a test statistic.

One-proportion z-test
A test of the null hypothesis that the proportion of a single sample equals a specified value ($H_0: p = p_0$) by comparing the statistic $z = \dfrac{\hat{p} - p_0}{SD(\hat{p})}$ to a standard Normal model.

One-sided alternative   An alternative hypothesis is one-sided (e.g., $H_A: p > p_0$ or $H_A: p < p_0$) when we are interested in deviations in *only one* direction away from the hypothesized parameter value.

P-value   The probability of observing a value for a test statistic at least as far from the hypothesized value as the statistic value actually observed if the null hypothesis is true. A small P-value indicates that the observation obtained is improbable given the null hypothesis and thus provides evidence against the null hypothesis.

Power   The probability that a hypothesis test will correctly reject a false null hypothesis. To find the power of a test, we must specify a particular alternative parameter value as the "true" value. For any specific value in the alternative, the power is $1 - \beta$.

Significance level   Another term for the alpha level, used most often in a phrase such as "at the 5% significance level."

Two-sided alternative   An alternative hypothesis is two-sided ($H_A: p \neq p_0$) when we are interested in deviations in *either* direction away from the hypothesized parameter value.

Type I error   The error of rejecting a null hypothesis when in fact it is true (also called a "false positive"). The probability of a Type I error is α.

Type II error   The error of failing to reject a null hypothesis when in fact it is false (also called a "false negative"). The probability of a Type II error is commonly denoted β and depends on the effect size.

Skills

- Be able to state the null and alternative hypotheses for a one-proportion z-test.
- Know how to think about the assumptions and their associated conditions. Examine your data for violations of those conditions.
- Be able to identify and use the alternative hypothesis when testing hypotheses. Understand how to choose between a one-sided and two-sided alternative hypothesis and be able to explain your choice.

- Know how to perform a one-proportion z-test.

- Be able to interpret the results of a one-proportion z-test.
- Be able to interpret the meaning of a P-value in nontechnical language, making clear that the probability claim is about computed values under the assumption that the null model is true and not about the population parameter of interest.

Technology Help

Hypothesis tests for proportions are so easy and natural that many statistics packages don't offer special commands for them. Most statistics programs want to know the "success" and "failure" status for each case. Usually these are given as 1 or 0, but they might be category names like "yes" and "no." Often we just know the proportion of successes, $\hat{p}$, and the total count, n. Computer packages don't usually deal naturally with summary data like this, but see below for a couple of important exceptions (Minitab and JMP).

In some programs you can reconstruct the original values. But even when you have reconstructed (or can reconstruct) the raw data values, often you won't get *exactly* the same test statistic from a computer package as you would find working by hand. The reason is that when the packages treat the proportion as a mean, they make some approximations. The result is very close, but not exactly the same. If you use a computer package, you may notice slight discrepancies between your answers and the answers in the back of the book, but they're not important.

Reports about hypothesis tests generated by technologies don't follow a standard form. Most will name the test and provide the test statistic value, its standard deviation, and the P-value. But these elements may not be labeled clearly. For example, the expression "Prob > |z|" means the probability (the "Prob") of observing a test statistic whose magnitude (the absolute value tells us this) is larger than that of the one (the "z") found in the data (which, because it is written as "z," we know follows a Normal model). That is a fancy (and not very clear) way of saying P-value. In some packages, you can specify that the test be one-sided. Others might report three P-values, covering the ground for both one-sided tests and two-sided tests.

Sometimes a confidence interval and hypothesis test are automatically given together. The confidence interval ought to be for the corresponding confidence level: $1 - \alpha$.

Often, the standard deviation of the statistic is called the "standard error," and usually that's appropriate because we've had to estimate its value from the data. That's not the case for proportions, however: We get the standard deviation for a proportion from the null hypothesis value. Nevertheless, you may see the standard deviation called a "standard error" even for tests with proportions.

It's common for statistics packages and calculators to report more digits of "precision" than could possibly have been found from the data. You can safely ignore them. Round values such as the standard deviation to one digit more than the number of digits reported in your data.

Here are the kind of results you might see in typical computer output.

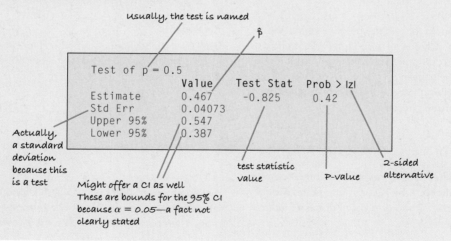

EXCEL

Inference methods for proportions are not part of the standard Excel tool set.

Comments

For summarized data, type the calculation into any cell and evaluate it.

You can use the DDXL add-in to compute hypothesis tests for proportions. Select the variables to test. From the **DDXL** menu, choose **Hypothesis Tests**. In the Hypothesis Tests dialog, choose **1 Var Prop Test**. Specify the variable to test and click **OK**.

MINITAB

Choose **Basic Statistics** from the **Stat** menu.

- Choose **1 Proportion** from the Basic Statistics submenu.
- If the data are category names in a variable, assign the variable from the variable list box to the **Samples in columns** box.
- If you have summarized data, click the **Summarized Data** button and fill in the number of trials and the number of successes.
- Click the **Options** button and specify the remaining details.

- If you have a large sample, check **Use test and interval based on Normal distribution.**
- Click the **OK** button.

Comments

When working from a variable that names categories, MINITAB treats the last category as the "success" category. You can specify how the categories should be ordered.

SPSS

SPSS does not find hypothesis tests for proportions.

JMP

For a **categorical** variable that holds category labels, the **Distribution** platform includes tests and intervals of proportions. For summarized data, put the category names in one variable and the frequencies in an adjacent variable. Designate the frequency column to have the **role** of **frequency**. Then use the **Distribution** platform.

Comments

JMP uses slightly different methods for proportion inferences than those discussed in this text. Your answers are likely to be slightly different.

DATA DESK

Data Desk does not offer built-in methods for inference with proportions. The **Replicate Y by X** command in the **Manip** menu will "reconstruct" summarized count data so that you can display it.

Comments

For summarized data, open a Scratchpad to compute the standard deviation and margin of error by typing the calculation. Then perform the test with the **z-test for individual s** found in the Test command.

Mini Case Study Projects

Metal Production

Ingots are huge pieces of metal, often weighing in excess of 20,000 pounds, made in a giant mold. They must be cast in one large piece for use in fabricating large structural parts for cars and planes. If they crack while being made, the crack may propagate into the zone required for the part, compromising its integrity. Airplane manufacturers insist that metal for their planes be defect-free, so the ingot must be made over if any cracking is detected.

Even though the metal from the cracked ingot is recycled, the scrap cost runs into the tens of thousands of dollars. Metal manufacturers would like to avoid cracking if at all possible. But the casting process is complicated, and not everything can be controlled completely. In one plant, only about 75% of the ingots have been free of cracks. In an attempt to reduce the cracking proportion, the plant engineers and chemists recently (January 2006) made changes to the casting process. The data from 5000 ingots produced since the changes are found in the file **ch11_MCSP_Ingots**. The variable *Crack* indicates whether a crack was found (1) or not (0). Select a random sample of 100 ingots and test the claim that the cracking rate has decreased from 25%. Find a confidence interval for the cracking rate as well. Now select a random sample of 1000 ingots and test the claim and find the confidence interval again. Compare the two tests and intervals and prepare a short report about your findings including the differences (if any) that you see from the two samples.

Loyalty Program

A marketing manager has sent out 10,000 mail pieces to a random sample of customers to test a new web-based loyalty program. The customers either received nothing (No Offer), a free companion airline ticket (Free Flight), or free flight insurance on their next flight (Free Insurance). The person in charge of selecting the 10,000 customers has assured the marketing manager that the sample is representative of the various marketing segments in the customer base. However, the manager is worried that the offer was not sent out to enough customers in the *Travel* segment which represents 25% of the entire customer base (variable *Spending.Segment*). In addition, he is worried that fewer than 1/3 of customers in that segment actually received no offer. Using the data found in **ch11_MCSP_Loyalty_Program** write a short report to the manager testing the appropriate hypotheses and summarizing your findings. Include in your report a 95% confidence interval for the proportion of customers who responded to the offer by signing up for the loyalty program. (The variable *Response* indicates a 1 for responders and 0 for non-responders.)

EXERCISES

1. Hypotheses. Write the null and alternative hypotheses to test each of the following situations.

a) An online clothing company is concerned about the timeliness of the delivery of their products. The VP of Operations and Marketing recently stated that she wanted the percentage of products delivered on time to be at least 90%, and she wants to know if the company has succeeded.

b) A realty company recently announced that the proportion of houses taking more than three months to sell is now greater than 50%.

c) A financial firm's accounting reports have an error rate below 2%.

2. More hypotheses. Write the null and alternative hypotheses to test each of the following situations.

a) A business magazine article reports that, in 1990, 35% of CEOs had an MBA degree. Has the percentage changed?

b) Recently, 20% of cars of a certain model have needed costly transmission work after being driven between 50,000 and 100,000 miles. The car manufacturer hopes that the redesign of a transmission component has solved this problem.

c) A market researcher for a cola company decides to field test a new flavor soft drink, planning to market it only if he is sure that over 60% of the people like the flavor.

3. Deliveries. The clothing company in Exercise 1a looks at a sample of delivery reports. They test the hypothesis that 90% of the deliveries are on time against the alternative that greater than 90% are on time and find a P-value of 0.22. Which of these conclusions is appropriate?

a) There's a 22% chance that 90% of the deliveries are on time.

b) There's a 78% chance than 90% of the deliveries are on time.

c) There's a 22% chance that the sample they drew shows the correct percentage of on-time deliveries

d) There's a 22% chance that natural sampling variation could produce a sample with an observed proportion of on-time deliveries such as the one they obtained if, in fact, 90% of deliveries are on time.

4. House sales. The realty company in Exercise 1b looks at a recent sample of houses that have sold. On testing the null hypothesis that 50% of the houses take more than three months to sell against the hypothesis that more than 50% of the houses take more than three months to sell, they find a P-value of 0.034. Which of these conclusions is appropriate?

a) There's a 3.4% chance that 50% of the houses take more than three months to sell.

b) If 50% of the houses take more than three months to sell, there's a 3.4% chance that a random sample would produce a sample proportion as high as the one they obtained.

c) There's a 3.4% chance that the null hypothesis is correct.

d) There's a 96.6% chance that 50% of the houses take more than three months to sell.

5. P-value. Have harsher penalties and ad campaigns increased seat belt use among drivers and passengers? Observations of commuter traffic have failed to find evidence of a significant change compared with three years ago. Explain what the study's P-value of 0.17 means in this context.

6. Another P-value. A company developing scanners to search for hidden weapons at airports has concluded that a new device is significantly better than the current scanner. The company made this decision based on a P-value of 0.03. Explain the meaning of the P-value in this context.

7. Ad campaign. An information technology analyst believes that they are losing customers on their website who find the checkout and purchase system too complicated. She adds a one-click feature to the website, to make it easier but finds that only about 10% of the customers are using it. She decides to launch an ad awareness campaign to tell customers about the new feature in the hope of increasing the percentage. She doesn't see much of a difference, so she hires a consultant to help her. The consultant selects a random sample of recent purchases, tests the hypothesis that the ads produced no change against the alternative that the percent who use the one-click feature is now greater than 10%, and finds a P-value of 0.22. Which conclusion is appropriate? Explain.

a) There's a 22% chance that the ads worked.

b) There's a 78% chance that the ads worked.

c) There's a 22% chance that the null hypothesis is true.

d) There's a 22% chance that natural sampling variation could produce poll results like these if the use of the one-click feature has increased.

e) There's a 22% chance that natural sampling variation could produce poll results like these if there's really no change in website use.

8. Mutual funds. A mutual fund manager claims that at least 70% of the stocks she selects will increase in price over the next year. We examined a sample of 200 of her selections over the past three years. Our P-value turns out to be 0.03. Test an appropriate hypothesis. Which conclusion is appropriate? Explain.

a) There's a 3% chance that the fund manager is correct.

b) There's a 97% chance that the fund manager is correct.

c) There's a 3% chance that a random sample could produce the results we observed, so it's reasonable to conclude that the fund manager is correct.

d) There's a 3% chance that a random sample could produce the results we observed if $p = .7$, so it's reasonable to conclude that the fund manager is not correct.

e) There's a 3% chance that the null hypothesis is correct.

9. Product effectiveness. A pharmaceutical company's old antacid formula provided relief for 70% of the people who used it. The company tests a new formula to see if it is better and gets a P-value of 0.27. Is it reasonable to conclude that the new formula and the old one are equally effective? Explain.

10. Car sales. A German automobile company is counting on selling more cars to the younger market segment—drivers under the age of 20. The company's market researchers survey to investigate whether or not the proportion of today's high school seniors who own their own cars is higher than it was a decade ago. They find a P-value of 0.017. Is it reasonable to conclude that more high school seniors have cars? Explain.

11. False claims? A candy company claims that in a large bag of holiday M&M's® half the candies are red and half the candies are green. You pick candies at random from a bag and discover that of the first 20 you eat, 12 are red.

a) If it were true that half are red and half are green, what is the probability you would have found that at least 12 out of 20 were red?

b) Do you think that half of the M&M's® candies in the bag are really red? Explain.

12. Scratch off. A retail company offers a "scratch off" promotion. Upon entering the store, you are given a card. When you pay, you may scratch off the coating. The company advertises that half the cards are winners and have immediate cashback savings of $5 (the others offer $1 off any future purchase of coffee in the cafe). You aren't sure the percentage is really 50% winners.

a) The first time you shop there, you get the coffee coupon. You try again and again get the coffee coupon. Do two failures in a row convince you that the true fraction of winners isn't 50%? Explain.

b) You try a third time. You get coffee again! What's the probability of not getting a cash savings three times in a row if half the cards really do offer cash savings?

c) Would three losses in a row convince you that the store is cheating?

d) How many times in a row would you have to get the coffee coupon instead of cash savings to be pretty sure that the company isn't living up to its advertised percentage of winners? Justify your answer by calculating a probability and explaining what it means.

13. Spike poll. In August 2004, *Time* magazine reported the results of a random telephone poll commissioned by the Spike network. Of the 1302 men who responded, only 39 said that their most important measure of success was their work.

a) Estimate the percentage of all American males who measure success primarily from their work. Use a 98% confidence interval. Don't forget to check the conditions first.

b) Some believe that few contemporary men judge their success primarily by their work. Suppose we wished to conduct a hypothesis test to see if the fraction has fallen below the 5% mark. What does your confidence interval indicate? Explain.

c) What is the significance level of this test? Explain.

14. Stocks. A young investor in the stock market is concerned that investing in the stock market is actually gambling, since the chance of the stock market going up on any given day is 50%. She decides to track her favorite stock for 250 days and finds that on 140 days the stock was "up."

a) Find a 95% confidence interval for the proportion of days the stock is "up." Don't forget to check the conditions first.

b) Does your confidence interval provide any evidence that the market is not random? Explain.

c) What is the significance level of this test? Explain.

15. Economy. In 2008, a Gallup Poll asked 2336 U.S. adults, aged 18 or over, how they rated economic conditions. In a poll conducted from January 27 through February 1, 2008, 24% rated the economy as Excellent/Good. A recent media outlet claimed that the percentage of Americans who felt the economy was in Excellent/Good shape was, in fact, 28%. Does the Gallup Poll support this claim?

a) Test the appropriate hypothesis. Find a 95% confidence interval for the sample proportion of U.S. adults who rated the economy as Excellent/Good. Check conditions.

b) Does your confidence interval provide evidence to support the claim?

c) What is the significance level of the test in b? Explain.

16. Economy, part 2. The same Gallup Poll data from Exercise 15 also reported that 33% of those surveyed rated the economy as Poor. The same media outlet claimed the true proportion to be 30%. Does the Gallup Poll support this claim?

a) Test the appropriate hypothesis. Find a 95% confidence interval for the sample proportion of U.S. adults who rated the economy as Poor. Check conditions.

b) Does your confidence interval provide evidence to support the claim.

c) What is the significance level of the test in b? Explain.

17. Convenient alpha. An enthusiastic junior executive has run a test of his new marketing program. He reports that it resulted in a "significant" increase in sales. A footnote on his report explains that he used an alpha level of 7.2% for his test. Presumably, he performed a hypothesis test against the null hypothesis of no change in sales.

a) If instead he had used an alpha level of 5%, is it more or less likely that he would have rejected his null hypothesis? Explain.

b) If he chose the alpha level 7.2% so that he could claim statistical significance, explain why this is not an ethical use of statistics.

18. Safety. The manufacturer of a new sleeping pill suspects that it may increase the risk of sleepwalking, which could be dangerous. A test of the drug fails to reject the null hypothesis of increased sleepwalking when tested at alpha = .01.

a) If the test had been performed at alpha = .05, would the test have been more or less likely to reject the null hypothesis of no increase in sleepwalking?

b) Which alpha level do you think the company should use? Why?

19. Product testing. Since many people have trouble programming their VCRs, an electronics company has developed what it hopes will be easier instructions. The goal is to have at least 96% of customers succeed at being able to program their VCRs. The company tests the new system on 200 people, 188 of whom were successful. Is this strong evidence that the new system fails to meet the company's goal? A student's test of this hypothesis is shown here. How many mistakes can you find?

$H_0: \hat{p} = 0.96$

$H_A: \hat{p} \neq 0.96$

SRS, $0.96(200) > 10$

$\dfrac{188}{200} = 0.94; SD(\hat{p}) = \sqrt{\dfrac{(0.94)(0.06)}{200}} = 0.017$

$z = \dfrac{0.96 - 0.94}{0.017} = 1.18$

$P = P(z > 1.18) = 0.12$

There is strong evidence that the new system does not work.

20. Marketing. In November 2001, the *Ag Globe Trotter* newsletter reported that 90% of adults drink milk. A regional farmers organization planning a new marketing campaign across its multicounty area polls a random sample of 750 adults living there. In this sample, 657 people said that they drink milk. Do these responses provide strong evidence that the 90% figure is not accurate for this region? Correct the mistakes you find in the following student's attempt to test an appropriate hypothesis.

$H_0: \hat{p} = 0.9$

$H_A: \hat{p} < 0.9$

SRS, $750 > 10$

$\dfrac{657}{750} = 0.876; SD(\hat{p}) = \sqrt{\dfrac{(0.88)(0.12)}{750}} = 0.012$

$z = \dfrac{0.876 - 0.94}{0.012} = -2$

$P = P(z > -2) = 0.977$

There is more than a 97% chance that the stated percentage is correct for this region.

21. Environment. In the 1980s, it was generally believed that congenital abnormalities affected about 5% of the nation's children. Some people believe that the increase in the number of chemicals in the environment has led to an increase in the incidence of abnormalities. A recent study examined 384 children and found that 46 of them showed signs of an abnormality. Is this strong evidence that the risk has increased? (We consider a P-value of around 5% to represent reasonable evidence.)

a) Write appropriate hypotheses.
b) Check the necessary assumptions.
c) Perform the mechanics of the test. What is the P-value?
d) Explain carefully what the P-value means in this context.
e) What's your conclusion?
f) Do environmental chemicals cause congenital abnormalities?

22. Billing company. A billing company that collects bills for doctors' offices in the area is concerned that the percentage of bills being paid by Medicare has risen. Historically, that percentage has been 31%. An examination of 8368 recent bills reveals that 32% of these bills are being paid by Medicare. Is this evidence of a change in the percent of bills being paid by Medicare?

a) Write appropriate hypotheses.
b) Check the assumptions and conditions.
c) Perform the test and find the P-value.
d) State your conclusion.
e) Do you think this difference is meaningful? Explain.

23. Education. The National Center for Education Statistics monitors many aspects of elementary and secondary education nationwide. Their 1996 numbers are often used as a baseline to assess changes. In 1996, 34% of students had not been absent from school even once during the previous month. In the 2000 survey, responses from 8302 students showed that this figure had slipped to 33%. Officials would be concerned if student attendance were declining. Do these figures give evidence of a decrease in student attendance?

a) Write appropriate hypotheses.
b) Check the assumptions and conditions.
c) Perform the test and find the P-value.
d) State your conclusion.
e) Do you think this difference is meaningful? Explain.

24. Consumer confidence. At various times in 2007, when asked if economic conditions were getter better or worse, consistently more than 20% of U.S. adults said better. On January 19–20, 2008, when Gallup polled 2590 U.S. adults, only 13% said that conditions were getting better. Do these responses give evidence that consumer confidence has decreased from the 2007 level?

a) Write appropriate hypotheses.
b) Check the assumptions and conditions.
c) Perform the test and find the P-value.
d) State your conclusion.
e) Do you think this difference is meaningful? Explain.

25. Retirement. A survey of 1000 workers indicated that approximately 520 have invested in an individual retirement account. National data suggests that 44% of workers invest in individual retirement accounts.

a) Create a 95% confidence interval for the proportion of workers who have invested in individual retirement accounts based on the survey.
b) Does this provide evidence of a change in behavior among workers? Using your confidence interval, test an appropriate hypothesis and state your conclusion.

26. Customer satisfaction. A company hopes to improve customer satisfaction, setting as a goal no more than 5% negative comments. A random survey of 350 customers found only 10 with complaints.

a) Create a 95% confidence interval for the true level of dissatisfaction among customers.
b) Does this provide evidence that the company has reached its goal? Using your confidence interval, test an appropriate hypothesis and state your conclusion.

27. Maintenance costs. A limousine company is concerned with increasing costs of maintaining their fleet of 150 cars. After testing, the company found that the emissions systems of 7 out of the 22 cars they tested failed to meet pollution control guidelines. They had forecasted costs assuming that a total of 30 cars would need updating to meet the latest guidelines. Is this strong evidence that more than 20% of the fleet might be out of compliance? Test an appropriate hypothesis and state your conclusion. Be sure the appropriate assumptions and conditions are satisfied before you proceed.

28. Damaged goods. An appliance manufacturer stockpiles washers and dryers in a large warehouse for shipment to retail stores. Sometimes in handling them the appliances get damaged. Even though the damage may be minor, the company must sell those machines at drastically reduced prices. The company goal is to keep the proportion of damaged machines below 2%. One day an inspector randomly checks 60 washers and finds that 5 of them have scratches or dents. Is this strong evidence that the warehouse is failing to meet the company goal? Test an appropriate hypothesis and state your conclusion. Be sure the appropriate assumptions and conditions are satisfied before you proceed.

29. Defective products. An internal report from a manufacturing company indicated that about 3% of all products were defective. Data from one batch found only 7 defective products out of 469 products. Is this consistent with the report? Test an appropriate hypothesis and state your conclusion. Be sure the appropriate assumptions and conditions are satisfied before you proceed.

30. Jobs. The accounting department of a major state university would like to advertise that more than 50% of its graduates obtained a job offer prior to graduation. A sample of 240 recent graduates indicated that 138 of these graduates had a job offer prior to graduation. Test an appropriate hypothesis and state your conclusion. Be sure the appropriate assumptions and conditions are satisfied before you proceed.

31. WebZine. A magazine called *WebZine* is considering the launch of an online edition. The magazine plans to go ahead only if it's convinced that more than 25% of current readers would subscribe. The magazine contacts a simple random sample of 500 current subscribers, and 137 of those surveyed expressed interest. What should the magazine do? Test an appropriate hypothesis and state your conclusion. Be sure the appropriate assumptions and conditions are satisfied before you proceed.

32. Truth in advertising. A garden center wants to store leftover packets of vegetable seeds for sale the following spring, but the center is concerned that the seeds may not germinate at the same rate a year later. The manager finds a packet of last year's green bean seeds and plants them as a test. Although the packet claims a germination rate of 92%, only 171 of 200 test seeds sprout. Is this evidence that the seeds have lost viability during a year in storage? Test an appropriate hypothesis and state your conclusion. Be sure the appropriate assumptions and conditions are satisfied before you proceed.

33. Women executives. A company is criticized because only 13 of 43 people in executive-level positions are women. The company explains that although this proportion is lower than it might wish, it's not surprising given that only 40% of their employees are women. What do you think? Test an appropriate hypothesis and state your conclusion. Be sure the appropriate assumptions and conditions are satisfied before you proceed.

34. Jury. Census data for a certain county shows that 19% of the adult residents are Hispanic. Suppose 72 people are called for jury duty, and only 9 of them are Hispanic. Does this apparent underrepresentation of Hispanics call into question the fairness of the jury selection system? Explain.

35. Nonprofit. A nonprofit company concerned with the school dropout rates in the United States has designed a tutoring program aimed at students between 16 to 18 years old. The National Center for Education Statistics reported that the high school dropout rate in the United States for the year 2000 was 10.9%. One school district, who adopted the use of the nonprofit's tutoring program and whose dropout rate has always been very close to the national average, reported in 2004 that 175 of their 1782 students dropped out. Is their experience evidence that the tutoring program has been effective? Explain.

36. Real estate. A national real estate magazine advertised that 15% of first home buyers had a family income below $40,000. A national real estate firm believes this percentage is too low and samples 100 of its records. The firm finds that 25 of its first home buyers did have a family income below $40,000. Does the sample suggest that the proportion of first home buyers with an income less than $40,000 is more than 15%? Comment and write up your own conclusions based on an appropriate confidence interval as well as a hypothesis test. Include any assumptions you made about the data.

37. Public relations. According to the U.S. Department of Transportation (DOT), passengers filed more complaints about airline service in 2007 than in 2006. One airline's public relations department says that their airline rarely loses luggage. Furthermore, it claims that when it does, 90% of the time the bags are recovered and delivered within 24 hours. A consumer group surveys a large number of air travelers and finds that 103 of 122 people who lost luggage were reunited with their missing items within 24 hours. Does this cast doubt on the airline's claim? Explain.

38. TV ads. A start-up company is about to market a new computer printer. It decides to gamble by running commercials during the Super Bowl. The company hopes that name recognition will be worth the high cost of the ads. The goal of the company is that over 40% of the public recognize its brand name and associate it with computer equipment. The day after the game, a pollster contacts 420 randomly chosen adults and finds that 181 of them know that this company manufactures printers. Would you recommend that the company continue to advertise during the Super Bowl? Explain.

39. Business ethics. One study reports that 30% of newly hired MBAs are confronted with unethical business practices during their first year of employment. One business school dean wondered if her MBA graduates had similar experiences. She surveyed recent graduates from her school's MBA program to find that 27% of the 120 graduates from the previous year claim to have encountered unethical business practices in the workplace. Can she conclude that her graduates' experiences are different?

40. Stocks, part 2. A young investor believes that he can beat the market by picking stocks that will increase in value. Assume that on average 50% of the stocks selected by a portfolio manager will increase over 12 months. Of the 25 stocks that the young investor bought over the last 12 months, 14 have increased. Can he claim that he is better at predicting increases than the typical portfolio manager?

41. U.S. politics. The national elections in 2008 are apparently drawing more interest and debate among voters than in prior U.S. elections. A national sample of 2020 U.S. adults, aged 18 and older, surveyed over the telephone (using both landlines and cell phones) between January 30 and February 2 in 2008 by Gallup revealed that 71% reported that they had given "quite a lot" of thought to the upcoming election for president. Is there any evidence that the percentage has changed from the historically reported benchmark of 58% during the same timeframe in 2004?
a) Find the z-score of the observed proportion.
b) Compare the z-score to the critical value for a 0.1% significance level using a two-sided alternative.
c) Explain your conclusion.

42. iPod reliability. MacInTouch reported that several versions of the iPod reported failure rates of 20% or more. From a customer survey, the color iPod, first released in 2004, showed 64 failures out of 517. Is there any evidence that the failure rate for this model may be lower than the 20% rate of previous models?
a) Find the z-score of the observed proportion.
b) Compare the z-score to the critical value for a 0.1% significance level using a one-sided alternative.
c) Explain your conclusion.

43. Testing cars. A clean air standard requires that vehicle exhaust emissions not exceed specified limits for various pollutants. Many states require that cars be tested annually to be sure they meet these standards. Suppose state regulators double-check a random sample of cars that a suspect repair shop has certified as okay. They will revoke the shop's license if they find significant evidence that the shop is certifying vehicles that do not meet standards.
a) In this context, what is a Type I error?
b) In this context, what is a Type II error?
c) Which type of error would the shop's owner consider more serious?
d) Which type of error might environmentalists consider more serious?

44. Quality control. Production managers on an assembly line must monitor the output to be sure that the level of defective products remains small. They periodically inspect a random sample of the items produced. If they find a significant increase in the proportion of items that must be rejected, they will halt the assembly process until the problem can be identified and repaired.
a) Write null and alternative hypothesis for this problem.
b) What is the Type I and Type II error in this context?
c) Which type of error would the factory owner consider more serious?
d) Which type of error might customers consider more serious?

45. Testing cars, again. As in Exercise 43, state regulators are checking up on repair shops to see if they are certifying vehicles that do not meet pollution standards.
a) In this context, what is meant by the power of the test the regulators are conducting?
b) Will the power be greater if they test 20 or 40 cars? Why?
c) Will the power be greater if they use a 5% or a 10% level of significance? Why?
d) Will the power be greater if the repair shop's inspectors are only a little out of compliance or a lot? Why?

46. Quality control, part 2. Consider again the task of the quality control inspectors in Exercise 44.
a) In this context, what is meant by the power of the test the inspectors conduct?
b) They are currently testing 5 items each hour. Someone has proposed they test 10 items each hour instead. What are the advantages and disadvantages of such a change?
c) Their test currently uses a 5% level of significance. What are the advantages and disadvantages of changing to a significance level of 1%?
d) Suppose that as a day passes one of the machines on the assembly line produces more and more items that are defective. How will this affect the power of the test?

47. Statistics software. A Statistics professor has observed that for several years about 13% of the students who initially enroll in his Introductory Statistics course withdraw before the end of the semester. A salesperson suggests that he try a statistics software package that gets students more involved with computers, predicting that it will cut the dropout rate. The software is expensive, and the salesperson offers to let the professor use it for a semester to see if the dropout rate goes down significantly. The professor will have to pay for the software only if he chooses to continue using it.
a) Is this a one-tailed or two-tailed test? Explain.
b) Write the null and alternative hypotheses.
c) In this context, explain what would happen if the professor makes a Type I error.
d) In this context, explain what would happen if the professor makes a Type II error.
e) What is meant by the power of this test?

48. Radio ads. A company is willing to renew its advertising contract with a local radio station only if the station can prove that more than 20% of the residents of the city have heard the ad and recognize the company's product. The radio station conducts a random phone survey of 400 people.

a) What are the hypotheses?
b) The station plans to conduct this test using a 10% level of significance, but the company wants the significance level lowered to 5%. Why?
c) What is meant by the power of this test?
d) For which level of significance will the power of this test be higher? Why?
e) They finally agree to use $\alpha = 0.05$, but the company proposes that the station call 600 people instead of the 400 initially proposed. Will that make the risk of Type II error higher or lower? Explain.

49. Statistics software, part 2. Initially, 203 students signed up for the Statistics course in Exercise 47. They used the software suggested by the salesperson, and only 11 dropped out of the course.

a) Should the professor spend the money for this software? Support your recommendation with an appropriate test.
b) Explain what your P-value means in this context.

50. Radio ads, part 2. The company in Exercise 48 contacts 600 people selected at random, and 133 can remember the ad.

a) Should the company renew the contract? Support your recommendation with an appropriate test.
b) Explain carefully what your P-value means in this context.

T 51. Customer spending, part 2. In Chapter 10, Exercise 51 constructed a confidence interval for the proportion of customers who qualified for a special offer by spending more than $1000 a month on their card. Historically, the percentage has been 11%, and the finance department wonders if it has increased. Test the appropriate hypothesis and write up a few sentences with your conclusions.

T 52. Fund-raising. In Chapter 10, Exercise 52 found a confidence interval for the proportion of donors that were 50 years old or older. The head of finance says that the American Association of Retired Persons (AARP) advertisement won't be worth the money unless at least 2/3 of the donors are 50 years old or older. Test the appropriate hypothesis and write up a few sentences with your conclusions.

JUST CHECKING ANSWERS

1 You can't conclude that the null hypothesis is true. You can conclude only that the experiment was unable to reject the null hypothesis. They were unable, on the basis of 12 patients, to show that aspirin was effective.
2 The null hypothesis is H₀: $p = 0.75$.
3 With a P-value of 0.0001, this is very strong evidence against the null hypothesis. We can reject H₀ and conclude that the improved version of the drug gives relief to a higher proportion of patients.
4 The parameter of interest is the proportion, p, of all delinquent customers who will pay their bills. H₀: $p = 0.30$ and Hₐ: $p > 0.30$.
5 At $\alpha = 0.05$, you can't reject the null hypothesis because 0.30 is contained in the 90% confidence interval—it's plausible that sending the DVDs is no more effective than sending letters.
6 The confidence interval is from 29% to 45%. The DVD strategy is more expensive and may not be worth it. We can't distinguish the success rate from 30% given the results of this experiment, but 45% would represent a large improvement. The bank should consider another trial, increasing the sample size to get a narrower confidence interval.
7 A Type I error would mean deciding that the DVD success rate is higher than 30%, when it isn't. The bank would adopt a more expensive method for collecting payments that's no better than its original, less expensive strategy.
8 A Type II error would mean deciding that there's not enough evidence to say the DVD strategy works when in fact it does. The bank would fail to discover an effective method for increasing revenue from delinquent accounts.
9 Higher; the larger the effect size, the greater the power. It's easier to detect an improvement to a 60% success rate than to a 32% rate.

12

Confidence Intervals and Hypothesis Tests for Means

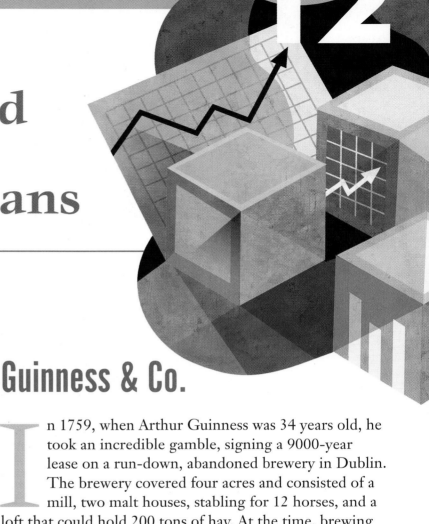

Guinness & Co.

I n 1759, when Arthur Guinness was 34 years old, he took an incredible gamble, signing a 9000-year lease on a run-down, abandoned brewery in Dublin. The brewery covered four acres and consisted of a mill, two malt houses, stabling for 12 horses, and a loft that could hold 200 tons of hay. At the time, brewing was a difficult and competitive market. Gin, whiskey, and the traditional London porter were the drinks of choice.

In addition to the lighter ales that Dublin was known for, Guinness began to brew dark porters to compete directly with those of the English brewers. Forty years later, Guinness stopped brewing light Dublin ales altogether to concentrate on his stouts and porters. Upon his death in 1803, his son Arthur Guinness II took over the business, and a few years later the company began to export Guinness stout to other parts of Europe. By the 1830s, the Guinness St. James's Gate Brewery, had become the largest in Ireland. In 1886, the Guinness Brewery, with an annual production of 1.2 million barrels, was the first major brewery to be incorporated as a public company on the London Stock Exchange. During the 1890s, the company began to employ scientists. One of those, William S. Gosset, was hired as a chemist to test the quality of the brewing process. Gosset was not only an early pioneer of quality control method in industry but his statistical work made modern statistical inference possible.[1]

A s a chemist at the Guinness Brewery in Dublin, William S. Gosset was in charge of quality control. His job was to make sure that the stout (a thick, dark beer) leaving the brewery was of high enough quality to meet the standards of the brewery's many discerning customers. It's easy to imagine, when testing stout, why testing a large amount of stout might be undesirable, not to mention dangerous to one's health. So to test for quality Gosset often used a sample of only 3 or 4 observations per batch. But he noticed that with samples of this size, his tests for quality weren't quite right. He knew this because when the batches that he rejected were sent back to the laboratory for more extensive testing, too often the test results turned out to be wrong. As a practicing statistician, Gosset knew he had to be wrong *some* of the time, but he hated being wrong more often than the theory predicted. One result of Gosset's frustrations was the development of a test to handle small samples, the main subject of this chapter.

12.1 The Sampling Distribution for the Mean

You've learned how to create confidence intervals and test hypotheses about proportions. Now we want to do the same thing for means. For proportions we found the confidence interval as

$$\hat{p} \pm ME.$$

The ME was equal to a critical value, z^*, times $SE(\hat{p})$. Our confidence interval for means will be

$$\bar{y} \pm ME,$$

And our ME will be a critical value times $SE(\bar{y})$. So let's put the pieces together. What the Central Limit Theorem told us back in Chapter 9 looks like what we need.

---

[1] Source: Guinness & Co. 2006, www.guinness.com/global/story/history.

> ### The Central Limit Theorem
> When a random sample is drawn from *any* population with mean μ and standard deviation σ, its sample mean, $\bar{y}$, has a sampling distribution whose *shape* is approximately Normal as long as the sample size is large enough. The larger the sample used, the more closely the Normal approximates the sampling distribution for the mean. The mean of the sampling distribution is μ, and its standard deviation is $SD(\bar{y}) = \dfrac{\sigma}{\sqrt{n}}$.

This gives us a sampling distribution and a standard deviation for the mean. All we need is a random sample of quantitative data and the true value of the population standard deviation σ.

But wait. That could be a problem. To compute $\sigma/\sqrt{n}$ we need to know σ. How are we supposed to know σ? Suppose we told you that for 25 young executives the mean value of their stock portfolios is $125,672. Would that tell you the value of σ? No, the standard deviation depends on how similarly the executives invest, not on how well they invested (the mean tells us that). But we need σ because it's the numerator of the standard deviation of the sample mean: $SD(\bar{y}) = \dfrac{\sigma}{\sqrt{n}}$. So what can we do? The obvious answer is to use the sample standard deviation, s, from the data instead of σ. The result is the standard error: $SE(\bar{y}) = \dfrac{s}{\sqrt{n}}$.

A century ago, people just plugged the standard error into the Normal model, assuming it would work. And for large sample sizes it *did* work pretty well. But they began to notice problems with smaller samples. The extra variation in the standard error was wreaking havoc with the P-values and margins of error.

William S. Gosset was the first to investigate this phenomenon. He realized that not only do we need to allow for the extra variation with larger margins of error and P-values, but we also need a new sampling distribution model. In fact, we need a whole *family* of models, depending on the sample size, n. These models are unimodal, symmetric, and bell-shaped, but the smaller our sample, the more we must stretch out the tails. Gosset's work transformed Statistics, but most people who use his work don't even know his name.

Gosset's *t*

Gosset checked the stout's quality by performing hypothesis tests. He knew that if he set $\alpha = 0.05$ the test would make some Type I errors by rejecting about 5% of the good batches of stout. However, the lab told him that he was in fact rejecting about 15% of the good batches. Gosset knew something was wrong, and it bugged him.

Gosset took time off from his job to study the problem and earn a graduate degree in the emerging field of Statistics. He figured out that when he used the standard error $\dfrac{s}{\sqrt{n}}$, the shape of the sampling model was no longer Normal. He even figured out what the new model was and called it a *t*-distribution.

The Guinness Company didn't give Gosset a lot of support for his work. In fact, it had a policy against publishing results. Gosset had to convince the company that he was not publishing an industrial secret and (as part of getting permission to publish) had to use a pseudonym. The pseudonym he chose was "Student," and ever since, the model he found has been known as **Student's *t***.

Gosset's model is always bell-shaped, but the details change with the sample sizes. So the Student's *t*-models form a family of related distributions that depend on a parameter known as **degrees of freedom**. We often denote degrees of freedom as df and the model as t_{df}, with the numerical value of the degrees of freedom as a subscript.

To find the sampling distribution of $\dfrac{\bar{y}}{s/\sqrt{n}}$, Gosset simulated it by hand. He drew paper slips of small samples from a hat hundreds of times and computed the means and standard deviations with a mechanically cranked calculator. Today you could repeat in seconds on a computer the experiment that took him over a year. Gosset's work was so meticulous that not only did he get the shape of the new histogram approximately right, but he even figured out the exact formula for it from his sample. The formula was not confirmed mathematically until years later by Sir R. A. Fisher.

NOTATION ALERT:
Ever since Gosset, the letter *t*
has been reserved in Statistics
for his distribution.

12.2 A Confidence Interval for Means

To make confidence intervals or to test hypotheses for means, we need to use
Gosset's model. Which one? Well, for means, it turns out the right value for
degrees of freedom is df $= n - 1$.

Practical Sampling Distribution Model for Means

When certain conditions are met, the standardized sample mean,

$$t = \frac{\bar{y} - \mu}{SE(\bar{y})}$$

follows a Student's *t*-model with $n - 1$ degrees of freedom. We find the standard
error from:

$$SE(\bar{y}) = \frac{s}{\sqrt{n}}.$$

When Gosset corrected the Normal model for the extra uncertainty, the margin
of error got bigger, as you might have guessed. When you use Gosset's model instead
of the Normal model, your confidence intervals will be just a bit wider and your
P-values just a bit larger (Figure 12.1). That's just the correction you need. By using
the *t*-model, you've compensated for the extra variability in precisely the right way.

One-sample *t*-interval

When the assumptions and conditions are met, we are ready to find the
confidence interval for the population mean, μ. The confidence interval is:

$$\bar{y} \pm t^{*}_{n-1} \times SE(\bar{y})$$

where the standard error of the mean is:

$$SE(\bar{y}) = \frac{s}{\sqrt{n}}.$$

The critical value t^{*}_{n-1} depends on the particular confidence level, C, that you
specify and on the number of degrees of freedom, $n - 1$, which we get from the
sample size.

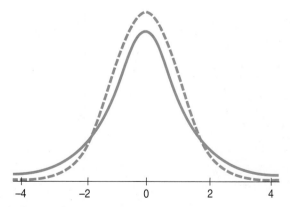

Figure 12.1   *The t-model (solid curve) with 2 degrees
of freedom has fatter tails than the Normal model (dashed
curve). So the 68-95-99.7 Rule doesn't work for t-models
with only a few degrees of freedom.*

Student's *t*-models are unimodal, symmetric, and bell-shaped, just like the
Normal model. But *t*-models with only a few degrees of freedom have a narrower

peak than the Normal model and have much fatter tails. (That's what makes the margin of error bigger.) As the degrees of freedom increase, the *t*-models look more and more like the Normal model. In fact, the *t*-model with infinite degrees of freedom is exactly Normal.[2] This is great news if you happen to have an infinite number of data values. Unfortunately, that's not practical. Fortunately, above a few hundred degrees of freedom it's very hard to tell the difference. Of course, in the rare situation that we *know* σ, it would be foolish not to use that information. If we don't have to estimate σ, we can use the Normal model. Typically that value of σ would be based on (lots of) experience, or on a theoretical model. Usually, however, we estimate σ by *s* from the data and use the *t*-model.

z or t?

If you know σ, use *z*. (That's rare!) Whenever you use *s* to estimate σ, use *t*.

12.3 Assumptions and Conditions

Gosset found the *t*-model by simulation. Years later, when Sir Ronald Fisher showed mathematically that Gosset was right, he needed to make some assumptions to make the proof work. These are the assumptions we need in order to use the Student's *t*-models.

Independence Assumption

Independence Assumption: The data values should be independent. There's really no way to check independence of the data by looking at the sample, but we should think about whether the assumption is reasonable.

Randomization Condition: The data arise from a random sample or suitably randomized experiment. Randomly sampled data—and especially data from a Simple Random Sample (SRS)—are ideal.

When a sample is drawn without replacement, technically we ought to confirm that we haven't sampled a large fraction of the population, which would threaten the independence of our selections.

10% Condition: The sample size should be no more than 10% of the population. In practice, though, we often don't mention the 10% Condition when estimating means. Why not? When we made inferences about proportions, this condition was crucial because we usually had large samples. But for means our samples are generally smaller, so this problem arises only if we're sampling from a small population (and then there's a correction formula we could use).

Normal Population Assumption

Student's *t*-models won't work for data that are badly skewed. How skewed is too skewed? Well, formally, we assume that the data are from a population that follows a Normal model. Practically speaking, there's no way to be certain this is true.

And it's almost certainly *not* true. Models are idealized; real data are, well, real. The good news, however, is that even for small samples, it's sufficient to check a condition.

Nearly Normal Condition. The data come from a distribution that is unimodal and symmetric. This is a much more practical condition and one we can check by making a histogram.[3] For small samples, it can be hard to see any

We Don't *Want* to Stop

We check conditions hoping that we can make a meaningful analysis of our data. The conditions serve as *disqualifiers*—we keep going unless there's a serious problem. If we find minor issues, we note them and express caution about our results. If the sample is not an SRS, but we believe it's representative of some populations, we limit our conclusions accordingly. If there are outliers, rather than stop, we perform the analysis both with and without them. If the sample looks bimodal, we try to analyze subgroups separately. Only when there's major trouble—like a strongly skewed small sample or an obviously nonrepresentative sample—are we unable to proceed at all.

---

[2] Formally, in the limit as the number of degrees of freedom goes to infinity.
[3] Or we could check a more sensitive display called a normal probability plot, discussed in Chapter 16.

distribution shape in the histogram. Unfortunately, the condition matters most when it's hardest to check.[4]

For very small samples ($n < 15$ or so), the data should follow a Normal model pretty closely. Of course, with so little data, it's rather hard to tell. But if you do find outliers or strong skewness, don't use these methods.

For moderate sample sizes (n between 15 and 40 or so), the t methods will work well as long as the data are unimodal and reasonably symmetric. Make a histogram to check.

When the sample size is larger than 40 or 50, the t methods are safe to use unless the data are extremely skewed. Make a histogram anyway. If you find outliers in the data and they aren't errors that are easy to fix, it's always a good idea to perform the analysis twice, once with and once without the outliers, even for large samples. The outliers may well hold additional information about the data, so they deserve special attention. If you find multiple modes, you may well have different groups that should be analyzed and understood separately.

If the data are extremely skewed (like the CEO data from Chapter 6—see Figure 12.2), the mean may not be the most appropriate summary. But in business we often are concerned with costs and profits. When our data consist of a collection of instances whose *total* is the business consequence—as when we add up the profits (or losses) from many transactions or the costs of many supplies—then the mean is just that total divided by n. And that's the value with a business consequence. Fortunately, in this instance, the Central Limit Theorem comes to our rescue. Even when we must sample from a very skewed distribution, the sampling distribution of our sample mean will be close to Normal, so we can use Student's t methods without much worry as long as the sample size is *large enough*.

How large is large enough? Here's the histogram of CEO compensations ($000).

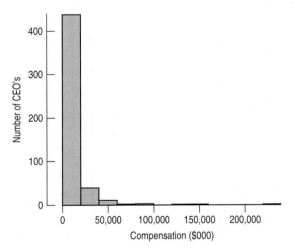

Figure 12.2   *It's hard to imagine a distribution more skewed than these annual compensations from the Fortune 500 CEO's.*

Although this distribution is very skewed, the Central Limit Theorem will make the sampling distribution of the means of samples from this distribution more and more Normal as the sample size grows. Here's a histogram and a Normal probability plot of the means of samples of size 100 CEO's:

---

[4] There are formal tests of Normality, but they don't really help. When we have a small sample—just when we really care about checking Normality—these tests have very little power. So it doesn't make much sense to use them in deciding whether to perform a t-test.

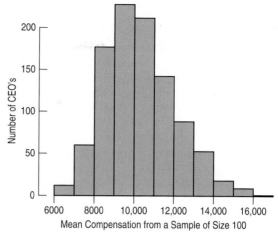

Figure 12.3 *Even samples as small as 100 from the CEO data set produce means whose sampling distribution is nearly normal. Larger samples will have sampling distributions even more Normal.*

Often, in modern business applications, we have samples of many hundreds, or thousands. We should still be on guard for outliers and multiple modes and we should be sure that the observations are independent. But if the mean is of interest, the Central Limit Theorem works quite well in insuring that the sampling distribution of the mean will be close to the Normal for samples of this size.

JUST CHECKING

Every 10 years, the United States takes a census that tries to count every resident. In addition, the census collects information on a variety of economic and social questions. Businesses of all types use the census data to plan sales and marketing strategies and to understand the underlying demographics of the areas that they serve.

There are two census forms: the "short form," answered by most people, and the "long form," sent only to about one in six or seven households chosen at random. According to the Census Bureau (factfinder.census.gov), " . . . each estimate based on the long form responses has an associated confidence interval."

1 Why does the Census Bureau need a confidence interval for long-form information, but not for the questions that appear on both the long and short forms?

2 Why must the Census Bureau base these confidence intervals on *t*-models?

The Census Bureau goes on to say, "These confidence intervals are wider . . . for geographic areas with smaller populations and for characteristics that occur less frequently in the area being examined (such as the proportion of people in poverty in a middle-income neighborhood)."

3 Why is this so? For example, why should a confidence interval for the mean amount families spend monthly on housing be wider for a sparsely populated area of farms in the Midwest than for a densely populated area of an urban center? How does the formula for the one-sample *t*-interval show this will happen?

To deal with this problem, the Census Bureau reports long-form data only for " . . . geographic areas from which about two hundred or more long forms were completed—which are large enough to produce good quality estimates. If smaller weighting areas had been used, the confidence intervals around the estimates would have been significantly wider, rendering many estimates less useful."

4 Suppose the Census Bureau decided to report on areas from which only 50 long forms were completed. What effect would that have on a 95% confidence interval for, say, the mean cost of housing? Specifically, which values used in the formula for the margin of error would change? Which values would change a lot, and which values would change only slightly? Approximately how much wider would that confidence interval based on 50 forms be than the one based on 200 forms?

GUIDED EXAMPLE | *Insurance Profits*

Insurance companies take risks. When they insure a property or a life, they must price the policy in such a way that their expected profit enables them to survive. They can base their projections on actuarial tables, but the reality of the insurance business often demands that they discount policies to a variety of customers and situations. Managing this risk is made even more difficult by the fact that until the policy expires, the company won't know if they've made a profit, no matter what premium they charge.

A manager wanted to see how well one of her sales representatives was doing, so she selected 30 matured policies that had been sold by the sales rep and computed the (net) profit (premium charged minus paid claims), for each of the 30 policies.

The manager would like you, as a consultant, to construct a 95% confidence interval for the mean profit of the policies sold by this sales rep.

Profit (in $) from 30 policies		
222.80	463.35	2089.40
1756.23	−66.20	2692.75
1100.85	57.90	2495.70
3340.66	833.95	2172.70
1006.50	1390.70	3249.65
445.50	2447.50	−397.70
3255.60	1847.50	−397.31
3701.85	865.40	186.25
−803.35	1415.65	590.85
3865.90	2756.34	578.95

PLAN

Setup State what we want to know. Identify the variables and their context.

Make a picture. Check the distribution shape and look for skewness, multiple modes, and outliers.

We wish to find a 95% confidence interval for the mean profit of policies sold by this sales rep. We have data for 30 matured policies.

Here's a boxplot and histogram of these values.

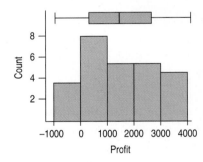

The sample appears to be unimodal and fairly symmetric with profit values between −$1000 and $4000 and no outliers.

Model Think about the assumptions and check the conditions.

✓ **Independence Assumption**

This is a random sample so observations should be independent.

✓ **Randomization Condition**

This sample was selected randomly from the matured policies sold by the sales representative of the company.

		✓ **Nearly Normal Condition**
	State the sampling distribution model for the statistic.	The distribution of profits is unimodal and fairly symmetric without strong skewness. We will use a Student's *t*-model with $n - 1 = 30 - 1 = 29$ degrees of freedom and find a one-sample *t*-interval for the mean.

DO

Mechanics Compute basic statistics and construct the confidence interval.	Using software, we obtain the following basic statistics:

$$n = 30$$
$$\bar{y} = \$1438.90$$
$$s = \$1329.60$$

Remember that the standard error of the mean is equal to the standard deviation divided by the square root of *n*.

The standard error of the mean is:

$$SE(\bar{y}) = \frac{s}{\sqrt{n}} = \frac{1329.60}{\sqrt{30}} = \$242.75$$

The critical value we need to make a 95% confidence interval comes from a Student's *t* table, a computer program, or a calculator. We have $30 - 1 = 29$ degrees of freedom. The selected confidence level says that we want 95% of the probability to be caught in the middle, so we exclude 2.5% in *each* tail, for a total of 5%. The degrees of freedom and 2.5% tail probability are all we need to know to find the critical value. Here it's 2.045.

There are $30 - 1 = 29$ degrees of freedom. The manager has specified a 95% level of confidence, so the critical value (from table T) is 2.045.

The margin of error is:

$$ME = 2.045 \times SE(\bar{y})$$
$$= 2.045 \times 242.75$$
$$= \$496.42$$

The 95% confidence interval for the mean profit is:

$$\$1438.90 \pm \$496.42$$
$$= (\$942.48, \$1935.32)$$

REPORT

Conclusion Interpret the confidence interval in the proper context.

MEMO

Re: Profit from Policies

From our analysis of the selected policies, we are 95% confident that the true mean profit of policies sold by this sales rep is contained in the interval from $942.48 to $1935.32.

When we construct confidence intervals in this way, we expect 95% of them to cover the true mean and 5% to miss the true value. That's what "95% confident" means.

Caveat: Insurance losses are notoriously subject to outliers. One very large loss could influence the average profit substantially. However, there were no such cases in this data set.

The critical value in the Guided Example was found in the Student's *t* Table in Appendix C. To find the critical value, locate the row of the table corresponding to the degrees of freedom and the column corresponding to the probability you want. Since a 95% confidence interval leaves 2.5% of the values on either side, we look for 0.025 at the top of the column or look for 95% confidence directly in the bottom row of the table. The value in the table at that intersection is the critical value we need. In the Guided Example, the number of degrees of freedom was $30 - 1 = 29$, so we located the value of 2.045.

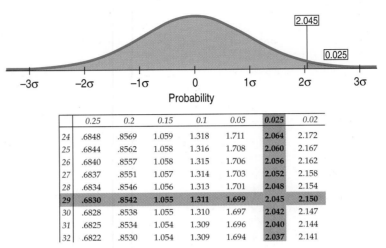

	0.25	0.2	0.15	0.1	0.05	0.025	0.02
24	.6848	.8569	1.059	1.318	1.711	2.064	2.172
25	.6844	.8562	1.058	1.316	1.708	2.060	2.167
26	.6840	.8557	1.058	1.315	1.706	2.056	2.162
27	.6837	.8551	1.057	1.314	1.703	2.052	2.158
28	.6834	.8546	1.056	1.313	1.701	2.048	2.154
29	.6830	.8542	1.055	1.311	1.699	2.045	2.150
30	.6828	.8538	1.055	1.310	1.697	2.042	2.147
31	.6825	.8534	1.054	1.309	1.696	2.040	2.144
32	.6822	.8530	1.054	1.309	1.694	2.037	2.141

Figure 12.4 *Using Table T to look up the critical value t* for a 95% confidence level with 29 degrees of freedom.*

12.4 Cautions About Interpreting Confidence Intervals

So What *Should* You Say?

Since 95% of random samples yield an interval that captures the true mean, you should say: "I am 95% confident that the interval from $942.48 to $1935.32 contains the mean profit of all policies sold by this sales representative." It's also okay to say something slightly less formal: "I am 95% confident that the mean profit for all policies sold by this sales rep is between $942.48 and $1935.32." Remember: *Your uncertainty is about the interval, not the true mean.* The interval varies randomly. The true mean profit is neither variable nor random—just unknown.

Confidence intervals for means offer new, tempting wrong interpretations. Here are some ways to keep from going astray:

◆ **Don't say,** "*95% of all the policies* sold by this sales rep have profits between $942.48 and $1935.32." The confidence interval is about the *mean*, not about the measurements of individual policies.

◆ **Don't say,** "We are 95% confident that *a randomly selected policy* will have a net profit between $942.48 and $1935.32." This false interpretation is also about individual policies rather than about the *mean* of the policies. We are 95% confident that the *mean* profit of all (similar) policies sold by this sales rep is between $942.48 and $1935.32

◆ **Don't say,** "The mean profit is $1438.90 95% *of the time*." That's about means, but still wrong. It implies that the true mean varies, when in fact it is the confidence interval that would have been different had we gotten a different sample.

◆ Finally, **don't say,** "95% *of all samples* will have mean profits between $942.48 and $1935.32." That statement suggests that *this* interval somehow sets a standard for every other interval. In fact, this interval is no more (or less) likely to be correct than any other. You could say that 95% of all possible samples would produce intervals that contain the true mean profit. (The problem is that because we'll never know what the true mean profit is, we can't know if our sample was one of those 95%.)

12.5 One-Sample *t*-Test

The manager has a more specific concern. Company policy states that if a sales rep's mean profit is below $1500, the sales rep has been discounting too much and will have to adjust his pricing strategy. Is there evidence from this sample that the mean is really less than $1500? This question calls for a hypothesis test called the **one-sample *t*-test for the mean**.

You already know enough to construct this test. The test statistic looks just like the others we've seen. We've always compared the difference between the observed statistic and a hypothesized value to the standard error. For means that looks like: $\dfrac{\bar{y} - \mu_0}{SE(\bar{y})}$. We already know that the appropriate probability model to use is Student's *t* with $n - 1$ degrees of freedom.

One-sample *t*-test for the mean

The conditions for the one-sample *t*-test for the mean are the same as for the one-sample *t*-interval. We test the hypothesis $H_0: \mu = \mu_0$ using the statistic

$$t_{n-1} = \frac{\bar{y} - \mu_0}{SE(\bar{y})},$$

where the standard error of $\bar{y}$ is: $SE(\bar{y}) = \dfrac{s}{\sqrt{n}}$.

When the conditions are met and the null hypothesis is true, this statistic follows a Student's *t*-model with $n - 1$ degrees of freedom. We use that model to obtain a P-value.

GUIDED EXAMPLE | *Insurance Profits Revisited*

Let's apply the one-sample *t*-test to the 30 mature policies sampled by the manager. From these 30 policies, the management would like to know if there's evidence that the mean profit of policies sold by this sales rep is less than $1500.

PLAN

Setup State what we want to know. Make clear what the population and parameter are. Identify the variables and context.	We want to test whether the mean profit of the sales rep's policies is less than $1500. We have a random sample of 30 mature policies from which to judge.
Hypotheses We give benefit of the doubt to the sales rep. The null hypothesis is that the true mean profit is equal to $1500. Because we're interested in whether the profit is less, the alternative is one sided.	$H_O: \mu = \$1500$ $H_A: \mu < \$1500$

Make a graph. Check the distribution for skewness, multiple modes, and outliers.

We checked the histogram of these data in the previous Guided Example and saw that it had a unimodal, symmetric distribution.

Model Check the conditions.

We checked the Randomization and Nearly Normal Conditions in the previous Guided Example.

State the sampling distribution model.

Choose your method.

The conditions are satisfied, so we'll use a Student's *t*-model with $n - 1 = 29$ degrees of freedom and a one-sample *t*-test for the mean.

DO

Mechanics Compute the sample statistics. Be sure to include the units when you write down what you know from the data.

Using software, we obtain the following basic statistics:

$$n = 30$$
$$\text{Mean} = \$1438.90$$
$$\text{StDev} = \$1329.60$$

The *t*-statistic calculation is just a standardized value. We subtract the hypothesized mean and divide by the standard error.

$$t = \frac{1438.90 - 1500}{1329.60/\sqrt{30}} = -0.2517$$

(The observed mean is less than one standard error below the hypothesized value.)

We assume the null model is true to find the P-value. Make a picture of the *t*-model, centered at μ_0. Since this is a lower-tail test, shade the region to the left of the observed average profit.

The P-value is the probability of observing a sample mean as small as \$1438.90 (or smaller) *if* the true mean were \$1500, as the null hypothesis states. We can find this P-value from a table, calculator, or computer program.

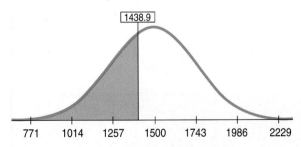

P-value = $P(t_{29} < -0.2517) = 0.4015$
(or from a table $0.1 < P < 0.5$)

**REPORT**

Conclusion Link the P-value to your decision about H_0, and state your conclusion in context.

MEMO

Re: Sales Performance

The mean profit on 30 sampled contracts closed by the sales rep in question has fallen below our standard of \$1500, but there is not enough evidence in this sample of policies to indicate that the true mean is below \$1500. If the mean were \$1500, we would expect a sample of size 30 to have a mean this low about 40.15% of the time.

Notice that the way this hypothesis was set up, the sales rep's mean profit would have to be well below $1500 to reject the null hypothesis. Because the null hypothesis was that the mean was $1500 and the alternative was that it was less, this set up gave some benefit of the doubt to the sales rep. There's nothing intrinsically wrong with that, but keep in mind that it's always a good idea to make sure that the hypotheses are stated in ways that will guide you make the right business decision.

Finding *t*-Values by Hand

The Student's *t*-model is different for each value of degrees of freedom. We might print a table like Table Z (in Appendix C) for each degrees of freedom value, but that's a lot of pages and not likely to be a bestseller. One way to shorten the book is to limit ourselves to 80%, 90%, 95% and 99% confidence levels. So Statistics books usually have one table of *t*-model critical values for a selected set of confidence levels. This one does, too; see Table T in Appendix C. (You can also find tables on the Internet.)

The *t*-tables run down the page for as many degrees of freedom as can fit, and they are much easier to use than the Normal tables (Figure 12.5). Then they get to the bottom of the page and run out of room. Of course, for *enough* degrees of freedom, the *t*-model gets closer and closer to the Normal, so the tables give a final row with the critical values from the Normal model and label it "∞ df."

| Two tail probability | | 0.20 | 0.10 | 0.05 |
One tail probability		0.10	0.05	0.025
Table T	df			
Values of t_a	1	3.078	6.314	12.706
	2	1.886	2.920	4.303
	3	1.638	2.353	3.182
	4	1.533	2.132	2.776
	5	1.476	2.015	2.571
	6	1.440	1.943	2.447
	7	1.415	1.895	2.365
	8	1.397	1.860	2.306
	9	1.383	1.833	2.262
	10	1.372	1.812	2.228
	11	1.363	1.796	2.201
	12	1.356	1.782	2.179
	13	1.350	1.771	2.160
	14	1.345	1.761	2.145
	15	1.341	1.753	2.131
	16	1.337	1.746	2.120
	17	1.333	1.740	2.110
	18	1.330	1.734	2.101
	19	1.328	1.729	2.093
	⋮	⋮	⋮	⋮
	∞	1.282	1.645	1.960
Confidence levels		80%	90%	95%

Figure 12.5 *Part of Table T in Appendix C.*

JUST CHECKING

In discussing estimates based on the long-form samples, the Census Bureau notes, "The disadvantage . . . is that . . . estimates of characteristics that are also reported on the short form will not match the [long-form estimates]."

The short-form estimates are values from a complete census, so they are the "true" values—something we don't usually have when we do inference.

5 Suppose we use long-form data to make 100 95% confidence intervals for the mean age of residents, one for each of 100 of the census-defined areas. How many of these 100 intervals should we expect will *fail* to include the true mean age (as determined from the complete short-form census data)?

6 Based on the long-form sample, we might test the null hypothesis that the mean household income in a region was the same as in the previous census. Would the standard error for such a test be likely to increase or decrease if we used an area with more long-form respondents?

For large degrees of freedom, the shape of Student's *t*-models changes more gradually. Table T in Appendix C includes degrees of freedom between 100 and 1000 so you can pin down the P-value for just about any df. If your df's aren't listed, take the cautious approach by using the next lower value or use technology to find the exact value.

For example, suppose we've performed a one-sample *t*-test finding for large degrees of freedom, with 19 df and want the upper tail P-value. From the table we see that 1.639 falls between 1.328 and 1.729. All we can say is that the P-value lies between the P-values of these two critical values, so $0.05 < P < 0.10$.

Or we can use technology. Calculators or statistics programs can give critical values for a *t*-model for any number of degrees of freedom and for any confidence level you need. And they can go straight to P-values when you test a hypothesis. With tables we can only approximate P-values by pinning them down between two of the columns. Usually that's good enough. More precision won't necessarily help make a good business decision.

Did we need to perform a one-sample *t*-test to know that we would fail to reject a null hypothesis that the mean was $1500?

After all, we saw that the interval $942.48 to $1935.32 contained all the plausible values for the mean profit at 95% confidence. Since $1500 was one of those plausible values, we have no evidence to suggest that the mean is not $1500.

Because we wanted a one-sided test, our α level from the 95% confidence interval would be 0.025, corresponding to only one side of the confidence interval. If we wanted an α level of 0.05 we could look at the narrower 90% confidence interval: ($1022.26, $1855.54). Because $1500 is also in this interval we would come to the same conclusion and fail to reject the hypothesis that the mean is $1500.

12.6 Sample Size

How large a sample do we need? The simple answer is always "larger." But more data cost money, effort, and time. So how much is enough? Suppose your computer took an hour to download a movie you wanted to watch. You wouldn't be happy. Then you hear about a program that claims to download movies in under a half hour. You're interested enough to spend $29.95 for it, but only if it really delivers. So you get the free evaluation copy and test it by downloading a movie 10 times. Of course, the mean download time is not exactly 30 minutes as claimed. Observations vary. If the margin of error were 8 minutes, though, you'd probably be able to decide whether the software was worth the money. Doubling the sample size would require another 5 or so hours of testing and would reduce your margin of error to a bit under 6 minutes. You'd need to decide whether that's worth the effort.

As we make plans to collect data, we should have some idea of how small a margin of error is required to be able to draw a conclusion or detect a difference we want to see. If the size of the effect we're studying is large, then we may be able to tolerate a larger ME. If we need great precision, however, we'll want a smaller ME, and, of course, that means a larger sample size.

Armed with the ME and confidence level, we can find the sample size we'll need. Almost.

We know that for a mean, $ME = t^*_{n-1} \times SE(\bar{y})$ and that $SE(\bar{y}) = \dfrac{s}{\sqrt{n}}$, so we can determine the sample size by solving this equation for *n*:

$$ME = t^*_{n-1} \times \frac{s}{\sqrt{n}}.$$

The good news is that we have an equation; the bad news is that we won't know most of the values we need to compute it. When we thought about sample size for proportions, we ran into a similar problem. There we had to guess a working value for *p* to compute a sample size. Here, we need to know *s*. We don't know *s* until we get some data, but we want to calculate the sample size *before* collecting the data. We

Sample size calculations by hand

Let's give the sample size formula a spin. Suppose we want an ME of 8 minutes and we think the standard deviation of download times is about 10 minutes. Using a 95% confidence interval and $z^* = 1.96$, we solve for n:

$$8 = 1.96\frac{10}{\sqrt{n}}$$

$$\sqrt{n} = \frac{1.96 \times 10}{8} = 2.45$$

$$n = (2.45)^2 = 6.0025$$

That's a small sample size, so we use $(6 - 1) = 5$ degrees of freedom to substitute an appropriate t^* value. At 95%, $t_5^* = 2.571$. Now we can solve the equation one more time:

$$8 = 2.571\frac{10}{\sqrt{n}}$$

$$\sqrt{n} = \frac{2.571 \times 10}{8} \approx 3.214$$

$$n = (3.214)^2 \approx 10.33$$

To make sure the ME is no larger than you want, you should always round *up*, which gives $n = 11$ runs. So, to get an ME of 8 minutes, we should find the downloading times for $n = 11$ movies.

might be able to make a good guess, and that is often good enough for this purpose. If we have no idea what the standard deviation might be or if the sample size really matters (for example, because each additional individual is very expensive to sample or experiment on), it might be a good idea to run a small *pilot study* to get some feeling for the size of the standard deviation.

That's not all. Without knowing n, we don't know the degrees of freedom, and we can't find the critical value, t_{n-1}^*. One common approach is to use the corresponding z^* value from the Normal model. If you've chosen a 95% confidence level, then just use 2, following the 68-95-99.7 Rule, or 1.96 to be more precise. If your estimated sample size is 60 or more, it's probably okay—z^* was a good guess. If it's smaller than that, you may want to add a step, using z^* at first, finding n, and then replacing z^* with the corresponding t_{n-1}^* and calculating the sample size once more.

Sample size calculations are *never* exact. The margin of error you find *after* collecting the data won't match exactly the one you used to find n. The sample size formula depends on quantities that you won't have until you collect the data, but using it is an important first step. Before you collect data, it's always a good idea to know whether the sample size is large enough to give you a good chance of being able to tell you what you want to know.

*12.7 Degrees of Freedom – Why $n - 1$?

The number of degrees of freedom $(n - 1)$ might have reminded you of the value we divide by to find the standard deviation of the data (since, after all, it's the same number). We promised back when we introduced that formula to say a bit more about why we divide by $n - 1$ rather than by n. The reason is closely tied to the reasoning of the t-distribution.

If only we knew the true population mean, μ, we would find the sample standard deviation using n instead of $n - 1$ as:

$$s = \sqrt{\frac{\sum(y - \mu)^2}{n}} \text{ and we'd call it } s.$$

We use $\bar{y}$ instead of μ, though, and that causes a problem. For any sample, $\bar{y}$ will be as close to the data values as possible. Generally the population mean, μ, will be farther away. Think about it. GMAT scores have a population mean of 525. If you took a random sample of 5 students who took the test, their sample mean wouldn't be 525. The five data values will be closer to their own $\bar{y}$ than to 525. So if we use $\sum(y - \bar{y})^2$ instead of $\sum(y - \mu)^2$ in the equation to calculate s, our standard deviation estimate will be too small. The amazing mathematical fact is that we can compensate for the fact that $\sum(y - \bar{y})^2$ is too small just by dividing by $n - 1$ instead of by n. So that's all the $n - 1$ is doing in the denominator of s. We call $n - 1$ the degrees of freedom.

WHAT CAN GO WRONG?

First, you must decide when to use Student's *t* methods.

- **Don't confuse proportions and means.** When you treat your data as categorical, counting successes and summarizing with a sample proportion, make inferences using the Normal model methods. When you treat your data as quantitative, summarizing with a sample mean, make your inferences using Student's *t* methods.

- **Be careful of interpretation when confidence intervals overlap.** If confidence intervals for the means from two groups overlap, don't jump to the conclusion that the means are equal. It can be the case that two means are significantly different, and yet their confidence intervals will overlap. We'll see in the next chapter how to test the difference between two means directly. If the confidence intervals don't overlap, we are safe in rejecting the null hypothesis, but the methods in the next chapter are more powerful.

Student's *t* methods work only when the Normal Population Assumption is true. Naturally, many of the ways things can go wrong turn out to be ways that the Normal Population Assumption can fail. It's always a good idea to look for the most common kinds of failure. It turns out that you can even fix some of them.

- **Beware of multimodality.** The Nearly Normal Condition clearly fails if a histogram of the data has two or more modes. When you see this, look for the possibility that your data come from two groups. If so, your best bet is to try to separate the data into groups. (Use the variables to help distinguish the modes, if possible. For example, if the modes seem to be composed mostly of men in one and women in the other, split the data according to the person's sex.) Then you can analyze each group separately.

- **Beware of skewed data.** Make a histogram of the data. If the data are severely skewed, you might try re-expressing the variable. Re-expressing may yield a distribution that is unimodal and symmetric, making it more appropriate for the inference methods for means. Re-expression cannot help if the sample distribution is not unimodal.

> As tempting as it is to get rid of annoying values, you can't just throw away outliers and not discuss them. It is not appropriate to lop off the highest or lowest values just to improve your results.

- **Investigate outliers.** The Nearly Normal Condition also fails if the data have outliers. If you find outliers in the data, you need to investigate them. Sometimes, it's obvious that a data value is wrong and the justification for removing or correcting it is clear. When there's no clear justification for removing an outlier, you might want to run the analysis both with and without the outlier and note any differences in your conclusions. Any time data values are set aside, you *must* report on them individually. Often they will turn out to be the most informative part of your report on the data.[5]

---

[5] This suggestion may be controversial in some disciplines. Setting aside outliers is seen by some as unethical because the result is likely to be a narrower confidence interval or a smaller P-value. But an analysis of data with outliers left in place is *always* wrong. The outliers violate the Nearly Normal Condition and also the implicit assumption of a homogeneous population, so they invalidate inference procedures. An analysis of the non-outlying points, along with a separate discussion of the outliers, is often much more informative, and can reveal important aspects of the data.

Of course, Normality issues aren't the only risks you face when doing inferences about means.

- **Watch out for bias.** Measurements of all kinds can be biased. If your observations differ from the true mean in a systematic way, your confidence interval may not capture the true mean. And there is no sample size that will save you. A bathroom scale that's 5 pounds off will be 5 pounds off even if you weigh yourself 100 times and take the average. We've seen several sources of bias in surveys, but measurements can be biased, too. Be sure to think about possible sources of bias in your measurements.

- **Make sure data are independent.** Student's *t* methods also require the sampled values to be mutually independent. We check for random sampling and the 10% Condition. You should also think hard about whether there are likely violations of independence in the data collection method. If there are, be very cautious about using these methods.

- **Make sure that data are from an appropriately randomized sample.** Ideally, all data that we analyze are drawn from a simple random sample or generated by a randomized experiment. When they're not, be careful about making inferences from them. You may still compute a confidence interval correctly or get the mechanics of the P-value right, but this can't save you from making a serious mistake in inference.

ETHICS IN ACTION

Recent reports have indicated that waiting times in hospital emergency rooms (ERs) across the United States are getting longer, with the average now at 30 minutes (WashingtonPost.com; January 2008). Several reasons have been cited for this rise in average ER waiting time including the closing of hospital emergency rooms in urban areas and problems with managing hospital flow. Tyler Hospital, located in rural Ohio, has recently joined the Joint Commission's Continuous Service Readiness program and consequently agreed to monitor its ER waiting times. After collecting data for a random sample of 30 ER patients arriving at Tyler's ER during the last month, they found an average waiting time of 26 minutes with a standard deviation of 8.25 minutes. Further statistical analysis yielded a 95% confidence interval of 22.92 to 29.08 minutes, clear indication that Tyler's ER patients wait less than 30 minutes to see a doctor. Tyler's administration was

not only pleased with the findings, but also sure that the Joint Commission would also be impressed. Their next step was to consider ways of including this message, "95% of Tyler's ER patients can expect to wait less than the national average to see a doctor," in their advertising and promotional materials.

ETHICAL ISSUE *Interpretation of the confidence interval is incorrect and misleading (related to Item C, ASA Ethical Guidelines). The confidence interval does not provide results for individual patients. So, it is incorrect to state that 95% of individual ER patients wait less (or can expect to wait less) than 30 minutes to see a doctor.*

ETHICAL SOLUTION *Interpret the results of the confidence interval correctly, in terms of the mean waiting time and not individual patients.*

What have we learned?

We first learned to create confidence intervals and test hypotheses about proportions. Now we've turned our attention to means and learned that statistical inference for means relies on the same concepts; only the mechanics and our model have changed.

- We've learned that what we can say about a population mean is inferred from data, using the mean and standard deviation of a representative random sample.
- We've learned to describe the sampling distribution of sample means using a new model we select from the Student's t family based on our degrees of freedom.
- We've learned that our ruler for measuring the variability in sample means is the standard error:

$$SE(\bar{y}) = \frac{s}{\sqrt{n}}.$$

- We've learned to find the margin of error for a confidence interval using that standard error ruler and a critical value based on a Student's t-model.
- We've also learned to use that ruler to test hypotheses about the population mean.

Above all, we've learned that the reasoning of inference, the need to verify that the appropriate assumptions are met, and the proper interpretation of confidence intervals and P-values all remain the same regardless of whether we are investigating means or proportions.

Terms

Degrees of freedom (df)   A parameter of the Student's t-distribution that depends upon the sample size. Typically, more degrees of freedom reflects increasing information from the sample.

One-sample t-interval for the mean   A one-sample t-interval for the population mean is:

$$\bar{y} \pm t^*_{n-1} \times SE(\bar{y}) \text{ where } SE(\bar{y}) = \frac{s}{\sqrt{n}}.$$

The critical value t^*_{n-1} depends on the particular confidence level, C, that you specify and on the number of degrees of freedom, $n - 1$.

One-sample t-test for the mean   The one-sample t-test for the mean tests the hypothesis $H_0: \mu = \mu_0$ using the statistic $t_{n-1} = \dfrac{\bar{y} - \mu_0}{SE(\bar{y})}$, where $SE(\bar{y}) = \dfrac{s}{\sqrt{n}}.$

Student's t   A family of distributions indexed by its degrees of freedom. The t-models are unimodal, symmetric, and bell-shaped, but generally have fatter tails and a narrower center than the Normal model. As the degrees of freedom increase, t-distributions approach the Normal model.

Skills

PLAN

- Be able to state the assumptions required for t-tests and t-based confidence intervals.
- Know to examine your data for violations of conditions that would make inference about the population mean unwise or invalid.

- Understand that a hypothesis test can be performed with an appropriately chosen confidence interval.

- Know how to compute and interpret a *t*-test for the population mean using a statistics software package or by working from summary statistics for a sample.

- Know how to compute and interpret a *t*-based confidence interval for the population mean using a statistics software package or by working from summary statistics for a sample.

REPORT

- Be able to explain the meaning of a confidence interval for a population mean. Make clear that the randomness associated with the confidence level is a statement about the interval bounds and not about the population parameter value.

- Understand that a 95% confidence interval does not trap 95% of the sample values.

- Be able to interpret the result of a test of a hypothesis about a population mean.

- Know that we do not "accept" a null hypothesis if we cannot reject it. We say that we fail to reject it.

- Understand that the P-value of a test does not give the probability that the null hypothesis is correct.

Technology Help: Inference for Means

Statistics packages offer convenient ways to make histograms of the data. That means you have no excuse for skipping the check that the data are nearly Normal.

Any standard statistics package can compute a hypothesis test. Here's what the package output might look like in general (although no package we know gives the results in exactly this form).

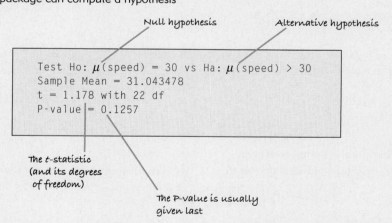

Null hypothesis   Alternative hypothesis

```
Test Ho: μ(speed) = 30 vs Ha: μ(speed) > 30
Sample Mean = 31.043478
t = 1.178 with 22 df
P-value = 0.1257
```

The *t*-statistic (and its degrees of freedom)

The P-value is usually given last

The package computes the sample mean and sample standard deviation of the variable and finds the P-value from the *t*-distribution based on the appropriate number of degrees of freedom. All modern statistics packages report P-values. The package may also provide additional information such as the sample mean, sample standard deviation, *t*-statistic value, and degrees of freedom. These are useful for interpreting the resulting P-value and telling the difference between a meaningful result and one that is merely statistically significant.

Statistics packages that report the estimated standard deviation of the sampling distribution usually label it "standard error" or "SE."

Inference results are also sometimes reported in a table. You may have to read carefully to find the values you need. Often, test results and the corresponding confidence interval bounds are given together. And often you must read carefully to find the alternative hypotheses. On the next page is an example of that kind of output.

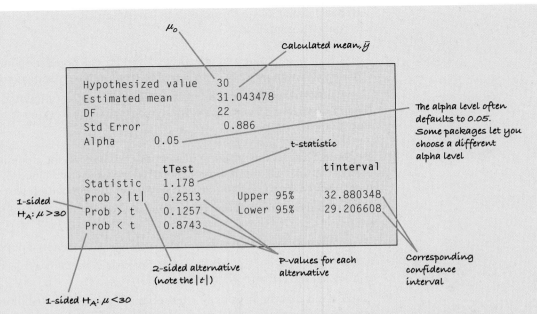

The commands to do inference for means on common statistics programs and calculators are not always obvious. (By contrast, the resulting output is usually clearly labeled and easy to read.) The guides for each program can help you start navigating.

EXCEL

Specify formulas. Find t^* with the TINV(alpha, df) function. The DDXL add-in offers both t-tests and confidence intervals.

Comments

Excel has no built in functions for finding P-values.

MINITAB

From the **Stat** menu, choose the **Basic Statistics** submenu. From that menu, choose **1-sample t....** Then fill in the dialog.

Comments

The dialog offers a clear choice between confidence interval and test.

SPSS

From the **Analyze** menu, choose the **Compare Means** submenu. From that, choose **One-Sample t-test** command.

Comments

The commands suggest neither a single mean nor an interval. But the results provide both a test and an interval.

JMP

From the **Analyze** menu, select **Distribution.** For a confidence interval, scroll down to the "Moments" section to find the interval limits. (Be sure that your variables are "Continuous" type so that this section will be available.) For a hypothesis test, click the red triangle next to the variable's name and choose **Test Mean** from the menu. Then fill in the resulting dialog.

Comments

"Moment" is a fancy statistical term for means, standard deviations, and other related statistics.

Mini Case Study Projects

Real Estate

A real estate agent is trying to understand the pricing of homes in her area, a region comprised of small to midsize towns and a small city. For each of 1200 homes recently sold in the region, the file **ch12_MCSP_Real_Estate** holds the following variables:

- *Sale Price* (in $)
- *Lot size* (size of the lot in acres)
- *Waterfront* (Yes, No)
- *Age* (in years)
- *Central Air* (Yes, No)
- *Fuel Type* (Wood, Oil, Gas, Electric, Propane, Solar, Other)
- *Condition* (1 to 5, 1 = Poor, 5 = Excellent)
- *Living Area* (living area in square feet)
- *Pct College* (% in zip code who attend a four-year college)
- *Full Baths* (number of full bathrooms)
- *Half Baths* (number of half bathrooms)
- *Bedrooms* (number of bedrooms)
- *Fireplaces* (number of fireplaces)

The agent has a family interested in a four bedroom house. Using confidence intervals, how should she advise the family on what the average price of a four bedroom house might be in this area? Compare that to a confidence interval for two bedroom homes. How does the presence of central air conditioning affect the mean price of houses in this area? Use confidence intervals and graphics to help answer that question.

Explore other questions that might be useful for the real estate agent in knowing how different categorical factors affect the sale price and write up a short report on your findings.

Donor Profiles

A philanthropic organization collects and buys data on their donor base. The full database contains about 4.5 million donors and over 400 variables collected on each, but the data set **ch12_MCSP_Donor_Profiles** is a sample of 916 donors and includes the variables:

- *Age* (in years)
- *Homeowner* (H = Yes, U = Unknown)
- *Gender* (F = Female, M = Male, U = Unknown)

- *Wealth* (Ordered categories of total household wealth from 1 = Lowest to 9 = Highest)
- *Children* (Number of children)
- *Donated Last* (0 = Did not donate to last campaign, 1 = Did donate to last campaign)
- *Amt Donated Last* ($ amount of contribution to last campaign)

The analysts at the organization want to know how much people donate on average to campaigns, and what factors might influence that amount. Compare the confidence intervals for the mean *Amt Donated Last* by those known to own their homes with those whose homeowner status is unknown. Perform similar comparisons for *Gender* and two of the *Wealth* categories. Write up a short report using graphics and confidence intervals for what you have found. (Be careful not to make inferences directly about the differences between groups. We'll discuss that in the next chapter. Your inference should be about single groups.)

(The distribution of *Amt Donated Last* is highly skewed to the right, and so the median might be thought to be the appropriate summary. But the median is $0.00 so the analysts must use the mean. From simulations, they have ascertained that the sampling distribution for the mean is unimodal and symmetric for samples larger than 250 or so. Note that small differences in the mean could result in millions of dollars of added revenue nationwide. The average cost of their solicitation is $0.67 per person to produce and mail.)

EXERCISES

1. *t*-models Using the *t* tables, software, or a calculator, estimate:

a) the critical value of *t* for a 90% confidence interval with df = 17.

b) the critical value of *t* for a 98% confidence interval with df = 88.

c) the P-value for $t \geq 2.09$ with 4 degrees of freedom.

d) the P-value for $|t| > 1.78$ with 22 degrees of freedom.

2. *t*-models, part 2. Using the *t* tables, software, or a calculator, estimate:

a) the critical value of *t* for a 95% confidence interval with df = 7.

b) the critical value of *t* for a 99% confidence interval with df = 102.

c) the P-value for $t \leq 2.19$ with 41 degrees of freedom.

d) the P-value for $|t| > 2.33$ with 12 degrees of freedom.

3. Confidence intervals. Describe how the width of a 95% confidence interval for a mean changes as the standard deviation (*s*) of a sample increases, assuming sample size remains the same.

4. Confidence intervals, part 2. Describe how the width of a 95% confidence interval for a mean changes as the sample size (*n*) increases, assuming the standard deviation remains the same.

5. Confidence intervals and sample size. A confidence interval for the price of gasoline from a random sample of 30 gas stations in a region gives the following statistics:

$$\bar{y} = \$4.49 \quad s = \$0.29$$

a) Find a 95% confidence interval for the mean price of regular gasoline in that region.

b) Find the 90% confidence interval for the mean.

c) If we had the same statistics from a sample of 60 stations, what would the 95% confidence interval be now?

6. Confidence intervals and sample size, part 2. A confidence interval for the price of gasoline from a random sample of 30 gas stations in a region gives the following statistics:

$$\bar{y} = \$4.49 \quad SE(\bar{y}) = \$0.06$$

a) Find a 95% confidence interval for the mean price of regular gasoline in that region.

b) Find the 90% confidence interval for the mean.

c) If we had the same statistics from a sample of 60 stations, what would the 95% confidence interval be now?

7. Marketing livestock feed. A feed supply company has developed a special feed supplement to see if it will promote weight gain in livestock. Their researchers report that the 77 cows studied gained an average of 56 pounds and that a 95% confidence interval for the mean weight gain this supplement produces has a margin of error of ±11 pounds. Staff in their marketing department wrote the following conclusions. Did anyone interpret the interval correctly? Explain any misinterpretations.

a) 95% of the cows studied gained between 45 and 67 pounds.

b) We're 95% sure that a cow fed this supplement will gain between 45 and 67 pounds.

c) We're 95% sure that the average weight gain among the cows in this study was between 45 and 67 pounds.

d) The average weight gain of cows fed this supplement is between 45 and 67 pounds 95% of the time.

e) If this supplement is tested on another sample of cows, there is a 95% chance that their average weight gain will be between 45 and 67 pounds.

8. Meal costs. A company is interested in estimating the costs of lunch in their cafeteria. After surveying employees, the staff calculated that a 95% confidence interval for the mean amount of money spent for lunch over a period of six months is ($780, $920). Now the organization is trying to write its report and considering the following interpretations. Comment on each.

a) 95% of all employees pay between $780 and $920 for lunch.
b) 95% of the sampled employees paid between $780 and $920 for lunch.
c) We're 95% sure that employees in this sample averaged between $780 and $920 for lunch.
d) 95% of all samples of employees will have average lunch costs between $780 and $920.
e) We're 95% sure that the average amount all employees pay for lunch is between $780 and $920.

9. CEO compensation. A sample of 20 CEOs from the Forbes 500 shows total annual compensations ranging from a minimum of $0.1 to $62.24 million. The average for these 20 CEOs is $7.946 million. The histogram and boxplot are as follows:

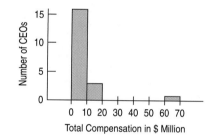

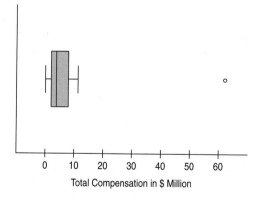

Based on these data, a computer program found that a 95% confidence interval for the mean annual compensation of all Forbes 500 CEOs is (1.69, 14.20) $M. Why should you be hesitant to trust this confidence interval?

10. Credit card charges. A credit card company takes a random sample of 100 cardholders to see how much they charged on their card last month. A histogram and boxplot are as follows:

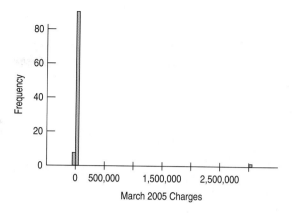

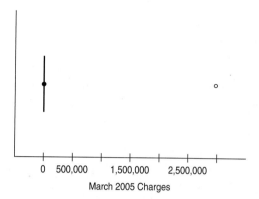

A computer program found that the 95% confidence interval for the mean amount spent in March 2005 is (−$28366.84, $90691.49). Explain why the analysts didn't find the confidence interval useful, and explain what went wrong.

11. Parking. Hoping to lure more shoppers downtown, a city builds a new public parking garage in the central business district. The city plans to pay for the structure through parking fees. For a random sample of 44 weekdays, daily fees collected averaged $126, with a standard deviation of $15.

a) What assumptions must you make in order to use these statistics for inference?
b) Find a 90% confidence interval for the mean daily income this parking garage will generate.
c) Explain in context what this confidence interval means.
d) Explain what 90% confidence means in this context.
e) The consultant who advised the city on this project predicted that parking revenues would average $128 per day. Based on your confidence interval, what do you think of the consultant's prediction? Why?

12. Housing 2008 was a difficult year for the economy. There were a large number of foreclosures of family homes. In one large community, realtors randomly sampled 36 bids from potential buyers to determine the average loss in home value. The sample showed the average loss was $11,560 with a standard deviation of $1500.

a) What assumptions and conditions must be checked before finding a confidence interval? How would you check them?

b) Find a 95% confidence interval for the mean loss in value per home.

c) Interpret this interval and explain what 95% confidence means.

d) Suppose nationally, the average loss in home values at this time was $10,000. Do you think the loss in the sampled community differs significantly from the national average? Explain.

13. Parking, part 2. Suppose that for budget planning purposes the city in Exercise 11 needs a better estimate of the mean daily income from parking fees.

a) Someone suggests that the city use its data to create a 95% confidence interval instead of the 90% interval first created. How would this interval be better for the city? (You need not actually create the new interval.)

b) How would the 95% confidence interval be worse for the planners?

c) How could they achieve a confidence interval estimate that would better serve their planning needs?

14. Housing, part 2. In Exercise 12, we found a 95% confidence interval to estimate the loss in home values.

a) Suppose the standard deviation of the losses was $3000 instead of the $1500 used for that interval. What would the larger standard deviation do to the width of the confidence interval (assuming the same level of confidence)?

b) Your classmate suggests that the margin of error in the interval could be reduced if the confidence level were changed to 90% instead of 95%. Do you agree with this statement? Why or why not?

c) Instead of changing the level of confidence, would it be more statistically appropriate to draw a bigger sample?

15. State budgets. States that rely on sales tax for revenue to fund education, public safety, and other programs often end up with budget surpluses during economic growth periods (when people spend more on consumer goods) and budget deficits during recessions (when people spend less on consumer goods). Fifty-one small retailers in a state with a growing economy were recently sampled. The sample showed a mean increase of $2350 in additional sales tax revenue collected per retailer compared to the previous quarter. The sample standard deviation = $425.

a) Find a 95% confidence interval for the mean increase in sales tax revenue.

b) What assumptions have you made in this inference? Do you think the appropriate conditions have been satisfied?

c) Explain what your interval means and provide an example of what it does not mean.

16. State budgets, part 2. Suppose the state in Exercise 15 sampled 16 small retailers instead of 51, and for the sample of 16, the sample mean increase again equaled $2350 in additional sales tax revenue collected per retailer compared to the previous quarter. Also assume the sample standard deviation = $425.

a) What is the standard error of the mean increase in sales tax revenue collected?

b) What happens to the accuracy of the estimate when the interval is constructed using the smaller sample size?

c) Find and interpret a 95% confidence interval.

d) How does the margin of error for the interval constructed in Exercise 15 compare with the margin of error constructed in this exercise? Explain statistically how sample size changes the accuracy of the constructed interval. Which sample would you prefer if you were a state budget planner? Why?

17. Departures. What are the chances your flight will leave on time? The U.S. Bureau of Transportation Statistics of the Department of Transportation publishes information about airline performance. Here are a histogram and summary statistics for the percentage of flights departing on time each month from 1995 through 2006.

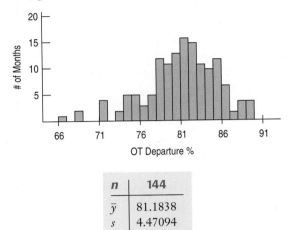

n	144
$\bar{y}$	81.1838
s	4.47094

There is no evidence of a trend over time. (The correlation of On Time Departure% with time is $r = -0.016$.)

a) Check the assumptions and conditions for inference.

b) Find a 90% confidence interval for the true percentage of flights that depart on time.

c) Interpret this interval for a traveler planning to fly.

18. Late arrivals. Will your flight get you to your destination on time? The U.S. Bureau of Transportation Statistics reported the percentage of flights that were late each month from 1995 through 2006. Here's a histogram, along with some summary statistics:

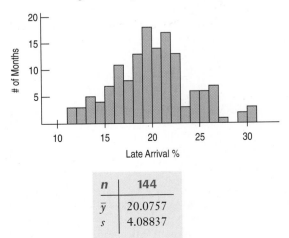

n	144
$\bar{y}$	20.0757
s	4.08837

We can consider these data to be a representative sample of all months. There is no evidence of a time trend ($r = -0.07$).

a) Check the assumptions and conditions for inference about the mean.

b) Find a 99% confidence interval for the true percentage of flights that arrive late.

c) Interpret this interval for a traveler planning to fly.

19. E-commerce. A market researcher at a major clothing company that has traditionally relied on catalog mail-order sales decides to investigate whether the amount of online sales has changed. She compares the mean monthly online sales of the past several months with a historical figure for mean monthly sales for online purchases. She gets a P-value of 0.01. Explain in this context what the 1% represents.

20. Performance standards. The United States Golf Association (USGA) sets performance standards for golf balls. For example, the initial velocity of the ball may not exceed 250 feet per second when measured by an apparatus approved by the USGA. Suppose a manufacturer introduces a new kind of ball and provides a randomly selected sample of balls for testing. Based on the mean speed in the sample, the USGA comes up with a P-value of 0.34. Explain in this context what the 34% represents.

21. Social Security payments. The average monthly Social Security benefit for widows and widowers in 2005 was $967 (*Statistical Abstract of the United States*, U.S. Census Bureau). Payments vary from region to region. In Texas, the 2005 average monthly benefit equaled $940. A social welfare advocate from a rural Texas county believed the 2005 average Social Security benefit for widows and widowers differed significantly from the overall Texas average. To test this assumption, the advocate randomly sampled 100 widow/widower monthly benefit payments and found the sample mean = $915 with a standard deviation of $90.

a) Find and interpret a 95% confidence interval estimate for the sample mean 2005 Social Security widow/widower benefit in the rural Texas county.

b) In a hypothesis test performed to determine whether the county's average payment was different, the test was rejected with a P-value = .007 (using the same sample results shown above and a level of significance = .05). Explain how the confidence interval constructed in part a is consistent with the hypothesis test results. Your discussion should include level of confidence, the interval bounds, the P-value, and the hypothesis test decision.

22. Social Security payments, part 2. In a neighboring county, a newspaper wrote that widow/widower Social Security benefits were significantly lower in the county than in all the other counties of Texas. The newspaper reported that the average 2005 monthly benefit for widows and widowers was $900, based on a random sample of 100 with standard deviation of $90.

a) Find and interpret a 95% confidence interval estimate for the sample mean Social Security widow/widower benefit in this county.

b) Is this county's average widow/widower benefit different from the Texas average of $940?

23. TV safety. The manufacturer of a metal stand for home TV sets must be sure that its product will not fail under the weight of a typical TV set. Since some larger sets weigh nearly 300 pounds, the company's safety inspectors have set a standard of ensuring that the stands can support an average of 500 pounds. Their inspectors regularly subject a random sample of the stands to increasing weight until they fail. They test the hypothesis H_0: $\mu = 500$ against H_A: $\mu > 500$, using the level of significance $\alpha = 0.01$. If the sample of stands fails this safety test, the inspectors will not certify the product for sale to the general public.

a) Is this an upper-tail or lower-tail test? In the context of the problem, why do you think this is important?

b) Explain what will happen if the inspectors commit a Type I error.

c) Explain what will happen if the inspectors commit a Type II error.

24. Quality control. During an angiogram, heart problems can be examined via a small tube (a catheter) threaded into the heart from a vein in the patient's leg. It's important that the company who manufactures the catheter maintain a catheter diameter of 2.00 mm. (The standard deviation is quite small.) Each day, quality control personnel make several measurements to test H_0: $\mu = 2.00$ against H_A: $\mu \neq 2.00$ at a significance level of $\alpha = 0.05$. If they discover a problem, they will stop the manufacturing process until it is corrected.

a) Is this a one-sided or two-sided test? In the context of the problem, why do you think this is important?

b) Explain in this context what happens if the quality control people commit a Type I error.

c) Explain in this context what happens if the quality control people commit a Type II error.

25. TV safety, revisited. The manufacturer of the metal TV stands in Exercise 23 is thinking of revising its safety test.

a) If the company's lawyers are worried about being sued for selling an unsafe product, should they increase or decrease the value of α? Explain.

b) In this context, what is meant by the power of the test?

c) If the company wants to increase the power of the test, what options does it have? Explain the advantages and disadvantages of each option.

26. Quality control, part 2. The catheter company in Exercise 24 is reviewing its testing procedure.

a) Suppose the significance level is changed to $\alpha = 0.01$. Will the probability of Type II error increase, decrease, or remain the same?

b) What is meant by the power of the test the company conducts?

c) Suppose the manufacturing process is slipping out of proper adjustment. As the actual mean diameter of the catheters produced gets farther and farther above the desired 2.00 mm, will the power of the quality control test increase, decrease, or remain the same?

d) What could they do to improve the power of the test?

27. E-commerce, part 2. The average age of online consumers a few years ago was 23.3 years. As older individuals gain confidence with the Internet, it is believed that the average age has increased. We would like to test this belief.

a) Write appropriate hypotheses.

b) We plan to test the null hypothesis by selecting a random sample of 40 individuals who have made an online purchase during 2007. Do you think the necessary assumptions for inference are satisfied? Explain.

c) The online shoppers in our sample had an average age of 24.2 years, with a standard deviation of 5.3 years. What's the P-value for this result?

d) Explain (in context) what this P-value means.

e) What's your conclusion?

28. Fuel economy. A company with a large fleet of cars hopes to keep gasoline costs down and sets a goal of attaining a fleet average of at least 26 miles per gallon. To see if the goal is being met, they check the gasoline usage for 50 company trips chosen at random, finding a mean of 25.02 mpg and a standard deviation of 4.83 mpg. Is this strong evidence that they have failed to attain their fuel economy goal?

a) Write appropriate hypotheses.

b) Are the necessary assumptions to perform inference satisfied?

c) Test the hypothesis and find the P-value.

d) Explain what the P-value means in this context.

e) State an appropriate conclusion.

T 29. Pricing for competitiveness. SLIX wax is developing a new high performance fluorocarbon wax for cross country ski racing designed to be used under a wide variety of conditions. In order to justify the price marketing wants, the wax needs to be very fast. Specifically, the mean time to finish their standard test course should be less than 55 seconds for the former Olympic champion who is now their consultant. To test it, the consultant will ski the course 8 times.

a) The champion's times are 56.3, 65.9, 50.5, 52.4, 46.5, 57.8, 52.2, and 43.2 seconds to complete the test course. Should they market the wax? Explain.

b) Suppose they decide not to market the wax after the test, but it turns out that the wax really does lower the champion's average time to less than 55 seconds. What kind of error have they made? Explain the impact to the company of such an error.

T 30. Popcorn. Pop's Popcorn, Inc. needs to determine the optimum power and time settings for their new licorice-flavored microwave popcorn. They want to find a combination of power and time that delivers high-quality popcorn with less than 10% of the kernels left unpopped, on average—a value that their market research says is demanded by their customers. Their research department experiments with several settings and determines that power 9 at 4 minutes is optimum. Their tests confirm that this setting meets the less than 10% requirement. They change the instructions on the box and promote a new money back guarantee of less than 10% unpopped kernels.

a) If, in fact, the setting results in more than 10% kernels unpopped, what kind of error have they made? What will the consequence be for the company?

b) To reduce the risk of making an error, the president (Pop himself) tells them to test 8 more bags of popcorn (selected at random) at the specified setting. They find the following percentage of unpopped kernels: 7, 13.2, 10, 6, 7.8, 2.8, 2.2, 5.2. Does this provide evidence that the setting meets their goal of less than than 10% unpopped? Explain.

31. False claims? A manufacturer claims that a new design for a portable phone has increased the range to 150 feet, allowing many customers to use the phone throughout their homes and yards. An independent testing laboratory found that a random sample of 44 of these phones worked over an average distance of 142 feet, with a standard deviation of 12 feet. Is there evidence that the manufacturer's claim is false?

32. False claims? part 2. The makers of *Abolator*, a portable exercise device that sells for $149.95, claim that using their machine for only 6 minutes a day results in an average weight loss of 8 pounds during the first week. A consumer organization recruits 30 volunteers to use the product according to the manufacturer's recommendations and finds an average weight loss of 4.7 pounds with a standard deviation of 6.1 pounds. Is there evidence that the makers of the *Abolator* claim is false?

T 33. Chips Ahoy. In 1998, as an advertising campaign, the Nabisco Company announced a "l000 Chips Challenge," claiming that every 18-ounce bag of their Chips Ahoy cookies contained at least 1000 chocolate chips. Dedicated Statistics students at the Air Force Academy (no kidding) purchased some randomly selected bags of cookies, and counted the chocolate chips. Some of their data are given below. (*Chance*, 12, no. 1 [1999])

```
1219 1214 1087 1200 1419 1121 1325 1345
1244 1258 1356 1132 1191 1270 1295 1135
```

a) Check the assumptions and conditions for inference. Comment on any concerns you have.

b) Create a 95% confidence interval for the average number of chips in bags of Chips Ahoy cookies.

c) What does this evidence say about Nabisco's claim? Use your confidence interval to test an appropriate hypothesis and state your conclusion.

T 34. Consumer Reports. *Consumer Reports* tested 14 brands of vanilla yogurt and found the following numbers of calories per serving: 160, 200, 220, 230, 120, 180, 140, 130, 170, 190, 80, 120, 100, and 170.

a) Check the assumptions and conditions for inference.

b) Create a 95% confidence interval for the average calorie content of vanilla yogurt.

c) A diet guide claims that you will get 120 calories from a serving of vanilla yogurt. What does this evidence indicate? Use your confidence interval to test an appropriate hypothesis and state your conclusion.

35. Investment. Investment style plays a role in constructing a mutual fund. Many individual stocks can be grouped into two distinct groups: Growth and Value. A Growth stock is one with high earning potential and often pays little or no dividends to shareholders. Conversely, Value stocks are commonly viewed as steady or more conservative stocks with a lower earning potential. A family is trying to decide what type of funds to invest in. An independent advisor claims that Value Mutual Funds provided an annualized return of greater than 8% over a recent 5-year period. Below are the summary statistics for the 5-year return for a random sample of such Value funds.

Variable	N	Mean	SE Mean	StDev
5 yr Return	35	8.418	0.493	2.916

	Minimum	Q1	Median	Q3	Maximum
	2.190	6.040	7.980	10.840	14.320

Test the hypothesis that the mean 5-year return for value funds is greater than 8%, assuming a significance level of 5%. What does this evidence say about the portfolio managers' claim that the annualized 5-year return was greater than 8%? State your conclusion.

36. Manufacturing. A tire manufacturer is considering a newly designed tread pattern for its all-weather tires. Tests have indicated that these tires will provide better gas mileage and longer tread life. The last remaining test is for braking effectiveness. The company hopes the tire will allow a car traveling at 60 mph to come to a complete stop within an average of 125 feet after the brakes are applied. They will adopt the new tread pattern unless there is strong evidence that the tires do not meet this objective. The distances (in feet) for 10 stops on a test track were 129, 128, 130, 132, 135, 123, 102, 125, 128, and 130. Should the company adopt the new tread pattern? Test an appropriate hypothesis and state your conclusion. Explain how you dealt with the outlier and why you made the recommendation you did.

37. Collections. Credit card companies lose money on cardholders who fail to pay their minimum payments. They use a variety of methods to encourage their delinquent cardholders to pay their credit card balances, such as letters, phone calls and eventually the hiring of a collection agency. To justify the cost of using the collection agency, the agency must collect an average of at least $200 per customer. After a trial period during which the agency attempted to collect from a random sample of 100 delinquent cardholders, the 90% confidence interval on the mean collected amount was reported as ($190.25, $250.75). Given this, what recommendation(s) would you make to the credit card company about using the collection agency?

38. Free gift. A philanthropic organization sends out "free gifts" to people on their mailing list in the hope that the receiver will respond by sending back a donation. Typical gifts are mailing labels, greeting cards, or post cards. They want to test out a new gift that costs $0.50 per item to produce and mail. They mail it to a "small" sample of 2000 customers and find a 90%

confidence interval of the mean donation to be ($0.489, $0.879). As a consultant, what recommendation(s) would you make to the organization about using this gift?

39. Collections, part 2. The owner of the collection agency in Exercise 37 is quite certain that they can collect more than $200 per customer on average. He urges that the credit card company run a larger trial. Do you think a larger trial might help the company make a better decision? Explain.

40. Free gift, part 2. The philanthropic organization of Exercise 38 decided to go ahead with the new gift. In mailings to 98,000 prospects, the new mailing yielded an average of $0.78. If they had decided based on their initial trial *not* to use this gift, what kind of error would they have made? What aspects of their initial trial might have suggested to you (as their consultant) that a larger trial would be worthwhile?

41. Batteries. A battery company claims that its batteries last an average of 100 hours under normal use. There have been several complaints that the batteries don't last that long, so an independent testing agency is engaged to test them. For the 16 batteries they tested, the mean lifetime was 97 hours with a standard deviation of 12 hours.

a) What are the null and alternative hypotheses?
b) A consumer advocate (who does not know statistics) says that 97 hours is a lot less than the advertised 100 hours, so we should reject the company's claim. Explain to him the problem with doing that.
c) What assumptions must we make in order to proceed with inference?
d) At a 5% level of significance, what do you conclude?
e) Suppose that, in fact, the average life of the company's batteries is only 98 hours. Has an error been made in part d? If so, what kind?

42. Pet breeding. Fancy pets are big business, so a young entrepreneur (age 12) decided to breed golden hamsters, a type well known to pet stores and collectors. (Oddly enough, nearly all the golden hamsters in captivity are descendants of one litter found in Syria in 1930.) Of 47 recent litters, there were an average of 7.27 baby hamsters with a standard deviation of 2.5 hamsters per litter.

a) Find and interpret a 90% confidence interval for the mean litter size.
b) How much smaller or larger would the margin of error be for a 99% confidence interval? Explain.
c) Based on these statistics, how many litters would we need to estimate the mean litter size to within one baby hamster with 95% confidence?

43. Fish production. Farmed salmon is much cheaper to bring to market than salmon caught in the wild, but consumers are concerned about several issues recently publicized about farmed salmon including the type of food they are fed and the contaminants found in their meat (see www.healthcastle.com). Among the contaminants are such compounds as polychlorinated biphenyls (PCBs), dioxins, toxaphene, dieldrin, hexachlorobenzene, lindane, heptachlor epoxide, cis-nonachlor, trans-nonachlor, gamma-chlordane, alpha-chlordane, Mirex, endrin, and DDT.

The EPA recommends that salmon contain no more than 0.08 ppm of the insecticide Mirex. A local environmental group is considering a boycott of salmon if it exceeds 0.08 ppm. A 95% confidence interval from a sample of farmed salmon from a random sample of 150 different salmon farms (*Science*, 9, January 2004) is found to be (0.0834 to 0.0992) ppm. The data were unimodal and symmetric with no outliers.

a) Is there evidence that the farms are producing salmon with mean Mirex contamination higher than the EPA recommended amount? Your explanation should discuss the confidence level, P-value and the decision.

b) Discuss the two types of errors that can be made in this decision in the context of a business decision of whether to prohibit the producers from selling their salmon.

44. Downloading speed. A student recently bought an anti-spyware program for her computer to help increase performance. Before she installed the program, her mean download speed was 480 kbps (kilobits per second). The program cost $29.95 and she wants to see if her speed has increased.

She tried a band width speed test (www.bandwidthplace.com) 16 different times and found a 90% confidence interval to be (482.6, 505.9) kbps. She plotted the 16 speeds and found the data to be unimodal and roughly symmetric with no obvious outliers.

a) Is there evidence to suggest that the computer's downloading speed is greater than 480 kbps? Your explanation should discuss the confidence level, P-value, and the decision.

b) If she finds out later that the mean is really 465 kbps, what kind of error has she made?

c) Do you think her decision to buy the software was the right one based on the data?

45. Computer lab fees. The technology committee has stated that the average time spent by students per lab visit has increased, and the increase supports the need for increased lab fees. To substantiate this claim, the committee randomly samples 12 student lab visits and notes the amount of time spent using the computer. The times in minutes are as follows:

Time	Time
52	74
57	53
54	136
76	73
62	8
52	62

a) Plot the data. Are any of the observations outliers? Explain.

b) The previous mean amount of time spent using the lab computer was 55 minutes. Test the hypothesis that the mean is now higher than 55 minutes at $\alpha = .05$. What is your conclusion?

c) If outliers exist, eliminate the outlier(s) and retest the hypothesis. What is your conclusion?

d) Discuss the statistical implications of eliminating outliers. Why might some researchers disagree with deleting outlying observations from a data set?

46. Cell phone batteries. A company that produces cell phones claims its standard phone battery lasts longer on average than other batteries in the market. To support this claim, the company publishes an ad reporting the results of a recent experiment showing that under normal usage, their batteries last at least 35 hours. To investigate this claim, a consumer advocacy group asked the company for the raw data. The company sends the group the following results:

35, 34, 32, 31, 34, 34, 32, 33, 35, 55, 32, 31

a) Test an appropriate hypothesis and state your conclusion.

b) Explain how you dealt with the outlier, and why you made the recommendation you did.

47. Growth and air pollution. Government officials have difficulty attracting new business to communities with troubled reputations. Nevada has been one of the fastest growing states in the country for a number of years. Accompanying the rapid growth are massive new construction projects. Since Nevada has a dry climate, the construction creates visible dust pollution. High pollution levels may paint a less than attractive picture of the area, and can also result in fines levied by the federal government. As required by government regulation, researchers continually monitor pollution levels. In the most recent test of pollution levels, 121 air samples were collected. The dust particulate levels must be reported to the federal regulatory agencies. In the report sent to the federal agency, it was noted that the mean particulate level = 57.6 micrograms/cubic liter of air, and the 95% confidence interval estimate is (52.06 mg to 63.07 mg). A graph of the distribution of the particulate amounts was also included and is shown below.

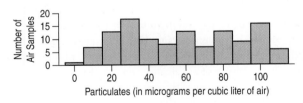

a) Discuss the assumptions and conditions for using Student's *t* inference methods with these data.

b) Do you think the confidence interval noted in the report is valid? Briefly explain why or why not.

48. Convention revenues. At one time, Nevada was the only U.S. state that allowed gambling. Although gambling continues to be one of the major industries in Nevada, the proliferation of legalized gambling in other areas of the country has required state and local governments to look at other growth possibilities. The convention and visitor's authorities in many Nevada cities actively recruit national conventions that bring thousands of visitors to the state. Various demographic and economic data are collected from surveys given to convention attendees. One statistic of interest is the amount visitors spend on slot machine gambling. Nevada often reports the slot machine expenditure as amount spent per hotel guest room. A recent survey of 500 visitors asked how much they spent on gambling. The average expenditure per room was $180.

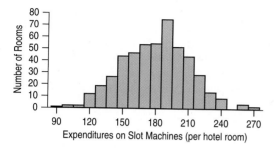

Casinos will use the information reported in the survey to estimate slot machine expenditure per hotel room. Do you think the estimates produced by the survey will accurately represent expenditures? Explain using the statistics reported and graph shown.

49. Traffic speed. Police departments often try to control traffic speed by placing speed-measuring machines on roads that tell motorists how fast they are driving. Traffic safety experts must determine where machines should be placed. In one recent test, police recorded the average speed clocked by cars driving on one busy street close to an elementary school. For a sample of 25 speeds, it was determined that the average amount over the speed limit for the 25 clocked speeds was 11.6 mph with a standard deviation of 8 mph. The 95% confidence interval estimate for this sample is 8.30 mph to 14.90 mph.

a) What is the margin of error for this problem?

b) The researchers commented that the interval was too wide. Explain specifically what should be done to reduce the margin of error to no more than ±2 mph.

50. Traffic speed II. The speed-measuring machines must measure accurately to maximize effectiveness in slowing traffic. The accuracy of the machines will be tested before placement on city streets. To ensure that error rates are estimated accurately, the researchers want to take a large enough sample that will insure usable and accurate interval estimates of how much the machines may be off in measuring actual speeds. Specially, the researchers want the margin of error for a single speed measurement to be no more than ±1.5 mph.

a) Discuss how the researchers may obtain a reasonable estimate of the standard deviation of error in the measured speeds.

b) Suppose the standard deviation for the error in the measured speeds equals 4 mph. At 95% confidence, what sample size should be taken to ensure that the margin of error is no larger than ±1.0 mph?

51. Tax audits I. Certified public accountants are often required to appear with clients if the IRS audits the client's tax return. Some accounting firms give the client an option to pay a fee when the tax return is completed that guarantees tax advice and support from the accountant if the client were audited. The fee is charged up front like an insurance premium and is less than the amount that would be charged if the client were later audited and then decided to ask the firm for assistance during the audit. A large accounting firm is trying to determine what fee to charge for next year's returns. In previous years, the actual mean cost to the firm for attending a client audit session was $650. To determine if this cost has changed, the firm randomly samples 32 client audit fees. The sample mean audit cost was $680 with a standard deviation of $75.

a) Develop a 95% confidence interval estimate for the mean audit cost.

b) Perform the appropriate test to determine if the mean audit cost is now different from the historical mean of $650. Use a .05 level of significance.

c) Comment on how the confidence interval estimate supports the results of the hypothesis test.

T 52. Tax audits II. While reviewing the sample of audit fees, a senior accountant for the firm notes that the fee charged by the firm's accountants depends on the complexity of the return. A comparison of actual charges therefore might not provide the information needed to set next year's fees. To better understand the fee structure, the senior accountant requests a new sample that measures the time the accountants spent on the audit. Last year, the average hours charged per client audit was 3.25 hours. A new sample of 10 audit times shows the following times in hours:

4.2, 3.7, 4.8, 2.9, 3.1, 4.5, 4.2, 4.1, 5.0, 3.4

a) Assume the conditions necessary for inference are met. Find a 90% confidence interval estimate for the mean audit time.

b) Perform the appropriate test to determine if the mean audit time for this year's audits is significantly different from last year's 3.25 hours. Use α = .10.

c) Comment on how the confidence interval estimate supports the results of the hypothesis test.

T 53. Wind power. Should you generate electricity with your own personal wind turbine? That depends on whether you have enough wind on your site. To produce enough energy, your site should have an annual average wind speed of at least 8 miles per hour, according to the Wind Energy Association. One candidate site was monitored for a year, with wind speeds recorded every 6 hours. A total of 1114 readings of wind speed averaged 8.019 mph with a standard deviation of 3.813 mph. You've been asked to make a statistical report to help the landowner decide whether to place a wind turbine at this site.

a) Discuss the assumptions and conditions for using Student's *t* inference methods with these data. Here are some plots that may help you decide whether the methods can be used:

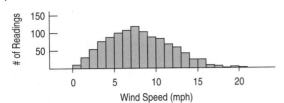

b) What would you tell the landowner about whether this site is suitable for a small wind turbine? Explain

54. Real estate crash? After the sub-prime crisis of late 2007, real estate prices fell almost everywhere in the U.S. In 2006–2007 before the crisis, the average selling price of homes in a region in upstate New York was $191,300. A real estate agency wants to know how much the prices have fallen since then. They collect a sample of 1231 homes in the region and find the average asking price to be $178,613.50 with a standard deviation of $92,701.56. You have been retained by the real estate agency to report on the current situation.

a) Discuss the assumptions and conditions for using *t*-methods for inference with these data. Here are some plots that may help you decide what to do.

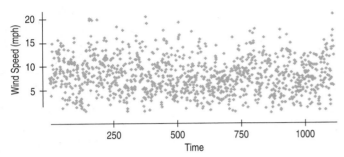

b) What would you report to the real estate agency about the current situation?

JUST CHECKING ANSWERS

1 Questions on the short form are answered by everyone in the population. This is a census, so means or proportions *are* the true population values. The long forms are just given to a sample of the population. When we estimate parameters from a sample, we use a confidence interval to take sample-to-sample variability into account.

2 They don't know the population standard deviation, so they must use the sample SD as an estimate. The additional uncertainty is taken into account by *t*-models.

3 The margin of error for a confidence interval for a mean depends, in part, on the standard error:

$$SE(\bar{y}) = \frac{s}{\sqrt{n}}$$

Since *n* is in the denominator, smaller sample sizes generally lead to larger SEs and correspondingly wider intervals. Because long forms are sampled at the same rate of one in every six or seven households throughout the country, samples will be smaller in less populous areas and result in wider confidence intervals.

4 The critical values for *t* with fewer degrees of freedom would be slightly larger. The $\sqrt{n}$ part of the standard error changes a lot, making the SE much larger. Both would increase the margin of error. The smaller sample is one fourth as large, so the confidence interval would be roughly twice as wide.

5 We expect 95% of such intervals to cover the true value, so 5 of the 100 intervals might be expected to miss.

6 The standard error would be likely to decrease if we have a larger sample size.

Comparing Two Means

Visa Global Organization

Today, more than one billion people and 24 million merchants use the Visa card in 170 countries worldwide. But back in the early 1950s when the idea of cashless transactions first took hold, only Diners Club and some retailers, notably oil companies, issued charge cards. The vast majority of purchases were made by cash or personal check. Bank of America pioneered its BankAmericard program in Fresno, California, in 1958, and American Express issued the first plastic card in 1959. The idea of a credit "card" really gained momentum a decade later when a group of banks formed a joint venture to create a centralized system of payment. National BankAmericard, Inc. (NBI) took ownership of the credit card system in 1970 and for simplicity and marketability changed its name to Visa in 1976. (The name Visa is pronounced nearly the same way in every language.)

That year, Visa processed 679,000 transactions—a volume that is processed on average every four minutes today. As technology changed, so did the credit card industry. By 1986, cardholders were able to use their Visa cards to get cash from ATMs.

Now a global organization, Visa is divided into six regional entities: Visa Asia Pacific, Visa Canada, Visa Central & Eastern Europe, Visa Middle East & Africa, Visa Europe, Visa Latin American & Caribbean, and Visa USA. The Visa system is currently able to handle a load of about 6800 transactions a second. That was nearly surpassed on December 23, 2005, during the height of the Christmas season, when Visa processed an average of nearly 6400 transaction messages every second.[1]

The credit card business can be extremely profitable. In 2003, card issuers worldwide made $2.5 billion per month before taxes. The typical American household has eight cards with a total of $7500 debt on them, usually at rates much higher than conventional loans and mortgages. Not surprisingly, the credit card business is also intensely competitive. Rival banks and lending agencies are constantly trying to create new products and offers to win new customers, keep current customers, and provide incentives for current customers to charge more on their cards.

Are some credit card promotions more effective than others? For example, do customers spend more using their credit card if they know they will be given "double miles" or "double coupons" toward flights, hotel stays, or store purchases? To answer questions such as this, credit card issuers often perform experiments on a sample of customers, making them an *offer* of an incentive, while other customers receive no offer. Promotions cost the company money, so the company needs to estimate the size of any increased revenue to judge whether it is sufficient to cover their expenses. By comparing the performance of the two offers on the sample, they can decide whether the new offer would provide enough potential profit if they were to "roll it out" and offer it to their entire customer base.

Experiments that compare two groups are common throughout both science and industry. Other applications include comparing the effects of a new drug with the traditional therapy, the fuel efficiency of two car engine designs, or the sales of new products on two different customer segments. Usually the experiment is carried out on a subset of the population, often a much smaller subset. Using statistics, we can make statements about whether the means of the two groups differ in the population at large, and how large that difference might be.

13.1 Testing Differences Between Two Means

The natural display for comparing the means of two groups is side-by-side boxplots (see Figure 13.1). For the credit card promotion, the company judges performance by comparing the *mean* spend lift (the change in spending from before receiving the promotion to after receiving it) for the two samples. If the difference in spend lift between the group that received the promotion and the group that didn't is high

---

[1] Source: © 2006 Visa International Service Association, www.corporate.visa.com.

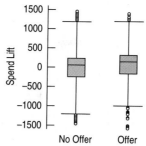

Figure 13.1 *Side-by-side boxplots show a small increase in spending for the group that received the promotion.*

enough, this will be viewed as evidence that the promotion worked. Looking at the two boxplots, it's not obvious that there's much of a difference. Can we conclude that the slight increase seen for those who received the promotion is more than just random fluctuation? We'll need statistical inference.

For two groups, the statistic of interest is the difference in the observed means of the offer and no offer groups: $\bar{y}_o - \bar{y}_n$. We've offered the promotion to a random sample of cardholders and used another sample of cardholders, who got no special offer, as a control group. We know what happened in our samples, but what we'd really like to know is the difference of the means in the population at large: $\mu_o - \mu_n$.

We compare two means in much the same way as we compared a single mean to a hypothesized value. But now the population model parameter of interest is the *difference* between the means. In our example, it's the true difference between the mean spend lift for customers offered the promotion and for customers for whom no offer was made. We estimate the difference with $\bar{y}_o - \bar{y}_n$. How can we tell if a difference we observe in the sample means indicates a real difference in the underlying population means? We'll need to know the sampling distribution model and standard deviation of the difference. Once we know those, we can build a confidence interval and test a hypothesis just as we did for a single mean.

We have data on 500 randomly selected customers who were offered the promotion and another randomly selected 500 who were not. It's easy to find the mean and standard deviation of the spend lift for each of these groups. From these, we can find the standard deviations of the means, but that's not what we want. We need the standard deviation of the *difference* in their means. For that, we can use a simple rule: *If the sample means come from independent samples, the variance of their sum or difference is the sum of their variances.*

Variances Add for Sums *and* Differences At first, it may seem that this can't be true for differences as well as for sums. Here's some intuition about why variation increases even when we subtract two random quantities. Grab a full box of cereal. The label claims that it contains 16 ounces of cereal. We know that's not exact. There's a random quantity of cereal in the box with a mean (presumably) of 16 ounces and some variation from box to box. Now pour a 2-ounce serving of cereal into a bowl. Of course, your serving isn't exactly 2 ounces. There's some variation there, too. How much cereal would you guess was left in the box? Can you guess as accurately as you could for the full box? The mean should be 14 ounces. But does the amount left in the box have *less* variation than it did before you poured your serving? Almost, certainly not! *After* you pour your bowl, the amount of cereal in the box is still a random quantity (with a smaller mean than

before), but you've made it *more variable* because of the uncertainty in the amount you poured. However, notice that we don't add the *standard deviations* of these two random quantities. As we'll see, it's the *variance* of the amount of cereal left in the box that's the sum of the two variances.

As long as the two groups are independent, we find the standard deviation of the *difference* between the two sample means by adding their variances and then taking the square root:

$$SD(\bar{y}_1 - \bar{y}_2) = \sqrt{Var(\bar{y}_1) + Var(\bar{y}_2)}$$

$$= \sqrt{\left(\frac{\sigma_1}{\sqrt{n_1}}\right)^2 + \left(\frac{\sigma_2}{\sqrt{n_2}}\right)^2}$$

$$= \sqrt{\frac{\sigma_1^2}{n_1} + \frac{\sigma_2^2}{n_2}}.$$

Of course, usually we don't know the true standard deviations of the two groups, σ_1 and σ_2, so we substitute the estimates, s_1 and s_2, and find a *standard error:*

$$SE(\bar{y}_1 - \bar{y}_2) = \sqrt{\frac{s_1^2}{n_1} + \frac{s_2^2}{n_2}}.$$

Just as we did for one mean, we'll use the standard error to see how big the difference really is. You shouldn't be surprised that, just as for a single mean, the ratio of the difference in the means to the standard error of that difference has a sampling model that follows a Student's *t* distribution.

An Easier Rule?

The formula for the degrees of freedom of the sampling distribution of the difference between two means is complicated. So some books teach an easier rule: The number of degrees of freedom is always at *least* the smaller of $n_1 - 1$ and $n_2 - 1$ and at most $n_1 + n_2 - 2$. The problem is that if you need to perform a two-sample *t*-test and don't have the formula at hand to find the correct degrees of freedom, you have to be conservative and use the lower value. And *that* approximation can be a poor choice because it can give less than *half* the degrees of freedom you're entitled to from the correct formula.

What else do we need? Only the degrees of freedom for the Student's *t*-model. Unfortunately, *that* formula isn't as simple as $n - 1$. The problem is that the sampling model isn't *really* Student's *t*, but only something close. The reason is that we estimated two different variances (s_1^2 and s_2^2) and they may be different. That extra variability makes the distribution even more variable than the Student's *t* for either of the means. But by using a special, adjusted degrees of freedom value, we can find a Student's *t*-model that is so close to the right sampling distribution model that nobody can tell the difference. The adjustment formula is straightforward but doesn't help our understanding much, so we leave it to the computer or calculator. (If you are curious and really want to see the formula, look in the footnote.[2])

---

[2] The result is due to Satterthwaite and Welch.

Satterthwaite, F. E. (1946). "An Approximate Distribution of Estimates of Variance Components," *Biometrics Bulletin* 2: 110–114.

Welch, B. L. (1947). "The Generalization of 'Student's' Problem when Several Different Population Variances are Involved," *Biometrika* 34: 28–35.

$$df = \frac{\left(\frac{s_1^2}{n_1} + \frac{s_2^2}{n_2}\right)^2}{\frac{1}{n_1 - 1}\left(\frac{s_1^2}{n_1}\right)^2 + \frac{1}{n_2 - 1}\left(\frac{s_2^2}{n_2}\right)^2}$$

This approximation formula usually doesn't even give a whole number. If you are using a table, you'll need a whole number, so round down to be safe. If you are using technology, the approximation formulas that computers and calculators use for the Student's *t*-distribution can deal with fractional degrees of freedom.

> **A Sampling Distribution for the Difference Between Two Means**
>
> When the conditions are met (see Section 13.3), the standardized sample difference between the means of two independent groups,
>
> $$t = \frac{(\bar{y}_1 - \bar{y}_2) - (\mu_1 - \mu_2)}{SE(\bar{y}_1 - \bar{y}_2)},$$
>
> can be modeled by a Student's t-model with a number of degrees of freedom found with a special formula. We estimate the standard error with
>
> $$SE(\bar{y}_1 - \bar{y}_2) = \sqrt{\frac{s_1^2}{n_1} + \frac{s_2^2}{n_2}}.$$

13.2 The Two-Sample *t*-Test

NOTATION ALERT:
Δ_0 (pronounced "delta naught") isn't so standard that you can assume everyone will understand it. We use it because it's the capital Greek letter "D" for "difference."

Now we've got everything we need to construct the hypothesis test, and you already know how to do it. It's the same idea we used when testing one mean against a hypothesized value. Here, we start by hypothesizing a value for the true difference of the means. We'll call that hypothesized difference Δ_0. (It's so common for that hypothesized difference to be zero that we often just assume $\Delta_0 = 0$.) We then take the ratio of the difference in the means from our samples to its standard error and compare that ratio to a critical value from a Student's t-model. The test is called the **two-sample *t*-test**.

> **Two-sample *t*-test**
>
> When the appropriate assumptions and conditions are met, we test the hypothesis:
>
> $$H_0: \mu_1 - \mu_2 = \Delta_0$$
>
> where the hypothesized difference Δ_0 is almost always 0. We use the statistic:
>
> $$t = \frac{(\bar{y}_1 - \bar{y}_2) - \Delta_0}{SE(\bar{y}_1 - \bar{y}_2)}.$$
>
> The standard error of $\bar{y}_1 - \bar{y}_2$ is:
>
> $$SE(\bar{y}_1 - \bar{y}_2) = \sqrt{\frac{s_1^2}{n_1} + \frac{s_2^2}{n_2}}.$$
>
> When the null hypothesis is true, the statistic can be closely modeled by a Student's t-model with a number of degrees of freedom given by a special formula. We use that model to compare our t ratio with a critical value for t or to obtain a P-value.

13.3 Assumptions and Conditions

Before we can perform a two-sample t-test, we have to check the assumptions and conditions.

Independence Assumption

The data in each group must be drawn independently and at random from each group's own homogeneous population or generated by a randomized comparative experiment. We can't expect that the data, taken as one big group, come from a homogeneous population because that's what we're trying to test. But without randomization of some sort, there are no sampling distribution models and no inference.

We should think about whether the independence assumption is reasonable. We can also check two conditions:

Randomization Condition: Were the data collected with suitable randomization? For surveys, are they a representative random sample? For experiments, was the experiment randomized?

10% Condition: We usually don't check this condition for differences of means. We'll check it only if we have a very small population or an extremely large sample. We needn't worry about it at all for randomized experiments.

Normal Population Assumption

As we did before with Student's *t*-models, we need the assumption that the underlying populations are *each* Normally distributed. So we check one condition.

Nearly Normal Condition: We must check this for *both* groups; a violation by either one violates the condition. As we saw for single sample means, the Normality Assumption matters most when sample sizes are small. When either group is small ($n < 15$), you should not use these methods if the histogram or Normal probability plot shows skewness. For *n*'s closer to 40, a mildly skewed histogram is OK, but you should remark on any outliers you find and not work with severely skewed data. When both groups are bigger than that, the Central Limit Theorem starts to work unless the data are severely skewed or there are extreme outliers, so the Nearly Normal Condition for the data matters less. Even in large samples, however, you should still be on the lookout for outliers, extreme skewness, and multiple modes.

Independent Groups Assumption

To use the two-sample *t* methods, the two groups we are comparing must be independent of each other. In fact, the test is sometimes called the two *independent samples t*-test. No statistical test can verify that the groups are independent. You have to think about how the data were collected. The assumption would be violated, for example, if one group were comprised of husbands and the other group, their wives. Whatever we measure on one might naturally be related to the other. Similarly, if we compared subjects' performances before some treatment with their performances afterward, we'd expect a relationship of each "before" measurement with its corresponding "after" measurement. Measurements taken for two groups over time when the observations are taken at the same time may be related—especially if they share, for example, the chance they were influenced by the overall economy or world events. In cases such as these, where the observational units in the two groups are related or matched, *the two-sample methods of this chapter can't be applied*. When this happens, we need a different procedure, discussed in the next chapter.

GUIDED EXAMPLE | *Credit Card Promotions and Spending*

Our preliminary market research has suggested that a new incentive may increase customer spending. However, before we invest in this promotion on the entire population of cardholders, let's test a hypothesis on a sample. To judge whether the incentive works, we will examine the change in spending (called the *spend lift*) over a six month period. We will see whether the *spend lift* for the group that received the offer was greater than the *spend lift* for the group that received no offer. If we observe differences, how will we know whether these differences are important (or real) enough to justify our costs?

PLAN

Setup State what we want to know.

Identify the *parameter* we wish to estimate. Here our parameter is the difference in the means, not the individual group means.

Identify the *population(s)* about which we wish to make statements.

Identify the variables and context.

We want to know if cardholders who are offered a promotion spend more on their credit card. We have the spend lift (in $) for a random sample of 500 cardholders who were offered the promotion and for a random sample of 500 customers who were *not*.

H_0: The mean *spend lift* for the group who received the offer is the same as for the group who did not:

H_0: $\mu_{Offer} = \mu_{No\ Offer}$

H_A: The mean *spend lift* for the group who received the offer is higher:

H_0: $\mu_{Offer} > \mu_{No\ Offer}$

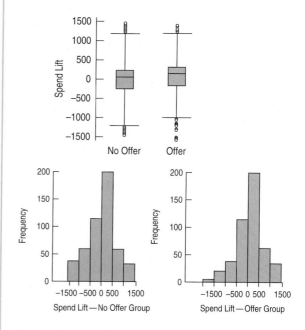

Make a graph to compare the two groups and check the distribution of each group. For completeness, we should report any outliers. If any outliers are extreme enough, we should consider performing the test both with and without the outliers and reporting the difference.

The boxplots and histograms show the distribution of both groups. It looks like the distribution for each group is fairly symmetric.

The boxplots indicate several outliers in each group, but we have no reason to delete them, and their impact is minimal.

Model Check the assumptions and conditions.

For large samples like these with quantitative data, we often don't worry about the 10% Condition.

✓ **Independence Assumption.** We have no reason to believe that the spending behavior of one customer would influence the spending behavior of another customer in the same group. The data report the "spend lift" for each customer for the same time period.

✓ **Randomization Condition.** The customers who were offered the promotion were selected at random.

✓ **Nearly Normal Condition.** The samples are large, so we are not overly concerned with this condition, and the boxplots and histograms show symmetric distributions for both groups.

✓ **Independent Groups Assumption.** Customers were assigned to groups at random. There's no reason to think that those in one group can affect the spending behavior of those in the other group.

State the sampling distribution model for the statistic. Here the degrees of freedom will come from the approximation formula in footnote 2.

Under these conditions, it's appropriate to use a Student's t-model.

Specify your method.

We will use a two-sample t-test.

Mechanics List the summary statistics. Be sure to include the units along with the statistics. Use meaningful subscripts to identify the groups.

We know $n_{\text{No Offer}} = 500$ and $n_{\text{Offer}} = 500$. From technology, we find:

$$\bar{y}_{\text{No Offer}} = \$7.69 \qquad \bar{y}_{\text{Offer}} = \$127.61$$

$$s_{\text{No Offer}} = \$611.62 \quad s_{\text{Offer}} = \$566.05$$

The observed difference in the two means is:

$$\bar{y}_{\text{Offer}} - \bar{y}_{\text{No Offer}} = \$127.61 - \$7.69 = \$119.92$$

Use the sample standard deviations to find the standard error of the sampling distribution.

The groups are independent, so:

$$SE(\bar{y}_{\text{Offer}} - \bar{y}_{\text{No Offer}}) = \sqrt{\frac{(611.62)^2}{500} + \frac{(566.05)^2}{500}}$$

$$= \$37.27$$

The best alternative is to let the computer use the approximation formula for the degrees of freedom and find the P-value.

The observed t-value is:

$$t = 119.92/37.27 = 3.218$$

with 992.0 df (from technology).

(To use critical values, we could find that the one-sided 0.01 critical value for a t with 992.0 df is $t^* = 2.33$.

Our observed t-value is larger than this, so we could reject the null hypothesis at the 0.01 level.)

Using software to obtain the P-value, we get:

```
Promotional Group   N     Mean    StDev
No                 500     7.69   611.62
Yes                500   127.61   566.05

Difference = mu (1) − mu (0)
Estimate for difference: 119.9231
t = 3.2178, df = 992.007
One-sided P-value = 0.0006669
```

REPORT

Conclusion Interpret the test results in the proper context.

MEMO:

Re: Credit Card Promotion

Our analysis of the credit card promotion experiment found that customers offered the promotion spent more than those not offered the promotion. The difference was statistically significant, with a P-value < 0.001. So we conclude that this promotion will increase spending. The difference in spend lift averaged $119.92, but our analyses so far have not determined how much income this will generate for the company and thus whether the estimated increase in spending is worth the cost of the offer.

JUST CHECKING

Many office "coffee stations" collect voluntary payments for the food consumed. Researchers at the University of Newcastle upon Tyne performed an experiment to see whether the image of eyes watching would change employee behavior.[3] They alternated pictures of eyes looking at the viewer with pictures of flowers each week on the cupboard behind the "honesty box." They measured the consumption of milk to approximate the amount of food consumed and recorded the contributions (in £) each week per liter of milk. The table summarizes their results.

	Eyes	**Flowers**
n(# weeks)	5	5
$\bar{y}$	0.417 £/liter	0.151 £/liter
s	0.1811	0.067

1 What null hypothesis were the researchers testing?

2 Check the assumptions and conditions needed to test whether there really is a difference in behavior due to the difference in pictures.

13.4 A Confidence Interval for the Difference Between Two Means

We rejected the null hypothesis that customers' mean spending would not change when offered a promotion. Because the company took a random sample of customers for each group, and our P-value was convincingly small, we concluded this difference is not zero for the population. Does this mean that we should offer the promotion to all customers?

---

[3] Melissa Bateson, Daniel Nettle, and Gilbert Roberts, "Cues of Being Watched Enhance Cooperation in a Real-World Setting," *Biol. Lett.* Doi:10.1098/rsbl.2006.0509.

A hypothesis test really says nothing about the size of the difference. All it says is that the observed difference is large enough that we can be confident it isn't zero. That's what the term "statistically significant" means. It doesn't say that the difference is important, financially significant, or interesting. Rejecting a null hypothesis simply says that the observed statistic is unlikely to have been observed if the null hypothesis were true.

So, what recommendations can we make to the company? Almost every business decision will depend on looking at a range of likely scenarios—precisely the kind of information a confidence interval gives. We construct the confidence interval for the difference in means in the usual way, starting with our observed statistic, in this case $(\bar{y}_1 - \bar{y}_2)$. We then add and subtract a multiple of the standard error $SE(\bar{y}_1 - \bar{y}_2)$ where the multiple is based on the Student's t distribution with the same df formula we saw before.

Confidence Interval for the Difference Between Two Means

When the conditions are met, we are ready to find a **two-sample t-interval** for the difference between means of two independent groups, $\mu_1 - \mu_2$. The confidence interval is:

$$(\bar{y}_1 - \bar{y}_2) \pm t^*_{df} \times SE(\bar{y}_1 - \bar{y}_2),$$

where the standard error of the difference of the means is:

$$SE(\bar{y}_1 - \bar{y}_2) = \sqrt{\frac{s_1^2}{n_1} + \frac{s_2^2}{n_2}}.$$

The critical value t^*_{df} depends on the particular confidence level, and on the number of degrees of freedom.

GUIDED EXAMPLE

Confidence Interval for Credit Card Spending

We rejected the null hypothesis that the mean spending in the two groups was equal. But, to find out whether we should consider offering the promotion nationwide, we need to estimate the magnitude of the spend lift.

PLAN

Setup State what we want to know.	We want to find a 95% confidence interval for the mean difference in spending between those who are offered a promotion and those who aren't.
Identify the *parameter* we wish to estimate. Here our parameter is the difference in the means, not the individual group means.	
Identify the *population(s)* about which we wish to make statements.	We looked at the boxplots and histograms of the groups and checked the conditions before. The same assumptions and conditions are appropriate here, so we can proceed directly to the confidence interval.
Identify the variables and context.	
Specify the method.	We will use a two-sample t-interval.

DO

Mechanics Construct the confidence interval. Be sure to include the units along with the statistics. Use meaningful subscripts to identify the groups.

Use the sample standard deviations to find the standard error of the sampling distribution.

The best alternative is to let the computer use the approximation formula for the degrees of freedom and find the confidence interval.

Ordinarily, we rely on technology for the calculations. In our hand calculations, we rounded values at intermediate steps to show the steps more clearly. The computer keeps full precision and is the one you should report. The difference between the hand and computer calculations is about $0.08.

In our previous analysis, we found:

$$\bar{y}_{\text{No Offer}} = \$7.69 \qquad \bar{y}_{\text{Offer}} = \$127.61$$
$$s_{\text{No Offer}} = \$611.62 \qquad s_{\text{Offer}} = \$566.05$$

The observed difference in the two means is:

$$\bar{y}_{\text{Offer}} - \bar{y}_{\text{No Offer}} = \$127.61 - \$7.69 = \$119.92,$$

and the standard error is:

$$SE(\bar{y}_{\text{Offer}} - \bar{y}_{\text{No Offer}}) = \$37.27$$

From technology, the df is 992.007, and the 0.025 critical value for t with 992.007 df is 1.96. So the 95% confidence interval is:

$$119.92 \pm 1.96(37.27) = (\$46.87, \$192.97)$$

Using software to obtain these computations, we get:

```
95 percent confidence interval:
 46.78784, 193.05837
sample means:
No Offer  Offer
7.690882  127.613987
```

REPORT

Conclusion Interpret the test results in the proper context.

MEMO:

Re: Credit Card Promotion Experiment

In our experiment, the promotion resulted in an increased spend lift of $119.92 on average. Further analysis gives a 95% confidence interval of ($46.79, $193.06). In other words, we expect with 95% confidence that under similar conditions, the mean spend lift that we achieve when we roll out the offer to all similar customers will be in this interval. We recommend that the company consider whether the values in this interval will justify the cost of the promotion program.

13.5 The Pooled *t*-Test

If you bought a used camera in good condition from a friend, would you pay the same as you would if you bought the same item from a stranger? A researcher at Cornell University[4] wanted to know how friendship might affect simple sales such as this. She randomly divided subjects into two groups and gave each group descriptions of items they might want to buy. One group was told to imagine buying from a friend whom they expected to see again. The other group was told to imagine buying from a stranger.

[4] J. J. Halpern (1997). "The Transaction Index: A Method for Standardizing Comparisons of Transaction Characteristics Across Different Contexts," *Group Decision and Negotiation*, 6, no. 6: 557–572.

Here are the prices they offered for a used camera in good condition.

PRICE OFFERED FOR A USED CAMERA ($)	
Buying from a Friend	**Buying from a Stranger**
275	260
300	250
260	175
300	130
255	200
275	225
290	240
300	

The researcher who designed the friendship study was interested in testing the impact of friendship on negotiations. Previous theories had doubted that friendship had a measurable effect on pricing, but she hoped to find such an effect. The usual null hypothesis is that there's no difference in means and that's what we'll use for the camera purchase prices.

When we performed the *t*-test earlier in the chapter, we used an approximation formula that adjusts the degrees of freedom to a lower value. When $n_1 + n_2$ is only 15, as it is here, we don't really want to lose any degrees of freedom. Because this is an experiment, we might be willing to make another assumption. The null hypothesis says that whether you buy from a friend or a stranger should have no effect on the mean amount you're willing to pay for a camera. If it has no effect on the means, should it affect the variance of the transactions?

If we're willing to *assume* that the variances of the groups are equal (at least when the null hypothesis is true), then we can save some degrees of freedom. To do that, we have to *pool* the two variances that we estimate from the groups into one common, or *pooled*, estimate of the variance:

$$s^2_{pooled} = \frac{(n_1 - 1)s_1^2 + (n_2 - 1)s_2^2}{(n_1 - 1) + (n_2 - 1)}.$$

(If the two sample sizes are equal, this is just the average of the two variances.)

Now we just substitute this pooled variance in place of each of the variances in the standard error formula. Remember, the standard error formula for the difference of two independent means is:

$$SE(\bar{y}_1 - \bar{y}_2) = \sqrt{\frac{s_1^2}{n_1} + \frac{s_2^2}{n_2}}.$$

We substitute the common pooled variance for each of the two variances in this formula, making the pooled standard error formula simpler:

$$SE_{pooled}(\bar{y}_1 - \bar{y}_2) = \sqrt{\frac{s^2_{pooled}}{n_1} + \frac{s^2_{pooled}}{n_2}} = s_{pooled}\sqrt{\frac{1}{n_1} + \frac{1}{n_2}}.$$

The formula for degrees of freedom for the Student's *t*-model is simpler, too. It was so complicated for the two-sample *t* that we stuck it in a footnote. Now it's just df $= (n_1 - 1) + (n_2 - 1)$.

Substitute the pooled-*t* estimate of the standard error and its degrees of freedom into the steps of the confidence interval or hypothesis test and you'll be using pooled-*t* methods. Of course, if you decide to use a pooled-*t* method, you must defend your assumption that the variances of the two groups are equal.

To use the pooled *t*-methods, you'll need to add the **Equal Variance Assumption** that the variances of the two populations from which the samples have been drawn are equal. That is, $\sigma_1^2 = \sigma_2^2$. (Of course, we can think about the standard deviations being equal instead.)

Pooled *t*-Test and Confidence Interval for the Difference Between Means

The conditions for the **pooled *t*-test** for the difference between the means of two independent groups (commonly called a pooled *t*-test) are the same as for the two-sample *t*-test with the additional assumption that the variances of the two groups are the same. We test the hypothesis:

$$H_0: \mu_1 - \mu_2 = \Delta_0,$$

where the hypothesized difference Δ_0 is almost always 0, using the statistic

$$t = \frac{(\bar{y}_1 - \bar{y}_2) - \Delta_0}{SE_{\text{pooled}}(\bar{y}_1 - \bar{y}_2)}.$$

The standard error of $\bar{y}_1 - \bar{y}_2$ is:

$$SE_{\text{pooled}}(\bar{y}_1 - \bar{y}_2) = s_{\text{pooled}}\sqrt{\frac{1}{n_1} + \frac{1}{n_2}},$$

where the pooled variance is:

$$s_{\text{pooled}}^2 = \frac{(n_1 - 1)s_1^2 + (n_2 - 1)s_2^2}{(n_1 - 1) + (n_2 - 1)}.$$

When the conditions are met and the null hypothesis is true, we can model this statistic's sampling distribution with a Student's *t*-model with $(n_1 - 1) + (n_2 - 1)$ degrees of freedom. We use that model to obtain a P-value for a test or a margin of error for a confidence interval.

The corresponding **pooled-*t* confidence interval** is:

$$(\bar{y}_1 - \bar{y}_2) \pm t_{\text{df}}^* \times SE_{\text{pooled}}(\bar{y}_1 - \bar{y}_2),$$

where the critical value t^* depends on the confidence level and is found with $(n_1 - 1) + (n_2 - 1)$ degrees of freedom.

GUIDED EXAMPLE | *Role of Friendship in Negotiations*

The usual null hypothesis in a pooled *t*-test is that there's no difference in means and that's what we'll use for the camera purchase prices.

PLAN

Setup State what we want to know.

Identify the *parameter* we wish to estimate. Here our parameter is the difference in the means, not the individual group means. Identify the variables and context.

We want to know whether people are likely to offer a different amount for a used camera when buying from a friend than when buying from a stranger. We wonder whether the difference between mean amounts is zero. We have bid prices from 8 subjects buying from a friend and 7 subjects buying from a stranger, found in a randomized experiment.

Hypotheses State the null and alternative hypotheses.

The research claim is that friendship changes what people are willing to pay.[5] The natural null hypothesis is that friendship makes no difference.

We didn't start with any knowledge of whether friendship might increase or decrease the price, so we choose a two-sided alternative.

Make a graph. Boxplots are the display of choice for comparing groups. We'll also want to check the distribution of each group. Histograms may do a better job.

H_0: The difference in mean price offered to friends and the mean price offered to strangers is zero:

$$\mu_F - \mu_S = 0,$$

H_A: The difference in mean prices is not zero:

$$\mu_F - \mu_S \neq 0.$$

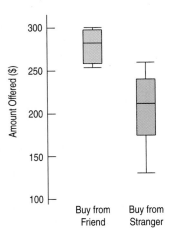

REALITY CHECK

Looks like the prices are higher if you buy from a friend. The two ranges barely overlap, so we'll be pretty surprised if we don't reject the null hypothesis.

Model Think about the assumptions and check the conditions. (Because this is a randomized experiment, we haven't sampled at all, so the 10% Condition doesn't apply.)

✓ **Independence Assumption.** There is no reason to think that the behavior of one person will influence the behavior of another.
✓ **Randomization Condition.** The experiment was randomized. Subjects were assigned to treatment groups at random.
✓ **Independent Groups Assumption.** Randomizing the experiment gives independent groups.
✓ **Nearly Normal Condition.** Histograms of the two sets of prices show no evidence of skewness or extreme outliers.

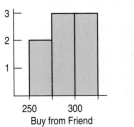 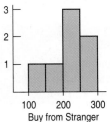

State the sampling distribution model.

Specify the method.

Because this is a randomized experiment with a null hypothesis of no difference in means, we can make the Equal Variance Assumption. If, as we are assuming from the null hypothesis, the treatment doesn't

---
[5] This claim is a good example of what is called a "research hypothesis" in many social sciences. The only way to check it is to deny that it's true and see where the resulting null hypothesis leads us.

change the means, then it is reasonable to assume that it also doesn't change the variances. Under these assumptions and conditions, we can use a Student's *t*-model to perform a pooled *t*-test.

Mechanics List the summary statistics. Be sure to use proper notation.

From the data:

$$n_F = 8 \qquad n_S = 7$$
$$\bar{y}_F = \$281.88 \qquad \bar{y}_S = \$211.43$$
$$s_F = \$18.31 \qquad s_S = \$46.43$$

Use the null model to find the P-value. First determine the standard error of the difference between sample means.

The pooled variance estimate is:

$$s_p^2 = \frac{(n_F - 1)s_F^2 + (n_S - 1)s_S^2}{n_F + n_S - 2}$$
$$= \frac{(8 - 1)(18.31)^2 + (7 - 1)(46.43)^2}{8 + 7 - 2}$$
$$= 1175.48$$

The standard error of the difference becomes:

$$SE_{pooled}(\bar{y}_F - \bar{y}_S) = \sqrt{\frac{s_p^2}{n_F} + \frac{s_p^2}{n_S}}$$
$$= 17.744$$

Make a graph. Sketch the *t*-model centered at the hypothesized difference of zero. Because this is a two-tailed test, shade the region to the right of the observed difference and the corresponding region in the other tail.

The observed difference in means is:

$$(\bar{y}_F - \bar{y}_S) = 281.88 - 211.43 = \$70.45$$

which results in a *t*-ratio

$$t = \frac{(\bar{y}_F - \bar{y}_S) - (0)}{SE_{pooled}(\bar{y}_F - \bar{y}_S)} = \frac{70.45}{17.744} = 3.97$$

Find the *t*-value.

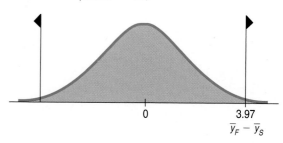

A statistics program can find the P-value.

The computer output for a pooled *t*-test appears here.

```
Pooled T Test for friend vs. stranger
              N    Mean    StDev    SE Mean
Friend        8    281.9   18.3     6.5
Stranger      7    211.4   46.4     18

t = 3.9699, df = 13, P-value = 0.001600
Alternative hypothesis: true difference in
means is not equal to 0

95 percent confidence interval:
 32.11047 108.78238
```

Conclusion Link the P-value to your decision about the null hypothesis and state the conclusion in context.

Be cautious about generalizing to items whose prices are outside the range of those in this study. The confidence interval can reveal more detailed information about the size of the difference. In the original article (referenced in footnote 5 in this chapter), the researcher tested several items and proposed a model relating the size of the difference to the price of the items.

MEMO:

Re: Role of friendship in negotiations.

Results of a small experiment show that people are likely to offer a different amount for a used camera when bargaining with a friend than when bargaining with a stranger. The difference in mean offers was statistically significant ($P = .0016$).

The confidence interval suggests that people tend to offer more to a friend than they would to a stranger. For the camera, the 95% confidence interval for the mean difference in price was $32.11 to $108.78, but we suspect that the actual difference may vary with the price of the item purchased.

JUST CHECKING

Recall the experiment to see whether pictures of eyes would improve compliance in voluntary contributions at an office coffee station.

3 What alternative hypothesis would you test?

4 The P-value of the test was less than 0.05. State a brief conclusion.

When Should You Use the Pooled *t*-Test?

When the variances of the two groups are in fact equal, the two methods give pretty much the same result. Pooled methods have a small advantage (slightly narrower confidence intervals, slightly more powerful tests) mostly because they usually have a few more degrees of freedom, but the advantage is slight. When the variances are *not* equal, the pooled methods are just not valid and can give poor results. You have to use the two-sample methods instead.

As the sample sizes get bigger, the advantages that come from a few more degrees of freedom make less and less difference. So the advantage (such as it is) of the pooled method is greatest when the samples are small—just when it's hardest to check the conditions. And the difference in the degrees of freedom is greatest when the variances are not equal—just when you can't use the pooled method anyway. Our advice is to use the two-sample *t* methods to compare means.

Why did we devote a whole section to a method that we don't recommend using? That's a good question. The answer is that pooled methods are actually very important in Statistics, especially in the case of designed experiments, where we start by assigning subjects to treatments at random. We know that at the start of the experiment each treatment group is a random sample from the same population,[6] so each treatment group begins with the same population variance. In this case, assuming the variances are equal after we apply the treatment is the same as

> Because the advantages of pooling are small, and you are allowed to pool only rarely (when the equal variances assumption is met), our advice is: ***don't.***
>
> **It's never wrong *not* to pool.**

---

[6] That is, the population of experimental subjects. Remember that to be valid, experiments do not need a representative sample drawn from a population because we are not trying to estimate a population model parameter.

assuming that the treatment doesn't change the variance. When we test whether the true means are equal, we may be willing to go a bit farther and say that the treatments made no difference *at all*. That's what we did in the friendship and bargaining experiment. Then it's not much of a stretch to assume that the variances have remained equal. We'll come back to pooled methods for experiments in Chapter 23.

The other reason to discuss the pooled *t*-test is historical. Until recently, many software packages offered the pooled *t*-test as the default for comparing means of two groups and required you to specifically request the *two-sample t-test* (or sometimes the misleadingly named "unequal variance *t*-test") as an option. That's changing, but be careful when using software to specify the right test.

There is also a hypothesis test that you could use to test the assumption of equal variances. However, it is sensitive to failures of the assumptions and works poorly for small sample sizes—just the situation in which we might care about a difference in the methods. When the choice between two-sample *t* and pooled *t* methods makes a difference (that is, when the sample sizes are small), the test for whether the variances are equal hardly works at all.

Even though pooled methods are important in Statistics, the ones for comparing two means have good alternatives that don't require the extra assumption. The two-sample methods apply to more situations and are safer to use.

*13.6 Tukey's Quick Test

If you think that the *t*-test is a lot of work for what seemed like an easy comparison, you're not alone. The famous statistician John Tukey[7] was once challenged to come up with a simpler alternative to the two-sample *t*-test that, like the 68-95-99.7 Rule, had critical values that could be remembered easily. The test he came up with asks you only to count and to remember three numbers: 7, 10, and 13.

When you first looked at the boxplots of the friendship data, you might have noticed that they didn't overlap very much. That's the basis for Tukey's test. To use Tukey's test, one group must have the highest value, and the other must have the lowest. We just count how many values in the high group are higher than *all* the values of the lower group. Add to this the number of values in the low group that are lower than *all* the values of the higher group. (You can count ties as $\frac{1}{2}$.) If the total of these exceedences is 7 or more, we can reject the null hypothesis of equal means at $\alpha = 0.05$. The "critical values" of 10 and 13 correspond to α's of 0.01 and 0.001.

Let's try it. The "Friend" group has the highest value ($300), and the "Stranger" group has the lowest value ($130). Six of the values in the Friend group are higher than the highest value of the Stranger group ($260), and one is a tie. Six of the Stranger values are lower than the lowest value for the Friend group. That's a total of $12\frac{1}{2}$ exceedences. That's more than 10, but less than 13. So the P-value is between 0.01 and 0.001—just what we found with the pooled *t*.

This is a remarkably good test. The only assumption it requires is that the two samples be independent. It's so simple to do that there's no reason not to do one to check your two-sample *t* results. If they disagree, check the assumptions. Tukey's quick test, however, is not as widely known or accepted as the two-sample *t*-test, so you still need to know and use the two-sample *t*.

---

[7] Famous seems like an understatement for John Tukey. The *New York Times* called Tukey "one of the most influential statisticians of the" 20th century and noted that he is credited with inventing the words "software" and "bit." He also invented both the stem-and-leaf display and the boxplot.

WHAT CAN GO WRONG?

- **Watch out for paired data.** The Independent Groups Assumption deserves special attention. Some researchers *deliberately* violate the Independent Groups Assumption. For example, suppose you wanted to test a diet program. You select 10 people at random to take part in your diet. You measure their weights at the beginning of the diet and after 10 weeks of the diet. So, you have two columns of weights, one for *before* and one for *after*. Can you use these methods to test whether the mean has gone down? No! The data are related; each "after" weight goes naturally with the "before" weight for the *same* person. If the samples are *not* independent, you can't use two-sample methods. This is probably the main thing that can go wrong when using these two-sample methods. Certainly, someone's weight before and after the 10 weeks will be related (whether the diet works or not). The methods of this chapter can be used *only* if the observations in the two groups are *independent*. (See Chapter 14 for methods that work on paired data.)

- **Don't use individual confidence intervals for each group to test the difference between their means.** If you make 95% confidence intervals for the means of two groups separately and you find that the intervals don't overlap, you can reject the hypothesis that the means are equal (at the corresponding α level). But, if the intervals do overlap, that doesn't mean that you *can't* reject the null hypothesis. The margin of error for the difference between the means is smaller than the sum of the individual confidence interval margins of error. Comparing the individual confidence intervals is like adding the standard deviations. But we know that it's the variances that we add, and when we do it right, we actually get a more powerful test. So, don't test the difference between group means by looking at separate confidence intervals. Always make a two sample *t*-interval or perform a two sample *t*-test.

- **Look at the plots.** The usual (by now) cautions about checking for outliers and non-Normal distributions apply. The simple defense is to make and examine boxplots. You may be surprised how often this simple step saves you from the wrong or even absurd conclusions that can be generated by a single undetected outlier. You don't want to conclude that two methods have very different means just because one observation is atypical.

- **Do what we say, not what we do.** Precision machines used in industry often have a bewildering number of parameters that have to be set, so experiments are performed in an attempt to try to find the best settings. Such was the case for a hole-punching machine used by a well-known computer manufacturer to make printed circuit boards. The data were analyzed by one of the authors, but because he was in a hurry, he didn't look at the boxplots first and just performed *t*-tests on the experimental factors. When he found extremely small P-values even for factors that made no sense, he plotted the data. Sure enough, there was one observation 1,000,000 times bigger than the others. It turns out that it had been recorded in microns (millionths of an inch), while all the rest were in inches.

ETHICS IN ACTION

Advocacy groups for equity and diversity in the workplace often cite middle managers as an obstacle to many organizations' efforts to be more inclusive. In response to this concern, Michael Schrute, the CEO for a large manufacturing company, asked Albert Fredericks, VP of Human Resources, to look into the possibility of instituting some type of diversity training for the company's middle managers. One option under consideration was an online education program that focused on cultural diversity, gender equity, and disability awareness. Although cost-effective, Albert suspected that an online program would not be as effective as traditional training for middle managers. In order to evaluate the online program under consideration, 20 middle managers were selected to participate. Before beginning, they were given a test to assess their knowledge of and sensitivity to various diversity and equity issues. Out of a possible perfect score of 100, the test average was 63.65. Each of the 20 managers then completed the 6-week online diversity education program and was retested. The average on the test after completing the online program was 69.15. Although the group achieved a higher mean test score after completing the program, the two-sample *t*-test revealed that this average test score was not significantly higher than the average prior to completing the online program (*t* = −0.94, P-value = 0.176). Albert was not surprised and began to explore more traditional diversity educational programs.

ETHICAL ISSUE: *The pretest and posttest design violates independence and therefore the two-sample t-test is not appropriate (related to Item A, ASA Ethical Guidelines).*

ETHICAL SOLUTION: *Use the correct test. The two-sample t-test is not appropriate for these data. (Chapter 14 discusses methods appropriate for such data.) Using the correct test shows that the online diversity education program was effective.*

What have we learned?

Are the means of two groups equal? If not, how different are they? We've learned to use statistical inference to compare the means of two independent groups and we have learned that:

- Confidence intervals and hypothesis tests about the difference between two means, like those for an individual mean, use *t*-models.
- Checking assumptions that tell us whether our method will work is important.
- The standard error for the difference in sample means relies on the assumption that our data come from independent groups. Unlike proportions, pooling is usually not the best choice here.
- We can add variances of independent random variables to find the standard deviation of the difference in two independent means.

Terms

Pooled *t*-interval   A confidence interval for the difference in the means of two independent groups used when we are willing and able to make the additional assumption that the variances of the groups are equal. It is found as:

$$(\bar{y}_1 - \bar{y}_2) \pm t_{df}^* \times SE_{pooled}(\bar{y}_1 - \bar{y}_2),$$

$$\text{where } SE_{pooled}(\bar{y}_1 - \bar{y}_2) = s_{pooled}\sqrt{\frac{1}{n_1} + \frac{1}{n_2}},$$

and the pooled variance is

$$s_{\text{pooled}}^2 = \frac{(n_1 - 1)s_1^2 + (n_2 - 1)s_2^2}{(n_1 - 1) + (n_2 - 1)}.$$

The number of degrees of freedom is $(n_1 - 1) + (n_2 - 1)$.

Pooled *t*-test A hypothesis test for the difference in the means of two independent groups when we are willing and able to assume that the variances of the groups are equal. It tests the null hypothesis

$$H_0: \mu_1 - \mu_2 = \Delta_0,$$

where the hypothesized difference Δ_0 is almost always 0, using the statistic

$$t_{\text{df}} = \frac{(\bar{y}_1 - \bar{y}_2) - \Delta_0}{SE_{\text{pooled}}(\bar{y}_1 - \bar{y}_2)},$$

where the pooled standard error is defined as for the pooled interval and the degrees of freedom is $(n_1 - 1) + (n_2 - 1)$.

Pooling Data from two or more populations may sometimes be combined, or *pooled*, to estimate a statistic (typically a pooled variance) when we are willing to assume that the estimated value is the same in both populations. The resulting larger sample size may lead to an estimate with lower sample variance. However, pooled estimates are appropriate only when the required assumptions are true.

Two-sample *t*-interval A confidence interval for the difference in the means of two independent groups found as

$$(\bar{y}_1 - \bar{y}_2) \pm t_{\text{df}}^* \times SE(\bar{y}_1 - \bar{y}_2), \text{ where}$$

$$SE(\bar{y}_1 - \bar{y}_2) = \sqrt{\frac{s_1^2}{n_1} + \frac{s_2^2}{n_2}}$$

and the number of degrees of freedom is given by the approximation formula in footnote 2 of this chapter, or with technology.

Two-sample *t*-test A hypothesis test for the difference in the means of two independent groups. It tests the null hypothesis

$$H_0: \mu_1 - \mu_2 = \Delta_0,$$

where the hypothesized difference Δ_0 is almost always 0, using the statistic

$$t_{\text{df}} = \frac{(\bar{y}_1 - \bar{y}_2) - \Delta_0}{SE(\bar{y}_1 - \bar{y}_2)},$$

with the number of degrees of freedom given by the approximation formula in footnote 2 of this chapter, or with technology.

Skills

PLAN
- Be able to recognize situations in which we want to do inference on the difference between the means of two independent groups.
- Know how to examine your data for violations of conditions that would make inference about the difference between two population means unwise or invalid.
- Be able to recognize when a pooled-*t* procedure might be appropriate and be able to explain why you decided to use a two-sample method anyway.

DO
- Be able to perform a two-sample *t*-test using a statistics package or calculator (at least for finding the degrees of freedom).

REPORT
- Be able to interpret a test of the null hypothesis that the means of two independent groups are equal. (If the test is a pooled *t*-test, your interpretation should include a defense of your assumption of equal variances.)

Technology Help: Two-Sample Methods

Here's some typical computer package output with comments:

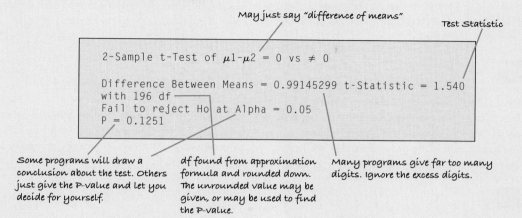

May just say "difference of means"

Test Statistic

```
2-Sample t-Test of μ1-μ2 = 0 vs ≠ 0

Difference Between Means = 0.99145299  t-Statistic = 1.540
with 196 df
Fail to reject Ho at Alpha = 0.05
P = 0.1251
```

Some programs will draw a conclusion about the test. Others just give the P-value and let you decide for yourself.

df found from approximation formula and rounded down. The unrounded value may be given, or may be used to find the P-value.

Many programs give far too many digits. Ignore the excess digits.

Most statistics packages compute the test statistic for you and report a P-value corresponding to that statistic. And statistics packages make it easy to examine the boxplots of the two groups, so you have no excuse for skipping the important check of the Nearly Normal Condition.

Some statistics software automatically tries to test whether the variances of the two groups are equal. Some automatically offer both the two-sample-*t* and pooled-*t* results. Ignore the test for the variances; it has little power in any situation in which its results could matter. If the pooled and two-sample methods differ in any important way, you should stick with the two-sample method. Most likely, the Equal Variance Assumption needed for the pooled method has failed.

The degrees of freedom approximation usually gives a fractional value. Most packages seem to round the approximate value down to the next smallest integer (although they may actually compute the P-value with the fractional value, gaining a tiny amount of power).

There are two ways to organize data when we want to compare two independent groups. The first, called **unstacked data**, lists the data in two columns, one for each group. Each list can be thought of as a variable. In this method, the variables in the credit card example would be "Offer" and "No Offer." Graphing calculators usually prefer this form, and some computer programs can use it as well.

The alternative way to organize the data is as **stacked data**. What is the response variable for the credit card experiment? It's the "Spend Lift"—the amount by which customers increased their spending. But the values of this variable in the unstacked lists are in both columns, and actually there's an experiment factor here, too—namely, whether the customer was offered the promotion or not. So we could put the data into two different columns, one with the "Spend Lift"s in it and one with a "Yes" for those who were offered the promotion and a "No" for those who weren't. The stacked data would look like this:

Spend Lift	Offer
969.74	Yes
915.04	Yes
197.57	No
77.31	No
196.27	Yes
...	...

This way of organizing the data makes sense as well. Now the factor and the response variables are clearly visible. You'll have to see which method your program requires. Some packages even allow you to structure the data either way.

The commands to do inference for two independent groups on common statistics technology are not always found in obvious places. Here are some starting guidelines.

EXCEL

From the Data Tab, Analysis Group, choose **Data Analysis**. Alternatively (if the Data Analysis Tool Pack is not installed), in the Formulas Tab, choose More functions > Statistical > TTEST, and specify Type=3 in the resulting dialog.
Fill in the cell ranges for the two groups, the hypothesized difference, and the alpha level.

Comments

Excel expects the two groups to be in separate cell ranges. Notice that, contrary to Excel's wording, we do not need to assume that the variances are *not* equal; we simply choose not to assume that they *are* equal.

MINITAB

From the **Stat** menu, choose the **Basic Statistics** submenu. From that menu, choose **2-sample t....** and select data in "one column" or "two columns" depending on whether the data are stacked or unstacked. You may also enter summarized data. Then fill in the dialog.

Comments

The **Graphs** button enables you to create boxplots of your two samples and the **options** button allows you to conduct either a one-sided or two-sided test.

SPSS

From the **Analyze** menu, choose the **Compare Means** submenu. From that, choose the **Independent-Samples t-test** command. Specify the data variable and "group variable." Then type in the labels used in the group variable. SPSS offers both the two-sample and pooled-*t* results in the same table.

Comments

SPSS expects the data in one variable and group names in the other. If there are more than two group names in the group variable, only the two that are named in the dialog box will be compared.

JMP

From the **Analyze** menu, select **Fit y by x.** Select variables: a **Y, Response** variable that holds the data and an **X, Factor** variable that holds the group names. JMP will make a dotplot. Click the **red triangle** in the dotplot title, and choose **Unequal variances.** The *t*-test is at the bottom of the resulting table. Find the P-value from the Prob>F section of the table (they are the same).

Comments

JMP expects data in one variable and category names in the other. Don't be misled: There is no need for the variances to be unequal to use two-sample *t* methods.

DATA DESK

Select variables.
From the **Calc** menu, choose **Estimate** for confidence intervals or **Test** for hypothesis tests. Select the interval or test from the drop-down menu and make other choices in the dialog.

Comments

Data Desk expects the two groups to be in separate variables.

Mini Case Study Project

Real Estate

In Chapter 12 we examined the regression of the sales price of a home on its size and saw that larger homes generally fetch a higher price. How much can we learn about a house from the fact that it has a fireplace or more than the average number of bedrooms? Data for a random sample of 1047 homes from the upstate New York area can be found in the file **ch13_MCSP_Real_Estate**. There are 6 quantitative variables: *Price* ($), *Living Area (sq.ft)*, *Bathrooms(#)*, *Bedrooms(#)*, *Lot Size(Acres)*, and *Age(years)*, and one categorical variable, *Fireplace?(1 = Yes;0 = No)* denoting whether the house has at least one fireplace. We can use *t*-methods to see, for example, whether homes with fireplaces sell for more, on average, and by how much. For the quantitative variables, create new categorical variables by splitting them at the median or some other splitting point of your choice, and compare home prices above and below this value. For example, the median number of *Bedrooms* of these homes is 2. You might compare the prices of homes with 1 or 2 bedrooms to those with more than 2. Write up a short report summarizing the differences in mean price based on the categorical variables that you created.

EXERCISES

1. Hot dogs and calories. Consumers increasingly make food purchases based on nutrition values. In the July 2007 issue, *Consumer Reports* examined the calorie content of two kinds of hot dogs: meat (usually a mixture of pork, turkey, and chicken) and all beef. The researchers purchased samples of several different brands. The meat hot dogs averaged 111.7 calories, compared to 135.4 for the beef hot dogs. A test of the null hypothesis that there's no difference in mean calorie content yields a P-value of 0.124. What would you conclude?

2. Hot dogs and sodium. The *Consumer Reports* article described in Exercise 1 also listed the sodium content (in mg) for the various hot dogs tested. A test of the null hypothesis that beef hot dogs and meat hot dogs don't differ in the mean amounts of sodium yields a P-value of 0.110. What would you conclude?

3. Learning math. The Core Plus Mathematics Project (CPMP) is an innovative approach to teaching mathematics that engages students in group investigations and mathematical modeling. After field tests in 36 high schools over a three-year period, researchers compared the performances of CPMP students with those taught using a traditional curriculum. In one test, students had to solve applied algebra problems using calculators. Scores for 320 CPMP students were compared with those of a control group of 273 students in a traditional math program. Computer software was used to create a confidence interval for the difference in mean scores (*Journal for Research in Mathematics Education*, 31, no. 3, 2000).

    Conf. level: 95%
    Variable: μ(CPMP) − μ(Ctrl)
    Interval: (5.573, 11.427)

a) What is the margin of error for this confidence interval?
b) If we had created a 98% confidence interval, would the margin of error be larger or smaller?
c) Explain what the calculated interval means in this context.
d) Does this result suggest that students who learn mathematics with CPMP will have significantly higher mean scores in applied algebra than those in traditional programs? Explain.

4. Sales performance. A chain that specializes in healthy and organic food would like to compare the sales performance of two of its primary stores in the state of Massachusetts. These stores are both in urban, residential areas with similar demographics. A comparison of the weekly sales randomly sampled over a period of nearly two years for these two stores yields the following information:

Store	N	Mean	StDev	Minimum	Median	Maximum
Store #1	9	242170	23937	211225	232901	292381
Store #2	9	235338	29690	187475	232070	287838

a) Create a 95% confidence interval for the difference in the mean store weekly sales.
b) Interpret your interval in context.
c) Does it appear that one store sells more on average than the other store?
d) What is the margin of error for this interval?
e) Would you expect a 99% confidence interval to be wider or narrower? Explain.
f) If you computed a 99% confidence interval, would your conclusion in part c change? Explain.

5. CPMP, again. During the study described in Exercise 3, students in both CPMP and traditional classes took another algebra test that did not allow them to use calculators. The table shows the results. Are the mean scores of the two groups significantly different? Assume that the assumptions for inference are satisfied.

Math Program	*n*	Mean	SD
CPMP	312	29.0	18.8
Traditional	265	38.4	16.2

a) Write an appropriate hypothesis.
b) Here is computer output for this hypothesis test. Explain what the P-value means in this context.

    2-Sample t-Test of μ1 − μ2 ≠ 0
    t-Statistic = −6.451 w/574.8761 df
    P < 0.0001

c) State a conclusion about the CPMP program.

6. IT training costs. An accounting firm is trying to decide between IT training conducted in-house and the use of third party consultants. To get some preliminary cost data, each type of training was implemented at two of the firm's offices located in different cities. The table below shows the average annual training cost per employee at each location. Are the mean costs significantly different? Assume that the assumptions for inference are satisfied.

IT Training	*n*	Mean	SD
In-House	210	$490.00	$32.00
Consultants	180	$500.00	$48.00

a) Write the appropriate hypotheses.
b) Below is computer output for this hypothesis test. Explain what the P-value means in this context.

    2-Sample t-Test of μ1 − μ2 ≠ 0
    t-Statistic = −2.38 w/303df
    P = .018

c) State a conclusion about IT training costs.

7. CPMP and word problems. The study of the new CPMP mathematics methodology described in Exercise 3 also tested students' abilities to solve word problems. This table shows how the CPMP and traditional groups performed. What do you conclude? (Assume that the assumptions for inference are met.)

Math Program	n	Mean	SD
CPMP	320	57.4	32.1
Traditional	273	53.9	28.5

8. Statistical training. The accounting firm described in Exercise 6 is interested in providing opportunities for its auditors to gain more expertise in statistical sampling methods. They wish to compare traditional classroom instruction with online self-paced tutorials. Auditors were assigned at random to one type of instruction, and the auditors were then given an exam. The table shows how the two groups performed. What do you conclude? (Assume the assumptions for inference are met.)

Program	n	Mean	SD
Traditional	296	74.5	11.2
Online	275	72.9	12.3

9. Trucking company. A trucking company would like to compare two different routes for efficiency. Truckers are randomly assigned to two different routes. Twenty truckers following Route A report an average of 40 minutes, with a standard deviation of 3 minutes. Twenty truckers following Route B report an average of 43 minutes, with a standard deviation of 2 minutes. Histograms of travel times for the routes are roughly symmetric and show no outliers.
a) Find a 95% confidence interval for the difference in average time for the two routes.
b) Will the company save time by always driving one of the routes? Explain.

10. Change in sales. Suppose the specialty food chain from Exercise 4 wants to now compare the change in sales across different regions. An examination of the difference in sales over a 37-week period in a recent year for 8 stores in the state of Massachusetts compared to 12 stores in nearby states reveals the following descriptive statistics for relative increase in sales. (If these means are multiplied by 100, they show % increase in sales.)

State	N	Mean	StDev
MA	8	0.0738	0.0666
Other	12	0.0559	0.0503

a) Find the 90% confidence interval for the difference in relative increase in sales over this time period.
b) Is there a significant difference in increase in sales between these two groups of stores? Explain.
c) What would you like to see to check the conditions?

⊤ 11. Cereal company. A food company is concerned about recent criticism of the sugar content of their children's cereals. The data show the sugar content (as a percentage of weight) of several national brands of children's and adults' cereals.

Children's cereals: 40.3, 55, 45.7, 43.3, 50.3, 45.9, 53.5, 43, 44.2, 44, 47.4, 44, 33.6, 55.1, 48.8, 50.4, 37.8, 60.3, 46.6
Adults' cereals: 20, 30.2, 2.2, 7.5, 4.4, 22.2, 16.6, 14.5, 21.4, 3.3, 6.6, 7.8, 10.6, 16.2, 14.5, 4.1, 15.8, 4.1, 2.4, 3.5, 8.5, 10, 1, 4.4, 1.3, 8.1, 4.7, 18.4

a) Write the null and alternative hypotheses.
b) Check the conditions.
c) Find the 95% confidence interval for the difference in means.
d) Is there a significant difference in mean sugar content between these two types of cereals? Explain.

⊤ 12. Foreclosure rates. According to recent reports, home foreclosures were up 47% in March 2008 compared to the previous year (realestate.msn.com; April 2008). The data show home foreclosure rates (as % change from the previous year) for a sample of cities in two regions of the U.S., the Northeast and the Southwest.

Northeast: 2.99, −2.36, 3.03, 1.01, 5.77, 9.95, −3.52, 7.16, −3.34, 4.75, 5.25, 6.21, 1.67, −2.45, −0.55, 3.45, 4.50, 1.87, −2.15, −0.75
Southwest: 10.15, 23.05, 18.95, 21.16, 17.45, 12.67, 13.75, 29.42, 11.45, 16.77, 12.67, 13.69, 25.81, 21.16, 19.67, 11.88, 13.67, 18.00, 12.88

a) Write the null and alternative hypotheses.
b) Check the conditions.
c) Test the hypothesis and find the P-value.
d) Is there a significant difference in the mean home foreclosure rates between these two regions of the U.S.? Explain.

13. Investment. Investment style plays a role in constructing a mutual fund. Each individual stock is grouped into two distinct groups: "Growth" and "Value." A Growth stock is one with high earning potential and often pays little or no dividends to shareholders. Conversely, Value stocks are commonly viewed as steady, or more conservative, with a lower earning potential. You are trying to decide what type of funds to invest in. Because you are saving toward your retirement, if you invest in a Value fund, you hope that the fund remains conservative. We would call such a fund "consistent." If the fund did not remain consistent and became higher risk, that could impact your retirement savings. The funds in this data set have been identified as either being "style consistent" or "style drifter." Portfolio managers wonder whether consistency provides the optimal chance for successful retirement, so they believe that style consistent funds outperform style drifters. Out of a sample of 140 funds, 66 were identified as style consistent, while 74 were identified as style drifters. Their statistics for their average return over 5 years are given on the next page:

Type	N	Mean	StDev	Minimum	Q1	Q3	Maximum 5-yr Return
Consistent	66	9.382	2.675	1.750	7.675	11.110	15.920
Drifter	74	8.563	3.719	−0.870	5.928	11.288	17.870

a) Write the null and alternative hypotheses.
b) Find the 95% confidence interval of the difference in mean return between style consistent and style drifter funds.
c) Is there a significant difference in 5-year return for these two types of funds? Explain.

14. Technology adoption. The Pew Internet & American Life Project (www.pewinternet.org/) conducts surveys to gauge how the internet and technology impact daily life of individuals, families, and communities. In a recent survey Pew asked respondents if they thought that computers and technology give people more or less control over their lives. Companies that are involved in innovative technologies use the survey results to better understand their target market. One might suspect that younger and older respondents might differ in their opinions of whether computers and technology give them more control over their lives. A subset of the data from this survey (*February-March 2007 Tracking Data Set*) shows the mean ages of two groups of respondents, those who reported that they believed that computers and technology give them "more" control and those that reported "less" control.

Group	N	Mean	StDev	Min	Q1	Med	Q3	Max
More	74	54.42	19.65	18	41.5	53.5	68.5	99.0
Less	29	54.34	18.57	20	41.0	58.0	70.0	84.0

a) Write the null and alternative hypotheses. b) Find the 95% confidence interval for the difference in mean age between the two groups of respondents.
c) Is there a significant difference in the mean ages between these two groups? Explain.

15. Product testing. A company is producing and marketing new reading activities for elementary school children that it believes will improve reading comprehension scores. A researcher randomly assigns third graders to an eight-week program in which some will use these activities and others will experience traditional teaching methods. At the end of the experiment, both groups take a reading comprehension exam. Their scores are shown in the back-to-back stem-and-leaf display. Do these results suggest that the new activities are better? Test an appropriate hypothesis and state your conclusion.

New Activities		Control
	1	07
4	2	068
3	3	377
96333	4	12222238
9876432	5	355
721	6	02
1	7	
	8	5

16. Product placement. The owner of a small organic food store was concerned about her sales of a specialty yogurt manufactured in Greece. As a result of increasing fuel costs, she recently had to increase its price. To help boost sales, she decided to place the product on a different shelf (near eye level for most consumers) and in a location near other popular international products. She kept track of sales (number of containers sold per week) for six months after she made the change. These values are shown below, along with the sales numbers for the six months prior to making the change, in stem and leaf displays.

After Change		Before Change	
3	2	2	0
3	9	2	899
4	23	3	224
4	589	3	7789
5	0012	4	0000223
5	55558	4	5567
6	00123	5	0
6	67	5	6
7	0		

Do these results suggest that sales are better after the change in product placement? Test an appropriate hypothesis and state your conclusion. Be sure to check assumptions and conditions.

17. Acid rain. Researchers collected samples of water from streams in the Adirondack Mountains to investigate the effects of acid rain. They measured the pH (acidity) of the water and classified the streams with respect to the kind of substrate (type of rock over which they flow). A lower pH means the water is more acidic. Here is a plot of the pH of the streams by substrate (limestone, mixed, or shale):

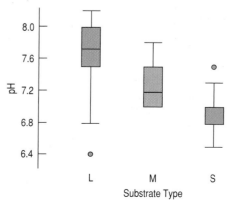

Here are selected parts of a software analysis comparing the pH of streams with limestone and shale substrates:

2-Sample t-Test of μ1 − μ2 = 0
Difference Between Means = 0.735
t-Statistic = 16.30 w/133 df
P ≤ 0.0001

a) State the null and alternative hypotheses for this test.
b) From the information you have, do the assumptions and conditions appear to be met?
c) What conclusion would you draw?

T 18. Hurricanes. It has been suggested that global warming may increase the frequency of hurricanes. The data show the number of hurricanes recorded annually before and after 1970. Has the frequency of hurricanes increased since 1970?

Before (1944–1969)	After (1970–2000)
3, 3, 1, 2, 4, 3, 8, 5, 3, 4, 2,	2, 1, 0, 1, 2, 3, 2, 1, 2, 2, 2,
6, 2, 2, 5, 2, 2, 7, 1, 2, 6, 1,	3, 1, 1, 1, 3, 0, 1, 3, 2, 1, 2,
3, 1, 0, 5	1, 1, 0, 5, 6, 1, 3, 5, 3, 4, 2, 3, 6, 7, 2

a) Write the null and alternative hypotheses.
b) Are the conditions for hypothesis testing satisfied?
c) If so, test the hypothesis.

T 19. Product testing, part 2. A pharmaceutical company is producing and marketing a ginkgo biloba supplement to enhance memory. In an experiment to test the product, subjects were assigned randomly to take ginkgo biloba supplements or a placebo. Their memory was tested to see whether it improved. Here are boxplots comparing the two groups and some computer output from a two-sample *t*-test computed for the data.

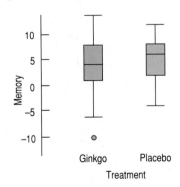

2-Sample t-Test of μ_G − μ_P > 0
Difference Between Means = −0.9914
t-Statistic = −1.540 w/196 df
P = 0.9374

a) Explain in this context what the P-value means.
b) State your conclusion about the effectiveness of ginkgo biloba.
c) Proponents of ginkgo biloba continue to insist that it works. What type of error do they claim your conclusion makes?

T 20. Baseball 2007. American League baseball teams play their games with the designated hitter rule, meaning that pitchers do not bat. The league believes that replacing the pitcher, traditionally a weak hitter, with another player in the batting order produces more runs and generates more interest among fans. The data provided in the file on the disk include the average numbers of runs scored per game (*Runs per game*) by American League and National League teams for almost the complete first half of the 2007 season.

a) Create an appropriate display of these data. What do you see?
b) With a 95% confidence interval, estimate the mean number of runs scored by American League teams.
c) With a 95% confidence interval, estimate the mean number of runs scored by National League teams.
d) Explain why you should not use two separate confidence intervals to decide whether the two leagues differ in average number of runs scored.

21. Productivity. A factory hiring people to work on an assembly line gives job applicants a test of manual agility. This test counts how many strangely shaped pegs the applicant can fit into matching holes in a one-minute period. The table summarizes the data by gender of the job applicant. Assume that all conditions necessary for inference are met.

	Male	Female
Number of subjects	50	50
Pegs placed:		
Mean	19.39	17.91
SD	2.52	3.39

a) Find 95% confidence intervals for the average number of pegs that males and females can each place.
b) Those intervals overlap. What does this suggest about any gender-based difference in manual agility?
c) Find a 95% confidence interval for the difference in the mean number of pegs that could be placed by men and women.
d) What does this interval suggest about any gender-based difference in manual agility?
e) The two results seem contradictory. Which method is correct: doing two-sample inference, or doing one-sample inference twice?
f) Why don't the results agree?

22. Online shopping. Online shopping statistics are routinely reported by www.shop.org. Of interest to many online retailers are gender based differences in shopping preferences and behaviors. The average monthly online expenditures are reported for males and females:

Group		Male	Female
	n	45	45
	Mean	$352	$310
	StDev	$95	$80

a) Find 95% confidence intervals for the average monthly online expenditures for males and females.

b) These intervals overlap. What does this suggest about any gender-based difference in monthly online expenditures?

c) Find a 95% confidence interval for the *difference* in average monthly online expenditures between males and females.

d) The two results seem contradictory. Which method is correct: doing two-sample inference, or doing one-sample inference twice?

T 23. Double header. Do the data in Exercise 20 suggest that the American League's designated hitter rule may lead to more runs?

a) Write the null and alternative hypothesis.

b) Find a 95% confidence interval for the difference in mean, Runs per game, and interpret your interval.

c) Test the hypothesis stated above in part a and find the P-value.

d) Interpret the P-value and state your conclusion. Does the test suggest that the American League scores more runs on average?

24. Online shopping again. In 2004, it was reported that the average male spends more money shopping online per month than the average female, $204 compared to $186 (www.shop.org; accessed April 2008). Do the data reported in Exercise 22 indicate that this is still true?

a) Write the null and alternative hypothesis.

b) Test the hypothesis stated in part (a) and find the P-value.

c) Interpret the P-value and state your conclusion. Does the test suggest that males continue to spend more online on average than females?

T 25. Drinking water. In an investigation of environmental causes of disease, data were collected on the annual mortality rate (deaths per 100,000) for males in 61 large towns in England and Wales. In addition, the water hardness was recorded as the calcium concentration (parts per million, ppm) in the drinking water. The data set also notes for each town whether it was south or north of Derby. Is there a significant difference in mortality rates in the two regions? Here are the summary statistics.

Summary of: mortality
For categories in: Derby

Group	Count	Mean	Median	StdDev
North	34	1631.59	1631	138.470
South	27	1388.85	1369	151.114

a) Test appropriate hypotheses and state your conclusion.

b) The boxplots of the two distributions show an outlier among the data north of Derby. What effect might that have had on your test?

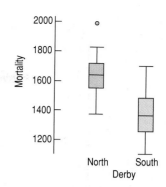

26. Sustainable stocks. The earnings per share ratio (EPS) is one of several important indicators of a company's profitability. There are several categories of "sustainable" stocks including natural foods/health and green energy/bio fuels. Below are earnings per share for a sample of stocks from both of these categories (Yahoo Financial, April 6, 2008). Is there a significant difference in earnings per share values for these two groups of sustainable stocks?

Group	Count	Mean	Median	StDev
Foods/Health	15	0.862	1.140	0.745
Energy/Fuel	16	−0.320	−0.545	0.918

a) Test appropriate hypotheses and state your conclusion.

b) Based upon the boxplots of the two distributions shown below, what might you suspect about your test? Explain.

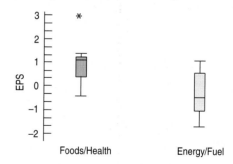

T 27. Job satisfaction. A company institutes an exercise break for its workers to see if this will improve job satisfaction, as measured by a questionnaire that assesses workers' satisfaction. Scores for 10 randomly selected workers before and after implementation of the exercise program are shown. The company wants to assess the effectiveness of the exercise program. Explain why you can't use the methods discussed in this chapter to do that. (Don't worry, we'll give you another chance to do this the right way.)

Worker Number	Job Satisfaction Index	
	Before	**After**
1	34	33
2	28	36
3	29	50
4	45	41
5	26	37
6	27	41
7	24	39
8	15	21
9	15	20
10	27	37

28. ERP effectiveness. When implementing a packaged Enterprise Resource Planning (ERP) system, many companies report that the module they first install is Financial Accounting. Among the measures used to gauge the effectiveness of their ERP system implementation is acceleration of the financial close process. Below is a sample of 8 companies that report their average time (in weeks) to financial close before and after the implementation of their ERP system.

Company	Before	After
1	6.5	4.2
2	7.0	5.9
3	8.0	8.0
4	4.5	4.0
5	5.2	3.8
6	4.9	4.1
7	5.2	6.0
8	6.5	4.2

Check the assumptions and conditions for using two-sample *t* methods. Can you proceed with them? If not, why not?

29. Delivery time. A small appliance company is interested in comparing delivery times of their product during two months. They are concerned that the summer slow-downs in August cause delivery times to lag during this month. Given the following delivery times (in days) of their appliances to the customer for a random sample of 6 orders each month, test if delivery times differ across these two months.

June	54	49	68	66	62	62
August	50	65	74	64	68	72

30. Branding. In June 2002, the *Journal of Applied Psychology* reported on a study that examined whether the content of TV shows influenced the ability of viewers to recall brand names of items featured in the commercials. The researchers randomly assigned volunteers to watch one of three programs, each containing the same nine commercials. One of the programs had violent content, another sexual content, and the third neutral content. After the shows ended, the subjects were asked to recall the brands of products that were advertised.

		Program Type		
		Violent	**Sexual**	**Neutral**
Brands Recalled	*n*	108	108	108
	Mean	2.08	1.71	3.17
	SD	1.87	1.76	1.77

a) Do these results indicate that viewer memory for ads may differ depending on program content? Test the hypothesis that there is no difference in ad memory between programs with sexual content and those with violent content. State your conclusion.

b) Is there evidence that viewer memory for ads may differ between programs with sexual content and those with neutral content? Test an appropriate hypothesis and state your conclusion.

31. Ad campaign. You are a consultant to the marketing department of a business preparing to launch an ad campaign for a new product. The company can afford to run ads during one TV show, and has decided not to sponsor a show with sexual content. You read the study described in Exercise 30 and then use a computer to create a confidence interval for the difference in mean number of brand names remembered between the groups watching violent shows and those watching neutral shows.

Two-Sample t
95% CI for $\mu_{viol} - \mu_{neut}$: (−1.578, −0.602)

a) At the meeting of the marketing staff, you have to explain what this output means. What will you say?

b) What advice would you give the company about the upcoming ad campaign?

32. Branding, part 2. In the study described in Exercise 30, the researchers also contacted the subjects again, 24 hours later, and asked them to recall the brands advertised. Results for the number of brands recalled are summarized in the table.

	Program Type		
	Violent	**Sexual**	**Neutral**
No. of subjects	101	106	103
Mean	3.02	2.72	4.65
SD	1.61	1.85	1.62

a) Is there a significant difference in viewers' abilities to remember brands advertised in shows with violent vs. neutral content?

b) Find a 95% confidence interval for the difference in mean number of brand names remembered between the groups watching shows with sexual content and those watching neutral shows. Interpret your interval in this context.

33. Ad recall. In Exercises 30 and 32, we see the number of advertised brand names people recalled immediately after watching TV shows and 24 hours later. Strangely enough, it appears that they remembered more about the ads the next day. Should we conclude this is true in general about people's memory of TV ads?

a) Suppose one analyst conducts a two-sample hypothesis test to see if memory of brands advertised during violent TV shows is higher 24 hours later. The P-value is 0.00013. What might she conclude?

b) Explain why her procedure was inappropriate. Which of the assumptions for inference was violated?

c) How might the design of this experiment have tainted these results?

d) Suggest a design that could compare immediate brand name recall with recall one day later.

34. Hybrid SUV's. The Chevy Tahoe Hybrid got a lot of attention in 2008. It is a relatively high-priced hybrid SUV that makes use of the latest technologies for fuel efficiency. One of the more popular hybrid SUV's on the market is the modestly priced Ford Escape Hybrid. A consumer group was interested in comparing the gas mileage of these two models. In order to do so, each vehicle was driven on the same 10 routes that combined both highway and city streets. The results showed that the mean mileage for the Chevy Tahoe was 29 mpg and for the Ford Escape it was 31 mpg. The standard deviations were 3.2 mpg and 2.5 mpg, respectively.

a) An analyst for the consumer group computed the two-sample *t* 95% confidence interval for the difference between the two means as (−.71, 4.71). What conclusion would he reach based on this analysis?

b) Why is this procedure inappropriate? What assumption is violated?

c) In what way do you think this may have impacted the results?

35. Science scores. Newspaper headlines recently announced a decline in science scores among high school seniors. In 2000, 15,109 seniors tested by the National Assessment in Education Program (NAEP) scored a mean of 147 points. Four years earlier, 7537 seniors had averaged 150 points. The standard error of the difference in the mean scores for the two groups was 1.22.

a) Have the science scores declined significantly? Cite appropriate statistical evidence to support your conclusion.

b) The sample size in 2000 was almost double that in 1996. Does this make the results more convincing or less? Explain.

36. Credit card debt. The average household credit card debt has been reported to be between $8000 and $10,000. Often of interest is the average credit card debt carried by college students. In 2008, the average credit card debt for college students was reported to be $2200 based on 12,500 responses. A year earlier it was reported to be $2190 based on survey of 8200 college students. The standard error of the difference in mean credit card balances was $1.75.

a) Has the average credit card balance carried by college students increased significantly? Cite appropriate statistical evidence to support your conclusion.

b) Is this a meaningful difference to the typical student? Is it meaningful to a credit card company?

c) The sample size in 2008 is one and a half times that in 2007. Does this make the results more or less convincing? Explain.

37. The Internet. The NAEP report described in Exercise 35 compared science scores for students who had home Internet access with the scores of those who did not, as shown in the graph. They report that the differences are statistically significant.

a) Explain what "statistically significant" means in this context.

b) If their conclusion is incorrect, which type of error did the researchers commit?

c) Does this prove that using the Internet at home can improve a student's performance in science?

d) What companies might be interested in this information?

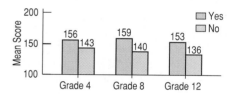

38. Credit card debt public or private. The average credit card debt carried by college students was compared at public versus private universities. It was reported that a significant difference existed between the two types of institutions and that students at private universities carried higher credit card debt.

a) Explain what "statistically significant" means in this context.

b) If this conclusion is incorrect, which type of error was committed?

c) Does this prove that students who choose to attend public institutions will carry lower credit card debt?

39. Pizza sales. A national food product company believes that it sells more frozen pizza during the winter months than during the summer months. Average weekly sales for a sample of stores in the Baltimore area over a three-year period provided the following data for sales volume (in pounds) during the two seasons.

Season	N	Mean	StDev	Minimum	Maximum
Winter	38	31234	13500	15312	73841
Summer	40	22475	8442	12743	54706

a) How much difference is there between the mean amount of this brand of frozen pizza sold (in pounds) between the two seasons? (Assume that this time frame represents typical sales in the Baltimore area.)

b) Construct and interpret a 95% confidence interval for the difference between weekly sales during the winter and summer months.

c) Suggest factors that might have influenced the sales of the frozen pizza during the winter months.

40. More pizza sales. Here's some additional information about the pizza sales data presented in Exercise 39. It is generally thought that sales spike during the weeks leading up to AFC and NFC football championship games, as well as leading up to the Super Bowl at the end of January each year. If we omit those 6 weeks of sales from this three-year period of weekly sales, the summary statistics look like this. Do sales appear to be higher during the winter months after omitting those weeks most influenced by football championship games?

Season	N	Mean	StDev	Minimum	Maximum
Winter	32	28995	9913	15312	48354
Summer	40	22475	8442	12743	54706

a) Write the null and alternative hypotheses.

b) Test the null hypotheses and state your conclusion.

c) Suggest additional factors that may influence pizza sales not accounted for in this exercise.

41. Olympic heats. In Olympic running events, preliminary heats are determined by random draw, so we should expect that the ability level of runners in the various heats to be about the same, on average. Here are the times (in seconds) for the 400-m women's run in the 2004 Olympics in Athens for preliminary heats 2 and 5. Is there any evidence that the mean time to finish is different for randomized heats? Explain. Be sure to include a discussion of assumptions and conditions for your analysis.

Country	Name	Heat	Time
USA	HENNAGAN Monique	2	51.02
BUL	DIMITROVA Mariyana	2	51.29
CHA	NADJINA Kaltouma	2	51.50
JAM	DAVY Nadia	2	52.04
BRA	ALMIRAO Maria Laura	2	52.10
FIN	MYKKANEN Kirsi	2	52.53
CHN	BO Fanfang	2	56.01
BAH	WILLIAMS-DARLING Tonique	5	51.20
BLR	USOVICH Svetlana	5	51.37
UKR	YEFREMOVA Antonina	5	51.53
CMR	NGUIMGO Mireille	5	51.90
JAM	BECKFORD Allison	5	52.85
TOG	THIEBAUD-KANGNI Sandrine	5	52.87
SRI	DHARSHA K V Damayanthi	5	54.58

42. Swimming heats. In Exercise 41 we looked at the times in two different heats for the 400-m women's run from the 2004 Olympics. Unlike track events, swimming heats are *not* determined at random. Instead, swimmers are seeded so that better swimmers are placed in later heats. Here are the times (in seconds) for the women's 400-m freestyle from heats 2 and 5. Do these results suggest that the mean times of seeded heats are not equal? Explain. Include a discussion of assumptions and conditions for your analysis.

Country	Name	Heat	Time
ARG	BIAGIOLI Cecilia Elizabeth	2	256.42
SLO	CARMAN Anja	2	257.79
CHI	KOBRICH Kristel	2	258.68
MKD	STOJANOVSKA Vesna	2	259.39
JAM	ATKINSON Janelle	2	260.00
NZL	LINTON Rebecca	2	261.58
KOR	HA Eun-Ju	2	261.65
UKR	BERESNYEVA Olga	2	266.30
FRA	MANAUDOU Laure	5	246.76
JPN	YAMADA Sachiko	5	249.10
ROM	PADURARU Simona	5	250.39
GER	STOCKBAUER Hannah	5	250.46
AUS	GRAHAM Elka	5	251.67
CHN	PANG Jiaying	5	251.81
CAN	REIMER Brittany	5	252.33
BRA	FERREIRA Monique	5	253.75

43. Tee tests. Does it matter what kind of tee a golfer places the ball on? The company that manufactures "Stinger" tees claims that the thinner shaft and smaller head will lessen resistance and drag, reducing spin and allowing the ball to travel farther. In August 2003, Golf Laboratories, Inc. compared the distance traveled by golf balls hit off regular wooden tees to those hit off Stinger tees. All the balls were struck by the same golf club using a robotic device set to swing the club head at approximately 95 miles per hour. Summary statistics from the test are shown in the table. Assume that 6 balls were hit off each tee and that the data were suitable for inference.

		Total Distance (yards)	Ball Velocity (mph)	Club Velocity (mph)
Regular tee	Avg.	227.17	127.00	96.17
	SD	2.14	0.89	0.41
Stinger tee	Avg.	241.00	128.83	96.17
	SD	2.76	0.41	0.52

Is there evidence that balls hit off the Stinger tees would have a higher initial velocity?

44. Tee tests, again. Given the test results on golf tees described in Exercise 43, is there evidence that balls hit off Stinger tees would travel farther? Again assume that 6 balls were hit off each tee and that the data were suitable for inference.

45. Marketing slogan. A company is considering marketing their classical music as "music to study by." Is this a valid slogan? In a study conducted by some Statistics students, 62 people were randomly assigned to listen to rap music, music by Mozart, or no music while attempting to memorize objects pictured on a page. They were then asked to list all the objects they could remember. Here are summary statistics for each group.

	Rap	Mozart	No Music
Count	29	20	13
Mean	10.72	10.00	12.77
SD	3.99	3.19	4.73

a) Does it appear that it is better to study while listening to Mozart than to rap music? Test an appropriate hypothesis and state your conclusion.

b) Create a 90% confidence interval for the mean difference in memory score between students who study to Mozart and those who listen to no music at all. Interpret your interval.

46. Marketing slogan, part 2. Using the results of the experiment described in Exercise 45, does it matter whether one listens to rap music while studying, or is it better to study without music at all?

a) Test an appropriate hypothesis and state your conclusion.

b) If you concluded there is a difference, estimate the size of that difference with a 90% confidence interval and explain what your interval means.

47. Mutual funds. You have heard that if you leave your money in mutual funds for a longer period of time, you will see a greater return. So you would like to compare the 3-year and 5-year returns of a random sample of mutual funds to see if indeed, your return is expected to be greater if you leave your money in the funds for 5 years.

a) Using the data provided, check the conditions for this test.

b) Write the null and alternative hypothesis for this test.

c) Test the hypothesis and find the P-value if appropriate.

d) What is your conclusion?

48. Mutual funds, part 2. An investor now tells you that if you leave your money in as long as 10 years, you will see an even greater return, so you would like to compare the 5-year and 10-year returns of a random sample of mutual funds to see if your return is expected to be greater if you leave your money in the funds for 10 years.

a) Using the data provided, check the conditions for this test.

b) Write the null and alternative hypothesis for this test.

c) Test the hypothesis and find the P-value if appropriate.

d) What is your conclusion?

49. Real estate. Residents of neighboring towns in a state in the United States have an ongoing disagreement over who lays claim to the higher average price of a single-family home. Since you live in one of these towns, you decide to obtain a random sample of homes listed for sale with a major local realtor to investigate if there is actually any difference in the average home price.

a) Using the data provided, check the conditions for this test.

b) Write the null and alternative hypothesis for this test.

c) Test the hypothesis and find the P-value.

d) What is your conclusion?

50. Real estate, part 2. Residents of one of the towns discussed in Exercise 49 claim that since their town is much smaller, the sample size should be increased. Instead of random sampling 30 homes, you decide to sample 42 homes from the database to test the difference in the mean price of single-family homes in these two towns.

a) Using the data provided, check the conditions for this test.

b) Write the null and alternative hypothesis for this test.

c) Test the hypothesis and find the P-value.

d) What is your conclusion? Did the sample size make a difference?

51. Home run. For the same reasons identified in Exercise 20, a friend of yours claims that the average number of home runs hit per game is higher in the American League than in the National League. Using the same 2007 data as in Exercises 20 and 23, you decide to test your friend's theory.

a) Using the data provided, check the conditions for this test.

b) Write the null and alternative hypothesis for this test.

c) Test the hypothesis and find the P-value.

d) What is your conclusion?

52. Statistics journals. When a professional statistician has information to share with colleagues, he or she will submit an article to one of several Statistics journals for publication. This can be a lengthy process; typically, the article must be circulated for "peer review" and perhaps edited before being accepted for publication. Then the article must wait in line with other articles before actually appearing in print. In the Winter 1998 issue of *Chance* magazine, Eric Bradlow and Howard Wainer reported on this delay for several journals between 1990 and 1994. For 288 articles published in *The American Statistician*, the mean length of time between initial submission and publication was 21 months, with a standard deviation of 8 months. For 209 *Applied Statistics* articles, the mean time to publication was 31 months, with a standard deviation of 12 months. Create and interpret a 90% confidence interval for the difference in mean delay, and comment on the assumptions that underlie your analysis.

JUST CHECKING ANSWERS

1 H_0: $\mu_{eyes} - \mu_{flowers} = 0$

2 ✓ **Independence Assumption:** The amount paid by one person should be independent of the amount paid by others.

   ✓ **Randomization Condition:** This study was observational. Treatments alternated a week at a time and were applied to the same group of office workers.

   ✓ **Nearly Normal Condition:** We don't have the data to check, but it seems unlikely there would be outliers in either group.

   ✓ **Independent Groups Assumptions:** The same workers were recorded each week, but week-to-week independence is plausible.

3 H_A: $\mu_{eyes} - \mu_{flowers} \neq 0$. An argument could be made for a one-sided test because the research hypothesis was that eyes would improve honest compliance.

4 Office workers' compliance in leaving money to pay for food at an office coffee station was different when a picture of eyes was placed behind the "honesty box" than when the picture was one of flowers.

Paired Samples and Blocks

The Coke/Pepsi Rivalry

Pepsi-Cola and Coca-Cola were introduced in the late 1800s as fountain beverages (Coke in 1886 and Pepsi a few years later), and they have dominated the worldwide soft drink industry virtually ever since then. The origins of the names have intrigued people for almost as long as the products have existed. Pepsi-Cola first came out as "Brad's drink," after inventor pharmacist Caleb Bradham, but was changed to Pepsi-Cola in 1898 and trademarked in 1903. There are many theories for the name, but the two discussed on the PepsiCo website are that Bradham bought the name "Pep Kola" from a rival or that Pepsi-Cola is an anagram of Episcopal, after the church across the street from the birthplace of Pepsi. Another theory is that it was initially intended to cure stomach pains and that the name came from the condition for indigestion (dyspepsia) or from the use of pepsin root as an ingredient. Coca-Cola got its name from the fact that it originally contained trace amounts of both cocaine and kola nuts. The amount of cocaine was miniscule, but it helped protect the name and its marketing campaign as "the intellectual drink."

Both companies grew in the early 20th century as bottling started overtaking fountain sales. Coke's popularity grew as it stepped up its bottling efforts and with the introduction of the distinctive 6-ounce contour bottle in 1916. In 1923 high sugar prices forced Pepsi-Cola into bankruptcy, and the Pepsi trademark was sold to Roy C. Megargel. Eight years later, the company went bankrupt again, resulting in a reformulation of the Pepsi-Cola syrup formula.

During the Great Depression, Pepsi gained ground on Coke after the introduction in 1934 of a 12-ounce bottle. With twelve ounces in a bottle instead of the six ounces sold by Coca-Cola, Pepsi turned the price difference to its advantage with a radio advertising campaign. This was the first use of a jingle in advertising: "Pepsi cola hits the spot/ Twelve full ounces, that's a lot/ Twice as much for a nickel, too/ Pepsi-Cola is the drink for you." The campaign succeeded as price-conscious consumers made the switch, and Pepsi-Cola's profits doubled. Meanwhile, Coca-Cola started expanding into the rest of the world and by 1940 was selling and being bottled in 44 countries.

After WWII, the war between the soft drink companies heated up, and marketing became increasingly used to distinguish the two products. In the 1940s and 1950s Pepsi specifically targeted African-American consumers. In the 1960s, Pepsi attempted to capture the newly financially empowered teen market with the slogan "The Pepsi Generation." After Royal Crown Cola's success with a diet cola, Diet Pepsi was introduced in 1964 and became the first diet drink popular across the entire United States. In 1966, Pepsi merged with the snack food company Frito-Lay to become PepsiCo.

Since Pepsi's merger with Frito-Lay, the two companies have followed different paths, with PepsiCo becoming a global food giant and Coke remaining in the drink business. Both have been enormously successful. Coca-Cola still outsells Pepsi-Cola in most countries, but PepsiCo is a more diversified company.

The diet soft drink market has grown steadily since artificially sweetened beverages were introduced in the 1960s. Much of the competition between Coke and Pepsi takes place in this arena. One of the challenges of formulating diet drinks is that aspartame (marketed as NutraSweet®) loses sweetness over time. One important consideration when companies such as Coke and Pepsi test new diet colas is how well flavor is retained during storage. When subtle changes in taste are the issue, companies often use carefully designed tasting experiments employing trained tasters.

Trained tasters practice matching and describing flavors. For this study, ten tasters were given three sugar solutions of increasing sweetness to establish a numerical scale of sweetness from 1 to 10. Then each taster tasted the cola and rated its sweetness on this scale. The cola samples were then stored at elevated temperature for one month to simulate four months of shelf storage.

One month later all ten tasters returned to the laboratory, where they again rated the cola on the same scale. Because the same ten tasters rated the cola before and after storage, the measurements are not independent, so we can't use a two-sample *t*-test. Not only isn't it appropriate in this situation, but the company testing the cola isn't interested in how sweet the tasters found the cola. They are interested in whether there was a perceptible loss in sweetness. That is, they are interested in any *difference* in sweetness as perceived by each taster.

14.1 Paired Data

Data such as these are called **paired**. Paired data arise in a number of ways. Perhaps the most common is a situation like this when we compare measurements at two different times. When pairs arise from an experiment, we compare measurements before and after a treatment, and the pairing is a type of *blocking*. When they arise from an observational study, it is called *matching*.

Pairing isn't a problem; it's an opportunity. If you know the data are paired, you can take advantage of the pairing—in fact, you *must* take advantage of it. You *should not* use the two-sample (or pooled two-sample) method when the data are paired. Be careful. There is no statistical test to determine whether the data are paired. You must decide whether the data are paired from understanding how they were collected and what they mean (check the W's).

Once we recognize that the taste test data are matched pairs, it makes sense to consider the change in each taster's perceptions. That is, we look at the collection of pairwise differences. The company is interested in the *difference* in the ratings before and after the storage time. Because it is the *differences* we care about, we can treat them as if there was a single variable of interest holding those differences. With only one variable to consider, we can use a simple one-sample *t*-test. A **paired *t*-test** is just a one-sample *t*-test for the mean of the pairwise differences. The sample size is the number of pairs.

14.2 Assumptions and Conditions

Paired Data Assumption

The data must actually be paired. You can't just decide to pair data from independent groups. When you have two groups with the same number of observations, it may be tempting to match them up, but that's not valid. You can't pair data just because they "seem to go together." To use paired methods you must determine from knowing how the data were collected whether the two groups are paired or independent. Usually the context will make it clear.

Be sure to recognize paired data when you have it. Remember, two-sample *t*-methods aren't valid unless the groups are independent, and paired groups aren't independent.

Independence Assumption

For these methods, it's the *differences* that must be independent of each other. This is just the one-sample *t*-test assumption of independence, now applied to the differences. In our example, one rater's opinion shouldn't affect how another person rated the colas. As always, randomization helps to ensure independence.

Randomization Condition. Randomness can arise in many ways. The *pairs* may be a random sample. For example, we may be comparing opinions of husbands and wives from a random selection of couples. In an experiment, the order of the two treatments may be randomly assigned, or the treatments may be randomly assigned to one member of each pair. In a before-and-after study, like this one, we may believe that the observed differences are a representative sample from a population of interest. If we have any doubts, we'll need to include a control group to be able to draw conclusions. What we want to know usually focuses our attention on where the randomness should be.

10% Condition. When we sample from a finite population, we should be careful not to sample more than 10% of that population. Sampling too large a fraction of the population calls the independence assumption into question. Here, we can regard our tasters as representative of the (potentially very large) population of trained tasters. As with other quantitative data situations, we don't usually explicitly check the 10% condition, but be sure to think about it.

Normal Population Assumption

We need to assume that the population of *differences* follows a Normal model. We don't need to check the data in each of the two individual groups. In fact, the data from each group can be quite skewed, but the differences can still be unimodal and symmetric.

Nearly Normal Condition. This condition can be checked with a histogram of the differences. As with the one-sample *t*-methods, this assumption matters less as we have more pairs to consider. You may be pleasantly surprised when you check this condition. Even if your original measurements are skewed or bimodal, the *differences* may be nearly Normal. After all, the individual who was way out in the tail on an initial measurement is likely to still be out there on the second one, giving a perfectly ordinary difference.

14.3 The Paired *t*-Test

The paired *t*-test is mechanically a one-sample *t*-test. We treat the differences as our variable. We simply compare the mean difference to its standard error. If the *t*-statistic is large enough, we reject the null hypothesis.

Paired *t*-test

When the conditions are met, we are ready to test whether the mean paired difference is significantly different from a hypothesized value (called Δ_0). We test the hypothesis:

$$H_0: \mu_d = \Delta_0,$$

where the *d*'s are the pairwise differences and Δ_0 is almost always 0.

We use the statistic:

$$t_{n-1} = \frac{\bar{d} - \Delta_0}{SE(\bar{d})},$$

where $\bar{d}$ is the mean of the pairwise differences, *n* is the number of *pairs*, and

$$SE(\bar{d}) = \frac{s_d}{\sqrt{n}},$$

where s_d is the standard deviation of the pairwise differences.

When the conditions are met and the null hypothesis is true, the sampling distribution of this statistic is a Student's *t*-model with $n - 1$ degrees of freedom and we use that model to obtain the P-value.

Similarly, we can construct a confidence interval for the true difference. As in a one-sample *t*-interval, we center our estimate at the mean difference in our data. The margin of error on either side is the standard error multiplied by a critical *t*-value (based on our confidence level and the number of pairs we have).

Paired *t*-interval

When the conditions are met, we are ready to find the confidence interval for the mean of the paired differences. The confidence interval is:

$$\bar{d} \pm t^*_{n-1} \times SE(\bar{d}),$$

where the standard error of the mean difference is $SE(\bar{d}) = \frac{s_d}{\sqrt{n}}$.

The critical value t^* from the Student's *t*-model depends on the particular confidence level that you specify and on the degrees of freedom, $n - 1$, which is based on the number of pairs, *n*.

GUIDED EXAMPLE | *Loss of Sweetness in a Cola*

PLAN

Setup State what we want to know.

Identify the variables and the parameter to estimate.

For a paired analysis, the parameter of interest is the mean of the differences.

Hypotheses State the null and alternative hypotheses.

We want to know whether the sweetness of the tested diet cola changed after one month of storage at elevated temperatures. We have sweetness ratings by ten trained taste testers before and after the storage, and we will examine sweetness loss, (d = sweetness before − sweetness after).

H_0: Storage did not change the sweetness; the mean difference is zero: $\mu_d = 0$.

A loss of sweetness would mean a decrease in the sweetness rating, so the *Before* minus *After* difference would be positive.

H_A: The colas lost sweetness: $\mu_d > 0$.

Model Check the conditions.

✓ **Paired Data Assumption:** The data are paired because they are for the same focus group participants.

✓ **Independence Assumption:** The tasters are professionals and we assume that their ratings are independent of each other.

✓ **Randomization Condition:** The taster's responses are assumed to be representative of all tasters.

✓ **Nearly Normal Condition:** Even though these data are not strictly quantitative, the histogram of the differences is roughly unimodal and symmetric. With only 10 values, it is difficult to assess Normality, but we can see that there are no outliers or extreme skewness.

Make a graph to check the nearly normal condition.

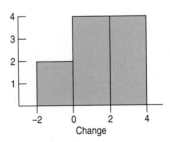

State the sampling distribution model.

Choose your method.

The conditions are met, so we can use a Student's t-model with $(n - 1) = 9$ degrees of freedom and perform a paired t-test.

Mechanics n is the number of *pairs*, here, the number of tasters.

$\bar{d}$ is the mean difference.

s_d is the standard deviation of the differences.

$$n = 10 \text{ tasters}$$
$$\bar{d} = 1.02$$
$$s_d = 1.196$$

We estimate the standard error of $\bar{d}$ as:

$$SE(\bar{d}) = \frac{s_d}{\sqrt{n}} = \frac{1.196}{\sqrt{10}} = 0.378$$

The df for the t-model is $n - 1 = 9$.

We find $t = \dfrac{\bar{d} - 0}{SE(\bar{d})} = \dfrac{1.02}{0.378} = 2.697$

A t-statistic with 9 degrees of freedom and a value of 2.697 has a one-sided P-value of 0.0123.

REALITY CHECK ▶	This result makes sense, since the majority of the differences in sweetness were positive.	On average, the difference in ratings was 1.02 on a 10-point scale of sweetness that ranges from not sweet at all to very sweet. That was a perceptible difference, at least to the trained taste testers.
REPORT	**Conclusion** Interpret the results of the hypothesis test in context.	**MEMO:** **Re: Diet Cola Formulation** In an industrial laboratory taste test employing ten trained tasters, the loss in sweetness of the diet cola formulation due to the degradation of aspartame was detectable. The laboratory estimates that their test simulated four months of storage. If the diet cola needs to preserve flavor for that long or longer, we may need to consider adjusting the formula. The mean sweetness loss was about one point in a scale from 1 to 10. We don't know if the average consumer would detect an effect of this size.

14.4 How the Paired *t*-Test Works

When data are paired, the *t*-test can sometimes see differences that a two-sample *t*-test can't. Even though they are trained, the judges will not all agree with each other. But we want to know about our product, not about the judges. By focusing on the *differences* of each taster's score, we see past that variation so that we can more easily detect changes in what we want to know.

A paired design has roughly half the degrees the freedom of the two sample *t*-test. Typically, we'd want *more* degrees of freedom, but usually pairing more than compensates for this by reducing the variation. We could have employed two independent teams (one group tasting before storage and one tasting it afterward), but using the *same* judges not only seems more sensible, it is more efficient.

Unfortunately, you can't take the benefit of pairing unless the data are actually paired. If you're designing the study, you may be able to arrange for the data to be paired (before vs. after; using the same people to test two different methods; tracking the same customers in two different months, etc.). Experimental designers call this general technique *blocking* (see Chapter 23). You may also be able to use pairing even when the data don't arise from an experiment if you can justify the matching. But, be careful. If the data from the two groups are independent, you may not pair them just because the groups have the same number of observations. There must be a link that you can identify and justify between the pairs. Data on husbands and wives, or observations made before and after some event on the same people, companies, or subjects are naturally paired.

✓ JUST CHECKING

Think about each of the following situations. Would you use a two-sample t or paired t-method (or neither)? Why?

1 Random samples of 50 men and 50 women are surveyed on the amount they invest on average in the stock market on an annual basis. We want to estimate any gender difference in how much they invest.

2 Random samples of students were surveyed on their perception of ethical and community service issues both in their first year and fourth year at a university. The university wants to know whether their required programs in ethical decision-making and service learning change student perceptions.

3 A random sample of work groups within a company was identified. Within each work group, one male and one female worker were selected at random. Each was asked to rate the secretarial support that their workgroup received. When rating the same support staff, do men and women rate them equally on average?

4 A total of 50 companies are surveyed about business practices. They are categorized by industry, and we wish to investigate differences across industries.

5 These same 50 companies are surveyed again one year later to see if their perceptions, business practices, and R&D investment have changed.

GUIDED EXAMPLE   *Seasonal Spending*

Economists and credit card banks know that people tend to spend more near the holidays in December. In fact, sales in the few days after Thanksgiving (the fourth Thursday of the month in the United States) provide an indication of the strength of the holiday season sales and an early look at the strength of the economy in general. After the holidays, spending decreases substantially. Because credit card banks receive a percentage of each transaction, they need to forecast how much the average spending will increase or decrease from month to month. How much less do people tend to spend in January than December? For any particular segment of cardholders, a credit card bank could select two random samples—one for each month—and simply compare the average amount spent in January with that in December. A more sensible approach might be to select a single random sample and compare the spending between the two months for *each cardholder*. Designing the study in this way and examining the paired differences gives a more precise estimate of the actual change in spending.

Here we have a sample of cardholders from a particular market segment and the amount they charged on their credit card in both December 2004 and January 2005. (There were 1000 cardholders in the original sample, but 89 of them had at least one month missing leaving a sample of $n = 911$.) We can create a paired *t*-confidence interval to estimate the true mean difference in spending between the two months.

WHO	Cardholders in a particular market segment of a major credit card issuer
WHAT	Amount charged on their credit card in December and January
WHEN	2004–2005
WHERE	United States
WHY	To estimate the amount of decrease in spending one could expect after the holiday shopping season

PLAN

Setup State what we want to know.

Identify the *parameter* we wish to estimate and the sample size.

We want to know how much we can expect credit card charges to change, on average, from December to January for this market segment. We have the total amount charged in December 2004 and January 2005 for *n* = 911 cardholders in this segment. We want to find a confidence interval for the true mean difference in charges between these two months for all cardholders in this segment. Because we know that people tend to spend more in December, we will look at the difference: *December spend − January spend*. A positive difference will mean a *decrease* in spending.

Model Check the conditions.

State why the data are paired. Simply having the same number of individuals in each group or displaying them in side-by-side columns, doesn't make them paired.

✓ **Paired Data Assumption:** The data are paired because they are measurements on the same cardholders in two different months.

✓ **Independence Assumption:** The behavior of any individual is independent of the behavior of the others, so the differences are mutually independent.

Think about what we hope to learn and where the randomization comes from.

✓ **Randomization Condition:** This was a random sample from a large market segment.

Make a picture of the differences. Don't plot separate distributions of the two groups—that would entirely miss the pairing. For paired data, it's the Normality of the differences that we care about. Treat those paired differences as you would a single variable, and check the Nearly Normal condition.

✓ **Nearly Normal Condition:** The distribution of the differences is unimodal and symmetric. Although there are many observations nominated by the boxplot as outliers, the distributions are symmetric. (This is typical of the behavior of credit card spending.) There are no isolated cases that would unduly dominate the mean difference, so we will leave all observations in the study.

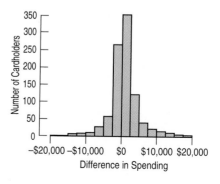

Specify the sampling distribution model.

Choose the method.

The conditions are met, so we'll use a Student's *t*-model with $(n - 1) = 910$ degrees of freedom and find a **paired *t*-confidence interval**.

DO

Mechanics *n* is the number of *pairs*, in this case, the number of cardholders.

$\bar{d}$ is the mean difference.

s_d is the standard deviation of the differences.

Make a picture. Sketch a *t*-model centered at the observed mean of 788.18.

The computer output tells us:

$$n = 911 \text{ pairs}$$
$$\bar{d} = \$788.18$$
$$s_d = \$3,740.22$$

Find the standard error and the *t*-score of the observed mean difference. There is nothing new in the mechanics of the paired *t*-methods. These are the mechanics of the *t*-interval for a mean applied to the differences.

We estimate the standard error of $\bar{d}$ using:

$$SE(\bar{d}) = \frac{s_d}{\sqrt{n}} = \frac{3740.22}{\sqrt{911}} = \$123.919$$

$$t^*_{910} = 1.96$$

The margin of error, $ME = t^*_{910} \times SE(\bar{d})$

$$= 1.96 \times 123.919 = 242.88$$

So a 95% CI is $\bar{d} \pm ME = (\$545.30, \$1031.06)$.

REPORT

Conclusion Link the results of the confidence interval to the context of the problem.

MEMO:

Re: Credit Card Expenditure Changes

In the sample of cardholders studied, the change in expenditures between December and January averaged $788.18, which means that, on average, cardholders spend $788.18 less in January than the month before. Although we didn't measure the change for all cardholders in the segment, we can be 95% confident that the true mean decrease in spending is between $545.30 to $1031.06.

WHAT CAN GO WRONG?

- **Don't use a paired *t*-method when the samples aren't paired.** When two groups don't have the same number of values, it's easy to see that they can't be paired. But just because two groups have the same number of observations doesn't mean they can be paired, even if they are shown side-by-side in a table. We might have 25 men and 25 women in our study, but they might be completely independent of one another. If they were siblings or spouses, we might consider them paired. Remember that you cannot *choose* which method to use based on your preferences. Only if the data are from an experiment or study in which observations were paired, can you use a paired method.

- **Don't forget outliers.** The outliers we care about now are in the differences. A subject who is extraordinary both before and after a treatment may still have a perfectly typical difference. But one outlying difference can completely distort your conclusions. Be sure to plot the differences (even if you also plot the data).

- **Don't look for the difference in side-by-side boxplots.** The point of the paired analysis is to remove extra variation. The boxplots of each group still contain that variation. Comparing them is likely to be misleading.

ETHICS IN ACTION

Boyd Casey, a scientist at a regional university, has made a name for himself by researching the health benefits of a variety of substances (e.g., caffeine, vitamin C, etc.). Recently he was approached by Nature's Plenty Inc., a company specializing in a limited line of health food supplements and products, about doing some research on the *acai* berry. Touted as a super food, rich in antioxidants, *acai* also contains a significant amount of essential fatty acids (omega-6 and omega-9) believed to reduce cholesterol levels. Nature's Plenty is considering adding some products made with *acai*, but would first like to establish its ability to reduce cholesterol. They have available a list of 100 customers of a health-food store who regularly consume *acai* and another 100 customers who do not consume *acai*.

Boyd collects cholesterol levels on both groups. He is disappointed to find that the difference in cholesterol isn't large enough for statistical significance. But then he tries another approach. He sorts the participants in each group in order according to their cholesterol level, lowest to highest. Then he matches them up, comparing the participant with the lowest cholesterol level in the *acai* group with the participant with the lowest cholesterol level in the non-*acai* group and computing a paired *t*-test. He is pleased that now he can find a statistically significant difference between the groups, and he prepares to report that result to Nature's Plenty.

ETHICAL ISSUE *Inappropriate analysis. Boyd's groups are not paired. Sorting and aligning them is inappropriate and leads to an incorrect analysis. (Related to Items II A2 and E5, ASA Ethical Guidelines.)*

ETHICAL SOLUTION *He should use a two-sample t-test and report a failure to reject the null hypothesis.*

What have we learned?

Pairing can be an effective strategy. Because pairing can help control variability between individual subjects, paired methods are usually more powerful than methods that compare independent groups. Now we've learned that analyzing data from matched pairs requires different inference procedures.

- We've learned that paired *t*-methods look at pairwise differences. Based on these differences, we test hypotheses and generate confidence intervals. Our procedures are mechanically identical to the one-sample *t*-methods.

- We've also learned to think about the design of the study that collected the data before we proceed with inference. We must be careful to recognize pairing when it is present but not assume it when it is not. Making the correct decision about whether to use independent *t*-procedures or paired *t*-methods is the first critical step in analyzing the data.

Terms

Paired data
Data are paired when the observations are collected in pairs or the observations in one group are naturally related to observations in the other. The simplest form of pairing is to measure each subject twice—often before and after a treatment is applied. Pairing in experiments is a form of blocking and arises in other contexts. Pairing in observational and survey data is a form of matching.

Paired *t*-test
A hypothesis test for the mean of the pairwise differences of two groups. It tests the null hypothesis $H_0: \mu_d = \Delta_0$, where the hypothesized difference is almost always 0, using the statistic $t = \dfrac{\bar{d} - \Delta_0}{SE(\bar{d})}$ with $n - 1$ degrees of freedom, where $SE(\bar{d}) = \dfrac{s_d}{\sqrt{n}}$ and n is the number of pairs.

Paired *t*-confidence interval
A confidence interval for the mean of the pairwise differences between paired groups found as $\bar{d} \pm t^*_{n-1} \times SE(\bar{d})$, where $SE(\bar{d}) = \dfrac{s_d}{\sqrt{n}}$ and n is the number of pairs.

Skills

 PLAN
- Recognize whether a design that compares two groups is paired or not.

DO
- Know how to find a paired confidence interval, recognizing that it is mechanically equivalent to doing a one-sample *t*-interval applied to the differences.

- Be able to perform a paired *t*-test, recognizing that it is mechanically equivalent to a one-sample *t*-test applied to the differences.

 REPORT
- Know how to interpret a paired *t*-test, recognizing that the hypothesis tested is about the mean of the differences between paired values rather than about the differences between the means of two independent groups.

- Know how to interpret a paired *t*-interval, recognizing that it gives an interval for the mean difference in the pairs.

Technology Help: Paired *t*

Most statistics programs can compute paired-*t* analyses. Some may want you to find the differences yourself and use the one-sample *t*-methods. Those that perform the entire procedure will need to know the two variables to compare. The computer, of course, cannot verify that the variables are naturally paired. Most programs will check whether the two variables have the same number of observations, but some stop there, and that can cause trouble.

Most programs will automatically omit any pair that is missing a value for either variable. You must look carefully to see whether that has happened.

As we've seen with other inference results, some packages pack a lot of information into a simple table, but you must locate what you want for yourself. Here's a generic example with comments

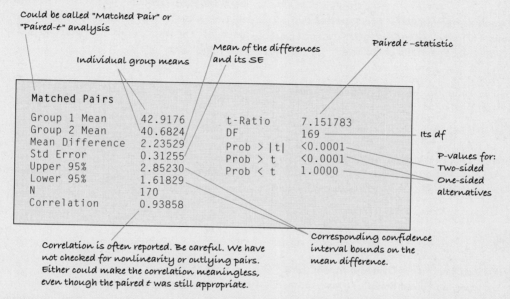

Could be called "Matched Pair" or "Paired-*t*" analysis

Individual group means

Mean of the differences and its SE

Paired *t*-statistic

```
Matched Pairs

Group 1 Mean       42.9176        t-Ratio      7.151783
Group 2 Mean       40.6824        DF           169
Mean Difference     2.23529       Prob > |t|   <0.0001
Std Error           0.31255       Prob > t     <0.0001
Upper 95%           2.85230       Prob < t      1.0000
Lower 95%           1.61829
N                 170
Correlation         0.93858
```

Its df

P-values for:
Two-sided
One-sided
alternatives

Correlation is often reported. Be careful. We have not checked for nonlinearity or outlying pairs. Either could make the correlation meaningless, even though the paired *t* was still appropriate.

Corresponding confidence interval bounds on the mean difference.

Other packages try to be more descriptive. It may be easier to find the results, but you may get less information from the output table.

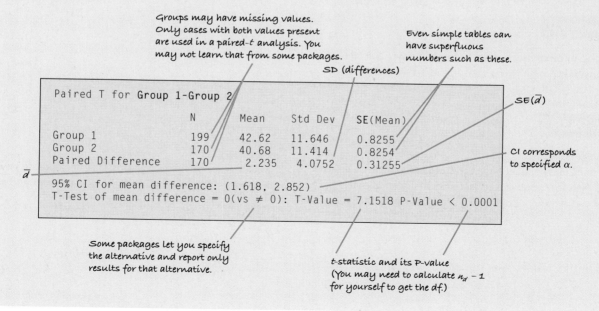

Groups may have missing values. Only cases with both values present are used in a paired-*t* analysis. You may not learn that from some packages.

SD (differences)

Even simple tables can have superfluous numbers such as these.

SE($\bar{d}$)

```
Paired T for Group 1-Group 2

                    N       Mean    Std Dev   SE(Mean)
Group 1            199     42.62     11.646    0.8255
Group 2            170     40.68     11.414    0.8254
Paired Difference  170      2.235     4.0752   0.31255

95% CI for mean difference: (1.618, 2.852)
T-Test of mean difference = 0(vs ≠ 0): T-Value = 7.1518 P-Value < 0.0001
```

$\bar{d}$

CI corresponds to specified α.

Some packages let you specify the alternative and report only results for that alternative.

t-statistic and its P-value (You may need to calculate $n_d - 1$ for yourself to get the df.)

Computers make it easy to examine the boxplots of the two groups and the histogram of the differences—both important steps. Some programs offer a scatterplot of the two variables. That can be helpful. In terms of the scatterplot, a paired *t*-test is about whether the points tend to be above or below the 45° line $y = x$. (Note that pairing says nothing about whether the scatterplot should be straight. That doesn't matter for our *t*-methods.)

EXCEL

In Excel 2003 and earlier, select **Data Analysis** from the **Tools** menu.
In Excel 2007, select **Data Analysis** from the **Analysis** Group on the **Data** Tab.
From the **Data Analysis** menu, choose **t-test: paired two-sample for Means**. Fill in the cell ranges for the two groups, the hypothesized difference, and the alpha level.

Warning: Do not compute this test in Excel without checking for missing values. If there are any missing values (empty cells), Excel will usually give a wrong answer. Excel compacts each list, pushing values up to cover the missing cells, and then checks only that it has the same number of values in each list. The result is mismatched pairs and an entirely wrong analysis.

Comments

Excel expects the two groups to be in separate cell ranges.

MINITAB

From the **Stat** menu, choose the **Basic Statistics** submenu. From that menu, choose **Paired t...** Then fill in the dialog boxes for the two paired samples, or fill in the summary data for the differences.

Comments

Minitab takes "First sample" minus "Second sample."

SPSS

From the **Analyze** menu, choose the **Compare Means** submenu. From that, choose the **Paired-Samples t-test** command. Select pairs of variables to compare, and click the arrow to add them to the selection box.

Comments

You can compare several pairs of variables at once. Options include the choice to exclude cases missing in any pair from all tests.

JMP

From the **Analyze** menu, select **Matched Pairs**. Specify the columns holding the two groups in the **Y Paired Response** Dialog. Click **OK**.

DATA DESK

Select variables.
From the **Calc** menu, choose **Estimate** for confidence intervals or **Test** for hypothesis tests. Select the interval or test from the drop-down menu, and make other choices in the dialog.

Comments

Data Desk expects the two groups to be in separate variables and in the same "Relation"—that is, about the same cases.

Mini Case Study Projects

A Taste Test (Data Collection and Analysis)

Assume that you are a marketing manager at PepsiCo and that you have been asked to gather taste data. Your biggest competitor for one of your leading soft drinks (Diet Pepsi) is Diet Coke. Invite a random sample of individuals to participate in a focus group. For each participant, present them with the colas in random order with a glass of water to sip in between. Ask them to rate each cola on a scale from 1–7.

Once you have the data collected, check the necessary assumptions and conduct the appropriate statistical tests to examine if the participants prefer one cola to another. State your methods and conclusions in a report from the perspective of a marketing manager at PepsiCo. You may need to get approval from your University's Institutional Research Board before conducting your own taste test. If that proves problematic, analyze data in **ch14_MCSP_Taste_Test** from a test already run by a professor of Marketing (the original form used for this experiment is found in **ch14_tastetestsurvey.doc**). The question of interest is to see whether participants switched preferences before and after the taste test. Look at any difference of Post − Pre (on Taste, Freshness, or Quality) to see if preferences changed.

Consumer Spending Patterns (Data Analysis)

You are on the financial planning team for monitoring a high spending segment of a credit card. You know that customers tend to spend more during December before the holidays, but you're not sure about the pattern of spending in the months after the holidays. Look at the data set **ch14_MCSP_Consumer_spending**. It contains the monthly credit card spending of 1200 customers during the months December, January, February, and March. Report on the spending differences between the months. If you had failed to realize that these are paired data, what difference would that have made in your reported confidence intervals and tests?

EXERCISES

1. Egg production. Can a food additive increase egg production? Egg producers want to design an experiment to find out. They have 100 hens available. They have two kinds of feed: the regular feed and the new feed with the additive. They plan to run their experiment for one month, recording the number of eggs each hen produces.

a) Design an experiment that will require a two-sample t procedure to analyze the results.

b) Design an experiment that will require a paired t procedure to analyze the results.

c) Which experiment would you consider the stronger design? Why?

2. Productivity and music. Some offices pipe in background music. The vendor claims this improves productivity, but might it cause more distraction? A firm's HR department wants to learn whether productivity is affected by background music. They hire a research firm to conduct an experiment. The researchers will time some volunteers to see how long it takes them to complete some relatively easy crossword puzzles. During some of the trials, the room will be quiet; during other trials in the same room, background music will be piped in.

a) Design an experiment that will require a two-sample t procedure to analyze the results.

b) Design an experiment that will require a paired t procedure to analyze the results.

c) Which experiment would you consider the stronger design? Why?

3. Advertisements. Ads for many products use sexual images to try to attract attention to the product, but do these ads bring people's attention to the item being advertised? A company wants to design an experiment to see if the presence of sexual images in an advertisement affects people's ability to remember the product.

a) Describe an experimental design that would require a paired *t* procedure to analyze the results.

b) Describe an experimental design that would require an independent sample procedure to analyze the results.

4. All you can eat. Some sports arenas and ballparks are offering "all you can eat" sections where, for a higher ticket price, fans can feast on all the hot dogs and popcorn they want. (Alcohol and desserts are extra.) But, of course, the teams want to price those tickets appropriately. They want to design an experiment to determine how much fans are likely to eat in an "all you can eat" section and whether it is more or less than they might ordinarily eat in similar regular seats.

a) Design an experiment that would require a two-sample *t* procedure for analysis.

b) Design an experiment that would require a paired *t* procedure for analysis.

5. Labor force. Values for the labor force participation rate (proportion) of women (LFPR) are published by the U.S. Bureau of Labor Statistics. We are interested in whether there was a difference between female participation in 1968 and 1972, a time of rapid change for women. We check LFPR values for 19 randomly selected cities for 1968 and 1972. Here is software output for two possible tests.

Paired t-Test of $\mu(1 - 2)$
Test Ho: $\mu(1972 - 1968) = 0$ vs Ha: $\mu(1972 - 1968) \neq 0$
Mean of Paired Differences = 0.0337
t-Statistic = 2.458 w/18 df
p = 0.0244

2-Sample t-Test of $\mu1 - \mu2$
Ho: $\mu1 - \mu2 = 0$ Ha: $\mu1 - \mu2 \neq 0$
Test Ho: $\mu(1972) - \mu(1968) = 0$ vs
Ha: $\mu(1972) - \mu(1968) \neq 0$
Difference Between Means = 0.0337
t-Statistic = 1.496 w/35 df
p = 0.1434

a) Which of these tests is appropriate for these data? Explain.

b) Using the test you selected, state your conclusion.

6. Rain. It has long been a dream of farmers to summon rain when it is needed for their crops. Crop losses to drought have significant economic impact. One possibility is cloud seeding in which chemicals are dropped into clouds in an attempt to induce rain. Simpson, Alsen, and Eden (*Technometrics*, 1975) report the results of trials in which clouds were seeded and the amount of

rainfall recorded. The authors report on 26 seeded (Group 2) and 26 unseeded (Group 1) clouds. Each group has been sorted in order of the amount of rainfall, largest amount first. Here are two possible tests to study the question of whether cloud seeding works.

Paired t-Test of $\mu(1 - 2)$
Mean of Paired Differences = -277.4
t-Statistic = -3.641 w/25 df p = 0.0012

2-Sample t-Test of $\mu1 - \mu2$
Difference Between Means = -277.4
t-Statistic = -1.998 w/33 df p = 0.0538

a) Which of these tests is appropriate for these data? Explain.

b) Using the test you selected, state your conclusion.

7. Friday the 13th. The *British Medical Journal* published an article titled, "Is Friday the 13th Bad for Your Health?" Researchers in Britain examined how Friday the 13th affects human behavior. One question was whether people tend to stay at home more on Friday the 13th. The data show the number of cars passing Junctions 9 and 10 on the M25 motorway for consecutive Fridays (6th and 13th) for five different time periods.

Year	Month	6th	13th
1990	July	134,012	132,908
1991	September	133,732	131,843
1991	December	121,139	118,723
1992	March	124,631	120,249
1992	November	117,584	117,263

Here are summaries of two possible analyses.

Paired t-Test of $\mu1 = \mu2$ vs. $\mu1 > \mu2$
Mean of Paired Differences: 2022.4
t-Statistic = 2.9377 w/4 df
P = 0.0212

2-Sample t-Test of $\mu1 = \mu2$ vs. $\mu1 > \mu2$
Difference Between Means: 2022.4
t-Statistic = 0.4273 w/7.998 df
P = 0.3402

a) Which of the tests is appropriate for these data? Explain.

b) Using the test you selected, state your conclusion.

c) Are the assumptions and conditions for inference met?

8. Friday the 13th, part 2. The researchers in Exercise 7 also examined the number of people admitted to emergency rooms for vehicular accidents on 12 Friday evenings (6 each on the 6th and 13th).

Year	Month	6th Group 1	13th Group 2
1989	October	9	13
1990	July	6	12
1991	September	11	14
1991	December	11	10
1992	March	3	4
1992	November	5	12

Based on these data, is there evidence that more people are admitted on average on Friday the 13th? Here are two possible analyses of the data.

Paired t-Test of $\mu 1 = \mu 2$ vs. $\mu 1 < \mu 2$
Mean of Paired Differences = 3.333
t-Statistic = 2.7116 w/5 df
P = 0.0211

2-Sample t-Test of $\mu 1 = \mu 2$ vs. $\mu 1 < \mu 2$
Difference Between Means = 3.333
t-Statistic = 1.6644 w/9.940 df
P = 0.0636

a) Which of these tests is appropriate for these data? Explain.
b) Using the test you selected, state your conclusion.
c) Are the assumptions and conditions for inference met?

T 9. Online insurance. After seeing countless commercials claiming one can get cheaper car insurance from an online company, a local insurance agent was concerned that he might lose some customers. To investigate, he randomly selected profiles (type of car, coverage, driving record, etc.) for 10 of his clients and checked online price quotes for their policies. The comparisons are shown in the table. His statistical software produced the following summaries (where *PriceDiff = Local − Online*):

Variable	Count	Mean	StdDev
Local	10	799.200	229.281
Online	10	753.300	256.267
PriceDiff	10	45.9000	175.663

Local	Online	PriceDiff
568	391	177
872	602	270
451	488	−37
1229	903	326
605	677	−72
1021	1270	−249
783	703	80
844	789	55
907	1008	−101
712	702	10

At first, the insurance agent wondered whether there was some kind of mistake in this output. He thought the Pythagorean Theorem of Statistics should work for finding the standard deviation of the price differences—in other words, that $SD(Local − Online) = \sqrt{SD^2(Local) + SD^2(Online)}$. But when he checked, he found that $\sqrt{(229.281)^2 + (256.267)^2} = 343.864$, not 175.663 as given by the software. Tell him where his mistake is.

T 10. Windy. Alternative sources of energy are of increasing interest throughout the energy industry. Wind energy has great potential. But appropriate sites must be found for the turbines. To select the site for an electricity-generating wind turbine, wind speeds were recorded at several potential sites every 6 hours for a year. Two sites not far from each other looked good. Each had a

mean wind speed high enough to qualify, but we should choose the site with a higher average daily wind speed. Because the sites are near each other and the wind speeds were recorded at the same times, we should view the speeds as paired. Here are the summaries of the speeds (in miles per hour):

Variable	Count	Mean	StdDev
site2	1114	7.452	3.586
site4	1114	7.248	3.421
site2 − site4	1114	0.204	2.551

Is there a mistake in this output? Why doesn't the Pythagorean Theorem of Statistics work here? In other words, shouldn't $SD(site2 − site4) = \sqrt{SD^2(site2) + SD^2(site4)}$? But $\sqrt{(3.586)^2 + (3.421)^2} = 4.956$, not 2.551 as given by the software. Explain why this happened.

T 11. Online insurance, part 2. In Exercise 9, we saw summary statistics for 10 drivers' car insurance premiums quoted by a local agent and an online company. Here are displays for each company's quotes and for the difference (*Local − Online*):

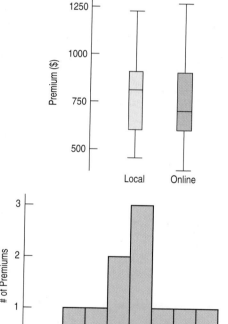

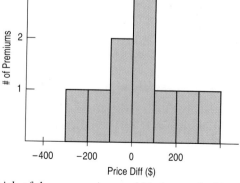

a) Which of the summaries would help you decide whether the online company offers cheaper insurance? Why?
b) The standard deviation of *PriceDiff* is quite a bit smaller than the standard deviation of prices quoted by either the local or online companies. Discuss why.
c) Using the information you have, discuss the assumptions and conditions for inference with these data.

12. Windy, part 2. In Exercise 10, we saw summary statistics for wind speeds at two sites near each other, both being considered as locations for an electricity-generating wind turbine. The data, recorded every 6 hours for a year, showed each of the sites had a mean wind speed high enough to qualify, but how can we tell which site is best? Here are some displays:

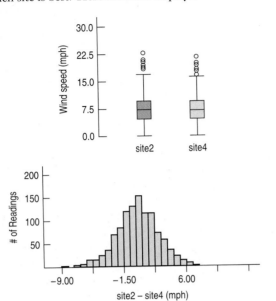

a) The boxplots show outliers for each site, yet the histogram shows none. Discuss why.

b) Which of the summaries would you use to select between these sites? Why?

c) Using the information you have, discuss the assumptions and conditions for paired *t* inference for these data. (*Hint:* Think hard about the independence assumption in particular.)

13. Online insurance, part 3. Exercises 9 and 11 give summaries and displays for car insurance premiums quoted by a local agent and an online company. Test an appropriate hypothesis to see if there is evidence that drivers might save money by switching to the online company.

14. Windy, part 3. Exercises 10 and 12 give summaries and displays for two potential sites for a wind turbine. Test an appropriate hypothesis to see if there is evidence that either of these sites has a higher average wind speed.

15. Wheelchair marathon. The Boston Marathon has had a wheelchair division since 1977. Who do you think is typically faster, the men's marathon winner on foot or the women's wheelchair marathon winner? Because the conditions differ year to year and speeds have improved over the years, it seems best to treat these as paired measurements. Here are summary statistics for the pairwise differences in finishing time (in minutes):

Summary of wheelchairF − runM
N = 31
Mean = −2.12097
SD = 33.4434

a) Comment on the assumptions and conditions.

b) Assuming that these times are representative of such races and the differences appeared acceptable for inference, construct and interpret a 95% confidence interval for the mean difference in finishing times.

c) Would a hypothesis test at $\alpha = 0.05$ reject the null hypothesis of no difference? What conclusion would you draw?

16. Boston start-up years. When we considered the Boston Marathon in Exercise 15 we were unable to check the Nearly Normal Condition. Here's a histogram of the differences.

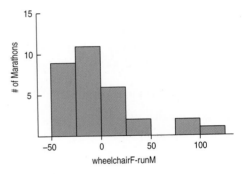

Those three large differences are the first three years of wheelchair competition, 1977, 1978, and 1979. Often the start-up years of new events are different; later on more athletes train and compete. If we omit those three years, the summary statistics change as follows.

Summary of wheelchairF − runM
N = 28
Mean = 12.1780
SD = 19.5116

a) Comment on the assumptions and conditions.

b) Assuming that these times are representative of such races, construct and interpret a 95% confidence interval for the mean difference in finishing time.

c) Would a hypothesis test at $\alpha = 0.05$ reject the null hypothesis of no difference? What conclusion would you draw?

17. Employee athletes. An ergonomics consultant is engaged by a large consumer products company to see what they can do to increase productivity. The consultant recommends an "employee athlete" program, encouraging every employee to devote 5 minutes an hour to physical activity. The company worries that the gains in productivity will be offset by the loss in time on the job. They'd like to know if the program increases or decreases productivity. To measure it, they monitor a random sample of 145 employees who word process, measuring their hourly key strokes both before and after the program is instituted. Here are the data:

Keystrokes per Hour			
	Before	**After**	**Difference (After − Before)**
Mean	1534.2	1556.9	22.7
SD	168.5	149.5	113.6
N	145	145	145

a) What are the null and alternative hypotheses?
b) What can you conclude? Explain.
c) Give a 95% confidence interval for the mean change in productivity (as measured by keystrokes per hour).

18. Employee athletes, part 2. A small company, on hearing about the employee athlete program (see Exercise 17) at the large company down the street, decides to try it as well. To measure the difference in productivity, they measure the average number of keystrokes per hour of 23 employees before and after the 5 minute an hour program is instituted. The data follow:

Keystrokes per Hour			
	Before	**After**	**Difference (After − Before)**
Mean	1497.3	1544.8	47.5
SD	155.4	136.7	122.8
N	23	23	23

a) Is there evidence to suggest that the program increases productivity?
b) Give a 95% confidence interval for the mean change in productivity (as measured by keystrokes per hour).
c) Given this information and the results of Exercise 17, what recommendations would you make to the company about the effectiveness of the program?

19. Productivity. A national fitness firm claims that a company may increase employee productivity by implementing one of the firm's fitness programs at the job site. As evidence of this, the fitness firm reports that one company was able to increase job productivity of a random sample of 30 employees from 57 to 70 (on a scale of 100). The standard deviation of the increases was 7.9. The national fitness firm wants to estimate the mean increase a company could expect after implementing the fitness program.

a) Check the assumptions and conditions for inference.
b) Find a 95% confidence interval.
c) Explain what your interval means in this context.

20. Productivity, part 2. After implementing the fitness program described in Exercise 19, another company found that a random sample of 48 employees increased their productivity score from 49 to 56, with a standard deviation of 6.2. This company believes that the fitness firm may have exaggerated the potential results of their program. Is there evidence that the mean improvement seen by this company is less than the one claimed by the fitness company in Exercise 19? Be sure to check the assumptions and conditions for inference.

21. BST. Many dairy cows now receive injections of BST, a hormone intended to spur greater milk production. After the first injection, a test herd of 60 Ayrshire cows increased their mean daily production from 47 pounds to 61 pounds of milk. The standard deviation of the increases was 5.2 pounds. We want to estimate the mean increase a farmer could expect in his own cows.

a) Check the assumptions and conditions for inference.
b) Write a 95% confidence interval.
c) Explain what your interval means in this context.
d) Given the cost of BST, a farmer believes he cannot afford to use it unless he is sure of attaining at least a 25% increase in milk production. Based on your confidence interval, what advice would you give him?

22. BST, part 2. In the experiment about hormone injections in cows described in Exercise 21, a group of 52 Jersey cows increased average milk production from 43 pounds to 52 pounds per day, with a standard deviation of 4.8 pounds. Is this evidence that the hormone may be more effective in one breed than the other? Test an appropriate hypothesis and state your conclusion. Be sure to discuss any assumptions you make.

T 23. European temperatures. The following table gives the average high temperatures in January and July for several European cities. Find a 90% confidence interval for the mean temperature difference between summer and winter in Europe. Be sure to check conditions for inference, and clearly explain what your interval means within the context of the situation.

Mean High Temperatures (°F)		
City	**January**	**July**
Vienna	34	75
Copenhagen	36	72
Paris	42	76
Berlin	35	74
Athens	54	90
Rome	54	88
Amsterdam	40	69
Madrid	47	87
London	44	73
Edinburgh	43	65
Moscow	21	76
Belgrade	37	84

T 24. Marathons 2007. Shown are the winning times (in minutes) for men and women in the New York City Marathon between 1978 and 2007. Assuming that performances in the Big Apple resemble performances elsewhere, we can think of these data as a sample of performance in marathon competitions. Create a 90% confidence interval for the mean difference in winning times for male and female marathon competitors. (www.nycmarathon.org)

Year	Men	Women	Year	Men	Women
1978	132.2	152.5	1993	130.1	146.4
1979	131.7	147.6	1994	131.4	147.6
1980	129.7	145.7	1995	131.0	148.1
1981	128.2	145.5	1996	129.9	148.3
1982	129.5	147.2	1997	128.2	148.7
1983	129.0	147.0	1998	128.8	145.3
1984	134.9	149.5	1999	129.2	145.1
1985	131.6	148.6	2000	130.2	145.8
1986	131.1	148.1	2001	127.7	144.4
1987	131.0	150.3	2002	128.1	145.9
1988	128.3	148.1	2003	130.5	142.5
1989	128.0	145.5	2004	129.5	143.2
1990	132.7	150.8	2005	129.5	144.7
1991	129.5	147.5	2006	130.0	145.1
1992	129.5	144.7	2007	129.1	143.2

T 25. Exercise equipment. A leading manufacturer of exercise equipment wanted to collect data on the effectiveness of their equipment. An August 2001 article in the journal *Medicine and Science in Sports and Exercise* compared how long it would take men and women to burn 200 calories during light or heavy workouts on various kinds of exercise equipment. The results summarized in the following table are the average times for a group of physically active young men and women whose performances were measured on a representative sample of exercise equipment:

AVERAGE MINUTES TO BURN 200 CALORIES

Machine Type	Hard Exertion Men	Hard Exertion Women	Light Exertion Men	Light Exertion Women
Treadmill	12	17	14	22
X-C skier	12	16	16	23
Stair climber	13	18	20	37
Rowing machine	14	16	21	25
Exercise rider	22	24	27	36
Exercise bike	16	20	29	44

a) On average, how many minutes longer than a man must a woman exercise at a light exertion rate in order to burn 200 calories? Find a 95% confidence interval.

b) Estimate the average number of minutes longer a woman must work out at light exertion than at heavy exertion to get the same benefit. Find a 95% confidence interval.

c) These data are actually averages rather than individual times. How might this affect the margins of error in these confidence intervals?

T 26. Market value. Real estate agents want to set the price of a house that's about to go on the real estate market correctly. They must choose a price that strikes a balance between one that is so high that the house takes too long to sell and one that's so low that not enough value will go to the homeowner. One appraisal method is the "Comparative Market Analysis" approach by which the market value of a house is based on recent sales of similar homes in the neighborhood. Because no two houses are exactly the same, appraisers have to adjust comparable homes for such features as extra square footage, bedrooms, fireplaces, upgrading, parking facilities, swimming pool, lot size, location and so on. Here are the appraised market values and the selling prices of 45 homes from the same region.

Market Value	Sale Price	Sale Price – Market value
296,700	420,000	123,300
273,333	285,500	12,167
124,778	165,000	40,222
170,000	190,400	20,400
181,793	276,000	94,207
154,400	200,000	45,600
261,300	292,000	30,700
153,625	159,695	6,070
140,000	137,000	−3,000
157,000	158,000	1,000
497,460	502,420	4,960
161,900	165,000	3,100
195,500	195,500	0
119,800	183,900	64,100
191,489	170,000	−21,489
206,900	226,500	19,600
193,000	215,000	22,000
118,000	116,000	−2,000
105,000	168,000	63,000
342,400	341,500	−900
106,300	151,000	44,700
96,250	110,000	13,750
52,000	90,000	38,000
262,500	332,750	70,250
270,100	379,900	109,800
104,900	109,900	5,000
283,200	335,000	51,800
251,250	305,000	53,750
145,600	123,000	−22,600
558,100	625,000	66,900
176,900	245,000	68,100
355,400	330,500	−24,900
262,500	240,000	−22,500
100,375	118,000	17,625
111,600	196,100	84,500
184,600	195,000	10,400

184,600	195,000	10,400
206,300	230,000	23,700
488,700	605,000	116,300
99,800	132,000	32,200
150,500	185,000	34,500
167,875	152,000	−15,875
298,500	420,000	121,500
155,000	195,000	40,000
220,800	231,000	10,200
194,247	260,000	65,753

a) Test the hypothesis that on average, the market value and the sale price of homes from this region are the same.
b) Find a 95% confidence interval for the mean difference.
c) Explain your findings in a sentence or two in context.

T 27. Job satisfaction. (When you first read about this exercise break plan in Chapter 13, you did not have an inference method that would work. Try again now.) A company institutes an exercise break for its workers to see if this will improve job satisfaction, as measured by a questionnaire that assesses workers' satisfaction. Scores for 10 randomly selected workers before and after the implementation of the exercise program are shown in the following table.

Worker Number	Job Satisfaction Index	
	Before	**After**
1	34	33
2	28	36
3	29	50
4	45	41
5	26	37
6	27	41
7	24	39
8	15	21
9	15	20
10	27	37

a) Identify the procedure you would use to assess the effectiveness of the exercise program and check to see if the conditions allow for the use of that procedure.
b) Test an appropriate hypothesis and state your conclusion.

T 28. Summer school. Having done poorly on their math final exams in June, six students repeat the course in summer school and take another exam in August.

If we consider these students to be representative of all students who might attend this summer school in other years, do these results provide evidence that the program is worthwhile?

June	54	49	68	66	62	62
August	50	65	74	64	68	72

a) Identify the procedure you would use to assess whether or not this program is worthwhile and check to see if the conditions allow for the use of that procedure.
b) Test an appropriate hypothesis and state your conclusion.

T 29. Efficiency. Many drivers of cars that can run on regular gas actually buy premium gas in the belief that they will get better gas mileage. To test that belief, a consumer research group evaluated the use of 10 cars in a company fleet in which all the cars run on regular gas. Each car was filled first with either regular or premium gasoline, decided by a coin toss, and the mileage for that tankful was recorded. Then the mileage was recorded again for the same cars for a tankful of the other kind of gasoline. The consumer research group did not let the drivers know about this experiment. Here are the results (miles per gallon):

Car #	1	2	3	4	5	6	7	8	9	10
Regular	16	20	21	22	23	22	27	25	27	28
Premium	19	22	24	24	25	25	26	26	28	32

a) Is there evidence that cars get better gas mileage on average with premium gasoline?
b) How big might that difference be? Check a 90% confidence interval.
c) Even if the difference is significant, why might the car fleet company choose to stick with regular gasoline?
d) Suppose you had mistakenly treated these data as two independent samples instead of matched pairs. What would the significance test have found? Carefully explain why the results are so different.

30. Advertising. A company developing an ad campaign for their cola is investigating the impact of caffeine on studying in the hope of finding evidence of their claim that caffeine helps memory. The firm asked 30 subjects, randomly divided into two groups, to take a memory test. The subjects then each drank two cups of regular (caffeinated) cola or caffeine-free cola. Thirty minutes later they each took another version of the memory test, and the changes in their scores were noted. Among the 15 subjects who drank caffeine, scores fell an average of −0.933 points with a standard deviation of 2.988 points. Among the no-caffeine group, scores went up an average of 1.429 points with a standard deviation of 2.441 points. Assumptions of Normality were deemed reasonable based on histograms of differences in scores.

a) Did scores change significantly for the group who drank caffeine? Test an appropriate hypothesis and state your conclusion.
b) Did scores change significantly for the no-caffeine group? Test an appropriate hypothesis and state your conclusion.
c) Does this indicate that some mystery substance in non-caffeinated soda may aid memory? What other explanation is plausible?

T 31. Quality control In an experiment on braking performance, a tire manufacturer measured the stopping distance for one of its tire models. On a test track, a car made repeated stops from 60 miles per hour. Twenty tests were run, 10 each on both dry and wet pavement, with results shown in the following table. (Note that actual *braking distance*, which takes into account the driver's reaction time, is much longer, typically nearly 300 feet at 60 mph!)

a) Find a 95% confidence interval for the mean dry pavement stopping distance. Be sure to check the appropriate assumptions and conditions, and explain what your interval means.

b) Find a 95% confidence interval for the mean increase in stopping distance on wet pavement. Be sure to check the appropriate assumptions and conditions, and explain what your interval means.

Stopping Distance (ft)	
Dry Pavement	**Wet Pavement**
145	211
152	191
141	220
143	207
131	198
148	208
126	206
140	177
135	183
133	223

T **32. Quality control, part 2.** For another test of the tires in Exercise 31, the company tried them on 10 different cars, recording the stopping distance for each car on both wet and dry pavement. Results are shown in the following table.

Stopping Distance (ft)		
Car #	**Dry Pavement**	**Wet Pavement**
1	150	201
2	147	220
3	136	192
4	134	146
5	130	182
6	134	173
7	134	202
8	128	180
9	136	192
10	158	206

a) Find a 95% confidence interval for the mean dry pavement stopping distance. Be sure to check the appropriate assumptions and conditions, and explain what your interval means.

b) Find a 95% confidence interval for the mean increase in stopping distance on wet pavement. Be sure to check the appropriate assumptions and conditions, and explain what your interval means.

T **33. Environment.** One major impact on the environment is the emission of CO_2 by power plants. Two states that produce the most CO_2 emissions are Texas and California. Both states claim that their power plants are improving. A random sample of power plants in the state of Texas allows us to compare their CO_2

emissions (in tons) between the years 2000 and 2007. Using the data provided in the computer file, test if there has been a significant change in CO_2 emissions in these power plants.

T **34. Student satisfaction.** Student surveys are often used to evaluate student satisfaction at the *end* of a course. In a recent paper in the *Journal of the Academy of Business Education* by C. Comm and D. Mathaisel, the authors suggested using "Gap Analysis," as used in marketing methodology to measure the expectation (or importance) and subsequent perception of a customer with regard to a specific product. If we regard the delivery of a college course as a "product," then we can measure the expectation of a student before the course begins and compare it to the perception of a student after the course has ended. The student survey consisted of 26 statements. A five-point Likert scale was used for each statement, where 1 = strongly agree, 2 = agree, 3 = neutral, 4 = disagree, and 5 = strongly disagree. The data in the computer file includes a subsample of a larger data set and represents the responses of 30 students in a quantitative course at a private institution to a question gauging "interest in the subject." Based on these data, assess any gap in the average student's interest before and after the course. Assuming that gap is calculated as the prescore minus postscore, what does a positive gap, or difference, suggest about the course?

T **35. Advertising claims.** Advertisements for an instructional video claim that the techniques will improve the ability of Little League pitchers to throw strikes and that, after undergoing the training, players will be able to throw strikes on at least 60% of their pitches. To test this claim, we have 20 Little Leaguers throw 50 pitches each, and we record the number of strikes. After the players participate in the training program, we repeat the test. The following table shows the number of strikes each player threw before and after the training.

a) Is there evidence that after training players can throw strikes more than 60% of the time?

b) Is there evidence that the training is effective in improving a player's ability to throw strikes?

Number of Strikes (out of 50)			
Before	**After**	**Before**	**After**
28	35	33	33
29	36	33	35
30	32	34	32
32	28	34	30
32	30	34	33
32	31	35	34
32	32	36	37
32	34	36	33
32	35	37	35
33	36	37	32

T **36. Drug costs.** In a full-page ad that ran in many U.S. newspapers in August 2002, a Canadian discount pharmacy listed costs of drugs that could be ordered from a website in Canada. The following table compares prices (in U.S. dollars) for commonly prescribed drugs.

Cost per 100 Pills

Drug Name	United States	Canada	Percent Savings
Cardizem	131	83	37
Celebrex	136	72	47
Cipro	374	219	41
Pravachol	370	166	55
Premarin	61	17	72
Prevacid	252	214	15
Prozac	263	112	57
Tamoxifen	349	50	86
Vioxx	243	134	45
Zantac	166	42	75
Zocor	365	200	45
Zoloft	216	105	51

a) Find a 95% confidence interval for the average savings in dollars.
b) Find a 95% confidence interval for the average savings in percent.
c) Which analysis do you think is more appropriate? Why?
d) In small print, the newspaper ad says, "Complete list of all 1500 drugs available on request." How does this comment affect your conclusions above?

T 37. Advertisements, part 2. In Exercise 3 you considered the question of whether sexual images in ads affected people's abilities to remember the item being advertised. To investigate, a group of Statistics students cut ads out of magazines. They were careful to find two ads for each of 10 similar items, one with a sexual image and one without. They arranged the ads in random order and had 39 subjects look at them for one minute. Then they asked the subjects to list as many of the products as they could remember. Their data are shown in the table. Is there evidence that the sexual images mattered?

Subject Number	Sexual Image	No Sex	Subject Number	Sexual Image	No Sex
	Ads Remembered			**Ads Remembered**	
1	2	2	21	2	3
2	6	7	22	4	2
3	3	1	23	3	3
4	6	5	24	5	3
5	1	0	25	4	5
6	3	3	26	2	4
7	3	5	27	2	2
8	7	4	28	2	4
9	3	7	29	7	6
10	5	4	30	6	7
11	1	3	31	4	3
12	3	2	32	4	5
13	6	3	33	3	0
14	7	4	34	4	3
15	3	2	35	2	3
16	7	4	36	3	3
17	4	4	37	5	5
18	1	3	38	3	4
19	5	5	39	4	3
20	2	2			

T 38. Freshman 15. Cornell Professor of Nutrition David Levitsky recruited students from two large sections of an introductory health course to test the validity of the "Freshman 15" theory that first year students gain 15 pounds their first year. Although they were volunteers, they appeared to match the rest of the freshman class in terms of demographic variables such as sex and ethnicity. The students were weighed during the first week of the semester, then again 12 weeks later. Based on Professor Levitsky's data, estimate the mean weight gain in first-semester freshmen and comment on the "freshman 15." (Weights are in pounds.)

Subject Number	Initial Weight	Terminal Weight	Subject Number	Initial Weight	Terminal Weight
1	171	168	35	148	150
2	110	111	36	164	165
3	134	136	37	137	138
4	115	119	38	198	201
5	150	155	39	122	124
6	104	106	40	146	146
7	142	148	41	150	151
8	120	124	42	187	192
9	144	148	43	94	96
10	156	154	44	105	105
11	114	114	45	127	130
12	121	123	46	142	144
13	122	126	47	140	143
14	120	115	48	107	107
15	115	118	49	104	105
16	110	113	50	111	112
17	142	146	51	160	162
18	127	127	52	134	134
19	102	105	53	151	151
20	125	125	54	127	130
21	157	158	55	106	108
22	119	126	56	185	188
23	113	114	57	125	128
24	120	128	58	125	126
25	135	139	59	155	158
26	148	150	60	118	120
27	110	112	61	149	150
28	160	163	62	149	149
29	220	224	63	122	121
30	132	133	64	155	158
31	145	147	65	160	161
32	141	141	66	115	119
33	158	160	67	167	170
34	135	134	68	131	131

T 39. Store sales. A company that owns a chain of specialty food stores would like to see if their sales have increased over the same time period from the previous year. A random sample of stores produced the average weekly sales for the current quarter compared to the average weekly sales for the same quarter one year ago for a sample of 15 stores. Using the data provided, determine if the average weekly sales has increased this past year for stores in this chain.

T 40. Store profits. The store managers for the sample of stores in Exercise 39 maintain that their stores are doing better this year, despite relatively flat sales. Their argument is that they have been able to reduce costs through more efficient staffing and inventory management. Using the data provided, determine if the average weekly profits for one quarter has increased for these stores over the past year (from year 1 to year 2). Do your results support the claim of the store managers?

T 41. Yogurt. Do these data suggest that there is a significant difference in calories between servings of strawberry and vanilla yogurt? Test an appropriate hypothesis and state your conclusion, including a check of assumptions and conditions.

| | | Calories per Serving | |
		Strawberry	Vanilla
Brand	America's Choice	210	200
	Breyer's Lowfat	220	220
	Columbo	220	180
	Dannon Light 'n Fit	120	120
	Dannon Lowfat	210	230
	Dannon la Crème	140	140
	Great Value	180	80
	La Yogurt	170	160
	Mountain High	200	170
	Stonyfield Farm	100	120
	Yoplait Custard	190	190
	Yoplait Light	100	100

T 42. Housing starts. The population in the United States has been shifting to the western and southern regions. Most states in these regions have experienced rapid population growth. One measure of economic growth compiled by the federal government is the number of housing starts for new privately–owned units. To determine if housing starts increased, on average, between 2000 and 2004, the number of starts was recorded for 6 western states.

Housing Starts in Western States		
State	**2000**	**2004**
Arizona	59400	64400
Colorado	52500	35900
Idaho	11300	13500
Nevada	31100	37800
Oregon	18800	19200
Utah	18100	20200

a) Was there an increase in housing starts between 2000 and 2004?
b) Were you surprised by the findings? To answer the question, compare the differences for each state in the sample. Which difference in particular might affect the results of the test? Can you think of reasons why one state had a large decrease in starts while many of the other states all experienced increases in starts?

Between 2004 and 2005, growth in housing units stalled. The 2005 housing starts were 600 less on average than the 2004 housing starts. To determine whether the decline in housing starts continued through 2006, the following comparisons were made:

Housing Starts in Western States		
State	**2005**	**2006**
Arizona	61900	60100
Colorado	36800	37700
Idaho	13100	12800
Nevada	36100	35000
Oregon	19600	20000
Utah	19900	19800

c) Is there evidence the number of housing starts in 2006 have decreased, on average, by more than 600 units from 2005? www.census.gov/prod/2007pubs/08abstract/construct.pdf

T 43. Small business. Many economists consider small businesses to be the backbone of the U.S. economy, but a majority of them fail within the first 5 years of opening. In an annual report issued to the President, differences in small business performance were compared between 2003 and 2004. One statistic that was investigated was the change in bankruptcies filed for each state. The table below shows the number of bankruptcies filed in New England states in 2003 and 2004.

Bankruptcies of Small Businesses		
State	**2003**	**2004**
Connecticut	187	132
Maine	105	138
Massachusetts	396	315
New Hampshire	178	158
Rhode Island	48	74
Vermont	78	65

Source: The Small Business Economy, 2005 Report to the President, United States Government Printing Office, Washington D.C. www.sba.gov/idc/groups/public/documents/ sba_homepage/ebeconomy2005.pdf

Was there a change in bankruptcies in New England between the years 2003 and 2004? Perform the test, and discuss whether the mean difference was statistically significant.

T 44. Auto repair shops. Certain businesses and professions have reputations for being somewhat dishonest when dealing with customers. One area of concern is the honesty of auto repair shops. Many states require smog checks. A vehicle that does not pass the check must be repaired. In one state, the department of motor vehicles has been receiving numerous complaints about a particular auto repair chain. The state decided to check the shops to determine whether they were unlawfully issuing "no pass" reports in order to charge customers unnecessary repair fees. The state procured 8 vehicles. Each vehicle was first tested on department of motor vehicle smog equipment. The eight vehicles were then randomly sent to auto repair shops for a smog check. As part of the check for accuracy, the hydrocarbon (HC) emission in parts per million (ppm) were compared:

Vehicle	1	2	3	4	5	6	7	8
DMV HC Level:	7	10	3	1	5	8	30	7
Auto Shop HC Level:	20	11	5	10	5	7	42	15

a) Is there a difference between the measured HC levels taken from the Auto Shop and the DMV measurements? Find a suitable confidence interval.

b) Do you think the department of motor vehicles has evidence the auto shop readings differ from the department readings? Perform the appropriate test.

c) If you found the test results to be significant, can the department of motor vehicles automatically assume the auto shop is cheating its customers? What other possible explanations could cause the differences in readings?

T 45. Airlines. In recent years, the airline industry has been severely criticized for a variety of service-related issues including poor on-time performance, canceled flights, and lost luggage. Some believe airline service is declining while the price of airline fares is increasing. A sample of 10 third quarter changes in airfares is shown below.

		Third Quarter 2006	Third Quarter 2007	Percent Change from 3rd Qtr 2006
Origin	Cincinnati, OH	511.11	575.67	12.6
	Salt Lake City, UT	319.29	344.48	7.9
	Dallas Love, TX	185.12	198.74	7.4
	New York JFK, NY	324.75	345.97	6.5
	Hartford, CT	341.05	363.17	6.5
	Charleston, SC	475.10	367.08	−22.7
	Columbus, OH	322.60	277.24	−14.1
	Kona, HI	206.50	180.40	−12.6
	Memphis, TN	418.70	382.29	−8.7
	Greensboro/ High Point, NC	411.95	377.41	−8.4

Source: Bureau of Transportation Statistics. Top Five Third Quarter U.S. Domestic Average Itinerary Fare Increases and Decreases, 3rd Qtr 2006– 3rd Qtr 2007—Top 100 Airports Based on 2006 U.S. Originating Domestic Passengers. Fares based on 2006 U.S. domestic itinerary fares, round-trip or one-way for which no return is purchased. Averages do not include frequent flyer fares. www.bts.gov/press_releases/2008

a) Does the percent change in airfare from the third quarter 2006 column represent paired data? Why or why not?

b) Was there an actual change, on average, in airline fares between the two quarters? Perform the test on both the actual and percentage differences. Discuss the results of the test and explain how you chose between the fares and the percent differences as the data to test.

T 46. Grocery prices. WinCo Foods, a large discount grocery retailer in the western United States promotes itself as the lowest priced grocery retailer. In newspaper ads printed and distributed during January 2008, WinCo Foods published a price comparison for products between WinCo and several competing grocery retailers. One of the retailers compared against WinCo was Wal-Mart, also known as a low price competitor. WinCo selected a variety of products, listed the price of the product charges at each retailer, and showed the sales receipt to prove the prices at WinCo were the lowest in the area. A sample of the product and their price comparison at both WinCo and Wal-Mart are shown in the following table:

Item	WinCo Price	Wal-Mart Price
Bananas (lb)	0.42	0.56
Red Onions (lb)	0.58	0.98
Mini Peeled Carrots (1 lb bag)	0.98	1.48
Roma Tomatoes (lb)	0.98	2.67
Deli Tater Wedges (lb)	1.18	1.78
Beef Cube Steak (lb)	3.83	4.118
Beef Top Round London Broil (lb)	3.48	4.12
Pillsbury Devils Food Cake Mix (18.25 oz)	0.88	0.88
Lipton Rice and Sauce Mix (5.6 oz)	0.88	1.06
Sierra Nevada Pale Ale (12 − 12 oz bottles)	12.68	12.84
GM Cheerios Oat Clusters (11.3 oz)	1.98	2.74
Charmin Bathroom Tissue (12 roll)	5.98	7.48
Bumble Bee Pink Salmon (14.75 oz)	1.58	1.98
Pace Thick & Chunky Salsa, Mild (24 oz)	2.28	2.78
Nalley Chili, Regular w/Beans (15 oz)	0.78	0.78
Challenge Butter (lb quarters)	2.18	2.58
Kraft American Singles (12 oz)	2.27	2.27
Yuban Coffee FAC (36 oz)	5.98	7.56
Totino's Pizza Rolls, Pepperoni (19.8 oz)	2.38	2.42
Rosarita Refried Beans, Original (16 oz)	0.68	0.73
Barilla Spaghetti (16 oz)	0.78	1.23
Sun-Maid Mini Raisins (14 − .5 oz)	1.18	1.36
Jif Peanut Butter, Creamy (28 oz)	2.54	2.72
Dole Fruit Bowl, Mixed Fruit (4 − 4 oz)	1.68	1.98
Progresso Chicken Noodle Soup (19 oz)	1.28	1.38
Precious Mozzarella Ball, Part Skim (16 oz)	3.28	4.23
Mrs. Cubbison Seasoned Croutons (6 oz)	0.88	1.12
Kellogg's Raisin Bran (20 oz)	1.98	2.50
Campbell's Soup at Hand, Cream of Tomato (10.75 oz)	1.18	1.26

a) Do the prices listed indicate that, on average, prices at WinCo are lower than prices at Wal-Mart?

b) At the bottom of the price list, the following statement appears: "Though this list is not intended to represent a typical weekly grocery order or a random list of grocery items, WinCo continues to be the area's low price leader." Why do you think WinCo added this statement?

c) What other comments could be made about the statistical validity of the test on price comparisons given in the ad?

JUST CHECKING ANSWERS

1 These are independent groups sampled at random, so use a two-sample *t* confidence interval to estimate the size of the difference.

2 If the same random sample of students was sampled both in the first year and again in the fourth year of their university experience, then this would be a paired *t*-test.

3 A male and female are selected from each work group. The question calls for a paired *t*-test.

4 Since the sample of companies is different in each of the industries, this would be a two-sample test.

5 Since the same 50 companies are surveyed twice to examine a change in variables over time, this would be a paired *t*-test.

15

Inference for Counts: Chi-Square Tests

SAC Capital

Hedge funds, like mutual funds and pension funds, pool investors' money in an attempt to make profits. Unlike these other funds, however, hedge funds are not required to register with the U.S. Securities and Exchange Commission (SEC) because they issue securities in "private offerings" only to "qualified investors" (investors with either $1 million in assets or annual income of at least $200,000).

Hedge funds don't necessarily "hedge" their investments against market moves. But typically these funds use multiple, often complex, strategies to exploit inefficiencies in the market. For these reasons, hedge fund managers have the reputation for being obsessive traders.

One of the most successful hedge funds is SAC Capital, which was founded by Steven (Stevie) A. Cohen in 1992 with nine employees and $25 million in assets under management (AUM). SAC Capital

401

returned annual gains of 40% or more through much of the 1990s and by 2007 had more than 800 employees and over $14 billion in assets under management.

Cohen, a legendary figure on Wall Street, is known for taking advantage of any information he can find and for turning that information into profit. SAC Capital is one of the most active trading organizations in the world. According to *Business Week* (7/21/2003), Cohen's firm "routinely accounts for as much as 3% of the NYSE's average daily trading, plus up to 1% of the NASDAQ's—a total of at least 20 million shares a day."

I n a business as competitive as hedge fund management, information is gold. Being the first to have information and knowing how to act on it can mean the difference between success and failure. Hedge fund managers look for small advantages everywhere, hoping to exploit inefficiencies in the market and to turn those inefficiencies into profit.

Wall Street has plenty of "wisdom" about market patterns. For example, investors are advised to watch for "calendar effects," certain times of year or days of the week that are particularly good or bad: "As goes January, so goes the year" and "Sell in May and go away." Some analysts claim that the "bad period" for holding stocks is from the sixth trading day of June to the fifth-to-last trading day of October. Of course, there is also Mark Twain's advice.

"October. This is one of the peculiarly dangerous months to speculate in stocks. The others are July, January, September, April, November, May, March, June, December, August, and February."

—Pudd'nhead Wilson's Calendar

One common claim is that stocks show a weekly pattern. For example, some argue that there is a *weekend effect* in which stock returns on Mondays are often lower than those of the immediately preceding Friday. Are patterns such as this real? We have the data, so we can check. Between October 1, 1928 and June 6, 2007, there were 19,755 trading sessions. Let's first see how many trading days fell on each day of the week. It's not exactly 20% for each day because of holidays. The distribution of days is shown in Table 15.1.

Day of Week	Count	% of days
Monday	3820	19.34%
Tuesday	4002	20.26
Wednesday	4024	20.37
Thursday	3963	20.06
Friday	3946	19.97

Table 15.1 *The distribution of days of the week among the 19,755 trading days from October 1, 1928 to June 6, 2007. We expect about 20% to fall in each day, with minor variations due to holidays and other events.*

Of these 19,755 trading sessions, 10,272, or about 52% of the days, saw a gain in the Dow Jones Industrial Average (DJIA). To test for a pattern, we need a model. The model comes from the supposition that any day is as likely to show a gain as any other. In any sample of positive or "up" days, we should expect to see the same distribution of days as in Table 15.1—in other words, about 19.34% of "up" days would be Mondays, 20.26% would be Tuesdays, and so on. Here is the distribution of days in one such random sample of 1000 "up" days.

Day of Week	Count	% of days in the sample of "up" days
Monday	192	19.2%
Tuesday	189	18.9
Wednesday	202	20.2
Thursday	199	19.9
Friday	218	21.8

Table 15.2 *The distribution of days of the week for a sample of 1000 "up" trading days selected at random from October 1, 1928 to June 6, 2007. If there is no pattern, we would expect the proportions here to match fairly closely the proportions observed among all trading days in Table 15.1.*

Of course, we expect some variation. We wouldn't expect the proportions of days in the two tables to match exactly. In our sample, the percentage of Mondays in Table 15.2 is slightly lower than in Table 15.1, and the proportion of Fridays is a little higher. Are these deviations enough for us to declare that there is a recognizable pattern?

15.1 Goodness-of-Fit Tests

To address this question, we test the table's **goodness-of-fit**, where *fit* refers to the null model proposed. Here, the null model is that there is no pattern, that the distribution of *up* days should be the same as the distribution of trading days overall. (If there were no holidays or other closings, that would just be 20% for each day of the week.)

Assumptions and Conditions

Data for a goodness-of-fit test are organized in tables, and the assumptions and conditions reflect that. Rather than having an observation for each individual, we typically work with summary counts in categories. Here, the individuals are trading days, but rather than list all 1000 trading days in the sample, we have totals for each weekday.

Counted Data Condition. The data must be counts for the categories of a categorical variable. This might seem a silly condition to check. But many kinds of values can be assigned to categories, and it is unfortunately common to find the methods of this chapter applied incorrectly (even by business professionals) to proportions or quantities just because they happen to be organized in a two-way table. So check to be sure that you really have counts.

Independence Assumption

Independence Assumption. The counts in the cells should be independent of each other. You should think about whether that's reasonable. If the data are a random sample you can simply check the randomization condition.

Randomization Condition.   The individuals counted in the table should be a random sample from some population. We need this condition if we want to generalize our conclusions to that population. We took a random sample of 1000 trading days on which the DJIA rose. That lets us assume that the market's performance on any one day is independent of performance on another. If we had selected 1000 consecutive trading days, there would be a risk that market performance on one day could affect performance on the next, or that an external event could affect performance for several consecutive days.

Sample Size Assumption

Sample Size Assumption.   We must have enough data for the methods to work. We usually just check the following condition:

Expected Cell Frequency Condition.   We should expect to see at least 5 individuals in each cell. The expected cell frequency condition should remind you of—and is, in fact, quite similar to—the condition that np and nq be at least 10 when we test proportions.

Chi-Square Model

We have observed a count in each category (weekday). We can compute the number of up days we'd *expect* to see for each weekday if the null model were true. For the trading days example, the expected counts come from the null hypothesis that the up days are distributed among weekdays just as trading days are. Of course, we could imagine almost any kind of model and base a null hypothesis on that model.

To decide whether the null model is plausible, we look at the differences between the expected values from the model and the counts we observe. We wonder: Are these differences so large that they call the model into question, or could they have arisen from natural sampling variability? We denote the *differences* between these observed and expected counts, $(Obs - Exp)$. As we did with variance, we square them. That gives us positive values and focuses attention on any cells with large differences. Because the differences between observed and expected counts generally get larger the more data we have, we also need to get an idea of the *relative* sizes of the differences. To do that, we divide each squared difference by the expected count for that cell.

The test statistic, called the **chi-square (or chi-squared) statistic**, is found by adding up the sum of the squares of the deviations between the observed and expected counts divided by the expected counts:

$$\chi^2 = \sum_{all\ cells} \frac{(Obs - Exp)^2}{Exp}.$$

The chi-square statistic is denoted χ^2, where χ is the Greek letter chi (pronounced ki). The resulting family of sampling distribution models is called the **chi-square models**.

The members of this family of models differ in the number of degrees of freedom. The number of degrees of freedom for a goodness-of-fit test is $k - 1$, where k is the number of cells—in this example, 5 weekdays.

We will use the chi-square statistic only for testing hypotheses, not for constructing confidence intervals. A small chi-square statistic means that our model fits the data well, so a small value gives us no reason to doubt the null hypothesis. If the observed counts don't match the expected counts, the statistic will be large. If the calculated statistic value is large enough, we'll reject the null hypothesis. So the chi-square test is always one-sided. What could be simpler? Let's see how it works.

Expected Cell Frequencies
Companies often want to assess the relative successes of their products in different regions. However, a company whose sales regions had 100, 200, 300, and 400 representatives might not expect equal sales in all regions. They might expect observed sales to be proportional to the size of the sales force. The null hypothesis in that case would be that the proportions of sales were 1/10, 2/10, 3/10, and 4/10, respectively. With 500 total sales, their expected counts would be 50, 100, 150, and 200.

NOTATION ALERT:
We compare the counts *observed* in each cell with the counts we *expect* to find. The usual notation uses *Obs* and *Exp* as we've used here. The expected counts are found from the null model.

NOTATION ALERT:
The only use of the Greek letter χ in Statistics is to represent the chi-square statistic and the associated sampling distribution. This is another violation of the general rule that Greek letters represent population parameters. Here we are using a Greek letter simply to name a family of distribution models and a statistic.

The Chi-Square Calculation

Here are the steps to calculate the chi-square statistic:

1. **Find the expected values.** These come from the null hypothesis model. Every null model gives a hypothesized proportion for each cell. The expected value is the product of the total number of observations times this proportion. (The result need not be an integer.)
2. **Compute the residuals.** Once you have expected values for each cell, find the residuals, $Obs - Exp$.
3. **Square the residuals.** $(Obs - Exp)^2$
4. **Compute the components.** Find $\dfrac{(Obs - Exp)^2}{Exp}$ for each cell.
5. **Find the sum of the components.** That's the chi-square statistic, $\chi^2 = \displaystyle\sum_{all\ cells} \dfrac{(Obs - Exp)^2}{Exp}$.
6. **Find the degrees of freedom.** It's equal to the number of cells minus one.
7. **Test the hypothesis.** Large chi-square values mean lots of deviation from the hypothesized model, so they give small P-values. Look up the critical value from a table of chi-square values such as table X in Appendix C, or use technology to find the P-value directly.

The steps of the chi-square calculations are often laid out in tables. Use one row for each category, and columns for observed counts, expected counts, residuals, squared residuals, and the contributions to the chi-square total:

	Observed	Expected	Residual = $(Obs - Exp)$	$(Obs - Exp)^2$	Component = $\dfrac{(Obs - Exp)^2}{Exp}$
Monday	192	193.369	−1.369	1.879	0.0097
Tuesday	189	202.582	−13.582	184.461	0.9105
Wednesday	202	203.695	−1.695	2.874	0.0141
Thursday	199	200.607	−1.607	2.584	0.0129
Friday	218	199.747	18.253	333.176	1.6680

Table 15.3 Calculations for the chi-square statistic in the trading days example.

GUIDED EXAMPLE — *Stock Market Patterns*

We have counts of the "up" days for each day of the week. The economic theory we want to investigate is whether there is a pattern in "up" days. So, our null hypothesis is that across all days in which the DJIA rose, the days of the week are distributed as they are across all trading days. (As we saw, the trading days are not quite *evenly* distributed because of holidays, so we use the *trading days* percentages as the null model.) We refer to this as *uniform*, accounting for holidays. The alternative hypothesis is that the observed percentages are *not* uniform. The test statistic looks at how closely the observed data match this idealized situation.

PLAN

Setup State what you want to know.

Identify the variables and context.

We want to know whether the distribution for "up" days differs from the null model (the trading days distribution). We have the number of times each weekday appeared among a random sample of 1000 "up" days.

Hypotheses State the null and alternative hypotheses. For χ^2 tests, it's usually easier to state the hypotheses in words than in symbols.

H_O: The days of the work week are distributed among the up days as they are among all trading days.

H_A: The trading days model does not fit the up days distribution.

Model Think about the assumptions and check the conditions.

✓ **Counted Data Condition** We have counts of the days of the week for all trading days and for the "up" days.

✓ **Independence Assumption** We have no reason to expect that one day's performance will affect another's, but to be safe we've taken a random sample of days. The randomization should make them far enough apart to alleviate any concerns about dependence.

✓ **Randomization Condition** We have a random sample of 1000 days from the time period.

✓ **Expected Cell Frequency Condition** All the expected cell frequencies are much larger than 5.

Specify the sampling distribution model.

Name the test you will use.

The conditions are satisfied, so we'll use a χ^2 model with $5 - 1 = 4$ degrees of freedom and do a **chi-square goodness-of-fit test**.

Mechanics Each cell contributes a value equal to $\dfrac{(Obs - Exp)^2}{Exp}$ to the chi-square sum.

To find the expected number of days, we take the fraction of each weekday from *all* days and multiply by the number of "up" days.

For example, there were 3,820/19,755 Mondays overall, or 19.34%.

So, we'd expect there would be 1000 × 19,34% or 193.4 Mondays among the 1000 "up" days.

The expected values are:

Monday: 193.4
Tuesday: 202.6
Wednesday: 203.7
Thursday: 200.6
Friday: 199.7

And we observe:

Monday: 192
Tuesday: 189
Wednesday: 202
Thursday: 199
Friday: 218

Add up these components. If you do it by hand, it can be helpful to arrange the calculation in a table.

$$\chi^2 = \frac{(192 - 193.4)^2}{193.4} + \cdots + \frac{(218 - 199.7)^2}{199.7} = 2.62$$

The P-value is the probability in the upper tail of the χ^2 model. It can be found using software or a table (see table X in Appendix C).

Using table X in Appendix C, we find that for a significance level of 5% and 4 degrees of freedom, we'd need a value of 9.488 or more to have a P-value less than .05. Our value of 2.62 is less than that.

Using a computer to generate the P-value, we find:

$$\text{P-value} = P(\chi^2_4 > 2.62) = 0.62$$

The χ^2 models are skewed to the high end. Large χ^2 statistic values correspond to small P-values, which would lead us to reject the null hypothesis.

Conclusion Link the P-value to your decision. Be sure to say more than a fact about the distribution of counts. State your conclusion in terms of what the data mean.

MEMO

Re: Stock Market Patterns

Our investigation of whether there are day-of-the-week patterns in the behavior of the DJIA in which one day or another is more likely to be an "up" day found no evidence of such a pattern. Our statistical test indicated that a pattern such as the one found in our sample of trading days would happen by chance about 62% of the time.

We conclude that there is, unfortunately, no evidence of a pattern that could be used to guide investment in the market. We were unable to detect a "weekend" or other day of the week effect in the market.

15.2 Interpreting Chi-Square Values

When we calculated χ^2 for the trading days example, we got 2.62. That value was not large for 4 degrees of freedom, so we were unable to reject the null hypothesis. In general, what *is* big for a χ^2 statistic?

Think about how χ^2 is calculated. In every cell any deviation from the expected count contributes to the sum. Large deviations generally contribute more, but if there are a lot of cells, even small deviations can add up, making the χ^2 value larger. So the more cells there are, the higher the value of χ^2 has to be before it becomes significant. For χ^2, the decision about how big is big depends on the number of degrees of freedom.

Unlike the Normal and *t* families, χ^2 models are skewed. Curves in the χ^2 family change both shape and center as the number of degrees of freedom grows. Here, for example, are the χ^2 curves for 5 and for 9 degrees of freedom.

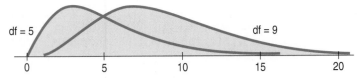

| **Figure 15.1**   The χ^2 curves for 5 and 9 degrees of freedom.

Notice that the value $\chi^2 = 10$ might seem somewhat extreme when there are 5 degrees of freedom, but appears to be rather ordinary for 9 degrees of freedom. Here are two simple facts to help you think about χ^2 models:

◆ The mode is at $\chi^2 = df - 2$. (Look at the curves; their peaks are at 3 and 7.)

◆ The expected value (mean) of a χ^2 model is its number of degrees of freedom. That's a bit to the right of the mode—as we would expect for a skewed distribution.

Goodness-of-fit tests are often performed by people who have a theory of what the proportions *should* be in each category and who believe their theory to be true. In some cases, unlike our market example, there isn't an obvious null hypothesis against which to test the proposed model. So, unfortunately, in those cases, the only null hypothesis available is that the proposed theory is true. And as we know, the hypothesis testing procedure allows us only to reject the null or fail to reject it. We can never confirm that a theory is in fact true; we can never confirm the null hypothesis.

At best, we can point out that the data are consistent with the proposed theory. But this doesn't prove the theory. The data *could* be consistent with the model even if the theory were wrong. In that case, we fail to reject the null hypothesis but can't conclude anything for sure about whether the theory is true.

Why Can't We Prove the Null?

A student claims that it really makes no difference to your starting salary how well you do in your Statistics class. He surveys recent graduates, categorizes them according to whether they earned an A, B, or C in Statistics, and according to whether their starting salary is above or below the median for their class. He calculates the proportion above the median salary for each grade. His null model is that in each grade category, 50% of students are above the median. With 40 respondents, he gets a P-value of .07 and declares that Statistics grades don't matter. But then more questionnaires are returned, and he finds that with a sample size of 70, his P-value is .04. Can he ignore the second batch of data? Of course not. If he could do that, he could claim almost any null model was true just by having too little data to refute it.

15.3 Examining the Residuals

Chi-square tests are always one-sided. The chi-square statistic is always positive, and a large value provides evidence against the null hypothesis (because it shows that the fit to the model is *not* good), while small values provide little evidence that the model doesn't fit. In another sense, however, chi-square tests are really many sided; a large statistic doesn't tell us *how* the null model doesn't fit. In our market theory example, if we had rejected the uniform model, we wouldn't have known *how* it failed. Was it because there were not enough Mondays represented, or was it that all five days showed some deviation from the uniform?

When we reject a null hypothesis in a goodness-of-fit test, we can examine the residuals in each cell to learn more. In fact, whenever we reject a null hypothesis, it's a good idea to examine the residuals. (We don't need to do that when we fail to reject because when the χ^2 value is small, all of its components must have been small.)

Because we want to compare residuals for cells that may have very different counts, we standardize the residuals. We know the mean residual is zero,[1] but we need to know each residual's standard deviation. When we tested proportions, we saw a link between the expected proportion and its standard deviation. For counts, there's a similar link. To standardize a cell's residual, we divide by the square root of its expected value:[2]

$$\frac{(Obs - Exp)}{\sqrt{Exp}}.$$

Notice that these **standardized residuals** are the square roots of the components we calculated for each cell, with the plus (+) or the minus (−) sign indicating whether we observed more or fewer cases than we expected.

The standardized residuals give us a chance to think about the underlying patterns and to consider how the distribution differs from the model. Now that we've divided each residual by its standard deviation, they are z-scores. If the null hypothesis was true, we could even use the 68-95-99.7 Rule to judge how extraordinary the large ones are.

Here are the standardized residuals for the trading days data:

	Standardized Residual $= \dfrac{(Obs - Exp)}{\sqrt{Exp}}$
Monday	−0.0984
Tuesday	−0.9542
Wednesday	−0.1188
Thursday	−0.1135
Friday	1.292

| **Table 15.4** *Standardized residuals.*

None of these values is remarkable. The largest, Friday, at 1.292, is not impressive when viewed as a z-score. The deviations are in the direction suggested by the "weekend effect," but they aren't quite large enough for us to conclude that they are real.

15.4 The Chi-Square Test of Homogeneity

Skin care products are big business. According to the American Academy of Dermatology, "the average adult uses at least seven different products each day," including moisturizers, skin cleansers, and hair cosmetics.[3] Growth in the skin care market in China during 2006 was 15%, fueled by its population size and massive economic growth. But not all cultures and markets are the same. Global companies must understand cultural differences in the importance of various skin care products in order to compete effectively.

The GfK Roper Reports® Worldwide Survey, which we first saw in Chapter 4, asked 30,000 consumers in 23 countries about their attitudes on health, beauty, and other personal values. One question participants were asked was how important is: "Seeking the utmost attractive appearance" to you? Responses were a scale with 1 = Not at all important and 7 = Extremely important. Is agreement with this

---

[1] Residual = observed − expected. Because the total of the expected values is the same as the observed total, the residuals must sum to zero.

[2] It can be shown mathematically that the square root of the expected value estimates the appropriate standard deviation.

[3] www.aad.org/public/Publications/pamphlets/Cosmetics.htm.

WHO	Respondents in the GfK Roper Reports Worldwide Survey
WHAT	Responses to questions relating to perceptions of food and health
WHEN	Fall 2005; published in 2006
WHERE	Worldwide
HOW	Data collected by GfK Roper Consulting using a multistage design
WHY	To understand cultural differences in the perception of the food and beauty products we buy and how they affect our health

question the same across the five countries for which we have data (China, France, India, U.K., and U.S.)? Here is a table with the counts.

| | Country | | | | | |
	China	France	India	U.K.	U.S.	Total
Appearance						
7—Extremely important	197	274	642	210	197	**1520**
6	257	405	304	252	203	**1421**
5	315	364	196	348	250	**1473**
4—Average importance	480	326	263	486	478	**2033**
3	98	82	41	125	100	**446**
2	63	46	36	70	58	**273**
1—Not at all important	92	38	53	62	29	**274**
Total	**1502**	**1535**	**1535**	**1553**	**1315**	**7440**

| **Table 15.5** *Responses to how important is: "Seeking the utmost attractive appearance."*

We can compare the countries more easily by examining the column percentages.

| | Country | | | | | |
	China	France	India	U.K.	U.S.	Row %
Appearance						
7—Extremely important	13.12%	17.85	41.82	13.52	14.98	**20.43%**
6	17.11	26.38	19.80	16.23	15.44	**19.10**
5	20.97	23.71	12.77	22.41	19.01	**19.80**
4—Average importance	31.96	21.24	17.13	31.29	36.35	**27.33**
3	6.52	5.34	2.67	8.05	7.60	**5.99**
2	4.19	3.00	2.35	4.51	4.41	**3.67**
1—Not at all important	6.13	2.48	3.45	3.99	2.21	**3.68**
Total	**1502**	**1535**	**1535**	**1553**	**1315**	**7440**

| **Table 15.6** *Responses as a percentage of respondents by country.*

The stacked barchart of the responses by country shows the patterns more vividly:

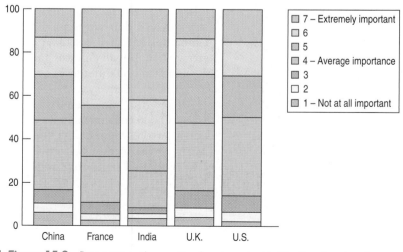

Figure 15.2 *Responses to the question how important is: "Seeking the utmost attractive appearance" by country. India stands out for the proportion of respondents who said Important or Extremely important.*

It seems that India stands out from the other countries. There is a much larger proportion of respondents from India who responded *Extremely Important*. But are the observed differences in the percentages real or just natural sampling variation? Our null hypothesis is that the proportions choosing each alternative are the same for each country. To test that hypothesis, we use a **chi-square test of homogeneity**. This is just another chi-square test. It turns out that the mechanics of the test of this hypothesis are nearly identical to the chi-square goodness-of-fit test we just saw in Section 15.1. The difference is that the goodness-of-fit test compared our observed counts to the expected counts from a *given* model. The test of homogeneity, by contrast, has a null hypothesis that the distributions are the same for all the groups. The test examines the differences between the observed counts and what we'd expect under that assumption of homogeneity.

For example, 20.43% (the row %) of *all* 7440 respondents said that looking good was extremely important to them. If the distributions were homogenous across the five countries (as the null hypothesis asserts), then that proportion should be the same for all five countries. So 20.43% of the 1315 U.S. respondents, or 268.66, would have said that looking good was extremely important. That's the number we'd *expect* under the null hypothesis.

Working in this way, we (or, more likely, the computer) can fill in expected values for each cell. The following table shows these expected values for each response and each country.

		Country					
		China	**France**	**India**	**U.K.**	**U.S.**	**Row %**
	7—Extremely important	306.86	313.60	313.60	317.28	268.66	**20.43%**
	6	286.87	293.18	293.18	296.61	251.16	**19.10**
	5	297.37	303.91	303.91	307.47	260.35	**19.80**
Appearance	**4—Average importance**	410.43	419.44	419.44	424.36	359.33	**27.33**
	3	90.04	92.02	92.02	93.10	78.83	**5.99**
	2	55.11	56.32	56.32	56.99	48.25	**3.67**
	1—Not at all important	55.32	56.53	56.53	57.19	48.43	**3.68**
	Total	**1502**	**1535**	**1535**	**1553**	**1315**	**7440**

Table 15.7 *Expected values for the responses. Because these are theoretical values, they don't have to be integers.*

The term *homogeneity* means that things are the same. Here, we ask whether the distribution of responses about the importance of looking good is the same across the five countries. The chi-square test looks for differences large enough to step beyond what we might expect from random sample-to-sample variation. It can reveal a large deviation in a single category or small but persistent differences over all the categories—or anything in between.

Assumptions and Conditions

The assumptions and conditions are the same as for the chi-square test for goodness-of-fit. The **Counted Data Condition** says that these data must be counts. You can never perform a chi-square test on a quantitative variable. For example, if Roper had recorded how much respondents spent on skin care products, you wouldn't be able to use a chi-square test to determine whether the mean expenditures in the five countries were the same.[4]

---

[4] To do that, you'd use a method called Analysis of Variance (see Chapter 23).

How to Find Expected Values

In a contingency table, to test for homogeneity, we need to find the expected values when the null hypothesis is true. To find the expected value for row *i* and column *j*, we take:

$$Exp_{ij} = \frac{Total_{Row\ i} \times Total_{Col\ j}}{Table\ Total}$$

Here's an example:

Suppose we ask 100 people, 40 men and 60 women, to name their magazine preference: *Sports Illustrated, Cosmopolitan,* or *The Economist* with the following result:

Magazine Preference

	Sports Illustrated	Cosmopolitan	Economist	Total
Men	25	5	10	40
Women	10	45	5	60
Total	35	50	15	100

Then, for example, the expected value under homogeneity for *Men* who prefer *The Economist* would be:

$$Exp_{13} = \frac{40 \times 15}{100} = 6$$

Performing similar calculations for all cells gives the expected values:

	Sports Illustrated	Cosmopolitan	Economist	
Men	14	20	6	40
Women	21	30	9	60
	35	50	15	100

Independence Assumption. So that we can generalize, we need the counts to be independent of each other. We can check the **Randomization Condition.** Here, we have random samples, so we *can* assume that the observations are independent and draw a conclusion comparing the populations from which the samples were taken.

We must be sure we have enough data for this method to work. The **Sample Size Assumption** can be checked with the **Expected Cell Frequency Condition,** which says that the expected count in each cell must be at least 5. Here, our samples are certainly large enough.

Following the pattern of the goodness-of-fit test, we compute the component for each cell of the table:

$$\textbf{Component} = \frac{(\textbf{\textit{Obs}} - \textbf{\textit{Exp}})^2}{\textbf{\textit{Exp}}}.$$

Summing these components across all cells gives the chi-square value:

$$\chi^2 = \sum_{all\ cells} \frac{(Obs - Exp)^2}{Exp}.$$

The degrees of freedom are different than they were for the goodness-of-fit test. For a test of homogeneity, there are $(R - 1) \times (C - 1)$ degrees of freedom, where *R* is the number of rows and *C* is the number of columns.

In our example, we have $6 \times 4 = 24$ degrees of freedom. We'll need the degrees of freedom to find a P-value for the chi-square statistic.

GUIDED EXAMPLE   *Attitudes on Appearance*

PLAN

Setup State what you want to know. Identify the variables and context.	We want to know whether the distribution of responses to how important is "Seeking the utmost attractive appearance" is the same for the five countries for which we have data: China, France, India, U.K., and U.S.
Hypotheses State the null and alternative hypotheses.	H_0: The responses are homogeneous (have the same distribution for all five countries). H_A: The responses are not homogeneous.

	Model Thick about the assumptions and check the conditions.	We have counts of the number of respondents in each country who choose each response.
		✓ **Counted Data Condition** The data are counts of the number of people choosing each possible response.
		✓ **Randomization Condition** The data were obtained from a random sample by a professional global marketing company.
		✓ **Expected Cell Frequency Condition** The expected values in each cell are all at least 5.
	State the sampling distribution model. Name the test you will use.	The conditions seem to be met, so we can use a χ^2 model with $(7-1) \times (5-1) = 24$ degrees of freedom and use a **chi-square test of homogeneity**.

Mechanics Show the expected counts for each cell of the data table. You could make separate tables for the observed and expected counts or put both counts in each cell. A segmented bar chart is often a good way to display the data.

The observed and expected counts are in Tables 15.5 and 15.7. The bar graph shows the column percentages:

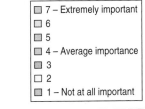

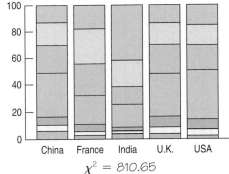

$$\chi^2 = 810.65$$

Use software to calculate χ^2 and the associated P-value.

Here, the calculated value of the χ^2 statistic is extremely high, so the P-value is quite small.

P-value $= P(\chi^2_{24} > 810.65) < 0.001$, so we reject the null hypothesis.

REPORT

Conclusion State your conclusion in the context of the data. Discuss whether the distributions for the groups appear to be different. For a small table, examine the residuals.

MEMO

Re: Importance of Appearance

Our analysis of the Roper data shows large differences across countries in the distribution of how important respondents say it is for them to look attractive. Marketers of cosmetics are advised to take notice of these differences, especially when selling products to India.

If you find that simply rejecting the hypothesis of homogeneity is a bit unsatisfying, you're in good company. It's hardly a shock that responses to this question differ from country to country. What we'd really like to know is where the differences were and how big they were. The test for homogeneity doesn't answer these interesting questions, but it does provide some evidence that can help us. A look at the standardized residuals can help identify cells that don't match the homogeneity pattern.

15.5 Comparing Two Proportions

Many employers require a high school diploma. In October 2000, the U.S. Department of Commerce researchers contacted more than 25,000 24-year-old Americans to see if they had finished high school and found that 84.9% of the 12,460 men and 88.1% of the 12,678 women reported having high school diplomas. Should we conclude that girls are more likely than boys to complete high school?

The U.S. Department of Commerce gives percentages, but it's easy to find the counts and put them in a table. It looks like this.

	Men	Women	Total
HS diploma	10,579	11,169	**21,748**
No diploma	1,881	1,509	**3,390**
Total	**12,460**	**12,678**	**25,138**

Table 15.8 Numbers of men and women earning a high school diploma in 2000 in a sample of 25,138 24-year-old Americans.

Overall, $\frac{21748}{25138} = 86.5144\%$ of the sample had high school diplomas. So, under the homogeneity assumption, we would expect the same percentage of the 12,460 men (or $0.865144 \times 12460 = 10,779.7$ men) to have diplomas. Completing the table, the expected counts look like this.

	Men	Women	Total
HS diploma	10,779.7	10,968.3	**21,748**
No diploma	1,680.3	1,709.7	**3,390**
Total	**12,460**	**12,678**	**25,138**

| *Table 15.9 The expected values.*

The chi-square statistic with $(2 - 1) \times (2 - 1) = 1$ df is:

$$\chi_1^2 = \frac{(10579 - 10779.7)^2}{10779.7} + \frac{(11169 - 10968.3)^2}{10968.3} + \frac{(1881 - 1680.3)^2}{1680.3}$$
$$+ \frac{(1509 - 1709.7)^2}{1709.7} = 54.941$$

This has a P-value < 0.001, so we reject the null hypothesis and conclude that the distribution of high school diplomas is different for men and women.

A chi-square test on a 2 × 2 table, which has only 1 df, is equivalent to testing whether two proportions (in this case, the proportions of men and women with diplomas) are equal. There is an equivalent way of testing the equality of two proportions that uses a *z*-statistic, and it gives exactly the same P-value. You may encounter the z-test for two proportions, so remember that it's the same as the chi-square test on the equivalent 2×2 table.

Even though the z-test and the chi-square test are equivalent for testing whether two proportions are the same, the z-test can also give a confidence interval. This is an important advantage, as the following example shows.

Confidence Interval for the Difference of Two Proportions

As we saw, 88.1% of the women and 84.9% of the men earned high school diplomas in the United States in the year 2000 according to the survey. That's a difference of 3.2%. If we knew the standard error of that quantity, we could use a z-statistic to construct a confidence interval for the true difference in the population. It's not hard to find the standard error. All we need is the formula[5]:

$$SE(\hat{p}_1 - \hat{p}_2) = \sqrt{\frac{\hat{p}_1\hat{q}_1}{n_1} + \frac{\hat{p}_2\hat{q}_2}{n_2}}$$

The confidence interval has the same form as the confidence interval for a single proportion, with this new standard error:

$$(\hat{p}_1 - \hat{p}_2) \pm z^*SE(\hat{p}_1 - \hat{p}_2).$$

Confidence Interval for the Difference of Two Proportions

When the conditions are met, we can find the confidence interval for difference of two proportions, $p_1 - p_2$. The confidence interval is

$$(\hat{p}_1 - \hat{p}_2) \pm z^*SE(\hat{p}_1 - \hat{p}_2),$$

where we find the standard error of the difference as

$$SE(\hat{p}_1 - \hat{p}_2) = \sqrt{\frac{\hat{p}_1\hat{q}_1}{n_1} + \frac{\hat{p}_2\hat{q}_2}{n_2}}$$

from the observed proportions.

The critical value z^* depends on the particular confidence level that you specify.

For the high school graduation, a 95% confidence interval for the true difference between women's and men's rates is:

$$(0.881 - 0.849) \pm 1.96 \times \sqrt{\frac{(0.881)(0.119)}{12678} + \frac{(0.849)(0.151)}{12460}}$$
$$= (0.0236, 0.0404), \text{ or } 2.36\% \text{ to } 4.04\%.$$

We can be 95% confident that women's graduation rates in 2000 were 2.36 to 4.04% higher than men's. With a sample size this large, we can be quite confident that the difference isn't zero. But is it a difference that matters? That, of course, depends on the *reason* we are asking the question. The confidence interval shows us the effect size—or at least, the interval of plausible values for the effect size. If we are considering changing hiring or recruitment policies, this difference may be too small to warrant much of an adjustment even though the difference is statistically "significant." Be sure to consider the effect size if you plan to make a business decision based on rejecting a null hypothesis using chi-square methods.

---

[5] The standard error of the difference is found from the general fact that the variance of a difference of two independent quantities is the *sum* of their variances. See Chapter 21 for details.

15.6 Chi-Square Test of Independence

We saw that the importance people place on their personal appearance varies a great deal from one country to another, a fact that might be crucial for the marketing department of a global cosmetics company. Suppose the marketing department wants to know whether the age of the person matters as well. That might affect the kind of media channels they use to advertise their products. Do older people feel as strongly as younger people that personal appearance is important?

		Age						
		13–19	**20–29**	**30–39**	**40–49**	**50–59**	**60+**	**Total**
	7—Extremely important	396	337	300	252	142	93	**1520**
	6	325	326	307	254	123	86	**1421**
	5	318	312	317	270	150	106	**1473**
Appearance	**4—Average importance**	397	376	403	423	224	210	**2033**
	3	83	83	88	93	54	45	**446**
	2	37	43	53	58	37	45	**273**
	1—Not at all important	40	37	53	56	36	52	**274**
	Total	**1596**	**1514**	**1521**	**1406**	**766**	**637**	**7440**

| **Table 15.10**   *Responses to the question about personal appearance by age group.*

When we examined the five countries, we thought of the countries as five different groups, rather than as levels of a variable. But here, we can (and probably should) think of *Age* as a second variable whose value has been measured for each respondent along with his or her response to the appearance question. Asking whether the distribution of responses changes with *Age* now raises the question of whether the variables personal *Appearance* and *Age* are independent.

Whenever we have two variables in a contingency table like this, the natural test is a **chi-square test of independence**. Mechanically, this chi-square test is identical to a test of homogeneity. The difference between the two tests is in how we think of the data and, thus, what conclusion we draw.

Here we ask whether the response to the personal appearance question is independent of age. Remember, that for any two events, **A** and **B**, to be independent, the probability of event **A** given that event **B** occurred must be the same as the probability of event **A**. Here, this means the probability that a randomly selected respondent thinks personal appearance is extremely important is the same for all age groups. That would show that the response to the personal *Appearance* question is independent of that respondent's *Age*. Of course, from a table based on data, the probabilities will never be exactly the same. But to tell whether they are different enough, we use a chi-square test of independence.

Now we have two categorical variables measured on a single population. For the homogeneity test, we had a single categorical variable measured independently on two or more populations. Now we ask a different question: "Are the variables independent?" rather than "Are the groups homogeneous?" These are subtle differences, but they are important when we draw conclusions.

> The only difference between the test for homogeneity and the test for independence is in the decision you need to make.

Assumptions and Conditions

Of course, we still need counts and enough data so that the expected counts are at least five in each cell.

If we're interested in the independence of variables, we usually want to generalize from the data to some population. In that case, we'll need to check that the data are a representative random sample from that population.

GUIDED EXAMPLE | *Personal Appearance and Age*

PLAN

Setup State what you want to know.

Identify the variables and context.

We want to know whether the categorical variables personal *Appearance* and *Age* are statistically independent. We have a contingency table of 7440 respondents from a sample of five countries.

Hypotheses State the null and alternative hypotheses.

We perform a test of independence when we suspect the variables may not be independent. We are making the claim that knowing the respondents age will change the distribution of their response to the question about personal *Appearance*, and testing the null hypothesis that it is *not* true.

H_0: personal *Appearance* and *Age* are independent.[6]

H_A: personal *Appearance* and *Age* are not independent.

Model Check the conditions.

✓ **Counted Data Condition** We have counts of individuals categorized on two categorical variables.

✓ **Randomization Condition** These data are from a randomized survey conducted in 30 countries. We have data from five of them. Although they are not an SRS, they were selected to avoid biases.

✓ **Expected Cell Frequency Condition** The expected values are all much larger than 5.

This table shows the expected counts below for each cell. The expected counts are calculated exactly as they were for a test of homogeneity; in the first cell, for example, we expect $\frac{1520}{7440} = 20.43\%$ of 1596 which is 326.06.

			Expected Values				
				Age			
		13–19	**20–29**	**30–39**	**40–49**	**50–59**	**60 +**
	7—Extremely important	326.065	309.312	310.742	287.247	156.495	130.140
	6	304.827	289.166	290.503	268.538	146.302	121.664
	5	315.982	299.748	301.133	278.365	151.656	126.116
Appearance	**4—Average importance**	436.111	413.705	415.617	384.193	209.312	174.062
	3	95.674	90.759	91.178	84.284	45.919	38.186
	2	58.563	55.554	55.811	51.591	28.107	23.374
	1—Not at all important	58.777	55.758	56.015	51.780	28.210	23.459

The stacked bar graph shows that the response seems to be dependent on *Age*. Older people tend

---

[6] As in other chi-square tests, the hypotheses are usually expressed in words, without parameters. The hypothesis of independence itself tells us how to find expected values for each cell of the contingency table. That's all we need.

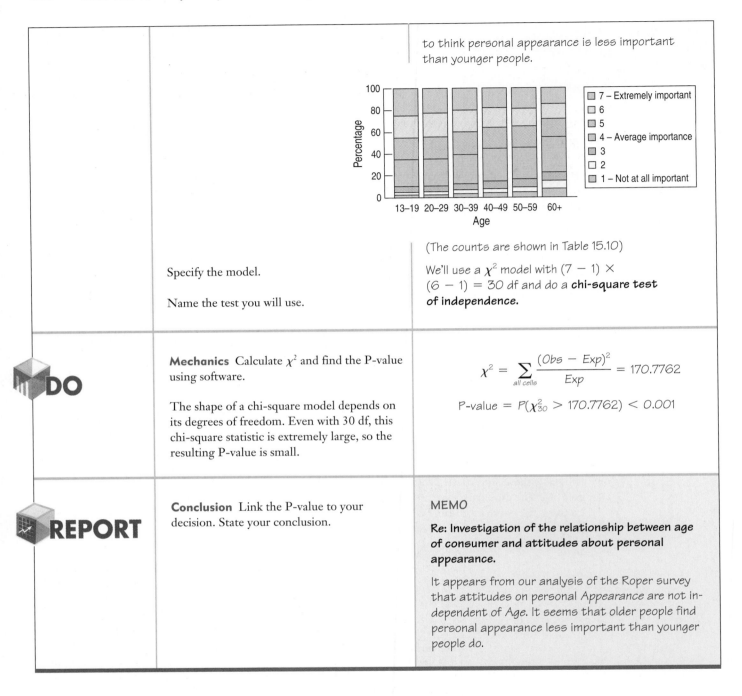

to think personal appearance is less important than younger people.

(The counts are shown in Table 15.10)

Specify the model.

Name the test you will use.

We'll use a χ^2 model with $(7 - 1) \times (6 - 1) = 30$ df and do a **chi-square test of independence**.

DO

Mechanics Calculate χ^2 and find the P-value using software.

The shape of a chi-square model depends on its degrees of freedom. Even with 30 df, this chi-square statistic is extremely large, so the resulting P-value is small.

$$\chi^2 = \sum_{all\ cells} \frac{(Obs - Exp)^2}{Exp} = 170.7762$$

$$\text{P-value} = P(\chi^2_{30} > 170.7762) < 0.001$$

REPORT

Conclusion Link the P-value to your decision. State your conclusion.

MEMO

Re: Investigation of the relationship between age of consumer and attitudes about personal appearance.

It appears from our analysis of the Roper survey that attitudes on personal Appearance are not independent of Age. It seems that older people find personal appearance less important than younger people do.

We rejected the null hypothesis of independence between *Age* and attitudes about personal *Appearance*. With a sample size this large, we can detect very small deviations from independence, so it's almost guaranteed that the chi-square test will reject the null hypothesis. Examining the residuals can help you see the cells that deviate farthest from independence. To make a meaningful business decision, you'll have to look at effect sizes as well.

Suppose the company was specifically interested in deciding how to split advertising resources between the teen market and the 30–39-year-old market. How much of a difference are the proportions of those in each age group that rated personal *Appearance* as very important (responding either 6 or 7)?

JUST CHECKING

Which of the three chi-square tests would you use in each of the following situations—goodness-of-fit, homogeneity, or independence?

1 A restaurant manager wonders whether customers who dine on Friday nights have the same preferences among the chef's four special entrées as those who dine on Saturday nights. One weekend he has the wait staff record which entrées were ordered each night. Assuming these customers to be typical of all weekend diners, he'll compare the distributions of meals chosen Friday and Saturday.

2 Company policy calls for parking spaces to be assigned to everyone at random, but you suspect that may not be so. There are three lots of equal size: lot A, next to the building; lot B, a bit farther away; and lot C on the other side of the highway. You gather data about employees at middle management level and above to see how many were assigned parking in each lot.

3 Is a student's social life affected by where the student lives? A campus survey asked a random sample of students whether they lived in a dormitory, in off-campus housing, or at home and whether they had been out on a date 0, 1–2, 3–4, or 5 or more times in the past two weeks.

For that we'll need to construct a confidence interval on the difference. From Table 15.10, we find that the percentages of those answering 6 and 7 are 45.17% and 39.91% for the teen and 30–39-year-old groups, respectively. The 95% confidence interval is:

$$(\hat{p}_1 - \hat{p}_2) \pm z^* SE(\hat{p}_1 - \hat{p}_2)$$

$$= (0.4517 - 0.3991) \pm 1.96 \times \sqrt{\frac{(0.4517)(0.5483)}{1596} + \frac{(0.3991)(0.6009)}{1521}}$$

$$= (0.018, 0.087), \text{ or } (1.8\% \text{ to } 8.7\%)$$

This is a statistically significant difference, but now we can see that the difference may be as small as 1.8%. When deciding how to allocate advertising expenditures, it is important to keep these estimates of the effect size in mind.

Chi-Square Tests and Causation

Chi-square tests are common. Tests for independence are especially widespread. Unfortunately, many people interpret a small P-value as proof of causation. We know better. Just as correlation between quantitative variables does not demonstrate causation, a failure of independence between two categorical variables does not show a cause-and-effect relationship between them, nor should we say that one variable *depends* on the other.

The chi-square test for independence treats the two variables symmetrically. There is no way to differentiate the direction of any possible causation from one variable to the other. While we can see that attitudes on personal *Appearance* and *Age* are related, we can't say that getting older *causes* you to change attitudes. And certainly it's not correct to say that changing attitudes on personal appearance makes you older.

Of course, there's never any way to eliminate the possibility that a lurking variable is responsible for the observed lack of independence. In some sense, a failure of independence between two categorical variables is less impressive than a strong, consistent, linear association between quantitative variables. Two categorical variables can fail the test of independence in many ways, including ways that show no consistent pattern of failure. Examination of the chi-square standardized residuals can help you think about the underlying patterns.

WHAT CAN GO WRONG?

- **Don't use chi-square methods unless you have counts.** All three of the chi-square tests apply only to counts. Other kinds of data can be arrayed in two-way tables. Just because numbers are in a two-way table doesn't make them suitable for chi-square analysis. Data reported as proportions or percentages can be suitable for chi-square procedures, *but only after they are converted to counts.* If you try to do the calculations without first finding the counts, your results will be wrong.

- **Beware large samples.** Beware *large* samples? That's not the advice you're used to hearing. The chi-square tests, however, are unusual. You should be wary of chi-square tests performed on very large samples. No hypothesized distribution fits perfectly, no two groups are exactly homogeneous, and two variables are rarely perfectly independent. The degrees of freedom for chi-square tests don't grow with the sample size. With a sufficiently large sample size, a chi-square test can always reject the null hypothesis. But we have no measure of how far the data are from the null model. There are no confidence intervals to help us judge the effect size except in the case of two proportions.

- **Don't say that one variable "depends" on the other just because they're not independent.** "Depend" can suggest a model or a pattern, but variables can fail to be independent in many different ways. When variables fail the test for independence, it may be better to say they are "associated."

ETHICS IN ACTION

eliberately Different specializes in unique accessories for the home such as hand-painted switch plates and hand-embroidered linens, offered through a catalog and a website. Its customers tend to be women, generally older, with relatively high household incomes. Although the number of customer visits to the site has remained the same, management noticed that the proportion of customers visiting the site who make a purchase has been declining. Megan Cally, the product manager for Deliberately Different, was in charge of working with the market research firm hired to examine this problem. In her first meeting with Jason Esgro, the firm's consultant, she directed the conversation toward website design. Jason mentioned several reasons for consumers abandoning online purchases, the two most common being concerns about transaction security and unanticipated shipping/handling charges. Because Deliberately Different's shipping charges are reasonable, Megan asked him to look further into the issue of security concerns. They developed a survey that randomly sampled customers who had visited the website. They contacted these customers by e-mail and asked them to respond to a brief survey, offering the chance of winning a prize, which would be awarded at random among the respondents. A total of 2450 responses were received. The analysis of the responses included chi-square tests for independence, checking to see if responses on the security question were independent of gender and income category. Both tests were significant, rejecting the null hypothesis of independence. Megan reported to management that concerns about online transaction security were dependent on gender, and income, so Deliberately Different began to explore ways in which they could assure their older female customers that transactions on the website are indeed secure. As product manager, Megan was relieved that the decline in purchases was not related to product offerings.

ETHICAL ISSUE: *The chance of rejecting the null hypothesis in a chi-square test for independence increases with sample size. Here the sample size is very large. In addition, it is misleading to state that concerns about security depend on gender, age, and income. Furthermore, patterns of association were not examined (for instance, with varying age categories). Finally, as product manager, Megan intentionally steered attention away from examining the product offerings, which could be a factor in declining purchases. Instead she reported to management that they have pinpointed the problem without noting that they had not explored other potential factors (related to Items A and H, ASA Ethical Guidelines).*

ETHICAL SOLUTION: *Interpret results correctly, cautioning about the large sample size and looking for any patterns of association, realizing that there is no way to estimate the effect size.*

What have we learned?

We've learned how to test hypotheses about categorical variables. We use one of three related methods. All look at counts of data in categories, and all rely on chi-square models, a new family indexed by degrees of freedom.

- Goodness-of-fit tests compare the observed distribution of a single categorical variable to an expected distribution based on a theory or model.
- Tests of homogeneity compare the distribution of several groups for the same categorical variable.
- Tests of independence examine counts from a single group for evidence of an association between two categorical variables.

We've seen that mechanically these tests are almost identical. Although the tests appear to be one-sided, we've learned that conceptually they are many-sided because there are many ways that a table of counts can deviate significantly from what we hypothesized. When that happens and we reject the null hypothesis, we've learned to examine standardized residuals in order to better understand the patterns in the table.

Terms

Cell
A cell of a two-way table is one element of the table corresponding to a specific row and a specific column. Table cells can hold counts, percentages, or measurements on other variables, or they can hold several values.

Chi-square models
Chi-square models are skewed to the right. They are parameterized by their degrees of freedom and become less skewed with increasing degrees of freedom.

Chi-square (or chi-squared) statistic
The chi-square statistic is found by summing the chi-square components. Chi-square tests can be used to test goodness-of-fit, homogeneity, or independence.

Contingency table
A two-way table that classifies individuals according to two categorical variables.

Goodness-of-fit
A test of whether the distribution of counts in one categorical variable matches the distribution predicted by a model. A chi-square test of goodness-of-fit finds

$$\chi^2 = \sum_{all\ cells} \frac{(Obs - Exp)^2}{Exp},$$

where the expected counts come from the predicting model. It finds a P-value from a chi-square model with $n - 1$ degrees of freedom, where n is the number of categories in the categorical variable.

Homogeneity
A test comparing the distribution of counts for two or more groups on the same categorical variable. A chi-square test of homogeneity finds

$$\chi^2 = \sum_{all\ cells} \frac{(Obs - Exp)^2}{Exp},$$

where the expected counts are based on the overall frequencies, adjusted for the totals in each group. We find a P-value from a chi-square distribution with $(R - 1) \times (C - 1)$ degrees of freedom, where R gives the number of categories (rows) and C gives the number of independent groups (columns).

Independence	A test of whether two categorical variables are independent. It examines the distribution of counts for one group of individuals classified according to both variables. A chi-square test of *independence* uses the same calculation as a test of homogeneity. We find a P-value from a chi-square distribution with $(R - 1) \times (C - 1)$ degrees of freedom, where R gives the number of categories in one variable and C gives the number of categories in the other.
Standardized residual	In each cell of a two-way table, a standardized residual is the square root of the chi-square component for that cell with the sign of the *Observed − Expected* difference:

$$\frac{(Obs - Exp)}{\sqrt{Exp}}$$

When we reject a chi-square test, an examination of the standardized residuals can sometimes reveal more about how the data deviate from the null model.

Skills

 PLAN

- Be able to recognize when a test of goodness-of-fit, a test of homogeneity, or a test of independence would be appropriate for a table of counts.

- Understand that the degrees of freedom for a chi-square test depend on the dimensions of the table and not on the sample size. Understand that this means that increasing the sample size increases the ability of chi-square procedures to reject the null hypothesis.

 DO

- Be able to display and interpret counts in a two-way table.

- Know how to use the chi-square tables to perform chi-square tests.

- Know how to perform a chi-square test using statistics software or a calculator.

- *Know how to find a confidence interval for the difference of two proportions.

- Be able to examine the standardized residuals to explain the nature of the deviations from the null hypothesis.

 REPORT

- Know how to communicate the results of chi-square tests, whether goodness-of-fit, homogeneity, or independence, in a few sentences.

Technology Help: Chi-Square

Most statistics packages associate chi-square tests with contingency tables. Often chi-square is available as an option only when you make a contingency table. This organization can make it hard to locate the chi-square test and may confuse the three different roles that the chi-square test can take. In particular, chi-square tests for goodness-of-fit may be hard to find or missing entirely. Chi-square tests for homogeneity are computationally the same as chi-square tests for independence, so you may have to perform the mechanics as if they were tests of independence and interpret them afterward as tests of homogeneity.

Most statistics packages work with data on individuals rather than with the summary counts. If the only information you have is the table of counts, you may find it more difficult to get a statistics package to compute chi-square. Some packages offer a way to reconstruct the data from the summary counts so that they can then be passed back through the chi-square calculation, finding the cell counts again. Many packages offer chi-square standardized residuals (although they may be called something else).

EXCEL

Excel offers the function **CHITEST (actual_range, expected_range),** which computes a chi-square P-value for homogeneity. Both ranges are of the form UpperLeftCell: LowerRightCell, specifying two rectangular tables that must hold counts (although Excel will not check for integer values). The two tables must be of the same size and shape. DDXL offers all 3 tests. Use the **Tables** command.

Comments

Excel's documentation claims this is a test for independence and labels the input ranges accordingly, but Excel offers no way to find expected counts, so the function is not particularly useful for testing independence. You can use this function only if you already know the expected values or are willing to program additional calculations.

MINITAB

From the **Start** menu, choose the **Tables** submenu. From that menu, choose **Chi Square Test . . . .** In the dialog, identify the columns that make up the table. Minitab will display the table and print the chi-square value and its P-value.

Comments

Alternatively, select the **Cross Tabulation . . .** command to see more options for the table, including expected counts and standardized residuals.

SPSS

From the **Analyze** menu, choose the **Descriptive Statistics** submenu. From that submenu, choose **Crosstabs . . . .** In the Crosstabs dialog, assign the row and column variables from the variable list. Both variables must be categorical. Click the **Cells** button to specify that standardized residuals should be displayed. Click the **Statistics** button to specify a chi-square test.

Comments

SPSS offers only variables that it knows to be categorical in the variable list for the Crosstabs dialog. If the variables you want are missing, check that they have the right type.

JMP

From the **Analyze** menu, select **Fit Y by X**. Choose one variable as the Y, response variable, and the other as the X, factor variable. Both selected variables must be Nominal or Ordinal. JMP will make a plot and a contingency table. Below the contingency table, **JMP** offers a **Tests** panel. In that panel, the Chi Square for independence is called a **Pearson ChiSquare**. The table also offers the P-value.

Click on the contingency Table title bar to drop down a menu that offers to include a **Deviation** and Cell **Chi square** in each cell of the table.

Comments

JMP will choose a chi-square analysis for a **Fit Y by X** if both variables are nominal or ordinal (marked with an N or O), but not otherwise. Be sure the variables have the right type. Deviations are the observed—expected differences in counts. Cell chi-squares are the squares of the standardized residuals. Refer to the deviations for the sign of the difference. Look under **Distributions** in the **Analyze** menu to find a chi-square test for goodness-of-fit.

DATA DESK

Select variables. From the **Calc** menu, choose **Contingency Table.** From the table's HyperView menu, choose **Table Options** (or choose **Calc > Calculation Options > Table Options**). In the dialog, check the boxes for **Chi Square** and for **Standardized Residuals.** Data Desk will display the chi-square and its P-value below the table and the standardized residuals within the table.

Comments

Data Desk automatically treats variables selected for this command as categorical variables even if their elements are numerals.

The **Compute Counts** command in the table's HyperView menu will make variables that hold the table contents (as selected in the Table Options dialog), including the standardized residuals.

Mini Case Study Projects

Health Insurance

With the rising costs of medical insurance and the declining interest and ability of employers to maintain proper medical coverage for their employees, business owners and employees alike are wondering: Who will insure future workers in the United States? The government has spent decades debating different initiatives to expand health insurance coverage, but no comprehensive bill has been passed.

Just how widespread is the lack of medical coverage? The media claims that the segments of the population most at risk are women, children, and the elderly. The tables give the number of uninsured (in thousands) by sex and by age in 2004.[7] Using the appropriate chi-square test and the statistics software of your choice, investigate the accuracy of the media's statement using these data. Be sure to discuss your assumptions, methods, results, and conclusions.

Sex			
	Male	**Female**	**Total**
Insured	86,176	93,329	**179,505**
Uninsured	16,026	15,117	**31,143**
Total	**102,202**	**108,446**	**210,648**

Age Groups						
	0–17	**18–24**	**25–44**	**45–64**	**65–80**	**Total**
Insured	57,375	12,755	47,850	41,176	20,349	**179,505**
Uninsured	6,755	5,464	12,105	6,607	212	**31,143**
Total	**64,130**	**18,219**	**59,955**	**47,783**	**20,561**	**210,648**

Loyalty Program

A marketing executive tested two incentives to see what percentage of customers would enroll in a new web-based loyalty program. The customers were asked to log on to their accounts on the Web and provide some demographic and spending information. As an incentive, they were offered either Nothing (No Offer), Free flight insurance on their next flight (Free Insurance), or a free companion Airline ticket (Free Flight). The customers were segmented according to their past year's spending patterns as spending primarily in one of five areas: *Travel, Entertainment, Dining, Household,* or *Balanced.* The executive wanted to know whether the incentives resulted in different enrollment rates (*Response*). Specifically, she wanted to know how much higher the enrollment rate for the free flight was compared to the free insurance. She also wanted to see whether *Spending Pattern* was associated with *Response.* Using the data **ch15_MCSP_Loyalty_Program**, write up a report for the marketing executive using appropriate graphics, summary statistics, statistical tests, and confidence intervals.

---

[7] Source: U.S. Census Bureau, Current Population Survey, Annual Social and Economic Supplement, 2005.

EXERCISES

1. Concepts. For each of the following situations, state whether you'd use a chi-square goodness-of-fit test, chi-square test of homogeneity, chi-square test of independence, or some other statistical test.

a) A brokerage firm wants to see whether the type of account a customer has (Silver, Gold, or Platinum) affects the type of trades that customer makes (in person, by phone, or on the Internet). It collects a random sample of trades made for its customers over the past year and performs a test.

b) That brokerage firm also wants to know if the type of account affects the size of the account (in dollars). It performs a test to see if the mean size of the account is the same for the three account types.

c) The academic research office at a large community college wants to see whether the distribution of courses chosen (Humanities, Social Science, or Science) is different for its residential and nonresidential students. It assembles last semester's data and performs a test.

2. Concepts, part 2. For each of the following situations, state whether you'd use a chi-square goodness-of-fit test, a chi-square test of homogeneity, a chi-square test of independence, or some other statistical test.

a) Is the quality of a car affected by what day it was built? A car manufacturer examines a random sample of the warranty claims filed over the past two years to test whether defects are randomly distributed across days of the work week.

b) A researcher for the American Booksellers Association wants to know if retail sales/sq. ft. is related to serving coffee or snacks on the premises. She examines a database of 10,000 independently owned bookstores testing whether retail sales (dollars/sq. ft.) is related to whether or not the store has a coffee bar.

c) A researcher wants to find out whether education level (some high school, high school graduate, college graduate, advanced degree) is related to the type of transaction most likely to be conducted using the Internet (shopping, banking, travel reservations, auctions). He surveys 500 randomly chosen adults and performs a test.

3. Dice. After getting trounced by your little brother in a children's game, you suspect that the die he gave you is unfair. To check, you roll it 60 times, recording the number of times each face appears. Do these results cast doubt on the die's fairness?

a) If the die is fair, how many times would you expect each face to show?

b) To see if these results are unusual, will you test goodness-of-fit, homogeneity, or independence?

c) State your hypotheses.

d) Check the conditions.

e) How many degrees of freedom are there?

f) Find χ^2 and the P-value.

g) State your conclusion.

Face	Count
1	11
2	7
3	9
4	15
5	12
6	6

4. Quality control. Mars, Inc. says that the colors of its M&M's® candies are 14% yellow, 13% red, 20% orange, 24% blue, 16% green and 13% brown. (www.mms.com/us/about/products/milkchocolate). On his way home from work the day he was writing these exercises, one of the authors bought a bag of plain M&M's. He got 29 yellow, 23 red, 12 orange, 14 blue, 8 green, and 20 brown. Is this sample consistent with the company's advertised proportions? Test an appropriate hypothesis and state your conclusion.

a) If the M&M's are packaged in the advertised proportions, how many of each color should the author have expected in his bag of M&M's?

b) To see if his bag was unusual, should he test goodness-of-fit, homogeneity, or independence?

c) State the hypotheses.

d) Check the conditions.

e) How many degrees of freedom are there?

f) Find χ^2 and the P-value.

g) State a conclusion.

5. Quality control, part 2. A company advertises that its premium mixture of nuts contains 10% Brazil nuts, 20% cashews, 20% almonds, 10% hazelnuts, and that the rest are peanuts. You buy a large can and separate the various kinds of nuts. Upon weighing them, you find there are 112 grams of Brazil nuts, 183 grams of cashews, 207 grams of almonds, 71 grams of hazelnuts, and 446 grams of peanuts. You wonder whether your mix is significantly different from what the company advertises.

a) Explain why the chi-square goodness-of-fit test is not an appropriate way to find out.

b) What might you do instead of weighing the nuts in order to use a χ^2 test?

6. Sales rep travel. A sales representative who is on the road visiting clients thinks that, on average, he drives the same distance each day of the week. He keeps track of his mileage for several weeks and discovers that he averages 122 miles on Mondays, 203 miles on Tuesdays, 176 miles on Wednesdays, 181 miles on Thursdays, and 108 miles on Fridays. He wonders if this evidence contradicts his belief in a uniform distribution of miles across the days of the week. Is it appropriate to test his hypothesis using the chi-square goodness-of-fit test? Explain.

7. Maryland lottery. For a lottery to be successful, the public must have confidence in its fairness. One of the lotteries in Maryland is Pick-3 Lottery, where 3 random digits are drawn each day.[8] A fair game depends on every value (0 to 9) being equally likely at each of the three positions. If not, then someone detecting a pattern could take advantage of that and beat the lottery. To investigate the randomness, we'll look at data collected over a recent 32-week period. Although the winning numbers look like three-digit numbers, in fact, each digit is a randomly drawn numeral. We have 654 random digits in all. Are each of the digits from 0 to 9 equally likely? Here is a table of the frequencies.

Group	Count	%
0	62	9.480
1	55	8.410
2	66	10.092
3	64	9.786
4	75	11.468
5	57	8.716
6	71	10.856
7	74	11.315
8	69	10.550
9	61	9.327

a) Select the appropriate procedure.
b) Check the assumptions.
c) State the hypotheses.
d) Test an appropriate hypothesis and state your results.
e) Interpret the meaning of the results and state a conclusion.

8. Employment discrimination? Census data for New York City indicate that 29.2% of the under-18 population is white, 28.2% black, 31.5% Latino, 9.1% Asian, and 2% are of other ethnicities. The New York Civil Liberties Union points out that of 26,181 police officers, 64.8% are white, 14.5% black, 19.1% Hispanic, and 1.4% Asian. Do the police officers reflect the ethnic composition of the city's youth?

a) Select the appropriate procedure.
b) Check the assumptions.
c) State the hypotheses.
d) Test an appropriate hypothesis and state your results.
e) Interpret the meaning of the results and state a conclusion.

T 9. Titanic. Here is a table showing who survived the sinking of the *Titanic* based on whether they were crew members or passengers booked in first-, second-, or third-class staterooms.

	Crew	First	Second	Third	Total
Alive	212	202	118	178	**710**
Dead	673	123	167	528	**1491**
Total	**885**	**325**	**285**	**706**	**2201**

a) If we draw an individual at random from this table, what's the probability that we will draw a member of the crew?
b) What's the probability of randomly selecting a third-class passenger who survived?
c) What's the probability of a randomly selected passenger surviving, given that the passenger was in a first-class stateroom?
d) If someone's chances of surviving were the same regardless of their status on the ship, how many members of the crew would you expect to have lived?
e) State the null and alternative hypotheses we would test here.
f) Give the degrees of freedom for the test.
g) The chi-square value for the table is 187.8, and the corresponding P-value is barely greater than 0. State your conclusions about the hypotheses.

T 10. Promotion discrimination? The table shows the rank attained by male and female officers in the New York City Police Department (NYPD). Do these data indicate that men and women are equitably represented at all levels of the department?

Rank		Male	Female
	Officer	21,900	4281
	Detective	4058	806
	Sergeant	3898	415
	Lieutenant	1333	89
	Captain	359	12
	Higher ranks	218	10

a) What's the probability that a person selected at random from the NYPD is a female?
b) What's the probability that a person selected at random from the NYPD is a detective?
c) Assuming no bias in promotions, how many female detectives would you expect the NYPD to have?
d) To see if there is evidence of differences in ranks attained by males and females, will you test goodness-of-fit, homogeneity, or independence?
e) State the hypotheses.
f) Test the conditions.
g) How many degrees of freedom are there?
h) Find the chi-square value and the associated P-value.
i) State your conclusion.
j) If you concluded that the distributions are not the same, analyze the differences using the standardized residuals of your calculations.

11. Birth order and college choice. Students in an Introductory Statistics class at a large university were classified by birth order and by the college they attend.

Birth Order (1 = oldest or only child)

College		1	2	3	4 or more	Total
	Arts and Sciences	34	14	6	3	57
	Agriculture	52	27	5	9	93
	Social Science	15	17	8	3	43
	Professional	13	11	1	6	31
	Total	114	69	20	21	224

Expected Values
Birth Order (1 = oldest or only child)

College		1	2	3	4 or more
	Arts and Sciences	29.0089	17.5580	5.0893	5.3438
	Agriculture	47.3304	28.6473	8.3036	8.7188
	Social Science	21.8839	13.2455	3.8393	4.0313
	Professional	15.7768	9.5491	2.7679	2.9063

a) What kind of chi-square test is appropriate—goodness-of-fit, homogeneity, or independence?

b) State your hypotheses.

c) State and check the conditions.

d) How many degrees of freedom are there?

e) The calculation yields $\chi^2 = 17.78$, with P $= 0.0378$. State your conclusion.

f) Examine and comment on the standardized residuals. Do they challenge your conclusion? Explain.

Standardized Residuals
Birth Order (1 = oldest or only child)

College		1	2	3	4 or more
	Arts and Sciences	0.92667	−0.84913	0.40370	−1.01388
	Agriculture	0.67876	−0.30778	−1.14640	0.09525
	Social Science	−1.47155	1.03160	2.12350	−0.51362
	Professional	−0.69909	0.46952	−1.06261	1.81476

12. Automobile manufacturers. *Consumer Reports* uses surveys given to subscribers of its magazine and website (www. ConsumerReports.org) to measure reliability in automobiles. This annual survey asks about problems that consumers have had with their cars, vans, SUVs, or trucks during the previous 12 months. Each analysis is based on the number of problems per 100 vehicles.

Origin of Manufacturer

	Asia	Europe	U.S.	Total
No Problems	88	79	83	250
Problems	12	21	17	50
Total	100	100	100	300

Expected Values

	Asia	Europe	U.S.
No Problems	83.33	83.33	83.33
Problems	16.67	16.67	16.67

a) State your hypotheses.

b) State and check the conditions.

c) How many degrees of freedom are there?

d) The calculation yields $\chi^2 = 2.928$, with P $= 0.231$. State your conclusion.

e) Would you expect that a larger sample might find statistical significance? Explain.

T 13. Cranberry juice. It's common folk wisdom that cranberries can help prevent urinary tract infections in women. A leading producer of cranberry juice would like to use this information in their next ad campaign, so they need evidence of this claim. In 2001, the *British Medical Journal* reported the results of a Finnish study in which three groups of 50 women were monitored for these infections over 6 months. One group drank cranberry juice daily, another group drank a lactobacillus drink, and the third group drank neither of those beverages, serving as a control group. In the control group, 18 women developed at least one infection compared with 20 of those who consumed the lactobacillus drink and only 8 of those who drank cranberry juice. Does this study provide supporting evidence for the value of cranberry juice in warding off urinary tract infections in women?

a) Select the appropriate procedure.

b) Check the assumptions.

c) State the hypotheses.

d) Test an appropriate hypothesis and state your results.

e) Interpret the meaning of the results and state a conclusion.

f) If you concluded that the groups are not the same, analyze the differences using the standardized residuals of your calculations.

T 14. Car company. A European manufacturer of automobiles claims that their cars are preferred by the younger generation and would like to target university students in their next ad campaign. Suppose we test their claim with our own survey. A random survey of autos parked in the student lot and the staff lot at a large university classified the brands by country of origin, as seen in the following table. Are there differences in the national origins of cars driven by students and staff?

	Driver	
	Student	**Staff**
American	107	105
European	33	12
Asian	55	47

(Origin)

a) Is this a test of independence or homogeneity?
b) Write appropriate hypotheses.
c) Check the necessary assumptions and conditions.
d) Find the P-value of your test.
e) State your conclusion and analysis.

15. Market segmentation. The Chicago Female Fashion Study[9] surveyed customers to determine characteristics of the "frequent" shoppers at different department stores in the Chicago area. Suppose you are a marketing manager at one of the department stores. You would like to know if a customer's shopping frequency and her age are related. Here are the data:

	Age			
Shopping Frequency	**18–24**	**25–44**	**45–54**	**55 or over**
Never/Hardly Ever	32	171	45	24
1–2 times/yr	18	134	40	37
3–4 times/yr	21	109	48	27
≥ 5 times/yr	39	134	71	50

	Standardized Residuals Age			
Shopping Frequency	**18–24**	**25–44**	**45–54**	**55 or over**
Never/Hardly Ever	0.3803	1.7974	−1.4080	−2.2094
1–2 times/yr	−1.4326	0.7595	−0.9826	0.9602
3–4 times/yr	−0.3264	−0.3151	0.9556	−0.2425
≥ 5 times/yr	1.1711	−2.1360	1.4235	1.4802

a) Is this a test of homogeneity or independence?
b) Write an appropriate hypothesis.
c) Are the conditions for inference satisfied?
d) The calculation yields $\chi^2 = 26.084$, P-value = 0.002. State your conclusion.
e) Given the standardized residuals in the table, state a complete conclusion.

[9] Original *Market Segmentation Exercise* prepared by K. Matsuno, D. Kopcso, and D. Tigert, Babson College in 1997 (Babson Case Series #133-C97A-U).

16. Seafood company. A large company in the northeastern United States that buys fish from local fishermen and distributes them to major companies and restaurants is considering launching a new ad campaign on the health benefits of fish. As evidence, they would like to cite the following study. Medical researchers followed 6272 Swedish men for 30 years to see if there was any association between the amount of fish in their diet and prostate cancer ("Fatty Fish Consumption and Risk of Prostate Cancer," *Lancet*, June 2001).

	Prostate Cancer	
Fish Consumption	**No**	**Yes**
Never/seldom	110	14
Small part of diet	2420	201
Moderate part	2769	209
Large part	507	42

a) Is this a survey, a retrospective study, a prospective study, or an experiment? Explain.
b) Is this a test of homogeneity or independence?
c) Do you see evidence of an association between the amount of fish in a man's diet and his risk of developing prostate cancer?
d) Does this study prove that eating fish does not prevent prostate cancer? Explain.

17. Shopping. A survey of 430 randomly chosen adults finds that 47 of 222 men and 37 of 208 women had purchased books online.

a) Is there evidence that the sex of the person and whether they buy books online are associated?
b) If your conclusion in fact proves to be wrong, did you make a Type I or Type II error?
c) Give a 95% confidence interval for the difference in proportions of buying online for men and women.

18. Information technology. A recent report suggests that Chief Information Officers (CIO's) who report directly to Chief Financial Officers (CFO's) rather than Chief Executive Officers (CEO's) are more likely to have IT agendas that deal with cost cutting and compliance (SearchCIO.com, March 14, 2006). In a random sample of 535 companies, it was found that CIO's reported directly to CFO's in 173 out of 335 service firms and in 95 out of 200 manufacturing companies.

a) Is there evidence that type of business (service versus manufacturing) and whether or not the CIO reports directly to the CFO are associated?
b) If your conclusion proves to be wrong, did you make a Type I or Type II error?
c) Give a 95% confidence interval for the difference in proportions of companies in which the CIO reports directly to the CFO between service and manufacturing firms.

19. Fast food. GfK Roper Consulting gathers information on consumer preferences around the world to help companies monitor attitudes about health, food, and health care products. They asked people in many different cultures how they felt about the following statement: *I try to avoid eating fast foods.*

In a random sample of 800 respondents, 411 people were 35 years old or younger, and, of those, 197 agreed (completely or somewhat) with the statement. Of the 389 people over 35 years old, 246 people agreed with the statement.

a) Is there evidence that the percentage of people avoiding fast food is different in the two age groups?
b) Give a 90% confidence interval for the difference in proportions.

20. Computer gaming. In order to effectively market electronic games, a manager wanted to know what age group of boys played more. A survey in 2006 found that 154 of 223 boys aged 12–14 said they "played computer or console games like Xbox or PlayStation . . . or games online." Of 248 boys aged 15–17, 154 also said they played these games.

a) Is there evidence that the percentage of boys who play these types of games is different in the two age groups?
b) Give a 90% confidence interval for the difference in proportions.

21. Foreclosure rates. The two states with the highest home foreclosure rates in March 2008 were Nevada and Colorado (realestate.msn.com, April 2008). In the second quarter of 2008, there were 8 foreclosures in a random sample of 1098 homes in Nevada, and 6 in a sample of 1460 homes in Colorado.

a) Is there evidence that the percentage of foreclosures is different in the two states?
b) Give a 90% confidence interval for the difference in proportions.

22. Labor force. Immigration reform has focused on dividing illegal immigrants into two groups: long-term and short-term. According to a recent report, short-term unauthorized workers make up nearly 6% of the U.S. labor force in construction (Pew Hispanic Center Fact Sheet, April 13, 2006). The regions of the country with the lowest percentage of unauthorized short-term immigrant construction workers are the Northeast and the Midwest. In a random sample of 958 construction workers from the Northeast, 66 are illegal short-term immigrants. In the Midwest, 42 out of a sample of 1070 are illegal short-term immigrants.

a) Is there evidence that the percentage of construction workers who are illegal short-term immigrants differs in the two regions?
b) Give a 90% confidence interval for the difference in proportions.

T 23. Market segmentation, part 2. The survey described in Exercise 15 also investigated the customers' marital status. Using the same definitions for *Shopping Frequency* as in Exercise 15, the calculations yielded the following table. Test an appropriate hypothesis for the relationship between marital status and the frequency of shopping at the same department store as in Exercise 15, and state your conclusions.

	Counts			
	Single	**Widowed**	**Married**	**Total**
Never/Hardly Ever	105	5	162	**272**
1–2 times/yr	53	15	161	**229**
3–4 times/yr	57	8	140	**205**
≥ 5 times/yr	72	15	207	**294**
Total	**287**	**43**	**670**	**1000**

24. Investment options. The economic slowdown in early 2008 and the possibility of future inflation prompted a full service brokerage firm to gauge the level of interest in inflation-beating investment options among its clients. It surveyed a random sample of 1200 clients asking them to indicate the likelihood that they would add inflation-linked annuities and bonds to their portfolios within the next year. The table below shows the distribution of responses by the investors' tolerance for risk. Test an appropriate hypothesis for the relationship between risk tolerance and the likelihood of investing in inflation linked options.

		Risk Tolerance			
		Averse	**Neutral**	**Seeking**	**Total**
Likelihood of Investing in Inflation-Linked Options	**Certain Will Invest**	191	93	40	**324**
	Likely to Invest	82	106	123	**311**
	Not Likely to Invest	64	110	101	**275**
	Certain Will not Invest	63	91	136	**290**
	Total	**400**	**400**	**400**	**1200**

25. Accounting. The Sarbanes Oxley (SOX) Act was passed in 2002 as a result of corporate scandals and in an attempt to regain public trust in accounting and reporting practices. Two random samples of 1015 executives were surveyed and asked their opinion about accounting practices in both 2000 and in 2006. The table below summarizes all 2030 responses to the question, "Which of the following do you consider most critical to establishing ethical and legal accounting and reporting practices?" Did the distribution of responses change from 2000 to 2006?

		2000	**2006**
Responses	**Training**	142	131
	IT Security	274	244
	Audit Trails	152	173
	IT Policies	396	416
	No Opinion	51	51

a) Select the appropriate procedure.
b) Check the assumptions.
c) State the hypotheses.
d) Test an appropriate hypothesis and state your results.
e) Interpret the meaning of the results and state a conclusion.

26. Entrepreneurial executives. A leading CEO mentoring organization offers a program for chief executives, presidents, and business owners with a focus on developing entrepreneurial skills. Women and men executives that recently completed the program rated its value. Are perceptions of the program's value the same for men and women?

		Men	Women
Perceived Value	**Excellent**	3	9
	Good	11	12
	Average	14	8
	Marginal	9	2
	Poor	3	1

a) Will you test goodness-of-fit, homogeneity, or independence?
b) Write appropriate hypotheses.
c) Find the expected counts for each cell, and explain why the chi-square procedures are not appropriate for this table.

T 27. Market segmentation, again. The survey described in Exercise 15 also investigated the customers' emphasis on *Quality* by asking them the question: "For the same amount of money, I will generally buy one good item rather than several of lower price and quality." Using the same definitions for *Shopping Frequency* as in Exercise 15, the calculations yielded the following table. Test an appropriate hypothesis for the relationship between a customer's emphasis on *Quality* and the *Shopping Frequency* at this department store.

a) Select the appropriate procedure.
b) Check the assumptions.
c) State the hypotheses.
d) Test an appropriate hypothesis and state your results.
e) Interpret the meaning of the results and state a conclusion.

	Counts			
	Disagree	**Moderately Disagree/Agree**	**Agree**	**Total**
Never/Hardly Ever	15	97	160	**272**
1–2 times/yr	28	107	94	**229**
3–4 times/yr	30	90	85	**205**
≥ 5 times/yr	35	140	119	**294**
Total	**108**	**434**	**458**	**1000**

28. Online shopping. A recent report concludes that while Internet users like the convenience of online shopping, they do have concerns about privacy and security (*Online Shopping*, Washington, DC, Pew Internet & American Life Project, February 2008). Respondents were asked to indicate their level of agreement with the statement "I don't like giving my credit card number or personal information online." The table gives a subset of responses. Test an appropriate hypothesis for the relationship between age and level of concern about privacy and security online.

		Strongly Agree	Agree	Disagree	Strongly Disagree	Total
Age Category	**Ages 18–29**	127	147	138	10	**422**
	Ages 30–49	141	129	78	55	**403**
	Ages 50–64	178	102	64	51	**395**
	Ages 65 +	180	132	54	14	**380**
	Total	**626**	**510**	**334**	**130**	**1600**

a) Select the appropriate procedure.
b) Check the assumptions.
c) State the hypotheses.
d) Test an appropriate hypothesis and state your results.
e) Interpret the meaning of the results and state a conclusion.

29. Entrepreneurial executives again. In some situations where the expected counts are too small, as in Exercise 26, we can complete an analysis anyway. We can often proceed after combining cells in some way that makes sense and also produces a table in which the conditions are satisfied. Here is a new table displaying the same data, but combining "Marginal" and "Poor" into a new category called "Below Average."

		Men	Women
Perceived Value	**Excellent**	3	9
	Good	11	12
	Average	14	8
	Below Average	12	3

a) Find the expected counts for each cell in this new table, and explain why a chi-square procedure is now appropriate.
b) With this change in the table, what has happened to the number of degrees of freedom?
c) Test your hypothesis about the two groups and state an appropriate conclusion.

30. Small business. The director of a small business development center located in a mid-sized city is reviewing data about its clients. In particular, she is interested in examining if the distribution of business owners across the various stages of the business life cycle is the same for white-owned and Hispanic-owned businesses. The data are shown below.

Stage in Business	White-Owned	Hispanic-Owned
Planning	11	9
Starting	14	11
Managing	20	2
Getting Out	15	1

a) Will you test goodness-of-fit, homogeneity, or independence?

b) Write the appropriate hypotheses.

c) Find the expected counts for each cell and explain why chi-square procedures are not appropriate for this table.

d) Create a new table by combining categories so that a chi-square procedure can be used.

e) With this change in the table, what has happened to the number of degrees of freedom?

f) Test your hypothesis about the two groups and state an appropriate conclusion.

T **31. Racial steering.** A subtle form of racial discrimination in housing is "racial steering." Racial steering occurs when real estate agents show prospective buyers only homes in neighborhoods already dominated by that family's race. This violates the Fair Housing Act of 1968. According to an article in *Chance* magazine (Vol. 14, no. 2, 2001), tenants at a large apartment complex recently filed a lawsuit alleging racial steering. The complex is divided into two parts: Section A and Section B. The plaintiffs claimed that white potential renters were steered to Section A, while African-Americans were steered to Section B. The following table displays the data that were presented in court to show the locations of recently rented apartments. Do you think there is evidence of racial steering?

New Renters

	White	Black	Total
Section A	87	8	95
Section B	83	34	117
Total	170	42	212

32. Titanic, again. Newspaper headlines at the time and traditional wisdom in the succeeding decades have held that women and children escaped the *Titanic* in greater proportion than men. Here's a table with the relevant data. Do you think that survival was independent of whether the person was male or female? Defend your conclusion.

	Female	Male	Total
Alive	343	367	710
Dead	127	1364	1491
Total	470	1731	2201

33. Racial steering, revisited. Find a 95% confidence interval for the difference in the proportions of Black renters in the two sections for the data in Exercise 31.

34. Titanic, one more time. Find a 95% confidence interval for the difference in the proportion of women who survived and the proportion of men who survived for the data in Exercise 32.

35. Industry sector and outsourcing. Many companies have chosen to outsource segments of their business to external providers in order to cut costs and improve quality and/or efficiencies. Common business segments that are outsourced include Information Technology (IT) and Human Resources (HR). The data below show the types of outsourcing decisions made (no outsourcing, IT only, HR only, both IT and HR) by a sample of companies from various industry sectors.

Industry Sector	No Outsourcing	IT Only	HR Only	Both IT and HR
Healthcare	810	6429	4725	1127
Financial	263	1598	549	117
Industrial Goods	1031	1269	412	99
Consumer Goods	66	341	305	197

Do these data highlight significant differences in outsourcing by industry sector?

a) Select the appropriate procedure.

b) Check the assumptions.

c) State the hypotheses.

d) Test an appropriate hypothesis and state your results.

e) Interpret the meaning of the results and state a conclusion.

36. Industry sector and outsourcing, part 2. Consider only the companies that have outsourced their IT and HR business segments. Do these data suggest significant differences between companies in the financial and industrial goods sectors with regard to their outsourcing decisions?

Industry Sector	IT Only	HR Only	Both IT and HR
Financial	1598	549	117
Industrial Goods	1269	412	99

a) Select the appropriate procedure.

b) Check the assumptions.

c) State the hypotheses.

d) Test an appropriate hypothesis and state your results.

e) Interpret the meaning of the results and state the conclusion.

	Management Styles				
Employee Job Satisfaction	**Exploitative Authoritarian**	**Benevolent Authoritarian**	**Laissez Faire**	**Consultative**	**Participative**
Very Satisfied	27	50	52	71	101
Satisfied	82	19	88	83	59
Somewhat Satisfied	43	56	26	20	20
Not Satisfied	48	75	34	26	20

37. Management styles. Use the survey results in the table at the top of the page to investigate differences in employee job satisfaction among organizations in the United States with different management styles.

a) Select the appropriate procedure.
b) Check the assumptions.
c) State the hypotheses.
d) Test an appropriate hypothesis and state your results.
e) Interpret the meaning of the results and state a conclusion.

38. Ranking companies. Every year Fortune Magazine lists the 100 best companies to work for, based on criteria such as pay, benefits, turnover rate, and diversity. In 2008, the top three were Google, Quicken Loans, and Wegmans Food Markets (*Fortune*, February 4, 2008). Of the best 100 companies to work for, 33 experienced double digit job growth (10%–68%), 49 experienced single digit job growth (1%–9%), and 18 experienced no growth or a decline. A closer examination of the top 30 showed that 15 had job growth in the double digits, 11 in the single digits, and only 4 had no growth or a decline. Is there anything unusual about job growth among the 30 top companies?

a) Select the appropriate procedure.
b) Check the assumptions.
c) State the hypotheses.
d) Test an appropriate hypothesis and state your results.
e) Interpret the meaning of the results and state a conclusion.

39. Businesses and blogs. The Pew Internet & American Life Project routinely conducts surveys to gauge the impact of the Internet and technology on daily life. A recent survey asked respondents if they read online journals or blogs, an Internet activity of potential interest to many businesses. A subset of the data from this survey (*February–March 2007 Tracking Data Set*) shows responses to this question. Test whether reading online journals or blogs is independent of generation.

	Read online journal or blog			
Generation	**Yes, Yesterday**	**Yes, but not Yesterday**	**No**	**Total**
Gen-Y (18–30)	29	35	62	**126**
Gen X (31–42)	12	34	137	**183**
Trailing Boomers (43–52)	15	34	132	**181**
Leading Boomers (53–61)	7	22	83	**112**
Matures (62 +)	6	21	111	**138**
Total	**69**	**146**	**525**	**740**

40. Businesses and blogs again. The Pew Internet & American Life Project survey described in Exercise 39 also asked respondents if they ever created or worked on their own online journal or blog. Again, a subset of the data from this survey (*February–March 2007 Tracking Data Set*) shows responses to this question. Test whether creating online journals or blogs is independent of generation.

	Create online journal or blog . . .			
Generation	**Yes/ Yesterday**	**Yes/ Not Yesterday**	**No**	**Total**
Gen Y (18–30)	18	24	85	**127**
Gen X (31–42)	6	15	162	**183**
Boomers (43–61)	5	15	273	**293**
Matures (62+)	3	3	132	**138**
Total	**32**	**57**	**652**	**741**

41. Information systems. In a recent study of enterprise resource planning (ERP) system effectiveness, researchers asked companies about how they assessed the success of their ERP systems. Out of 335 manufacturing companies surveyed, they found that 201 used return on investment (ROI), 100 used reductions in inventory levels, 28 used improved data quality, and 6 used on-time delivery. In a survey of 200 service firms, 40 used ROI, 40 used inventory levels, 100 used improved data quality, and 20 used on-time delivery. Is there evidence that the measures used to assess ERP system effectiveness differ between service and manufacturing firms? Perform the appropriate test and state your conclusion.

	West (Far West, Southwest, and Rocky Mtn.)	Midwest (Great Lakes and Plains States)	Southeast	Northeast (Mideast and New England States)	Total
Top two quintiles (top 40%)	5	5	5	5	**20**
Bottom three quintiles (bottom 60%)	10	7	7	7	**31**
Total	**15**	**12**	**12**	**12**	**51**

42. U.S. Gross Domestic Product. The U.S. Bureau of Economic Analysis provides information on the Gross Domestic Product (GDP) in the United States by state (see the website www.bea.gov). The Bureau recently released figures that showed the real GDP by state for 2007. Using the data in the table at the top of the page, examine if there is independence of the GDP and region of the country. (Note that both Alaska and Hawaii are considered to be part of the West Region and D.C is included in the Mideast Region.)

43. Economic growth. The U.S. Bureau of Economic Analysis also provides information on the growth of the U.S. economy (see the website www.bea.gov). The Bureau recently released figures that they claimed showed a growth spurt in the western region of the United States. Using the table and map below, determine if the percent change in real GDP by state for 2005–2006 was independent of region of the country. (Note that both Alaska and Hawaii are considered to be part of the West Region and D.C is included in the Mideast Region.)

Percent Change in Real GDP by State, 2005–2006

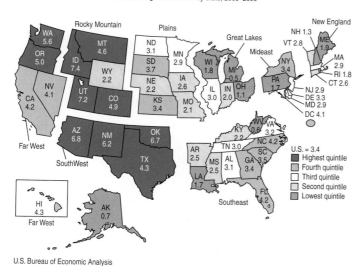

U.S. Bureau of Economic Analysis

	West (Far West, Southwest, and Rocky Mtn.)	Midwest (Great Lakes and Plains States)	Southeast	Northeast (Mideast and New England States)	Total
Top two quintiles (top 40%)	13	2	4	2	**21**
Bottom three quintiles (bottom 60%)	2	10	8	10	**30**
Total	**15**	**12**	**12**	**12**	**51**

44. Economic growth, revisited. The U.S. Bureau of Economic Analysis provides information on the GDP in the United States by metropolitan area (see the website www.bea.gov). The Bureau recently released figures that showed the percent change in real GDP by metropolitan area for 2004–2005. Using the data in the following table, examine if there is independence of the growth in metropolitan GDP and region of the country. (Note that both Alaska and Hawaii are considered part of the West Region, and some of the metropolitan areas may have been combined for this analysis.)

		West (Far West, Southwest, and Rocky Mtn.)	Midwest (Great Lakes and Plains States)	Southeast	Northeast (Mideast and New England States)	Total
Quintile	Top two quintiles (top 40%)	62	9	38	12	**121**
	Bottom three quintiles (bottom 60%)	46	87	58	36	**227**
	Total	**108**	**96**	**96**	**48**	**348**

Percent Change in Real GDP by Metropolitan Area, 2004–2005

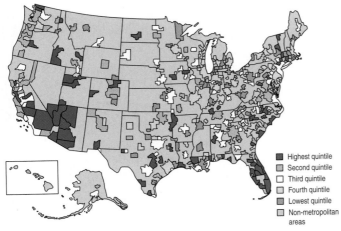

Highest quintile
Second quintile
Third quintile
Fourth quintile
Lowest quintile
Non-metropolitan areas

U.S. Bureau of Economic Analysis

JUST CHECKING ANSWERS

1 This is a test of homogeneity. The clue is that the question asks whether the distributions are alike.

2 This is a test of goodness-of-fit. We want to test the model of equal assignment to all lots against what actually happened.

3 This is a test of independence. We have responses on two variables for the same individuals.

PART III

Exploring Relationships Among Variables

Inference for Regression

Nambé Mills

Nambé (nam 'bei) Mills, Inc. was founded in 1951 near the tiny village of Nambé Pueblo, about 10 miles north of Santa Fe, New Mexico. Known for its elegant, functional cooking and tableware, Nambé Mills now sells its products in luxury stores throughout the world. Many of its products are made from an eight-metal alloy created at the Los Alamos National Laboratory (where the atomic bomb was developed during World War II) and now used exclusively by Nambé Mills. The alloy has the luster of silver and the solidity of iron, but its main component is aluminum. In fact, it does not contain silver, lead, or pewter (a tin and copper alloy), and it does not tarnish. Because it's a trade secret, Nambé Mills does not divulge the rest of the formula. Up to 15 craftsmen may be involved in the production process of an item, which includes molding, pouring, grinding, polishing, and buffing.

Because Nambé Mills's metal products are sand-cast, they must go through a lengthy production process. To rationalize its production schedule, management examined the total polishing times of 59 tableware items. Here's a scatterplot showing the retail price of the items and the amount of time (in minutes) spent in the polishing phase (Figure 16.1).

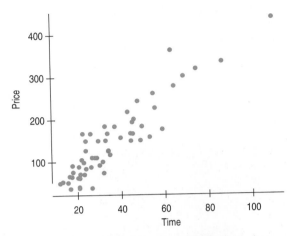

Figure 16.1 *A scatterplot of Price ($) against polishing Time (minutes) for Nambé tableware products shows that the items that take longer to polish cost more, on average.*

Back in Chapter 8 we modeled relationships like this by fitting a straight line. The equation of the least squares line for these data is:

$$\widehat{Price} = -4.871 + 4.200 \times Time$$

The slope says that, on average, the price increases by $4.20 for every extra minute of polishing time.

How useful is this model? When we fit linear models before, we used them to describe the relationship between the variables, and we interpreted the slope and intercept as descriptions of the data. Now we'd like to know what the regression model can tell us beyond the sample we used to generate this regression. To do that, we'll want to make confidence intervals and test hypotheses about the slope and intercept of the regression line.

16.1 The Population and the Sample

Our data are a sample of 59 items. If we take another sample, we hope the regression line will be similar to the one we found here, but we know it won't be exactly the same. Observations vary from sample to sample. But we can imagine a true line that summarizes the relationship between *Price* and *Time*. Following our usual conventions, we write the idealized line with Greek letters and consider the coefficients (slope and intercept) to be parameters: β_0 is the intercept, and β_1 is the slope. Corresponding to our fitted line of $\hat{y} = b_0 + b_1x$, we write $\mu_y = \beta_0 + \beta_1x$. We write μ_y instead of y because the regression line assumes that the *means* of the y values for each value of x fall exactly on the line. We can picture the relationship as in Figure 16.2. The means fall exactly on the line (for our idealized model), and the y values at each x are distributed around them.

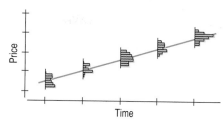

Figure 16.2 *There's a distribution of Prices for each value of polishing Time. The regression model assumes that the means line up perfectly like this.*

Now, if only we had all the values in the population, we could find the slope and intercept of this *idealized regression line* explicitly by using least squares.

Of course, not all the individual y's are at these means. In fact, the line will miss most—and usually all—of the plotted points. Some y's lie above the line and some below the line, so like all models, this one makes errors. If we want to account for each individual value of y in our model, we have to include these errors, which we denote by ε:

$$y = \beta_0 + \beta_1 x + \varepsilon.$$

This equation has an ε to soak up the deviation at each point, so the model gives a value of y for each value of x.

We estimate the β's by finding a regression line, $\hat{y} = b_0 + b_1 x$, as we did in Chapter 8. The residuals, $e = y - \hat{y}$, are the sample-based versions of the errors, ε. We'll use them to help us assess the regression model.

We know that least squares regression will give us reasonable estimates of the parameters of this model from a random sample of data. We also know that our estimates won't be equal to the parameters in the idealized or "true" model. Our challenge is to account for the uncertainty in our estimates by making confidence intervals as we've done for means and proportions. For that, we need to make some assumptions about the model and the errors.

16.2 Assumptions and Conditions

Back in Chapter 8 when we fit lines to data, we needed both the **Linearity** and the **Equal Variance Assumptions**, and so we checked four conditions. Now, when we want to make inferences about the coefficients of the line, we'll have to assume more, so we'll add more conditions.

Also, we need to be careful about the order in which we check conditions. So we number the assumptions, and check conditions for each in order: (1) Linearity Assumption, (2) Independence Assumption, (3) Equal Variance Assumption, and (4) Normal Population Assumption.

1. Linearity Assumption

If the true relationship of two quantitative variables is far from linear and we use a straight line to fit the data, our entire analysis will be useless, so we always check linearity first (and we check the **Quantitative Variable Condition** for both variables as well).

The **Linearity Condition** is satisfied if a scatterplot looks straight. It's generally not a good idea to draw a line through the scatterplot when checking. That can fool your eye into seeing the plot as straighter than it really is. Recall the errors, or residuals, we computed in Chapter 8 for each observation. Sometimes it's easier to see violations of this condition by looking at a scatterplot of the residuals against x or against the predicted values, $\hat{y}$. That plot should have no pattern if the condition is satisfied.

If the scatterplot is straight enough, we can go on to some assumptions about the errors. If not, we stop here, or consider transforming the variables to make the scatterplot more linear.[1]

2. Independence Assumption

The errors in the true underlying regression model (the ε's) must be independent of each other. As usual, there's no way to be sure that the Independence Assumption is true.

When we care about inference for the regression parameters, it's often because we think our regression model might apply to a larger population. In such cases, we can check the **Randomization Condition** that the individuals are a random sample from that population.

We can also check displays of the regression residuals for evidence of patterns, trends, or clumping, any of which would suggest a failure of independence. In the special case when we have a time series, a common violation of the Independence Assumption is for the errors to be correlated with each other (autocorrelated). (The error our model makes today may be similar to the one it made yesterday.) We can check this violation by plotting the residuals against time (usually the x-variable for a time series) and looking for patterns.

3. Equal Variance Assumption

The variability of y should be about the same for all values of x. In Chapter 8, we looked at the standard deviation of the residuals (s_e) to measure the size of the scatter. Now we'll need this standard deviation to build confidence intervals and test hypotheses. The standard deviation of the residuals is the building block for the standard errors of all the regression parameters. But it only makes sense if the scatter of the residuals is the same everywhere. In effect, the standard deviation of the residuals "pools" information across all of the individual distributions of y at each x-value, and pooled estimates are appropriate only when they combine information for groups with the same variance. A scatterplot of residuals against predicted values can help us see if the spread changes in any way. (You can also plot the residuals against x.)

We always check the **Equal Spread Condition** by looking at a scatterplot of residuals against x or $\hat{y}$. Make sure the spread around the line is nearly constant. Be alert for a "fan" shape or other tendency for the variation to grow or shrink in one part of the scatterplot.

If the plot is straight enough, the data are independent, and the spread doesn't change, we can move on to the final assumption and its associated condition.

4. Normal Population Assumption

We assume the errors around the idealized regression line at each value of x follow a Normal model. We need this assumption so that we can use a Student's t-model for inference.

---

[1] We've seen that re-expressing variables can help us understand relationships in several other contexts back in Chapters 7 and 8. Chapter 17 discusses transformations for regression more thoroughly.

As with other times when we've used Student's *t*, we'll settle for the residuals satisfying the **Nearly Normal Condition**.[2] As we have noted before, the Normality Assumption becomes less important as the sample size grows because the model is about means and the Central Limit Theorem takes over. A histogram of the residuals is one way to check whether they are nearly Normal. Alternatively, we can look at a **Normal probability plot** of the residuals (see Figure 16.3). It finds deviations from the Normal model more efficiently than a histogram. If the distribution of the data is Normal, the Normal probability plot will look roughly like a diagonal straight line. Deviations from a straight line indicate that the distribution is not Normal. This plot is usually able to show deviations from Normality more clearly than the corresponding histogram, but it's usually easier to understand *how* a distribution fails to be Normal by looking at its histogram. Another common failure of Normality is the presence of an outlier. So, we still check the **Outlier Condition** to ensure that no point is exerting too much influence on the fitted model.

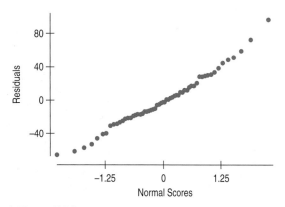

Figure 16.3 *A Normal probability plot graphs the actual standardized residuals against those expected (Normal Scores) for a sample from a standard Normal containing the same number of observations.*

◆ **How does the Normal probability plot work?**   A Normal probability plot compares each value (in our case each of the 59 residuals) with the value we would have *expected* to get if we'd just drawn a sample of 59 values from a Standard Normal model. The key is to match our numbers in order to the expected Normal values in order.

It helps to think in terms of standardized values. For example, the lowest (most negative) residual in our example has a value of −$69.48. Standardizing, we find that it is 2.03 standard deviations below the mean giving a *z*-score of −2.03. We can learn from theory that if we draw a sample of 59 values at random from a standard Normal model, we'd expect the smallest of them to have a value of −2.39. We're drawing from a standard Normal, so that's already a *z*-score. We can see, then, that our lowest residual isn't quite as far from the mean as we might have expected (had the residuals been perfectly Normal).

We can continue in this way, comparing each observed value with the value we'd expect from a Normal model. The easiest way to make the comparison, of course, is to graph it. If our observed values look like a sample from a Normal model, then the probability plot stretches out in a straight line from lower left to

---

[2] *This* is why we check the conditions in order. We check that the residuals are independent and that the variation is the same for all *x*'s before we can lump all the residuals together to check the Normal Condition.

upper right. But if our values deviate from what we'd expect, the plot will bend or have jumps in it.

The values we'd expect from a Normal model are called **Normal scores**, or sometimes nscores. Statistics programs haven't agreed on whether to plot the normal scores on the *x*-axis or the *y*-axis, so you need to look to be sure. But since you usually just want to check whether the plot is straight or not, it really doesn't matter.

A Normal probability plot is a great way to check whether the distribution is nearly Normal. But when it isn't straight, it is often a good idea to make a histogram of the values as well to get a sense of just how the data are distributed.

The best advice on using a Normal probability plot is to check whether it is straight. If it is, then your data look like data from a Normal model. If not, make a histogram to understand how they differ from the model.

Summary of Assumptions and Conditions

If all four assumptions were true, the idealized regression model would look like Figure 16.4.

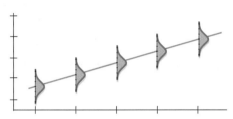

Figure 16.4 *The regression model has a distribution of y-values for each x-value. These distributions follow a Normal model with means lined up along the line and the same standard deviations.*

At each value of *x*, there is a distribution of *y*-values that follows a Normal model, and each of these Normal models is centered on the line and has the same standard deviation. Of course, we don't expect the assumptions to be exactly true. As George Box said, "all models are wrong." But the linear model is often close enough to be useful.

In regression, there's a little catch. The best way to check many of the conditions is with the residuals, but we get the residuals only *after* we compute the regression. Before we compute the regression, however, we should check at least one of the conditions.

So we work in this order:

1. **Make a scatterplot of the data** to check the Linearity Condition (and always check that the variables are quantitative as well). (This checks the **Linearity Assumption.**)

2. If the data are straight enough, **fit a regression and find the residuals, *e*, and predicted values, $\hat{y}$.**

3. If you know when the measurements were made, **plot the residuals against time** to check for evidence of patterns that suggest they may not be independent (**Independence Assumption**).

4. **Make a scatterplot of the residuals against *x* or the predicted values.** This plot should have no pattern. Check in particular for any bend (which would suggest that the data weren't that straight after all), for any thickening (or thinning), and, of course, for any unusual observations. (If you discover any errors,

"Truth will emerge more readily from error than from confusion."
—Francis Bacon (1561–1626)

correct them or omit those points, and go back to step 1. Otherwise, consider performing two regressions—one with and one without the unusual observations.) (**Equal Variance Assumption**)

5. If the scatterplots look OK, then **make a histogram and Normal probability plot of the residuals** to check the **Nearly Normal** and **Outlier Conditions** (**Normal Population Assumption**).

16.3 The Standard Error of the Slope

There's only one regression model for the population. Sample regressions try to estimate the parameters, β_0 and β_1. We expect the estimated slope for any sample, b_1, to be close to—but not actually equal to—the model slope, β_1. If we could see the collection of slopes from many samples (imagined or real) we would see a distribution of values around the true slope. That's the sampling distribution of the slope.

What is the standard deviation of this distribution? What aspects of the data affect how much the slope vary from sample to sample?

◆ **Spread around the line.** Figure 16.5 shows samples from two populations. Which underlying population would give rise to the more consistent slopes?

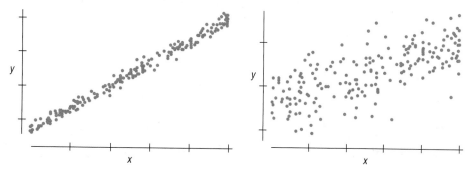

Figure 16.5 *Which of these scatterplots would give the more consistent regression slope estimate if we were to sample repeatedly from its underlying population?*

Less scatter around the line means the slope will be more consistent from sample to sample. Recall that we measure the spread around the line with the **residual standard deviation:**

$$s_e = \sqrt{\frac{\sum (y - \hat{y})^2}{n - 2}}.$$

The less scatter around the line, the smaller the residual standard deviation and the stronger the relationship between x and y.

◆ **Spread of the x's:** Here are samples from two more populations (Figure 16.6). Which of these would yield more consistent slopes?

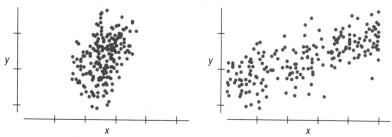

Figure 16.6 *Which of these scatterplots would give the more consistent regression slope estimate if we were to sample repeatedly from the underlying population?*

A plot like the one on the right has a broader range of *x*-values, so it gives a more stable base for the slope. We might expect the slopes of samples from situations like that to vary less from sample to sample. A large standard deviation of *x*, s_x, as in the figure on the right, provides a more stable regression.

◆ **Sample size.** What about the two scatterplots in Figure 16.7?

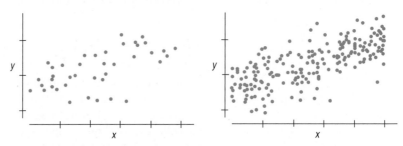

Figure 16.7 *Which of these scatterplots would give the more consistent regression slope estimate if we were to sample repeatedly from the underlying population?*

It shouldn't shock you that a larger sample size (scatterplot on the right), *n*, gives more consistent estimates from sample to sample.

Let's summarize what we've seen in these three figures:

The standard error of the regression slope

Three aspects of the scatterplot that affect the standard error of the regression slope are:

- Spread around the line: s_e
- Spread of *x* values: s_x
- Sample size: *n*

These are in fact the *only* things that affect the standard error of the slope. The formula for the standard error of the slope is:

$$SE(b_1) = \frac{s_e}{s_x \sqrt{n-1}}.$$

The error standard deviation, s_e, is in the *numerator*, since a larger spread around the line *increases* the slope's standard error. On the other hand, the *denominator* has both a sample size term ($\sqrt{n-1}$) and s_x because increasing either of these *decreases* the slope's standard error.

16.4 A Test for the Regression Slope

We know b_1 varies from sample to sample. As you'd expect, its sampling distribution model is centered at β_1, the slope of the idealized regression line. Now we can estimate its standard deviation with $SE(b_1)$. What about its shape? Here the Central Limit Theorem and Gosset come to the rescue again. When we standardize the slopes by subtracting the model mean and dividing by their standard error, we get a Student's *t*-model, this time with $n - 2$ degrees of freedom:

$$\frac{b_1 - \beta_1}{SE(b_1)} \sim t_{n-2}.$$

> **The sampling distribution for the regression slope**
>
> When the conditions are met, the standardized estimated regression slope,
>
> $$t = \frac{b_1 - \beta_1}{SE(b_1)},$$
>
> follows a Student's t-model with $n - 2$ degrees of freedom. We estimate the standard error with $SE(b_1) = \dfrac{s_e}{s_x \sqrt{n-1}}$, where $s_e = \sqrt{\dfrac{\sum (y - \hat{y})^2}{n - 2}}$, n is the number of data values, and s_x is the standard deviation of the x-values.

The same reasoning applies for the intercept. We write:

$$\frac{b_0 - \beta_0}{SE(b_0)} \sim t_{n-2}.$$

We could use this statistic to construct confidence intervals and test hypotheses, but often the value of the intercept isn't interesting. Most hypothesis tests and confidence intervals for regression are about the slope. But in case you really want to see the formula for the standard error of the intercept, we've parked it in a footnote.[3]

Now that we have the standard error of the slope and its sampling distribution, we can test a hypothesis about it and make confidence intervals. The usual null hypothesis about the slope is that it's equal to 0. Why? Well, a slope of zero would say that y doesn't tend to change linearly when x changes—in other words, that there is no linear association between the two variables. If the slope were zero, there wouldn't be much left of our regression equation.

A null hypothesis of a zero slope questions the entire claim of a linear relationship between the two variables, and often that's just what we want to know. In fact, every software package or calculator that does regression simply assumes that you want to test the null hypothesis that the slope is really zero.

> **The t-test for the regression slope**
>
> When the assumptions and conditions are met, we can test the hypothesis $H_0: \beta_1 = 0$ vs. $H_A: \beta_1 \neq 0$ (or a one-sided alternative hypothesis) using the standardized estimated regression slope,
>
> $$t = \frac{b_1 - \beta_1}{SE(b_1)},$$
>
> which follows a Student's t-model with $n - 2$ degrees of freedom. We can use the t-model to find the P-value of the test.

What if the slope is 0?

If $b_1 = 0$, our prediction is $\hat{y} = b_0 + 0x$, and the equation collapses to just $\hat{y} = b_0$. Now x is nowhere in sight, so y doesn't depend on x at all.

In this case, b_0 would turn out to be $\bar{y}$. Why? Because we know that $b_0 = \bar{y} - b_1 \bar{x}$, and when $b_1 = 0$, that becomes simply $b_0 = \bar{y}$. It turns out, that when the slope is 0, the entire regression equation is just $\hat{y} = \bar{y}$, so for every value of x, we predict the mean value ($\bar{y}$) for y.

This is just like every other t-test we've seen: a difference between the statistic and its hypothesized value divided by its standard error. This test is the *t-test* that the regression slope is 0, usually referred to as the **t-test for the regression slope**.

Another use of these values might be to make a confidence interval for the slope. We can build a confidence interval in the usual way, as an estimate plus or minus a margin of error. As always, the margin of error is just the product of the standard error and a critical value.

---

[3] $SE(b_0) = s_e \sqrt{\dfrac{1}{n} + \dfrac{\bar{x}^2}{\sum (x - \bar{x})^2}}$

> **The confidence interval for the regression slope**
>
> When the assumptions and conditions are met, we can find a confidence interval for β_1 from
>
> $$b_1 \pm t^*_{n-2} \times SE(b_1),$$
>
> where the critical value t^* depends on the confidence level and has $n - 2$ degrees of freedom.

GUIDED EXAMPLE	*Nambé Mills*

Now that we have a method to draw inferences from our regression equation, let's try it out on the Nambé Mills data. The slope of the regression gives the impact of *Time* on *Price*. Let's test the hypothesis that the slope is different from zero.

PLAN

Setup State the objectives.

Identify the parameter you wish to estimate. Here our parameter is the slope.

Identify the variables and their context.

We want to test the theory that the price of a particular item at Nambé Mills is related to the time it takes to polish it. We have data for 59 items sold by Nambé Mills. The slope of this relationship will indicate the impact of *Time* on *Price*. Our null hypothesis will be that the slope of the regression is 0.

Hypotheses Write the null and alternative hypotheses.

H_0: The *Price* of an item is not related to the polishing *Time*: $\beta_1 = 0$.
H_A: The *Price* is, in fact, related to the *Time*: $\beta_1 \neq 0$.

Model Check the assumptions and conditions.

✓ **Linearity Condition:** There is no obvious curve in the scatterplot of y versus x.

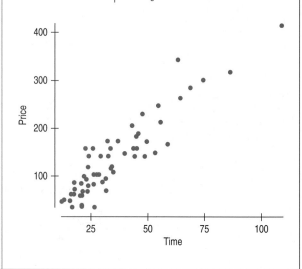

Make graphs. Because our scatterplot of *y* versus *x* seems straight enough, we can find the least squares regression and plot the residuals.

Usually, we check for suggestions that the Independence Assumption fails by plotting the residuals against time. Patterns or trends in that plot raise our suspicions.

✓ **Independence Assumption:** These data are on 59 different items manufactured by the company. There is no reason to suggest that the error in price of one item should be influenced by another.

✓ **Randomization Condition:** The data are not a random sample, but we assume they are representative of the prices and polishing times of Nambé Mills items.

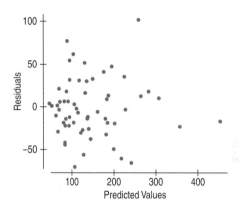

✓ **Equal Spread Condition:** The plot of residuals against the predicted values shows no obvious patterns. The spread is about the same for all predicted values, and the scatter appears random.

✓ **Nearly Normal Condition:** A histogram of the residuals is unimodal and symmetric, and the normal probability plot is reasonably straight.

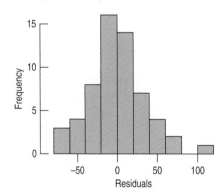

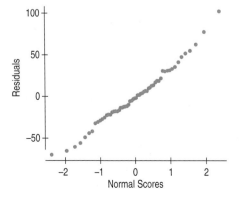

	State the sampling distribution model.	Under these conditions, the sampling distribution of the regression slope can be modeled by a Student's t-model with $(n - 2) = 59 - 2 = 57$ degrees of freedom, so we'll proceed with a regression slope t-test.
	Choose the method.	

DO

Mechanics The regression equation can be found from the formulas in Chapter 8, but regressions are almost always found from a computer program or calculator.

Here's the computer output for this regression.

Variable	Coefficient	SE(coeff)	t-ratio	P-value
Intercept	−4.871	9.654	−0.50	0.6159
Time	4.200	0.2385	17.61	<0.0001

$S = 32.54$ $R\text{-}Sq = 84.5\%$

The P-values given in the regression output table are from the Student's t-distribution on $(n - 2) = 57$ degrees of freedom. They are appropriate for two-sided alternatives.

The P-value < 0.0001 means that the association we see in the data is unlikely to have occurred by chance. Therefore, we reject the null hypothesis and conclude that there is strong evidence that the Price is linearly related to the polishing Time.

Create a confidence interval for the true slope. To obtain the t-value for 57 degrees of freedom, use the t-table at the back of your textbook. The estimated slope and SE for the slope are obtained from the regression output.

A 95% confidence interval for β_1 is:

$$b_1 \pm t^*_{n-2} \times SE(b_1) = (3.722, 4.678) \text{ \$/minute}$$

Interpret the interval.

Simply rejecting the standard null hypothesis doesn't guarantee that the size of the effect is large enough to be important.

I am 95% confident that the price increases, on average, between \$3.72 and \$4.68 for each additional minute of polishing time. (Technically: I am 95% confident that the interval from \$3.72 to \$4.68 per minute captures the true rate at which the Price increases with polishing Time.)

REPORT

Conclusion State the conclusion in the proper context.

MEMO:

Re: Nambé Mills Pricing

We investigated the relationship between polishing time and pricing of 59 Nambé Mills items. The regression analysis showed that, on average, the price increased \$4.20 for every additional minute of polishing time.

Assuming that these items are representative, we are 95% confident that the actual price of a metal item manufactured by Nambé Mills increases, on average, between \$3.72 and \$4.68 for each additional minute of polishing work required.

16.5 A Hypothesis Test for Correlation

We just tested whether the slope, β_1, was 0. To test it, we estimated the slope from the data and then, using its standard error and the t-distribution, measured how far the slope was from 0: $t = \dfrac{b_1 - 0}{SE(b_1)}$. What if we wanted to test whether the *correlation* between x and y is 0? We write ρ for the parameter (true population value) of the correlation, so we're testing $H_0: \rho = 0$. Remember that the regression slope estimate is $b_1 = r\dfrac{s_y}{s_x}$. The same is true for the parameter versions of these statistics: $\beta_1 = \rho\dfrac{\sigma_y}{\sigma_x}$. That means that if the slope is really 0, then the correlation has to be 0, too. So if we test $H_0: \beta_1 = 0$, that's really the same as testing $H_0: \rho = 0$. Sometimes a researcher, however, might want to test correlation without fitting a regression, so you'll see the test of correlation as a separate test (it's also slightly more general), but the results are mathematically the same even though the form looks a little different. Here's the t-test for the correlation coefficient.

> **The t-test for the correlation coefficient**
>
> When the conditions are met, we can test the hypothesis $H_0: \rho = 0$ vs. $H_A: \rho \neq 0$ using the test statistic:
>
> $$t = r\sqrt{\frac{n-2}{1-r^2}},$$
>
> which follows a Student's t-model with $n - 2$ degrees of freedom. We can use the t-model to find the P-value of the test.

JUST CHECKING

General economic theory suggests that as unemployment rises and jobs become harder to find, more students will enroll in universities. Researchers analyzed college enrollment at the University of New Mexico and unemployment data in New Mexico to determine whether or not there is any statistical relationship between the two variables. The data were collected by the University of New Mexico over a period of 29 years, starting with 1961 and ending with 1989. The variable *Enrollment* is in number of students and the variable *Unemp* is a percentage. Here is some regression output for these data.

Predictor	Coeff	SE(Coeff)	t-ratio	P-value
Intercept	3957	4000	0.99	0.331
Unemp	1133.8	513.1	2.21	0.036

S = 3049.50   R-Sq = 15.3%

1 What would you like to see before proceeding with inference on this regression? Why?

2 Assuming the assumptions and conditions for regression are met, find the 95% confidence interval for the slope.

3 Clearly state the null and alternative hypothesis for the slope. Interpret the P-value.

4 Is there a strong relationship between enrollment and unemployment?

5 Interpret the value of R-Sq in the output.

6 The correlation between enrollment and unemployment for this sample is 0.391, which gives a t value of 2.21 with 27 degrees of freedom and a two-sided P-value of 0.036. What does this say about the true correlation between enrollment and unemployment? Does this give you any new information?

16.6 Standard Errors for Predicted Values

We've seen how to construct the confidence interval for a slope or intercept, but we're often interested in prediction. We know how to compute predicted values of *y* for any value of *x*. We first did that in Chapter 8. This predicted value would be our best estimate, but it's still just an informed guess. Now, however, we have standard errors. We can use those SE's to construct confidence intervals for the predictions and to report our uncertainty honestly.

From our model of Nambé Mills items, we can use polishing *Time* to get a reasonable estimate of *Price*. Suppose we want to predict the *Price* of an item that takes 40 minutes of *Time* to polish. A confidence interval can tell us how precise that prediction is. The precision depends on the question we ask, however, and there are two different questions we could ask:

> Do we want to know the mean *Price* for *all items* that have a polishing *Time* of 40 minutes?

or,

> Do we want to estimate the *Price* for a *particular* item whose polishing *Time* is 40 minutes?

What's the difference between the two questions? If we were the manufacturer, we might be more naturally interested in the *mean Price* of all items that take a certain *Time* to polish. On the other hand, if we're interested in purchasing an item, we might be more interested in knowing how much an *individual* item's *Price* will vary at that polishing *Time*. Both questions are interesting. The predicted *Price* value is the same for both, but one question leads to a much more precise interval than the other. If your intuition says that it's easier to be more precise about the mean than about the individuals, you're on the right track. Because individual items vary much more than means, we can predict the *mean Price* for all items with a lot more precision than we can predict the *Price* of a particular item with the same polishing *Time*.

Let's start by predicting the *Price* for a new *Time*, one that was not necessarily part of the original data set. To emphasize this, we'll call this *x*-value "*x* sub new" and write it x_ν. As an example, we'll take x_ν to be 40 minutes. The regression equation predicts *Price* by $\hat{y}_\nu = b_0 + b_1 x_\nu$. Now that we have the predicted value, we can construct intervals around this number. Both intervals take the form:

$$\hat{y}_\nu \pm t^*_{n-2} \times SE.$$

Even the t^* value is the same for both. It's the critical value (from Table T or technology) for $n - 2$ degrees of freedom and the specified confidence level. The difference between the two intervals is in the standard errors.

The confidence interval for the predicted mean value

When the conditions are met, we find the confidence interval for the predicted mean value μ_ν at a value x_ν as

$$\hat{y}_\nu \pm t^*_{n-2} \times SE,$$

where the standard error is

$$SE(\hat{\mu}_\nu) = \sqrt{SE^2(b_1) \times (x_\nu - \overline{x})^2 + \frac{s_e^2}{n}}.$$

The details behind the standard error can be found in the Math Box on page 452, but the ideas behind the interval are best understood by looking at an example. Figure 16.8 shows the confidence interval for the mean predictions. In this plot, the intervals for all the mean *Prices* at all values of *Time* are shown together as confidence bands. Notice that the bands get wider as we attempt to predict values that lie farther away from the mean *Time* (35.82 minutes). (That's the $(x_v - \bar{x})^2$ term in the SE formula.) As we move away from the mean x value, there is more uncertainty associated with our prediction. We can see, for example, that a 95% confidence interval for the mean *Price* of an item that takes 40 minutes to polish would go from about $150 to $170. (It's actually $153.90 to $172.34.) The interval is much wider for items that take 100 minutes to polish.

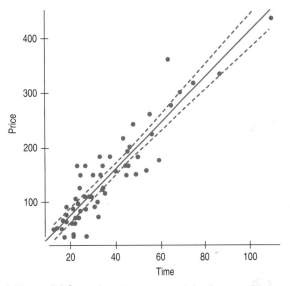

Figure 16.8 *The confidence intervals for the mean* Price *at a given polishing* Time *are shown as the green dotted lines. Near the mean* Time *(35.8 minutes) our confidence interval for the mean* Price *is much narrower than for values far from the mean, like 100 minutes.*

Like all confidence intervals, the width of these confidence intervals varies with the sample size. A sample larger than 59 items would result in narrower intervals. A regression on 10,000 items would have much narrower bands. The last factor affecting our confidence intervals is the spread of the data around the line. If there is more spread around the line, predictions are less certain, and the confidence interval bands are wider.

From Figure 16.8, it's easy to see that most *points* don't fall within the confidence interval bands—and we shouldn't expect them to. These bands show confidence intervals for the *mean*. An even larger sample would have given even narrower bands. Then we'd expect an even smaller percentage of the points to fall within them.

If we want to capture an individual price, we need to use a wider interval, called a **prediction interval**. Figure 16.9 shows these prediction intervals for the Nambé Mills data. Prediction intervals are based on the same quantities as the confidence intervals, but in order to capture a percentage of all the future predictions, they include an extra term for the spread around the line. As we can see in Figure 16.9, these bands also widen as we move from the mean of x, but it's less obvious because the extra width across the entire range of x makes the change harder to see.

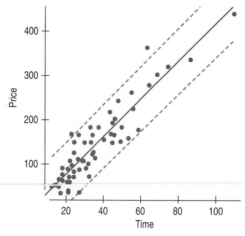

Figure 16.9 *Prediction intervals (in red) estimate the interval that contains say, 95% of the distribution of the y values that might be observed at a given value of x. If the assumptions and conditions hold, then there's about a 95% chance that a particular y-value at x_ν will be covered by the interval.*

The standard errors for prediction depend on the same kinds of things as the coefficients' standard errors. If there is more spread around the line, we'll be less certain when we try to predict the response. Of course, if we're less certain of the slope, we'll be less certain of our prediction. If we have more data, our estimate will be more precise. And there's one more piece. If we're farther from the center of our data, our prediction will be less precise. It's a lot easier to predict a data point near the middle of the data set than to extrapolate far from the center.

The prediction interval for an individual value

When the conditions are met, we can find the prediction interval for all values of y at a value x_ν as

$$\hat{y}_\nu \pm t^*_{n-2} \times SE,$$

where the standard error is

$$SE(\hat{y}_\nu) = \sqrt{SE^2(b_1) \times (x_\nu - \bar{x})^2 + \frac{s_e^2}{n} + s_e^2}.$$

The critical value t^* depends on the confidence level that you specify.

Remember to keep the distinction between the two kinds of intervals when looking at computer output. The narrower ones are confidence intervals for the *mean* and the wider ones are prediction intervals for *individual* values.

MATH BOX

Some insight into the differences between the two intervals can be gained by looking at the formulas for their standard errors and how they're derived.

To predict a y-value for a new value of x, x_ν, we'd have

$\hat{y}_\nu = b_0 + b_1 x_\nu$, which, because $b_0 = \bar{y} - b_1\bar{x}$, can be written $\hat{y}_\nu = b_1(x_\nu - \bar{x}) + \bar{y}$.

We use $\hat{y}_\nu$ in two ways. First, we use it to estimate the mean value of all the y's at x_ν, in which case we *call* it $\hat{\mu}_\nu$.

To create a confidence interval for the mean value, we need to measure the variability in this prediction

$$Var(\hat{\mu}_v) = Var(b_1(x_v - \bar{x}) + \bar{y}).$$

We now call on the Pythagorean Theorem of Statistics: The slope, b_1, and mean, $\bar{y}$, are independent, so their variances add.

$$Var(\hat{\mu}_v) = Var(b_1(x_v - \bar{x})) + Var(\bar{y}).$$

The horizontal distance from our specific x-value to the mean, $x_v - \bar{x}$, is a constant, so it comes "out" of the variance.

$$Var(\hat{\mu}_v) = (Var(b_1))(x_v - \bar{x})^2 + Var(\bar{y}).$$

Let's write that equation in terms of standard deviations.

$$SD(\hat{\mu}_v) = \sqrt{(SD^2(b_1))(x_v - \bar{x})^2 + SD^2(\bar{y})}.$$

Because we'll need to estimate these standard deviations using sample statistics, we're really dealing with standard errors.

$$SE(\hat{\mu}_v) = \sqrt{(SE^2(b_1))(x_v - \bar{x})^2 + SE^2(\bar{y})}.$$

We know that the standard deviation of a mean, $\bar{y}$, is $\dfrac{\sigma}{\sqrt{n}}$. Here we'll estimate σ using s_e, which describes the variability in how far the line we drew through our sample mean may lie above or below the true mean.

$$SE(\hat{\mu}_v) = \sqrt{(SE^2(b_1))(x_v - \bar{x})^2 + \left(\frac{s_e}{\sqrt{n}}\right)^2}.$$

$$= \sqrt{(SE^2(b_1))(x_v - \bar{x})^2 + \frac{s_e^2}{n}}.$$

And there it is—the standard error we need to create a confidence interval for a predicted mean value.[4]

When we try to predict an *individual* value of y, we also must worry about how far the true point may lie above or below the regression line. We represent that uncertainty by adding another term, e, to the original equation to get:

$$y = \hat{\mu}_v + e = b_1(x_v - \bar{x}) + \bar{y} + e.$$

To make a long story short (and the equation just a bit longer), that additional term simply adds one more standard error to the sum of the variances.

$$SE(\hat{y}_v) = \sqrt{(SE^2(b_1))(x_v - \bar{x})^2 + \frac{s_e^2}{n} + s_e^2}.$$

We've written the predicted value as $\hat{y}_v$ instead of $\hat{\mu}_v$, this time not because it's a different value, but to emphasize that we're *using* it to predict an individual now and not the mean of all y values at x_v.

---

[4] You may see the standard error expressions written in other, equivalent ways. The most common alternatives are

$$SE(\hat{\mu}_v) = s_e\sqrt{\frac{1}{n} + \frac{(x_v - \bar{x})^2}{\sum(x - \bar{x})^2}} \quad \text{and} \quad SE(\hat{y}_v) = s_e\sqrt{1 + \frac{1}{n} + \frac{(x_v - \bar{x})^2}{\sum(x - \bar{x})^2}}.$$

16.7 Using Confidence and Prediction Intervals

Now that we have standard errors, we can ask how well our analysis can predict the mean price for objects that take 25 minutes to polish. The regression output table provides most of the numbers we need.

Variable	Coefficient	SE(coeff)	t-ratio	P-value
Intercept	−4.871	9.654	−0.50	0.6159
Time	4.200	0.2385	17.61	<0.0001

S = 32.54 R-Sq = 84.5%

The regression model gives a predicted value at $x_\nu = 25$ minutes of:

$$-4.871 + 4.200\,(25) = \$100.13$$

Using this, we'll first find the 95% confidence interval for the mean *Price* for all objects whose polishing *Time* is 25 minutes. We find the standard error from the formula using the values in the regression output.

$$SE(\hat{\mu}_\nu) = \sqrt{(SE^2(b_1))\,(x_\nu - \bar{x})^2 + \left(\frac{s_e}{\sqrt{n}}\right)^2}$$

$$= \sqrt{(0.2385)^2\,(25 - 335.82)^2 + \left(\frac{32.54}{\sqrt{59}}\right)^2} = \$4.96$$

The t^* value that excludes 2.5% in either tail with $59 - 2 = 57$ df is (according to the tables) 2.002.

Putting it all together, we find the margin of error as:

$$ME = 2.002(4.96) = \$9.93$$

So, we are 95% confident that the interval

$$\$100.13 \pm 9.93 = (\$90.20, \$110.06)$$

includes the true mean *Price* of objects whose *Time* is 25 minutes.

Suppose, however, that instead of the mean price, we want to know how much an item that needs 25 minutes of polishing will cost. The confidence interval we just found is too narrow. It may contain the mean price, but it's unlikely to cover many of the individual price values. To make a prediction interval for an *individual* item's price with a polishing time of 25 minutes, we need the larger standard error formula to account for the greater variability. Using the formula

$$SE(\hat{y}_\nu) = \sqrt{(SE^2(b_1))\,(x_\nu - \bar{x})^2 + \frac{s_e^2}{n} + s_e^2} = \$32.92,$$

we find the ME to be

$$ME = t^*SE(\hat{y}_\nu) = 2.002 \times 32.92 = \$65.91,$$

and so the prediction interval is

$$\hat{y} \pm ME = 100.13 \pm 65.91 = (\$34.22, \$166.04).$$

Notice how much wider this interval is than the 95% confidence interval. Most of the time we will use a software package to compute and display these intervals. Most packages generate displays that show the regression line along with both the 95% confidence and prediction intervals (combining what we've shown in Figures 16.8 and 16.9). This makes it easier to see how much wider the prediction intervals are than the corresponding confidence intervals (see Figure 16.10).

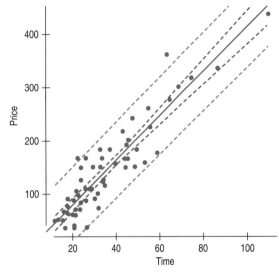

Figure 16.10 A scatterplot of Price versus Time with a least squares regression line. The inner lines (green) near the regression line show the extent of the 95% confidence intervals, and the outer lines (red) show the prediction intervals. Most of the points are contained within the prediction intervals (as they should be), but not within the confidence interval for the means.

WHAT CAN GO WRONG?

In this chapter we've added inference to the regression explorations that we did in Chapter 8. Everything covered in that chapter that could go wrong with regression can still go wrong.

With inference, we've put numbers on our estimates and predictions, but these numbers are only as good as the model. Here are the main things to watch out for:

- **Don't fit a linear regression to data that aren't straight.** This is the most fundamental assumption. If the relationship between *x* and *y* isn't approximately linear, there's no sense in fitting a straight line to it.

- **Watch out for changing spread.** The common part of confidence and prediction intervals is the estimate of the error standard deviation, the spread around the line. If it changes with *x*, the estimate won't make sense. Imagine making a prediction interval for these data:

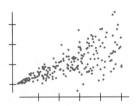

(continued)

When x is small, we can predict y precisely, but as x gets larger, it's much harder to pin y down. Unfortunately, if the spread changes, the single value of s_e won't pick that up. The prediction interval will use the average spread around the line, with the result that we'll be too pessimistic about our precision for low x-values and too optimistic for high x-values. A re-expression of y (see Chapter 17) is often a good fix for changing spread.

- **Watch out for non-Normal errors.** When we make a prediction interval for an individual y-value, the Central Limit Theorem can't come to our rescue. For us to believe the prediction interval, the errors must follow the Normal model. Check the histogram and Normal probability plot of the residuals to see if this assumption looks reasonable.

- **Watch out for extrapolation.** It's tempting to think that because we have prediction *intervals*, they'll take care of all of our uncertainty so we don't have to worry about extrapolating. Wrong. The interval is only as good as the model. The uncertainty our intervals predict is only correct if our model is true. There's no way to adjust for wrong models. That's why it's always dangerous to predict for x-values that lie far from the center of the data.

- **Watch out for high-influence points and unusual observations.** We should always be on the lookout for a few points that have undue influence on our estimates, and regression analysis is certainly no exception.

- **Watch out for one-tailed tests.** Because tests of hypotheses about regression coefficients are usually two-tailed, software packages report two-tailed P-values. If you are using that type of software to conduct a one-tailed test about the slope, you'll need to divide the reported P-value by two.

ETHICS IN ACTION

The need for senior care businesses that offer companionship and non-medical home services is increasing as the U.S. population continues to age. One such franchise, Independent Senior Care, tries to set itself apart from its competitors by offering an additional service to prospective franchisees. In addition to standard information packets that provide tools, training, and mentorship opportunities, Independent Senior Care has an analyst on staff, Allen Ackman, to help prospective franchisees evaluate the feasibility of opening an elder care business in their area. Allen was contacted recently by Kyle Sennefeld, a recent business school graduate with a minor in gerontology, who is interested in starting a senior care franchise in northeastern Pennsylvania. Allen decides to use a regression model that relates annual profit to the number of residents over the age of 65 that live within a 100-mile radius of a franchise location. Even though the R^2 for this model is small, the variable is statistically significant, and the model is easy to explain to prospective franchisees. Allen sends Kyle a report that estimates the annual profit at Kyle's proposed location. Kyle was excited to see that opening an Independent Senior Care franchise in northeastern Pennsylvania would be a good business decision.

ETHICAL ISSUE *The regression model has a small R^2, so its predictive ability is questionable. Related to Items A and B, ASA Ethical Guidelines.*

ETHICAL SOLUTION *Disclose the value of R^2 along with the prediction results and disclose if the regression is being used to extrapolate outside the range of x values. Allen should provide a prediction interval as well as an estimate of the profit. Because Kyle will be assessing his franchise's chances for profit from this interval, Allen should make sure it is a prediction interval and not a confidence interval for the mean profit at all similar locations.*

What have we learned?

In this chapter, we have extended our study of inference methods by applying them to regression models. We've found that the same methods we used for means—Student's *t*-models—work for regression in much the same way they did for means. And we've seen that although this makes the mechanics familiar, we need to check new conditions and be careful when describing the hypotheses we test and the confidence intervals we construct.

- We've learned that under certain assumptions, the sampling distribution for the slope of a regression line can be modeled by a Student's *t*-model with $n - 2$ degrees of freedom.

- We've learned to check four conditions before we proceed with inference. We've learned the importance of checking these conditions in order, and we've seen that most of the checks can be made by graphing the data and the residuals.

- We've learned to use the appropriate *t*-model to test a hypothesis about the slope. If the slope of our regression line is significantly different from zero, we have strong evidence that there is an association between the two variables.

- We've also learned to create and interpret a confidence interval for the true slope.

Terms

Confidence interval for the regression slope

When the assumptions are satisfied, we can find a confidence interval for the slope parameter from $b_1 \pm t^*_{n-2} \times SE(b_1)$. The critical value, t^*_{n-2}, depends on the confidence interval specified and on the Student's *t*-model with $n - 2$ degrees of freedom.

Confidence interval for the predicted mean value

Different samples will give different estimates of the regression model and, so, different predicted values for the same value of x. We find a confidence interval for the mean of these predicted values at a specified x-value, x_ν, as

$$\hat{y}_\nu \pm t^*_{n-2} \times SE(\hat{\mu}_\nu),$$

where

$$SE(\hat{\mu}_\nu) = \sqrt{SE^2(b_1) \times (x_\nu - \bar{x})^2 + \frac{s_e^2}{n}}.$$

The critical value, t^*_{n-2}, depends on the specified confidence level and the Student's *t*-model with $n - 2$ degrees of freedom.

Residual standard deviation

The measure, denoted s_e, of the spread of the data around the regression line:

$$s_e = \sqrt{\frac{\sum(y - \hat{y})^2}{n - 2}} = \sqrt{\frac{\sum e^2}{n - 2}}.$$

Prediction interval for a future observation

A confidence interval for individual values. Prediction intervals are to observations as confidence intervals are to parameters. They predict the distribution of individual values, while confidence intervals specify likely values for a true parameter. The prediction interval takes the form

$$\hat{y}_\nu \pm t^*_{n-2} \times SE(\hat{y}_\nu),$$

where

$$SE(\hat{y}_\nu) = \sqrt{SE^2(b_1) \times (x_\nu - \bar{x})^2 + \frac{s_e^2}{n} + s_e^2}.$$

The critical value, t^*_{n-2}, depends on the specified confidence level and the Student's t-model with $n - 2$ degrees of freedom. The extra s^2_e in $SE(\hat{y}_\nu)$ makes the interval wider than the corresponding confidence interval for the mean.

t-test for the regression slope

The usual null hypothesis is that the true value of the slope is zero. The alternative is that it is not. A slope of zero indicates a complete lack of linear relationship between y and x.

To test $H_0: \beta_1 = 0$ we find

$$t = \frac{b_1 - 0}{SE(b_1)},$$

where $SE(b_1) = \dfrac{s_e}{s_x \sqrt{n - 1}}$, $s_e = \sqrt{\dfrac{\sum(y - \hat{y})^2}{n - 2}}$, n is the number of cases, and s_x is the standard deviation of the x-values. We find the P-value from the Student's t-model with $n - 2$ degrees of freedom.

t-test for the correlation coefficient

When the conditions are met, we can test the hypothesis $H_0: \rho = 0$ vs. $H_A: \rho \neq 0$ using the test statistic

$$t = r\sqrt{\frac{n - 2}{1 - r^2}},$$

which follows a Student's t-model with $n - 2$ degrees of freedom. We can use the t-model to find the P-value of the test.

Skills

PLAN

- Understand that the "true" regression line does not fit the population data perfectly, but rather is an idealized summary of that data.

- Know how to examine your data and a scatterplot of y vs. x for violations of assumptions that would make inference for regression unwise or invalid.

- Know how to examine displays of the residuals from a regression to double-check that the conditions required for regression have been met. In particular, know how to judge linearity and constant variance from a scatterplot of residuals against predicted values. Know how to judge Normality from a histogram and Normal probability plot.

- Remember to be especially careful to check for failures of the Independence Assumption when working with data recorded over time. To search for patterns, examine scatterplots both of x against time and of the residuals against time.

DO

- Know how to test the standard hypothesis that the true regression slope is zero. Be able to state the null and alternative hypotheses. Know where to find the relevant numbers in standard computer regression output.

- Be able to find a confidence interval for the slope of a regression based on the values reported in a standard regression output table.

REPORT

- Be able to summarize a regression in words. In particular, be able to state the meaning of the true regression slope, the standard error of the estimated slope, and the standard deviation of the errors.

- Be able to interpret the P-value of the t-statistic for the slope to test the standard null hypothesis.

- Be able to interpret a confidence interval for the slope of a regression.

Technology Help: Regression Analysis

All statistics packages make a table of results for a regression. These tables differ slightly from one package to another, but all are essentially the same. We've seen two examples of such tables already.

All packages offer analyses of the residuals. With some, you must request plots of the residuals as you request the regression. Others let you find the regression first and then analyze the residuals afterward. Either way, your analysis is not complete if you don't check the residuals with a histogram or Normal probability plot and a scatterplot of the residuals against x or the predicted values.

You should, of course, always look at the scatterplot of your two variables before computing a regression.

Regressions are almost always found with a computer or calculator. The calculations are too long to do conveniently by hand for data sets of any reasonable size. No matter how the regression is computed, the results are usually presented in a table that has a standard form. Here's a portion of a typical regression results table, along with annotations showing where the numbers come from.

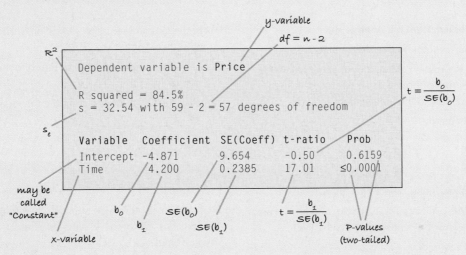

The regression table gives the coefficients (once you find them in the middle of all this other information). This regression predicts *Price* from *Time*. The regression equation is

$$\widehat{Price} = -4.871 + 4.200\ Time$$

and the R^2 for the regression is 84.5%.

The column of *t*-ratios gives the test statistics for the respective null hypotheses that the true values of the coefficients are zero. The corresponding P-values are also usually reported.

EXCEL

- In Excel 2003 and earlier, select **Data Analysis** from the **Tools** menu. In Excel 2007, select **Data Analysis** from the **Analysis Group** on the Data Tab.
- Select Regression from the **Analysis Tools** list.
- Click the **OK** button.
- Enter the data range holding the Y-variable in the box labeled "Y-range".
- Enter the range of cells holding the X-variable in the box labeled "X-range."
- Select the **New Worksheet Ply** option.
- Select **Residuals** options. Click the **OK** button.

Comments

The Y and X ranges do not need to be in the same rows of the spreadsheet, although they must cover the same number of cells. But it is a good idea to arrange your data in parallel columns as in a data table.

Although the dialog offers a Normal probability plot of the residuals, the data analysis add-in does not make a correct probability plot, so don't use this option.

MINITAB

- Choose **Regression** from the **Stat** menu.
- Choose **Regression...** from the **Regression** submenu.
- In the Regression dialog, assign the Y-variable to the Response box and assign the X-variable to the Predictors box.
- Click the **Graphs** button.
- In the Regression-Graphs dialog, select **Standardized residuals,** and check **Normal plot of residuals**, **Residuals versus fits** and **Residuals versus order.**

- Click the **OK** button to return to the Regression dialog.
- Click the **OK** button to compute the regression.

Comments

You can also start by choosing a Fitted Line plot from the **Regression** submenu to see the scatterplot first—usually good practice.

SPSS

- Choose **Regression** from the **Analyze** menu.
- Choose **Linear** from the **Regression** submenu.
- In the Linear Regression dialog that appears, select the Y-variable and move it to the dependent target. Then move the X-variable to the independent target.
- Click the **Plots** button.

- In the Linear Regression Plots dialog, choose to plot the *SRESIDs against the *ZPRED values.
- Click the **Continue** button to return to the Linear Regression dialog.
- Click the **OK** button to compute the regression.

JMP

- From the **Analyze** menu, select **Fit Y by X.**
- Select variables: a Y, Response variable, and an X, Factor variable. Both must be continuous (quantitative).
- JMP makes a scatterplot.
- Click on the red triangle beside the heading labeled **Bivariate Fit...** and choose **Fit Line.** JMP draws the least squares regression line on the scatterplot and displays the results of the regression in tables below the plot.
- The portion of the table labeled "Parameter Estimates" gives the coefficients and their standard errors, *t*-ratios, and P-values.

Comments

JMP chooses a regression analysis when both variables are "Continuous." If you get a different analysis, check the variable types.

The Parameter table does not include the residual standard deviation s_e. You can find that as Root Mean Square Error in the Summary of Fit panel of the output.

DATA DESK

- Select Y- and X-variables.
- From the **Calc** menu, choose **Regression.**
- Data Desk displays the regression table.
- Select plots of residuals from the Regression table's **HyperView** menu.

Comments

You can change the regression by dragging the icon of another variable over either the Y- or X-variable name in the table and dropping it there. The regression will recompute automatically.

Mini Case Study Projects

Frozen Pizza

The product manager at a subsidiary of Kraft Foods, Inc. is interested in learning how sensitive sales are to changes in the unit price of a frozen pizza in Dallas, Denver, Baltimore, and Chicago. The product manager has been provided data on both *Price* and *Sales* volume every fourth week over a period of nearly four years for the four cities (**ch16_MCSP_Frozen_Pizza**).

Examine the relationship between *Price* and *Sales* for each city. Be sure to discuss the nature and validity of this relationship. Is it linear? Is it negative? Is it significant? Are the conditions of regression met? Some individuals in the product manager's division suspect that frozen pizza sales are more sensitive to price in some cities than in others. Is there any evidence to suggest that? Write up a short report on what you find. Include 95% confidence intervals for the mean *Sales* if the *Price* is $2.50 and discuss how that interval changes if the *Price* is $3.50.

Global Warming?

Every spring, Nenana, Alaska, hosts a contest in which participants try to guess the exact minute that a wooden tripod placed on the frozen Tanana River will fall through the breaking ice. The contest started in 1917 as a diversion for railroad engineers, with a jackpot of $800 for the closest guess. It has grown into an event in which hundreds of thousands of entrants enter their guesses on the Internet and vie for more than $300,000.

Because so much money and interest depends on the time of the ice breakup, it has been recorded to the nearest minute with great accuracy ever since 1917 (**ch16_MCSP_Global_Warming**). And because a standard measure of breakup has been used throughout this time, the data are consistent. An article in *Science* ("Climate Change in Nontraditional Data Sets," *Science* 294, October 2001) used the data to investigate global warming. Researchers are interested in the following questions. What is the rate of change in the date of breakup over time (if any)? If the ice is breaking up earlier, what is your conclusion? Does this necessarily suggest global warming? What could be other reasons for this trend? What is the predicted breakup date for the year 2015? (Be sure to include an appropriate prediction or confidence interval.) Write up a short report with your answers.

EXERCISES

1. Online shopping. Several studies have found that the frequency with which shoppers browse Internet retailers is related to the frequency with which they actually purchase products and/or services online. Here are data showing the age of respondents and their answer to the question "how many minutes do you browse online retailers per week?"

Age	Browsing Time (min/wk)
22	492
50	186
44	180
32	384
55	120
60	120
38	276
22	480
21	510
45	252
52	126
33	360
19	570
17	588
21	498

a) Make a scatterplot for these data.
b) Do you think a linear model is appropriate? Explain.
c) Find the equation of the regression line.
d) Check the residuals to see if the conditions for inference are met.

T 2. El Niño. Concern over the weather associated with El Niño has increased interest in the possibility that the climate on Earth is getting warmer. The most common theory relates an increase in atmospheric levels of carbon dioxide (CO_2), a greenhouse gas, to increases in temperature. Here is part of a regression analysis of the mean annual CO_2 concentration in the atmosphere, measured in parts per thousand (ppt), at the top of Mauna Loa in Hawaii and the mean annual air temperature over both land and sea across the globe, in degrees Celsius. The scatterplots and residuals plots indicated that the data were appropriate for inference and the response variable is *Temp*.

Variable	Coeff	SE(Coeff)
Intercept	16.4328	0.0557
CO_2	0.0405	0.0116

R squared = 25.8%
s = 0.0854 with 37 − 2 = 35 degrees of freedom

a) Write the equation of the regression line.
b) Find the value of the correlation and test whether the true correlation is zero. Is there evidence of an association between CO_2 level and global temperature?
c) Find the *t*-value and P-value for the slope. Is there evidence of an association between CO_2 level and global temperature? What do you know from the slope and *t*-test that you might not have known from testing the correlation?
d) Do you think predictions made by this regression will be very accurate? Explain.

T 3. Movie budgets. How does the cost of a movie depend on its length? Data on the cost (millions of dollars) and the running time (minutes) for major release films of 2005 are summarized in these plots and computer output:

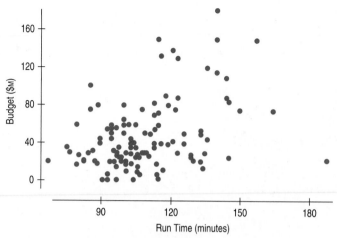

Dependent variable is: Budget($M)
R squared = 15.4%
s = 32.95 with 120 − 2 = 118 degrees of freedom

Variable	Coefficient	SE(Coeff)	t-ratio	P-value
Intercept	−31.39	17.12	−1.83	0.0693
Run Time	0.71	0.15	4.64	≤0.0001

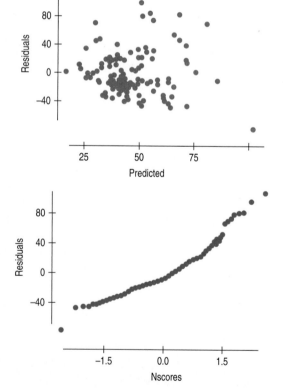

a) Explain in words and numbers what the regression says.
b) The intercept is negative. Discuss its value, taking note of the P-value.
c) The output reports s = 32.95. Explain what that means in this context.
d) What's the value of the standard error of the slope of the regression line?
e) Explain what that means in this context.

T 4. House prices. How does the price of a house depend on its size? Data from Saratoga, New York, on 1064 randomly selected houses that had been sold include data on price ($1000's) and size (1000's ft²), producing the following graphs and computer output:

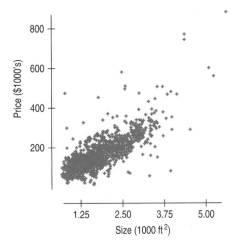

Dependent variable is: Price
R squared = 59.5%
s = 53.79 with 1064 − 2 = 1062 degrees of freedom

Variable	Coefficient	SE(Coeff)	t-ratio	P-value
Intercept	−3.1169	4.688	−0.665	0.5063
Size	94.4539	2.393	39.465	≤0.0001

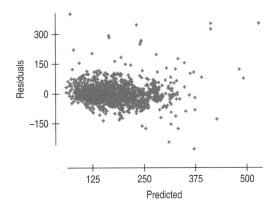

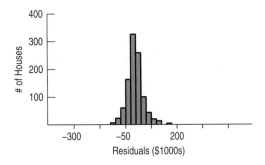

a) Explain in words and numbers what the regression says.
b) The intercept is negative. Discuss its value, taking note of its P-value.

c) The output reports *s* = 53.79. Explain what that means in this context.
d) What's the value of the standard error of the slope of the regression line?
e) Explain what that means in this context.

T 5. Movie budgets: the sequel. Exercise 3 shows computer output examining the association between the length of a movie and its cost.

a) Check the assumptions and conditions for inference.
b) Find a 95% confidence interval for the slope and interpret it.

T 6. Second home. Exercise 4 shows computer output examining the association between the sizes of houses and their sale prices.

a) Check the assumptions and conditions for inference.
b) Find a 95% confidence interval for the slope and interpret it.

T 7. Water hardness. In an investigation of environmental causes of disease, data were collected on the annual mortality rate (deaths per 100,000) for males in 61 large towns in England and Wales. In addition, the water hardness was recorded as the calcium concentration (parts per million, or ppm) in the drinking water. Here are the scatterplot and regression analysis of the relationship between mortality and calcium concentration, where the dependent variable is *Mortality*.

Variable	Coeff	SE(Coeff)
Intercept	1676.36	29.30
Calcium	−3.226	0.485

R squared = 42.9%
s = 143.0 with 61 − 2 = 59 degrees of freedom

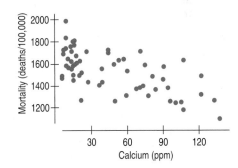

a) Is there an association between the hardness of the water and the mortality rate? Write the appropriate hypothesis.
b) Assuming the assumptions for regression inference are met, what do you conclude?
c) Create a 95% confidence interval for the slope of the true line relating calcium concentration and mortality.
d) Interpret your interval in context.

T 8. Mutual funds. In March 2002, *Consumer Reports* listed the rate of return for several large cap mutual funds over the previous 3-year and 5-year periods. (Here, "large cap" refers to companies worth over $10 billion.) It's common for advertisements to carry the disclaimer that "past returns may not be indicative of future performance." Do these data indicate that there was an association between 3-year and 5-year rates of return?

Fund Name	Annualized Returns (%)	
	3-year	**5-year**
Ameristock	7.9	17.1
Clipper	14.1	18.2
Credit Suisse Strategic Value	5.5	11.5
Dodge & Cox Stock	15.2	15.7
Excelsior Value	13.1	16.4
Harbor Large Cap Value	6.3	11.5
ICAP Discretionary Equity	6.6	11.4
ICAP Equity	7.6	12.4
Neuberger Berman Focus	9.8	13.2
PBHG Large Cap Value	10.7	18.1
Pelican	7.7	12.1
Price Equity Income	6.1	10.9
USAA Cornerstone Strategy	2.5	4.9
Vanguard Equity Income	3.5	11.3
Vanguard Windsor	11.0	11.0

T 9. Youth unemployment. The United Nations has developed a set of millennium goals for countries, and the United Nations Statistics Division (UNSD) maintains databases to measure economic progress toward these goals (unstats. un.org/unsd/mdg). Data extracted from this source are on the disk. One measure that is tracked is the youth (ages 15–24) unemployment rate in different countries. Is the unemployment rate for the male youth related to the unemployment rate for the female youth?

a) Find a regression model predicting *Male Rate* from the *Female Rate* in 2005 for the sample of 57 countries provided by UNSD.

b) Examine the residuals to determine if a linear regression is appropriate.

c) Test an appropriate hypothesis to determine if the association is significant.

d) What percentage of the variability in the *Male Rate* is accounted for by the regression model?

T 10. Male unemployment. Using the unemployment data provided by the United Nations, investigate the association between the male unemployment rate in 2004 and 2005 for a sample of 52 countries. (The sample is smaller than in Exercise 9, since not all countries reported rates in both years.)

a) Find a regression model predicting the *2005-Male* rate from the *2004-Male* rate.

b) Examine the residuals to determine if a linear regression is appropriate.

c) Test an appropriate hypothesis to determine if the association is significant.

d) What percentage of the variability in the *2005-Male* rate is accounted for by the *regression model*?

T 11. Used cars 2007. Classified ads in a newspaper offered several used Toyota Corollas for sale. Listed below are the ages of the cars and the advertised prices.

Age (yr)	Prices Advertised ($)
1	13990
1	13495
3	12999
4	9500
4	10495
5	8995
5	9495
6	6999
7	6950
7	7850
8	6999
8	5995
10	4950
10	4495
13	2850

a) Make a scatterplot for these data.

b) Do you think a linear model is appropriate? Explain.

c) Find the equation of the regression line.

d) Check the residuals to see if the conditions for inference are met.

T 12. Property assessments. The following software outputs provide information about the size (in square feet) of 18 homes in Ithaca, New York, and the city's assessed value of those homes, where the response variable is *Assessment*.

Predictor	Coeff	SE(Coeff)	t-ratio	P-value
Intercept	37108.85	8664.33	4.28	0.0006
Size	11.90	4.29	2.77	0.0136

s = 4682.10   R-Sq = 32.5%

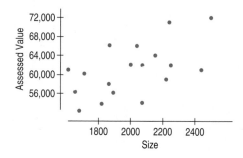

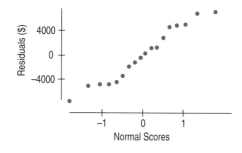

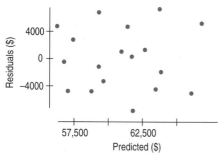

a) Explain why inference for linear regression is appropriate with these data.

b) Is there a significant linear association between the *Size* of a home and its *Assessment*? Test an appropriate hypothesis and state your conclusion.

c) What percentage of the variability in assessed value is accounted for by this regression?

d) Give a 90% confidence interval for the slope of the true regression line, and explain its meaning in the proper context.

e) From this analysis, can we conclude that adding a room to your house will increase its assessed value? Why or why not?

f) The owner of a home measuring 2100 square feet files an appeal, claiming that the $70,200 assessed value is too high. Do you agree? Explain your reasoning.

T 13. Used cars 2007, again. Based on the analysis of used car prices you did for Exercise 11, if appropriate, create a 95% confidence interval for the slope of the regression line and explain what your interval means in context.

T 14. Assets and sales. A business analyst is looking at a company's assets and sales to determine the relationship (if any) between the two measures. She has data (in $million) from a random sample of 79 Fortune 500 companies, and obtained the linear regression below:

The regression equation is Assets = 1867.4 + 0.975 Sales

Predictor	Coeff	SE(Coeff)	t-ratio	P-value
Constant	1867.4	804.5	2.32	0.0230
Sales	0.975	0.099	9.84	≤0.0001

S = 6132.59   R-Sq = 55.7%   R-Sq(adj) = 55.1%

Use the data provided to find a 95% confidence interval, if appropriate, for the slope of the regression line and interpret your interval in context.

T 15. Fuel economy. A consumer organization has reported test data for 50 car models. We will examine the association

between the weight of the car (in thousands of pounds) and the fuel efficiency (in miles per gallon). Use the data provided on the disk to answer the following questions, where the response variable is *Fuel Efficiency* (mpg).

a) Create the scatterplot and obtain the regression equation.

b) Are the assumptions for regression satisfied?

c) Write the appropriate hypotheses for the slope.

d) Test the hypotheses and state your conclusion.

16. Consumer Reports. In October 2002, *Consumer Reports* listed the price (in dollars) and power (in cold cranking amps) of auto batteries. We want to know if more expensive batteries are generally better in terms of starting power. Here are the regression and residual output, where the response variable is *Power*.

Dependent variable is: Power
R squared = 25.2%
s = 116.0 with 33 − 2 = 31 degrees of freedom

Variable	Coefficient	SE(Coeff)	t-ratio	P-value
Intercept	384.594	93.55	4.11	0.0003
Cost	4.146	1.282	3.23	0.0029

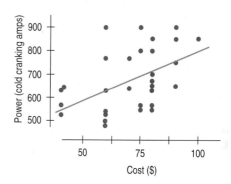

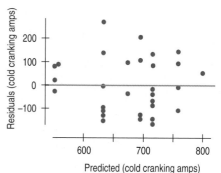

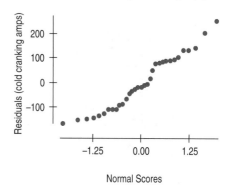

a) How many batteries were tested?

b) Are the conditions for inference satisfied? Explain.

c) Is there evidence of a linear association between the cost and cranking power of auto batteries? Test an appropriate hypothesis and state your conclusion.

d) Is the association strong? Explain.

e) What is the equation of the regression line?

f) Create a 90% confidence interval for the slope of the true line.

g) Interpret your interval in this context.

T 17. SAT scores. How strong was the association between student scores on the Math and Verbal sections of the old SAT? Scores on this exam ranged from 200 to 800 and were widely used by college admissions offices. Here are summary statistics, regression analysis, and plots of the scores for a graduating class of 162 students at Ithaca High School, where the response variable is *Math Score*.

Predictor	Coeff	SE(Coeff)	t-ratio	P-value
Intercept	209.55	34.35	6.10	<0.0001
Verbal	0.675	0.057	11.88	<0.0001

s = 71.75   R-Sq = 46.9%

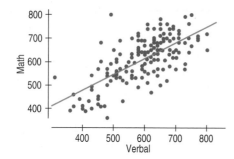

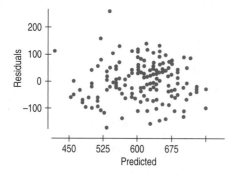

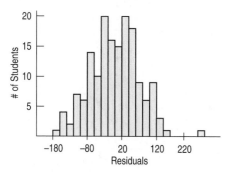

a) Is there evidence of a linear association between *Math* and *Verbal* scores? Write an appropriate hypothesis.

b) Discuss the assumptions for inference.

c) Test your hypothesis and state an appropriate conclusion.

18. Productivity. How strong is the association between labor productivity and labor costs? Data from the Bureau of Labor Statistics, specifically 2006 index values (computed using 1997 as the base year) for labor productivity and unit labor costs across 53 industries, are used to examine this relationship (ftp://ftp.bls.gov; accessed May 2008). Here are the results of a regression analysis where the response variable is *Labor Productivity*.

Predictor	Coeff	SE(Coeff)	t-ratio	P-value
Intercept	212.67	12.18	17.46	<0.0001
Unit Labor Cost	−0.768	0.111	−6.92	<0.0001

s = 14.39   R-Sq = 48.4%

a) Is there evidence of a linear association between *Labor Productivity* and Unit Labor Costs? Write appropriate hypotheses.

b) Test your null hypothesis and state an appropriate conclusion (assume that assumptions and conditions are met).

T 19. Football salaries. Football owners are constantly in competition for good players. The better the winning percentage, the more likely that the team will provide good business returns for the owners. Of course, the resources that each of the 32 teams has in the National Football League (NFL) vary. Does the size of the payroll matter? Here is a scatterplot and regression showing the association between team salaries in the NFL in 2006 and winning percentage.

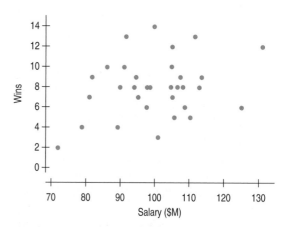

Predictor	Coeff	SE(Coeff)	t-ratio	P-value
Intercept	1.783	3.964	0.45	0.6560
Salary ($M)	0.062	0.039	1.58	0.1244

s = 2.82   R-Sq = 7.7%

a) State the hypotheses about the slope.

b) Perform the hypothesis test and state your conclusion in context.

T 20. Gallup poll. The Gallup organization has, over six decades periodically asked the following question:

If your party nominated a generally well-qualified person for president who happened to be a woman, would you vote for that person?

We wonder if the proportion of the public who have "no opinion" on this issue has changed over the years. Here is a regression for the proportion of those respondents whose response to this question about voting for a woman president was "no opinion." Assume that the conditions for inference are satisfied and that the response variable is *No Opinion*.

Predictor	Coeff	SE(Coeff)	t-ratio	P-value
Intercept	7.693	2.445	3.15	0.0071
year	−0.042	0.035	−1.21	0.2460

s = 2.28   R-Sq = 9.5%

a) State the hypotheses about the slope (both numerically and in words) that describes how voters' thoughts have changed about voting for a woman.
b) Assuming that the assumptions for inference are satisfied, perform the hypothesis test and state your conclusion.
c) Examine the scatterplot corresponding to the regression for No Opinion. How does it change your opinion of the trend in "no opinion" responses? Do you think the true slope is negative as shown in the regression output?

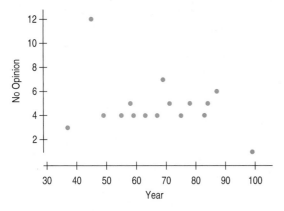

T 21. Fuel economy, part 2. Consider again the data in Exercise 15 about the gas mileage and weights of cars.

a) Create a 95% confidence interval for the slope of the regression line.
b) Explain in this context what your confidence interval means.

T 22. SAT scores, part 2. Consider the high school SAT scores data from Exercise 17.

a) Find a 90% confidence interval for the slope of the true line describing the association between Math and Verbal scores.
b) Explain in this context what your confidence interval means.

T 23. Mutual funds. It is common economic theory that the money flowing into and out of mutual funds (fund flows) is related to the performance of the stock market. Another way of stating this is that investors are more likely to place money into mutual funds when the market is performing well. One way to measure market performance is the Wilshire 5000 Total Market Return (%), which is a value-weighted return. (The return of each stock in the index is weighted by its percent of market value for all stocks.) Here are the scatterplot and regression analysis, where the response variable is *Fund Flows* ($ million) and the explanatory variable is *Market Return* (%), using data from January 1990 to October 2002.

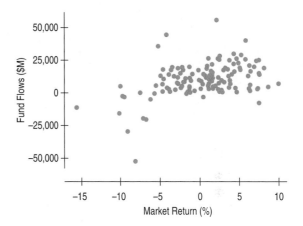

Dependent variable is: Fund Flows
R squared = 17.9%
s = 10999 with 154 − 2 = 152 degrees of freedom

Variable	Coeff	SE(Coeff)	t-ratio	P-value
Intercept	9599.10	896.9	10.7	≤ 0.0001
Market Return	1156.40	201.1	5.75	≤ 0.0001

a) State the null and alternative hypotheses under investigation.
b) Assuming that the assumptions for regression inference are reasonable, test the hypothesis.
c) State your conclusion.

24. Marketing managers. Are wages for various marketing managerial positions related? One way to determine this is to examine the relationship between the mean hourly wages for two managerial occupations in marketing: sales managers and advertising managers. The average hourly wage for both occupations across all U.S. states and territories are analyzed (data.bls.gov/oes; Occupational Employment Statistics; accessed May 2008). Here are the regression analysis results.

Predictor	Coeff	SE(Coeff)	t-ratio	P-value
Constant	10.317	4.382	2.35	0.0227
Sales Mgr Avg Hourly Wage	0.56349	0.09786	5.76	< 0.0001

a) State the null and alternative hypothesis under investigation.
b) Assuming that the assumptions for regression inference are reasonable, test the null hypothesis.
c) State your conclusion.

T 25. Cost index. Recall the *Worldwide Cost of Living Survey* from Chapter 8 that determined the cost of living in the most expensive cities in the world as an index. This index scales New York City as 100 and expresses the cost of living in other cities as a percentage of the New York cost. For example, in 2007 the cost of living index in Tokyo was 122.1, which means that it was 22% higher than New York. The output shows the regression of the 2006 on the 2007 index for the most expensive cities in 2007, where *Index 2007* is the response variable.

Predictor	Coeff	SE(Coeff)	t-ratio	P-value
Intercept	12.02	12.25	0.98	0.3446
Index 2006	0.943	0.115	8.17	< 0.0001

s = 4.45   R-Sq = 83.7%

a) State the hypotheses about the slope (both numerically and in words).
b) Perform the hypothesis test and state your conclusion in context.
c) Explain what the *R*-squared in this regression means.
d) Do these results indicate that, in general, cities with a higher index in 2006 will also have a higher index in 2007? Explain.

26. Job growth. Fortune Magazine publishes the top 100 companies to work for every year. Among the information listed is the percentage growth in jobs at each company. The output below shows the regression of the 2008 job growth (%) on the 2006 job growth for a sample of 29 companies. Note that *Job Growth 2008* is the response variable. (money.cnn.com/magazines/fortune/bestcompanies/full_list; accessed May, 2008).

Dependent variable is: Job Growth 2008

R squared = 25.6%   R squared (adjusted) = 22.9%

s = 6.129 with 29 − 2 = 27 degrees of freedom

Variable	Coefficient	SE(Coeff)	t-ratio	P-value
Intercept	2.993	1.441	2.08	0.0475
Job Growth 2006	0.399	0.131	3.05	0.0051

a) State the hypotheses about the slope (both numerically and in words).
b) Assuming that the assumptions for inference are satisfied, perform the hypothesis test and state your conclusion in context.
c) Explain what the *R*-squared in this regression means.
d) Do these results indicate that, in general, companies with a higher job growth in 2006 will also have a higher job growth in 2008? Explain.

T 27. Oil prices. The Organization of Petroleum Exporting Countries (OPEC) is a cartel, so it artificially sets prices. But are prices related to production? Using the data provided on the disk for crude oil prices ($/barrel) and oil production (thousand barrels per day) between 2001 and 2007, answer the following questions. Use the *Crude price* as the response variable and *Production* as the predictor variable.

a) Examine a scatterplot for the two variables and test the conditions for regression.
b) Do you think there is a linear association between oil prices and production? Explain.

T 28. Attendance 2006. Traditionally, athletic teams that perform better grow their fan base and generate greater attendance at games or matches. This should hold true regardless of the sport—whether it's soccer, football, or American baseball. Using the data provided for the number of wins and attendance in 2006 for the 14 American League baseball teams, answer the following questions. Use *Home Attendance* as the dependent variable and *Wins* as the explanatory variable.

a) Examine a scatterplot for the two variables and test the conditions for regression.
b) Do you think there is a linear association between *Home Attendance* and *Wins*? Explain.

T 29. Printers. In March 2002, *Consumer Reports* reviewed several models of inkjet printers. The following table shows the speed of the printer (in pages per minute) and the cost per page printed. Is there evidence of an association between *Speed* and *Cost*? Test an appropriate hypothesis and state your conclusion.

Speed (ppm)	Cost (cents/page)
4.6	12.0
5.5	8.5
4.5	6.2
3.8	3.4
4.6	2.6
3.7	4.0
4.7	5.8
4.7	8.1
4.0	9.4
3.1	14.9
1.9	2.6
2.2	4.3
1.8	4.6
2.0	14.8
2.0	4.4

T 30. Product testing. Remember the Little League instructional video discussed in Chapter 14? Ads claimed that the techniques would improve the performances of Little League pitchers. To test this claim, 20 Little Leaguers threw 50 pitches each, and we recorded the number of strikes. After the players participated in the training program, we repeated the test. The following table shows the number of strikes each player threw before and after the training. A test of paired differences failed to show that this training was effective in improving a player's ability to throw strikes. Is there any evidence that the *Effectiveness* (*After* − *Before*) of the video depends on the player's *Initial Ability* (*Before*) to throw strikes? Test an appropriate hypothesis and state your conclusion. Propose an explanation for what you find.

Number of Strikes (out of 50)			
Before	**After**	**Before**	**After**
28	35	33	33
29	36	33	35
30	32	34	32
32	28	34	30
32	30	34	33
32	31	35	34
32	32	36	37
32	34	36	33
32	35	37	35
33	36	37	32

T 31. Fuel economy, revisited. Consider again the data in Exercise 15 about the fuel economy and weights of cars.

a) Create a 95% confidence interval for the average fuel efficiency among cars weighing 2500 pounds, and explain what your interval means.

b) Create a 95% prediction interval for the gas mileage you might get driving your new 3450-pound SUV, and explain what that interval means.

T 32. SAT scores, again. Consider the high school SAT scores data from Exercise 17 once more. The mean Verbal score was 596.30.

a) Find a 90% confidence interval for the mean SAT Math score for all students with an SAT Verbal score of 500.

b) Find a 90% prediction interval for the Math score of the senior class president, if you know she scored 710 on the Verbal section.

T 33. Mutual funds part 2. Using the same mutual fund flow data provided in Exercise 23, answer the following questions.

a) Find the 95% prediction interval for month that reports a market return of 8%.

b) Do you think predictions made by this regression will be very accurate? Explain.

c) Would your prediction be more or less precise if you were to omit the points noted in Exercise 23?

T 34. Assets and sales, revisited. A business analyst was interested in the relationship between a company's assets and its sales. She collected data (in millions of dollars) from a random sample of 79 Fortune 500 companies and created the following regression analysis. Economists commonly take the logarithm of these variables to make the relationship more nearly linear, and she did too. The dependent variable is *LogSales*. The assumptions for regression inference appeared to be satisfied.

Dependent variable is: LogSales

R squared = 33.9%

s = 0.4278 with 79 − 2 = 77 degrees of freedom

Variable	Coefficient	SE(Coeff)	t-ratio	P-value
Intercept	1.303	0.3211	4.06	0.0001
LogAssets	0.578	0.0919	6.28	≤ 0.0001

a) Is there a significant linear association between *LogAssets* and *LogSales*? Find the *t*-value and P-value to test an appropriate hypothesis and state your conclusion in context.

b) Do you think that a company's assets serve as a useful predictor of their sales?

T 35. All the efficiency money can buy. A sample of 84 model-2004 cars from an online information service was examined to see how fuel efficiency (as highway mpg) relates to the cost (Manufacturer's Suggested Retail Price in dollars) of cars. Here are displays and computer output:

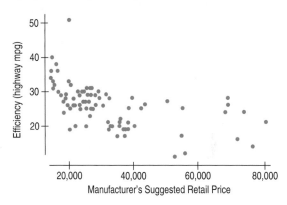

Dependent variable is: Highway MPG

R squared = 30.1%

s = 5.298 with 84 − 2 = 82 degrees of freedom

Variable	Coefficient	SE(Coeff)	t-ratio	P-value
Constant	33.06	1.299	25.5	≤ 0.0001
MSRP	−2.165e-4	3.639e-5	−5.95	≤ 0.0001

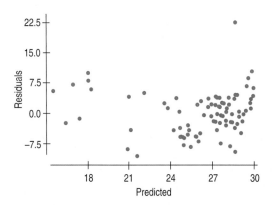

a) State what you want to know, identify the variables, and give the appropriate hypotheses.

b) Check the assumptions and conditions.

c) If the conditions are met, complete the analysis.

T 36. Energy use. Based on data collected from the United Nations Millennium Indicators Database related to measuring the goal of *ensuring environmental sustainability*, investigate the association between energy use (kg oil equivalent per $1000 GDP) in 1990 and 2004 for a sample of 96 countries (unstats.un.org/unsd/mi/mi_goals.asp; accessed May 2008).

a) Find a regression model showing the relationship between *2004 Energy Use* (reponse variable) and *1990 Energy Use* (predictor variable).

b) Examine the residuals to determine if a linear regression is appropriate.

c) Test an appropriate hypothesis to determine if the association is significant.

d) What percentage of the variability in *2004 Energy Use* is explained by *1990 Energy Use*?

37. Youth unemployment, part 2. Refer to the United Nations data referenced in Exercise 9. Here is a scatterplot showing the regression line, 95% confidence interval, and 95% prediction interval, using 2005 youth unemployment data for a sample of 57 nations. The response variable is the *Male Rate*, and the predictor variable is the *Female Rate*.

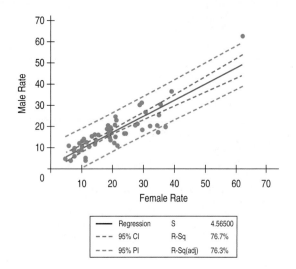

	Regression	S	4.56500
---	95% CI	R-Sq	76.7%
---	95% PI	R-Sq(adj)	76.3%

a) Explain the meaning of the 95% prediction interval in this context.

b) Explain the meaning of the 95% confidence interval in this context.

c) Identify the unusual observation, and discuss its potential impact on the regression.

38. Male unemployment, part 2. Refer to the United Nations data referenced in Exercise 10. Here is a scatterplot showing the regression line, 95% confidence interval, and 95% prediction interval, using 2005 and 2004 male unemployment data for a sample of 52 nations. The response variable is the *2005-Male Rate*, and the predictor variable is the *2004-Male Rate*.

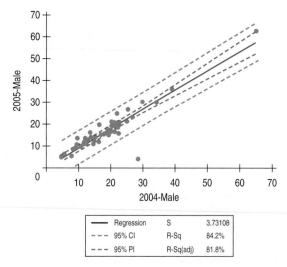

	Regression	S	3.73108
---	95% CI	R-Sq	84.2%
---	95% PI	R-Sq(adj)	81.8%

a) Explain the meaning of the 95% prediction interval in this context.

b) Explain the meaning of the 95% confidence interval in this context.

c) Identify the unusual observation, and discuss its potential impact on the regression.

39. Energy use again. Examine the regression and scatterplot showing the regression line, 95% confidence interval, and 95% prediction interval using *1990* and *2004 energy use* (kg oil equivalent per $1000 GDP) for a sample of 96 countries. The response variable is *2004 Energy Use*.

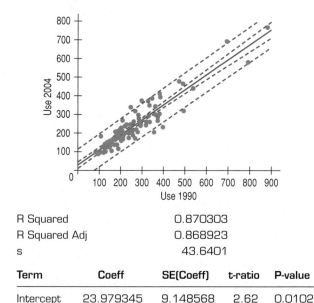

R Squared	0.870303
R Squared Adj	0.868923
s	43.6401

Term	Coeff	SE(Coeff)	t-ratio	P-value
Intercept	23.979345	9.148568	2.62	0.0102
Use 1990	0.8073999	0.032148	25.12	<.0001

a) Explain the meaning of the 95% prediction interval in this context.

b) Explain the meaning of the 95% confidence interval in this context.

T **40. Global reach.** The Internet has revolutionized business and offers unprecedented opportunities for globalization. However, the ability to access the Internet varies greatly among different regions of the world. One of the variables the United Nations collects data on each year is *Personal Computers per 100 Population* (unstats.un.org/unsd/cdb/cdb_help/cdb_quick_start.asp) for various countries. Below is a scatterplot showing the regression line, 95% confidence interval, and 95% prediction interval using 2000 and 2004 computer adoption (personal computers per 100 population) for a sample of 85 countries. The response variable is *PC/100 2004*.

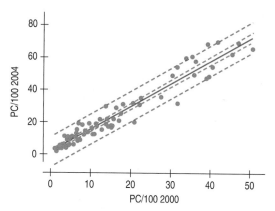

a) Find a regression model showing the relationship between personal computer adoption in 2004 *PC/100 2000* (the response variable) and personal computer adoption in 2000 *PC/100 2000* (the predictor variable).
b) Explain the meaning of the 95% prediction interval in this context.
c) Explain the meaning of the 95% confidence interval in this context.

T **41. Seasonal spending revisited.** Spending on credit cards decreases after the Christmas spending season (as measured by amount charged on a credit card in December). The data set on the DVD contains the monthly credit card charges of a random sample of 99 cardholders.

a) Build a regression model to predict January spending from December's spending.
b) How much, on average, will cardholders who charged $2000 in December charge in January?
c) Give a 95% confidence interval for the average January charges of cardholders who charged $2000 in December.
d) From c), give a 95% confidence interval for the average decrease in the charges of cardholders who charged $2000 in December.
e) What reservations, if any, do you have about the confidence intervals you made in c) and d)?

T **42. Seasonal spending revisited part 2.** Financial analysts know that January credit card charges will generally be much lower than those of the month before. What about the difference between January and the next month? Does the trend continue? The data set on the DVD contains the monthly credit card charges of a random sample of 99 cardholders.

a) Build a regression model to predict February charges from January's charges.
b) How much, on average, will cardholders who charged $2000 in January charge in February?
c) Give a 95% confidence interval for the average February charges of cardholders who charged $2000 in January.
d) From c), give a 95% confidence interval for the average decrease in the charges of cardholders who charged $2000 in January.
e) What reservations, if any, do you have about the confidence intervals you made in c) and d)?

43. Environment. The Environmental Protection Agency is examining the relationship between the ozone level (in parts per million) and the population (in millions) of U.S. cities. Part of the regression analysis is shown, where the dependent variable is *Ozone*.

Dependent variable is: Ozone
R squared = 84.4%
s = 5.454 with 16 − 2 = 14 df

Variable	Coefficient	SE(Coeff)
Intercept	18.892	2.395
Pop	6.650	1.910

a) We suspect that the greater the population of a city, the higher its ozone level. Is the relationship significant? Assuming the conditions for inference are satisfied, find the *t*-value and P-value to test the appropriate hypothesis. State your conclusion in context.
b) Do you think that the population of a city is a useful predictor of ozone level? Use the values of both R^2 and *s* in your explanation.

44. Environment part 2. Consider again the relationship between the population and ozone level of U.S. cities that you analyzed in Exercise 43.

a) Give a 90% confidence interval for the approximate increase in *Ozone* level associated with each additional million city inhabitants.
b) For the cities studied, the mean population was 1.7 million people. The population of Boston is approximately 0.6 million people. Predict the mean ozone level for cities of that size with an interval in which you have 90% confidence.

JUST CHECKING ANSWERS

1 I would need to see a scatterplot to see if the linearity assumption is reasonable, to make sure that there are no outliers, and a residual plot to check the equal spread condition. I'd also like to see a histogram or Normal probability plot of the residuals to make sure the nearly normal condition is satisfied. Finally, I'd like to see the residuals plotted in time to see if the residuals appear independent. Without verifying these conditions, I wouldn't know whether my analysis is valid.

2 The 95% CI for the slope is $1133.8 \pm 2.052 (513.1)$, or (80.9, 2186.7).

3 H_0: The slope $\beta_1 = 0$. H_A: The slope $\beta_1 \neq 0$. Since the P-value $= 0.036$, we reject the null hypothesis (at $\alpha = 0.05$) and conclude that there is a linear relationship between enrollment and unemployment.

4 Strength is a judgment call, but I'd be hesitant to call a relationship with an R^2 value of only 15% strong.

5 Approximately 15% of the variation in enrollment at the University of New Mexico is accounted for by variation in the unemployment rate in New Mexico.

6 The test says that we can reject the hypothesis that the correlation is 0 (at $\alpha = 0.05$) and conclude that there is a linear relationship between enrollment and unemployment. This is exactly what the test of the slope in part 3 told us. The correlation is 0, if and only if the slope is 0. There is no new information here.

Understanding Residuals

Kellogg's

J ohn Harvey Kellogg was a physician who ran the Battle Creek Sanitarium. He was an advocate of vegetarian diets and of living without caffeine, alcohol, or sex. He was an early advocate of peanut butter, patenting one of the first devices for making it. But he is best remembered for his work with his brother Will Keith Kellogg. Together, in 1897, they founded the Sanitas Food Company to manufacture whole grain cereals.

At the start of the 20th century, breakfast was typically a large, high-fat meal of eggs and meat for the wealthy and a less nutritious meal of porridge or gruel for the poor. The Kellogg brothers introduced toasted corn flakes as a healthy and affordable alternative. But in 1906, they argued when Will wanted to add sugar to the recipe—an idea that horrified John Harvey. Will founded the Battle Creek Toasted Corn Flake Company, which eventually became the Kellogg Company, using its founder's "W. K. Kellogg" signature as a logo—a marketing concept that survives in the script "Kellogg's" on their boxes to this day.

True to its roots in healthy nutrition, Kellogg's brands include Kashi® health foods, Morningstar Farms® vegetarian foods, and Nutri-Grain® cereal bars. (However, they also make Cocoa Krispies®, Froot Loops®, Cheez-Its®, and Keebler® Cookies.) In 1923, Kellogg hired the first dietitian to work in the food industry, and in the 1930s, Kellogg was the first company to print nutrition information on their boxes. The W. K. Kellogg Institute for Food and Nutrition Research, a world-class research facility, opened in 1997. The company continues to advocate for healthy nutrition, offering advice and education on their website and partnering with organizations such as the American Heart Association.

Regression may be the most widely used Statistics method. It is used every day throughout the world to help make good business decisions. The applications of linear regression are limitless. It can be used to predict customer loyalty, staffing needs at hospitals, sales of automobiles, and almost anything that can be quantified. Because regression is so widely used, it's also widely abused and misinterpreted. This chapter presents examples of regressions in which things are not quite as simple as they may seem at first and shows how you can still use regression to discover what the data have to say.

The residuals from the fitting of a linear regression hold an incredible amount of information about the model. Remember that a residual is the difference between the actual data and the value we predict for it: $e = y - \hat{y}$. Residuals can help tell you how well the model is performing and provide clues for fixing it if it's not working as well as it could. In this chapter, we'll show a variety of ways in which detecting patterns in the residuals can help you improve the model. Examining residuals can reveal more about the data than was apparent at first, or even second, glance. That's why no regression analysis is ever complete without a display of the residuals and a thorough examination of what they have to say.

17.1 Examining Residuals for Groups

It seems that ever since the Kellogg brothers fought over sugar in breakfast cereals, it has been a concern. Using data from those nutrition labels introduced by Kellogg, we can examine the relationship between the calories in a serving and the amount of sugar (in grams). Figure 17.1 appears to satisfy the conditions for regression; the relationship is linear with no outliers.

The least squares regression model,

$$\widehat{calories} = 89.5 + 2.50 \, sugar$$

has an R^2 of 32%. Figure 17.2 shows the residuals.

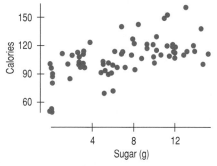

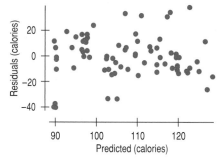

Figure 17.1 Calories *versus* Sugar content (grams) per serving of breakfast cereal.

Figure 17.2 Residuals for the regression plotted against predicted Calories.

At first glance, the scatterplot seems to have no particular structure, and as you may remember from Chapter 8, that's exactly what we hope to see. But let's check a histogram of the residuals.

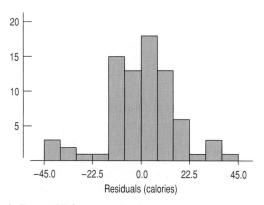

Figure 17.3 The distribution of the regression residuals shows modes above and below the central large mode. These may be worth a second look.

How would you describe the shape of this histogram? It looks like there might be small modes on either side of the central body of the data. A few cereals stand out with larger negative residuals—that is, fewer calories than we might have predicted. And a few stand out with larger positive residuals. Of course, the sample size here is not very large. We can't say for sure that there are three modes, but it's worth a closer look.

Let's look more carefully at the residuals. Figure 17.4 repeats the scatterplot of Figure 17.2, but with the points in those modes marked. Now we can see that those two groups stand away from the central pattern in the scatterplot. Doing a little more work and examining the data set, we find that the high-residual cereals (green x's) are *Just Right Fruit & Nut*; *Muesli Raisins, Dates & Almonds*; *Peaches & Pecans*; *Mueslix Crispy Blend*; and *Nutri-Grain Almond Raisin*. Do these cereals have something in common? These high-calorie cereals all market themselves as "healthy." This might be surprising, but in fact, "healthy" cereals often contain more fat. They often contain nuts and oil which are "natural" and don't necessarily contain sugar, but are higher in fat than grain and sugar. So, they may have more calories than we might expect from looking at their sugar content alone.

The low-residual (red) cereals are *Puffed Rice*, *Puffed Wheat*, three bran cereals, and *Golden Crisps*. These cereals have fewer calories than we would expect based on their sugar content. We might not have grouped these cereals together before.

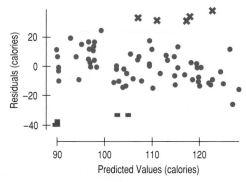

Figure 17.4 *A scatterplot of the residuals vs. predicted values for the cereal regression. The green x's are cereals whose calorie content is higher than the linear model predicts. The red –'s show cereals with fewer calories than the model predicts. Is there something special about these cereals?*

What they have in common is a low calorie count *relative to their sugar content*—even though their sugar contents are quite different. (They are low calorie because of their shape and structure.)

These observations may not lead us to question the overall linear model, but they do help us understand that other factors may be part of the story. An exploration of residuals often leads us to discover more about individual cases. When we discover groups in our data, we may decide to analyze them separately, using a different model for each group.

Often, more research can help us discover why certain cases tend to behave similarly. Here, certain cereals group together in the residual plot because cereal manufacturers aim cereals at different segments of the market. A common technique used to attract different customers is to place different types of cereals on certain shelves. Cereals for kids tend to be on the "kid's shelf," at their eye level. Toddlers aren't likely to grab a box from this shelf and beg, "Mom, can we please get this *All-Bran with Extra Fiber*?"

How can we take this extra information into account in our analysis? Figure 17.5 shows a scatterplot of *Calories* and *Sugar,* colored according to the shelf on which the

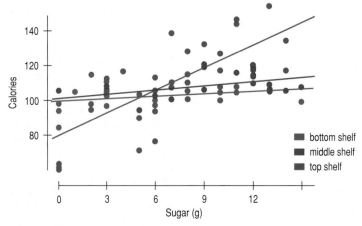

Figure 17.5 *Calories and Sugars colored according to the shelf on which the cereal was found in a supermarket, with regression lines fit for each shelf individually. Do these data appear homogeneous? That is, do the cereals seem to all be from the same population of cereals? Or are there kinds of cereals that we might want to consider separately?*

cereals were found, with a separate regression line fit for each shelf. Now we can see that the top shelf is unlike the bottom two shelves. We might want to report two regressions, one for the top shelf and one for the bottom two shelves.[1]

17.2 Extrapolation and Prediction

Linear models give a predicted value for each case in the data. Put a new x-value into the equation, and it gives a predicted value, $\hat{y}$, to go with it. But when the new x-value lies far from the data we used to build the regression, how trustworthy is the prediction?

The simple answer is that the farther the new x-value is from $\bar{x}$, the center of the x-values, the less trust we should place in the predicted value. Once we venture into new x territory, such a prediction is called an **extrapolation**. Extrapolations are dangerous because they require the additional—and questionable—assumption that nothing about the relationship between x and y changes, even at extreme values of x and beyond. Extrapolations can get us into deep trouble, especially if we try to predict far into the future.

As a cautionary example, let's examine oil prices from 1972 to 1981 in constant (2005) dollars.[2] In the mid 1970s, in the midst of an energy crisis, oil prices surged, and long lines at gas stations were common. In 1970, the price of oil was about $3 a barrel. A few years later, it had surged to $15. In 1975, a survey of 15 top econometric forecasting models (built by groups that included Nobel prize-winning economists) found predictions for 1985 oil prices that ranged from $50 to $200 a barrel (or $181 to $726(!) dollars a barrel in 2005 dollars). How close were these forecasts? Let's look at Figure 17.6.

When the Data Are Years
We usually don't enter them as four-digit numbers. Here, we used 0 for 1970, 10 for 1980, and so on. It's common to assign 0 to the date of the first observation in our data set if we are working with a time series. Another option is to enter two digits for the year, using 88 for 1988, for instance. Rescaling years like this often makes calculations easier and equations simpler. But be careful; if 1988 is 88, then 2004 is 104 (not 4).

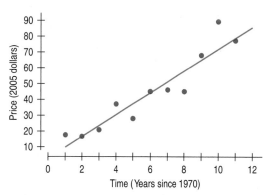

Figure 17.6 *The price of oil per barrel in constant (2005) dollars from 1971 to 1982 shows a linear trend increasing at about $7 a year.*

The regression model for the *Price* of oil against *Time* (Years since 1970) for these data is

$$\widehat{Price} = -0.85 + 7.39\ Time,$$

which says that prices increased, on average, $7.39 per year, or nearly $75 in 10 years. If they continued to increase linearly, it would have been easy to predict oil prices. And indeed, many forecasters made that assumption. So, how well did

---

[1] Another alternative is to fit a multiple regression model by adding variables (called dummy or indicator variables) that distinguish the groups. This method will be discussed in Chapter 19.

[2] We will discuss special models for fitting data when x is time in Chapter 20, but simple regression models are often used. Even when using more sophisticated methods, the dangers of extrapolation don't disappear.

they do? Well, in the period from 1982 to 1998, oil prices didn't exactly continue that steady increase. In fact, they went down so much that by 1998, prices (adjusted for inflation) were the lowest they'd been since before World War II (Figure 17.7).

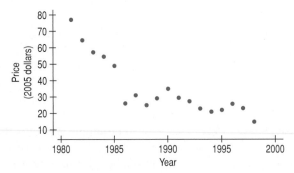

Figure 17.7   *Time series plot of price of oil in constant (2005) dollars shows a fairly constant decrease over time.*

For example, the average price of oil in 1985 turned out to be less than $30 per barrel—not quite the $100 predicted by the model. Extrapolating out beyond the original data by just four years produced some vastly inaccurate forecasts. While the time series plot in Figure 17.7 shows a fairly steady decline, this pattern clearly didn't continue (or oil would be free by now).

In the 1990s, the U.S. government decided to include scenarios in their forecasts. The result was that the Energy Information Administration (EIA) offered *two* 20-year forecasts for oil prices after 1998 in their Annual Energy Outlook (AEO). Both of these scenarios, however, called for relatively modest increases in oil prices (Figure 17.8).

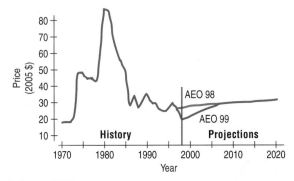

Figure 17.8   *This graph, adapted from one by the Energy Information Administration, shows oil prices from 1970 to 1998 with two sets of forecasts for the period 1999 to 2020.*

So, how accurate have these forecasts been? Let's compare these predictions to the actual prices in constant (2005) dollars (Figure 17.9).

Figure 17.9   *Here are the same EIA forecasts as in Figure 17.8, together with the actual prices from 1981 to 2007. Neither forecast predicted the sharp run-up in the past few years.*

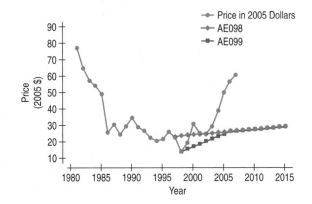

The experts seem to have missed the sharp run-up in oil prices in the first decade of the 21st century. Where do you think oil prices will go in the *next* decade? Your guess may be as good as anyone's. Clearly, these forecasts did not take into account many of the unforeseen global and economic events that occurred since 2000. Providing accurate long-term forecasts is extremely difficult.

Extrapolation far from the data is dangerous. Linear models are based on the *x*-values of the data at hand and cannot be trusted beyond that span. Some phenomena do exhibit a kind of inertia that allows us to guess that the currently observed systematic behavior will continue outside this range. When *x* is time, you should be especially wary. Such regularity can't be counted on in phenomena such as stock prices, sales figures, hurricane tracks, or public opinion.

Extrapolating from current trends is a mistake made not only by regression beginners or the naïve. Professional forecasters are prone to the same mistakes, and sometimes the errors are striking. However, because the temptation to predict the future is so strong, our more realistic advice is this:

> *If you extrapolate far into the future, be prepared for the actual values to be (possibly quite) different from your predictions.*

> *"Prediction is difficult, especially about the future."*
>
> —Niels Bohr, Danish physicist

17.3 Unusual and Extraordinary Observations

Your credit card company makes money each time you use your card. To encourage you to use your card, the card issuer may offer you an incentive such as airline miles, rebates, or gifts.[3] Of course, this is profitable to the company only if the increased use brings in enough revenue to offset the cost of the incentives. New ideas for offers (referred to as "creatives") are typically tested on a sample of cardholders before they are rolled out to the entire segment or population, a process referred to as a "campaign." Typically, the new offer (the "challenger") is tested against a control group who may be offered nothing or the current best offer ("the champion").

One campaign offered one of the highest-performing market segments an incentive for three months: one redeemable anytime air mile for each dollar spent. They hoped that the cardholders would increase their spending enough to pay for the campaign, but they feared that some cardholders would move

---

[3] There are websites dedicated to finding credit card "deals." Search "credit card rewards."

spending forward into the incentive period, with a resulting drop in spending afterward.

For this particular segment, the typical cardholder charged about $1700 a month. During the campaign period, the group averaged around $1919.61 a month, a difference that was both statistically and financially significant. But analysts were suprised to see that the increase in spending continued well beyond the offer period. To investigate it, they made a scatterplot like the one shown in Figure 17.10.

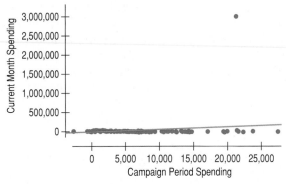

Figure 17.10 *Spending after the campaign plotted against spending during the campaign period reveals a surprising value and a positive regression slope.*

The outlying point, at the top of the graph, represents a cardholder who charged nearly $3 million in the month after the free miles period ended. Remarkably, the point was verified to be a real purchase! Nevertheless, this cardholder is clearly not typical of the rest of the segment. To answer the company's question, we need to examine the plot without the outlying point (Figure 17.11).

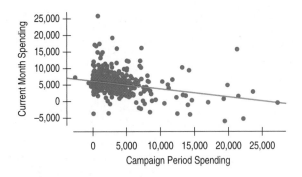

Figure 17.11 *A plot of current spending against the spending during the campaign period, with the outlier set aside. Now the slope is negative, and significantly so.*

The plot does show that those with the largest charges during the campaign spent less in the month after the campaign. Just one outlier was capable of changing the slope's direction from strongly negative to strongly positive. On the basis of this finding, the analysts decided to focus only on people whose spending during *both* periods was less than $10,000 a month, figuring that if someone decides to spend more than $10,000 on their credit card, their primary motivation is probably not the airline miles incentive.

"Give me a place to stand and I will move the Earth."

—Archimedes (287–211 BCE)

> Influence depends on both the leverage and the residual; a case with high leverage whose *y*-value sits right on the line fit to the rest of the data is not influential. A case with low leverage but a very large residual can be influential. The only way to be sure is to fit the regression with and without the potential influential point.

"For whoever knows the ways of Nature will more easily notice her deviations; and, on the other hand, whoever knows her deviations will more accurately describe her ways."

—Francis Bacon (1561–1626)

Outliers, Leverage, and Influence

By providing a simple description of how data behave, models help us see when and how data values are unusual. In regression, a point can stand out in two ways. A case can have a large residual, as our $3 million spender certainly did. Because they are not like the other cases, points with large residuals always deserve special attention and are called **outliers**.

A data point can also be unusual if its *x*-value is far from the mean of the *x*-values. Such a point is said to have high **leverage**. The physical image of a lever is exactly right. The least squares line must pass through $(\bar{x}, \bar{y})$, so you can picture that point as the fulcrum of the lever. Just as sitting farther from the center of a seesaw gives you more leverage, points with values far from $\bar{x}$ pull more strongly on the regression line.

A point with high leverage has the potential to change the regression line but it doesn't always use that potential. If the point lines up with the pattern of the other points, it doesn't change our estimate of the line. By sitting so far from $\bar{x}$ though, it may appear to strengthen the relationship, inflating the correlation and R^2.

How can you tell if a high-leverage point changes the model? Just fit the linear model twice, both with and without the point in question. We say that a point is **influential** if omitting it from the analysis gives a very different model (as the high spender did in our example).[4]

Unusual points in a regression often tell us more about the data and the model than any other cases. Whenever you have—or suspect that you have—influential points, you should fit the linear model to the other cases alone and then compare the two regression models to understand how they differ. A model dominated by a single point is unlikely to be useful for understanding the rest of the cases. The best way to understand unusual points is against the background of the model established by the other data values. Don't give in to the temptation to delete points simply because they don't fit the line. That can give a false picture of how well the model fits the data. But often the best way to identify interesting cases and subgroups is to note that they are influential and to find out what makes them special.

Not all points with large influence have large residuals. Sometimes, their influence pulls the regression line so close that it makes the residual deceptively small. Influential points like that can have a shocking effect on the regression. Figure 17.12 shows IQ plotted against shoe size from a fanciful study of intelligence and foot size. The outlier is Bozo the clown, known for his large feet and hailed as a comic genius.

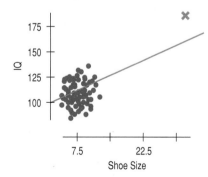

Figure 17.12 *Bozo the clown's extraordinarily large shoes give his data point high-leverage in the regression of:* $\widehat{IQ} = 93.3 + 2.08$ *shoe size, even though the R^2 is 25%. Wherever Bozo's IQ happens to be, the regression line will follow.*

---

[4] Some textbooks use the term *influential point* for any observation that influences the slope, intercept, or R^2. We'll reserve the term for points that influence the slope.

Although this is a silly example, it illustrates an important and common potential problem. Almost all of the variance accounted for ($R^2 = 25\%$) is due to *one* point, namely, Bozo. Without Bozo, there is little correlation between shoe size and IQ. If we run the regression after omitting Bozo, we get an R^2 of only 0.7%—a weak linear relationship (as one might expect). One single point exhibits a great influence on the regression analysis.

JUST CHECKING

Each of these scatterplots shows an unusual point. For each, tell whether the point is a high-leverage point, would have a large residual, and/or is influential.

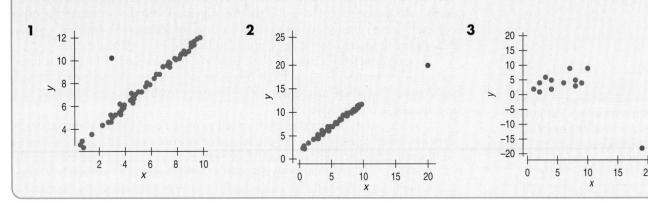

What should you do with a high-leverage point? Sometimes these values are important (they may be customers with extremely high incomes or employees with unusually long service to the company), and they may say more about the relationship between y and x than any of the other data values. However, at other times, high-leverage points are values that really don't belong with the rest of the data. Such points should probably be omitted, and a linear model found without them for comparison. When in doubt, it's usually best to fit regressions both with and without the points and compare the two models.

◆ **Warning:** Influential points can hide in plots of residuals. Points with high-leverage pull the line close to them, so they often have small residuals. You'll see influential points more easily in scatterplots of the original data, and you'll see their effects by finding a regression model with and without the points.

17.4 Working with Summary Values

Scatterplots of statistics summarized over groups tend to show less variability than we would see if we measured the same variables on individuals. This is because the summary statistics themselves vary less than the data on the individuals.

Wind power is getting increasing attention as an alternative, carbon-free method of generating electricity. Of course, there must be enough wind to make it cost-effective. In a study to find a site for a wind generator, wind speeds were collected four times a day (at 6:00 a.m., noon, 6:00 p.m., and midnight) for a year at several possible sites. Figure 17.13 plots the wind speeds for two of these sites. The correlation is 0.736.

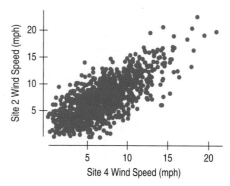

Figure 17.13 *The wind speed at sites 2 and 4 are correlated.*

What would happen to the correlation if we used only one measurement per day? If, instead of plotting four data points for each day, we record an average speed for each day, the resulting scatterplot shows less variation, as Figure 17.14 shows. The correlation for these values increases to 0.844.

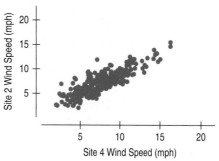

Figure 17.14 *Daily average wind speeds show less variation.*

Let's average over an even longer time period. Figure 17.15 shows *monthly* averages for the year (plotted on the same scale). Now the correlation is 0.942.

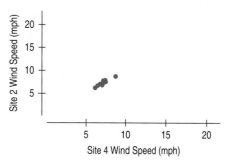

Figure 17.15 *Monthly averages are even less variable.*

What these scatterplots show is that summary statistics exhibit less scatter than the data on individuals on which they're based and can give us a false impression of how well a line summarizes the data. There's no simple correction for this phenomenon. If we're given summary data, we usually can't get the original values back. You should be a bit suspicious of conclusions based on regressions of summary data. They may look better than they really are.

Another way to reduce the number of points in a data set is to select or sample points rather than average them. This can be especially important with data, such as the wind speeds, that are measured over time. For example, if instead of finding the *mean* for each day, we select just one of the four daily measurements—say the one made at noon on each day. We would have just as many points as in Figure 17.14, but the correlation is 0.730—essentially the same as for the full data. Figure 17.16 shows the relationship.

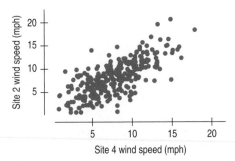

Figure 17.16 *Selecting only the noon measurements doesn't reduce the variation. Compare this scatterplot to Figures 17.13 and 17.14.*

17.5 Autocorrelation

Time series data that are collected at regular time points often have the property that points near each other in time will be related. When values at time *t* are correlated with values at time *t*-1 we say the values are **autocorrelated** in the first order. If values are correlated with values two time periods back, we say second-order autocorrelation is present, and so on.

A regression model applied to autocorrelated data will have errors that are not independent, and that violates an assumption for regression. The statistical tests and confidence intervals for the slope depend on independence, and its violation can render these tests and intervals invalid. Fortunately, there is a statistic called the Durbin-Watson statistic that can detect first-order autocorrelation from the residuals of a regression analysis.

The Mini Case Study of Chapter 16 provided data on *Price* and *Sales* volume of frozen pizza for several cities for one week each month. We actually have the data for *every* week in the same three-year period. Here is the regression of *Sales* volume on *Price* for Dallas.

	Coeff	SE(Coeff)	t-value	P-value
Intercept	139547	11302	12.347	<0.0001
Price	−33527	4308	−7.783	<0.0001

A plot of the residuals against predicted values shows nothing particularly unusual.

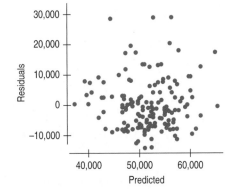

Figure 17.17 *A scatterplot of the residuals vs. predicted values for the 156 weeks of pizza sales data reveals no obvious patterns.*

But because these data points are consecutive weekly data, we should investigate the residuals *vs.* time. Here we've plotted the *Residuals* against *Week*, consecutively from week 1 to week 156.

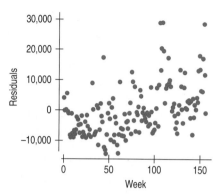

Figure 17.18 *A scatterplot of the residuals against* Week *for the 156 weeks of pizza sales data seems to show some trend.*

It may not be obvious that there's a pattern here. Autocorrelation can be difficult to see in residuals. It does seem, however, that there is a tendency in Figure 17.18 for the residuals to be related to nearby points. Notice the overall positive trend. We shouldn't see such a trend in residuals that are independent of each other. The **Durbin-Watson** statistic estimates the autocorrelation by summing squares of consecutive differences and comparing the sum with its expected value under the null hypothesis of no autocorrelation. The Durbin-Watson statistic is computed as follows:

$$D = \frac{\sum_{t=2}^{n}(e_t - e_{t-1})^2}{\sum_{t=1}^{n}e_t^2}$$

where e_t is the residual at time t. The statistic always falls in the interval from 0 to 4. When the null hypothesis of no autocorrelation is true, the value of D should be 2. Values of D below 2 are evidence of positive autocorrelation, while values of D above 2 indicate possible negative autocorrelation. Positive autocorrelation is more common than negative autocorrelation. How far below or above 2 does D need to be to show "strong" or significant autocorrelation? It may be surprising, but the answer to this question depends only on the sample size, n, and the number of predictors in the regression model, k, which for simple regression is equal to 1.

A standard Durbin-Watson table (see Appendix C) shows the sample size down the left-hand column, so that each row corresponds to a different sample size n, with the number of predictors k across the top. For each k there are two columns: d_L and d_U. (The significance level of the table is also shown at the top of the page.) The test has several possible outcomes:

If $D < d_L$ (lower critical value), then there is evidence of positive autocorrelation.

If $d_L < D < d_U$, then the test is inconclusive.

If $D > d_U$ (upper critical value), then there is no evidence of positive autocorrelation.

To test negative autocorrelation, we use the same values of d_L and d_U, but we subtract them from 4:

If $D > 4 - d_L$ (lower critical value), then there is evidence of negative autocorrelation.

Why 0 and 4?
Notice that if the adjacent residuals are equal (perfectly correlated), then the numerator and the value of D equals 0. If, on the other hand, the residuals are equal but have opposite signs (perfectly negatively correlated), then each difference is twice the residual. Then after squaring, the numerator will be 2^2, or four times the denominator.

If $4 - d_L < D < 4 - d_U$, then the test is inconclusive.

If $D < 4 - d_U$ (upper critical value), then there is no evidence of negative autocorrelation.

We usually rely on technology to compute the statistic. For the pizza example, we have $n = 156$ weeks and one predictor (*Price*), so $k = 1$. The value of D is $D = 0.8812$. Using the table in Appendix C, we find the largest value of n listed is $n = 100$, and at $\alpha = 0.05$, $d_L = 1.65$. Because our value is less than that, we conclude that there is evidence of positive autocorrelation. (A software package would find the P-value to be < 0.0001.) We conclude that the residuals are *not* independent but that residuals from one week have a positive correlation with the residuals from the preceding week. The standard errors and test for the slope are not valid since we don't have independence.

Time series methods (see Chapter 20) attempt to deal with the problem of autocorrelation by modeling the errors. Another solution is to find a predictor variable that accounts for some of the autocorrelation and removes the dependence in the residuals (see Chapter 19). A simple solution that often works is to sample from the time series so that the values are more distant in time and thus less likely to be correlated. If we take every fourth week of data (as we did in Chapter 16), starting at week 4, from the Dallas pizza data, our regression becomes:

	Coeff	SE(Coeff)	t-ratio	P-value
Intercept	148350	22266	6.663	8.01e-08
Price	−36762	8583	−4.283	0.000126

Now, $D = 1.617$. With $n = 39$, the upper critical value d_U is 1.54. Since our new value of D is larger than that, we see no evidence of autocorrelation. Output from technology shows the P-value:

```
Durbin-Watson test
```

$D = 1.6165,\ P = 0.098740$

We should feel more comfortable basing our confidence and prediction intervals on this model.

17.6 Linearity

Increasing gas prices and concern for the environment have lead to increased attention to automobile fuel efficiency. The most important factor in fuel efficiency is the weight of the car.

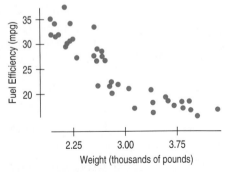

Figure 17.19 Fuel Efficiency (mpg) vs. Weight (thousands of pounds) shows a strong, apparently linear, negative trend.

The relationship is strong ($R^2 = 81.6\%$), clearly negative, and apparently linear. The regression equation

$$\widehat{Fuel\ Efficiency} = 48.7 - 8.4\ Weight$$

says that fuel efficiency drops by 8.4 mpg per 1000 pounds, starting from a value of 48.7 mpg. We check the **Linearity Condition** by plotting the residuals versus either the *x* variable or the predicted values.

The scatterplot of the residuals against *Weight* (Figure 17.20) holds a surprise. Residual plots should have no pattern, but this one has a bend. Look back at the original scatterplot. The scatter of points isn't really straight. There's a slight bend to the plot, but the bend is much easier to see in the residuals.

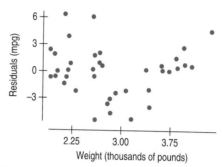

Figure 17.20 *Plotting residuals against weight reveals a bend. The bend can be seen if you look carefully at the original scatterplot, but here it's easier to see.*

When the relationship isn't straight, we shouldn't fit a regression or summarize the strength of the association with correlation. But often we can make the relationship straighter. All we have to do is re-express (or transform) one or both of the variables with a simple function. In this case, there's a natural function. In the U.S., automobile fuel efficiency is measured in miles per gallon. But throughout the rest of the world, things are different. Not only do other countries use metric measures, and thus kilometers and liters, but they measure fuel efficiency in liters per 100 kilometers. That's the *reciprocal* of miles per gallon (times a scale constant). That is, the gas amount (gallons or liters) is in the numerator, and the distance (miles or kilometers) is now in the denominator.

There's no reason to prefer one form or the other, so let's try the (negative) reciprocal form.

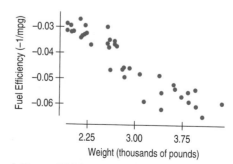

Figure 17.21 *The reciprocal of* Fuel Efficiency vs. Weight *is straighter.*

The residuals look better as well.

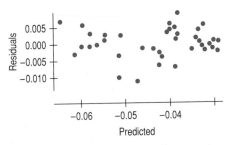

Figure 17.22 *Residuals from the regression of* Fuel Efficiency (−1/mpg) *on* Weight *show less bend.*

There's a clear improvement using the reciprocal, so we should use the reciprocal as the response in our regression model.

17.7 Transforming (Re-expressing) Data

Are we allowed to re-express fuel efficiency as its reciprocal? For these data, it seems like a good idea. After all, we chose a form of the variable used by most people in the world. The general idea of transforming data to improve and simplify its structure extends beyond straightening scatterplots. In fact, you use re-expressions in everyday life. How fast can you go on a bicycle? If you measure your speed, you probably do it in distance per time (miles per hour or kilometers per hour). In 2005, during a 12-mile-long time trial in the Tour de France, Dave Zabriskie averaged nearly 35 mph (54.7 kph), beating Lance Armstrong by 2 seconds. You probably realize that's a tough act to follow. It's fast. You can tell that at a glance because you have no trouble thinking in terms of distance covered per time.

If you averaged 12.5 mph (20.1 kph) for a mile run, would that be fast? Would it be fast for a 100-meter dash? Even if you run the mile often, you probably have to stop and calculate. Although we measure speed of bicycles in distance per time, we don't usually measure running speed that way. Instead, we re-express it as the *reciprocal*—time per distance (minutes per mile, seconds per 100 meters, etc.). Running a mile in under 5 minutes (12 mph) is fast. A mile at 16 mph would be a world record (that's a 3-minute 45-second mile).

The point is that there is no single natural way to measure speed. In some cases, we use distance traveled per time, and in other cases, we use the reciprocal. It's just because we're used to thinking that way in each case, not because one way is correct. It's important to realize that the way these quantities are measured is not sacred. It's usually just convenience or custom. When we re-express a quantity to make it satisfy certain conditions, we may either leave it in those new units when we explain the analysis to others or convert it back to the original units.

Goals of Re-expression

We re-express data for several reasons. Each of these goals helps make the data more suitable for analysis by our methods. We'll illustrate each goal by looking at data about large companies.

Goal 1 *Make the distribution of a variable (as seen in its histogram, for example) more symmetric.* It's easier to summarize the center of a symmetric distribution, and for nearly symmetric distributions, we can use the mean and standard deviation. If the distribution is unimodal, then the resulting distribution may be closer to the Normal model, allowing us to use the 68-95-99.7 Rule.

WHO	77 large companies
WHAT	*Assets, Sales,* and *Market Sector*
UNITS	$100,000
WHEN	1986
WHY	To examine distribution of *Assets* for the top *Forbes* 500 companies

Here are the *Assets* of these companies we first saw in Chapter 4.

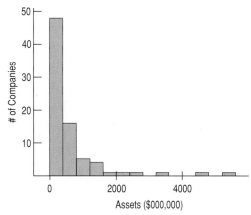

Figure 17.23 *The distribution of the Assets of large companies is skewed to the right. Data on wealth often look like this.*

The skewed distribution is made much more symmetric by taking logs.

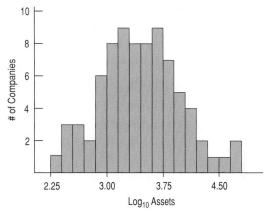

Figure 17.24 *Taking logs makes the distribution more symmetric.*

Goal 2 *Make the spread of several groups (as seen in side-by-side boxplots) more alike,* even if their centers differ. Groups that share a common spread are easier to compare. We'll see methods later in the book that can be applied only to groups with a common standard deviation. We saw an example of re-expression for comparing groups with boxplots in Chapter 6.

Here are the *Assets* of these companies by *Market Sector.*

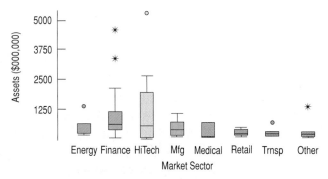

Figure 17.25 *Assets of large companies by Market Sector. It's hard to compare centers or spreads, and there seem to be a number of high outliers.*

Taking logs makes the individual boxplots more symmetric and gives them spreads that are more nearly equal.

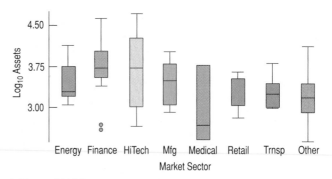

Figure 17.26 *After re-expressing using logs, it's much easier to compare across Market Sectors. The boxplots are more symmetric, most have similar spreads, and the companies that seemed to be outliers before are no longer extraordinary. Two new outliers have appeared in the finance sector. They are the only companies in that sector that are not banks.*

This makes it easier to compare *Assets* across *Market Sectors*. It can also reveal problems in the data. Some companies that looked like outliers on the high end turned out to be more typical. But two companies in the Finance sector now stick out. They are not banks. Unlike the rest of the companies in that sector, they may have been placed in the wrong sector, but we couldn't see that in the original data.

Goal 3 *Make the form of a scatterplot more nearly linear.* Linear scatterplots are easier to describe. We saw an example of scatterplot straightening in the opening example of this chapter. The value of re-expressing data to straighten a relationship is that we can fit a linear model once the relationship is straight.

Here are *Assets* plotted against the logarithm of *Sales*.

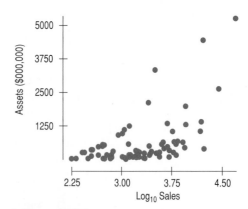

Figure 17.27 *Assets vs. log Sales shows a positive association (bigger sales goes with bigger Assets) with a bent shape.*

Note that the plot of *Assets* versus log *Sales* shows that the points go from tightly bunched at the left to widely scattered at the right—a "fan" shape. The plot's shape is bent. Taking logs makes the relationship much more linear (see Figure 17.28). If we re-express the company *Assets* using logarithms, we get a graph that shows a more linear association. Also note that the "fan" shape has disappeared and the variability at each value of x is about the same.

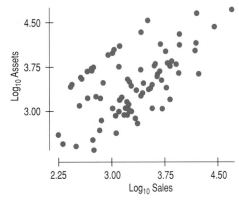

Figure 17.28 Log Assets *vs.* log Sales *shows a positive linear association.*

Goal 4 *Make the scatter in a scatterplot or residual plot spread out evenly rather than following a fan shape.* Having an even scatter is a condition of many methods of Statistics, as we'll see in later chapters. This goal is closely related to Goal 2, but it often comes along with Goal 3. Indeed, a glance back at the scatterplot (Figure 17.27) shows that the plot for *Assets* is much more spread out on the right than on the left, while the plot for log *Assets* (Figure 17.28) has roughly the same variation in log *Assets* for any *x*-value.

17.8 The Ladder of Powers

We've seen that taking logs or reciprocals can improve an analysis of relationships. Other transformations can be helpful too, but how do we know which re-expression to use? We could use trial and error to choose a re-expression, but there's an easier way. We can choose our re-expressions from a family of simple mathematical expressions that move data toward our goals in a consistent way. This family includes the most common ways to re-express data. More important, the members of the family line up in order, so that the farther you move away from the original data (the "1" position), the greater the effect on the data. This fact lets you search systematically for a transformation that works—stepping a bit farther from "1" or taking a step back toward "1" as you see the results.

Where to start? It turns out that certain kinds of data are more likely to be helped by particular re-expressions. Knowing that gives you a good place to start your search for a mathematical expression. We call this collection of re-expressions the **Ladder of Powers.** The following table shows some of the most useful powers with each one specified as a single value.

Power	Name	Comment
2	The square of the data values, y^2	Try this for unimodal distributions that are skewed to the left.
1	The raw data—no change at all. This is "home base." The farther you step from here up or down the ladder, the greater the effect.	Data that can take on both positive and negative values with no bounds are less likely to benefit from re-expression.
1/2	The square root of the data values $\sqrt{y}$	Counts often benefit from a square root re-expression. For counted data, start here.

Continued

Power	Name	Comment
"0"	Although mathematicians define the "0-th" power differently, for us the place is held by the logarithm.	Measurements that cannot be negative, and especially values that grow by percentage increases such as salaries or populations, often benefit from a log re-expression. When in doubt, start here. If your data have zeros, try adding a small constant to all values before finding the logs.
−1/2	The (negative) reciprocal square root $-1/\sqrt{y}$	An uncommon re-expression, but sometimes useful. Changing the sign to take the *negative* of the reciprocal square root preserves the direction of relationships, which can be a bit simpler.
−1	The (negative) reciprocal, $-1/y$	Ratios of two quantities (miles per hour, for example) often benefit from a reciprocal. (You have about a 50-50 chance that the original ratio was taken in the "wrong" order for simple statistical analysis and would benefit from re-expression.) Often, the reciprocal will have simple units (hours per mile). Change the sign if you want to preserve the direction of relationships. If your data have zeros, try adding a small constant to all values before finding the reciprocal.

The Ladder of Powers orders the *effects* that the re-expressions have on data. If you try, say, taking the square roots of all the values in a variable and it helps, but not enough, then moving farther down the ladder to the logarithm or reciprocal root will have a similar effect on your data, but even stronger. If you go too far, you can always back up. But don't forget—when you take a negative power, the *direction* of the relationship will change. You can always change the sign of the response if you want to keep the same direction.

JUST CHECKING

4 You want to model the relationship between prices for various items in Paris and Hong Kong. The scatterplot of Hong Kong prices *vs.* Paris prices shows a generally straight pattern with a small amount of scatter. What re-expression (if any) of the Hong Kong prices might you start with?

5 You want to model the population growth of the United States over the past 200 years with a percentage growth that's nearly constant. The scatterplot shows a strongly upwardly curved pattern. What re-expression (if any) of the population might you start with?

WHAT CAN GO WRONG?

This entire chapter has warned about things that can go wrong in a regression analysis. So let's just recap. When you make a linear model:

- **Make sure the relationship is straight enough to fit a regression model.** Check the Linearity Condition on the scatterplot of y against x and always examine the residuals for evidence that the Linearity Assumption has failed. It's often easier to see deviations from a straight line in the residuals plot than in the scatterplot of the original data. Pay special attention to the most extreme residuals because they may have something to add to the story told by the linear model.

- **Be on guard for different groups.** Check for evidence that the data consist of separate subsets. If you find subsets that behave differently, consider fitting a different linear model to each subset.

- **Beware of extrapolating.** Beware of extrapolation beyond the x-values that were used to fit the model. Although it's common to use linear models to extrapolate, be cautious.

- **Beware of extrapolating far into the future.** Be especially cautious about extrapolating far into the future with linear models. A linear model assumes that changes over time will continue forever at the same rate you've observed in the past. Predicting the future is particularly tempting and particularly dangerous.

- **Look for unusual points.** Unusual points always deserve attention and may well reveal more about your data than the rest of the points combined. Always look for them and try to understand why they stand apart. Making a scatterplot of the data is a good way to reveal high-leverage and influential points. A scatterplot of the residuals against the predicted values is a good tool for finding points with large residuals.

- **Beware of high-leverage points, especially of those that are influential.** Influential points can alter the regression model a great deal. The resulting model may say more about one or two points than about the overall relationship.

- **Consider setting aside outliers and re-running the regression.** To see the impact of outliers on a regression, try running two regressions, one with and one without the extraordinary points, and then discuss the differences.

- **Treat unusual points honestly.** If you remove enough carefully selected points, you will eventually get a regression with a high R^2. But it won't get you very far. Some data are not simple enough for a linear model to fit very well. When that happens, report the failure and stop.

- **Be alert for autocorrelation.** Data measured over time may fail the Independence Assumption. A Durbin-Watson test can check for that.

- **Watch out when dealing with data that are summaries.** Be cautious in working with data values that are themselves summaries, such as means or medians. Such statistics are less variable than the data on which they are based, so they tend to inflate the impression of the strength of a relationship.

- **Re-express your data when necessary.** When the data don't have the right form for the model you are fitting, your analysis can't be valid. Be alert for opportunities to re-express data to achieve simpler forms.

ETHICS IN ACTION

Certain types of fish and shellfish, such as sword-fish and shark, are known to contain high levels of mercury. The FDA and EPA publish guidelines on the recommended consumption per week for various types of fish to protect public health, particularly the health of children and pregnant women. There has been a recent resurgence of media attention on the potentially high levels of mercury in tuna. A few years ago the state of California tried unsuccessfully to include mercury warnings on canned tuna labels. More recently, the *New York Times* (January 23, 2008) published a study that showed the levels of mercury in high-grade sushi tuna are above FDA recommended cutoffs. This prompted James Halibut, director for the U.S. Pacific Coast Tuna Association, to check on the progress being made by researcher Gary Waters. Gary, a biologist at a regional university, is examining factors that affect mercury levels in tuna. His work is being funded, in part, by a grant from the association. Since most of the tuna caught off the U.S. Pacific Coast is relatively small in size, James told Gary that he is particularly interested in the relationship between the weight of tuna and mercury levels. Gary had found a significant linear relationship between mercury levels and tuna weight, but it was not very strong. Moreover, when he analyzed the residuals from the regression with tuna weight, he discovered an influential outlier. In fact, Gary had performed the regression analysis twice, including and then excluding this point. Omitting this point further reduced the strength of the relationship. Feeling a bit pressured by James, Gary decided to only discuss with him the regression results including this point.

ETHICAL ISSUE *An influential outlier needs to be examined further; including it to make a linear relationship appear stronger than it would otherwise be is unethical. Related to Item H, ASA Ethical Guidelines.*

ETHICAL SOLUTION *The results, even though not as favorable as expected by the funding agency, need to be discussed honestly. Gary must disclose the outlier and its effects on the relationship. James should not place any pressure on the researcher.*

What have we learned?

We've learned that there are many ways in which a data set may be unsuitable for a regression analysis.

- Watch out for more than one group hiding in your regression analysis. If you find subsets of the data that behave differently, consider fitting a different regression model to each subset.

- The **Linearity Condition** says that the relationship should be reasonably straight to fit a regression. Paradoxically, it may be easier to see that the relationship is not straight *after* you fit the regression and examine the residuals.

- The **Outlier Condition** refers to two ways in which cases can be extraordinary. They can have large residuals or high leverage (or, of course, both). Cases with either kind of extraordinary behavior can influence the regression model significantly.

Terms

Extrapolation Although linear models provide an easy way to predict values of y for a given value of x, it is unsafe to predict for values of x far from the ones used to find the linear model equation. Be cautious when extrapolating.

Influential If omitting a point from the data changes the regression model substantially, that point is considered influential.

Leverage Data points whose x-values are far from the mean of x are said to exert leverage on a linear model. High-leverage points pull the line close to them, so they can have a large effect on the line, sometimes completely determining the slope and intercept. Points with high enough leverage can have deceptively small residuals.

Outlier
Any data point that stands away from the regression line by having a large residual is called an outlier.

Transformation (or Re-expression)
A function—typically a simple power or root—applied to the values of a quantitative variable to make its distribution more symmetric and/or to simplify its relationship with other variables.

Skills

 PLAN

- Understand that we cannot fit linear models or use linear regression if the underlying relationship between the variables is not itself linear.

- Understand that data used to find a model must be homogeneous. Look for subgroups in data before you find a regression, and analyze each separately.

- Know the danger of extrapolating beyond the range of the *x*-values used to find the linear model, especially when the extrapolation tries to predict into the future.

- Understand that points can be unusual by having a large residual or by having high leverage.

- Understand that an influential point can change the slope and intercept of the regression line.

- Be able to identify variables that might benefit from a re-expression to make them more symmetric, equalize their spread across groups, or make them more nearly linear when plotted against another variable.

 DO

- Know how to look for high-leverage and influential points by examining a scatterplot of the data. Know how to look for points with large residuals by examining a scatterplot of the residuals against the predicted values or against the *x*-variable. Understand how fitting a regression line with and without influential points can add to understanding of the regression model.

- Know how to look for high leverage points by examining the distribution of the *x*-values or by recognizing them in a scatterplot of the data, and understand how they can affect a linear model.

- Be alert to subgroups in your data.

- Know how to search for an apt re-expression from the Ladder of Powers, moving up and down the ladder to achieve the best improvement in the form of the variable and its relationship with other variables.

 REPORT

- Include diagnostic information such as plots of residuals and leverages as part of your report of a regression.

- Report any high-leverage points.

- Report any outliers. Consider reporting analyses with and without outliers included to assess their influence on the regression.

- Include appropriate cautions about extrapolation when reporting predictions from a linear model.

- Be able to describe a model that includes re-expressed variables.

Technology Help

Most statistics technology offers simple ways to check whether your data satisfy the conditions for regression and to re-express the data when that is called for. We have already seen that these programs can make a simple scatterplot. They can also help us check the conditions by plotting residuals. Most statistics packages offer a way to re-express and compute with variables. Some packages permit you to specify the power of a re-expression with a slider or other moveable control, possibly while watching the consequences of the re-expression on a plot or analysis. This is an effective way to find a good re-expression.

EXCEL

The Data Analysis add-in for Excel includes a Regression command.

The dialog box it shows offers to make plots of residuals. Excel is an excellent place to re-expres data. Just use Excel's built-in functions as you would for any calculation. Changing a value in the original column will change the re-expressed value.

Comments

Do not use the Normal probability plot offered in the regression dialog. It is not what it claims to be and is wrong.

MINITAB

From the **Stat** menu, choose **Regression.** From the Regression submenu, select **Regression** again. In the Regression dialog, enter the response variable name in the "Response" box and the predictor variable name in the "Predictor" box. To specify saved results, in the Regression dialog, click **Storage.** Check "Residuals" and "Fits." Click **OK.** To specify displays, in the Regression dialog, click **Graphs.** Under "Residual Plots," select "Individual plots" and check "Residuals versus fits." Click **OK.**

Now back in the Regression dialog, click **OK.** Minitab computes the regression and the requested saved values and graphs.

To re-express a variable in MINITAB, choose **Calculator** from the **Calc** menu. In the Calculator dialog, specify a name for the new re-expressed variable. Use the **Functions List,** the calculator buttons, and the **Variables list** box to build the expression. Click **OK.**

SPSS

From the **Analyze** menu, choose **Regression.** From the Regression submenu, choose **Linear.** After assigning variables to their roles in the regression, click the "**Plots . . .**" button.

In the Plots dialog, you can specify a Normal probability plot of residuals and scatterplots of various versions of standardized residuals and predicted values.

To re-express a variable in SPSS, choose **Compute** from the **Transform** menu. Enter a name in the Target Variable field. Use the calculator and Function List to build the expression.

Move a variable to be re-expressed from the source list to the Numeric Expression field. Click the **OK** button.

Comments

A plot of ***ZRESID** against ***PRED** will look most like the residual plots we've discussed. SPSS standardizes the residuals by dividing by their standard deviation. (There's no need to subtract their mean; it must be zero.) The standardization doesn't affect the scatterplot.

JMP

From the **Analyze** menu, choose **Fit Y by X**. Select **Fit Line.** Under Linear Fit, select **Plot Residuals.** You can also choose to **Save Residuals.**
Subsequently, from the **Distribution** menu, choose **Normal quantile plot** or **histogram** for the residuals.
To re-express a variable in JMP, double-click to the right of the last column of data to create a new column. Name the new column and select it. Choose **Formula** from the **Cols**

menu. In the Formula dialog, choose the transformation and variable that you wish to assign to the new column. Click the **OK** button.
JMP places the re-expressed data in the new column.

Comments

The log and square root re-expressions are found in the **Transcendental** menu of functions in the formula dialog.

DATA DESK

Click on the **HyperView** menu on the **Regression** output table. A menu drops down to offer scatterplots of residuals against predicted values, Normal probability plots of residuals, or just the ability to save the residuals and predicted values.
Click on the name of a predictor in the regression table to be offered a scatterplot of the residuals against that predictor.
To re-express a variable in Data Desk, select the variable and choose the function to re-express it from the **Manip > Transform** menu. Square root, log, reciprocal, and reciprocal root are immediately available. For others, make a derived variable and type the function. Data Desk makes a new derived variable that holds the re-expressed values. Any value

changed in the original variable will immediately be re-expressed in the derived variable.

Comments

If you change any of the variables in the regression analysis, Data Desk will offer to update the plots of residuals.
An alternative way to re-express a variables is to select it and choose **Manip > Transform > Dynamic > Box-Cox** to generate a continuously changeable variable and a slider that specifies the power. Set plots to **Automatic Update** in their HyperView menus and watch them change dynamically as you drag the slider.

Mini Case Study Projects

Gross Domestic Product

The Gross Domestic Product (GDP) per capita is a widely used measure of a country's (or state's) economy. It is defined as the total market value of all goods and services produced within a country (or state) in a specified period of time. The most common computation of GDP includes five items: consumption, gross investment, government spending, exports, and imports (which negatively impact the total). The Census Bureau reports the GDP for each state in the United States quarterly. The government also reports annual personal income totals (seasonally adjusted in $millions) by state and each state's population. Let's examine how personal income is related to GDP at the state level. Use the data in the file **ch17_MCSP_GDP** to investigate the relationship between GDP and personal income.

Find a model to predict *Personal Income* from *GDP*. Write a short report detailing what you find. Be sure to include appropriate plots, look for influential points, and consider transforming either or both variables.

Repeat the analysis after dividing both variables by state population in 2005 to create per capita versions of the variables. Discuss which regression you think best helps to describe how personal income and GDP are related. Be sure to examine the residuals and discuss the regression assumptions.

Energy Sources

Renewable sources of energy are of growing importance in the economy. The U.S. government (www.stat-usa.gov) reports the amount of renewable energy generated (in thousands of kilowatt-hours) in each of the states for each of several renewable sources. Some states do not have reports, so there are missing values in the data. The data for 2004 are in the file **ch17_MCSP_Alternative_Energy**.

Consider the relationship of hydroelectric power with energy from wind. Find a model for this relationship. Be sure to deal with any extraordinary or influential points. You may want to transform one or both variables. Discuss your residual analysis and the regression assumptions.

Now graph the relationship between (re-expressed) hydroelectric power and wind-generated power. Locate subgroups of states within this plot and discuss how they differ. Do you think a single model for the relationship of these renewable sources is appropriate? Summarize your conclusions in a report.

EXERCISES

T 1. Marriage age 2003. Weddings are one of the fastest growing businesses; about $40 billion is spent on weddings in the United States each year. But demographics may be changing, and this could affect wedding retailers' marketing plans. Is there evidence that the age at which women get married has changed over the past 100 years? The scatterplot shows the trend in age at first marriage for American women. (www.census.gov)

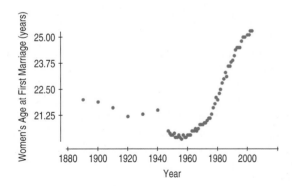

a) Do you think there is a clear pattern? Describe the trend.
b) Is the association strong?
c) Is the correlation high? Explain.
d) Do you think a linear model is appropriate for these data? Explain.

T 2. Smoking 2004. Even with campaigns to reduce smoking, Americans still consume more than four packs of cigarettes per month per adult. (ssdc.ucsd.edu/tobacco/sales/) The Centers for Disease Control and Prevention track cigarette smoking in the United States. How has the percentage of people who smoke changed since the danger became clear during the last half of the 20th century? The scatterplot shows percentages of smokers among men 18–24 years of age, as estimated by surveys, from 1965 through 2004. (www.cdc.gov/nchs/)

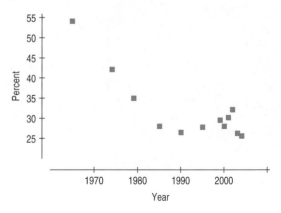

a) Do you think there is a clear pattern? Describe the trend.
b) Is the association strong?
c) Is a linear model appropriate for these data? Explain.

3. Human Development Index. The United Nations Development Programme (UNDP) collects data in the developing world to help countries solve global and national development challenges. In the UNDP annual Human Development Report, you can find data on over 100 variables for each of 177 countries worldwide. One summary measure used by the agency is the Human Development Index (HDI), which attempts to summarize in a single number the progress in health, education, and economics of a country. In 2006, the HDI was as high as 0.965 for Norway and as low as 0.331 for Niger. The gross domestic product per capita (GDPPC), by contrast, is often used to summarize the *overall* economic strength of a country. Is the HDI related to the GDPPC? Here is a scatterplot of *HDI* against *GDPPC*.

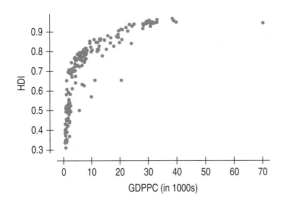

a) Explain why fitting a linear model to these data would be misleading.

b) If you fit a linear model to the data, what do you think a scatterplot of residuals versus predicted *HDI* will look like?

c) There is an outlier (Luxembourg) with a *GDPPC* of around $70,000. Will setting this point aside improve the model substantially? Explain.

4. HDI, part 2. The United Nations Development Programme (UNDP) uses the Human Development Index (HDI) in an attempt to summarize in one number the progress in health, education, and economics of a country. The number of cell phone subscribers per 1000 people is positively associated with economic progress in a country. Can the number of cell phone subscribers be used to predict the HDI? Here is a scatterplot of HDI against cell phone subscribers:

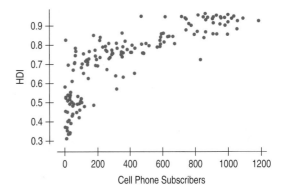

a) Explain why fitting a linear model to these data might be misleading.

b) If you fit a linear model to the data, what do you think a scatterplot of residuals versus predicted *HDI* will look like?

5. Good model? In justifying his choice of a model, a consultant says "I know this is the correct model because $R^2 = 99.4\%$."

a) Is this reasoning correct? Explain.

b) Does this model allow the consultant to make accurate predictions? Explain.

6. Bad model? An intern who has created a linear model is disappointed to find that her R^2 value is a very low 13%.

a) Does this mean that a linear model is not appropriate? Explain.

b) Does this model allow the intern to make accurate predictions? Explain.

7. Movie dramas. Here's a scatterplot of the production budgets (in millions of dollars) vs. the running time (in minutes) for major release movies in 2005. Dramas are plotted in red and all other genres are plotted in blue. A separate least squares regression line has been fitted to each group. For the following questions, just examine the plot:

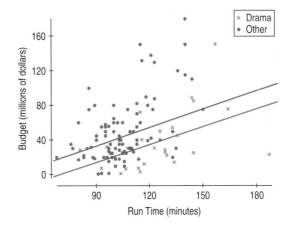

a) What are the units for the slopes of these lines?

b) In what way are dramas and other movies similar with respect to this relationship?

c) In what way are dramas different from other genres of movies with respect to this relationship?

8. Movie ratings. Does the cost of making a movie depend on its audience? Here's a scatterplot of the same data we examined in Exercise 7. Movies with an R rating are colored blue, those with a PG-13 rating are purple, and those with a PG rating are green. Regression lines have been found for each group. (The red points are G-rated, but there were too few to fit a line reliably.)

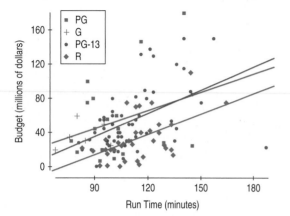

a) In what ways is the relationship between run times and budgets similar for the three different ratings groups?
b) How do the costs of R-rated movies differ from those of PG-13 and PG rated movies? Discuss both the slopes and the intercepts.
c) The film *King Kong*, with a run time of 187 minutes, is the purple point sitting at the lower right. If it were omitted from this analysis, how might that change your conclusions about PG-13 movies?

9. Oakland passengers. Much attention has been paid to the challenges faced by the airline industry. Patterns in customer demand are an important variable to watch. The scatterplot below shows the number of passengers departing from Oakland (CA) airport month by month since the start of 1997. Time is shown as years since 1990, with fractional years used to represent each month. (Thus, June of 1997 is 7.5—halfway through the 7th year after 1990.) www.oaklandairport.com

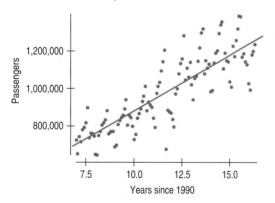

Here's a regression and the residuals plotted against *Years since 1990*:

Dependent variable is: Passengers
R-squared = 71.1% s = 104330

Variable	Coeff
Constant	282584
Year-1990	59704.4

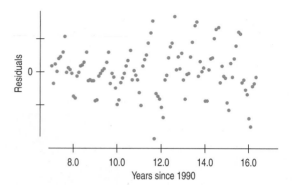

a) Interpret the slope and intercept of the regression model.
b) What does the value of R^2 say about how successful the model is?
c) Interpret s_e in this context.
d) Compute the Durbin-Watson statistic and comment.
e) Would you use this model to predict the numbers of passengers in 2010 (*YearsSince1990* = 20)? Explain.
f) There's a point near the middle of this time span with a large negative residual. Can you explain this outlier?

10. Tracking hurricanes. Like many businesses, The National Hurricane Center also participates in a program to improve the quality of data and predictions by government agencies. They report their errors in predicting the path of hurricanes. The following scatterplot shows the trend in 48-hour tracking errors since 1970. (www.nhc.noaa.gov)

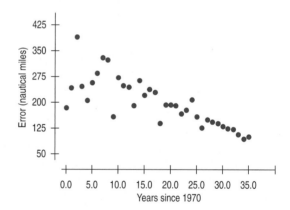

Dependent variable is: Error
R-squared = 63.0% s = 42.87

Variable	Coeff
Intercept	292.089
Years-1970	−5.22924

a) Interpret the slope and intercept of the regression model.
b) Interpret s_e in this context.
c) The Center had a stated goal of achieving an average tracking error of 125 nautical miles in 2009. Will they make it? Explain.
d) Compute the Durbin-Watson statistic and comment.
e) What if their goal were an average tracking error of 90 nautical miles?
f) What cautions would you state about your conclusion?

11. Unusual points. Each of the four scatterplots a–d that follow shows a cluster of points and one "stray" point. For each, answer questions 1–4:

1) In what way is the point unusual? Does it have high leverage, a large residual, or both?

2) Do you think that point is an influential point?

3) If that point were removed from the data, would the correlation become stronger or weaker? Explain.

4) If that point were removed from the data, would the slope of the regression line increase, decrease, or remain the same? Explain.

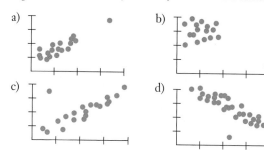

12. More unusual points. Each of the following scatterplots a–d shows a cluster of points and one "stray" point. For each, answer questions 1–4:

1) In what way is the point unusual? Does it have high leverage, a large residual, or both?

2) Do you think that point is an influential point?

3) If that point were removed from the data, would the correlation become stronger or weaker? Explain.

4) If that point were removed from the data, would the slope of the regression line increase, decrease, or remain the same? Explain.

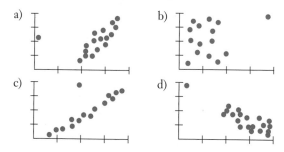

13. The extra point. The scatterplot shows five blue data points at the left. Not surprisingly, the correlation for these points is $r = 0$. Suppose *one* additional data point is added at one of the five positions suggested below in green. Match each point (a–e) with the correct new correlation from the list given.

1) −0.90

2) −0.40

3) 0.00

4) 0.05

5) 0.75

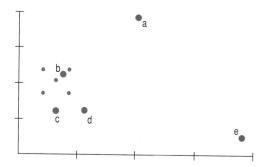

14. The extra point, part 2. The original five points in Exercise 13 produce a regression line with slope 0. Match each of the green points (a–e) with the slope of the line after that one point is added:

1) −0.45

2) −0.30

3) 0.00

4) 0.05

5) 0.85

15. What's the cause? A researcher gathering data for a pharmaceutical firm measures blood pressure and the percentage of body fat for several adult males and finds a strong positive association. Describe three different possible cause-and-effect relationships that might be present.

16. What's the effect? Published reports about violence in computer games have become a concern to developers and distributors of these games. One firm commissioned a study of violent behavior in elementary-school children. The researcher asked the children's parents how much time each child spent playing computer games and had their teachers rate each child's level of aggressiveness when playing with other children. The researcher found a moderately strong positive correlation between computer game time and aggressiveness score. But does this mean that playing computer games increases aggression in children? Describe three different possible cause-and-effect explanations for this relationship.

17. Heating cost. Small businesses must track every expense. A flower shop owner tracked her costs for heating and related it to the average daily Fahrenheit temperature, finding the model $\widehat{Cost} = 133 - 2.13\, Temp$. The residuals plot for her data is shown.

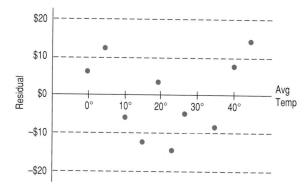

a) Interpret the slope of the line in this context.
b) Interpret the *y*-intercept of the line in this context.
c) During months when the temperature stays around freezing, would you expect cost predictions based on this model to be accurate, too low, or too high? Explain.
d) What heating cost does the model predict for a month that averages 10°?
e) During one of the months on which the model was based, the temperature did average 10°. What were the actual heating costs for that month?
f) Do you think the home owner should use this model? Explain.
g) Would this model be more successful if the temperature were expressed in degrees Celsius? Explain.

18. Fuel economy. How does the speed at which a car drives affect fuel economy? Owners of a taxi fleet, watching their bottom line sink beneath fuel costs, hired a research firm to tell them the optimal speed for their taxis to drive. Researchers drove a compact car for 200 miles at speeds ranging from 35 to 75 miles per hour. From their data, they created the model $\overline{Fuel\ Efficiency} = 32 - 0.1\ Speed$ and created this residual plot:

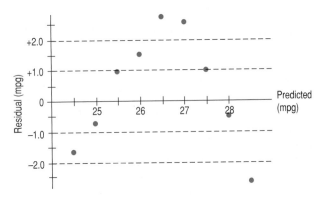

a) Interpret the slope of this line in context.
b) Explain why it's silly to attach any meaning to the *y*-intercept.
c) When this model predicts high *Fuel Efficiency*, what can you say about those predictions?
d) What *Fuel Efficiency* does the model predict when the car is driven at 50 mph?
e) What was the actual *Fuel Efficiency* when the car was driven at 45 mph?
f) Do you think there appears to be a strong association between *Speed* and *Fuel Efficiency*? Explain.
g) Do you think this is the appropriate model for that association? Explain.

T 19. Interest rates. Here's a plot showing the federal rate on 3-month Treasury bills from 1950 to 1980, and a regression model fit to the relationship between the *Rate* (in %) and *Years since 1950*. (www.gpoaccess.gov/eop/)

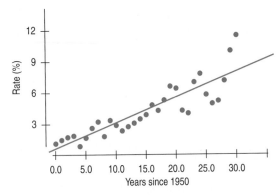

Dependent variable is: Rate
R-squared = 77.4%   s = 1.239

Variable	Coeff
Intercept	0.640282
Year-1950	0.247637

a) What is the correlation between *Rate* and *Year*?
b) Interpret the slope and intercept.
c) What does this model predict for the interest rate in the year 2000?
d) Compute the Durbin-Watson statistic and comment.
e) Would you expect this prediction to have been accurate? Explain.

T 20. Ages of couples 2003. In Exercise 1 we looked at the age at which women married as one of the variables considered by those selling wedding services. Another variable of concern is the *difference* in age of the two partners. The graph shows the ages of both men and women at first marriage. (www.census.gov)

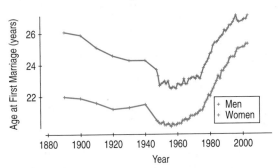

Clearly, the pattern for men is similar to the pattern for women. But are the two lines getting closer together?

On the next page is a timeplot showing the *difference* in average age (men's age − women's age) at first marriage, the regression analysis, and the associated residuals plot.

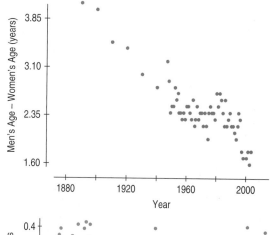

Dependent variable is: Age Difference
R-squared = 75.1%   s = 0.2333

Variable	Coeff
Constant	35.0167
Year	−0.016585

a) What is the correlation between *Age Difference* and *Year*?
b) Interpret the slope of this line.
c) Predict the average age difference in 2015.
d) Compute the Durbin-Watson statistic and comment.
e) Describe reasons why you might not place much faith in that prediction.

21. Interest rates, part 2. In Exercise 19 you investigated the federal rate on 3-month Treasury bills between 1950 and 1980. The scatterplot below shows that the trend changed dramatically after 1980, so we've built a new regression model that includes only the data since 1980 (from $x = 30$ and on in the plot below).

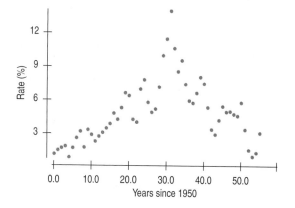

Dependent variable is: Rate
R-squared = 74.5%   s = 1.630

Variable	Coeff
Intercept	21.0688
Year-50	−0.356578

a) How does this model compare to the one in Exercise 19?
b) What does this model estimate the interest rate to have been in 2000? How does this compare to the rate you predicted in Exercise 19?
c) Do you trust this newer predicted value? Explain.
d) Given these two models, what would you predict the interest rate on 3-month Treasury bills will be in 2020?

22. Ages of couples, part 2. Has the trend of decreasing difference in age at first marriage seen in Exercise 20 gotten stronger recently? Here are the scatterplot and residual plot for the data from 1975 through 2003, along with a regression for just those years.

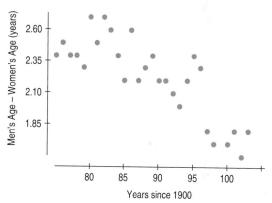

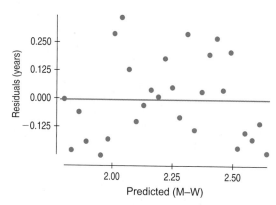

Dependent variable is: Men − Women
R-squared = 65.6%   s = 0.1869

Variable	Coeff
Intercept	4.88424
Year	−0.029959

a) Why is R^2 higher for the first model (in Exercise 20)?
b) Is this linear model appropriate for the post-1975 data? Explain.
c) What does the slope say about marriage ages since 1975?
d) Explain why it's not safe to interpret the *y*-intercept.

23. Colorblind. Although some women are colorblind, this condition is found primarily in men. An advertisement for socks marked so they were easy for someone who was colorblind to match started out "There's a strong correlation between sex and colorblindness." Explain in statistics terms why this isn't a correct statement (whether or not it might be a good ad).

24. New homes. A real estate agent collects data to develop a model that will use the *Size* of a new home (in square feet) to predict its *Sale Price* (in thousands of dollars). Which of these is most likely to be the slope of the regression line: 0.008, 0.08, 0.8, or 8? Explain.

25. Residuals. Suppose you have fit a linear model to some data and now take a look at the residuals. For each of the following possible residuals plots, tell whether you would try a re-expression and, if so, why.

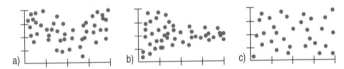

26. Residuals, part 2. Suppose you have fit a linear model to some data and now take a look at the residuals. For each of the following possible residuals plots, tell whether you would try a re-expression and, if so, why.

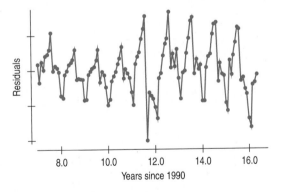

27. Oakland passengers, part 2. In Exercise 9, we created a linear model describing the trend in the number of passengers departing from Oakland (CA) airport each month since the start of 1997. Here's the residual plot, but with lines added to show the order of the values in time:

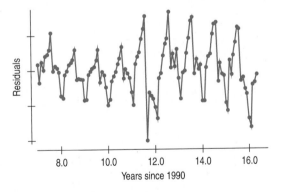

a) Can you account for the pattern shown here?
b) Would a re-expression help us deal with this pattern? Explain.

28. Home Depot sales. The home retail industry has experienced relatively consistent annual growth over the past few decades. Here is a scatterplot of the *Net Sales* ($B) of *The Home Depot* from 1995 through 2004, along with a regression and a time series plot of the residuals.

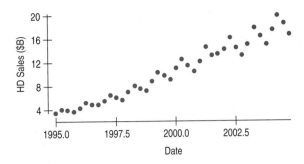

Dependent variable is: HDSales
R-squared = 95.5%
s = 1.044 with 40 − 2 = 38 degrees of freedom

Variable	Coeff	SE(Coeff)	t-ratio	P-value
Constant	−3234.87	114.4	−28.3	≤ 0.0001
Date	1.62283	0.0572	28.4	≤ 0.0001

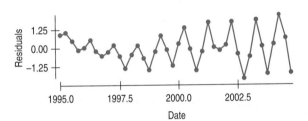

a) What does the R^2 value in the regression mean?
b) What features of the residuals should be noted with regards to this regression?
c) What features of the residuals might be dealt with by a re-expression? Which ones would not be helped by a re-expression?
d) Can you use the regression model to help in your understanding of the growth of this market?

29. Models. For each of the models listed below, predict *y* when *x* = 2.

a) $\hat{y} = 1.2 + 0.8x$
b) $\ln \hat{y} = 1.2 + 0.8x$
c) $\sqrt{\hat{y}} = 1.2 + 0.8x$
d) $\dfrac{1}{\hat{y}} = 1.2 + 0.8x$
e) $\hat{y} = 1.2x^{0.8}$

30. More models. For each of the models listed below, predict *y* when *x* = 2.

a) $\hat{y} = 1.2 + 0.8 \log x$
b) $\log \hat{y} = 1.2 + 0.8x$
c) $\hat{y} = 1.2 + 0.8\sqrt{x}$
d) $\hat{y} = 1.2(0.8^x)$
e) $\hat{y} = 0.8x^2 + 1.2x + 1$

31. Models, again. Find the predicted value of *y*, using each model for *x* = 10.

a) $\hat{y} = 2 + 0.8 \ln x$ b) $\log \hat{y} = 5 - 0.23x$

c) $\dfrac{1}{\sqrt{\hat{y}}} = 17.1 - 1.66x$

32. Models, last time. Find the predicted value of y, using each model when $x = 4$.

a) $\hat{y} = 10 + \sqrt{x}$

b) $\dfrac{1}{y} = 14.5 - 3.45x$

c) $\sqrt{y} = 3.0 + 0.5x$

Ⓣ 33. Lobster industry. According to the Maine Department of Marine Resources, in 2004 more than 72,666,846 pounds of lobster were landed in Maine—a catch worth more than $297,164,000. The lobster fishing industry is carefully controlled and licensed, and facts about it have been recorded for more than a century, so it is an important industry that we can examine in detail. We'll look at annual data (available at www.maine.gov/dmr) from 1950 through 2006.

The value of the annual lobster catch has grown. Here's a scatterplot of the value in millions of dollars over time:

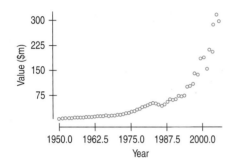

a) Which regression assumptions and conditions appear to be violated according to this plot?

Here's a scatterplot of the *log* of the value:

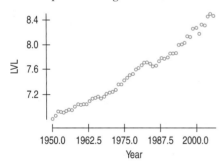

b) Discuss the same assumptions as in part a. Does taking logs make these data suitable for regression?

After performing a regression on the log values, we obtain the following plot of residuals:

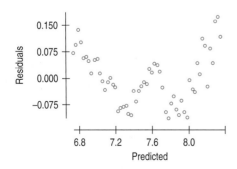

c) Discuss what this plot shows. Would a different transformation be likely to do better than the *log*? Explain.

Ⓣ 34. Lobster traps. Lobster are caught in traps, which are baited and left in the open ocean. Licenses to fish for lobster are limited, there is a small additional fee for each trap in use, and there are limits on the numbers of traps that can be placed in each of seven fishing zones. But those limits have changed over time. Here's a scatterplot of the number of traps per licensed lobster fisher over time:

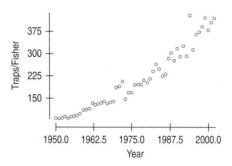

a) Does this plot satisfy the regression assumptions and conditions? Explain.

A regression of *Traps/Fisher* vs. *Year* yields the following plot of residuals:

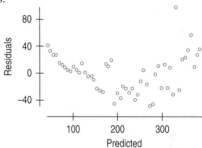

b) What can you see in the plots of residuals that may not have been clear in the original scatterplot of the data?

Ⓣ 35. Lobster value. Here's a regression model relating the *logValue* of the annual Maine lobster catch to the number of licensed lobster *Fishers*:

Dependent variable is: LogValue
R-squared = 18.9%   R-squared (adjusted) = 17.3%
s = 0.4306 with 56 − 2 = 54 degrees of freedom

Variable	Coeff	SE(Coeff)	t-ratio	P-value
Intercept	6.43619	0.3138	20.5	≤ 0.0001
Fishers	1.56021e-4	0.0000	3.54	0.0008

a) The number of licensed lobster fishers has fluctuated over the years between roughly 5000 and 10,000. Recently the number has been just over 7000. But licenses are in demand (and tightly restricted). What does this model predict the value of the catch would be in a year if there were 10,000 licensed fishers? (Take care to interpret the coefficient correctly and to compute the inverse of the log transformation.)

b) Interpret the slope coefficient. Do more fishers cause a higher valued harvest? Suggest alternative explanations.

T **36. Lobster price.** Of course, what matters most to the individual entrepreneur—the licensed commercial lobster fisher—is the price of lobster. Here's an analysis relating that price ($/lb) to the number of traps (millions):

Dependent variable is: Price/lb
R-squared = 93.7%   R-squared (adjusted) = 93.6%
s = 0.2890 with 56 − 2 = 54 degrees of freedom

Variable	Coeff	SE(Coeff)	t-ratio	P-value
Intercept	−0.276454	0.0812	−3.40	0.0013
Traps(M)	1.25210	0.0441	28.4	≤ 0.0001

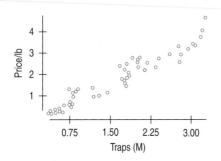

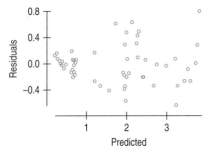

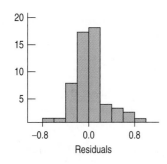

a) Are the assumptions and conditions for regression inference satisfied?
b) What does the coefficient of *Traps* mean in this model? Does it predict that licensing more traps would cause an increase in the price of lobster? Suggest some alternative explanations.

37. GDP. The scatterplot shows the gross domestic product (GDP) of the United States in billions of dollars plotted against years since 1950.

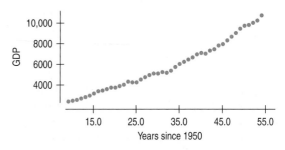

A linear model fit to the relationship looks like this:

Dependent variable is: GDP
R-squared = 97.2%   s = 406.6

Variable	Coeff
Intercept	240.171
Year-1950	177.889

a) Does the value 97.1% suggest that this is a good model? Explain.
b) Here's a scatterplot of the residuals. Now do you think this is a good model for these data? Explain?

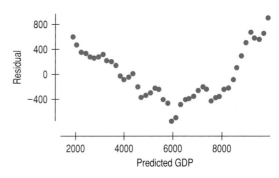

38. Better GDP model? Consider again the post-1950 trend in U.S. GDP we examined in Exercise 35. Here are a regression and residual plot when we use the log of GDP in the model. Is this a better model for GDP? Explain.

Dependent variable is: LogGDP
R-squared = 99.4%   s = 0.0150

Variable	Coeff
Intercept	3.29092
Year-1950	0.013881

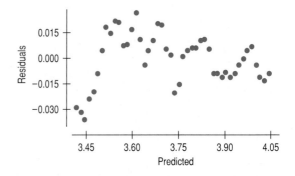

39. Logs (not logarithms). Many professions use tables to determine key quantities. The value of a log is based on the number of board feet of lumber the log may contain. (A board foot is the equivalent of a piece of wood 1 inch thick, 12 inches wide, and 1 foot long. For example, a 2″ × 4″ piece that is 12 feet long contains 8 board feet.) To estimate the amount of lumber in a log, buyers measure the diameter inside the bark at the smaller end. Then they look in a table based on the Doyle Log Scale. The table below shows the estimates for logs 16 feet long.

Diameter of Log	8″	12″	16″	20″	24″	28″
Board Feet	16	64	144	256	400	576

a) What transformation of *Board Feet* makes this relationship linear?
b) How much lumber would you estimate that a log 10 inches in diameter contains?
c) What does this model suggest about logs 36 inches in diameter?

T 40. Life expectancy. Life insurance rates are based on life expectancy values compiled for large demographic groups. But with improvements in medical care and nutrition, life expectancies have been changing. Here is a table from the National Vital Statistics Report that gives the Life Expectancy for white males in the United States every decade during the last century (1 = 1900 to 1910, 2 = 1911 to 1920, etc.). Consider a linear model to predict future increases in life expectancy. Would re-expressing either variable make a better model?

Decade	1	2	3	4	5	6	7	8	9	10
Life exp.	48.6	54.4	59.7	62.1	66.5	67.4	68.0	70.7	72.7	74.9

T 41. OECD GDP. The Organization for Economic Cooperation and Development (OECD) is an organization comprised of thirty countries. To belong, a country must support the principles of representative democracy and a free market economy. How have these countries grown in the decade from 1988 to 1998–2000? Here are the GDP per capita for 24 of the OECD members (both in year 2000 dollars). (www.sba.gov/idc/groups/public/documents/sbahomepage/rs264tot.pdf)

Country	1988 GDP/Capita	1998–2000 GDP/Capita
Australia	18,558	23,713
Austria	25,626	31,192
Belgium	24,204	30,506
Canada	19,349	22,605
Denmark	31,517	38,136
Finland	25,682	31,246
France	24,663	29,744
Germany	27,196	32,256
Greece	10,606	13,181
Ireland	13,050	27,282
Italy	17,339	20,710
Japan	36,301	44,154
Korea	7038	12,844
Mexico	3024	3685
Netherlands	23,159	30,720
New Zealand	15,480	17,979
Norway	28,241	37,934
Portugal	8935	12,756
Spain	12,879	17,197
Sweden	26,634	30,873
Switzerland	43,375	46,330
Turkey	2457	2947
United Kingdom	17,676	22,153
United States	25,324	31,296

Make a model of 1998–2000 GDP/Capita in terms of 1988 GDP/Capita. Plot the residuals and discuss any concerns you may have.

T 42. Orange production. Orange growers know that the larger an orange the higher the price it will bring. But as the number of oranges on a tree increases, the fruit tends to be smaller. Here's a table of that relationship. Create a model for this relationship, and express any concerns you may have.

Number of Oranges/Tree	Average Weight/Fruit (lb)
50	0.60
100	0.58
150	0.56
200	0.55
250	0.53
300	0.52
350	0.50
400	0.49
450	0.48
500	0.46
600	0.44
700	0.42
800	0.40
900	0.38

T 43. Human Development Index, again. In Exercise 3 we saw that the United Nations Development Programme (UNDP) uses the Human Development Index (HDI) in an attempt to summarize the progress in health, education, and economics of a country with one number. The gross domestic product per capita (GDPPC) attempts to summarize the wealth produced by a country in one number. Here is a plot of *GDPPC* against *HDI* for 172 countries throughout the world:

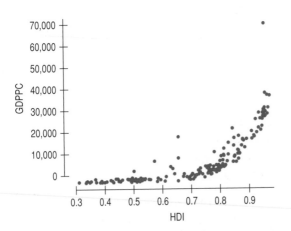

GDPPC is measured in dollars. Incomes and other economic measures tend to be highly right skewed. Taking logs often makes the distribution more unimodal and symmetric. Compare the histogram of *GDPPC* to the histogram of *log(GDPPC)*.

T 44. Human Development Index, last time. In Exercise 41 we examined the relationship between *log(GDPPC)* and *HDI* for 172 countries. The number of cell phone subscribers (per 1000 people) is also positively associated with economic progress in a country. Here's a scatterplot of *CellPhones* (subscribers per 1000 people) against *HDI* for 154 countries:

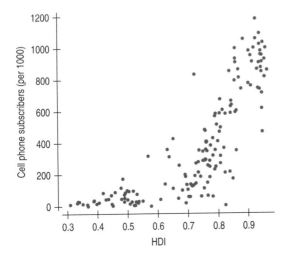

a) *CellPhones* is a count of subscribers (per 1000). What re-expression is often useful for counts? Examine the histogram of *CellPhones* and the histogram of *Cellphones* using the re-expression you suggested. Comment.
b) Use the re-expression in a) for the scatterplot against *HDI*. Comment.
c) Why might you be skeptical of using this relationship to predict the number of cell phone users based on the *HDI*?

T 45. Lobster fishers. How has the number of licensed lobster fishers changed? Here's a plot of the number of *Fishers vs. Year*:

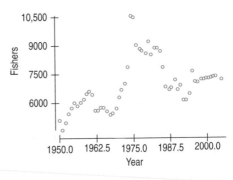

This plot isn't straight. Would a transformation help? If so, which one? If not, why not?

46. Lobster price. How has the price of a lobster changed? Here's a plot tracking the price of lobster per pound in constant year 2000 dollars.

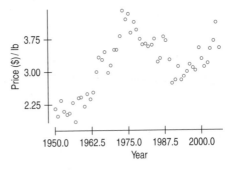

This plot is not straight. Would a transformation help? If so, which one? If not, why not?

 JUST CHECKING ANSWERS
1 Not high-leverage, not influential, large residual
2 High-leverage, not influential, small residual
3 High-leverage, influential, not large residual
4 None
5 Logarithm

CHAPTER 18

Multiple Regression

Zillow.com

Zillow.com is a real estate research site, founded in 2005 by Rich Barton and Lloyd Frink. Both are former Microsoft executives and founders of Expedia.com, the Internet-based travel agency. Zillow collects publicly available data and provides an estimate (called a Zestimate®) of the home's worth. The estimate is based on a model of the data that Zillow has been able to collect on a variety of predictor variables, including the past history of the home's sales, the location of the home, and characteristics of the house such as its size and number of bedrooms and bathrooms.

The site is enormously popular among both potential buyers and sellers of homes. According to Rismedia.com, Zillow is one of the most-visited U.S. real estate sites on the Web, with approximately 5 million unique users each month. These users include more than one-third of all mortgage professionals in the U.S.—or approximately 125,000—in any given month. Additionally, 90 percent of Zillow users are homeowners, and two-thirds are either buying and selling now, or plan to in the near future.

WHO:	Houses
WHAT:	Sale price (2002 dollars) and other facts about the houses
WHEN:	2002–2003
WHERE:	Upstate New York near Saratoga Springs
WHY:	To understand what influences housing prices and how to predict them

How exactly does Zillow figure the worth of a house? According to the Zillow.com site, "We compute this figure by taking zillions of data points—much of this data is public—and entering them into a formula. This formula is built using what our statisticians call 'a proprietary algorithm'—big words for 'secret formula.' When our statisticians developed the model to determine home values, they explored how homes in certain areas were similar (i.e., number of bedrooms and baths, and a myriad of other details) and then looked at the relationships between actual sale prices and those home details." These relationships form a pattern, and they use that pattern to develop a model to estimate a market value for a home. In other words, the Zillow statisticians use a model, most likely a regression model, to predict home value from the characteristics of the house. We've seen how to predict a response variable based on a single predictor. That's been useful, but the types of business decisions we'll want to make are often too complex for simple regression.[1] In this chapter, we'll expand the power of the regression model to take into account many predictor variables into what's called a multiple regression model. With our understanding of simple regression as a base, getting to multiple regression isn't a big step, but it's an important and worthwhile one. Multiple regression is probably the most powerful and widely used statistical tool today.

As anyone who's ever looked at house prices knows, house prices depend on the local market. To control for that, we will restrict our attention to a single market. We have a random sample of 1057 home sales from the public records of sales in Upstate New York, in the region around the city of Saratoga Springs. The first thing often mentioned in describing a house for sale is the number of bedrooms. Let's start with just one predictor variable. Can we use *Bedrooms* to predict home *Price*?

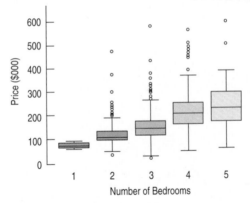

Figure 18.1 *Side-by-side boxplots of* Price *against* Bedrooms *show that price increases, on average, with more bedrooms.*

The number of *Bedrooms* is a quantitative variable, but it holds only a few values (from 1 to 5 in this data set). So a scatterplot may not be the best way to examine the relationship between *Bedrooms* and *Price*. In fact, at each value for *Bedrooms* there is a whole distribution of prices. Side-by-side boxplots of *Price* against *Bedrooms* (Figure 18.1) show a general increase in price with more bedrooms, and an approximately linear growth.

Figure 18.1 also shows a clearly increasing spread from left to right, violating the Equal Spread Condition, and that's a possible sign of trouble. For now, we'll

---

[1] When we need to note the difference, a regression with a single predictor is called a **simple regression**.

proceed cautiously. We'll fit the regression model, but we will be cautious about using inference methods for the model. Later we'll add more variables to increase the power and usefulness of the model.

The output from a linear regression model of *Price* on *Bedrooms* shows:

```
Response variable: Price

R² = 21.4%
s = 68432.21 with 1057 − 2 = 1055 degrees of freedom

Variable     Coeff       SE(Coeff)   t-ratio   P-value
Intercept    14349.48    9297.69     1.54      0.1230
Bedrooms     48218.91    2843.88     16.96     ≤ 0.0001
```

| **Table 18.1** *Linear regression of Price on Bedrooms.*

Apparently, just knowing the number of bedrooms gives us some useful information about the sale price. The model tells us that, on average, we'd expect the price to increase by almost $50,000 for each additional bedroom in the house, as we can see from the slope value of $48,219.90:

$$\widehat{Price} = 14349.48 + 48218.91 \times Bedrooms.$$

Even though the model does tell us something, notice that the R^2 for this regression is only 21.4%. The variation in the number of bedrooms accounts for only 21% of the variation in house prices. Perhaps some of the other facts about these houses can account for portions of the remaining variation.

18.1 The Multiple Regression Model

For simple regression, we wrote the predicted values in terms of one predictor variable:

$$\hat{y} = b_0 + b_1 x.$$

To include more predictors in the model, we just write the regression model with more predictor variables. The resulting **multiple regression** looks like this:

$$\hat{y} = b_0 + b_1 x_1 + b_2 x_2 + \ldots + b_k x_k$$

where b_0 is still the intercept and each b_k is the estimated coefficient of its corresponding predictor x_k. Although the model doesn't look much more complicated than a simple regression, it isn't practical to determine a multiple regression by hand. This is a job for a statistics program on a computer. Remember that for simple regression, we found the coefficients for the model using the least squares solution, the one whose coefficients made the sum of the squared residuals as small as possible. For multiple regression, a statistics package does the same thing and can find the coefficients of the least squares model easily.

If you know how to find the regression of *Price* on *Bedrooms* using a statistics package, you can probably just add another variable to the list of predictors in your program to compute a multiple regression. A multiple regression of *Price* on the two variables *Bedrooms* and *Living Area* generates a multiple regression table like this one.

```
Response variable: Price

R² = 57.8%
s = 50142.4 with 1057 − 3 = 1054 degrees of freedom

Variable        Coeff       SE(Coeff)    t-ratio    P-value
Intercept       20986.09    6816.3       3.08       0.0021
Bedrooms        −7483.10    2783.5       −2.69      0.0073
Living Area     93.84       3.11         30.18      ≤ 0.0001
```

Table 18.2 *Multiple regression output for the linear model predicting* Price *from* Bedrooms *and* Living Area.

You should recognize most of the numbers in this table, and most of them mean what you expect them to. The value of R^2 for a regression on two variables gives the fraction of the variability of *Price* accounted for by both predictor variables together. With *Bedrooms* alone predicting *Price*, the R^2 value was 22.1%, but this model accounts for 57.8% of the variability in *Price*. We shouldn't be surprised that the variability explained by the model has gone up. It was for this reason—the hope of accounting for some of that leftover variability—that we tried a second predictor. We also shouldn't be surprised that the size of the house, as measured by *Living Area*, also contributes to a good prediction of house prices. Collecting the coefficients of the multiple regression of *Price* on *Bedrooms* and *Living Area* from Table 18.2, we can write the estimated regression as:

$$\widehat{Price} = 20,986.09 - 7,483.10 Bedrooms + 93.84 Living\ Area.$$

As before, we define the residuals as:

$$e = y - \hat{y}.$$

The standard deviation of the residuals is still denoted as s (or also sometimes as s_e as in simple regression—for the same reason—to distinguish it from the standard deviation, s_y, of y). The degrees of freedom calculation comes right from our definition. The degrees of freedom is the number of observations ($n = 1057$) minus one for each coefficient estimated:

$$df = n - k - 1,$$

where k is the number of predictor variables and n is the number of cases. For this model, we subtract 3 (the two coefficients and the intercept). To find the standard deviation of the residuals, we use that number of degrees of freedom in the denominator:

$$s_e = \sqrt{\frac{\sum(y - \hat{y})^2}{n - k - 1}}.$$

For each predictor, the regression output shows a coefficient, its standard error, a *t*-ratio, and the corresponding P-value. As with simple regression, the *t*-ratio measures how many standard errors the coefficient is away from 0. Using a Student's *t*-model, we can use its P-value to test the null hypothesis that the true value of the coefficient is 0.

What's different? With so much of the multiple regression looking just like simple regression, why devote an entire chapter to the subject?

There are several answers to this question. First, and most important, is that the meaning of the coefficients in the regression model has changed in a subtle, but important, way. Because that change is not obvious, multiple regression coefficients are often misinterpreted. We'll show some examples to explain this change in meaning.

Second, multiple regression is an extraordinarily versatile model, underlying many widely used statistics methods. A sound understanding of the multiple regression model will help you to understand these other applications as well.

Third, multiple regression offers you a first glimpse into statistical models that use more than two quantitative variables. The real world is complex. Simple models of the kind we've shown so far are a great start, but they're not detailed enough to be useful for understanding, predicting, and making business decisions in many real-world situations. Models that use several variables can be a big step toward realistic and useful modeling of complex phenomena and relationships.

18.2 Interpreting Multiple Regression Coefficients

It makes sense that both the number of bedrooms and the size of the living area would influence the price of a house. We'd expect both variables to have a positive effect on price—houses with more bedrooms typically sell for more money, as do larger houses. But look at the coefficient for *Bedrooms* in the multiple regression equation. It's negative: −7483.09. How can it be that the coefficient of *Bedrooms* in the multiple regression is negative? And not just slightly negative, its *t*-ratio is large enough for us to be quite confident that the true value is really negative. Yet from Table 18.1, we saw the coefficient was equally clearly positive when *Bedrooms* was the sole predictor in the model (see Figure 18.2).

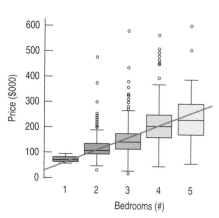

Figure 18.2 *The slope of Bedrooms is positive. For each additional bedroom, we would predict an additional $48,000 in the price of a house from the simple regression model of Table 18.1.*

The explanation of this apparent paradox is that in a multiple regression, coefficients have a more subtle meaning. Each coefficient takes into account the other predictor(s) in the model.

Think about a group of houses of about the same size. For the *same size* living area, a house with more bedrooms is likely to have smaller rooms. That might actually make it *less* valuable. To see this in the data, let's look at a group of similarly sized homes from 2500 to 3000 square feet of living area, and examine the relationship between *Bedrooms* and *Price* just for houses in this size range (see Figure 18.3).

For houses with between 2500 and 3000 square feet of living area, it appears that homes with *fewer* bedrooms have a higher price, on average, than those with more bedrooms. When we think about houses in terms of *both* variables, we can see that this makes sense. A 2500 square foot house with five bedrooms would have either relatively small, cramped bedrooms or not much common living space. The same size house with only three bedrooms could have larger, more

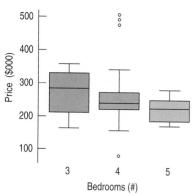

Figure 18.3 *For the 96 houses with Living Area between 2500 and 3000 square feet, the slope of Price on Bedrooms is negative. For each additional bedroom, restricting data to homes of this size, we would predict that the house's Price was about $17,800 lower.*

JUST CHECKING

Body fat percentage is an important health indicator, but it is difficult to measure accurately. One way to do so is to take an MRI (magnetic resonance image) at a cost of about $1000 per image. Insurance companies want to know if body fat percentage can be estimated from easier to measure characteristics such as *Height* and *Weight*. A scatterplot of *Percent Body Fat* against *Height* shows no pattern, and the correlation is −0.03 and is not statistically significant. A multiple regression using *Height (inches)*, *Age (years)*, and *Weight (pounds)* finds the following model:

```
             Coeff        SE(Coeff)      t-ratio      P-value
Intercept    57.27217     10.39897        5.507       < 0.0001
Height       -1.27416      0.15801       -8.064       < 0.0001
Weight        0.25366      0.01483       17.110       < 0.0001
Age           0.13732      0.02806        4.895       < 0.0001

s = 5.382 on 246 degrees of freedom
Multiple R-squared: 0.584,
F-statistic: 115.1 on 3 and 246 DF, P-value:  < 0.0001
```

1 Interpret the R^2 of this regression model.

2 Interpret the coefficient of *Age*.

3 How can the coefficient of *Height* have such a small P-value in the multiple regression when the correlation between *Height* and *Percent Body Fat* was not statistically distinguishable from zero?

attractive bedrooms and still have adequate common living space. What the coefficient of *Bedrooms* is saying in the multiple regression is that, after accounting for living area, houses with more bedrooms tend to sell for a *lower* price. In other words, what we saw by *restricting* our attention to homes of a certain size and seeing that additional bedrooms had a negative impact on price was generally true across all sizes. What seems confusing at first is that without taking *Living Area* into account, *Price* tends to go *up* with more bedrooms. But that's because *Living Area* and *Bedrooms* are also related. Multiple regression coefficients must always be interpreted in terms of the other predictors in the model. That can make their interpretation more subtle, more complex, and more challenging than when we had only one predictor. This is also what makes multiple regression so versatile and effective. The interpretations are more sophisticated and more appropriate.

There's a second common pitfall in interpreting coefficients. Be careful not to interpret the coefficients causally. For example, this analysis cannot tell a homeowner how much the price of his home will change if he combines two of his four bedrooms into a new master bedroom. And it can't be used to predict whether adding a 100 square foot child's bedroom onto the house would increase or decrease its value. The model simply reports the relationship between the number of *Bedrooms* and *Living Area* and *Price* for existing houses. As always with regression, we should be careful not to assume causation between the predictor variables and the response.

18.3 Assumptions and Conditions for the Multiple Regression Model

We can write the multiple regression model like this, numbering the predictors arbitrarily (the order doesn't matter), writing betas for the model coefficients (which we will estimate from the data), and including the errors in the model:

$$y = \beta_0 + \beta_1 x_1 + \beta_2 x_2 + \ldots + \beta_k x_k + \varepsilon.$$

The assumptions and conditions for the multiple regression model are nearly the same as for simple regression, but with more variables in the model, we'll have to make a few changes, as described in the following sections.

Linearity Assumption

We are fitting a linear model.[2] For that to be the right kind of model for this analysis, we need to verify an underlying linear relationship. But now we're thinking about several predictors. To confirm that the assumption is reasonable, we'll check the Linearity Condition for *each* of the predictors.

Linearity Condition. Scatterplots of y against each of the predictors are reasonably straight. The scatterplots need not show a strong (or any) slope; we just check to make sure that there isn't a bend or other nonlinearity. For the real estate data, the scatterplot is linear in both *Bedrooms* and *Living Area* as we saw in Chapter 16.

As in simple regression, it's a good idea to check the residual plot for any violations of the linearity condition. We can fit the regression and plot the residuals against the predicted values (Figure 18.4), checking to make sure we don't find patterns—especially bends or other nonlinearities.

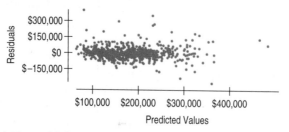

Figure 18.4 *A scatterplot of residuals against the predicted values shows no obvious pattern.*

Independence Assumption

As with simple regression, the errors in the true underlying regression model must be independent of each other. As usual, there's no way to be sure that the Independence Assumption is true, but we should think about how the data were collected to see if that assumption is reasonable. We should check the randomization condition as well.

Randomization Condition. Ideally, the data should arise from a random sample or randomized experiment. Randomization assures us that the data are representative of some identifiable population. If you can't identify the population, you can interpret the regression model as a description of the data you have, but you can't interpret the hypothesis tests at all because such tests are about a regression model for a specific population. Regression methods are often applied to data that were not collected with randomization. Regression models fit to such data may still do a good job of modeling the data at hand, but without some reason to believe that the data are representative of a particular population, you should be reluctant to believe that the model generalizes to other situations.

---

[2] By *linear* we mean that each x appears simply multiplied by its coefficient and added to the model, and that no x appears in an exponent or some other more complicated function. That ensures that as we move along any x-variable, our prediction for y will change at a constant rate (given by the coefficient) if nothing else changes.

We also check the regression residuals for evidence of patterns, trends, or clumping, any of which would suggest a failure of independence. In the special case when one of the *x*-variables is related to time (or *is* itself *Time*), be sure that the residuals do not have a pattern when plotted against that variable. In addition to checking the plot of residuals against the predicted values, we recommend that you check the individual plots of the residuals against each of the explanatory, or *x*, variables in the model. These individual plots can yield important information on necessary transformations, or re-expressions, for the predictor variables.

The real estate data were sampled from a larger set of public records for sales during a limited period of time. The houses were not related in any way, so we can be fairly confident that their measurements are independent.

Equal Variance Assumption

The variability of the errors should be about the same for all values of *each* predictor. To see whether this assumption is valid, we look at scatterplots and check the Equal Spread Condition.

Equal Spread Condition.   The same scatterplot of residuals against the predicted values (Figure 18.4) is a good check of the consistency of the spread. We saw what appeared to be a violation of the equal spread condition when *Price* was plotted against *Bedrooms* (Figure 18.2.) But here in the multiple regression, the problem has dissipated when we look at the residuals. Apparently, much of the tendency of houses with more bedrooms to have greater variability in prices was accounted for in the model when we included *Living Area* as a predictor.

If residual plots show no pattern, if the data are plausibly independent, and if the plots don't thicken, we can feel good about interpreting the regression model. Before we test hypotheses, however, we must check one final assumption: the normality assumption.

Normality Assumption

We assume that the errors around the idealized regression model at any specified values of the *x*-variables follow a Normal model. We need this assumption so that we can use a Student's *t*-model for inference. As with other times when we've used Student's *t*, we'll settle for the residuals satisfying the Nearly Normal Condition. As with means, the assumption is less important as the sample size grows. Our inference methods will work well even when the residuals are moderately skewed, if the sample size is large. If the distribution of residuals is unimodal and symmetric, there is little to worry about.[3]

Nearly Normal Condition.   Because we have only one set of residuals, this is the same set of conditions we had for simple regression. Look at a histogram or Normal probability plot of the residuals.

---

[3] The only procedure that needs strict adherence to Normality of the errors is finding prediction intervals for individuals in multiple regression. Because they are based on Normal probabilities, the errors must closely follow a Normal model.

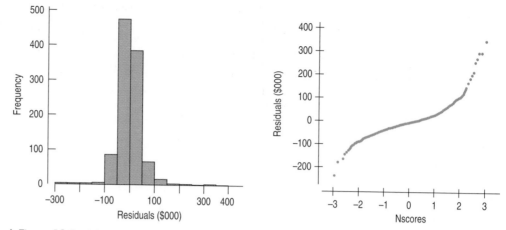

Figure 18.5 *A histogram of the residuals shows a unimodal, symmetric distribution, but the tails seem a bit longer than one would expect from a Normal model. The Normal probability plot confirms that.*

The histogram of residuals in the real estate example certainly looks unimodal and symmetric. The Normal probability plot has some bend on both sides, which indicates that there are more residuals in the tails than Normally distributed data would have. However, as we have said before, the Normality Assumption becomes less important as the sample size grows, and here we have no skewness and more than 1000 cases. (The Central Limit Theorem will help our confidence intervals and tests based on the *t*-statistic with large samples.)

Let's summarize all the checks of conditions that we've made and the order in which we've made them.

1. Check the Linearity Condition with scatterplots of the *y*-variable against each *x*-variable.

2. If the scatterplots are straight enough, fit a multiple regression model to the data. (Otherwise, either stop or consider re-expressing an *x*-variable or the *y*-variable.)

3. Find the residuals and predicted values.

4. Make a scatterplot of the residuals against the predicted values (and ideally against each predictor variable separately). These plots should look pattern-less. Check, in particular, for any bend (which would suggest that the data weren't all that straight after all) and for any thickening. If there's a bend, consider re-expressing the *y* and/or the *x* variables. If the variation in the plot grows from one side to the other, consider re-expressing the *y*-variable. If you re-express a variable, start the model fitting over.

5. Think about how the data were collected. Should they be independent? Was suitable randomization used? Are the data representative of some identifiable population? If the data are measured over time, check for evidence of patterns that might suggest they're not independent by plotting the residuals against time to look for patterns.

6. If the conditions check out this far, feel free to interpret the regression model and use it for prediction.

7. Make a histogram and Normal probability plot of the residuals to check the Nearly Normal Condition. If the sample size is large, the Normality is less important for inference, but always be on the lookout for skewness or outliers.

GUIDED EXAMPLE | *Housing Prices*

Zillow.com attracts millions of users each month who are interested in finding out how much their house is worth. Let's see how well a multiple regression model can do. The variables available include:

Price   The price of the house as sold in 2002

Living Area The size of the living area of the house in square feet

Bedrooms   The number of bedrooms

Bathrooms   The number of bathrooms (a half bath is a toilet and sink only)

Age   Age of the house in years

Fireplaces Number of fireplaces in the house

PLAN

Setup State the objective of the study. Identify the variables.

We want to build a model to predict house prices for a region in Upstate New York. We have data on *Price* ($), *Living Area* (sq ft), *Bedrooms* (#), *Bathrooms* (#), *Fireplaces* (#), and *Age* (in years).

Model Think about the assumptions and check the conditions.

Linearity Condition

To fit a regression model, we first require linearity. Scatterplots (or side-by-side boxplots) of *Price* against all potential predictor variables are shown.

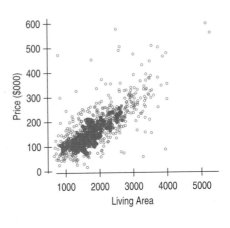

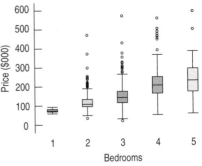

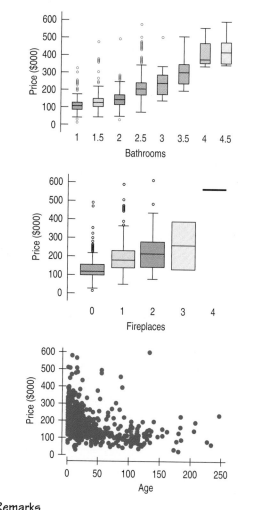

Remarks

There are a few anomalies in the plots that deserve discussion. The plot of *Price* against *Bathrooms* shows a positive relationship, but it is not quite linear. There seem to be two slopes, one from 1–2 bathrooms and then a steeper one from 2–4. For now, we'll proceed cautiously, realizing that any slope we find will average these two. The plot of *Price* against *Fireplaces* shows an outlier—an expensive home with four fireplaces. We tried setting this home aside and running the regression without it, but its influence on the slopes was not great, so we decided to include it in the model. The plot of *Price* against *Age* shows that there may be some curvature. We should be cautious in interpreting the slope, especially for newer homes.

✓ **Independence Assumption.** We can regard the house prices as being independent of one another since they are from a fairly large geographic area.

✓ **Randomization Condition.** These 1057 houses are a random sample of a much larger set.

✓ **Equal Spread Condition.** A scatterplot of residuals *vs.* predicted values shows no evidence of changing spread. There is a group of homes whose residuals are larger (both negative and positive) than the vast majority. This is also seen in the long tails of the histogram of residuals.

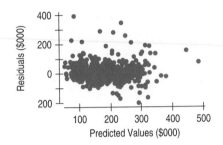

We need the Nearly Normal Condition only if we want to do inference and the sample size is not large. If the sample size is large, we need the distribution to be Normal only if we plan to produce prediction intervals.

✓ **Nearly Normal Condition, Outlier Condition.** The histogram of residuals is unimodal and symmetric, but long tailed. The Normal probability plot supports that.

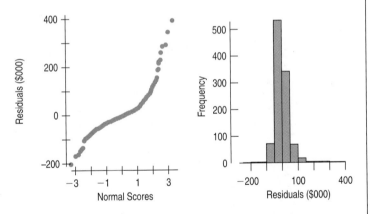

Under these conditions, we can proceed with caution to a multiple regression analysis. We will return to some of our concerns in the discussion.

 DO

Mechanics We always fit multiple regression models with computer software. An output table like this one isn't exactly what any of the major packages produce, but it is enough like all of them to look familiar.

Here is the computer output for the multiple regression, using all five predictors.

	Coeff	SE(Coeff)	t-ratio	P-value
Intercept	15712.702	7311.427	2.149	0.03186
Living Area	73.446	4.009	18.321	< 0.0001
Bedrooms	−6361.311	2749.503	−2.314	0.02088
Bathrooms	19236.678	3669.080	5.243	< 0.0001
Fireplaces	9162.791	3194.233	2.869	0.00421
Age	−142.740	48.276	−2.957	0.00318

Residual standard error: 48615.95 on 1051 degrees of freedom
Multiple R-squared: 0.6049.
F-statistic: 321.8 on 5 and 1051 DF. P-value: < 0.0001

The estimated equation is:

$$\widehat{Price} = 15{,}712.70 + 73.45Living\ Area - 6361.31Bedrooms + 19{,}236.68Bathrooms + 9162.79Fireplaces - 142.74Age$$

All of the P-values are small which indicates that even with five predictors in the model, all are contributing. The R^2 value of 60.49% indicates that more than 60% of the overall variation in house prices has been accounted for by this model. The residual standard error of $48,620 gives us a rough indication that we can predict the price of a home to within about 2 × $48,620 = $97,240. If that's close enough to be useful, then our model is potentially useful as a price guide.

 **REPORT**

Summary and Conclusions Summarize your results and state any limitations of your model in the context of your original objectives.

MEMO:

Re: Regression Analysis of Home Price Predictions

A regression model of *Price* on *Living Area, Bedrooms, Bathrooms, Fireplaces,* and *Age* accounts for 60.5% of the variation in the price of homes in Upstate New York. A statistical test of each coefficient shows that each one is almost certainly not zero, so each of these variables appears to be a contributor of the price of a house.

This model reflects the common wisdom in real estate about the importance of various aspects of a home. An important variable not included is the location, which every real estate agent knows is crucial to pricing a house. This is ameliorated by the fact that all these houses are in the same general area. However, knowing more specific information about where they are located would almost certainly help the model. The price found from this model is to be used as a starting point for comparing a home with comparable homes in the area.

The model may be improved by re-expressing one or more of the predictors, especially *Age* and *Bathrooms*. We recommend caution in interpreting the slopes across the entire range of these predictors.

18.4 Testing the Multiple Regression Model

There are several hypothesis tests in the multiple regression output, but all of them talk about the same thing. Each is concerned with whether the underlying model parameters (the slopes and intercept) are actually zero. The first of these hypotheses is one we skipped over for simple regression (for reasons that will be clear in a minute).

Now that we have more than one predictor, there's an overall test we should perform before we consider inference for the coefficients. We ask the global question: Is this multiple regression model any good at all? If home prices were set randomly or based on other factors than those we have as predictors, then the best estimate would just be the mean price.

To address the overall question, we'll test the null hypothesis that all the slope coefficients are zero:

$$H_0: \beta_1 = \ldots = \beta_k = 0 \text{ } vs \text{ } H_A: \text{at least one } \beta \neq 0.$$

We can test this hypothesis with an **F-test**. (It's the generalization of the t-test to more than one predictor.) The sampling distribution of the statistic is labeled with the letter F (in honor of Sir Ronald Fisher). The F-distribution has two degrees of freedom, k, the number of predictors, and $n - k - 1$. In our Guided Example, we have $k = 5$ predictors and $n = 1057$ homes, which means that the F-value of 321.8 has 5 and $1057 - 5 - 1 = 1051$ degrees of freedom. The regression output shows that it has a P-value < 0.0001. The null hypothesis is that the regression model predicts no better than the mean. The alternative is that it does. The test is one-sided—bigger F-values mean smaller P-values. If the null hypothesis were true, the F-statistic would be near 1. The F-statistic here is quite large, so we can easily reject the null hypothesis and conclude that the multiple regression model for predicting house prices with these five variables is better than just using the mean.[4]

Once we check the F-test and reject its null hypothesis—and, if we are being careful, *only* if we reject that hypothesis—we can move on to checking the test statistics for the individual coefficients. Those tests look like what we did for the slope of a simple regression in Chapter 16. For each coefficient, we test the null hypothesis that the slope is zero against the (two-sided) alternative that it isn't zero. The regression table gives a standard error for each coefficient and the ratio of the estimated coefficient to its standard error. If the assumptions and conditions are met (and now we need the Nearly Normal Condition or a large sample), these ratios follow a Student's t-distribution:

$$t_{n-k-1} = \frac{b_j - 0}{SE(b_j)}.$$

Where did the degrees of freedom $n - k - 1$ come from? We have a rule of thumb that works here. The degrees of freedom value is the number of data values minus the number of estimated parameters (including the intercept). For the house price regression on five predictors, that's $n - 5 - 1$. Almost every regression report includes both the t-statistics and their corresponding P-values.

We can build a confidence interval in the usual way, with an estimate plus or minus a margin of error. As always, the margin of error is the product of the standard error and a critical value. Here the critical value comes from the t-distribution on $n - k - 1$ degrees of freedom, and the standard errors are in the regression table. So a confidence interval for each slope β_j is:

$$b_j \pm t^*_{n-k-1} \times SE(b_j).$$

The tricky parts of these tests are that the standard errors of the coefficients now require harder calculations (so we leave it to technology), and the meaning of a coefficient, as we have seen, depends on all the other predictors in the multiple regression model.

That last point is important. If we fail to reject the null hypothesis for a multiple regression coefficient, it does *not* mean that the corresponding predictor variable has no linear relationship to y. It means that the corresponding predictor contributes nothing to modeling y *after allowing for all the other predictors*.

The multiple regression model looks so simple and straightforward. It *looks* like each β_j tells us the effect of its associated predictor, x_j, on the response variable, y.

---

[4] There are F tables in the back of the book, and most regression tables include a P-value for the F-statistic.

But that is not true. This is, without a doubt, the most common error that people make with multiple regression. In fact:

♦ The coefficient β_j in a multiple regression can be quite different from zero even when it is possible there is no simple linear relationship between y and x_j.

♦ It is even possible that the multiple regression slope changes sign when a new variable enters the regression. We saw this for the *Price* on *Bedrooms* real estate example when *Living Area* was added to the regression.

So we'll say it once more: The coefficient of x_j in a multiple regression depends as much on the *other* predictors as it does on x_j. Failing to interpret coefficients properly is the most common error in working with regression models.

18.5 Adjusted R^2, and the F-statistic

In Chapter 16, for simple linear regression, we interpreted R^2 as the variation in y accounted for by the model. The same interpretation holds for multiple regression, where now the model contains more than one predictor variable. The R^2 value tells us how much (as a fraction or percentage) of the variation in y is accounted for by the model with all the predictor variables included.

There are some relationships among the standard error of the residuals, s_e, the F-ratio, and R^2 that are useful for understanding how to assess the value of the multiple regression model. To start, we can write the standard error of the residuals as:

$$s_e = \sqrt{\frac{SSE}{n - k - 1}},$$

where $SSE = \sum e^2$ is called the **Sum of Squared Residuals (the E is for error)**. As we know, a larger SSE (and thus s_e) means that the residuals are more variable and that our predictions will be correspondingly less precise.

We can look at the total variation of the response variable, y, which is called the **Total Sum of Squares** and is denoted SST: $SST = \sum (y - \bar{y})^2$. For any regression model, we have no control over SST, but we'd like SSE to be as small as we can make it by finding predictor variables that account for as much of that variation as possible. In fact, we can write an equation that relates the total variation SST to SSE:

$$SST = SSR + SSE,$$

where $SSR = \sum (\hat{y} - \bar{y})^2$ is called the **Regression Sum of Squares** because it comes from the predictor variables and tells us how much of the total variation in the response is due to the regression model. For a model to account for a large portion of the variability in y, we need SSR to be large and SSE to be small. In fact, R^2 is just the ratio of SSR to SST:

$$R^2 = \frac{SSR}{SST} = 1 - \frac{SSE}{SST}.$$

When the SSE is nearly 0, the R^2 value will be close to 1.

In Chapter 16, we saw that for the relationship between two quantitative variables, testing the standard null hypothesis about the correlation coefficient, $H_0: \rho = 0$, was equivalent to testing the standard null hypothesis about the slope, $H_0: \beta_1 = 0$. A similar result holds here for multiple regression. Testing the overall hypothesis tested by the F-statistic, $H_0: \beta_1 = \beta_2 = \ldots = \beta_k = 0$, is equivalent to testing whether the true multiple regression R^2 is zero. In fact, the F-statistic for testing that all the slopes are zero can be found as:

$$F = \frac{R^2/k}{(1 - R^2)/(n - k - 1)} = \frac{\dfrac{SSR}{SST}\dfrac{1}{k}}{\dfrac{SSE}{SST}\dfrac{1}{n - k - 1}} = \frac{SSR/K}{SSE/(n - k - 1)} = \frac{MSR}{MSE}.$$

Mean Squares

Whenever a sum of squares is divided by its degrees of freedom, the result is called a mean square. For example, the Mean Square for Error, which you may see written as MSE, is found as $SSE/(n - k - 1)$. It estimates the variance of the errors.

Similarly $SST/(n - 1)$ divides the total sum of square by *its* degrees of freedom. That is sometimes called the Mean Square for Total and denoted MST. We've see this one before; the MST is just the variance of y.

And SSR/k is the Mean Square for Regression.

In other words, using an *F*-test to see whether any of the true coefficients is different from 0 is equivalent to testing whether the R^2 value is different from zero. A rejection of either hypothesis says that at least one of the predictors accounts for enough variation in *y* to distinguish it from noise. Unfortunately, the test doesn't say which slope is responsible. You need to look at individual *t*-tests on the slopes to determine that. Because removing one predictor variable from the regression equation can change any number of slope coefficients, it is not straightforward to determine the right subset of predictors to use. We'll return to that problem when we discuss model selection in Chapter 19.

R^2 and Adjusted R^2

Adding a predictor variable to a multiple regression equation does not always increase the amount of variation accounted for by the model, but it can never reduce it. Adding new predictor variables will always keep the R^2 value the same or increase it. It can never decrease it. But, even if the R^2 value grows, that doesn't mean that the resulting model is a better model or that it has greater predictive ability. If we have a model with *k* predictors (all of which have statistically significant coefficients at some *α* level) and want to see if including a new variable, x_{k+1}, is warranted, we could fit the model with all $k + 1$ variables and simply test the slope of the added variable with a *t*-test of the slope.

This method can test whether the last added variable adds significantly to the model, but choosing the "best" subset of predictors is not necessarily straightforward. We'll spend time discussing strategies for model selection in Chapter 19. The trade-off between a small (parsimonious) model and one that fits the data well is one of the great challenges of any serious model-building effort. Various statistics have been proposed to provide guidance for this search, and one of the most common is called adjusted R^2. **Adjusted R^2** imposes a "penalty" for each new term that's added to the model in an attempt to make models of different sizes (numbers of predictors) comparable. It differs from R^2 because it can shrink when a predictor is added to the regression model or grow when a predictor is removed if the predictor in question doesn't contribute usefully to the model. In fact, it can even be negative.

For a multiple regression with *k* predictor variables and *n* cases, it is defined as

$$R_{adj}^2 = 1 - (1 - R^2)\frac{n - 1}{n - k - 1} = 1 - \frac{\text{SSE}/(n - k - 1)}{\text{SST}/(n - 1)}.$$

In the Guided Example, we saw that the regression of *Price* on *Bedrooms, Bathrooms, Living Area, Fireplaces,* and *Age* resulted in an R^2 of 0.6049. All the coefficients had P-values well below 0.05. The adjusted R^2 value for this model is 0.6030. If we add the variable *Lot Size* to the model, we get the following regression model:

	Coeff	SE(Coeff)	t-ratio	P-value
Intercept	15360.011	7334.804	2.094	0.03649
Living Area	73.388	4.043	18.154	< 0.00001
Bedrooms	−6096.387	2757.736	−2.211	0.02728
Bathrooms	18824.069	3676.582	5.120	< 0.00001
Fireplaces	9226.356	3191.788	2.891	0.00392
Age	−152.615	48.224	−3.165	0.00160
Lot Size	847.764	1989.112	0.426	0.67005

```
Residual standard error: 48440 on 1041 degrees of freedom
Multiple R-squared: 0.6081, Adjusted R-squared: 0.6059
F-statistic: 269.3 on 6 and 1041 DF, P-value: < 0.0001
```

The most striking feature of this output, as compared to the output in the Guided Example on page 520, is that although most of the coefficients have changed very little, the coefficient of *Lot Size* is far from significant, with a P-value of 0.670. Yet, the adjusted R^2 value is actually higher than for the previous model. This is why we warn against putting too much faith in this statistic. Especially for large samples, the adjusted R^2 does not always adjust downward enough to make sensible model choices. The other problem with comparing these two models is that 9 homes had missing values for *Lot Size*, which means that we're not comparing the models on exactly the same data set. When we matched the two models on the smaller data set, the adjusted R^2 value actually did "make the right decision" but just barely—0.6059 versus 0.6060 for the model without *Lot Size*. One might expect a larger difference considering we added a variable whose *t*-ratio is much less than 1.

The lesson to be learned here is that there is no "correct" set of predictors to use for any real business decision problem, and finding a reasonable model is a process that takes a combination of science, art, business knowledge, and common sense. Look at the adjusted R^2 value for any multiple regression model you fit, but be sure to think about all the other reasons for including or not including any given predictor variable. We will have much more to say about this important subject in Chapter 19.

*18.6 The Logistic Regression Model

Business decisions often depend on whether something will happen or not. Will my customers leave my wireless service at the end of their subscription? How likely is it that my customers will respond to the offer I just mailed? The response variable is either Yes or No—a dichotomous response. By definition, that's a categorical response, so we can't use linear regression methods to predict it.

If we coded this categorical response variable by giving the Yes values the value 1 and the No values the value 0, we could "pretend" that it's quantitative and try to fit a linear model. Let's imagine that we're trying to model whether someone will respond to an offer based on how much they spent in the last year at our company. The regression of *Purchase (1 = Yes: 0 = No)* on *Spending* might look like this.

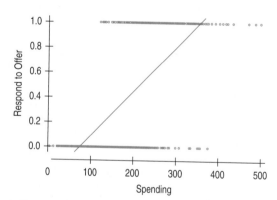

Figure 18.6 *A linear model of* Respond to Offer *shows that as* Spending *increases, the proportion of customers responding to the offer increases as well. Unfortunately, the line predicts values outside the interval (0,1), but those are the only sensible values for proportions or probabilities.*

The model shows that as past spending increases, something having to do with responding to the offer increases. Look at *Spending* near $500. Nearly all the customers who spend that much responded to the offer, while for those customers who spent less than $100 last year, almost none responded. The line seems to show the proportion of those responding for different values of *Spending*. Looking at the plot, what would you predict the proportion of responders to be for those that spent about $225? You might say that the proportion was about 0.50.

What's wrong with this interpretation? We know proportions (or probabilities) must lie between 0 and 1. But the linear model has no such restrictions. The line crosses 1.0 at about $350 in *Spending* and goes below 0 at about $75 in *Spending*. However, by transforming the probability, we can make the model behave better and get more sensible predictions for all values of *Spending*. We could just cut off all values greater than 1 at 1 and all values less than 0 at 0. That would work fine in the middle, but we know that things like customer behavior don't change that abruptly. So we'd prefer a model that curves at the ends to approach 0 and 1 gently. That's likely to be a better model for what really happens. A simple function is all we need. There are several that can do the job, but one common model is the logistic regression model. Here is a plot of the logistic regression model predictions for the same data.

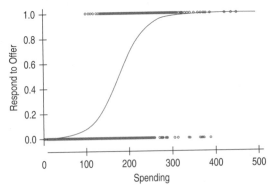

Figure 18.7   *A logistic model of Respond to Spending predicts values between 0 and 1 for the proportion of customers who Respond for all values of Spending.*

For many values of *Spending*, especially those near the mean value of *Spending*, the predictions of the probability of responding is similar, but the logistic transformation approaches the limits at 0 and 1 smoothly. A logistic regression is an example of a nonlinear regression model. The computer finds the coefficients by solving a system of equations that in some ways mimic the least squares calculations we did for linear regression. The **logistic regression** model looks like this:

$$\log\left(\frac{p}{1-p}\right) = \beta_0 + \beta_1 x_1 + \ldots + \beta_k x_k.$$

In other words, it's not the probabilities themselves that are fit to a (multiple) regression model, but a transformation (the logistic function) of the probabilities that are fit using a linear regression in the predictor variables. When the probabilities are transformed back and plotted, we get an S-shaped curve (in each predictor), as seen in Figure 18.7.

As with multiple regression, computer output for a logistic regression model provides estimates for each coefficient, standard errors, and a test for whether the coefficient is zero. Unlike multiple regression, the coefficients for each variable are tested with a Chi-square statistic[5] rather than a *t*-statistic, and the overall R^2 is no

---

[5] Yes, the same sampling distribution as we used in Chapter 15.

longer available. (Some computer programs use a *z*-statistic to test individual slope coefficients. The tests are equivalent.) Because the probabilities are not a linear function of the predictors, it is more difficult to interpret the coefficients. The fitted logistic regression equation can be written:

$$\log\left(\frac{\hat{p}}{1 - \hat{p}}\right) = b_0 + b_1 x_1 + \ldots + b_k x_k.$$

Some researchers try to interpret the logistic regression equation directly by realizing that the expression on the left of the logistic regression equation, $\log\left(\frac{\hat{p}}{1 - \hat{p}}\right)$[6], can be thought of as the predicted log odds of the probability, *p*. The higher the log odds, the higher the probability. Negative log odds indicate that the probability is less than 0.5 because then $\frac{p}{1 - p}$ will be less than 1 and thus have a negative logarithm. Positive log odds indicate a probability greater than 0.5. If the coefficient of a particular predictor is positive, it means that higher values of it are associated with higher log odds and thus a higher probability of the response. So, the *direction* is interpretable in the same way as a multiple regression coefficient. But an increase of one unit of the predictor *increases* the *log odds* by an amount equal to the coefficient of that predictor, after allowing for the effects of all the other predictors. It does not increase the *probability* by that amount.

The transformation back to the probabilities is straightforward, but nonlinear. Once we have fit the logistic regression equation

$$\log\left(\frac{\hat{p}}{1 - \hat{p}}\right) = b_0 + b_1 x_1 + \ldots + b_k x_k,$$

we can find the individual probability estimates from the equation:

$$\hat{p} = \frac{1}{1 + e^{-(b_0 + b_1 x_1 + \ldots + b_k x_k)}} = \frac{e^{(b_0 + b_1 x_1 + \ldots + b_k x_k)}}{1 + e^{(b_0 + b_1 x_1 + \ldots + b_k x_k)}}.$$

The assumptions and conditions for fitting a logistic regression are similar to multiple regression. We still need the independence assumption and randomization condition. We no longer need the linearity or equal spread condition. However, it's a good idea to plot the response variable against each predictor variable to make sure that there are no outliers that could unduly influence the model. A customer who both spent $10,000 and responded to the offer could possibly change the shape of the curve shown in Figure 18.7. However, residual analysis for logistic regression models is beyond the scope of this book.

GUIDED EXAMPLE   *Time on Market*

A real estate agent used information on 1115 houses, such as we used in our multiple regression Guided Example. She wants to predict whether each house sold in the first 3 months it was on the market based on other variables. The variables available include:

Sold — 1 = Yes—the house sold within the first 3 months it was listed; 0 = No, it did not sell within 3 months.

Price — The price of the house as sold in 2002.

[6] The function $\log\left(\frac{\hat{p}}{1 - \hat{p}}\right)$ is also called the *logit* function.

Living Area	The size of the living area of the house in square feet	*Age*	Age of the house in years
Bedrooms	The number of bedrooms	*Fireplaces*	Number of fireplaces in the house
Bathrooms	The number of bathrooms (a half bath is a toilet and sink only)		

PLAN

Setup State the objective of the study. Identify the variables.

Model Think about the assumptions and check the conditions.

We want to build a model to predict whether a house will sell within the first 3 months it's on the market based on *Price* ($), *Living Area* (sq ft), *Bedrooms* (#), *Bathrooms* (#), *Fireplaces* (#), and *Age* (years). Notice that now *Price* is a predictor variable in the regression.

Outlier Condition

To fit a logistic regression model, we check that there are no outliers in the predictors that may unduly influence the model.

Here are scatterplots of *Sold* (1 = Yes; 0 = No) against each predictor. (Plots of *Sold* against the variables *Bathrooms*, *Bedrooms*, and *Fireplaces* are uninformative because these predictors are discrete.)

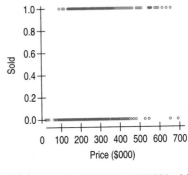

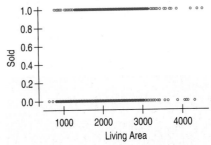

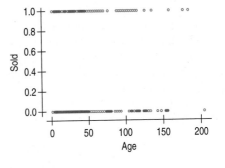

✓ **Outlier Condition.** There do not seem to be any outliers in the predictors. Of course, there can't be any in the response variable; it is only 0 or 1.

✓ **Independence Assumption.** We can regard the house prices as being independent of one another since they come from a fairly large geographic area.

✓ **Randomization Condition.** These 1115 houses are a random sample from a larger collection of houses.

We can fit a logistic regression predicting *Sold* from the six predictor variables.

Mechanics We always fit logistic regression models with computer software. An output table like this one isn't exactly what any of the major packages produce, but it is enough like all of them to look familiar.

Here is the computer output for the logistic regression using all six predictors.

	Coeff	SE(Coeff)	z-value	P-value
Intercept	−3.222e + 00	3.826e − 01	−8.422	< 0.0001
Living Area	−1.444e − 03	2.518e − 04	−5.734	< 0.0001
Age	4.900e − 03	2.823e − 03	1.736	0.082609
Price	1.693e − 05	1.444e − 06	11.719	< 0.0001
Bedrooms	4.805e − 01	1.366e − 01	3.517	0.000436
Bathrooms	−1.813e − 01	1.829e − 01	−0.991	0.321493
Fireplaces	−1.253e − 01	1.633e − 01	−0.767	0.442885

The estimated equation is:

$$\text{logit}(\hat{p}) = -3.22 - 0.00144 \, Living\ Area \\ + 0.0049 Age + 0.00000169 Price \\ + 0.481 Bedrooms - 0.181 Bathrooms \\ - 0.125 Fireplaces$$

Three of the P-values are quite small, two are large (*Bathrooms* and *Fireplaces*), and one is marginal (*Age*). After examining several alternatives, we chose the following model.

Strategies for model selection will be discussed in Chapter 19.

	Coeff	SE(Coeff)	z-value	P-value
Intercept	−3.351e + 00	3.601e − 01	−9.305	< 0.0001
Living Area	−1.574e − 03	2.342e − 04	−6.719	< 0.0001
Age	6.106e − 03	2.668e − 03	2.289	0.022102
Bedrooms	4.631e − 01	1.354e − 01	3.421	0.000623
Price	1.672e − 05	1.428e − 06	11.704	< 0.0001

The estimated logit equation is:

$$\text{logit}(\hat{p}) = -3.351 - 0.00157 \, Living\ Area \\ + 0.00611 Age + 0.00000167 Price \\ + 0.463 Bedrooms$$

While interpretation is difficult, it appears that for a house of a given size (and age and bedrooms), higher priced homes may have a higher chance of selling within 3 months. For a house of given size and price and age, having more bedrooms may be associated with a greater chance of selling within 3 months.

**REPORT**

Summary and Conclusions Summarize your results and state any limitations of your model in the context of your original objectives.

MEMO:

Re: Logistic Regression Analysis of Selling

A logistic regression model of *Sold* on various predictors was fit, and a model based on *Living Area*, *Bedrooms*, *Price*, and *Age* found that these four predictors were statistically significant in predicting the probability that a house will sell within 3 months. More thorough analysis to understand the meaning of the coefficients is needed, but each of these variables appears to be an important predictor of whether the house will sell quickly.

However, knowing more specific information about other characteristics of a house and where it is located would almost certainly help the model.

 WHAT CAN GO WRONG?

Interpreting Coefficients

- **Don't claim to "hold everything else constant" for a single individual.** It's often meaningless to say that a regression coefficient says what we expect to happen if all variables but one were held constant for an individual and the predictor in question changed. While it's mathematically correct, it often just doesn't make any sense. For example, in a regression of salary on years of experience, years of education, and age, subjects can't gain a year of experience or get another year of education without getting a year older. Instead, we *can* think about all those who fit given criteria on some predictors and ask about the conditional relationship between y and one x for those individuals.

- **Don't interpret regression causally.** Regressions are usually applied to observational data. Without deliberately assigned treatments, randomization, and control, we can't draw conclusions about causes and effects. We can never be certain that there are no variables lurking in the background, causing everything we've seen. Don't interpret b_1, the coefficient of x_1 in the multiple regression, by saying: "If we were to change an individual's x_1 by 1 unit (holding the other x's constant), it would change his y by b_1 units." We have no way of knowing what applying a change to an individual would do.

- **Be cautious about interpreting a regression model as predictive.** Yes, we do call the x's predictors, and you can certainly plug in values for each of the x's and find a corresponding *predicted value*, $\hat{y}$. But the term "prediction" suggests extrapolation into the future or beyond the data, and we know that we can get into trouble when we use models to estimate $\hat{y}$ values for x's not in the range of the data. Be careful not to extrapolate very far from the span of your data. In simple regression, it was easy to tell when you extrapolated. With many predictor variables, it's often harder to know when you are outside the bounds of your original data.[7]

---

[7] With several predictors we can wander beyond the data because of the *combination* of values even when individual values are not extraordinary. For example, houses with 1 bathroom and houses with 5 bedrooms can both be found in the real estate records, but a single house with 5 bedrooms and only 1 bathroom would be quite unusual. The model we found is not appropriate for predicting the price of such an extraordinary house.

We usually think of fitting models to the data more as modeling than as prediction, so that's often a more appropriate term.

- **Don't think that the sign of a coefficient is special.** Sometimes our primary interest in a predictor is whether it has a positive or negative association with *y*. As we have seen, though, the sign of the coefficient also depends on the other predictors in the model. Don't look at the sign in isolation and conclude that "the direction of the relationship is positive (or negative)." Just like the value of the coefficient, the sign is about the relationship after allowing for the linear effects of the other predictors. The sign of a variable can change depending on which other predictors are in or out of the model. For example, in the regression model for house prices, we saw the coefficient of *Bedrooms* change sign when *Living Area* was added to the model as a predictor. It isn't correct to say either that houses with more bedrooms sell for more on average or that they sell for less. The truth is more subtle and requires that we understand the multiple regression model.

- **If a coefficient's *t*-statistic is not significant, don't interpret it at all.** You can't be sure that the value of the corresponding parameter in the underlying regression model isn't really zero.

WHAT ∧*Else* CAN GO WRONG?

- **Don't fit a linear regression to data that aren't straight.** This is the most fundamental regression assumption. If the relationship between the *x*'s and *y* isn't approximately linear, there's no sense in fitting a linear model to it. What we mean by "linear" is a model of the form we have been writing for the regression. When we have two predictors, this is the equation of a plane, which is linear in the sense of being flat in all directions. With more predictors, the geometry is harder to visualize, but the simple structure of the model is consistent; the predicted values change consistently with equal size changes in any predictor.

  Usually we're satisfied when plots of *y* against each of the *x*'s are straight enough. We'll also check a scatterplot of the residuals against the predicted values for signs of nonlinearity.

- **Watch out for the plot thickening.** The estimate of the error standard deviation shows up in all the inference formulas. But that estimate assumes that the error standard deviation is the same throughout the range of the *x*'s so that we can combine all the residuals when we estimate it. If s_e changes with any *x*, these estimates won't make sense. The most common check is a plot of the residuals against the predicted values. If plots of residuals against several of the predictors all show a thickening and especially if they also show a bend, then consider re-expressing *y*. If the scatterplot against only one predictor shows thickening, consider re-expressing that predictor.

- **Make sure the errors are nearly Normal.** All of our inferences require that the true errors be modeled well by a Normal model. Check the histogram and Normal probability plot of the residuals to see whether this assumption looks reasonable.

- **Watch out for high-influence points and outliers.** We always have to be on the lookout for a few points that have undue influence on our model, and regression is certainly no exception. Chapter 19 discusses this issue in greater depth.

ETHICS IN ACTION

Alpine Medical Systems, Inc. is a large provider of medical equipment and supplies to hospitals, doctors, clinics, and other health care professionals. Alpine's VP of Marketing and Sales, Kenneth Jadik, asked one of the company's analysts, Nicole Haly, to develop a model that could be used to predict the performance of the company's sales force. Based on data collected over the past year, as well as records kept by Human Resources, she considered five potential independent variables: (1) gender, (2) starting base salary, (3) years of sales experience, (4) personality test score, and (5) high school grade point average. The dependent variable (sales performance) is measured as the sales dollars generated per quarter. In discussing the results with Nicole, Kenneth asks to see the full regression model with all five independent variables included. Kenneth notes that a *t*-test for the coefficient of gender shows no significant effect on sales performance and recommends that it be eliminated from the model. Nicole reminds him of the company's history of offering lower starting base salaries to women, recently corrected under court order. If instead, starting base salary is removed from the model, gender is statistically significant, and its coefficient indicates that women on the sales force outperform men (taking into account the other variables). Kenneth argues that because gender is not significant when all predictors are included, it is the variable that should be omitted.

ETHICAL ISSUE *The choice of predictors for the regression model is politically motivated. Because gender and base salary are related, it is impossible to separate their effects on sales performance, and inappropriate to conclude that one or the other is irrelevant. Related to Item A, ASA Ethical Guidelines.*

ETHICAL SOLUTION *The situation is more complex than a single model can explain. Both the model with gender but not base salary and the one with base salary but not gender should be reported. Then the discussion of these models should point out that the two variables are related because of previous company policy and note that the conclusion that those with lower base salary have better sales and the conclusion that women tend to have better sales performance are equivalent as far as these data are concerned.*

What have we learned?

In Chapter 16, we learned to apply our inference methods to linear regression models. Now we've seen that much of what we know about those models is also true for multiple regression.

- The assumptions and conditions are the same: linearity (checked now with scatterplots of *y* against each *x*), independence (think about it), constant variance (checked with the scatterplot of residuals against predicted values), and nearly Normal residuals (checked with a histogram or probability plot).

- R^2 is still the fraction of the variation in *y* accounted for by the regression model.

- s_e is still the standard deviation of the residuals—a good indication of the precision of the model.

- The degrees of freedom (in the denominator of s_e and for each of the *t*-tests) follows the same rule: *n* minus the *number of parameters estimated.*

- The regression table produced by any statistics package shows a row for each coefficient, giving its estimate, a standard error, a *t*-statistic, and a P-value.

- If all the conditions are met, we can test each coefficient against the null hypothesis that its parameter value is zero with a Student's *t*-test.

And we've learned some new things that are useful now that we have multiple predictors.

- We can perform an overall test of whether the multiple regression model provides a better summary for y than its mean by using the F-distribution.
- We learned that R^2 may not be appropriate for comparing multiple regression models with different numbers of predictors. Adjusted R^2 is one approach to this problem.

Finally, we've learned that multiple regression models extend our ability to model the world to many more situations but that we must take great care when we interpret its coefficients. To interpret a coefficient of a multiple regression model, remember that it estimates the linear relationship between y and that predictor *after accounting for two things:* 1) the linear effects of all the other predictors on y and 2) the linear relationship between that predictor and all other x's.

Terms

Adjusted R^2

An adjustment to the R^2 statistic that attempts to allow for the number of predictors in the model. It is sometimes used when comparing regression models with different numbers of predictors:

$$R^2_{adj} = 1 - (1 - R^2)\frac{n - 1}{n - k - 1} = 1 - \frac{\text{SSE}/(n - k - 1)}{\text{SST}/(n - 1)}.$$

F-test

The F-test is used to test the null hypothesis that the overall regression is no improvement over just modeling y with its mean:

$$H_0: \beta_1 = \ldots = \beta_k = 0. \quad \textit{vs } H_0: \text{at least one } \beta \neq 0$$

If this null hypothesis is not rejected, then you should not proceed to test the individual coefficients.

Least squares

We still fit multiple regression models by choosing the coefficients that make the sum of the squared residuals as small as possible. This is called the *method of least squares.*

Multiple regression

A linear regression with two or more predictors whose coefficients are found by least squares. When the distinction is needed, a least squares linear regression with a single predictor is called a *simple regression.* The multiple regression model is: $y = \beta_0 + \beta_1 x_1 + \ldots + \beta_k x_k + \varepsilon.$

Regression Sum of Squares, SSR

A measure of the total variation in the response variable due to the model. $SSR = \sum(\hat{y} - \bar{y})^2.$

Sum of Squared Residuals, SSE

A measure of the variation in the residuals. $SSE = \sum(y - \hat{y})^2.$

Total Sum of Squares, SST

A measure of the variation in the response variable. $SST = \sum(y - \bar{y})^2.$ Note that $\frac{SST}{n - 1} = Var(y).$

t-ratios for the coefficients

The t-ratios for the coefficients can be used to test the null hypotheses that the true value of each coefficient is zero against the alternative that it is not. The t distribution is also used in the construction of confidence intervals for each slope coefficient.

Skills

When you complete this lesson you should:

PLAN

- Understand that the "true" regression model is an idealized summary of the data.
- Know how to examine scatterplots of *y* vs. each *x* for violations of assumptions that would make inference for regression unwise or invalid.
- Know how examine displays of the residuals from a multiple regression to check that the conditions have been satisfied. In particular, know how to judge linearity and constant variance from a scatterplot of residuals against predicted values. Know how to judge Normality from a histogram and Normal probability plot.
- Remember to be especially careful to check for failures of the independence assumption when working with data recorded over time. Examine scatterplots of the residuals against time and look for patterns.

DO

- Be able to use a statistics package to perform the calculations and make the displays for multiple regression, including a scatterplot of the response against each predictor, a scatterplot of residuals against predicted values, and a histogram and Normal probability plot of the residuals.
- Know how to use the *F*-test to check that the overall regression model is better than just using the mean of *y*.
- Know how to test the standard hypotheses that each regression coefficient is really zero. Be able to state the null and alternative hypotheses. Know where to find the relevant numbers in standard computer regression output.

REPORT

- Be able to summarize a regression in words. In particular, be able to state the meaning of the regression coefficients, taking full account of the effects of the other predictors in the model.
- Be able to interpret the *F*-statistic and R^2 for the overall regression.
- Be able to interpret the P-value of the *t*-statistics for the coefficients to test the standard null hypotheses.

Technology Help: Regression Analysis

All statistics packages make a table of results for a regression. The table for multiple regression looks very similar to the table for simple regression. You'll want to look at the Analysis of Variance (ANOVA) table, and you'll see information for each of the coefficients.

Most packages offer to plot residuals against predicted values. Some will also plot residuals against the *x*'s. With some packages, you must request plots of the residuals when you request the regression. Others let you find the regression first and then analyze the residuals afterward. Either way,

your analysis is not complete if you don't check the residuals with a histogram or Normal probability plot and a scatterplot of the residuals against the *x*'s or the predicted values.

One good way to check assumptions before embarking on a multiple regression analysis is with a scatterplot matrix. This is sometimes abbreviated SPLOM (or Matrix Plot) in commands.

Multiple regressions are always found with a computer or programmable calculator. Before computers were available, a full multiple regression analysis could take months or even years of work.

EXCEL

- In Excel 2003 and earlier, select **Data Analysis** from the **Tools** menu.
- In Excel 2007, select **Data Analysis** from the **Analysis Group** on the Data Tab.
- Select **Regression** from the **Analysis Tools** list.
- Click the **OK** button.
- Enter the data range holding the Y-variable in the box labeled "Y-range."
- Enter the range of cells holding the X-variables in the box labeled "X-range."
- Select the **New Worksheet Ply** option.
- Select **Residuals** options. Click the **OK** button.

Comments

The Y and X ranges do not need to be in the same rows of the spreadsheet, although they must cover the same number of cells. But it is a good idea to arrange your data in parallel columns as in a data table. The X-variables must be in adjacent columns. No cells in the data range may hold non-numeric values or be left blank.

Although the dialog offers a Normal probability plot of the residuals, the data analysis add-in does not make a correct probability plot, so don't use this option.

MINITAB

- Choose **Regression** from the **Stat** menu.
- Choose **Regression . . .** from the **Regression** submenu.
- In the Regression dialog, assign the Y-variable to the Response box and assign the X-variables to the Predictors box.
- Click the **Graphs** button.

- In the Regression-Graphs dialog, select **Standardized residuals**, and check **Normal plot of residuals** and **Residuals versus fits**.
- Click the **OK** button to return to the Regression dialog.
- Click the **OK** button to compute the regression.

SPSS

- Choose **Regression** from the **Analyze** menu.
- Choose **Linear** from the **Regression** submenu.
- When the Linear Regression dialog appears, select the Y-variable and move it to the dependent target. Then move the X-variables to the independent target.
- Click the **Plots** button.

- In the Linear Regression Plots dialog, choose to plot the *SRESIDs against the *ZPRED values.
- Click the **Continue** button to return to the Linear Regression dialog.
- Click the **OK** button to compute the regression.

JMP

- From the **Analyze** menu, select **Fit Model**.
- Specify the response, Y. Assign the predictors, X, in the **Construct Model Effects** dialog box.
- Click on **Run Model**.

Comments

JMP chooses a regression analysis when the response variable is "Continuous." The predictors can be any combination of quantitative or categorical. If you get a different analysis, check the variable types.

DATA DESK

- Select Y- and X-variable icons.
- From the **Calc** menu, choose **Regression**.
- Data Desk displays the regression table.
- Select plots of residuals from the Regression table's HyperView menu.

Comments

You can change the regression by dragging the icon of another variable over either the Y- or an X-variable name in the table and dropping it there. You can add a predictor by dragging its icon into that part of the table. The regression will recompute automatically.

Mini Case Study Project

Golf Success

Professional sports, like many other professions, require a variety of skills for success. That makes it difficult to evaluate and predict success. Fortunately, sports provide examples we can use to learn about modeling success because of the vast amount of data which are available. Here's an example.

What makes a golfer successful? The game of golf requires many skills. Putting well or hitting long drives will not, by themselves, lead to success. Success in golf requires a combination of skills. That makes multiple regression a good candidate for modeling golf achievement.

A number of Internet sites post statistics for the current PGA players. We have data for 204 top players of 2006 in the file **ch18_MCSP_Golfers**.

All of these players earned money on the tour, but they didn't all play the same number of events. And the distribution of earnings is quite skewed. (Tiger Woods earned $662,000 per event. In second place, Jim Furyk earned only $300,000 per event. Median earnings per event were $36,600.) So it's a good idea to take logs of Earnings/Event as the response variable.

The variables in the data file include:

Log$/E	The logarithm of earnings per event
GIR	Greens in Regulation. Percentage of holes played in which the ball is on the green with two or more strokes left for par.
Putts	Average number of putts per hole in which the green was reached in regulation.
Save%	Each time a golfer hits a bunker by the side of a green but needs only one or two additional shots to reach the hole, he is credited with a save. This is the percentage of opportunities for saves that are realized.
DDist	Average Drive Distance (yards). Measured as averages over pairs of drives in opposite directions (to account for wind).
DAcc	Drive Accuracy. Percent of drives landing on the fairway.

Investigate these data. Find a regression model to predict golfers' success (measured in log earnings per event). Write a report presenting your model including an assessment of its limitations. Note: Although you may consider several intermediate models, a good report is about the model you think best, not necessarily about all the models you tried along the way while searching for it.

EXERCISES

The first 12 exercises consist of two sets of 6 (one even-numbered, one odd-numbered). Each set guides you through a multiple regression analysis. We suggest that you do all 6 exercises in a set. Remember that the answers to the odd-numbered exercises can be found in the back of the book.

T 1. Police salaries. Is the amount of violent crime related to what police officers are paid? The U.S. Bureau of Labor Statistics publishes data on occupational employment and wage estimates

(www.bls.gov/oes/). Here are data on those states for which 2006 data were available. The variables are:

Violent Crime (crimes per 100,000 population)

Police Officer Wage (mean $/hr)

Graduation Rate (%)

One natural question to ask of these data is how police officer wages are related to violent crime across these states.

First, here are plots and background information.

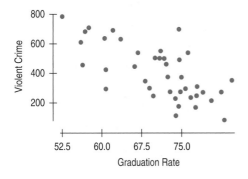

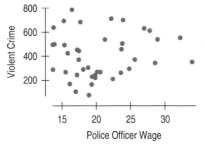

Correlations			
	Violent Crime	**Graduation Rate**	**Police Officer Wage**
Violent Crime	1.000		
Graduation Rate	−0.682	1.000	
Police Officer Wage	0.103	0.213	1.000

a) Name and check (to the extent possible) the regression assumptions and their corresponding conditions.

b) If we found a regression to predict *Violent Crime* just from *Police Officer Wage*, what would the R^2 of that regression be?

T 2. Ticket prices. On a typical night in New York, about 25,000 people attend a Broadway show, paying an average price of more than $75 per ticket. *Variety* (www.variety.com), a news weekly that reports on the entertainment industry, publishes statistics about the Broadway show business. The data file on the disk holds data about shows on Broadway for most weeks of 2006–2008. (A few weeks are missing data.) The following variables are available for each week:

Receipts ($ million)

Paid Attendance (thousands)

 # Shows

Average Ticket Price ($)

Viewing this as a business, we'd like to model *Receipts* in terms of the other variables.

First, here are plots and background information.

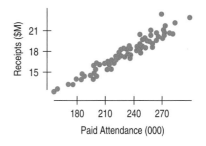

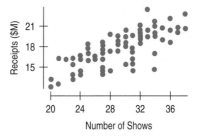

Correlations				
	Receipts	**Paid Attendance**	**# Shows**	**Average Ticket Price**
Receipts ($M)	1.000			
Paid Attendance	0.961	1.000		
# Shows	0.745	0.640	1.000	
Average Ticket Price	0.258	0.331	−0.160	1.000

a) Name and check (to the extent possible) the regression assumptions and their corresponding conditions.

b) If we found a regression to predict *Receipts* only from *Paid Attendance*, what would the R^2 of that regression be?

T 3. Police salaries, part 2. Here's a multiple regression model for the variables considered in Exercise 1.

Dependent variable is: Violent Crime
R squared = 53.0% R squared (adjusted) = 50.5%
s = 129.6 with 37 degrees of freedom

Source	Sum of Squares	df	Mean Square	F-ratio
Regression	701648	2	350824	20.9
Residual	621060	37	16785.4	

Variable	Coeff	SE(Coeff)	*t*-ratio	P-value
Intercept	1390.83	185.9	7.48	< 0.0001
Police Officer Wage	9.33	4.125	2.26	0.0297
Graduation Rate	−16.64	2.600	−6.40	< 0.0001

a) Write the regression model.
b) What does the coefficient of *Police Officer Wage* mean in the context of this regression model?
c) In a state in which the average police officer wage is $20/hour and the high school graduation rate is 70%, what does this model estimate the violent crime rate would be?
d) Is this likely to be a good prediction? Why do you think that?

T 4. **Ticket prices, part 2.** Here's a multiple regression model for the variables considered in Exercise 2:[8]

Dependent variable is: Receipts($M)
R squared = 99.9% R squared (adjusted) = 99.9%
s = 0.0931 with 74 degrees of freedom

Source	Sum of Squares	df	Mean Square	F-ratio
Regression	484.789	3	161.596	18634
Residual	0.641736	74	0.008672	

Variable	Coeff	SE(Coeff)	*t*-ratio	P-value
Intercept	−18.320	0.3127	−58.6	< 0.0001
Paid Attendance	0.076	0.0006	126.7	< 0.0001
# Shows	0.0070	0.0044	1.59	0.116
Average Ticket Price	0.24	0.0039	61.5	< 0.0001

a) Write the regression model.
b) What does the coefficient of *Paid Attendance* mean in this regression? Does that make sense?
c) In a week in which the paid attendance was 200,000 customers attending 30 shows at an average ticket price of $70, what would you estimate the receipts would be?
d) Is this likely to be a good prediction? Why do you think that?

T 5. **Police salaries, part 3.** Using the regression table in Exercise 3, answer the following questions.
a) How was the *t*-ratio of 2.26 found for *Police Officer Wage*? (Show what is computed using numbers from the table.)
b) How many states are used in this model. How do you know?
c) The *t*-ratio for *Graduation Rate* is negative. What does that mean?

T 6. **Ticket prices, part 3.** Using the regression table in Exercise 4, answer the following questions.
a) How was the *t*-ratio of 126.7 found for *Paid Attendance*? (Show what is computed using numbers found in the table.)
b) How many weeks are included in this regression? How can you tell?
c) The *t*-ratio for the intercept is negative. What does that mean?

---

[8] Some values are rounded to simplify the exercises. If you recompute the analysis with your statistics software you may see slightly different numbers.

T 7. **Police salaries, part 4.** Consider the coefficient of *Police Officer Wage*.
a) State the standard null and alternative hypotheses for the true coefficient of *Police Officer Wage*.
b) Test the null hypothesis (at $\alpha = 0.05$) and state your conclusion.
c) A state senate aide challenges your conclusion. She points out that we can see from the scatterplot and correlation (see Exercise 1) that there is almost no linear relationship between police officer wages and violent crime. Therefore, she claims, your conclusion in part (a) must be mistaken. Explain to her why this is not a contradiction.

T 8. **Ticket prices, part 4.** Consider the coefficient of *# Shows*.
a) State the standard null and alternative hypotheses for the true coefficient of *# Shows*.
b) Test the null hypothesis (at $\alpha = 0.05$) and state your conclusion.
c) A Broadway investor challenges your analysis. He points out that the scatterplot of *Receipts* vs. *# Shows* in Exercise 2 shows a strong linear relationship and claims that your result in part (a) can't be correct. Explain to him why this is not a contradiction.

T 9. **Police salaries, part 5.** The Senate aide in Exercise 7 now accepts your analysis but claims that it demonstrates that if the state pays police more, it will actually *increase* the rate of violent crime. Explain why this interpretation is not a valid use of this regression model. Offer some alternative explanations.

T 10. **Ticket prices, part 5.** The investor in Exercise 8 now accepts your analysis but claims that it demonstrates that it doesn't matter how many shows are playing on Broadway; receipts will be essentially the same. Explain why this interpretation is not a valid use of this regression model. Be specific.

T 11. **Police salaries, part 6.** Here are some plots of residuals for the regression of Exercise 3.

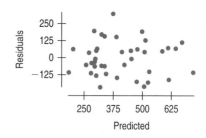

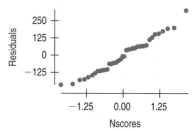

Which of the regression conditions can you check with these plots? Do you find that those conditions are met?

T 12. **Ticket prices, part 6.** Here are some plots of residuals for the regression of Exercise 4.

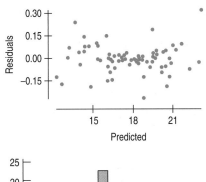

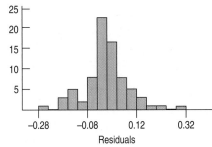

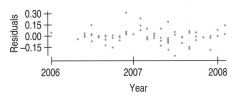

Which of the regression conditions can you check with these plots? Do you find that those conditions are met?

13. Real estate prices. A regression was performed to predict selling price of houses based on *Price* in dollars, *Area* in square feet, *Lotsize* in square feet, and *Age* in years. The R^2 is 92%. The equation from this regression is given here.

$$Price = 169,328 + 35.3 \, Area + 0.718 \, Lotsize - 6543 \, Age$$

One of the following interpretations is correct. Which is it? Explain what's wrong with the others.

a) Each year a house ages, it is worth $6543 less.

b) Every extra square foot of area is associated with an additional $35.50 in average price, for houses with a given lot size and age.

c) Every additional dollar in price means lot size increases 0.718 square feet.

d) This model fits 92% of the data points exactly.

14. Wine prices. Many factors affect the price of wine, including such qualitative characteristics as the variety of grape, location of winery, and label. Researchers developed a regression model considering two quantitative variables: the tasting score of the wine and the age of the wine (in years) when released to market. They found the following regression equation, with an R^2 of 65%, to predict the price (in dollars) of a bottle of wine.

$$Price = 6.25 + 1.22 \, Tasting \, Score + 0.55 \, Age$$

One of the following interpretations is correct. Which is it? Explain what's wrong with the others.

a) Each year a bottle of wine ages, its price increases about $.55.

b) This model fits 65% of the points exactly.

c) For a unit increase in tasting score, the price of a bottle of wine increases about $1.22.

d) After allowing for the age of a bottle of wine, a wine with a one unit higher tasting score can be expected to cost about $1.22 more.

15. Appliance sales. A household appliance manufacturer wants to analyze the relationship between total sales and the company's three primary means of advertising (television, magazines, and radio). All values were in millions of dollars. They found the following regression equation.

$$Sales = 250 + 6.75 \, TV + 3.5 \, Radio + 2.3 \, Magazine$$

One of the following interpretations is correct. Which is it? Explain what's wrong with the others.

a) If they did no advertising, their income would be $250 million.

b) Every million dollars spent on radio makes sales increase $3.5 million, all other things being equal.

c) Every million dollars spent on magazines increases TV spending $2.3 million.

d) Sales increase on average about $6.75 million for each million spent on TV, after allowing for the effects of the other kinds of advertising.

16. Wine prices, part 2. Here are some more interpretations of the regression model to predict the price of wine developed in Exercise 14. One of these interpretations is correct. Which is it? Explain what is wrong with the others.

a) The minimum price for a bottle of wine that has not aged is $6.25.

b) The price for a bottle of wine increases on average about $.55 for each year it ages, after allowing for the effects of tasting score.

c) Each year a bottle of wine ages, its tasting score increases by 1.22.

d) Each dollar increase in the price of wine increases its tasting score by 1.22.

17. Cost of pollution. What is the financial impact of pollution abatement on small firms? The U.S. government's Small Business Administration studied this and reported the following model.

$$Pollution \, abatement/employee = -2.494 - 0.431 \, ln(Number \, of \, Employees) + 0.698 \, ln(Sales)$$

Pollution abatement is in dollars per employee (Crain, M. W., *The Impact of Regulatory Costs on Small Firms*, available at www.sba.gov/idc/groups/public/documents/sba_homepage/rs264tot.pdf).

a) The coefficient of *ln(Number of Employees)* is negative. What does that mean in the context of this model? What does it mean that the coefficient of *ln(Sales)* is positive?

b) The model uses the (natural) logarithms of the two predictors. What does the use of this transformation say about their effects on pollution abatement costs?

18. OECD economic regulations. A study by the U.S. Small Business Administration modeled the GDP per capita of 24 of the countries in the Organization for Economic Cooperation and Development (OECD) (Crain, M. W., *The Impact of Regulatory*

Costs on Small Firms, available at www.sba.gov/idc/groups/public/ documents/sba_homepage/rs264tot.pdf). One analysis estimated the effect on GDP of economic regulations, using an index of the degree of OECD economic regulation and other variables. They found the following regression model.

GDP(1998–2002) = 10487 − 1343 OECD Economic Regulation Index + 1.078 GDP/Capita(1988) − 69.99 Ethno-linguistic Diversity Index + 44.71 Trade as share of GDP (1998–2002) − 58.4 Primary Education(%Eligible Population)

All *t*-statistics on the individual coefficients have P-values < 0.05, except the coefficient of *Primary Education.*

a) Does the coefficient of the OECD Economic Regulation Index indicate that more regulation leads to lower GDP? Explain.

b) The *F*-statistic for this model is 129.61 (5, 17 *df*). What do you conclude about the model?

c) If *GDP/Capita(1988)* is removed as a predictor, then the *F*-statistic drops to 0.694 and none of the *t*-statistics is significant (all P-values > 0.22). Reconsider your interpretation in (a).

19. Home prices. Many variables have an impact on determining the price of a house. A few of these are size of the house (square feet), lot size, and number of bathrooms. Information for a random sample of homes for sale in the Statesboro, Georgia, area was obtained from the Internet. Regression output modeling the asking price with square footage and number of bathrooms gave the following result.

Dependent Variable is: Asking Price
s = 67013 R-Sq = 71.1% *R*-Sq(adj) = 64.6%

Predictor	Coeff	SE(Coeff)	t-ratio	P-value
Intercept	−152037	85619	−1.78	0.110
Baths	9530	40826	0.23	0.821
Area	139.87	46.67	3.00	0.015

Analysis of Variance

Source	DF	SS	MS	F	P-value
Regression	2	99303550067	49651775033	11.06	0.004
Residual	9	40416679100	4490742122		
Total	11	1.39720E + 11			

a) Write the regression equation.

b) How much of the variation in home asking prices is accounted for by the model?

c) Explain in context what the coefficient of Area means.

d) The owner of a construction firm, upon seeing this model, objects because the model says that the number of bathrooms has no effect on the price of the home. He says that when *he* adds another bathroom, it increases the value. Is it true that the number of bathrooms is unrelated to house price? (*Hint:* Do you think bigger houses have more bathrooms?)

20. Home prices, part 2. Here are some diagnostic plots for the home prices data from Exercise 19. These were generated by a computer package and may look different from the plots generated by the packages you use. (In particular, note that the axes of the Normal probability plot are swapped relative to the plots we've made in the text. We only care about the pattern of this plot, so it shouldn't affect your interpretation.) Examine these plots and discuss whether the assumptions and conditions for the multiple regression seem reasonable.

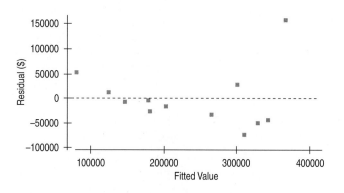

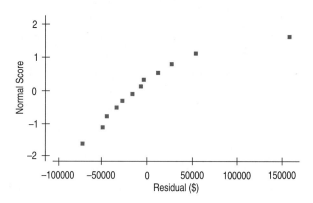

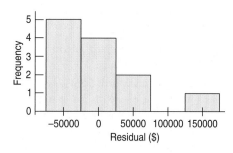

21. Secretary performance. The AFL-CIO has undertaken a study of 30 secretaries yearly salaries (in thousands of dollars). The organization wants to predict salaries from several other variables. The variables to be considered potential predictors of salary are:

X1 = months of service
X2 = years of education
X3 = score on standardized test
X4 = words per minute (wpm) typing speed
X5 = ability to take dictation in words per minute

A multiple regression model with all five variables was run on a computer package, resulting in the following output.

Variable	Coeff	Std. Error	t-value
Intercept	9.788	0.377	25.960
X1	0.110	0.019	5.178
X2	0.053	0.038	1.369
X3	0.071	0.064	1.119
X4	0.004	0.0307	0.013
X5	0.065	0.038	1.734

$s = 0.430$ $R\text{-sq} = 0.863$

Assume that the residual plots show no violations of the conditions for using a linear regression model.

a) What is the regression equation?

b) From this model, what is the predicted salary (in thousands of dollars) of a secretary with 10 years (120 months) of experience, 9th grade education (9 years of education), 50 on the standardized test, 60 wpm typing speed, and the ability to take 30 wpm dictation?

c) Test whether the coefficient for words per minute of typing speed (X4) is significantly different from zero at $\alpha = 0.05$.

d) How might this model be improved?

e) A correlation of age with salary finds $r = 0.682$, and the scatterplot shows a moderately strong positive linear association. However, if X6 = Age is added to the multiple regression, the estimated coefficient of age turns out to be $b_6 = -0.154$. Explain some possible causes for this apparent change of direction in the relationship between age and salary.

22. Wal-Mart revenue. Here's a regression of monthly revenue of Wal-Mart Corp, relating that revenue to the Total U.S. Retail Sales, the Personal Consumption Index, and the Consumer Price Index.

Dependent variable is: Wal-Mart_Revenue
R squared = 66.7% R squared (adjusted) = 63.8%
s = 2.327 with 39 − 4 = 35 degrees of freedom

Source	Sum of Squares	df	Mean Square	F-ratio
Regression	378.749	3	126.250	23.3
Residual	189.474	35	5.41354	

Variable	Coeff	SE(Coeff)	t-ratio	P-value
Intercept	87.0089	33.60	2.59	0.0139
Retail Sales	0.000103	0.000015	6.67	< 0.0001
Persnl Consmp	0.00001108	0.000004	2.52	0.0165
CPI	−0.344795	0.1203	−2.87	0.0070

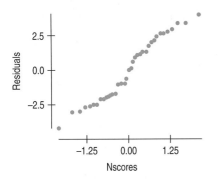

a) Write the regression model.

b) Interpret the coefficient of the Consumer Price Index (CPI). Does it surprise you that the sign of this coefficient is negative? Explain.

c) Test the standard null hypothesis for the coefficient of CPI and state your conclusions.

23. Mutual funds returns. In Chapter 16, Exercise 23 considered the relationship between the Wilshire 5000 Total Market Return and the amount of money flowing into and out of mutual funds (fund flows) monthly from January 1990 through October of 2002. The data file included data on the unemployment rate in each of the months. The original model looked like this.

Dependent variable is: Fund_Flows
R squared = 17.9% R squared (adjusted) = 17.3%
s = 10999 with 154 − 2 = 152 degrees of freedom

Source	Sum of Squares	df	Mean Square	F-ratio
Regression	4002044231	1	4002044231	33.1
Residual	18389954453	152	120986542	

Variable	Coeff	SE(Coeff)	t-ratio	P-value
Intercept	9599.10	896.9	10.7	< 0.0001
Wilshire	1156.40	201.1	5.75	< 0.0001

Adding the *Unemployment Rate* to the model yields:

Dependent variable is: Fund_Flows
R squared = 28.8% R squared (adjusted) = 27.8%
s = 10276 with 154 − 3 = 151 degrees of freedom

Source	Sum of Squares	df	Mean Square	F-ratio
Regression	6446800389	2	3223400194	30.5
Residual	15945198295	151	105597340	

Variable	Coeff	SE(Coeff)	t-ratio	P-value
Intercept	30212.2	4365	6.92	≤ 0.0001
Wilshire	1183.29	187.9	6.30	< 0.0001
Unemployment	−3719.55	773.0	−4.81	< 0.0001

a) Interpret the coefficient of the *Unemployment Rate*.

b) The *t*-ratio for the *Unemployment Rate* is negative. Explain why.

c) State and complete the standard hypothesis test for the *Unemployment Rate*.

24. Lobster value, revisited. In Chapter 17, Exercise 33 predicted the annual value of the Maine lobster industry catch from the number of licensed lobser fishers. The lobster industry is an important one in Maine, with annual landings worth about $300,000,000 and employment consequences that extend throughout the state's economy. We saw in chapter 17 that it was best to transform Value by logarithms. Here's a more sophisticated multiple regression to predict the logValue from other variables published by the Maine Department of Marine Resources (maine.gov/dmr). The predictors are number of *Traps* (millions), number of licensed *Fishers*, and *Water Temperature* (°F).

Dependent variable is: LogValue
R squared = 97.4% R squared (adjusted) = 97.3%
s = 0.0782 with 56 − 4 = 52 degrees of freedom

Source	Sum of Squares	df	Mean Square	F-ratio
Regression	12.0200	3	4.00667	656
Residual	0.317762	52	0.006111	

Variable	Coeff	SE(Coeff)	t-ratio	P-value
Intercept	7.70207	0.3706	20.8	< 0.0001
Traps(M)	0.575390	0.0152	37.8	< 0.0001
Fishers	−0.0000532	0.00001	−5.43	< 0.0001
Water Temp	−0.015185	0.0074	−2.05	0.0457

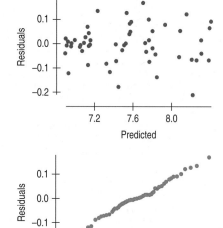

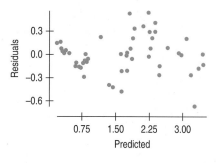

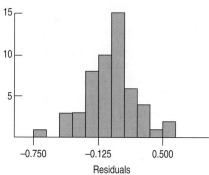

a) Write the regression model.

b) Are the assumptions and conditions met?

c) Interpret the coefficient of *Fishers*. Would you expect that restricting the number of lobstering licenses to even fewer fishers would increase the value of the harvest?

d) State and test the standard null hypothesis for the coefficient of *Water Temperature*. Scientists claim that this is an important predictor of the harvest. Do you agree?

T 25. Lobster price, revisited 2. In Chapter 17, Exercise 34 predicted the price ($/lb) of lobster harvested in the Maine lobster fishing industry. Here's a multiple regression to predict the *Price* from the number of *Traps* (millions), the number of *Fishers*, and the *Catch/Trap* (metric tonnes).

```
Dependent variable is:    Price/lb
R squared = 94.4%   R squared (adjusted) = 94.1%
s = 0.2462 with 53 − 4 = 49 degrees of freedom
```

Source	Sum of Squares	df	Mean Square	F-ratio
Regression	50.4337	3	16.8112	277
Residual	2.96960	49	0.060604	

Variable	Coeff	SE(Coeff)	t-ratio	P-value
Intercept	0.845123	0.3557	2.38	0.0215
Traps(M)	1.20094	0.0619	19.4	≤ 0.0001
Fishers	−0.0001218	0.00004	−3.30	0.0018
Ctch/Trp	−22.0207	10.96	−2.01	0.0500

a) Write the regression model.

b) Are the assumptions and conditions met?

c) State and test the standard null hypothesis for the coefficient of *Catch/Trap*. Use the standard α-level of .05 and state your conclusion.

d) Does the coefficient of *Catch/Trap* mean that when the catch per trap declines the price will increase?

e) This model has an adjusted R^2 of 94.1%. The previous model of Chapter 17 Exercise 34 had an adjusted R^2 of 93.6%. Explain why this is an appropriate statistic to use to compare these models.

T 26. Price of beef. How is the price of beef related to other factors? The data in Table 18.1 give information on the price of beef (*PBE*) and the possible explanatory variables: consumption of beef per capita (*CBE*), retail food price index (*PFO*), food consumption per capita index (*CFO*), and an index of real disposable income per capita (*RDINC*) for the years 1925 to 1941 in the United States.

a) Use computer software to find the regression equation for predicting the price of beef based on all of the given explanatory variables. What is the regression equation?

b) Produce the appropriate residual plots to check the assumptions. Is this inference appropriate for this model? Explain.

c) How much variation in the price of beef can be explained by this model?

d) Consider the coefficient of beef consumption per capita (*CBE*). Does it say that that price of beef goes up when people eat less beef? Explain.

Year	PBE	CBE	PFO	CFO	RDINC
1925	59.7	58.6	65.8	90.9	68.5
1926	59.7	59.4	68	92.1	69.6
1927	63	53.7	65.5	90.9	70.2
1928	71	48.1	64.8	90.9	71.9
1929	71	49	65.6	91.1	75.2
1930	74.2	48.2	62.4	90.7	68.3
1931	72.1	47.9	51.4	90	64
1932	79	46	42.8	87.8	53.9
1933	73.1	50.8	41.6	88	53.2
1934	70.2	55.2	46.4	89.1	58
1935	82.2	52.2	49.7	87.3	63.2
1936	68.4	57.3	50.1	90.5	70.5
1937	73	54.4	52.1	90.4	72.5
1938	70.2	53.6	48.4	90.6	67.8
1939	67.8	53.9	47.1	93.8	73.2
1940	63.4	54.2	47.8	95.5	77.6
1941	56	60	52.2	97.5	89.5

| *Table 18.1*

T 27. Wal-Mart revenue, part 2. Wal-Mart is the second largest retailer in the world. The data file on the disk holds monthly data on Wal-Mart's revenue, along with several possibly related economic variables.

a) Using computer software, find the regression equation predicting Wal-Mart revenues from the *Retail Index*, the Consumer Price index (*CPI*), and *Personal Consumption*.

b) Does it seem that Wal-Mart's revenue is closely related to the general state of the economy?

T 28. Wal-Mart revenue, part 3. Consider the model you fit in Exercise 27 to predict Wal-Mart's revenue from the Retail Index, CPI, and Personal Consumption index.

a) Plot the residuals against the predicted values and comment on what you see.

b) Identify and remove the four cases corresponding to December revenue and find the regression with December results removed.

c) Does it seem that Wal-Mart's revenue is closely related to the general state of the economy?

*29. Right-to-work laws. Are state right-to-work laws related to the percent of public sector employees in unions and the percent of private sector employees in unions? This data set looks at these percentages for the states in the United States in 1982. The dependent variable is whether the state had a right-to-work law or not. The computer output for the logistic regression is given here. (Source: N. M. Meltz, "Interstate and Interprovincial Differences in Union Density," *Industrial Relations*, 28:2 (Spring 1989), 142–158 by way of DASL.)

Logistic Regression Table

Predictor	Coeff	SE(Coeff)	z	P
Intercept	6.19951	1.78724	3.47	0.001
publ	−0.106155	0.0474897	−2.24	0.025
pvt	−0.222957	0.0811253	−2.75	0.006

a) Write out the estimated regression equation.

b) The following are scatterplots of the response variable against each of the explanatory variables. Examine them for the conditions required by logistic regression. Does logistic regression seem appropriate here? Explain.

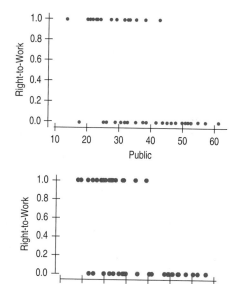

*30. Cost of higher education. Are there fundamental differences between liberal arts colleges and universities? In this case, we have information on the top 25 liberal arts colleges and the top 25 universities in the Unites States. We will consider the type of school as our response variable and will use the percent of students who were in the top 10% of their high school class and the amount of money spent per student by the college or university as our explanatory variables. The output from this logistic regression is given here.

Logistic Regression Table

Predictor	Coeff	SE(Coeff)	z	P
Intercept	−13.1461	3.98629	−3.30	0.001
Top 10%	0.0845469	0.0396345	2.13	0.033
$/Student	0.0002594	0.0000860	3.02	0.003

a) Write out the estimated regression equation.

b) Is percent of students in the top 10% of their high school class statistically significant in predicting whether or not the school is a university? Explain.

c) Is the amount of money spent per student statistically significant in predicting whether or not the school is a university? Explain.

T 31. Motorcycles. More than one million motorcycles are sold annually (www.webbikeworld.com). Off-road motorcycles (often called "dirt bikes") are a market segment (about 18%) that is highly specialized and offers great variation in features. This makes it a good segment to study to learn about which features account for the cost (manufacturer's suggested retail price, MSRP) of a dirt bike. Researchers collected data on 2005 model dirt bikes (lib.stat.cmu.edu/datasets/dirtbike_aug.csv). Their original goal was to study market differentiation among brands (*The Dirt on Bikes: An Illustration of CART Models for Brand Differentiation*, Jiang Lu, Joseph B. Kadane, and Peter Boatwright, server1.tepper.cmu.edu/gsiadoc/WP/2006-E57.pdf), but we can use these to predict msrp from other variables.

Here are scatterplots of three potential predictors, *Wheelbase (in)*, *Displacement (cu in)*, and *Bore (in)*.

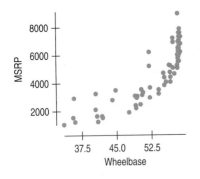

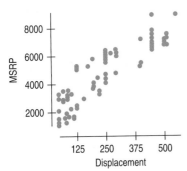

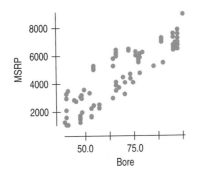

Which of these variables would you choose first as a predictor in a regression to model *MSRP*? Explain.

T 32. Motorcycles, part 2. In Exercise 31, we saw data on off-road motorcycles and examined scatterplots. Review those scatterplots. Here's a regression of *MSRP* on both *Displacement* and *Bore*. Both of the predictors are measures of the size of the engine. The displacement is the total volume of air and fuel mixture that an engine can draw in during one cycle. The bore is the diameter of the cylinders.

Dependent variable is: MSRP
R squared = 77.0% R squared (adjusted) = 76.5%
s = 979.8 with 98 − 3 = 95 degrees of freedom

Variable	Coeff	SE(Coeff)	*t*-ratio	P-value
Intercept	318.352	1002	0.318	0.7515
Bore	41.1650	25.37	1.62	0.1080
Displacement	6.57069	3.232	2.03	0.0449

a) State and test the standard null hypothesis for the coefficient of *Bore*.
b) Both of these predictors seem to be linearly related to *MSRP*. Explain what your result in (a) means.

33. Motorcycles, part 3. Here's another model for the *MSRP* of off-road motorcycles.

Dependent variable is: MSRP
R squared = 90.9% R squared (adjusted) = 90.6%
s = 617.8 with 95 − 4 = 91 degrees of freedom

Source	Sum of Squares	df	Mean Square	F-ratio
Regression	346795061	3	115598354	303
Residual	34733372	91	381685	

Variable	Coeff	SE(Coeff)	*t*-ratio	P-value
Intercept	−2682.38	371.9	−7.21	< 0.0001
Bore	86.5217	5.450	15.9	< 0.0001
Clearance	237.731	30.94	7.68	< 0.0001
Engine strokes	−455.897	89.88	−5.07	< 0.0001

a) Would this be a good model to use to predict the price of an off-road motorcycle if you knew its bore, clearance, and engine strokes? Explain.
b) The Suzuki DR650SE had an *MSRP* of $4999 and a 4-stroke engine, with a bore of 100 inches. Can you use this model to estimate its *Clearance*? Explain.

T 34. Demographics. Here is a data set on various measures of the 50 United States. The *Murder* rate is per 100,000, *HS Graduation* rate is in %, *Income* is per capita income in dollars, *Illiteracy* rate is per 1000, and *Life Expectancy* is in years. Find a regression model for *Life Expectancy* with three predictor variables by trying all four of the possible models.

State name	Murder	HS grad	Income	Illiteracy	Life expectancy
Alabama	15.1	41.3	3624	2.1	69.05
Alaska	11.3	66.7	6315	1.5	69.31
Arizona	7.8	58.1	4530	1.8	70.55
Arkansas	10.1	39.9	3378	1.9	70.66
California	10.3	62.6	5114	1.1	71.71
Colorado	6.8	63.9	4884	0.7	72.06
Connecticut	3.1	56	5348	1.1	72.48
Delaware	6.2	54.6	4809	0.9	70.06
Florida	10.7	52.6	4815	1.3	70.66
Georgia	13.9	40.6	4091	2	68.54
Hawaii	6.2	61.9	4963	1.9	73.6
Idaho	5.3	59.5	4119	0.6	71.87
Illinois	10.3	52.6	5107	0.9	70.14
Indiana	7.1	52.9	4458	0.7	70.88
Iowa	2.3	59	4628	0.5	72.56
Kansas	4.5	59.9	4669	0.6	72.58
Kentucky	10.6	38.5	3712	1.6	70.1
Louisiana	13.2	42.2	3545	2.8	68.76
Maine	2.7	54.7	3694	0.7	70.39
Maryland	8.5	52.3	5299	0.9	70.22
Massachusetts	3.3	58.5	4755	1.1	71.83
Michigan	11.1	52.8	4751	0.9	70.63
Minnesota	2.3	57.6	4675	0.6	72.96
Mississippi	12.5	41	3098	2.4	68.09
Missouri	9.3	48.8	4254	0.8	70.69
Montana	5	59.2	4347	0.6	70.56
Nebraska	2.9	59.3	4508	0.6	72.6
Nevada	11.5	65.2	5149	0.5	69.03
New Hampshire	3.3	57.6	4281	0.7	71.23
New Jersey	5.2	52.5	5237	1.1	70.93
New Mexico	9.7	55.2	3601	2.2	70.32
New York	10.9	52.7	4903	1.4	70.55
North Carolina	11.1	38.5	3875	1.8	69.21
North Dakota	1.4	50.3	5087	0.8	72.78
Ohio	7.4	53.2	4561	0.8	70.82
Oklahoma	6.4	51.6	3983	1.1	71.42
Oregon	4.2	60	4660	0.6	72.13
Pennsylvania	6.1	50.2	4449	1	70.43
Rhode Island	2.4	46.4	4558	1.3	71.9
South Carolina	11.6	37.8	3635	2.3	67.96
South Dakota	1.7	53.3	4167	0.5	72.08
Tennessee	11	41.8	3821	1.7	70.11
Texas	12.2	47.4	4188	2.2	70.9
Utah	4.5	67.3	4022	0.6	72.9
Vermont	5.5	57.1	3907	0.6	71.64
Virginia	9.5	47.8	4701	1.4	70.08
Washington	4.3	63.5	4864	0.6	71.72
West Virginia	6.7	41.6	3617	1.4	69.48
Wisconsin	3	54.5	4468	0.7	72.48
Wyoming	6.9	62.9	4566	0.6	70.29

a) Which model appears to do the best?

b) Would you leave all three predictors in this model?

c) Does this model mean that by changing the levels of the predictors in this equation, we could affect life expectancy in that state? Explain.

d) Be sure to check the conditions for multiple regression. What do you conclude?

 35. Burger King nutrition. Like many fast-food restaurant chains, Burger King (BK) provides data on the nutrition content of its menu items on its website. Here's a multiple regression predicting calories for Burger King foods from *Protein* content *(g)*, *Total Fat (g)*, *Carbohydrate (g)*, and *Sodium (mg)* per serving.

Dependent variable is: Calories
R-squared = 100.0% *R*-squared [adjusted] = 100.0%
s = 3.140 with 31 − 5 = 26 degrees of freedom

Source	Sum of Squares	df	Mean Square	F-ratio
Regression	1419311	4	354828	35994
Residual	256.307	26	9.85796	

Variable	Coeff	SE(Coeff)	t-ratio	P-value
Intercept	6.53412	2.425	2.69	0.0122
Protein	3.83855	0.0859	44.7	<0.0001
Total fat	9.14121	0.0779	117	<0.0001
Carbs	3.94033	0.0338	117	<0.0001
Na/Serv.	−0.69155	0.2970	−2.33	0.0279

a) Do you think this model would do a good job of predicting calories for a new BK menu item? Why or why not?

b) The mean of *Calories* is 455.5 with a standard deviation of 217.5. Discuss what the value of *s* in the regression means about how well the model fits the data.

c) Does the R^2 value of 100.0% mean that the residuals are all actually equal to zero?

✓ JUST CHECKING ANSWERS

1 58.4% of the variation in *Percent Body Fat* can be accounted for by the multiple regression model using *Height*, *Age*, and *Weight* as predictors.

2 For a given *Height* and *Weight*, an increase of one year in *Age* is associated with an increase of 0.137% in *Body Fat* on average.

3 The multiple regression coefficient is interpreted for *given* values of the other variables. That is, for people of the *same Weight* and *Age*, an increase of one inch of *Height* is associated with, on average, a *decrease* of 1.274% in *Body Fat*. The same cannot be said when looking at people of all *Weights* and *Ages*.

CHAPTER 19

Building Multiple Regression Models

Bolliger and Mabillard

The story is told that when John Wardley, the award-winning concept designer of theme parks and roller coasters, including *Nitro* and *Oblivion*, was about to test the roller coaster *Nemesis* for the first time, he asked Walter Bolliger, the president of the coaster's manufacturer, B&M, "What if the coaster stalls? How will we get the trains back to the station?" Bolliger replied, "Our coasters never stall. They always work perfectly the first time." And, of course, it did work perfectly.

Roller coaster connoisseurs know that Bolliger & Mabillard Consulting Engineers, Inc (B&M) is responsible for some of the most innovative roller coasters in the business. The company was founded in the late 1980s when Walter Bolliger and Claude Mabillard left Intamin AG, where they had designed the company's first stand-up coaster. B&M has built its reputation on innovation. They developed the first "inverted" roller coaster in which the train runs under the track with the seats attached to the wheel carriage and pioneered "diving machines," which feature a vertical drop, introduced first with *Oblivion*.

B&M coasters are famous among enthusiasts for particularly smooth rides and for their reliability, easy maintainability, and excellent safety record. Unlike some other manufacturers, B&M does not use powered launches, preferring, as do many coaster connoisseurs, gravity-powered coasters. B&M is an international leader in the roller coaster design field, having designed 24 of the top 50 steel roller coasters on the 2006 Golden Ticket Awards list and 4 of the top 10.

Theme parks are big business. In the United States alone, there are nearly 500 theme and amusement parks which generate over $10 billion a year in revenue. The U.S. industry is fairly mature, but parks in the rest of the world are still growing. Europe now generates more than $1 billion a year from their theme parks, and Asia's industry is growing fast. Although theme parks have started to diversify to include water parks and zoos, rides are still the main attraction at most parks, and at the center of the rides is the roller coaster. Engineers and designers compete to make them bigger and faster. For a two-minute ride on the fastest and best roller coasters, fans will wait for hours.

Can we learn what makes a roller coaster fast? What are the most important design considerations in getting the fastest coaster? Here are data on some of the fastest roller coasters in the world today:

Name	Park	Country	Type	Duration (sec)	Speed (mph)	Height (ft)	Drop (ft)	Length (ft)	Inversion?
New Mexico Rattler	Cliff's Amusement Park	USA	Wooden	75	47	80	75	2750	No
Fujiyama	Fuji-Q Highlands	Japan	Steel	216	80.8	259.2	229.7	6708.67	No
Goliath	Six Flags Magic Mountain	USA	Steel	180	85	235	255	4500	No
Great American Scream Machine	Six Flags Great Adventure	USA	Steel	140	68	173	155	3800	Yes
Hangman	Wild Adventures	USA	Steel	125	55	115	95	2170	Yes
Hayabusa	Tokyo SummerLand	Japan	Steel	108	60.3	137.8	124.67	2559.1	No
Hercules	Dorney Park	USA	Wooden	135	65	95	151	4000	No
Hurricane	Myrtle Beach Pavilion	USA	Wooden	120	55	101.5	100	3800	No

A small selection of coasters from the larger data set available on the CD.

| **Table 19.1** *Facts about some roller coasters. Source: The Roller coaster database at www.rcdb.com.*

<table>
<tr><td>WHO</td><td>Roller coasters</td></tr>
<tr><td>WHAT</td><td>See Table 19.1 for the variables and their units.</td></tr>
<tr><td>WHERE</td><td>Worldwide</td></tr>
<tr><td>WHEN</td><td>All were in operation in 2003.</td></tr>
<tr><td>WHY</td><td>To understand characteristics that affect speed and duration</td></tr>
</table>

◆ *Type* indicates what kind of track the roller coaster has. The possible values are "wooden" and "steel." (The frame usually is of the same construction as the track, but doesn't have to be.)

◆ *Duration* is the duration of the ride in seconds.

◆ *Speed* is top speed in miles per hour.

◆ *Height* is maximum height above ground level in feet.

◆ *Drop* is greatest drop in feet.

◆ *Length* is total length of the track in feet.

◆ *Inversions* reports whether riders are turned upside down during the ride. It has the values "yes" or "no."

Customers not only want the ride to be fast; they also want it to last. It makes sense that the longer the track is, the longer the ride will last. Let's have a look at *Duration* and *Length* to see what the relationship is.

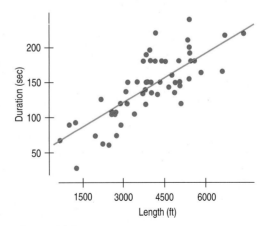

Figure 19.1 *The relationship between* Duration *and* Length *looks strong and positive. On average,* Duration *increases linearly with* Length.

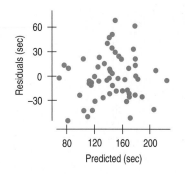

```
Dependent variable is: Duration
R-squared = 62.0%  R-squared (adjusted) = 61.4%
s = 27.23 with 63 - 2 = 61 degrees of freedom

Source       Sum of Squares    DF    Mean Square    F-ratio
Regression   73901.7           1     73901.7        99.6
Residual     45243.7           61    741.700

Variable    Coefficient    SE(Coeff)    t-ratio    P-value
Intercept   53.9348        9.488        5.68       <0.0001
Length      0.0231         0.0023       9.98       <0.0001
```

| **Table 19.2** *The regression of* Duration *on* Length *looks strong, and the conditions seem to be met.*

As the scatterplots of the variables (Figure 19.1) and residuals (Figure 19.2) show, the regression conditions seem to be met, and the regression makes sense. We'd expect longer tracks to give longer rides. Starting from the intercept at about 53.9 seconds, the duration of the ride increases, on average, by 0.0231 seconds per foot of track—or 23.1 seconds more for each 1000 additional feet of track.

19.1 Indicator (or Dummy) Variables

Of course, there's more to these data. One interesting variable might not be one you'd naturally think of. Many modern coasters have "inversions" where riders get turned upside down, with loops, corkscrews, or other devices. These inversions add excitement, but they must be carefully engineered, and that enforces some speed limits on that portion of the ride. Riders like speed, but they also like inversions which affect both the speed and the duration of the ride. We'd like to add the information of whether the roller coaster has an inversion to our model. Until now, all our predictor variables have been quantitative. Whether or not a roller coaster has an inversion is a categorical variable ("yes" or "no"). Let's see how to introduce the categorical variable *Inversions* as a predictor in our regression model. Figure 19.2 shows the same scatterplot of *Duration* against *Length*, but now with the roller coasters that have inversions shown as red x's and those without shown as blue dots. There's a separate regression line for each type of roller coaster.

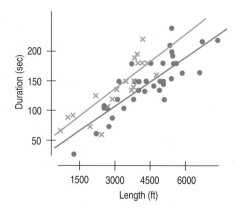

Figure 19.2 *The two lines fit to coasters with and without inversions are roughly parallel.*

It's easy to see that, for a given length, the roller coasters with inversions take a bit longer and that for each type of roller coaster, the slopes of the relationship between duration and length are not quite equal but are similar. If we split the data into two groups—coasters without inversions and those with inversions—and compute the regression for each group, the output looks like this:

```
Dependent variable is: Duration
Cases selected according to: No inversions
R-squared = 69.4%  R-squared (adjusted) = 68.5%
s = 25.12 with 38 − 2 = 36 degrees of freedom
```

Variable	Coefficient	SE(Coeff)	t-ratio	P-value
Intercept	25.9961	14.10	1.84	0.0734
Length	0.0274	0.003	9.03	<0.0001

```
Dependent variable is: Duration
Cases selected according to: Inversions
R-squared = 70.5%  R-squared (adjusted) = 69.2%
s = 23.20 with 25 − 2 = 23 degrees of freedom
```

Variable	Coefficient	SE(Coeff)	t-ratio	P-value
Intercept	47.6454	12.50	3.81	0.0009
Length	0.0299	0.004	7.41	<0.0001

Table 19.3 *The regressions computed separately for the two types of roller coasters show similar slopes but different intercepts.*

As the scatterplot showed, the slopes are very similar, but the intercepts are different. When we have a situation like this with roughly parallel regressions for each group,[1] there's an easy way to add the group information to a single regression model. We create a new variable that *indicates* what type of roller coaster we have, giving it the value 1 for roller coasters that have inversions and the value 0 for those that don't. (We could have reversed the coding; it's an arbitrary choice.[2]) Such variables are called **indicator variables** or *indicators* because they indicate which category each case is in. They are also often called **dummy variables**. When we add our new indicator, *Inversions*, to the regression model as a second variable, the multiple regression model looks like this.

Inversions = 1 if a coaster has inversions.
Inversions = 0 if not.

```
Dependent variable is: Duration
R-squared = 70.4%  R-squared (adjusted) = 69.4%
s = 24.24 with 63 − 3 = 60 degrees of freedom
```

Variable	Coefficient	SE(Coeff)	t-ratio	P-value
Intercept	22.3909	11.39	1.97	0.0539
Length	0.0282	0.0024	11.7	<0.0001
Inversions	30.0824	7.290	4.13	<0.0001

Table 19.4 *The regression model with a dummy, or indicator, variable for* Inversions.

This looks like a better model than the simple regression for all the data. The R^2 is larger, the *t*-ratios of both coefficients are large, and now we can understand the effect of inversions with a single model without having to compare two regressions. (The residuals look reasonable as well.) But what does the coefficient for *Inversions* mean? Let's see how an indicator variable works when we calculate predicted values for two of the roller coasters listed in Table 19.1.

---

[1] The fact that the individual regression lines are nearly parallel is a part of the Linearity Condition. You should check that the lines are nearly parallel before using this method or read on to see what to do if they are not parallel enough.
[2] Some implementations of indicator variables use 1 and −1 for the levels of the categories.

Name	Park	Country	Type	Duration	Speed	Height	Drop	Length	Inversion?
Hangman	Wild Adventures	USA	Steel	125	55	115	95	2170	Yes
Hayabusa	Tokyo SummerLand	Japan	Steel	108	60.3	137.8	124.67	2559.1	No

Ignoring the variable *Inversions* for the moment, the model (in Table 19.4) says that for all coasters, the predicted *Duration* is:

$$22.39 + 0.0282 \text{ } Length + 30.08 \text{ } Inversions.$$

Now remember that for this indicator variable, the value 1 means that a coaster has an inversion, while a 0 means it doesn't. For *Hayabusa*, with no inversion, the value of *Inversions* is 0, so the coefficient of *Inversions* doesn't affect the prediction at all. With a length of 2259.1 feet, we predict its duration as:[3]

$$22.39 + 0.0282 \text{ } (2559.1) + 30.08 \times 0 = 94.56 \text{ seconds},$$

which is close to its actual duration of 108 seconds. The *Hangman* (with a length of 2170 feet) has an inversion, and so the model predicts an "additional" 30.0824 seconds for its duration:

$$22.39 + 0.0282 \text{ } (2170.0) + 30.08 \times 1 = 113.66 \text{ seconds}.$$

That compares well with the actual duration of 125 seconds.

Notice how the indicator works in the model. When there is an inversion (as in *Hangman*), the value 1 for the indicator causes the amount of the indicator's coefficient, 30.08, to be *added* to the prediction. When there is no inversion (as in *Hayabusa*), the indicator is 0, so nothing is added. Looking back at the scatterplot, we can see that this is exactly what we need. The difference between the two lines is a vertical shift of about 30 seconds. This may seem a bit confusing at first because we usually think of the coefficients in a multiple regression as slopes. For indicator variables, however, they act differently. They're vertical shifts that keep the slopes for the other variables apart.

An indicator variable that is 0 or 1 can only shift the line up and down. It can't change the slope, so it works only when we have lines with the same slope and different intercepts.

19.2 Adjusting for Different Slopes—Interaction Terms

Consumers, even of fast food, are increasingly concerned with nutrition. So, like many restaurant chains, Burger King publishes the nutrition details of its menu items on its website (www.bk.com/Nutrition/). Many customers count calories or carbohydrates. Of course, these are likely to be related to each other. We can examine that relationship in Burger king foods by looking at a scatterplot.

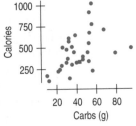

Figure 19.3 Calories of Burger King foods plotted against Carbs seems to fan out.

---

[3] We round coefficient values when we write the model but calculate with the full precision, rounding at the end of the calculation.

It's not surprising to see that an increase in *Carbs* is associated with more *Calories*, but the plot seems to thicken as we move from left to right. Could there be something else going on?

We divide Burger King foods into two groups, coloring those with meat (including chicken and fish) in orange and those without meat in blue. Looking at the regressions for each group, we see a different picture:

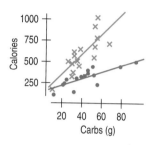

Figure 19.4 *Plotting the meat-based and non-meat items separately, we see two distinct linear patterns.*

Clearly, meat-based items contribute more calories from their carbohydrate content than do other Burger King foods. But unlike Figure 19.2, when the lines were parallel, we can't account for the kind of difference we see here by just including a dummy variable in a regression. It isn't just the heights of the lines that are different; they have entirely different slopes.

We'll start, as before, by constructing the indicator for the two groups, *Meat*, which is 1 for foods that contain meat and 0 for the others. The variable *Meat* can adjust the intercepts of the two lines. To adjust the slopes, we have to construct another variable—the *product* of *Meat* and the predictor variable *Carbs*. The coefficient of this **interaction term** in a multiple regression gives an adjustment to the slope for the cases in the indicated group. The resulting variable *Carbs*Meat* has the value of *Carbs* for foods containing meat (those coded 1 in the *Meat* indicator) and the value 0 for the others. By including this interaction variable in the model, we can adjust the slope of the line fit to the meat-containing foods. Here's the resulting analysis:

```
Dependent variable is: Calories
R-squared = 78.1%  R-squared (adjusted) = 75.7%
s = 106.0 with 32 − 4 = 28 degrees of freedom
```

Source	Sum of Squares	DF	Mean Square	F-ratio
Regression	1119979	3	373326	33.2
Residual	314843	28	11244.4	

Variable	Coefficient	SE(Coeff)	t-ratio	P-value
Intercept	137.395	58.72	2.34	0.0267
Carbs(g)	3.93317	1.113	3.53	0.0014
Meat	−26.1567	98.48	−0.266	0.7925
Carbs*Meat	7.87530	2.179	3.61	0.0012

Table 19.5 *The regression model with both an indicator variable and an interaction term.*

What does the coefficient for the indicator *Meat* do? It provides a different intercept to separate the meat and non-meat items at the origin (where *Carbs* = 0). Each group has its own slope, but the two lines nearly meet at the origin, so there seems to be no need for an additional intercept adjustment. The difference of 26.16

calories is small. That's why the coefficient for the indicator variable *Meat* has a small *t*-statistic (−0.266).

By contrast, the coefficient of the interaction term, *Carbs*Meat*, says that the slope relating calories to carbohydrates is steeper by 7.88 calories per carbohydrate gram for meat-containing foods than for meat-free foods. Its small P-value suggests that this difference is real. Overall, the regression model predicts calories to be:

$$137.40 + 3.93 \ Carbs - 26.16 \ Meat + 7.88 \ Carbs*Meat.$$

Let's see how these adjustments work. A BK Whopper has 53 grams of carbohydrates and is a meat dish. The model predicts its *Calories* as:

$$137.40 + 3.93 \times 53 - 26.16 \times 1 + 7.88 \times 53 \times 1 = 737.2 \text{ calories},$$

not far from the measured calorie count of 680. By contrast, the Veggie Burger, with 43 grams of carbohydrates, has value 0 for *Meat* and so has a value of 0 for *Carbs*Meat* as well. Those indicators contribute nothing to its predicted calories:

$$137.40 + 3.93 \times 43 - 26.16 \times 0 + 7.88 \times 0 \times 43 = 306.4 \text{ calories},$$

close to the 330 measured officially.

Indicators for Three or More Categories

It's easy to construct indicators for a variable with two categories; we just assign 0 to one level and 1 to the other. But business and economic variables such as *Month* or *Socioeconomic Class* may have several levels. You can construct indicators for a categorical variable with several levels by constructing a separate indicator for each of these levels. There's just one thing to keep in mind. If a variable has *k* levels, you can create only *k-1* indicators. You have to choose one of the *k* categories as a "baseline" and *leave out* its indicator. Then the coefficients of the other indicators can be interpreted as the amount by which their categories differ from the baseline, after allowing for the linear effects of the other variables in the model.

For the two-category variable *Inversions*, we used "no inversion" as the baseline, and coasters with an inversion got a 1. We needed only one variable for two levels. If we wished to represent *Month* with indicators, we would need 11 of them. We might, for example, define *January* as the baseline and make indicators for *February, March, . . . , November*, and *December*. Each of these indicators would be 0 for all cases except for the ones that had that value for the variable *Month*.

Why couldn't we use a single variable with "1" for *January*, "2" for *February*, and so on? That would require the pretty strict assumption that the responses to the months are linear and equally spaced—that is, that the change in our response variable from January to February is the same in both direction and amount as the change from July to August. That's a pretty severe restriction and usually isn't true. Using 11 indicators releases the model from that restriction even though it adds complexity to the model.

For the real estate data we analyzed in Chapter 18, we might want to introduce a variable for the *Fuel Type* used to heat the house. That variable is reported with values: 1 = None; 2 = Gas; 3 = Electric; 4 = Oil; 5 = Wood; 6 = Solar; 7 = Unknown/Other.

Because there are 7 levels, we need 6 indicator variables. We'll use 1 = None as the *baseline*. Fuel2 would have the value 1 for Gas and 0 for all other types, Fuel3 would have value 1 for Electric and 0 for all other types, etc. However, when we made a bar chart of the variables, we found that although there were 7 possible levels, only 6 houses actually had values other than 2, 3, and 4. So we decided to leave only 2 indicators, *Fuel2* and *Fuel3* indicating Gas and Electric respectively, leaving Oil as the baseline. We had to set aside 6 houses (out of over 1700) that used other fuels, but their impact on the model was small.

Once you've created multiple indicator variables (up to $k - 1$) for a categorical variable with *k* levels, it often helps to combine levels with similar characteristics and with similar relationships with the response. This can help keep the number of variables in a multiple regression from exploding.

19.3 Multiple Regression Diagnostics

We often use regression analyses to make important business decisions. By working with the data and creating models, we can learn a great deal about the relationships among variables. As we saw with simple regression, sometimes we can learn as much from the cases that *don't* fit the model as from the bulk of cases that do. Extraordinary cases often tell us more just by the ways in which they fail to conform and the reasons we can discover for those deviations. If a case doesn't conform to the others, we should identify it and, if possible, understand why it is different. In simple regression, a case can be extraordinary by standing away from the model in the *y* direction or by having unusual values in an *x*-variable. In multiple regression, it can also be extraordinary by having an unusual *combination* of values in the *x*-variables. Just as in simple regression, large deviations in the *y* direction show up in the residuals as outliers. Deviations in the *x*'s show up as *leverage*.

Leverage

In a regression of a single predictor and a response, it's easy to see if a value has high leverage, because it's far from the mean of the *x* values in a scatterplot. In a multiple regression with *k* predictor variables, things are more complicated. A point might actually not be far from any of the *x* means and yet still exert large leverage because it has an unusual *combination* of predictor values. Even a graphics program designed to display points in high dimensional spaces may not make it obvious. Fortunately, there are values of leverage that can be calculated and are standard for most multiple regression programs.

The **leverage** is defined as follows. For any case, add (or subtract) 1 to its *y*-value. Recompute the regression, and see how much the *predicted* value of the case changes. The amount of the change is the leverage. It can never be greater than 1 or less than 0. A point with zero leverage has no effect at all on the regression slope, although it does participate in the calculations of R^2, *s*, and the *F*- and *t*-statistics. (The leverage of the ith point in a data set is often denoted by h_i.)

A point with high leverage may not actually influence the regression coefficients if it follows the pattern of the model set by the other points, but it's worth examining simply because of its *potential* to do so. Looking at leverage values can be an effective way to discover cases that are extraordinary on a combination of *x*-variables. In business, such cases often deserve special attention.

There are no tests for whether the leverage of a case is too large. The average leverage value among all cases in a regression is $1/n$, but that doesn't give us much of a guide. Some packages use rules of thumb to indicate high leverage values,[4] but another common approach is to just make a histogram of the leverages. Any case whose leverage stands out in a histogram of leverages probably deserves special attention. You may decide to leave the case in the regression or to see how the regression model changes when you delete the case, but you should be aware of its potential to influence the regression.

We've already seen that the *Duration* of a roller coaster ride depends linearly on its *Length*. But even more than a long ride, roller coaster customers like speed. So, rather than predict the duration of a roller coaster ride, let's build a model for how fast it travels. A multiple regression in two variables shows that both the total

---

[4] One common rule for determining when a leverage is large is to indicate any leverage value greater than $3(k + 1)/n$, where *k* is the number of predictors.

Height and the *Drop* (the maximum distance from the top to the bottom of the largest drop in the ride) are important factors:

Variable	Coeff	SE(Coeff)	t-ratio	P-Value
Intercept	37.01333	1.47723	25.056	<.0001
Height	0.06581	0.01911	3.444	0.000953
Drop	0.12540	0.01888	6.643	<.0001

Multiple R-squared: 0.855   Adjusted R-squared: 0.851
F-statistic: 215.2 on 2 and 73 DF, P-value < 0.0001

Table 19.6 *Regression of Speed on Height and Drop shows both predictor variables to be highly significant.*

The regression certainly seems reasonable. The R^2 value is high, and the residual plot looks patternless:

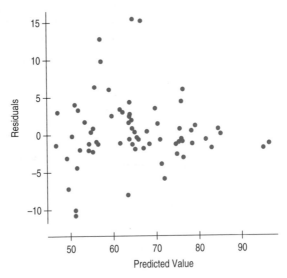

Figure 19.5 *The scatterplot of residuals against predicted values shows nothing unusual for the regression of Speed on Height and Drop.*

A histogram of the leverage values, however, shows something interesting:

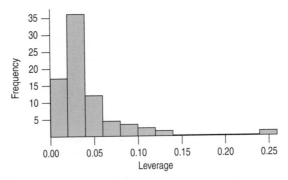

Figure 19.6 *The distribution of leverage values shows a few high values and one extraordinarily high-leverage point.*

The case with very high leverage is a coaster called *Oblivion*, a steel roller coaster in England that opened as the world's first "vertical drop coaster" in 1998. What's unusual about *Oblivion* is that its *Height* is only about 65 feet above ground (placing it below the median), and yet it drops 180 feet to achieve a top speed of 68 mph. The unique feature of *Oblivion* is that it plunges *underground* nearly 120 feet.

Leverage points can affect not only the coefficients of the model, but also our choice of whether to include a predictor in a regression model as well. The more complex the regression model, the more important it is to look at high-leverage values and their effects.

Residuals and Standardized Residuals

Residuals are not all alike. Consider a point with leverage 1.0. That's the highest a leverage can be, and it means that the line follows the point perfectly. So, a point like that must have a zero residual. And since we know the residual exactly, that residual has zero standard deviation. This tendency is true in general: The larger the leverage, the smaller the standard deviation of its residual. When we want to compare values that have differing standard deviations, it's a good idea to standardize them. We can do that with the regression residuals, dividing each one by an estimate of its own standard deviation. When we do that, the resulting values follow a Student's *t*-distribution. In fact, these standardized residuals are called **Studentized residuals**.[5] It's a good idea to examine the Studentized residuals (rather than the simple residuals) to check the Nearly Normal Condition and the Equal Spread Condition. Any Studentized residual that stands out from the others deserves your attention.

It may occur to you that we've always plotted the *unstandardized* residuals when we made regression models. We treated them as if they all had the same standard deviation when we checked the Nearly Normal Condition. It turns out that this was a simplification. It didn't matter much for simple regression, but for multiple regression models, it's a better idea to use the Studentized residuals when checking the Nearly Normal Condition and when making scatterplots of residuals against predicted values.

Influence Measures

A case that has *both* high leverage and a large Studentized residual is likely to have changed the regression model substantially all by itself. Such a case is said to be **influential**. An influential case cries out for special attention because removing it is likely to give a very different regression model. The surest way to tell whether a case is influential is to try leaving it out[6] and see how much the regression model changes. You should call a case "influential" if omitting it changes the regression model by enough to matter for *your* purposes.

To identify possibly influential cases, check the leverage and Studentized residuals. Two statistics that combine leverage and Studentized residuals into a single measure of

---

[5] There's more than one way to Studentize residuals according to how you estimate *s*. You may find statistics packages referring to *externally Studentized residuals* and *internally Studentized residuals*. It is the *externally Studentized* version that follows a *t*-distribution, so those are the ones we recommend.

[6] Or, equivalently, include an indicator variable that selects only for that case. See the discussion in the next section.

influence, **Cook's Distance** (Cook's D) and DFFITs, are offered by many statistics programs. If either of these measures is unusually large for a case, that case should be checked as a possible influential point. Cook's D is found from the leverage, the residual, the number of predictors, and the residual standard error:

$$D_i = \frac{e_i^2}{ks_e^2}\left[\frac{h_i}{(1 - h_i)^2}\right].$$

A histogram of the Cook's distances from the model in Table 19.6 shows a few influential values:

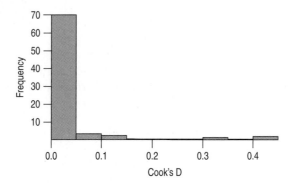

Here are coasters with the four highest values of Cook's D.

Name	Type	Duration	Speed	Height	Drop	Length	Inversions	Cook's D
HyperSonic XLC	Steel	NA	80	165	133	1560	0	0.1037782
Oblivion	Steel	NA	68	65	180	1222	0	0.1124114
Volcano, The								
Blast Coaster	Steel	NA	70	155	80	2757	1	0.3080218
Xcelerator	Steel	62	82	205	130	2202	0	0.4319336

| **Table 19.7** *Coasters with high Cook's D*

In addition to the *Oblivion*, Cook's Distance singles out three other coasters: *HyperSonic XLC*; *Volcano, The Blast Coaster*; and *Xcelerator*. A little research finds that these three coasters *are* different as well. Our model found that *Height* and *Drop* significantly influence a coaster's *Speed*. But these three coasters have something extra— a hydraulic catapult that accelerates the coasters more than gravity alone. In fact, the *Xcelerator* reaches 82 mph in 2.3 seconds, using only 157 feet of track to launch it. Removing these three accelerator coasters from the model has a striking effect.

```
Variable     Coeff       SE(Coeff)    t-ratio    P-value
Intercept    36.47453    1.06456      34.262     <0.0001
Drop         0.17519     0.01493      11.731     <0.0001
Height       0.01600     0.01507       1.062      0.292

Residual standard error: 3.307 on 70 degrees of freedom
Multiple R-squared: 0.9246   Adjusted R-squared: 0.9225
F-statistic: 429.5 on 2 and 70 DF, p-value: <0.0001
```

Table 19.8 *Removing the three blast coasters has made Height no longer important to the model. A simple regression model, such as in Table 19.9, may be a more effective summary than the model with two predictor variables.*

The Height of the coaster is no longer a statistically significant predictor, so we might choose to omit that variable.

Variable	Coeff	SE(Coeff)	t-ratio	P-value
Intercept	36.743925	1.034798	35.51	<0.0001
Drop	0.189474	0.006475	29.26	<0.0001

Residual standard error: 3.31 on 71 degrees of freedom
Multiple R-squared: 0.9234   Adjusted R-squared: 0.9224
F-statistic: 856.3 on 1 and 71 DF, p-value: <0.0001

Table 19.9 *A simple linear regression model without the three blast coasters and with Height deleted.*

Indicators for Influence

One good way to examine the effect of an extraordinary case on a regression is to construct a special indicator variable that is zero for all cases *except* the one we want to isolate. Including such an indicator in the regression model has the same effect as removing the case from the data, but it has two special advantages. First, it makes it clear to anyone looking at the regression model that we have treated that case specially. Second, the *t*-statistic for the indicator variable's coefficient can be used as a test of whether the case is influential. If the P-value is small, then that case really didn't fit well with the rest of the data. Typically, we name such an indicator with the identifier of the case we want to remove. Here's the last roller coaster model in which we have removed the influence of the three blast coasters by constructing indicators for them instead of by removing them from the data. Notice that the coefficient for drop is just the same as the ones we found by omitting the cases.

Dependent variable is: Speed
R squared = 92.7% R squared (adjusted) = 92.3%
s = 3.310 with 76 − 5 = 71 degrees of freedom

Variable	Coeff	SE(Coeff)	t-ratio	P-value
Intercept	36.7439	1.035	35.5	<0.0001
Drop	0.189474	0.0065	29.3	<0.0001
Xcelerator	20.6244	3.334	6.19	<0.0001
HyperSonic	18.0560	3.334	5.42	<0.0001
Volcano	18.0981	3.361	5.38	<0.0001

Table 19.10 *The P-values for the three indicator variables confirm that each of these roller coasters doesn't fit with the others.*

Diagnosis Wrapup

What have we learned from diagnosing the regression? We've discovered four roller coasters that may be strongly influencing the model. And for each of them, we've been able to understand why and how it differed from the others. The oddness of *Oblivion* plunging into a hole in the ground may cause us to value *Drop* as a predictor of *Speed* more than *Height*. The three influential cases with high Cook's D values turned out to be different from the other roller coasters because they are "blast coasters" that don't rely only on gravity for their acceleration. Although we can't count on always discovering why influential cases are special, diagnosing influential cases raises the question of what about them might be different and can help us understand our model better.

When a regression analysis has cases that have both high leverage and large Studentized residuals, it would be irresponsible to report only the regression on all the data. You should also compute and discuss the regression found with such cases removed, and discuss the extraordinary cases individually if they offer additional insight. If your interest is to understand the world, the extraordinary cases may tell you more than the rest of the model. If your only interest is in the model (for example, because you hope to use it for prediction), then you'll want to be certain that the model wasn't determined by only a few influential cases, but instead was built on the broader base of the body of your data.

19.4 Building Regression Models

When many possible predictors are available, we will naturally want to select only a few of them for a regression model. But which ones? The first and most important thing to realize is that often there is no such thing as the "best" regression model. In fact, no regression model is "right." Often, several alternative models may be useful or insightful. The "best" for one purpose may not be best for another, and the one with the highest R^2 may not be best for many purposes.

Multiple regressions are subtle. The coefficients often don't mean what they may appear to mean at first. The choice of which predictors to use determines almost everything about the regression. Predictors interact with each other, which complicates interpretation and understanding. So it is usually best to build a parsimonious model, using as few predictors as you can. On the other hand, we don't want to leave out predictors that are theoretically or practically important. Making this trade-off is the heart of the challenge of selecting a good model.[7] The best regression models, in addition to satisfying the assumptions and conditions of multiple regression, have:

◆ Relatively few predictors, to keep the model simple.

◆ A relatively high R^2, indicating that much of the variability in y is accounted for by the regression model.

◆ A relatively small value of s_e, the standard deviation of the residuals, indicating that the magnitude of the errors is small.

◆ Relatively small P-values for the F- and t-statistics, showing that the overall model is better than a simple summary with the mean and that the individual coefficients are reliably different from zero.

◆ No cases with extraordinarily high leverage that might dominate and alter the model.

◆ No cases with extraordinarily large residuals, and Studentized residuals that appear to be nearly Normal. Outliers can alter the model and certainly weaken the power of any test statistics, and the Nearly Normal Condition is required for inference.

◆ Predictors that are reliably measured and relatively unrelated to each other.

The term "relatively" in this list is meant to suggest that you should favor models with these attributes over others that satisfy them less, but of course, there are many trade-offs and no absolute rules. In addition to favoring predictors that can be measured reliably, you may want to favor those that are less expensive to measure, especially if your model is intended for prediction with values not yet measured.

It should be clear from this discussion that the selection of a regression model calls for judgment. This is yet another of those decisions in Statistics that just can't be made automatically. Indeed, it is one that we shouldn't want to make automatically; there are so many aspects of what makes a model useful that human judgment is necessary to make a final choice. Nevertheless, there are tools that can help by identifying potentially interesting models.

Best Subsets and Stepwise Regression

How can we find the best multiple regression model? The list of desirable features we just looked at should make it clear that there is no simple definition of the "best" model. The choice of a multiple regression model always requires judgment to choose among potential models. Sometimes it can help to look at models that are "good" in some arbitrary sense to understand some possibilities, but such models should never be accepted blindly.

---

[7] This trade-off is sometimes referred to as Occam's Razor after the medieval philosopher William of Occam.

If we choose a single criterion such as finding a model with the highest adjusted R^2, then, for modest size data sets and a modest number of potential predictors, it is actually possible for computers to search through *all* possible models. The method is called a **Best Subsets Regression.** Often the computer reports a collection of "best" models: the best with three predictors, the best with four, and so on.[8] Of course, as you add predictors, the R^2 can never decrease, but the improvement may not justify the added complexity of the additional predictors. One criterion that might help is to use adjusted R^2. Best subsets regression programs usually offer a choice of criteria, and of course, different criteria usually lead to different "best" models.

Although best subsets programs are quite clever about computing far fewer than all the possible alternative models, they do become overwhelmed by more than a few dozen possible predictors or very many cases. So, unfortunately, they aren't useful in many data mining applications. (We'll discuss those more in Chapter 24.)

Another alternative is to have the computer build a regression "stepwise." In a **stepwise regression,** at each step, a predictor is either added to or removed from the model. The predictor chosen to add is the one whose addition increases the adjusted R^2 the most (or similarly improves some other measure). The predictor chosen to remove is the one whose removal reduces the adjusted R^2 least (or similarly loses the least on some other measure). The hope is that by following this path, the computer can settle on a good model. The model will gain or lose a predictor only if that change in the model makes a big enough change in the performance measure. The changes stop when no more changes pass the criterion.

Best subsets and stepwise methods offer both a final model and information about the paths they followed. The intermediate stage models can raise interesting questions about the data and suggest relationships that you might not have thought about. Some programs offer the chance for you to make choices as the process progresses. By interacting with the process at each decision stage, you can exclude a variable that you judge inappropriate for the model (even if including it would help the statistic being optimized) or include a variable that wasn't the top choice at the next step, if you think it is important for your purposes. Don't let a variable that doesn't make sense enter the model just because it has a high correlation, but at the same time, don't exclude a predictor just because you didn't initially think it was important. (That would be a good way to make sure that you never learn anything new.) Finding the balance between these two choices underlies the art of successful model building and makes it challenging.

Unlike best subset methods, stepwise methods can work even when the number of potential predictors is so large that you can't examine them individually. In such cases, using a stepwise method can help you identify potentially interesting predictors, especially when you use it sequentially.

Both methods are powerful. But as with many powerful tools, they require care when you use them. You should be aware of what the automated methods *fail* to do: They don't check the Assumptions and Conditions. Some, such as the independence of the cases, you can check before performing the analyses. Others, such as the Linearity condition and concerns over outliers and influential cases, must be checked for each model. There's a risk that automated methods will be influenced by nonlinearities, by outliers, by high leverage points, by clusters, and by the need for constructed dummy variables to account for subgroups.[9] And these influences affect not just the coefficients in the final model, but the *selection* of the predictors themselves. If there is a case that is influential for even one possible multiple regression model, a best

---

[8] Best subsets regressions don't actually compute every regression. Instead, they cleverly exclude models they know to be worse than some they've already examined. Even so, there are limits to the size of the data and number of variables they can deal with comfortably.

[9] This risk grows dramatically with larger and more complex data sets—just the kind of data for which these methods can be most helpful.

subsets search is guaranteed to consider that model (because it considers *all* possible models) and have its decision influenced by that one case.

◆ **Choosing the Wrong "Best" Model.** Here's a simple example of how stepwise and best subsets regressions can go astray. We might want to find a regression to model *Horsepower* in a sample of cars from the car's engine size (*Displacement*) and its *Weight*. The simple correlations are as follows:

	HP	Disp	Wt
Horsepower	1.000		
Displacement	0.872	1.000	
Weight	0.917	0.951	1.000

Because *Weight* has a slightly higher correlation with *Horsepower*, stepwise regression will choose it first. Then, because *Weight* and engine size (*Displacement*) are so highly correlated, once *Weight* is in the model, *Displacement* won't be added to the model. And a best subsets regression will prefer the regression on *Weight* because it has a higher R^2 and adjusted R^2. But *Weight* is, at best, a lurking variable leading to both the need for more horsepower and a larger engine. Don't try to tell an engineer that the best way to increase horsepower is to add weight to the car and that the engine size isn't important! From an engineering standpoint, *Displacement* is a far more appropriate predictor of *Horsepower*, but stepwise regression can't find that model.

Challenges in Building Regression Models

The data set used to construct the regression in the Guided Example of Chapter 18 originally contained more than 100 variables on more than 10,000 houses in the upstate New York area. Part of the challenge in constructing models is simply preparing the data for analysis. A simple scatterplot can often reveal a data value mistakenly coded, but with hundreds of potential variables, the task of checking the data for accuracy, missing values, consistency, and reasonableness can become the major part of the effort. We will return to this issue when we discuss data mining in Chapter 24.

Another challenge in building large models is Type I error. Although we've warned against using 0.05 as an unquestioned guide to statistical significance, we have to start somewhere, and this critical value is often used to test whether a variable can enter (or leave) a regression model. Of course, using 0.05 means that about 1 in 20 times, a variable whose contribution to the model may be negligible will appear as significant. Using something more stringent than 0.05 means that potentially valuable variables may be overlooked. Whenever we use automatic methods (stepwise, best subsets or others), the actual number of different models considered becomes huge, and the probability of a Type I error grows with it. There is no easy remedy for this problem. Building a model that includes predictors that actually contribute to reducing the variation of the response and avoiding predictors that simply add noise to the predictions is the challenge of modern model building. Much current research is devoted to criteria and automatic methods to make this search easier and more reliable, but for the foreseeable future, you will need to use your own judgment and wisdom in addition to your statistical knowledge to build sensible useful regressions.

GUIDED EXAMPLE | *Housing Prices*

Let's return to the upstate New York data set to predict house prices, this time using a sample of 1734 houses and 16 variables.

The variables available include:

Price The price of the house as sold in 2002

Lot Size The size of the land in *acres*

Waterfront An indicator variable coded as 1 if the property contains waterfront, 0 otherwise

Age The age of the house in *years*

Land Value The assessed value of the property without the structures

New Construct An indicator variable coded as 1 if the house is new construction, 0 otherwise

Central Air An indicator variable coded as 1 if the house has central air conditioning, 0 otherwise

Fuel Type A categorical variable describing the main type of fuel used to heat the house:

    1 = None; 2 = Gas; 3 = Electric; 4 = Oil;
    5 = Wood; 6 = Solar; 7 = Unknown/Other

Heat Type A categorical variable describing the heating system of the house:

    1 = None; 2 = Forced Hot Air; 3 = Hot Water;
    4 = Electric

Sewer Type A categorical variable describing the sewer system of the house:

    1 = None/Unknown; 2 = Private (Septic System);
    3 = Commercial/Public

Living Area The size of the living area of the house in *square feet*

Pct College The percent of the residents of the zip code that attended four-year college (from the U.S. Census Bureau)

Full Baths The *number* of full bathrooms

Half Baths The *number* of half bathrooms

Bedrooms The *number* of bedrooms

Fireplaces The *number* of fireplaces

PLAN

Setup State the objective of the study.

Identify the variables.

Model Think about the assumptions and check the conditions. A scatterplot matrix is a good way to examine the relationships for the quantitative variables.

We want to build a model to predict house prices for a region in Upstate New York. We have data on *Price* ($), and 15 potential predictor variables selected from a much larger list.

✓ **Linearity Condition.** To fit a regression model, we first require linearity. (Scatterplots of *Price* against *Living Area, Age, Bedrooms, Bathrooms,* and *Fireplaces* are similar to the plots shown in the regression of Chapter 18 and are not shown here.)

✓ **Independence Assumption.** We can regard the house prices as being independent of one another since they are from a fairly large geographic area.

✓ **Randomization Condition.** These 1728 houses are a random sample of a much larger set. That supports the idea that these houses are independent.

To check equal variance and Normality, we usually find a regression and examine the residuals. Linearity is all we need for that.

Remarks

Examination of *Fuel Type* showed that there were only 6 houses that did not have categories 2, 3 or 4.

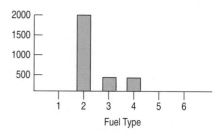

Fuel Type

Two of those had unknown *Heat Type*. We decided to set these 6 houses aside, leaving three categories. So we can use two dummy variables for each. We combined the two *Bathroom* variables into a single variable *Bathroom* equal to the sum of the full baths plus 0.5*half baths. We now have 17 potential predictor variables.

We started by fitting a model to all of them.

```
Dependent variable is: Price
R squared = 65.1%R squared (adjusted) = 64.8%
s = 58408 with 1728 - 18 = 1710 degrees of freedom
```

Variable	Coeff	SE(Coeff)	t-Ratio	P-value
Intercept	18794.0	23333	0.805	0.4207
Lot.Size	7774.34	2246	3.46	0.0006
Waterfront	119046	15577	7.64	≤ 0.0001
Age	−131.642	58.54	−2.25	0.0246
Land.Value	0.9258	0.048	19.4	≤ 0.0001
New.Construct	−45234.8	7326	−6.17	≤ 0.0001
Central.Air	9864.30	3487	2.83	0.0047
Fuel Type[2]	4225.35	5027	0.840	0.4008
Fuel Type[3]	−8148.11	12906	−0.631	0.5279
Heat Type[2]	−1185.54	12345	−0.096	0.9235
Heat Type[3]	−11974.4	12866	−0.931	0.3521
Sewer Type[2]	4051.84	17110	0.237	0.8128
Sewer Type[3]	5571.89	17165	0.325	0.7455
Living.Area	75.769	4.24	17.9	≤ 0.0001
Pct.College	−112.405	151.9	−0.740	0.4593
Bedrooms	−4963.36	2405	−2.06	0.0392
Fireplaces	768.058	2992	0.257	0.7975
Bathrooms	23077.4	3378	6.83	≤ 0.0001

✓ **Equal Spread Condition.** A scatterplot of the Studentized residuals against predicted values shows no thickening or other patterns. There is a group of homes whose residuals are larger (both negative and positive) than the vast majority, whose Studentized residual values are larger than 3 or 4 in absolute value. We'll revisit them after we've selected our model.

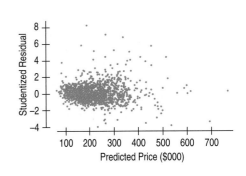

We need the Nearly Normal Condition only if we want to do inference and the sample size is not large. If the sample size is large, we need the distribution to be Normal only if we plan to produce prediction intervals.

✓ **Nearly Normal Condition, Outlier Condition.**
The histogram of residuals is unimodal and symmetric, but slightly long tailed. The Normal Probability plot supports that.

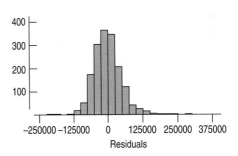

Under these conditions, we can proceed to search for a suitable multiple regression model using a subset of the predictors. We will return to some of our concerns in the discussion.

Mechanics We first let the stepwise program proceed backward from the full model on all 14 predictors.

Here is the computer output for the multiple regression, starting with all 14 predictor variables and proceeding backward until no more candidates were nominated for exclusion.

Dependent variable is: Price
R squared = 65.1%   R squared (adjusted) = 64.9%
s = 58345 with 1728 − 12 = 1716 degrees of freedom

Variable	Coeff	SE(Coeff)	t-ratio	P-value
Intercept	9643.14	6546	1.47	0.1409
Lot.Size	7580.42	2049	3.70	0.0002
Waterfront	119372	15365	7.77	≤ 0.0001
Age	−139.704	57.18	−2.44	0.0147
Land.Value	0.921838	0.0463	19.9	≤ 0.0001
New.Construct	−44172.8	7159	−6.17	≤ 0.0001
Central.Air	9501.81	3402	2.79	0.0053
Heat Type[2]	10099.9	4048	2.50	0.0127
Heat Type[3]	−791.243	5215	−0.152	0.8794
Living.Area	75.9000	4.124	18.4	≤ 0.0001
Bedrooms	−4843.89	2387	−2.03	0.0426
Bathrooms	23041.0	3333	6.91	≤ 0.0001

The estimated equation is:

$$\widehat{Price} = -9643.14 + 7580.42 LotSize$$
$$+ 119{,}372 Waterfront - 139.70 Age$$
$$+ 0.922 LandValue$$
$$- 44{,}172.8 NewConstuction$$
$$+ 9501.81 CentralAir$$
$$+ 10099.9 Heat\ Type2$$
$$- 791.24 Heat\ Type3$$
$$+ 75.90 Living\ Area$$
$$- 4843.89 Bedrooms$$
$$+ 23041 Bathrooms$$

All of the P-values are small which indicates that even with 11 predictors in the model, all are contributing. The R^2 value of 65.1% indicates that more than 65% of the overall variation in house prices has been accounted for by this model and the fact that the adjusted R^2 has actually increased, suggest that we haven't removed any important predictors from the model. The residual standard error of $58,345 gives us a rough indication that we can predict the price of a home to within about 2 × $58,345 = $116,690. If that's close enough to be useful, then our model is potentially useful as a price guide.

Remarks

We also tried running the stepwise regression program forward and obtained the same model. There are some houses that have large Studentized residual values and some that have somewhat large leverage, but omitting them from the model did not significantly change the coefficients. We will use this model as a starting basis for pricing homes in the area.

REPORT

Summary and Conclusions Summarize your results and state any limitations of your model in the context of your original objectives.

MEMO

Re: Regression Analysis of Home Price Predictions

A regression model of *Price* on 11 predictors accounts for about 65% of the variation in the price of homes in these data from Upstate New York. Tests of each coefficient show that each of these variables appears to be an aspect of the price of a house.

This model reflects the common wisdom in real estate about the importance of various aspects of a home. An important variable not included is the location, which every real estate agent knows is crucial to pricing a house. This is ameliorated by the fact that all these houses are in the same general area. However, knowing more specific information about where they are located would probably improve the model. The price found from this model can be used as a starting point for comparing a home with comparable homes in the area.

As with all multiple regression coefficients, when we interpret the effect of a predictor, we must take account of all the other predictors and be careful not to suggest a causal relationship between the predictor and the response variable. Here are some interesting features of the model. It appears that among houses with the same values of the other variables, those with waterfront access are worth on average about $119,000 more. Among houses with the same values of the other variables, those with more bedrooms have lower sales prices by, on average, $4843 for each bedroom, while among those with the same values of the other variables, those with more bathrooms have higher prices by, on average, $23,000 per bathroom. Not surprisingly, the value of the land is positively associated with the sale price, accounting for, on average, about $0.92 of the sale price for each $1 of assessed land value among houses that are otherwise alike on the other variables.

This model reflects the prices of 1728 homes in a random sample of homes taken in this area.

19.5 Collinearity

From Chapter 18, we know that houses with more rooms generally cost more than houses with fewer rooms. A simple regression of *Price* on *Rooms* showed:

Variable	Coeff	SE(Coeff)	t-ratio	P-value
Intercept	53015.6	6424.3	8.252	<0.0001
Rooms	22572.9	866.7	26.046	<0.0001

An additional room seems to be "worth" about $22,500 on average to these homes. We also know that the *Living Area* of a house is an important predictor, which associates each extra square foot with an average increase of $113.12:

Variable	Coeff	SE(Coeff)	t-ratio	P-value
Intercept	13439.394	4992.353	2.692	0.00717
Living Area	113.123	2.682	42.173	<0.0001

Finally, we saw that a simple regression on *Bedrooms* also shows increasing price with number of *Bedrooms*, with an additional *Bedroom* associated with an increase, on average, of $48,218:

Variable	Coeff	SE(Coeff)	t-ratio	P-value
Intercept	59863	8657	6.915	<0.0001
Bedrooms	48218	2656	18.151	<0.0001

But, when we put more than one of these variables into a regression equation simultaneously, things can change. Here's a regression with both *Living Area* and *Bedrooms*:

Variable	Coeff	SE(Coeff)	t-ratio	P-value
Intercept	36667.895	6610.293	5.547	<0.0001
Living Area	125.405	3.527	35.555	<0.0001
Bedrooms	−14196.769	2675.159	−5.307	<0.0001

Now, it appears that an extra bedroom is associated with a *lower* sale *Price*.

This type of coefficient change often happens in multiple regression and can seem counterintuitive. When two predictor variables are correlated, their coefficients in a multiple regression (with both of them present) can be quite different from their simple regression slopes. In fact, the coefficient can change from being significantly positive to significantly negative with the inclusion of one correlated predictor, as is the case here with *Bedrooms* and *Living Area*. The problem arises when one of the predictor variables can be predicted well from the others. This phenomenon is called **collinearity**.[10]

Collinearity in the predictors can have other consequences in a multiple regression. If instead of adding *Bedrooms* to the model, we add *Rooms*, we see a different outcome:

Variable	Coeff	SE(Coeff)	t-ratio	P-value
Intercept	11691.586	5521.253	2.118	0.0344
Living Area	110.974	3.948	28.109	<0.0001
Rooms	783.579	1056.568	0.742	0.4584

The coefficient for *Living Area* has hardly changed at all. It still shows an increase of about $111 per square foot, but the coefficient for *Rooms* is indistinguishable from 0. With the addition of *Living Area* to the model the coefficient for *Rooms* changed from having a *t*-statistic over 25, with a very small P-value (in the simple regression) to having a P-value of 0.458. Notice also that the standard errors of the coefficients have increased. The standard error of *Living Area* increased from 2.68 to 3.95. That may not seem like much, but it's an increase of nearly 50%.

This variance inflation of the coefficients is another consequence of collinearity. The stronger the correlation between predictors, the more the variance of their coefficients increases when both are included in the model. Sometimes this effect can change a coefficient from statistically significant to indistinguishable from zero.

Data sets in business often have related predictor variables. General economic variables, such as interest rates, unemployment rates, GDP, and other productivity measures are highly correlated. The choice of which subsets to include in the model can significantly change the coefficients, their standard errors, and their P-values, making both selecting the models and interpreting them difficult.

How can we detect and deal with collinearity? Let's look at a regression among just the predictor variables. If we regress *Rooms* on *Bedrooms* and *Living Area* we find:

Variable	Coeff	SE(Coeff)	t-ratio	P-value
Intercept	0.680	0.141	4.821	<0.0001
Bedrooms	0.948	0.057	16.614	<0.0001
Living Area	0.009	0.000	25.543	<0.0001

```
Residual standard error: 1.462 on 1725 degrees of freedom
Multiple R-squared: 0.602   Adjusted R-squared: 0.6015
F-statistic: 1304 on 2 and 1725 DF, P-value: <0.0001
```

Look at the R^2 for that regression. What does it tell us? Since R^2 is the fraction of variability accounted for by the regression, in this case, that's the fraction of the variability in *Rooms* accounted for by the other two predictors.

Now we can be precise about collinearity. If that R^2 were 100%, we'd have perfect collinearity. *Rooms* would then be perfectly predictable from the other two

Sometimes we can understand what the coefficients are telling us even in such paradoxical situations. Here, it seems that a house that allocates more of its living area to bedrooms (and correspondingly less to other functions) will be worth less.

In the second example, we see that more rooms don't make a house worth more if they just carve up the existing living area. The value of more rooms we saw before was probably because houses with more rooms tend to have more living area as well.

---

[10] You may see also this problem called "multicollinearity."

predictors and so could tell us nothing new about *Price* because it didn't vary in any way not already accounted for by the predictors already in the model. In fact, we couldn't even perform the calculation. Its coefficient would be indeterminate, and its standard error would be infinite. (Statistics packages usually print warnings when this happens.[11]) Conversely, if the R^2 were 0%, then *Rooms* would bring entirely new information to the model, and we'd have no collinearity at all.

Clearly, there's a range of possible collinearities for each predictor. The statistic that measures the degree of collinearity of the jth predictor with the others is called the **Variance Inflation Factor (VIF)** and is found as:

$$VIF_j = \frac{1}{1 - R_j^2}.$$

The R_j^2 here shows how well the jth predictor can be predicted by the other predictors. The $1 - R_j^2$ term measures what that predictor has left to bring to the regression model. If R_j^2 is high, then not only is that predictor superfluous, but it can damage the predictor model. The VIF tells how much the variance of the coefficient has been inflated due to this collinearity. The higher the VIF, the higher the standard error of its coefficient and the less it can contribute to the regression model. Since R_j^2 can't be less than zero, the minimum value of the VIF is 1.0. The VIF takes into account all the other predictors. Nevertheless, any time you see a high correlation among two predictors, be on the lookout for collinearity.

As a final blow, when a predictor is collinear with the other predictors, it's often difficult to figure out what its coefficient means in the multiple regression. We've blithely talked about "removing the effects of the other predictors," but now when we do that, there may not be much left. What is left is not likely to be about the original predictor, but more about the fractional part of that predictor not associated with the others. In a regression of *Horsepower* on *Weight* and *Engine Size*, once we've removed the effect of *Weight* on *Horsepower*, *Engine Size* doesn't tell us anything *more* about *Horsepower*. That's certainly not the same as saying that *Engine Size* doesn't tell us anything at all about *Horsepower*. It's just that most cars with big engines also weigh a lot.

To summarize, when a predictor is collinear with the other predictors in the model, two things can happen:

1. Its coefficient can be surprising, taking on an unanticipated sign or being unexpectedly large or small.

2. The standard error of its coefficient can be large, leading to a smaller *t*-statistic and correspondingly large P-value.

One telltale sign of collinearity is the paradoxical situation in which the overall *F*-test for the multiple regression model is significant, showing that at least one of the coefficients is significantly different from zero, and yet most or all of the individual coefficients have small *t*-values, each in effect, denying that *it* is the significant one.

What should you do about a collinear regression model? The simplest cure is to remove some of the predictors. That both simplifies the model and generally improves the *t*-statistics. If several predictors give pretty much the same information, removing some of them won't hurt the model. Which should you remove? Keep the predictors that are most reliably measured, least expensive to find, or even those that are politically important. Another alternative that may make sense is to construct a new predictor by combining variables. For example, several different measures of a product's durability (perhaps for different parts of it) could be added together to create a single durability measure.

---

[11] Excel does not. It gives 0 as the estimate of most values and a NUM! warning for the standard error of the coefficient.

Facts about Collinearity

◆ The collinearity of any predictor with the others in the model can be measured with its Variance Inflation Factor.

◆ High collinearity leads to the coefficient being poorly estimated and having a large standard error (and correspondingly low *t*-statistic). The coefficient may seem to be the wrong size or even the wrong sign.

◆ Consequently, if a multiple regression model has a high R^2 and large F, but the individual *t*-statistics are not significant, you should suspect collinearity.

◆ Collinearity is measured in terms of the R_i^2 between a predictor and *all* of the other predictors in the model. It is not measured in terms of the correlation between any two predictors. Of course, if two predictors are highly correlated, then the R_i^2 with even more predictors must be at least that large and will usually be even higher.

19.6 Quadratic Terms

After the women's downhill ski racing event at the 2002 Winter Olympic Games in Salt Lake City, Picabo Street of the U.S. team was disappointed with her 16th place finish after she'd posted the fastest practice time. Changing snow conditions can affect finish times, and in fact, the top seeds can choose their starting positions and try to guess when the conditions will be best. But how much impact was there? On the day of the women's downhill race, it was unusually sunny. Skiers expect conditions to improve and then, as the day wears on, to deteriorate, so they try to pick the optimum time. But their calculations were upset by a two-hour delay. Picabo Street chose to race in 26th position. By then conditions had turned around, and the slopes had begun to deteriorate. Was that the reason for her disappointing finish?

The regression in Table 19.11 seems to support her point. Times did get slower as the day wore on.

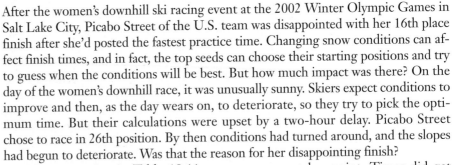

```
Dependent variable is:  Time

R squared = 37.9%  R squared (adjusted) = 36.0%
s = 1.577 with 35 - 2 = 33 degrees of freedom

Variable      Coeff        SE(Coeff)      t-ratio    P-value
Intercept     100.069      0.5597         179        <0.0001
StartOrder      0.108563   0.0242           4.49     <0.0001
```

| *Table 19.11* *Time to ski the women's downhill event at the 2002 Winter Olympics depended on starting position.*

But a plot of the residuals warns us that the linearity assumption isn't met.

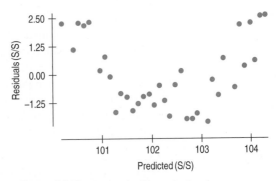

| *Figure 19.7*   *The residuals reveal a bend.*

If we return to plot the data, we can see that re-expression can't help us because the times first trend down and then turn around and increase.

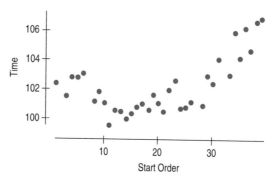

Figure 19.8 *The original data trend down and then up. That kind of bend can't be improved with re-expression.*

How can we use regression here? We can introduce a squared term to the model:

$$\hat{y} = b_0 + b_1 startorder + b_2 startorder^2.$$

The fitted function is a *quadratic*, which can follow bends like the one in these data. Table 19.12 has the regression table.

```
Dependent variable is:  Time

R squared = 83.3%  R squared (adjusted) = 82.3%
s = 0.8300 with 35 - 3 = 32 degrees of freedom
```

Source	Sum of Squares	df	Mean Square	F-ratio
Regression	110.139	2	55.0694	79.9
Residual	22.0439	32	0.688871	

Variable	Coeff	SE(Coeff)	t-ratio	P-value
Intercept	103.547	0.4749	218	<0.0001
StartOrder	−0.367408	0.0525	−6.99	<0.0001
SartOrder^2	0.011592	0.0012	9.34	<0.0001

Table 19.12 *A regression model with a quadratic term fits these data better.*

This model fits the data better. Adjusted R^2 is 82.3%, up from 36.0% for the linear version. And the residuals look generally unstructured, as Figure 19.9 shows.

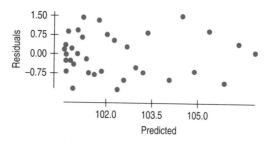

Figure 19.9 *The residuals from the quadratic model show no structure.*

However, one problem remains. In the new model, the coefficient of *Start Order* has changed from significant and positive to significant and negative. As we've just seen, that's a signal of possible collinearity. Quadratic models often have collinearity problems because for many variables, x and x^2 are highly correlated. In these data, *Start Order* and *Start Order*^{2} have a correlation of 0.97.

There's a simple fix for this problem. Instead of using *Start Order*^{2}, we can use $(Start\ Order - \overline{Start\ Order})^2$. The form with the mean subtracted has a *zero* correlation with the linear term. Table 19.13 shows the resulting regression.

```
Dependent variable is: Time
R squared = 83.3%  R squared (adjusted) = 82.3%
s = 0.8300 with 35 - 3 = 32 degrees of freedom
```

Source	Sum of Squares	df	Mean Square	F-ratio
Regression	110.139	2	55.0694	79.9
Residual	22.0439	32	0.688871	

Variable	Coeff	SE(Coeff)	t-ratio	P-value
Intercept	98.7493	0.3267	302	<0.0001
StartOrder	0.104239	0.0127	8.18	<0.0001
(SO-mean)^2	0.011592	0.0012	9.34	<0.0001

| *Table 19.13* *Using a centered quadratic term alleviates the collinearity.*

The predicted values and residuals are the same for these two models,[12] but the coefficients of the second one are easier to interpret.

So, did Picabo Street have a valid complaint? Well, times did increase with start order, but (from the quadratic term), they decreased before they turned around and increased. Picabo's start order of 26 has a predicted time of 101.83 seconds. Her performance at 101.17 was better than predicted, but her residual of −0.66 is not large in magnitude compared to that for some of the other skiers. Skiers who had much later starting positions *were* disadvantaged, but Picabo's start position was only slightly later than the best possible one (about 16th according to this model), and her performance was not extraordinary by Olympic standards.

One final note: Quadratic models can do an excellent job of fitting curved patterns such as this one. But they are particularly dangerous to extrapolate beyond the range of the x-values. So you should use them with care.

Regression Roles

We build regression models for a number of reasons. One reason is to model how variables are related to each other in the hope of understanding the relationships. Another is to build a model that might be used to predict values for a response variable when given values for the predictor variables. When we hope to understand, we are often particularly interested in simple, straightforward models in which predictors are as unrelated to each other as possible. We are especially happy when the *t*-statistics are large, indicating that the predictors each contribute to the model.

By contrast, when prediction is our goal, we are more likely to care about the overall R^2. Good prediction occurs when much of the variability in y is accounted for by the model. We might be willing to keep variables in our model that have relatively small *t*-statistics simply for the stability that having several predictors can provide. We care less whether the predictors are related to each other because we don't intend to interpret the coefficients anyway, so collinearity is less of a concern.

In both roles, we may include some predictors to "get them out of the way." Regression offers a way to approximately control for factors when we have observational data because each coefficient estimates a relationship *after removing the effects of the other predictors*. Of course, it would be better to control for factors in a randomized experiment, but in the real world of business that's often just not possible.

---

[12] This can be shown algebraically for any quadratic model with a centered squared term.

WHAT CAN GO WRONG?

Removing a high-influence point may surprise you with unexpected collinearity. Alternatively, a single case that is extreme on several predictors can make them appear to be collinear when in fact they would not be if you removed that point. Removing that point may make apparent collinearities disappear (and would probably result in a more useful regression model).

- **Beware missing data.** Values may be missing or unavailable for any case in any variable. In simple regression, when the cases are missing for reasons that are unrelated to the variable we're trying to predict, that's not a problem. We just analyze the cases for which we have data. But when several variables participate in a multiple regression, any case with data missing on any of the variables will be omitted from the analysis. You can unexpectedly find yourself with a much smaller set of data than you started with. Be especially careful, when comparing regression models with different predictors, that the cases participating in the models are the same.

- **Don't forget linearity.** The **Linearity Assumption** requires linear relationships among the variables in a regression model. As you build and compare regression models, be sure to plot the data to check that it is straight. Violations of this assumption make everything else about a regression model invalid.

- **Check for parallel regression lines.** When you introduce an indicator variable for a category, check the underlying assumption that the other coefficients in the model are essentially the same for both groups. If not, consider adding an interaction term.

ETHICS IN ACTION

Fred Barolo heads a travel company that offers, among other services, customized travel packages. These packages provide a relatively high profit margin for his company, but Fred worries that a weakened economic outlook will adversely affect this segment of his business. He read in a recent Travel Industry Association report that there is increasing interest among U.S. leisure travelers in trips focused on a unique culinary and wine-related experience. To gain an understanding of travel trends in this niche market, he seeks advice from a market analyst, Smith Nebbiolo, whose firm also handles promotional campaigns. Smith has access to several databases, some through membership with the Travel Industry Association and others through the U.S. Census Bureau and Bureau of Economic Analysis. Smith suggests developing a model to predict demand for culinary wine-related travel, and Fred agrees. As the dependent variable, Smith uses the monthly dollar amount spent on travel packages advertised as culinary wine experiences. He considers a number of monthly economic indicators such as gross domestic product (GDP) and personal consumption expenditures (PCE); variables related to the travel industry (e.g., the American Consumer Satisfaction Index (ACSI) for airlines and hotels, etc.); and factors specific to culinary and wine travel experiences such as advertising expenditure and price. With so many variables, Smith uses an automatic stepwise procedure to select among them. The final model Smith presents to Fred does not include any monthly economic indicators; ACSI for airlines and hotels are included as was advertising expenditure for these types of trips. Fred and Smith discuss how it appears that the economy has little effect on this niche travel market and how Fred should start thinking of how he will promote his new culinary wine-related travel packages.

ETHICAL ISSUE: *Although using an automatic stepwise procedure is useful in narrowing down the number of independent variables to consider for the model, usually more thoughtful analysis is required. In this case, many of the potential independent variables are highly correlated with each other. It turns out that the ACSI is a strong predictor of economic indicators such as GDP and PCE. Its presence in the model would preclude the entry of GDP and PCE, but saying that economic factors don't affect the dependent variable is misleading. Further, these data are time-dependent. No variables capturing trend or potential seasonality are considered. Related to Item A, ASA Ethical Guidelines.*

ETHICAL SOLUTION: *The interrelationships between the independent variables need to be examined. More expertise is required on the part of the model builder; residuals need to be examined to determine if there are any time-dependent patterns (i.e., seasonality).*

What have we learned?

In Chapter 18, we learned that multiple regression is a natural way to extend what we knew about linear regression models to include several predictors. Now we've learned that multiple regression is both more powerful and more complex than it may appear at first. As with other chapters in this book whose titles spoke of greater "wisdom," this chapter has drawn us deeper into the uses and cautions of multiple regression. We've glimpsed the power of the multiple regression model. We can incorporate categorical data by using indicator variables, modeling relationships that have parallel slopes but at different intercepts for different groups. With interaction terms, we can allow for different slopes as well. We can create identifier variables that isolate individual cases to remove their influence from the model while exhibiting how they differ from the other points and testing whether that difference is statistically significant.

We've learned to beware unusual cases. A single case can have high leverage, allowing it to unduly influence the entire regression. Such cases should be treated specially, possibly by fitting the model both with and without them or by including indicator variables to isolate their influence.

We've learned that in complex models one has to be careful in interpreting the coefficients. Associations among the predictors can change the coefficients to values that can be quite different from the coefficient in the simple regression of a predictor and the response, even changing the sign. And we've learned that building multiple regression models is an art that speaks to the central goal of statistical analysis: understanding the world with data. The graphical methods are the same ones we learned in the early chapters, and the inference methods are those we originally developed for means. In short, there's been a consistent tale of how we understand data to which we've added more and more detail and richness, but which has been consistent throughout.

Terms

Best Subsets Regression   A regression method that checks all possible combinations of the available predictors to identify the combination that optimizes an arbitrary measure of regression success.

Collinearity   When one (or more) of the predictors can be fit closely by a regression on the other predictors, we have collinearity. When collinear predictors are in a regression model, they may have unexpected coefficients and often have inflated standard errors (and correspondingly small t-statistics).

Cook's Distance   A measure of the influence of a single case on the coefficients in a multiple regression.

Dummy variable   An indicator variable.

Indicator variable   A variable constructed to indicate for each case whether it is in a designated group or not. Usually the values are 0 and 1, where 1 indicates group membership.

Influential case   A case is *influential* on a multiple regression model if, when it is omitted, the model changes by enough to matter for your purposes. (There is no specific amount of change defined to declare a case influential.) Cases with high leverage and large Studentized residual are likely to be influential.

Interaction term	A variable constructed by multiplying a predictor variable by an indicator variable. An interaction term adjusts the slope of that predictor for the cases identified by the indicator.
Leverage	A measure of the amount of influence an individual case has on the regression. Moving a case in the *y* direction by 1 unit (while changing nothing else) will move its predicted value by the leverage, denoted as h.
Stepwise regression	An automated method of building regression models in which predictors are added to or removed from the model one at a time in an attempt to optimize a measure of the success of the regression. Stepwise methods rarely find the best model and are easily affected by influential cases, but they can be valuable in winnowing down a large collection of candidate predictors.
Studentized residual	When a residual is divided by an independent estimate of its standard deviation, the result is a Studentized residual. The type of Studentized residual that has a *t*-distribution is an *externally Studentized residual*.
Variance inflation factor (VIF)	A measure of the degree to which a predictor in a multiple regression model is collinear with other predictors. It is based on the R^2 of the regression of that predictor on all the other predictors in the model:

$$VIF_j = \frac{1}{1 - R_j^2}$$

Skills

 PLAN

- Understand how individual cases can influence a regression model.
- Know how to define and use indicator variables to introduce categorical variables as predictors in a multiple regression model.
- Know how to examine histograms of leverages and of Studentized residuals to identify extraordinary cases that deserve special attention.
- Know how to recognize when a regression model may suffer from collinearity.

 DO

- Know how to check for high-leverage cases by identifying cases whose leverage stands apart from the others.
- Know how to check for cases with large Studentized residuals.
- Be able to use a statistics package to diagnose a multiple regression model.
- Know how to build a multiple regression model, selecting predictors from a larger collection of potential predictors.

 REPORT

- Be able to interpret the coefficients found for indicator variables in a multiple regression.
- Be able to discuss the influence that a case with high leverage or a large Studentized residual may have in a regression.
- Be able to recognize when collinearity among the predictors may be present. Be able to check for it and discuss its consequences.
- Be careful in interpreting regression coefficients when the predictors are collinear. Avoid the pitfalls of interpreting the sign of the coefficient as if it were special. If you can't interpret the first digit of the coefficient, you probably can't interpret the sign either.

Technology Help: Regression Analysis on the Computer

Statistics packages differ in how much information they provide to diagnose a multiple regression. Most packages provide leverage values. Many provide far more, including statistics that we have not discussed. But for all, the principle is the same. We hope to discover any cases that don't behave like the others in the context of the regression model and then to understand why they are special.

Many of the ideas in this chapter rely on the concept of examining a regression model and then finding a new one based on your growing understanding of the model and the data. Regression diagnosis is meant to provide steps along that road. A thorough regression analysis may involve finding and diagnosing several models.

EXCEL

Excel does not offer diagnostic statistics with its regression function.

Comments

Although the dialog offers a Normal probability plot of the residuals, the data analysis add-in does not make a correct probability plot, so don't use this option. The "stan-

dardized residuals" are just the residuals divided by their standard deviation (with the wrong df), so they too should be ignored.

The DDXL add-in provides most of the diagnostic statistics and displays discussed in the chapter, but does not provide either stepwise or best subsets regression.

MINITAB

- Choose **Regression** from the **Stat** menu.
- Choose **Regression...** from the **Regression** submenu.
- In the Regression dialog, assign the Y variable to the Response box and assign the X-variables to the Predictors box.
- Click on the **Options** button to obtain the VIF in the regression output.
- In the Regression **Storage** dialog, you can select a variety of diagnostic statistics. They will be stored in the columns of your worksheet.
- Click the **OK** button to return to the Regression dialog.
- To specify displays, click **Graphs**, and check the displays you want.

- Click the **OK** button to return to the Regression dialog.
- Click the **OK** button to compute the regression.

Comments

Your will probably want to make displays of the stored diagnostic statistics. Use the usual Minitab methods for creating displays.

Minitab also offers both stepwise and best subsets regression from the **Regression** dialog. Indicate the response variable, the predictors eligible for inclusion, and any predictors that you wish to force into the model.

SPSS

- Choose **Regression** from the **Analyze** menu.
- Choose **Linear** from the **Regression** submenu.
- When the Linear Regression dialog appears, select the Y-variable and move it to the dependent target. Then move the X-variables to the independent target.
- Click the **Save** button.
- In the Linear Regression Save dialog, choose diagnostic statistics. These will be saved in your worksheet along with your data.
- Click the **Continue** button to return to the Linear Regression dialog.
- Click the **OK** button to compute the regression.

Comments

SPSS offers stepwise methods (use the **Method** drop-down menu), but not best subsets (in the student version). Click on the **Statistics** button to find collinearity diagnostics and on the **Save** button for influential point diagnostics. (The residuals SPSS calls "Studentized deleted" are the externally studentized residuals that we've recommended in this chapter.) You may want to plot the saved diagnostics using SPSS's standard graphics methods.

JMP

- From the **Analyze** menu select **Fit Model.**
- Specify the response, Y. Assign the predictors, X, in the **Construct Model Effects** dialog box.
- Click on **Run Model.**
- Click on the red triangle in the title of the Model output to find a variety of plots and diagnostics available.

Comments

JMP chooses a regression analysis when the response variable is "Continuous."

In JMP, stepwise regression is a *personality* of the Model Fitting platform; it is one of the selections in the Fitting Personality popup menu on the Model Specification dialog. Stepwise provides best subsets with an **All Possible Models** command, accessible from the red triangle drop-down menu on the stepwise control panel after you've computed a stepwise regression analysis.

DATA DESK

Request diagnostic statistics and graphs from the HyperView menus in the regression output table. Most will update and can be set to update automatically when the model or data change.

Comments

You can add a predictor to the regression by dragging its icon into the table or replace variables by dragging the icon over their name in the table. Click on a predictor's name to drop down a menu that lets you remove it from the model.

Data desk does not provide an automated stepwise regression, but you can use its automatic updating and drag-and-drop features to build models. Compute correlations of y and all candidate x's. Compute the regression of y on one promising x. From the regression table HyperView menu, compute residuals and drop them into the correlation table. Now the remaining candidate x variables that are most highly correlated with the residuals are good predictors to investigate. Make a scatterplot of the residuals against any candidate by clicking on the correlation value. Drag the predictor you select into the regression table. As you add variables to the model, you can update the correlations.

Mini Case Study Project

The Paralyzed Veterans of America (PVA) is a philanthropic organization sanctioned by the U.S. government to represent the interests of those veterans who are disabled. (For more information on the PVA see the opening of Chapter 24). To generate donations, the PVA sends out greeting cards and mailing address labels periodically with their requests for donations. To increase their efficiency, they would like to be able to model the amount of donations based on past giving and demographic variables on their donors. The data set **ch19_MCSP_pva.txt** contains data on 3,648 donors who gave to a recent solicitation. There are 26 predictor variables and 1 response variable. The response variable (GIFTAMNT) is the amount of money donated by the donor to the last solicitation. Find a model from the 26 predictor variables using any model selection procedure you like to predict this amount. The variables include:

Variables based on the donor's zipcode

> MALEVET (% Male veterans)
> VIETVETS (% Vietnam Veterans)
> WWIIVETS (% WWII veterans)
> LOCALGOV (% Employed by local government)

STATEGOV (% Employed by state government)

FEDGOV (% Employed by federal government)

Variables specific to the individual donor

CARDPROM (Number of card promotions received lifetime)

MAXADATE (Date of most recent promotion received in YYMM Year Month format)

NUMPROM (Number of promotions received lifetime)

CARDPRM12 (Number of card promotions received in last 12 months)

NUMPRM12 (Number of promotions received in last 12 months)

NGIFTALL (Number of gifts given lifetime to date)

CARDGIFT (Number of gifts to card promotions given lifetime to date)

MINRAMNT (Amount of smallest gift to date in $)

MINRDATE (Date associated with the smallest gift to date—YYMM format)

MAXRAMNT (Amount of largest gift to date in $)

MAXRDATE (Date associated with the largest gift to date—YYMM format)

LASTGIFT (Amount of most recent gift in $)

AVGGIFT (Average amount of gifts to date in $)

CONTROLN (Control number—unique record identifier)

HPHONE_D (Indicator variable for presence of a published home phone number: 1 = Yes; 0 = No)

CLUSTER2 (Classic Cluster Code—nominal field)

CHILDREN (Number of children living at home)

Response variable

GIFTAMNT (Response variable—amount of last gift in $)

Be sure to include exploratory data analysis and evaluate the relationship among these variables using graphical and correlation analysis to guide you in building your regression models. Write a report summarizing your analysis.

EXERCISES

T 1. Pizza ratings. Manufacturers of frozen foods often reformulate their products to maintain and increase customer satisfaction and sales. So they pay particular attention to evaluations of their products in comparison to their competitors' products. Frozen pizzas are a major sector of the frozen food market, accounting for $2.84 Billion in sales in 2007 (www.aibonline.org/resources/statistics/2007pizza.htm). The prestigious Consumer's Union rated frozen pizzas for flavor and quality, assigning an overall score to each brand tested. A regression model to predict the Consumer's Union score from Calories, Type (1 = cheese, 0 = pepperoni), and fat content gives the following result:

Dependent variable is: Score
R squared = 28.7%   R squared (adjusted) = 20.2%
s = 19.79 with 29 − 4 = 25 degrees of freedom

Source	Sum of Squares	df	Mean Square	F-ratio
Regression	3947.34	3	1315.78	3.36
Residual	9791.35	25	391.654	

Variable	Coeff	SE(Coeff)	t-ratio	P-value
Intercept	−148.817	77.99	−1.91	0.0679
Calories	0.743023	0.3066	2.42	0.0229
Type	15.6344	8.103	1.93	0.0651
Fat	−3.89135	2.138	−1.82	0.0807

a) What is the interpretation of the coefficient of *Type* in this regression? According to these results, what type would you expect to sell better—cheese or pepperoni?

b) What displays would you like to see to check assumptions and conditions for this model?

T 2. Traffic delays. The Texas Transportation Institute (tti.tamu.edu) studies traffic delays. They estimate that in the year 2000, the 75 largest metropolitan areas experienced 3.6 billion vehicle hours of delay, resulting in 5.7 billion gallons of wasted fuel and $67.5 billion in lost productivity. That's about 0.7% of the nation's GDP that year. Data the institute published for the year 2001 include information on the *Total Delay per Person* (hours per year spent delayed by traffic), the *Average Arterial Road Speed* (mph), the *Average Highway Road Speed* (mph), and the *Size* of the city (small, medium, large, very large). The regression model based on these variables looks like this. The variables *Small, Large,* and *Very Large* are indicators constructed to be 1 for cities of the named size and 0 otherwise.

Dependent variable is: Delay/person
R squared = 79.1%   R squared (adjusted) = 77.4%
s = 6.474 with 68 − 6 = 62 degrees of freedom

Source	Sum of Squares	df	Mean Square	F-ratio
Regression	9808.23	5	1961.65	46.8
Residual	2598.64	62	41.9135	

Variable	Coeff	SE(Coeff)	t-ratio	P-value
Intercept	139.104	16.69	8.33	≤0.0001
HiWay MPH	−1.07347	0.2474	−4.34	≤0.0001
Arterial MPH	−2.04836	0.6672	−3.07	0.0032
Small	−3.58970	2.953	−1.22	0.2287
Large	5.00967	2.104	2.38	0.0203
Very Large	3.41058	3.230	1.06	0.2951

a) Why is there no coefficient for *Medium*?

b) Explain how the coefficients of *Small, Large, and Very Large* account for the size of the city in this model.

T 3. Pizza ratings, part 2. Here's a scatterplot of the residuals against predicted values for the regression model found in Exercise 1.

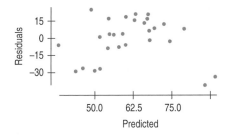

a) The two extraordinary points in the lower right are *Reggio's* and *Michelina's,* two gourmet brands. Interpret these points.

b) Do you think these two pizzas are likely to be influential in the regression. Would setting them aside be likely to change the coefficients? What other statistics might help you decide?

T 4. Traffic delays, part 2. Here's a scatterplot of the residuals from the regression in Exercise 2 plotted against mean *Highway mph.*

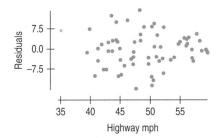

a) The point plotted with an "x" is Los Angeles. Read the graph and explain what it says about traffic delays in Los Angeles and about the regression model.

b) Is Los Angeles likely to be an influential point in this regression?

T 5. Wal-Mart revenue. Each week about 100 million customers—nearly one-third of the U.S. population—visit one of Wal-Mart's U.S. stores. How does Wal-Mart's revenue relate to the state of the economy in general? Here's a regression table predicting Wal-Mart's monthly revenue ($Billion) from the end of 2003 through the start of 2007 from the consumer price index (*CPI*), and a scatterplot of the relationship.

Dependent variable is: WM_Revenue

R squared = 11.4%   R squared (adjusted) = 9.0%
s = 3.689 with 39 − 2 = 37 degrees of freedom

Variable	Coeff	SE(Coeff)	t-ratio	P-value
Intercept	−24.4085	19.25	−1.27	0.2127
CPI	0.071792	0.0330	2.18	0.0358

a) The points plotted with "x" are the four December values. We can construct a variable that is "1" for those four values and "0" otherwise. What is such a variable called?

Here's the resulting regression.

Dependent variable is: WM_Rev

R squared = 80.3%   R squared (adjusted) = 79.2%
s = 1.762 with 39 − 3 = 36 degrees of freedom

Variable	Coeff	SE(Coeff)	t-ratio	P-value
Intercept	−34.7755	9.238	−3.76	0.0006
CPI	0.087707	0.0158	5.55	≤0.0001
December	10.4905	0.9337	11.2	≤0.0001

b) What is the interpretation of the coefficient of the constructed variable *December*?

c) What additional assumption is required to include the variable *December* in this model? Is there reason to believe that it is satisfied?

T 6. Baseball attendance. Baseball team owners want to attract fans to games. The New York Mets acquired the pitcher Pedro Martinez in 2005. Martinez is considered one of the best pitchers of his era, having won the Cy Young Award three times. Martinez had his own fans. Possibly, he attracted more fans to the ballpark when he pitched at home, helping to justify his 4-year, $53 million contract. Is there really a "Pedro effect" in attendance? We have data for the Mets home games of the 2005 season. The regression has the following predictors:

Weekend   1 if game is on Saturday or Sunday, 0 otherwise

Yankees   1 if game is against the Yankees (a hometown rivalry), 0 otherwise

Rain Delay   1 if the game was delayed by rain (which might have depressed attendance), 0 otherwise

Opening Day   1 for opening day, 0 for the others

Pedro Start   1 if Pedro was the starting pitcher, 0 otherwise

Here's the regression.

Dependent variable is: Attendance

R squared = 53.9% R squared (adjusted) = 50.8%
s = 6998 with 80 − 6 = 74 degrees of freedom

Variable	Coeff	SE(Coeff)	t-ratio	P-value
Intercept	28896.9	1161	24.9	≤0.0001
Weekend	9960.50	1620	6.15	≤0.0001
Yankees	15164.3	4218	3.59	0.0006
Rain Delay	−17427.9	7277	−2.39	0.0192
Opening Day	24766.1	7093	3.49	0.0008
Pedro Start	5428.02	2017	2.69	0.0088

a) All of these predictors are of a special kind. What are they called?

b) What is the interpretation of the coefficient for *Pedro Start*?

c) If we're primarily interested in Pedro's effect on attendance, why is it important to have the other variables in the model?

d) Could Pedro's agent claim, based on this regression, that his man attracts more fans to the ballpark? What statistics should he cite?

T 7. Pizza ratings, part 3. In Exercise 3, we raised questions about two gourmet pizzas. After removing them, the resulting regression looks like this.

Dependent variable is: Score

R squared = 64.4% R squared (adjusted) = 59.8%
s = 14.41 with 27 − 4 = 23 degrees of freedom

Source	Sum of Squares	df	Mean Square	F-ratio
Regression	8649.29	3	2883.10	13.9
Residual	4774.56	23	207.590	

Variable	Coeff	SE(Coeff)	t-ratio	P-value
Intercept	−363.109	72.15	−5.03	≤0.0001
Calories	1.56772	0.2824	5.55	≤0.0001
Type	25.1540	6.214	4.05	0.0005
Fat	−8.82748	1.887	−4.68	0.0001

A plot of the residuals against the predicted values for this regression looks like this. It has been colored according to the *Type* of pizza.

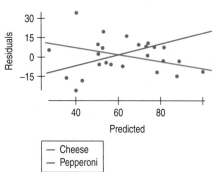

a) What does this plot say about how the regression model deals with these two types of pizza?

We constructed another variable consisting of the indicator variable *Type* multiplied by *Calories*. Here's the resulting regression.

Dependent variable is: Score

R squared = 73.1% R squared (adjusted) = 68.2%
s = 12.82 with 27 − 5 = 22 degrees of freedom

Source	Sum of Squares	df	Mean Square	F-ratio
Regression	9806.53	4	2451.63	14.9
Residual	3617.32	22	164.424	

Variable	Coeff	SE(Coeff)	t-ratio	P-value
Intercept	−464.498	74.73	−6.22	≤0.0001
Calories	1.92005	0.2842	6.76	≤0.0001
Type	183.634	59.99	3.06	0.0057
Fat	−10.3847	1.779	−5.84	≤0.0001
Type*Cals	−0.461496	0.1740	−2.65	0.0145

b) Interpret the coefficient of *Type*Cals* in this regression model.

c) Is this a better regression model than the one in Exercises 1 and 3?

T 8. Traffic delays, part 3. Here's a plot of the Studentized residuals from the regression model of Exercise 2 plotted against *ArterialMPH*. The plot is colored according to *City Size* (Small, Medium, Large, and Very Large), and regression lines are fit for each city size.

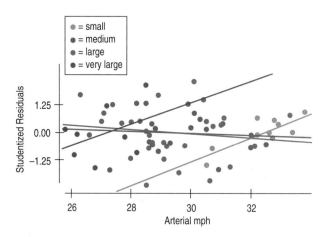

a) The model in Exercise 2 includes indicators for city size. Considering this display, have these indicator variables accomplished what is needed for the regression model? Explain.

Here is another model that adds two new constructed variables to the model in Exercise 2. They are the product of *ArterialMPH* and *Small* and the product of *ArterialMPH* and *VeryLarge*.

Dependent variable is: Delay/person
R squared = 80.7% R squared (adjusted) = 78.5%
s = 6.316 with 68 − 8 = 60 degrees of freedom

Source	Sum of Squares	df	Mean Square	F-ratio
Regression	10013.0	7	1430.44	35.9
Residual	2393.82	60	39.8970	

Variable	Coeff	SE(Coeff)	t-ratio	P-value
Intercept	153.110	17.42	8.79	≤0.0001
HiWayMPH	−1.02104	0.2426	−4.21	≤0.0001
ArterialMPH	−2.60848	0.6967	−3.74	0.0004
Small	−125.979	66.92	−1.88	0.0646
Large	4.89837	2.053	2.39	0.0202
VeryLarge	−89.4993	63.25	−1.41	0.1623
AM*Sml	3.81461	2.077	1.84	0.0712
AM*VLg	3.38139	2.314	1.46	0.1491

b) What does the predictor *AM*Sml* (*ArterialMPH* by *Small*) do in this model? Interpret the coefficient.

c) Does this model improve on the model in Exercise 2? Explain.

T 9. Insurance (Life expectancy). Insurance companies base their premiums on many factors, but basically all the factors are variables that predict life expectancy. Life expectancy varies from place to place. Here's a regression that models *Life Expectancy* in terms of other demographic variables that we saw in Exercise 34 of Chapter 18. (Refer to that exercise for variable definitions and units.)

Dependent variable is: Life exp
R squared = 67.0%   R squared (adjusted) = 64.0%
s = 0.8049 with 50 − 5 = 45 degrees of freedom

Source	Sum of Squares	df	Mean Square	F-ratio
Regression	59.1430	4	14.7858	22.8
Residual	29.1560	45	0.647910	

Variable	Coeff	SE(Coeff)	t-ratio	P-value
Intercept	69.4833	1.325	52.4	≤0.0001
Murder	−0.261940	0.0445	−5.89	≤0.0001
HSGrad	0.046144	0.0218	2.11	0.0403
Income	1.24948e-4	0.0002	0.516	0.6084
Illiteracy	0.276077	0.3105	0.889	0.3787

a) The state with the highest leverage and largest Cook's distance is Alaska. It is plotted with an "x" in the residuals plot. What evidence do you have from these diagnostic plots that Alaska might be an influential point?

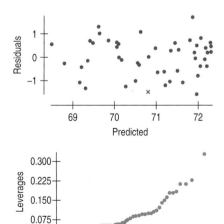

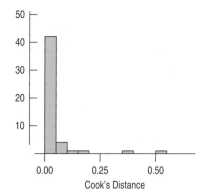

Here's another regression with a dummy variable for Alaska added to the regression model.

Dependent variable is: Life exp

R squared = 70.8%   R squared (adjusted) = 67.4%
s = 0.7660 with 50 − 6 = 44 degrees of freedom

Source	Sum of Squares	df	Mean Square	F-ratio
Regression	62.4797	5	12.4959	21.3
Residual	25.8193	44	0.586802	

Variable	Coeff	SE(Coeff)	t-ratio	P-value
Intercept	67.6377	1.480	45.7	≤0.0001
Murder	−0.250395	0.0426	−5.88	≤0.0001
HSGrad	0.055792	0.0212	2.63	0.0116
Illiteracy	0.458607	0.3053	1.50	0.1401
Income	3.68218e-4	0.0003	1.46	0.1511
Alaska	−2.23284	0.9364	−2.38	0.0215

b) What does the coefficient for the dummy variable for Alaska mean? Is there evidence that Alaska is an outlier in this model?

c) Which model would you prefer for understanding or predicting *Life Expectancy*? Explain.

T 10. Cereal nutrition. Breakfast cereal manufacturers publish nutrition information on each box of their product. As we saw in Chapter 17, there is a long history of cereals being associated with nutrition. Here's a regression to predict the number of *Calories* in breakfast cereals from their *Sodium*, *Potassium*, and *Sugar* content, and some diagnostic plots.

Dependent variable is: Calories

R squared = 38.4%   R squared (adjusted) = 35.9%
s = 15.60 with 77 − 4 = 73 degrees of freedom

Variable	Coeff	SE(Coeff)	t-ratio	P-value
Intercept	83.0469	5.198	16.0	≤0.0001
Sodium	0.057211	0.0215	2.67	0.0094
Potassium	−0.019328	0.0251	−0.769	0.4441
Sugar	2.38757	0.4066	5.87	≤0.0001

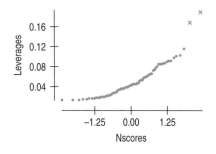

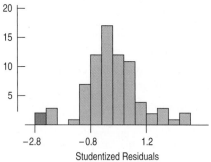

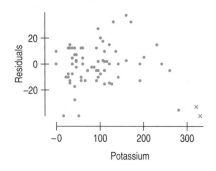

The shaded part of the histogram corresponds to the two cereals plotted with x's in the normal probability plot of the leverages. These are *All-Bran with Extra Fiber* and *All-Bran*.

a) What do the displays say about the influence of these two cereals on this regression? (The histogram is of the Studentized residuals.)

Here's another regression with dummy variables defined for each of the two bran cereals.

Dependent variable is: Calories
R squared = 50.7%   R squared (adjusted) = 47.3%
s = 14.15 with 77 − 6 = 71 degrees of freedom

Variable	Coeff	SE(Coeff)	t-ratio	P-value
Intercept	79.0874	4.839	16.3	≤0.0001
Sodium	0.068341	0.0198	3.46	0.0009
Potassium	0.043063	0.0272	1.58	0.1177
Sugar	2.03202	0.3795	5.35	≤0.0001
All-Bran	−50.7963	15.84	−3.21	0.0020
All-Bran Extra	−52.8659	16.03	−3.30	0.0015

b) Explain what the coefficients of the bran cereal dummy variables mean.

c) Which regression would you select for understanding the interplay of these nutrition components. Explain. (Note: Both are defensible.)

d) As you can see from the scatterplot, there's another cereal with high potassium. Not too surprisingly, it is *100% Bran*. But it does not have leverage as high as the other two bran cereals.

Do you think it should be treated like them (i.e., removed from the model, fit with its own dummy, or left in the model with no special attention, depending on your answer to (c))? Explain.

T 11. Lobster value diagnosed. In Chapter 18, Exercise 24 we constructed the following regression model to predict the log of the annual value of Maine's lobster harvest.

Dependent variable is: LogValue
R squared = 97.4%   R squared (adjusted) = 97.3%
s = 0.0782 with 56 − 4 = 52 degrees of freedom

Variable	Coeff	SE(Coeff)	t-ratio	P-value
Intercept	7.70207	0.3706	20.8	≤0.0001
Traps(M)	0.575390	0.0152	37.8	≤0.0001
Fishers	−5.32221e-5	0.0000	−5.43	≤0.0001
Water Temp	−0.015185	0.0074	−2.05	0.0457

At that time we also examined plots of the residuals which appeared to satisfy the assumptions and conditions for regression inference. Let's look a bit deeper. Here's a histogram of the Cook's distances for this model.

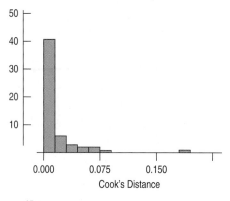

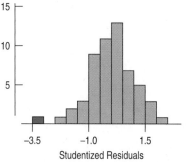

The case with high Cook's distance is 1994. (You can find a scatterplot of the *logValue* over time in Exercise 31 of Chapter 17.) What does this suggest about this model? What would you recommend?

T 12. Lobster price diagnosed. In Chapter 18, Exercise 25, we constructed the following regression model to predict the *Price* of lobster harvested in Maine's lobster fishing industry.

Dependent variable is: Price/lb
R squared = 94.4%   R squared (adjusted) = 94.1%
s = 0.2462 with 53 − 4 = 49 degrees of freedom

Variable	Coeff	SE(Coeff)	t-ratio	P-value
Intercept	0.845123	0.3557	2.38	0.0215
Traps(M)	1.20094	0.0619	19.4	≤0.0001
Fishers	−1.21820e-4	0.0000	−3.30	0.0018
Catch/Trap	−22.0207	10.96	−2.01	0.0500

At that time we also examined plots of the residuals which appeared to satisfy the assumptions and conditions for regression inference. Let's look a bit deeper. Here's a histogram of the Cook's distances for this model.

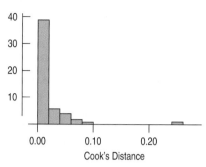

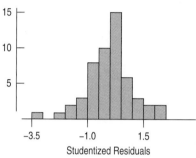

The case with large Cook's distance is 1994, which was also the year with the lowest Studentized residual. What does this suggest about this model? What would you recommend?

T 13. Economic regulation model building. A study by the U.S. Small Business Administration modeled the GDP per capita of 24 of the countries in the Organization for Economic Cooperation and Development (OECD) (Crain, M. W., *The Impact of Regulatory Costs on Small Firms*, available at www.sba.gov/idc/groups/public/documents/sba_homepage/rs264tot.pdf). One analysis estimated the effect on GDP of economic regulations using an index of the degree of OECD economic regulation and other variables. (We considered this model in Exercise 18 of Chapter 18.) They found the following regression model.

Dependent variable is: GDP/Capita

R squared = 97.4%   R squared (adjusted) = 96.6%
s = 2084 with 24 − 6 = 18 degrees of freedom

Source	Sum of Squares	df	Mean Square	F-ratio
Regression	2895078376	5	579015675	133
Residual	78158955	18	4342164	

Variable	Coeff	SE(Coeff)	t-ratio	P-value
Intercept	10487.3	9431	1.11	0.2808
OECD Reg Index	−1343.28	626.4	−2.14	0.0459
Ethnolinguistic Diversity	−69.9875	23.26	−3.01	0.0075
Int'l Trade/ GDP	44.7096	14.00	3.19	0.0050
Primary Education(%)	−58.4084	86.11	−0.678	0.5062
1988 GDP/ Capita	1.07767	0.0448	24.1	≤0.0001

a) If we remove *Primary Education* from the model, the R^2 decreases to 97.3%, but the adjusted R^2 *increases* to 96.7%. How can that happen? What does it mean? Would you include *Primary Education* in this model?

Here's a part of that regression.

Dependent variable is: GDP/Capita
R squared = 97.3%   R squared (adjusted) = 96.7%
s = 2054 with 24 − 5 = 19 degrees of freedom

Variable	Coeff	SE(Coeff)	t-ratio	P-value
Intercept	4243.21	2022	2.10	0.0495
OECD Reg Index	−1244.20	600.4	−2.07	0.0521
Ethnolinguistic Diversity	−64.4200	21.45	−3.00	0.0073
Int'l Trade/ GDP	40.3905	12.29	3.29	0.0039
1988 GDP/ Capita	1.08492	0.0429	25.3	≤0.0001

b) Consider the *t*-statistic for *OECD Regulation* in the reduced model. That was the predictor of interest to this author. Do you agree with his conclusion that OECD regulation reduced GDP/Capita in these countries? Why do you think he chose to include *Primary Education* as a predictor? Explain.

T 14. Dirt bikes. Off-road motorcycles (often called "dirt bikes") are a segment (about 18%) of the growing motorcycle market. Because dirt bikes offer great variation in features, they are a good market segment to study to learn about which features account for the cost (manufacturer's suggested retail price, *MSRP*) of a bike. Researchers collected data on 2005-model dirt bikes (lib.stat.cmu.edu/datasets/dirtbike_aug.csv). Their original goal was to study market differentiation among brands. (*The Dirt on Bikes: An Illustration of CART Models for Brand Differentiation*, Jiang Lu, Joseph B. Kadane, and Peter Boatwright). In Chapter 18, Exercises 31, 32, and 33 dealt with these data, but several bikes were removed from those data to simplify the analysis. Now we'll take on the full set.[13]

Here's a regression model and some associated graphs.

Dependent variable is: MSRP

R squared = 91.0%   R squared (adjusted) = 90.5%
s = 606.4 with 100 − 6 = 94 degrees of freedom

Source	Sum of Squares	df	Mean Square	F-ratio
Regression	349911096	5	69982219	190
Residual	34566886	94	367733	

Variable	Coeff	SE(Coeff)	t-ratio	P-value
Intercept	−5514.66	826.2	−6.67	≤0.0001
Bore	83.7950	6.145	13.6	≤0.0001
Clearance	152.617	52.02	2.93	0.0042
Engine Strokes	−315.812	89.83	−3.52	0.0007
Total Weight	−13.8502	3.017	−4.59	≤0.0001
Wheelbase	119.138	34.26	3.48	0.0008

---

[13] Well, in honesty, we've removed one luxury, handmade bike whose MSRP was $19,500 as a clearly identified outlier.

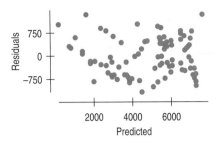

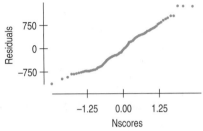

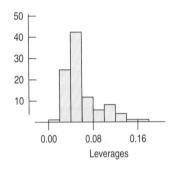

a) List aspects of this regression model that lead to the conclusion that it is likely to be a useful model.

b) What aspects of the displays indicate that the model is a good one?

T 15. Airlines on time. Airlines strive to be on time, in part, because customers can refer to government-published statistics to select flights that are most often on time. We have data for 19 airlines for March 2006 reporting the following variables:

On Time	(number of on-time arrivals)
Cancelled	(number of cancelled flights)
Diverted	(number of diverted flights)
Carrier	(number of delays due to the carrier)
Weather	(number of delays due to weather)
NAS Delay	(Delays due to the National Airspace System (traffic control))
Late Arrival	(number of delays due to late arrival of equipment or crew)

Here's a regression model.

Dependent variable is: On Time

R squared = 93.9%   R squared (adjusted) = 90.8%
s = 5176 with 19 − 7 = 12 degrees of freedom

Source	Sum of Squares	df	Mean Square	F-ratio
Regression	4947151273	6	824525212	30.8
Residual	321546284	12	26795524	

Variable	Coeff	SE(Coeff)	t-ratio	P-value
Intercept	1357.10	2316	0.586	0.5687
Cancelled	−18.5514	6.352	−2.92	0.0128
Diverted	39.5623	95.59	0.414	0.6863
Carrier	10.9620	3.104	3.53	0.0041
Weather	10.4637	9.462	1.11	0.2905
NAS Delay	2.46727	1.091	2.26	0.0431
Late Arrival	4.64874	1.445	3.22	0.0074

a) Interpret the coefficient of *Diverted*. (Hint: This is a trick question.)

Here's a scatterplot of *On Time* vs. *Diverted* and another regression. Note that *Diverted* is the response variable in this second regression.

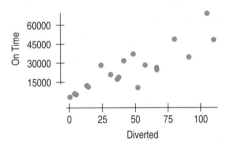

Dependent variable is: Diverted
R squared = 85.6%   R squared (adjusted) = 80.0%
s = 15.02 with 19 − 6 = 13 degrees of freedom

Variable	Coeff	SE(Coeff)	t-ratio	P-value
Intercept	−1.65088	6.703	−0.246	0.8093
Cancelled	−0.016896	0.0178	−0.948	0.3604
Carrier	0.013823	0.0081	1.70	0.1136
Weather	0.050714	0.0236	2.15	0.0509
Late Arrival	6.33917e-3	0.0038	1.67	0.1196
NAS Delay	7.09830e-3	0.0025	2.86	0.0133

b) It seems from the scatterplot that *Diverted* would be a good predictor of *On Time*, but that seems not to be the case. Why do you think the coefficient of *Diverted* is not significant in the first regression?

c) How does the second regression explain this apparent contradiction?

d) Find the value of the Variance Inflation Factor statistic for *Diverted* in the first regression.

T 16. Dirt bikes, part 2. The model in Exercise 14 is missing one predictor that we might have expected to see. *Engine Displacement* is highly correlated ($r = 0.783$) with *MSRP*, but that variable has not entered the model (and, indeed, would have a P-value of 0.54 if it were added to the model). Here is some evidence to explain why that may be. (Hint: Notice that *Displacement* is the response variable in this regression.)

Dependent variable is: Displacement
R squared = 95.9%   R squared (adjusted) = 95.7%
s = 35.54 with 100 − 6 = 94 degrees of freedom

Variable	Coeff	SE(Coeff)	t-ratio	P-value
Intercept	−8.05901	48.42	−0.166	0.8682
Bore	9.10890	0.3601	25.3	≤0.0001
Clearance	3.55912	3.048	1.17	0.2460
Engine Stroke	−27.3943	5.264	−5.20	≤0.0001
Total Weight	1.03749	0.1768	5.87	≤0.0001
Wheelbase	−10.0612	2.008	−5.01	≤0.0001

a) What term describes the reason *Displacement* doesn't contribute to the regression model for *MSRP*?

b) Find the value of the Variance Inflation Factor for *Displacement* in the regression on *MSRP*.

T 17. Show business diagnosed. Exercises in Chapter 18 considered weekly receipts from shows on Broadway in New York City. To simplify matters, we omitted a few weeks from those data. Here is the same regression with all the data present.

Dependent variable is: Receipts($M)
R squared = 93.3%   R squared (adjusted) = 93.1%
s = 0.9589 with 92 − 4 = 88 degrees of freedom

Source	Sum of Squares	df	Mean Square	F-ratio
Regression	1130.43	3	376.811	410
Residual	80.9067	88	0.919395	

Variable	Coeff	SE(Coeff)	t-ratio	P-value
Intercept	−22.1715	2.221	−9.98	≤0.0001
Paid Attendance	0.087031	0.0046	19.0	≤0.0001
Number of Shows	−0.024934	0.0338	−0.737	0.4632
Average Ticket Price	0.265756	0.0286	9.29	≤0.0001

Here are some diagnostic plots.

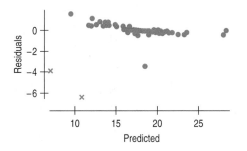

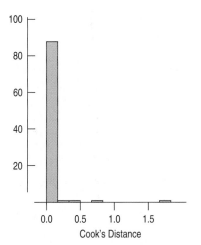

a) The two points plotted with "x" in the residuals vs. predicted plot are the two with the highest Cook's distances. What does this information say about how those points might be affecting the analysis?

In fact, those points are reports published the last week of November and the first week of December in 2007—a time when the Stagehands Union Local One was on strike, closing down most Broadway shows. It seems that these are not representative weeks for business on Broadway. Removing them results in the following regression.

Dependent variable is: Receipts($M)

R squared = 98.5%   R squared (adjusted) = 98.5%
s = 0.3755 with 90 − 4 = 86 degrees of freedom

Source	Sum of Squares	df	Mean Square	F-ratio
Regression	809.793	3	269.931	1915
Residual	12.1250	86	0.140988	

Variable	Coeff	SE(Coeff)	t-ratio	P-value
Intercept	−21.3165	0.8984	−23.7	≤0.0001
Paid Attendance	0.071955	0.0019	37.4	≤0.0001
Number of Shows	0.045793	0.0137	3.35	0.0012
Average Ticket Price	0.274503	0.0115	23.8	≤0.0001

b) What changes in these estimated models support the conclusion that those two weeks were influential points? (Hint: The increase in adjusted R^2 isn't one of the reasons.)

c) Which model would be the best to use in analyzing the business of Broadway shows—the one with all the data or the one with the two influential points removed? Explain.

T 18. Economic regulation model building, part 2. Exercise 13 raised some questions about the regression model built to understand the effect of OECD regulation on GDP/Capita in 24 OECD countries. Let's look more deeply. Here's a histogram of the Cook's distances for that model.

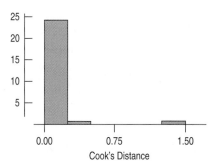

The country with high Cook's distance is Ireland.

a) What does the Cook's distance value suggest about Ireland in this model?

Of the predictors available for this model, by far the best (highest R^2 predictor) is *1988 GDP/Capita*. In a scatterplot of *GDP/Capita* vs. *1988 GDP/Capita*, Ireland stands away from the overall linear trend.

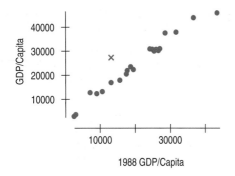

We can define a dummy variable that is "1" for Ireland and "0" for all other countries. The resulting model looks like this.

Dependent variable is: GDP/Capita
R squared = 98.3%   R squared (adjusted) = 97.7%
s = 1713 with 24 − 7 = 17 degrees of freedom

Variable	Coeff	SE(Coeff)	t-ratio	P-value
Intercept	13609.0	7818	1.74	0.0998
1988 GDP/ Capita	1.10397	0.0378	29.2	≤0.0001
OECD Reg Index	−520.181	579.2	−0.898	0.3816
Int'l Trade/ GDP	21.0171	13.81	1.52	0.1463
Ethnolinguistic Diversity	−49.8210	20.19	−2.47	0.0245
Primary Education	−99.3369	72.00	−1.38	0.1856
Ireland	8146.64	2624	3.10	0.0064

b) Explain what the dummy variable for Ireland accomplishes in this model.

c) What do you conclude now about the effects of OECD regulation in these data?

T 19. *Lobster catch.** In Chapter 17, we saw data on the lobster fishing industry in Maine. We have annual data from 1950 through 2006. The annual catch (metric tons) has increased, but is this due to more licensed lobster fishers or to more efficient fishing? Because both fishers and traps are individually licensed, we have detailed data. Here's a scatterplot of *Catch/Fisher* (tons) vs. the number of *Traps* (in millions).

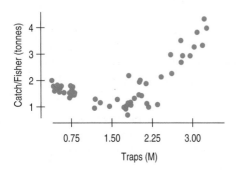

a) Consider the assumptions and conditions for regression. Would re-expressing either variable help? Explain.

Here's a model that includes a quadratic term.

Dependent variable is: Catch/Fisher
R squared = 85.1%   R squared (adjusted) = 84.5%
s = 0.3343 with 56 − 3 = 53 degrees of freedom

Source	Sum of Squares	df	Mean Square	F-ratio
Regression	33.8265	2	16.9132	151
Residual	5.92352	53	0.111765	

Variable	Coeff	SE(Coeff)	t-ratio	P-value
Intercept	2.82971	0.1699	16.7	≤0.0001
Traps	−2.45241	0.2286	−10.7	≤0.0001
Traps^2	0.874513	0.0647	13.5	≤0.0001

b) In 2001, almost 3 million lobster traps were licensed in Maine to 7327 fishers. What would this model estimate as the total catch for that year?

c) The coefficient of *Traps* in this model is negative, but we can see from the scatterplot that the catch has increased substantially as the number of traps increased. What accounts for this apparent anomaly?

20. *Water temperature.** Like many businesses, the Maine lobster fishing industry is seasonal. But the peak lobstering season has been shifting later in the fall. According to scientists, the biggest factor in the time of the peak is water temperature. The data at www.maine.gov/dmr include water temperature. How has it changed over the past half century? Here's a scatterplot of Water Temperature(°F) vs. Year.

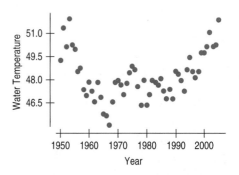

a) Consider the assumptions and conditions for regression. Would re-expressing either variable help? Explain.

A regression model for *Water Temperature* by *Year* looks like this.

Dependent variable is: Water Temperature
R squared = 2.7%   R squared (adjusted) = 0.9%
s = 1.522 with 56 − 2 = 54 degrees of freedom

Variable	Coeff	SE(Coeff)	t-ratio	P-value
Intercept	17.7599	24.88	0.714	0.4784
Year	0.015437	0.0126	1.23	0.2251

b) Do the statistics associated with this model indicate that there is no relationship between *Water Temperature* and *Year*? Explain.

Here's a regression model including a quadratic term.

Dependent variable is: Water Temperature
57 total cases of which 1 is missing

R squared = 60.0%   R squared (adjusted) = 58.5%
s = 0.9852 with 56 − 3 = 53 degrees of freedom

Variable	Coeff	SE(Coeff)	t-ratio	P-value
Intercept	19208.2	2204	8.71	≤0.0001
Year	−19.3947	2.229	−8.70	≤0.0001
Year^2	4.90774e-3	0.0006	8.71	≤0.0001

c) This model fits better, but the coefficient of *Year* is now strongly negative. We can see from the scatterplot that temperatures have not been dropping at 19 degrees per year. Explain this apparent anomaly.

Here's the same regression model[14] but with the quadratic term in the form of $(Year - \overline{Year})^2$.

Dependent variable is: Water Temperature
R squared = 60.0%   R squared (adjusted) = 58.5%
s = 0.9852 with 56 − 3 = 53 degrees of freedom

Variable	Coeff	SE(Coeff)	t-ratio	P-value
Intercept	6.77149	16.16	0.419	0.6768
Year	0.020345	0.0082	2.49	0.0159
(Year-mean)^2	4.90774e-3	0.0006	8.71	≤0.0001

d) Using the second model, predict what the water temperature will be in 2015, and then explain why you don't trust that prediction.

---

[14] Notice that, even though mathematically the P-value of the coefficient for *Year* should be the same in this table as in the previous regression, it is not. This is because *Year* and $Year^2$ are so highly correlated (their correlation is 0.99999309) that the collinearity has caused a large enough rounding error to affect the calculation in the previous regression.

Time Series Analysis

Whole Foods Market®

In 1978, twenty-five-year-old John Mackey and twenty-one-year-old Rene Lawson Hardy borrowed $45,000 from family and friends to open the doors of a small natural foods store they named SaferWay in Austin, Texas. Two years later, they joined forces with Clarksville Natural Grocery to open the first Whole Foods Market. With 10,500 square feet of floor space and a staff of 19, their store was large in comparison to other health food stores of the time.

In the mid 1980s, Whole Foods Market began expanding outside of Austin and opened stores elsewhere in the southern United States and California. During the next 15 years, Whole Foods Market grew rapidly, in part due to several mergers and acquisitions. Among the stores they acquired were Wellspring Grocery, Bread & Circus, Mrs. Gooch's Natural Foods, and Fresh Fields Markets. Since 2000, Whole Foods has expanded outside of North America with the purchase of seven Fresh & Wild stores in the United Kingdom. Today, Whole Foods Market, Inc. has 53,000 employees and more than 270 stores. In 2005 they ranked fifth in the Fortune 100 best places to work. In 2007, they merged with Wild Oats of Boulder, Colorado. The firm has continued to grow both by opening new stores and by acquiring related firms.[1]

T he decade of the 1990s was a period of growth for most companies, but unlike firms in many other industries, Whole Foods Market continued to grow into the next decade. Here is a time series plot showing the quarterly Sales ($M) plotted by quarter since 1995. If you were asked to summarize the trend in *Sales* over this decade what would you say?

WHAT	Quarterly *Sales*
UNITS	Millions of U.S. dollars
WHEN	1995–2007
WHERE	United States, Canada, and United Kingdom
WHY	To forecast sales for Whole Foods Market

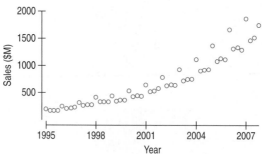

Figure 20.1 *Quarterly Sales (in $M) for Whole Foods from 1995 to 2007.*

Clearly, sales at Whole Foods Market, Inc. grew between 1995 and 2007, starting at below $250 million and reaching nearly $2 billion per quarter. But we'd like to be able to say something more than that, and ultimately, we'd like to *model* the growth. Time series models are primarily built for predicting into the near future. Some also offer interpretable coefficients.

Corporations often look at time series plots to examine prior patterns in the data and to forecast future values. The executives at Whole Foods Market, Inc. might be interested in understanding patterns in *Sales* to:

◆ Plan inventory and distribution of goods
◆ Schedule hiring and staffing
◆ Understand the impact of seasons, or time of the year, on sales
◆ Develop advertising campaigns
◆ Forecast profits and plan corporate strategy

Suppose that you are an analyst at Whole Foods Market, Inc. and you have been asked to forecast *Sales* for the next four quarters. How can you analyze the time series to produce accurate forecasts of *Sales*? How can you measure accuracy and compare your different forecasting models?

---

[1] Source: Official Whole Foods Market website: www.wholefoodsmarket.com.

20.1 What Is a Time Series?

The Whole Foods sales are recorded each financial quarter, and we're interested in the growth of sales over time. Whenever we have data recorded sequentially over time, and we consider *Time* to be an important aspect of the data, we have a **time series**.[2] Most time series are equally spaced at roughly regular intervals, such as monthly, quarterly, or annually.[3] The Whole Foods sales data in Figure 20.1 are a time series measured quarterly. The fiscal year at Whole Foods Market, Inc. starts on or near September 30, so, unlike many companies, the first fiscal quarter reports sales for the end of the calendar year, and includes the holiday season.

Because a time series is likely to be revealed to us sequentially and only up to the present, most time series methods use prior values of a series to predict future values. Some methods describe the overall pattern in the series and predict that it will continue, at least for a short while. Others methods estimate **time series models** that identify structured components like trends and seasonal fluctuation in the series and model those. Some introduce external variables to help predict the response, as in regression, while others simply look at the prior values in the series to try to discern patterns.[4] The objective of most time series analyses is to provide **forecasts** of future values of the time series. Of course, predicting the future is valuable to business decision makers, so time series methods are widely used.

20.2 Components of a Time Series

When we examined the distribution of a single variable, we looked at its shape, center, and spread. When we looked at scatterplots of two variables, we asked about the direction, form, and strength. For a time series, we look for the trend, seasonal patterns, and long-term cycles. Some time series exhibit some of these components, some show all of them, and others have no particular large-scale structure at all.

The Trend Component

Look at the time series plot of the Whole Foods *Sales* data in Figure 20.1. What overall pattern are the *Sales* following? Not only have sales at Whole Foods Market been increasing, but they seem to be accelerating. Of course, there are fluctuations around this pattern, but viewed over more than a decade, the overall trend is clear. We'd describe the direction as positive, and the shape as curving upwards (see Figure 20.2). This overall pattern is the **trend component** of the time series. This is often the most interesting aspect of a time series. For example, it is what an investor would want to know about.

Most series have an increasing or decreasing trend with other fluctuations around the trend. Some, however, just fluctuate randomly, much like residual plots from a successful regression. If a series shows no particular trend over time and has a relatively consistent mean, it is said to be **stationary in the mean**.

---

[2] Actually, the methods of this chapter will work for any values that are ordered.

[3] Some series, such as those recording values for trading days or on the first day of each month, are not exactly equally spaced. If there are actual gaps in a time series, researchers use a variety of methods to fill in the missing observations before analyzing the data.

[4] Advanced models such as dynamic regression models or distributed lag models are outside the scope of this text.

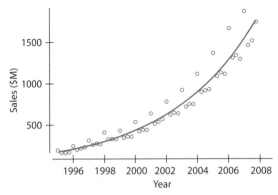

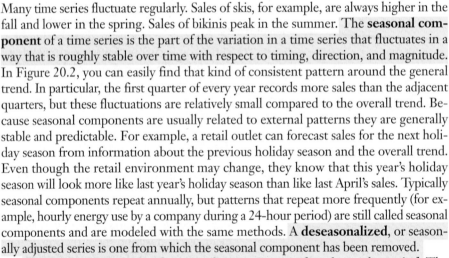

Figure 20.2 *The trend in the Whole Foods data looks nonlinear. In fact, it is increasing and curved.*

If the trend grows roughly linearly, we can use the linear regression methods of Chapter 16 to estimate the linear component. We'll do that later in this chapter. (You may wonder if we can use regression for data in which successive values are not likely to be independent. But if we don't plan to test the coefficients against the standard regression null hypothesis, we don't need any regression assumptions other than linearity.)

The Seasonal Component

Many time series fluctuate regularly. Sales of skis, for example, are always higher in the fall and lower in the spring. Sales of bikinis peak in the summer. The **seasonal component** of a time series is the part of the variation in a time series that fluctuates in a way that is roughly stable over time with respect to timing, direction, and magnitude. In Figure 20.2, you can easily find that kind of consistent pattern around the general trend. In particular, the first quarter of every year records more sales than the adjacent quarters, but these fluctuations are relatively small compared to the overall trend. Because seasonal components are usually related to external patterns they are generally stable and predictable. For example, a retail outlet can forecast sales for the next holiday season from information about the previous holiday season and the overall trend. Even though the retail environment may change, they know that this year's holiday season will look more like last year's holiday season than like last April's sales. Typically seasonal components repeat annually, but patterns that repeat more frequently (for example, hourly energy use by a company during a 24-hour period) are still called seasonal components and are modeled with the same methods. A **deseasonalized**, or seasonally adjusted series is one from which the seasonal component has been removed.

The time between peaks of a seasonal component is referred to as the **period**. The period is independent of the actual timing of peaks and valleys. For example, many retail companies' sales spike during the holiday season in December, and sales of water skis peak in early summer, but both have a seasonal period of one year.

Cyclical Component

Regular cycles in the data with periods longer than one year are referred to as **cyclical components**. Economic and business cycles can sometimes be modeled, but often we do little more than describe them. When a cyclical component can be related to a predictable phenomenon, then it can be modeled based on some regular behavior or by introducing a variable that represents that predictable phenomenon, and added it to whatever model we are building for the time series.

Irregular Component

We will see a number of ways to model time series. Just as we did with linear models and regression in Chapter 8, in this chapter we'll find that the residuals—the part of the data *not* fit by the model—can be informative. In time series modeling, these residuals are called the **irregular component**. As with regression residuals, it's a good idea to plot the irregular component to look for extraordinary cases or other unexpected patterns. Often our interest in the irregular component is in how variable it is, whether that variability changes over time, and whether there are any outliers or spikes that may deserve special attention. A time series that has a relatively constant variance is said to be **stationary in the variance**. If a time series is simply said to be stationary, most often what is meant is that it has stationary variance.

To summarize, we identify four *components* of a time series:

1. Trend component (T)
2. Seasonal component (S)
3. Cyclical component (C)
4. Irregular component (I)

Table 20.1 provides a summary of the components as applied to the Whole Foods *Sales* data.

Component	Description	Rationale	Period Length
Trend	Positive and nonlinear	Overall increase in sales, with a change in the rate of increase of sales in the last 7 years	
Seasonal	Peaks every first quarter in 1995 through 2007	Larger sales in first quarter; reason unknown at present	4 quarters—yearly
Cyclical	Insufficient data to observe peaks every 2 to 10 years	Could be due to economic cycles and factors, such as inflation, interest rates, and employment that might impact consumer spending	Insufficient data to observe cyclical period
Irregular	Random fluctuations in data	Due to irregular or unpredictable events, such as mergers and acquisitions of other companies, or natural disasters, such as floods	No regular repeating pattern; no period

| **Table 20.1** *Time series components terms applied to the series Sales for Whole Foods Market, Inc. from 1995 to 2007.*

Even though some time series exhibit only two or one of these components, or even none, an understanding of them can help us structure our understanding of a time series. Just as we look at the direction, form, and strength of a scatterplot, remembering to think about the trend, seasonal, and cyclical parts of a time series can clarify our view of a time series.

Modeling Time Series

Methods for forecasting a time series fall into two general classes. Smoothing methods work from the bottom up. They try to "smooth out" the irregular component so any underlying patterns will be easier to see. They have the advantage that they don't assume that there is a trend or seasonal component—and indeed, they'll work even when no seasonal component is present and the trend is complex. For example, we can use smoothing methods on a time series that has only a general cyclical component but no clear trend or seasonal fluctuations to model. The disadvantage of smoothing methods is that they can forecast only the immediate future. Lacking a model of behaviors that can be trusted to continue (such as a seasonal component based on calendar shopping patterns or temperature variation over the year), they don't have a basis for long-term forecasting. Instead, they rely on the assumption that most time series show patterns that vary more slowly than each successive observation, so the next value in the future will resemble the most recent ones.

When we can discern a trend or both a trend and seasonal component, we'll often prefer regression-based modeling methods. These use the methods of multiple regression we've learned in Chapters 18 and 19 to estimate each component's contribution to the time series and to build a model for the time series. As with any regression-based model, models of this kind can be used to forecast for any value of *Time* and thus can generate forecasts further into the future than one time period. However, as always, we'll need to be cautious with such extrapolations.

The next sections discuss several kinds of smoothing methods, followed by a discussion of regression-based models. Although the smoothing methods don't explicitly use the time series components, it is a good idea to keep them in mind. The regression models explicitly estimate the components as a basis for building the models.

20.3 Smoothing Methods

Most time series contain some random fluctuations that vary up and down rapidly—often for consecutive observations. But, precisely because they are random, these fluctuations provide no help in forecasting. Even if we believe that a time series will continue to fluctuate randomly, we can't predict *how* it will do so. The only aspects of a time series that we have any hope of predicting are those that vary either regularly or slowly. One way to identify these aspects is to smooth away the rapid random fluctuations.[5] To forecast the value of a time series in the future, we want to identify the underlying, consistent behavior of the series. In many time series, these slower changes have a kind of inertia. They change and fluctuate, but recent behavior is often a good indication of behavior in the near future. Smoothing methods damp down random fluctuations and try to reveal the underlying behavior so we can use it to forecast values in the immediate future.

The time series for the daily Intel stock price in 2002 (Figure 20.3) shows no regular repeating patterns and no evidence of a regular seasonal effect. But it does show rapid fluctuations and some evidence of longer-term movements.

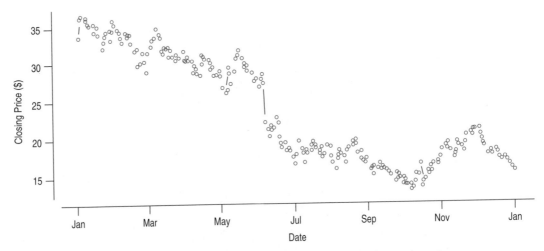

| *Figure 20.3*   *Daily closing prices for Intel stock in 2002 show no seasonal or other regular patterns.*

---

[5] To an engineer, this would be separating the signal from the rapidly varying noise.

Methods for smoothing out the apparently random fluctuations generally work by averaging adjacent values in the series. We know from Chapter 12 that means vary less than the underlying data. We can use that principle to find values that are typical of the local value of the series while varying less than the original data.

20.4 Simple Moving Average Methods

The most commonly used smoothing method is the method of **moving averages**. A moving average replaces each value in a time series by an average of the adjacent values. The number of values we use to construct the averages is called the *length* of the moving average (*L*). Almost every stock tracking service on the Internet offers a variety of moving averages (often with 50-, 100-, and 200-day lengths) to help track stock movements.

A moving average simply uses the mean of the previous *L* values as the fitted value at each time. Because it focuses on only the recent values, a moving average with a short length can respond to rapid changes in a time series. A moving average with a longer length will respond more slowly. The general form of a moving average is:

$$\tilde{y}_t = \frac{\sum_{i=t-L+1}^{t} y_i}{L}$$

The length, *L*, of a moving average is a subjective choice, but it must be specified when discussing a moving average. We write MA(*L*) for a moving average of length *L* and use the tilde to denote a moving average calculated from a sequence of data values.

Let's begin by using a moving average of length of 5 on the Intel stock series to illustrate the calculation. The data are in Table 20.2 and the formulas to calculate the smoothed stock price for the fifth and sixth day in the series are:

$$\widetilde{\text{Price}}_5 = \frac{\sum_{1}^{5} \text{Price}_t}{5} = \frac{33.00 + 35.52 + 35.79 + 35.27 + 35.58}{5} = 35.03$$

and

$$\widetilde{\text{Price}}_6 = \frac{\sum_{2}^{6} \text{Price}_t}{5} = \frac{35.52 + 35.79 + 35.27 + 35.58 + 35.36}{5} = 35.50$$

The MA(5) smoothed stock price for each day in the series in 2002 is computed from that day's closing price and the preceding four daily closing prices using similar formulas. If we instead select *L* = 15 for our moving averages, then the calculations will average the 15 previous closing prices (including today's price). Table 20.2 shows the computed values for the two moving averages using *L* = 5 and *L* = 15 and the actual closing price of Intel daily stock for the first 30 days when the market was open in 2002.

There are no moving averages for the first (*L* − 1) days in the series for each moving average model. What happens to the moving average as the length, *L*, increases? The two smoothed series produced by computing moving averages for the daily Intel stock price using *L* = 5 and *L* = 15 in Figure 20.4 shows that the moving average series with the greater length is smoother. That should be what you expected. Of course, a smoother series is not

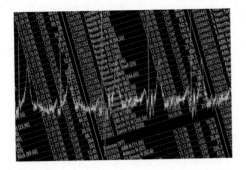

Date	Price	MA(5)	MA(15)
2-Jan-02	$33.00	*	*
3-Jan-02	$35.52	*	*
4-Jan-02	$35.79	*	*
7-Jan-02	$35.27	*	*
8-Jan-02	$35.58	$35.03	*
9-Jan-02	$35.36	$35.50	*
10-Jan-02	$34.65	$35.33	*
11-Jan-02	$34.55	$35.08	*
14-Jan-02	$34.84	$35.00	*
15-Jan-02	$34.68	$34.82	*
16-Jan-02	$33.71	$34.49	*
17-Jan-02	$34.53	$34.46	*
18-Jan-02	$33.48	$34.25	*
22-Jan-02	$31.70	$33.62	*
23-Jan-02	$32.45	$33.17	$34.34
24-Jan-02	$33.20	$33.07	$34.35
25-Jan-02	$33.68	$32.90	$34.23
28-Jan-02	$33.92	$32.99	$34.11
29-Jan-02	$32.68	$33.19	$33.93
30-Jan-02	$33.86	$33.47	$33.82
31-Jan-02	$35.04	$33.84	$33.80
1-Feb-02	$34.67	$34.03	$33.80
4-Feb-02	$33.98	$34.05	$33.76
5-Feb-02	$33.80	$34.27	$33.69
6-Feb-02	$32.92	$34.08	$33.57
7-Feb-02	$32.31	$33.54	$33.48
8-Feb-02	$32.52	$33.11	$33.35
11-Feb-02	$33.57	$33.02	$33.35
12-Feb-02	$32.97	$32.86	$33.44
13-Feb-02	$33.38	$32.95	$33.50

Table 20.2 *Moving Averages for L = 5, or MA(5) and L = 15, or MA(15) for the closing price for Intel stock during the first 30 trading days of 2002.*

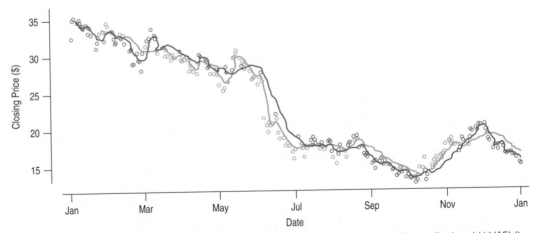

Figure 20.4 *A time series plot of the daily Intel stock closing prices in 2002 with MA(5) (in yellow) and MA(15) (in blue) forecasts superimposed. The moving average with the smaller length, L = 5, follows the data more closely while the moving average with the longer length is smoother.*

necessarily a better model for the data because it has a hard time following the series when it changes rapidly. Look, for example, at the two series during June of the year. The stock price fell rapidly, but the MA(15) smooth changed too slowly, running above the data for several weeks before it "caught up."

To obtain a forecast for a new time point, analysts use the last average in the series:

$$\hat{y}_{t+1} = \widetilde{y}_t$$

This is the **simple moving average forecast**. Of course, this method can forecast only one time period into the future, for *Time* $= t + 1$. (You can repeat that value as a forecast beyond period $t + 1$, but unless the time series is essentially unstructured and horizontal, it won't be a very good forecast.) If the length of the moving average is 1 ($L = 1$), then the forecast is simply that the next value will be the same as the previous one, $\hat{y}_{t+1} = y_t$. As the simplest forecast, this is called the **naive forecast**.

Often, moving averages are used primarily as summaries of how a time series is changing. The length selected depends on the purpose of the analysis. If the focus is on long-term behavior, a longer moving average is appropriate. But an analyst interested in shorter-term changes would choose a shorter length. Sometimes (as you can see in the sidebar) analysts compare a shorter length moving average and one with a longer length, hoping to learn something from how they compare.

One potential problem with a moving average is that, as we know from Chapter 6, means can be affected by outliers. An outlier in a time series would be a spike in the series far from the adjacent values. Such a spike will contaminate all of the averages in which it participates, spreading its influence over a number of values.

> **Investors and Moving Averages**
>
> Moving averages can help investors identify short- and long-term trends for a particular stock or mutual fund. Many analysts will consider it a bull market when the 50-day moving average is above the 200-day moving average, and a bear market if the opposite is true. When the 50-day moving average crosses above the 200-day average, this indicates that recent values have increased above the level established by the longer-term average and is taken as a signal for investors to buy. It is referred to as a "golden cross." When the 50-day moving average moves below the 200-day average, then this is a "sell" signal and is called a "death cross."
>
> —*Fidelity Outlook*, August 2002, p. 6

20.5 Weighted Moving Averages

In a simple moving average, we just average the most recent L values. But we can benefit from a more sophisticated averaging scheme. We can assign a *weight* to each value according to how far it is before the current value. The result is a *weighted* average. In a weighted average, each value is multiplied by a *weight* before they are added up, and the total is divided by the sum of the weights:

$$\widetilde{y}_t = \frac{\sum w_i y_{t-i}}{\sum w_i}$$

The weights might be specified, or they might be found as part of the smoothing process.

Weighted moving averages form a very general class of smoothers.[6] We will consider two types of weighted moving average smoothers that are commonly used on time series data, exponential smoothers and autoregressive moving averages.

---

[6] In this general form, these smoothers are known as *linear smoothers*. They are important in engineering and finance.

20.6 Exponential Smoothing Methods

Smoothing methods summarize each value of a time series with an average of recent values. In many time series, recent values of the series are more relevant for modeling than older ones. So a weighted moving average that weights the more recent values more heavily than the older ones makes sense. Exponential smoothing does just that. **Exponential smoothing** is a weighted moving average with weights that decline exponentially into the past. The most recent data are weighted the most and the most distant data are weighted the least. This model is the **single-exponential smoothing (SES) model**:

$$\tilde{y}_t = \alpha y_t + (1 - \alpha)\tilde{y}_{t-1}$$

The choice of the weight α is up to the data analyst, although it is usually restricted to $0 < \alpha < 1$. When $\alpha = 0.50$, the current data point and the entire set of historical data (all points before the current one) are weighted equally. If $\alpha = 0.75$, then historical data are weighted only 25% and the current value has more weight at 75%. If the objective is to produce forecasts that are stable and smoother, then choose a smoothing coefficient closer to zero. If, however, the objective is to react to volatile events rapidly, then choose a smoothing coefficient close to one.[7]

Unlike a simple moving average, exponential smoothing uses *all* previous values, although distant ones typically get very small weight. If we expand the calculation, we can see that the smoothed value at time t is a *weighted* average of the current value and all the previous values, with the weights depending on a smoothing coefficient, α:

$$\tilde{y}_t = \alpha y_t + \alpha(1 - \alpha)y_{t-1} + \alpha(1 - \alpha)^2 y_{t-2} + \alpha(1 - \alpha)^3 y_{t-3} + \cdots.$$

As with the moving average model, we use $\tilde{y}_t$ as our prediction for time $t + 1$.

Figure 20.5 shows the Intel stock prices again, this time with exponentially smoothed values using $\alpha = 0.75$ and $\alpha = 0.10$. You can see that the curve computed using the larger α follows the original series closely. By contrast, the curve computed using the smaller α is smoother, but doesn't follow rapid changes in the series such as the sharp price drop in June.

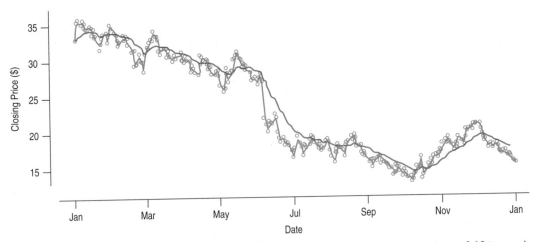

Figure 20.5   *The Intel stock prices along with exponential smoothing models ($\alpha = 0.75$ in red, $\alpha = 0.10$ in green). The model with the larger alpha follows the data more closely, and the model with the smaller alpha is smoother.*

---

[7] The initial smoothed value to get the algorithm started is either the initial observed value ($\tilde{y}_1 = y_1$) or the average of some of the initial values. In Minitab, for example, the initial smoothed value is equal to the average of the first six observations.

20.7 Summarizing Forecast Error

Whenever we model a time series, it is natural to ask how closely the model fits the series. A closely related question is how well the model *forecasts* the series. For smoothing models, we use the most recent model value as the forecast for the next time period. When we make a plot of the data and the smoothed series we usually plot the smoothed value $\widetilde{y}_t$ at time t. But if we are interested in the error the model makes when trying to forecast each value, we must compare the data value y_t not to $\widetilde{y}_t$, but rather to $\hat{y}_t$, which is actually equal to $\widetilde{y}_{t-1}$. We can find a forecast error for each time in the series for which we have such a forecast:

$$e_t = (y_t - \hat{y}_t)$$

When there is a particular forecast of interest, it makes sense to look at the forecast error, e_t. More often, we ask about the overall success of a model at forecasting for a time series. That calls for a summary of the forecast errors and, as often happens in statistics, we have several summaries to choose among.

In Chapter 18, we found the **mean squared error (MSE)** to summarize the magnitude of the errors. We can do the same with the forecast errors:

$$MSE = \frac{1}{n}\Sigma(y_t - \hat{y}_t)^2$$

The MSE penalizes large forecast errors because the errors are squared. It also has the problem that it is not in the same units as the data, but rather in the square of those units. We can address both of these problems by summing the *absolute values* of the errors. That gives the **mean absolute deviation (MAD)**:

$$MAD = \frac{1}{n}\Sigma|y_t - \hat{y}_t|$$

The most common approach to measuring forecast error compares the absolute errors to the magnitude of the estimated quantities to sum the *proportions* of the values that are in error. Multiplying the proportions by 100 gives the **absolute percentage error (APE)**. If we average the APE for all the forecasts, we have the **mean absolute percentage error (MAPE)**:

$$MAPE = 100 \times \frac{1}{n}\Sigma\frac{|y_t - \hat{y}_t|}{|y_t|}$$

The MAPE is expressed in percent, so it is independent of the units of the y variable. If you choose to rescale y, both the MSE and MAD will change, but the MAPE will remain the same.

In summary, MSE resembles the error measures we've used for regression models, but it isn't in the same units as the data. MAD is in the same units as the data, but that means that it will be rescaled if the measurements are rescaled. MAPE is a percentage relating the size of the errors to the magnitudes of the data values.

Table 20.3 shows forecast error for two different single-exponential smoothing (SES) models over the final 30 days of the Intel stock series along with the *MAD, MSE,* and *MAPE* computed for the entire time series. The smoothing model that uses a larger smoothing coefficient ($\alpha = 0.75$) more accurately forecasts the daily stock price over this time period.

Date	SES α = 0.75	Forecast error	SES α = 0.10	Forecast error	Price
31-Oct-02	17.19	0.11	15.74	1.56	$17.30
1-Nov-02	18.02	0.28	16.00	2.30	$18.30
4-Nov-02	18.58	0.19	16.27	2.50	$18.77
5-Nov-02	18.41	−0.06	16.48	1.87	$18.35
6-Nov-02	18.96	0.19	16.75	2.40	$19.15
7-Nov-02	18.57	−0.13	16.92	1.52	$18.44
8-Nov-02	18.26	−0.11	17.04	1.11	$18.15
11-Nov-02	17.57	−0.23	17.07	0.27	$17.34
12-Nov-02	17.73	0.06	17.14	0.65	$17.79
13-Nov-02	18.02	0.10	17.24	0.88	$18.12
14-Nov-02	18.91	0.30	17.44	1.77	$19.21
15-Nov-02	18.83	−0.03	17.57	1.23	$18.80
18-Nov-02	18.62	−0.07	17.67	0.88	$18.55
19-Nov-02	18.27	−0.12	17.72	0.43	$18.15
20-Nov-02	18.93	0.22	17.86	1.29	$19.15
21-Nov-02	19.89	0.32	18.10	2.11	$20.21
22-Nov-02	20.01	0.04	18.29	1.76	$20.05
25-Nov-02	20.36	0.12	18.51	1.97	$20.48
26-Nov-02	20.24	−0.04	18.68	1.52	$20.20
27-Nov-02	20.74	0.16	18.90	2.00	$20.90
29-Nov-02	20.84	0.04	19.10	1.78	$20.88
2-Dec-02	21.00	0.05	19.29	1.76	$21.05
3-Dec-02	20.48	−0.17	19.40	0.91	$20.31
4-Dec-02	19.87	−0.20	19.42	0.25	$19.67
5-Dec-02	19.19	−0.23	19.38	−0.42	$18.96
6-Dec-02	18.83	−0.12	19.31	−0.60	$18.71
9-Dec-02	17.97	−0.29	19.15	−1.47	$17.68
10-Dec-02	18.09	0.04	19.05	−0.92	$18.13
11-Dec-02	18.14	0.02	18.96	−0.80	$18.16
12-Dec-02	18.18	0.01	18.88	−0.69	$18.19
MAD		$0.135		$1.321	
MSE		0.026		2.162	
MAPE		0.714%		6.91%	

Table 20.3 *The Intel stock prices along with smoothed values reported as forecasts one period ahead for two models. For this series, the single-exponential smoothing (SES) model with the larger coefficient (α = 0.75) has a lower forecast error.*

20.8 Autoregressive Models

Simple moving averages and exponential smoothing methods are good choices for time series with no regular long-term patterns. But if some patterns are present—even if they don't rise to the level of a well-structured seasonal fluctuation—we may want to choose weights that facilitate modeling that structure—something that exponential smoothing can find difficult to do. Such weights might even be negative. (Imagine a series with values that alternated up and down for successive times. A good weighted average would give a negative weight to the most recent value and a positive weight to the one before that.)

But how could we find appropriate weights and how could we choose among the huge number of possible weights? It turns out that we can use the methods of multiple regression we saw in Chapter 19, along with the fact that the data come to us in time sequence order, to discover weights for a weighted moving average smoother. We shift the data by a time period or a few time periods. This shift is known as *lagging*, and the resulting variables are called *lagged variables*. For example, Table 20.4 shows the first 15 values of the daily Intel stock prices in 2002 along with the lagged values for lags of one, two, three, and four days.

Date	Price	Price$_{lag1}$	Price$_{lag2}$	Price$_{lag3}$	Price$_{lag4}$
2-Jan-02	$33.00	*	*	*	*
3-Jan-02	$35.52	$33.00	*	*	*
4-Jan-02	$35.79	$35.52	$33.00	*	*
7-Jan-02	$35.27	$35.79	$35.52	$33.00	*
8-Jan-02	$35.58	$35.27	$35.79	$35.52	$33.00
9-Jan-02	$35.36	$35.58	$35.27	$35.79	$35.52
10-Jan-02	$34.65	$35.36	$35.58	$35.27	$35.79
11-Jan-02	$34.55	$34.65	$35.36	$35.58	$35.27
14-Jan-02	$34.84	$34.55	$34.65	$35.36	$35.58
15-Jan-02	$34.68	$34.84	$34.55	$34.65	$35.36
16-Jan-02	$33.71	$34.68	$34.84	$34.55	$34.65
17-Jan-02	$34.53	$33.71	$34.68	$34.84	$34.55
18-Jan-02	$33.48	$34.53	$33.71	$34.68	$34.84
22-Jan-02	$31.70	$33.48	$34.53	$33.71	$34.68
23-Jan-02	$32.45	$31.70	$33.48	$34.53	$33.71

Table 20.4 *The lagged values for the first 15 days in the Intel daily stock time series for one, two, three, and four days.*

If we fit a regression to predict a time series from its lag1 and lag2 versions,

$$\hat{y} = b_0 + b_1 y_{lag1} + b_2 y_{lag2},$$

each predicted value, $\hat{y}_t$, is just a sum of the two previous values, y_{lag1} and y_{lag2} (plus a constant) weighted by the fitted coefficients b_1 and b_2. That's just a weighted moving average with weights found by the regression.

But wait. Regression methods assume that the data values are mutually independent. And this method works only if the data values are *not* independent—that is, if recent values can help predict current ones. Isn't this a violation of the regression model? Well, yes and no. The independence assumption is certainly required for inference on the coefficients, for example, to test the standard null hypothesis that the true coefficient is zero. But we're not doing inference here; we're just building a model. And for that purpose, the failure of independence is really more an opportunity than a problem.

In fact, we can specifically account for the association of cases with previous ones. The correlation between a series and a (lagged) version of the same series that is offset by a fixed number of time periods is called **autocorrelation**.[8] Table 20.5 shows some autocorrelations for the Intel series.

	Price	Lag1	Lag2	Lag3	Lag4
Price	1.000				
Lag1	0.992	1.000			
Lag2	0.984	0.992	1.000		
Lag3	0.978	0.984	0.992	1.000	
Lag4	0.973	0.977	0.984	0.992	1.000

Table 20.5 *Autocorrelations of the daily Intel closing stock price for the entire year 2002 for lags 1, 2, and 3.*

---

[8] Recall that we evaluated the presence of autocorrelation using the Durbin-Watson statistic in Chapter 17.

A regression model that is based on an average of prior values in the series weighted according to a regression on lagged versions of the series, is called an **autoregressive model**. A model based on only the first lagged variable is called a first-order autoregressive model, often abbreviated as AR(1).

A pth-order autoregressive model has the form:

$$\hat{y} = b_0 + b_1 y_{lag1} + \ldots + b_p y_{lag\,p}.$$

For the Intel stock price series, we find the coefficients for a fourth-order autoregressive model from a multiple regression of the series on its first four lagged values, as shown in Table 20.6.

```
Dependent variable is:   Price
R squared = 98.4%    R squared (adjusted) = 98.4%
s = 0.8793 with 236 − 5 = 231 degrees of freedom
```

Variable	Coef	SE(Coeff)	t-ratio	P-value
Intercept	0.126434	0.2032	0.622	0.5344
Lag1	0.963981	0.0655	14.7	≤ 0.0001
Lag2	−0.046396	0.0911	−0.509	0.6110
Lag3	−0.056936	0.0905	−0.629	0.5300
Lag4	0.134714	0.0641	2.10	0.0368

Table 20.6 *A fourth-order autoregressive model for the Intel stock prices. Note that there are 240 values in the series but, because lagged variables have missing values at the beginning of the series, there are only 236 complete cases.*

The resulting fourth-order autoregressive model is

$$\hat{y}_t = 0.126434 + 0.963981 y_{lag1} - 0.046396 y_{lag2} - 0.056936 y_{lag3} + 0.134714 y_{lag4}$$

Looking at the coefficients, we can see that the model puts most of its weight on the data value just preceding the one we're estimating (the lag 1 value), some on the lag 4, and relatively little on the other two lags.

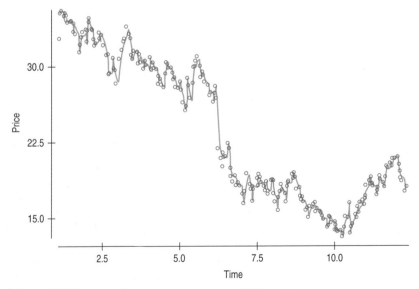

| **Figure 20.6** *The Intel stock price series with an AR(4) model.*

We can use an autoregressive moving average model to predict the next value in a time series, but, we can only predict as far as the model permits. A model that uses lag1 data can only predict one time period into the future. A model in which the shortest lag is lag 4 can predict four periods ahead.

The forecasts, forecast errors, and APE produced by the two moving average models (MA(5) and MA(15)), the exponential smoothing models ($\alpha = 0.10$) and ($\alpha = 0.75$), and the fourth-order autoregressive model, or AR(4), are shown in Table 20.7.

Model	Forecast ($\hat{y}_t$)	Actual Price (y_t)	Forecast Error ($y_t - \hat{y}_t$)	Absolute Percent Error (APE) $100 \times |y_t - \hat{y}_t|/|y_t|$
MA-5	$17.95	$17.58	−$0.37	2.10%
MA-15	$19.40	$17.58	−$1.82	10.35%
SES ($\alpha = 0.10$)	$18.75	$17.58	−$1.17	6.66%
SES ($\alpha = 0.75$)	$17.45	$17.58	$0.13	0.85%
AR(4)	$18.17	$17.58	−$0.59	0.77%

Table 20.7 *The forecasts and forecast error for each of the moving average, single exponential smoothing, and autoregressive models for the Intel daily stock price on December 13, 2002.*

JUST CHECKING

J. Crew is a clothing company known for its preppy fashions, including jeans, khakis, and other basic items sold to young professionals through its catalogs, websites, and some 260 retail and outlet stores in the United States. It was recently made prominent because First Lady Michelle Obama shops there. We have their reported quarterly revenue from Q1 2003 through 2007. Here's a time series plot:

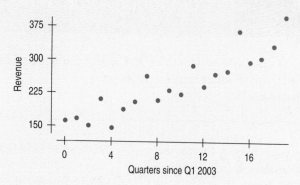

1 Which time series components do you see in this plot?

2 The final four values of this series are

Q1	2007	297.312
Q2	2007	304.731
Q3	2007	332.744
Q4	2007	399.936

Without calculating anything, roughly where on the graph would the final value of a 4-point simple moving average be? Do you think it would be a good prediction for Q1 of 2008?

3 If the exponential smooth value for Q3 2007 is 319.3, what is the exponential smooth value of Q4 when an α of 0.5 is used?

4 If you wished to fit an autoregressive model, how many terms should you include? Why?

Unlike simple moving average models and single exponential smoothing models, AR models can follow time series that have seasonal fluctuations. The AR method will assign a larger weight to lags that correspond to the period of the fluctuation. For example, AR models will tend to predict quarterly sales that show a seasonal cycle by assigning a large weight to the lag4 version of the series, so sales in the same quarter of the previous year are counted heavily.

20.9 Random Walks

All moving average models include the special naïve model that predicts that the next value will be the same as the current one. For an autoregressive model, when the coefficient of y_{lag1} is close to one and the estimated intercept is close to zero, the first-order autoregressive model is approximately

$$\hat{y}_{t+1} = y_t$$

The naïve forecast model is sometimes called a **random walk** because each new value can be thought of as a random step away from the previous value. Time series that are modeled by a random walk can have rapid and sudden changes in direction, but they also may have long periods of runs up or down that can be mistaken for cycles.[9] Random walks include series such as the assets of a gambler over time, the location of a molecule in a gas, and the path taken by a foraging animal. The random walk hypothesis in finance predicts that in an efficient market, stock prices should follow a random walk.[10, 11] There is a large literature dealing with random walks. They show up in mathematics, economics, finance, and physics, as well as other fields.

GUIDED EXAMPLE	*Comparing Time Series Methods*

In Chapter 8, we first looked at sales at The Home Depot. At that time, we related them to other extraneous variables. But those values are a time series. So let's return to them and consider ways to model sales over time.

PLAN

Setup State your objective. Identify the quantitative variables you wish to examine. Report the time frame over which the data have been collected and define each variable.

We want to build time series models for quarterly sales at The Home Depot from 1995 through 2004. We have quarterly sales ($Billion).

---

[9] This is one reason we recommended that the identification of cycles be based on theory, on established patterns, or on other variables.

[10] In the case of first-order autoregressive models, or in cases where autocorrelation continues to present a problem, one solution is to model the *differenced* time series. For example, in this case, we might model the *change* in Intel daily stock price $(y_t - y_{t-1})$ as the response variable.

[11] Princeton economist Burton Malkiel made the random walk theory of the stock market famous in his book *A Random Walk Down Wall Street: The Time-Tested Strategy for Successful Investing*, first published in 1973. The theory originated in the 1950s with Maurice Kendall, a British statistician.

Plot Plot the time series and clearly label the axes to identify the scale and units.

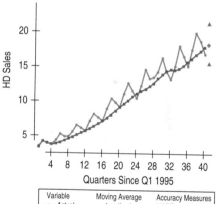

Model Think about the assumptions and check the conditions.

These are quantitative data measured over time at consistent intervals. So it is appropriate to use time series methods.

There is a consistent increasing trend with fluctuations around it that appear to be seasonal because they repeat every four quarters. Some smoothing methods may have difficulty with this kind of series.

Mechanics, Part 1 Try a moving average. For data with a strong seasonal component, such as these, a moving average length that is a multiple of the period is a good idea. But series with a strong trend, such as this one, won't be fit well by an uncentered moving average.

Here is a simple moving average of length 4 as computed by a statistics program:

Variable	Moving Average	Accuracy Measures	
—— Actual	length 4	MAPE	10.9425
—•— Fits		MAD	1.1999
—•— Forecasts		MSD	2.1150
—▲— 95.0% PI			

Evaluate how well this method fits the series.

The program reports:

MAD = 1.1999
MSD = 2.115
MAPE = 10.9425

Make a forecast for Q1 2005 ($t = 40$).

The program offers a forecast of 18.2735 \$B for the first quarter of 2005.

Mechanics, Part 2 Exponential smoothing can be a good compromise between a simple moving average smoother and a fit with a seasonal component.

Let's try an exponential smooth. Now we have to choose a smoothing weight. We'll use $\alpha = 0.5$, which weights the current data value equally with

In series with a strong trend such as this one, exponential smooths will inevitably lag behind the data.

all the rest in the past. Here's the result from a computer-generated smooth:

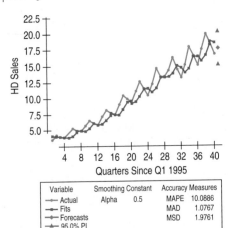

The statistics package reports:

MAD = 1.0767
MSE = 1.9761
MAPE = 10.0886

Evaluate how well this model fits.

The package generates a prediction of 17.6770 $B for the first quarter of 2005.

Forecast sales for Q1 2005.

Mechanics, Part 3 An autoregressive model is a multiple regression on the series itself lagged, or offset, by 1, 2, or more time periods. When we know that there is a seasonal component, it is important to include the corresponding lag—here lag4. This model has terms for each lag.

Now let's fit an autoregressive model. Because we know that the seasonal pattern is 4 quarters long, we'll fit four terms, using multiple regression to find the smoothing weights:

```
Dependent variable is HDSales
R squared = 98.7%  R squared (adjusted) = 98.6%
s = 0.5412 with 36 − 5 = 31 degrees of freedom
```

Variable	Coeff	SE(Coeff)	t-ratio	P-value
Intercept	0.947082	0.2469	3.84	0.0006
Lag1	0.428426	0.1455	2.94	0.0061
Lag2	−0.396168	0.1548	−2.56	0.0156
Lag3	0.353743	0.1612	2.19	0.0358
Lag4	0.646243	0.1552	4.16	0.0002

The AR(4) model is

$$\hat{y}_t = 0.947 + 0.428y_{t-1} - 0.396y_{t-2} + 0.354y_{t-3} + 0.646y_{t-4}$$

Plot the fit.

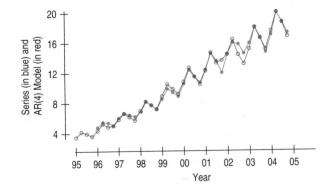

Plot the residuals.

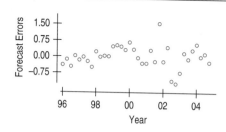

The forecast errors show some disturbance near and just after Q4 2001. We've already noticed this quarter as a high outlier. Notice here how its effects propagate through the fit and residuals. A lack of resistance to outliers is one disadvantage of autoregressive models.

Calculate fit measures and a prediction.

MAD = 0.32667
MSE = 0.227
MAPE = 3.46

A predicted value for the first quarter of 2005 is:

$$0.9470 + 0.4284 \times 16.812 - 0.39617 \times 18.772 + 0.3537 \times 19.96 + 0.6462 \times 17.55 = 19.11$$

REPORT

Conclusion Compare the advantages and disadvantages of the time series methods. Be sure to state your interpretations in the proper context.

MEMO:

Re: Time Series Analyses of The Home Depot Quarterly Sales

We compared several time series methods to fit quarterly data on sales at The Home Depot for the period from 1995 through 2004. The actual sales in Q1 of 2005 were $19B.

The different methods had differing strengths and weaknesses. The moving average method smoothed out all of the seasonal effects, but provided a good description of the trend and a good prediction.

The exponential smoothing method followed more of the seasonal pattern, but that didn't improve its prediction.

The autoregressive method worked well and made a good prediction, but it was clearly affected by the single outlying quarter, which contaminated the fit for several quarters afterward.

20.10 Multiple Regression-based Models

We noted earlier that some time series have identifiable components; a trend, a seasonal component, and possibly a cyclical component. Simple moving average models work best on time series that lack any consistent structures like these. Exponential smoothing models don't usually follow a seasonal component well either. Autoregressive moving averages can follow all of these components provided the length of the smoother is at least as long as the period of the seasonal or cyclical component.

When some or all of these components are present, we can gain two distinct advantages by modeling them directly. First, we may be able to forecast beyond the immediate next time period—something the smoothing models can't do easily. Second, we may be able to understand the components themselves and reach a deeper understanding of the time series itself.

Modeling the Trend Component

When a time series has a linear trend, the natural thing to do is to model it with a regression. If the trend is linear, a linear regression of y_t on *Time* can model the trend. The residuals would then be a *detrended* version of the time series.

The Home Depot sales data that we examined in the Guided Example seem to have a roughly linear trend. The regression to estimate the trend is in Table 20.8.

```
Dependent variable is:  HDSales
R squared = 95.5%   R squared (adjusted) = 95.4%
s = 1.044 with 40 - 2 = 38 degrees of freedom

Variable   Coeff     SE(Coeff)  t-ratio  P-value
Intercept  2.67102   0.3241      8.24    ≤0.0001
Time       0.405707  0.0143     28.4     ≤0.0001
```

Table 20.8 *Estimating the trend component in the Home Depot sales data by regression.*

One attractive feature of a regression-based model is that the coefficient of *Time* can be interpreted directly as the change in *y* (here, Home Depot Sales) *per* time unit (here, quarters). The trend in Home Depot sales was that they increased by 0.405 billion dollars per quarter.

By contrast, we saw in Figure 20.1 that the Whole Foods sales data *don't* have a linear trend. As we learned in Chapter 17, we can often improve the linearity of a relationship with a re-expression. The re-expression that most often works for time series is the logarithm. That's because many time series grow or shrink exponentially. Typically, the bigger you are, the larger the absolute increment in your profits. Growth by a consistent percentage is *exponential growth*. For example, if sales each year are 5% higher than the previous year, then the overall growth will

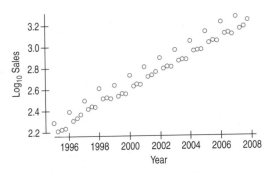

Figure 20.7 *The logarithm of Whole Foods quarterly Sales is linear over time.*

be exponential. And the logarithm makes exponential growth linear. Figure 20.7 shows the result of taking logs in the Whole Foods data. The corresponding model for the trend is

```
Dependent variable is:  Log WFSales
R squared = 96.5%   R squared (adjusted) = 96.4%
s = 0.0574 with 52 − 2 = 50 degrees of freedom

Variable    Coeff       SE(Coeff)  t-ratio  P-value
Intercept   −154.663    4.243      −36.4    ≤ 0.0001
Year          0.078650  0.0021      37.1    ≤ 0.0001
```

Now the interpretation of the trend coefficient is different. Adding 0.07865 to the logarithm of sales each year is the same as multiplying sales by $10^{0.07865} \approx 1.20$. And that's an increase of 20 percent. So we can say that Whole Foods Market's sales were increasing by 20 percent per year.[12]

Re-expressing the Whole Foods quarterly sales data by logarithms reveals a second advantage of the re-expression. The seasonal fluctuations evident in Figure 20.1 grew in magnitude as the sales themselves grew. But in Figure 20.7, those fluctuations are nearly constant in size. That will make them much easier to model.

Modeling the Seasonal Component

Figure 20.7 shows that the Whole Foods data have a strong seasonal component. Every fourth quarter there's a spike. That's not unusual in time series related to retail sales. The simplest version of a seasonal component is one that adds a different value to the series (in addition to the trend) for each season. We can see that this is a good description of the Whole Foods data; the first quarter is above the overall trend by roughly the same amount each year. Figure 20.8 shows the pattern.

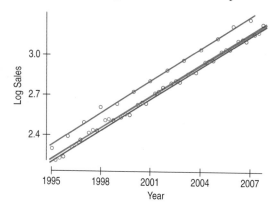

Figure 20.8 *The logarithm of Whole Foods quarterly sales. Each quarter is displayed in its own color and has regression line fit to it. The lines differ almost exclusively by a shift up or down.*

---

[12] Of course, if you use natural logarithms rather than base-10 logarithms, the function that returns to the original units is e^b where b is the Trend coefficient. But the interpretation of the result (and, allowing for rounding error, the value itself) will be the same.

As we learned in Chapter 19, a pattern such as the one shown in Figure 20.8 can be modeled by introducing an indicator or dummy variable for each season. For example, we can define our dummy variables to be:

$$Q_1 = 1 \text{ in quarter 1 and 0, otherwise}$$

$$Q_2 = 1 \text{ in quarter 2 and 0, otherwise, and}$$

$$Q_3 = 1 \text{ in quarter 3 and 0, otherwise.}$$

Recall from Chapter 19 that for a categorical variable with k levels to enter a regression model we use $k - 1$ dummy variables. We can't use all k because that would create a collinearity. So we leave out one of them. It doesn't really matter which one we choose to leave out. The intercept coefficient will estimate a level for the period "left out," and the coefficient of each dummy variable estimates the shift up or down in the series relative to that base level. With four quarters, we use three dummy variables. For this example, we'll arbitrarily choose to leave out the dummy for Q4. Then cases in Q4 will have the value zero for all three of our dummy variables (Q_1, Q_2 and Q_3) and the mean adjustment relative to the trend will be estimated by the intercept (b_0). In any other quarter, the adjustment relative to Q4 will be the coefficient for that quarter's dummy variable.

20.11 Additive and Multiplicative Models

Adding dummy variables to the regression of a time series on *Time* turns what was a simple one-predictor regression, such as we dealt with in Chapter 8, into a multiple regression, such as those we learned about in Chapter 18. That, combined with the question of whether to work with the original time series or the logarithm of the series, raises a new question. If we model the original values, we have added the seasonal component (in the form of dummy variables) to the trend component (in the form of an intercept coefficient and a regression with the *Time* variable as a predictor). We can write $\hat{y}_t = T + S$.

This is an **additive model** because the components are added up in the model. For example, we've seen that Sales from The Home Depot seem to grow linearly with a seasonal pattern. Table 20.9 shows the regression that models those sales in terms of a trend component and three quarterly dummy variables.

```
Dependent variable is:   HDSales
R squared = 98.7%    R squared (adjusted) = 98.5%
s = 0.5915 with 40 − 5 = 35 degrees of freedom
```

Variable	Coeff	SE(Coeff)	t-ratio	P-value
Intercept	1.38314	0.2534	5.46	≤0.0001
Time	0.410336	0.0081	50.4	≤0.0001
Q1	1.22191	0.2657	4.60	≤0.0001
Q2	2.41907	0.2650	9.13	≤0.0001
Q3	1.14944	0.2647	4.34	0.0001

Table 20.9 A regression to model Home Depot sales with a trend component and three dummy variables representing a seasonal component in an additive model.

The model contains a trend component that predicts growth at about $0.41B per quarter with adjustments for each quarter that are consistent over the entire time period. For example, because Q4 is the quarter without a dummy variable, sales in Q4 are predicted to be on average $1.38 + 0.41 Time$ billion dollars. The seasonal dummy variable coefficients adjust the predictions for each quarter by adding the value of their coefficients to the intercept. For example, sales in Q1 are predicted to be $1.38 + 0.41\,Time + 1.22 = \$2.60B + 0.41\,Time$ (see Figure 20.9). But you can see from Figure 20.9 that the seasonal fluctuations are small early in the series and grow larger later in the series—a pattern this model doesn't fit.

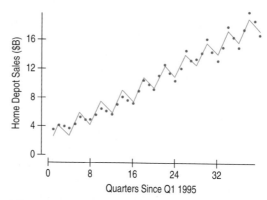

Figure 20.9 *Sales at The Home Depot with predictions from the additive model. The model predicts a consistent seasonal component although the seasonal component of the data varies.*

When we examined the Whole Foods sales data, we saw that we could straighten the trend and make the seasonal fluctuations more nearly the same size by finding the *logarithm*. We can still find a multiple regression, but now the response variable is re-expressed by logarithms. The model is in Table 20.10.

```
Dependent variable is:  Log WFSales
R squared = 99.6%   R squared (adjusted) = 99.6%
s = 0.0188 with 52 − 5 = 47 degrees of freedom

Variable   Coeff      SE(Coeff)   t-ratio   P-value
Constant  −156.488    1.396       −112      ≤0.0001
Year         0.079541 0.0007       114      ≤0.0001
Q1           0.132511 0.0074        17.9    ≤0.0001
Q2           0.016969 0.0074         2.30   0.0261
Q3           0.012885 0.0074         1.75   0.0874
```

Table 20.10 *A regression model for the logarithm of the Whole Foods quarterly Sales data with trend and seasonal components.*

Now, however, there's a difference in the model. Because we are modeling the logarithm of sales, when we think in terms of the sales themselves, the model components are multiplied rather than added, so we have a **multiplicative model**,

$$\hat{y} = T \times S.$$

Although we acknowledge that the terms in a multiplicative model are multiplied, we always fit the multiplicative model by taking logs, changing the form to an additive model that can be fit by multiple regression.

As we observed earlier, seasonal fluctuations are often proportional to the overall level of the values in the series; the Q1 lift in sales in a multiplicative model is a *proportion* of overall sales, not a fixed additive increment. Specifically, it is

> The terms *exponential model* and *multiplicative model* are equivalent for these models and can be used interchangeably.

$10^{0.1325} \approx 1.36$—about 36% higher than the sales at that time. Because the sales themselves were growing (at 20% per year), this 36% lift grew as well, in dollar terms. But after taking logs, it is a constant lift, and easier to model.

Look back at the Whole Foods sales in Figure 20.1. You can see this growth in the size of the seasonal component in the plot as well as in the trend. Taking logs not only turns exponential growth into linear growth, but it also tends to stabilize the size of the seasonal fluctuations. In fact, if you look back at the Home Depot data in Figure 20.9, you may notice a similar increase in seasonal fluctuations. Although the trend is already linear and we can fit an additive model to those data, a multiplicative model is likely to work even better.

20.12 Cyclical and Irregular Components

Many time series are more complex than the trend and seasonal components can model. The Intel Stock price data of Figure 20.3 is one example. Models of time series components are usually said to include two additional components, a cyclical component and an irregular component. Consistent with their form for the trend and seasonal components, we write for additive models:

$$\hat{y}_t = T + S + C + I,$$

and for multiplicative models:

$$\hat{y}_t = T \times S \times C \times I.$$

The Cyclical Component

Long-term business cycles may influence economic and financial time series. Other time series may be influenced by other long-term fluctuations. Whenever there is a business, economic, or physical cycle whose cause is understood and that can be relied on, we should look for an external or **exogenous** variable to model the cycle. The regression models we've been considering can accommodate such additional predictors naturally.

Just calling a long-term fluctuation in the data that isn't fit by the trend or seasonal component a "cyclical component" doesn't add much to our understanding. Cyclical patterns may not be immediately evident in the data, so it is wise to compute and plot the residuals, known for time series models as the irregular component.

Irregular Components

The irregular components are the residuals—what's left over after we fit all the other components. We should examine them to check any assumptions and also to see if there might be other patterns apparent in the residuals that we might model. For multiple regression, most statistics programs plot the residuals against the predicted values, but for time series models, it is essential to plot them against *Time*. Figure 20.10 shows the residuals of the Whole Foods data plotted against year. Two quarters stand out. Both extraordinary quarters are fourth quarters of their years, which suggests that our seasonal model may need to be improved.[13] We also see a possible cyclical pattern with a period of about 4 years. This might be something worth investigating to see if we can add a component to our model.

---

[13] It's possible to include a dummy variable to model a specific event in time. Doing so is known as intervention analysis.

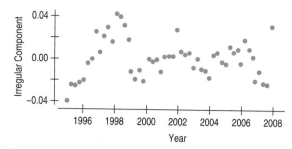

Figure 20.10 *The irregular component, or residuals, of the multiplicative model for the Whole Foods sales. In the beginning years, growth was apparently faster than after 1999 or so. There are also a couple of quarters that deserve attention: Q4 of 2001 and Q4 of 2007 which were underestimated.*

JUST CHECKING

Continuing our analysis of the J. Crew revenue data form the previous Just Checking, here is a regression fitting a linear model to predict the *Revenue* ($M):

```
Dependent variable is:  Revenue
R squared = 96.1%   R squared (adjusted) = 95.0%
s = 15.83 with 20 − 5 = 15 degrees of freedom
```

Variable	Coeff	SE(Coeff)	t-ratio	P-value
Intercept	193.037	9.87	19.6	≤ 0.0001
Q since '03	10.2338	0.6255	16.4	≤ 0.0001
Q1	−64.0281	10.18	−6.29	≤ 0.0001
Q2	−52.8349	10.09	−5.24	0.0001
Q3	−57.5810	10.03	−5.74	≤ 0.0001

5 Locate and interpret the trend coefficient.

6 Why is there no term for Q4? Would that term have a positive or negative coefficient?

20.13 Forecasting with Regression-based Models

Regression models are easy to use for forecasting because they give us a formula. It is easy to substitute future values for *Time*, for the appropriate dummy variables, and for any exogenous variables and calculate a predicted value. Unlike smoothing models, regression-based models can forecast beyond the next time period. But any kind of forecast is uncertain, and the uncertainty grows the further we extrapolate, so it's wise to limit forecasts to the near term. The forecast error measures we discussed in Section 20.7 apply equally well to regression models and are computed from the residuals (or Irregular component.)

The most reliable part of a regression time series model is the seasonal component, because it is probably driven by fairly regular economic or environmental phenomena. We can reasonably expect those patterns to continue into the future, so we can feel comfortable basing longer term forecasts on them. A company that

Car Makers' Boom Years Now Look Like a Bubble
By Neal E. Boudette and Norihiko Shirouzu

. . . Through most of the 1990s, auto makers sold a little over 15 million cars and light trucks a year in the U.S. market. That changed in the late 1990s: With gasoline prices low and many U.S. consumers feeling flush from the tech-stock boom, auto sales surged. Sales peaked at 17.4 million in 2000 and remained near 17 million for another five years. Heads of General Motors Corp. and Toyota said the U.S. was entering a golden age of the automobile. In 2003, Toyota's head of North American sales predicted the industry would soon be selling 20 million vehicles a year.

They were wrong. Sales started falling in 2006 and this year are expected to be right back where they were in the 1990s, at just over 15 million.[14]

Source: Wall Street Journal, May 20, 2008, page 1.

sees large fourth-quarter holiday season sales can probably count on higher fourth-quarter sales even several years in the future.

The trend component is less reliable. As much as we might like to think that a growing company will continue to grow, it should be clear that no company can grow forever. Exponential growth is even harder to maintain. Changes in the trend can be quite sudden. The business news is filled with stories of companies whose growth "suddenly" stopped or slowed, of products whose sales unexpectedly shot up and then, just as unexpectedly fell back, and of economic predictions by experts that, in retrospect, look ill-informed. While long-term government forecasts can be found for such reliable indicators as the Gross Domestic Product (GDP) and Disposable Income (DI), long-term forecasts should be made with great care. Changes in the economy or the market cannot be anticipated by the trend component, and can change a company's business quite suddenly.

The reliability of cyclical components for forecasting is something you must simply judge for yourself based on your knowledge and understanding of the underlying economic cycles. An empirical cycle that is not understood makes a risky basis for prediction. A cycle that is understood and is due to underlying phenomena that are stable over time would be a more reliable component of a forecast.

Table 20.11 compares the forecasts from four regression-based models for the Whole Foods Markets quarterly *Sales* to the actual sales in the first (calendar) quarter of 2008. When the forecast error is negative, the model provides an *over*estimate. When the error is positive, the model provides an *under*estimate for the forecast. The first model is the simple linear trend model (additive model). The second model is the additive model with quarterly seasonal components. The third model is the multiplicative trend model, and the fourth model is the multiplicative model with trend and quarterly seasonal components.

Model	Forecast ($M) $\hat{y}_t$	Actual Sales ($M) Q1 2008 ($y_t$)	Forecast Error ($y_t - \hat{y}_t$)	Absolute Percent Error $100 \times \frac{|y_t - \hat{y}_t|}{y_t}$
Simple Linear Trend Model	1489.717	2457.258	967.541	39.4%
Additive Trend plus Seasonal Component Model	1652.760	2457.258	804.498	32.7%
Multiplicative Trend Model	1930.640	2457.258	526.618	21.4%
Multiplicative Trend plus Seasonal Component Model	2419.637	2457.258	37.621	1.5%

Table 20.11   *The forecasts and forecast error for the first quarter of 2008 using additive and multiplicative models with and without seasonal components.*

From the original scatterplot, we saw that the trend wasn't linear, so it's no surprise that the two additive models grossly underestimate Q1 of 2008. The multiplicative trend model without seasonal components doesn't take into account that Q1 is

---

[14]The May estimate was also wrong. Sales fell to about $13.7 million

typically a big quarter for Whole Foods, so it still underestimates it by more than 21%. The multiplicative trend model with seasonal components comes closest, getting within 1.5% of the actual sales.

| **GUIDED EXAMPLE** | *Comparing Time Series Methods, Part 2* |

In the first Guided Examples, we looked at smoothing models for sales at The Home Depot. But that time series shows a clear linear trend and a seasonal component. So let's try the regression models and see how well they do by comparison.

PLAN

Setup State your objective. Identify the quantitative variables you wish to examine. Report the time frame over which the data have been collected and define each variable.

We want to build regression-based time series models for quarterly sales at The Home Depot from 1995 through 2004. We have quarterly sales ($Billion).

Plot Plot the time series and clearly label the axes to identify the scale and units.

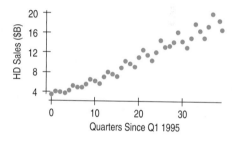

Model Think about the assumptions and check the conditions.

This is a time series recording quantitative values.

There is a consistent, increasing trend and fluctuations around it that may be seasonal.

We can start by estimating the trend.

DO

Mechanics, Part 1 Fit an additive trend model using regression.

```
Dependent variable is: HDSales
R squared = 95.5%    R squared (adjusted) = 95.4%
s = 1.044 with 40 − 2 = 38 degrees of freedom

Variable   Coeff     SE(Coeff)  t-ratio  P-value
Intercept  2.67102   0.3241     8.24     ≤ 0.0001
time       0.405707  0.0143     28.4     ≤ 0.0001
```

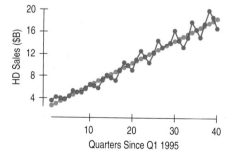

The blue series tracks the data. The model is a straight line, shown here in red.

Plot the residuals.	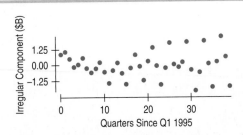
Evaluate how well this method fits the series.	There's a bend in the irregular component and increasing spread. A seasonal fluctuation is easier to see in the irregular component.
Make a forecast for Q1 of 2005 ($t = 40$).	We can examine various measures of how well this model fits the series. These summaries are mostly useful to compare models, as we will here. MAD = 0.7706 MSE = 1.0356 MAPE = 8.1697 Forecast = 2.671 + 0.4057 × 40 = 18.899 ($B)
Mechanics, Part 2 Fit a multiplicative model with trend and seasonal terms.	We fit a multiplicative model by multiple regression of the log series on both *Time* and dummy variables for three of the four quarters.[15]

```
Dependent variable is: log HDsales
R squared = 97.2%    R squared (adjusted) = 96.9%
s = 0.0393 with 40 - 5 = 35 degrees of freedom

Variable   Coeff      SE(Coeff)  t-ratio  P-value
Intercept  0.556994   0.0168     33.1     ≤ 0.0001
Time       0.018850   0.0005     34.8     ≤ 0.0001
Q1         0.049673   0.0177     2.81     0.0080
Q2         0.096707   0.0176     5.49     ≤ 0.0001
Q3         0.046880   0.0176     2.67     0.0116
```

Plot the fit vs. the data (here on the log scale) and the residuals.	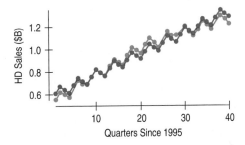

Again, the data are the blue series and the model is the more regular red pattern.

This long-term pattern could be a cycle, but we'd need to see it repeat over the next 40 quarters to be sure of that and to be able to fit it. In fact, it probably is not a business cycle.	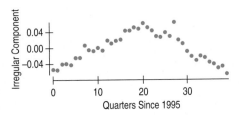

---

[15] If you don't recall why we must leave out one of the quarters or how dummy variables work, see Chapter 19.

To make a prediction with a multiplicative model, we use the inverse function of the logarithm.

The Irregular component shows that the growth in Home Depot sales slowed in the middle of this period. There is also one point that stands out above the pattern in the Irregular component. This is Q4 of 2001. Possibly, the events of 9/11/01 encouraged people to fix up their homes.

MAD = 1.077
MSE = 1.0144
MAPE = 12.0386

Note that for comparability, these must be computed on the residuals transformed back to original units: $10^{residuals}$.

Make a forecast for Q1 of 2005.

$$\text{Forecast} = 10^{(0.556994 + 0.018850*40 + 0.049673)}$$
$$= 22.94 \text{ \$B}$$

 **REPORT**

Conclusion Compare the advantages and disadvantages of the time series models. Be sure to state your interpretations in the proper context.

MEMO:

Re: Time Series Analyses of The Home Depot Quarterly Sales

We compared several time series methods to fit quarterly data on sales at The Home Depot for the period from 1995 through 2004. The actual sales in Q1 of 2005 were \$19B, so the best prediction was made by the simplest model—the additive linear trend model.

The different methods had differing strengths and weaknesses. The linear trend model is simple to find and, for these data, was the best predictor. This may be in part because average sales in Q1 are close to the annual average. If we were forecasting either Q2 or Q4, we would expect the multiplicative model with seasonal dummy variables to do better.

The multiplicative component model fit and predicted least well, but by providing a model, it revealed both a slowing of growth in Home Depot sales and an outlying quarter of unexpectedly high sales in Q4 of 2001, possibly related to the events of 9/11/01. Both of these observations may be important in understanding the series.

20.14 Choosing a Time Series Forecasting Method

We've considered several methods for modeling and forecasting time series. How can you choose among them the ones that fit your needs? Your choice depends both on the nature of your data and on what you hope to learn from the data.

Simple moving averages demand the least of the data. You can apply them to almost any time series. However:

◆ They can forecast well only for the next time period. Longer forecasts just repeat the single forecast value.

◆ They are sensitive to spikes or outliers in the series, and can smear the shock of a spike across several adjacent time periods.

◆ They don't do well on series that have a strong trend, tending to lag behind the trend.

Exponential smoothing methods also make few assumptions about the data. They have the advantage of controlling the relative importance of recent values relative to older ones, and do a good job of following the movement of a time series. However:

◆ They can forecast well only for the next time period. Longer forecasts just repeat the single forecast value.

◆ They are sensitive to spikes or outliers in the series.

◆ They don't do well on series that have a strong trend, tending to lag behind the trend.

Autoregressive moving average models use automatically determined weights to allow them to follow time series that may have regular fluctuations, such as a seasonal component, or series with a consistent trend. However:

◆ They can forecast for a limited span, depending on the shortest lag in the model.

◆ They are sensitive to spikes or outliers in two different ways. Those outliers will influence the regression that determines the smoothing weights. And then again, when the smoothing is done, the effect of spikes in the series can be spread out, contaminating several points.

Regression-based models estimate the trend and seasonal components by regression against *Time* and using dummy variables to represent the seasonal period. They can incorporate exogenous variables to help model business cycles and other phenomena. And, unlike moving average models, they can forecast into the future. However:

◆ You must decide whether to fit an additive model to the time series (if the trend is linear and the seasonal fluctuations have a consistent size) or to re-express the series by logarithms and fit the resulting multiplicative model.

◆ Because they are based on regression, these models are sensitive to outliers and failures of linearity. Because they use dummy variables to estimate the seasonal effects, those effects must be consistent in magnitude during the time covered by the data.

◆ Forecasts depend on the continuation of the trend and seasonal patterns. Although seasonal patterns may be reliable, trends are difficult to forecast and hazardous to assume beyond the near future. Cycles are best forecast when they are based on some identifiable (and predictable) phenomenon.

20.15 Interpreting Time Series Models: The Whole Foods Data Revisited

When you use a time series model based on moving averages, you can (and should) summarize the patterns seen in the series and any patterns noticed in the residuals around the smooth trend. But time series models based on regression encourage us to interpret the coefficients.

Many time series of retail sales have a strong seasonal component. We've seen two in this chapter. But one of those, the quarterly sales at Whole Foods, is problematic. No, there's no problem with the models we've fit. But there is a problem if we interpret them without thinking.

Why should there be a seasonal spike in Whole Foods' sales? Food isn't a seasonal item. To be honest, it took the authors a while before we asked these questions. But that sent us back to the data. It turns out that Whole Foods Market divides its financial year into three quarters of 12 weeks and one of 16 weeks. The spike is entirely due to this bookkeeping anomaly. You can check for yourself; the seasonal peaks are $16/12 = 1.33$ times as big as the other quarters—almost exactly what the multiplicative model estimated them to be.

That doesn't invalidate any of our models. We'd still have to allow for the seasonal peak to model Whole Foods sales. But it is a cautionary tale that warns us not to jump to conclusions when interpreting our models.

WHAT CAN GO WRONG?

- **Don't use a linear trend model to describe a nonlinear trend.** Be sure to examine the scatter of the observations around the linear trend line to determine if there is a pattern. A curved pattern may indicate a need to transform the series. Plot the residuals. Variation in the residuals that increases when the central value is higher is a sign of a need to take logarithms. Series that grow or shrink by a constant percentage each time period are exponential and should be transformed before fitting a model.

- **Don't use a trend model for short-term forecasting.** Trend models are most effective for long-term forecasting and are rarely the best models to forecast one or two time periods ahead. Be aware that forecast errors are greater for long-term forecasting.

- **Don't use a moving average, an exponential smoothing, or an autoregressive model for long-term forecasting.** Smoothing models are most effective for short-term forecasting because they require continuous updating as new values in the time series are observed. However, it is less important to transform an exponentially trending series for these models because they base their predictions on recent values.

- **Don't ignore autocorrelation if it is present.** Look at correlations among the lagged versions of the time series. Compute the Durbin-Watson statistic, and make residual plots to identify autocorrelation. If present, then try using lagged variables to model the time series.

ETHICS IN ACTION

Kevin Crammer, a broker for a large financial services firm, is getting ready to meet with a new client, Sally. Sally just inherited a rather large sum of money and is looking to invest it long term. She is not an experienced investor, and even though Kevin sent her information about various financial investment options, she is confused and still wants to meet with him to get his advice. Kevin's company has several of its own mutual fund products they like their brokers to push. Kevin selected one of these funds and, based on historical returns, prepared a graph showing how an initial amount of $10,000 invested 20 years would have grown over time. While some cyclical fluctuations tied to economic conditions were noticeable, the underlying trend was upward. Kevin decided to fit a linear trend line to the graph. Checking out the residuals, he did notice a pattern indicating a curvature. Transforming the data by logs yielded a better model. Upon closer examination, he was able to see that the straight line didn't fit as well to the raw data as to the transformed data. Although past pattern is no indication of future behavior, he did use both models to project future values for an initial investment of the size Sally was considering. The linear trend line provided higher values. As he prepared for his meeting with Sally, he thought it would be a good idea to keep the linear trend analysis handy when discussing this investment option.

ETHICAL ISSUE *A linear trend should not be used to model a nonlinear trend. In this case, the linear trend produced forecasts that suited Kevin's purpose and he ignored the observed cyclical variation in the data. Related to Item B, ASA Ethical Guidelines.*

ETHICAL SOLUTION *Present all relevant models. Caution about forecasting future values based on past patterns.*

What have we learned?

Terms

Absolute percentage error (APE)	A measure of the error of a forecast: $$APE = \frac{|y_t - \hat{y}_t|}{|y_t|}$$
Additive model	A model for a time series that models the times series with a sum of all or some of the following terms: a trend, a seasonal pattern, and a cyclical pattern.
Autocorrelation	The correlation between a data sequence, such as a time series, and that same sequence offset (or lagged) by one or more positions. Autocorrelation is one measure of lack of independence of the individual cases.
Autoregressive model	A pth-order autoregressive model has the form: $$\hat{y} = b_0 + b_1 y_{lag1} + \ldots + b_p y_{lag\,p}.$$
Cyclical component	The part of a model for a time series that describes regular repeating fluctuations with a period of 2 to 10 years.
Deseasonalized	A time series that has had a seasonal component estimated and subtracted.
Exogenous	Variables that are not part of a time series but nevertheless might be helpful in modeling it.

Exponential model	A model for a time series that grows exponentially. It is estimated by transforming the response variable by the logarithm:

$$\widehat{log}(y) = b_0 + b_1 Time + other\ terms.$$

Exponential smoothing, Single exponential smoothing (SES model)	An exponential smoother has the form

$$\widetilde{y}_t = \alpha y_t + (1 - \alpha)\widetilde{y}_{t-1}$$

or equivalently,

$$\widetilde{y}_t = \alpha y_t + \alpha(1 - \alpha)y_{t-1} + \alpha(1 - \alpha)^2 y_{t-2} + \alpha(1 - \alpha)^3 y_{t-3} \dots$$

The parameter α determines how the smoother behaves. Larger values of α give more weight to recent values of the series. Smaller values give more weight to more distant values.

Forecast	Many analyses of time series attempt to forecast future values. We denote a forecast value $\hat{y}_t$
Forecast error	The difference between the observed value and the forecasted value for a particular time in a time series:

$$e_t = y_t - \hat{y}_t.$$

Irregular component	The part of a time series model that describes random, or unpredictable, behavior; the residuals.
Linear trend model	A time series model that assumes a constant rate of increase (or decrease) over time:

$$\hat{y} = b_0 + b_1 t.$$

Mean absolute deviation (MAD)	A measure of forecast error of the form:

$$MAD = \frac{1}{n}\sum |y_t - \hat{y}_t|.$$

Mean squared error (MSE)	A measure of forecast error of the form:

$$MSE = \frac{1}{n}\sum (y_t - \hat{y}_t)^2.$$

Mean absolute percentage error (MAPE)	A measure of forecast error of the form:

$$MAPE = 100 \times \frac{1}{n}\sum \frac{|y_t - \hat{y}_t|}{|y_t|}.$$

Moving average	An estimate that uses the arithmetic average of the prior L values in a time series to forecast the next value

$$\widetilde{y}_t = \frac{\displaystyle\sum_{i=t-L+1}^{t} y_i}{L}.$$

Multiplicative model	A classical time series model consisting of four components,

$$\hat{y} = T \times S \times C \times I,$$

where T is the trend component, S is the seasonal component, C is the cyclic component, and I is the irregular component.

Naive forecast	Forecasting that the next value in a time series will be equal to the current one.

$$\hat{y}_{t+1} = y_t.$$

Period	The time between peaks of a regular oscillation in a time series.
Random walk	A time series that exhibits random periods of upturns and downturns and is best modeled using a naive forecast.
Seasonal component	The part of a model for a time series that fits a regular pattern that has a period of less than or equal to 12 months.
Single exponential smoothing (SES) model	(See Exponential smoothing.)
Simple moving average forecast	A simple moving average averages L consecutive values of a time series, where L is the length of the moving average. When the final L values have been averaged, the resulting smooth value for time t can be used to forecast the series for time $t + 1$. (See Moving average.)
Stationary in the mean	A time series that has a relatively constant mean value over the time frame of the series is said to be stationary in the mean.
Stationary in the variance	A time series that has a relatively constant variance is said to be stationary in the variance. This is equivalent to homoskedasticity. (If a time series is simply said to be stationary, most often what is meant is that it has stationary variance.)
Time series	A time series is data recorded sequentially over time at roughly equally spaced intervals.
Trend component	The part of a model for a time series that fits long-term changes in the mean of the series.

Skills

When you complete this chapter, you should:

- Be able to recognize when data are a time series.
- Know how to identify whether a linear or nonlinear trend model is appropriate.
- Be able to judge whether a time series is stationary in the mean and/or variance.
- Be able to recognize when a seasonal or cyclical component is present in a time series.
- Understand the impact of changing lengths on a moving average model.
- Understand the impact of changing smoothing coefficients on a single exponential smoothing model.

- Know how to develop linear and nonlinear trend models.
- Be able to compute moving averages of different lengths to obtain forecasts.
- Be able to compute a single exponential smoothing model using different coefficients.
- Know how to calculate forecast error, including MAD, MSE, and MAPE.
- Know how to determine if autocorrelation is present.
- Be able to develop autoregressive models for seasonal and nonseasonal models.
- Know how and when to use dummy variables to develop seasonal models.

REPORT
- Know how to use forecast error to compare alternative time series models.
- Know how to determine which lagged variables are significant in an autoregressive model.
- Know how to identify different advantages for alternative time series models depending on the objective.

Technology Help

EXCEL

Excel offers some, but not all, time series methods in this chapter.

Exponential Smoothing:
Tools > **Data Analysis** > **Exponential Smoothing** and enter the input and output range, as well as the smoothing coefficient (Damping factor in Excel).

Moving Averages:
Tools > **Data Analysis** > **Moving Average** and enter the input and output range, as well as the Length (Interval in Excel).

MINITAB

The time series commands are in the **Time Series** submenu of the **Stat** menu. They are generally self-explanatory. Most commands open a dialog in which you specify the series to analyze, specify parameters (e.g., for smoothers), and request predictions.

SPSS

The commands related to time series are found in the **Time Series** submenu of the **Analyze** menu. Choose **Create Models** for a choice of methods.

To make a time series plot, choose **Analyze** > **Time Series** > **Create Models** and specify the variable you wish to plot. Click on the **Plots Tab**, and check **Series**.

Trend Analysis:
Transform > **Date/Time** > **Assign Periodicity** > **Days**, to create the time variable. Then choose **Analyze** > **Regression** > **Linear**. Enter the variable you wish to analyze as the dependent variable and days as the independent variable.

Moving Averages:
Analyze > **Time Series** > **Create Models** and enter the variable you wish to analyze. Then select **ARIMA** > **Criteria** and enter the moving average period length in the criteria box.

Exponential Smoothing:
Analyze > **Time Series** > **Create Models** and enter the variable you wish to analyze. Then select **Exponential Smoothing** > **Criteria** and enter the weight (smoothing coefficient) to use.

Lagged Variables:
Transform > **Create Time Series** and enter the variable for which you wish to create a lag function. Select **Lag** and enter the length of the new lagged variable.

DATA DESK

Use the **Lineplot** command to display a time series and the **Multiple Lineplot** command to compare series.

The **Hyperview** menu on a scatterplot or lineplot offers smoothing options, although they are more sophisticated smoothers than those discussed here.

To lag a series, select its icon and choose **Manip** > **Transform** > **Dynamic** > **Lag**. Then slide the slider to specify the lag amount. You can adjust the lag even after the variable is in use in a regression model and view the consequences to the fit or residuals dynamically.

Mini Case Study Projects

Intel Corporation

Intel Corporation, located in Santa Clara, California, was founded by three engineers in 1968 to develop technology for silicon-based electronic chips. Intel has a reputation as an innovator, and their net revenues more than tripled between 1993 and 2002. Although Intel remains the world's largest maker of chips, the semiconductor industry is impacted by volatility in sales of personal computers and other electronic and computing devices. As PC sales leveled off in the United States, modest semiconductor growth resulted from sales in video games, wireless and smart phones, networking equipment, and handheld devices. In 2002, worldwide semiconductor sales were over $150 billion.

The daily closing prices for Intel stock in the first four months of 2007 are in the file **Ch20_MCSP_Intel**. Examine the time series and identify appropriate times series models to forecast the daily stock price for Intel. Download the stock prices for Intel in the month of May 2007. Compare the forecasts for May 2007 obtained using your time series models with the actual closing price using the measures of forecast error discussed in this chapter. Which models are more appropriate and/or more accurate for forecasting Intel daily stock prices in 2007? Explain.

Tiffany & Co.

Tiffany was founded in 1837, when Charles Lewis Tiffany opened his first store in downtown Manhattan. Tiffany retails and distributes a selection of Tiffany & Co. brand jewelry at a range of prices. Today, more than 150 Tiffany & Co. stores sell to customers in U.S. and international markets. In addition to jewelry, it sells Tiffany & Co. brand merchandise in the following categories: timepieces and clocks; sterling silver merchandise; stainless steel flatware; crystal, glassware, china, and other tableware; custom engraved stationery; writing instruments; and fashion accessories. Fragrance products are sold under the trademarks Tiffany, Pure Tiffany, and Tiffany for Men. Tiffany also sells other brands of timepieces and tableware in its U.S. stores.[16]

Tiffany's quarterly sales from 1995 to 2002 are in the file **Ch20_MCSP_Tiffany**. Tiffany's quarterly sales (in $million) from 1995 to 2002 are shown here.

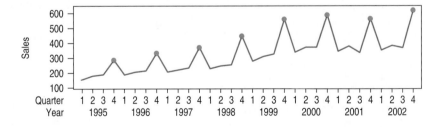

Build alternative time series models for Tiffany's sales and forecast future sales. In addition, develop causal, or multiple regression, models to forecast Tiffany's

---

[16] Source: www.shareholder.com/tiffany.

sales by using economic data, such as gross domestic product (GDP), consumer price index (CPI), disposable income, unemployment, and interest rates over this same time period. These economic data are also provided in the data file. Given that Tiffany is known for its high-end quality and price, it was of interest for the corporate executives to see how sensitive Tiffany sales were to economic indicators. Develop a hypothesis before developing your models. Compare the forecasts from your time series models and regression models to quarterly sales in 2003. (These are also included in the data file.) Use alternative measures of forecast error and recommend a model for the executives at Tiffany to use for forecasting their quarterly sales.

EXERCISES

1. Concepts.

a) Which will be smoother, a 50-day or a 200-day moving average?

b) Which will be smoother, a single exponential smoothing (SES) model using $\alpha = 0.10$ or a model using $\alpha = 0.80$?

c) What is the difference in how historical data are used when the smoothing coefficient in a single exponential smoothing (SES) model is raised from 0.10 to 0.80?

2. Concepts, again. We are trying to forecast monthly sales for a company that sells ski equipment and clothing. Assume that the company's sales peak each December and that the monthly sales have been growing at the rate of 1% each month. Answer the following questions.

a) Based on the description of these data, what time series components can you identify?

b) If you identified a seasonal component, what is the period?

c) If you use seasonal dummy variables, specify the dummy variables you would use.

d) After examining the residuals and using the information provided, you decide to transform the sales data. What transformation are you likely to suggest? Why?

3. More concepts. For each of the following time series, suggest an appropriate model:

a) Weekly stock prices that reveal erratic periods of up and down swings.

b) Annual sales that reveal a consistent percentage annual increase.

c) Quarterly sales for a bicycle shop that reveal a seasonal pattern where sales peak in Q2 of each year.

4. Final concepts. For each of the following time series, suggest an appropriate model:

a) Daily stock prices that reveal erratic periods of up and down swings.

b) Monthly sales that reveal a consistent percentage increase from month to month.

c) Quarterly sales for a woman's clothing company that reveal an annual peak each December.

5. Liquid assets. The Bank of New York Company was founded by Alexander Hamilton in 1784 and was a major commercial bank until its merger with the Mellon Financial Corporation in 2007. Their year-end financial reports for the final five years of independent operation give the following values for their liquid assets (Source: *The Financial Times*).

Year	Liquid Assets ($M)
2002	18,546
2003	22,364
2004	22,413
2005	19,881
2006	26,670

a) Use a 3-year moving average to predict what liquid assets would have been in 2007.

b) Predict the value for 2007 using a single exponential smooth with smoothing parameter $\alpha = 0.2$

6. Baking profits. Sara Lee Corp., maker of food, beverage, and household products, is known especially for its baked products, marketed under its corporate name. For the five years ending July 1 of each year from 2002 to 2006, their bakery division reported the following profits.

Fiscal Year	Profits ($M)
2002	97
2003	98
2004	156
2005	−4
2006	−197

a) Use a 4-year moving average to predict profits for 2007.

b) Predict the profits for 2007 using a single exponential smooth with smoothing parameter $\alpha = 0.5$

c) Think about the exponential smoother. If the parameter were .8 would you expect the prediction to be higher or lower? What if it were .2? Explain.

7. Banana prices. The price of bananas fluctuates on the world market. Here are the prices ($/tonne) for the years 2000–2004. (Source: *Holy See Country Review*, 2008).

2000	2001	2002	2003	2004
422.27	584.70	527.61	375.19	524.84

a) Find a 3-year moving average prediction for the price in 2005.

b) Find a prediction for 2005 with an exponential smoothing model with $\alpha = 0.4$.

c) The actual price of bananas in 2005 was 577 $/tonne (www.imf.org/external/np/res/commod/table3.pdf.) Compute the absolute percentage error for each prediction.

8. Target earnings. Target Corp. operates "big box" stores that sell everyday essentials and fashionable differentiated merchandise. It also operates an online business at target.com. Target's reported gross earnings per share for the years 2003–2006 are given here.

2003	2004	2005	2006
$1.82	2.02	2.17	2.73

a) Find a prediction for 2007 based on a 3-year moving average and one for a 4-year moving average.

b) Find a prediction for 2007 based on an exponential smoothing model with $\alpha = 0.8$.

c) Earnings per share in 2007 were, in fact, $3.18. Compute the absolute percentage error for each prediction.

T **9. Toyota stock prices.** The following time series graph shows daily closing stock prices (adjusted for splits and dividends) for Toyota Motor Manufacturing from April 1, 2008, through July 3, 2008 (Source: Yahoo! Finance).

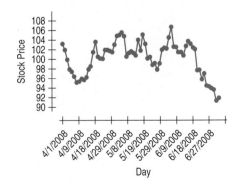

a) Which time series components seem to be present?

The method of moving averages was applied to these data. Here are time series graphs showing moving average results using two different lengths.

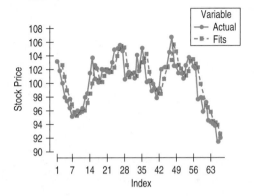

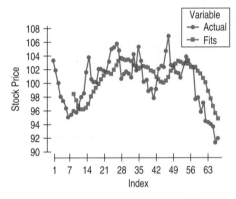

b) In which application is a larger length used?

10. Exponential smoothing. The following time series graph shows daily closing stock prices (adjusted for splits and dividends) for Google, Inc. from February 1, 2008, through June 30, 2008 (Source: Yahoo! Finance).

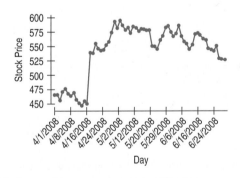

a) Which time series components are evident?

Single exponential smoothing (SES) models were found for these data. Examine the time series graphs on the next page showing two different smoothing coefficients values ($\alpha = 0.2$ and $\alpha = 0.8$).

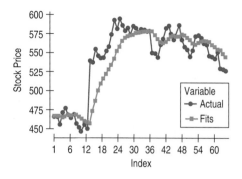

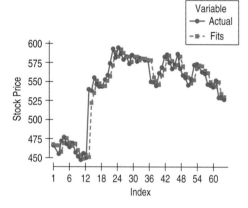

b) In which graph is a larger value of α used?

T 11. Men's weekly earnings. This graph shows the quarterly median weekly earnings from the first quarter of 2000 through the fourth quarter of 2007 for men, 25 years of age or older, in the United States (www.bls.gov).

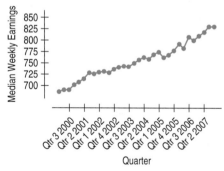

Here are time series plots showing a 2-quarter moving average and an 8-quarter moving average.

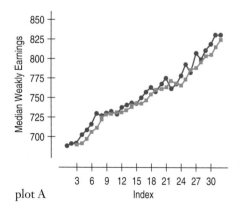

plot A

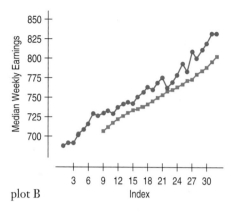

plot B

Identify which plot is the 2-quarter moving average and which is the 8-quarter moving average. Explain why the better-fitting model fits better.

T 12. Women's weekly earnings. The following graph shows the quarterly median weekly earnings for U.S. women 25 years of age or older (www.bls.gov). Data are provided from the first quarter of 2000 through the fourth quarter of 2007.

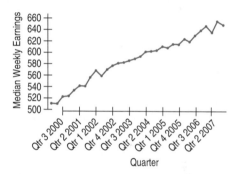

Here is single exponential smoothing model to these data using $\alpha = 0.2$.

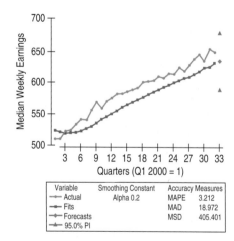

Here is the exponential smoothing model using $\alpha = 0.8$:

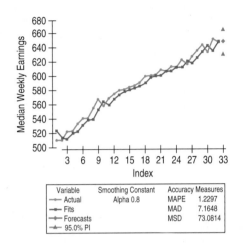

Variable	Smoothing Constant	Accuracy Measures	
Actual	Alpha 0.8	MAPE	1.2297
Fits		MAD	7.1648
Forecasts		MSD	73.0814
95.0% PI			

Why does the exponential smoother with the higher coefficient fit the series better? What about this series is important for this result? What does this suggest about using exponential smoothers on such time series?

13. Autoregressive model. Suppose an autoregressive model is used for data in which quarterly sales in 2005 were: 1.9, 1.7, 2.2, and 2.3 ($Billion).

a) If a first-order autoregressive model is developed with estimated parameters of $b_0 = 0.100$ and $b_1 = 1.12$, compute the forecast for Q4 of 2006.

b) Compare this forecast to the actual value ($2.9B) by computing the absolute percentage error (APE). Did you over-forecast or under-forecast?

c) Assuming these quarterly sales have a seasonal component of length 4, use the following model to compute a forecast for Q4 of 2006: $y_t = 0.410 + 1.35\, y_{t-4}$. Compare the APE for this forecast to that in question a. Compare the appropriateness of the different models.

14. Another autoregressive model. Suppose an autoregressive model is used to model sales for a company that peaks twice per year (in June and December).

a) What lagged variables would you try in a regression to forecast sales? Explain.

b) How would you determine which of your lagged variables should remain in the model? Explain.

T 15. Coffee prices. Coffee is the world's second largest legal export commodity (after oil) and is the second largest source of foreign exchange for developing nations. The United States consumes about one-fifth of the world's coffee. The International Coffee Organization (ICO) computes a coffee price index using Colombian, Brazilian, and a mixture of other coffee data. Data are

provided for the ICO price index (in $US) from January 2001 to December 2005.

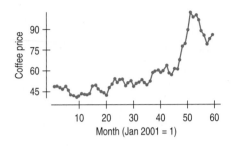

Here's an autoregressive model for the composite prices.

Dependent variable is: Composite
R squared = 94.8% R squared (adjusted) = 94.6%
s = 3.892 with 57 − 3 = 54 degrees of freedom

Variable	Coeff	SE(Coeff)	t-ratio	P-value
Intercept	1.59175	1.941	0.820	0.4157
lag1	1.22856	0.1327	9.26	≤ 0.0001
lag2	−0.247256	0.1350	−1.83	0.0725

a) Here are the last several values of the series: 88.48, 85.31, 78.79, 82.55, and 85.93. What price does this model predict for the next value in the series?

b) The next value in the series was, in fact, 86.85. Compute the APE.

c) Find a prediction based on a 2-point moving average. How does it compare with the AR model?

T 16. Gas prices smoothed. We have data on the weekly average retail price (cents per gallon) of regular gas nationwide from 2002 through the early part of 2007. Here's a time series plot.

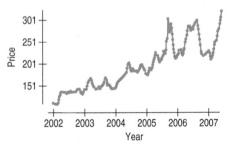

Here are a 3-point moving average and a 3-term autoregressive model fit to these data.

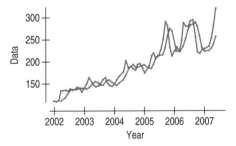

Dependent variable is: Price
R squared = 92.5%   R squared (adjusted) = 92.2%
s = 14.60 with 63 − 4 = 59 degrees of freedom

Variable	Coeff	SE(Coeff)	t-ratio	P-value
Intercept	7.93591	7.504	1.06	0.2946
Price1	1.47044	0.1250	11.8	≤ 0.0001
Price2	−0.896360	0.2003	−4.47	≤ 0.0001
Price3	0.400086	0.1270	3.15	0.0026

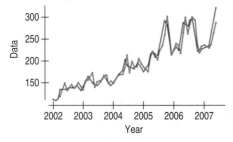

Which model is the better fit for the data?

17. Gallup poll. The Gallup organization periodically asks the following question:

If your party nominated a generally well-qualified person for president who happened to be a woman, would you vote for that person?

Here is a time series plot of the percentage answering "yes" versus the year of the (20th) century. The least squares trend line is given by: $\hat{y}_t = 5.58 + 0.999\,Year$, where $Year = 37, 45, \ldots 99$ to represent the years during which the survey was given.

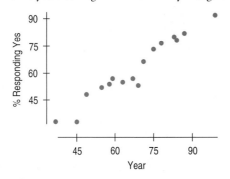

a) The R^2 for this trend line is 94%. A student decided to use this linear model to obtain a forecast for the percent who will respond "yes" in 2012. What value should the student use for *Year*?
b) Find the predicted value for the year 2012. Is it realistic?

T 18. Consumer price index. The Consumer Price Index (CPI) represents changes in prices of all goods and services purchased for consumption by urban households.[17] The most common use

---

[17] User fees (such as water and sewer service) and sales and excise taxes paid by the consumer are also included. Income taxes and investment items (such as stocks, bonds, and life insurance) are not included. Most of the specific CPI indexes have a 1982–1984 reference base. That is, BLS sets the average index level (representing the average price level) for the 36-month period covering the years 1982, 1983, and 1984 equal to 100 and then measures changes in relation to that figure. An index of 110, for example, means there has been a 10-percent increase in price since the reference period. See www.bls.gov/cpi for more information.

of the CPI is as an economic indicator to forecast inflation and evaluate the effectiveness of government policies. Following is the time series plot for the monthly CPI (not seasonally adjusted) from January 2001 to March 2007. The linear trend line is: CPI = 173 + 0.414 t, where $t = 0, 1, \ldots 74$ to represent the months in the series.

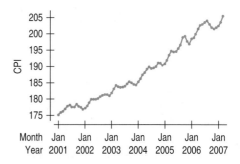

a) What does the intercept 173 represent in this trend line? What does the slope represent?
b) The R^2 for this trend line is 97%. Use this linear model to obtain a forecast for the CPI in April 2007.
c) The actual CPI in April 2007 was 206.686. Compute the absolute percent error for your forecast.

T 19. Retail gas prices. We have data on the weekly average retail price (cents per gallon) of regular gas nationwide from 2002 through the early part of 2007. Here's a time series plot:

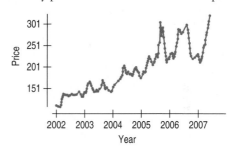

a) What components can you see in this plot?
Here's a linear trend model fit to these data.

Dependent variable is: Price
R squared = 82.3%   R squared (adjusted) = 82.2%
s = 21.88 with 282 − 2 = 280 degrees of freedom

Variable	Coeff	SE(Coeff)	t-ratio	P-value
Intercept	−60025.3	1668	−36.0	≤ 0.0001
Year	30.0388	0.8322	36.1	≤ 0.0001

b) Interpret the trend coefficient in this model.
c) How would you interpret the intercept coefficient?
Here's a time series plot of the residuals.

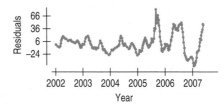

d) Comment on any patterns you see.

T 20. Interest rates. Average annual interest rates (banks prime lending) in the United States from 1980 through 2006 are shown in the following time series graph (unstats.un.org).

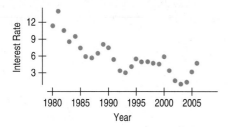

a) What components do you see in this series?

Here's a linear trend model fit to these data.

```
Dependent variable is:     Interest Rate
R squared = 69.9%   R squared (adjusted) = 68.7%
s = 1.732 with 27 − 2 = 25 degrees of freedom
```

Variable	Coeff	SE(Coeff)	t-ratio	P-value
Intercept	656.539	85.30	7.70	≤ 0.0001
Year	−0.326490	0.0428	−7.63	≤ 0.0001

b) Interpret the trend coefficient mean in this model.
c) Predict the interest rate for 2006. Do you trust the prediction? Why or why not?

Here's a time series plot of the residuals from this model.

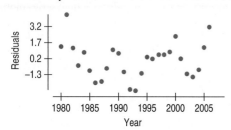

d) Discuss any patterns you see in this plot.
e) Would an exponential model be likely to do better for these data? Explain.

21. Seasonal model. Use the following model to forecast quarterly sales ($Million) for a company (where time is rescaled to begin at zero and Q_2, Q_3, and Q_4 are dummy variables for the indicated quarters.), and answer the following questions.

$$\hat{y} = 1.1 + 0.2\,t - 0.1\,Q_2 - 0.5\,Q_3 + 0.5\,Q_4$$

a) For the first quarter of the time series, what are the sales?
b) What is the quarter that on average has the lowest level of sales over the time frame of the series?
c) What is the quarter that on average has the highest level of sales over the time frame of the series?
d) Interpret the coefficient of the dummy variable named Q_4.

22. Another seasonal model. Use the following model to forecast quarterly sales ($000) for a start-up (where time is rescaled to begin at zero and Q_2, Q_3, and Q_4 are dummy variables for the indicated quarters.), and answer the following questions.

$$\hat{y} = 15.1 + 10.5\,t - 5.0\,Q_2 - 7.2\,Q_3 + 7.5\,Q_4$$

a) For the first quarter of the time series, what are the sales?
b) What is the quarter that on average has the lowest level of sales over the time frame of the series?
c) What is the quarter that on average has the highest level of sales over the time frame of the series?
d) Interpret the coefficient of the dummy variable named Q_4.

T 23. Wal-Mart revenue. Wal-Mart has grown rapidly in recent years. Here is the monthly revenue ($Billion) for Wal-Mart from November 2003 to January 2007.

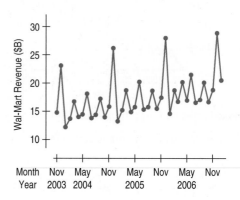

a) What components of a time series do you see in this timeplot?

Here's a regression model fit using dummy variables for months and a *Time* variable that counts from 1 for the first data value in the series.

```
Dependent variable is:     WM Rev
R squared = 94.3%   R squared (adjusted) = 91.6%
s = 1.121 with 39 − 13 = 26 degrees of freedom
```

Variable	Coeff	SE(Coeff)	t-ratio	P-value
Intercept	12.0322	0.6562	18.3	≤ 0.0001
Time	0.145241	0.0163	8.93	≤ 0.0001
Feb	1.46096	0.8599	1.70	0.1013
Mar	2.84671	0.8585	3.32	0.0027
Apr	1.67981	0.8574	1.96	0.0609
May	0.870232	0.8567	1.02	0.3191
Jun	4.99999	0.8562	5.84	≤ 0.0001
Jul	0.106417	0.8560	0.124	0.9020
Aug	0.434176	0.8562	0.507	0.6164
Sep	3.25327	0.8567	3.80	0.0008
Oct	−0.219640	0.8574	−0.256	0.7998
Nov	1.87023	0.7932	2.36	0.0262
Dec	11.5625	0.7927	14.6	≤ 0.0001

b) Interpret the coefficient of *Time*.
c) Interpret the coefficient of *Dec*.
d) What revenue would you predict for Wal-Mart in February 2007 (the 40th month in this series)?
e) What does it mean that the coefficient for *Oct* is the only negative coefficient in the model?

T 24. Movie revenue. The movie *Harry Potter and the Sorcerer's Stone* opened as a great success. But every movie sees declining revenue over time. Here are the daily revenues for the movie during its first 17 days.

Day	Date	Earnings ($M)
Friday	11/16/01	35
Saturday	11/17/01	30
Sunday	11/18/01	25
Monday	11/19/01	9
Tuesday	11/20/01	10
Wednesday	11/21/01	10
Thursday	11/22/01	10
Friday	11/23/01	22
Saturday	11/24/01	23
Sunday	11/25/01	13
Monday	11/26/01	3
Tuesday	11/27/01	3
Wednesday	11/28/01	3
Thursday	11/29/01	2
Friday	11/30/01	9
Saturday	12/1/01	9
Sunday	12/2/01	6

a) Without plotting the data, what components can you see in this series? Be specific.

For some series, a "seasonal" effect repeats weekly rather than annually. Here's a regression model fit to these data with dummy variables for days of the week. (*Day#* counts days starting at 1.)

Dependent variable is: Earnings
R squared = 96.9%   R squared (adjusted) = 94.6%
s = 2.365 with 17 − 8 = 9 degrees of freedom

Variable	Coeff	SE(Coeff)	t-ratio	P-value
Intercept	21.0000	2.090	10.0	≤ 0.0001
Day#	−1.42857	0.1194	−12.0	≤ 0.0001
Friday	12.4286	2.179	5.70	0.0003
Saturday	12.5238	2.166	5.78	0.0003
Sunday	7.95238	2.160	3.68	0.0051
Monday	−4.28571	2.392	−1.79	0.1068
Tuesday	−2.35714	2.377	−0.992	0.3473
Wednesday	−0.928571	2.368	−0.392	0.7041

b) Interpret the coefficient of *Day#*.

c) Interpret the coefficient of *Saturday* in this model.

d) Predict what earnings probably were for Monday 12/3/01. What does this say about the model?

e) What probably happened to earnings after the initial 17 days?

T 25. Hawaii tourism. Much of the public and private industry in Hawaii depends on tourism. The following time series plot shows the total number of visitors to Hawaii by air from the rest of the United States per month from January 2002 to December 2006.

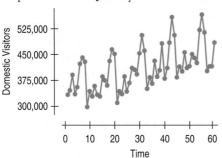

Here is a regression model fit to these data with dummy variables for months and a *Time* variable that starts at 0 and counts months:

Dependent variable is: Domestic Visitors
R squared = 95.5%   R squared (adjusted) = 94.4%
s = 14336 with 60 − 13 = 47 degrees of freedom

Variable	Coeff	SE(Coeff)	t-ratio	P-value
Intercept	312,312	6925	45.1	≤ 0.0001
Time	2016.84	109.1	18.5	≤ 0.0001
Feb	8924.36	9068	0.984	0.3301
Mar	58,693.7	9070	6.47	≤ 0.0001
Apr	20,035.3	9073	2.21	0.0321
May	19,501.0	9078	2.15	0.0369
Jun	90,440.8	9084	9.96	≤ 0.0001
Jul	132,893	9091	14.6	≤ 0.0001
Aug	96,037.3	9099	10.6	≤ 0.0001
Sep	−27,919.7	9109	−3.07	0.0036
Oct	1244.42	9120	0.136	0.8921
Nov	−12,181.4	9133	−1.33	0.1887
Dec	39,201.9	9146	4.29	≤ 0.0001

a) Interpret the P-value for the trend coefficient.

b) You are planning to visit Hawaii and hope to avoid the crowds. When should you go? (Name the two best months.)

c) You are planning to advertise your product (a new sunscreen/massage oil combination) to tourists visiting Hawaii. What two months would be the best times to advertise?

d) How many tourists would you predict for Hawaii in April 2007 (month 63 of this series)?

T 26. Hawaii tourism, part 2. In Exercise 25 we examined domestic tourists who visit Hawaii. Now, let's consider international tourism. Here's a time series plot of international visitors.

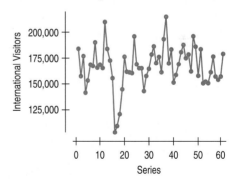

Here's the linear trend model with dummy variables for months.

Dependent variable is: International
R squared = 62.9%   R squared (adjusted) = 53.4%
s = 13684 with 60 − 13 = 47 degrees of freedom

Variable	Coeff	SE(Coeff)	t-ratio	P-value
Intercept	185963	6610	28.1	≤ 0.0001
Time	73.1937	104.1	0.703	0.4854
Feb	−23134.8	8655	−2.67	0.0103
Mar	−14961.8	8657	−1.73	0.0905
Apr	−50026.4	8660	−5.78	≤ 0.0001
May	−41912.4	8664	−4.84	≤ 0.0001
Jun	−33370.4	8670	−3.85	0.0004
Jul	−21521.2	8677	−2.48	0.0168
Aug	−5021.56	8685	−0.578	0.5659
Sep	−22294.3	8694	−2.56	0.0136
Oct	−20751.5	8705	−2.38	0.0212
Nov	−27338.9	8717	−3.14	0.0029
Dec	6075.27	8730	0.696	0.4899

a) Interpret the P-value for the *Time* coefficient.

b) The R^2 for this model is lower than for the model fit to domestic visitors in exercise 25. Does that mean that an exponential trend model would do better?

c) International tourists often visit Hawaii in January. How can you tell that from this model?

d) Even though the R^2 for this model is lower than the corresponding R^2 for the model fit in Exercise 25 to domestic tourist visits, you might feel more comfortable predicting the number of international visitors for April 2007 with this model than you did predicting the number of domestic visitors with the previous model. Explain why.

T **27. Oakland passengers.** The Port of Oakland airport reports the number of passengers passing through each month. At first glance, this is just simple growth, but by recognizing the series as a time series, we may learn more.

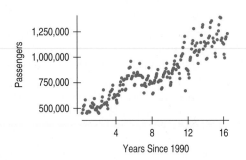

The increasing spread of the data suggests that a multiplicative model may fit better.

Here's an exponential trend model with seasonal (monthly) dummy variables.

Dependent variable is: LogPassengers
R squared = 92.4% R squared (adjusted) = 91.9%
s = 0.0358 with 194 − 13 = 181 degrees of freedom

Variable	Coeff	SE(Coeff)	t-ratio	P-value
Intercept	5.64949	0.0101	558	≤ 0.0001
Years since '90	0.024929	0.0006	45.2	≤ 0.0001
Feb	−0.013884	0.0127	−1.10	0.2747
Mar	0.048429	0.0127	3.82	0.0002
Apr	0.049028	0.0125	3.93	0.0001
May	0.066227	0.0125	5.30	≤ 0.0001
Jun	0.082862	0.0127	6.54	≤ 0.0001
Jul	0.095295	0.0127	7.52	≤ 0.0001
Aug	0.121779	0.0127	9.61	≤ 0.0001
Sep	0.029382	0.0127	2.32	0.0216
Oct	0.049596	0.0127	3.91	0.0001
Nov	0.039649	0.0127	3.13	0.0020
Dec	0.043545	0.0127	3.44	0.0007

a) Interpret the slope.

b) Interpret the intercept.

c) Which months have the lowest traffic at Oakland airport? (Hint: Consider all 12 months.)

Here's a plot of the residuals from the model fit to the Oakland airport passengers:

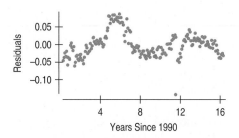

d) What components would you now say are in this series?

T **28. Retail gas prices, part 2.** In Exercise 16 we saw data on the U.S. average regular retail gasoline price for one week of each month from January 2002 to May 2007.

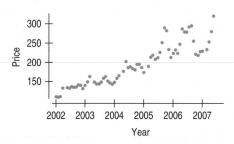

The bend in the plot and the increasing variation suggest a multiplicative model. Here is a multiplicative model with seasonal dummy variables for months.

Dependent variable is: Log price
R squared = 92.0% R squared (adjusted) = 90.2%
s = 0.0377 with 66 − 13 = 53 degrees of freedom

Variable	Coeff	SE(Coeff)	t-ratio	P-value
Intercept	2.04566	0.0164	125	≤ 0.0001
Year since 2002	0.071754	0.0030	24.2	≤ 0.0001
Feb	0.002682	0.0210	0.128	0.8988
Mar	0.030088	0.0210	1.43	0.1574
Apr	0.053455	0.0210	2.55	0.0138
May	0.058316	0.0210	2.78	0.0075
Jun	0.044912	0.0221	2.03	0.0471
Jul	0.055008	0.0221	2.49	0.0159
Aug	0.061186	0.0221	2.77	0.0077
Sep	0.057329	0.0221	2.60	0.0122
Oct	0.039903	0.0221	1.81	0.0764
Nov	0.012609	0.0221	0.571	0.5704
Dec	−0.085243	0.0221	−0.386	0.7011

a) Interpret the slope.

b) Interpret the intercept.

c) In what month of the year are gas prices highest?

29. Oakland outlier. The plot of residuals in Exercise 27 shows an outlier that wasn't as evident in the data. The outlier is September 2001. Clearly, this wasn't a typical month for air travel. Here are three models fit to this series, a single exponential smooth, a 12-point moving average, and the fitted values from the seasonal regression model of Exercise 27. Discuss how each deals with the outlier.

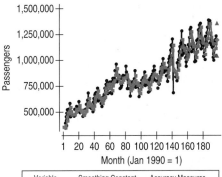

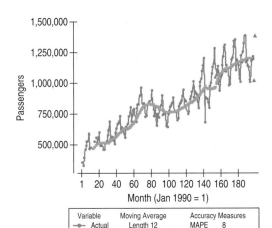

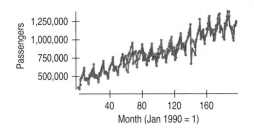

The remaining exercises require the use of statistics software. Statistics packages vary in their capabilities and in the default decisions that some make. As a result, depending upon which package you choose, your answers may differ from those in the back of the book.

T 30. E-commerce. Quarterly e-commerce retail sales (in millions of dollars) in the United States are provided. (Source: www.census.gov.) Use this time series to answer the following questions.

a) Fit a linear trend model to this series but do not use the last two quarters (Q4 2007 and Q1 2008).

b) Fit an exponential trend model to this series but do not use the last two quarters.

c) Use both models to forecast the quarterly values for Q4 2007 and Q1 2008. Which model produces better forecasts?

d) What other time series components (besides trend) are likely present in this series?

T 31. E-commerce, part 2.

a) Fit a linear trend model with dummy variables for the seasonal effect to the e-commerce data in Exercise 30.

b) Fit an exponential trend (multiplicative) model with dummy variables to these data.

c) Which model fits better?

T 32. J. Crew revenue. J. Crew launched its first catalog in 1983, and the first J. Crew store opened in 1989 at the South Street Seaport in New York City. Since then, J. Crew has expanded to 198 retail and 65 outlet stores across the country. Quarterly *Revenues* ($ Million) for J. Crew follow; use the data provided to answer the questions.

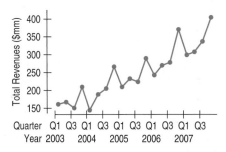

a) Identify the time series components you see in the graph. If you see seasonality, what is the period?

b) Use your answers to part (a) to develop an autoregressive model.

c) Examine the residuals over time to evaluate the Independence Assumption.

T 33. Retail gas prices, part 3. Using the data from Exercise 28, develop and compare the following models.

a) Fit an appropriate autoregressive model by testing for the significance of each autoregressive term.

b) Obtain a forecast for the same week you did in Exercise 28 (May 28, 2007).

c) Compare your forecast to the actual value (by computing APE) and to the forecasts obtained in Exercise 28.

d) Recommend an appropriate model for forecasting this time series.

T 34. Oil prices. Recall the crude oil price data from Chapter 16. A time series plot of monthly crude oil price ($/barrel) from January 2001 to March 2007 is shown here.

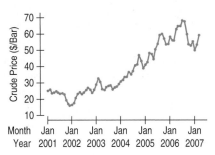

Using these data,

a) Fit an appropriate autoregressive model by testing for the significance of each autoregressive term.

b) Obtain a forecast for March 2007.

35. Oil prices again. Return to the oil price data of Exercise 34.

a) Find a linear model for this series.

b) Find an exponential (multiplicative) model for this series.

c) Use these methods to forecast the crude oil price for March 2007.

d) The actual price for March 2007 was $58.70. Compute measures of forecast error (e.g., MAD and MAPE) and compare the accuracy of the forecasts for the models of Exercises 34 and 35.

36. U.S. unemployment rate. Following is the time series plot for the monthly U.S. *Unemployment rate* (%) from January 1997 to March 2007. These data have been seasonally adjusted (meaning that the seasonal component has already been removed).

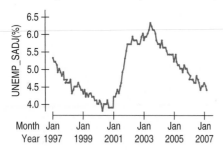

a) What time series components do you observe in this series?

b) Develop a 6-month and 12-month moving average model for this series.

c) Fit a single exponential smoothing model to these series.

d) Use these models to forecast the *Unemployment rate* for March 2007.

e) Compute measures of forecast error (e.g., MAD and MAPE) and compare the accuracy of the forecasts for these two models.

37. Mutual funds. The money that is invested in mutual funds is often referred to as "Fund Flows." The following time series plot shows the monthly *Fund Flows* ($million) from January 1990 to October 2002.

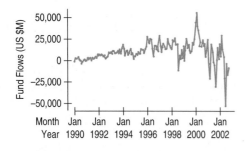

a) What time series components do you observe in this series?

b) Develop a 6-month and 12-month moving average model for this series.

c) Fit a single exponential smoothing model to these series.

d) Use these models to forecast the *Fund Flows* for October 2002.

e) Compute measures of forecast error (e.g., MAD and MAPE) and compare the accuracy of the forecasts for these two models.

38. U.S. unemployment rate, part 2. Using the data from Exercise 36, develop and compare the following models.

a) Fit an appropriate autoregressive model by testing for the significance of each autoregressive term (assume $\alpha = 0.05$).

b) Obtain a forecast for March 2007.

c) Compare your forecast for March to the actual value (by computing APE) and to the forecasts obtained in Exercise 36.

d) Recommend an appropriate model for forecasting this time series.

39. Hotel occupancy rates. The monthly occupancy rates for hotels in Honolulu, Hawaii, from January 2000 to December 2004 are available for the same time period as we examined in Exercises 25 and 26.

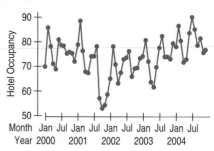

Use the data to answer the following questions.

a) Identify the time series components you see in the graph. If you see seasonality, what is the period? Use your answers to develop an autoregressive model.

b) Examine the residuals over time to evaluate the Independence Assumption.

c) Obtain a forecast for January 2005.

40. OPEC production. Recall the data for OPEC oil production from Chapter 16, Exercise 27. A time series plot of monthly OPEC production (thousand barrels/day) is shown here.

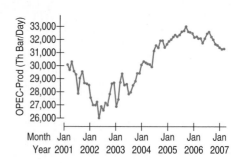

a) What time series components do you observe in this series?

b) Develop a 6-month and 12-month moving average model for this series.

c) Fit a single exponential smoothing model to these series.

d) Use these models to forecast the OPEC oil production for March 2007.

e) Compute measures of forecast error (e.g., MAD and MAPE) and compare the accuracy of the forecasts for these two models.

T **41. OPEC production, part 2.** Using the data from Exercise 40, develop and compare the following models.

a) Fit an appropriate autoregressive model by testing for the significance of each autoregressive term.

b) Obtain a forecast for March 2007.

c) Compare your forecast for March to the actual value (by computing APE) and to the forecasts obtained in Exercise 40.

d) Recommend an appropriate model for forecasting this time series.

T **42. Interest rates, part 2.** Consider the average annual interest rates we saw in Exercise 20. That exercise showed a linear model fit to these data.

a) Fit a linear trend model to this series but do not use the values for 2005 and 2006.

b) Fit an exponential trend model to this series but do not use the values for 2005 and 2006.

c) Use both models to forecast the average interest rate for 2005 and 2006. Which model produces better forecasts?

d) What other time series components (besides trend) are likely present in this series?

JUST CHECKING ANSWERS

1 Trend and seasonal.

2 The value is the mean of the final 4 values, about 333. It might be a bit low as a prediction.

3 $(319.3 + 399.9)/2 = 359.6$.

4 Four terms because there is a strong seasonal component with period 4.

5 10.2338. Revenue grew at about 10.2 million dollars per quarter.

6 Dummy variables require that we leave one out. Q4 coefficient will be positive.

PART IV

Building Models for Decision Making

Random Variables and Probability Models

Metropolitan Life Insurance Company

In 1863, at the height of the U.S. Civil War, a group of businessmen in New York City decided to form a new company to insure Civil War soldiers against disabilities and injuries suffered from the war. After the war ended, they changed direction and decided to focus on selling life insurance. The new company was named Metropolitan Life (MetLife) because the bulk of the company's clients were in the "metropolitan" area of New York City.

Although an economic depression in the 1870s put many life insurance companies out of business, MetLife survived, modeling their business on similar successful programs in England. Taking advantage of spreading industrialism and the selling methods of British insurance agents, the company soon was enrolling as many as 700 new policies per day. By 1909, MetLife was the nation's largest life insurer in the United States.

During the Great Depression of the 1930s, MetLife expanded their public service by promoting public health campaigns, focusing on educating the urban poor in major U.S.

cities about the risk of tuberculosis. Because the company invested primarily in urban and farm mortgages, as opposed to the stock market, they survived the crash of 1929 and ended up investing heavily in the post-war U.S. housing boom. They were the principal investors in both the Empire State Building (1929) and Rockefeller Center (1931). During World War II, the company was the single largest contributor to the Allied cause, investing more than half of their total assets in war bonds.

Today, in addition to life insurance, MetLife manages pensions and investments. In 2000, the company held an initial public offering and entered the retail banking business in 2001 with the launch of MetLife Bank. The company's public face is well known because of their use of Snoopy, the dog from the cartoon strip "Peanuts."

Insurance companies make bets all the time. For example, they bet that you're going to live a long life. Ironically, you bet that you're going to die sooner. Both you and the insurance company want the company to stay in business, so it's important to find a "fair price" for your bet. Of course, the right price for *you* depends on many factors, and nobody can predict exactly how long you'll live. But when the company averages its bets over enough customers, it can make reasonably accurate estimates of the amount it can expect to collect on a policy before it has to pay out the benefit. In order to do that effectively, it must model the situation with a probability model. Using the resulting probabilities, the company can find the fair price of almost any situation involving risk and uncertainty.

Here's a simple example. An insurance company offers a "death and disability" policy that pays $100,000 when a client dies or $50,000 if the client is permanently disabled. It charges a premium of only $500 per year for this benefit. Is the company likely to make a profit selling such a plan? To answer this question, the company needs to know the *probability* that a client will die or become disabled in any year. From actuarial information such as this and the appropriate model, the company can calculate the expected value of this policy.

21.1 Expected Value of a Random Variable

To model the insurance company's risk, we need to define a few terms. The amount the company pays out on an individual policy is an example of a **random variable**, called that because its value is based on the outcome of a random event. We use a capital letter, in this case, X, to denote a random variable. We'll denote a particular *value* that it can have by the corresponding lowercase letter, in this case, x. For the insurance company, x can be $100,000 (if you die that year), $50,000 (if you are disabled), or $0 (if neither occurs). Because we can list all the outcomes, we call this random variable a **discrete random variable**. A random variable that can take on any value between two values is called a **continuous random variable**. Continuous random variables are common in business applications for modelling physical quantities like heights and weights, and monetary quantities such as profits, revenues, and spending.

Sometimes it is obvious whether to treat a random variable as discrete or continuous, but at other times the choice is more subtle. Age, for example, might be viewed as discrete if it is measured only to the nearest decade with possible values 10, 20, 30, ... In a scientific context, however, it might be measured more precisely and treated as continuous.

For both discrete and continuous variables, the collection of all the possible values and the probabilities associated with them is called the **probability model** for the random variable. For a discrete random variable, we can list the probability of all possible values in a table, or describe it by a formula. For example, to model the possible outcomes of a fair die, we can let X be the number showing on the face. The probability model for X is simply:

$$P(X = x) = \begin{cases} 1/6 & \text{if } x = 1, 2, 3, 4, 5, \text{ or } 6 \\ 0 & \text{otherwise} \end{cases}$$

Suppose in our insurance risk example that the death rate in any year is 1 out of every 1000 people and that another 2 out of 1000 suffer some kind of disability. The loss, which we'll denote as X, is a discrete random variable because it takes on only 3 possible values. We can display the probability model for X in a table, as in Table 21.1.

Policyholder Outcome	Payout x (cost)	Probability $P(X = x)$
Death	100,000	$\frac{1}{1000}$
Disability	50,000	$\frac{2}{1000}$
Neither	0	$\frac{997}{1000}$

| **Table 21.1** *Probability model for an insurance policy.*

Of course, we can't predict exactly what *will* happen during any given year, but we can say what we *expect* to happen—in this case, what we expect the profit of a policy will be. The expected value of a policy is a parameter of the probability model. In fact, it's the mean. We'll signify this with the notation $E(X)$, for expected value (or sometimes EV, or sometimes μ). We use the term "mean" for this quantity just as we did for data, but be careful. This isn't an average of data values, so we won't estimate it. Instead, we calculate it directly from the probability model for the random variable. Because it comes from a model and not data, we use the parameter μ to denote it (and *not* $\bar{y}$ or $\bar{x}$.)

To see what the insurance company can expect, think about some (convenient) number of outcomes. For example, imagine that they have exactly 1000 clients and that the outcomes in one year followed the probability model exactly: 1 died, 2 were disabled, and 997 survived unscathed. Then our total payout would be:

$$E(X) = \frac{100,000(1) + 50,000(2) + 0(997)}{1000} = 200$$

So our total payout (cost) comes to $200 per policy.

Instead of writing the expected value as one big fraction, we can rewrite it as separate terms, each divided by 1000.

$$E(X) = \$100,000\left(\frac{1}{1000}\right) + \$50,000\left(\frac{2}{1000}\right) + \$0\left(\frac{997}{1000}\right)$$
$$= \$200$$

Writing it this way, we can see that for each policy, there's a 1/1000 chance that we'll have to pay $100,000 for a death and a 2/1000 chance that we'll have to pay $50,000 for a disability. Of course, there's a 997/1000 chance that we won't have to pay anything.

So the **expected value** of a (discrete) random variable is found by multiplying each possible value of the random variable by the probability that it occurs and then summing all those products. This gives the general formula for the expected value of a discrete random variable:[1]

$$E(X) = \Sigma x \cdot P(x).$$

Be sure that *every* possible outcome is included in the sum. Verify that you have a valid probability model to start with—the probabilities should each be between 0 and 1 and should sum to one. (Recall the rules of probability in Chapter 5.)

21.2 Standard Deviation of a Random Variable

Of course, this expected value (or mean) is not what actually happens to any *particular* policyholder. No individual policy actually costs the company $200. We are dealing with random events, so some policyholders receive big payouts and others nothing. Because the insurance company must anticipate this variability, it needs to know the standard deviation of the random variable.

For data, we calculate the standard deviation by first computing the deviation of each data value from the mean and squaring it. We perform a similar calculation when we compute the **standard deviation** of a (discrete) random variable as well. First, we find the deviation of each payout from the mean (expected value). (See Table 21.2.)

Policyholder Outcome	Payout x (cost)	Probability $P(X = x)$	Deviation $(x - EV)$
Death	100,000	$\dfrac{1}{1000}$	$(100{,}000 - 200) = 99{,}800$
Disability	50,000	$\dfrac{2}{1000}$	$(50{,}000 - 200) = 49{,}800$
Neither	0	$\dfrac{997}{1000}$	$(0 - 200) = -200$

| **Table 21.2** *Deviations between the expected value and each payout (cost).*

Next, we square each deviation. The **variance** is the expected value of those squared deviations. To find it, we multiply each by the appropriate probability and sum those products:

$$Var(X) = 99{,}800^2\left(\frac{1}{1000}\right) + 49{,}800^2\left(\frac{2}{1000}\right) + (-200)^2\left(\frac{997}{1000}\right)$$
$$= 14{,}960{,}000.$$

Finally, we take the square root to get the standard deviation:

$$SD(X) = \sqrt{14{,}960{,}000} \approx \$3{,}867.82$$

---

[1] The concept of expected values for continuous random variables is similar, but the calculation requires calculus and is beyond the scope of this text.

The insurance company can expect an average payout of $200 per policy, with a standard deviation of $3,867.82.

Think about that. The company charges $500 for each policy and expects to pay out $200 per policy. Sounds like an easy way to make $300. (In fact, most of the time—probability 997/1000—the company pockets the entire $500.) But would you be willing to take on this risk yourself and sell all your friends policies like this? The problem is that occasionally the company loses big. With a probability of 1/1000, it will pay out $100,000, and with a probability of 2/1000, it will pay out $50,000. That may be more risk than you're willing to take on. The standard deviation of $3,867.82 gives an indication of the uncertainty of the profit, and that seems like a pretty big spread (and risk) for an average profit of $300.

Here are the formulas for these arguments. Because these are parameters of our probability model, the variance and standard deviation can also be written as σ^2 and σ, respectively (sometimes with the name of the random variable as a subscript). You should recognize both kinds of notation:

$$\sigma^2 = Var(X) = \sum(x - EV)^2 P(x) = \sum(x - \mu)^2 P(x), \text{ and}$$
$$\sigma = SD(X) = \sqrt{Var(X)}.$$

GUIDED EXAMPLE | *Computer Inventory*

As the head of inventory for a computer company, you've had a challenging couple of weeks. One of your warehouses recently had a fire, and you had to flag all the computers stored there to be recycled. On the positive side, you were thrilled that you had managed to ship two computers to your biggest client last week. But then you discovered that your assistant hadn't heard about the fire and had mistakenly transported a whole truckload of computers from the damaged warehouse into the shipping center. It turns out that 30% of all the computers shipped last week were damaged.

You don't know whether your biggest client received two damaged computers, two undamaged ones, or one of each. Computers were selected at random from the shipping center for delivery.

If your client received two undamaged computers, everything is fine. If the client gets one damaged computer, it will be returned at your expense—$100—and you can replace it. However, if both computers are damaged, the client will cancel all other orders this month, and you'll lose $10,000. What is the expected value and the standard deviation of your loss under this scenario?

PLAN

| **Setup** State the problem. | We want to analyze the potential consequences of shipping damaged computers to a large client. We'll look at the expected value and standard deviation of the amount we'll lose.

Let X = amount of loss. We'll denote the receipt of an undamaged computer by **U** and the receipt of a damaged computer by **D.** The three possibilities are: two undamaged computers (**U** and **U**), two damaged computers (**D** and **D**), and one of each (**UD** or **DU**). Because the computers were selected randomly and the number in the warehouse is large, we can assume independence. |

DO

Model List the possible values of the random variable, and compute all the values you'll need to determine the probability model.

Because the events are independent, we can use the multiplication rule (Chapter 5) and find:

$$P(UU) = P(U) \times P(U)$$
$$= 0.7 \times 0.7 = 0.49$$
$$P(DD) = P(D) \times P(D)$$
$$= 0.3 \times 0.3 = 0.09$$

So, $P(UD \text{ or } DU) = 1 - (0.49 + 0.09) = 0.42$

We have the following model for all possible values of X.

Outcome	x	$P(X = x)$
Two damaged	10000	$P(DD) = 0.09$
One damaged	100	$P(UD \text{ or } DU) = 0.42$
Neither damaged	0	$P(UU) = 0.49$

Mechanics Find the expected value.

$$E(X) = 0(0.49) + 100(0.42) + 10000(0.09)$$
$$= \$942.00$$

Find the variance.

$$Var(X) = (0 - 942)^2 \times (0.49)$$
$$+ (100 - 942)^2 \times (0.42)$$
$$+ (10000 - 942)^2 \times (0.09)$$
$$= 8{,}116{,}836$$

Find the standard deviation.

$$SD(X) = \sqrt{8{,}116{,}836} = \$2849.01$$

REPORT

Conclusion Interpret your results in context.

MEMO:

Re: Damaged Computers

The recent shipment of two computers to our large client may have some serious negative impact. Even though there is about a 50% chance that they will receive two perfectly good computers, there is a 9% chance that they will receive two damaged computers and will cancel the rest of their monthly order. We have analyzed the expected loss to the firm as $942 with a standard deviation of $2,849.01. The large standard deviation reflects the fact that there is real possibility of losing $10,000 from the mistake.

Both numbers seem reasonable. The expected value of $942 is between the extremes of $0 and $10,000, and there's great variability in the outcome values.

REALITY CHECK

21.3 Properties of Expected Values and Variances

Our example insurance company expected to pay out an average of $200 per policy, with a standard deviation of about $3,868. The expected profit then was $500 − $200 = $300 per policy. Suppose that the company decides to lower the price of the premium by $50 to $450. It's pretty clear that the expected profit would drop an average of $50 per policy, to $450 − $200 = $250.

What about the standard deviation? We know that adding or subtracting a constant from data shifts the mean but doesn't change the variance or standard deviation. The same is true of random variables:[2]

$$E(X \pm c) = E(X) \pm c,$$
$$Var(X \pm c) = Var(X), \text{ and}$$
$$SD(X \pm c) = SD(X).$$

What if the company decides to *double* all the payouts—that is, pay $200,000 for death and $100,000 for disability? This would double the average payout per policy and also increase the variability in payouts. In general, multiplying each value of a random variable by a constant multiplies the mean by that constant and multiplies the variance by the *square* of the constant:

$$E(aX) = aE(X), \text{ and}$$
$$Var(aX) = a^2 Var(X).$$

Taking square roots of the last equation shows that the standard deviation is multiplied by the absolute value of the constant:

$$SD(aX) = |a|SD(X).$$

This insurance company sells policies to more than just one person. We've just seen how to compute means and variances for one person at a time. What happens to the mean and variance when we have a collection of customers? The profit on a group of customers is the *sum* of the individual profits, so we'll need to know how to find expected values and variances for sums. To start, consider a simple case with just two customers who we'll call Mr. Ecks and Ms. Wye. With an expected payout of $200 on each policy, we might expect a total of $200 + $200 = $400 to be paid out on the two policies—nothing surprising there. In other words, we have the **Addition Rule for Expected Values of Random Variables**: *The expected value of the sum (or difference) of random variables is the sum (or difference) of their expected values:*

$$E(X \pm Y) = E(X) \pm E(Y).$$

The variability is another matter. Is the risk of insuring two people the same as the risk of insuring one person for twice as much? We wouldn't expect both clients to die or become disabled in the same year. In fact, because we've spread the risk, the standard deviation should be smaller. Indeed, this is the fundamental principle behind insurance. By spreading the risk among many policies, a company can keep the standard deviation quite small and predict costs more accurately. It's much less risky to insure thousands of customers than one customer when the total expected payout is the same, assuming that the events are independent. Catastrophic events such as hurricanes or earthquakes that affect large numbers of customers at the same time destroy the independence assumption, and often the insurance company along with it.

But how much smaller is the standard deviation of the sum? It turns out that, if the random variables are independent, we have the **Addition Rule for**

---

[2] The rules in this section are true for both discrete *and* continuous random variables.

Variances of (independent) Random Variables: *The variance of the sum or difference of two independent random variables is the sum of their individual variances:*

$$Var(X \pm Y) = Var(X) + Var(Y)$$

if X and Y are independent.

MATH BOX

Pythagorean Theorem of Statistics

We often use the standard deviation to measure variability, but when we add independent random variables, we use their variances. Think of the Pythagorean Theorem. In a right triangle (only), the *square* of the length of the hypotenuse is the sum of the *squares* of the lengths of the other two sides:

$$c^2 = a^2 + b^2.$$

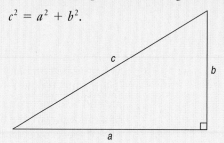

For independent random variables (only), the *square* of the standard deviation of their sum is the sum of the *squares* of their standard deviations:

$$SD^2(X + Y) = SD^2(X) + SD^2(Y).$$

It's simpler to write this with *variances*:

$$Var(X + Y) = Var(X) + Var(Y),$$

but we'll use the standard deviation formula often as well:

$$SD(X + Y) = \sqrt{Var(X) + Var(Y)}.$$

For Mr. Ecks and Ms. Wye, the insurance company can expect their outcomes to be independent, so (using X for Mr. Ecks's payout and Y for Ms. Wye's):

$$Var(X + Y) = Var(X) + Var(Y)$$
$$= 14{,}960{,}000 + 14{,}960{,}000$$
$$= 29{,}920{,}000.$$

Let's compare the variance of writing two independent policies to the variance of writing only one for twice the size. If the company had insured only Mr. Ecks for twice as much, the variance would have been

$$Var(2X) = 2^2 Var(X) = 4 \times 14{,}960{,}000 = 59{,}840{,}000, \text{or}$$

twice as big as with two independent policies, even though the expected payout is the same.

Of course, variances are in squared units. The company would prefer to know standard deviations, which are in dollars. The standard deviation of the payout for two independent policies is $SD(X + Y) = \sqrt{Var(X + Y)} = \sqrt{29{,}920{,}000} = \$5{,}469.92$. But the standard deviation of the payout for a single policy of twice the size is twice the standard deviation of a single policy: $SD(2X) = 2SD(X) = 2(3867.82) = 7{,}735.64$, or about 40% more than the standard deviation of the sum of the two independent policies, \$5469.92.

If the company has two customers, then it will have an expected annual total payout (cost) of $400 with a standard deviation of about $5470. If they write one policy with an expected annual payout of $400, they increase the standard deviation by about 40%. Spreading risk by insuring many independent customers is one of the fundamental principles in insurance and finance.

Let's review the rules of expected values and variances for sums and differences.

◆ *The expected value of the sum of two random variables is the sum of the expected values.*

◆ *The expected value of the difference of two random variables is the difference of the expected values:*

$$E(X \pm Y) = E(X) \pm E(Y).$$

◆ *If the random variables are independent, the variance of their sum or difference is always the sum of the variances:*

$$Var(X \pm Y) = Var(X) + Var(Y).$$

Do we always *add* variances? Even when we take the *difference* of two random quantities? Yes! Think about the two insurance policies. Suppose we want to know the mean and standard deviation of the *difference* in payouts to the two clients. Since each policy has an expected payout of $200, the expected difference is $200 − $200 = $0. If we computed the variance of the difference by subtracting variances, we would get $0 for the variance. But that doesn't make sense. Their difference won't always be exactly $0. In fact, the difference in payouts could range from $100,000 to −$100,000, a spread of $200,000. The variability in differences *increases* as much as the variability in sums. If the company has two customers, the difference in payouts has a mean of $0 and a standard deviation of about $5,470.

◆ **For random variables, does X + X + X = 3X?**   Maybe, but be careful. As we've just seen, insuring one person for $300,000 is not the same risk as insuring three people for $100,000 each. When each instance represents a different outcome for the same random variable, though, it's easy to fall into the trap of writing all of them with the same symbol. Don't make this common mistake. Make sure you write each instance as a *different* random variable. Just because each random variable describes a similar situation doesn't mean that each random outcome will be the same. What you really mean is $X_1 + X_2 + X_3$. Written this way, it's clear that the sum shouldn't necessarily equal 3 times *anything*.

JUST CHECKING

1 Suppose that the time it takes a customer to get and pay for seats at the ticket window of a baseball park is a random variable with a mean of 100 seconds and a standard deviation of 50 seconds. When you get there, you find only two people in line in front of you.

   a) How long do you expect to wait for your turn to get tickets?
   b) What's the standard deviation of your wait time?
   c) What assumption did you make about the two customers in finding the standard deviation?

21.4 Discrete Probability Models

We've seen how to compute means and standard deviations of random variables. But plans based just on averages are, on average, wrong. At least that's what Sam Savage, Professor at Stanford University says in his book, *The Flaw of Averages*.

Unfortunately, many business owners make decisions based solely on averages—the average amount sold last year, the average number of customers seen last month, etc. Instead of relying on averages, the business decision maker can incorporate much more by modeling the situation with a probability model. Probability models can play an important and pivotal role in helping decision makers better predict both the outcome and the consequences of their decisions. In this section we'll see that some fairly simple models provide a framework for thinking about how to model a wide variety of business phenomena.

The Uniform Model

When we first studied probability in Chapter 5, we saw that equally likely events were the simplest case. For example, a single die can turn up 1, 2, . . . , 6 on one toss. A probability model for the toss is Uniform because each of the outcomes has the same probability (1/6) of occurring. Similarly if X is a random variable with possible outcomes 1, 2, . . . , n and $P(X = i) = 1/n$ for each value of i, then we say X has a **discrete Uniform distribution, U[1, . . . , n]**.

Bernoulli Trials

In September, 2008, Google Inc. announced the release of their web browser *Chrome*, designed to compete with Microsoft's *Internet Explorer*, Apple's *Safari*, and others. One of the goals of *Chrome* was to insulate the browser from websites that fail to display. The developers of *Chrome* worked hard to minimize the probability that their browser will have trouble displaying a website. Before releasing the product, they had to test many websites to discover those that might fail. Although web browsers are relatively new, *quality control inspection* such as this is common throughout manufacturing worldwide and has been in use in industry for nearly 100 years.

The developers of *Chrome* sampled websites, recording whether the browser displayed the website correctly or had a problem. We call the act of inspecting a website a trial. There are two possible outcomes—either the website renders correctly or it doesn't. By convention, one of the outcomes is denoted a "success" and the other a "failure." Which one is called a success is arbitrary, but often the less common outcome or the one that calls for action is called a success. Here, that would mean that a website that *doesn't* work would be a "success." (That may seem strange, but if you're a quality inspector, you'll want to find problems so you can fix them.) At least early in this work, the probability of a success didn't change from trial to trial. Situations like this occur often and are called **Bernoulli trials**. To summarize, trials are Bernoulli if:

◆ There are only two possible outcomes (called *success* and *failure*) for each **trial**. For example, either you get a website that fails to display correctly (success), or you don't (failure).

◆ The probability of success, denoted p, is the same on every trial. (The probability of failure, $1 - p$ is often denoted q.)

◆ The trials are independent. Finding that one website does not display correctly does not change what might happen with the next website.

Common examples of Bernoulli trials include tossing a coin, collecting responses on Yes/No questions from surveys or even shooting free throws in a basketball game. Bernoulli trials are remarkably versatile and can be used to model a wide variety of real life situations. The specific question you might ask in different situations will give rise to different random variables that, in turn, have different probability models.

The Geometric Model

What's the probability that the first website that fails to display is the second one that we test? Let X denote the number of trials (websites) until the first such "success." For X to be 2, the first website must have displayed correctly (which has probability $1 - p$), and then the second one must have failed to display correctly—a success, with probability p. Since the trials are independent, these probabilities can be multiplied, and so $P(X = 2) = (1 - p)(p)$ or qp. Maybe you won't find a success until the fifth trial. What are the chances of that? You'd have to fail 4 times in a row and then succeed, so $P(X = 5) = (1 - p)^4(p) = q^4p$. See the Math Box for an extension and more explanation.

Whenever we want to know how long (how many trials) it will take us to achieve the first success, the model that tells us this probability is called the **geometric probability model**. Geometric models are completely specified by one parameter, p, the probability of success, and are denoted Geom(p).

The geometric model can tell Google something important about its software. No large complex program is entirely free of bugs. So before releasing a program or upgrade, developers typically ask not whether it is free of bugs, but how long it is likely to be until the next bug is discovered. If the expected number of pages displayed until the next failure is high enough, then the program is ready to ship.

Daniel Bernoulli (1700–1782) was the nephew of Jakob, whom you saw in Chapter 5. He was the first to work out the mathematics for what we now call Bernoulli trials.

! NOTATION ALERT:

Now we have two more reserved letters. Whenever we deal with Bernoulli trials, p represents the probability of success, and q represents the probability of failure. (Of course, $q = 1 - p$.)

Geometric probability model for Bernoulli trials: Geom(p)

p = probability of success (and $q = 1 - p$ = probability of failure)
X = number of trials until the first success occurs

$$P(X = x) = q^{x-1}p$$

Expected value: $\mu = \dfrac{1}{p}$

Standard deviation: $\sigma = \sqrt{\dfrac{q}{p^2}}$

MATH BOX

We want to find the mean (expected value) of random variable X, using a geometric method with probability of success p.

First write the probabilities:

x	1	2	3	4	$\ldots$
$P(X = x)$	p	qp	q^2p	q^3p	$\ldots$

The expected value is: $E(X) = 1p + 2qp + 3q^2p + 4q^3p + \cdots$

Let $p = 1 - q$: $\qquad = (1 - q) + 2q(1 - q) + 3q^2(1 - q) + 4q^3(1 - q) + \cdots$

Simplify: $\qquad = 1 - q + 2q - 2q^2 + 3q^2 - 3q^3 + 4q^3 - 4q^4 + \cdots$

That's an infinite geometric series, with first term 1 and common ratio q:

$\qquad = 1 + q + q^2 + q^3 + \cdots$

$\qquad = \dfrac{1}{1 - q}$

So, finally $\qquad E(X) = \dfrac{1}{p}.$

Independence

One of the important requirements for Bernoulli trials is that the trials be independent. Sometimes that's a reasonable assumption. Is it true for our example? It's easy to imagine that related sites might have similar problems, but if the sites are selected at random, whether one has a problem should be independent of others.

The 10% Condition: Bernoulli trials must be independent. In theory, we need to sample from a population that's infinitely big. However, if the population is finite, it's still okay to proceed as long as the sample is smaller than 10% of the population. In Google's case, they just happened to have a directory of millions of websites, so most samples would easily satisfy the 10% condition.

The Binomial Model

Suppose Google tests 5 websites. What's the probability that *exactly* 2 of them have problems (2 "successes")? When we studied the geometric model, we asked how long it would take until our first success. Now we want to find the probability of getting exactly 2 successes among the 5 trials. We are still talking about Bernoulli trials, but we're asking a different question.

This time we're interested in the *number of successes* in the 5 trials, which we'll denote by Y. We want to find $P(Y = 2)$. Whenever the random variable of interest is the number of successes in a series of Bernoulli trials, it's called a **Binomial random variable**. It takes two parameters to define this **Binomial probability model**: the number of trials, n, and the probability of success, p. We denote this model $Binom(n, p)$.

Suppose that in this phase of development, 10% of the sites exhibited some sort of problem so that $p = 0.10$. (Early in the development phase of a product, it is not uncommon for the number of defects to be much higher than it is when the product is released.) Exactly 2 successes in 5 trials means 2 successes and 3 failures. It seems logical that the probability should be $(p)^2(1 - p)^3$. Unfortunately, it's not *quite* that easy. That calculation would give you the probability of finding two successes and then three failures—*in that order*. But you could find the two successes in a lot of other ways, for example in the 2nd and 4th website you test. The probability of that sequence is $(1 - p)p(1 - p)(p)(1 - p)$ which is also $p^2(1 - p)^3$. In fact, as long as there are two successes and three failures, the probability will always be the same, regardless of the order of the sequence of successes and failures. The probability will be $(p)^2(1 - p)^3$. To find the probability of getting 2 successes in 5 trials in any order, we just need to know how many ways that outcome can occur.

Fortunately, all the possible sequences that lead to the same number of successes are *disjoint*. (For example, if your successes came on the first two trials, they couldn't come on the last two.) So once we find all the different sequences, we can add up their probabilities. And since the probabilities are all the same, we just need to find how many sequences there are and multiply $(p)^2(1 - p)^3$ by that number.

Each different order in which we can have k successes in n trials is called a "combination." The total number of ways this can happen is written $\binom{n}{k}$ or $_nC_k$ and pronounced "n choose k:"

$$\binom{n}{k} = {}_nC_k = \frac{n!}{k!(n - k)!} \text{ where } n! = n \times (n - 1) \times \cdots \times 1.$$

For 2 successes in 5 trials,

$$\binom{5}{2} = \frac{5!}{2!(5 - 2)!} = \frac{(5 \times 4 \times 3 \times 2 \times 1)}{(2 \times 1 \times 3 \times 2 \times 1)} = \frac{(5 \times 4)}{(2 \times 1)} = 10.$$

So there are 10 ways to get 2 successes in 5 websites, and the probability of each is $(p)^2(1 - p)^3$. To find the probability of exactly 2 successes in 5 trials, we multiply the probability of any particular order by this number:

$$P(exactly\ 2\ successes\ in\ 5\ trials) = 10p^2(1 - p)^3 = 10(0.10)^2(0.90)^3 = 0.0729$$

In general, we can write the probability of exactly k successes in n trials as

$$P(Y = k) = \binom{n}{k}p^k q^{n-k}.$$

If the probability that any single website has a display problem is 0.10, what's the expected number of websites with problems if we test 100 sites? You probably said 10. We suspect you didn't use the formula for expected value that involves multiplying each value times its probability and adding them up. In fact, there is an easier way to find the expected value for a Binomial random variable. You just multiply the probability of success by n. In other words, $E(Y) = np$. We prove this in the next Math Box.

The standard deviation is less obvious and you can't just rely on your intuition. Fortunately, the formula for the standard deviation also boils down to something simple: $SD(Y) = \sqrt{npq}$. If you're curious to know where that comes from, it's in the Math Box, too.

In our website example, with $n = 100$, $E(Y) = np = 100(0.10) = 10$ so we expect to find 10 successes out of the 100 trials. The standard deviation is

$$\sqrt{100 \times 0.10 \times 0.90} = 3\ websites.$$

To summarize, a Binomial probability model describes the distribution of the number of successes in a specified number of trials.

Binomial model for Bernoulli trials: Binom(n, p)

n = number of trials
p = probability of success (and $q = 1 - p$ = probability of failure)
X = number of successes in n trials

$$P(X = x) = \binom{n}{x} p^x q^{n-x}, where\ \binom{n}{x} = \frac{n!}{x!(n - x)!}$$

Mean: $\mu = np$
Standard deviation: $\sigma = \sqrt{npq}$

MATH BOX

To derive the formulas for the mean and standard deviation of the Binomial model, we start with the most basic situation.

Consider a single Bernoulli trial with probability of success p. Let's find the mean and variance of the number of successes.

Here's the probability model for the number of success:

x	0	1
$P(X = x)$	q	p

Find the expected value:

$$E(X) = 0q + 1p$$
$$E(X) = p$$

Now the variance:

$$Var(X) = (0 - p)^2 q + (1 - p)^2 p$$
$$= p^2 q + q^2 p$$
$$= pq(p + q)$$
$$= pq(1)$$
$$Var(X) = pq$$

What happens when there is more than one trial? A Binomial model simply counts the number of successes in a series of n independent Bernoulli trials. That makes it easy to find the mean and standard deviation of a binomial random variable, Y.

$$\text{Let } Y = X_1 + X_2 + X_3 + \cdots + X_n$$
$$E(Y) = E(X_1 + X_2 + X_3 + \cdots + X_n)$$
$$= E(X_1) + E(X_2) + E(X_3) + \cdots + E(X_n)$$
$$= p + p + p + \cdots + p \text{ (There are } n \text{ terms.)}$$

So, as we thought, the mean is $E(Y) = np$.

And since the trials are independent, the variances add:

$$Var(Y) = Var(X_1 + X_2 + X_3 + \cdots + X_n)$$
$$= Var(X_1) + Var(X_2) + Var(X_3) + \cdots + Var(X_n)$$
$$= pq + pq + pq + \cdots + pq \text{ (Again, } n \text{ terms.)}$$
$$Var(Y) = npq$$

Voila! The standard deviation is $SD(Y) = \sqrt{npq}$.

GUIDED EXAMPLE   *The American Red Cross*

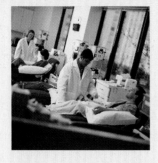

Every two seconds someone in America needs blood.

The American Red Cross is a nonprofit organization that runs like a large business. It serves over 3,000 hospitals around the United States, providing a wide range of high quality blood products and blood donor and patient testing services. It collects blood from over 4 million donors and provides blood to millions of patients with a dedication to meeting customer needs.[3]

The balancing of supply and demand is complicated not only by the logistics of finding donors that meet health criteria, but by the fact that the blood type of donor and patient must be matched. People with O-negative blood are called "universal donors" because O-negative blood can be given to patients with any blood type. Only about 6% of people have O-negative blood, which presents a challenge in managing and planning. This is especially true, since, unlike a manufacturer who can balance supply by planning to produce or to purchase more or less of a key item, the Red Cross gets its supply from volunteer donors who show up more-or-less at random (at least in terms of blood type). Modeling the arrival of samples with various blood types helps Red Cross managers to plan their blood allocations.

Here's a small example of the kind of planning required. In the next 20 donors to arrive at a blood donation center, how many universal donors can be expected? Specifically, what are the mean and standard deviation of the number of universal donors? What is the probability that there are 2 or 3 universal donors?

---

[3] Source: www.redcross.org

Question 1: What are the mean and standard deviation of the number of universal donors?

Question 2: What is the probability that there are exactly 2 or 3 universal donors out of the 20 donors?

PLAN

Setup State the question.

Check to see that these are Bernoulli trials.

Variable Define the random variable.

Model Specify the model.

We want to know the mean and standard deviation of the number of universal donors among 20 people and the probability that there are 2 or 3 of them.

✓ There are two outcomes:
success = O-negative
failure = other blood types

✓ $p = 0.06$

✓ **10% Condition:** Fewer than 10% of all possible donors have shown up.

Let X = number of O-negative donors among $n = 20$ people.

We can model X with a Binom(20, 0.06) model.

DO

Mechanics Find the expected value and standard deviation.

Calculate the probability of 2 or 3 successes.

$E(X) = np = 20(0.06) = 1.2$

$SD(X) = \sqrt{npq} = \sqrt{20(0.06)(0.94)} \approx 1.06$

$P(X = 2 \text{ or } 3) = P(X = 2) + P(X = 3)$

$$= \binom{20}{2}(0.06)^2(0.94)^{18}$$

$$+ \binom{20}{3}(0.06)^3(0.94)^{17}$$

$$\approx 0.2246 + 0.0860$$

$$= 0.3106$$

REPORT

Conclusion Interpret your results in context.

MEMO:

Re: Blood Drive

In groups of 20 randomly selected blood donors, we'd expect to find an average of 1.2 universal donors, with a standard deviation of 1.06. About 31% of the time, we'd expect to find exactly 2 or 3 universal donors among the 20 people.

Simeon Denis Poisson was a French mathematician interested in rare events. He originally derived his model to approximate the Binomial model when the probability of a success, *p*, is very small and the number of trials, *n*, is very large. Poisson's contribution was providing a simple approximation to find that probability. When you see the formula, however, you won't necessarily see the connection to the Binomial.

W. S. Gosset, the quality control chemist at the Guinness brewery in the early 20th century who developed the methods of Chapter 12, was one of the first to use the Poisson in industry. He used it to model and predict the number of yeast cells so he'd know how much to add to the stock. The Poisson is a good model to consider whenever your data consist of counts of occurrences. It requires only that the events be independent and that the mean number of occurrences stays constant.

The Poisson Model

Not all discrete events can be modeled as Bernoulli trials. Sometimes we're interested simply in the number of events that occur over a given interval of time or space. For example, we might want to model the number of customers arriving in our store in the next ten minutes, the number of visitors to our website in the next minute, or the number of defects that occur in a computer monitor of a certain size. In cases like these, the number of occurrences can be modeled by a **Poisson random variable.** The Poisson's parameter, the mean of the distribution, is usually denoted by λ.

For example, data show an average of about 4 hits per minute to a small business website during the afternoon hours from 1:00 to 5:00 P.M. We can use the Poisson model to find the probability that any

---

Poisson probability model for occurrences: Poisson (λ)

λ = mean number of occurrences.
X = number of occurrences.

$$P(X = x) = \frac{e^{-\lambda}\lambda^x}{x!}$$

Expected value: $E(X) = \lambda$

Standard deviation: $SD(X) = \sqrt{\lambda}$

---

number of hits will arrive. For example, if we let X be the number of hits arriving in the next minute, then $P(X = x) = \dfrac{e^{-\lambda}\lambda^x}{x!} = \dfrac{e^{-4}4^x}{x!}$, using the given average rate of 4 per minute. So, the probability of no hits during the next minute would be $P(X = 0) = \dfrac{e^{-4}4^0}{0!} = e^{-4} = 0.0183$ (recall that $e \approx 2.71828$).

One interesting and useful feature of the Poisson model is that it scales according to the interval size. For example, suppose we want to know the probability of no hits to our website in the next 30 seconds. Since the mean rate is 4 hits per minute, it's 2 hits per 30 seconds, so we can use the model with $\lambda = 2$ instead. If we let Y be the number of hits arriving in the next 30 seconds, then:

$$P(Y = 0) = \frac{e^{-2}2^0}{0!} = e^{-2} = 0.1353.$$

The Poisson model has been used to model phenomenon such as customer arrivals, hot streaks in sports, and disease clusters.

Whenever or wherever rare events happen closely together, people want to know whether the occurrence happened by chance or whether an underlying change caused the unusual occurrence. The Poisson model can be used to find the probability of the occurrence and can be the basis for making the judgment.

The Poisson distribution was the model used in the famous 1982 Woburn toxic waste trial when eight families from Woburn, Massachusetts, sued W. R. Grace & Company, alleging that the company contaminated the public water supplies by dumping toxic materials near city wells. The families argued that eight recent cases of leukemia were the result of the company's actions. The resulting trial was the basis for the book and movie *A Civil Action*. For the Woburn case, the probability (based on national averages) for eight leukemia cases in a town of that size in the given time period was determined to be about 0.04. That's a small chance, but not strikingly unusual.

21.5 Continuous Random Variables

Discrete random variables are great for modeling occurrences, categories, or small counts. But in industry we often measure quantities that a discrete variable just can't handle. For example, the time until a computer battery needs to be charged might take on any value between two and four hours.

When a random variable can take on any value in an interval, we can no longer model it using a discrete probability model and must use a continuous probability model instead. For any continuous random variable, the distribution of its probability can be shown with a curve. That curve is called the **probability density function (pdf)**, usually denoted as $f(x)$. You've seen the Normal or bell-shaped curve. Technically, that is known as the Normal probability density function.

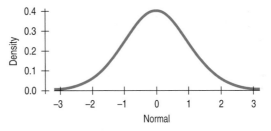

Figure 21.1 *The standard Normal density function (a normal with mean 0 and standard deviation 1). The probability of finding a z-score in any interval is simply the area over that interval under the curve. For example, the probability that the z-score falls between −1 and 1 is about 68%, which can be seen from the density function or found more precisely from a table or technology.*

Density functions must satisfy two requirements. They must stay nonnegative for every possible value, and the total area under the curve must be exactly 1.0. This last requirement corresponds to the Probability Assignment Rule of Chapter 5, which said that the total probability (equal to 1.0) must be assigned somewhere.

Any density function can give the probability that the random variable lies in an interval. But remember, the probability that X lies in the interval from a to b is the *area* under the density function, $f(x)$, between the values a and b and not the value $f(a)$ or $f(b)$. In general, finding that area requires calculus or numerical analysis, and is beyond the scope of this text. But for the models we'll discuss, the probabilities are found either from tables (the Normal and Exponential) or simple computations (Uniform).

There are many (in fact, there are an infinite number of) possible continuous distributions, but we'll explore only three of the most commonly used to model business phenomena: the **Uniform model**, the **Normal model**, and the **Exponential model**.

The Uniform Model

We've already seen the discrete version of the uniform model. A continuous uniform shares the principle that all events should be equally likely, but with a continuous model, we can't talk about the probability of a particular value because each value has probability zero. Instead, for a continuous random variable X, we say that the probability that X lies in any interval depends only on the length of that interval. Not surprisingly the density function of a continuous uniform random variable looks flat (see Figure 21.2).

The density function of a continuous uniform random variable defined on the interval a to b can be defined by the formula (see also Figure 21.2).

$$f(x) = \begin{cases} \dfrac{1}{b-a} & \text{if} \quad a \le x \le b \\ 0 & \text{otherwise} \end{cases}$$

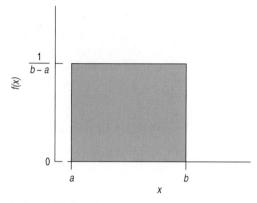

Figure 21.2 *The density function of a continuous uniform random variable on the interval from a to b.*

From Figure 21.2, it's easy to see that the probability that X lies in any interval between a and b is the same as any other interval of the same length. In fact, the probability is just the ratio of the length of the interval to the total length: $b - a$. In other words:

For values c and d ($c \le d$) both within the interval [a,b]:

$$P(c \le X \le d) = \frac{(d-c)}{(b-a)}$$

As an example, suppose you arrive at a bus stop and want to model how long you'll wait for the next bus. The sign says that busses arrive about every 20 minutes, but no other information is given. You might assume that the arrival is equally likely to be anywhere in the next 20 minutes, and so the density function would be

$$f(x) = \begin{cases} \dfrac{1}{20} & \text{if} \quad 0 \le x \le 20 \\ 0 & \text{otherwise} \end{cases}$$

and would look as shown in Figure 21.3.

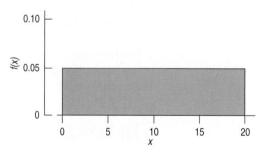

Figure 21.3 *The density function of a continuous uniform random variable on the interval [0,20]. Notice that the mean (the balancing point) of the distribution is at 10 minutes.*

Just as the mean of a data distribution is the balancing point of a histogram, the mean of any continuous random variable is the balancing point of the density function. Looking at Figure 21.3, we can see that the balancing point is halfway between the end points at 10 minutes. In general, the expected value is:

$$E(X) = \frac{a + b}{2}$$

for a uniform distribution on the interval (a, b). With $a = 0$ and $b = 20$ the expected value would be 10 minutes.

The variance and standard deviation are less intuitive:

$$Var(X) = \frac{(b - a)^2}{12}; \ SD(X) = \sqrt{\frac{(b - a)^2}{12}}.$$

Using these formulas, our bus wait will have an expected value of 10 minutes with a standard deviation of $\sqrt{\frac{(20 - 0)^2}{12}} = 5.77$ minutes.

How can every value have probability 0?

At first it may seem illogical that every value of a continuous random variable has probability 0. Let's look at the standard Normal random variable, Z. The Normal model is a continuous model that you're already familiar with from Chapter 9. We could find (from a table, website, or computer program) that the probability that Z lies between 0 and 1 is:

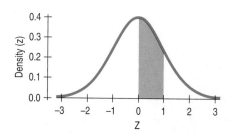

0.3413, which is the area under the Normal pdf (in red) between the values 0 and 1.

So, what's the probability that Z is between 0 and 1/10?

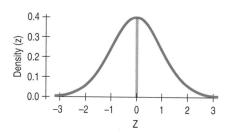

That area is only 0.0398. What is the chance then that Z will fall between 0 and 1/100? There's not much area—the probability is only 0.0040. If we kept going, the probability would keep getting smaller. The probability that Z is between 0 and 1/100,000 is less than 0.0001.

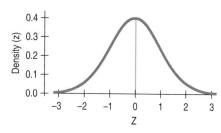

So, what's the probability that Z is *exactly* 0? Well, there's *no* area under the curve right at $x = 0$, so the probability is 0. It's only intervals that have positive probability, but that's OK. In real life we never mean exactly 0.0000000000 or any other value. If you say "exactly 164 pounds," you might really mean between 163.5 and 164.5 pounds or even between 163.99 and 164.01 pounds, but realistically not 164.000000000 . . . pounds.

The Normal Model

Normal models show up throughout Statistics and are often referred to more generally. When you hear of a "bell-shaped curve," the formal definition is usually a Normal model. We find probabilities related to Normal models using tables (such as Table Z in the back of the book) or technology. (Chapter 9 discusses how to do this specifically.)

Normal models show up so often because they have some special properties. An important one is that the sum or difference of two independent Normal random variables is also Normal.

A company manufactures small stereo systems. At the end of the production line, the stereos are packaged and prepared for shipping. Stage 1 of this process is called "packing." Workers must collect all the system components (a main unit, two speakers, a power cord, an antenna, and some wires), put each in plastic bags, and then place everything inside a protective form. The packed form then moves on to Stage 2, called "boxing," in which workers place the form and a packet of instructions in a cardboard box and then close, seal, and label the box for shipping.

Because the times required for packing and boxing can take on any value, they must be modeled by a continuous random variable. In particular, the company says that times required for the packing stage are unimodal and symmetric and can be described by a Normal model with a mean of 9 minutes and standard deviation of 1.5 minutes. (See Figure 21.4.) The times for the boxing stage can also be modeled as Normal, with a mean of 6 minutes and standard deviation of 1 minute.

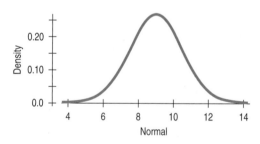

Figure 21.4 *The Normal model for the packing stage with a mean of 9 minutes and standard deviation of 1.5 minutes.*

The company is interested in the total time that it takes to get a system through both packing and boxing, so they want to model the sum of the two random variables. Fortunately, the special property that adding independent Normals yields another Normal model allows us to apply our knowledge of Normal probabilities to questions about the sum or difference of independent random variables. To use this property of Normals, we'll need to check the Independence Assumption as well as the Normal Model Assumption for each variable.

GUIDED EXAMPLE *Packaging Stereos*

Consider the company that manufactures and ships small stereo systems that we discussed previously.

If the time required to pack the stereos can be described by a Normal model, with a mean of 9 minutes and standard deviation of 1.5 minutes, and the times for the boxing stage can also be modeled as Normal, with a mean of 6 minutes and standard deviation of 1 minute, what is the probability that packing an order of two systems takes over 20 minutes? What percentage of the stereo systems takes longer to pack than to box?

Question 1: What is the probability that packing an order of two systems takes over 20 minutes?

 **PLAN**

Setup State the problem.

We want to estimate the probability that packing an order of two systems takes more than 20 minutes.

Variables Define your random variables.

Let P_1 = time for packing the first system
P_2 = time for packing the second system
T = total time to pack two systems
$T = P_1 + P_2$

Write an appropriate equation for the variables you need.

Think about the model assumptions.

✓ **Normal Model Assumption.** We are told that packing times are well modeled by a Normal model, and we know that the sum of two Normal random variables is also Normal.

✓ **Independence Assumption.** There is no reason to think that the packing time for one system would affect the packing time for the next, so we can reasonably assume the two are independent.

 DO

Mechanics Find the expected value. (Expected values always add.)

$$E(T) = E(P_1 + P_2)$$
$$= E(P_1) + E(P_2)$$
$$= 9 + 9 = 18 \text{ minutes}$$

Find the variance.

For sums of independent random variables, variances add. (In general, we don't need the variables to be Normal for this to be true—just independent.)

Since the times are independent,

$$Var(T) = Var(P_1 + P_2)$$
$$= Var(P_1) + Var(P_2)$$
$$= 1.5^2 + 1.5^2$$
$$Var(T) = 4.50$$

Find the standard deviation.

$$SD(T) = \sqrt{4.50} \approx 2.12 \text{ minutes}$$

Now we use the fact that both random variables follow Normal models to say that their sum is also Normal.

Sketch a picture of the Normal model for the total time, shading the region representing over 20 minutes.

We can model the time, *T*, with a *N* (18, 2.12) model.

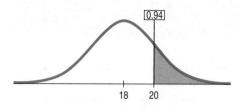

Find the z-score for 20 minutes.

$$z = \frac{20 - 18}{2.12} = 0.94$$

Use technology or a table to find the probability.

$$P(T > 20) = P(z > 0.94) = 0.1736$$

REPORT

Conclusion Interpret your result in context.

MEMO:

Re: Computer Systems Packing

Using past history to build a model, we find slightly more than a 17% chance that it will take more than 20 minutes to pack an order of two stereo systems.

Question 2: What percentage of stereo systems take longer to pack than to box?

PLAN

Setup State the question.

We want to estimate the percentage of the stereo systems that takes longer to pack than to box.

Variables Define your random variables.

Let *P* = time for packing a system
  B = time for boxing a system
  D = difference in times to pack and box a system

Write an appropriate equation.

D = *P* − *B*

What are we trying to find? Notice that we can tell which of two quantities is greater by subtracting and asking whether the difference is positive or negative.

A system that takes longer to pack than to box will have *P* > *B*, and so *D* will be positive. We want to find *P*(*D* > 0).

✓ **Normal Model Assumption.** We are told that both random variables are well modeled by Normal models, and we know that the difference of two Normal random variables is also Normal.

Don't forget to think about the assumptions.

✓ **Independence Assumption.** There is no reason to think that the packing time for a system will affect its boxing time, so we can reasonably assume the two are independent.

DO

Mechanics Find the expected value.	$E(D) = E(P - B)$ $\quad = E(P) - E(B)$ $\quad = 9 - 6 = 3 \text{ minutes}$
For the difference of independent random variables, the variance is the sum of the individual variances.	Since the times are independent, $\quad Var(D) = Var(P - B)$ $\quad\quad = Var(P) + Var(B)$ $\quad\quad = 1.5^2 + 1^2$ $Var(D) = 3.25$ $SD(D) = \sqrt{3.25} \approx 1.80 \text{ minutes}$
Find the standard deviation. State what model you will use. Sketch a picture of the Normal model for the difference in times and shade the region representing a difference greater than zero.	We can model D with a N (3, 1.80) model. 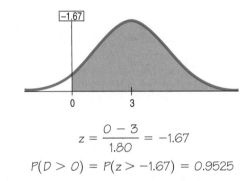
Find the z-score. Then use a table or technology to find the probability.	$z = \dfrac{0 - 3}{1.80} = -1.67$ $P(D > 0) = P(z > -1.67) = 0.9525$

REPORT

Conclusion Interpret your result in context.	MEMO: **Re: Computer Systems Packing** In our second analysis, we found that just over 95% of all the stereo systems will require more time for packing than for boxing.

The Normal Model for the Binomial

Even though the Normal is a continuous model, it is often used as an approximation for discrete events when the number of possible events is large. In particular, it is a good model for sums of independent random variables of which a Binomial random variable is a special case. Here's an example of how the Normal can be used to calculate binomial probabilities. Suppose that the Tennessee Red Cross anticipates the need for at least 1850 units of O-negative blood this year. It estimates that it will collect blood from 32,000 donors. How likely is the Tennessee Red Cross to meet its need? We've just learned how to calculate such probabilities. We could use the binomial model with $n = 32{,}000$ and $p = 0.06$. The probability of getting *exactly* 1850 units of O-negative blood from 32,000 donors is $\binom{32000}{1850} \times 0.06^{1850} \times 0.94^{30150}$. No calculator on earth can calculate that first term (it has more than 100,000 digits).[4] And that's just the beginning. The problem

---

[4] If your calculator *can* find Binom(32000, 0.06), then apparently it's smart enough to use an approximation.

said *at least* 1850, so we would have to do it again for 1851, for 1852, and all the way up to 32,000. (No thanks.)

When we're dealing with a large number of trials like this, making direct calculations of the probabilities becomes tedious (or outright impossible). But the Normal model can come to the rescue.

The Binomial model has mean $np = 1920$ and standard deviation $\sqrt{npq} \approx 42.48$. We could try approximating its distribution with a Normal model, using the same mean and standard deviation. Remarkably enough, that turns out to be a very good approximation. Using that mean and standard deviation, we can find the *probability*:

$$P(X \geq 1850) = P\left(z \geq \frac{1850 - 1920}{42.48}\right) \approx P(z \geq -1.65) \approx 0.95$$

There seems to be about a 95% chance that this Red Cross chapter will have enough O-negative blood.

The Continuity Correction

When we use a continuous model to model a set of discrete events, we may need to make an adjustment called the **continuity correction**. We approximated the Binomial distribution (50, 0.2) with a Normal model. But what does the Normal model say about the probability that $X = 10$? Every specific value in the Normal probability model has probability 0. That's not the answer we want.

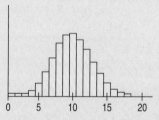

Because X is really discrete, it takes on the exact values 0, 1, 2, ... , 50, each with positive probability. The histogram holds the secret to the correction. Look at the bin corresponding to $X = 10$ in the histogram. It goes from 9.5 to 10.5. What we really want is to find the area under the normal curve *between* 9.5 and 10.5. So when we use the Normal model to approximate discrete events, we go halfway to the next value on the left and/or the right. We approximate $P(X = 10)$ by finding $P(9.5 \leq X \leq 10.5)$. For a Binomial (50, 0.2), $\mu = 10$ and $\sigma = 2.83$.

So $P(9.5 \leq X \leq 10.5) \approx P\left(\dfrac{9.5 - 10}{2.83} \leq z \leq \dfrac{10.5 - 10}{2.83}\right)$
$= P(-0.177 \leq z \leq 0.177)$
$= 0.1405$

By comparison, the *exact* Binomial probability is 0.1398.

Can we always use a Normal model to make estimates of Binomial probabilities? No. It depends on the sample size. Suppose we are searching for a prize in cereal boxes, where the probability of finding a prize is 20%. If we buy five boxes, the actual Binomial probabilities that we get 0, 1, 2, 3, 4, or 5 prizes are 33%, 41%, 20%, 5%, 1%, and 0.03%, respectively. The histogram just below shows that this probability model is skewed. That makes it clear that we should not try to estimate these probabilities by using a Normal model.

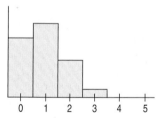

If we open 50 boxes of this cereal and count the number of prizes we find, we get the histogram below. It is centered at $np = 50(0.2) = 10$ prizes, as expected, and it appears to be fairly symmetric around that center.

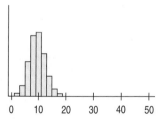

Let's have a closer look. The third histogram (in the side bar above) shows the same distribution, but this time magnified somewhat and centered at the expected value of 10 prizes. It looks close to Normal for sure. With this larger sample size, it appears that a Normal model might be a useful approximation.

A Normal model is a close enough approximation to the Binomial only for a large enough number of trials. And what we mean by "large enough" depends on the probability of success. We'd need a larger sample if the probability of success were very low (or very high). It turns out that a Normal model works pretty well if we expect to see at least 10 successes and 10 failures. That is, we check the Success/Failure Condition.

Success/Failure Condition: A Binomial model is approximately Normal if we expect at least 10 successes and 10 failures:

$$np \geq 10 \text{ and } nq \geq 10.$$

Why 10? Well, actually it's 9, as revealed in the following Math Box.

MATH BOX

It's easy to see where the magic number 10 comes from. You just need to remember how Normal models work. The problem is that a Normal model extends infinitely in both directions. But a Binomial model must have between 0 and n successes, so if we use a Normal to approximate a Binomial, we have to cut off its tails. That's not very important if the center of the Normal model is so far from 0 and n that the lost tails have only a negligible area. More than three standard deviations should do it because a Normal model has little probability past that.

So the mean needs to be at least 3 standard deviations from 0 and at least 3 standard deviations from n. Let's look at the 0 end.

We require:	$\mu - 3\sigma > 0$
Or, in other words:	$\mu > 3\sigma$
For a Binomial that's:	$np > 3\sqrt{npq}$
Squaring yields:	$n^2p^2 > 9npq$
Now simplify:	$np > 9q$
Since $q \leq 1$, we require:	$np > 9$

For simplicity we usually demand that np (and nq for the other tail) be at least 10 to use the Normal approximation which gives the Success/Failure Condition.[5]

The Exponential Model

We saw earlier that the Poisson model is a good model for the arrival of, or occurrence, of events. We found, for example, the probability that x visits to our website will occur within the next minute. The exponential model with parameter λ can be used to model the time *between* those events. Its density function has the form:

$$f(x) = \lambda e^{-\lambda x} \text{ for } x \geq 0 \text{ and } \lambda > 0$$

The use of the parameter λ again is not coincidental. It highlights the relationship between the exponential and the Poisson.

---

[5] Looking at the final step, we see that we need $np > 9$ in the worst case, when q (or p) is near 1, making the Binomial model quite skewed. When q and p are near 0.5—for example, between 0.4 and 0.6—the Binomial model is nearly symmetric, and $np > 5$ ought to be safe enough. Although we'll always check for 10 expected successes and failures, keep in mind that for values of p near 0.5, we can be somewhat more forgiving.

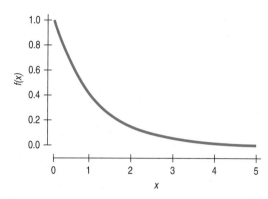

Figure 21.5 *The exponential density function (with $\lambda = 1$).*

If a discrete random variable can be modeled by a Poisson model with rate λ, then the times between events can be modeled by an exponential model with the same parameter λ. The mean of the exponential is $1/\lambda$. The inverse relationship between the two means makes intuitive sense. If λ increases and we expect *more* hits per minute, then the expected time between hits should go down. The standard deviation of an exponential random variable is $1/\lambda$.

Like any continuous random variable, probabilities of an exponential random variable can be found only through the density function. Fortunately, the area under the exponential density between any two values, s and t $(s \le t)$, has a particularly easy form:

$$P(s \le X \le t) = e^{-\lambda s} - e^{-\lambda t}.$$

In particular, by setting s to be 0, we can find the probability that the waiting time will be less than t from

$$P(X \le t) = P(0 \le X \le t) = e^{-\lambda 0} - e^{-\lambda t} = 1 - e^{-\lambda t}.$$

The function $P(X \le t) = F(t)$ is called the **cumulative distribution function (cdf)** of the random variable X. If arrivals of hits to our website can be well modeled by a Poisson with $\lambda = 4/\text{minute}$, then the probability that we'll have to wait less than 20 seconds (1/3 of a minute) is $F(1/3) = P(0 \le X \le 1/3) = 1 - e^{-4/3} = 0.736$. That seems about right. Arrivals are coming about every 15 seconds on average, so we shouldn't be surprised that nearly 75% of the time we won't have to wait more than 20 seconds for the next hit.

JUST CHECKING

Roper Worldwide reports that they are able to contact 76% of the randomly selected households drawn for a telephone survey.

2 Explain why these phone calls can be considered Bernoulli trials.

3 Which of the models of this chapter (Geometric, Binomial, Normal, Poisson, Exponential, or Uniform) would you use to model the number of successful contacts from a list of 1000 sampled households?

4 Roper also reports that even after they contacted a household, only 38% of the contacts agreed to be interviewed. So the probability of getting a completed interview from a randomly selected household is only 0.29. Which of the models of this chapter would you use to model the number of households Roper has to call before they get the first completed interview?

WHAT CAN GO WRONG?

- **Probability models are still just models.** Models can be useful, but they are not reality. Think about the assumptions behind your models. Question probabilities as you would data.

- **If the model is wrong, so is everything else.** Before you try to find the mean or standard deviation of a random variable, check to make sure the probability model is reasonable. As a start, the probabilities in your model should all be between 0 and 1 and they should add up to 1. If not, you may have calculated a probability incorrectly or left out a value of the random variable.

- **Don't assume everything's Normal.** Just because a random variable is continuous or you happen to know a mean and standard deviation doesn't mean that a Normal model will be useful. You must think about whether the **Normality Assumption** is justified. Using a Normal model when it really does not apply will lead to wrong answers and misleading conclusions.

- **Watch out for variables that aren't independent.** You can add expected values of *any* two random variables, but you can only add variances of independent random variables. Suppose a survey includes questions about the number of hours of sleep people get each night and also the number of hours they are awake each day. From their answers, we find the mean and standard deviation of hours asleep and hours awake. The expected total must be 24 hours; after all, people are either asleep or awake. The means still add just fine. Since all the totals are exactly 24 hours, however, the standard deviation of the total will be 0. We can't add variances here because the number of hours you're awake depends on the number of hours you're asleep. Be sure to check for independence before adding variances.

- **Don't write independent instances of a random variable with notation that looks like they are the same variables.** Make sure you write each instance as a different random variable. Just because each random variable describes a similar situation doesn't mean that each random outcome will be the same. These are *random* variables, not the variables you saw in Algebra. Write $X_1 + X_2 + X_3$ rather than $X + X + X$.

- **Don't forget:** Variances of independent random variables add. Standard deviations don't.

- **Don't forget:** Variances of independent random variables add, even when you're looking at the difference between them.

- **Be sure you have Bernoulli trials.** Be sure to check the requirements first: two possible outcomes per trial ("success" and "failure"), a constant probability of success, and independence. Remember that the 10% Condition provides a reasonable substitute for independence.

- **Don't confuse Geometric and Binomial models.** Both involve Bernoulli trials, but the issues are different. If you are repeating trials until your first success, that's a Geometric probability. You don't know in advance how many trials you'll need—theoretically, it could take forever. If you are counting the number of successes in a specified number of trials, that's a Binomial probability.

- **Don't use the Normal approximation with small *n*.** To use a Normal approximation in place of a Binomial model, there must be at least 10 expected successes and 10 expected failures.

ETHICS IN ACTION

While e-government services are available on-line, many Americans, especially those who are older, prefer to deal with government agencies in person. For this reason, the U.S. Social Security Administration (SSA) has local offices distributed across the country. Pat Mennoza is the office manager for one of the larger SSA offices in Phoenix. Since the initiation of the SSA website, his staff has been severely reduced. Yet, because of the number of retirees in the area, his office is one of the busiest. Although there have been no formal complaints, Pat expects that customer waiting times have increased. He decides to keep track of customer wait times for a one month period in the hopes of making a case for hiring additional staff. He finds that the average wait time is 5 minutes with a standard deviation of 6 minutes. He reasons that 50% of customers who visit his office wait longer than 5 minutes for service. The target wait time is 10 minutes or less. Applying the Normal probability model, Pat finds that more than 20% of customers will have to wait longer than 10 minutes! He has uncovered what he suspected. His next step is to request additional staff based on his findings.

ETHICAL ISSUE *Waiting times are generally skewed and therefore not usually modeled using the Normal distribution. Pat should have checked the data to see if a Normal model was appropriate. Using the Normal for data that are highly skewed to the right will inflate the probability a customer will have to wait longer than 10 minutes. Related to Item A, ASA Ethical Guidelines.*

ETHICAL SOLUTION *Check reasonableness of applying the Normal probability model.*

What have we learned?

We've learned to work with random variables. We can use the probability model for a discrete random variable to find its expected value and its standard deviation.

We've learned that the mean of the sum or difference of two random variables, discrete or continuous, is just the sum or difference of their means. And we've learned the Pythagorean Theorem of Statistics: *For independent random variables, the variance of their sum or difference is always the sum of their variances.* We have also learned that Normal models are once again special: Sums or differences of Normally distributed random variables also follow Normal models.

Finally, we've learned that Bernoulli trials show up in lots of places. Depending on the random variable of interest, we can use one of three models to estimate probabilities for Bernoulli trials:

- a Geometric model, when we're interested in the number of Bernoulli trials until the next success;

- a Binomial model, when we're interested in the number of successes in a certain number of Bernoulli trials;

- a Normal model can approximate a Binomial model when we expect at least 10 successes and 10 failures.

Terms

Addition Rule for Expected Values of Random Variables

$E(X \pm Y) = E(X) \pm E(Y)$

Addition Rule for Variances of Random Variables

(Pythagorean Theorem of Statistics)

If X and Y are *independent*: $Var(X \pm Y) = Var(X) + Var(Y)$,

and $SD(X \pm Y) = \sqrt{Var(X) + Var(Y)}$.

Bernoulli trials	A sequence of trials are called Bernoulli trials if:
	1. There are exactly two possible outcomes (usually denoted *success* and *failure*).
	2. The probability of success is constant.
	3. The trials are independent.
Binomial probability model	A Binomial model is appropriate for a random variable that counts the number of successes in a fixed number of Bernoulli trials.
Changing a random variable by a constant	$$E(X \pm c) = E(X) \pm c \qquad Var(X \pm c) = Var(X) \qquad SD(X \pm c) = SD(X)$$ $$E(aX) = aE(X) \qquad Var(aX) = a^2 Var(X) \qquad SD(aX) = |a|SD(X)$$
Continuous random variable	A random variable that can take any numeric value within a range of values. The range may be infinite or bounded at either or both ends.
Cumulative distribution function (cdf)	For any random variable X, and any value x, the cumulative distribution function $F(x) = P(X \le x)$.
Discrete random variable	A random variable that can take one of a finite number[6] of distinct outcomes.
Expected value	The expected value of a random variable is its theoretical long-run average value, the center of its model. Denoted μ or $E(X)$, it is found (if the random variable is discrete) by summing the products of variable values and probabilities: $$\mu = EV = E(X) = \sum x \cdot P(x).$$
Exponential probability model	A model often used to model waiting times between events, especially when the number of arrivals of those events are well modeled by a Poisson model.
Geometric probability model	A model appropriate for a random variable that counts the number of Bernoulli trials until the first success.
Normal model	The most famous continuous probability model, the Normal is used to model a wide variety of phenomena whose distributions are unimodal and symmetric. The Normal model is also used as an approximation to the Binomial model for large n, when np and $nq \ge 10$, and is used as the model for sampling distributions of sums and means under a wide variety of conditions.
Poisson model	A discrete model often used to model the number of arrivals of events such as customers arriving in a queue or calls arriving into a call center.
Probability density function (pdf)	A function $f(x)$ that represents the probability distribution of a random variable X. The probability that X is in an interval A is the area under the curve $f(x)$ over A.
Probability model	A function that associates a probability P with each value of a discrete random variable X, denoted $P(X = x)$, or with any interval of values of a continuous random variable.
Random variable	Assumes any of several different values as a result of some random event. Random variables are denoted by a capital letter, such as X.
Standard deviation of a random variable	Describes the spread in the model and is the square root of the variance.

---

[6] Technically, there could be an infinite number of outcomes as long as they're *countable*. Essentially, that means we can imagine listing them all in order, like the counting numbers 1, 2, 3, 4, 5, ...

Uniform model	For a discrete uniform model over a set of *n* values, each value has probability $1/n$. For a continuous uniform random variable over an interval $[a, b]$, the probability that *X* lies in any subinterval within $[a, b]$ is the same and is just equal to the length of the interval divided by the length of $[a, b]$ which is $b - a$.
Variance	The variance of a random variable is the expected value of the squared deviations from the mean. For discrete random variables, it can be calculated as:

$$\sigma^2 = Var(X) = \sum (x - \mu)^2 P(x).$$ |

Skills

When you complete this lesson, you should:

PLAN

- Be able to recognize random variables.
- Understand that random variables must be independent in order to determine the variability of their sum or difference by adding variances.

DO

- Be able to find the probability model for a discrete random variable.
- Know how to find the mean (expected value) and the variance of a random variable.
- Always use the proper notation for these population parameters: μ or $E(X)$ for the mean and σ, $SD(X)$, σ^2, or $Var(X)$ when discussing variability.
- Know how to determine the new mean and standard deviation after adding a constant, multiplying by a constant, or adding or subtracting two independent random variables.

REPORT

- Be able to interpret the meaning of the expected value and standard deviation of a random variable in the proper context.

Mini Case Study Project

A young entrepreneur has just raised a sum of money ($30,000) from investors, and she would like to invest it while she continues her fundraising in hopes of starting her company one year from now. She wants to do due diligence and understand the risk of each of her investment options. After speaking with her colleagues in finance, she believes that she has three choices: (1) she can purchase a $30,000 certificate of deposit (CD); (2) she can invest in a mutual fund with a balanced portfolio; or (3) she can invest in a growth stock that has a greater potential payback but also has greater volatility. Each of her options will yield a different payback on her $30,000, depending on the state of the economy.

During the next year, she knows that the CD yields a constant annual percentage rate, regardless of the state of the economy. If she invests in a balanced mutual fund, she estimates that she will earn 9% if the economy remains strong, but earn only 3% if the economy takes a downturn. Finally, if she invests all $30,000 in a growth stock, experienced investors tell her that she can earn as much as 50% in a strong economy, but may *lose* as much as 50% in a poor economy.

Estimating these returns, along with the likelihood of a strong economy, is challenging. Therefore, often a "sensitivity analysis" is conducted, where figures are computed using a range of values for each of the uncertain parameters in the problem. Following this advice, this investor decides to compute measures for a range of interest rates for CDs, a range of returns for the mutual fund, and a range of returns for the growth stock. In addition, the likelihood of a strong economy is unknown, so she will vary these probabilities as well.

Assume that the probability of a strong economy over the next year is either 0.5, 0.3 or 0.7. To help this investor make an informed decision, evaluate the expected value and volatility of each of her investments using the following ranges of rates of growth:

CD: Look up the current annual rate for the return on a 3-year CD and use this value $\pm$ 0.5%.
Mutual Fund: Use values of 8%, 10%, and 12% for a strong economy and values of 2%, 0%, and −4% for a weak economy.
Growth Stock: Use values of 15%, 20%, and 25% in a strong economy and values of −25%, −15%, and −10% in a weak economy.

Discuss the expected returns and uncertainty of each of the alternative investment options for this investor in each of the scenarios you analyzed. Be sure to compare the volatility of each of her options.

EXERCISES

1. New website. You have just launched the website for your company that sells nutritional products online. Suppose X = the number of different pages that a customer hits during a visit to the website.

a) Assuming that there are n different pages in total on your website, what are the possible values that this random variable may take on?
b) Is the random variable discrete or continuous?

2. New website, part 2. For the website described in Exercise 1, let Y = the total time (in minutes) that a customer spends during a visit to the website.

a) What are the possible values of this random variable?
b) Is the random variable discrete or continuous?

3. Job interviews. Through the career services office, you have arranged preliminary interviews at four companies for summer jobs. Each company will either ask you to come to their site for a follow-up interview or not. Let X be the random variable equal to the total number of follow-up interviews that you might have.

a) List all the possible values of X.
b) Is the random variable discrete or continuous?
c) Do you think a uniform distribution might be appropriate as a model for this random variable? Explain briefly.

4. Help desk. The computer help desk is staffed by students during the 7:00 P.M. to 11:00 P.M. shift. Let Y denote the random variable that represents the number of students seeking help during the 15-minute time slot 10:00 to 10:15 P.M.

a) What are the possible values of Y?
b) Is the random variable discrete or continuous?

5. Orthodontist. An orthodontist has three financing packages, and each has a different service charge. He estimates that 30% of patients use the first plan which has a $10 finance charge; 50%

use the second plan which has a $20 finance charge; and 20% use the third plan which has a $30 finance charge.

a) Find the expected value of the service charge.
b) Find the standard deviation of the service charge.

6. Timeshare. A marketing agency has developed three vacation packages to promote a timeshare plan at a new resort. They estimate that 20% of potential customers will choose the Day Plan which does not include overnight accommodations; 40% will choose the Overnight Plan which includes one night at the resort; and 40% will choose the Weekend Plan which includes two nights.

a) Find the expected value of the number of nights potential customers will need.
b) Find the standard deviation of the number of nights potential customers will need.

7. Concepts I. Given independent random variables, X and Y, with means and standard deviations as shown, find the mean and standard deviation of each of the variables in parts a) to d).

a) $3X$
b) $Y + 6$
c) $X + Y$
d) $X - Y$

	Mean	SD
X	10	2
Y	20	5

8. Concepts II. Given independent random variables, X and Y, with means and standard deviations as shown, find the mean and standard deviation of each of the variables in parts a) to d).

a) $X - 20$
b) $0.5Y$
c) $X + Y$
d) $X - Y$

	Mean	SD
X	80	12
Y	12	3

9. Lottery. Iowa has a lottery game called Pick 3 in which customers buy a ticket for $1 and choose three numbers, each from zero to nine. They also must select the play type, which determines what combinations are winners. In one type of play, called the "Straight/Box," they win if they match the three numbers in any order, but the payout is greater if the order is exact. For the case where all three of the numbers selected are different, the probabilities and payouts are:

	Probability	Payout
Straight/Box Exact	1 in 1000	$350
Straight/Box Any	5 in 1000	$50

a) Find the amount a Straight/Box player can expect to win.
b) Find the standard deviation of the player's winnings.
c) Tickets to play this game cost $1 each. If you subtract $1 from the result in part a, what is the expected result of playing this game?

10. Software company. A small software company will bid on a major contract. It anticipates a profit of $50,000 if it gets it, but thinks there is only a 30% chance of that happening.

a) What's the expected profit?
b) Find the standard deviation for the profit.

11. Commuting to work. A commuter must pass through five traffic lights on her way to work and will have to stop at each one that is red. After keeping record for several months, she developed the following probability model for the number of red lights she hits:

X = # of red	0	1	2	3	4	5
P(X = x)	0.05	0.25	0.35	0.15	0.15	0.05

a) How many red lights should she expect to hit each day?
b) What's the standard deviation?

12. Defects. A consumer organization inspecting new cars found that many had appearance defects (dents, scratches, paint chips, etc.). While none had more than three of these defects, 7% had three, 11% had two, and 21% had one defect.

a) Find the expected number of appearance defects in a new car.
b) What is the standard deviation?

13. Fishing tournament. A sporting goods manufacturer was asked to sponsor a local boy in two fishing tournaments. They claim the probability that he will win the first tournament is 0.4. If he wins the first tournament, they estimate the probability that he will also win the second is 0.2. They guess that if he loses the first tournament, the probability that he will win the second is 0.3.

a) According to their estimates, are the two tournaments independent? Explain your answer.
b) What's the probability that he loses both tournaments?
c) What's the probability he wins both tournaments?

d) Let random variable X be the number of tournaments he wins. Find the probability model for X.
e) What are the expected value and standard deviation of X?

14. Contracts. Your company bids for two contracts. You believe the probability that you get contract #1 is 0.8. If you get contract #1, the probability that you also get contract #2 will be 0.2, and if you do not get contract #1, the probability that you get contract #2 will be 0.3.

a) Are the outcomes of the two contract bids independent? Explain.
b) Find the probability you get both contracts.
c) Find the probability you get neither contract.
d) Let X be the number of contracts you get. Find the probability model for X.
e) Find the expected value and standard deviation of X.

15. Battery recall. A company has discovered that a recent batch of batteries had manufacturing flaws, and has issued a recall. You have 10 batteries covered by the recall, and 3 are dead. You choose 2 batteries at random from your package of 10.

a) Has the assumption of independence been met? Explain.
b) Create a probability model for the number of good batteries chosen.
c) What's the expected number of good batteries?
d) What's the standard deviation?

16. Grocery supplier. A grocery supplier believes that the mean number of broken eggs per dozen is 0.6, with a standard deviation of 0.5. You buy 3 dozen eggs without checking them.

a) How many broken eggs do you expect to get?
b) What's the standard deviation?
c) Is it necessary to assume the cartons of eggs are independent? Why?

17. Commuting, part 2. A commuter finds that she waits an average of 14.8 seconds at each of five stoplights, with a standard deviation of 9.2 seconds. Find the mean and the standard deviation of the total amount of time she waits at all five lights. What, if anything, did you assume?

18. Repair calls. A small engine shop receives an average of 1.7 repair calls per hour, with a standard deviation of 0.6. What is the mean and standard deviation of the number of calls they receive for an 8-hour day? What, if anything, did you assume?

19. Insurance company. An insurance company estimates that it should make an annual profit of $150 on each homeowner's policy written, with a standard deviation of $6000.

a) Why is the standard deviation so large?
b) If the company writes only two of these policies, what are the mean and standard deviation of the annual profit?
c) If the company writes 1000 of these policies, what are the mean and standard deviation of the annual profit?
d) What is the probability that the company will make money if they write 1000 policies?
e) What circumstances could violate the assumption of independence of the policies?

20. Casino. At a casino, people play the slot machines in hopes of hitting the jackpot, but most of the time, they lose their money. A certain machine pays out an average of $0.92 (for every dollar played), with a standard deviation of $120.

a) Why is the standard deviation so large?
b) If a gambler plays 5 times, what are the mean and standard deviation of the casino's profit?
c) If gamblers play this machine 1000 times in a day, what are the mean and standard deviation of the casino's profit?

21. Bike sale. A bicycle shop plans to offer 2 specially priced children's models at a sidewalk sale. The basic model will return a profit of $120 and the deluxe model $150. Past experience indicates that sales of the basic model will have a mean of 5.4 bikes with a standard deviation of 1.2, and sales of the deluxe model will have a mean of 3.2 bikes with a standard deviation of 0.8 bikes. The cost of setting up for the sidewalk sale is $200.

a) Define random variables and use them to express the bicycle shop's net profit.
b) What's the mean of the net profit?
c) What's the standard deviation of the net profit?
d) Do you need to make any assumptions in calculating the mean? How about the standard deviation?

22. Farmers' market. A farmer has 100 lbs of apples and 50 lbs of potatoes for sale. The market price for apples (per pound) each day is a random variable with a mean of 0.5 dollars and a standard deviation of 0.2 dollars. Similarly, for a pound of potatoes, the mean price is 0.3 dollars and the standard deviation is 0.1 dollars. It also costs him 2 dollars to bring all the apples and potatoes to the market. The market is busy with eager shoppers, so we can assume that he'll be able to sell all of each type of produce at that day's price.

a) Define your random variables, and use them to express the farmer's net income.
b) Find the mean of the net income.
c) Find the standard deviation of the net income.
d) Do you need to make any assumptions in calculating the mean? How about the standard deviation?

23. NASCAR. For a new type of tire, a NASCAR team found the average distance a set of tires would run during a race is 168 miles, with a standard deviation of 14 miles. Assume that tire mileage is independent and follows a Normal model.

a) If the team plans to change tires twice during a 500-mile race, what is the expected value and standard deviation of miles remaining after two changes?
b) What is the probability they won't have to change tires a third time before the end of a 500 mile race?

24. Swimming medley. In the 4 × 100 medley relay event, four swimmers swim 100 yards, each using a different stroke. A college team preparing for the conference championship looks at the times their swimmers have posted and creates a model based on the following assumptions:

• The swimmers' performances are independent.

• Each swimmer's times follow a Normal model.
• The means and standard deviations of the times (in seconds) are as shown here.

Swimmer	Mean	SD
1 (backstroke)	50.72	0.24
2 (breaststroke)	55.51	0.22
3 (butterfly)	49.43	0.25
4 (freestyle)	44.91	0.21

a) What are the mean and standard deviation for the relay team's total time in this event?
b) The team's best time so far this season was 3:19.48. (That's 199.48 seconds.) What is the probability that they will beat that time in the next event?

25. Movie rentals. To compete with Netflix, the owner of a movie rental shop decided to try sending DVDs through the mail. In order to determine how many copies of newly released titles he should purchase, he carefully observed turnaround times. Since nearly all of his customers were in his local community, he tested delivery times by sending DVDs to his friends. He found the mean delivery time was 1.3 days, with a standard deviation of 0.5 days. He also noted that the times were the same whether going to the customer or coming back to the shop.

a) Find the mean and standard deviation of the round-trip delivery times for a DVD (mailed to the customer and then mailed back to the shop).
b) The shop owner tries to process a DVD that is returned to him and get it back in the mail in one day, but circumstances sometimes prevent it. His mean turnaround time is 1.1 days, with a standard deviation of 0.3 days. Find the mean and standard deviation of the turnaround times combined with the round-trip times in part a.
c) The complete rent cycle is the time from when a DVD is placed in the mail until it is returned, processed, and placed back in the mail. Initially, the shop owner estimated the rent cycle would take 9 days. If the time customers hold DVDs has a mean of 3.7 days and a standard deviation of 2.0 days, combine customer times with round-trip and process times in part b and determine what proportion of DVD rentals would take longer than 9 days to complete the cycle. (Assume the distribution of rent cycle time has a Normal model.)

26. Online applications. Researchers for an online marketing company suggest that new customers who have to become a member before they can check out on the website are very intolerant of long applications. One way to rate an application is by the total number of keystrokes required to fill it out.

a) One common frustration is having to enter an e-mail address twice. If the mean length of e-mail addresses is 13.3 characters, with a standard deviation of 2.8 characters, what is the mean and standard deviation of total characters typed if entered twice?
b) The company found the mean and standard deviation of the length of customers' names (including spaces) were 13.4 and 2.4

characters, respectively, and for addresses, 30.8 and 6.3 characters. What is the mean and standard deviation of the combined lengths of entering the e-mail addresses twice and then the name and the address?

c) The store's researchers suggested the frustration limit is 80 characters, beyond which a potential customer is likely to close the application without completing the purchase. What proportion of applications found in part b will exceed that? (Assume the distribution of application lengths has a Normal model.)

27. eBay. A collector purchased a quantity of action figures and is going to sell them on eBay. He has 19 Hulk figures. In recent auctions, the mean selling price of similar figures has been $12.11, with a standard deviation of $1.38. He also has 13 Iron Man figures which have had a mean selling price of $10.19, with a standard deviation of $0.77. His insertion fee will be $0.55 on each item, and the closing fee will be 8.75% of the selling price. He assumes all will sell without having to be relisted.

a) Define your random variables, and use them to create a random variable for the collector's net income.

b) Find the mean (expected value) of the net income.

c) Find the standard deviation of the net income.

d) Do you have to assume independence for the sales on eBay? Explain.

28. Real estate. A real estate broker purchased 3 two-bedroom houses in a depressed market for a combined cost of $71,000. He expects the cleaning and repair costs on each house to average $3700, with a standard deviation of $1450. When he sells them, after subtracting taxes and other closing costs, he expects to realize an average of $39,000 per house, with a standard deviation of $1100.

a) Define your random variables, and use them to create a random variable for the broker's net profit.

b) Find the mean (expected value) of the net profit.

c) Find the standard deviation of the net profit.

d) Do you have to assume independence for the repairs and sale prices of the houses? Explain.

29. Bernoulli. Can we use probability models based on Bernoulli trials to investigate the following situations? Explain.

a) Each week a doctor rolls a single die to determine which of his six office staff members gets the preferred parking space.

b) A medical research lab has samples of blood collected from 120 different individuals. How likely is it that the majority of them are Type A blood, given that Type A is found in 43% of the population?

c) From a workforce of 13 men and 23 women, all five promotions go to men. How likely is that, if promotions are based on qualifications rather than gender?

d) We poll 500 of the 3000 stockholders to see how likely it is that the proposed budget will pass.

e) A company realizes that about 10% of its packages are not being sealed properly. In a case of 24 packages, how likely is it that more than 3 are unsealed?

30. Bernoulli, part 2. Can we use probability models based on Bernoulli trials to investigate the following situations? Explain.

a) You are rolling 5 dice. How likely is it to get at least two 6's to win the game?

b) You survey 500 potential customers to determine their color preference.

c) A manufacturer recalls a doll because about 3% have buttons that are not properly attached. Customers return 37 of these dolls to the local toy store. How likely are they to find any buttons not properly attached?

d) A city council of 11 Republicans and 8 Democrats picks a committee of 4 at random. How likely are they to choose all Democrats?

e) An executive reads that 74% of employees in his industry are dissatisfied with their jobs. How many dissatisfied employees can he expect to find among the 481 employees in his company?

31. Closing sales. A salesman normally makes a sale (closes) on 80% of his presentations. Assuming the presentations are independent, find the probability of each of the following.

a) He fails to close for the first time on his fifth attempt.

b) He closes his first presentation on his fourth attempt.

c) The first presentation he closes will be on his second attempt.

d) The first presentation he closes will be on one of his first three attempts.

32. Computer chip manufacturer. Suppose a computer chip manufacturer rejects 2% of the chips produced because they fail presale testing. Assuming the bad chips are independent, find the probability of each of the following.

a) The fifth chip they test is the first bad one they find.

b) They find a bad one within the first 10 they examine.

c) The first bad chip they find will be the fourth one they test.

d) The first bad chip they find will be one of the first three they test.

33. Side effects. Researchers testing a new medication find that 7% of users have side effects. To how many patients would a doctor expect to prescribe the medication before finding the first one who has side effects?

34. Credit cards. College students are a major target for advertisements for credit cards. At a university, 65% of students surveyed said they had opened a new credit card account within the past year. If that percentage is accurate, how many students would you expect to survey before finding one who had not opened a new account in the past year?

35. Missing pixels. A company that manufactures large LCD screens knows that not all pixels on their screen light, even if they spend great care when making them. In a sheet 6 ft by 10 ft (72 in. by 120 in.) that will be cut into smaller screens, they find an average of 4.7 blank pixels. They believe that the occurrences of blank pixels are independent. Their warranty policy states that they will replace any screen sold that shows more than 2 blank pixels.

a) What is the mean number of blank pixels per square foot?

b) What is the standard deviation of blank pixels per square foot?

c) What is the probability that a 2 ft by 3 ft screen will have at least one defect?

d) What is the probability that a 2 ft by 3 ft screen will be replaced because it has too many defects?

36. Bean bags. Cellophane that is going to be formed into bags for items such as dried beans or bird seed is passed over a light sensor to test if the alignment is correct before it passes through the heating units that seal the edges. Small adjustments can be made by the machine automatically. But if the alignment is too bad, the process is stopped and an operator has to manually adjust it. These misalignment stops occur randomly and independently. On one line, the average number of stops is 52 per 8-hour shift.

a) What is the mean number of stops per hour?

b) What is the standard deviation of stops per hour?

c) When the machine is restarted after a stop, what is the probability it will run at least 15 minutes before the next stop?

37. Hurricane insurance. An insurance company needs to assess the risks associated with providing hurricane insurance. Between 1990 and 2006, Florida was hit by 22 tropical storms or hurricanes. If tropical storms and hurricanes are independent and the mean has not changed, what is the probability of having a year in Florida with each of the following. (Note that 1990 to 2006 is 17 years.)

a) No hits?

b) Exactly one hit?

c) More than three hits?

38. Hurricane insurance, part 2. Between 1965 and 2007, there were 95 major hurricanes (category 3 or more) in the Atlantic basin. Assume that hurricanes are independent and the mean has not changed.

a) What is the mean number of major hurricanes per year? (There are 43 years from 1965 to 2007.)

b) What is the standard deviation of the frequency of major hurricanes?

c) What is the probability of having a year with no major hurricanes?

d) What is the probability of going three years in a row without a major hurricane?

39. Professional tennis. Serena Williams made a successful first serve 67% of the time in a Wimbledon finals match against her sister Venus. If she continues to serve at the same rate the next time they play and serves 6 times in the first game, determine the following probabilities. (Assume that each serve is independent of the others.)

a) All 6 first serves will be in.

b) Exactly 4 first serves will be in.

c) At least 4 first serves will be in.

40. American Red Cross. Only 4% of people have Type AB blood. A bloodmobile has 12 vials of blood on a rack. If the distribution of blood types at this location is consistent with the general population, what's the probability they find AB blood in:

a) None of the 12 samples?

b) At least 2 samples?

c) 3 or 4 samples?

41. Professional tennis, part 2. Suppose Serena continues to make 67% of her first serves as in Exercise 39 and serves 80 times in a match.

a) What's the mean and standard deviation of the number of good first serves expected?

b) Justify why you can use a Normal model to approximate the distribution of the number of good first serves.

c) Use the 68-95-99.7 Rule to describe this distribution.

d) What's the probability she makes at least 65 first serves in the 80 attempts?

42. American Red Cross, part 2. The bloodmobile in Exercise 40 received 300 donations in one day.

a) Assuming the frequency of AB blood is 4%, determine the mean and standard deviation of the number of donors who are AB.

b) Justify why you can use a Normal model to approximate the distribution of Type AB blood.

c) How likely would it be to find 10 or more samples with type AB blood in 300 samples?

43. No-shows. Because many passengers who make reservations do not show up, airlines often overbooks flights (sell more tickets than there are seats). A Boeing 767-400ER holds 245 passengers. If the airline believes the rate of passenger no-shows is 5% and sells 255 tickets, is it likely they will not have enough seats and someone will get bumped?

a) Use the Normal model to approximate the Binomial to determine the probability of at least 246 passengers showing up.

b) Should the airline change the number of tickets they sell for this flight? Explain.

44. Euro. Shortly after the introduction of the Belgian euro coin, newspapers around the world published articles claiming the coin is biased. The stories were based on reports that someone had spun the coin 250 times and gotten 140 heads—that's 56% heads.

a) Use the Normal model to approximate the Binomial to determine the probability of spinning a fair coin 250 times and getting at least 140 heads.

b) Do you think this is evidence that spinning a Belgian euro is unfair? Would you be willing to use it at the beginning of a sports event? Explain.

45. Satisfaction survey. A cable provider wants to contact customers in a particular telephone exchange to see how satisfied they are with the new digital TV service the company has provided. All numbers are in the 452 exchange, so there are 10,000 possible numbers from 452-0000 to 452-9999. If they select the numbers with equal probability:

a) What distribution would they use to model the selection?

b) What is the probability the number selected will be an even number?

c) What is the probability the number selected will end in 000?

46. Manufacturing quality. In an effort to check the quality of their cell phones, a manufacturing manager decides to take a random sample of 10 cell phones from yesterday's production run, which produced cell phones with serial numbers ranging (according to when they were produced) from 43005000 to 43005999. If each of the 1000 phones is equally likely to be selected:

a) What distribution would they use to model the selection?
b) What is the probability that a randomly selected cell phone will be one of the last 100 to be produced?
c) What is the probability that the first cell phone selected is either from the last 200 to be produced or from the first 50 to be produced?
d) What is the probability that the first two cell phones are both from the last 100 to be produced?

47. Web visitors. A website manager has noticed that during the evening hours, about 3 people per minute check out from their shopping cart and make an online purchase. She believes that each purchase is independent of the others and wants to model the number of purchases per minute.

a) What model might you suggest to model the number of purchases per minute?
b) What is the probability that in any one minute at least one purchase is made?
c) What is the probability that no one makes a purchase in the next 2 minutes?

48. Quality control. The manufacturer in Exercise 46 has noticed that the number of faulty cell phones in a production run of cell phones is usually small and that the quality of one day's run seems to have no bearing on the next day.

a) What model might you use to model the number of faulty cell phones produced in one day.

b) If the mean number of faulty cell phones is 2 per day, what is the probability that no faulty cell phones will be produced tomorrow?
c) If the mean number of faulty cell phones is 2 per day, what is the probability that 3 or more faulty cell phones were produced in today's run?

49. Web visitors, part 2. The website manager in Exercise 47 wants to model the time between purchases. Recall that the mean number of purchases in the evening is 3 per minute.

a) What model would you use to model the time between events?
b) What is the mean time between purchases?
c) What is the probability that the time to the next purchase will be between 1 and 2 minutes?

50. Quality control, part 2. The cell phone manufacturer in Exercises 46 and 48 wants to model the time between events. The mean number of defective cell phones is 2 per day.

a) What model would you use to model the time between events?
b) What would the probability be that the time to the next failure is 1 day or less?
c) What is the mean time between failures?

JUST CHECKING ANSWERS

1 a) $100 + 100 = 200$ seconds
b) $\sqrt{50^2 + 50^2} = 70.7$ seconds
c) The times for the two customers are independent.
2 There are two outcomes (contact, no contact), the probability of contact stays constant at 0.76, and random calls should be independent.
3 Binomial (or Normal approximation)
4 Geometric

22

Decision Making and Risk

Data Description, inc.

Small businesses are 99.7 percent of all U.S. firms. They generate more than half of the private non-farm gross domestic product and create 60 to 80 percent of net new jobs. (www.sba.gov). Here's the story of one such firm.

Data Description was founded in 1985 by Paul Velleman to develop, market, and support the Data Desk® statistics program. Personal computers were becoming more widely available, and graphical desktop interfaces offered new and exciting opportunities for data analysis and graphics. Data Desk was introduced in 1986 for the newly released Macintosh computer and released for the PC in 1997, with expanded capabilities for analysis, and the ability to work efficiently with large (several million case) data sets. Data Description then launched *ActivStats*, a multimedia education product on DVD, which pioneered multimedia teaching in statistics and led to the development of MediaDX, a development platform that deploys a full array of media, narration, animation, video, sound, and interactive tools for teaching.

Data Description employs both a local staff and programmers who work remotely from other parts of the country and "telecommute." In these ways, it is typical of many technologically-centered small businesses relying on the Internet for development, marketing, and customer support and competing with larger firms by staying fast and flexible.

Company president John Sammis notes that the Internet has enabled small businesses such as Data Description to compete with larger international firms and that decisions about how best to use the Internet are key to competing successfully. By its twentieth anniversary, Data Description (www.datadesk.com), based in Ithaca, New York, was providing software and services to education and business clients throughout the world.

We make decisions every day, often without knowing how they will turn out. Most of our daily decisions have no serious consequences. But the decisions a business makes determine the success or failure of the venture. The consequences of business decisions can often be quantified in monetary terms, but the costs and benefits themselves often depend on events beyond the control or foresight of decisionmakers. Decisions must be made nevertheless. How can people make intelligent decisions when they don't know for certain what the future holds?

One decision facing Data Description (as with many high-tech companies) is how best to provide technical support for its customers. In the past, Data Description has relied on its thorough documentation, a built-in help system, and free telephone support. But as its user base has grown, telephone-based customer support has become a significant expense. Data Description's president, John Sammis, must decide whether to invest in developing an online FAQ (Frequently Asked Questions) help system as the first line of defense for customer support, to continue free telephone support through the help desk, or to hire and train additional telephone support staff and charge customers for it. For a small business, decisions such as these carry significant costs, but making the right decision can be the difference between keeping customers happy and losing them to larger competitors.

22.1 Actions, States of Nature, and Outcomes

We refer to the alternative choices we can make for a decision as **actions**. Actions are mutually exclusive; if you choose one course of action, you haven't chosen the others. We'll call Data Descriptions' choices:

◆ Online FAQ

◆ Free Help Desk

◆ Pay Help Desk

The facts about the world that affect the consequences of each action are called **states of nature**, or sometimes just **states**. We often simplify the states to make it easier to understand the decision alternatives. We know, for example, that the economy isn't just "good" or "bad," and the number of customers can be any value in some reasonable range. But we'll simplify by grouping the states of nature into a few categories and treating them as discrete possibilities.

For Data Description, the *states of nature* concern the type of question coming into the Help Desk. Management classifies them into two basic types:

◆ Simple
or
◆ Complex

A simple question is one that can be answered with an online FAQ system alone. A complex question requires either human technical support by telephone or backup e-mail support for FAQ users.

Each *action* has consequences that depend on the *state of nature* that actually occurs. These consequences are called the **outcomes** or **payoffs** (because usually in business the consequence can be measured in money).

To make informed decisions, Data Description president John Sammis must estimate the costs and benefits of each action under each state of nature. According to his estimates:

◆ The FAQ system can answer a simple question for about $3.50. A complex question can't be answered by the FAQ system and needs additional resources. Unfortunately the back-up system is slow and the marketing manager estimates that this would cost about $15 in goodwill and future business. Adding that to the FAQ system cost gives a total cost of $18.50 for a complex question.

◆ A live telephone support person costs about $10 per question whether it is simple or complex.

◆ Charging the customer $3 for the telephone support could defray some of the cost, but it might upset a customer with only a simple question. That resulting ill will will cost the company an estimated $15 for each simple question. On the other hand, customers with a complex question might be happy to pay for the telephone support, so the net cost for them would be $7.

22.2 Payoff Tables and Decision Trees

We can summarize the actions, states of nature, and corresponding payoffs in a **payoff table**. Table 22.1 shows the payoff table for Data Description's decision about customer support. Customer support is an expense, so all the entries in the payoff table are costs, which Data Description hopes to keep to a minimum. The table shows the costs for a single tech support request.

		State of Nature	
		Simple Question	**Complex Question**
Action	**FAQ**	$3.50	$18.50
	Free Help Desk	$10.00	$10.00
	Pay Help Desk	$15.00	$7.00

Table 22.1 *A payoff table showing costs of actions for two states of nature. Note that these are costs and not profits. As "payoffs," they might be written as negative values, but it is simpler to just remember that we want to minimize costs. Payoff tables of profits look to maximize values.*

The payoff table is easy to read, but it doesn't display the sequential nature of the decision process. Another way to display this information that shows the dynamics of the decision making process is with a **decision tree**. The decision tree mimics the actual decision making process by showing actions first followed by the possible states of nature and finally the outcomes that would result from each combination (Figure 22.1).

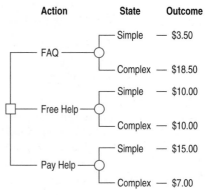

Action	State	Outcome
FAQ	Simple	$3.50
	Complex	$18.50
Free Help	Simple	$10.00
	Complex	$10.00
Pay Help	Simple	$15.00
	Complex	$7.00

Figure 22.1 *A decision tree for the customer support decision.*

The decision process moves from left to right through the decision tree. The decisionmaker chooses an action (indicated by a square node), determining which branch of the tree to follow. The circle nodes indicate branches due to the state of nature, which lie outside the control of the decisionmaker. The tree can expand to show decisions with any (reasonable) number of actions and states of nature.

> The square node, or decision node, indicates that a decision must be made. The circular node, or chance node, indicates a choice made by nature.

22.3 Minimizing Loss and Maximizing Gain

A conservative approach to decision-making would look at the worst possible loss and try to minimize it. Here are the outcomes for each decision and the greatest cost for each.

		State of Nature		
		Simple Question	**Complex Question**	**Greatest Cost**
Action	**FAQ**	$3.50	$18.50	$18.50
	Free Help Desk	$10.00	$10.00	$10.00
	Pay Help Desk	$15.00	$7.00	$15.00

Table 22.2 *The greatest cost for each decision.*

The decision to go with a FAQ could cost the company as much as $18.50 when complex calls come in, and the decision for a pay help desk could cost up to $15.00. The safe choice—the one that minimizes the maximum cost—is the free help desk. This is known as the **minimax** choice.

Of course, one could take the other extreme position and try for the maximum gain (or, equivalently here, the minimum cost). In this table, that would call for committing to a FAQ system (and hoping that the calls were simple). Such a choice is known as the **maximax** (when maximizing return) or the **minimin** (when minimizing cost) choice.

Choosing actions based on worst and best case scenarios rarely leads to the best business decision. Instead, successful small business executives rely on knowing their market so they can make less absolute decisions. A more realistic modelling approach takes into account how frequently the decision maker expects to experience each of the states of nature, and finds the optimum under that model.

22.4 The Expected Value of an Action

How can Data Description choose an action to maximize profit (or minimize loss)? They can be pretty sure that they'll get a mix of both simple and complex questions. Decision makers often estimate the *probability* of a state of nature based on their understanding of their business. Such probabilities can be *subjective* probabilities or can be grounded in data, and viewed as the *frequentist* probabilities we've used earlier in this book. In either case they express the expert's opinion of and belief in the relative likelihood of the states. We'll write $P(s_j)$ for the probability of the jth state of nature.

As with all probabilities, we need to check that the probabilities are *legitimate*. If there are N states of nature, then we require that:

$$P(s_j) \geq 0 \text{ for all } j,$$

$$\sum_{j=1}^{N} P(s_j) = 1.$$

If the probabilities are legitimate, then we can find the **expected value (EV)** of action a_i in essentially the same way as we found the expected value of a discrete random variable in Chapter 21[1]:

$$EV(a_i) = \sum_{j=1}^{N} o_{ij} P(s_j),$$

where there are N possible states of nature and o_{ij} is the outcome or payoff of action i when the state of nature is s_j.

At Data Description, about 40% of the questions are simple, so managers assigned a probability of 0.4 that a support question will be simple and a probability of 0.6 that it will be complex. To compute the expected value, we start by placing the probabilities on the decision tree in the appropriate place. With these values in place, we can find the expected value of each of the actions. The probabilities are associated with each state of nature and are the same regardless of the action. We place these values on the branches associated with each state of nature, repeated for each one of the possible actions (the circle nodes).

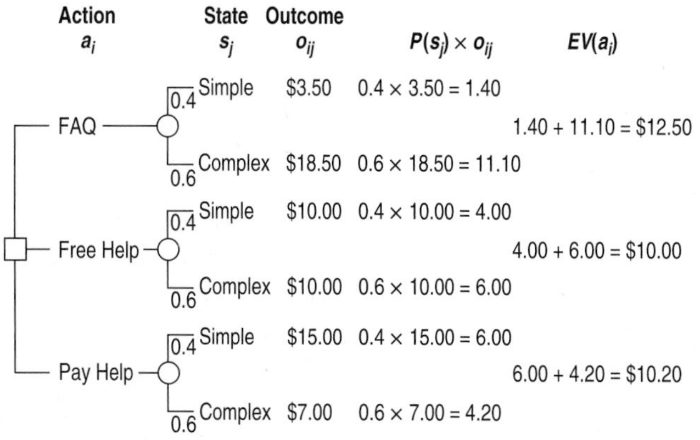

There are three general types of decisionmakers:
- a risk-averse individual will sacrifice EV for lower variation
- a risk-neutral individual will try to maximize EV
- a risk-tolerant individual may be willing to absorb greater variation in return for the chance of a higher payoff

Action a_i	State s_j	Outcome o_{ij}	$P(s_j) \times o_{ij}$	$EV(a_i)$
FAQ	Simple (0.4)	$3.50	$0.4 \times 3.50 = 1.40$	
	Complex (0.6)	$18.50	$0.6 \times 18.50 = 11.10$	$1.40 + 11.10 = \$12.50$
Free Help	Simple (0.4)	$10.00	$0.4 \times 10.00 = 4.00$	
	Complex (0.6)	$10.00	$0.6 \times 10.00 = 6.00$	$4.00 + 6.00 = \$10.00$
Pay Help	Simple (0.4)	$15.00	$0.4 \times 15.00 = 6.00$	
	Complex (0.6)	$7.00	$0.6 \times 7.00 = 4.20$	$6.00 + 4.20 = \$10.20$

Figure 22.2 *Calculating the expected value of each action using a decision tree.*

---

[1] In Chapter 21 we computed the EV for one action, such as investing in a CD. This chapter extends this concept by computing and comparing EVs for different actions.

For each combination of outcome and probability, we calculate its contribution to the expected value $P(s_j) \times o_{ij}$. Because only one state of nature can occur for each call, the states of natures are disjoint. We can therefore find the expected value for each action by summing the contributions over all the states of nature, in this case Simple and Complex (see Figure 22.2).

From the expected values in Figure 22.2, it appears that expanding free telephone support may be the best action because it costs Data Description the *least*. The expected value of that action is a cost of $10.00.

22.5 Expected Value with Perfect Information

Unfortunately, we can't predict the state of nature with certainty. But it can be informative to consider what it costs us to live with the uncertainty. If we knew the true state of nature—rather than just the probable state of nature—how much might that be worth to us? To help quantify that, let's consider a single customer support question. Knowing whether it was simple or complex would enable us to take the optimal action in each case. For a simple question we'd offer online FAQ support and the "payoff" would be $3.50. For a complex question we'd provide paid telephone support costing the company $7.00.

Using the probabilities of the two states of nature, we can calculate the expected value of this optimal strategy as:

$$0.6 \times \$7.00 + 0.4 \times \$3.50 = \$5.60$$

This is called the **expected value with perfect information** and is sometimes denoted *EVwPI*.

By contrast, the expected value of the optimal strategy that we calculated without knowing the state of nature was $10.00. The absolute value of the difference between these two quantities is called the **expected value of perfect information**.

$$EVPI = |EVwPI - EV|$$

In our example, this difference is $|\$5.60 - \$10.00| = \$4.40$. (Notice the absolute value in that calculation. Information should increase profit or, as in our example, reduce costs. Either way, the *value* of information is positive.) Our lack of perfect knowledge about the state of nature costs us $4.40 per call. The expected value of information, EVPI, gives the maximum amount the company might be willing to pay for perfect information about the outcome.

22.6 Decisions Made with Sample Information

Usually the costs and profits—the payoffs—of alternative actions under various states of nature can be estimated fairly well, because they depend on business-related events and actions that are well understood. By contrast, the probabilities assigned to the states of nature may be based only on expert judgment. These probabilities are sometimes called **prior probabilities** because they are determined before obtaining any additional information on the states of nature. A survey of customers, however, or a planned experiment might provide information that could bring the probabilities closer to reality and make for a more informed decision. Revised probabilities are also called **posterior probabilities**. Should the decisionmaker first gather data to help estimate the probabilities of the states of nature? That's a decision in itself. Surveys and experiments cost money, and the resulting information may not be worth the expense. We can incorporate the decision to gather data into the overall

decision process. First, it is worthwhile to ask whether useful information is even possible. If the states of nature in the decision process are about the future health of the economy or the future value of the Dow Jones Industrial Average, it is not likely that we can learn very much new. But market research surveys of customers and trial marketing incentives can often provide useful data.

To improve their estimates of the probabilities making up the mix of simple and complex questions, Data Description could collect data about their customers. Would the added expense of that work be worthwhile? It depends on whether the information about customers helps to predict the likelihood of each type of question. Suppose they could classify their users into those who are *technically sophisticated* users and those who are *beginners*. That information might be useful because for *technically sophisticated* users, Data Description thinks that 75% of the questions will be complex but for *beginners*, that percentage is only 40%. Knowing how to classify their customers will help them make the right decision.

These probabilities are *conditional* probabilities of the kind we first saw in Chapter 5. We write:

$$P(\text{Complex}|\text{Sophisticated}) = 0.75; \; P(\text{Simple}|\text{Sophisticated}) = 0.25$$
$$P(\text{Complex}|\text{Beginner}) = 0.40; \; P(\text{Simple}|\text{Beginner}) = 0.60$$

In each case, we know the missing probability from the requirement that the probabilities sum to one.

Now our decision tree must include both the decision of whether or not to conduct the study and the possible outcomes of the study (if it is conducted), each with its appropriate conditional probability. Figure 22.3 shows the results. The costs in Figure 22.3 are the same ones we've used all along from Table 22.1. The expected values when there is no study are those worked out previously in Figure 22.2.

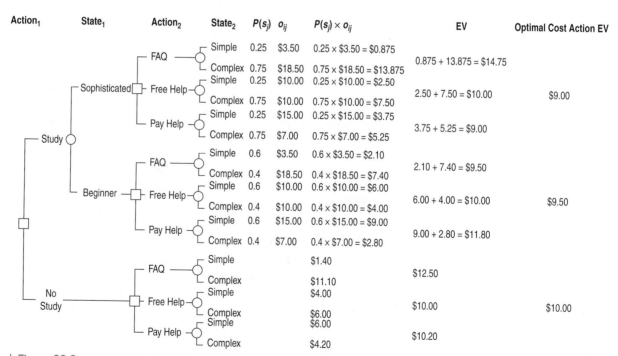

Figure 22.3 *A decision tree including a customer history study. For each decision node under each circumstance, the optimal (in this example, lowest cost) alternative is indicated in color, and the corresponding consequence is assigned as the expected value for that alternative.*

This decision process has two sets of decision nodes. The first decision is whether to conduct the study. The second is what action to take given the information from the study (if available). The outcome of the study is a chance node because we don't know *a priori* what the study will tell us.

To determine a decision strategy we compute the expected payoff at the end of each branch of the tree following the same methods we used before and now using the estimated probabilities that might come from the study. In this example, if we perform the study and it shows that sophisticated users are more common, we'll choose a pay help desk at a cost of $9.00. If we perform the study but it shows that beginners are the norm, we'll choose the FAQ system, at a cost of $9.50. If we don't perform the study at all, our optimal choice was the free phone help at a cost of $10.00. In this example, both alternatives following the study are less costly than the nonstudy choice, so it appears that any information the study may provide will save the company money. To be more precise, however, the company can go further and estimate the likelihood of the study's outcome. This will give the company an estimate of the **expected value with sample information** and help them estimate the potential worth of the study.

For example, if they think there's a probability of 0.8 that the study will show that most users are now technically sophisticated (and, correspondingly, a probability of 0.2 that it would find that beginners are the norm), then the expected value of the decision with that sample information would be:

$$EVwSI = 0.8 \times \$9.00 + 0.2 \times \$9.50 = \$9.10$$

The absolute value of the difference between this payoff and the payoff without the study is the **expected value of sample information.**

$$EVSI = |EVwSI - EV| = |\$9.10 - \$10.00| = \$0.90$$

An analysis like this can help us understand the value of sample information. It tells us that a study of this sort is likely to be worth about $0.90 per technical support question in its ability to help the company choose the optimal action.

22.7 Estimating Variation

In Chapter 6, we said that means should always be reported with an associated standard deviation, but thus far we've found expected values (which are, after all, means) without finding the corresponding standard deviations. The expected values tell the decisionmaker which action is likely to be worth the most. But to assess the *risk* of the decision, a decisionmaker must understand whether the likely outcome of the decision will always be close to that expected value or whether it might be quite different. We assess that risk by finding the standard deviation of the outcome.

Recall that the calculation of the expected value of an action is:

$$EV(a_i) = \sum_{j=1}^{N} o_{ij} P(s_j),$$

where there are N possible states of nature, o_{ij} is the outcome or payoff of action i when the state of nature is s_j, and $P(s_j)$ is the probability of the jth state of nature. We saw the equivalent formula for expected value in Chapter 21. The corresponding formulas for variance and standard deviation (also seen in Chapter 21) are:

$$Var(a_i) = \sum_{i=1}^{N} (o_{ij} - EV(a_i))^2 P(s_j),$$

$$\text{and } SD(a_i) = \sqrt{Var(a_i)}.$$

To calculate some values, we'll need the outcomes and probabilities we saw earlier and the expected values we calculated for the actions. Recall that Data Description thinks the probability of a simple question is 0.40 and of a complex question is 0.60. The payoffs for each state of nature and action were first given in Table 22.1. Now we also have expected values found from those payoffs and probabilities. Table 22.3 summarizes them.

		State of Nature		
		Simple Question $P=0.40$	**Complex Question** $P=0.60$	$EV(a_i)$
Action	**FAQ**	$3.50	$18.50	$12.50
	Free Help Desk	$10.00	$10.00	$10.00
	Pay Help Desk	$15.00	$7.00	$10.20

Table 22.3   *The payoff table for the customer support decision along with the expected values computed earlier. Recall that these are costs and thus could be written as negative values. Of course, for finding a standard deviation, writing these as negative numbers would give the same result.*

Now, for example, the standard deviation of the outcome for the FAQ action is

$$SD(FAQ) = \sqrt{(3.50 - 12.50)^2 \times 0.40 + (18.50 - 12.50)^2 \times 0.60} = \$7.35$$

and the standard deviation of the Pay Help Desk option is

$$SD(Pay\,Help) = \sqrt{(15.00 - 10.20)^2 \times 0.4 + (7.00 - 10.20)^2 \times 0.6} = \$3.92$$

The standard deviation of the Free Help Desk option is zero. (Can you see why?)

In the customer support example, the choice to rely on an online FAQ system has higher variability. That may make it a less desirable alternative even in situations where its expected value seems to be optimal.

One way to combine the expected value and standard deviation is to find their ratio, the **coefficient of variation**.

$$CV(a_i) = \frac{SD(a_i)}{EV(a_i)}.$$

For the customer support example, the CVs are 0.588 for online FAQ, 0 for Free Help, and 0.384 for Pay Help. Actions with smaller CVs are generally seen as less risky, and may be preferred for that reason by some people. You can see why Data Description may prefer Free Help even though it may be more expensive; the company can predict its cost regardless of the state of nature.

One action may dominate another if its worst outcome is still better than the best outcome for the other. In such cases, the better action should be chosen regardless of how much variation it may have. **Dominance** is defined as the situation in which one decision alternative is never an optimal action regardless of the state of nature. A dominated action can be eliminated as a variable option or strategy.

Some discussions of risk prefer to concentrate on the expected value rather than the standard deviation, so they look at the reciprocal of the CV and give it the name **return to risk ratio**

$$RRR(a_i) = \frac{1}{CV(a_i)} = \frac{EV(a_i)}{SD(a_i)}.$$

The RRR can't be computed for actions with zero standard deviation, because they have no risk. Generally, actions with higher RRR are favored because they are less risky for their expected return. The units of RRR are dollars returned per dollar at risk.

22.8 Sensitivity

One weakness of the methods we've discussed here is that they require that you estimate—subjectively or by research—probabilities of states of nature. You may wish to assess the *sensitivity* of your conclusions about the expected value and standard deviation of alternative actions to the probabilities in your model. One way to do this is to re-compute those values with slightly (or even substantially) different probabilities. If the decision advised by the model doesn't change, you can be more confident in it. But if small changes in the probabilities result in large differences in the estimated value of the action alternatives, you should take great care and not rely too heavily on your decision analysis.[2]

For example, if the probability of a complex question were 0.8 rather than 0.6, would that change the decision? Table 22.4 shows the calculations.

	State of Nature		
	Simple Question $P = 0.20$	**Complex Question** $P = 0.80$	$EV(a_i)$
FAQ	$3.50	$18.50	$15.50
Free Help Desk	$10.00	$10.00	$10.00
Pay Help Desk	$15.00	$7.00	$8.60

(Row label: **Action**)

Table 22.4   Alternative calculations (compare with Table 22.3) with different probabilities for the states of nature.

It seems that the best decision would change if the probability of a complex question were as large as 0.8. The decisionmakers at Data Description should consider how sensitive these expected values are to their probabilities by finding the SD and EV for a plausible range of probabilities and then think about how certain they are about those estimated probabilities.

22.9 Simulation

Another alternative to assessing the sensitivity of a decision model to the choice of probabilities is to simulate the model for a variety of plausible probability values. Rather than specify single probabilities, you can specify a distribution of plausible values. You can then run a *simulation* (such as those we discussed in Chapter 5) in which the computer draws states of nature at random according to your distribution of probabilities and evaluates the consequences. This approach can deal with much more complex decision models than those we've discussed here. The result of the simulation is not a single decision but a distribution of outcomes. Such a distribution may be a more appropriate description for a decisionmaker than any single expected value.

---

[2] If you completed the Mini Case Studies at the end of Chapter 21, you have already conducted a small sensitivity analysis.

Programs such as @*Risk* (www.palisade.com/) and Crystal Ball (www.oracle.com/appserver/business-intelligence/crystalball/index.html) provide ways for you to specify actions, states of nature, and the associated probabilities and outcomes and then use simulation methods to generate distributions of outcomes.

GUIDED EXAMPLE | *Insurance Services*

 InterCon Travel Health is a Toronto-based company that provides services to health insurers of foreign tourists who travel to the United States and Canada.[3] The primary focus of InterCon is to act as an interface between local health providers and the overseas insurance company who insured the tourist should the traveler find himself or herself in need of medical attention. The customer base of sick or injured tourists is potentially profitable because most tourists who fall ill require only minor treatment and rarely come back for expensive follow-up visits. So InterCon can pass savings along to the foreign insurer by facilitating claim management and payment and can collect processing fees from these insurance companies in return.

Currently, InterCon charges a processing fee of 9.5%, collected partly from the medical providers and partly from the foreign insurers. It has experienced an average annual growth of 3%. However, to help with rising costs, they may consider increasing the processing fee from 9.5% to 10.5%. Although this would generate additional revenue, they would also incur the risk of losing contracts with insurers and health care providers. Table 22.5 gives the company's estimates of the impact of a change in rates on the annual growth in claims depending on the strength of foreign tourism

	State of Nature	
	Weak Tourism $P = 0.40$	**Strong Tourism** $P = 0.60$
9.5% Fee	$54.07M	$56.23M
10.5% Fee	$52.97M	$54.56M

(Action)

Table 22.5 *InterCon's estimates of the impact on the annual growth in claims for two different actions (changes in rate) under two possible states of nature (strong or weak tourism).*

What would be the best choice for the company?

 **PLAN**

Setup State the objective of the study.	We want to evaluate the two alternative actions the company is considering.
Identify the variables.	We have estimates of the outcomes and the probabilities of two states of nature: weak or strong tourism.
Model Think about the assumptions and check the conditions.	The only condition to check is that the probabilities are legitimate; 0.40 + 0.60 = 1.0, and 0.4 and 0.6 are both between 0 and 1.

---

[3] This example is based on the case by G. Truman, D. Pachamanova, and M. Goldstein entitled "InterCon Travel Health Case Study" in *Journal of the Academy of Business Education*, Vol. 8, Summer 2007, pp. 17–32.

DO

Mechanics The calculations of expected value and standard deviation are straightforward.

P	Action	State	Outcome	$o_{ij}P(s)$	$EV(a_i)$
0.4		Weak	$54.07M	$21.63M	
	9.5% fee				$55.37M
0.6		Strong	$56.23M	$33.74M	
0.4		Weak	$52.97M	$21.188M	
	10.5% fee				$53.92M
0.6		Strong	$54.56M	$32.736M	

It can help to diagram the decision process with a tree diagram.

For example:

$$0.4 \times \$54.07M = \$21.63M$$

$$\$21.63M + \$33.74M = \$55.37M$$

Standard deviations:

For 9.5% fee:

$$(o_{ij} - EV(a_i)) \qquad\qquad (o_{ij} - EV(a_i))^2 P(s_j)$$
$$(54.07 - 55.37) = -1.30 \quad (-1.30)^2 \times 0.4 = 0.676$$
$$(56.23 - 55.37) = 0.86 \quad (0.86^2) \times 0.6 = 0.444$$
$$SD = \sqrt{0.676 + 0.444}$$
$$= \$1.06M$$

For 10.5% fee:

$$SD = \$0.779M$$

RRR (9.5% fee) = 55.37/1.06 = 52.24

RRR (10.5% fee) = 53.92/0.779 = 69.22

REPORT

Summary and Conclusions Summarize your results and state any limitations of your model in the context of your original objectives.

MEMO:

Re: Recommendations for Processing Fee Rate

The expected value of the alternative with the lower processing fee rate is slightly higher ($55.37 million compared to $53.92 million), but it is also a riskier decision since resulting revenue will be less if tourism remains strong. The return to risk ratio (RRR) at the higher processing fee rate of 10.5% is 69.22; while at the lower rate, it is only 52.24.

Because one action has the better expected value but the other seems less risky, the company should determine the impact of the downside of their decision and evaluate the probability estimates for the strength of the foreign tourism industry.

22.10 Probability Trees

Some decisions involve more subtle evaluation of probabilities. Given the probabilities of various states of nature, an analyst can use trees similar to the ones presented earlier to calculate the likelihood of complex combinations of events. This can enable the analyst to compare several possible scenarios. Here's a manufacturing example.

Personal electronic devices, such as personal digital assistants (PDAs) and MP3 players, are getting smaller all the time. Manufacturing components for these devices is a challenge, and at the same time, consumers are demanding more and more functionality and increasing sturdiness. Microscopic and even submicroscopic flaws can develop during their fabrication that can blank out pixels on the screens or cause intermittent performance failures. Defects will always occur, so the quality engineer in charge of the production process must monitor the number of defects and take action if the process seems out of control.

Let's suppose that the engineer is called down to the production line because the number of defects has crossed a threshold and the process has been declared to be out of control. She must decide between two possible actions. She knows that a small adjustment to the robots that assemble the components can fix a variety of problems, but for more complex problems, the entire production line needs to be shut down in order to pinpoint the problem. The adjustment requires that production be stopped for about an hour. But shutting down the line takes at least an entire shift (8 hours). Naturally, her boss would prefer that she make the simple adjustment. But without knowing the source or severity of the problem, she can't be sure whether that will be successful.

If the engineer wants to predict whether the smaller adjustment will work, she can use a probability tree to help make the decision. This tree is much like the decision tree that we saw before, except that now each node corresponds to a state of nature that occurs randomly, whereas in the decision tree, some of the nodes depicted decisions made to select an action. Based on her experience, the engineer thinks that there are three possible problems: (1) the motherboards could have faulty connections, (2) the memory could be the source of the faulty connections, or (3) some of the cases may simply be seating incorrectly in the assembly line. She knows from past experience how often these types of problem crop up and how likely it is that just making an adjustment will fix each type of problem. *Motherboard* problems are rare (10%), *memory* problems have been showing up about 30% of the time, and *case* alignment issues occur most often (60%). We can put those probabilities on the first set of branches.

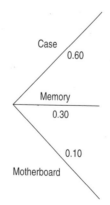

Case
0.60

Memory
0.30

0.10
Motherboard

Figure 22.4 *Possible problems and their probabilities.*

Notice that we've covered all the possibilities, and so the probabilities sum to one. To this diagram we can now add the *conditional* probabilities that a minor adjustment will fix each type of problem. Most likely the engineer will rely on her experience or assemble a team to help determine these probabilities. For example, the engineer knows that motherboard connection problems are not likely to be fixed with a simple adjustment: P(Fix|Motherboard) = 0.10. After some discussion, she and her team determine that P(Fix|Memory) = 0.50 and P(Fix|Case alignment) = 0.80. At the end of each branch representing the problem type, we draw two possibilities (*Fix* or *Not Fixed*) and write the conditional probabilities on the branches.

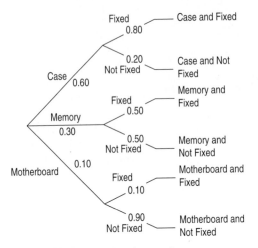

Figure 22.5 *Extending the tree diagram, we can show both the problem class and the outcome probabilities. The outcome (Fixed or Not fixed) probabilities are conditional on the problem type, and they change depending on which branch we follow.*

At the end of each second branch, we write the *joint event* corresponding to the combination of the two branches. For example, the top branch is the combination of the problem being Case alignment, and the outcome of the small adjustment is that the problem is now Fixed. For each of the joint events, we can use the general multiplication rule to calculate their joint probability. For example:

$$P(Case \text{ and } Fixed) = P(Case) \times P(Fixed|Case)$$
$$= 0.60 \times 0.80 = 0.48$$

We write this probability next to the corresponding event. Doing this for all branch combinations gives the Figure 22.6.

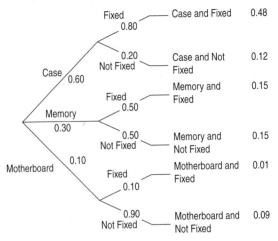

Figure 22.6 *We can find the probabilities of compound events by multiplying the probabilities along the branch of the tree that leads to the event, just the way the General Multiplication Rule specifies.*

All the outcomes at the far right are disjoint because at every node, all the choices are disjoint alternatives. And those alternatives are *all* the possibilities, so the probabilities on the far right must add up to one.

Because the final outcomes are disjoint, we can add up any combination of probabilities to find probabilities for compound events. In particular, the engineer can answer her question: What's the probability that the problem will be fixed by a simple adjustment? She finds all the outcomes on the far right in which the problem was fixed. There are three (one corresponding to each type of problem), and she adds their probabilities: $0.48 + 0.15 + 0.01 = 0.64$. So 64% of all problems are fixed by the simple adjustment. The other 36% require a major investigation.

*22.11 Reversing the Conditioning: Bayes's Rule

The engineer in our story decided to try the simple adjustment and, fortunately, it worked. Now she needs to report to the quality engineer on the next shift what she thinks the problem was. Was it more likely to be a case alignment problem or a motherboard problem? We know the probabilities of those problems beforehand, but they change now that we have more information. What are the likelihoods that each of the possible problems was, in fact, the one that occurred?

Unfortunately, we can't read those probabilities from the tree in Figure 22.6. For example, the tree gives us $P(\textit{Fixed and Case}) = 0.48$, but we want $P(\textit{Case}|\textit{Fixed})$. We know $P(\textit{Fixed}|\textit{Case}) = 0.80$, but that's not the same thing. It isn't valid to reverse the order of conditioning in a conditional probability statement. To "turn" the probability around, we need to go back to the definition of conditional probability.

$$P(\textit{Case}|\textit{Fixed}) = \frac{P(\textit{Case and Fixed})}{P(\textit{Fixed})}$$

We can read the probability in the numerator from the tree, and we've already calculated the probability in the denominator by adding all the probabilities on the final branches that correspond to the event *Fixed*. Putting those values in the formula, the engineer finds:

$$P(\textit{Case}|\textit{Fixed}) = \frac{0.48}{0.48 + 0.15 + 0.01} = 0.75$$

She knew that 60% of all problems were due to case alignment, but now that she knows the problem has been fixed, she knows more. Given the additional information that a simple adjustment was able to fix the problem, she now can increase the probability that the problem was case alignment to 0.75.

It's usually easiest to solve problems like this by reading the appropriate probabilities from the tree. However, we can write a general formula for finding the reverse conditional probability. To understand it, let's review our example again. Let $A_1 = \{Case\}$, $A_2 = \{Memory\}$, and $A_3 = \{Motherboard\}$ represent the three types of problems. Let $B = \{Fixed\}$, meaning that the simple adjustment fixed the problem. We know $P(B|A_1) = 0.80$, $P(B|A_2) = 0.50$, and $P(B|A_3) = 0.10$. We want to find the reverse probabilities, $P(A_i|B)$, for the three possible problem types. From the definition of conditional probability, we know (for any of the three types of problems):

$$P(A_i|B) = \frac{P(A_i \textit{ and } B)}{P(B)}$$

We still don't know either of these quantities, but we use the definition of conditional probability again to find $P(A_i \textit{ and } B) = P(B|A_i)P(A_i)$, both of which we know. Finally, we find P(B) by adding up the probabilities of the three events.

$$P(B) = P(A_1 \textit{ and } B) + P(A_2 \textit{ and } B) + P(A_3 \textit{ and } B) =$$
$$P(A_1|B)P(A_1) + P(A_2|B)P(A_2) + P(A_3|B)P(A_3)$$

In general, we can write this for *n* events A_i that are mutually exclusive (each pair is disjoint) and exhaustive (their union is the whole space). Then:

$$P(A_i|B) = \frac{P(B|A_i)P(A_i)}{\sum\limits_j P(B|A_j)P(A_j)}$$

This formula is known as Bayes's rule, after the Reverend Thomas Bayes (1702–1761), even though historians doubt that it was actually Bayes who first came up with the reverse conditioning probability. When you need to find reverse conditional probabilities, we recommend drawing a tree and finding the appropriate probabilities as we did at the beginning of the section, but the formula gives the general rule.

22.12 More Complex Decisions

The methods of this chapter can be extended to decisions that are more complex and that may have second or third stages conditional on the outcomes of the first decisions. For example, customer support questions might be handled best by an initial online FAQ system that then offers customers with complex questions a choice of free e-mail support or paid telephone support. To work out the expected values for these choices, we'd need to estimate the probabilities of a customer with a complex question choosing free e-mail or paid telephone support—something with which the company has no prior experience.

Or the probabilities may be changed by a change in circumstances. When the company releases an upgraded version, technical support calls increase, but there's a higher probability of simple calls. As the customer base learns the new version, their questions generally become more sophisticated.

WHAT CAN GO WRONG?

- **Decision trees do not tell the decisionmaker the correct decision.** Models can be useful, but they are based on assumptions. Decision trees model the expected outcome and risk for each decision. The best decision depends on the level of risk the decisionmaker is willing to accept, as well as the state of nature that occurs.

- **The computation of the expected values and standard deviations are sensitive to the probabilities.** The probabilities required by the calculations of expected value and standard deviation may be reliably based on past experience or data, or they may just represent the subjective judgment of the decisionmaker. Don't trust them blindly. Using a decision model and ignoring the sensitivity of the model to different probabilities can lead to poor decisions and misleading conclusions. One way to address the problem is to solve the model with a range of plausible probabilities to assess how sensitive the conclusions are to the choice of probabilities. Another way is to use software that simulates the results for randomly generated probabilities from an appropriate distribution.

- **Be sure that the probabilities are legitimate and that the action choices are disjoint.** The calculations in this chapter depend upon both of these being true. Methods for conditional probabilities exist but are beyond the scope of this chapter.

ETHICS IN ACTION

Nelson Greene develops real estate in the southwestern United States. His next planned project is to build a housing development on a parcel of land in New Mexico, not far from Santa Fe. However, with the recent decline in housing prices, particularly in areas of the southwest like Las Vegas, Nelson is reconsidering his decision. He set up a meeting with some key individuals, including Tom Barnick, a representative from the township supervisors. Tom and the township council were positive about the proposed development since it would provide increased tax revenue. So Tom wanted to influence Nelson to go forward with the development. Tom had some analysis prepared, which he presented at the meeting. Based on several assumptions, including the probabilities associated with different housing market conditions in New Mexico (improving, worsening, or staying the same), the analysis showed a positive expected payoff for going forward with the project. Tom mentioned that the probabilities were obtained from a small group of housing market experts, although they had difficulty reaching consensus. Nelson noticed that not all the payoffs associated with going forward with the project were positive for each possible market condition, but the positive expected payoff made him view the project more favorably.

ETHICAL ISSUE *The expected payoff is a long run average and not an actual outcome for a decision. Although the expected payoff was positive, there was no mention of the expected payoff for not going forward with the development.*

ETHICAL SOLUTION *If the experts couldn't agree on the probabilities, the sensitivity analysis should be done for a range of probabilities. Related to Item A, ASA Ethical Guidelines.*

What have we learned?

We've learned to evaluate the likely outcomes of decisions when faced with uncertain outcomes. We can use payoff tables or tree diagrams to help structure thinking about the decision process. And we've learned that decision models aid the decisionmaker by providing information regarding the expected payoff (or cost) and relative risk of each decision; the actual decision that is made depends on the level of risk that each decisionmaker is willing to accept.

Terms

Action　Alternative choices that one can make for a decision.

Coefficient of variation　The coefficient of variation (CV) of each decision shows the "$ at risk" for each "$ returned":
$$CV(a_i) = SD(a_i)/EV(a_i).$$

Decision tree　A decision tree organizes the actions, states of nature, and outcomes to display the sequence of the elements in a decision.

Dominance　An alternative dominates when it pays off at least as much as (or costs less than) the alternative it dominates.

Expected value　The expected value of an action under uncertain states of nature is:
$$EV(a_i) = \sum_{j=1}^{N} o_{ij}P(s_j)$$

where there are N possible states of nature and o_{ij} is the outcome or payoff of action i when the state of nature is s_j.

Expected value of perfect information (EVPI)	The difference between the expected value with perfect information (EVwPI) and the expected value (without perfect information) of the strategy (EV): $$EVPI = |EVwPI - EV|$$
Expected value with perfect information (EVwPI)	The expected value of a hypothetical strategy assuming that probabilities of the states of nature are known.
Expected value of sample information (EVSI)	The difference between the expected value with sample information (EVwSI) and the expected value (without sample information) of the strategy (EV): $$EVSI = |EVwSI - EV|$$
Expected value with sample information (EVwSI)	The expected value of a hypothetical strategy assuming that probabilities of the states of nature are estimated.
Maximax choice (or strategy)	The action that maximizes the maximum possible return across all states of nature in a payoff table.
Minimax choice (or strategy)	The action that minimizes the maximum possible cost across all states of nature in a payoff table.
Minimin choice (or strategy)	The action that minimizes the minimum possible cost across all states of nature in a payoff table.
Outcome (or payoff)	The consequence of an action combined with a state of nature, usually expressed in monetary units (positive or negative).
Payoff table	A table containing the actions, states of nature, and the corresponding payoffs for each combination of the two.
Return to risk ratio	The return to risk ratio (RRR) measures the expected return of an action relative to the risk, represented by its standard deviation: $$RRR(a_i) = 1/CV = EV(a_i)/SD(a_i).$$
Standard deviation of an action, Variance of an action	The standard deviation of an action measures the variability of possible outcomes, calculated in terms of their estimated probabilities of occurring. The standard deviation is the square root of the variance: $$Var(a_i) = \sum_{i=1}^{N} (o_{ij} - EV(a_i))^2 \, P(s_j)$$ $$SD(a_i) = \sqrt{Var(a_i)}$$
States of nature	The facts about the world that affect the consequences of an action, usually summarized into a few possible values.

Skills

PLAN

When you complete this lesson, you should:

- Be able to identify alternative actions and states of nature for a decision process.

- Be able to assign probabilities to states of nature based on judgment or data and be able to check that these probabilities are legitimate (i.e., that each is between 0 and 1 and that they sum to 1.0).

- Know how to draw the decision tree for a set of alternative actions.
- Be able to find the expected value of an action under uncertain states of nature.
- Know how to find the coefficient of variation and risk to return ratio for an action.

- Know how to evaluate the relative risk of each alternative decision and recommend a decision for decisionmakers who are risk averse, risk neutral, and risktakers.

Mini Case Study Projects

Texaco-Pennzoil

"Oil is big business." A classic example of this is the Texaco-Pennzoil court case, which appeared in the book *Making Hard Decisions*[4] and in a subsequent case study by T. Reilly and N. Sharpe (2001). In 1984, a merger was hammered out between two oil giants, Pennzoil and Getty Oil. Before the specifics had been agreed to in a written and binding form, another oil giant—Texaco—offered Getty Oil more money. Ultimately, Getty sold out to Texaco.

Pennzoil immediately sued Texaco for illegal interference, and in late 1985 was awarded $11.1 billion—an enormous award at the time. (A subsequent appeal reduced the award to $10.3 billion.) The CEO of Texaco threatened to fight the judgment all the way to the U.S. Supreme Court, citing improper negotiations held between Pennzoil and Getty. Concerned about bankruptcy if forced to pay the required sum of money, Texaco offered Pennzoil $2 billion to settle the case. Pennzoil considered the offer, analyzed the alternatives, and decided that a settlement price closer to $5 billion would be more reasonable.

The CEO of Pennzoil had a decision to make. He could make the low-risk decision of accepting the $2 billion offer, or he could decide to make the counteroffer of $5 billion. If Pennzoil countered with $5 billion, what are the possible outcomes? First, Texaco could accept the offer. Second, Texaco could refuse to negotiate and demand settlement in the courts. Assume that the courts could order one of the following:

- Texaco must pay Pennzoil $10.3 billion.
- Texaco must pay Pennzoil's figure of $5 billion.
- Texaco wins and pays Pennzoil nothing.

The award associated with each outcome—whether ordered by the court or agreed upon by the two parties—is what we will consider to be the "payoff" for Pennzoil. To simplify Pennzoil's decision process, we make a few assumptions. First, we assume that Pennzoil's objective is to maximize the amount of the settlement. Second, the likelihood of each of the outcomes in this high-profile case is based on similar cases. We will assume that there is an even chance (50%) that Texaco will refuse the counteroffer and go to court. In a *Fortune* article,[5] the CEO of Pennzoil reportedly believed that should the offer be refused, Texaco had a chance to win the case with appeals, which would leave Pennzoil with high legal

[4] Clemen, R. T. and Reilly, T. (2001), *Making Hard Decisions*, New York: Brooks/Cole
[5] *Fortune*, May 11, 1987, pp. 50–58.

fees and no payoff. Based on prior similar court cases and expert opinion, assume that there is also a 50% probability that the court will order a compromise and require Texaco to pay Pennzoil the suggested price of $5 billion. What are the remaining options for the court? Assume that the other two alternatives—Pennzoil receiving the original total award ($10.3 billion) or Pennzoil getting nothing—are almost equal, with the likelihood of the original verdict being upheld slightly greater (30%) than the likelihood of reversing the decision (20%).

Evaluate the expected payoff and risk of each decision for Pennzoil.

Insurance Services, Revisited

InterCon Travel Health is a Toronto-based company that provides services to health insurers of foreign tourists who travel to the United States and Canada.[6] As described in the Guided Example in this chapter, the primary focus of InterCon is to act as an interface between local health providers and the overseas insurance company who insured the tourist should the traveler find himself or herself in need of medical attention. The customer base of sick or injured tourists is potentially profitable because most tourists who fall ill require only minor treatment and rarely come back for expensive follow-up visits. Therefore, InterCon can pass savings along to the foreign insurer by facilitating claim management and payment and can collect processing fees from these insurance companies in return.

Currently, this Canadian company charges a processing fee of 9.5% (collected partly from the medical providers and partly from the foreign insurers) and has experienced an average annual growth rate of 3%. However, a backlog of claims and a cyclical pattern to the claims (due to high tourist seasons) has caused a delay in filing claims with the foreign insurers, resulting in a number of noncollectible claims. In addition, while the growing company has been trying to keep costs to a minimum, they are considering adding a new information technology system to help them streamline the process. Most of the company's revenue comes from claims for inpatient hospital stays, which average approximately $10,000. These claims represent 20% of InterCon's claims, but over 80% of its claim revenue. The remaining claims arise from clinic and emergency room visits. Currently, the company has about 20,000 annual claims, so the associated revenue is estimated to be as follows.

Claim Class	# Claims by Class	% of Total	Average Claim Amount	Claim Revenue ($M)	% of Total	Fee Rate	Fees ($M)
Class A	4000	20%	$10000	$40.0	85%	9.5%	$3.8
Class B	6000	30%	$1000	$6.0	13%	9.5%	$0.57
Class C	10000	50%	$100	$1.0	2%	9.5%	$0.095
Total (Annual)	**20000**			**$47.0**			**$4.465**

To help with rising costs, this firm may consider raising the processing fee as high as 11.0%. While this would generate additional revenue, the firm would also incur the risk of losing contracts with insurers and health care providers. Here are estimates of the impact of a change in rates on the annual growth in claims depending on the strength of foreign tourism (reverse of the strength of the U.S. economy) for the 9.5% processing fee rate. Find the payoffs (total revenue) for the 11.0% fee rate based on the preceding table and the growth rates provided.

---

[6] This case is based on the case by G. Truman, D. Pachamanova, and M. Goldstein entitled "InterCon Travel Health Case Study" in *Journal of the Academy of Business Education*, Vol. 8, Summer 2007, pp. 17–32.

Foreign Tourism	Probability	Processing Fee Rate at 9.5%		Processing Fee Rate at 11.0%	
		Growth Rate	Payoff ($)	Growth Rate	Payoff ($)
Weak	0.40	3%	54069129	1%	??
Strong	0.60	4%	56231894	3%	??

What is the expected revenue for the respective fee alternatives based on the growth rate and total income from claims and fees? Suppose the manager gets new information and revises the probabilities for the strength of tourism to be even at 50-50. What are the new values for expected revenue, CV, and return to risk ratio?

EXERCISES

1. Flight decision. You are planning a trip home at the end of the semester and need to make plane reservations soon. However, you've just had a preliminary interview with a consulting firm that seemed to go very well. There's a chance that they will want you to stay for a few days at the end of the semester for a round of interviews at their offices, which will mean you'll have to change the date of the flight if you make the reservation now. Suppose that you can purchase a changeable fare for $750 or a non-refundable fare for $650 for which a change costs $150. Construct a payoff table for this set of actions using total cost as the "payoff."

2. Product introduction. A small company has the technology to develop a new personal data assistant (PDA), but it worries about sales in the crowded market. They estimate that it will cost $600,000 to develop, launch, and market the product. Analysts have produced revenue estimates for three scenarios: If sales are high, they will sell $1.2M worth of the phones, if sales are moderate, they will sell $800,000 worth, and if sales are low, they will sell only $300,000 worth. Construct a payoff table for this set of actions using net profit as the "payoff." Don't forget the possible action of doing nothing.

3. Advertising strategies. After a series of extensive meetings, several of the key decisionmakers for a small marketing firm have produced the following payoff table (expected profit per customer) for various advertising strategies and two possible states of the economy.

		Consumer Confidence	
		Rising	Falling
Action	Prime Time Ad	$20.00	$2.00
	Targeted Web Marketing	$12.00	$10.00
	Direct Mail Piece	$10.00	$15.00

Construct a decision tree for this payoff table.

4. Energy investment. An investment bank is thinking of investing in a start-up alternative energy company. They can become a major investor for $6M, a moderate investor for $3M, or a small investor for $1.5M. The worth of their investment in 12 months will depend on how the price of oil behaves between then and now. A financial analyst produces the following payoff table with the net worth of their investment (predicted worth— initial investment) as the payoff.

		Price of Oil		
		Substantially Higher	About the Same	Substantially Lower
Action	Major investment	$5,000,000	$3,000,000	−$2,000,000
	Moderate investment	$2,500,000	$1,500,000	−$1,000,000
	Small investment	$1,000,000	$500,000	−$100,000

Construct a decision tree for this payoff table.

5. Flight decision tree. Construct a decision tree for the payoff table in Exercise 1.

6. Product introduction tree. Construct a decision tree for the payoff table in Exercise 2.

7. Flight decision expected value. If you think the probability of being called for an interview is 0.30, calculate the expected value of each action in Exercise 1. Which is the better action in this case?

8. Product introduction expected value. An analyst for the company in Exercise 2 thinks the probabilities of high, moderate, and low sales are 0.2, 0.5, and 0.3, respectively. In this case calculate the expected value of each action. Which is the best action in this case?

9. Flight decision, part 4. For the decision of Exercise 1, you've just learned that you are on the short list and now estimate the chance that you'll be called for an interview is 0.70. Does that change your choice of actions?

10. Product introduction, part 4. For the product launch decision of Exercise 2, the economy isn't looking that good. Your very cautious boss says that he thinks there's a 60% chance of low sales, and a 30% chance of moderate sales. Which course should the company follow?

11. Advertising strategies decisions. For the payoff table in Exercise 3, find the action with the highest expected value.
a) If forecasters think the probability of rising consumer confidence is 0.70, what is its expected value?
b) What action would have the highest expected value if they think the probability of rising consumer confidence is only 0.40?

12. Energy investment decisions.
a) For the payoff table in Exercise 4, find the investment strategy under the assumption that the probability that the price of oil goes substantially higher is 0.4 and that the probability that it goes substantially lower is 0.2.
b) What if those two probabilities are reversed?

13. Advertising strategies EVPI.
a) For the advertising strategies of Exercise 11 and using the probability of 0.70 for rising consumer confidence, what is the Expected Value of Perfect Information (EVwPI)?
b) What is the EVPI if the probability of rising consumer confidence is only 0.40?

14. Energy investment EVPI. For the energy investment of Exercise 12 and using both of the probabilities considered in that exercise, find the Expected Value of Perfect Information.

15. Advertising strategies with information. The company from Exercises 3, 11, and 13 has the option of hiring an economics consulting firm to predict consumer confidence. The company has already considered that the probability of rising consumer confidence could be as high as 0.70 or as low as 0.40. They could ask the consultants for their choice between those two probabilities, or they could just pick a probability in the middle, such as .50, and choose a strategy based on that.
a) Draw the decision tree including the decision to hire the consultants.
b) Would the consultants' information be useful? Explain.
c) The company thinks there's an equal chance of either of the consulting alternatives being what the consultants report. What's the value to the company (per customer) of the extra information?

16. Energy investment with information. The company in Exercises 4, 12, and 14 could send a team to Saudi Arabia to obtain additional information about the probabilities that oil will increase or decrease in price. They hope that the fact-finding trip would choose between the two alternatives considered in Exercise 12, or they could just estimate that the probabilities are equal.

a) Make a decision tree for these decisions.
b) Should the company send the fact-finding trip? Explain.
c) The company's experts estimate that if they send the fact-finding mission, there's a 70% chance that they'll conclude there's a 0.4 probability of higher oil prices. What would the value of the additional information be to the company?

17. Investing in equipment. *KickGrass Lawncare* is a service that cares for lawns in a large, affluent community. Shawn Overgrowth, the owner, is considering the purchase of new zero-turn riding lawn tractors, which would allow him to expand his business. The tractors cost $6300 each, and he would purchase two of them. Another alternative is to purchase three additional mowers of the current type to add to his current equipment. Those would cost $475 apiece. Or he could face the coming gardening season with his existing equipment. Shawn estimates that in a good growing season, the tractors would allow him to expand his business by $40,000. But if the summer is hot and dry (so lawns don't grow) or cold and wet (ditto), he'd only be able to add about $15,000 in contracts. If he purchases the mowers, he could expand his business by $10,000 in a good year or by just $5,000 in a bad one. And if he spends nothing, he won't expand his business. In a bad year, his income would contract by about $1,000.
   Construct a payoff table and tree diagram for Shawn's decision. Don't forget to include his expenses in the calculations.

18. Market segmentation. Demand and price are related; raising prices typically lowers demand. Many companies understand that if they can *segment* their market and offer different prices to different segments, they can often capture more revenue. *Aaron'sAir* is a small commuter airline. They typically charge $150 for a one-way flight between a resort island they serve and the mainland. In times of low, medium, and high demand, Aaron (the owner and pilot) estimates that he'll sell 100, 200, or 500 seats per week, respectively. He is considering offering two different fares based on whether his customers stay over a Saturday night on the island. He thinks that business travelers coming to the island for conferences and retreats would typically not stay, but vacationers would. He expects that the low fare will attract additional customers. However, he anticipates that some of his regular customers will also pay less. The two fares would be $90 and $210. Aaron estimates that in times of low demand he'd sell 30 high-fare and 80 low-fare tickets—revenue of $30 \times \$210 + 80 \times \$90 = \$13,500$. In times of medium demand, he estimates 110 high-fare and 250 low-fare tickets, for an estimated revenue of $45,600. And in times of high demand, he expects 500 low-fare customers and 250 high-fare customers, yielding $97,500.
   Make a payoff table and decision tree for this decision.

19. Investing in equipment, maxes and mins. Shawn Overgrowth, whom we met in Exercise 17, is an entrepreneur who is optimistic about the growing season. What choice should he make to maximize his return? His assistant, Lance Broadleaf, is very conservative, and argues that *KickGrass* should minimize their potential downside results. Which alternative decision does he argue for?

20. Market segmentation minimax. Aaron, who we met in Exercise 18, tends to be optimistic about business conditions. What is his maximax strategy that would maximize his results?

21. Weather or not to invest in equipment. Shawn Overgrowth, from Exercise 17, estimates that the probability of a good growing season is 0.70. Based on that:

a) Find the EV for his actions.
b) Find the standard deviations.
c) Compute the RRRs. Which action is preferred based on the RRRs?

22. Market segmentation and chance. *Aaron'sAir* (see Exercises 18 and 20) estimates that high-demand periods (which depend on the weather and on bookings for conferences) occur with probability .3 and medium demand periods occur with probability .5. The rest are low-demand periods.

a) What's the expected value of each of Aaron's alternative actions?
b) What are the standard deviations for each action?
c) What are the RRRs? Based on the RRRs, what action is best?

23. Equipment and data. Shawn, of Exercises 17, 19, and 21, could obtain long-range predictions of the growing conditions for next summer. He thinks that those might show a probability of good growing conditions as low as 50% or as high as 80%. If he doesn't obtain those predictions, he'll make a decision based on his previous estimates (see Exercise 21).

a) Draw the decision tree.
b) If Shawn thinks there's a 60% chance the long-range predictions will predict a 50% chance of good conditions, find the corresponding EVwSI.
c) Should Shawn purchase the long-range predictions?

24. Segments and surveys. *Aaron'sAir* (see Exercises 18, 20, and 22), could purchase a market survey from a firm that has advised the island tourist and conference bureau. He thinks their projections would help him determine whether the probability of high demand might be as high as .5 or as low as .2, with the corresponding probabilities for medium demand being .3 and .4. If he doesn't purchase the market survey, he'll make a decision based on his previous best estimates (see Exercise 22).

a) Draw the decision tree.
b) Aaron thinks the market survey is likely to be optimistic. He'd estimate a 65% probability that it would predict the higher (.5) probability of high demand. What would be the EVwSI?
c) If the consultant's report costs $2,000, should Aaron pay for it?

25. Investment strategy. An investor is considering adding a stock to her portfolio. Assuming she buys 100 shares, here is an estimated payoff table for the alternative stocks if she holds onto them for six months. The value of the stock depends on whether or not an acquisition is approved for one of the companies, since the companies are actually competitors. She estimates the probability of an acquisition to be 0.3.

	Acquisition?	
	Yes (0.3)	**No (0.7)**
Stock A	$5000	−$1000
Stock B	−$500	$3500

a) Compute the EV for each alternative decision.
b) Compute the SD for each decision.
c) Compute the CV and RRR for each decision.
d) Which stock would you choose and why?

26. Mutual fund investing. An investor is considering how to invest her money. She has two options—either a domestic mutual fund that only invests in blue chip stocks or an international aggressive mutual fund that invests in young technical firms. The payoff (profit) after one year for these investments depends on the state of the economy.

	Economy Improves (0.5)	Declines (0.5)
Domestic Mutual Fund	$1500	$1000
International Mutual Fund	$3500	−$1000

a) Compute the EV for each alternative decision.
b) Compute the SD for each decision.
c) Compute the CV and RRR for each decision.
d) Which mutual fund would you invest in and why?

27. Bicycle sales. A bicycle shop owner is deciding which products to stock. His distributor will give him a deal if he buys more of the same kind of bike. The payoff table shows monthly sales for a high-end bike (selling at $950) or a moderately priced bike (selling at $500). Based on past experience, the shop owner makes the following assumption about the demand for the high-end bike. Demand will be low, moderate, or high with probabilities 0.3, 0.5, and 0.2, respectively. He also assumes that if demand is low for the high-end bike, it will be higher for the moderately-priced bike.

	Demand for High-End Bike		
	Low Demand (0.3)	**Moderate Demand (0.5)**	**High Demand (0.2)**
High-End Bike	$1900	$4750	$7600
Moderately-Priced Bike	$4000	$2500	$1000

a) Compute the EV for each alternative product (decision).
b) Compute the SD for each decision.

c) Compute the CV and RRR for each decision.

d) Which bike would you stock and why?

28. Bicycle sales, part 2. The bike shop owner has now done a bit more research and believes that the demand for high-end bikes has shifted, so that now the low demand is 50% likely and the high demand is only 10% likely. How does this change your responses to Exercise 27? Find the new RRR. Does your recommendation to the shop owner change?

29. Website experiment. Summit Projects provides marketing services and website management for many companies that specialize in outdoor products and services (www.summitprojects.com). To understand customer Web behavior, the company experiments with different offers and website design. The results of such experiments can help to maximize the probability that customers purchase products during a visit to a website. Possible actions by the website include offering the customer an instant discount, offering the customer free shipping, or doing nothing. A recent experiment found that customers make purchases 6% of the time when offered the instant discount, 5% when offered free shipping, and 2% when no special offer was given. Suppose 20% of the customers are offered the discount and an additional 30% are offered free shipping.

a) Construct a probability tree for this experiment.

b) What percent of customers who visit the site made a purchase?

c) Given that a customer made a purchase, what is the probability that they were offered free shipping?

30. Website experiment, part 2. The company in Exercise 29 performed another experiment in which they tested three website designs to see which one would lead to the highest probability of purchase. The first (design A) used enhanced product information, the second (design B) used extensive iconography, and the third (design C) allowed the customer to submit their own product ratings. After 6 weeks of testing, the designs delivered probabilities of purchase of 4.5%, 5.2%, and 3.8%, respectively. Equal numbers of customers were sent randomly to each website design.

a) Construct a probability tree for this experiment.

b) What percent of customers who visited the site made a purchase?

c) What is the probability that a randomly selected customer was sent to design C?

d) Given that a customer made a purchase, what is the probability that the customer had been sent to design C?

31. Contract bidding. As manager for a construction firm, you are in charge of bidding on two large contracts. You believe the probability you get contract #1 is 0.8. If you get contract #1, the probability you also get contract #2 will be 0.2, and if you do not get #1, the probability you get #2 will be 0.4.

a) Sketch the probability tree.

b) What is the probability you will get both contracts?

c) Your competitor hears that you got the second contract but hears nothing about the first contract. Given that you got the second contract, what is the probability that you also got the first contract?

32. Extended warranties. A company that manufactures and sells consumer video cameras sells two versions of their popular hard disk camera, a basic camera for $750 and a deluxe version for $1250. About 75% of customers select the basic camera. Of those, 60% purchase the extended warranty for an additional $200. Of the people who buy the deluxe version, 90% purchase the extended warranty.

a) Sketch the probability tree for total purchases.

b) What is the percentage of customers who buy an extended warranty?

c) What is the expected revenue of the company from a camera purchase (including warranty if applicable)?

d) Given that a customer purchases an extended warranty, what is the probability that he or she bought the deluxe version?

33. Computer reliability. Laptop computers have been growing in popularity according to a study by Current Analysis Inc. Laptops now represent more than half the computer sales in the United States. A campus bookstore sells both types and in the last semester sold 56% laptops and 44% desktops. Reliability rates for the two types of machines are quite different, however. In the first year, 5% of desktops require service, while 15% of laptops have problems requiring service.

a) Sketch a probability tree for this situation.

b) What percentage of computers sold by the bookstore last semester required service?

c) Given that a computer required service, what is the probability that it was a laptop?

34. Titanic survival. Of the 2201 passengers on the RMS *Titanic*, only 711 survived. The practice of "women and children first" was first used to describe the chivalrous actions of the sailors during the sinking of the HMS *Birkenhead* in 1852, but became popular after the sinking of the *Titanic*, during which 53% of the children and 73% of the women survived, but only 21% of the men survived. Part of the protocol stated that passengers enter lifeboats by ticket class as well. Here is a table showing survival by ticket class.

		Class				
		First	**Second**	**Third**	**Crew**	**Total**
Survival	**Alive**	203 / 28.6%	118 / 16.6%	178 / 25.0%	212 / 29.8%	711 / 100%
	Dead	122 / 8.2%	167 / 11.2%	528 / 35.4%	673 / 45.2%	1490 / 100%

a) Find the conditional probability of survival for each type of ticket.

b) Draw a probability tree for this situation.

c) Given that a passenger survived, what is the probability they had a first-class ticket?

Design and Analysis of Experiments and Observational Studies

Capital One

Not everyone graduates first in their class at a prestigious business school. But even doing that won't guarantee that the first company you start will become a Fortune 500 company within a decade. Richard Fairbank managed to do both. When he graduated from Stanford Business School in 1981, he wanted to start his own company, but, as he said in an interview with the *Stanford Business Magazine*, he had no experience, no money, and no business ideas. So he went to work for a consulting firm. Wanting to be on his own, he left in 1987 and landed a contract to study the operations of a large credit card bank in New York. It was then that he realized that the secret lay in data. He and his partner, Nigel Morris asked themselves, "Why not use the mountains of data that credit cards produce to design cards with prices and terms to satisfy different customers?" But they had a hard time selling this idea to the large credit card issuers. At the time all cards carried the same interest rate—19.8% with a $20 annual fee, and almost half of the population didn't qualify for a card. And credit issuers were naturally resistant to new ideas.

Finally, Fairbank and Morris signed on with Signet, a regional bank that hoped to expand its modest credit card operation. Using demographic and financial data about Signet's customers, they designed and tested combinations of card features that allowed them to offer credit to customers who previously didn't qualify. Signet's credit card business grew and, by 1994, was spun off as Capital One with a market cap of $1.1B. By 2000, Capital One was the ninth largest issuer of credit cards with $29.5B in cardholder balances.

Fairbank also introduced "scientific testing." Capital One designs experiments to gather data about customers. For example, customers who hear about a better deal than the one their current card offers may phone, threatening to switch to another bank unless they get a better deal. To help identify which potential card-hoppers were serious, Fairbank designed an experiment. When a card-hopper called, the customer service agent's computer randomly ordered one of three actions: match the claimed offer, split the difference in rates or fees, or just say no. In that way the company could gather data on who switched, who stayed, and how they behaved. Now when a potential card-hopper phones, the computer can give the operator a script specifying the terms to offer—or instruct the operator to bid the card-hopper a pleasant good-bye.

Fairbank attributes the phenomenal success of Capital One to their use of such experiments. According to Fairbank, "Anyone in the company can propose a test and, if the results are promising, Capital One will rush the new product or approach into use immediately." Why does this work for Capital One? Because, as Fairbank says, "We don't hesitate because our testing has already told us what will work."

In 2002, Capital One won the Wharton Infosys Business Transformation Award, which recognizes enterprises that have transformed their businesses by leveraging information technology.

23.1 Observational Studies

Fairbank started by analyzing the data that had already been collected by the credit card company. These data weren't from designed studies of customers. He simply *observed* the behavior of customers from the data that were already there. Such studies are called **observational studies**. Many companies collect data from customers with "frequent shopper" cards, which allow the companies to record each purchase. A company might study that data to identify associations between customer behavior and demographic information. For example, customers with pets might tend to spend more. The company can't conclude that owning a pet *causes* these customers to spend. People who have pets may also have higher incomes on average or be more likely to own their own homes. Nevertheless, the company may decide to make special offers targeted at pet owners.

Observational studies are used widely in public health and marketing because they can reveal trends and relationships. Observational studies that study an outcome in the present by delving into historical records are called **retrospective studies**. When Fairbank looked at the accumulated experiences of Signet bank's credit card customers, he started with information about which customers earned the bank the most money and sought facts about these customers that could identify others like them, so he was performing a retrospective study. Retrospective studies can often generate testable hypotheses because they identify interesting relationships although they can't demonstrate a causal link.

When it is practical, a somewhat better approach is to observe individuals over time, recording the variables of interest and seeing how things turn out. For example, if we thought pet ownership might be a way to identify profitable customers, we might start by selecting a random sample of new customers and ask whether they have a pet. We could then track their performance and compare those who own pets to those who don't. Identifying subjects in advance and collecting data as events unfold would make this a **prospective study**. Prospective studies are often used in public health, where by following smokers or runners over a period of time we may find that one group or the other develops emphysema or arthritic knees (as you might expect), or dental cavities (which you might not anticipate).

Although an observational study may identify important variables related to the outcome we are interested in, there is no guarantee that it will find the right or the most important related variables. People who own pets may differ from the other customers in ways that we failed to observe. It may be this difference—whether we know what it is or not—rather than owning a pet in itself that leads pet owners to be more profitable customers. It's just not possible for observational studies, whether prospective or retrospective, to demonstrate a causal relationship. That's why we need experiments.

JUST CHECKING

In early 2007, a larger-than-usual number of cats and dogs developed kidney failure; many died. Initially, researchers didn't know why, so they used an observational study to investigate.

1 Suppose that, as a researcher for a pet food manufacturer, you are called on to plan a study seeking the cause of this problem. Specify how you might proceed. Would your study be prospective or retrospective?

23.2 Randomized, Comparative Experiments

Experiments are the only way to show cause-and-effect relationships convincingly, so they are a critical tool for understanding what products and ideas will work in the marketplace. An **experiment** is a study in which the experimenter *manipulates* attributes of what is being studied and observes the consequences. Usually, the attributes, called **factors**, are manipulated by being set to particular **levels** and then allocated or assigned to individuals. An experimenter identifies at least one factor to manipulate and at least one response variable to measure. Often the observed **response** is a quantitative measurement such as the amount of a product sold. However, responses can be categorical ("customer purchased"/ "customer didn't purchase"). The combination of factor levels assigned to a subject is called that subject's **treatment**.

The individuals on whom or which we experiment are known by a variety of terms. Humans who are experimented on are commonly called **subjects** or **participants**. Other individuals (rats, products, fiscal quarters, company divisions) are commonly referred to by the more generic term **experimental units**.

You've been the subject of marketing experiments. Every credit card offer you receive is actually a combination of various factors that specify your "treatment," the specific offer you get. For example, the factors might be *Annual Fee, Interest Rate,* and *Communication Channel* (e-mail, direct mail, phone, etc.). The particular treatment you receive might be a combination of *no Annual Fee* and a *moderate Interest Rate* with the offer being sent by *e-mail*. Other customers receive different treatments. The response might be categorical (do you accept the offer of that card?) or quantitative (how much do you spend with that card during the first three months you have it?).

Two key features distinguish an experiment from other types of investigations. First, the experimenter actively and deliberately manipulates the factors to specify the treatment. Second, the experiment assigns the subjects to those treatments at *random*. The importance of **random assignment** may not be immediately obvious. Experts, such as business executives and physicians, may think that they know how different subjects will respond to various treatments. In particular, marketing executives may want to send what they consider the best offer to the their best customers, but this makes fair comparisons of treatments impossible and invalidates the inference from the test. Without random assignment, we can't perform the hypothesis tests that allow us to conclude that differences among the treatments were responsible for any differences we observed in the responses. By using random assignment to ensure that the groups receiving different treatments are comparable, the experimenter can be sure that these differences are *due* to the differences in treatments. There are many stories of experts who were certain they knew the effect of a treatment and were proven wrong by a properly designed study. In business, it is important to get the facts rather than to just rely on what you may think you know from experience.

23.3 The Four Principles of Experimental Design

There are four **principles of experimental design**.

1. **Control.** We control sources of variation other than the factors we are testing by making conditions as similar as possible for all treatment groups. In a test of a new credit card, all alternative offers are sent to customers at the same time and in the same manner. Otherwise, if gas prices soar, the stock market drops, or interest rates spike dramatically during the study, those events could influence customers' responses, making it difficult to assess the effects of the treatments. So an experimenter tries to make any other variables that are not manipulated as alike as possible. Controlling extraneous sources of variation reduces the variability of the responses, making it easier to discern differences among the treatment groups.

There is a second meaning of control in experiments. A bank testing the new creative idea of offering a card with special discounts on chocolate to attract more customers will want to compare its performance against one of their standard cards. Such a baseline measurement is called a control treatment, and the group that receives it is called the **control group**.

2. **Randomize.** In any true experiment, subjects are assigned treatments at random. Randomization allows us to equalize the effects of unknown or uncontrollable sources of variation. Although randomization can't eliminate the effects of these sources, it spreads them out across the treatment levels so that we can see past them. Randomization also makes it possible to use the powerful methods of inference to draw conclusions from your study. Randomization protects us even from effects we didn't know about. Perhaps women are more likely to respond to the chocolate benefit card. We don't need to test equal numbers of men and women—our mailing list may not have that information. But if we randomize, that tendency won't contaminate our results. There's an adage that says "Control what you can, and randomize the rest."

3. **Replicate.** Replication shows up in different ways in experiments. Because we need to estimate the variability of our measurements, we must make more than one observation at each level of each factor. Sometimes that just means making repeated observations. But, as we'll see later, some experiments combine two or more factors in ways that may permit a single observation for each *treatment*— that is, each combination of factor levels. When such an experiment is repeated in its entirety, it is said to be *replicated*. Repeated observations at each treatment are called **replicates**. If the number of replicates is the same for each treatment combination, we say that the experiment is **balanced**.

   A second kind of replication is to repeat the entire experiment for a different group of subjects, under different circumstances, or at a different time. Experiments do not require, and often can't obtain, representative random samples from an identified population. Experiments study the consequences of different levels of their factors. They rely on the random assignment of treatments to the subjects to generate the sampling distributions and to control for other possibly contaminating variables. When we detect a significant difference in response among treatment groups, we can conclude that it is due to the difference in treatments. However, we should take care in generalizing that result too broadly if we've only studied a specialized population. A special offer of accelerated checkout lanes for regular customers may attract more business in December, but it may not be effective in July. Replication in a variety of circumstances can increase our confidence that our results apply to other situations and populations.

4. **Blocking.** Sometimes we can identify a factor not under our control whose effect we don't care about, but which we suspect might have an effect either on our response variable or on the ways in which the factors we are studying affect that response. Perhaps men and women will respond differently to our chocolate offer. Or maybe customers with young children at home behave differently than those without. Platinum card members may be tempted by a premium offer much more than standard card members. Factors like these can account for some of the variation in our observed responses because subjects at different

JUST CHECKING

Following concerns over the contamination of its pet foods by melamine, which had lead to kidney failure, a manufacturer now claims its products are safe. You are called on to design the study to demonstrate the safety of the new formulation.

2 Identify the treatment and response.

3 How would you implement control, randomization, and replication?

levels respond differently. But we can't *assign* them at random to subjects. So we deal with them by grouping, or **blocking**, our subjects together and, in effect, analyzing the experiment separately for each block. Such factors are called **blocking factors**, and their levels are called **blocks**. Blocking in an experiment is like stratifying in a survey design. Blocking reduces variation by comparing subjects within these more homogenous groups. That makes it easier to discern any differences in response due to the factors of interest. In addition, we may want to study the effect of the blocking factor itself. Blocking is an important compromise between randomization and control. However, unlike the first three principles, blocking is not required in all experiments.

23.4 Experimental Designs

Completely Randomized Designs

When each of the possible treatments is assigned to at least one subject at random, the design is called a **completely randomized design**. This design is the simplest and easiest to analyze of all experimental designs. A diagram of the procedure can help in thinking about experiments. In this experiment, the subjects are assigned at random to the two treatments.

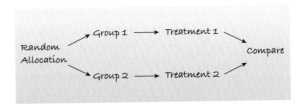

Figure 23.1 *The simplest randomized design has two groups randomly assigned two different treatments.*

Randomized Block Designs

When one of the factors is a blocking factor, complete randomization isn't possible. We can't randomly assign factors based on people's behavior, age, sex, and other attributes. But we may want to block by these factors in order to reduce variability and to understand their effect on the response. When we have a blocking factor, we randomize the subject to the treatments *within each block*. This is called a **randomized block design**. In the following experiment, a marketer wanted to know the effect of two types of offers in each of two segments: a high spending group and a low spending group. The marketer selected 12,000 customers *from each group* at random and then randomly assigned the three treatments to the 12,000 customers *in each group* so that 4000 customers in each segment received each of the three treatments. A display makes the process clearer.

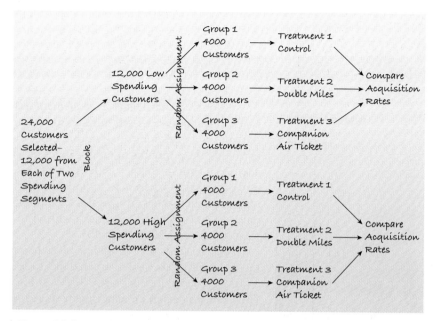

Figure 23.2 *This example of a randomized block design shows that customers are randomized to treatments within each segment, or block.*

Factorial Designs

An experiment with more than one manipulated factor is called a **factorial design**. A full factorial design contains treatments that represent all possible combinations of all levels of all factors. That can be a lot of treatments. With only three factors, one at 3 levels, one at 4, and one at 5, there would be $3 \times 4 \times 5 = 60$ different treatment combinations. So researchers typically limit the number of levels to just a few.

It may seem that the added complexity of multiple factors is not worth the trouble. In fact, just the opposite is true. First, if each factor accounts for some of the variation in responses, having the important factors in the experiment makes it *easier* to discern the effects of each. Testing multiple factors in a single experiment makes more efficient use of the available subjects. And testing factors together is the only way to see what happens at *combinations* of the levels.

An experiment to test the effectiveness of offering a $50 coupon for free gas may find that the coupon increases customer spending by 1%. Another experiment finds that lowering the interest rate increases spending by 2%. But unless some customers were offered *both* the $50 free gas coupon *and* the lower interest rate, the analyst can't learn whether offering both together would lead to still greater spending or less.

When the combination of two factors has a different effect than you would expect by adding the effects of the two factors together, that phenomenon is called an **interaction**. If the experiment does not contain both factors, it is impossible to see interactions. That can be a major omission because such effects can have the most important and surprising consequences in business.

GUIDED EXAMPLE *Designing a Direct Mail Experiment*

At a major credit card bank, management has been pleased with the success of a recent campaign to cross-sell Silver card customers with the new SkyWest Gold card. But you, as a marketing analyst, think that the revenue of the card can be increased by adding three months of double miles on SkyWest to the offer, and you think the additional gain in charges will offset the cost of the double miles. You want to design a marketing experiment to find out what the difference will be in revenue if you offer the double miles. You've also been thinking about offering a new version of the miles called "use anywhere miles," which can be transferred to other airlines, so you want to test that version as well.

You also know that customers receive so many offers that they tend to disregard most of their direct mail. So, you'd like to see what happens if you send the offer in a shiny gold envelope with the SkyWest logo prominently displayed on the front. How can we design an experiment to see whether either of these factors has an effect on charges?

PLAN

State the problem.

We want to study two factors to see their effect on the revenue generated for a new credit card offer.

Response Specify the response variable.

Revenue is a percentage of the amount charged to the card by the cardholder. To measure the success, we will use the monthly charges of customers who receive the various offers. We will use the three months after the offer is sent out as the collection period and use the total amount charged per customer during this period as the response.

Factors Identify the factors you plan to test.

Levels Specify the levels of the factors you will use.

We will offer customers three levels of the factor **miles** for the SkyWest Gold card: no miles, double miles, and double "use anywhere miles." Customers will receive the offer in the standard envelope or the new SkyWest logo envelope (factor **envelope**).

Experimental Design Observe the principles of design:

Control any sources of variability you know of and can control.

Randomly assign experimental units to treatments to equalize the effects of unknown or uncontrollable sources of variation.

Replicate results by placing more than one customer (usually many) in each treatment group.

We will send out all the offers to customers at the same time (in mid September) and evaluate the response as total charges in the period October through December.

A total of 30,000 current Silver card customers will be randomly selected from our customer records to receive one of the six offers.

✓ No miles with standard envelope
✓ Double miles with standard envelope
✓ Double "anywhere miles" with standard envelope
✓ No miles with Logo envelope
✓ Double miles with Logo envelope
✓ Double "anywhere miles" with Logo envelope

Make a Picture A diagram of your design can help you think about it.

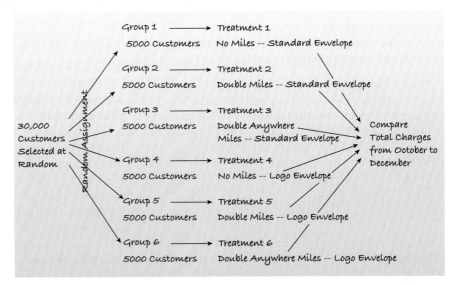

Specify any other details about the experiment. You must give enough details so that another experimenter could exactly replicate your experiment.

It's generally better to include details that might seem irrelevant because they may turn out to make a difference.

Specify how to measure the response.

On January 15, we will examine the total card charges for each customer for the period October 1 through December 31.

DO

Once you collect the data, you'll need to display them (if appropriate) and compare the results for the treatment groups. (Methods of analysis for factorial designs will be covered later in the chapter.)

REPORT

To answer the initial question, we ask whether the differences we observe in the means (or proportions) of the groups are meaningful.

Because this is a randomized experiment, we can attribute significant differences to the treatments. To do this properly, we'll need methods from the analysis of factorial designs covered later in the chapter.

MEMO:

Re: Test Mailing for Creative Offer and Envelope

The mailing for testing the Double Miles and Logo envelope ideas went out on September 17. On January 15, once we have total charges for everyone in the treatment groups, I would like to call the team back together to analyze the results to see:

✓ Whether offering Double Miles is worth the cost of the miles

✓ Whether the "use anywhere miles" are worth the cost

✓ Whether the Logo envelope increased spending enough to justify the added expense

Blinding by Misleading

Social science experiments can sometimes blind subjects by misleading them about the purpose of a study. One of the authors participated as an undergraduate volunteer in a (now infamous) psychology experiment using such a blinding method. The subjects were told that the experiment was about three-dimensional spatial perception and were assigned to draw a model of a horse. While they were busy drawing, a loud noise and then groaning were heard coming from the room next door. The *real* purpose of the experiment was to see how people reacted to the apparent disaster. The experimenters wanted to see whether the social pressure of being in groups made people react to the disaster differently. Subjects had been randomly assigned to draw either in groups or alone. The experimenter had no interest in how well the subjects could draw the horse, but the subjects were blind to the treatment because they were misled.

23.5 Blinding and Placebos

Humans are notoriously susceptible to errors in judgment—all of us. When we know what treatment is assigned, it's difficult not to let that knowledge influence our response or our assessment of the response, even when we try to be careful.

Suppose you were trying to sell your new brand of cola to be stocked in a school's vending machines. You might hope to convince the committee designated to make the choice that students prefer your less expensive cola, or at least that they can't taste the difference. You could set up an experiment to see which of the three competing brands students prefer (or whether they can tell the difference at all). But people have brand loyalties. If they know which brand they are tasting, it might influence their rating. To avoid this bias, it would be better to disguise the brands as much as possible. This strategy is called **blinding** the participants to the treatment. Even professional taste testers in food industry experiments are blinded to the treatment to reduce any prior feelings that might influence their judgment.

But it isn't just the subjects who should be blind. Experimenters themselves often subconsciously behave in ways that favor what they believe. It wouldn't be appropriate for you to run the study yourself if you have an interest in the outcome. People are so good at picking up subtle cues about treatments that the best (in fact, the only) defense against such biases in experiments on human subjects is to keep anyone who could affect the outcome or the measurement of the response from knowing which subjects have been assigned to which treatments. So, not only should your cola-tasting subjects be blinded, but also you, as the experimenter, shouldn't know which drink is which—at least until you're ready to analyze the results.

There are two main classes of individuals who can affect the outcome of the experiment:

◆ Those who could influence the results (the subjects, treatment administrators, or technicians)

◆ Those who evaluate the results (judges, experimenters, etc.)

When all the individuals in either one of these classes are blinded, an experiment is said to be **single-blind**. When everyone in both classes is blinded, we call the experiment **double-blind**. Double-blinding is the gold standard for any experiment involving both human subjects and human judgment about the response.

Often simply applying *any* treatment can induce an improvement. Every parent knows the medicinal value of a kiss to make a toddler's scrape or bump stop hurting. Some of the improvement seen with a treatment—even an effective treatment—can be due simply to the act of treating. To separate these two effects, we can sometimes use a control treatment that mimics the treatment itself. A "fake" treatment that looks just like the treatments being tested is called a **placebo**. Placebos are the best way to blind subjects so they don't know whether they have received the treatment or not. One common version of a placebo in drug testing is a "sugar pill." Especially when psychological attitude can affect the results, control group subjects treated with a placebo may show an improvement.

The fact is that subjects treated with a placebo sometimes improve. It's not unusual for 20% or more of subjects given a placebo to report reduction in pain, improved movement, or greater alertness or even to demonstrate improved health or performance. This **placebo effect** highlights both the importance of effective blinding and

The placebo effect is stronger when placebo treatments are administered with authority or by a figure who appears to be an authority. "Doctors" in white coats generate a stronger effect than salespeople in polyester suits. But the placebo effect is not reduced much, even when subjects know that the effect exists. People often suspect that they've gotten the placebo if nothing at all happens. So, recently, drug manufacturers have gone so far in making placebos realistic that they cause the same side effects as the drug being tested! Such "active placebos" usually induce a stronger placebo effect. When those side effects include loss of appetite or hair, the practice may raise ethical questions.

JUST CHECKING

The pet food manufacturer we've been following hires you to perform the experiment to test whether their new formulation is safe and nutritious for cats and dogs.

4 How would you establish a control group?

5 Would you use blinding? How? (Can or should you use double-blinding?)

6 Both cats and dogs are to be tested. Should you block? Explain.

the importance of comparing treatments with a control. Placebo controls are so effective that you should use them as an essential tool for blinding whenever possible.

The best experiments are usually:

◆ Randomized

◆ Double-blind

◆ Comparative

◆ Placebo-controlled

23.6 Confounding and Lurking Variables

A credit card bank wanted to test the sensitivity of the market to two factors: the annual fee charged for a card and the annual percentage rate charged. The bank selected 100,000 people at random from a mailing list and sent out 50,000 offers with a low rate and no fee and 50,000 offers with a higher rate and a $50 annual fee. They discovered that people preferred the low-rate, no-fee card. No surprise. In fact, customers signed up for that card at over twice the rate as the other offer. But the question the bank really wanted to answer was: "How much of the change was due to the rate, and how much was due to the fee?" Unfortunately, there's simply no way to separate out the two effects with that experimental design.

If the bank had followed a factorial design in the two factors and sent out all four possible different treatments—low rate with no fee; low rate with $50 fee; high rate with no fee, and high rate with $50 fee—each to 25,000 people, it could have learned about both factors and could have also learned about the interaction between rate and fee. But we can't tease apart these two effects because the people who were offered the low rate were also offered the no-fee card. When the levels of one factor are associated with the levels of another factor, we say that the two factors are **confounded**.

Confounding can also arise in well-designed experiments. If some other variable not under the experimenter's control but associated with a factor has an effect on the response variable, it can be difficult to know which variable is really responsible for the effect. A shock to the economic or political situation that occurs during a marketing experiment can overwhelm the effects of the factors being tested. Randomization will usually take care of confounding by distributing uncontrolled factors over the treatments at random. But be sure to watch out for potential confounding effects even in a well-designed experiment.

Confounding may remind you of the problem of lurking variables that we discussed in Chapter 7. Confounding variables and lurking variables are alike in that they interfere with our ability to interpret our analyses simply. Each can mislead us, but they are not the same. A lurking variable is associated with two variables in such a way that it creates an apparent, possibly causal relationship between them. By contrast, confounding arises when a variable associated with a factor has an effect on the response variable, making it impossible to separate the effect of the factor from the effect of the confounder. Both confounding and lurking variables are outside influences that make it harder to understand the relationship we are modeling.

23.7 Analyzing a Design in One Factor— The One-Way Analysis of Variance

The most common experimental design used in business is the single factor experiment with two levels. Often these are known as champion/challenger designs because typically they're used to test a new idea (the challenger) against the current version (the champion). In this case, the customers offered the champion are the control group, and the customers offered the challenger (a special deal, a new offer, a new service, etc.) are the test group. As long as the customers are randomly assigned to the two groups, we already know how to analyze data from experiments like these. When the response is quantitative, we can test whether the means are equal with a two-sample *t*-test, and if the response is 0-1 (yes/no), we would test whether the two proportions are equal using a two proportion *z*-test.

But those methods can compare only two groups. What happens when we introduce a third level into our single factor experiment? Suppose an associate in a percussion music supply company, *Tom's Tom-Toms*, wants to test ways to increase the amount purchased from the catalog the company sends out every three months. He decides on three treatments: a coupon for free drum sticks with any purchase, a free practice pad, and a $50 discount on any purchase. The response will be the dollar amount of sales per customer. He decides to keep some customers as a control group by sending them the catalog without any special offer. The experiment is a single factor design with four levels: no coupon, coupon for free drum sticks, coupon for the practice pad, and $50 coupon. He assigns the same number of customers to each treatment randomly.

Now the hypothesis to test isn't quite the same as when we tested the difference in means between two independent groups. To test whether all *k* means are equal, the hypothesis becomes:

$$H_0: \mu_1 = \mu_2 = \ldots = \mu_k$$

H_A: *at least one mean is different*

The test statistic compares the variance of the means to what we'd expect that variance to be based on the variance of the individual responses. Figure 23.3 illustrates the concept. The differences among the means are the same for the two sets of boxplots, but it's easier to see that they are different when the underlying variability is smaller.

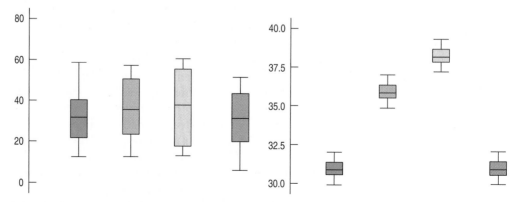

Figure 23.3   *The means of the four groups in the left display are the same as the means of the four groups in the right display, but the differences are much easier to see in the display on the right because the variation within each group is less.*

Why is it easier to see that the means[1] of the groups in the display on the right are different and much harder to see it in the one on the left? It is easier because we naturally compare the differences *between* the group means to the variation *within* each group. In the bottom picture, there is much less variation within each group so the differences among the group means are evident.

This is exactly what the test statistic does. It's the ratio of the variation among the group means to the variation within the groups. When the numerator is large enough, we can be confident that the differences among the group means are greater than we'd expect by chance, and reject the null hypothesis that they are equal. The test statistic is called the *F*-statistic in honor of Sir Ronald Fisher, who derived the sampling distribution for this statistic. The *F*-statistic showed up in multiple regression (Chapter 18) to test the null hypothesis that all slopes were zero. Here, it tests the null hypothesis that the means of all the groups are equal.

The ***F*-statistic** compares two quantities that measure variation, called *mean squares*. The numerator measures the variation *between* the groups (treatments) and is called the *Mean Square due to Treatments (MST)*. The denominator measures the variation *within* the groups, and is called the *Mean Square due to Error (MSE)*. The *F*-statistic is their ratio:

$$F_{k-1, N-k} = \frac{MST}{MSE}$$

We reject the null hypothesis that the means are equal if the *F*–statistic is too big. The critical value for deciding whether *F* is too big depends both on its degrees of freedom and the α-level you choose. Here, the degrees of freedom are $k - 1$ (for the MST) and $N - k$ (for the MSE) where k is the number of groups and N is the total number of observations. Alternatively, we could find the P-value of this statistic and reject the null hypothesis when that value is small.

This analysis is called an **Analysis of Variance (ANOVA)**, but the hypothesis is actually about *means*. The null hypothesis is that the means are all equal. The collection of statistics—the sums of squares, mean squares, *F*-statistic, and P-value—are usually presented in a table, called the **ANOVA table**, like this one:

Source	DF	Sum of Squares	Mean Square	F-Ratio	Prob > *F*
Treatment (Between)	$k - 1$	SST	MST	MST/MSE	P-value
Error (Within)	$N - k$	SSE	MSE		
Total	$N - 1$	SSTotal			

| *Table 23.1* *An ANOVA table displays the treatment and error sums of squares, mean squares, F-ratio, and P-value.*

◆ **How does the Analysis of Variance work?** When looking at side-by-side boxplots to see whether we think there are real differences between treatment means, we naturally compare the variation *between* the groups to the variation *within* the groups. The variation between the groups indicates how large an effect the treatments have. The variation within the groups shows the underlying variability. To model those variations, the one way ANOVA decomposes the data into several parts: the grand average, the treatment effects, and the residuals.

$$y_{ij} = \bar{\bar{y}} + (\bar{y}_i - \bar{\bar{y}}) + (y_{ij} - \bar{y}_i).$$

We can write this as we did for regression as

$$data = predicted + residual.$$

---

[1] Of course the boxplots show medians at their centers, and we're trying to find differences among means. But for roughly symmetric distributions like these, the means and medians are very close.

To estimate the variation *between* the groups we look at how much their means vary. The SST (sometimes called the *between* sum of squares) captures it like this:

$$SST = \sum_{i=1}^{k} n_i (\bar{y}_i - \bar{\bar{y}})^2$$

where $\bar{y}_i$ is the mean of group i, n_i is the number of observations in group i and $\bar{\bar{y}}$ is the overall mean of all observations.

We compare the *SST* to how much variation there is *within* each group. The SSE captures that like this:

$$SSE = \sum_{i=1}^{k} (n_i - 1) s_i^2$$

where s_i^2 is the sample variance of group i.

To turn these estimates of variation into variances, we divide each sum of squares by its associated degrees of freedom:

$$MST = \frac{SST}{k - 1}$$

$$MSE = \frac{SSE}{N - k}$$

Remarkably (and this is Fisher's real contribution), these two variances estimate the *same* variance when the null hypothesis is true. When it is false (and the group means differ), the MST gets larger.

The *F*-statistic tests the null hypothesis by taking the ratio of these two mean squares:

$$F_{k-1,\, N-k} = \frac{MST}{MSE},$$ and rejecting the hypothesis if the ratio is too large.

The critical value and P-value depend on the two degrees of freedom $k - 1$ and $N - k$.

Let's look at an example. For the summer catalog of the percussion supply company *Tom's Tom-Toms*, 4000 customers were selected at random to receive one of four offers[2]: *No Coupon*, *Free Sticks* with purchase, *Free Pad* with purchase, or $50 off next purchase. All the catalogs were sent out on March 15 and sales data for the month following the mailing were recorded.

The first step is to plot the data. Here are boxplots of the spending of the four groups for the month after the mailing:

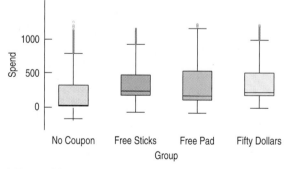

Figure 23.4 *Boxplots of the spending of the four groups show that the coupons seem to have stimulated spending.*

---

[2] Realistically, companies often select equal (and relatively small sizes) for the treatment groups and consider all other customers as the control. To make the analysis easier, we'll assume that this experiment just considered 4000 "control" customers. Adding more controls wouldn't increase the power very much.

Here are summary statistics for the four groups:

	Group			
	No Coupon	**Free Sticks**	**Free Pad**	**Fifty Dollars**
Median	$0.00	$233.00	$157.50	$232.00
Mean	$216.68	$385.87	$339.54	$399.95
SD	$390.58	$331.10	$364.17	$337.07

The ANOVA table shows the components of the calculation of the *F*-test.

Source	DF	Sum of Squares	Mean Square	*F*-Ratio	Prob > *F*
Groups (Between)	3	20,825,966	6,941,988.66	54.6169	< 0.0001
Error (Within)	3996	507,905,263	127,103.42		
Total	3999	528,731,229			

Table 23.2 *The ANOVA table shows that the F-statistic has a very small P-value, so we can reject the null hypothesis that the means of the four treatments are equal.*

The very small P-value is an indication that the differences we saw in the box-plots are not due to chance, so we reject the null hypothesis of equal means and conclude that the four means are not equal.

23.8 Assumptions and Conditions for ANOVA

Whenever we compute P-values and make inferences about a hypothesis, we need to make assumptions and check conditions to see if the assumptions are reasonable. The ANOVA is no exception. Because it's an extension of the two-sample *t*-test, many of the same assumptions apply.

Independence Assumption

The groups must be independent of each other. No test can verify this assumption. You have to think about how the data were collected. The individual observations must be independent as well.

We check the **Randomization Condition**. Did the experimental design incorporate suitable randomization? We were told that the customers were assigned to each treatment group at random.

Equal Variance Assumption

ANOVA assumes that the true variances of the treatment groups are equal. We can check the corresponding **Similar Variance Condition** in various ways:

◆ Look at side-by-side boxplots of the groups to see whether they have roughly the same spread. It can be easier to compare spreads across groups when they have the same center, so consider making side-by-side boxplots of the residuals. If the groups have differing spreads, it can make the pooled variance—the

MSE—larger, reducing the *F*-statistic value and making it less likely that we can reject the null hypothesis. So the ANOVA will usually fail on the "safe side," rejecting H_0 less often than it should. Because of this, we usually require the spreads to be quite different from each other before we become concerned about the condition failing. If you've rejected the null hypothesis, this is especially true.

◆ Look at the original boxplots of the response values again. In general, do the spreads seem to change systematically with the centers? One common pattern is for the boxes with bigger centers to have bigger spreads. This kind of systematic trend in the variances is more of a problem than random differences in spread among the groups and should not be ignored. Fortunately, such systematic violations are often helped by re-expressing the data. If, in addition to spreads that grow with the centers, the boxplots are skewed with the longer tail stretching off to the high end, then the data are pleading for a re-expression. Try taking logs of the dependent variable for a start. You'll likely end up with a much cleaner analysis.

◆ Look at the residuals plotted against the predicted values. Often, larger predicted values lead to larger magnitude residuals. This is another sign that the condition is violated. If the residual plot shows more spread on one side or the other, it's usually a good idea to consider re-expressing the response variable. Such a systematic change in the spread is a more serious violation of the equal variance assumption than slight variations of the spreads across groups.

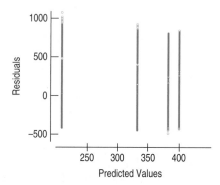

Figure 23.5   *A plot of the residuals against the predicted values from the ANOVA shows no sign of unequal spread.*

Normal Population Assumption

Like Student's *t*-tests, the *F*-test requires that the underlying errors follow a Normal model. As before when we faced this assumption, we'll check a corresponding **Nearly Normal Condition**.

Technically, we need to assume that the Normal model is reasonable for the populations underlying each treatment group. We can (and should) look at the side-by-side boxplots for indications of skewness. Certainly, if they are all (or mostly) skewed in the same direction, the Nearly Normal Condition fails (and re-expression is likely to help). However, in many business applications, samples sizes are quite large, and when that is true, the Central Limit Theorem implies that the sampling distribution of the means may be nearly Normal in spite of skewness. Fortunately, the *F*-test is conservative. That means that if you see a small P-value it's probably safe to reject the null hypothesis for large samples even when the data are nonnormal.

Check Normality with a histogram or a Normal probability plot of all the residuals together. Because we really care about the Normal model within each group, the Normal Population Assumption is violated if there are outliers in any of the groups. Check for outliers in the boxplots of the values for each treatment.

The Normal Probability plot for the *Tom's Tom-Toms* residuals holds a surprise.

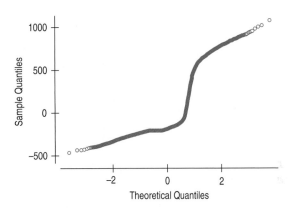

Figure 23.6 *A normal probability plot shows that the residuals from the ANOVA of the* Tom's Tom-Toms *data are clearly not normal.*

Investigating further with a histogram, we see the problem.

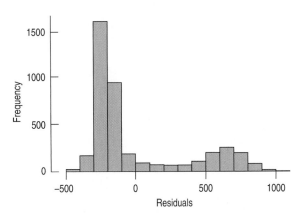

Figure 23.7 *A histogram of the residuals reveals bimodality.*

The histogram shows clear bimodality of the residuals. If we look back to histograms of the spending of each group, we can see that the boxplots failed to reveal the bimodal nature of the spending.

The manager of the company wasn't surprised to hear that the spending is bimodal. In fact, he said, "We typically have customers who either order a complete new drum set, or who buy accessories. And, of course, we have a large group of customers who choose not to purchase anything during a given quarter."

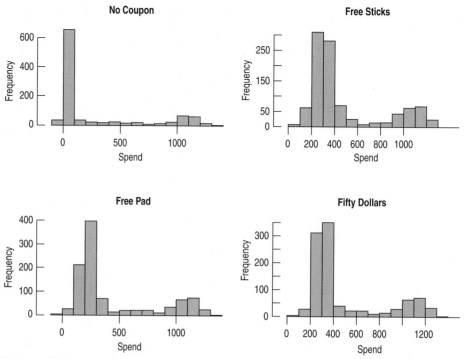

Figure 23.8 *The spending appears to be bimodal for all the treatment groups. There is one mode near $1000 and another larger mode between $0 and $200 for each group.*

These data (and the residuals) clearly violate the Nearly Normal Condition. Does that mean that we can't say anything about the null hypothesis? No. Fortunately, the sample sizes are large, and there are no individual outliers that have undue influence on the means. With sample sizes this large, we can appeal to the Central Limit Theorem and still make inferences about the means. In particular, we are safe in rejecting the null hypothesis. When the Nearly Normal Condition is not satisfied, the *F*-test will tend to fail on the safe side and be less likely to reject the null. Since we have a very small P-value, we can be fairly sure that the differences we saw were real.

JUST CHECKING

Your experiment to test the new pet food formulation has been completed. One hypothesis you have tested is whether the new formulation is different in nutritional value (measured by having veterinarians evaluate the test animals) from a standard food known to be safe and nutritious. The ANOVA has an *F*-statistic of 1.2, which (for the degrees of freedom in your experiment) has a P-value of 0.87. Now you need to make a report to the company.

7 Write a brief report. Can you conclude that the new formulation is safe and nutritious?

*23.9 Multiple Comparisons

Simply rejecting the null hypothesis is almost never the end of an Analysis of Variance. Knowing that the means are different leads to the question of which ones are different and by how much. Tom, the owner of *Tom's Tom-Toms*, would hardly be satisfied with a consultant's report that told him that the offers generated different amounts of spending, but failed to indicate which offers did better and by how much.

We'd like to know more, but the *F*-statistic doesn't offer that information. What can we do? If we can't reject the null hypothesis, there's no point in further testing. But if we can reject the null hypothesis, we can do more. In particular, we can test whether any pairs or combinations of group means differ. For example, we might want to compare treatments against a control or against the current standard treatment.

We could do *t*-tests for any pair of treatment means that we wish to compare. But each test would have some risk of a Type I error. As we do more and more tests, the risk that we'll make a Type I error grows. If we do enough tests, we're almost sure to reject one of the null hypotheses by mistake—and we'll never know which one.

There is a solution to this problem. In fact, there are several solutions. As a class, they are called methods for **multiple comparisons**. All multiple comparisons methods require that we first reject the overall null hypothesis with the ANOVA's *F*-test. Once we've rejected the overall null, we can think about comparing several—or even all—pairs of group means.

One such method is called the **Bonferroni method**. This method adjusts the tests and confidence intervals to allow for making many comparisons. The result is a wider margin of error (called the **minimum significant difference, or MSD**) found by replacing the critical *t*-value *t** with a slightly larger number. That makes the confidence intervals wider for each pairwise difference and the corresponding Type I error rates lower for each test, and it keeps the overall Type I error rate at or below α.

The Bonferroni method distributes the error rate equally among the confidence intervals. It divides the error rate among J confidence intervals, finding each interval at confidence level $1 - \dfrac{\alpha}{J}$ instead of the original $1 - \alpha$. To signal this adjustment, we label the critical value *t*** rather than *t**. For example, to make the six confidence intervals comparing all possible pairs of offers at our overall α risk of 5%, instead of making six 95% confidence intervals, we'd use

$$1 - \frac{0.05}{6} = 1 - .0083 = .9917$$

instead of 0.95. So we'd use a critical *t*** value of 2.64 instead of 1.96. The ME would then become:

$$ME = 2.642 \times 356.52 \sqrt{\frac{1}{1000} + \frac{1}{1000}} = 42.12$$

This change doesn't affect our decision that each offer increases the mean sales compared to the *No Coupon* group, but it does adjust the comparison of average sales for the *Free Sticks* offer and the *Free Pad* offer. With a margin of error of $42.12, the difference between average sales for those two offers is now $(385.87 - 339.54) \pm 42.12 = (\$4.21, \$88.45)$.

The confidence interval says that the *Free Sticks* offer generated between $4.21 and $88.45 more sales per customer on average than the *Free Pad* offer. In order to make a valid business decision, the company should now calculate their expected *profit* based on the confidence interval. Suppose they make 8% profit on sales. Then, multiplying the confidence interval

$$0.08 \times (\$4.21, \$88.45) = (\$0.34, \$7.08)$$

we find that the *Free Sticks* generate between $0.34 and $7.08 profit per customer on average. So, if the *Free Sticks* cost $1.00 more than the pads, the confidence interval for profit would be:

$$(\$0.34 - \$1.00, \$7.08 - \$1.00) = (-\$0.66, \$6.08)$$

Carlo Bonferroni (1892–1960) was a mathematician who taught in Florence. He wrote two papers in 1935 and 1936 setting forth the mathematics behind the method that bears his name.

There is a possibility that the *Free Sticks* may actually be a less profitable offer. The company may decide to take the risk or to try another test with a larger sample size to get a more precise confidence interval.

Many statistics packages assume that you'd like to compare all pairs of means. Some will display the result of these comparisons in a table such as this.

Fifty Dollars	$399.95	A		
Free Sticks	$385.87	A		
Free Pad	$339.54		B	
No Coupon	$216.68			C

Table 23.3   The output shows that the two top-performing offers are indistinguishable in terms of mean spending, but that the Free Pad is distinguishable from both those two and from No Coupon.

This table indicates that the top two are indistinguishable, that all are distinguishable from *No Coupon*, and that *Free Pad* is also distinguishable from the other three.

The subject of multiple comparisons is a rich one because of the many ways in which groups might be compared. Most statistics packages offer a choice of several methods. When your analysis calls for comparing all possible pairs, consider Tukey's HSD method. If one of the groups is a control group, and you wish to compare all the other groups to it, consider Dunnett's method. But whenever you look at differences after rejecting the null hypothesis of equal means, you should consider using a multiple comparisons method that attempts to preserve the *overall* α risk.

23.10 ANOVA on Observational Data

So far we've applied ANOVA only to data from designed experiments. That application is appropriate for several reasons. The primary one is that randomized comparative experiments are specifically designed to compare the results for different treatments. The overall null hypothesis, and the subsequent tests on pairs of treatments in ANOVA, address such comparisons directly. In addition, the **Equal Variance Assumption** (which we need for all of the ANOVA analyses) is often plausible in a randomized experiment because when we randomly assign subjects to treatments, all the treatment groups start out with the same underlying variance of the experimental units.

Sometimes, though, we just can't perform an experiment. When ANOVA is used to test equality of group means from observational data, there's no *a priori* reason to think the group variances might be equal at all. Even if the null hypothesis of equal means were true, the groups might easily have different variances. But you can use ANOVA on observational data if the side-by-side boxplots of responses for each group show roughly equal spreads and symmetric, outlier-free distributions.

Observational data tend to be messier than experimental data. They are much more likely to be unbalanced. If you aren't assigning subjects to treatment groups, it's harder to guarantee the same number of subjects in each group. And because you are not controlling conditions as you would in an experiment, things tend to be, well, less controlled. The only way we know to avoid the effects of possible

Recall that a design is called *balanced* if it has an equal number of observations for each treatment level.

lurking variables is with control and randomized assignment to treatment groups, and for observational data, we have neither.

ANOVA is often applied to observational data when an experiment would be impossible or unethical. (We can't randomly break some subjects' legs, but we *can* compare pain perception among those with broken legs, those with sprained ankles, and those with stubbed toes by collecting data on subjects who have already suffered those injuries.) In such data, subjects are already in groups, but not by random assignment.

Be careful; if you have not assigned subjects to treatments randomly, you can't draw *causal* conclusions even when the *F*-test is significant. You have no way to control for lurking variables or confounding, so you can't be sure whether any differences you see among groups are due to the grouping variable or to some other unobserved variable that may be related to the grouping variable.

Because observational studies often are intended to estimate parameters, there is a temptation to use pooled confidence intervals for the group means for this purpose. Although these confidence intervals are statistically correct, be sure to think carefully about the population that the inference is about. The relatively few subjects that you happen to have in a group may not be a simple random sample of any interesting population, so their "true" mean may have only limited meaning.

23.11 Analysis of Multifactor Designs

In our direct mail example, we looked at two factors: *Miles* and *Envelope*. *Miles* had three levels: *No Miles*, *Double Miles*, and *Double Anywhere Miles*. The factor *Envelope* had two levels: *Standard* and new *Logo*. The three levels of *Miles* and the two levels of *Envelope* resulted in six treatment groups. Because this was a completely randomized design, the 30,000 customers were selected at random, and 5,000 were assigned at random to each treatment.

Three months after the offer was mailed out, the total charges on the card were recorded for each of the 30,000 cardholders in the experiment. Here are boxplots of the six treatment groups' responses, plotted against each factor.

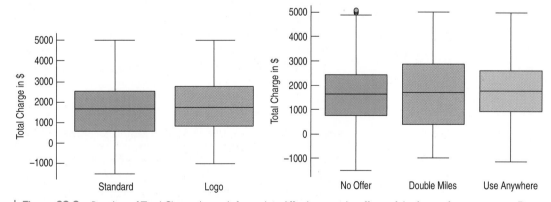

Figure 23.9 *Boxplots of Total Charge by each factor. It is difficult to see the effects of the factors for two reasons. First, the other factor hasn't been accounted for, and second, the effects are small compared to the overall variation in charges.*

If you look closely, you may be able to discern a very slight increase in the *Total Charges* for some levels of the factors, but it's very difficult to see. There are two reasons for this. First, the variation due to each factor gets in the way of seeing the effect of the other factor. For example, some of the customers in the boxplot for the *Logo Envelope* got each of the offers. If those offers had an effect on spending, then that increased the variation within the *Logo* treatment group. Second, as is typical in a marketing experiment of this kind, the effects are very small compared to the variability in people's spending. That's why companies use such a large sample size.

The analysis of variance for two factors removes the effects of each factor from consideration of the other. It can also model whether the factors interact, increasing or decreasing the effect. In our example, it will separate out the effect of changing the levels of *Miles* and the effect of changing the levels of *Envelope*. It will also test whether the effect of the *Envelope* is the same for the three different *Miles* levels. If the effect is different, that's called an interaction effect between the two factors.

The details of the calculations for the two-way ANOVA with interaction are less important than understanding the summary, the model, and the assumptions and conditions under which it's appropriate to use the model. For a one-way ANOVA, we calculated three sums of squares (SS): the Total SS, the Treatment SS, and the Error SS. For this model, we'll calculate five: the Total SS, the SS due to Factor A, the SS due to Factor B, the SS due to the interaction, and the Error SS.

Let's suppose we have a levels of factor A, b levels of factor B, and r replicates at each treatment combination. In our case, $a = 2$, $b = 3$, $r = 5000$, and $a \times b \times r = N$ is 30,000. Then the ANOVA table will look like this.

Source	DF	Sum of Squares	Mean Square	F-Ratio	Prob > F
Factor A	$a - 1$	SSA	MSA	MSA/MSE	P-value
Factor B	$b - 1$	SSB	MSB	MSB/MSE	P-value
Interaction	$(a - 1) \times (b - 1)$	SSAB	MSAB	MSAB/MSE	P-value
Error	$ab(r - 1)$	SSE	MSE		
Total (Corrected)	$N - 1$	SSTotal			

Table 23.4 *An ANOVA table for a replicated two factor design with a row for each factor's sum of squares, interaction sum of square, error, and total.*

There are now three null hypotheses—one that asserts that the means of the levels of factor A are equal, one that asserts that the means of the levels of factor B are all equal, and one that asserts that the effects of factor A are *constant* across the levels of factor B (or vice versa). Each P-value is used to test the corresponding hypothesis.

Here is the ANOVA table for the marketing experiment.

Source	DF	Sum of Squares	Mean Square	F-Ratio	Prob > F
Miles	2	201,150,000	100,575,000	66.20	< .0001
Envelope	1	203,090,000	203,090,000	133.68	< .0001
Miles × Envelope	2	1,505,200	752,600	0.50	0.61
Error	29994	45,568,000,000	1,519,237		

Table 23.5 *The ANOVA table for the marketing experiment. Both the effect of Miles and Envelope are highly significant, but the interaction term is not.*

From the ANOVA table, we can see that both the *Miles* and the *Envelope* effects are highly significant, but that the interaction term is not. An **interaction plot**, a plot of means for each treatment group, is essential for sorting out what these P-values mean.

Figure 23.10 *An interaction plot of the* Miles *and* Envelope *effects. The parallel lines show that the effects of the three* Miles *offers are roughly the same over the two different* Envelopes *and therefore that the interaction effect is small.*

The interaction plot shows the mean *Charges* at all six treatment groups. The levels of one of the factors, in this case *Miles*, are shown on the *x*-axis, and the mean *Charges* of the groups for each *Envelope* level are shown at each *Miles* level. The means of each level of *Envelope* are connected for ease of understanding. Notice that the effect of *Double Miles* over *No Offer* is about the same for both the *Standard* and *Logo Envelopes*. And the same is true for the *Use Anywhere* miles. This indicates that the effect of *Miles* is constant for the two different *Envelopes*. The lines are parallel, which indicates that there is no interaction effect.

We reject the null hypothesis that the mean *Charges* at the three different levels of *Miles* are equal (with P-values < 0.0001), and also we reject that the mean *Charges* for *Standard* and *Logo* are the same (with P-value < 0.0001). We have no evidence, however, to suggest that there is an interaction between the factors.

After rejecting the null hypotheses, we can create a confidence interval for any particular treatment mean or perform a hypothesis test for the difference between any two means. If we want to do several tests or confidence intervals, we will need to use a multiple comparisons method that adjusts the size of the confidence interval or the level of the test to keep the *overall* Type I error rate at the level we desire.

When the interaction term is not significant, we can talk about the overall effect of either factor. Because the effect of *Envelope* is roughly the same for all three *Miles* offers (as we know by virtue of not rejecting the hypothesis that the interaction effect is zero), we can calculate and interpret an overall *Envelope* effect. The means of the two *Envelope* levels are:

Logo $1871.75   *Standard* $1707.19

and so the *Logo* envelope generated a difference in average charge of $1871.75 − $1707.19 = $164.56. A confidence interval for this difference is ($136.66, $192.45), which the analysts can use to decide whether the added cost of the *Logo* envelope is worth the expense.

But when an interaction term *is* significant, we must be very careful not to talk about the effect of a factor, *on average*, because the effect of one factor *depends* on the level of the other factor. In that case, we always have to talk about the factor effect at a specific level of the other factor, as we'll see in the next example.

◆ **How does Two-Way Analysis of Variance work?** In Two-Way ANOVA, we have two factors. Each treatment consists of a level of each of the factors, so we write the individual responses as y_{ij}, to indicate the ith level of the first factor and the jth level of the second factor. Both factors could be manipulated by the experimenter or one could be a blocking factor. The more general formulas are no more informative; just more complex. We will start with an unreplicated design with one observation in each treatment group. We'll call the factors A and B, each with *a* and *b* levels, respectively. Then the total number of observations is $n = a \times b$.

For the first factor (factor A), the *Treatment Sum of Squares*, SSA, is the same as we calculated for one-way ANOVA:

$$SSA = \sum_{i=1}^{a} b(\bar{y}_i - \bar{\bar{y}})^2 = \sum_{j=1}^{b} \sum_{i=1}^{a} (\bar{y}_i - \bar{\bar{y}})^2$$

where *b* is the number of levels of factor B, $\bar{y}_i$ is the mean of all subjects assigned level *i* of factor A (regardless of which level of factor B they were assigned), and $\bar{\bar{y}}$ is the *overall* mean of all observations. The mean square for treatment A (MSA) is

$$MSA = \frac{SSA}{a-1}.$$

The treatment sum of squares for the second factor (B) is computed in the same way, but of course the treatment means are now the means for each level of this second factor:

$$SSB = \sum_{j=1}^{b} a(\bar{y}_j - \bar{\bar{y}})^2 = \sum_{i=1}^{a} \sum_{j=1}^{b} (\bar{y}_j - \bar{\bar{y}})^2, \text{ and}$$

$$MSB = \frac{SSB}{b-1}$$

where *a* is the number of levels of factor *A*, and $\bar{\bar{y}}$, as before, is the overall mean of all observations. $\bar{y}_j$ is the mean of all subjects assigned the jth level of factor B.

The SSE can be found by subtraction:

$$SSE = SSTotal - (SSA + SSB)$$

where

$$SSTotal = \sum_{i=1}^{a} \sum_{j=1}^{b} (y_{ij} - \bar{\bar{y}})^2.$$

The mean square for error is $MSE = \dfrac{SSE}{N - (a + b - 1)}$, where $N = a \times b$.

There are now two *F*-statistics, the ratio of each of the treatment mean squares to the MSE, which are associated with each null hypothesis.

To test whether the means of all the levels of factor A are equal, we would find a P-value for $F_{a-1, N-(a+b-1)} = \dfrac{MSA}{MSE}$. For factor B, we would find a P-value for $F_{b-1, N-1(a-b+1)} = \dfrac{MSB}{MSE}$.

Provided the experiment is replicated (say *r* times), we can also estimate the interaction between the two factors, and test whether it is zero. The sums of squares for each factor have to be multiplied by *r*. Alternatively they can be written as a (triple) sum where the sum is now over factor A, factor B, and the replications:

$$SSA = \sum_{k=1}^{r} \sum_{j=1}^{b} \sum_{i=1}^{a} (\bar{y}_i - \bar{\bar{y}})^2 \text{ and } SSB = \sum_{k=1}^{r} \sum_{i=1}^{a} \sum_{j=1}^{b} (\bar{y}_j - \bar{\bar{y}})^2.$$

We find the sum of squares for the interaction effect AB as:

$$SSAB = \sum_{k=1}^{r} \sum_{j=1}^{b} \sum_{i=1}^{a} (\bar{y}_{ij} - \bar{y}_i - \bar{y}_j - \bar{\bar{y}})^2, \text{ and}$$

$$MSAB = \frac{SSAB}{(a-1)(b-1)}.$$

The SSE is the sum of the squared residuals:

$$SSE = \sum_{k=1}^{r} \sum_{j=1}^{b} \sum_{i=1}^{a} (y_{ijk} - \bar{y}_{ij})^2 \text{ and } MSE = \frac{SSE}{ab(r-1)}.$$

There are now three *F*-statistics associated with the three hypotheses (factor A, factor B, and the interaction). They are the ratios of each of these mean squares with the MSE:

$$F_{a-1,\,ab(r-1)} = \frac{MSA}{MSE}, \; F_{b-1,\,ab(r-1)} = \frac{MSB}{MSE}, \text{ and } F_{(a-1)(b-1),\,ab(r-1)} = \frac{MSAB}{MSE}.$$

Note that $N = r \times a \times b$ is the total number of observations in the experiment.

GUIDED EXAMPLE | *A Follow Up Experiment*

After analyzing the data, the bank decided to go with the *Logo* envelope, but a marketing specialist thought that more *Miles* might increase spending even more.

A new test was designed to test both the type of *Miles* and the amount. Again, total *Charge* in three months is the response.

PLAN

State the problem.

We want to study the two factors *Miles* and *Amount* to see their effect on the revenue generated for a new credit card offer.

Response Specify the response variable.

To measure the success, we will use the monthly charges of customers who receive the various offers. We will use the three months after the offer is sent out as the collection period and the total amount charged per customer during this period as the response.

Factors Identify the factors you plan to test.

Levels Specify the levels of the factors you will use.

We will offer each customer one of the two levels of the factor *Miles* for the SkyWest Gold card: Sky-West miles or *Use Anywhere Miles*. Customers are offered three levels of *Miles*: *Regular Miles*, *Double Miles*, and *Triple Miles*.

We will send out all the offers to customers at the same time (in mid March) and evaluate the response as total charges in the period April through June.

Experimental Design Specify the design.

A total of 60,000 current Gold card customers will be randomly selected from our customer records to receive one of the six offers.

✓ Regular SkyWest miles
✓ Double SkyWest miles
✓ Triple SkyWest miles
✓ Regular *Use Anywhere* Miles
✓ Double *Use Anywhere* Miles
✓ Triple *Use Anywhere* Miles

Make a Picture A diagram of your design can help you think about it. We could also draw this diagram like the one on page 707 with 6 treatment groups, but now we are thinking of the design as having two distinct factors that we wish to evaluate individually, so this form gives the right impression.

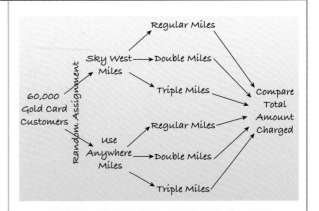

On June 15, we will examine the total card charges for each customer for the period April 1 through June 30.

Specify any other experimental details. You must give enough details so that another experimenter could exactly replicate your experiment.

We want to measure the effect of the two types of *Miles* and the three award *Amounts*.

It's generally better to include details that might seem irrelevant than to leave out matters that could turn out to make a difference.

Specify how to measure the response and your hypotheses.

The three null hypotheses are:

H_0: The mean charges for *Sky West Miles* and *Use Anywhere Miles* are the same (the means for *Miles* are equal).

H_0: The mean charges for *No Miles*, *Double Miles*, and *Triple Miles* are the same (the means for *Amount* are equal).

H_0: The effect of *Miles* is the same for all levels of *Amount* (and vice-versa) (no interaction effect).

The alternative for the first hypothesis is that the mean *Charges* for the two levels of *Miles* are different.

The alternative for the second hypothesis is that at least one of the mean charges for the three levels of *Amount* is different.

The alternative for the third hypothesis is that there is an interaction effect.

Plot Examine the boxplots and interaction plots.

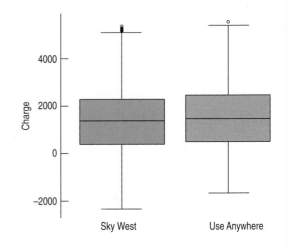

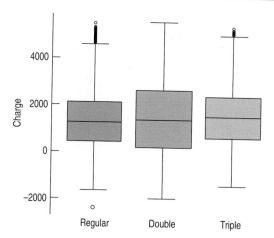

Boxplots by each factor show that there may be a slight increase in charges due to the *Use Anywhere* miles and the *Amount* of miles offered, but the differences are hard to see because of the intrinsic variation in *Charges*.

There are some outliers apparent in the boxplots, but none exerts a large influence on its group mean, so we will leave them in.

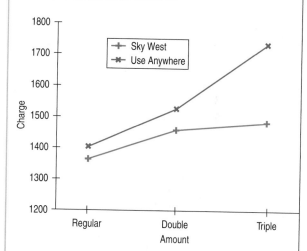

The interaction plot shows that offering *Triple* miles may have a much larger effect for *Use Anywhere* miles than for *Sky West* miles.

Assumptions and Conditions Think about the assumptions and check the conditions.

✓ **Independence Assumption, Randomization Condition.** The experiment was randomized to current cardholders.

✓ **Similar Variance Condition.** The boxplots show that the variances across all groups are similar. (We can recheck with a residual plot after fitting the ANOVA model.)

✓ **Outlier Condition.** There are some outliers, but none appear to be exerting undue influence on the group means.

DO

Show the ANOVA table.

Source	Df	SS	MS	*F*-Ratio	P-Value
Miles	1	103576768	103576768	61.6216	<0.0001
Amount	2	253958660.1	126979330	75.5447	<0.0001
Miles × Amount	2	64760963.01	32380481.51	19.2643	<0.0001
Error	29994	50415417459	1680850		
Total	29999	50837713850			

Check the remaining conditions on the residuals.

✓ **Nearly Normal Condition.** A histogram of the residuals shows that they are reasonably unimodal and symmetric.

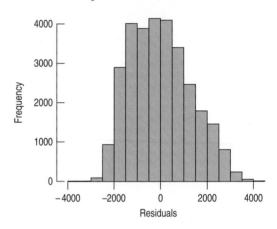

Under these conditions, it is appropriate to interpret the F-ratios and their P-values.

Discuss the results of the ANOVA table.

The F-ratios are all large, and the P-values are all very small, so we reject all three null hypotheses. Because the interaction effect is significant, we cannot talk about the overall effect of the amount of miles but must make the discussion specific to the type of *Miles* offered.

Show a table of means, possibly with confidence intervals or tests from an appropriate Multiple Comparisons method.

Level					Mean
Use Anywhere, Triple	A				1732.21
Use Anywhere, Double		B			1526.93
Sky West, Triple		B			1484.34
Sky West, Double		B	C		1460.20
Use Anywhere, Regular			C	D	1401.89
Sky West, Regular				D	1363.94

 **REPORT**

To answer the initial question, we ask whether the differences we observe in the means of the groups are meaningful.

MEMO:

Re: Test Mailing for Creative Offer and Envelope

The mailing for testing the Triple Miles initiative went out in March, and results on charges from

Because this is a randomized experiment, we can attribute significant differences to the treatments.

April through June were available in early July. We found that *Use Anywhere* miles performed better than the standard *Sky West* miles, but that the amount they increased charges depended on the amount offered.

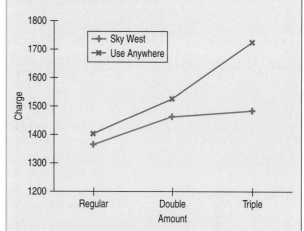

As we can see, *Triple Miles* for the *Sky West* miles didn't increase *Charge* significantly and is probably not worth the added expense. However, *Triple Miles* for the *Use Anywhere* miles generated an average $205 more in average *Charge* (with a confidence interval from $131 to $279). Even at the low end of this interval, we feel that the added revenue of the *Triple Miles* justifies their cost.

Be sure to make recommendations based on the context of your business decision.

In summary, we recommend offering Triple Miles for the Use Anywhere miles offers but would keep the Double miles offer for the Sky West miles.

WHAT CAN GO WRONG?

- **Don't give up just because you can't run an experiment.** Sometimes we can't run an experiment because we can't identify or control the factors. Sometimes it would simply be unethical to run the experiment. (Consider randomly assigning employees to two environments—one where workers were exposed to massive amounts of cigarette smoke and one a smoke-free environment—to see differences in health and productivity.) If we can't perform an experiment, often an observational study is a good choice.

- **Beware of confounding.** Use randomization whenever possible to ensure that the factors not in your experiment are not confounded with your treatment levels. Be alert to confounding that cannot be avoided, and report it along with your results.

- **Bad things can happen even to good experiments.** Protect yourself by recording additional information. An experiment in which the air-conditioning failed for two weeks, affecting the results, was saved by recording the temperature (although that was not originally one of the factors) and estimating the

(continued)

effect the higher temperature had on the response.[3] It's generally good practice to collect as much information as possible about your experimental units and the circumstances of the experiment. For example, in the direct mail experiment, it would be wise to record details of the general economy and any global events (such as a sharp downturn in the stock market) that might affect customer behavior.

- **Don't spend your entire budget on the first run.** Just as it's a good idea to pretest a survey, it's always wise to try a small pilot experiment before running the full-scale experiment. You may learn, for example, how to choose factor levels more effectively, about effects you forgot to control, and about unanticipated confounding.

- **Watch out for outliers.** One outlier in a group can change both the mean and the spread of that group. It will also inflate the Error Mean Square, which can influence the *F*-test. The good news is that ANOVA fails on the safe side by losing power when there are outliers. That is, you are less likely to reject the overall null hypothesis if you have (and leave) outliers in your data, so they are not likely to cause you to make a Type I error.

- **Watch out for changing variances.** The conclusions of the ANOVA depend crucially on the assumptions of independence and constant variance and (somewhat less seriously as *n* increases) on Normality. If the conditions on the residuals are violated, it may be necessary to re-express the response variable to approximate these conditions more closely. ANOVA benefits so greatly from a judiciously chosen re-expression that the choice of a re-expression might be considered a standard part of the analysis.

- **Be wary of drawing conclusions about causality from observational studies.** ANOVA is often applied to data from randomized experiments for which causal conclusions are appropriate. If the data are not from a designed experiment, however, the Analysis of Variance provides no more evidence for causality than any other method we have studied. Don't get into the habit of assuming that ANOVA results have causal interpretations.

- **Be wary of generalizing to situations other than the one at hand.** Think hard about how the data were generated to understand the breadth of conclusions you are entitled to draw.

- **Watch for multiple comparisons.** When rejecting the null hypothesis, you can conclude that the means are not *all* equal. But you can't start comparing every pair of treatments in your study with a *t*-test. You'll run the risk of inflating your Type I error rate. Use a multiple comparisons method when you want to test many pairs.

- **Be sure to fit an interaction term when it exists.** When the design is replicated, it is always a good idea to fit an interaction term. If it turns out not to be statistically significant, you can then fit a simpler two-factor main effects model instead.

- **When the interaction effect is significant, don't interpret the main effects.** Main effects can be very misleading in the presence of interaction terms. Look at this interaction plot:

---

[3] R. D. DeVeaux and M. Szelewski, "Optimizing Automatic Splitless Injection Parameters for Gas Chromatographic Environmental Analysis," *Journal of Chromatographic Science* 27, no. 9 (1989): 513–518.

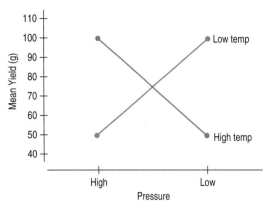

Figure 23.11 *An interaction plot of* Yield *by* Temperature *and* Pressure. *The main effects are misleading. There is no (main) effect of Pressure because the average Yield at the two pressures is the same. That doesn't mean that Pressure has no effect on the Yield. In the presence of an interaction effect, be careful when interpreting the main effects.*

The experiment was run at two temperatures and two pressure levels. High amounts of material were produced at high pressure with high temperature and at low pressure with low temperature. What's the effect of *Temperature?* Of *Pressure?* Both main effects are 0, but it would be silly (and wrong) to say that neither *Temperature* nor *Pressure* was important. The real story is in the interaction.

ETHICS IN ACTION

Professors at many state universities belong to a faculty union. The unionized faculty in one state's university system are preparing for contract negotiations. Cheryl McCrady, recently elected union president at one of the state university campuses, has long been concerned about the salary differential between male and female faculty. As union president, she now has access to faculty salary information, and she decides to run some analyses. After consulting with a few colleagues who regularly use statistics, she settles on using analysis of variance to determine if differences in salary can be attributed to gender accounting for faculty rank (assistant professor, associate professor, and full professor). She's not surprised by the results. While there is no significant interaction effect of gender and rank, she does find that both gender and rank are significant factors in explaining salary differences. Given that discrimination based on gender is a serious issue, she is wondering how she should proceed.

ETHICAL ISSUE *This is an observational study lacking the control of an experimental study. Confounding variables are likely to exist, but are not discussed. For instance, lower paid disciplines (e.g., Education) tend to have more female faculty than higher paid disciplines (e.g., Business). Related to Item A, ASA Guidelines. She should also check for outliers. Special cases, such as a star football coach or Nobel prize winner, may command unusually large salaries but not be relevant to the pay of ordinary faculty members.*

ETHICAL SOLUTION *Make all caveats explicit. This is a complex issue that should not be treated simply.*

What have we learned?

We've learned to recognize sample surveys, observational studies, and randomized comparative experiments. We know that these methods collect data in different ways and lead us to different conclusions.

We've learned to identify retrospective and prospective observational studies and understand the advantages and disadvantages of each.

We've learned that only well-designed experiments can allow us to reach cause-and-effect conclusions. We manipulate levels of treatments to see if the factor we have identified produces changes in our response variable.

We've learned the principles of experimental design:

- We want to be sure that variation in the response variable can be attributed to our factor, so we identify as many other sources of variability as possible.

- We control the sources of variability we can and consider blocking to reduce variability from sources we recognize but cannot control.

- Because there are many possible sources of variability that we cannot identify, we try to equalize those by randomly assigning experimental units to treatments.

- We replicate the experiment on as many subjects as possible. We've learned the value of having a control group and of using blinding and placebo controls.

We learned, in Chapter 13, how to test whether the means of two groups are equal. Now in this chapter, we've extended that to testing whether the means of several groups are equal. We first learned in Chapter 4 that a good first step in looking at the relationship between a quantitative response and a categorical grouping variable is to look at side-by-side boxplots. We've seen that it's still a good first step before formally testing the null hypothesis.

We've learned that the F-test is a generalization of the t-test that we used for testing two groups. We've seen that although this makes the mechanics familiar, there are new conditions to check. We've also learned that when the null hypothesis is rejected and we conclude that there are differences, we need to adjust the confidence intervals for the pairwise differences between means. We also need to adjust the alpha levels of tests we perform once we've rejected the null hypothesis.

- We've learned that under certain assumptions, the statistic used to test whether the means of k groups are equal is distributed as a F-statistic with $k - 1$ and $N - k$ degrees of freedom.

- We've learned to check four conditions to verify the assumptions before we proceed with inference, and we've seen that most of the checks can be made by graphing the data and the residuals with the methods we learned in earlier chapters.

- We've learned that if the F-statistic is large enough, we reject the null hypothesis that all the means are equal.

- We've also learned to create and interpret confidence intervals for the differences between each pair of group means, recognizing that we need to adjust the confidence interval for the number of comparisons we make.

- We've learned that sometimes factors can interact with each other. When we have at least two observations at each combination of factor levels, we can add an interaction term to our model to account for the possible interaction.

- Finally, we've learned to recognize the problems posed by confounding variables in experiments and lurking variables in observational studies.

Terms

Analysis of Variance (ANOVA)	An analysis method for testing equality of means across treatment groups.
ANOVA table	The ANOVA table is convenient for showing the degrees of freedom, treatment mean square, error mean square, their ratio, F-statistic, and its P-value. There are usually other quantities of lesser interest included as well.
Blind, Blinding	Any individual associated with an experiment who is not aware of how subjects have been allocated to treatment groups is said to be blinded.
Block	When groups of experimental units are similar, it is often a good idea to gather them together into blocks. By blocking we isolate the variability attributable to the differences between the blocks so that we can see the differences in the means due to the treatments more clearly.
Bonferroni method	One of many methods for adjusting the margin of error to control the overall risk of making a Type I error when testing many pairwise differences between group means.
Confounded	When the levels of one factor are associated with the levels of another factor so that their effects cannot be separated, we say that these two factors are confounded.
Control	When we limit the levels of a factor not explicitly part of the experiment design, we have controlled that factor. (By contrast, the factors we are testing are said to be *manipulated*.)
Control group	The experimental units assigned to a baseline treatment level, typically either the default treatment, which is well understood, or a null, placebo treatment. Their responses provide a basis for comparison.
Designs	• **Randomized block design:** The randomization occurs only within blocks. • **Completely randomized design:** All experimental units have an equal chance of receiving any treatment. • **Factorial design:** Includes more than one factor in the same design and includes every combination of all the levels of each factor.

Double-blind, Single-blind	There are two classes of individuals who can affect the outcome of an experiment:
	
	    those who could *influence the results* (subjects, treatment administrators, or technicians)
	    those who *evaluate the results* (judges, treating physicians, etc.)
	
	When every individual in *either* of these classes is blinded, an experiment is said to be single-blind.
	When everyone in *both* classes is blinded, we call the experiment double-blind.

Experiment	An experiment *manipulates* factor levels to create treatments, *randomly assigns* subjects to these treatment levels, and then *compares* the responses of the subject groups across treatment levels.
Experimental units	Individuals on whom an experiment is performed. Usually called subjects or participants when they are human.
F-distribution	The F-distribution is the sampling distribution of the F-statistic when the null hypothesis that the treatment means are equal is true. It has two degrees of freedom parameters, one for the numerator, $(k-1)$, and one for the denominator, $N-k$, where N is the total number of observations and k is the number of groups.

F-statistic	The *F*-statistic is the ratio MST/MSE. When the *F*-statistic is sufficiently large, we reject the null hypothesis that the group means are equal.
F-test	The *F*-test tests the null hypothesis that all the group means are equal against the one-sided alternative that they are not all equal. We reject the hypothesis of equal means if the *F*-statistic exceeds the critical value from the *F*-distribution corresponding to the specified significance level and degrees of freedom.
Factor	A variable whose levels are controlled by the experimenter. Experiments attempt to discover the effects that differences in factor levels may have on the responses of the experimental units.
Interaction	When the effects of the levels of one factor change depending on the level of the other factor, the two factors are said to interact. When interaction terms are present, it is misleading to talk about the main effect of one factor because how large it is *depends* on the level of the other factor.
Interaction plot	A plot that shows the means at each treatment combination, highlighting the factor effects and their behavior at all the combinations.
Level	The specific values that the experimenter chooses for a factor are called the levels of the factor.
Mean Square	A sum of squares divided by its associated degrees of freedom. • **Mean Square due to Error (MSE)** The estimate of the error variance obtained by pooling the variance of each treatment group. The square root of the MSE is the estimate of the error standard deviation, s_p. • **Mean Square due to Treatment (MST)** The estimate of the error variance under the null hypothesis that the treatment means are all equal. If the null hypothesis is not true, the MST will be larger than the error variance.
Multiple comparisons	If we reject the null hypothesis of equal means, we often then want to investigate further and compare pairs of treatment group means to see if they differ. If we want to test several such pairs, we must adjust for performing several tests to keep the overall risk of a Type I error from growing too large. Such adjustments are called methods for multiple comparisons.
Observational study	A study based on data in which no manipulation of factors has been employed.
Placebo	A treatment known to have no effect, administered so that all groups experience the same conditions. Many subjects respond to such a treatment (a response known as a *placebo effect*). Only by comparing with a placebo can we be sure that the observed effect of a treatment is not due simply to the placebo effect.
Placebo effect	The tendency of many human subjects (often 20% or more of experiment subjects) to show a response even when administered a placebo.
Principles of experimental design	• **Control** aspects of the experiment that we know may have an effect on the response, but that are not the factors being studied. • **Randomize** subjects to treatments to even out effects that we cannot control. • **Replicate** over as many subjects as possible. Results for a single subject are just anecdotes. • **Block** to reduce the effects of identifiable attributes of the subjects that cannot be controlled.
Prospective study	An observational study in which subjects are followed to observe future outcomes. Because no treatments are deliberately applied, a prospective study is not an

experiment. Nevertheless, prospective studies typically focus on estimating differences among groups that might appear as the groups are followed during the course of the study.

Random assignment	To be valid, an experiment must assign experimental units to treatment groups at random. This is called random assignment.
Residual standard deviation	The residual standard deviation gives an idea of the underlying variability of the response values.
Response	A variable whose values are compared across different treatments. In a randomized experiment, large response differences can be attributed to the effect of differences in treatment level.
Retrospective study	An observational study in which subjects are selected and then their previous conditions or behaviors are determined. Because retrospective studies are not based on random samples, they usually focus on estimating differences between groups or associations between variables.
Subjects or **Participants**	When the experimental units are people, they are usually referred to as Subjects or Participants.
Treatment	The process, intervention, or other controlled circumstance applied to randomly assigned experimental units. Treatments are the different levels of a single factor or are made up of combinations of levels of two or more factors.

Skills

- Recognize when an observational study would be appropriate.
- Be able to identify observational studies as retrospective or prospective and understand the strengths and weaknesses of each method.
- Know the four basic principles of sound experimental design: control, randomize, replicate, and block, and be able to explain each.
- Be able to recognize the factors, treatments, and response variable in a description of a designed experiment.
- Understand the essential importance of randomization in assigning treatments to experimental units.
- Understand the importance of replication to move from anecdotes to general conclusions.
- Understand the value of blocking so that variability due to differences in attributes of the subjects can be removed.
- Understand the importance of a control group and the need for a placebo treatment in some studies.
- Understand the importance of blinding and double-blinding in studies on human subjects and be able to identify blinding and the need for blinding in experiments.
- Understand the value of a placebo in experiments with human participants.
- Recognize situations for which ANOVA is the appropriate analysis.
- Know how to examine your data for violations of conditions that would make ANOVA unwise or invalid.
- Recognize when a further analysis of differences between group means would be appropriate.
- Understand the advantages of an experiment in two factors.

DO

- Be able to design a completely randomized experiment to test the effect of a single factor.
- Be able to design an experiment in which blocking is used to reduce variation.
- Know how to use graphical displays to compare responses for different treatment groups.
- Be able to perform an ANOVA using a statistics package or calculator for one response variable and one factor with any number of levels.
- Be able to perform several subsequent tests using a multiple comparisons procedure.
- Be able to use a statistics package to compute a two-way ANOVA.
- Know how to interpret an interaction plot for replicated data with two factors.

REPORT

- Know how to report the results of an observational study. Identify the subjects, how the data were gathered, and any potential biases or flaws you may be aware of. Identify the factors known and those that might have been revealed by the study.
- Know how to report the results of an experiment. Tell who the subjects are and how their assignment to treatments was determined. Report on how the response variable was measured and in what measurement units.
- Understand that your description of an experiment should be sufficient for another researcher to replicate the study with the same methods.
- Be able to explain the contents of an ANOVA table, in particular the role of the MST, MSE, and the pooled standard deviation, s_p.
- Be able to interpret a test of the null hypothesis that the true means of several independent groups are equal. (Your interpretation should include a defense of your assumption of equal variances.)
- Be able to interpret the results of tests that use multiple comparisons methods.
- Be able to interpret main effects in a two-way ANOVA.
- Be able to use an interaction plot to explain an interaction effect.
- Be able to distinguish when a discussion of main effects is appropriate in the presence of a significant interaction.

Technology Help

Most analyses of variance are performed with computers, and all statistics packages present the results in an ANOVA table much like the ones in the chapter. Technology also makes it easy to examine the side-by-side boxplots and check the residuals for violations of the assumptions and conditions. Statistics packages offer different choices among possible multiple comparisons methods. This is a specialized area. Get advice or read further if you need to choose a multiple comparisons method. As we saw in Chapter 13, there are two ways to organize data recorded for several groups. We can put all the response values in a single variable and use a second, "factor," variable to hold the group identities. This is sometimes called *stacked format*. The alternative is an unstacked format, placing the data for each group in its own column or variable. Then the variable identities become the group identifiers. Stacked format is necessary for experiments with more than one factor. Each factor's levels are named in a variable. Some packages can work with either format for simple one-factor designs, and some use one format for some things and the other for others. (Be careful, for example, when you make side-by-side boxplots; be sure to give the appropriate version of that command to correspond to the structure of your data.) Most packages offer to save residuals and predicted values and make them available for further tests of conditions. In some packages, you may have to request them specifically.

Some statistics packages have different commands for models with one factor and those with two or more factors. You must be alert to these differences when analyzing a two-factor ANOVA. It's not unusual to find ANOVA models in several different places in the same package. (Look for terms like "Linear Models.")

EXCEL

To compute a single-factor ANOVA:

- From the tools menu (or the Data Ribbon in Office 2007), select **Data Analysis**.
- Select **Anova Single Factor** from the list of analysis tools.
- Click the **OK** button.
- Enter the data range in the box provided.
- Check the **Labels in First Row** box, if applicable.
- Enter an alpha level for the *F*-test in the box provided.
- Click the **OK** button.

Comments

The data range should include two or more columns of data to compare. Unlike statistics packages, Excel expects each column of the data to represent a different level of the factor. However, it offers no way to label these levels. The columns need not have the same number of data values, but the selected cells must make up a rectangle large enough to hold the column with the most data values.

Excel cannot compute multiway ANOVA, but the Excel Data Analysis Add-in offers a two-way ANOVA "with and without replication." That command requires the data to be in a special format and cannot deal with unbalanced (i.e., unequal counts in treatment groups) data.

MINITAB

- Choose **ANOVA** from the Stat menu.
- Choose **One-way...** or **Two-way...** from the **ANOVA** submenu.
- In the dialog, assign a quantitative Y variable to the Response box and assign the categorical X factor(s) to the Factor box.
- In a two-way ANOVA, specify interactions.
- Check the **Store Residuals** check box.
- Click the **Graphs** button.

- In the ANOVA-Graphs dialog, select **Standardized residuals,** and check **Normal plot of residuals** and **Residuals versus fits**.
- Click the **OK** button to return to the ANOVA dialog.
- Click the **OK** button to compute the ANOVA.

Comments

If your data are in unstacked format, with separate columns for each treatment level, Minitab can compute a one-way ANOVA directly. Choose **One-way (unstacked)** from the **ANOVA** submenu. For two-way ANOVA, you must use the stacked format.

SPSS

To compute a one-way ANOVA:

- Choose **Compare Means** from the Analyze menu.
- Choose **One-way ANOVA** from the **Compare Means** submenu.
- In the One-Way ANOVA dialog, select the Y-variable and move it to the dependent target. Then move the X-variable to the independent target.
- Click the **OK** button.

To compute a two-way ANOVA:

- Choose **Analyze > General Linear Model > Univariate**.
- Assign the response variable to the **Dependent Variable** box.

- Assign the two factors to the **Fixed Factor(s)** box. This will fit the model with interactions by default.
- To omit interactions, click on **Model.** Select **Custom**. Highlight the factors. Select **Main Effects** under the **Build Terms** arrow and click the arrow.
- Click **Continue** and **OK** to compute the model.

Comments

SPSS expects data in stacked format. The **Contrasts** and **Post Hoc** buttons offer ways to test contrasts and perform multiple comparisons. See your SPSS manual for details.

JMP

To compute a one-way ANOVA:

- From the **Analyze** menu, select **Fit Y by X**.
- Select variables: a quantitative Y, Response variable, and a categorical X, Factor variable.
- JMP opens the **Oneway** window.
- Click on the red triangle beside the heading, select **Display Options**, and choose **Boxplots**.
- From the same menu, choose the **Means/ANOVA t-test** command.
- JMP opens the oneway ANOVA output.

To compute a two-way ANOVA:

- From the **Analyze** menu, select **Fit Model**.
- Select variables and **Add** them to the **Construct Model Effects** box.

- To specify an interaction, select both factors and press the **Cross** button.
- Click **Run Model**.
- JMP opens a **Fit Least Squares** window.
- Click on the red triangle beside each effect to see the means plots for that factor. For the interaction term, this is the interaction plot.
- Consult JMP documentation for information about other features.

Comments

JMP expects data in "stacked" format with one continuous response and two nominal factor variables.

DATA DESK

To compute a one-way or two-way ANOVA:

- Select the response variable as Y and the factor variable as X.
- From the **Calc** menu, choose **ANOVA > ANOVA** or (for two-way ANOVA with interactions) **ANOVA > ANOVA with Interactions**.
- Data Desk displays the ANOVA table.

- Select plots of residuals from the ANOVA table's Hyper-View menu.

Comments

Data Desk expects data in "stacked" format. You can change the ANOVA by dragging the icon of another variable over either the Y or the X variable name in the table and dropping it there. The analysis and any displays will recompute automatically.

Mini Case Study Project

Design, carry out, and analyze your own multifactor experiment. The experiment doesn't have to involve human subjects. In fact, an experiment designed to find the best settings for microwave popcorn, the best paper airplane design, or the optimal weight and placement of coins on a toy car to make it travel farthest and fastest down an incline are all fine ideas. Be sure to define your response variable of interest before you start the experiment and detail how you'll perform the experiment, specifically including the elements you control, how you use randomization, and how many times you replicate the experiment. Analyze the results of your experiment and write up your analysis and conclusions including any recommendations for further testing.

EXERCISES

1. Laundry detergents. A consumer group wants to test the efficacy of a new laundry detergent. They take 16 pieces of white cloth and stain each with the same amount of grease. They decide to try it using both hot and cold water settings and at both short and long washing times. Half of the 16 pieces will get the new detergent, and half will get a standard detergent. They'll compare the shirts by using an optical scanner to measure whiteness.

a) What are the factors they are testing?
b) Identify all the factor levels.
c) What is/are the response(s)?

2. Sales scripts. An outdoor products company wants to test a new website design where customers can get information about their favorite outdoor activity. They randomly send half of the customers coming to the website to the new design. They want to see whether the Web visitors spend more time at the site and whether they make a purchase.

a) What are the factors they are testing?
b) Identify all the factor levels.
c) What is/are the response(s)?

3. Laundry detergents, part 2. One member of the consumer group in Exercise 1 is concerned that the experiment will take too long and makes some suggestions to shorten it. Comment briefly on each idea.

a) Cut the runs to 8 by testing only the new detergent. Compare the results to results on the standard detergent published by the manufacturer.
b) Cut the runs to 8 by testing only in hot water.
c) Keep the number of runs at 16, but save time by running all the standard detergent runs first to avoid swapping detergents back and forth.

4. Swimsuits. A swimsuit manufacturer wants to test the speed of its newly designed $550 suit. They design an experiment by having 6 randomly selected Olympic swimmers swim as fast as they can with their old swimsuit first and then swim the same event again with the new, expensive swim suit. They'll use the difference in times as the response variable. Criticize the experiment and point out some of the problems with generalizing the results.

5. Mozart. Will listening to a Mozart piano sonata make you smarter? In a 1995 study, Rauscher, Shaw, and Ky reported that when students were given a spatial reasoning section of a standard IQ test, those who listened to Mozart for 10 minutes improved their scores more than those who simply sat quietly.

a) These researchers said the differences were statistically significant. Explain what that means in this context.
b) Steele, Bass, and Crook tried to replicate the original study. The subjects were 125 college students who participated in the experiment for course credit. Subjects first took the test. Then they were assigned to one of three groups: listening to a Mozart piano sonata, listening to music by Philip Glass, and sitting for 10 minutes in

silence. Three days after the treatments, they were retested. Draw a diagram displaying the design of this experiment.
c) The boxplots show the differences in score before and after treatment for the three groups. Did the Mozart group show improvement?
d) Do you think the results prove that listening to Mozart is beneficial? Explain.

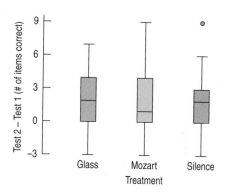

6. More Mozart. An advertisement selling specially designed CDs of Mozart's music specifically because they will "strengthen your mind, heal your body, and unlock your creative spirit" claims that "in Japan, a brewery actually reported that their best sake is made when Mozart is played near the yeast." Suppose you wished to design an experiment to test whether this is true. Assume you have the full cooperation of the sake brewery. Specify how you would design the experiment. Indicate factors and response and how they would be measured, controlled, or randomized.

7. Cereal marketing. The makers of Frumpies, "the breakfast of rug rats," want to improve their marketing, so they consult you.

a) They first want to know what fraction of children, ages 10 to 13, like their celery-flavored cereal. What kind of study should they perform?
b) They are thinking of introducing a new flavor, maple-marshmallow Frumpies and want to know whether children will prefer the new flavor to the old one. Design a completely randomized experiment to investigate this question.
c) They suspect that children who regularly watch the Saturday morning cartoon show starring Frump, the flying teenage warrior rabbit who eats Frumpies in every episode, may respond differently to the new flavor. How would you take that into account in your design?

8. Wine marketing. A 2001 Danish study published in the *Archives of Internal Medicine* casts significant doubt on suggestions that adults who drink wine have higher levels of "good" cholesterol and fewer heart attacks. These researchers followed a group of individuals born at a Copenhagen hospital between 1959 and 1961 for 40 years. Their study found that in this group the adults who drank wine were richer and better educated than those who did not.

a) What kind of study was this?

b) It is generally true that people with high levels of education and high socioeconomic status are healthier than others. How does this call into question the supposed health benefits of wine?

c) Can studies such as these prove causation (that wine helps prevent heart attacks, that drinking wine makes one richer, that being rich helps prevent heart attacks, etc.)? Explain.

9. SAT prep courses. Can special study courses actually help raise SAT scores? One organization says that the 30 students they tutored achieved an average gain of 60 points when they retook the test.

a) Explain why this does not necessarily prove that the special course caused the scores to go up.

b) Propose a design for an experiment that could test the effectiveness of the tutorial course.

c) Suppose you suspect that the tutorial course might be more helpful for students whose initial scores were particularly low. How would this affect your proposed design?

10. Safety switch. An industrial machine requires an emergency shutoff switch that must be designed so that it can be easily operated with either hand. Design an experiment to find out whether workers will be able to deactivate the machine as quickly with their left hands as with their right hands. Be sure to explain the role of randomization in your design.

11. Fuel economy. These boxplots show the relationship between the number of cylinders in a car's engine and its fuel economy from a study conducted by a major car manufacturer.

a) What are the null and alternative hypotheses? Talk about cars and fuel efficiency, not symbols.

b) Do the conditions for an ANOVA seem to be met here? Why or why not?

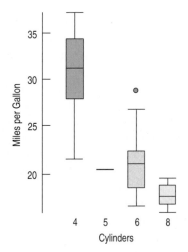

12. Wine production. The boxplots display case prices (in dollars) of wine produced by wineries along three of the Finger Lakes in upstate New York.

a) What are the null and alternative hypotheses? Talk about prices and location, not symbols.

b) Do the conditions for an ANOVA seem to be met here? Why or why not?

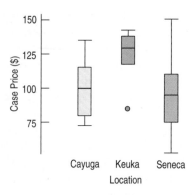

13. Cell phone adoption. Cell phone adoption rates are available for various countries in the United Nations Database (unstats. un.org). Countries were randomly selected from three regions (Africa, Asia, and Europe), and cell phone adoption (per 100 inhabitants) rates retrieved. The boxplots display the data.

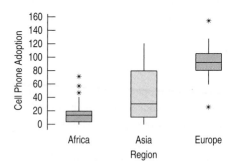

a) What are the null and alternative hypotheses (in words, not symbols)?

b) Are the conditions for ANOVA met? Why or why not?

14. Marketing managers' salaries. A sample of eight states was selected randomly from each of three regions in the United States (Northeast, Southeast, and West). Mean annual salaries for marketing managers were retrieved from the U.S. Bureau of Labor Statistics (data.bls.gov/oes). The boxplots display the data.

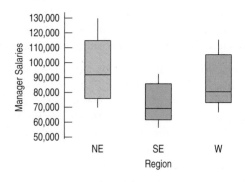

a) What are the null and alternative hypotheses (in words, not symbols)?

b) Are the conditions for ANOVA met? Why or why not?

15. Bank tellers. A bank is studying the average time that it takes 6 of its tellers to serve a customer. Customers line up in the queue and are served by the next available teller. Here is a boxplot of the times it took to serve the last 140 customers.

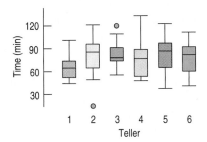

Source	DF	Sum of Squares	Mean Square	F-ratio	P-value
Teller	5	3315.32	663.064	1.508	0.1914
Error	134	58919.1	439.695		
Total	139	62234.4			

a) What are the null and alternative hypotheses?
b) What do you conclude?
c) Would it be appropriate to run a multiple comparisons test (for example, a Bonferroni test) to see which tellers differ from each other? Explain.

16. Product development. Vendors of hearing aids test them by having patients listen to lists of words and repeat what they hear. The word lists are supposed to be equally difficult to hear accurately. But the challenge of hearing aids is perception when there is background noise. A researcher investigated four different word lists used in hearing assessment (Loven, 1981). She wanted to know whether the lists were equally difficult to understand in the presence of a noisy background. To find out, she tested 24 subjects with normal hearing and measured the number of words perceived correctly in the presence of background noise. Here are the boxplots of the four lists.

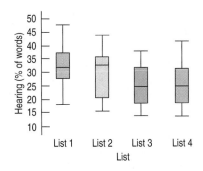

Source	DF	Sum of Squares	Mean Square	F-ratio	P-value
List	3	920.4583	306.819	4.9192	0.0033
Error	92	5738.1667	62.371		
Total	95	6658.6250			

a) What are the null and alternative hypotheses?
b) What do you conclude?
c) Would it be appropriate to run a multiple comparisons test (for example, a Bonferroni test) to see which lists differ from each other in terms of mean percent correct? Explain.

17. E-security. A report released by the Pew Internet & American Life Project entitled *The Internet & Consumer Choice* focused on current online issues (www.pewinternet.org/data.asp). Respondents were asked to indicate their level of agreement (1 = strongly agree to 4 = strongly disagree) with a variety of statements including "I don't like giving my credit card number or personal information online." A part of the data set was used to determine whether the type of community in which the individual resides (Urban, Suburban, or Rural) affected responses. Here are the results in the form of a partially completed Analysis of Variance table.

Source	DF	Sum of Squares	Mean Square	F-ratio	P-value
Community	2	6.615			
Error	183	96.998			
Total	185	103.613			

a) Is this an experimental or observational study? Explain.
b) Is this a prospective or retrospective study? Explain.
c) State the null and alternative hypothesis.
d) Calculate the *F*-statistic.
e) The P-value for this statistic turns out to be 0.002. State the conclusion. Can a causal link be established? Explain.

18. Internet usage. Internet usage rates are available for various countries in the United Nations Common Database (unstats.un.org). Countries were randomly selected from three regions (Africa, Asia, and Europe), and Internet usage (per 100 inhabitants) data from 2005 were retrieved. The data were analyzed to determine if Internet usage rates were the same across regions. The partially completed Analysis of Variance table is shown here.

Source	DF	Sum of Squares	Mean Square	F-ratio	P-value
Region	2	21607			
Error	93	20712			
Total	95	42319			

a) Is this an experimental or observational study? Explain.
b) Is this a prospective or retrospective study? Explain.
c) State the null and alternative hypotheses.
d) Calculate the *F*-statistic.
e) The P-value for this statistic turns out to be < 0.001. State the conclusion. Can a causal link be established? Explain.

19. Colorful experiment? In a recent article published in *Quality Progress*, each student in a statistics class had a randomly assigned bag of peanut M&M's® and counted the number of each color (*Blue, Red, Orange, Green, Brown, Yellow*). The bags were all the same size (1.74 ounces). The investigators claimed to use a randomized block design, with *Bag* as the blocking factor. They counted the number of candies of each color in each bag. Their results are reproduced here (Lin, T. and Sanders, M.S., "A Sweet Way to Learn DOE," *Quality Progress*, Feb. 2006, p. 88).

Source	Degrees of Freedom	Sum of Squares	Mean Square	F-ratio	P-value
Bag	13	4.726	0.364	0.10	1.000
Color	5	350.679	70.136	18.72	<0.001
Error	65	243.488	3.746		
Total	83	598.893			

a) Was this an observational or experimental study?
b) What was the treatment? What factors were manipulated?
c) What was the response variable?

20. Six Sigma training. A large financial institution is interested in training its college educated workforce in Six Sigma principles and methods. One part of the training involves basic statistical concepts and tools. Management is considering three approaches: *online, traditional classroom,* and *hybrid (a mix of both).* Prior to launching the program throughout the entire organization, they decided to pilot test the three approaches. Because they believed that educational background may affect the results, they selected 3 employees from each of 10 different college major programs of study (*liberal arts, accounting, economics, management, marketing, finance, information systems, computer science, operations, other*), and randomly assigned each to one of the three approaches. At the end of training, each participant took an exam. The results are shown here.

Source	Degrees of Freedom	Sum of Squares	Mean Square	F-ratio	P-value
Major	9	2239.47	248.830	21.69	<0.001
Training	2	171.47	85.735	7.47	0.004
Error	18	206.53	11.474		
Total	29	2617.47			

a) Was this an observational study or an experiment?
b) What was the purpose of using Major as a blocking factor?
c) Given the results, was it necessary to use Major as a blocking factor? Explain.
d) State the conclusion from this analysis.

21. E-trust. Online retailers want customers to trust their websites and want to alleviate any concerns potential customers may have about privacy and security. In a study investigating the factors that affect e-trust, participants were randomly assigned to carry out online transactions on fictitious retailers' websites. The sites were configured in one of three ways: (1) *with a third-party assurance seal (e.g., BBBOnLine) displayed;* (2) *a self-proclaimed assurance displayed;* or (3) *no assurance.* In addition, participants made a transaction involving one of three products (*book, camera,* or *insurance*). These products represent varying degrees of risk. After completing the transaction, they rated how "trustworthy" the website was on a scale of 1 (not at all) to 10 (extremely trustworthy).

a) Is this an experiment or an observational study? Explain.
b) What is the response variable?

c) How many factors are involved?
d) How many treatments are involved?
e) State the hypotheses (in words, not symbols).

22. Injection molding. In order to improve the quality of molded parts, companies often test different levels of parameter settings in order to find the best combinations. Injection molding machines typically have many adjustable parameters. One company used three different mold temperatures (25, 35, and 45 degrees Celsius) and four different cooling times (10, 15, 20, and 25 minutes) to examine how they affect the tensile strength of the resulting molded parts. Five parts were randomly sampled and measured from each treatment combination.

a) Is this an experiment or an observational study? Explain.
b) What is the response variable?
c) What are the factors?
d) How many treatments are involved?
e) State the hypotheses (in words, not symbols).

23. Stock returns. Companies that are ISO 9000 certified have met standards that ensure they have a quality management system committed to continuous improvement. Going through the certification process generally involves a substantial investment that includes the hiring of external auditors. A group of such auditors, wishing to "prove" that ISO 9000 certification pays off, randomly selected a sample of small and large companies with and without ISO 9000 certification. Size was based on the number of employees. They computed the % change in closing stock price from August 2006 to August 2007. The two-way ANOVA results are presented here (data obtained from *Yahoo! Finance*).

Source	DF	Sum of Squares	Mean Square	F-ratio	P-value
ISO 9000	1	2654.4	2654.41	5.78	0.022
Size	1	0.2	0.18	0.004	0.984
Interaction	1	1505.5	1505.49	3.28	0.079
Error	36	16545.9	459.61		
Total	39	20705.9			

a) Is this an experiment or an observational study?
b) State the hypotheses.
c) Given the small P-value associated with the ISO 9000 factor and that the mean annual return for the companies with ISO 9000 is 30.7% compared to 14.4% for those without, the auditors state that achieving ISO 9000 certification results in higher stock prices. Do you agree with their statement? Explain.

24. Company bonuses. After complaints about gender discrimination regarding bonus incentive pay, a large multinational firm collected data on bonuses awarded during the previous year (% of base pay). Human Resources (HR) randomly sampled male and female managers from three different levels: *senior, middle,* and *supervisory.* The two-way ANOVA results are presented here.

Source	DF	Sum of Squares	Mean Square	F-ratio	P-value
Gender	1	32.033	32.033	9.76	0.005
Level	2	466.200	233.100	70.99	0.000
Interaction	2	20.467	10.233	3.12	0.063
Error	24	78.800	3.283		
Total	29	597.500			

a) Is this an experiment or an observational study?

b) State the hypotheses.

c) Given the small P-value associated with the gender and that the mean annual bonus percent for females is 12.5% compared to 14.5% for males, HR concludes that gender discrimination exists. Do you agree? Explain.

25. Managers' hourly wages. What affects marketing managers' hourly wages? In order to find out, mean hourly wages were retrieved from the U.S. Bureau of Labor Statistics for two managerial occupations in marketing (*Sales managers, Advertising managers*) for a random sample of states from three regions (*Midwest, Southeast, West*) (data.bls.gov/oes). Here are boxplots showing mean hourly wages for the two marketing occupations and the three regions as well as the results for a two-way ANOVA.

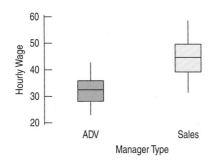

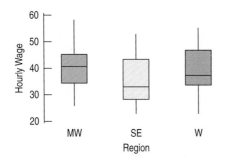

Source	DF	Sum of Squares	Mean Square	F-ratio	P-value
Manager Type	1	1325.93	1325.93	31.84	0.000
Region	2	153.55	76.78	1.84	0.176
Interaction	2	32.74	16.37	0.39	0.678
Error	30	1249.32	41.64		
Total	35	2761.55			

a) Is this an experiment or an observational study? Explain.

b) Are the conditions for two-way ANOVA met?

c) If so, perform the hypothesis tests and state your conclusions in terms of hourly wages, occupational type, and region.

d) Is it appropriate to interpret the main effects in this case? Explain.

26. Concrete testing. A company that specializes in developing concrete for construction strives to continually improve the properties of its materials. In order to increase the compressive strength of one of its new formulations, they varied the amount of alkali content (*low, medium, high*). Since the type of sand used may also affect the strength of concrete, they used three different types of sand (Types I, II, III). Four samples were randomly selected from each treatment combination to be tested. The boxplots show the test results on compressive strength (in psi) for the three levels of alkali content and three types of sand. Two-way ANOVA results are also given.

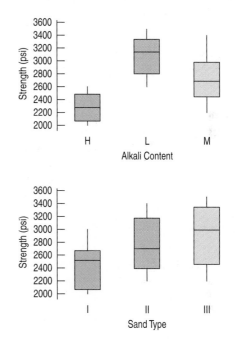

Source	DF	Sum of Squares	Mean Square	F-ratio	P-value
Alkali Content	2	4016600	2008300	46.38	0.000
Sand Type	2	1547817	773908	17.87	0.000
Interaction	4	177533	44383	1.02	0.412
Error	27	1169250	43306		
Total	35	6911200			

a) Is this an experiment or an observational study? Explain.

b) Are the conditions for two-way ANOVA met?

c) If so, perform the hypothesis tests and state your conclusions in terms of compressive strength, alkali content, and sand type.

d) Is it appropriate to interpret the main effects in this case? Explain.

T 27. Production problems. A manufacturing company that makes dental drills was experiencing problems with a specific part on the production line. Management suspected a machining problem that resulted in the length of the part to vary outside of target specification. Two factors were examined: the machine setting (at three levels) and the shift (morning, afternoon, and night). New hires were typically scheduled for night shift, and management believed that their relative inexperience may also be contributing to the variation. Three parts were randomly selected and measured from each treatment combination. The deviation from specified size was measured in microns. The data and two-way ANOVA results are shown.

Size Error	Machine Setting	Shift
1.1	1	Morning
3.6	2	Morning
3.3	3	Morning
2.1	1	Morning
0.9	2	Morning
2.6	3	Morning
0.6	1	Morning
2.3	2	Morning
3.2	3	Morning
2	1	Afternoon
2.4	2	Afternoon
5	3	Afternoon
1.8	1	Afternoon
4.3	2	Afternoon
3.2	3	Afternoon
2.5	1	Afternoon
5	2	Afternoon
2.3	3	Afternoon
3.8	1	Night
5.5	2	Night
5	3	Night
2.9	1	Night
6.7	2	Night
5.8	3	Night
2.8	1	Night
3	2	Night
5.3	3	Night

Source	DF	Sums of Squares	Mean Square	F-ratio	P-value
MachSet	2	17.1119	8.55593	7.3971	0.0045
Shift	2	24.9607	12.4804	10.790	0.0008
Interaction	4	1.4970	0.374259	0.32357	0.8585
Error	18	20.8200	1.15667		
Total	26	64.3896			

a) Is this an experiment or an observational study? Explain.
b) What is the response variable?
c) How many treatments are involved?
d) Based on the two-way ANOVA results, management concluded that shift has a significant impact on the length of the part and that consequently operator inexperience is the root cause of the part problems. Do you agree with this conclusion? Explain.

28. Process improvements. One way to improve a process is to eliminate nonvalue added activities (e.g., extra movements) and wasted effort (e.g., looking for materials). A consultant was hired to improve the efficiency in a large shop floor operation. She tested three different workspace designs and two different storage/retrieval systems. She measured process flow time for three randomly selected operations through each of the combinations of workspace design and storage/retrieval systems. The data and two-way ANOVA results are shown here.

Workspace Design	Storage System	Flow Time (Days)
1	1	4.5
2	1	3.3
3	1	3.4
1	1	4.0
2	1	3.0
3	1	2.9
1	1	4.2
2	1	3.0
3	1	3.2
1	1	4.5
2	1	3.5
3	1	3.2
1	1	3.8
2	1	2.8
3	1	3.0
1	2	3.0
2	2	3.8
3	2	3.6
1	2	2.8
2	2	4.0
3	2	3.5
1	2	3.0
2	2	3.5
3	2	3.8
1	2	4.0
2	2	4.2
3	2	4.2
1	2	3.0
2	2	3.6
3	2	3.8

Source	DF	Sum of Squares	Mean Square	F-ratio	P-value
Workspace Design	2	0.30867	0.15433	1.56	0.230
Storage System	1	0.07500	0.07500	0.76	0.392
Interaction	2	4.87800	2.43900	24.72	<0.001
Error	24	2.36800	0.09867		
Total	29	7.62967			

a) Is this an experiment or observational study? Explain.
b) What is the response variable?

c) How many treatments are involved?

d) Based on the two-way ANOVA results, management concludes that neither the workspace design nor the storage/retrieval system impacts process flow time (and that the consultant wasn't worth the money). Do you agree with this conclusion? Explain.

29. Yogurt research. An experiment to determine the effect of several methods of preparing cultures for use in commercial yogurt was conducted by a food science research group. Three batches of yogurt were prepared using each of three methods: traditional, ultrafiltration, and reverse osmosis. A trained expert then tasted each of the 9 samples, presented in random order, and judged them on a scale from 1 to 10. A partially complete Analysis of Variance table of the data follows.

Source	Sum of Squares	Degrees of Freedom	Mean Square	F-ratio
Treatment	17.300			
Residual	0.460			
Total	17.769			

a) Calculate the mean square of the treatments and the mean square of the error.

b) Form the *F*-statistic by dividing the two mean squares.

c) The P-value of this *F*-statistic turns out to be 0.000017. What does this say about the null hypothesis of equal means?

d) What assumptions have you made in order to answer part c?

e) What would you like to see in order to justify the conclusions of the *F*-test?

f) What is the average size of the error standard deviation in the judge's assessment?

30. Smokestack scrubbers. Particulate matter is a serious form of air pollution often arising from industrial production. One way to reduce the pollution is to put a filter, or scrubber, at the end of the smokestack to trap the particulates. An experiment to determine which smokestack scrubber design is best was run by placing four scrubbers of different designs on an industrial stack in random order. Each scrubber was tested 5 times. For each run, the same material was produced, and the particulate emissions coming out of the scrubber were measured (in parts per billion). A partially complete Analysis of Variance table of the data is shown here.

Source	Sum of Squares	Degrees of Freedom	Mean Square	F-ratio
Treatment	81.2			
Residual	30.8			
Total	112.0			

a) Calculate the mean square of the treatments and the mean square of the error.

b) Form the *F*-statistic by dividing the two mean squares.

c) The P-value of this *F*-statistic turns out to be 0.00000949. What does this say about the null hypothesis of equal means?

d) What assumptions have you made in order to answer part c?

e) What would you like to see in order to justify the conclusions of the *F*-test?

f) What is the average size of the error standard deviation in particulate emissions?

31. Cereal shelf placement. Supermarkets often place similar types of cereal on the same supermarket shelf. The shelf placement for 77 cereals was recorded as their sugar content. Does sugar content vary by shelf? Here's a boxplot and an ANOVA table.

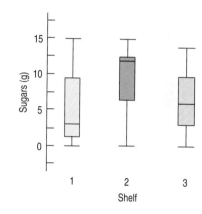

Source	DF	Sum of Squares	Mean Square	F-ratio	P-value
Shelf	2	248.4079	124.204	7.3345	0.0012
Error	74	1253.1246	16.934		
Total	76	1501.5325			

Level	n	Mean	StdDev
1	20	4.80000	4.57223
2	21	9.61905	4.12888
3	36	6.52778	3.83582

a) What kind of design or study is this?

b) What are the null and alternative hypotheses?

c) What does the ANOVA table say about the null hypothesis? (Be sure to report this in terms of sugar content and shelf placement.)

d) Can we conclude that cereals on shelf 2 have a different mean sugar content than cereals on shelf 3? Can we conclude that cereals on shelf 2 have a different mean sugar content than cereals on shelf 1? What can we conclude?

e) To check for significant differences between the shelf means, we can use a Bonferroni test, whose results are shown here. For each pair of shelves, the difference is shown along with its standard error and significance level. What does it say about the questions in part d?

Dependent Variable: SUGARS (Exercise 31)

	(I) SHELF	(J) SHELF	Mean Difference (I-J)	Std. Error	P-value	95% Confidence Interval	
						Lower Bound	Upper Bound
Bonferroni	1	2	−4.819	1.2857	0.001	−7.969	−1.670
		3	−1.728	1.1476	0.409	−4.539	1.084
	2	1	4.819	1.2857	0.001	1.670	7.969
		3	3.091	1.1299	0.023	0.323	5.859
	3	1	1.728	1.1476	0.409	−1.084	4.539
		2	−3.091	1.1299	0.023	−5.859	−0.323

T **32. Cereal shelf placement, part 2.** We also have data on the protein content on the 77 cereals in Exercise 31. Does protein content vary by shelf? Here's a boxplot and an ANOVA table.

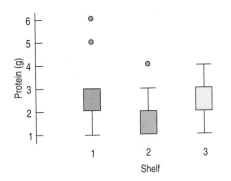

Source	DF	Sum of Squares	Mean Square	F-ratio	P-value
Shelf	2	12.4258	6.2129	5.8445	0.0044
Error	74	78.6650	1.0630		
Total	76	91.0909			

Level	n	Mean	StdDev
1	20	2.65000	1.46089
2	21	1.90476	0.99523
3	36	2.86111	0.72320

a) What kind of design or study is this?

b) What are the null and alternative hypotheses?

c) What does the ANOVA table say about the null hypothesis? (Be sure to report this in terms of protein content and shelf placement.)

d) Can we conclude that cereals on shelf 2 have a lower mean protein content than cereals on shelf 3? Can we conclude that cereals on shelf 2 have a lower mean protein content than cereals on shelf 1? What can we conclude?

e) To check for significant differences between the shelf means, we can use a Bonferroni test, whose results are shown here. For each pair of shelves, the difference is shown along with its standard error and significance level. What does it say about the questions in part d?

Dependent Variable: PROTEIN
Bonferroni

(I) SHELF	(J) SHELF	Mean Difference (I-J)	Std. Error	P-value	95% Confidence Interval	
					Lower Bound	Upper Bound
1	2	0.75	0.322	0.070	−0.04	1.53
	3	−0.21	0.288	1.000	−0.92	0.49
2	1	−0.75	0.322	0.070	−1.53	0.04
	3	−0.96	0.283	0.004	−1.65	−0.26
3	1	0.21	0.288	1.000	−0.49	0.92
	2	0.96	0.283	0.004	0.26	1.65

33. Automotive safety. The National Highway Transportation Safety Administration runs crash tests in which stock automobiles are crashed into a wall at 35 mph with dummies in both the passenger and the driver's seats. The THOR Alpha crash dummy is capable of recording 134 channels of data on the impact of the crash at various sites on the dummy. In this test, 335 cars are crashed. The response variable is a measure of head injury. Researchers want to know whether the seat the dummy is sitting in affects head injury severity, as well as whether the type of car affects severity. Here are boxplots for the 2 different *Seats* (*driver, passenger*) and the 6 different *Size* classifications (*compact, light, medium, mini, pickup, van*).

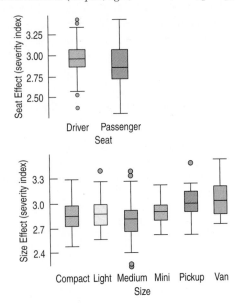

An interaction plot shows:

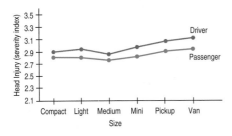

A scatterplot of residuals vs. predicted values shows:

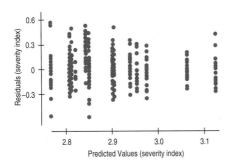

The ANOVA table follows:

Source	DF	Sum of Squares	Mean Square	F-ratio	P-value
Seat	1	0.88713	0.88713	25.501	< 0.0001
Size	5	1.49253	0.29851	8.581	< 0.0001
Seat × Size	5	0.07224	0.01445	0.415	0.838
Error	282	9.8101	0.03479		
Total	293	12.3853			

a) State the null hypotheses about the main effects (in words, not symbols).

b) Are the conditions for two-way ANOVA met?

c) If so, perform the hypothesis tests and state your conclusion. Be sure to state it in terms of head injury severity, seats, and vehicle types.

34. Gas additives. An experiment to test a new gasoline additive, *Gasplus*, was performed on three different cars: a sports car, a minivan, and a hybrid. Each car was tested with both *Gasplus* and regular gas on 10 different occasions, and their gas mileage was recorded. Here are the boxplots.

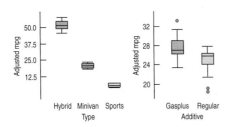

A two-way ANOVA with interaction model was run, and the following ANOVA table resulted.

Source	DF	Sum of Squares	Mean Square	F-ratio	P-value
Type	2	23175.4	11587.7	2712.2	< 0.0001
Additive	1	92.1568	92.1568	21.57	< 0.0001
Interaction	2	51.8976	25.9488	6.0736	0.0042
Error	54	230.711	4.27242		
Total	59	23550.2			

A plot of the residuals vs. predicted values showed:

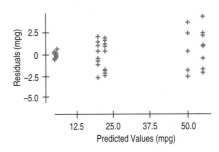

What conclusions about the additive and car types do you draw? Do you see any potential problems with the analysis?

✔ **JUST CHECKING ANSWERS**

1 Gather reports from veterinarians and pet hospitals. Look into the histories of sick animals. This would be a retrospective observational study.

2 Treatment: Feed the new food and a standard food to pets.

Response: Judge the health of the animals, possibly by having a veterinarian examine them before and after the feeding trials.

3 Control by choosing similar animals. Perhaps choose just one breed and test animals of the same age and health. Treat them otherwise the same in terms of exercise, attention, and so on.

Randomize by assigning animals to treatments at random.

Replicate by having more than one animal fed each formulation.

4 A control group could be fed a standard laboratory food, if we have one known to be safe. Otherwise we could prepare a special food in our test kitchens to be certain of its safety.

5 The veterinarian evaluating the animals should be blind to the treatments. For double-blinding, all technicians handling the animals should also be blinded. That would require making the control food look as much like the test food as possible.

6 Yes. Test dogs and cats separately.

7 No. We have failed to reject the hypothesis of a difference, but that's all we can conclude. There is insufficient evidence to discern any difference. But that should be sufficient for the company's purposes.

24

Introduction to Data Mining

Paralyzed Veterans of America

The Paralyzed Veterans of America (PVA) is a philanthropic and service organization chartered by the U.S. government to serve the needs of U.S. veterans who suffer from spinal cord injury or disease. Since 1946, PVA has raised money to support a variety of activities, including advocacy for veterans' health care, research and education in spinal cord injury and disease, and support for veterans' benefits and rights.

PVA raises money in a variety of ways, but the majority of their fund-raising comes from direct mail campaigns, in which they send free address labels or greeting cards to potential donors on their mailing list and request a donation for receiving these gifts.

PVA sends out regular solicitations to a list of more than four million donors. In 2006, PVA received about $86 million in donations, but the effort cost them over $34 million in postage, administrative, and gift expenses. Most of that money was spent on mailings to people who never responded. In fact, in any one solicitation, groups like PVA are lucky to get responses from more than a few percent of the people they contact. Response rates for commercial companies such as large credit card banks are so low that they often measure responses not in percentage points, but in basis points—hundredths of a percent. If the PVA could avoid mailing to just half of the people who won't respond, they could save $17 million a year and produce only half as much wasted paper. Can statistical methods help them decide who should get what mail?

24.1 Direct Marketing

Rather than advertise in various media, companies sometimes try to appeal directly to the consumer. The appeal is usually in the form of an offer to buy something, to enroll in a special program, or, in the case of a philanthropic group, to donate money or time. You've undoubtedly seen these efforts in the form of mail, e-mail, or phone calls. Direct mail, often called "junk mail," generates about four million tons of paper waste per year. Direct e-mail, often called "spam," accounts for a sizeable proportion of all e-mails.

Companies use direct mail and e-mail because, compared to other options, it is inexpensive and effective. To make it more effective, companies want to identify people who are most likely to respond. In other words, the companies would like to *target* their promotions and offers.

To help them decide how likely a customer might be to respond to a particular offer, a company might build a model to estimate the probability. The data used to build the model is usually a combination of data on the customer that the company has collected and other data that it has bought. **Data mining** is the name for a process that uses a variety of data analysis tools to discover patterns and relationships in data to help build useful models and make predictions. A company's goal with data mining might be to restrict their mailings to the people most likely to respond to the solicitations. With data mining, they would achieve that by looking at past customer behavior and other demographic information, and build a model (or models) to predict who is most likely to respond. Many of the modeling techniques that we've covered in this book—especially multiple regression and logistic regression—are used in data mining. But because data mining has benefited from work in machine learning, computer science, and artificial intelligence, as well as statistics, it has a much richer set of tools than those we've discussed in this book.

24.2 The Data

Companies and philanthropic organizations collect an incredible amount of information about their customers. You've contributed to this data gathering yourself, perhaps without even knowing it. Every time you purchase something with a credit card, use a loyalty card at a supermarket, order by phone or on the Internet, or call an 800 number, your transaction is recorded. For a credit card company, this **transactional data** might contain dozens or even hundreds of entries a year for each customer. For a philanthropic organization, the data might also include a history of when solicitations were sent, whether the donor responded to each solicitation, and how much they contributed. For a credit card bank, the transactional data would include every purchase the customer made. Every time a transaction is attempted, information is sent back to the transactional database to check that the card is valid, has not been reported stolen, and that the credit limit of the customer has not been exceeded—that's why there's a small delay before your transaction is approved every time you charge something. Although companies collect transactional data to facilitate the transactions themselves, they recognize that these data hold a wealth of information about their business. The challenge is to find ways to extract—to mine—that information.

In addition to transactional data, companies often have separate databases containing information about customers and about products (inventory, price, and shipping costs, for example). The databases are indexed and can be linked to each other in a relational database. The properties of relational databases and some simple examples were discussed in Chapter 2.

The variables in the customer database are of two types: individual and regional. The individual variables are typically first gathered when the customer opens an account, registers on a website, or fills out a warranty card and are specific to that customer. They might include **demographic variables** such as a customer's age, income, and number of children. The company then adds to these variables others that arise from the customer's interactions with the company, including some that may summarize variables in the transactional database. For example, the total amount spent each month might be a variable in the customer database that is updated from the customer's individual purchases, stored in the transactional database. The company might also purchase additional demographic data. Some demographic data are based on zip codes and can be obtained from agencies like the U.S. Census Bureau. These data can provide information on the average income, education, home value, and ethnic composition of the neighborhood in which a customer lives, but are not customer specific, and are only generic to the entire zip code in which the customer lives. Customer-specific data can also be purchased from a variety of commercial organizations. For example, a credit card company may want to send out an offer of free flight insurance to customers who travel frequently. To help know which customers those are, the company may buy information about customers' magazine subscriptions to identify customers who subscribe to travel or leisure magazines. The sharing and selling of individual information is controversial and raises privacy concerns, especially when the purchased data involve health records and personal information. In fact, concerns over the sharing of health data in the United States led to a set of strict guidelines known as HIPAA (Health Insurance Portability and Accountability Act) regulations. The ability to collect and share information about customers varies widely from country to country.

The PVA customer database[1] is a typical mix of variables and contains 481 variables on each donor. There are 479 potential predictor variables and two responses: *TARGET_B*, a 0/1 variable which indicates whether the donor contributed to the most recent campaign and *TARGET_D* which gives the dollar amount of the contribution. Table 24.1 shows the first 18 records for a subset of the 481 variables found in the PVA data. We can guess what some of the variables mean from their names, but others are more mysterious.

Information about the variables, including their definitions, how they are collected, the date of collection, etc. is collectively called **metadata**. The variables in the PVA data set presented below are typical of the types of data found in the customer records of many companies. We shouldn't be surprised to find out that *AGE* is the donor's age measured in years and that *ZIP* is the

---

[1] The PVA made some of their data available for the 1998 Knowledge Discovery and Data Mining (KDD) contest. The object of the contest was to build a model to predict which donors should receive the next solicitation based on demographic and past giving information. The results were presented at the KDD conference (www.kdnuggets.com). The variables discussed in this chapter are the ones that PVA made available.

donor's postal zip code. But without the metadata, it would be hard to know that *TCODE* is the code used before the title on the address label (0 = blank; 1 = Mr.; 2 = Mrs.; 28 = Ms., and so on) or that *RFA_2A* is a summary of past giving.

ODATEDW	OSOURCE	TCODE	STATE	ZIP	DOB	RFA_2A	AGE	OWN	INC	SEX	WEALTH	AVGGIFT	TARGET_B	TARGET_D
9401	L16	2	GA	30738	6501	F	33	U	5	F	2	11.66667	0	0
9001	L01	1	MI	49028	2201	F	76	H	1	M	2	8.777778	0	0
8601	DNA	1	TN	37079	0	E		U	1	M		8.619048	1	10
8601	AMB	1	WI	53719	3902	G	59			M		16.27273	0	0
8601	EPL	2	TX	79925	1705	E	81	H	2	F	6	10.15789	0	0
8701	LIS	1	IN	46771	0	F				M		8.871333	0	0
9201	GRI	1	IL	60016	1807	F	79	H	4	M	6	13.8	0	0
9401	HOS	0	KS	67218	5001	G	48	U	7	F	7	18.33333	0	0
8901	DUR	0	MI	48304	1402	F	84	H	7	M	9	12.90909	0	0
8601	AMB	0	FL	34746	1412	F	83	H	2	F	3	9.090909	0	0
9501	CWR	2	LA	70582	0	D		U	5	F		5.8	0	0
9501	ARG	0	MI	48312-	4401	E	54			F		8	0	0
8601	ASC	0	TX	75644	2401	G	74	U	7	F		13.20833	0	0
9501	DNA	28	CA	90059	2001	E	78	H	1	F		10	0	0
9201	SYN	0	FL	33167	1906	F	79	U	2	M	3	10.09091	0	0
9401	MBC	2	MO	63084	3201	F	66	H	5	F		10	0	0
9401	HHH	28	WI	54235	0	F		H	7	F		20	0	0
9101	L02	28	AL	36108	4006	F	58	H	5	F		10.66667	0	0

Table 24.1   *Part of the customer records from the PVA data set. Shown here are 15 of the 481 variables and 18 of the nearly 100,000 customer records used in the 1998 KDD (Knowledge Discovery and Data Mining) data mining competition.[2] The actual PVA customer database contains several million customer records.*

About 10% of the PVA variables describe past giving behavior collected by the PVA itself. More than half of the variables are regional data (based on zip codes), most likely purchased from the Census Bureau, and the rest of the variables are donor-specific, either gathered by the PVA or purchased from other organizations.

Sometimes disparate databases are gathered or merged in a central repository called a **data warehouse**. Maintaining the data warehouse is a huge job, and companies spend millions of dollars each year for the software, hardware, and technical personnel to do just that. Once a company has invested in creating and maintaining a data warehouse, it's only natural that they would want to get as much value as they can from it. For example, at PVA, analysts might use the data to build models to predict who will respond to a direct mail campaign; a credit card company, similarly, might want to predict who is most likely to accept an offer for a new credit card or services.

24.3 The Goals of Data Mining

The purpose of data mining is to extract useful information hidden in these large databases. With a database as large as a typical data warehouse, that search can be like looking for a needle in a haystack. How can analysts hope to find what they're looking for? They may start their search by using a sequence of queries to deduce facts about customer behavior, asking specific questions based on the data. Guided by their knowledge of the specific business, they may try a query-driven approach,

---

[2] The KDD cup is the leading Data Mining competition in the world, and is organized by the SIGKDD interest group of the ACM (Association for Computing Machinery).

asking a series of specific questions to deduce patterns. Such an approach is **online analytical processing** or **OLAP**. Analysts in sales, marketing, budgeting, inventory, and finance often use OLAP to answer specific questions involving many variables. An OLAP question for the PVA might be: "How many customers under the age of 65 with incomes between $40,000 and $60,000 in the Western region who have not donated in the previous two years gave more than $25 to the most recent solicitation?" Although OLAP is efficient in answering such multivariable queries, it is question-specific. OLAP produces answers for specific queries, typically as tables, but does not build a predictive model, and so it is not appropriate to generalize using OLAP.

In contrast to an OLAP query, the outcome of a data mining analysis is a **predictive model**—a model that uses predictor variables to predict a response. For a quantitative response variable (as in linear regression), the model will predict the *value* of the response, whereas for a categorical response, the model estimates the probability that the response variable takes on a certain value (as in logistic regression). Both linear and logistic multiple regressions are common tools of a data miner. Like all statistical models and unlike query-based methods like OLAP, data mining generalizes to other similar situations through its predictive model. For example, an analyst using an OLAP query might find that customers in a certain age group responded to a recent product promotion. But without building a model, the analyst can't understand the relationship between the customer's age and the success of the promotion, and thus may be unable to predict how the product will do with a larger set of customers. The goal of a data mining project is to increase business understanding and knowledge by building a model to answer a specific set of questions raised at the beginning of the project.

Data mining is similar to traditional statistical analysis in that it involves exploratory data analysis and modeling. However, there are several aspects of data mining that distinguish it from more traditional statistical analysis. Although there is no consensus on exactly what data mining is and how it differs from statistics, some of the most important differences include:

- ◆ **The size of the databases.** Although no particular size is required for an analysis to be considered data mining, an analysis involving only a few hundred cases or only a handful of variables would not usually be considered data mining.

- ◆ **The exploratory nature of data mining.** Unlike a statistical analysis that might test a hypothesis or produce a confidence interval, the outcome of a data mining effort is typically a model used for prediction. Usually, the data miner is not interested in the values of the parameters of a specific model or in testing hypotheses.

- ◆ **The data are "happenstance."** In contrast to data arising from a designed experiment or survey, the typical data on which data mining is performed have not been collected in a systematic way. Thus, warnings not to confuse association with causation are especially pertinent for data mining. Moreover, the sheer number of variables involved makes any search for relationships among variables prone to Type I errors.

- ◆ **The results of a data mining effort are "actionable."** For business applications, there should be a consensus of what the problem of interest is and how the model will help solve the problem. There should be an action plan in place for a variety of possible outcomes of the model. Exploring large databases out of curiosity or just to see what they contain is not likely to be productive.

- ◆ **The modeling choices are automatic.** Typically the data miner will try several different types of models to see what each has to say, but doesn't want to spend a lot of time choosing which variables to include in the model or making the kinds of choices one might make in a more traditional statistical analysis. Unlike the analyst who wants to understand the terms in a stepwise regression, the data miner is more concerned with the predictive ability of the model. As long as the resulting model can help make decisions about who should receive the next offer, who is best suited for an online coupon, or who is most likely to switch cable providers next month, the data miner is likely to be satisfied.

24.4 Data Mining Myths

Data mining software usually contains a variety of exploratory and model-building tools and a graphical user interface designed to guide the user through the data mining process. Some people buy data mining software hoping that the tools will find information in their databases and write reports that spew knowledge with little or no effort or input from the user. Software vendors often capitalize on this hope by exaggerating the capabilities and the automatic nature of data mining to increase sales of their software. Data mining can often *assist* the analyst to find meaningful patterns and help predict future customer behavior, but the more the analyst *knows* about his or her business, the more likely he or she will be successful using data mining. Data mining is not a magic wand that can overcome poor data quality or collection. A product may have tools for detecting outliers and may be able to impute (assign a value for) missing data values, but all the issues you learned about good statistical analysis are still relevant for data mining.

Here are some of the more common myths about what data mining can do.

◆ **Myth 1:** Find answers to unasked questions

Even though data mining can build a model to help answer a specific question, it does not have the ability to answer questions that haven't been asked. In fact, formulating a precise question is a key first step in any data mining project.

◆ **Myth 2:** Automatically monitor a database for interesting patterns

Data mining techniques build predictive models and can answer queries, but do not find interesting patterns on their own any more than regression models do.

◆ **Myth 3:** Eliminate the need to understand the business

In fact, the more an analyst understands his or her business, the more effective the data mining effort will be.

◆ **Myth 4:** Eliminate the need to collect good data

Good data is as important for a data mining model as for any other statistical model that you've encountered. While some data mining software contains tools to help with missing data and data transformations, there is no substitute for quality data.

◆ **Myth 5:** Eliminate the need for good data analysis skills

The better the data analysis skills of the miner—the skills you've learned in every chapter of this book—the better the analysis will be when using data mining tools. Data mining tools are more powerful and flexible than a statistical tool like regression but are similar in the way they work and in how they're implemented.

24.5 Successful Data Mining

The size of a typical data warehouse makes any analysis challenging. The ability to store data is growing faster than the ability to use it effectively. Commercial data warehouses often contain terabytes (TB)—more than 1,000,000,000,000 (1 trillion) bytes—of data (one TB is equivalent to about 260,000 digitized songs), and warehouses containing petabytes (PB—one PB = 1000 TB) are now common. The tracking database of the United Parcel Service (UPS) is estimated to be on the order of 16 TB, or roughly the digital size of all the books in the U.S. Library of Congress. According to *Wired* magazine, about 20 TB of photos are uploaded to Facebook every month. All the U.S. Census data from 1790 to 2000 would take about 600 TB. However, it's estimated that the servers at *Google* process a petabyte of data every 72 minutes.[3] Data miners hope to uncover some important strategic information lying hidden within these massive collections of data.

To have a successful data mining outcome, the first step is to have a well-defined business problem. With 500 variables, there are over 100,000 possible two-way relationships between pairs of variables. The number of pairs that will be related just by chance is likely to be large. And it's human nature to find many of these relationships interesting and even to posit plausible reasons why two variables might be associated. Some of these variables may appear to provide a useful predictive model, when in fact they do not. A well-defined business objective can help you avoid going down a lot of blind paths.

As in painting a house, much of the effort of a data mining project is in the preparation, cleaning, and checking of the data. It is estimated that for any data mining project, 65% to 90% of the time spent is spent in such **data preparation**. Data preparation involves investigating missing values, correcting wrong and inconsistent entries, reconciling data definitions, and possibly creating new variables from the original ones. The data may need to be extracted from several databases and combined. Errors must be corrected or eliminated, and outliers must be identified. For example, let's look at a histogram and boxplot for the variable *AGE* for all 94,649 records of the PVA data set.

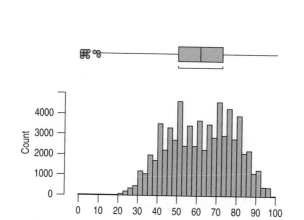

Quantiles		
100.0%	maximum	98.000
99.5%		95.000
97.5%		90.000
90.0%		83.000
75.0%	quartile	75.000
50.0%	median	62.000
25.0%	quartile	48.000
10.0%		39.000
2.5%		31.000
0.5%		26.000
0.0%	minimum	1.000
Mean		61.598019
Std Dev		16.666865
Std Err Mean		0.0624749
upper 95% Mean		61.720469
lower 95% Mean		61.475568
N		71170

| **Figure 24.1** *Graphical output and statistical summaries of the variable AGE help to identify questionable values.*

While most of the values seem reasonable, there is a group of cases with ages below 20. A closer examination reveals that there are a handful of cases whose ages are below 15, of whom 17 are younger than 5 years old and 9 have age 1 year old. Clearly some of these values are just wrong. There are also 23,479 missing values for the variable *AGE*. Our domain knowledge tells us that values below a certain

---

[3] *Wired*, Issue 16.07, June 2008.

age (probably below 19, but certainly below 10) or above a certain age (100?, 110?) are wrong, but, for other variables it is harder to know whether the data are correct. And, while checking one or two variables to see if their values make sense is possible, examining bar charts, histograms, and boxplots for 481 variables is a daunting task.

A successful data mining effort often requires a substantial amount of time devoted to basic examination, cleaning, and preparation of the data before any modeling is performed. It will have a clear objective, agreed to by a team of people who will share the work and the responsibility of the data mining. It will have an action plan once the results of the data mining are known, whether or not those results are what the team expected or desired. Finally, the data mining should be accompanied by as much knowledge about both the data and the business question as possible. Blind searches for patterns in large databases are rarely fruitful and will waste valuable analytic resources.

24.6 Data Mining Problems

Data mining can address different types of problems, some of which we have already encountered. When the goal is to predict a continuous response variable, the problem is generically called a **regression problem,** regardless of whether linear regression is used (or even considered) as one of the models. When the response variable is categorical, the problem is referred to as a **classification problem** because the model will either guess the most likely category for each or assign a probability to each class for that code. For example, for the classification problem of predicting *whether or not* a particular donor will give to the next campaign, a model will produce either the most probable class (*donate or not*) or the probabilities of each of these. Predicting the *amount* of money a donor will donate is a regression problem.

Both of the previous problems are referred to as **supervised problems**. In a supervised setting, we are given a set of data for which we know the response. That is, for the PVA data, we *know* for at least one group of donors the values of the responses *TARGET-B* and *TARGET-D*. We know whether they gave to the last campaign and how much they donated. The data miner would construct a model based on a portion of the original data, called the **training set**. In order to assess how well the model will work in the future on data that the model hasn't encountered, the modeler uses the original data set that was withheld from the model building and then tests the predictions of the model on these data. This second data set is called the **test set**.

By contrast, there are problems for which there is no particular response variable. In these **unsupervised problems**, the goal may be to build clusters of cases with similar attributes. For example, a company may want to cluster their customers into groups with similar buying behaviors and tastes. Such analyses resemble segmentation analyses performed by marketing analysts. In this case, there is no response variable. All the predictors (or a subset of them) are used to build clusters based on an index that measures the similarity between the customers. Many different algorithms are available to find clusters.

24.7 Data Mining Algorithms

The methods used for the problems discussed in the previous section are often referred to as **algorithms**, a term that describes a sequence of steps with a specific purpose. You may hear a method (even one like linear regression) referred to as a model, an algorithm, a tool, or generically as a method. The terms seem to be used interchangeably. This section touches on only a few of the models used in data mining. Some of the most common methods used for prediction are decision trees and neural networks (discussed below), support vector machines, belief nets, the regression methods discussed in Chapters 18 and 19, and random forests. It's a lively research field and new algorithms are appearing all the time.

Tree Models

Chapter 22 discussed decision trees in which the analyst examined a small number of actions and possible states of nature in a sequence to identify the action with the largest payoff. In data mining, the term **decision tree** is used to describe a *predictive model.* Data mining decision trees are superficially similar in form to the decision trees we saw in Chapter 22, but that's where the similarity ends. These tree models use *data* to select predictor variables that give predictions of the response. The models are entirely driven by the data with no input from the user.

The way a tree model works is fairly simple. To illustrate the process, imagine that we want to predict whether someone will default on a mortgage. To build the tree, we use past data from a group of customers on whom we have similar information and for whom we know whether they defaulted or not. For this simplified example, let's assume we'll predict it only from the following variables:

◆ Age (years)

◆ Household Income ($)

◆ Years on the Job (years)

◆ Debt ($)

◆ Homeowner (Yes/No)

The tree model tries to find predictors that can distinguish those people who will default on their mortgages from those who won't. To do that, it first examines *every* potential predictor variable and every possible way of *splitting* that variable into two groups. For example, it looks at *Age,* and for every value of *Age,* it calculates the default rate of those *above* that value and *below* that value. Keeping track of the difference in default rates for the two groups defined by every possible predictor variable split at every possible way, it chooses that pair of predictors and *split points* that produce the greatest difference in default rates.[4] The tree algorithm decides which variables to split and where to split them by trying (essentially) all combinations of variables and split points until it finds the best splits. For a categorical predictor, the model considers every possible way of putting the categories into two groups. There are several criteria used by different algorithms to define "best," but they all have in common some attempt to find two groups whose default rates are best separated the most by the split. After the algorithm finds the first split, it continues searching again on the two resulting groups, finding the next best variable and split point (possibly using the same variable as the previous split again). It continues in this way until one of several criteria is met (for example, number of customers too small, not a large enough difference in rates is found) and then stops at what are called **terminal nodes,** where the model produces a prediction. The predictions at the terminal nodes are simply the average (if the response is quantitative) or the proportions of each category (for a classification problem) of the cases at that node.

---

[4] This is a slight simplification. There are actually several possible variants on the criteria for splitting. Readers who want to know the details can read more advanced books on data mining or decision trees.

The tree for our hypothetical mortgage example is shown in Figure 24.3. To understand the tree, start at the top and imagine being presented with a new customer. The first question the tree asks is "Is the *Household Income* more than $40,000?". If the answer is yes, move down to the right. For these cases, *Debt* is the next variable split, this time at $10,000. If this customer's *Debt* exceeded $10,000 move down and to the right again. The tree estimates that customers like this (both *Household Income* > $40,000 and *Debt* > $10,000) defaulted at a rate of 5% (0.05). Just to the left are customers whose *Household Income* was > $40,000 but whose *Debt* was ≤ $10,000. They defaulted at a rate of 1%. For customers with incomes less than $40,000 (the left branch of the first split), the next variable split was not *Debt*, but *Job*. For those who have been at their job more than 5 years, the default rate was 6%, but was 11% for those with less time at their present job. The modeler may, at this point, label the outcomes as *risk* categories, for example by calling 1% *Very low*, 5 and 6% *Moderate*, and 11% *High*.

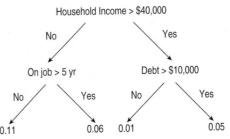

Figure 24.3 *Part of a tree model on a hypothetical examination of mortgage default. The tree selects Household Income as the most important variable to split on and selects $40,000 as the cut point. For those customers whose income is more than $40,000, Debt is the next most important variable, while for those whose income is less than $40,000, how long they've held their current job is more important.*

Tree models are very easy to implement and, in principle, are easy to interpret. They have the advantage of showing their logic clearly and are easy to explain to people who don't have a deep background in statistics. Even when they aren't used as a final model in a data mining project, they can be very useful for selecting a smaller subset of variables on which to do further analysis. Figure 24.4 shows a tree model run on the PVA data using *TARGET_B* as a response variable, which equals "YES" if the donor contributed to the most recent solicitation or "NO" if not.

The PVA tree model starts at the root node with all 94,649 customers (listed under **Count** in the top box of the figure). The single most important variable for predicting *TARGET_B* is the variable *RFA_4*, a variable that summarizes past giving. So, the first split occurs by splitting the levels of that variable into two groups, with 29,032 customers split to the left and the remaining 65,617 customers to the right. The variable *RFA_4* is comprised of codes that contain information on how recently the last gift solicitation was received, how frequently the donor has given and how large the last gift was. (Examples of these codes are A3C, S4B etc. seen in the output display). The percentages next to the levels 0 and 1 indicate the proportion of those who donated or not (*TARGET_B = 1 or 0*, respectively) in each terminal node. The terminal nodes are found in the bottom row. Notice the terminal node at the far left of the figure showing that 12.4% of the donors in this subgroup donated. This represents a substantial improvement over the average of 5.06% shown in the root node at the top.

It's interesting to see that of the 479 potential predictor variables, 3 of the top 4 variables involve past giving history and are not demographics. *RFA_4* is a summary of past giving, *LASTGIFT* measures the amount of the last gift, and

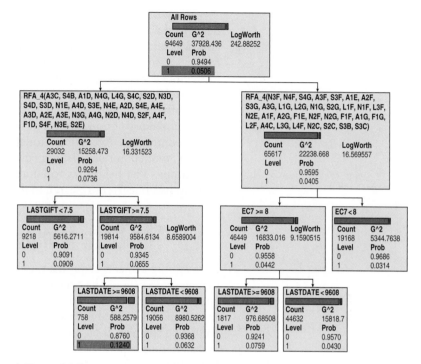

| **Figure 24.4** Part of a tree model run on the PVA data using JMP® software.

LASTDATE reports how recently the donor made the last gift. Only *EC*7 is a demographic variable, measuring the percentage of residents in the donor's zip code that have at least a bachelor's degree. By letting the tree grow further, we could grow the list of potential predictors to 20 or 30, a much larger number than shown here, but still a manageable subset of the original 479 variables. The tree, unlike many other data mining algorithms, can handle a great number of potential predictor variables, both categorical and quantitative. This makes it a great model for starting a data mining project. The modeler can then choose whether to use this model as a final model for predicting giving, or to use the variables suggested by the tree as inputs to other models that are less able to handle very large numbers of predictors.

Neural Networks

Another popular data mining tool is the multilayer perceptron or (artificial) **neural network**. This algorithm sounds more impressive than it really is. Although it was inspired by models that tried to mimic the function of the brain, it's really just an automatic, flexible, nonlinear regression tool. That is, it models a single response variable with a function of some number of predictor variables but, unlike multiple regression, it constructs a more complex function for the relationship. This can have the advantage of fitting the data better but has the disadvantage that complex functions are harder to interpret and understand. Indeed, modelers often don't even look at the models themselves. There are certainly things to be learned by studying how neural networks build models, but, for our purposes, we'll view them as "black box" models—models that don't produce an equation or graphical display that we can examine, but simply predict the response without much information about *how* it's doing it.

Even though they are black boxes, neural network algorithms do leave some clues. A listing of the most important variables is a common output feature, and some neural network software even provide plots of the predicted relationships between the response and the most important variables as measured by how much the predictions change when the variable is deleted. In the margin is a list of the top fifteen variables in the PVA problem ranked by "importance" from a neural network output in a data mining software package called Clementine®.

RFA_2A	2.31%
RFA_2	1.64%
RFA_2F	1.27%
LASTDATE	1.18%
INCOME	1.03%
PEPSTRFL	1.00%
ADATE_4	0.83%
ADATE_3	0.74%
RDATE_7	0.73%
LIFESRC	0.73%
HVP1	0.69%
RDATE_6	0.61%
DMA	0.55%
GENDER	0.54%
RDATE_3	0.51%

Table 24.3 *The top fifteen variables ranked in importance by the neural network node in Clementine®. The importance percentages indicate relative importance only and are not meant to be interpreted in an absolute sense.*

Many other algorithms are used by data miners, and new ones are being developed all the time. Typically, a data miner will build several different models and then test them before selecting one to use, or the data miner might decide to form a "committee" of models, combining the output from several models, much in the way a CEO might take input from a group of advisors in order to make a decision. For a classification problem, the final prediction for a case might be the class that's predicted most often by the models in the committee. For a regression problem, the prediction might simply be the average prediction of all the models in the committee. Averaging models provides protection against choosing the "wrong" model, but usually at a cost of not being able to interpret the resulting predictions. The best ways to combine many different models is an active and exciting area of current research in data mining.

24.8 The Data Mining Process

Data mining projects require a number of different skills. For that reason, a data mining project should be a team effort. No one person is likely to have the business knowledge, computer and database management skills, software expertise, and statistical training needed for all the steps in the process. Because data mining projects tend to be complex, it's useful to map out the steps for a successful project. A group of data mining experts have shared their combined expertise in a project called the Cross Industry Standard Process for Data Mining (CRISP-DM). A CRISP-DM schematic of the data mining cycle appears in Figure 24.5.

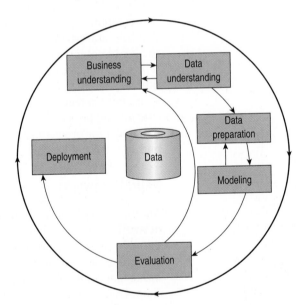

| **Figure 24.5** *A picture of the CRISP-DM data mining process.*

In this schematic, the process starts with the *Business understanding* phase. This is where the problem to be addressed is carefully articulated. It is best to have a specific problem before starting. A goal of understanding how to best manage customers sounds good but is not precise enough for a data mining project. A better, more specific question might be to understand which customers are most likely to switch cell phone providers in the next three months. It's important to have all the members of the data mining team involved at this stage, and the team should be representative of all parts of the business that may be affected by the resulting business decisions. If key constituents aren't represented, the model might not answer the right question. It can't be overemphasized how important it is to have consensus on a precise, correctly-formulated problem to be addressed before continuing with the data mining effort.

The phase of *Data understanding* is central to the entire data mining project. If you want to know which customers are likely to switch providers, you have to understand your data and have data that can support such an exploration. For example, you'll need a sample that contains both customers who have recently switched and those who have been loyal customers for some time. There should also be variables in the database that could reasonably explain or predict the behavior. The choice of those variables has to rely on the business knowledge of the team. At the beginning of the project, it's best to include all the variables that might be useful, but keep in mind that having *too* many variables can make the model selection phase more difficult. It is crucial to understand the data warehouse, what it contains, and what its limitations are at this stage.

Once the variables have been selected and the response variable(s) agreed upon, it's time to begin the *Data preparation* phase for modeling. As mentioned earlier, this can be a time-consuming part of the process and is likely to be a team effort. Investigating missing values, correcting wrong and inconsistent entries, reconciling data definitions, and merging data sources are all challenging issues. Some of these can be handled automatically, while others require painstaking detailed analysis. The team will have to decide how much effort is reasonable to make the data set as complete and reliable as possible, given the time and resource constraints of the project.

Once the data have been prepared, the analysts begin the *Modeling* phase by exploring and developing models. If the number of variables is very large, a preliminary model (such as a tree) might be considered in order to narrow the candidate predictor variables down to a reasonable number. If the number of predictors is small enough, modelers can use traditional graphical analysis (histograms, bar charts) for each variable to start and then investigate the relationship between each predictor and the response with bivariate graphs (scatterplots, boxplots, or segmented bar charts, depending on the variable types). The more knowledge of the data and variables that goes into the model, the higher the chance of success for the project.

The analysts should now have several models that fit the response variable with differing levels of accuracy in the training set. Once the analyst has several models that seem reasonable (based on domain knowledge and performance on the training set), the *Evaluation* phase can begin. Looking at the structure of the models and deciding what predictors are important for each model should give the analysts information about how the predictors can predict the response variable of interest. In the *Evaluation* phase, the candidate models are tested against the test set, and various criteria are used to judge the models. For example, for a regression problem with a quantitative response, the sum of squares of the residuals from predicting the response on the test set might be compared. For a (two-level) classification problem, the two types of errors that arise (predicting YES when the true response is NO and vice versa) are weighed. Different costs on the two types of errors might be warranted, and the total cost of misclassification should reflect that. At this point, the business question that motivated the project should be revisited. Does the model help to answer the question? If not, it may be necessary to go back to one of the previous steps to investigate why that happened.

If the model (or average of several models) seems to give insight into the business problem, then it's time for the *Deployment* phase. Usually, that means using the model to predict an outcome on a larger data set than the original, or on more recent data than were used to build the model. The reason that so many phases of the CRISP-DM diagram have arrows both to and from them is that this process is not a straightforward movement through the phases—it is an iterative and interconnected process. Knowledge gained at any phase may trigger a reexamination of an earlier phase. Even the "final" phase, deployment, is not really final. In many business situations, the environment changes rapidly, so models can become stale quickly. Although a data mining project is complex and involves the efforts of many people, mapping out the different phases can help assure that the project is as successful as possible.

24.9 Summary

There are many similarities between the modeling process of data mining and the basic approach to modeling that you've been learning throughout this book. What makes data mining different really has to do with the large number of algorithms and types of models available to the data miner and with the size and complexity of the data sets. But most of what makes a data mining project successful would make any statistical analysis successful. The same principles of understanding and exploring variables and their relationships are key to both processes. A famous statistician, Jerry Friedman, was once asked if there was a difference between statistics and data mining. Before answering, he asked if the questioner wanted the long answer or the short answer. Because the response was "the short answer," Jerry said simply, "No." We never got to hear the long answer, but we suspect it may have contained some of the differences discussed in this chapter.

Having a good set of statistical and data analysis tools is a great beginning to becoming a successful data miner. Being willing to learn new techniques, whether they come from statistics, computer science, machine learning, or other disciplines, is essential. Learning to work with other people whose skills complement yours will not only make the task more pleasant, but also is key to a successful data mining project. The need to understand the information contained in large databases will increase in the years ahead, so it will continue to be important to know about this rapidly growing field.

WHAT CAN GO WRONG?

- **Be sure that the question to be answered is specific.** Make sure that the business question to be addressed is specific enough so that a model can help to answer it. A goal as vague as "improving the business" is not likely to lead to a successful data mining project.

- **Be sure that the data have the potential to answer the question.** Check the variables to see whether a model can reasonably be built to predict the response. For example, if you want to know which type of customers are going to a particular Web page, make sure that data are being collected that link the Web page to the customer visiting the site.

- **Beware of overfitting to the data.** Because data mining tools are powerful and because the data sets used to train them are usually large, it is easy to think you're fitting the data well. Make sure you validate the model on a test set—a data set not used to fit the model.

- **Make sure that the data are ready to use in the data mining model.** Typically, data warehouses contain data from several different sources. It's important to confirm that variables with the same name actually measure the same thing in two different databases. Missing values, incorrect entries, and different time scales are all challenges to be overcome before the data can be used in the model building phase.

- **Don't try it alone.** Data mining projects require a variety of skills and a lot of work. Assembling the right team of people to carry out the effort is crucial.

ETHICS IN ACTION

With U.S. consumers becoming more environmentally conscious, there has been an explosion of eco-friendly products on the market. One notable entry has been the gas-electric hybrid car. Recent data show, however, that the rate of growth in U.S. sales of hybrid vehicles is starting to slow. A large nonprofit environmental group concerned about this trend would like to target customers likely to purchase hybrid cars in the future with a message expressing the urgency to do so sooner than later. They understand that direct mailings are very effective in this regard, but staying true to their environmental concerns, they want to avoid doing a nontargeted mass mailing. The executive team met to discuss the possibility of using data mining to help identify their target audience. The initial discussion revolved around data sources. Although they have several databases on demographics and transactional information for consumers who have purchased green products and donated to organizations that promote sustainability, someone suggested that they get data on political party affiliation. After all, there was a Green party, and Democrats tend to be more concerned about environmental issues than Republicans. Another member of the team was genuinely surprised that this was even possible. She wondered how ethical it is to use information about individuals that they may assume is being kept confidential.

ETHICAL ISSUE *Should information about individuals be collected without their knowledge? Data mining raises general concerns about privacy and confidentiality. Related to Item D, ASA Ethical Guidelines.*

ETHICAL SOLUTION *Data should never be used without the consent of the person to whom the data refer unless care has been taken to disguise the identity or otherwise ensure that the identity of the person cannot be inferred from the data themselves.*

What have we learned?

We've learned that the techniques we've studied in this book are being used in new and novel ways by science, industry, and government to help understand complex phenomena for which massive amounts of data are available.

We've seen that all the principles we've learned about data collection, exploration, and inference are just as relevant to massive data sets and complex problems.

We've learned, as in the multiple regression chapters, that model building is an art that speaks to the central goal of statistical analysis: understanding the world with data. We've learned that there is no "right" model and that each model reveals different aspects of the data, aspects that can help us understand the business phenomena at hand.

Terms

Algorithm
A set of instructions used for calculation and data processing. The algorithm specifies how the model is built from the data.

Classification problem
A prediction problem that involves a categorical response variable. (See also regression problem.)

Data mining
A process that uses a variety of data analysis tools to discover patterns and relationships in data that are useful for making predictions.

Data preparation
The process of cleaning data and checking its accuracy prior to modeling. Data preparation includes investigating missing values, correcting wrong and inconsistent entries, and reconciling data definitions.

Data warehouse	A digital repository for several large databases.
Decision tree (data mining version)	A model that predicts either a categorical or quantitative response in which the branches represent variable splits and the terminal nodes provide the predicted value.
Demographic variable	A variable containing information about a customer's personal characteristics or the characteristics of the region in which the customer lives. Commonly used demographics include age, income, race, and education.
Metadata	Information about the data including when and where the data were collected.
Neural network	A model based on analogies with human brain processing that uses combinations of predictor variables and nonlinear regression to predict either a categorical or quantitative response variable.
Online analytical processing (OLAP)	An approach for providing answers to queries that typically involve many variables simultaneously.
Predictive model	A model that provides predictions for the response variable.
Regression problem	A prediction problem that has a quantitative response variable. (See classification problem.)
Supervised problem	A classification or regression problem in which the analyst is given a set of data for which the response is known to use to build a model.
Terminal node	The final leaves of a decision tree where the predictions for the response variable are found.
Test set	The data set used in a supervised classification or regression problem *not used* to build the predictive model. The test set is withheld from the model building stage of the process, and the model's predictions on the test set are used to evaluate the model performance.
Training set	The data set used in a supervised classification or regression problem to build the predictive model.
Transactional data	Data describing an event involving a transaction, usually an exchange of money for goods or services.
Unsupervised problem	A problem, unlike classification or regression problems, where there is no response variable. The goal of an unsupervised problem is typically to put similar cases together into homogeneous groups or clusters.

Answers

Chapter 2

1. Answers will vary.
3. *Who*—50 recent oil spills; *What*—date, spillage amount (no specified unit) and cause of puncture; *When*—Recent years; *Where*—United States; *Why*—To determine whether or not spillage amount per oil spill has decreased since Congress passed the 1990 Oil Pollution Act and use that information in the design of new tankers; *How*—not specified; *Variables*—There are 3 variables. Spillage amount and date are quantitative variables, and cause of puncture is a categorical variable.
5. *Who*—existing stores; *What*—sales ($), town population (000), median age (years), median income ($), and whether or not they sell beer/wine; *When*—not specified; *Where*—United States; *Why*—The food retailer is interested in understanding any association of these variables to help them determine where to open their next store; *How*—collection from their stores; *Variables*—Sales ($), population (000), median age (years), and median income ($) are all quantitative variables. Whether or not the store sells beer/wine is categorical.
7. *Who*—Arby's sandwiches; *What*—type of meat, number of calories (in calories), and serving size (in ounces); *When*—not specified; *Where*—Arby's restaurants; *Why*—These data might be used to assess the nutritional value of the different sandwiches; *How*—Information was gathered from each of the sandwiches on the menu at Arby's, resulting in a census; *Variables*—There are three variables. Number of calories and serving size are quantitative variables, and type of meat is a categorical variable.
9. *Who*—385 species of flowers; *What*—date of first flowering (in days); *When*—over a period of 47 years; *Where*—Southern England; *Why*—The researchers believe that this indicates a warming of the overall climate; *How*—not specified; *Variables*—Date of first flowering is a quantitative variable; *Concerns*—Hopefully, date of first flowering was measured in days from January 1, or some other convention, to avoid problems with leap years.
11. *Who*—students; *What*—age (probably in years, though perhaps in years and months), race or ethnicity, number of absences, grade level, reading score, math score, and disabilities/special needs; *When*—current; *Where*—not specified; *Why*—Keeping this information is a state requirement; *How*—The information is collected and stored as part of school records; *Variables*—There are 7 variables. Race or ethnicity, grade level, and disabilities/special needs are categorical variables. Number of absences, age, reading score, and math score are quantitative variables; *Concerns*—What tests are used to measure reading and math ability, and what are the units of measure for the tests?
13. *Who*—customers of a start-up company; *What*—customer name, ID number, region of the country, date of last purchase, amount of purchase (probably in dollars), and item purchased; *When*—present time; *Where*—not specified; *Why*—The company is building a database of sales information; *How*—Presumably, the company records the information from each new customer; *Variables*—There are 6 variables. Name, ID number, region of the country, and item purchased are categorical variables. Date and amount of purchase are quantitative variables; *Concerns*—Region is a categorical variable, and it is potentially confusing to record it as a number.
15. *Who*—vineyards; *What*—size of vineyard (in acres), number of years in existence, state, varieties of grapes grown, average case price (in dollars), gross sales (probably in dollars), and percent profit; *When*—not specified; *Where*—not specified; *Why*—Business analysts hoped to provide information that would be helpful to producers of U.S. wines; *How*—not specified; *Variables*—There are 5 quantitative variables and two categorical variables. Size of vineyard, number of years in existence, average case price, gross sales, and percent profit are quantitative variables. State and variety of grapes grown are categorical variables.

17. *Who*—1,180 Americans; *What*—region, age (in years), political party affiliation, whether or not the person owned any shares of stock, and their attitude toward unions; *When*—not specified; *Where*—United States; *Why*—The information was gathered for presentation in a Gallup public opinion poll; *How*—phone survey; *Variables*—There are 5 variables. Region, political party affiliation, and stock ownership are categorical variables. Age and opinion about unions are quantitative variables.

19. *Who*—every model of automobile in the United States; *What*—vehicle manufacturer, vehicle type, weight (probably in pounds), horsepower (in horsepower), and gas mileage (in miles per gallon) for city and highway driving; *When*—This information is collected currently; *Where*—United States; *Why*—The Environmental Protection Agency uses the information to track fuel economy of vehicles; *How*—The data is collected from the manufacturer of each model; *Variables*—There are 6 variables. City mileage, highway mileage, weight, and horsepower are quantitative variables. Manufacturer and type of car are categorical variables.

21. *Who*—states in the United States; *What*—state name, whether or not the state sponsors a lottery, the number of numbers in the lottery, the number of matches required to win, and the probability of holding a winning ticket; *When*—1998; *Where*—United States; *Why*—It is likely that this study was performed in order to compare the chances of winning the lottery in each state; *How*—Although not specified, the researchers probably simply gathered data from a number of different sources, such as state lottery website and publications; *Variables*—There are 5 variables. State name and whether or not the state sponsors a lottery are categorical variables, and number, matches, and probability of winning are quantitative variables.

23. *Who*—students in an MBA statistics class; *What*—total personal investment in stock market ($), number of different stocks held, total invested in mutual funds ($), and name of each mutual fund; *When*—not specified; *Where*—United States; *Why*—The information was collected for use in classroom illustrations; *How*—An online survey was conducted. Presumably, participation was required for all members of the class; *Variables*—There are 4 variables. Name of mutual fund is a categorical variable. Number of stocks held, total amount invested in market ($) and in mutual funds ($) are quantitative variables.

25. *Who*—Indy 500 races; *What*—year, winner, car, time (hours), speed (mph) and car #. *When*—1911–2007; *Where*—Indianapolis, Indiana; *Why*—It is interesting to examine the trends in Indy 500 races; *How*—Official statistics are kept for the race every year; *Variables*—There are 6 variables. Winner, car and car # are categorical variables. Year, time, and speed are quantitative variables.

27. Each row should be a single mortgage loan. Columns hold the borrower name (which identifies the rows) and amount.

29. Each row is a week. Columns hold week number (to identify the row), sales prediction, sales, and difference.

31. Cross-sectional

33. Time series

Chapter 3

1. a) No. It would be nearly impossible to get exactly 500 males and 500 females by random chance.
   b) A stratified sample, stratified by whether the respondent is male or female.

3. a) Voluntary response.
   b) We have no confidence at all in estimates from such studies.

5. a) The population of interest is all adults in the United States aged 18 and older.
   b) The sampling frame is U.S. adults with landline telephones.
   c) Some members of the population (e.g., many college students) don't have landline phones, which could create a bias.

7. a) Population—Human resources directors of Fortune 500 companies.
   b) Parameter—Proportion who don't feel surveys intruded on their work day.
   c) Sampling Frame—List of HR directors at Fortune 500 companies.
   d) Sample—23% who responded.
   e) Method—Questionnaire mailed to all (nonrandom).
   f) Bias—Nonresponse. Hard to generalize because who responds is related to the question itself.

9. a) Population—Consumer Union subscribers.
   b) Parameter—Proportion who have used and benefited from alternative medicine.
   c) Sampling Frame—All Consumers Union subscribers.
   d) Sample—Those who responded (random).
   e) Method—Questionnaire to all (nonrandom).
   f) Bias—Nonresponse. Those who respond may have strong feelings one way or another.

11. a) Population—Adults.
    b) Parameter—Proportion who think drinking and driving is a serious problem.
    c) Sampling Frame—Bar patrons.
    d) Sample—Every 10th person leaving the bar.
    e) Method—Systematic sampling.
    f) Bias—Those interviewed had just left a bar. They probably think drinking and driving is less of a problem than do adults in general.

13. a) Population—Soil around a former waste dump.
    b) Parameter—Concentrations of toxic chemicals.
    c) Sampling Frame—Accessible soil around the dump.
    d) Sample—16 soil samples.
    e) Method—Not clear.
    f) Bias—Don't know if soil samples were randomly chosen. If not, may be biased toward more or less polluted soil.

15. a) Population—Snack food bags.
    b) Parameter—Weight of bags, proportion passing inspection.
    c) Sampling Frame—All bags produced each day.
    d) Sample—10 randomly selected cases, 1 bag from each case for inspection.
    e) Method—Multistage sampling.
    f) Bias—Should be unbiased.

17. Bias. Only people watching the news will respond, and their preference may differ from that of other voters. The sampling method may systematically produce samples that don't represent the population of interest.

19. a) Voluntary response. Only those who both see the ad *and* feel strongly enough will respond.
    b) Cluster sampling. One town may not be typical of all.
    c) Attempted census. Will have nonresponse bias.
    d) Stratified sampling with follow-up. Should be unbiased.

21. a) This is a multistage design, with a cluster sample at the first stage and a simple random sample for each cluster.
    b) If any of the three churches you pick at random is not representative of all churches, then you'll introduce sampling error by the choice of that church.

23. a) This is a systematic sample.
    b) It is likely to be representative of those waiting for the roller coaster. Indeed, it may do quite well if those at the front of the line respond differently (after their long wait) than those at the back of the line.
    c) The sampling frame is patrons willing to wait for the roller coaster on that day at that time. It should be representative of the people in line, but not of all people at the amusement park.

25. Answers will vary. Question 1 is the more neutrally worded question. Question 2 is biased in its wording.

27. Only those who think it worth the wait are likely to be in line. Those who don't like roller coasters are unlikely to be in the sampling frame, so the poll won't get a fair picture of whether park patrons overall would favor still more roller coasters.

29. a) Biased toward yes because of "pollute." "Should companies be responsible for any costs of environmental cleanup?"
    b) Biased toward no because of "enforce" and "strict". "Should companies have dress codes?"

31. a) Not everyone has an equal chance. People with unlisted numbers, people without phones, and those at work cannot be reached.
    b) Generate random numbers and call at random times.
    c) Under the original plan, those families in which one person stays home are more likely to be included. Under the second plan, many more are included. People without phones are still excluded.
    d) It improves the chance of selected households being included.
    e) This takes care of phone numbers. Time of day may be an issue. People without phones are still excluded.

33. a) Answers will vary.
    b) The amount of change you typically carry. Parameter is the true mean amount of change. Population is the amount on each day around noon.
    c) Population is now the amount of change carried by your friends. The average estimates the mean of these amounts.
    d) Possibly for your class. Probably not for larger groups. Your friends are likely to have similar needs for change during the day.

35. a) Assign numbers 001 to 120 to each order. Use random numbers to select 10 transactions to examine.
    b) Sample proportionately within each type. (Do a stratified random sample.)

37. a) Select three cases at random; then select one jar randomly from each case.
    b) Use random numbers to choose 3 cases from numbers 61 through 80; then use random numbers between 1 and 12 to select the jar from each case.
    c) No. Multistage sampling.

39. a) Depends on the Yellow Page listings used. If from regular (line) listings, this is fair if all doctors are listed. If from ads, probably not, as those doctors may not be typical.
    b) Not appropriate. This cluster sample will probably contain listings for only one or two business types.

Chapter 4

1. Answers will vary.
3. Answers will vary.
5. a) Yes, the categories divide a whole.
   b) Coca-Cola
7. a) The pie chart does a better job of showing portions of a whole
   b) There is no bar for "Other."
9. a) Yes, it is reasonable to assume that heart and respiratory disease caused approximately 38% of U.S. deaths in this year, since there is no possibility for overlap. Each person could only have one cause of death.
   b) Since the percentages listed add up to 73.7%, other causes must account for 26.3% of U.S. deaths.
   c) A bar graph or pie chart would be appropriate if a category for Other with 26.3% were added.
11. WebEx Communications, Inc. has the majority of the market share for web conferencing (58.4%), and Microsoft has approximately a quarter of the market share. There appears to be room for both to grow, because other companies comprise

about 15% of market share. A pie chart or bar chart would be appropriate.

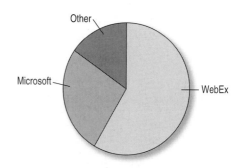

13. a) They total more than 100%; overlapping categories.
    b)

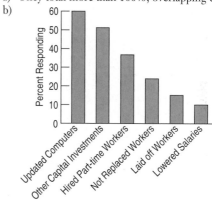

    c) No because the percentages do not total 100%.
    d) (Answers will vary). More than 50% of business owners say that they have either updated their computers or made other non-computer capital investments (or both). Smaller percentages of business owners (from 10 to 37%) made changes in either their hiring or salary structure.

15. The bar chart shows that grounding is the most frequent cause of oil spillage for these 312 spills, and allows the reader to rank the other types as well. If being able to differentiate between these close counts is required, use the bar chart. The pie chart is also acceptable as a display, but it's difficult to tell whether, for example, there is a greater percentage of spills caused by grounding or collisions. To showcase the causes of oil spills as a fraction of all 312 spills, use the pie chart.

17. a) 31%
    b) It looks like India's percentage is about 6 times as big, but it's not even twice as big.
    c) Start the percentages at 0% on the vertical axis, not 40%.
    d)

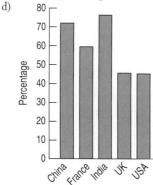

e) The percentage of people who say that wealth is important to them is highest in China and India (around 70%), followed by France (around 60%) and then the U.S. and U.K. where the percentage was only about 45%.

19. a) They must be column percentages because the sums are greater than 100% across the rows and all the columns add to 100%.

b)

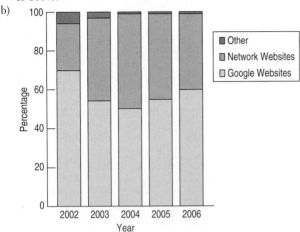

c) The main source of revenue for Google is from their own websites, which in 2002 was 70%, dropped down to 50% in 2004, and by 2006 was back up to 60%. The second largest source is from other network website. Licensing and other revenue was 6% in 2002, but since 2004 has been only 1%.

21. a) 62.5%
    b) 35%
    c) 15%
    d) 50%
    e) 61%
    f) 65%
    g) There does not appear to be any relationship between the performance of a stock on a single day and its performance over the prior year.

23. a) 10.2%; 13.2%
    b) 1.8%; 1.5%
    c) Sales decreased by 1.3%

25. a) 11.4% G; 14.3% PG; 48.6% PG-13; 25.7% R
    b) 0% G; 0% PG%; 57.9% PG-13; 42.1% R
    c)

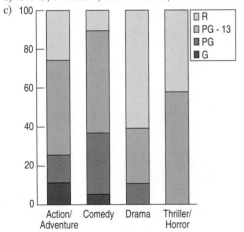

d) *Genre* and *Rating* are not independent. Thriller/Horror movie are all PG-13 or R and Drama are nearly so. Comedy moves are nearly 40% G and PG and only 10% R. Action/Adventure movies are nearly 15% G and 15% PG.

27. a) 62.7%
    b) 62.8%
    c) 62.5%
    d) 23.9% from Asia, 1.9% Europe, 7.8% Latin America, 3.7% Middle East, and 62.7% North America.
    e) The column percentages are given in the table.

	MBA program		
Origin	**Two-Yr**	**Evening**	**Total**
Asia	18.90	31.73	**23.88**
Europe	3.05	0.00	**1.87**
Latin America	12.20	0.96	**7.84**
Middle East	3.05	4.81	**3.73**
North America	62.80	62.50	**62.69**
Total	**100.00**	**100.00**	**100.00**

f) No. The distributions appear to be different. For example, the percentage from Latin America among those in Two-Yr. programs is nearly 20% while for those in Evening programs it is less than 1%.

29. a) 7%
    b) 5%
    c) 3.5%
    d) 57.5%
    e) 36.3%
    f) Here are row percentages:

					Total
2000–2005	5.0%	21.7%	57.5%	15.8%	100.0%
1996–1999	10.0%	17.5%	36.3%	36.3%	100.0%

PG-13 films increased from 36.3% in 1996–1999 to 57.5% in 2000–2005 and R-rated films decreased from 36.3% to 15.8%.

31. The study by the University of Texas Southwestern Medical Center provides evidence of an association between having a tattoo and contracting hepatitis C. Around 33% of the subjects who were tattooed in a commercial parlor had hepatitis C, compared with 13% of those tattooed elsewhere, and only 3.5% of those with no tattoo. If having a tattoo and having hepatitis C were independent, we would have expected these percentages to be roughly the same.

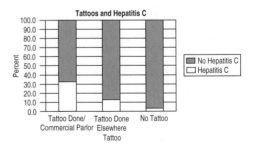

33. a) 8%
    b) No, because we're not given counts or totals.
    c) 92%
    d) There appears to be little, if any, relationship between revenue category and education level of the women CEOs.

35. a) 14.5% Hispanic, 12.5% African-American, and 73.0% Caucasian.

b) For 2006, 15.7% Hispanic, 12.0% African-American, 72.3% Caucasian.

c)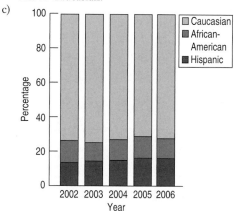

d) The (conditional) distribution of *Ethnicity* is almost the same across the five *Years*, however there seems to be a slight increase in the percentage of Hispanics who go to the movies from 13.1% in 2002 to 15.7% in 2006.

37. a) Row percentages.

b) A slightly higher percentage of urban women's business centers are established (at least 5 years old).

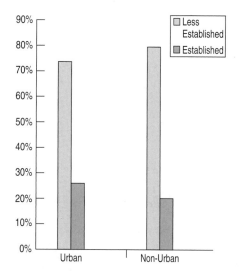

39. a) row percentages

b) No. We are given only the conditional distributions. We have no idea how much are sold in either Europe or America.

c)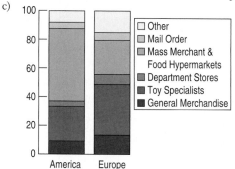

d) In America more than 50% of all toys are sold by large mass merchant discounters and food hypermarkets and only 25% are sold in toy specialty stores. In Europe 36% of all toys are sold in toy specialty stores while a relatively small 24% are sold in the large discount and hypermarket chains.

41. a) The marginal totals have been added to the table:

		Hospital Size		
		Large	**Small**	**Total**
Procedure	**Major surgery**	120 of 800	10 of 50	**130 of 850**
	Minor surgery	10 of 200	20 of 250	**30 of 450**
	Total	**130 of 1000**	**30 of 300**	**160 of 1300**

160 of 1300, or about 12.3% of the patients had a delayed discharge.

b) Major surgery patients were delayed 15.3% of the time. Minor Surgery patients were delayed 6.7% of the time.

c) Large Hospital had a delay rate of 13%. Small Hospital had a delay rate of 10%. The small hospital has the lower overall rate of delayed discharge.

d) Large Hospital: Major Surgery 15% and Minor Surgery 5%. Small Hospital: Major Surgery 20% and Minor Surgery 8%. Even though small hospital had the lower overall rate of delayed discharge, the large hospital had a lower rate of delayed discharge for each type of surgery.

e) Yes. While the overall rate of delayed discharge is lower for the small hospital, the large hospital did better with *both* major surgery and minor surgery.

f) The small hospital performs a higher percentage of minor surgeries than major surgeries. 250 of 300 surgeries at the small hospital were minor (83%). Only 200 of the large hospital's 1000 surgeries were minor (20%). Minor surgery had a lower delay rate than major surgery (6.7% to 15.3%), so the small hospital's overall rate was artificially inflated. The larger hospital is the better hospital when comparing discharge delay rates.

43. a) 1284 applicants were admitted out of a total of 3014 applicants. 1284/3014 = 42.6%

		Males Accepted (of applicants)	Females Accepted (of applicants)	Total
Program	**1**	511 of 825	89 of 108	**600 of 933**
	2	352 of 560	17 of 25	**369 of 585**
	3	137 of 407	132 of 375	**269 of 782**
	4	22 of 373	24 of 341	**46 of 714**
	Total	**1022 of 2165**	**262 of 849**	**1284 of 3014**

b) 1022 of 2165 (47.2%) of males were admitted. 262 of 849 (30.9%) of females were admitted.

c) Since there are four comparisons to make, the table below organizes the percentages of males and females accepted in each program. Females are accepted at a higher rate in every program.

Program	Males	Females
1	61.9%	82.4%
2	62.9%	68.0%
3	33.7%	35.2%
4	5.9%	7.0%

d) The comparison of acceptance rate within each program is most valid. The overall percentage is an unfair average. It fails to take the different numbers of applicants and different acceptance rates of each program. Women tended to apply to the programs in which gaining acceptance was difficult for everyone. This is an example of Simpson's Paradox.

Chapter 5

1. a) Outcomes are equally likely and independent.
   b) This is likely a personal probability expressing his degree of belief that there will be a rate cut.
3. a) There is no such thing as the "law of averages." The over-all probability of an airplane crash does not change due to recent crashes.
   b) There is no such thing as the "law of averages." The overall probability of an airplane crash does not change due to a period in which there were no crashes.
5. a) It would be foolish to insure your neighbor's house for $300. Although you would probably simply collect $300, there is a chance you could end up paying much more than $300. That risk is not worth the $300.
   b) The insurance company insures many people. The over-whelming majority of customers pay and never have a claim. The few customers who do have a claim are offset by the many who simply send their premiums without a claim. The relative risk to the insurance company is low.
7. a) yes
   b) yes
   c) no, probabilities sum to more than 1
   d) yes
   e) no, sum isn't 1 and one value is negative
9. 0.078
11. The events are disjoint. Use the addition rule.
   a) 0.72
   b) 0.89
   c) 0.28
13. a) 0.5184
   b) 0.0784
   c) 0.4816
15. a) The repair needs for the two cars must be independent of one another.
   b) This may not be reasonable. An owner may treat the two cars similarly, taking good (or poor) care of both. This may decrease (or increase) the likelihood that each needs to be repaired.
17. a) 0.68
   b) 0.32
   c) 0.04
19. a) 0.340
   b) 0.080
21. a) 0.4712
   b) 0.7112
   c) $1 - P(\text{interview}) = 1 - 0.2888 = 0.7112$
23. a) The events are disjoint (an M&M can't be two colors at once), so use the addition rule where applicable.
      i) 0.30
      ii) 0.30
      iii) 0.90
      iv) 0
   b) The events are independent (picking out one M&M doesn't affect the outcome of the next pick), so use the multiplication rule.
      i) 0.027
      ii) 0.128
      iii) 0.512
      iv) 0.271

25. a) disjoint
   b) independent
   c) No. Once you know that one of a pair of disjoint events has occurred, the other one cannot occur, so its probability has become zero.
27. a) 0.125
   b) 0.125
   c) 0.875
   d) independence
29. a) 0.0225
   b) 0.092
   c) 0.00008
   d) 0.556
31. a) Your thinking is correct. There are 47 cards left in the deck, 26 black and only 21 red.
   b) This is not an example of the Law of Large Numbers. The card draws are not independent.
33. a) 0.550
   b) 0.792
   c) 0.424
   d) 0.918
35. a) 0.333
   b) 0.429
   c) 0.667
37. a) 0.11
   b) 0.27
   c) 0.407
   d) 0.344
39. No. 28.8% of men with OK blood pressure have high choles-terol, but 40.7% of men with high blood pressure have high cholesterol.
41. a) 0.086
   b) 0.437
   c) 0.156
   d) 0.174
   e) 0.177
   f) No.
43. a) 0.47
   b) 0.266
   c) Having a garage and a pool are not independent events.
   d) Having a garage and a pool are not disjoint events.
45. a) 96.5%
   b) The probability of U.S. adults having a landline, given that they have a cell phone, is 58.2/(58.2 + 2.8) or about 95.4%. About 96.6% of U.S. adults have a landline. It appears that having a cell phone and having a landline are independent, since the probabilities are roughly the same.
47. No. 12.5% of the cars were of European origin, but about 16.9% of the students drive European cars.
49. a) 15.4%
   b) 11.4%
   c) 73.9%
   d) 18.5%

Chapter 6

1. Answers will vary.
3. This distribution is nearly symmetric and unimodal, centered at around $2500. The range is about $6000. Most of the tuitions lie between $1000 and $4000.
5. a) The distribution is skewed to the right. There are a few neg-ative values. The range is about $6000.
   b) The mean will be larger because the distribution is right skewed.
   c) Because of the skewness, the median is a better summary.

7. The distribution is unimodal and skewed to the right with two high outliers. The median is near 10%.

9. a) Five-Number Summary (Answers may vary depending on software.)

Min.	1st Qu.	Median	3rd Qu.	Max.
−10.820	7.092	11.270	17.330	94.940

b) Median = 11.275%; IQR = 10.24%

c)

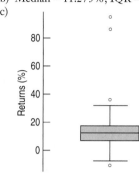

d) The histogram makes the skewness of the distribution clear.

11. a) Skewed to the right.

b) Yes, one high outlier

c) We don't know how far the high whisker should go because we don't know the largest value inside the fence.

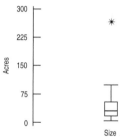

13. The stem-and-leaf display shows that many of the acreage values end in 0 or 5. Perhaps this is evidence that they are rounding or estimating the value.

```
24 | 0
22 |
20 |
18 |
16 |
14 | 0
12 | 0
10 | 0
 8 | 0
 6 | 920
 4 | 553500
 2 | 8655210987520
 0 | 751000086
```

Key: |8 0 = 80 acres

15. a) **Wayne Gretzky—Games played per season**

```
8 | 000000122
7 | 8899
7 | 0344
6 |
6 | 4
5 |
5 |
4 | 58
4 |
```

Key: 7 | 8 = 78 games

b)

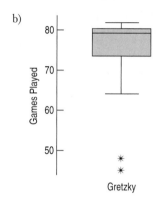

c) The distribution of the number of games played per season by Wayne Gretzky is skewed to the low end and has low outliers. The median is 79, and the range is 37 games.

d) There are two outlier seasons with 45 and 48 games. He may have been injured. The season with 64 games is also separated by a gap.

17. a) The median because the distribution is skewed.

b) lower

c) That display is not a histogram. It's a time series plot using bars to represent each point. The histogram should split up the range of the games into bins, not display the number of games over time.

19. a) Descriptive Statistics: Price ($)

Minimum	Q1	Median	Q3	Maximum
2.21	2.51	2.61	2.72	3.05

b) Range = max − min = 3.05 − 2.21 = $0.84; IQR = Q3 − Q1 = 2.72 − 2.51 = $0.21

c)

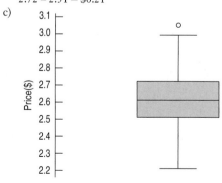

d) Symmetric with one high outlier. The mean is $2.62, with a standard deviation of $0.156.

e) There is one unusually high price that is greater than $3.00 per frozen pizza.

21. As we can see from the histogram and boxplot, the distribution of gasoline use is unimodal and skewed to the left with two low outliers, the District of Columbia and New York State. D.C. is a city, and New York may be dominated by New York City. Because of public transportation, gasoline usage is lower per capita in cities. The median usage is 485.7 gal./yr. per capita with an IQR of about 81.75 gal./yr (Values from different software may vary slightly.). The minimum is D.C. with 209.5, and the maximum is Wyoming with 589.18 gal./yr.

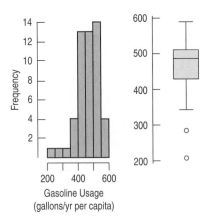

23. a) 1611 yards.
b) Between Quartile 1 = 5585.75 yards, and Quartile 3 = 6131 yards.
c) The distribution of golf course lengths appears roughly symmetric, so the mean and sd are appropriate.
d) The distribution of the lengths of all the golf courses in Vermont is roughly unimodal and symmetric. The mean length of the golf courses is approximately 5900 yards and the standard deviation is 386.6 yd.

25. a) A boxplot is shown. A histogram would also be appropriate.

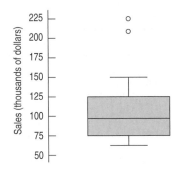

b) Descriptive Statistics: Sales ($) (Different statistics software may yield different results.)

Variable	N	N*	Mean	SE Mean	StDev	Minimum
Sales ($)	18	0	107845	11069	46962	62006

Q1	Median	Q3	Variable	Maximum
73111	95975	124439	Sales ($)	224504

The mean sale is $107,845, and the median is $95,975. The mean is higher because the outliers pull it up.
c) The median because the distribution has outliers.
d) The standard deviation of the distribution is $46,962, and the IQR is $51,328.
e) The IQR because the outliers inflate the standard deviation.
f) The mean would decrease. The standard deviation would decrease. The median and IQR would be relatively unaffected.

27. (To calculate the failure rate, divide the number failed by the sum of the number failed and the number OK for each model and then multiply by 100.) A histogram shows that the distribution is unimodal and skewed to the left. There do not appear to be any outliers. The median failure rate for these 17 models is 16.2%. The middle 50% of the models have failure rates between 10.87% and 21.2%. The best rate is 3.17% for the 60GB Video model, and the worst is the 40GB Click Wheel at 29.85%.

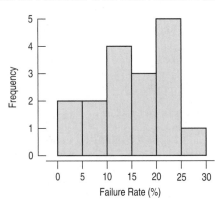

29. Sales in Location #1 were higher than sales in Location #2 in every week. The company might want to compare other stores in locations like these to see if this phenomenon holds true for other locations.

31. a) Gas prices increased over the three-year period, and the spread increased as well. The distribution of prices in 2002 was skewed to the left with several low outliers. Since then, the distribution has been increasingly skewed to the right. There is a high outlier in 2004, although it appears to be pretty close to the upper fence.
b) The distribution of gas prices in 2004 shows the greatest range and the biggest IQR, so the prices varied a great deal.

33. a) Seneca Lake.
b) Seneca Lake.
c) Keuka Lake.
d) Cayuga Lake vineyards and Seneca Lake vineyards have approximately the same average case price, of about $200, while a typical Keuka Lake vineyard has a case price of about $260. Keuka Lake vineyards have consistently high case prices, between $240 and $280, with one low outlier at about $170 per case. Cayuga Lake vineyards have case prices from $140 to $270, and Seneca Lake vineyards have highly variable case prices, from $100 to $300.

35. a) The median speed is the speed at which 50% of the winning horses ran slower. Find 50% on the left, move straight over to the graph and down to a speed of about 36 mph.
b) Q1 = 34.5 mph, and Q3 = 36.5 mph.
c) Range = 7 mph
IRQ = 2 mph.
d)

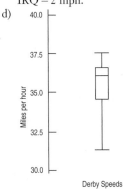

e) The distribution of winning speeds in the Kentucky Derby is skewed to the left. The lowest winning speed is just under 31 mph, and the fastest speed is about 37.5 mph. The median speed is approximately 36 mph, and 75% of winning speeds are above 34.5 mph. Only a few percent of winners have had speeds below 33 mph.

37. a) Class 3
    b) Class 3
    c) Class 3
    d) Class 1
    e) Probably Class 1 has. But without the actual scores, it is impossible to calculate the exact IQRs.

39. There is an extreme outlier for the slow speed drilling. One hole was drilled almost an inch away from the center of the target! If that distance is correct, the engineers at the computer production plant should investigate the slow speed drilling process closely. It may be plagued by extreme, intermittent inaccuracy. The outlier in the slow speed drilling process is so extreme that no graphical display can display the distribution in a meaningful way while including that outlier. That distance should be removed before looking at a plot of the drilling distances.

   With the outlier removed, we can see that the slow drilling process is more accurate. The greatest distance from the target for the slow drilling process, 0.000098 inches, is still more accurate than the smallest distance for the fast drilling process, 0.000100 inches.

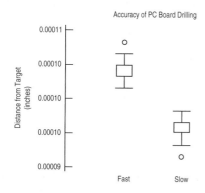

41. a) The mean of 54.41 is meaningless. These are categorical values.
    b) Typically, the mean and standard deviation are influenced by outliers and skewness.
    c) No. Summary statistics are only appropriate for quantitative data.

43. Over this 3-month period, International Funds generally outperformed the other two. Almost half of the International Funds outperformed all the funds in the other two categories. U.S. Domestic Large Cap Funds did better than U.S. Domestic Small/Mid Cap Funds in general. Large Cap funds had the least variation of the three types.

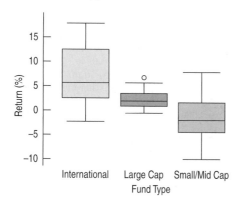

45. a) Even though MLS ID numbers are categorical identifiers, they are assigned sequentially, so this graph has some information. Most of the houses listed long ago have sold and are no longer listed.
    b) A histogram is generally not an appropriate display for categorical data.

47. a)

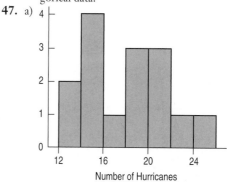

    b) The distribution is fairly uniform. There do not appear to be any decades that would be considered to be outliers.
    c)

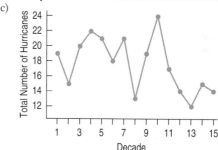

    d) This graph does not support the claim that the number of hurricanes has increased in recent decades.

49. What is the x-axis? If it is time, what are the units? Months? Years? Decades? How is "productivity" measured?

51. a) The distribution is skewed. That makes it difficult to estimate anything meaningful from the graph.
    b) Transform these data using either square roots or logs.

53. The house that sells for $400,000 has a z-score of $(400000 - 167900)/77158 = 3.01$, but the house with 4000 sq. ft. of living space has a z-score of $(4000 - 1819)/663 = 3.29$. So it's even more unusual.

55. U.S. z-scores are −0.04 and 1.63, total = 1.59. Ireland z-scores are 0.25 and 2.77, total 3.02. So Ireland "wins" the consumption battle.

57. a) The histogram shows that the distribution of prices is strongly skewed to the right.

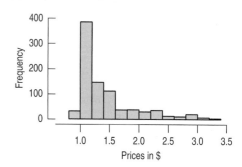

    b) Prices were relatively stable until the late 1990s, when they started increasing. Since 2005, the prices have been higher and unstable.

c) The time series plot.

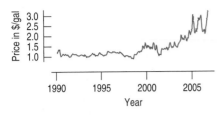

59. a) The bimodality of the distribution.
   b) The trend over time.
   c) The time series plot.
   d) Unemployment decreased steadily from about 5.6% in 1995 to below 4.0% by 2000, and then increased sharply to between 5.5 and 6.0% from 2002 to 2004.

Chapter 7

1. a) Number of text messages: explanatory; cost: response. To predict cost from number of text messages. Positive direction. Linear shape. Possibly an outlier for contracts with fixed cost for texting.
   b) Fuel efficiency: explanatory; sales volume: response. To predict sales from fuel efficiency. There may be no association between mpg and sales volume. Environmentalists hope that a higher mpg will encourage higher sales, which would be a positive association. We have no information about the shape of the relationship.
   c) Neither variable is explanatory. Both are responses to the lurking variable of temperature.
   d) Price: explanatory variable; demand: response variable. To predict demand from price. Negative direction. Linear shape in a narrow range, but curved over a larger range of prices.
3. a) None
   b) 3 and 4
   c) 2, 3, and 4
   d) 1 and 2
   e) 3 and 1
5. Not linear form. Moderately strong. The rate of increase from the beginning to about 1950 is steeper than from 1950 to the present. Horses may no longer be getting faster.
7. a)

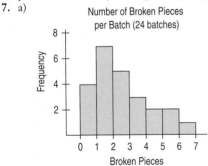

   b) Unimodal, skewed to the right. The skewness.
   c) The positive, somewhat linear relation between batch number and broken pieces.
9. a) 0.006
   b) 0.777
   c) −0.923
   d) −0.487
11. "Packaging" isn't a variable. At best, it is a category. There's no basis for computing a correlation.

13. The correlation will be near zero. There is a strong association, but it isn't linear. Sales are high at the start of each semester and lower in between.
15. a) The variables are both quantitative (measured in carats and dollars); the plot is straight enough and has no outliers. Conditions are met.
   b) Among diamonds of best color (D) and very good (VS1) clarity, there is a strong positive, linear association between the weight of a diamond and its price.
17. a) There is a strong negative linear association between Carbon Footprint and Highway mpg.
   b) Quantitative variables, straight enough. The Prius is far from the rest of the data. But it is in line with the linear pattern. It is correct to regard it as an outlier or not; that's a matter of judgment.
   c) $r = -0.94$; Removing the Prius reduces the correlation. Data values far from the main body of the data and in line with the linear trend tend to increase correlation and may make it misleading.
19. a) Yes, the variables are quantitative, and the plot is linear with no outliers.
   b) There is a strong positive, linear association. The correlation is 0.80.
21. a) Two quantitative variables. However, the scatterplot shows two points standing away from the others. Those outliers make correlation problematic.
   b) With the exception of the two oldest vineyards, no.
   c) This graph only includes vineyards in New York state, so that is the only geographic region about which we can draw conclusions.
23. a) Two quantitative variables. But the sctterplot shows a non-linear relationship, so correlation is not appropriate.
   b) For homes with between 2 and 6 bedrooms, there is a positive relationship between number of bedrooms and price. The few larger homes in the sample don't follow that pattern.
25. The variable *State* is not a quantitative variable, so the correlation between *Sales* and *State* is meaningless. The ordering of the states in alphabetical order is arbitrary. A bar chart might be a better way to display these data.
27. a) The scatterplot shows a moderately strong, positive, roughly linear relationship.
   b) Yes, states where community colleges are more expensive tend to charge more for four-year education as well.
   c) Yes, the correlation is 0.66, a moderately strong correlation.
29. a) It is negative, moderately strong, and roughly linear.
   b) Yes, given the answer in a) and both variables are quantitative.
   c) First, these data are all from one year. Second, there may be many reasons why the relationship looks as it does. We cannot conclude causation from correlation.
31. a) The relationship is positive, with moderate scatter and roughly linear.
   b) Yes, given the answer in a) and the fact that both variables are quantitative.
   c) A large correlation does not imply a causal relationship.
33. a) Assuming the relationship is linear, a correlation of −0.772 shows a strong relationship in a negative direction.
   b) Continent is a categorical variable.
35. If the data had an outlier that could dominate the correlation and if the relationship is curved, there could be an association but small correlation.
37. a) No, correlation does not imply causation.
   b) It is likely that more affluent countries consume more oil and have greater life expectancy due to improved nutrition, education, and medical attention.

39. a)

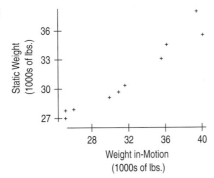

b) Positive, linear, and very strong.
c) The new scale is able to predict the static weight fairly well, except possibly at the high end. It may be possible to predict the static weight from the new scale accurately enough to be useful. But the weight-in-motion measurements seem a bit too high.
d) 0.965
e) The correlation would not change.
f) At the higher end of the scale, there is one point where the weight-in-motion is much higher than the static weight. The new scale may have to be recalibrated.

41. a) There are two quantitative variables, and the scatterplot is straight enough with no outliers. Assumptions and conditions are met.
b) -0.574
c) Yes, but it does not imply causation.
d) No. Change of units does not change the value of the correlation.

43. a) There is an outlier at the bottom of the graph, so the conditions are not met.
b) July, 2008; $r = 0.396$

45. a) Yes, the scatterplot is straight enough, variables are quantitative and there are no outliers.
b) Teams that score more runs generally have higher attendance.
c) There is a positive association, but correlation doesn't imply causation.

47. Conclusion not justified. Both Marital status and Giving are categorical and not suitable for correlation. They should examine a two-way table.

49. a) -0.14
b) No, the low correlation is because the relationship isn't linear.

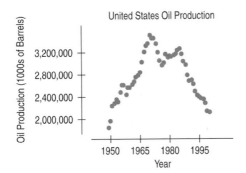

Chapter 8

1. a) *Price*
b) *Sales*
c) Sales decrease by 24,310 pounds per dollar.
d) It is just a base value. It means nothing because stores won't set their price to $0.

e) 56,572.32 pounds
f) 3427.68 pounds

3. a) *Salary*
b) *Wins*
c) On average team's win 0.62 more games per $10 million in salary.
d) Number of wins predicted for a team that spends $0 on salaries. This is not meaningful here.
e) 0.62 games more
f) 4.883 games. Better.
g) 3.117 games.

5. 47,084.23 pounds

7. The model is meaningless because the variable Region is not quantitative. The slope makes no sense because Region has no units. The boxplot comparisons are fine, but the regression is meaningless.

9. a) The variables are both quantitative (with units % of GDP), the plot is reasonably straight, there are no outliers, and the spread is roughly constant (although the spread is large).
b) About 21% of the variation in the growth rates of developing countries is accounted for by the growth rates of developed countries.
c) Years 1970–2007

11. a) $\widehat{Growth\ (Developing\ Countries)} = 3.46 + 0.433\ Growth$ *(Developed Countries)*
b) The predicted growth of developing countries in years of 0 growth in developed countries. Yes, this makes sense.
c) On average, GDP in developed countries increased 0.433% for a 1% increase in growth in developed countries.
d) 5.192%
e) More; we would predict 4.61%.
f) 1.48%

13. a) Positive, roughly straight (slight bending at the upper right), and moderately strong.
b) Yes (although answers may vary). It makes sense that a state's finances would influence both amounts in the same way.
c) $\widehat{Public.4yr} = 2826.00 + 1.216\ Public.2yr$
d) Yes
e) A public four-year education costs, on average, $2826 + 1.216 \times$ average tuition at two-year colleges.
f) 44.29% of the variation in the average tuition of four-yr state colleges is accounted for by the regression on two-yr college tuition.

15. a) The predicted value of the money *Flow* if the *Return* was 0%.
b) An increase of 1% in mutual fund return was associated with an increase of $771 million in money flowing into mutual funds.
c) $9747 million
d) $-$4747 million; Overestimated

17. a) Billions of dollars per thousand housing starts.
b) 0.49.
c) 0.70 standard deviations below the mean in *Sales*.

19. a) 88.3% of the variation in quarterly *Sales* can be accounted for by variation in U.S. unemployment *Rate*.
b) -0.94.
c) Its *Sales* would be lower by about $2.99 billion.

21. a) Model seems appropriate. Residual plot looks fine.
b) Model not appropriate. Relationship is nonlinear.
c) Model not appropriate. Spread is increasing.

23. a) *Sales* increase, on avearge by $0.0535 billion (or $53.5 million) per 1000 *Housing Starts*.
b) $15.25 billion.
c) A residual.

25. There are two outliers that inflate the R^2 value and affect the slope and intercept. Without those two points, the R^2 drops from 79% to about 31%. The analyst should set aside those two customers and refit the model.

27. a) A linear model is marginally appropriate. The variables are quantitative, the relationship is roughly straight, there is a possible outlier, and the spread is influened by the outlier. Moreover, there are only 10 data points, and more data for more store locations would provide information that might change this answer.

  b) 0.75.

  c) 56.9% of the variability in annual *Sales in 2000* is accounted for by variability in the *Population* of the town where the store is located.

29. 0.03

31. a) R^2 is an indication of the strength of the model, not the appropriateness of the model.

  b) The student should have said, "The model predicts that quarterly sales will be $10 million when $1.5 million is spent on advertising."

33. a) Quantitative variable condition: Both variables are quantatitive (*GPA* and *Starting Salary*).

  b) Linearity condition: Examine a scatterplot of *Starting Salary* by *GPA*.

  c) Outlier condition: Examine the scatterplot.

  d) Equal spread condition: A plot the regression residuals versus predicted values.

35. a) Linear, moderate in strength, and positive.

  b) One student got a 500 Verbal and 800 Math. That set of scores doesn't seem to fit the pattern.

  c) A moderate, positive association.

  d) $\widehat{Math} = 209.6 + 0.675(Verbal)$

  e) For each additional Verbal SAT, we expect the Math SAT score to be 0.675 points higher, on average.

  f) 547.1 points

  g) 50.4 points

37. a) $r = 0.685$

  b) $\widehat{Verbal} = 171.03 + 0.695\,Math$ (or $171.33 + 0.694\,Math$ directly from raw data).

  c) The observed *Verbal* score is higher than predicted.

  d) 518.5

  e) 559.6

  f) Regression to the mean; because we always predict a fraction (the correlation) of the SD from the mean, the successive predictions will get closer and closer to the mean.

39. a)

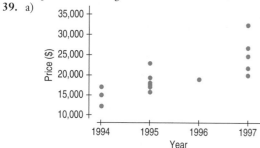

  b) There is a strong, positive, linear association between *Price* and *Year* of used BMW 840s.

  c) Yes

  d) 0.757

  e) 57.4% of the variability in *Price* of a used BMW 840 can be accounted for by the *Year* the car was made.

  f) The relationship is not perfect. Other factors, such as options, condition, and mileage, may account for some of the variability in price.

41. a)

  b) A linear model is possible, but with so few points and the two outliers, we should be very cautious with its use. The variables are quantitative, and the spread is roughly the same except for the two outliers.

  c) $\widehat{Sales}\,(million\ \$) = 24.114 - 0.4828\,Median\ Age$

  d) $8,664 million

  e) Yes, the only data between median age 30 and 36 years are two outliers, and the model is not appropriate.

43. a) The association between cost of living in 2007 and 2006 is linear, positive, and strong. The linearity of the scatterplot indicates that the linear model is appropriate.

  b) 83.7% of the variability in cost of living in 2007 can be explained by variability in cost of living in 2006.

  c) 0.915

  d) Moscow had a cost of living of 123.9% of New York's in 2006. According to the model, Moscow is predicted to have a cost of living in 2007 that is about 128.8% of New York's. Moscow actually had a cost of living in 2007 that was 134.4% of New York's, so its residual was about 5.6%.

45. The scatterplot of *Sales* by *Year* shows an association that is positive and strong. In general, *Sales* have increased over time. The scatterplot is straight enough to justify the use of the linear model.

  a) The linear model for predicting sales from the year has an R^2 of 99.8% and is *Sales* $= 178,165 + 25,701\,Time$

  b) *Sales* grew by $25.7 billion per year on average.

  c) The intercept is the value of net sales at the beginning of the time frame; *Sales* in 2001 were $178.165 billion according to the model (using 2001 = 0 for time).

47. a) $332,371 million

  b) The danger in using this model to predict net sales out to 2010 is the extrapolation involved and the impact of too many other unforeseen factors on Wal-Mart sales.

49. a) 0.578

  b) CO_2 levels account for 33.4% of the variation in mean temperature.

  c) $\widehat{Mean\ Temperature} = 15.3066 + 0.004\,CO_2$

  d) The predicted mean temperature has been increasing at an average rate of 0.004 degrees (C)/ppm of CO_2.

  e) One *could* say that with no CO_2 in the atmosphere, there would be a temperature of 15.3066 degrees Celsius, but this is extrapolation to a nonsensical point.

  f) No

  g) Predicted 16.7626 degrees C.

51. a) $\widehat{HighJump} = 2.681 - 0.00671\,800mTime$. High jump height is lower, on average, by 0.00671 meters per second of 800-m race time.

  b) 16.4%

  c) Yes, the slope is negative. Faster runners tend to jump higher as well.

d) There is a slight tendency for less variation in high jump height among the slower runners than among the faster runners.

e) Not especially; the residual standard deviation is 0.060 meters, which is not much smaller than the SD of all high jumps (0.066 meters). The model doesn't appear to do a very good job of predicting.

Chapter 9

1. a) 16% b) 50% c) 95% d) 0.15%

3. a) 2.4%
b) 8.0%
c) −8.8%
d) (−3.2% < x < 8.0%)

5. a) 50%
b) 16%
c) 2.5%
d) More than 1.542 is more unusual.

7. a) $x > 1.492$
b) $x < 1.459$
c) $(1.393 < x < 1.525)$
d) $x < 1.393$

9. a) 21.6% (using technology)
b) 48.9%
c) 59.9%
d) 33.4%

11. a) $x > 9.58\%$
b) $x < -2.31\%$
c) $(-0.54\% < x < 5.34\%)$
d) $x > -2.31\%$

13. a) 0.98%
b) 15.4%
c) 7.56%

15. a) 79.58
b) 18.50
c) 95.79
d) −2.79

17. $z_{SAT} = 1.30$; $z_{ACT} = 2$. The ACT score is the better score because it is farther above the mean in standard deviation units than the SAT score.

19. a) To know about their consistency and how long they might last. Standard deviation measures variability, which translates to consistency in everyday use. A type of battery with a small standard deviation would be more likely to have life spans close to their mean life span than a type of battery with a larger standard deviation.
b) The second company's batteries have a higher mean life span, but a larger standard deviation, so they have more variability. The decision is not clear-cut. The first company's batteries are not likely to fail in less than 21 months, but that wouldn't be surprising for the second company. But the second company's batteries could easily last longer than 39 months—a span very unlikely for the first company.

21. CEOs can have between 0 and maybe 40 (or possibly 50) years experience. A standard deviation of 1/2 year is impossible because many CEOs would be 10 or 20 SDs away from the mean, whatever it is. An SD of 16 years would mean that 2 SDs on either side of the mean is plus or minus 32, for a range of 64 years. That's too high. So, the SD must be 6 years.

23. a)

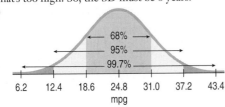

25. Any Job Satisfaction score more than 2 standard deviations below the mean or less than $100 - 2(12) = 76$ might be considered unusually low. We would expect to find someone with a Job Satisfaction score less than $100 - 3(12) = 64$ very rarely.

27. a) About 16%
b) One standard deviation below the mean is −1.27 hours, which is impossible.
c) Because the standard deviation is larger than the mean, the distribution is strongly skewed to the right, not symmetric.

29. a)

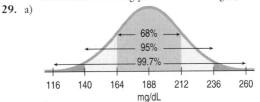

b) 30.85%
c) 17.00%
d) IQR = Q3 − Q1 = 32.38
e) Above 212.87 points

31. a) 0.657
b) 0.584
c) 0.507
d) 0.275

33. a) 5.3 grams
b) 6.4 grams
c) Since 5.3 < 6.4, the younger hens lay eggs that have more consistent weights than the eggs laid by the older hens.
d) According to the Normal model, the mean weight of the eggs is 62.7 grams, with a standard deviation of 6.2 grams.

35. a) $\mu_{\hat{p}} = p = 7\%$ and $\sigma(\hat{p}) \approx 1.8\%$
b) N(0.07, 0.018) Assume that these new clients are a random sample from same population on which the default percentage is based. This is not necessarily true. Assume independence—seems reasonable. Sample size condition is met.
c) 0.048

37. 0.212; reasonable that those polled are independent of each other and represent less than 10% of all potential voters. We assume the sample was selected at random. Success/Failure condition met: $np = 208 \geq 10$ and $nq = 194 \geq 10$.

39. 0.088 using N(0.08, 0.022) model

41. Answers will vary. Using $\mu + 3\sigma$ for "very sure," the restaurant should have 89 nonsmoking seats. It assumes customers at any time are independent of each other, a random sample, and represent less than 10% of all potential customers. $np = 72$, $nq = 48$, so N(0.60, 0.0447) model is reasonable.

43. a) $N\left(\mu, \dfrac{\sigma}{\sqrt{n}}\right)$
b) Standard deviation will be smaller. Center will remain the same.

45.

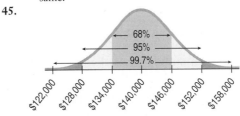

The sampling distribution model for the sample mean home values is approximately N(140,000, 6000).

47. a) Some people work far longer than the mean plus 2 or 3 SDs.
  b) The CLT says $\bar{y}$ is approximately Normal for large sample sizes, but not for samples of size 1 (individuals). Besides, people can't work fewer than 0 hours.
49. a) 0.0478
  b) 0.863
  c) 0.0019
  d) Essentially 0

Chapter 10

1. He believes the true proportion is within 4% of his estimate, with some (probably 95%) degree of confidence.
3. a) *Population*—all cars in the local area; *sample*—134 cars actually stopped at the checkpoint; *p*—population proportion of all cars with safety problems; $\hat{p}$—proportion of cars in the sample that actually have safety problems (10.4%); if sample (a cluster sample) is representative, then the methods of Chapter 10 will apply.
  b) *Population*—general public; *sample*—602 viewers that logged on to the website; *p*—population proportion of the general public that think corporate corruption is "worse"; $\hat{p}$—proportion that logged onto the website and voted that corporate corruption is "worse" (81.1%); can't use the methods of Chapter 10—sample is biased and nonrandom.
5. a) *Population*—all customers who recently bought new cars; *sample*—167 people surveyed about their experience; *p*—proportion of all new car buyers who are dissatisfied with the salesperson; $\hat{p}$—proportion of new car buyers surveyed who are dissatisfied with the salesperson (3%); can't use the methods of Chapter 10 because only 5 people were dissatisfied. Could use pseudo observation method of 10.5, but sample may not be representative.
  b) *Population*—all college students; *sample*—2883 who were asked about their cell phones at the football stadium; *p*—proportion of all college students with cell phones; $\hat{p}$—proportion of college students at the football stadium with cell phones (77.8%); be cautious—students entering the football stadium may not represent all students. Sample may be biased.
7. a) Not correct. This implies certainty.
  b) Not correct. Different samples will give different results. Most likely, none of the samples will have *exactly* 88% on-time orders.
  c) Not correct. A confidence interval says something about the unknown population proportion, not the sample proportion in different samples.
  d) Not correct. In this sample, we *know* that 88% arrived on time.
  e) Not correct. The interval is about the parameter, not about the days.
9. a) False
  b) True
  c) True
  d) False
11. We are 90% confident that between 29.9% and 47.0% of U.S. cars are made in Japan.
13. a) 0.025
  b) The pollsters are 90% confident that the true proportion of adults who do not use e-mail is within 2.5% of the estimated 38%.
  c) A 99% confidence interval requires a larger margin of error. In order to increase confidence, the interval must be wider.
  d) 0.039 or 3.9%.
  e) Smaller margins of error will give us less confidence in the interval.
15. a) (12.7%, 18.6%)
  b) We are 95% confident that between 12.7% and 18.6% of all accidents involve teenage drivers.

  c) About 95% of all random samples of size 582 will produce intervals that contain the true proportion of accidents involving teenage drivers.
  d) Contradicts—the interval is completely below 20%.
17. Probably nothing. Those who bothered to fill out the survey may be a biased sample.
19. This was a random sample of less than 10% of all Internet users; there were $703 \times 0.18 = 127$ successes and 576 failures, both at least 10. We are 95% confident that between 15.2% and 20.8% of Internet users have downloaded music from a site that was not authorized. (Answer could be 15.2% to 20.9% if $n = 127$ is used instead of 0.18).
21. a) $385/550 = 0.70$; 70% of U.S. chemical companies in the sample are certified.
  b) This was a random sample, but we don't know if it is less than 10% of all U.S. chemical companies; there were $550(0.70) = 385$ successes and 165 failures, both at least 10. We are 95% confident that between 66.2% and 73.8% of the chemical companies in the United States are certified. It appears that the proportion of companies certified in the United States is less than in Canada.
23. a) There may be response bias based on the wording of the question.
  b) (45.5%, 51.5%)
  c) The margin of error based on the pooled sample is smaller, since the sample size is larger.
25. a) The interval based on the survey conducted by the college Statistics class will have the larger margin of error, since the sample size is smaller.
  b) Both samples are random and are probably less than 10% of the city's voters (provided the city has more than 12,000 voters); there were 636 successes and 564 failures for the newspaper, both at least 10; there were 243 successes and 207 failures for the Statistics class, both at least 10; Newspaper poll: (50.2%, 55.8%); Statistics class: (49.4%, 58.6%).
  c) The Statistics class should conclude that the outcome is too close to call because 50% is in their interval.
27. a) This was a random sample of less than 10% of all English children; there were $2700(0.20) = 540$ successes and 2160 failures, both at least 10; (18.2%, 21.8%).
  b) We are 98% confident that between 18.2% and 21.8% of English children are deficient in vitamin D.
  c) About 98% of random samples of size 2700 will produce confidence intervals that contain the true proportion of English children that are deficient in vitamin D.
  d) No. The interval says nothing about causation.
29. a) This is not a random sample; even though it's representative we are not sure if we have less than 10% of all WBCs; there are 8 successes which is not greater than 10, so the sample is not large enough.
  b) Since the conditions are not met but the sample is representative, we could perform a pseudo observation confidence interval adding two success and two failures to the data. Then a 90% CI would be (0.251, 0.582).
31. a) This was a random sample of less than 10% of all companies in Vermont; there were 12 successes and 0 failures, which is not greater than 10, so the sample is not large enough.
  b) Using the pseudo observation confidence interval, $\widetilde{p} = \frac{14}{16} = 0.875$; (0.713, 1.037); so we can say that with 95% confidence the true proportion is at least 71.3%.
33. a) This was a random sample of less than 10% of all self-employed taxpayers; there were 20 successes and 206 failures, both at least 10.
  b) (5.2%, 12.6%)
  c) We are 95% confident that between 5.1% and 12.6% of all self-employed individuals had their tax returns audited in the past year.

d) If we were to select repeated samples of 226 individuals, we'd expect about 95% of the confidence intervals we created to contain the true proportion of all self-employed individuals who were audited.

35. a) This was a random sample of less than 10% of all U.S. adults; there were $703 \times 0.13 = 91$ successes and 612 failures, both at least 10.

b) (10.5%, 15.5%)

37. a) The 95% confidence interval for the true proportion of all 18- to 29-year-olds who believe the U.S. is ready for a woman president will be about twice as wide as the confidence interval for the true proportion of all U.S. adults, since it is based on a sample about one-fourth as large. (Assuming approximately equal proportions)

b) This was a random sample of less than 10% of all U.S. 15- to 29-year-old-adults; there were $250 \times 0.62 = 155$ successes and 95 failures, both at least 10. We are 95% confident that between 56.0% and 68.0% of 18- to 29-year-olds believe the U.S. is ready for a woman president.

39. a) The parameter is the proportion of digital songs in student libraries that are legal. The population is all songs held in digital libraries. The sample size is 117,079 songs, not 168 students.

b) This was a cluster sample (168 clusters of songs in the digital libraries); 117,079 is less than 10% of all digital songs; the number of legal songs and illegal songs in the sample are both much greater than 10.

c) We are 95% confident that between 22.9% and 23.3% of digital songs were legally purchased.

d) The very large sample size has made the confidence interval unreasonably narrow. It is hard to believe that such a narrow interval really captures the parameter of interest. Additionally, these data were collected in a cluster sample of only 168 students. This gives us less certainty about our ability to capture the true parameter.

41. a) This was a random sample of less than 10% of all Internet users; there were $703(0.64) = 450$ successes and 253 failures, both at least 10. We are 90% confident that between 61.0% and 67.0% of Internet users would still buy a CD.

b) In order to cut the margin of error in half, they must sample 4 times as many users; $4 \times 703 = 2812$ users.

43. a) 141
b) 318
c) 564

45. 1801

47. 384 total, using $p = 0.15$

49. Since $z^* \approx 1.634$, which is close to 1.645, the pollsters were probably using 90% confidence.

51. This was a random sample of less than 10% of all customers; there were 67 successes and 433 failures, both at least 10. From the data set, $\hat{p} = 67/500 = 0.134$. We are 95% confident that the true proportion of customers who spend $1000 per month or more is between 10.4% and 16.4%.

53. a) This was a random sample of less than 10% of all MA males; there were 2662 successes and 398 failures, both at least 10.

b) (0.858, 0.882)

c) We are 95% confident that between 85.8% and 88.2% of MA males have health insurance.

Chapter 11

1. a) Let p be the percentage of products delivered on time. $H_0: p = 0.90$ vs. $H_A: p > 0.90$

b) Let p be the proportion of houses taking more than 3 months to sell. $H_0: p = 0.50$ vs. $H_A: p > 0.50$

c) Let p be the error rate. $H_0: p = 0.02$ vs. $H_A: p < 0.02$

3. Statement d is correct.

5. If the rate of seat belt usage after the campaign is the same as the rate of seat belt usage before the campaign, there is a 17% chance of observing a rate of seat belt usage after the campaign this large or larger in a sample of the same size by natural sampling variation alone.

7. Statement e is correct.

9. No, we can say only that there is a 27% chance of seeing the observed effectiveness just from natural sampling if $p = 0.7$. There is no *evidence* that the new formula is more effective, but we can't conclude that they are equally effective.

11. a) 0.186 (using the normal model) 0.252 using exact probabilities

b) It seems reasonable to think there really may have been half of each. We would expect to get 12 or more reds out of 20 more than 15% of the time, so there's no real evidence that the company's claim is not true. The two sided P-value is greater than 0.30.

13. a) Conditions are satisfied: random sample; less than 10% of population; more than 10 successes and failures; (1.9%, 4.1%)

b) Since 5% is not in the interval, there is strong evidence that fewer than 5% of all men use work as their primary measure of success.

c) $\alpha = 0.01$; it's a lower tail test based on a 98% confidence interval.

15. a) Conditions are satisfied: random sample; less than 10% of population; more than 10 successes and failures; (0.223, 0.257); we are 95% confident that the true proportion of U.S. adults who rate the economy as Excellent/Good is between 0.223 and 0.257.

b) No. Since 0.28 is not within the interval, there is evidence that the proportion is not 28%.

c) $\alpha = 0.05$; it's a two-tail test based on a 95% confidence interval.

17. a) Less likely

b) Alpha levels must be chosen *before* examining the data. Otherwise the alpha level could always be selected to reject the null hypothesis.

19. 1. Use p, not $\hat{p}$, in hypotheses.

2. The question is about *failing* to meet the goal. H_A should be $p < 0.96$.

3. Did not check $nq = (200)(0.04) = 8$. Since $nq < 10$, the Success/Failure condition is violated. Didn't check the 10% condition.

4. $\hat{p} = \dfrac{188}{200} = 0.94$; $SD(\hat{p}) = \sqrt{\dfrac{pq}{n}} = \sqrt{\dfrac{(0.96)(0.04)}{200}} \approx 0.014$. The student used $\hat{p}$ and $\hat{q}$.

5. z is incorrect; should be $z = \dfrac{0.94 - 0.96}{0.014} \approx -1.43$

6. $P = P(z < -1.43) = 0.076$

7. There is only weak evidence that the new system has failed to meet the goal.

21. a) Let p = the percentage of children with genetic abnormalities. $H_0: p = 0.05$ vs. $H_A: p > 0.05$

b) SRS (not clear from information provided); $384 < 10\%$ of all children; $np = (384)(0.05) = 19.2 > 10$ and $nq = (384)(0.95) = 364.8 > 10$.

c) $z = 6.28$, $P < 0.0001$.

d) If 5% of children have genetic abnormalities, the chance of observing 46 children with genetic abnormalities in a random sample of 384 children is essentially 0.

e) Reject H_0. There is strong evidence that more than 5% of children have genetic abnormalities.

f) We don't know that environmental chemicals cause genetic abnormalities, only that the rate is higher now than in the past.

23. a) Let p = the percentage of students in 2000 with perfect attendance the previous month. $H_0: p = 0.34$ vs. $H_A: p < 0.34$

b) Although not specifically stated, assume that the National Center for Education Statistics used random sampling; $8302 < 10\%$ of all students; $np = (8302)(0.34) = 2822.68 > 10$ and $nq = (8302)(0.66) = 5479.32 > 10$.

c) $z = -1.92, P = 0.027$

d) Reject H_0 at $\alpha = 0.05$. There is evidence to suggest that the percentage of students with perfect attendance in the previous month has decreased in 2000.

e) This result is statistically significant at $\alpha = 0.05$, but it's not clear that it has any practical significance since the percentage dropped only from 34% to 33%.

25. a) SRS (not clear from information provided); $1000 < 10\%$ of all workers; $n\hat{p} = 520 > 10$ and $n\hat{q} = 480 > 10$; $(0.489, 0.551)$; we are 95% confident that between 48.9% and 55.1% of workers have invested in individual retirement accounts.

b) Let p = the percentage of workers who have invested. $H_0: p = 0.44$ vs. $H_A: p \neq 0.44$; since 44% is not in the 95% confidence interval, we reject H_0 at $\alpha = 0.05$. There is strong evidence that the percentage of workers who have invested in individual retirement accounts was not 44%. In fact, our sample indicates an increase in the percentage of adults who invest in individual retirement accounts.

27. Let p = the percentage of cars with faulty emissions. $H_0: p = 0.20$ vs. $H_A: p > 0.20$; two conditions are not satisfied: $22 > 10\%$ of the population of 150 cars and $np = (22)(0.20) = 4.4 < 10$. It's not a good idea to proceed with a hypothesis test.

29. Let p = the percentage of defective products. $H_0: p = 0.03$ vs. $H_A: p \neq 0.03$; SRS (not clear from information provided); $469 < 10\%$ of all products; $np = (469)(0.03) = 14.07 > 10$ and $nq = (469)(0.97) = 454.93 > 10$; $z = -1.91, P = 0.0556$; since the P-value = 0.0556 is technically greater than 0.05, we do not reject H_0.

31. Let p = the percentage of readers interested in an online edition. $H_0: p = 0.25$ vs. $H_A: p > 0.25$; SRS; $500 < 10\%$ of all potential subscribers; $np = (500)(0.25) = 125 > 10$ and $nq = (500)(0.75) = 375 > 10$; $z = 1.24, P = 0.1076$. Since the P-value is high, we fail to reject H_0. There is insufficient evidence to suggest that the proportion of interested readers is greater than 25%. The magazine should not publish the online edition.

33. Let p = the proportion of female executives. $H_0: p = 0.40$ vs. $H_A: p < 0.40$; data are for all executives in this company and may not be able to be generalized to all companies; $np = (43)(0.40) = 17.2 > 10$ and $nq = (43)(0.60) = 25.8 > 10$; $z = -1.31, P = 0.0951$. Since the P-value is high, we fail to reject H_0. There is insufficient evidence to suggest proportion of female executives is any different from the overall proportion of 40% female employees at the company.

35. Let p = the proportion of dropouts at this high school. $H_0: p = 0.109$ vs. $H_A: p < 0.109$; assume that the students at this high school are representative of all students nationally; $1792 < 10\%$ of all students nationally; $np = (1782)(0.109) = 194.238 > 10$ and $nq = (1782)(0.891) = 1587.762 > 10$; $z = -1.46, P = 0.072$. Since the P-value = 0.072 > 0.05, we fail to reject H_0. There is insufficient evidence of a decrease in dropout rate from 10.9%.

37. Let p = the proportion of lost luggage returned the next day. $H_0: p = 0.90$ vs. $H_A: p < 0.90$; it is reasonable to think that the people surveyed were independent with regard to their luggage woes; although not stated, we will hope that the survey was conducted randomly, or at least that these air travelers are representative of all air travelers for that airline; $122 < 10\%$ of all air travelers on the airline; $np = (122)(0.90) = 109.8 > 10$ and $nq = (122)(0.10) = 12.2 > 10$; $z = -2.05, P = 0.0201$. Since the P-value is low, we reject H_0. There is evidence that the proportion of lost luggage returned the next day is lower than the 90% claimed by the airline.

39. H_0: These MBA students are exposed to unethical practices at a similar rate to others in the program ($p = 0.30$).
H_A: These students are exposed to unethical practices at a different rate than other students ($p \neq 0.30$).
There is no reason to believe that students' rates would influence others; the professor considers this class typical of other classes; $120 < 10\%$ of all students in the MBA program; 27% of $120 = 32.4$—use 32 graduates; $np = 36 > 10$ and $nq = 84 > 10$; $z = -0.717, P = 0.4733$. Since the P-value is > 0.05, we fail to reject the null hypothesis. There is little evidence that the rate at which these students are exposed to unethical business practices is different from that reported in the study.

41. a) $z = 11.8$

b) $11.8 > 3.29$, if we assume a two-sided 0.1% significance level.

c) We conclude that the percent of U.S. adults giving "quite a lot" of thought to the upcoming election is significantly different in 2008 than it was 2004.

43. a) The regulators decide that the shop is not meeting standards when it actually is.

b) The regulators certify the shop when it is not meeting the standards.

c) Type I

d) Type II

45. a) The probability of detecting that the shop is not meeting standards when they are not

b) 40 cars; larger n

c) 10%; more chance to reject H_0

d) A lot; larger problems are easier to detect.

47. a) One-tailed; we are testing to see if a decrease in the dropout rate is associated with the software.

b) H_0 : The dropout rate does not change following the use of the software ($p = 0.13$).
H_A: The dropout rate decreases following the use of the software ($p < 0.13$).

c) The professor buys the software when the dropout rate has not actually decreased.

d) The professor doesn't buy the software when the dropout rate has actually decreased.

e) The probability of buying the software when the dropout rate has actually decreased.

49. a) H_0: The dropout rate does not change following the use of the software ($p = 0.13$).
H_A: The dropout rate decreases following the use of the software ($p < 0.13$).
One student's decision about dropping out should not influence another's decision; this year's class of 203 students is probably representative of all statistics students; $203 < 10\%$ of all students; $np = (203)(0.13) = 26.39 > 10$ and $nq = (203)(0.87) = 176.61 > 10$; $z = -3.21, P = 0.0007$. Since the P-value is very low, we reject H_0. There is strong evidence that the dropout rate has dropped since use of the software program was implemented. As long as the professor feels confident that this class of statistics students is representative of all potential students, then he should buy the program.

b) The chance of observing 11 or fewer dropouts in a class of 203 is only 0.07% if the dropout rate is really 13%.

51. $\hat{p} = \dfrac{67}{500} \approx 0.134; z = 1.715, P = 0.043$; reject H_0. However, the finance department might also look at the 95% confidence interval (10.4%, 16.4%) and make calculations based on this range of possible proportions to see the potential financial impact.

Chapter 12

1. a) 1.74   b) 2.37   c) 0.0524   d) 0.0889

3. As the variability of a sample increases, the width of a 95% confidence interval increases, assuming that sample size remains the same.

5. a) ($4.382, $4.598)
   b) ($4.400, $4.580)
   c) ($4.415, $4.565)

7. a) Not correct. A confidence interval is for the mean weight gain of the population of all cows. It says nothing about individual cows.
   b) Not correct. A confidence interval is for the mean weight gain of the population of all cows, not individual cows.
   c) Not correct. We don't need a confidence interval about the average weight gain for cows in this study. We are certain that the mean weight gain of the cows in this study is 56 pounds.
   d) Not correct. This statement implies that the average weight gain varies. It doesn't.
   e) Not correct. This statement implies that there is something special about our interval, when this interval is actually one of many that could have been generated, depending on the cows that were chosen for the sample.

9. The assumptions and conditions for a *t*-interval are not met. With a sample size of only 20, the distribution is too skewed. There is also a large outlier that is pulling the mean higher.

11. a) The data are a random sample of all days; the distribution is unimodal and symmetric with no outliers.
   b) ($122.20, $129.80)
   c) We are 90% confident that the interval $122.20 to $129.80 contains the true mean daily income of the parking garage.
   d) 90% of all random samples of size 44 will produce intervals that contain the true mean daily income of the parking garage.
   e) $128 is a plausible value.

13. a) We can be more confident that our interval contains the mean parking revenue.
   b) Wider (and less precise) interval
   c) By collecting a larger sample, they could create a more precise interval without sacrificing confidence.

15. a) $2350 \pm 2.009 (59.51)$ Interval: (2230.4, 2469.6)
   b) The assumptions and conditions that must be satisfied are:
      1) Independence: probably OK.
      2) Nearly Normal condition: can't tell.
      3) sample size of 51 is large enough.
   c) We are 95% confident the interval $2230.4 to $2469.6 contains the true mean increase in sales tax revenue.

   Examples of what the interval <u>does not</u> mean: The mean increase in sales tax revenue is $2350 95% of the time. 95% of all increases in sales tax revenue increases will be between $2230.4 and $2469.6. There's 95% confidence the next small retailer will have an increase in sales tax revenue between $2230.4 and $2469.6.

17. a) Given no time trend, the monthly on-time departure rates should be independent. Though not a random sample, these months should be representative, and they're fewer than 10% of all months. The histogram looks unimodal, but slightly left-skewed; not a concern with this large sample.
   b) (80.57%, 81.80%)
   c) We can be 90% confident that the interval from 80.57% to 81.80% holds the true mean monthly percentage of on-time flight departures.

19. If in fact the mean monthly sales due to online purchases has not changed, then only 1 out of every 100 samples would be expected to have mean sales as different from the historical figure as the mean sales observed in the sample.

21. a) $897.14 to $932.86
      We are 95% confident the interval $897.14 to $932.86 contains the true mean Social Security benefit for widows and widowers in the Texas county.
   b) With a P-value of .007, the hypothesis test results are significant and we reject the null hypothesis. We conclude the mean benefit payment for the Texas county is different from the $940 for the state. The 95% confidence interval estimate is $897.14 to $932.86. Since the interval does not contain the hypothesized mean of $940, we have evidence that the mean is unlikely to be $940.

23. a) Upper-tail. They need to prove that the stands will support 500 pounds (or more) easily.
   b) The inspectors certify the stands as safe, when they are not.
   c) The inspectors decide the stands are not safe when they are.

25. a) Decrease α. This means a smaller chance of declaring the stands safe if they are not.
   b) The probability of correctly detecting that the stands can safely hold over 500 pounds
   c) Decrease the standard deviation—probably costly. Increase the number of stands tested—takes more time for testing and is costly. Increase α—more Type I errors. Make the stands stronger—costly.

27. a) $H_0: \mu = 23.3$; $H_A: \mu > 23.3$
   b) **Randomization condition:** The 40 online shoppers were selected randomly. **Nearly Normal condition:** We should examine the distribution of the sample to check for serious skewness and outliers, but with a large sample of 40 shoppers, it should be safe to proceed.
   c) 0.145
   d) If the mean age of shoppers is still 23.3 years, there is a 14.5% chance of getting a sample mean of 24.2 years or older simply from natural sampling variation.
   e) There is no evidence to suggest that the mean age of online shoppers has increased from the mean of 23.3 years.

29. a) $H_0: \mu = 55$; $H_A: \mu < 55$; **Independence assumption:** Since the times are not randomly selected, we will assume that the times are independent and representative of all the champion's times. **Nearly Normal condition:** The histogram of the times is unimodal and roughly symmetric; P-value = 0.235; fail to reject H_0. There is insufficient evidence to conclude the mean time is less than 55 seconds. They should not market the new ski wax.
   b) Type II error. They won't market a competitive wax and thus lose the potential profit from having done so.

31. $H_0: \mu = 150$; $H_A: \mu < 150$; **Randomization condition:** The 44 phones in the sample were randomly selected. **Nearly Normal condition:** We don't have the actual data, so we cannot look at a graphical display. But since the sample is fairly large, it is safe to proceed. P-value < 0.001; reject H_0. There is strong evidence that the mean range of this type of phone is not 150 feet. Our evidence suggests that the mean range is actually less than 150 feet.

33. a) Random sample; the Nearly Normal Condition seems reasonable from a Normal probability plot. The histogram is roughly unimodal and symmetric with no outliers.
   b) (1187.9, 1288.4) chips
   c) Based on this sample, the mean number of chips in an 18-ounce bag is between 1187.9 and 1288.4, with 95% confidence. The *mean* number of chips is clearly greater than 1000. However, if the claim is about individual bags, then it's not necessarily true. If the mean is 1188 and the SD is near 94, then 2.5% of the bags will have fewer than 1000 chips, using the Normal model. If in fact the mean is 1288, the proportion below 1000 will be less than 0.1%, but the claim is still false.

35. **Independence assumption:** We assume that these mutual funds were selected at random and that 35 funds are less than 10% of all value funds. **Nearly Normal condition:** A histogram shows a nearly normal distribution.
    $H_0: \mu = 8$; $H_A: \mu > 8$; P-value = 0.201; fail to reject H_0. There is insufficient evidence that the mean 5-year return is greater than 8% for value funds.

37. Given this confidence interval, we cannot reject the null hypothesis of a mean $200 collection using $\alpha = 0.05$. However, the confidence interval suggests that there may be a large upside potential. The collection agency may be collecting as much as $250 per customer on average, or as little as $190 on average. If the possibility of collecting $250 on average is of interest to them, they may want to collect more data.

39. Yes, there is a large ($50) upside potential. The larger trial will likely narrow the confidence interval and make the decision clearer.

41. a) $H_0: \mu = 100$; $H_A: \mu < 100$
    b) Different samples give different means; this is a fairly small sample. The difference may be due to natural sampling variation.
    c) Batteries selected are a SRS (representative); fewer than 10% of the company's batteries; lifetimes are approximately Normal.
    d) $t = -1.0$; P-value = 0.167; do not reject H_0. This sample does not show that the average life of the batteries is significantly less than 100 hours.
    e) Yes; Type II.

43. a) The sample is random and the data are alleged to be nearly normal, so at $\alpha = 0.025$ we can reject the null hypothesis that the mean is 0.08 ppm and conclude that it is greater.
    b) A type I error would be deciding (as we did) that the mean Mirex contamination level is greater than 0.08 ppm when in fact it isn't. The boycott might harm the salmon producers needlessly. A type II error would be failing to reject the null hypothesis when it's false. In this case, the boycott would likely not take place, but the public would be exposed to the risk of eating salmon with elevated levels of Mirex.

45. a) The histogram of the lab fees shows 2 extreme outliers, so the conditions for inference are violated.
    b) $H_0: \mu = 55$
       $H_A: \mu > 55$
       $t = (63.25 - 55)/8.35 = 0.99$ $P = 0.172$
       Do not reject H_0 because $0.172 > 0.05$. We do not have evidence (at $\alpha = 0.05$) to conclude that the average time spent by students in the lab is greater than 55 minutes.
    c) $t = (61.5 - 55)/3.03 = 2.14$ $P = 0.030$
       When the 2 outliers are deleted, the decision is reject H_0 because $.03 < 0.05$ We have evidence at $\alpha < 0.05$ that the average time spent by students in the lab is greater than 55 minutes.
    d) Outliers, especially extreme outliers, are troublesome for inference because their presence violates the Nearly Normal Condition and assumption of a homogeneous population. When extreme outliers are present, the results of estimation and hypothesis testing can change dramatically. Testing and estimation should therefore be conducted both with and without outliers to see if changes do occur in the statistical results. When an outlier is deleted, an observation from the population is eliminated, so some researchers question whether it is appropriate to delete outliers. Addressing outliers and performing the analysis both with and without those outliers provides a more thorough analysis of the data.

47. a) The assumptions and conditions that must be satisfied are:
       The data come from a nearly normal distribution.
       The air samples were selected randomly, and there is no bias present in the sample.

    b) The histogram of air samples is not nearly normal, but the sample size is large, so inference is OK.

49. a) $14.90 - 11.6$ or ± 3.3 miles per hour
    b) The sample size for ME = 2 should be increased to
       $1.96 \times 8/2 = 7.84$
       $(7.84)^2 = 61.466 = 62$

51. a) Interval: $653 to $707
    b) $H_0: \mu = 650$
       $H_A: \mu \neq 650$
       $P = 0.031$
       Reject H_0. There is strong evidence that the mean audit cost is significantly different from $650.
    c) The confidence interval does not contain the hypothesized mean of $650. This provides evidence that the current year's mean audit cost is significantly different from $650.

53. a) The timeplot shows no pattern, so it seems that the measurements are independent. Although this is not a random sample, an entire year is measured, so it is likely that we have representative values. We certainly have fewer than 10% of all possible wind readings. The histogram appears nearly normal.
    b) Testing $H_0: \mu = 8$ mph vs. $H_A: \mu > 8$ mph with 1113 df gives $t = 0.1663$ for a P-value of about 0.43. Even though the observed mean wind speed is over 8 mph, we can't be confident that the true annual mean wind speed exceeds 8 mph. We would not recommend building a turbine at this site.

Chapter 13

1. The P-value is too high to reject H_0 at any reasonable α-level.

3. a) 2.927 points
   b) Larger
   c) We are 95% confident that the mean score for the CPMP math students will be between 5.573 and 11.427 points higher on this assessment than the mean score of the traditional students.
   d) Since the entire interval is above 0, there is strong evidence that students who learn with CPMP will have higher mean scores in applied algebra than those in traditional programs.

5. a) $H_0: \mu_C - \mu_T = 0$; $H_A: \mu_C - \mu_T \neq 0$
   b) If the mean scores for the CPMP and traditional students are really equal, there is less than a 1 in 10,000 chance of seeing a difference as large or larger than the observed difference of 9.4 points just from natural sampling variation.
   c) There is strong evidence that the CPMP students have a different mean score than the traditional students. The evidence suggests that the CPMP students have a higher mean score.

7. $H_0: \mu_C - \mu_T = 0$; $H_A: \mu_C - \mu_T \neq 0$.
   $P = 0.1602$; fail to reject H_0. There is no evidence that the CPMP students have a different mean score on the word problems test than the traditional students.

9. a) (1.36, 4.64); df = 33.1
   b) Since the CI does not contain 0, there is evidence that Route A is faster on average.

11. a) $H_0: \mu_C - \mu_A = 0$; $H_A: \mu_C - \mu_A \neq 0$
    b) Independent groups assumption: The percentage of sugar in the children's cereals is unrelated to the percentage of sugar in adult cereals. Randomization condition: It is reasonable to assume that the cereals are representative of all children's cereals and adult cereals, in regard to sugar content. Nearly Normal condition: The histogram of adult cereal sugar content is skewed to the right, but the sample sizes are reasonably large. The Central Limit Theorem allows us to proceed.
    c) (32.15, 40.82)%

d) Since the 95% confidence interval does not contain 0, we can conclude that the mean sugar content for the two cereals is significantly different at the 5% level of significance.

13. a) $H_0: \mu_C - \mu_D = 0$; $H_A: \mu_C - \mu_D \neq 0$
    b) $(-0.256, 1.894)$
    c) Since the confidence interval contains 0, there is insufficient evidence to conclude that the mean return over a 5-year period is different for consistent style funds as opposed to style drifters.

15. $H_0: \mu_N - \mu_C = 0$; $H_A: \mu_N - \mu_C > 0$ Independent groups assumption: Student scores in one group should not have an impact on the scores of students in the other group. Randomization condition: Students were randomly assigned to classes. Nearly Normal condition: The histograms of the scores are unimodal and symmetric. $P = 0.023$; reject H_0. There is evidence that the students taught using the new activities have a higher mean score on the reading comprehension test than the students taught using traditional methods.

17. a) $H_0: \mu_L - \mu_S = 0$; $H_A: \mu_L - \mu_S \neq 0$
    b) Independent groups assumption: pH levels from the two types of streams are independent. Independence assumption: Since we don't know if the streams were chosen randomly, assume that the pH level of one stream does not affect the pH of another stream. This seems reasonable. Nearly Normal condition: The boxplots provided show that the pH levels of the streams may be skewed (since the median is either the upper or lower quartile for the shale streams and the lower whisker of the limestone streams is stretched out), and there are outliers. However, since there are 133 degrees of freedom, we know that the sample sizes are large. It should be safe to proceed.
    c) $P \leq 0.0001$; reject H_0. There is strong evidence that the streams with limestone substrates have mean pH levels different than those of streams with shale substrates. The limestone streams are less acidic on average.

19. a) If the mean memory scores for people taking ginkgo biloba and people not taking it are the same, there is a 93.74% chance of seeing a difference in mean memory score this large or larger simply from natural sampling variability.
    b) Since the P-value is so high, there is no evidence that the mean memory test score for ginkgo biloba users is higher than the mean memory test score for non-users.
    c) Type II

21. a) Males: $(18.67, 20.11)$ pegs; females: $(16.95, 18.87)$ pegs
    b) It may appear to suggest that there is no difference in the mean number of pegs placed by males and females, but a two-sample t-interval should be constructed to assess the difference in mean number of pegs placed.
    c) $(0.29, 2.67)$ pegs
    d) We are 95% confident that the mean number of pegs placed by males is between 0.29 and 2.67 pegs higher than the mean number of pegs placed by females.
    e) Two-sample t-interval.
    f) If you attempt to use two confidence intervals to assess a difference in means, you are actually adding standard deviations. But it's the variances that add, not the standard deviations. The two-sample difference of means procedure takes this into account.

23. a) $H_0: \mu_A - \mu_N = 0$; $H_A: \mu_A - \mu_N > 0$
    b) We are 95% confident that the mean number of runs scored by American League teams is between 0.62 and 0.40 runs higher than the mean number of runs scored by National League teams.
    c) $t = 2.42$; $P = 0.013$; Reject H_0.
    d) There is evidence that the mean number of runs scored per game by the AL is greater than in the National League.

25. a) $H_0: \mu_N - \mu_S = 0$; $H_A: \mu_N - \mu_S \neq 0$. $t = 6.47$, df $= 53.49$, $P < 0.001$
    Since the P-value is low, we reject H_0. There is strong evidence that the mean mortality rate is different for towns north and south of Derby. There is evidence that the mortality rate north of Derby is higher.
    b) Since there is an outlier in the data north of Derby, the conditions for inference are not satisfied, and it is risky to use the two-sample t-test. The outlier should be removed, and the test should be performed again.

27. A two-sample t-procedure is not appropriate for these data because the two groups are not independent. They are before and after satisfaction scores for the same workers.

29. Independent groups assumption: assume that orders in June are independent of orders in August. Independence assumption: orders were a random sample. Nearly normal condition hard to check with small sample, but no outliers. $H_0: \mu_J - \mu_A = 0$; $H_A: \mu_J - \mu_A \neq 0$; $t = -1.17$; $P = 0.274$; fail to reject H_0. Thus, although the mean delivery time during August is higher, the difference in delivery time from June is not significant. A larger sample may produce a different result.

31. a) We are 95% confident that the mean number of ads remembered by viewers of shows with violent content will be between 1.6 and 0.6 lower than the mean number of brand names remembered by viewers of shows with neutral content.
    b) If they want viewers to remember their brand names, they should consider advertising on shows with neutral content, as opposed to shows with violent content.

33. a) She might attempt to conclude that the mean number of brand names recalled is greater after 24 hours.
    b) The groups are not independent. They are the same people, asked at two different time periods.
    c) A person with high recall right after the show might tend to have high recall 24 hours later as well. Also, the first interview may have helped the people to remember the brand names for a longer period of time than they would have otherwise.
    d) Randomly assign half of the group watching that type of content to be interviewed immediately after watching, and assign the other half to be interviewed 24 hours later.

35. a) Using df ≥ 7536, we are 95% confident that the mean score in 2000 was between 0.61 and 5.39 points lower than the mean score in 1996. Since 0 is not contained in the interval, this provides evidence that the mean score has decreased from 1996 to 2000.
    b) Both sample sizes are very large, which will make the standard errors of these samples very small. They are both likely to be very accurate. The difference in sample size shouldn't make you any more certain or any less certain.

37. a) The differences that were observed between the group of students with Internet access and those without were too great to be attributed to natural sampling variation.
    b) Type I
    c) No. There may be many other factors.
    d) It might be used to market computer services to parents.

39. a) 8759 pounds
    b) Independent groups assumption: sales in different seasons should be independent. Randomization Condition: Not a random sample of weeks, but it is of stores. Nearly Normal condition: can't verify, but we will proceed cautiously. We are 95% confident that the interval 3630.54 to 13,887.39 pounds contains the true difference in mean sales between winter and summer.
    c) Weather and sporting events may impact pizza sales.

41. $H_0: \mu_2 - \mu_5 = 0$; $H_A: \mu_2 - \mu_5 \neq 0$ Independent groups assumption: The two heats were independent. Randomization condition: Runners were randomly assigned. Nearly Normal

condition: Boxplots show an outlier in the distribution of times in heat 2. Perform the test twice, once with the outlier and once without. With outlier: $t = 0.035$; df $= 10.82$; P $= 0.972$; fail to reject H_0. Without the outlier in heat 2: $t = -1.141$; df $= 8.83$; P $= 0.287$; fail to reject H_0. Regardless of whether the outlier is included or excluded, there is no evidence that the mean time to finish differs between the two heats.

43. H_0: $\mu_S - \mu_R = 0$; H_A: $\mu_S - \mu_R > 0$. Assuming the conditions are satisfied, it is appropriate to model the sampling distribution of the difference in means with a Student's t-model, with 7.03 degrees of freedom (from the approximation formula). $t = 4.57$; P $= 0.0013$; reject H_0. There is strong evidence that the mean ball velocity for Stinger tees is higher than the mean velocity for regular tees.

45. a) H_0: $\mu_M - \mu_R = 0$; H_A: $\mu_M - \mu_R > 0$. Independent groups assumption: The groups are not related in regards to memory score. Randomization condition: Subjects were randomly assigned to groups. Nearly Normal condition: We don't have the actual data. We will assume that the distributions of the populations of memory test scores are Normal. $t = -0.70$; df $= 45.88$; P $= 0.7563$; fail to reject H_0. There is no evidence that the mean number of objects remembered by those who listen to Mozart is higher than the mean number of objects remembered by those who listen to rap music.

b) We are 90% confident that the mean number of objects remembered by those who listen to Mozart is between 0.189 and 5.352 objects lower than the mean of those who listened to no music.

47. a) Independent groups assumption: 3 and 5 year returns are not independent. These data are paired and not suited for a two-sample t-test. Randomization Condition: random sample of funds. Nearly Normal condition: histograms are unimodal and symmetric with no outliers.

b) H_0: $\mu_5 - \mu_3 = 0$; H_A: $\mu_5 - \mu_3 > 0$

c) Not appropriate for a two-sample t-test.

d) None. Can't use the methods of this chapter on these data.

49. a) Independent groups assumption: the prices in the two towns are not related. Randomization condition: each sample was a random sample of prices. Nearly Normal condition: both histograms are reasonably unimodal and symmetric with no outliers, so will use the two-sample t-test.

b) H_0: $\mu_1 = \mu_2$; H_A: $\mu_1 \neq \mu_2$

c) $t = -0.58$; P $= 0.567$; fail to reject H_0

d) We conclude that the mean price of homes in these two towns is not significantly different.

51. a) Independent groups assumption: the home runs hit in different leagues are independent. Randomization condition: not a random sample, but we will assume it's representative. Nearly Normal condition: both histograms are reasonably symmetric with no outliers. Unimodality is questionable, but we will proceed with caution. We will use the two-sample t-test.

b) H_0: $\mu_{AL} - \mu_{NL} = 0$; H_A: $\mu_{AL} - \mu_{NL} > 0$

c) $t = 0.90$; P $= 0.376$

d) There is insufficient evidence to conclude that the mean number of home runs is different in the two leagues. We fail to reject the null hypothesis.

Chapter 14

1. a) Randomly assign 50 hens to each of the two kinds of feed. Compare production at the end of the month.

b) Give all 100 hens the new feed for 2 weeks and the old feed for 2 weeks, randomly selecting which feed the hens get first. Analyze the differences in production for all 100 hens.

c) Matched pairs; Hens vary in egg production, and the matched-pairs design will control for that.

3. a) Show the same people ads with and without sexual images and record how many products they remember in each group. Randomly decide which ads a person sees first. Examine the differences for each person.

b) Randomly divide volunteers into two groups. Show one group ads with sexual images and the other group ads without. Compare how many products each group remembers.

5. a) The paired t-test is appropriate. The labor force participation rate for two different years was paired by city.

b) Since the P-value $= 0.0244$, there is evidence of a difference in the average labor force participation rate for women between 1968 and 1972. The evidence suggests an increase in the participation rate for women.

7. a) The paired t-test is appropriate since we have pairs of Fridays in 5 different months. Data from adjacent Fridays within a month may be more similar than randomly chosen Fridays.

b) Since the P-value $= 0.0212$, there is evidence that the mean number of cars on the M25 motorway on Friday the 13th is less than the mean number of cars on the previous Friday.

c) We don't know if these Friday pairs were selected at random. Obviously, if these are the Fridays with the largest differences, this will affect our conclusion. The Nearly Normal condition appears to be met by the differences, but the sample size of five pairs is small.

9. Adding variances requires that the variables be independent. These price quotes are for the same cars, so they are paired. Drivers quoted high insurance premiums by the local company will be likely to get a high rate from the online company, too.

11. a) The histogram—we care about differences in price.

b) Insurance cost is based on risk, so drivers are likely to see similar quotes from each company, making the differences relatively smaller.

c) The price quotes are paired; they were for a random sample of fewer than 10% of the agent's customers; the histogram of differences looks approximately Normal.

13. H_0: $\mu(\text{Local} - \text{Online}) = 0$ vs. H_A: $\mu(\text{Local} - \text{Online}) > 0$ $t = 0.826$ with 9 df. With a P-value of 0.215, we cannot reject the null hypothesis. These data don't provide evidence that online premiums are lower, on average.

15. a) Even if the individual times show a trend of improving speed over time, the differences may well be independent of each other. They are subject to random year-to-year fluctuations, and we may believe that these data are representative of similar races. We don't have any information with which to check the Nearly Normal condition.

b) We are 95% confident that the interval -15.49 (men are faster) to $+11.25$ minutes (women are faster) contains the true mean time difference between women's wheelchair times and men's running times.

c) No. The interval contains zero, so we would not reject a null hypothesis of no mean difference at a significance level of 0.05.

17. a) H_0: $\mu_d = 0$ vs. H_A: $\mu_d \neq 0$

b) $t_{144} = \dfrac{22.7}{113.6/\sqrt{145}} = 2.406$ 2-sided P $= 0.017$

We are able to reject the null hypothesis (with a P-value of 0.017) and conclude that mean number of keystrokes per hour has changed.

c) 95% CI for mean keystrokes per hour $22.7 \pm t_{0.025, 144} s/\sqrt{n} = (4.05, 41.35)$.

19. a) **Paired data assumption:** We are testing the employees, before and after implementation of the fitness program. **Randomization condition:** These employees are likely to be representative of all employees. **Normal population assumption:** We don't have the list of individual differences, so we can't look at a histogram. The sample is reasonably large, so we may proceed.

b) (10.05, 15.95)

c) We are 95% confident that the mean increase in productivity for employees after the fitness program is between 10.05 and 15.95 points.

21. a) Same cows before and after injection; the cows should be representative of others of their breed; cows are independent of each other; don't know about nearly Normal differences.

b) (12.66, 15.34)

c) Based on this sample, with 95% confidence, the average increase in milk production for Ayrshire cows given BST is between 12.66 and 15.34 pounds per day.

d) A 25% increase is an increase of 11.75 lbs. Since the confidence interval lies above this value, the farmer should pay for the BST.

23. **Paired data assumption:** The data are paired by city. **Randomization condition:** These cities might not be representative of all European cities, so be cautious in generalizing the results. **Normal population assumption:** A histogram of differences between January and July mean temperature is roughly unimodal and symmetric. We are 90% confident that the average high temperature in European cities in July is an average of between 32.3°F to 41.3°F higher than in January.

25. a) **Paired data assumption:** The data are paired by type of exercise machine. **Randomization condition:** Assume that the men and women participating are representative of all men and women in terms of number of minutes of exercise required to burn 200 calories. **Nearly Normal condition:** The histogram of differences between women's and men's times is roughly unimodal and symmetric. We are 95% confident that women take an average of 4.8 to 15.2 minutes longer to burn 200 calories than men when exercising at a light exertion rate.

b) **Nearly Normal condition:** There is no reason to think that this histogram does not represent differences drawn from a Normal population. We are 95% confident that women exercising with light exertion take an average of 4.9 to 20.4 minutes longer to burn 200 calories than women exercising with hard exertion.

c) Since these data are averages, we expect the individual times to be more variable. Our standard error would be larger, resulting in a larger margin of error.

27. a) **Paired data assumption:** The data are before and after job satisfaction rating for the same workers. **Randomization condition:** The workers were randomly selected to participate. **Nearly Normal condition:** A histogram of differences between before and after job satisfaction ratings is roughly unimodal and symmetric.

b) $H_0: \mu_d = 0$; $H_A: \mu_d > 0$; $t = 3.60$; df $= 9$; P-value $= 0.0029$; reject H_0. There is evidence that the mean job satisfaction rating has increased since the implementation of the exercise program.

29. a) $H_0: \mu_d = 0$; $H_A: \mu_d > 0$
Paired data assumption: The mileage is paired by car. **Randomization condition:** We randomized the order in which the different types of gasoline were used in each car. **Normal population assumption:** A histogram of differences between premium and regular is roughly unimodal and symmetric. $t = 4.47$; df $= 9$; P-value $= 0.0008$; reject H_0. There is strong evidence of a mean increase in gas mileage between regular and premium.

b) We are 90% confident that the mean increase in gas mileage when using premium rather than regular gasoline is between 1.18 and 2.82 miles per gallon.

c) Premium gasoline costs more than regular gasoline. The increase in price might outweigh the increase in mileage.

d) With $t = 1.25$ and a P-value $= 0.1144$, we would have failed to reject the null hypothesis and concluded that there was no evidence of a mean difference in mileage. The variation in performance of individual cars is greater than the variation related to the type of gasoline. This masked the true difference in mileage due to the gasoline. (Not to mention the fact that the two-sample test is not appropriate because we don't have independent samples!)

31. a) **Randomization condition:** These stops are probably representative of all such stops for this type of car, but not for all cars. **Nearly Normal condition:** A histogram of the stopping distances is roughly unimodal and symmetric. We are 95% confident that the mean dry pavement stopping distance for this type of car is between 133.6 and 145.2 feet.

b) **Independent groups assumption:** The wet pavement stops and dry pavement stops were made under different conditions and not paired in any way. **Randomization condition:** These stops are probably representative of all such stops for this type of car, but not for all cars. **Nearly Normal condition:** The histogram of wet pavement stopping distances is more uniform than unimodal, but no outliers. We are 95% confident that the mean stopping distance on wet pavement is between 51.4 and 74.6 feet longer than the mean stopping distance on dry pavement.

33. $H_0: \mu_d = 0$; $H_A: \mu_d \neq 0$
Paired data assumption: The data are paired by power plant. **Randomization condition:** The power plants were randomly sampled. **Normal population assumption:** The histogram of differences is roughly unimodal and symmetric. There are several outliers, but they are symmetric and removing them does not change the conclusion.
$t = 1.14$; df $= 21$; P-value $= 0.267$; fail to reject H_0. We have no evidence of any significant change in the CO_2 emissions for the power plants in Texas.

35. a) $H_0: \mu_A = 30$; $H_A: \mu_A > 30$
Randomization condition: Assume that these players are representative of all Little League pitchers. **Nearly Normal condition:** The histogram of the number of strikes thrown after the training is roughly unimodal and symmetric. $t = 6.06$; df $= 19$; P-value < 0.0001; reject H_0. There is strong evidence that the mean number of strikes that Little Leaguers can throw after the training is more than 30. (This test says nothing about the effectiveness of the training, just that Little Leaguers can throw more than 60% strikes on average after completing the training. This might not be an improvement.)

b) $H_0: \mu_d = 0$; $H_A: \mu_d > 0$
Paired data assumption: The data are paired by pitcher. **Randomization condition:** Assume that these players are representative of all Little League pitchers. **Normal population assumption:** The histogram of differences is roughly unimodal and symmetric. $t = 0.135$; df $= 19$; P-value $= 0.4472$; fail to reject H_0. There is no evidence of a mean difference in number of strikes thrown before and after the training. The training does not appear to be effective.

37. Assumptions and conditions for inference are satisfied. Using a t-test for paired differences, $t = -0.86$ and two-tailed $P = 0.396$. With a P-value so high, we fail to reject the null hypothesis of no mean difference. There is no evidence that sexual images in ads affects people's ability to remember the product being advertised.

39. $H_0: \mu_d = 0$; $H_A: \mu_d > 0$
Paired data assumption: The data are paired by store. **Randomization condition:** Assume the stores are representative of stores in the chain. **Normal population assumption:** The distribution of differences is slightly bimodal, but sample size

is small. We will proceed with caution. $t = -0.43$; df = 14; P-value = 336; fail to reject H_0. We cannot provide evidence that sales have increased.

41. H_0: $\mu_D = 0$ vs. H_A: $\mu_D \neq 0$. Data are paired by brand; brands are independent of each other; boxplot of differences shows an outlier (100) for Great Value:

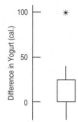

With the outlier included, the mean difference (Strawberry – Vanilla) is 12.5 calories with a *t*-stat of 1.332, with 11 df, for a P-value of 0.2098. Deleting the outlier, the difference is even smaller, 4.55 calories with a *t*-stat of only 0.833 and a P-value of 0.4241. With P-values so large, we do not reject H_0. We conclude that the data do not provide evidence of a difference in mean calories.

43. H_0: $\mu_d = 0$; H_A: $\mu_d \neq 0$
Data are paired. Differences are roughly unimodal and symmetric. Mean difference is -18.33; $t = -1.01$; P = 0.359. Can not reject null hypothesis that means bankruptcies are the same.

45. a) Yes. The percent change is computed using paired data. Same cities both years.
b) Data are paired. Differences (both actual and percent change) are roughly unimodal and symmetric. Actual differences: H_0: $\mu_d = 0$; H_A: $\mu_d \neq 0$
Mean difference: (3Q2007 − 3Q2006) = −10.10; $t = -0.67$; P = 0.535; do not reject H_0. There is no evidence the third quarter 2006 airfares are significantly different from the third quarter 2007 airfares.
t-test on the percent change:
H_0: $\mu = 0$; H_A: $\mu \neq 0$
The average percent change in airfare between third quarter 2006 and third quarter 2007 is −2.56%. Airfares were lower by 2.56% in 2007 vs 2006. $t = -0.67$; P = 0.520; do not reject H_0. There is no evidence the average percent change in airfare differs between the third quarter 2006 and third quarter 2007.
Both tests provide the same statistical conclusion. However, in general if there were extreme or outlying paired observations that skewed the mean difference, then the mean percent change might be a more appropriate test because it would eliminate the variability due to extreme differences.

Chapter 15

1. a) Chi-square test of independence; One sample, two variables. We want to see if the variable *Account type* is independent of the variable *Trade type*.
b) Some other statistical test; the variable *Account size* is quantitative, not counts.
c) Chi-square test of homogeneity; we have two samples (residential and nonresidential students) and one variable, *Courses*. We want to see if the distribution of *Courses* is the same for the two groups.

3. a) 10
b) Goodness-of-fit
c) H_0: The die is fair. (All faces have $p = 1/6$.)
H_A: The die is not fair. (Some faces are more or less likely to come up than others.)

d) Count data; rolls are random and independent of each other; expected frequencies are all greater than 5.
e) 5
f) $\chi^2 = 5.600$; P = 0.3471
g) Since P = 0.3471 is high, fail to reject H_0. There is not enough evidence to conclude that the die is unfair.

5. a) Weights are quantitative, not counts.
b) Count the number of each type of nut, assuming the company's percentages are based on counts rather than weights (which is not clear).

7. a) Goodness-of-fit
b) Count data; assume the lottery mechanism uses randomization and guarantees independence; expected frequencies are all greater than 5.
c) H_0: Likelihood of drawing each numeral is equal.
H_A: Likelihood of drawing each numeral is *not* equal.
d) $\chi^2 = 6.46$; df = 9; P = 0.693; fail to reject H_0.
e) The P-value says that if the drawings were in fact fair, an observed chi-square value of 6.46 or higher would occur about 69% of the time. This is not unusual at all, so we won't reject the null hypothesis that the values are uniformly distributed. The variation that we observed seems typical of that expected if the digits were drawn equally likely.

9. a) 40.2%   b) 8.1%   c) 62.2%   d) 285.48
e) H_0: Survival was independent of status on the ship.
H_A: Survival was not independent of the status.
f) 3
g) We reject the null hypothesis. Survival depended on status. We can see that first-class passengers were more likely to survive than passengers of any other class.

11. a) Independence
b) H_0: College enrollment is independent of birth order.
H_A: There is an association between college enrollment and birth order.
c) Count data; not a random sample of students, but assume that it is representative; expected counts are low for both the Social Science and Professional Colleges for both third and fourth or higher birth order. We'll keep an eye on these when we calculate the standardized residuals.
d) 9
e) With a P-value this low, we reject the null hypothesis. There is some evidence of an association between birth order and college enrollment.
f) Unfortunately, 3 of the 4 largest standardized residuals are in cells with expected counts less than 5. We should be very wary of drawing conclusions from this test.

13. a) Chi-square test for homogeneity
b) Count data; assume random assignment to treatments (although not stated); expected counts are all greater than 5.
c) H_0: The proportion of infection is the same for each group.
H_A: The proportion of infection is different among the groups.
d) $\chi^2 = 7.776$; df = 2; P = 0.02; reject H_0.
e) Since the P-value is low, we reject the null hypothesis. There is strong evidence of difference in the proportion of urinary tract infections for cranberry juice drinkers, lactobacillus drinkers, and women that drink neither of the two beverages.
f) The standardized residuals are:

	Cranberry	Lactobacillus	Control
Infection	−1.87276	1.191759	0.681005
No infection	1.245505	−0.79259	−0.45291

The significant difference appears to be primarily due to the success of cranberry juice.

15. a) Independence
    b) H_0: Age is independent of frequency of shopping at this department store.
       H_A: Age is not independent of frequency of shopping at this department store.
    c) Count data; assume survey was conducted randomly (not specifically stated); expected counts are all greater than 5.
    d) Since the P-value is low, we reject the null hypothesis. There is evidence of an association between age and frequency of shopping at this department store.
    e) Given the negative residuals for the low frequency categories among the older women and the positive residuals for the higher frequency categories among the older women, we conclude that older women in this survey shop more frequently at this department store than expected.

17. a) $P = 0.3767$. With a P-value this high, we fail to reject. There is not enough evidence to conclude that either men or women are more likely to make online purchases of books.
    b) Type II.
    c) $(-4.09\%, 10.86\%)$

19. a) $P < 0.001$. There is strong evidence that the proportions of the two groups are not equal.
    b) $(0.096, 0.210)$ (proportion of old $-$ proportion of young)

21. a) No, the P-value $= 0.300$.
    b) 90% confidence interval is $(-0.0019, 0.0082)$.

23. H_0: Marital status is independent of frequency of shopping.
    H_A: Marital status is not independent of frequency of shopping.
    Count data; assume survey was conducted randomly (not specifically stated); expected counts are all greater than 5.
    $\chi^2 = 23.858$; df $= 6$; P-value $= 0.001$
    Since the P-value is low, reject the null hypothesis. There is strong evidence of an association between marital status and frequency of shopping at this department store. Based on the residuals, married customers shopped at this store more frequently than expected, and more single women shopped never/hardly ever than expected.

25. a) Chi-square test for homogeneity.
    b) Count data; executives were surveyed randomly; expected counts are all greater than 5.
    c) H_0: The distribution of attitudes about critical factors affecting ethical and legal accounting practices was the same in 2000 and 2006.
       H_A: The distribution of attitudes about critical factors affecting ethical and legal accounting practices was not the same in 2000 and 2006.
    d) $\chi^2 = 4.030$; df $= 4$; $P = 0.4019$
    e) Since P-value is high, we fail to reject the null hypothesis. There is no evidence of a change in the distribution of attitudes about factors affecting ethical and legal accounting practices between 2000 and 2006.

27. a) Chi-square test of independence
    b) Count data; assume survey was conducted randomly (not specifically stated); expected counts are all greater than 5.
    c) H_0: Emphasis on quality is independent of frequency of shopping.
       H_A: Emphasis on quality is not independent of frequency of shopping.
    d) $\chi^2 = 30.007$; df $= 6$; $P < 0.001$
    e) Since the P-value is low, reject the null hypothesis. There is strong evidence of an association between emphasis on quality and frequency of shopping at this department store.

29. a)

	Men	Women
Excellent	6.667	5.333
Good	12.778	10.222
Average	12.222	9.778
Below Average	8.333	6.667

   Count data; assume that these executives are representative of all executives that have ever completed the program; expected counts are all greater than 5.
    b) Decreased from 4 to 3.
    c) $\chi^2 = 9.306$; $P = 0.0255$. Since the P-value is low, we reject the null hypothesis. There is evidence that the distributions of responses about the value of the program for men and women executives are different.

31. H_0: There is no association between race and the section of the apartment complex in which people live.
    H_A: There is an association between race and the section of the apartment complex in which people live.
    Count data; assume that the recently rented apartments are representative of all apartments in the complex; expected counts are all greater than 5.
    $\chi^2 = 14.058$; df $= 1$; $P < 0.001$
    Since the P-value is low, we reject the null hypothesis. There is strong evidence of an association between race and the section of the apartment complex in which people live. An examination of the components shows us that whites are more likely to rent in Section A (component $= 6.2215$) and that blacks are more likely to rent in Section B (component $= 5.0517$).

33. $\hat{p}_B - \hat{p}_A = 0.206$
    95% CI $= (0.107, 0.306)$

35. a) Chi-square test for independence.
    b) Count data; assume that the sample was taken randomly; expected counts are all greater than 5.
    c) H_0: Outsourcing is independent of industry sector.
       H_A: There is an association between outsourcing and industry sector.
    d) $\chi^2 = 2815.968$; df $= 9$; P-value is essentially 0.
    e) Since the P-value is so low, we reject the null hypothesis. There is strong evidence of an association between outsourcing and industry sector.

37. a) Chi-square test for homogeneity. (Could be independence if the categories are considered exhaustive.)
    b) Count data; assume that the sample was taken randomly; expected counts are all greater than 5.
    c) H_0: The distribution of employee job satisfaction level attained is the same for different management styles.
       H_A: The distribution of employee job satisfaction level attained is different for different management styles.
    d) $\chi^2 = 178.453$; df $= 12$; P-value is essentially 0.
    e) Since the P-value is so low, we reject the null hypothesis. There is strong evidence that the distribution of employee job satisfaction level attained is different across management styles. Generally, exploitative authoritarian management is more likely to have lower levels of employee job satisfaction than consultative or participative styles.

39. Assumptions and conditions for test of independence satisfied. H_0: Reading online journals or blogs is independent of generation. H_A: There is an association between reading online journals or blogs and generation. $\chi^2 = 48.408$; df $= 8$; $P < .001$. We reject the null hypothesis and conclude that reading online journals or blogs is not independent of generational age.

41. Chi-square test of homogeneity (unless these two types of firms are considered the only two types in which case it's a test of independence).

Count data; assume that the sample was random; expected counts are all greater than 5.

H_0: Systems used have same distribution for both types of industry.

H_A: Distributions of type of system differs in the two industries.

$\chi^2 = 157.256$; df = 3; P-value is essentially 0.

Since the P-value is low, we can reject the null hypothesis and conclude that the type of ERP system used differs across industry type. Those in manufacturing appear to use more of the inventory management and ROI systems.

43. Chi-square test of independence.

Count data; we assume that this time period is representative; expected counts are less than 5 in three cells, but very close.

H_0: Economic growth is independent of region of the United States.

H_A: Economic growth is not independent of region of the United States.

$\chi^2 = 19.0724$; df = 3; P < 0.001.

The P-value is low so we reject the null hypothesis and conclude that economic growth is not independent of region. The result seems clear enough even though the expected counts of 3 cells were slightly below 5.

Chapter 16

1. a)

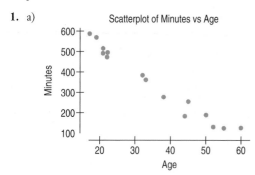

b) This scatterplot appears to have curvature at both ends of the age distribution so a linear regression may not be completely appropriate.

c) The regression equation is $\widehat{Minutes} = 750 - 11.5$

d) The residual plots are:

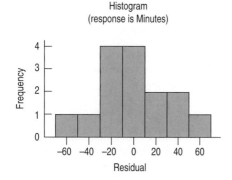

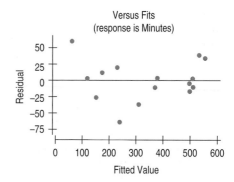

The nearly normal condition is satisfied. There may be some curvature to the residual plot.

3. a) $\widehat{Budget} = -31.39 + 0.71\ Run\ Time$. The model suggests that movies cost about \$710,000 per minute to make.

b) A negative starting value makes no sense, but the P-value of 0.07 indicates that we can't discern a difference between our estimated value and zero. The statement that a movie of zero length should cost \$0 makes sense.

c) Amounts by which movie costs differ from predictions made by this model vary, with a standard deviation of about \$33 million.

d) 0.15 \$m/min

e) If we constructed other models based on different samples of movies, we'd expect the slopes of the regression lines to vary, with a standard deviation of about \$150,000 per minute.

5. a) The scatterplot looks straight enough, the residuals look random and roughly normal, and the residuals don't display any clear change in variability although there may be some increasing spread.

b) I'm 95% confident that the cost of making longer movies increases at a rate of between 0.41 and 1.01 million dollars per minute. (CI is 0.41 to 1.02 using raw data.)

7. a) H_0: There is no linear relationship between calcium concentration in water and mortality rates for males. ($\beta_1 = 0$)

H_A: There is a linear relationship between calcium concentration in water and mortality rates for males. ($\beta_1 \neq 0$)

b) $t = -6.65$, P < 0.0001; reject the null hypothesis. There is strong evidence of a linear relationship between calcium concentration and mortality. Towns with higher calcium concentrations tend to have lower mortality rates.

c) For 95% confidence, use $t^*_{59} \approx 2.001$, or estimate from the table $t^*_{50} \approx 2.009$; (−4.19, −2.27).

d) We are 95% confident that the average mortality rate decreases by between 2.27 and 4.19 deaths per 100,000 for each additional part per million of calcium in drinking water.

9. a) $\widehat{male\ rate} = 2.376 + 0.75564 \times (female\ rate)$.

b) The scatterplot shows a high-leverage point. The residual plot suggests that it may not fit with the other points, so the regression should be run both with and without that point to examine its impact.

c) H_0: There is no linear relationship between male and female unemployment rate. ($\beta_1 = 0$)

H_A: There is a linear relationship between male and female unemployment rate. ($\beta_1 \neq 0$)

$t = 13.46$, df = 55, P < 0.001; reject the null hypothesis. There is strong evidence of a positive linear relationship between the male and female unemployment rate.

d) 76.7%

11. a)

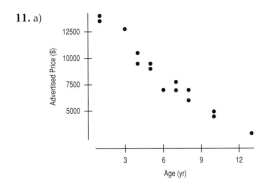

b) Yes, the plot seems linear.

c) $\widehat{Advertised\ Price} = 14286 - 959 \times Age$

d)

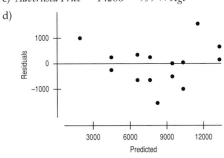

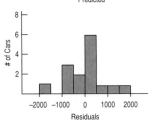

The residual plot shows some possible curvature. Inference may not be valid here, but we will proceed (with caution).

13. Based on these data, we are 95% confident that a used car's *Price* decreases between $819.50 and $1098.50 per year.

15. a)

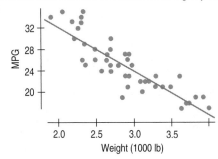

$\widehat{mpg} = 48.7393 - 8.21362 \times (weight)$

b) Yes, the conditions seem satisfied. Histogram of residuals is unimodal and symmetric; residual plot looks okay, but some "thickening" of the plot with increasing values. There may be one possible outlier.

c) H_0: There is no linear relationship between the weight of a car and its fuel efficiency. ($\beta_1 = 0$)
   H_A: There is a linear relationship between the weight of a car and its mileage. ($\beta_1 \neq 0$)

d) $t = -12.2$, df = 48, P < 0.0001; reject the null hypothesis. There is strong evidence of a linear relationship between weight of a car and its mileage. Cars that weigh more tend to have lower gas mileage.

17. a) H_0: There is no linear relationship between SAT Verbal and Math scores. ($\beta_1 = 0$)
   H_A: There is a linear relationship between SAT Verbal and Math scores. ($\beta_1 \neq 0$)

b) Assumptions seem reasonable, since conditions are satisfied. Residual plot shows no patterns (one outlier); histogram is unimodal and roughly symmetric.

c) $t = 11.9$, df = 160, P < 0.0001; reject the null hypothesis. There is strong evidence of a linear relationship between SAT Verbal and Math scores. Students with higher SAT Verbal scores tend to have higher SAT Math scores.

19. a) H_0: There is no linear relationship between team salaries and number of wins. ($\beta_1 = 0$)
   H_A: There is a linear relationship between team salaries and wins. ($\beta_1 \neq 0$)

b) $t = 1.58$; P = 0.124; we fail to reject the null hypothesis. There is no evidence of a linear relationship between team salary and team wins in 2006.

21. a) $(-9.57, -6.86)$ mpg per 1000 pounds

b) We are 95% confident that the mean mileage of cars decreases by between 6.86 and 9.57 miles per gallon for each additional 1000 pounds of weight.

23. a) H_0: There is no linear relationship between market return and fund flows. ($\beta_1 = 0$)
   H_A: There is a linear relationship between market return and fund flows. ($\beta_1 \neq 0$)

b) $t = 5.75$, P < 0.001; reject the null hypothesis. There is strong evidence of a linear relationship between money invested in mutual funds and market performance.

c) Greater investment in mutual funds tends to be associated with higher market return. I would suggest investigating the unusual observations and finding out when and why they occurred.

25. a) H_0: There is no linear association between the Index in 2006 and 2007. ($\beta_1 = 0$)
   H_A: There is a linear association between the Index in 2006 and 2007. ($\beta_1 \neq 0$)

b) $t = 8.17$; P < 0.001. The association is not likely to occur by chance. We reject the null hypothesis. There is strong evidence of a linear relationship between the Index in 2006 and 2007.

c) 83.7% of the variation in the 2007 Index is accounted for by the 2006 Index.

d) On average, as one increases, so does the other. However, this does not mean if the index is high in one year, it will necessarily be high in the other.

27. a)

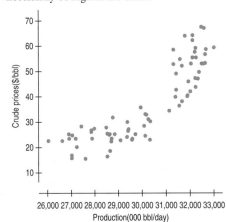

The conditions for regression fail, since the scatterplot bends.

b) No, the relationship does not appear to be linear.

29. H_0: There is no linear relationship between speed and cost of printers. ($\beta_1 = 0$)
H_A: There is a linear relationship between speed and cost of printers. ($\beta_1 \neq 0$)

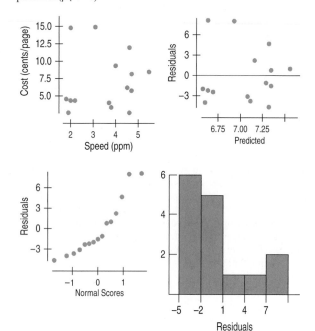

Assumptions do not seem reasonable. Scatterplot isn't straight. The histogram of the residuals is skewed to the right.
The conditions are not met, so inference is not appropriate.

31. a) We are 95% confident that the mean fuel efficiency of cars that weigh 2500 pounds is between 27.34 and 29.07 miles per gallon.

b) We are 95% confident that a car weighing 3450 pounds will have fuel efficiency between 15.44 and 25.37 miles per gallon.

33. a) $(-3145, 40845)$

b) Since the SE for the slope is relatively large and the R^2 is relatively small (18%), predictions using this regression will be imprecise.

c) Omitting outlying values makes the SE smaller and the R^2 larger, so predictions should be more precise.

35. a) We'd like to know if there is a linear association between *Price* and *Fuel Efficiency* in cars. We have data on 2004 model-year cars giving their highway mpg and retail price. H_0: $\beta_1 = 0$ (no linear relationship between *Price* and *Fuel Efficiency*); H_A: $\beta_1 \neq 0$.

b) The scatterplot fails the Straight Enough Condition. It shows a bend and it has an outlier. There is also some spreading from right to left, which would violate the Plot Thickens Condition. We cannot continue the analysis.

c) The conditions are not met; the regression equation should not be interpreted.

37. a) The 95% prediction interval shows the interval of uncertainty for a single predicted male unemployment rate, given a specific female unemployment rate.

b) The 95% confidence interval shows the interval of uncertainty for the mean male unemployment rate given a sample of female unemployment rates. Because this is an interval for an average, the variation or uncertainty is less, so the interval is narrower.

c) The unusual observation is the former Yugoslav Republic of Macedonia. Besides being an outlier, it is also a potential leverage value because its female rate is so removed from the average female rate for this sample of countries. Without this

leverage value, $\widehat{male\ rate} = 4.459 + 0.62951 \times (female\ rate)$. Note that the slope remains significant ($P < 0.001$), and the R^2 decreases slightly compared to the output in Exercise 9 without this leverage value.

39. a) The 95% prediction interval shows the interval of uncertainty for the predicted energy use in 2004 based on energy use in 1999 for a single country.

b) The 95% confidence interval shows the interval of uncertainty for the mean energy use in 2004 based on the same energy use in 1999 for a sample of countries. Because this is an interval for an average, the variation or uncertainty is less, so the interval is narrower.

41. a) $\widehat{Jan} = 120.73 + 0.6695 \times Dec$. We are told this is an SRS. One cardholders' spending should not affect another's. These are quantitative data with no apparent bend in the scatterplot. The residual plot shows some increased spread for larger values of January charges. A histogram of residuals is unimodal and slightly skewed to the right with several high outliers. We will proceed cautiously.

b) $1519.73

c) ($1330.24, $1709.24)

d) ($290.76, $669.76)

e) The residuals show increasing spread, so the confidence intervals may not be valid. I would be skeptical of interpreting them too literally.

43. a) H_0: There is no linear relationship between population and ozone level. ($\beta_1 = 0$)
H_A: There is a positive linear relationship between population and ozone level. ($\beta_1 > 0$)
$t = 3.48$, $P = 0.0018$; reject the null hypothesis. There is strong evidence of a positive linear relationship between ozone level and population. Cities with larger populations tend to have higher ozone levels.

b) City population is a good predictor of ozone level. Population explains 84% of the variability in ozone level and s is just over 5 parts per million.

Chapter 17

1. a) The trend appears to be somewhat linear up to about 1940, but from 1940 to about 1970 the trend appears to be nonlinear. From 1975 or so to about 1995, the trend appears to be linear.

b) Relatively strong for certain periods.

c) No, as a whole the graph is clearly nonlinear. Within certain periods (1975 to about 1995) the correlation is high.

d) Overall, no. You could fit a linear model to the period from 1975 to about 1995, but why? You don't need to interpolate, since every year is reported, and extrapolation seems dangerous.

3. a) The relationship is not straight.

b) It will be curved downward.

c) No. The relationship will still be curved.

5. a) No. We need to see the scatterplot first to see if the conditions are satisfied.

b) No, the linear model might not fit the data at all.

7. a) Millions of dollars per minute of run time.

b) Budgets for dramas increase at the same rate per minute as other movies.

c) Regardless of run time, dramas cost less by about $20 million.

9. a) The use of Oakland airport has been growing at about 59,700 passengers/year, starting from about 283,000 in 1990.

b) 71% of the variation in passengers is accounted for by this model.

c) Errors in predictions based on this model have a standard deviation of 104,330 passengers.

d) DW = 0.912; P-value < 0.001. Strong evidence of auto-correlation.

e) No, that would extrapolate too far from the years we've observed, and the series is autocorrelated.

f) The negative residual is September 2001. Air traffic was artificially low following the attacks on 9/11.

11. a) 1) High leverage, small residual.
   2) No, not influential for the slope.
   3) Correlation would decrease because outlier has large z_x and z_y, increasing correlation.
   4) Slope wouldn't change much because the outlier is in line with other points.

   b) 1) High leverage, probably small residual.
   2) Yes, influential.
   3) Correlation would weaken and become less negative because scatter would increase.
   4) Slope would increase toward 0, since outlier makes it negative.

   c) 1) Some leverage, large residual.
   2) Yes, somewhat influential.
   3) Correlation would strengthen, since scatter would decrease.
   4) Slope would increase slightly.

   d) 1) Little leverage, large residual.
   2) No, not influential.
   3) Correlation would become stronger and become more negative because scatter would decrease.
   4) Slope would change very little.

13. 1) e 2) d 3) c 4) b 5) a

15. Perhaps high blood pressure causes high body fat, high body fat causes high blood pressure, or both could be caused by a lurking variable such as a genetic or lifestyle issue.

17. a) *Cost* decreases by $2.13 per degree of average daily *Temp*. So warmer temperatures indicate lower costs.

   b) For an avg. monthly temperature of 0°F, the cost is predicted to be $133.

   c) Too high; the residuals (observed − predicted) around 32°F are negative, showing that the model overestimates the costs.

   d) $111.7

   e) About $105.7

   f) No, the residuals show a definite curved pattern. The data are probably not linear.

   g) No, there would be no difference. The relationship does not depend on the units.

19. a) 0.88

   b) Interest rates during this period grew at about 0.25% per year, starting from an interest rate of about 0.64%.

   c) Substituting 50 in the model yields a predicted value of about 13%.

   d) DW = 0.9527; P-value = 0.002; Strong evidence of auto-correlation.

   e) Not really. Extrapolating 20 years beyond the end of these data would be dangerous and unlikely to be accurate. Errors are autocorrelated as well.

21. a) The two models both fit well, but they have very different slopes.

   b) This model predicts the interest rate in 2000 to be 3.24%, much lower than the other model predicts.

   c) We can trust the new predicted value because it is in the middle of the data used for the regression.

   d) The best answer is "I can't predict that."

23. *Sex* and *Colorblindness* are both categorical variables, not quantitative. Correlation is meaningless for them, but we can say that the variables are associated.

25. a) No re-expression needed.
   b) Re-express to straighten the relationship.
   c) Re-express to equalize spread.

27. a) There's an annual pattern, so the residuals cycle up and down.

   b) No, this kind of pattern can't be helped by re-expression.

29. a) 2.8   b) 16.44   c) 7.84   d) 0.36   e) 2.09

31. a) 3.842   b) 501.187   c) 4.0

33. a) Linearity condition violated. Possibly increased variation at higher values. Independence may be in doubt since this is a time series.

   b) This plot looks both straighter and more consistent in variation.

   c) The residuals show a pattern, but it tracks up and down and could not be improved by an alternative transformation.

35. a) About $100,000,000

   b) *logValue* grows at .00015 per licensed *Fisher*. Possibly more fishers cause a more valuable harvest, but greater value would likely attract more fishers, so the causation might run the other way. Or improving technology could both lead to more value and attract more fishers.

37. a) Although more than 97% of the variation in GDP can be accounted for by this model, we should examine a scatterplot of the residuals to see if it's appropriate.

   b) No. The residuals show clear curvature.

39. a)

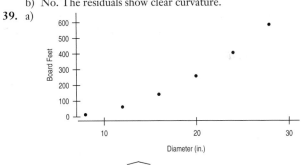

$$\widehat{\sqrt{Bdft}} = -4 + diam$$

The model is exact.

   b) 36 board feet.

   c) 1024 board feet.

41. Dependent variable is: 1998 GDP/Cap
R-squared = 94.2%   R-squared (adjusted) = 93.9%
s = 2803 with 24 − 2 = 22 degrees of freedom

Source	Sum of Squares	df	Mean Square	F-ratio
Regression	2800375675	1	2800375675	356
Residual	172861656	22	7857348	

Variable	Coeff	SE(Coeff)	t-ratio	P-value
Intercept	3295.80	1307	2.52	0.0194
88 GDP CAP	1.090	0.058	18.9	<0.0001

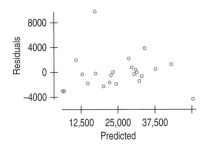

The regression has a high R^2 but the residuals show two influential points, Ireland with the largest positive residual and Switzerland with the largest negative residual. Assuming the data for these countries are correct, they should be set aside from the analysis and further work done to understand why each is unusual. Reports both with and without the outliers should be compared.

43. The histogram is much more unimodal and symmetric after we take logs.

45. The plot cannot be straightened by a transformation because it is not a simple monotonically increasing or decreasing shape, but rather it oscillates up and down.

Chapter 18

1. a) Linearity: The scatterplots appear linear.
Independence: States are not a random sample, but they may be independent of each other.
Equal variance: The scatterplot of *Violent Crime* vs. *Police Officer Wage* may become less spread to the right, or maybe that's just fewer data points.
Normality: To check the Nearly Normal condition, we'll need to look at the residuals; we can't check it with these plots.
  b) 1.1%

3. a) $\widehat{Violent\ Crime} = 1390.83 + 9.33\ Police\ Officer\ Wage - 16.64\ Graduation\ Rate$
  b) After allowing for the effects of graduation rate (or, alternatively, among states with similar graduation rates), states with higher police officer wages have more crime at the rate of 9.3 crimes per 100,000 for each dollar per hour of average wage.
  c) 412.63 crimes per 100,000
  d) Not very good; R^2 is only 53%.

5. a) $2.26 = \dfrac{9.33}{4.125}$
  b) 40, the degrees of freedom is 37, and that's equal to $n - k - 1$. With two predictors, $40 - 2 - 1 = 37$.
  c) The *t*-ratio is negative because the coefficient is negative.

7. a) $H_0: \beta_{Officer} = 0\ H_A: \beta_{Officer} \neq 0$
  b) $P = 0.0297$; that's small enough to reject the null hypothesis at $\alpha = .05$ and conclude that the coefficient is discernibly different from zero.
  c) The coefficient of *Police Officer Wage* reports the relationship *after allowing for the effects* of *Graduation Rate*. The scatterplot and correlation were only concerned with the two-variable relationship.

9. This is a causal interpretation, which is not supported by regression. For example, among states with high graduation rates, it may be that those with higher violent crime rates choose (or are obliged) to spend more to hire police officers, or that states with higher costs of living must pay more to attract qualified police officers but also have higher crime rates.

11. Constant variance condition: met by the residuals vs. predicted plot
Nearly Normal condition: met by the Normal probability plot

13. a) Doesn't mention other predictors; suggests direct relationship
  b) Correct
  c) Can't predict *x* from *y*
  d) Incorrect interpretation of R^2

15. a) Extrapolates far from the data
  b) Suggests a perfect relationship
  c) Can't predict one explanatory variable from another
  d) Correct

17. a) The sign of the coefficient for *ln(number of employees)* is negative. This means that for businesses that have the same amount of sales, those with more employees spend less per employee on pollution abatement on average. The sign of the coefficient for *ln(sales)* is positive. This means that for businesses with the same number of employees, those with larger sales spend more on pollution abatement on average.
  b) The logarithms mean that the effects become less severe (in dollar terms) as companies get larger either in sales or in number of employees.

19. a) $\widehat{Price} = -152{,}037 + 9530\ Baths + 139.87\ Area$
  b) $R^2 = 71.1\%$
  c) For houses with the same number of bathrooms, each square foot of area is associated with an increase of $139.87 in the price of the house, on average.
  d) The regression model says that for houses of the same size, those with more bathrooms are not priced higher. It says nothing about what would happen if a bathroom were added to a house. That would be a predictive interpretation, which is not supported by regression.

21. a) The regression equation is: $\widehat{Salary} = 9.788 + 0.110\ Service + 0.053\ Education + 0.071\ Test\ Score + 0.004\ Typing\ wpm + 0.065\ Dictation\ wpm$
  b) $Salary = 29.205$ or $29,205$
  c) The *t*-value is 0.013 with 24 df. P-value $= .4949$, which is not significant at $\alpha = 0.05$.
  d) Take out the explanatory variable for typing speed since it is not significant.
  e) *Age* is likely to be collinear with several of the other predictors already in the model. For example, secretaries with longer terms of *Service* will naturally also be older.

23. a) After allowing for the effects measured by the Wilshire 5000 index, an increase of one point in the *Unemployment Rate* is associated with a decrease on average of about $3719.55 million in the *Funds flow*.
  b) The *t*-ratio divides the coefficient by its standard error. The coefficient here is negative.
  c) $H_0: \beta = 0$, $H_A: \beta \neq 0$; P-value < 0.0001 is very small, so reject the null hypothesis.

25. a) $\widehat{Price} = 0.845 + 1.2\ Traps - 0.00012\ Fishers - 22.0\ Catch/Trap$
  b) Residuals show no pattern and have equal spread. Histogram of residuals is nearly Normal. We might wonder whether values from year to year are mutually independent.
  c) $H_0: \beta_{catch/trap} = 0$, $H_A: \beta_{catch/trap} \neq 0$; the P-value is exactly .0500. With a coefficient this large, it appears that *Catch/Trap* does contribute to the model. Answers may vary, but we'd reject the null hypothesis.
  d) No, it means that in years when the catch per trap is lower (allowing for the other variables) the price tends to be higher on average. It does not predict what might happen in a future year.
  e) Adjusted R^2 accounts for the number of predictors. These models have different numbers of predictors, so this is an appropriate comparison.

27. a) R squared $= 66.7\%$ R squared (adjusted) $= 63.8\%$
$s = 2.327$ with $39 - 4 = 35$ degrees of freedom

Variable	Coeff	SE(Coeff)	t-ratio	P-value
Intercept	87.0089	33.60	2.59	0.0139
CPI	−0.344795	0.1203	−2.87	0.0070
Personal Consumption	1.10842e−5	0.0000	2.52	0.0165
Retail Sales	1.03152e−4	0.0000	6.67	≤0.0001

$\widehat{Revenue} = 87.0 - 0.344\ CPI + 0.000011\ Personal$
$Consumption + 0.0001\ Retail\ Sales$
  b) R^2 is 66.7%, and all *t*-ratios are significant. It looks like these variables can account for much of the variation in Wal-Mart revenue.

29. a) $\widehat{Logit(Right\text{-}to\text{-}work)} = 6.19951 - 0.106155\ publ - 0.222957\ pvt$
  b) Yes

31. *Displacement* and *Bore* would be good predictors. Relationship with *Wheelbase* isn't linear.

33. a) Yes, R^2 of 90.9% says that most of the variability of *MSRP* is accounted for by this model.
  b) No, a regression model may not be inverted in this way.

35. a) Yes, R^2 is very large.
   b) The value of s, 3.140 calories, is very small compared with the initial standard deviation of *Calories*. This indicates that the model fits the data quite well, leaving very little variation unaccounted for.
   c) A true value of 100% would indicate zero residuals, but with real data such as these, it is likely that the computed value of 100% is rounded up from a slightly lower value.

Chapter 19

1. a) Cheese pizzas (*Type* 1) scored about 15.6 points higher, after allowing for the effects of calories and fat. Cheese pizzas can be expected to sell better than pepperoni pizzas based on these results.
   b) Scatterplots of the data and of residuals to check linearity and constant variance; a histogram or probability plot of residuals to check normality.

3. a) Both pizzas are predicted to have high scores. But both have scores more than 30 points lower than predicted.
   b) Yes, their unusually large predicted values indicate that they must be different from the other pizzas in some ways, and their large (negative) residuals indicate that they would pull on the regression. It might be helpful to see leverage values or Cook's Distance values for these points.

5. a) An indicator or "dummy" variable
   b) Sales are about 10.5 billion dollars higher in December, after accounting for the *CPI*.
   c) Must assume that the slope is the same for the December points as for the others. That appears to be true in the scatterplot.

7. a) There are different slopes for the two types.
   b) This is an interaction term. It says that cheese pizzas (*Type* 1) have scores that grow less rapidly with *Calories* (a slope 0.46 smaller) than do pepperoni pizzas.
   c) The adjusted R^2 for this model is higher than the adjusted R^2 for the previous model. Also, the *t*-ratios are larger. Overall, this looks like a more successful regression model.

9. a) Large Cook's Distance suggests that Alaska is influential. Its leverage is high, and its residual is one of the lowest.
   b) After allowing for the other predictors, Alaska's life expectancy is 2.23 years lower than the model predicts. The P-value of 0.02 says that Alaska is an outlier because the coefficient of the dummy variable is significantly not zero.
   c) The model is better (higher adjusted R^2, smaller standard deviation of residuals) when Alaska is removed from the model with a dummy variable.

11. The *Value* for 1994 is not particularly low, but apparently, judging from its studentized residual, it was unexpectedly low for the number of traps and fishers and the water temperature. Any point with a large Cook's distance is potentially influential, so we might want to set 1994 aside to see whether that changes the model.

13. a) R^2 adjusted accounts for the number of predictors. If removing a predictor causes it to increase, then that predictor contributed little to the model. *Primary Education* would best be omitted from the model.
   b) Whether the P-value for *OECD Regulation* is technically less than 0.05 or greater than 0.05 depends on including *Primary Education* in the model. We might be suspicious that the author, in his desire to claim that regulation harms *GDP/ Capita*, has found an irrelevant predictor that has the effect of nudging the P-value just below .05.

15. a) With a P-value of 0.686, we would not conclude that the true coefficient is not zero. It cannot be interpreted.
   b, c) The R^2 of the second regression shows that *Diverted* is linearly related to the other predictors and thus suffers from collinearity.
   d) 6.9

17. a) High Cook's distance values indicate that the points may be influential. The points have both unusually low predicted values and large residuals, so this seems likely.
   b) Two of the coefficients have not changed very much, but the coefficient of *#Shows* has changed from negative but not significant to positive and strongly significant. A point is influential if removing it from the model changes any coefficients in important ways.
   c) The second model would be better. The first one is influenced by extraordinary events, which are not the kinds of weeks we'd want to model.

19. a) The linearity condition is violated. Because the curve is not monotonic (consistently rising or falling), no re-expression can help.
   b) About 24,494.8 tonnes (24,745 by hand calculator).
   c) *Traps* and *Traps^2* are highly correlated. The collinearity accounts for the change in the coefficient of *Traps*.

Chapter 20

1. a) The 200-day moving average will be smoother.
   b) The SES model using $\alpha = 0.10$ is smoother.
   c) When the α is raised from 0.10 to 0.80, the most recent data point is weighted more heavily, so the model responds more quickly and is not as smooth.

3. a) Random walk, so use AR(1) or naive forecast.
   b) Exponential model
   c) Seasonal dummy variable model, or AR model that uses sales from the same quarter in prior years (lag4)

5. a) 22,988 $M
   b) 21,239 $M

7. a) 475.88 $/tonne
   b) 448.37 $/tonne
   c) 0.1753, 0.2229

9. a) Irregular
   b) Second graph

11. The 2-quarter moving average is shown in A and fits better. This series has a strong, consistent trend component, so a longer-span moving average is dragged down below the series.

13. a) $2.676 B
   b) APE = 7.7% (underforecast)
   c) Forecast is $2.975B; APE = 2.6% (overforecast).

15. a) 86.75
   b) 0.00115
   c) 84.24; the AR model prediction is closer.

17. a) 112
   b) The forecast will be over 100%, which is not possible.

19. a) Linear trend and irregular components
   b) Prices have increased at about 30 cents per year.
   c) This intercept has no interpretation because year zero has no meaning for this series.
   d) There is increased variability starting in the middle of 2005.

21. a) $1.1 million
   b) Q3
   c) Q4
   d) Sales on average are $0.5 million greater in Q4 than in Q1.

23. a) A positive trend component and a seasonal component
   b) Wal-Mart revenues have been increasing at about $0.145 billion per month.

c) Revenues in December tend to be about $11.56 billion more than in January (the base month).

d) $19.3028 billion

e) October is the only month in which Wal-Mart revenues are typically lower than they are in January, after allowing for the overall growth trend in revenue.

25. a) The P-value indicates that the time trend is significantly different from zero.

b) September and November

c) June and August

d) 459,408.22

27. a) The log of the number of passengers grew at 0.0249 per month.

b) In January 1990 there were about $10^{5.64949} = 446,159$ passengers.

c) January and February; we can tell that January is low because it is the base for the dummy variables and the other coefficients are positive.

d) Trend, seasonal, and possibly an 8-year cyclic component.

29. The single exponential smooth is pulled down by the outlier. The moving average spreads out the effect of the outlier across several months. The seasonal regression is not noticeably affected. (The September dummy coefficient absorbs the effect and spreads it out over other Septembers, but that isn't evident in the plot.)

31. a) Dependent variable is: E-Commerce Sales (Millions $)
   R squared = 96.9%   R squared (adjusted) = 96.4%
   s = 1740 with 34 − 5 = 29 degrees of freedom

Variable	Coeff	SE(Coeff)	t-ratio	P-value
Intercept	2022.8	777.4	2.60	0.014
Time	900.138	30.45	29.6	≤ 0.0001
Q2	−486.223	845.9	−0.575	0.5699
Q3	−709.611	845.4	−0.839	0.4081
Q4	3288.03	820.7	4.01	0.0004

Note: Choosing a different quarter as the base quarter will yield a different model, but one with the same R^2.

b) Dependent variable is: logE-commerce Sales
   R squared = 98.7%   R squared (adjusted) = 98.5%
   s = 0.0297 with 34 − 5 = 29 degrees of freedom

Variable	Coeff	SE(Coeff)	t-ratio	P-value
Intercept	3.73222	0.0136	274	≤ 0.0001
Time	0.024502	0.0005	47.1	≤ 0.0001
Q2	−2.63785e-3	0.0145	−0.182	0.8566
Q3	−5.43049e-3	0.0145	−0.376	0.7099
Q4	0.072158	0.0140	5.14	≤ 0.0001

c) The multiplicative model fits better judging from R^2 or adjusted R^2.

33. a) The regression equation is:
   U.S. Retail Gas Price (cents/gal) = 0.253 + 1.40 lag1 − 0.400 lag2

b) Forecast is $3.27.

c)

Model	Forecast ($)	Actual	APE (%)
Linear	2.38	3.19	25.4
Exponential	2.43	3.19	23.8
AR(2)	3.27	3.19	2.5

d) The AR(2) model does the best of those compared here.

35. a–c)

Model	Forecast ($)	Actual	APE (%)
Linear	60.95	58.70	3.8
Exponential	65.94	58.70	12.3
AR(1)	53.07	58.70	9.6

d) The linear model seems to fit best.

37. a) Only the irregular component; this is most likely a random walk, which is one reason these models are ineffective.

b) See sample computer output in Solutions.

c) See sample computer output in Solutions.

d) and e)

Model	MAPE	Forecast ($M)	Actual
MA(6)	79%	−12342.3	−7709
MA(12)	81%	−27.67	−7709
SES*	98%	−15886.0	−7709

*Using optimal alpha of 0.47; SES models using different coefficients will produce different values of MAPE.

39. a) Hotel Occupancy = 11.6 + 0.550 lag1 + 0.293 lag6

b) Residuals over time appear random, except for September 2001.

c) 80.04 %

41. a) The regression equation is:
   OPEC-Prod (Th Bar/Day) = 1148 + 0.962 lag1

b) Forecast is 31245.6.

c)

Model	Forecast (thousand barrels/day)	Actual	APE (%)
MA(6)	31689.8	31286.5	1.29
MA(12)	31937.5	31286.5	2.08
SES	31300.2	31286.5	0.04
AR(1)	31245.6	31286.5	0.13

d) The single exponential smoothing model seems to fit best.

Chapter 21

1. a) $1, 2, \ldots, n$

b) Discrete

3. a) 0,1,2,3,4

b) Discrete

c) No, the outcomes are not equally likely.

5. a) $19

b) $7

7. a) $\mu = 30; \sigma = 6$

b) $\mu = 26; \sigma = 5$

c) $\mu = 30; \sigma = 5.39$

d) $\mu = -10; \sigma = 5.39$

9. a) $0.60

b) $11.60

c) −$0.40

11. a) 2.25 lights

b) 1.26 lights

13. a) No, the probability he wins the second changes depending on whether he won the first.
   b) 0.42
   c) 0.08
   d)

x	0	1	2
P(X = x)	0.42	0.50	0.08

   e) $E(X) = 0.66$ tournaments; $\sigma = 0.62$ tournaments

15. a) Yes, because the probability of one battery being dead doesn't affect another.
   b)

Number good	0	1	2
P(number good)	$\left(\dfrac{3}{10}\right)\left(\dfrac{2}{9}\right) = \dfrac{6}{90}$	$\left(\dfrac{3}{10}\right)\left(\dfrac{7}{9}\right) + \left(\dfrac{7}{10}\right)\left(\dfrac{3}{9}\right) = \dfrac{42}{90}$	$\left(\dfrac{7}{10}\right)\left(\dfrac{6}{9}\right) = \dfrac{42}{90}$

   c) $\mu = 1.4$ batteries
   d) $\sigma = 0.61$ batteries

17. $\mu = E(\text{total wait time}) = 74.0$ seconds
   $\sigma = SD(\text{total wait time}) \approx 20.57$ seconds
   (Answers to standard deviation may vary slightly due to rounding of the standard deviation of the number of red lights each day.) The standard deviation may be calculated only if the days are independent of each other. This seems reasonable.

19. a) The standard deviation is large because the profits on insurance are highly variable. Although there will be many small gains, there will occasionally be large losses when the insurance company has to pay a claim.
   b) $\mu = E(\text{two policies}) = \$300$
      $\sigma = SD(\text{two policies}) \approx \$8,485.28$
   c) $\mu = E(1000 \text{ policies}) = \$150,000$
      $\sigma = SD(1000 \text{ policies}) = \$189,736.66$
   d) 0.785
   e) A natural disaster affecting many policyholders such as a large fire or hurricane

21. a) $B = $ number basic; $D = $ number deluxe
      Net Profit $= 120B + 150D - 200$
   b) \$928.00
   c) \$187.45
   d) Mean—no; SD—yes (sales are independent)

23. a) $\mu = E(\text{miles remaining}) = 164$ miles
      $\sigma = SD(\text{miles remaining}) \approx 19.799$ miles
   b) 0.566

25. a) $\mu = E(\text{time}) = 2.6$ days
      $\sigma = SD(\text{time}) \approx 0.707$ days
   b) $\mu = E(\text{combined time}) = 3.7$ days
      $\sigma = SD(\text{combined time}) \approx 0.768$ days
   c) 22.76% (22.66% from tables)

27. a) Let $X_i = $ price of ith Hulk figure sold; $Y_i = $ price of ith Iron Man figure sold; Insertion Fee $= \$0.55$; $T = $ Closing Fee $= 0.875(X_1 + X_2 + \ldots + X_{19} + Y_1 + \ldots + Y_{13})$
      Net Income $= (X_1 + X_2 + \ldots + X_{19} + Y_1 + \ldots + Y_{13}) - 32(0.55) - 0.0875(X_1 + X_2 + \ldots + X_{19} + Y_1 + \ldots + Y_{13})$
   b) $\mu = E(\text{net income}) = \$313.24$
   c) $\sigma = SD(\text{net income}) = \$6.05$
   d) Yes, to compute the standard deviation

29. a) No, these are not Bernoulli trials. The possible outcomes are 1, 2, 3, 4, 5, and 6. There are more than two possible outcomes.
   b) Yes, these may be considered Bernoulli trials. There are only two possible outcomes: Type A and not Type A. Assuming the 120 donors are representative of the population, the probability of having Type A blood is 43%. The trials are not independent because the population is finite, but the 120 donors represent less than 10% of all possible donors.

   c) No, these are not Bernoulli trials. The probability of choosing a man changes after each promotion and the 10% condition is violated.
   d) No, these are not Bernoulli trials. We are sampling without replacement, so the trials are not independent. Samples without replacement may be considered Bernoulli trials if the sample size is less than 10% of the population, but 500 is more than 10% of 3000.
   e) Yes, these may be considered Bernoulli trials. There are only two possible outcomes: sealed properly and not sealed properly. The probability that a package is unsealed is constant at about 10%, as long as the packages checked are a representative sample of all.

31. a) 0.0819
   b) 0.0064
   c) 0.16
   d) 0.992

33. $E(X) = 14.28$, so 15 patients

35. a) 0.078 pixels
   b) 0.280 pixels
   c) 0.375
   d) 0.012

37. a) 0.274
   b) 0.355
   c) 0.043

39. a) 0.090
   b) 0.329
   c) 0.687

41. a) $\mu = 53.6$ serves
      $\sigma = 4.2$ serves
   b) $np \geq 10$; $nq \geq 10$; serves assumed to be independent.
   c) According to the Normal model, in matches with 80 serves, she is expected to make between 49.4 and 57.8 first serves approximately 68% of the time, between 45.2 and 62.0 first serves approximately 95% of the time, and between 41.0 and 66.2 first serves approximately 99.7% of the time.
   d) 0.0034 (0.0048 with continuity correction)

43. a) 0.141 (0.175 with continuity correction)
   b) Answers may vary. That's a fairly high proportion, but the decision depends on the relative costs of not selling seats and bumping passengers.

45. a) A uniform; all numbers should be equally likely to be selected.
   b) 0.5
   c) 0.001

47. a) The Poisson model
   b) 0.9502
   c) 0.0025

49. a) The exponential model
   b) 1/3 minutes
   c) 0.0473

Chapter 22

1.

		Stay for Interview	
		No	**Yes**
Action	**Fully Changeable Fare**	$750	$750
	Non-refundable Fare	$650	$800

3.

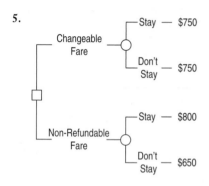

5.

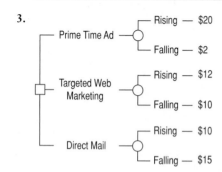

7. EV(Fully Changeable Fare) = $750
EV(Non-refundable Fare) = $695
Choose the non-refundable fare because we want to minimize the expected cost.

9. EV(Fully Changeable Fare) = $525, EV(not changeable) is now $755, so the changeable fare is a better choice.

11. If P(rising confidence) = 0.70, EV(Prime Time) = $14.60.
E(Targeted Web Marketing) = $11.40; EV(Direct Mail) = $11.50
Choose Prime Time since expected payoff is highest.
If P(rising confidence) = 0.40, EV(Prime Time) = $9.20;
EV(Targeted Web Marketing) = $10.80; EV(Direct Mail) = $13.
Choose Direct Mail in this case since expected payoff is highest.

13. a) $3.90
b) $4.00

15. a)

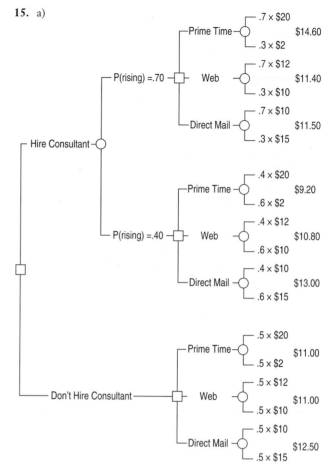

b) Yes, both alternatives with information have a better return.
c) $1.30

17.

		Growing Season	
		Good	**Bad**
Purchase Decision	**2 Tractors**	$27,400	$2,400
	3 Mowers	$8575	$3575
	No Purchase	$0	−$1000

19. Shawn favors Tractors, with an upside gain of $27,400. Lance prefers the Mowers, whose worst result is $3575.

21. a) and b)

		EV	SD	RRR
Purchase Decision	**2 Tractors**	$19,900.00	11,456.44	1.737
	3 Mowers	$7075	2291.28	3.088
	No Purchase	−$300.00	458.26	−0.655

c) Purchasing Mowers has the highest RRR, which is preferred.

23. a)

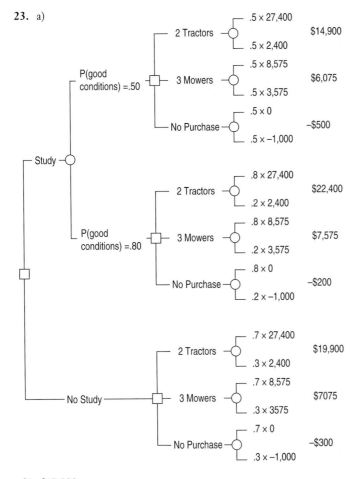

b) $17,900
c) No. In all scenarios, the Tractors investment seems the best course to take.

25. a) EV(Stock A) = $800; EV(Stock B) = $2,300
b) SD(Stock A) = $2,749.55; SD(Stock B) = $1,833.03
c) CV(Stock A) = 3.437; CV(Stock B) = 0.797; RRR(Stock A) = 0.291; RRR(Stock B) = 1.255
d) Stock B; it has a higher expected value and only a slightly higher SD. Its RRR is much higher which shows it has a better reward to risk ratio.

27. a) EV(High-End Bike) = $4465; EV(Moderately-Priced Bike) = $2650
b) SD(High-End Bike) = $1995.00; SD(Moderately-Priced Bike) = $1050.00
c) CV(High-End Bike) = 0.447; CV(Moderately-Priced Bike) = 0.396; RRR(High-End Bike) = 2.238; RRR(Moderately-Priced Bike) = 2.524
d) Answers may vary. The high-end bike has a much higher EV, but also a higher SD. Its RRR is lower but is very close to the moderately-priced bike. The decision will depend on the store owner's appetite for risk.

29. a)

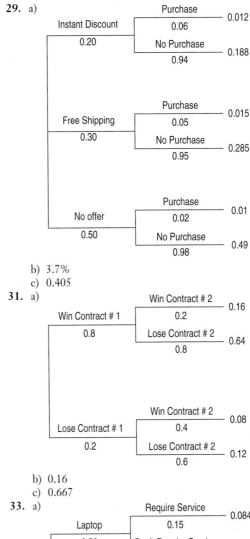

b) 3.7%
c) 0.405

31. a)

b) 0.16
c) 0.667

33. a)

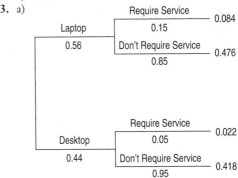

b) 10.6%
c) 0.792

Chapter 23

1. a) Temperature, Washing Time, Detergent
   b) Temperature (Hot, Cold), Washing time (Short, Long), Detergent (Standard, New).
   c) Whiteness as measured by an optical scanner
3. a) Both detergents should be tested under the same conditions to ensure that the results are comparable.
   b) It will be impossible to generalize to cold water washing.
   c) Treatments should be run in random order to ensure that no other unforeseen influences could affect the response in a systematic way.
5. a) The differences among the Mozart and quiet groups were more than would have been expected from ordinary sampling variation.
   b)

   c) The Mozart group seems to have the smallest median difference and thus the *least* improvement, but there does not appear to be a significant difference.
   d) No, the difference does not seem significant compared with the usual variation.
7. a) Observational; randomly select a group of children, ages 10 to 13, have them taste the cereal, and ask if they like the cereal.
   b) Answers may vary. Get volunteers ages 10 to 13. Each volunteer will taste one cereal, randomizing the assignment. Compare the percentage of favorable ratings for each cereal.

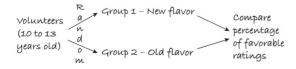

   c) Answers may vary. From the volunteers, identify the children who watch Frump and identify the children who do not watch Frump. Use a blocked design to reduce variation in cereal preference that may be associated with watching the Frump cartoon.

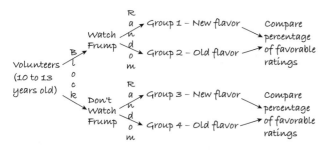

9. a) The students were not randomly assigned. Those who signed up for the prep course may be a special group whose scores would have improved anyway.
   b) Answers may vary. Find a group of volunteers who are willing to participate. Give all volunteers the SAT exam. Randomly assign the subjects to the review or no review group. Give the tutoring to one group. After a reasonable time, retest both groups. Check to see that the tutored group had a significant improvement in scores when compared with the no review group.
   c) After the volunteers have taken the first SAT exam, separate the volunteers into blocks of low, average, and high SAT exam score performance. Now assign half of each block to the review and half to the no review groups. Give the tutoring. Now retest all groups. Compare the differences between treatments for each block.
11. a) The null hypothesis is that the mean *Mileage* for all Cylinder levels is the same. The alternative is that not all means are equal.
   b) Randomization not specified, so inference is not clear. The spreads of the four groups look very different. The response variable probably should be re-expressed before proceeding with an Analysis of Variance. Consider omitting the single 5-cylinder car from the study since it is the only one in its group. Boxplots appear symmetric, but other plots should be used to check the Nearly Normal Condition.
13. a) H_0: Mean cell phone adoption rates are equal in the three regions.
   H_A: Mean cell phone adoption rates are not all equal in the three regions.
   b) Countries were selected randomly from the three regions, so the randomization condition is met. However, boxplots suggest the spreads may differ, and show several outliers, which may require attention. Other plots should be used to check the Nearly Normal Condition.
15. a) H_0: $\mu_1 = \mu_2 = \mu_3 = \mu_4 = \mu_5 = \mu_6$ (That is, each of the tellers takes, on average, the same amount of time to serve a customer.) vs. the alternative that not all means are equal. Here μ_k refers to the mean *Time* it takes *Teller k* to serve a customer.
   b) With a P-value of 0.19, we can't reject the null hypothesis.
   c) No, because we do not reject the null hypothesis, we cannot perform multiple comparisons.
17. a) Observational; the factor was not deliberately manipulated.
   b) Retrospective; using previously collected data.
   c) H_0: Mean responses to the e-security questions are all equal for the three types of community.
   H_A: Mean responses to the e-security questions are not all equal for the three types of community.
   d) $F = 6.24$
   e) With such a small P-value, we can reject the null hypothesis and conclude that mean responses to the e-security question are not all equal for the three types of community. A causal link cannot be established because this is an observational study.
19. a) Observational
   b) There is no treatment. No factors were manipulated.
   c) Numbers of candies of each color.
21. a) Experiment; the factors are deliberately manipulated to specify treatments, and the subjects are assigned to the treatments randomly.
   b) Website trustworthiness measured on a 10-point scale.
   c) There are two factors: (1) website configuration with respect to assurance and (2) type of product purchased.
   d) There are nine treatments.
   e) H_0: The mean *trustworthiness ratings* of websites are the same for those *with a third-party assurance seal, a self-proclaimed assurance displayed*, and *no assurance seal*.
   H_0: The mean *trustworthiness* of websites is the same for online purchases of *books, cameras,* and *insurance*.
   H_0: There is no interaction effect. (The effect of each factor is the same at all levels of the other factor.)
23. a) Observational study.
   b) H_0: The mean *percentage changes in stock prices* are the same for companies *with and without ISO 9000 certification*.
   H_0: The mean *percentage changes in stock prices* are the same for *large and small* companies.
   H_0: There is no interaction effect. (The effect of each factor is the same at all levels of the other factor.)

c) It is difficult to establish a causal link not only because this is an observational study, but also because of the possibility of confounding or lurking variables. For instance, industry sector may be related to percentage changes in stock prices and confounded with ISO 9000 certification but is unaccounted for in this study.

25. a) Observational study; the factors are not deliberately manipulated to specify treatments.

b) Boxplots of the hourly wages by both manager type and region indicate that the equal variance assumption is reasonable. Also, the boxplots indicate fairly symmetric distributions. However, other plots should be used to check the Nearly Normal Condition once the model has been fit. The states were randomly selected from the three regions, so the randomization condition is satisfied.

c) Based on the P-values, we can conclude that there is no significant interaction effect. There is no evidence that the mean hourly wages are different across the three regions but are different for sales and advertising managers.

d) Yes, because the interaction effect is not significant.

27. a) Experiment; the machine setting factor was deliberately manipulated. Shift is a blocking factor.

b) Size error of the part is the response variable.

c) There are nine treatments.

d) There is no significant interaction effect so we can conclude that both the Shift and the Machine Setting appear to have a significant effect. However, from this table alone, we cannot determine which of these effects is more important.

29. a) MST = 8.65; MSE = 0.0767

b) $F = 112.78$

c) The data provide very strong evidence that the means are not equal.

d) We have assumed that the experimental runs were performed in random order, that the variances of the treatment groups are similar, and that the residuals are nearly Normal.

e) A boxplot of the *Scores* by *Method*, a plot of residuals vs. predicted values, a Normal probability plot, and a histogram of the residuals.

f) $s_p = \sqrt{0.0767} = 0.277$ points

31. a) An observational study

b) The null hypothesis is that the mean *Sugar Content* is the same for the cereals on each *Shelf*. The alternative is that not all the means are equal.

c) The P-value of 0.0012 provides strong evidence that the mean *Sugar Content* is not the same for each *Shelf*.

d) We cannot conclude that cereals on *Shelf* 2 have a different mean *Sugar Content* than cereals on *Shelf* 3 or that cereals on *Shelf* 2 have a different mean *Sugar Content* than cereals on *Shelf* 1. We can conclude only that the means are not all equal.

e) The Bonferroni test shows that at $\alpha = .05$ the shelves are different, except for Shelf 1 and Shelf 3, which are not discernibly different. Now we can conclude that the mean *Sugar Content* of cereals on *Shelf* 2 is *not* equal to the mean *Sugar Content* on *Shelf* 1 and *Shelf* 3. In other words, we can conclude from this test what we wanted to conclude in part d.

33. a) H_0: Head injuries are, on average, about the same regardless of the size of the car

H_0: Which seat you sit in doesn't affect the degree of head injury.

b) The conditions appear to be met. We assume the data were collected independently, the boxplots show that the variance is roughly constant, and there are no patterns in the scatterplot of residuals vs. predicted values.

c) There is no significant interaction. The P-values for both *Seat* and *Size* are < 0.0001. Thus, we reject both the null hypotheses and conclude that both the *Seat* and the *Size* of car affect the severity of head injury.

Photo Acknowledgments

Chapter 1

Page 3 (Financial Times screenshot), *Financial Times* screenshot dated April 3, 2008, has been reprinted with permission from the *Financial Times*; (Wall Street), ©Bob Jacobson/Corbis; (charting graph), ©Comstock/Corbis.

Chapter 2

Page 9 (keyboard button), ©iStockphoto: (screen capture), ©Gregory Bajor/iStockphoto; (shopping bag), ©Feng Yu/iStockphoto; (Amazon.com distribution of *Harry Potter*), ©Justin Sullivan/Getty Images News; **Page 13** (dial telephone), ©Shutterstock; (punch card), public domain; **Page 18** (insurance company statistics), ©Shutterstock; **Page 22** (screen capture), ©Gregory Bajor/iStockphoto; (credit card), ©Konstantin Inozemtsev/iStockphoto.

Chapter 3

Page 27 (woman at desk), ©Digital Vision; (spectators waving American flags), ©Stockbyte; (Dewey Defeats Truman newspaper), public domain; **Page 30** (pot of soup), ©PhotoDisc/Getty Images; **Page 35** (group of similar business people), ©Digital Vision; **Page 36** (jigsaw puzzle college campus), ©Shutterstock; **Page 38** (*Calvin and Hobbes* 9.20.1993), ©Universal Press Syndicate; **Page 42** (*The Wizard of Id* 2.23.1991), ©Creators Syndicate; **Page 47** (woman at desk), ©Digital Vision; (market survey), ©iStockphoto.

Chapter 4

Page 53 (male jogger), ©image100/Corbis; (foot with orange shoe), ©Woods Wheatcroft, Keen, Inc.; (girl running in water), ©Mike Watson Images/Corbis; (green shoe silloutte/Barcelona shoe), ©Keen, Inc.; orange shoe sillouette/Apricot Shoe ©Keen, Inc.; **Page 55** (Google Analytics screenshot), public domain; **Page 56** (apricot flower sandal), ©Keen, Inc.; **Page 74** (foot with orange shoe), ©Woods Wheatcroft, Keen, Inc.; (couple squatting at river), ©Tetra Images/Corbis.

Chapter 5

Page 85 (large detached house), ©Phillip Spears/Digital Vision; (bills with a pen and a calculator) ©Sean Russell/fStop (Getty); (African American businessman), ©DreamPictures (Getty); (crumpled dollar bills), ©Frank Bean/UpperCut Images (Getty); **Page 89** (Jacob Bernouilli), public domain/St Andrews Univ MacTutor Archives; (Keno), ©Jean-Loup Gautreau/AFP/Getty Images; **Page 98** (person in bath tub) ©Digital Vision; **Page 103** (clothing rack), ©Martin Mcelligott/iStockphoto; (femaie shoppers), ©Jacob Wackerhausen/iStockphoto.

Chapter 6

Page 111 (man holding head), ©Getty; (LCD stock board), ©Andrey Volodin/iStockphoto; (Enron sign), ©James Nielsen/Stringer/Getty Images; **Page 123** (shopping and spending), ©Shutterstock; **Page 127** (2007 best place to work: Google), ©Kimberly White/Corbis; **Page 132** (wealth), ©Photodisc; **Page 143** (Luxury hotel pool), ©Klaas Lingbeek-van Kranen/iStockphoto; (Downtown Honolulu), ©Aimin Tang/iStockphoto.

Chapter 7

Page 159 (Home Depot), ©Tim Boyle/Getty Images; (metal background), ©KPT Power Photos; (construction tools), ©iStockphoto; **Page 173** (child reading), ©Shutterstock; **Page 181** (paint cans and brushes), ©Kutay Tanir/iStockphoto; (set of small wrenches), ©Jan Paul Schrage /iStockphoto.

Chapter 8

Page 193 (Best Buy), ©Scott Olson/Getty Images News; (flat LCD TV), ©Tomasz Pietryszek/iStockphoto (rock guitarist), ©Robert Kohlhuber/iStockphoto; (computer and LCD in a box), ©Rafal Zdeb/iStockphoto; **Page 199** (shoe size/height correlation), ©Shutterstock; **Page 200** (Sir Francis Galton), public domain/St Andrews Univ MacTutor Archives; **Page 213** (New York skyline), ©Klaas Lingbeek-van Kranen/iStockphoto; (financial material), ©Pali Rao/iStockphoto.

Chapter 9

Page 223 (MBNA logo), ©Scott Boehm/Getty Images Sport; (neoclassical columns), ©Rena Schild/iStockphoto; (stack of credit cards), ©Ilya Genkin/iStockphoto; **Page 230** (cereal), ©Blend Images/Getty RF; **Page 236** (central California neighborhood), ©Shutterstock; **Page 247** (living room), ©M. Eric Honeycutt/iStockphoto; (front porch), ©iStockphoto.

Chapter 10

Page 255 (news ticker board in Times Square), © iStockphoto; (business graph), ©José Luis Gutiérrez/iStockphoto; (inauguration of Franklin D. Roosevelt), ©National Archives/Handout/Getty Images; **Page 260** (*Garfield*) See text page for credit; **Page 262** (concerned employee), ©Photodisc; **Page 267** (pile of junk mail), ©Digital Vision; **Page 272** (watch and money), ©Clint Hild/iStockphoto; (stock market), ©iStockphoto.

Chapter 11

Page 279 (NYSE floor), ©Mario Tama/Staff/Getty Images; (market rates), ©Mike Bentley/iStockphoto; (watch), ©Alan Crosthwaite/iStockphoto; **Page 282** (jury trial), ©Digital Vision; **Page 286** (switching cell phone plans), ©Digital Vision; **Page 287** (2006 World Series), ©Getty Images Sport; **Page 305** (jet), ©Mark Evans/iStockphoto; (molten metal), ©iStockphoto.

Chapter 12

Page 313 (Arthur Guiness historical portrait), Arthur_Guinness.jpg, Wikipedia:Copyrights for more details; (bottle and glass), ©Roulier/Turiot/photocuisine/Corbis; (Eugene Hackett, brewery worker), ©Bert Hardy/Hulton Archive/Getty Images; **Page 315** (William Gossett), University of York Department of Mathematics; **Page 333** (for sale sign), ©Andy Dean/iStockphoto; (San Fernando Valley), © Scott Leigh/iStockphoto.

Chapter 13

Page 343 (Bank Of America), ©Stephen Chernin/Stringer/Getty Images; (Visa goes public), ©Chris Hondros/Staff/Getty Images News; (credit card), ©iStockphoto; **Page 345** (child and cereal), ©Digital Vision; **Page 351** (eyes and flowers), ©Shutterstock; **Page 353** (photo shoot), ©Shutterstock; **Page 355** (friendship negotiation), ©Shutterstock; **Page 364** (blueprint), ©Nicholas Belton/iStockphoto; (home at night), ©M. Eric Honeycutt/iStockphoto.

Chapter 14

Page 375 (Pepsi cans), ©Tim Boyle/Staff/Getty Images News; (Coke bottles), ©Justin Sullivan/Staff/Getty Images News; (Cola in cup), ©Christine Balderas/iStockphoto; **Page 382** (holiday shoppers), ©Blend Image/Getty RF; **Page 389** (blind taste test), ©Andy Reynolds/Stone/Getty Images; (soft drink), ©Andy Hwang/iStockphoto; **Page 392** (wheelchair racer), ©AP Wideworld Photos.

Chapter 15

Page 401 (investor's crystal ball), ©Paul Cowan/iStockphoto; (stock trader), ©Jonathan Kirn/Stone/Getty Images; (Wall Street sign), ©Christine Balderas/iStockphoto; **Page 411** (India bride beauty ritual), ©Stone/Getty Images; **Page 414** (graduating high school girls), ©Digital Vision; **Page 424** (medical records), ©Mark Kostich/iStockphoto; (elderly care), ©Alexander Raths/iStockphoto.

Chapter 16

Page 437 (molten metal), ©iStockphoto; (distressed metal surface), ©Bill Noll/iStockphoto; (Nambé products), ©Nambé Mills, Santa Fe, New Mexico; **Page 449** (older students in university), ©Digital Vision/Getty Images; **Page 456** (active seniors and caregiver), ©Photodisc/Getty Images; **Page 461** (frozen pizza), ©Associated Photo; (melting river ice), ©Josh Webb/iStockphoto.

Chapter 17

Page 473 (bowl of cereal), ©Christine Balderas/iStockphoto; (wheat field), ©Hougaard Malan/iStockphoto; (Kellogg's logo with Tony the Tiger), ©Lawrence Lucier/Stringer/Getty Images; **Page 477** (gas prices with a humorous slant), ©Shutterstock; **Page 479** (*Foxtrot* 5.10.2002), ©2002 Bill Amend. Reprinted with permission from Universal Press Syndicate. All rights reserved; **Page 482** (wind farm), ©Digital Vision/Getty Images; **Page 488** (Tour d'France), ©Getty Images Sport; **Page 497** (cargo ship), ©iStockphoto/Dan Barnes; (windmills), ©iStockphoto/Nicolas Skaanild; **Page 499** (class in developing country), ©PhotoDisc.

Chapter 18

Page 509 (selling a house), ©Sean Locke/iStockphoto; (new home), ©M. Eric Honeycutt/iStockphoto; (Zillow.com logo), ©Zillow; **Page 513** (house and for-sale sign), ©Shutterstock; **Page 526** (customers at appliance store), ©Digital Vision/Getty Images; **Page 536** (Golf ball on grass with driver), ©iStockphoto; (golf swing), ©Andrew Penner/iStockphoto.

Chapter 19

Page 547 (thinking businessman), ©Suprijono Suharjoto/iStockphoto; (construction of coaster), ©Manfred Steinbach/iStockphoto; **Page 548** (roller coaster), ©Marcio Silva/iStockphoto; **Page 550** (roller coaster), ©Shutterstock; **Page 552** (cheeseburger), ©Shutterstock; **Page 557** (Oblivion drop coaster), public domain, Wikipedia; **Page 570** (Picabo Street), Photographer's Choice/Getty Images; **Page 577** (Old Glory), ©RichVintage/Exclusive to iStockphoto; (stack of junk mail and unpaid bills), ©ShaneKato/Exclusive to iStockphoto.

Chapter 20

Page 589 (Whole Foods Market/Glendale California), ©Whole Foods Market; (fork with fresh salad), ©Ranplett/iStockphoto; (Japanese eggplant), ©Alison Stieglitz/iStockphoto; (leeks), ©Alison Stieglitz/iStockphoto; **Page 592** (seasonal shoppng/plants), ©Shutterstock; **Page 596** (stock market), ©Shutterstock; **Page 619** (grocery shopping), ©Shutterstock; **Page 624** (diamond necklace), ©Andrew Manley/Exclusive to iStockphoto; (computer chip), ©Andrey Volodin/Exclusive to iStockphoto.

Chapter 21

Page 639 (Met Life Building, Walter Gropius), ©Howard Architectural Models, Inc.; (pens for business men and women), ©AlbertSmirnov/iStockphoto; (contract series), ©Pali Rao/iStockphoto; (balancing accounts), ©Julien Bastide/iStockphoto; **Page 648** (*Calvin and Hobbes* 4.13.1993), Universal Press Syndicate; **Page 649** (Bernouille), public domain/St. Andrews Univ MacTutor Archive; **Page 652** (blood donors), ©PhotoDisc; **Page 654** (movie scene from A Civil Action), ©Touchstone/The Kobal Collection/ James, David; **Page 658** (stacking boxes on conveyor belt), ©Getty RF; **Page 668** (U.S. savings bonds), ©Richard Cano/Exclusive to iStockphoto; (stock market), ©Frank van den Bergh/iStockphoto.

Chapter 22

Page 675 (various images and logos), ©Data Description, Inc.; (computer studies), ©iStockphoto/Chris Schmidt; **Page 681** (technology support staff), ©Shutterstock; **Page 685** (tourists and health coverage), ©Shutterstock; **Page 687** (PDAs and/or MP3s manufacturing), ©Photodisc; **Page 693** (gas station), ©Exclusive to iStockphoto; (gavel and scale), ©iStockphoto/Christine Balderas.

Chapter 23

Page 699 (money), ©Exclusive to iStockphoto; (business graph), ©Exclusive to iStockphoto; (credit card), ©Shutterstock; **Page 705** (direct mail credit card offers), ©Getty RF; **Page 712** (drum shopping or playing), ©Shutterstock; **Page 717** (Carlo Bonferroni), public domain/St. Andrew's University History of Mathematics Archive; **Page 720** (plane travel), ©Digital Vision/Getty Images; **Page 736** (paper airplane), ©iStockphoto/Iain Sarjeant; (bowl of popcorn), ©iStockphoto/Martin Trebbin.

Chapter 24

Page 747 (envelopes), ©Exclusive to iStockphoto/Jarek Szymanski; (stamp honoring veterans), ©Exclusive to iStockphoto/Clifford Mueller; (U.S. medals on camouflage), ©Exclusive to iStockphoto/Laura Young; **Page 751** (business people reviewing data), ©Photodisc/Getty Images.

Tables and Selected Formulas

Row				TABLE OF RANDOM DIGITS					
1	96299	07196	98642	20639	23185	56282	69929	14125	38872	94168
2	71622	35940	81807	59225	18192	08710	80777	84395	69563	86280
3	03272	41230	81739	74797	70406	18564	69273	72532	78340	36699
4	46376	58596	14365	63685	56555	42974	72944	96463	63533	24152
5	47352	42853	42903	97504	56655	70355	88606	61406	38757	70657
6	20064	04266	74017	79319	70170	96572	08523	56025	89077	57678
7	73184	95907	05179	51002	83374	52297	07769	99792	78365	93487
8	72753	36216	07230	35793	71907	65571	66784	25548	91861	15725
9	03939	30763	06138	80062	02537	23561	93136	61260	77935	93159
10	75998	37203	07959	38264	78120	77525	86481	54986	33042	70648
11	94435	97441	90998	25104	49761	14967	70724	67030	53887	81293
12	04362	40989	69167	38894	00172	02999	97377	33305	60782	29810
13	89059	43528	10547	40115	82234	86902	04121	83889	76208	31076
14	87736	04666	75145	49175	76754	07884	92564	80793	22573	67902
15	76488	88899	15860	07370	13431	84041	69202	18912	83173	11983
16	36460	53772	66634	25045	79007	78518	73580	14191	50353	32064
17	13205	69237	21820	20952	16635	58867	97650	82983	64865	93298
18	51242	12215	90739	36812	00436	31609	80333	96606	30430	31803
19	67819	00354	91439	91073	49258	15992	41277	75111	67496	68430
20	09875	08990	27656	15871	23637	00952	97818	64234	50199	05715
21	18192	95308	72975	01191	29958	09275	89141	19558	50524	32041
22	02763	33701	66188	50226	35813	72951	11638	01876	93664	37001
23	13349	46328	01856	29935	80563	03742	49470	67749	08578	21956
24	69238	92878	80067	80807	45096	22936	64325	19265	37755	69794
25	92207	63527	59398	29818	24789	94309	88380	57000	50171	17891
26	66679	99100	37072	30593	29665	84286	44458	60180	81451	58273
27	31087	42430	60322	34765	15757	53300	97392	98035	05228	68970
28	84432	04916	52949	78533	31666	62350	20584	56367	19701	60584
29	72042	12287	21081	48426	44321	58765	41760	43304	13399	02043
30	94534	73559	82135	70260	87936	85162	11937	18263	54138	69564
31	63971	97198	40974	45301	60177	35604	21580	68107	25184	42810
32	11227	58474	17272	37619	69517	62964	67962	34510	12607	52255
33	28541	02029	08068	96656	17795	21484	57722	76511	27849	61738
34	11282	43632	49531	78981	81980	08530	08629	32279	29478	50228
35	42907	15137	21918	13248	39129	49559	94540	24070	88151	36782
36	47119	76651	21732	32364	58545	50277	57558	30390	18771	72703
37	11232	99884	05087	76839	65142	19994	91397	29350	83852	04905
38	64725	06719	86262	53356	57999	50193	79936	97230	52073	94467
39	77007	26962	55466	12521	48125	12280	54985	26239	76044	54398
40	18375	19310	59796	89832	59417	18553	17238	05474	33259	50595

Table Z	Second decimal place in z										
Areas under the standard Normal curve	0.09	0.08	0.07	0.06	0.05	0.04	0.03	0.02	0.01	0.00	z
										0.0000†	−3.9
	0.0001	0.0001	0.0001	0.0001	0.0001	0.0001	0.0001	0.0001	0.0001	0.0001	−3.8
	0.0001	0.0001	0.0001	0.0001	0.0001	0.0001	0.0001	0.0001	0.0001	0.0001	−3.7
	0.0001	0.0001	0.0001	0.0001	0.0001	0.0001	0.0001	0.0001	0.0002	0.0002	−3.6
	0.0002	0.0002	0.0002	0.0002	0.0002	0.0002	0.0002	0.0002	0.0002	0.0002	−3.5
	0.0002	0.0003	0.0003	0.0003	0.0003	0.0003	0.0003	0.0003	0.0003	0.0003	−3.4
	0.0003	0.0004	0.0004	0.0004	0.0004	0.0004	0.0004	0.0005	0.0005	0.0005	−3.3
	0.0005	0.0005	0.0005	0.0006	0.0006	0.0006	0.0006	0.0006	0.0007	0.0007	−3.2
	0.0007	0.0007	0.0008	0.0008	0.0008	0.0008	0.0009	0.0009	0.0009	0.0010	−3.1
	0.0010	0.0010	0.0011	0.0011	0.0011	0.0012	0.0012	0.0013	0.0013	0.0013	−3.0
	0.0014	0.0014	0.0015	0.0015	0.0016	0.0016	0.0017	0.0018	0.0018	0.0019	−2.9
	0.0019	0.0020	0.0021	0.0021	0.0022	0.0023	0.0023	0.0024	0.0025	0.0026	−2.8
	0.0026	0.0027	0.0028	0.0029	0.0030	0.0031	0.0032	0.0033	0.0034	0.0035	−2.7
	0.0036	0.0037	0.0038	0.0039	0.0040	0.0041	0.0043	0.0044	0.0045	0.0047	−2.6
	0.0048	0.0049	0.0051	0.0052	0.0054	0.0055	0.0057	0.0059	0.0060	0.0062	−2.5
	0.0064	0.0066	0.0068	0.0069	0.0071	0.0073	0.0075	0.0078	0.0080	0.0082	−2.4
	0.0084	0.0087	0.0089	0.0091	0.0094	0.0096	0.0099	0.0102	0.0104	0.0107	−2.3
	0.0110	0.0113	0.0116	0.0119	0.0122	0.0125	0.0129	0.0132	0.0136	0.0139	−2.2
	0.0143	0.0146	0.0150	0.0154	0.0158	0.0162	0.0166	0.0170	0.0174	0.0179	−2.1
	0.0183	0.0188	0.0192	0.0197	0.0202	0.0207	0.0212	0.0217	0.0222	0.0228	−2.0
	0.0233	0.0239	0.0244	0.0250	0.0256	0.0262	0.0268	0.0274	0.0281	0.0287	−1.9
	0.0294	0.0301	0.0307	0.0314	0.0322	0.0329	0.0336	0.0344	0.0351	0.0359	−1.8
	0.0367	0.0375	0.0384	0.0392	0.0401	0.0409	0.0418	0.0427	0.0436	0.0446	−1.7
	0.0455	0.0465	0.0475	0.0485	0.0495	0.0505	0.0516	0.0526	0.0537	0.0548	−1.6
	0.0559	0.0571	0.0582	0.0594	0.0606	0.0618	0.0630	0.0643	0.0655	0.0668	−1.5
	0.0681	0.0694	0.0708	0.0721	0.0735	0.0749	0.0764	0.0778	0.0793	0.0808	−1.4
	0.0823	0.0838	0.0853	0.0869	0.0885	0.0901	0.0918	0.0934	0.0951	0.0968	−1.3
	0.0985	0.1003	0.1020	0.1038	0.1056	0.1075	0.1093	0.1112	0.1131	0.1151	−1.2
	0.1170	0.1190	0.1210	0.1230	0.1251	0.1271	0.1292	0.1314	0.1335	0.1357	−1.1
	0.1379	0.1401	0.1423	0.1446	0.1469	0.1492	0.1515	0.1539	0.1562	0.1587	−1.0
	0.1611	0.1635	0.1660	0.1685	0.1711	0.1736	0.1762	0.1788	0.1814	0.1841	−0.9
	0.1867	0.1894	0.1922	0.1949	0.1977	0.2005	0.2033	0.2061	0.2090	0.2119	−0.8
	0.2148	0.2177	0.2206	0.2236	0.2266	0.2296	0.2327	0.2358	0.2389	0.2420	−0.7
	0.2451	0.2483	0.2514	0.2546	0.2578	0.2611	0.2643	0.2676	0.2709	0.2743	−0.6
	0.2776	0.2810	0.2843	0.2877	0.2912	0.2946	0.2981	0.3015	0.3050	0.3085	−0.5
	0.3121	0.3156	0.3192	0.3228	0.3264	0.3300	0.3336	0.3372	0.3409	0.3446	−0.4
	0.3483	0.3520	0.3557	0.3594	0.3632	0.3669	0.3707	0.3745	0.3783	0.3821	−0.3
	0.3859	0.3897	0.3936	0.3974	0.4013	0.4052	0.4090	0.4129	0.4168	0.4207	−0.2
	0.4247	0.4286	0.4325	0.4364	0.4404	0.4443	0.4483	0.4522	0.4562	0.4602	−0.1
	0.4641	0.4681	0.4721	0.4761	0.4801	0.4840	0.4880	0.4920	0.4960	0.5000	−0.0

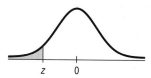

† For $z \leq -3.90$ the areas are 0.0000 to four decimal places.

Table Z (cont.)

Areas under the standard
Normal curve

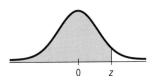

					Second decimal place in z					
z	0.00	0.01	0.02	0.03	0.04	0.05	0.06	0.07	0.08	0.09
0.0	0.5000	0.5040	0.5080	0.5120	0.5160	0.5199	0.5239	0.5279	0.5319	0.5359
0.1	0.5398	0.5438	0.5478	0.5517	0.5557	0.5596	0.5636	0.5675	0.5714	0.5753
0.2	0.5793	0.5832	0.5871	0.5910	0.5948	0.5987	0.6026	0.6064	0.6103	0.6141
0.3	0.6179	0.6217	0.6255	0.6293	0.6331	0.6368	0.6406	0.6443	0.6480	0.6517
0.4	0.6554	0.6591	0.6628	0.6664	0.6700	0.6736	0.6772	0.6808	0.6844	0.6879
0.5	0.6915	0.6950	0.6985	0.7019	0.7054	0.7088	0.7123	0.7157	0.7190	0.7224
0.6	0.7257	0.7291	0.7324	0.7357	0.7389	0.7422	0.7454	0.7486	0.7517	0.7549
0.7	0.7580	0.7611	0.7642	0.7673	0.7704	0.7734	0.7764	0.7794	0.7823	0.7852
0.8	0.7881	0.7910	0.7939	0.7967	0.7995	0.8023	0.8051	0.8078	0.8106	0.8133
0.9	0.8159	0.8186	0.8212	0.8238	0.8264	0.8289	0.8315	0.8340	0.8365	0.8389
1.0	0.8413	0.8438	0.8461	0.8485	0.8508	0.8531	0.8554	0.8577	0.8599	0.8621
1.1	0.8643	0.8665	0.8686	0.8708	0.8729	0.8749	0.8770	0.8790	0.8810	0.8830
1.2	0.8849	0.8869	0.8888	0.8907	0.8925	0.8944	0.8962	0.8980	0.8997	0.9015
1.3	0.9032	0.9049	0.9066	0.9082	0.9099	0.9115	0.9131	0.9147	0.9162	0.9177
1.4	0.9192	0.9207	0.9222	0.9236	0.9251	0.9265	0.9279	0.9292	0.9306	0.9319
1.5	0.9332	0.9345	0.9357	0.9370	0.9382	0.9394	0.9406	0.9418	0.9429	0.9441
1.6	0.9452	0.9463	0.9474	0.9484	0.9495	0.9505	0.9515	0.9525	0.9535	0.9545
1.7	0.9554	0.9564	0.9573	0.9582	0.9591	0.9599	0.9608	0.9616	0.9625	0.9633
1.8	0.9641	0.9649	0.9656	0.9664	0.9671	0.9678	0.9686	0.9693	0.9699	0.9706
1.9	0.9713	0.9719	0.9726	0.9732	0.9738	0.9744	0.9750	0.9756	0.9761	0.9767
2.0	0.9772	0.9778	0.9783	0.9788	0.9793	0.9798	0.9803	0.9808	0.9812	0.9817
2.1	0.9821	0.9826	0.9830	0.9834	0.9838	0.9842	0.9846	0.9850	0.9854	0.9857
2.2	0.9861	0.9864	0.9868	0.9871	0.9875	0.9878	0.9881	0.9884	0.9887	0.9890
2.3	0.9893	0.9896	0.9898	0.9901	0.9904	0.9906	0.9909	0.9911	0.9913	0.9916
2.4	0.9918	0.9920	0.9922	0.9925	0.9927	0.9929	0.9931	0.9932	0.9934	0.9936
2.5	0.9938	0.9940	0.9941	0.9943	0.9945	0.9946	0.9948	0.9949	0.9951	0.9952
2.6	0.9953	0.9955	0.9956	0.9957	0.9959	0.9960	0.9961	0.9962	0.9963	0.9964
2.7	0.9965	0.9966	0.9967	0.9968	0.9969	0.9970	0.9971	0.9972	0.9973	0.9974
2.8	0.9974	0.9975	0.9976	0.9977	0.9977	0.9978	0.9979	0.9979	0.9980	0.9981
2.9	0.9981	0.9982	0.9982	0.9983	0.9984	0.9984	0.9985	0.9985	0.9986	0.9986
3.0	0.9987	0.9987	0.9987	0.9988	0.9988	0.9989	0.9989	0.9989	0.9990	0.9990
3.1	0.9990	0.9991	0.9991	0.9991	0.9992	0.9992	0.9992	0.9992	0.9993	0.9993
3.2	0.9993	0.9993	0.9994	0.9994	0.9994	0.9994	0.9994	0.9995	0.9995	0.9995
3.3	0.9995	0.9995	0.9995	0.9996	0.9996	0.9996	0.9996	0.9996	0.9996	0.9997
3.4	0.9997	0.9997	0.9997	0.9997	0.9997	0.9997	0.9997	0.9997	0.9997	0.9998
3.5	0.9998	0.9998	0.9998	0.9998	0.9998	0.9998	0.9998	0.9998	0.9998	0.9998
3.6	0.9998	0.9998	0.9999	0.9999	0.9999	0.9999	0.9999	0.9999	0.9999	0.9999
3.7	0.9999	0.9999	0.9999	0.9999	0.9999	0.9999	0.9999	0.9999	0.9999	0.9999
3.8	0.9999	0.9999	0.9999	0.9999	0.9999	0.9999	0.9999	0.9999	0.9999	0.9999
3.9	1.0000[†]									

[†] For $z \geq 3.90$, the areas are 1.0000 to four decimal places.

	df	0.20	0.10	0.05	0.02	0.01		df
Two-tail probability		0.20	0.10	0.05	0.02	0.01		
One-tail probability		0.10	0.05	0.025	0.01	0.005		
Table T								
Values of t_α	1	3.078	6.314	12.706	31.821	63.657		1
	2	1.886	2.920	4.303	6.965	9.925		2
	3	1.638	2.353	3.182	4.541	5.841		3
	4	1.533	2.132	2.776	3.747	4.604		4
	5	1.476	2.015	2.571	3.365	4.032		5
	6	1.440	1.943	2.447	3.143	3.707		6
	7	1.415	1.895	2.365	2.998	3.499		7
	8	1.397	1.860	2.306	2.896	3.355		8
	9	1.383	1.833	2.262	2.821	3.250		9
	10	1.372	1.812	2.228	2.764	3.169		10
	11	1.363	1.796	2.201	2.718	3.106		11
	12	1.356	1.782	2.179	2.681	3.055		12
	13	1.350	1.771	2.160	2.650	3.012		13
	14	1.345	1.761	2.145	2.624	2.977		14
	15	1.341	1.753	2.131	2.602	2.947		15
	16	1.337	1.746	2.120	2.583	2.921		16
	17	1.333	1.740	2.110	2.567	2.898		17
	18	1.330	1.734	2.101	2.552	2.878		18
	19	1.328	1.729	2.093	2.539	2.861		19
	20	1.325	1.725	2.086	2.528	2.845		20
	21	1.323	1.721	2.080	2.518	2.831		21
	22	1.321	1.717	2.074	2.508	2.819		22
	23	1.319	1.714	2.069	2.500	2.807		23
	24	1.318	1.711	2.064	2.492	2.797		24
	25	1.316	1.708	2.060	2.485	2.787		25
	26	1.315	1.706	2.056	2.479	2.779		26
	27	1.314	1.703	2.052	2.473	2.771		27
	28	1.313	1.701	2.048	2.467	2.763		28
	29	1.311	1.699	2.045	2.462	2.756		29
	30	1.310	1.697	2.042	2.457	2.750		30
	32	1.309	1.694	2.037	2.449	2.738		32
	35	1.306	1.690	2.030	2.438	2.725		35
	40	1.303	1.684	2.021	2.423	2.704		40
	45	1.301	1.679	2.014	2.412	2.690		45
	50	1.299	1.676	2.009	2.403	2.678		50
	60	1.296	1.671	2.000	2.390	2.660		60
	75	1.293	1.665	1.992	2.377	2.643		75
	100	1.290	1.660	1.984	2.364	2.626		100
	120	1.289	1.658	1.980	2.358	2.617		120
	140	1.288	1.656	1.977	2.353	2.611		140
	180	1.286	1.653	1.973	2.347	2.603		180
	250	1.285	1.651	1.969	2.341	2.596		250
	400	1.284	1.649	1.966	2.336	2.588		400
	1000	1.282	1.646	1.962	2.330	2.581		1000
	∞	1.282	1.645	1.960	2.326	2.576		∞
Confidence levels		80%	90%	95%	98%	99%		

Two tails

One tail

Right-tail probability		0.10	0.05	0.025	0.01	0.005
Table X Values of χ_α^2	df					
	1	2.706	3.841	5.024	6.635	7.879
	2	4.605	5.991	7.378	9.210	10.597
	3	6.251	7.815	9.348	11.345	12.838
	4	7.779	9.488	11.143	13.277	14.860
	5	9.236	11.070	12.833	15.086	16.750
	6	10.645	12.592	14.449	16.812	18.548
	7	12.017	14.067	16.013	18.475	20.278
	8	13.362	15.507	17.535	20.090	21.955
	9	14.684	16.919	19.023	21.666	23.589
	10	15.987	18.307	20.483	23.209	25.188
	11	17.275	19.675	21.920	24.725	26.757
	12	18.549	21.026	23.337	26.217	28.300
	13	19.812	22.362	24.736	27.688	29.819
	14	21.064	23.685	26.119	29.141	31.319
	15	22.307	24.996	27.488	30.578	32.801
	16	23.542	26.296	28.845	32.000	34.267
	17	24.769	27.587	30.191	33.409	35.718
	18	25.989	28.869	31.526	34.805	37.156
	19	27.204	30.143	32.852	36.191	38.582
	20	28.412	31.410	34.170	37.566	39.997
	21	29.615	32.671	35.479	38.932	41.401
	22	30.813	33.924	36.781	40.290	42.796
	23	32.007	35.172	38.076	41.638	44.181
	24	33.196	36.415	39.364	42.980	45.559
	25	34.382	37.653	40.647	44.314	46.928
	26	35.563	38.885	41.923	45.642	48.290
	27	36.741	40.113	43.195	46.963	49.645
	28	37.916	41.337	44.461	48.278	50.994
	29	39.087	42.557	45.722	59.588	52.336
	30	40.256	43.773	46.979	50.892	53.672
	40	51.805	55.759	59.342	63.691	66.767
	50	63.167	67.505	71.420	76.154	79.490
	60	74.397	79.082	83.298	88.381	91.955
	70	85.527	90.531	95.023	100.424	104.213
	80	96.578	101.879	106.628	112.328	116.320
	90	107.565	113.145	118.135	124.115	128.296
	100	118.499	124.343	129.563	135.811	140.177

Table F

Numerator df

$\alpha = .01$

Denominator df	1	2	3	4	5	6	7	8	9	10	11	12	13	14	15	16	17	18	19	20	21	22
1	4052.2	4999.3	5403.5	5624.3	5764.0	5859.0	5928.3	5981.0	6022.4	6055.9	6083.4	6106.7	6125.8	6143.0	6157.0	6170.0	6181.2	6191.4	6200.7	6208.7	6216.1	6223.1
2	98.50	99.00	99.16	99.25	99.30	99.33	99.36	99.38	99.39	99.40	99.41	99.42	99.42	99.43	99.43	99.44	99.44	99.44	99.45	99.45	99.45	99.46
3	34.12	30.82	29.46	28.71	28.24	27.91	27.67	27.49	27.34	27.23	27.13	27.05	26.98	26.92	26.87	26.83	26.79	26.75	26.72	26.69	26.66	26.64
4	21.20	18.00	16.69	15.98	15.52	15.21	14.98	14.80	14.66	14.55	14.45	14.37	14.31	14.25	14.20	14.15	14.11	14.08	14.05	14.02	13.99	13.97
5	16.26	13.27	12.06	11.39	10.97	10.67	10.46	10.29	10.16	10.05	9.96	9.89	9.82	9.77	9.72	9.68	9.64	9.61	9.58	9.55	9.53	9.51
6	13.75	10.92	9.78	9.15	8.75	8.47	8.26	8.10	7.98	7.87	7.79	7.72	7.66	7.60	7.56	7.52	7.48	7.45	7.42	7.40	7.37	7.35
7	12.25	9.55	8.45	7.85	7.46	7.19	6.99	6.84	6.72	6.62	6.54	6.47	6.41	6.36	6.31	6.28	6.24	6.21	6.18	6.16	6.13	6.11
8	11.26	8.65	7.59	7.01	6.63	6.37	6.18	6.03	5.91	5.81	5.73	5.67	5.61	5.56	5.52	5.48	5.44	5.41	5.38	5.36	5.34	5.32
9	10.56	8.02	6.99	6.42	6.06	5.80	5.61	5.47	5.35	5.26	5.18	5.11	5.05	5.01	4.96	4.92	4.89	4.86	4.83	4.81	4.79	4.77
10	10.04	7.56	6.55	5.99	5.64	5.39	5.20	5.06	4.94	4.85	4.77	4.71	4.65	4.60	4.56	4.52	4.49	4.46	4.43	4.41	4.38	4.36
11	9.65	7.21	6.22	5.67	5.32	5.07	4.89	4.74	4.63	4.54	4.46	4.40	4.34	4.29	4.25	4.21	4.18	4.15	4.12	4.10	4.08	4.06
12	9.33	6.93	5.95	5.41	5.06	4.82	4.64	4.50	4.39	4.30	4.22	4.16	4.10	4.05	4.01	3.97	3.94	3.91	3.88	3.86	3.84	3.82
13	9.07	6.70	5.74	5.21	4.86	4.62	4.44	4.30	4.19	4.10	4.02	3.96	3.91	3.86	3.82	3.78	3.75	3.72	3.69	3.66	3.64	3.62
14	8.86	6.51	5.56	5.04	4.69	4.46	4.28	4.14	4.03	3.94	3.86	3.80	3.75	3.70	3.66	3.62	3.59	3.56	3.53	3.51	3.48	3.46
15	8.68	6.36	5.42	4.89	4.56	4.32	4.14	4.00	3.89	3.80	3.73	3.67	3.61	3.56	3.52	3.49	3.45	3.42	3.40	3.37	3.35	3.33
16	8.53	6.23	5.29	4.77	4.44	4.20	4.03	3.89	3.78	3.69	3.62	3.55	3.50	3.45	3.41	3.37	3.34	3.31	3.28	3.26	3.24	3.22
17	8.40	6.11	5.19	4.67	4.34	4.10	3.93	3.79	3.68	3.59	3.52	3.46	3.40	3.35	3.31	3.27	3.24	3.21	3.19	3.16	3.14	3.12
18	8.29	6.01	5.09	4.58	4.25	4.01	3.84	3.71	3.60	3.51	3.43	3.37	3.32	3.27	3.23	3.19	3.16	3.13	3.10	3.08	3.05	3.03
19	8.18	5.93	5.01	4.50	4.17	3.94	3.77	3.63	3.52	3.43	3.36	3.30	3.24	3.19	3.15	3.12	3.08	3.05	3.03	3.00	2.98	2.96
20	8.10	5.85	4.94	4.43	4.10	3.87	3.70	3.56	3.46	3.37	3.29	3.23	3.18	3.13	3.09	3.05	3.02	2.99	2.96	2.94	2.92	2.90
21	8.02	5.78	4.87	4.37	4.04	3.81	3.64	3.51	3.40	3.31	3.24	3.17	3.12	3.07	3.03	2.99	2.96	2.93	2.90	2.88	2.86	2.84
22	7.95	5.72	4.82	4.31	3.99	3.76	3.59	3.45	3.35	3.26	3.18	3.12	3.07	3.02	2.98	2.94	2.91	2.88	2.85	2.83	2.81	2.78
23	7.88	5.66	4.76	4.26	3.94	3.71	3.54	3.41	3.30	3.21	3.14	3.07	3.02	2.97	2.93	2.89	2.86	2.83	2.80	2.78	2.76	2.74
24	7.82	5.61	4.72	4.22	3.90	3.67	3.50	3.36	3.26	3.17	3.09	3.03	2.98	2.93	2.89	2.85	2.82	2.79	2.76	2.74	2.72	2.70
25	7.77	5.57	4.68	4.18	3.85	3.63	3.46	3.32	3.22	3.13	3.06	2.99	2.94	2.89	2.85	2.81	2.78	2.75	2.72	2.70	2.68	2.66
26	7.72	5.53	4.64	4.14	3.82	3.59	3.42	3.29	3.18	3.09	3.02	2.96	2.90	2.86	2.81	2.78	2.75	2.72	2.69	2.66	2.64	2.62
27	7.68	5.49	4.60	4.11	3.78	3.56	3.39	3.26	3.15	3.06	2.99	2.93	2.87	2.82	2.78	2.75	2.71	2.68	2.66	2.63	2.61	2.59
28	7.64	5.45	4.57	4.07	3.75	3.53	3.36	3.23	3.12	3.03	2.96	2.90	2.84	2.79	2.75	2.72	2.68	2.65	2.63	2.60	2.58	2.56
29	7.60	5.42	4.54	4.04	3.73	3.50	3.33	3.20	3.09	3.00	2.93	2.87	2.81	2.77	2.73	2.69	2.66	2.63	2.60	2.57	2.55	2.53
30	7.56	5.39	4.51	4.02	3.70	3.47	3.30	3.17	3.07	2.98	2.91	2.84	2.79	2.74	2.70	2.66	2.63	2.60	2.57	2.55	2.53	2.51
32	7.50	5.34	4.46	3.97	3.65	3.43	3.26	3.13	3.02	2.93	2.86	2.80	2.74	2.70	2.65	2.62	2.58	2.55	2.53	2.50	2.48	2.46
35	7.42	5.27	4.40	3.91	3.59	3.37	3.20	3.07	2.96	2.88	2.80	2.74	2.69	2.64	2.60	2.56	2.53	2.50	2.47	2.44	2.42	2.40
40	7.31	5.18	4.31	3.83	3.51	3.29	3.12	2.99	2.89	2.80	2.73	2.66	2.61	2.56	2.52	2.48	2.45	2.42	2.39	2.37	2.35	2.33
45	7.23	5.11	4.25	3.77	3.45	3.23	3.07	2.94	2.83	2.74	2.67	2.61	2.55	2.51	2.46	2.43	2.39	2.36	2.34	2.31	2.29	2.27
50	7.17	5.06	4.20	3.72	3.41	3.19	3.02	2.89	2.78	2.70	2.63	2.56	2.51	2.46	2.42	2.38	2.35	2.32	2.29	2.27	2.24	2.22
60	7.08	4.98	4.13	3.65	3.34	3.12	2.95	2.82	2.72	2.63	2.56	2.50	2.44	2.39	2.35	2.31	2.28	2.25	2.22	2.20	2.17	2.15
75	6.99	4.90	4.05	3.58	3.27	3.05	2.89	2.76	2.65	2.57	2.49	2.43	2.38	2.33	2.29	2.25	2.22	2.18	2.16	2.13	2.11	2.09
100	6.90	4.82	3.98	3.51	3.21	2.99	2.82	2.69	2.59	2.50	2.43	2.37	2.31	2.27	2.22	2.19	2.15	2.12	2.09	2.07	2.04	2.02
120	6.85	4.79	3.95	3.48	3.17	2.96	2.79	2.66	2.56	2.47	2.40	2.34	2.28	2.23	2.19	2.15	2.12	2.09	2.06	2.03	2.01	1.99
140	6.82	4.76	3.92	3.46	3.15	2.93	2.77	2.64	2.54	2.45	2.38	2.31	2.26	2.21	2.17	2.13	2.10	2.07	2.04	2.01	1.99	1.97
180	6.78	4.73	3.89	3.43	3.12	2.90	2.74	2.61	2.51	2.42	2.35	2.28	2.23	2.18	2.14	2.10	2.07	2.04	2.01	1.98	1.96	1.94
250	6.74	4.69	3.86	3.40	3.09	2.87	2.71	2.58	2.48	2.39	2.32	2.26	2.20	2.15	2.11	2.07	2.04	2.01	1.98	1.95	1.93	1.91
400	6.70	4.66	3.83	3.37	3.06	2.85	2.68	2.56	2.45	2.37	2.29	2.23	2.17	2.13	2.08	2.05	2.01	1.98	1.95	1.92	1.90	1.88
1000	6.66	4.63	3.80	3.34	3.04	2.82	2.66	2.53	2.43	2.34	2.27	2.20	2.15	2.10	2.06	2.02	1.98	1.95	1.92	1.90	1.87	1.85

Table F (cont.)

Numerator df

Denominator df

$\alpha = .01$	23	24	25	26	27	28	29	30	32	35	40	45	50	60	75	100	120	140	180	250	400	1000
1	6228.7	6234.3	6239.9	6244.5	6249.2	6252.9	6257.1	6260.4	6266.9	6275.3	6286.4	6295.7	6302.3	6313.0	6323.7	6333.9	6339.5	6343.2	6347.9	6353.5	6358.1	6362.8
2	99.46	99.46	99.46	99.46	99.46	99.46	99.46	99.47	99.47	99.47	99.48	99.48	99.48	99.48	99.48	99.49	99.49	99.49	99.49	99.50	99.50	99.50
3	26.62	26.60	26.58	26.56	26.55	26.53	26.52	26.50	26.48	26.45	26.41	26.38	26.35	26.32	26.28	26.24	26.22	26.21	26.19	26.17	26.15	26.14
4	13.95	13.93	13.91	13.89	13.88	13.86	13.85	13.84	13.81	13.79	13.75	13.71	13.69	13.65	13.61	13.58	13.56	13.54	13.53	13.51	13.49	13.47
5	9.49	9.47	9.45	9.43	9.42	9.40	9.39	9.38	9.36	9.33	9.29	9.26	9.24	9.20	9.17	9.13	9.11	9.10	9.08	9.06	9.05	9.03
6	7.33	7.31	7.30	7.28	7.27	7.25	7.24	7.23	7.21	7.18	7.14	7.11	7.09	7.06	7.02	6.99	6.97	6.96	6.94	6.92	6.91	6.89
7	6.09	6.07	6.06	6.04	6.03	6.02	6.00	5.99	5.97	5.94	5.91	5.88	5.86	5.82	5.79	5.75	5.74	5.72	5.71	5.69	5.68	5.66
8	5.30	5.28	5.26	5.25	5.23	5.22	5.21	5.20	5.18	5.15	5.12	5.09	5.07	5.03	5.00	4.96	4.95	4.93	4.92	4.90	4.89	4.87
9	4.75	4.73	4.71	4.70	4.68	4.67	4.66	4.65	4.63	4.60	4.57	4.54	4.52	4.48	4.45	4.41	4.40	4.39	4.37	4.35	4.34	4.32
10	4.34	4.33	4.31	4.30	4.28	4.27	4.26	4.25	4.23	4.20	4.17	4.14	4.12	4.08	4.05	4.01	4.00	3.98	3.97	3.95	3.94	3.92
11	4.04	4.02	4.01	3.99	3.98	3.96	3.95	3.94	3.92	3.89	3.86	3.83	3.81	3.78	3.74	3.71	3.69	3.68	3.66	3.64	3.63	3.61
12	3.80	3.78	3.76	3.75	3.74	3.72	3.71	3.70	3.68	3.65	3.62	3.59	3.57	3.54	3.50	3.47	3.45	3.44	3.42	3.40	3.39	3.37
13	3.60	3.59	3.57	3.56	3.54	3.53	3.52	3.51	3.49	3.46	3.43	3.40	3.38	3.34	3.31	3.27	3.25	3.24	3.23	3.21	3.19	3.18
14	3.44	3.43	3.41	3.40	3.38	3.37	3.36	3.35	3.33	3.30	3.27	3.24	3.22	3.18	3.15	3.11	3.09	3.08	3.06	3.05	3.03	3.02
15	3.31	3.29	3.28	3.26	3.25	3.24	3.23	3.21	3.19	3.17	3.13	3.10	3.08	3.05	3.01	2.98	2.96	2.95	2.93	2.91	2.90	2.88
16	3.20	3.18	3.16	3.15	3.14	3.12	3.11	3.10	3.08	3.05	3.02	2.99	2.97	2.93	2.90	2.86	2.84	2.83	2.81	2.80	2.78	2.76
17	3.10	3.08	3.07	3.05	3.04	3.03	3.01	3.00	2.98	2.96	2.92	2.89	2.87	2.83	2.80	2.76	2.75	2.73	2.72	2.70	2.68	2.66
18	3.02	3.00	2.98	2.97	2.95	2.94	2.93	2.92	2.90	2.87	2.84	2.81	2.78	2.75	2.71	2.68	2.66	2.65	2.63	2.61	2.59	2.58
19	2.94	2.92	2.91	2.89	2.88	2.87	2.86	2.84	2.82	2.80	2.76	2.73	2.71	2.67	2.64	2.60	2.58	2.57	2.55	2.54	2.52	2.50
20	2.88	2.86	2.84	2.83	2.81	2.80	2.79	2.78	2.76	2.73	2.69	2.67	2.64	2.61	2.57	2.54	2.52	2.50	2.49	2.47	2.45	2.43
21	2.82	2.80	2.79	2.77	2.76	2.74	2.73	2.72	2.70	2.67	2.64	2.61	2.58	2.55	2.51	2.48	2.46	2.44	2.43	2.41	2.39	2.37
22	2.77	2.75	2.73	2.72	2.70	2.69	2.68	2.67	2.65	2.62	2.58	2.55	2.53	2.50	2.46	2.42	2.40	2.39	2.37	2.35	2.34	2.32
23	2.72	2.70	2.69	2.67	2.66	2.64	2.63	2.62	2.60	2.57	2.54	2.51	2.48	2.45	2.41	2.37	2.35	2.34	2.32	2.30	2.29	2.27
24	2.68	2.66	2.64	2.63	2.61	2.60	2.59	2.58	2.56	2.53	2.49	2.46	2.44	2.40	2.37	2.33	2.31	2.30	2.28	2.26	2.24	2.22
25	2.64	2.62	2.60	2.59	2.58	2.56	2.55	2.54	2.52	2.49	2.45	2.42	2.40	2.36	2.33	2.29	2.27	2.26	2.24	2.22	2.20	2.18
26	2.60	2.58	2.57	2.55	2.54	2.53	2.51	2.50	2.48	2.45	2.42	2.39	2.36	2.33	2.29	2.25	2.23	2.22	2.20	2.18	2.16	2.14
27	2.57	2.55	2.54	2.52	2.51	2.49	2.48	2.47	2.45	2.42	2.38	2.35	2.33	2.29	2.26	2.22	2.20	2.18	2.17	2.15	2.13	2.11
28	2.54	2.52	2.51	2.49	2.48	2.46	2.45	2.44	2.42	2.39	2.35	2.32	2.30	2.26	2.23	2.19	2.17	2.15	2.13	2.11	2.10	2.08
29	2.51	2.49	2.48	2.46	2.45	2.44	2.42	2.41	2.39	2.36	2.33	2.30	2.27	2.23	2.20	2.16	2.14	2.12	2.10	2.08	2.07	2.05
30	2.49	2.47	2.45	2.44	2.42	2.41	2.40	2.39	2.36	2.34	2.30	2.27	2.25	2.21	2.17	2.13	2.11	2.10	2.08	2.06	2.04	2.02
32	2.44	2.42	2.41	2.39	2.38	2.36	2.35	2.34	2.32	2.29	2.25	2.22	2.20	2.16	2.12	2.08	2.06	2.05	2.03	2.01	1.99	1.97
35	2.38	2.36	2.35	2.33	2.32	2.30	2.29	2.28	2.26	2.23	2.19	2.16	2.14	2.10	2.06	2.02	2.00	1.98	1.96	1.94	1.92	1.90
40	2.31	2.29	2.27	2.26	2.24	2.23	2.22	2.20	2.18	2.15	2.11	2.08	2.06	2.02	1.98	1.94	1.92	1.90	1.88	1.86	1.84	1.82
45	2.25	2.23	2.21	2.20	2.18	2.17	2.16	2.14	2.12	2.09	2.05	2.02	2.00	1.96	1.92	1.88	1.85	1.84	1.82	1.79	1.77	1.75
50	2.20	2.18	2.17	2.15	2.14	2.12	2.11	2.10	2.08	2.05	2.01	1.97	1.95	1.91	1.87	1.82	1.80	1.79	1.76	1.74	1.72	1.70
60	2.13	2.12	2.10	2.08	2.07	2.05	2.04	2.03	2.01	1.98	1.94	1.90	1.88	1.84	1.79	1.75	1.73	1.71	1.69	1.66	1.64	1.62
75	2.07	2.05	2.03	2.02	2.00	1.99	1.97	1.96	1.94	1.91	1.87	1.83	1.81	1.76	1.72	1.67	1.65	1.63	1.61	1.58	1.56	1.53
100	2.00	1.98	1.97	1.95	1.93	1.92	1.91	1.89	1.87	1.84	1.80	1.76	1.74	1.69	1.65	1.60	1.57	1.55	1.53	1.50	1.47	1.45
120	1.97	1.95	1.93	1.92	1.90	1.89	1.87	1.86	1.84	1.81	1.76	1.73	1.70	1.66	1.61	1.56	1.53	1.51	1.49	1.46	1.43	1.40
140	1.95	1.93	1.91	1.89	1.88	1.86	1.85	1.84	1.81	1.78	1.74	1.70	1.67	1.63	1.58	1.53	1.50	1.48	1.46	1.43	1.40	1.37
180	1.92	1.90	1.88	1.86	1.85	1.83	1.82	1.81	1.78	1.75	1.71	1.67	1.64	1.60	1.55	1.49	1.47	1.45	1.42	1.39	1.35	1.32
250	1.89	1.87	1.85	1.83	1.82	1.80	1.79	1.77	1.75	1.72	1.67	1.64	1.61	1.56	1.51	1.46	1.43	1.41	1.38	1.34	1.31	1.27
400	1.86	1.84	1.82	1.80	1.79	1.77	1.76	1.75	1.72	1.69	1.64	1.61	1.58	1.53	1.48	1.42	1.39	1.37	1.33	1.30	1.26	1.22
1000	1.83	1.81	1.79	1.77	1.76	1.74	1.73	1.72	1.69	1.66	1.61	1.58	1.54	1.50	1.44	1.38	1.35	1.33	1.29	1.25	1.21	1.16

Table F (cont.)

Numerator df

α = .05	1	2	3	4	5	6	7	8	9	10	11	12	13	14	15	16	17	18	19	20	21	22
1	161.4	199.5	215.7	224.6	230.2	234.0	236.8	238.9	240.5	241.9	243.0	243.9	244.7	245.4	245.9	246.5	246.9	247.3	247.7	248.0	248.3	248.6
2	18.51	19.00	19.16	19.25	19.30	19.33	19.35	19.37	19.38	19.40	19.40	19.41	19.42	19.42	19.43	19.43	19.44	19.44	19.44	19.45	19.45	19.45
3	10.13	9.55	9.28	9.12	9.01	8.94	8.89	8.85	8.81	8.79	8.76	8.74	8.73	8.71	8.70	8.69	8.68	8.67	8.67	8.66	8.65	8.65
4	7.71	6.94	6.59	6.39	6.26	6.16	6.09	6.04	6.00	5.96	5.94	5.91	5.89	5.87	5.86	5.84	5.83	5.82	5.81	5.80	5.79	5.79
5	6.61	5.79	5.41	5.19	5.05	4.95	4.88	4.82	4.77	4.74	4.70	4.68	4.66	4.64	4.62	4.60	4.59	4.58	4.57	4.56	4.55	4.54
6	5.99	5.14	4.76	4.53	4.39	4.28	4.21	4.15	4.10	4.06	4.03	4.00	3.98	3.96	3.94	3.92	3.91	3.90	3.88	3.87	3.86	3.86
7	5.59	4.74	4.35	4.12	3.97	3.87	3.79	3.73	3.68	3.64	3.60	3.57	3.55	3.53	3.51	3.49	3.48	3.47	3.46	3.44	3.43	3.43
8	5.32	4.46	4.07	3.84	3.69	3.58	3.50	3.44	3.39	3.35	3.31	3.28	3.26	3.24	3.22	3.20	3.19	3.17	3.16	3.15	3.14	3.13
9	5.12	4.26	3.86	3.63	3.48	3.37	3.29	3.23	3.18	3.14	3.10	3.07	3.05	3.03	3.01	2.99	2.97	2.96	2.95	2.94	2.93	2.92
10	4.96	4.10	3.71	3.48	3.33	3.22	3.14	3.07	3.02	2.98	2.94	2.91	2.89	2.86	2.85	2.83	2.81	2.80	2.79	2.77	2.76	2.75
11	4.84	3.98	3.59	3.36	3.20	3.09	3.01	2.95	2.90	2.85	2.82	2.79	2.76	2.74	2.72	2.70	2.69	2.67	2.66	2.65	2.64	2.63
12	4.75	3.89	3.49	3.26	3.11	3.00	2.91	2.85	2.80	2.75	2.72	2.69	2.66	2.64	2.62	2.60	2.58	2.57	2.56	2.54	2.53	2.52
13	4.67	3.81	3.41	3.18	3.03	2.92	2.83	2.77	2.71	2.67	2.63	2.60	2.58	2.55	2.53	2.51	2.50	2.48	2.47	2.46	2.45	2.44
14	4.60	3.74	3.34	3.11	2.96	2.85	2.76	2.70	2.65	2.60	2.57	2.53	2.51	2.48	2.46	2.44	2.43	2.41	2.40	2.39	2.38	2.37
15	4.54	3.68	3.29	3.06	2.90	2.79	2.71	2.64	2.59	2.54	2.51	2.48	2.45	2.42	2.40	2.38	2.37	2.35	2.34	2.33	2.32	2.31
16	4.49	3.63	3.24	3.01	2.85	2.74	2.66	2.59	2.54	2.49	2.46	2.42	2.40	2.37	2.35	2.33	2.32	2.30	2.29	2.28	2.26	2.25
17	4.45	3.59	3.20	2.96	2.81	2.70	2.61	2.55	2.49	2.45	2.41	2.38	2.35	2.33	2.31	2.29	2.27	2.26	2.24	2.23	2.22	2.21
18	4.41	3.55	3.16	2.93	2.77	2.66	2.58	2.51	2.46	2.41	2.37	2.34	2.31	2.29	2.27	2.25	2.23	2.22	2.20	2.19	2.18	2.17
19	4.38	3.52	3.13	2.90	2.74	2.63	2.54	2.48	2.42	2.38	2.34	2.31	2.28	2.26	2.23	2.21	2.20	2.18	2.17	2.16	2.14	2.13
20	4.35	3.49	3.10	2.87	2.71	2.60	2.51	2.45	2.39	2.35	2.31	2.28	2.25	2.22	2.20	2.18	2.17	2.15	2.14	2.12	2.11	2.10
21	4.32	3.47	3.07	2.84	2.68	2.57	2.49	2.42	2.37	2.32	2.28	2.25	2.22	2.20	2.18	2.16	2.14	2.12	2.11	2.10	2.08	2.07
22	4.30	3.44	3.05	2.82	2.66	2.55	2.46	2.40	2.34	2.30	2.26	2.23	2.20	2.17	2.15	2.13	2.11	2.10	2.08	2.07	2.06	2.05
23	4.28	3.42	3.03	2.80	2.64	2.53	2.44	2.37	2.32	2.27	2.24	2.20	2.18	2.15	2.13	2.11	2.09	2.08	2.06	2.05	2.04	2.02
24	4.26	3.40	3.01	2.78	2.62	2.51	2.42	2.36	2.30	2.25	2.22	2.18	2.15	2.13	2.11	2.09	2.07	2.05	2.04	2.03	2.01	2.00
25	4.24	3.39	2.99	2.76	2.60	2.49	2.40	2.34	2.28	2.24	2.20	2.16	2.14	2.11	2.09	2.07	2.05	2.04	2.02	2.01	2.00	1.98
26	4.23	3.37	2.98	2.74	2.59	2.47	2.39	2.32	2.27	2.22	2.18	2.15	2.12	2.09	2.07	2.05	2.03	2.02	2.00	1.99	1.98	1.97
27	4.21	3.35	2.96	2.73	2.57	2.46	2.37	2.31	2.25	2.20	2.17	2.13	2.10	2.08	2.06	2.04	2.02	2.00	1.99	1.97	1.96	1.95
28	4.20	3.34	2.95	2.71	2.56	2.45	2.36	2.29	2.24	2.19	2.15	2.12	2.09	2.06	2.04	2.02	2.00	1.99	1.97	1.96	1.95	1.93
29	4.18	3.33	2.93	2.70	2.55	2.43	2.35	2.28	2.22	2.18	2.14	2.10	2.08	2.05	2.03	2.01	1.99	1.97	1.96	1.94	1.93	1.92
30	4.17	3.32	2.92	2.69	2.53	2.42	2.33	2.27	2.21	2.16	2.13	2.09	2.06	2.04	2.01	1.99	1.98	1.96	1.95	1.93	1.92	1.91
32	4.15	3.29	2.90	2.67	2.51	2.40	2.31	2.24	2.19	2.14	2.10	2.07	2.04	2.01	1.99	1.97	1.95	1.94	1.92	1.91	1.90	1.88
35	4.12	3.27	2.87	2.64	2.49	2.37	2.29	2.22	2.16	2.11	2.07	2.04	2.01	1.99	1.96	1.94	1.92	1.91	1.89	1.88	1.87	1.85
40	4.08	3.23	2.84	2.61	2.45	2.34	2.25	2.18	2.12	2.08	2.04	2.00	1.97	1.95	1.92	1.90	1.89	1.87	1.85	1.84	1.83	1.81
45	4.06	3.20	2.81	2.58	2.42	2.31	2.22	2.15	2.10	2.05	2.01	1.97	1.94	1.92	1.89	1.87	1.86	1.84	1.82	1.81	1.80	1.78
50	4.03	3.18	2.79	2.56	2.40	2.29	2.20	2.13	2.07	2.03	1.99	1.95	1.92	1.89	1.87	1.85	1.83	1.81	1.80	1.78	1.77	1.76
60	4.00	3.15	2.76	2.53	2.37	2.25	2.17	2.10	2.04	1.99	1.95	1.92	1.89	1.86	1.84	1.82	1.80	1.78	1.76	1.75	1.73	1.72
75	3.97	3.12	2.73	2.49	2.34	2.22	2.13	2.06	2.01	1.96	1.92	1.88	1.85	1.83	1.80	1.78	1.76	1.74	1.73	1.71	1.70	1.69
100	3.94	3.09	2.70	2.46	2.31	2.19	2.10	2.03	1.97	1.93	1.89	1.85	1.82	1.79	1.77	1.75	1.73	1.71	1.69	1.68	1.66	1.65
120	3.92	3.07	2.68	2.45	2.29	2.18	2.09	2.02	1.96	1.91	1.87	1.83	1.80	1.78	1.75	1.73	1.71	1.69	1.67	1.66	1.64	1.63
140	3.91	3.06	2.67	2.44	2.28	2.16	2.08	2.01	1.95	1.90	1.86	1.82	1.79	1.76	1.74	1.72	1.70	1.68	1.66	1.65	1.63	1.62
180	3.89	3.05	2.65	2.42	2.26	2.15	2.06	1.99	1.93	1.88	1.84	1.81	1.77	1.75	1.72	1.70	1.68	1.66	1.64	1.63	1.61	1.60
250	3.88	3.03	2.64	2.41	2.25	2.13	2.05	1.98	1.92	1.87	1.83	1.79	1.76	1.73	1.71	1.69	1.66	1.65	1.63	1.61	1.60	1.58
400	3.86	3.02	2.63	2.39	2.24	2.12	2.03	1.96	1.90	1.85	1.81	1.78	1.74	1.72	1.69	1.67	1.65	1.63	1.61	1.60	1.58	1.57
1000	3.85	3.00	2.61	2.38	2.22	2.11	2.02	1.95	1.89	1.84	1.80	1.76	1.73	1.70	1.68	1.65	1.63	1.61	1.60	1.58	1.57	1.55

Denominator df

Table F (cont.)

									Numerator df													
$\alpha = .05$	23	24	25	26	27	28	29	30	32	35	40	45	50	60	75	100	120	140	180	250	400	1000
1	248.8	249.1	249.3	249.5	249.6	249.8	250.0	250.1	250.4	250.7	251.1	251.5	251.8	252.2	252.6	253.0	253.3	253.4	253.6	253.8	254.0	254.2
2	19.45	19.45	19.46	19.46	19.46	19.46	19.46	19.46	19.46	19.47	19.47	19.47	19.48	19.48	19.48	19.49	19.49	19.49	19.49	19.49	19.49	19.49
3	8.64	8.64	8.63	8.63	8.63	8.62	8.62	8.62	8.61	8.60	8.59	8.59	8.58	8.57	8.56	8.55	8.55	8.55	8.54	8.54	8.53	8.53
4	5.78	5.77	5.77	5.76	5.76	5.75	5.75	5.75	5.74	5.73	5.72	5.71	5.70	5.69	5.68	5.66	5.66	5.65	5.65	5.64	5.64	5.63
5	4.53	4.53	4.52	4.52	4.51	4.50	4.50	4.50	4.49	4.48	4.46	4.45	4.44	4.43	4.42	4.41	4.40	4.39	4.39	4.38	4.38	4.37
6	3.85	3.84	3.83	3.83	3.82	3.82	3.81	3.81	3.80	3.79	3.77	3.76	3.75	3.74	3.73	3.71	3.70	3.70	3.69	3.69	3.68	3.67
7	3.42	3.41	3.40	3.40	3.39	3.39	3.38	3.38	3.37	3.36	3.34	3.33	3.32	3.30	3.29	3.27	3.27	3.26	3.25	3.25	3.24	3.23
8	3.12	3.12	3.11	3.10	3.10	3.09	3.08	3.08	3.07	3.06	3.04	3.03	3.02	3.01	2.99	2.97	2.97	2.96	2.95	2.95	2.94	2.93
9	2.91	2.90	2.89	2.89	2.88	2.87	2.87	2.86	2.85	2.84	2.83	2.81	2.80	2.79	2.77	2.76	2.75	2.74	2.73	2.73	2.72	2.71
10	2.75	2.74	2.73	2.72	2.72	2.71	2.70	2.70	2.69	2.68	2.66	2.65	2.64	2.62	2.60	2.59	2.58	2.57	2.57	2.56	2.55	2.54
11	2.62	2.61	2.60	2.59	2.59	2.58	2.58	2.57	2.56	2.55	2.53	2.52	2.51	2.49	2.47	2.46	2.45	2.44	2.43	2.43	2.42	2.41
12	2.51	2.51	2.50	2.49	2.48	2.48	2.47	2.47	2.46	2.44	2.43	2.41	2.40	2.38	2.37	2.35	2.34	2.33	2.33	2.32	2.31	2.30
13	2.43	2.42	2.41	2.41	2.40	2.39	2.39	2.38	2.37	2.36	2.34	2.33	2.31	2.30	2.28	2.26	2.25	2.25	2.24	2.23	2.22	2.21
14	2.36	2.35	2.34	2.33	2.33	2.32	2.31	2.31	2.30	2.28	2.27	2.25	2.24	2.22	2.21	2.19	2.18	2.17	2.16	2.15	2.15	2.14
15	2.30	2.29	2.28	2.27	2.27	2.26	2.25	2.25	2.24	2.22	2.20	2.19	2.18	2.16	2.14	2.12	2.11	2.11	2.10	2.09	2.08	2.07
16	2.24	2.24	2.23	2.22	2.21	2.21	2.20	2.19	2.18	2.17	2.15	2.14	2.12	2.11	2.09	2.07	2.06	2.05	2.04	2.03	2.02	2.02
17	2.20	2.19	2.18	2.17	2.17	2.16	2.15	2.15	2.14	2.12	2.10	2.09	2.08	2.06	2.04	2.02	2.01	2.00	1.99	1.98	1.98	1.97
18	2.16	2.15	2.14	2.13	2.13	2.12	2.11	2.11	2.10	2.08	2.06	2.05	2.04	2.02	2.00	1.98	1.97	1.96	1.95	1.94	1.93	1.92
19	2.12	2.11	2.11	2.10	2.09	2.08	2.08	2.07	2.06	2.05	2.03	2.01	2.00	1.98	1.96	1.94	1.93	1.92	1.91	1.90	1.89	1.88
20	2.09	2.08	2.07	2.07	2.06	2.05	2.05	2.04	2.03	2.01	1.99	1.98	1.97	1.95	1.93	1.91	1.90	1.89	1.88	1.87	1.86	1.85
21	2.06	2.05	2.05	2.04	2.03	2.02	2.02	2.01	2.00	1.98	1.96	1.95	1.94	1.92	1.90	1.88	1.87	1.86	1.85	1.84	1.83	1.82
22	2.04	2.03	2.02	2.01	2.00	2.00	1.99	1.98	1.97	1.96	1.94	1.92	1.91	1.89	1.87	1.85	1.84	1.83	1.82	1.81	1.80	1.79
23	2.01	2.01	2.00	1.99	1.98	1.97	1.97	1.96	1.95	1.93	1.91	1.90	1.88	1.86	1.84	1.82	1.81	1.81	1.79	1.78	1.77	1.76
24	1.99	1.98	1.97	1.97	1.96	1.95	1.95	1.94	1.93	1.91	1.89	1.88	1.86	1.84	1.82	1.80	1.79	1.78	1.77	1.76	1.75	1.74
25	1.97	1.96	1.96	1.95	1.94	1.93	1.93	1.92	1.91	1.89	1.87	1.86	1.84	1.82	1.80	1.78	1.77	1.76	1.75	1.74	1.73	1.72
26	1.96	1.95	1.94	1.93	1.92	1.91	1.91	1.90	1.89	1.87	1.85	1.84	1.82	1.80	1.78	1.76	1.75	1.74	1.73	1.72	1.71	1.70
27	1.94	1.93	1.92	1.91	1.90	1.90	1.89	1.88	1.87	1.86	1.84	1.82	1.81	1.79	1.76	1.74	1.73	1.72	1.71	1.70	1.69	1.68
28	1.92	1.91	1.91	1.90	1.89	1.88	1.88	1.87	1.86	1.84	1.82	1.80	1.79	1.77	1.75	1.73	1.71	1.71	1.69	1.68	1.67	1.66
29	1.91	1.90	1.89	1.88	1.88	1.87	1.86	1.85	1.84	1.83	1.81	1.79	1.77	1.75	1.73	1.71	1.70	1.69	1.68	1.67	1.66	1.65
30	1.90	1.89	1.88	1.87	1.86	1.85	1.85	1.84	1.83	1.81	1.79	1.77	1.76	1.74	1.72	1.70	1.68	1.68	1.66	1.65	1.64	1.63
32	1.87	1.86	1.85	1.85	1.84	1.83	1.82	1.82	1.80	1.79	1.77	1.75	1.74	1.71	1.69	1.67	1.66	1.65	1.64	1.63	1.61	1.60
35	1.84	1.83	1.82	1.82	1.81	1.80	1.79	1.79	1.77	1.76	1.74	1.72	1.70	1.68	1.66	1.63	1.62	1.61	1.60	1.59	1.58	1.57
40	1.80	1.79	1.78	1.77	1.77	1.76	1.75	1.74	1.73	1.72	1.69	1.67	1.66	1.64	1.61	1.59	1.58	1.57	1.55	1.54	1.53	1.52
45	1.77	1.76	1.75	1.74	1.73	1.73	1.72	1.71	1.70	1.68	1.66	1.64	1.63	1.60	1.58	1.55	1.54	1.53	1.52	1.51	1.49	1.48
50	1.75	1.74	1.73	1.72	1.71	1.70	1.69	1.69	1.67	1.66	1.63	1.61	1.60	1.58	1.55	1.52	1.51	1.50	1.49	1.47	1.46	1.45
60	1.71	1.70	1.69	1.68	1.67	1.66	1.66	1.65	1.64	1.62	1.59	1.57	1.56	1.53	1.51	1.48	1.47	1.46	1.44	1.43	1.41	1.40
75	1.67	1.66	1.65	1.64	1.63	1.63	1.62	1.61	1.60	1.58	1.55	1.53	1.52	1.49	1.47	1.44	1.42	1.41	1.40	1.38	1.37	1.35
100	1.64	1.63	1.62	1.61	1.60	1.59	1.58	1.57	1.56	1.54	1.52	1.49	1.48	1.45	1.42	1.39	1.38	1.36	1.35	1.33	1.31	1.30
120	1.62	1.61	1.60	1.59	1.58	1.57	1.56	1.55	1.54	1.52	1.50	1.47	1.46	1.43	1.40	1.37	1.35	1.34	1.32	1.30	1.29	1.27
140	1.61	1.60	1.58	1.57	1.57	1.56	1.55	1.54	1.53	1.51	1.48	1.46	1.44	1.41	1.38	1.35	1.33	1.32	1.30	1.29	1.27	1.25
180	1.59	1.58	1.57	1.56	1.55	1.54	1.53	1.52	1.51	1.49	1.46	1.44	1.42	1.39	1.36	1.33	1.31	1.30	1.28	1.26	1.24	1.22
250	1.57	1.56	1.55	1.54	1.53	1.52	1.51	1.50	1.49	1.47	1.44	1.42	1.40	1.37	1.34	1.31	1.29	1.27	1.25	1.23	1.21	1.18
400	1.56	1.54	1.53	1.52	1.51	1.50	1.50	1.49	1.47	1.45	1.42	1.40	1.38	1.35	1.32	1.28	1.26	1.25	1.23	1.20	1.18	1.15
1000	1.54	1.53	1.52	1.51	1.50	1.49	1.48	1.47	1.46	1.43	1.41	1.38	1.36	1.33	1.30	1.26	1.24	1.22	1.20	1.17	1.14	1.11

Denominator df

Critical Values d_L and d_U of the Durbin-Watson Statistic D (Critical Values are One-Sided)[a]

$\alpha = 0.05$

n	k = 1 d_L	k = 1 d_U	k = 2 d_L	k = 2 d_U	k = 3 d_L	k = 3 d_U	k = 4 d_L	k = 4 d_U	k = 5 d_L	k = 5 d_U
15	1.08	1.36	.95	1.54	.82	1.75	.69	1.97	.56	2.21
16	1.10	1.37	.98	1.54	.86	1.73	.74	1.93	.62	2.15
17	1.13	1.38	1.02	1.54	.90	1.71	.78	1.90	.67	2.10
18	1.16	1.39	1.05	1.53	.93	1.69	.82	1.87	.71	2.06
19	1.18	1.40	1.08	1.53	.97	1.68	.86	1.85	.75	2.02
20	1.20	1.41	1.10	1.54	1.00	1.68	.90	1.83	.79	1.99
21	1.22	1.42	1.13	1.54	1.03	1.67	.93	1.81	.83	1.96
22	1.24	1.43	1.15	1.54	1.05	1.66	.96	1.80	.86	1.94
23	1.26	1.44	1.17	1.54	1.08	1.66	.99	1.79	.90	1.92
24	1.27	1.45	1.19	1.55	1.10	1.66	1.01	1.78	.93	1.90
25	1.29	1.45	1.21	1.55	1.12	1.66	1.04	1.77	.95	1.89
26	1.30	1.46	1.22	1.55	1.14	1.65	1.06	1.76	.98	1.88
27	1.32	1.47	1.24	1.56	1.16	1.65	1.08	1.76	1.01	1.86
28	1.33	1.48	1.26	1.56	1.18	1.65	1.10	1.75	1.03	1.85
29	1.34	1.48	1.27	1.56	1.20	1.65	1.12	1.74	1.05	1.84
30	1.35	1.49	1.28	1.57	1.21	1.65	1.14	1.74	1.07	1.83
31	1.36	1.50	1.30	1.57	1.23	1.65	1.16	1.74	1.09	1.83
32	1.37	1.50	1.31	1.57	1.24	1.65	1.18	1.73	1.11	1.82
33	1.38	1.51	1.32	1.58	1.26	1.65	1.19	1.73	1.13	1.81
34	1.39	1.51	1.33	1.58	1.27	1.65	1.21	1.73	1.15	1.81
35	1.40	1.52	1.34	1.58	1.28	1.65	1.22	1.73	1.16	1.80
36	1.41	1.52	1.35	1.59	1.29	1.65	1.24	1.73	1.18	1.80
37	1.42	1.53	1.36	1.59	1.31	1.66	1.25	1.72	1.19	1.80
38	1.43	1.54	1.37	1.59	1.32	1.66	1.26	1.72	1.21	1.79
39	1.43	1.54	1.38	1.60	1.33	1.66	1.27	1.72	1.22	1.79
40	1.44	1.54	1.39	1.60	1.34	1.66	1.29	1.72	1.23	1.79
45	1.48	1.57	1.43	1.62	1.38	1.67	1.34	1.72	1.29	1.78
50	1.50	1.59	1.46	1.63	1.42	1.67	1.38	1.72	1.34	1.77
55	1.53	1.60	1.49	1.64	1.45	1.68	1.41	1.72	1.38	1.77
60	1.55	1.62	1.51	1.65	1.48	1.69	1.44	1.73	1.41	1.77
65	1.57	1.63	1.54	1.66	1.50	1.70	1.47	1.73	1.44	1.77
70	1.58	1.64	1.55	1.67	1.52	1.70	1.49	1.74	1.46	1.77
75	1.60	1.65	1.57	1.68	1.54	1.71	1.51	1.74	1.49	1.77
80	1.61	1.66	1.59	1.69	1.56	1.72	1.53	1.74	1.51	1.77
85	1.62	1.67	1.60	1.70	1.57	1.72	1.55	1.75	1.52	1.77
90	1.63	1.68	1.61	1.70	1.59	1.73	1.57	1.75	1.54	1.78
95	1.64	1.69	1.62	1.71	1.60	1.73	1.58	1.75	1.56	1.78
100	1.65	1.69	1.63	1.72	1.61	1.74	1.59	1.76	1.57	1.78

$\alpha = 0.01$

n	k = 1 d_L	k = 1 d_U	k = 2 d_L	k = 2 d_U	k = 3 d_L	k = 3 d_U	k = 4 d_L	k = 4 d_U	k = 5 d_L	k = 5 d_U
15	.81	1.07	.70	1.25	.59	1.46	.49	1.70	.39	1.96
16	.84	1.09	.74	1.25	.63	1.44	.53	1.66	.44	1.90
17	.87	1.10	.77	1.25	.67	1.43	.57	1.63	.48	1.85
18	.90	1.12	.80	1.26	.71	1.42	.61	1.60	.52	1.80
19	.93	1.13	.83	1.26	.74	1.41	.65	1.58	.56	1.77
20	.95	1.15	.86	1.27	.77	1.41	.68	1.57	.60	1.74
21	.97	1.16	.89	1.27	.80	1.41	.72	1.55	.63	1.71
22	1.00	1.17	.91	1.28	.83	1.40	.75	1.54	.66	1.69
23	1.02	1.19	.94	1.29	.86	1.40	.77	1.53	.70	1.67
24	1.04	1.20	.96	1.30	.88	1.41	.80	1.53	.72	1.66
25	1.05	1.21	.98	1.30	.90	1.41	.83	1.52	.75	1.65
26	1.07	1.22	1.00	1.31	.93	1.41	.85	1.52	.78	1.64
27	1.09	1.23	1.02	1.32	.95	1.41	.88	1.51	.81	1.63
28	1.10	1.24	1.04	1.32	.97	1.41	.90	1.51	.83	1.62
29	1.12	1.25	1.05	1.33	.99	1.42	.92	1.51	.85	1.61
30	1.13	1.26	1.07	1.34	1.01	1.42	.94	1.51	.88	1.61
31	1.15	1.27	1.08	1.34	1.02	1.42	.96	1.51	.90	1.60
32	1.16	1.28	1.10	1.35	1.04	1.43	.98	1.51	.92	1.60
33	1.17	1.29	1.11	1.36	1.05	1.43	1.00	1.51	.94	1.59
34	1.18	1.30	1.13	1.36	1.07	1.43	1.01	1.51	.95	1.59
35	1.19	1.31	1.14	1.37	1.08	1.44	1.03	1.51	.97	1.59
36	1.21	1.32	1.15	1.38	1.10	1.44	1.04	1.51	.99	1.59
37	1.22	1.32	1.16	1.38	1.11	1.45	1.06	1.51	1.00	1.59
38	1.23	1.33	1.18	1.39	1.12	1.45	1.07	1.52	1.02	1.58
39	1.24	1.34	1.19	1.39	1.14	1.45	1.09	1.52	1.03	1.58
40	1.25	1.34	1.20	1.40	1.15	1.46	1.10	1.52	1.05	1.58
45	1.29	1.38	1.24	1.42	1.20	1.48	1.16	1.53	1.11	1.58
50	1.32	1.40	1.28	1.45	1.24	1.49	1.20	1.54	1.16	1.59
55	1.36	1.43	1.32	1.47	1.28	1.51	1.25	1.55	1.21	1.59
60	1.38	1.45	1.35	1.48	1.32	1.52	1.28	1.56	1.25	1.60
65	1.41	1.47	1.38	1.50	1.35	1.53	1.31	1.57	1.28	1.61
70	1.43	1.49	1.40	1.52	1.37	1.55	1.34	1.58	1.31	1.61
75	1.45	1.50	1.42	1.53	1.39	1.56	1.37	1.59	1.34	1.62
80	1.47	1.52	1.44	1.54	1.42	1.57	1.39	1.60	1.36	1.62
85	1.48	1.53	1.46	1.55	1.43	1.58	1.41	1.60	1.39	1.63
90	1.50	1.54	1.47	1.56	1.45	1.59	1.43	1.61	1.41	1.64
95	1.51	1.55	1.49	1.57	1.47	1.60	1.45	1.62	1.42	1.64
100	1.52	1.56	1.50	1.58	1.48	1.60	1.46	1.63	1.44	1.65

[a] n = number of observations; k = number of independent variables.

Source: This table is reproduced from Biometrika, 41 (1951): 173 and 175, with the permission of the Biometrika Trustees.

Selected Formulas

$Range = Max - Min$

$IQR = Q3 - Q1$

Outlier Rule-of-Thumb: $y < Q1 - 1.5 \times IQR$ or $y > Q3 + 1.5 \times IQR$

$$\bar{y} = \frac{\sum y}{n}$$

$$s = \sqrt{\frac{\sum (y - \bar{y})^2}{n - 1}}$$

$z = \dfrac{y - \mu}{\sigma}$ (model based) $\qquad$ $z = \dfrac{y - \bar{y}}{s}$ (data based)

$$r = \frac{\sum z_x z_y}{n - 1}$$

$\hat{y} = b_0 + b_1 x \qquad$ where $b_1 = r \dfrac{s_y}{s_x}$ and $b_0 = \bar{y} - b_1 \bar{x}$

$P(\mathbf{A}) = 1 - P(\mathbf{A}^C)$

$P(\mathbf{A} \; or \; \mathbf{B}) = P(\mathbf{A}) + P(\mathbf{B}) - P(\mathbf{A} \; and \; \mathbf{B})$

$P(\mathbf{A} \; and \; \mathbf{B}) = P(\mathbf{A}) \times P(\mathbf{B}|\mathbf{A})$

$$P(\mathbf{B}|\mathbf{A}) = \frac{P(\mathbf{A} \; and \; \mathbf{B})}{P(\mathbf{A})}$$

If $\mathbf{A}$ and $\mathbf{B}$ are independent, $P(\mathbf{B}|\mathbf{A}) = P(\mathbf{B})$

$E(X) = \mu = \sum x \cdot P(x) \qquad\qquad Var(X) = \sigma^2 = \sum (x - \mu)^2 P(x)$

$E(X \pm c) = E(X) \pm c \qquad\qquad\quad Var(X \pm c) = Var(X)$

$E(aX) = aE(X) \qquad\qquad\qquad\quad Var(aX) = a^2 Var(X)$

$E(X \pm Y) = E(X) \pm E(Y) \qquad\quad Var(X \pm Y) = Var(X) + Var(Y)$

$\qquad\qquad\qquad\qquad\qquad\qquad\qquad\quad$ if X and Y are independent

Geometric: $\qquad P(x) = q^{x-1}p \qquad \mu = \dfrac{1}{p} \qquad \sigma = \sqrt{\dfrac{q}{p^2}}$

Binomial: $\qquad P(x) = {}_nC_x p^x q^{n-x} \qquad \mu = np \qquad \sigma = \sqrt{npq}$

$\hat{p} = \dfrac{x}{n} \qquad \mu(\hat{p}) = p \qquad SD(\hat{p}) = \sqrt{\dfrac{pq}{n}}$

Poisson probability model for successes: Poisson (λ)

λ = mean number of successes.

X = number of successes.

$$P(X = x) = \frac{e^{-\lambda} \lambda^x}{x!}$$

$$\text{Expected value:} \qquad E(X) = \lambda$$

$$\text{Standard deviation:} \qquad SD(X) = \sqrt{\lambda}$$

Sampling distribution of $\bar{y}$:

(CLT) As n grows, the sampling distribution approaches the Normal model with

$$\mu(\bar{y}) = \mu_y \qquad SD(\bar{y}) = \frac{\sigma}{\sqrt{n}}$$

Inference:

Confidence interval for parameter = **statistic ± critical value × SE(statistic)**

$$\text{Test statistic} = \frac{statistic - parameter}{SD(statistic)}$$

Parameter	Statistic	SD(statistic)	SE(statistic)
p	$\hat{p}$	$\sqrt{\dfrac{pq}{n}}$	$\sqrt{\dfrac{\hat{p}\hat{q}}{n}}$
μ	$\bar{y}$	$\dfrac{\sigma}{\sqrt{n}}$	$\dfrac{s}{\sqrt{n}}$
$\mu_1 - \mu_2$	$\bar{y}_1 - \bar{y}_2$	$\sqrt{\dfrac{\sigma_1^2}{n_1} + \dfrac{\sigma_2^2}{n_2}}$	$\sqrt{\dfrac{s_1^2}{n_1} + \dfrac{s_2^2}{n_2}}$
μ_d	$\bar{d}$	$\dfrac{\sigma_d}{\sqrt{n}}$	$\dfrac{s_d}{\sqrt{n}}$
σ_ε	$s_e = \sqrt{\dfrac{\sum (y - \hat{y})^2}{n - 2}}$	(divide by $n - k - 1$ in multiple regression)	
β_1	b_1	(in simple regression)	$\dfrac{s_e}{s_x \sqrt{n - 1}}$
μ_ν	$\hat{y}_\nu$	(in simple regression)	$\sqrt{SE^2(b_1) \cdot (x_\nu - \bar{x})^2 + \dfrac{s_e^2}{n}}$
y_ν	$\hat{y}_\nu$	(in simple regression)	$\sqrt{SE^2(b_1) \cdot (x_\nu - \bar{x})^2 + \dfrac{s_e^2}{n} + s_e^2}$

Pooling: For testing difference between proportions: $\hat{p}_{pooled} = \dfrac{y_1 + y_2}{n_1 + n_2}$

For testing difference between means: $s_p = \sqrt{\dfrac{(n_1 - 1)s_1^2 + (n_2 - 1)s_2^2}{n_1 + n_2 - 2}}$

Substitute these pooled estimates in the respective SE formulas for both groups when assumptions and conditions are met.

Chi-square: $\chi^2 = \sum \dfrac{(Obs - Exp)^2}{Exp}$

Assumptions for Inference	And the Conditions That Support or Override Them

Proportions (z)

- **One sample**
  1. Individuals are independent.
  2. Sample is sufficiently large.

1. SRS and $n < 10\%$ of the population.
2. Successes and failures each ≥ 10.

Means (t)

- **One Sample** (df $= n - 1$)
  1. Individuals are independent.
  2. Population has a Normal model.

1. SRS and $n < 10\%$ of the population.
2. Histogram is unimodal and symmetric.*

- **Matched pairs** (df $= n - 1$)
  1. Data are matched.
  2. Individuals are independent.
  3. Population of differences is Normal.

1. (Think about the design.)
2. SRS and $n < 10\%$ OR random allocation.
3. Histogram of differences is unimodal and symmetric.*

- **Two independent samples** (df from technology)
  1. Groups are independent.
  2. Data in each group are independent.
  3. Both populations are Normal.

1. (Think about the design.)
2. SRSs and $n < 10\%$ OR random allocation.
3. Both histograms are unimodal and symmetric.*

Distributions/Association (χ^2)

- **Goodness of fit** (df $=$ # of cells $- 1$; one variable, one sample compared with population model)
  1. Data are counts.
  2. Data in sample are independent.
  3. Sample is sufficiently large.

1. (Are they?)
2. SRS and $n < 10\%$ of the population.
3. All expected counts ≥ 5.

- **Homogeneity** [df $= (r - 1)(c - 1)$; many groups compared on one variable]
  1. Data are counts.
  2. Data in groups are independent.
  3. Groups are sufficiently large.

1. (Are they?)
2. SRSs and $n < 10\%$ OR random allocation.
3. All expected counts ≥ 5.

- **Independence** [df $= (r - 1)(c - 1)$; sample from one population classified on two variables]
  1. Data are counts.
  2. Data are independent.
  3. Sample is sufficiently large.

1. (Are they?)
2. SRSs and $n < 10\%$ of the population.
3. All expected counts ≥ 5.

Regression with k predictors (t, df $= n - k - 1$)

- **Association** of each quantitative predictor with the response variable
  1. Form of relationship is linear.

  2. Errors are independent.
  3. Variability of errors is constant.

  4. Errors follow a Normal model.

1. Scatterplots of y against each x are straight enough. Scatterplot of residuals against predicted values shows no special structure.
2. No apparent pattern in plot of residuals against predicted values.
3. Plot of residuals against predicted values has constant spread, doesn't "thicken."
4. Histogram of residuals is approximately unimodal and symmetric, or Normal probability plot is reasonably straight.*

Analysis of Variance (F, df depends on number of factors and number of levels in each.)

- **Equality** of the mean response across levels of categorical predictors
  1. Additive Model (if there are 2 factors with no interaction term).
  2. Independent errors.
  3. Equal variance across treatment levels.

  4. Errors follow a Normal model.

1. Interaction plot shows parallel lines (otherwise include an interaction term if possible).
2. Randomized experiment or other suitable randomization.
3. Plot of residuals against predicted values has constant spread. Boxplots (partial boxplots for 2 factors) show similar spreads.
4. Histogram of residuals is unimodal and symmetric, or Normal probability plot is reasonably straight.

(*Less critical as n increases)

Quick Guide to Inference

Plan				Do				Report
Inference about?	**One group or two?**	**Procedure**	**Model**	**Parameter**	**Estimate**	**SE**		**Chapter**
Proportions	One sample	1-Proportion z-Interval	z	p	$\hat{p}$	$\sqrt{\dfrac{\hat{p}\hat{q}}{n}}$		10
		1-Proportion z-Test				$\sqrt{\dfrac{p_0 q_0}{n}}$		11
Means	One sample	t-Interval t-Test	t df $= n-1$	μ	$\bar{y}$	$\dfrac{s}{\sqrt{n}}$		12
	Two independent groups	2-Sample t-Test 2-Sample t-Interval	t df from technology	$\mu_1 - \mu_2$	$\bar{y}_1 - \bar{y}_2$	$\sqrt{\dfrac{s_1^2}{n_1} + \dfrac{s_2^2}{n_2}}$		13
	Matched pairs	Paired t-Test Paired t-Interval	t df $= n-1$	μ_d	$\bar{d}$	$\dfrac{s_d}{\sqrt{n}}$		14
Distributions (one categorical variable)	One sample	Goodness of-Fit	χ^2 df $= cells - 1$					
	Many independent groups	Homogeneity χ^2 Test	χ^2 df $= (r-1)(c-1)$			$\sum \dfrac{(Obs - Exp)^2}{Exp}$		15
Independence (two categorical variables)	One sample	Independence χ^2 Test						
Association (two quantitative variables)	One sample	Linear Regression t-Test or Confidence Interval for β	t df $= n-2$	β_1	b_1	$\dfrac{s_e}{s_x \sqrt{n-1}}$ (compute with technology)		16
		*Confidence Interval for μ_v		μ_v	$\hat{y}_v$	$\sqrt{SE^2(b_1)\cdot(x_v - \bar{x})^2 + \dfrac{s_e^2}{n}}$		
		*Prediction Interval for y_v		y_v	$\hat{y}_v$	$\sqrt{SE^2(b_1)\cdot(x_v - \bar{x})^2 + \dfrac{s_e^2}{n} + s_e^2}$		
Association (one quantitative variable fit modeled by k quantitative variables)	One sample	Multiple Regression t-test or Confidence interval for each β_j	t $df = n - (k+1)$	β_j	b_j	(from technology)		17, 18, 19
		F test for regression model	F $df = k$ and $n - (k+1)$			MST/MSE		18, 19
Association (one quantitative and two or more categorical variables)	Two or more	ANOVA	F $df = k - 1$ and $N - k$			MST/MSE		23

Note: Page numbers in **boldface** indicate chapter-level topics; FE indicates For Example references; n indicates a footnote.